AF690987

# Treatise on Heavy-Ion Science

## Volume 8
## Nuclei Far From Stability

# Treatise on Heavy-Ion Science

EDITED BY D. ALLAN BROMLEY

Volume 1: Elastic and Quasi-Elastic Phenomena

Volume 2: Fusion and Quasi-Fusion Phenomena

Volume 3: Compound System Phenomena

Volume 4: Extreme Nuclear States

Volume 5: High-Energy Atomic Physics

Volume 6: Astrophysics, Chemistry, and Condensed Matter

Volume 7: Instrumentation and Techniques

Volume 8: Nuclei Far From Stability

# Treatise on Heavy-Ion Science

## Volume 8
## Nuclei Far From Stability

EDITED BY
D. ALLAN BROMLEY
*Henry Ford II Professor of Physics*
*Yale University*
*New Haven, Connecticut*

WITHDRAWN
IOWA STATE UNIVERSITY
LIBRARY

PLENUM PRESS • NEW YORK AND LONDON

Library of Congress Cataloging in Publication Data

(Revised for volume 8)

Treatise on heavy-ion science.

Includes bibliographies and indexes.
Contents: v. 1. Elastic and quasi-elastic phenomena—v. 2. Fusion and quasi-fusion phenomena—[etc.]—v. 8. Nuclei far from stability.
1. Heavy ions. I. Bromley, D. Allan (David Allan), 1926–
QC702.7.H42T74 1984 539.7 84-8384
ISBN 0-306-42949-7

© 1989 Plenum Press, New York
A Division of Plenum Publishing Corporation
233 Spring Street, New York, N.Y. 10013

All rights reserved

No part of this book may be reproduced, stored in a retrieval system, or transmitted in any form or by any means, electronic, mechanical, photocopying, microfilming, recording, or otherwise, without written permission from the Publisher

Printed in the United States of America

For Pat, Lynn, and David

# Preface to Treatise on Heavy-Ion Science

After a long gestation period, heavy-ion physics and chemistry is now, worldwide, the most rapidly growing area of nuclear science, and the concepts, techniques, and instrumentation of this heavy-ion work are finding ever-widening application in other areas of science and technology. Although there remain broad regions at higher energies, with heavier projectiles, and at higher excitations and angular momenta where heavy ions still provide gateways into the totally unknown, intensive studies over the past two decades have provided a sound framework of understanding of many of the salient features of interactions induced by these new heavier projectiles and a basis for coherent planning of future studies.

It thus seemed appropriate, at this point in the history of the field, to pull together in one place and in as coherent a fashion as possible, an overview of what has been accomplished and some enlightened speculation about where we go next. It is my hope that these volumes will provide a definitive reference for those working in this and neighboring fields—both students and professional scientists; beyond that I would also hope that they will make accessible to a much wider audience in other sciences and technologies some of the richness of heavy-ion research, and perhaps help to stimulate the transfer of techniques and concepts that I have already mentioned.

I have been singularly fortunate in planning these volumes in being able to persuade internationally recognized authorities to write in their areas of special expertise and interest, and also fortunate that I have not had to restrict them to any artificial or externally imposed procrustean restrictions on the scope or length of their chapters. I have asked each author to include sufficient background to make the chapter accessible to students and to the nonspecialist, to provide a broad selection of illustrations, and to feel free to extrapolate and to speculate about future directions.

In inviting contributions to these volumes I have made arbitrary decisions concerning both topics and contributors, and I make no claim to completeness. Indeed, a few chapters that I would have liked to include do not appear because of illness or other reasons which prevented their authors from completing them.

I should like to take this opportunity to thank all the authors represented for taking time in already full schedules to share with a wider audience their special experience and expertise in heavy-ion science. As was inevitable from the outset in a multiple-author venture of this scope—with over 65 different authors—my original scheduling and deadlines proved unrealistic. To all those authors who responded to them and produced manuscripts on or before the original deadline—in many cases, I am aware, at substantial personal cost—my most sincere thanks and appreciation. To them, too, however, go my apologies for the delay in bringing their work into print. I have delayed publication for over a year so that I might include a number of very important chapters without which the work would have been obviously incomplete.

Volumes 1-4 of the *Treatise on Heavy-Ion Science* are devoted to aspects of heavy-ion nuclear science, beginning with an overview of the historical development of the science and some of its simpler interactions—elastic and quasi-elastic, fusion and quasi-fusion phenomena—and moving from them to compound system phenomena and to much more complex and less well understood phenomena involving very heavy nuclear species and very high energies. Volume 5 is devoted to high-energy atomic physics, an entirely new field of science first made accessible by the availability of a broad range of heavy-ion beams, and one still very much in its infancy. Volume 6 considers the impact of heavy-ion studies on other sciences including astrophysics, chemistry, surface physics, condensed matter physics, materials science, and heavy-ion-induced fusion power. Volume 7 is devoted to some of the instrumentation peculiar to heavy-ion science and its applications. Volume 8, prepared substantially after the earlier volumes in this treatise, covers the relatively new studies of exotic nuclei and those far from stability to which heavy-ion beams give unique access.

Special thanks go to Amelia McNamara-Farrelly, Ellis Rosenberg, Bennett K. Ragnauth, and Kenneth Schubach of Plenum Press with whom it has been a pleasure to work on these volumes, and to Mary Anne Schulz for all her help in producing them. And I would also acknowledge my indebtedness to the Alexander von Humboldt Stiftung for a Humboldt Fellowship that I was privileged to hold during part of the time these volumes were in preparation.

D. Allan Bromley

New Haven, Connecticut

# Contributors to Volume 8

**Brian W. Allardyce**, European Center for Nuclear Research (CERN), Geneva, Switzerland

**Juha Äystö**, Department of Physics, University of Jyväskylä, Jyväskylä, Finland

**Joseph Cerny**, Department of Chemistry and Lawrence Berkeley Laboratory, University of California at Berkeley, Berkeley, California 94720

**Walter Greiner**, Institute for Theoretical Physics, University of Frankfurt, Frankfurt am Main, Federal Republic of Germany

**Joseph H. Hamilton**, Department of Physics, Vanderbilt University, Nashville, Tennessee 37235

**Martin Ivascu**, Central Institute of Physics and Nuclear Engineering, Bucharest, Romania

**Alexei A. Ogloblin**, Department of Nuclear Interaction, Kurchatov Institute of Atomic Energy, 123436 Moscow, USSR

**Ernst W. Otten**, Institute of Physics, University of Mainz, Mainz, Federal Republic of Germany

**Yuri E. Penionzhkevich**, Laboratory of Nuclear Reactions, Joint Institute for Nuclear Research, 101000 Moscow, USSR

**Dorin N. Poenaru**, Central Institute of Physics and Nuclear Engineering, Bucharest, Romania

**Helge L. Ravn**, European Center for Nuclear Research (CERN), Geneva, Switzerland

**Aurel Sandulescu**, Joint Institute for Nuclear Research, Dubna, USSR

**Isao Tanihata**, Institute of Physical and Chemical Research (RIKEN), Wako, Japan

**Vadim V. Volkov**, Laboratory of Nuclear Reactions, Joint Institute for Nuclear Research, 101000 Moscow, USSR

# Preface to Volume 8

After the completion of the first seven volumes of this Treatise, a growing number of readers pointed out the lack of definitive discussions of the use of heavy ions in the study of nuclei far removed from beta stability. Finding these arguments persuasive, and with the approval of Plenum Press, I undertook the assembly of this volume to redress this lack and was gratified by the immediate response from an outstanding group of authors who agreed to write chapters in their areas of special expertise.

Unhappily Georg Leander, who had agreed to write the opening theoretical chapter to put the entire volume in context, was prevented by a serious illness from doing so; at his recommendation I invited Jan Blomqvist to prepare this chapter but again, unhappily, it became impossible for him to complete the task. This also was the case for David Scott and James Symons who had agreed to write a chapter on neutron-rich nuclei far from stability.

Fortunately, the chapters included herein still provide an excellent overview of the volume topic and I am much indebted to the authors for their willingness to devote time in busy schedules to prepare these chapters. My apologies go to those authors who met my original optimistic deadlines and, in consequence, have watched their contributions become somewhat dated while I waited for the completion of other chapters without which this volume would have been incomplete.

This has been a most exciting three years since the first seven volumes of this Treatise appeared. New gamma-ray detection systems have allowed doubling of the angular momentum of discrete nuclear quantum states accessible to precision study. This has worked a renaissance in nuclear spectroscopy comparable to that for which the development of the laser was responsible in atomic spectroscopy.

The availability of ultrarelativistic beams of oxygen, silicon, and sulfur nuclei at Brookhaven and at CERN has opened up the new frontier of quark matter in nuclear physics. While early results are very encouraging, it is already clear that higher-energy beams of even heavier nuclei will be required before we can hope to delineate and understand the transition between hadronic and quark matter—a transition traversed in the earliest moments of our universe and in those most dramatic of all natural phenomena, the supernovae.

Whether or not additional volumes in this Treatise, covering these and other new topics in heavy-ion science, will be added will depend in very large measure on reader response and demand.

I again take this opportunity to express my indebtedness to all the authors who have contributed to the Treatise and to my Executive Assistant, Mrs. Mary Anne Schulz, for all her help in assembling it. Finally, let me only say that it has been a pleasure working with Plenum Press and its representatives over the years during which this work has been in production.

D. Allan Bromley

New Haven, Connecticut

# Contents

## 1. Structures of Nuclei Far From Stability

*Joseph H. Hamilton*



## 2. Production of Nuclei Far From Stability

*Vadim V. Volkov*

## 3. Proton-Rich Light Nuclei

*Juha Äystö and Joseph Cerny*



## 4. Very Neutron Rich Very Light Nuclei

*Alexei A. Ogloblin and Yuri E. Penionzhkevich*

## 5. On-Line Mass Separators

*Helge L. Ravn and Brian W. Allardyce*

## 6. On the Possible Use of Secondary Radioactive Beams

*Isao Tanihata*

## 7. Nuclear Radii and Moments of Unstable Isotopes

*Ernst W. Otten*



## 8. Cluster Radioactivities

*Walter Greiner, Martin Ivascu, Dorin N. Poenaru, and Aurel Sandulescu*

# Contents of Volumes 1–7

## Volume 1: Elastic and Quasi-Elastic Phenomena



## Volume 2: Fusion and Quasi-Fusion Phenomena

## Volume 3: Compound System Phenomena



## Volume 4: Extreme Nuclear States

## Volume 5: High-Energy Atomic Physics



## Volume 6: Astrophysics, Chemistry, and Condensed Matter

## Volume 7: Instrumentation and Techniques

# 1

JOSEPH H. HAMILTON is Landon C. Garland Professor of Physics at Vanderbilt University and Director of the Joint Institute for Heavy-Ion Research in Oak Ridge, TN. He received his Ph.D. in Physics from Indiana University in 1958. He received the Southeastern Section of the American Physical Society's Jesse Beams Award for Research in 1975 and the Alexander von Humboldt Prize from the Federal Republic of Germany in 1979. He has held an NSF Postdoctoral Fellowship at the University of Uppsala, Sweden, 1958–59; a research appointment at the Institute for Nuclear Research, Amsterdam, 1962–63; a Guest Professorship at the University of Frankfurt, Germany, 1979–80; and is an Adjunct Professor at Tsinghua University, Beijing, P.R. China, 1986–. He has served on many editorial boards and advisory committees.

# Structures of Nuclei Far From Stability

JOSEPH H. HAMILTON

## 1. Introduction

Studies of nuclei which are very far from the valley of beta stability are revealing an unexpected richness and diversity of nuclear shapes, structures, and decay modes which are transforming and challenging our understanding of the nucleus. Bergström (Be 66, Be 66b) drew major attention to the unique opportunities in this new field by organizing the first international conference on why and how to study nuclides far off the stability line (Fo 66). As one moves out of the valley of beta stability the energies for beta decay go from less than one MeV to over 10 MeV and the corresponding half-lives decrease down to milliseconds. Such studies required the development of major specialized techniques which employed the continuous production, separation by mass number, and isolation of specific nuclei for study. The experimental techniques are reviewed elsewhere (Ha 82, Ra 88).

Both the quantity and quality of the experimental data on the properties of nuclear-energy levels and nuclear-decay modes have increased tremendously over the last decade. Systematic studies of long isotonic and isotopic sequences of nuclei across major shell closures are providing stringent new tests of our theoretical understanding. To illustrate, Figure 1 traces the energies of the $2_1^+$ states in the Sr nuclei from one newly-discovered "deformed" shell gap at $N = 60$ for very large deformation across the well-known $N = 50$ "spherical" shell gap toward another newly-discovered "deformed" shell gap at $N = 38$ for similar, very large deformation. Not only are structures characteristic of all the individual nuclear models developed for particular mass regions, from harmonic vibrators to both oblate and prolate axially symmetric rotors on to triaxial rotors, being found in long isotopic chains, but many of these different shapes are being found even in one nucleus such as $^{185}$Au, where four different shape-coexisting bands are now identified at low energies (Zg 86). Models must be able to trace over long sequences of nuclides such changing (often rapidly changing) structures and to treat the multiple nuclear shapes seen in one to several nuclei in the chain. For example, the generalized collective model of the Frankfurt group (He 80,

JOSEPH H. HAMILTON • Department of Physics, Vanderbilt University, Nashville, Tennessee 37235.

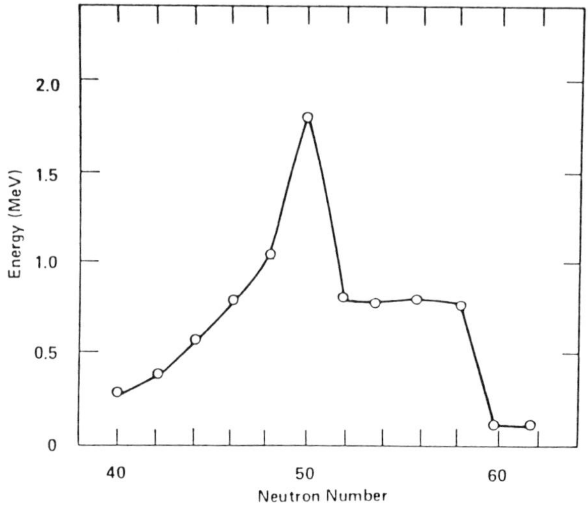


**Figure 1.** $2_1^+$ energies in $^{78-100}Sr$. Note smooth increase in deformation in going from the $N = 50$ spherical closed shell toward the new $N = 38$ deformed closed shell and sudden onset of deformation at the new $Z = 38$ and $N = 60$ deformed double closed shells. The suddenness of the onset is related to the spherical subshell closure at $N = 56$ reinforcing a spherical shape farther away from $N = 50$ than would have occurred otherwise.

Se 82), based on a description of collective states as quadrupole surface excitations, and the Lund–Oak Ridge version of calculating potential-energy surfaces (Be 86) along with the Interacting Boson Model (IBA) developed by Arima and Iachello (Ar 74, 81) with its various symmetries have been especially attractive because they can trace such changes. These two approaches are equivalent descriptions with the beauty of the generalized collective model related to its vivid physical pictures of the changing and multiple structures, as illustrated in Figure 2, and the beauty of IBA to its mathematical simplicity and elegance. More detailed models have been developed with remarkable

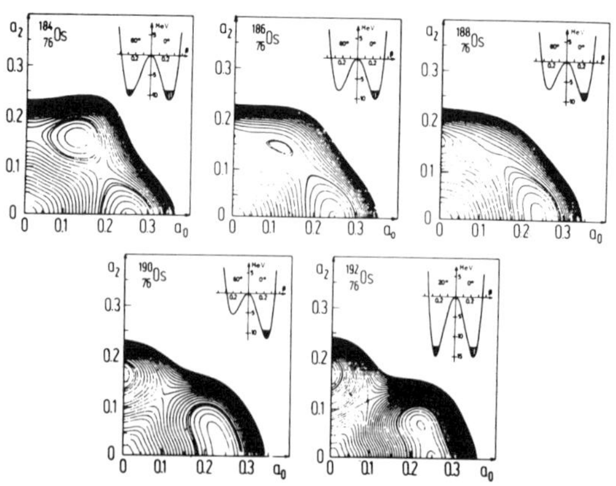

**Figure 2.** Potential-energy surfaces for a sequence of Os isotopes to illustrate the clear picture of the changing shapes one sees (He 80).

success to trace both the low-spin and high-spin structures, such as the Cranked Shell Model (Be 79, Be 79b) and Excited VAMPIR (Variational After Mean-Field Projections in Realistic Model Spaces, Sc 86). Finally, an exciting extension of searches for supersymmetry in collective nuclear structures is the recently developed Fermion dynamical-symmetry model (Wu 86, Wu 87).

In any review only selected highlights can be presented. Here, these will include: the coexistence of multiple shapes and structures associated with quite different deformations in the same nucleus at low excitations; unexpected, new deformed regions with the largest ground-state deformations observed; the importance of reinforcing of shell gaps for both neutrons and protons on the nuclear shape; new magic numbers now associated with shell gaps at nonzero deformations in addition to the magic spherical shell gap numbers; the role of intruder states from across the spherical shell gaps; and the possible existence of pairing free states at or near the ground state. These new results illustrate how studies of nuclei far from stability continue to give us significant, new insights which could not have been gained from those near stability.

More of the vast wealth of new nuclear-structure information and the breadth of the field opening up the discovery of new elements and new research in atomic, particle, and solid-state physics can be found in the published proceedings of the recent international conferences on the field in Nashville 1979 (Ha 80), Helsingør 1981 (Ha 81), Leuven 1984 (De 85), Chicago 1985 (Me 86), and Ontario 1987 (To 88), which was held after this review was completed. However, a few references from that conference have been added here. Some other details may be found in other recent reviews (Ha 85, Ha 85b, Ha 86, Ha 87).

## 2. Shapes and Structures of Nuclei: Views of a Decade Ago

To see how the new insights have significantly changed our views, let us look back to our understanding of the shapes and structures of nuclei a decade ago. In their review Baranger and Sorensen (Ba 69) classified nuclei into three types: "A) spherical, B) 'hard' deformed, and C) 'soft' transitional nuclei," and noted "spherical and hard deformed nuclei maintain their shapes through time," i.e., both these classes of nuclei have "permanent shapes."

The regions where these three types of nuclei are found can be seen by looking at Figure 3, which has been modified from their version to include new magic numbers for deformed shapes as discussed later. As they summarized, spherical, type "A" nuclei occurred around closed neutron and proton shells ($N$, $Z$ of 2, 8, 20, 28, . . . ) as calculated in the spherical-shell model developed independently by Mayer (Ma 50) and by Jensen and co-workers (Ha 50) and the hard deformed, type "B" nuclei, treated in the collective model of Bohr and Mottelson (Bo 53), in the regions well removed from the closed shells. The then known deformed regions were in the rare earths between $Z = 50$ and 82 and $N = 82$ and 126 and the actinides with $Z > 82$ and $N > 126$. The predicted new region of deformation far from stability, where $50 \leqslant N \leqslant 82$ and $50 \leqslant Z \leqslant 82$, is now well verified and is being extensively explored; for example, ${}^{124}_{58}Ce_{66}$ is well out into the region and has a measured ground-state deformation $\beta_2 = 0.31$ (Yi 86). In the regions between the deformed areas, shaded gray in Figure 3, and the spherical magic numbers, nuclei fell into type "C," the "soft" transitional nuclei which had no well-defined shape but varied through different shapes with small deformation $\leqslant 0.15$. They had no regular systematic level structure as had groups A and B. At that time, a nucleus had one of these three shapes, fixed for types A and B and soft for type C, and its shape determined the general features of its low-lying energy levels. We shall see how this picture has dramatically changed by looking at nuclei far from stability.

Earlier there had been isolated references to other possibilities. Hill and Wheeler (Hi 53) first suggested the possibility of nuclear shape isomers. Morinaga (Mo 56) included states with both deformed and spherical shapes to explain the first few, very high energy levels of double magic ${}^{16}_{8}O_8$. In a nonclosed-shell nucleus, ${}^{110}_{48}Cd_{62}$, Greiner (Gr 65) included nuclear-shape coexistence to explain its low-lying levels. Other theorists, Soloviev (So 66) and Kumar (Ku 75), extended the call for experimentalists to make concentrated efforts to find nuclei with low-lying excited levels with deformations quite different from their ground states. However, these isolated works received very little, if any, attention.

## 3. Shape Coexistence in Tl–Hg–Au–Pt Nuclei

### 3.1. Even–Even Nuclei

The discovery of the coexistence of complete bands of energy levels built on quite different deformations in ${}^{72}_{34}Se_{38}$ and ${}^{184,186,188}_{80}Hg_{104-108}$ [see reviews of Hamilton (Ha 76, Ha 79)] in two quite different mass regions confirmed the long-standing theoretical expectations of nuclear shape coexistence. These nuclei helped to completely change our picture that each nucleus had only one permanent shape (either type "A," "B," or "C" as described above) to characterize its ground and low-lying states. In striking contrast, nuclear-shape coexistence with various types of shapes is now found in the low-energy levels in many nuclei spread throughout the periodic table, including even "spherical" closed-shell nuclei with $Z = 50$ (Sn) and 82 (Pb). Additional examples to those presented here are found in the literature (Ha 81b, Ha 85, Ha 85b, He 83) and the Helsingør, Leuven, Chicago, and Ontario Conference proceedings (Ha 81, De 85, Me 86, To 88). Thus nuclear-shape coexistence, instead of being a rarity, is an important, widespread phenomenon which is providing many new challenges to our theoretical understanding.

The light-mass mercury isotopes, far from stability ($A = 184$–$188$), provide classic examples of the coexistence of overlapping bands of levels built on well-deformed ($|\beta| \simeq 0.25$) and near-spherical ($|\beta| \sim 0.12$) shapes as shown in Figure 4 [see (Ha 79) for more details and references]. Theoretical calculations indicate that the large deformation is associated with a prolate shape and the small deformation with an oblate shape [for example see (Be 86)]. These shapes are assumed in this paper. More direct evidence of these prolate and oblate assignments is important. As $N$ decreases from ${}^{198}_{80}Hg_{118}$ down to ${}^{190}Hg$, the low-lying levels are remarkably constant and characteristic of near-spherical nuclei (see Figure 5). Then suddenly there appears at low energy in ${}^{184,186,188}_{80}Hg$ a full band of states characteristic of a well-deformed shape which coexist with and overlap the near-spherical states seen in all the even–even mercury isotopes (as also seen in Figure 5). The two bands have high purity of shape above and below the mixed states where the bands cross, as established by the absence of crossing transitions between the bands except for the mixed states. The discovery of shape coexistence in these nuclei was surprising for two reasons. First, the Hg nuclei (see Figure 3) are very near the strong $Z = 82$ spherical closed shell where spherical and

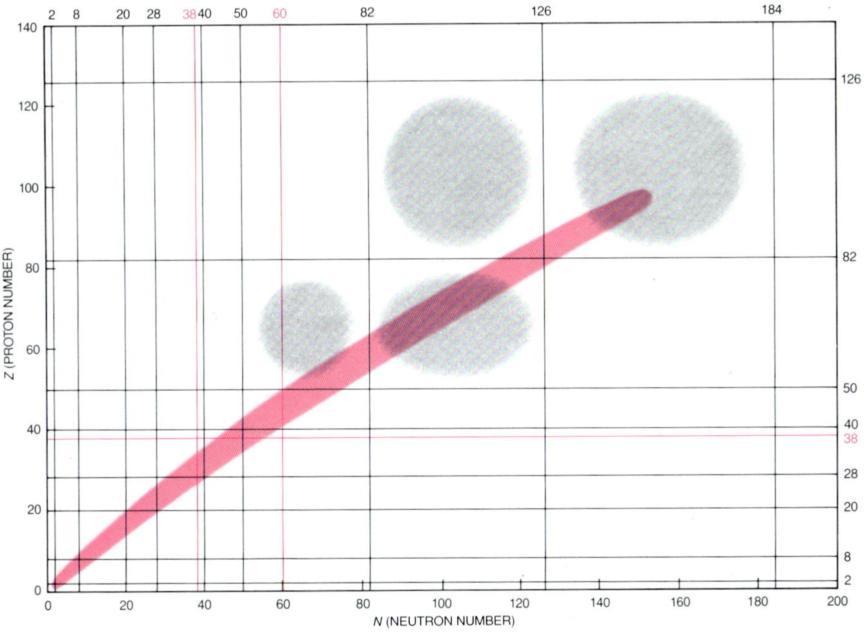


**Figure 3.** Chart of nuclides as a function of *N* and *Z* with spherical closed-shell magic numbers as given by Baranger and Sorenson (Ba 69) shown by vertical and horizontal lines and regions of deformed nuclei by ovals. The new magic numbers 38 and 60 for deformed shapes, as discussed in Section 4 in this chapter, are shown in color. The magenta band represents the stable nuclei.

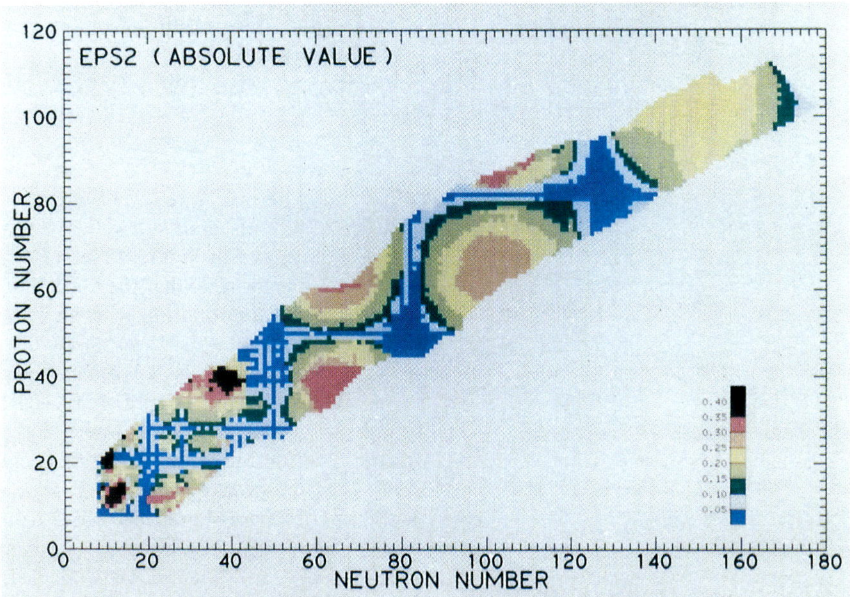


**Figure 35.** A *Z–N* chart of the nuclides where the colors indicate the magnitude of the $\epsilon_2$ deformations, as given in the color code (Mö 82, Be 84). For example, the light blue areas highlight the spherical magic numbers.

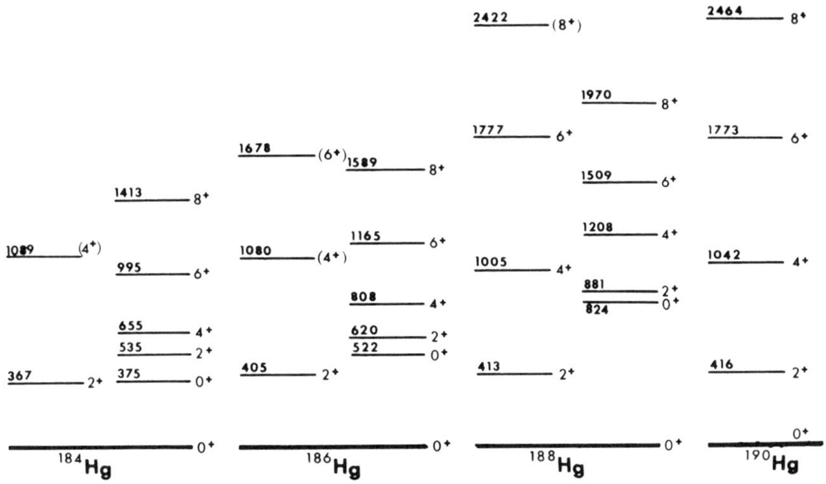


**Figure 4.** Coexistence of spherical and deformed bands in $^{184-188}$Hg [(Ha 79) and references cited therein].

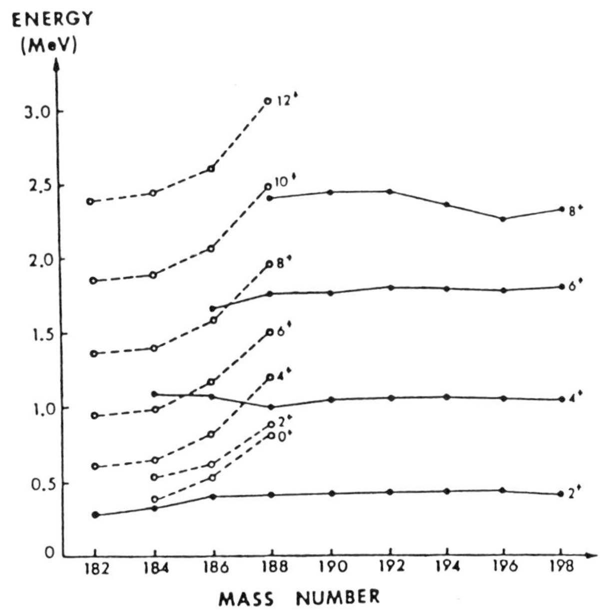


**Figure 5.** Energies of spherical and deformed states in $^{182}$Hg (Ma 84) compared with $^{184-198}$Hg (Ha 79).

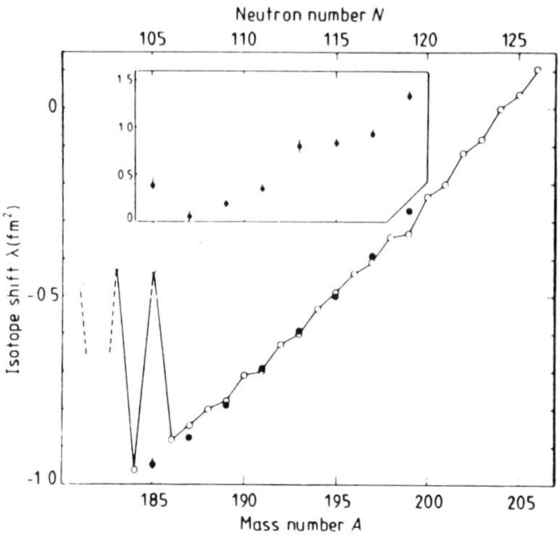


**Figure 6.** Changes of charge radii of mercury isotopes relative to $^{204}$Hg. The size parameter is defined as $\lambda = \delta\langle r^2\rangle - 1.1\times10^{-3}\delta\langle r^4\rangle + \text{higher terms}$. Open circles indicate ground states and full circles isomers. With the exception of $^{185}$Hg$^m$ the experimental error is smaller than the diameter of the circle. The line connects ground states of neighboring isotopes. The insert shows the even–odd staggering parameter for the $13/2^+$ isomers, defined as $\gamma = 2(\langle r^2\rangle_{N+1} - \langle r^2\rangle_N)/(\langle r^2\rangle_{N+2} - \langle r^2\rangle_N)$, for even neutron number $N$ (Ha 79b).

near-spherical structures were expected to dominate, as had been found experimentally earlier for $^{192-204}_{80}$Hg. Second, even if there were two minima in the potential-energy surfaces, one at near-spherical deformation and one at large deformation, we would not have expected at all that the barrier could be so high between these two minima that complete bands built on two different shapes each with high purity would be found.

The first evidence that something unusual was happening in this region came from the pioneering optical pumping work of the Mainz group (Bo 72) where, in the odd-$A$ Hg nuclei, a sudden large change in the isotope shift between $^{187}_{80}$Hg and $^{185}_{80}$Hg had been discovered. One interpretation of this shift was that it represented a sudden large change in ground-state deformation between $^{187}$Hg and $^{185}$Hg. Subsequent work has confirmed this interpretation. Their early (Bo 72) and more recent data (Kü 77, Da 79) are shown in Figure 6. However, the first in-beam studies of $^{186,184}_{80}$Hg (Pr 73, Ru 73) revealed that their ground states were still near-spherical. However, these studies and the first UNISOR studies of $^{188}_{80}$Hg (Ha 74, 75) revealed that in all three cases there was a shift around spins of $4^+$ to $6^+$ from near-spherical ground states to well-deformed structures. At UNISOR, the new isotopes $^{188,186,184}_{81}$Tl were discovered and used to study the levels of $^{188-184}_{80}$Hg (Ha 75, Co 76). Subsequently and independently the $^{188,186}$Hg results were confirmed in France (Bo 76, Bé 77). The decay studies went beyond the in-beam studies, where shifts from a near-spherical to well-deformed structures were established to identify the low spin states in the deformed band below the crossing point and to identify the higher spin states in the near-spherical ground bands above the crossing and to determine the purity of the states above and below the crossings.

In Figure 4 the energy of the $0_2^+$ deformed bandhead is seen to drop rapidly from 825 keV in $^{188}_{80}$Hg (Ha 75) to 372 keV in $^{184}_{80}$Hg (Co 76). This raised the question of what happens for lighter $N$. Do the ground states become deformed? Since in $^{184}_{80}$Hg$_{104}$ the neutrons are exactly midway between 82 and 126, one could expect symmetry around $N = 104$ and then the deformed band should rise in energy for $N < 104$. The yrast levels in $^{182}_{80}$Hg were established by Ma and co-workers (Ma 84). As seen in Figure 5, the $4^+$ to $12^+$ members of the deformed band continue to drop in energy relative to the ground-state near-spherical band (Ma 84). We will return to this question later.

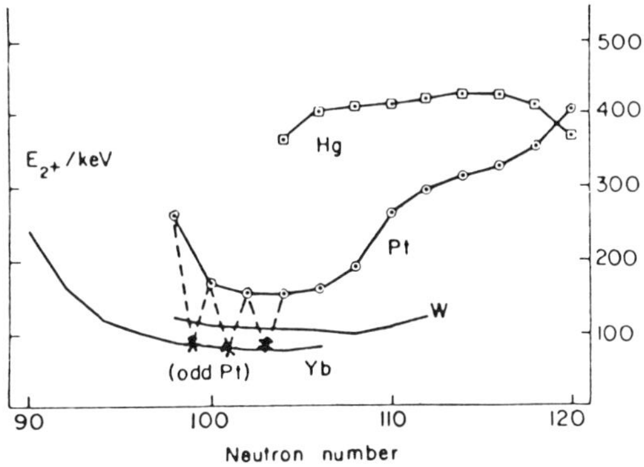


**Figure 7.** Systematics of $2_1^+$ energies for even-even Yb ($Z=70$), W ($Z=74$), Pt ($Z=78$), and Hg ($Z=80$) nuclei. The inertial parameters for the $1/2^-[521]$ bands in $^{177,179,181}$Pt were deduced and plotted as $6\hbar^2/2\mathcal{J}$ to show the odd-even staggering of the moment of inertia (Ha 79b).

The even-even Pt nuclei also are observed to undergo a variety of different shape changes with decreasing $N$. It has long been known that there is a shift from a small oblate deformed shape for the ground state of $^{192}_{78}$Pt, through a $\gamma$-soft ground state shape in $^{188}_{78}$Pt, to a well-deformed prolate shape for $^{186}_{78}$Pt. The increase in deformation is seen in the decrease in the $2_1^+$ energies for the even-even and analogous transitions in the odd-$A$ Pt isotopes, as shown in Figure 7 (Ha 78, Ha 79b). Wood (Wo 81) has reviewed the evidence for shape coexisting structures in $^{182-186}_{78}\mathrm{Pt}_{104-108}$ where, in contrast to the analogous $N = 106, 108$, $^{186,188}_{80}$Hg nuclei, the Pt ground states are well deformed (presumably prolate) and the excited bands near-spherical (presumably small oblate). The prolate well-deformed ground-state structures are present down to $^{178}_{78}$Pt (Dr 86).

Recently, evidence for a shift from a well-deformed to a small ground-state deformation and for shape coexistence was reported in $^{176}_{78}$Pt (Dr 86). The shift in ground-state deformation from large to small deformation between $^{178}$Pt and $^{176}$Pt and shape coexistence in $^{176}$Pt can be seen in Figure 8. The $2_1^+$ energy in $^{176}_{78}\mathrm{Pt}_{98}$ is much higher than the $2_1^+$ energy in $^{178}$Pt and than the $2_1^+$ energies in the well-deformed $^{172}_{74}\mathrm{W}_{98}$ and $^{174}_{76}\mathrm{Os}_{98}$, which are $N = 98$ isotones. However, above $4^+$ the yrast states in $^{176}_{78}$Pt, $^{174}_{76}$Os, and $^{172}_{74}$W all have essentially identical large deformation (Dr 78, Dr 86, Du 82). Thus, there is a clear shift from a small deformation for the ground state to large deformation above $4^+$ in $^{176}$Pt. In $^{178}$Pt, the $2_1^+$ energy is slightly shifted upward compared to its well-deformed $N = 100$ neighbors.

The experimental data seen in Figure 8 are very nicely reproduced by Potential Energy Surface (PES) calculations (Be 86) for $^{176-180}_{78}$Pt, as seen in Figure 9. The $^{180}$Pt PES contains a single minimum at a sizable prolate deformation, $\beta_2 \approx 0.24$. Thus, it is expected to have a highly collective, rotational ground-state band as is observed (Ri 85, Ko 86). The $^{178}$Pt PES is significantly different from the $^{180}$Pt PES. The lowest energy still occurs at a large prolate deformation almost identical to that in $^{180}$Pt, but the minimum is flat and extends far out into the $\gamma$ plane. As soon as the nucleus starts to rotate, however, the minimum is well localized at a large prolate shape because the rotation favors shapes with a large moment of inertia. The ground band of $^{178}$Pt is therefore expected to be nearly identical to that in $^{180}$Pt, again in agreement with experiment (Dr 86).

The $^{176}$Pt PES is quite different from those of $^{178,180}$Pt (Be 86). As seen in Figure 9, its ground state is triaxial. However, about 0.25 MeV above the triaxial ground state,

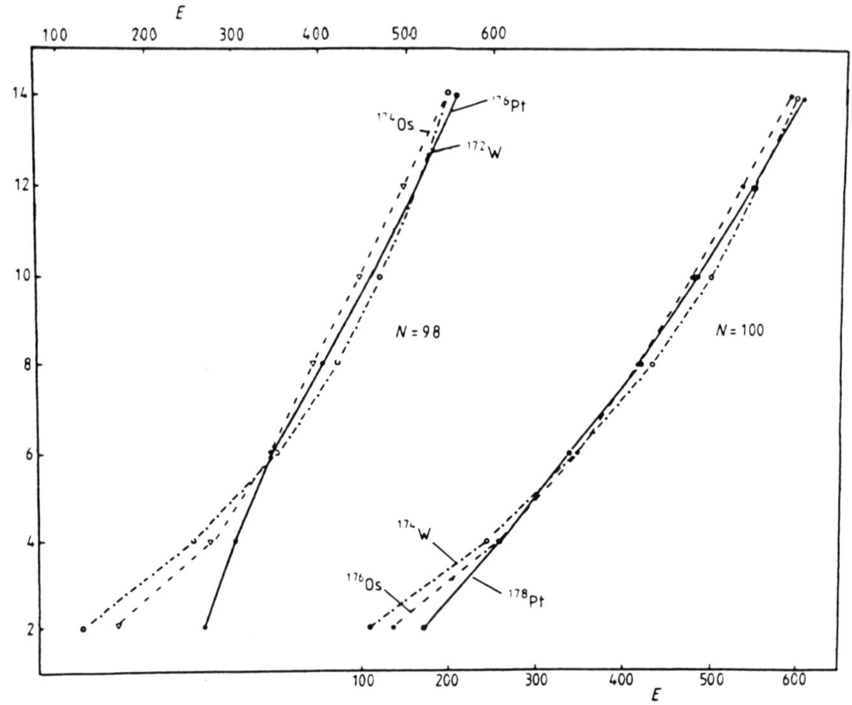


**Figure 8.** Energies (in keV) of transitions depopulating the yrast states as a function of spin for $^{172,174}_{74}$W, $^{174,176}_{76}$Os, and $^{176,178}_{78}$Pt (Be 86).

the PES has a flat region at a larger prolate deformation, $\beta_2 \approx 0.23$, which is almost as large as that in $^{178}$Pt and $^{180}$Pt. The large prolate minimum ($\beta_2 \approx 0.23$) is energetically formed in $^{176}$Pt when $I > 4\hbar$.

The small upward shift of the $^{178}$Pt $2^+$ energy seen in Figure 8 is also expected from the PES calculations. The extended region of minimal energy in the $^{178}$Pt PES contains two structures: one responsible for the prolate minimum in $^{180}$Pt and another corresponding to the triaxial minimum in $^{176}$Pt. But in $^{178}$Pt the $\beta_2 \approx 0.15$, $\gamma = 30°$ one (which is approximately the ground state of $^{176}$Pt) is only 55 keV higher in energy than the one at $\beta_2 \approx 0.24$, $\gamma = 0°$ (the ground state of $^{178,180}$Pt). The band-mixing calculations of Dracoulis and co-workers (Dr 86) likewise indicate that the unperturbed $0^+$ bandheads are nearly degenerate.

In summary, the shape changes and shape coexistence seen in $^{176,178}$Pt (Dr 86) and the earlier known prolate yrast band of $^{180}$Pt are in very nice agreement with the theoretical PES calculations (Be 86). However, while shape coexistence in $^{182-188}_{80}$Hg and in $^{182-186}_{78}$Pt is understood as resulting from competition between small oblate and large prolate axially-symmetric deformed shapes, in $^{178,176}$Pt the competition is predicted to be between triaxial and prolate shapes.

Bengtsson and co-workers (Be 86b) have completed extensive calculations of the potential-energy surfaces for the even–even Pt and Hg isotopes and their neighbors.

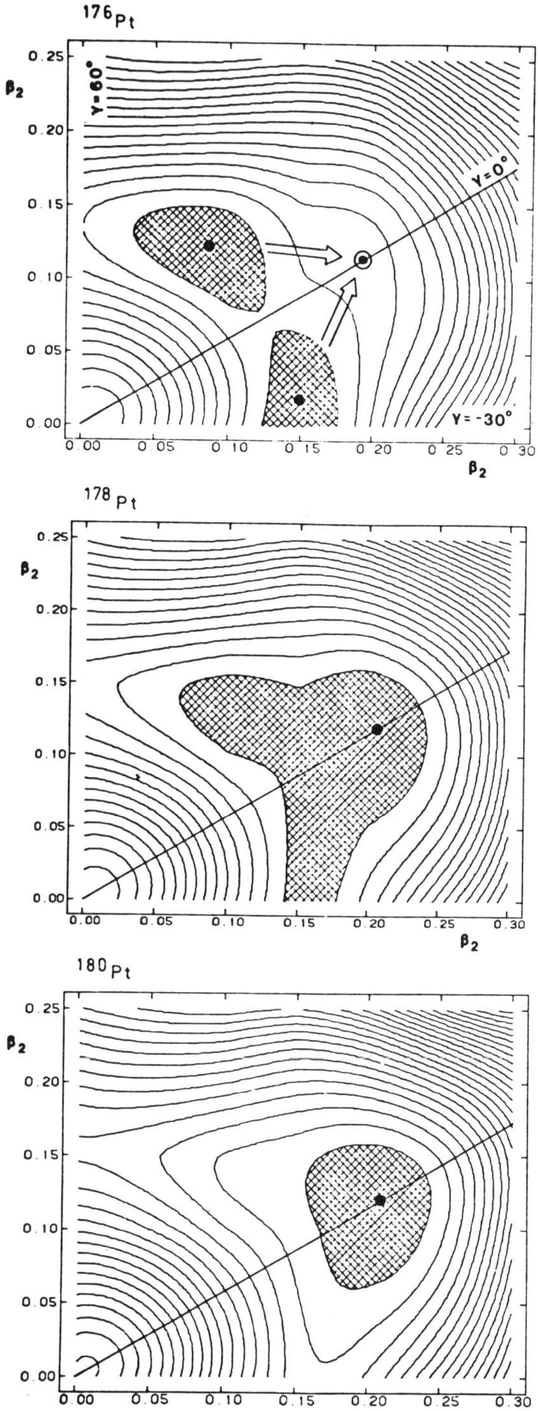


**Figure 9.** Potential-energy surfaces in the $(\beta_2,\gamma)$ plane for $I^\pi = 0^+$. The energy is minimized with respect to the hexadecapole deformation $\beta_4$. The energy difference between the contour lines is 0.1 MeV. The diagonal line is the $\gamma = 0$ prolate axis (Be 86).

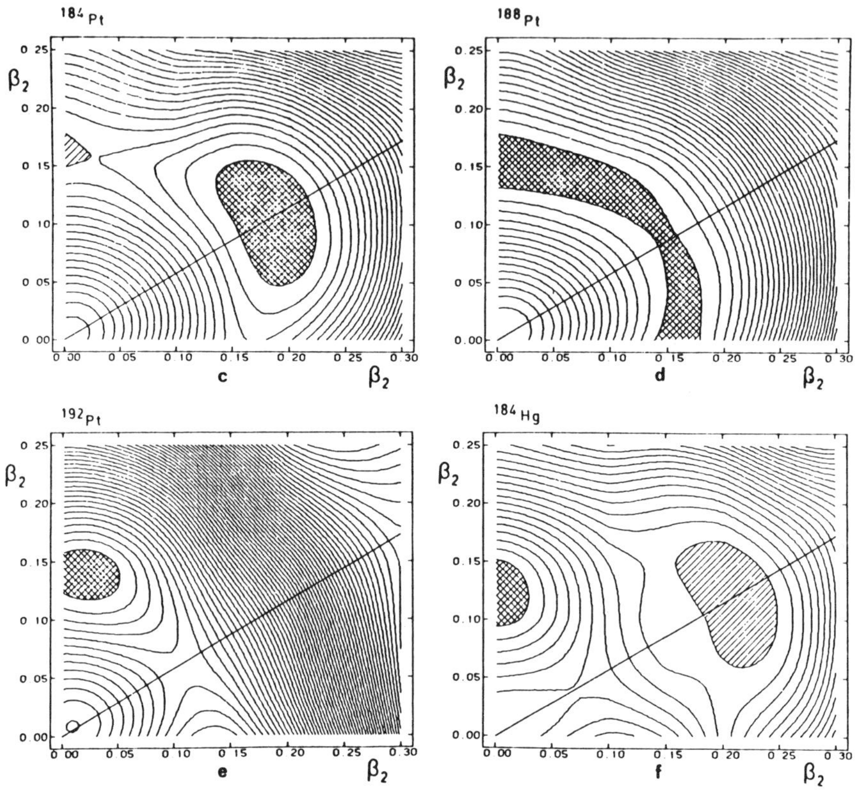


**Figure 10.** Potential-energy surfaces in the $(\beta_2, \gamma)$ plane for $I^\pi = 0^+$. The energy is minimized with respect to the hexadecapole deformation $\beta_4$. The energy difference between the contour lines is 0.1 MeV. The diagonal line is the $\gamma = 0$ prolate axis (Be 86b).

Calculations were made with both the Woods–Saxon (Na 85) and Modified Harmonic Oscillator (Nilsson) potential (Be 85, Zh 86). In both cases the Strutinsky renormalization plus BCS pairing approach was used, and the liquid-drop model was employed in calculating the macroscopic energy.

Some further results of their extensive calculations are shown in Figure 10. The $^{182,186}_{78}$Pt, $^{186,188}_{80}$Hg PES are like those of $^{184}$Pt, $^{184}$Hg shown in Figure 10. It is seen how by comparing the PES from $^{176}$Pt to $^{192}$Pt one has a vivid picture of the changing nuclear landscape from one type of shape coexistence in $^{176}$Pt, to a single prolate minimum in $^{180}$Pt, to a different shape coexistence in $^{182-186}$Pt, to $\gamma$ unstable $^{188}$Pt, and back to a single, now oblate, minimum in $^{192}$Pt. A summary of the calculations is as follows: a *single prolate minimum* $(0.20 \leq \beta_2 \leq 0.25)$ for W with $N \leq 114$, $^{174-178}_{76}$Os, $^{186,188}_{76}$Os, and $^{180}_{78}$Pt; a *single oblate minimum* $(-0.16 \leq \beta_2 \leq -0.10)$ for $_{80}$Hg $(N \geq 110)$, $_{78}$Pt $(N \geq 114)$, $_{76}$Os $(N \geq 118)$ and $_{74}$W $(N \geq 118)$; a *single triaxial minimum* for $^{176}$Pt. Two coexisting minima at $\leq$1.5 MeV in PES occur as follows: a *prolate ground state* $(0.20 \leq \beta_2 \leq 0.25)$ and an *excited secondary minima* at oblate shape $(\beta_2 \approx -0.16)$ in $^{180-184}_{76}$Os and $^{182-186}_{78}$Pt, however, the excited minimum is usually very shallow; an *oblate*

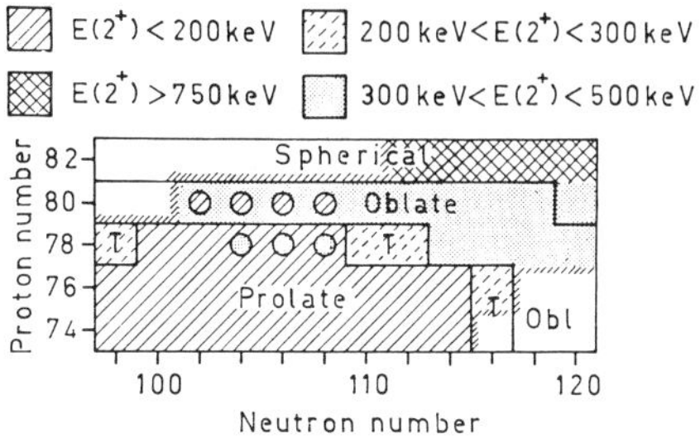


**Figure 11.** Deformation diagram in the $(Z, N)$-plane. The solid lines separate nuclei, which in the calculations based on the Woods-Saxon model are predicted to have spherical, oblate, prolate, or triaxial (T) deformation in the ground state. The various shadings show the experimental energy of the lowest $2^+$ state. The rings indicate nuclei for which the experimental data support the assumption of an oblate-prolate shape coexistence (Be 86b).

*ground state* $(-0.13 \leqslant \beta_2 \leqslant -0.10)$ and an *excited secondary minimum at prolate shape* $(0.20 \leqslant \beta_2 \leqslant 0.25)$ in $^{178\text{-}188}_{80}$Hg. In $^{188}_{78}$Pt and $^{190}_{74}$W, the PES are very $\gamma$ soft with approximately the same energy for oblate and prolate shapes. The changing nuclear landscape as predicted in their calculations (Be 86b) also is illustrated in Figure 11.

The two minima in $^{182\text{-}186}_{78}$Pt and $^{178\text{-}188}_{80}$Hg have structurally very different configurations: the prolate configuration corresponds to an excitation of at least two protons to the $h_{9/2}$ shell, which is not occupied in the oblate configuration. The importance of this intruder orbit has been proposed elsewhere (Wo 81, Zg 86) and discussed below. Furthermore, the prolate configuration contains two or four (depending on the neutron number) more $i_{13/2}$ neutrons than the oblate one.

The energy difference between the oblate and prolate minimum is shown in Figure 12 for the Hg isotopes. The theoretical prolate-oblate energy difference calculated with the Woods-Saxon potential is lowest at $N = 102$, 104 and increases sharply for higher and lower $N$. The experimental data clearly reproduce the increase for $N > 104$. However, the small continued drop in the energy separation in $^{182}$Hg goes against the predicted trend. Recently Simon and co-workers (Si 86) have assigned from NaI data three gamma rays to $^{180}$Hg by using the velocity filter SHIP to select mass 180 recoils in the reaction $^{90}$Zr + $^{90}$Zr as a coincidence gate. Based on systematics and relative $\gamma$-ray intensities, they suggested an ordering 194 $(1.0 \pm 0.3)$, 282 $(0.9 \pm 0.3)$, and 342 $(0.8 \pm 0.3)$ keV as transitions from the $2^+$, $4^+$, and $6^+$ yrast states where the relative

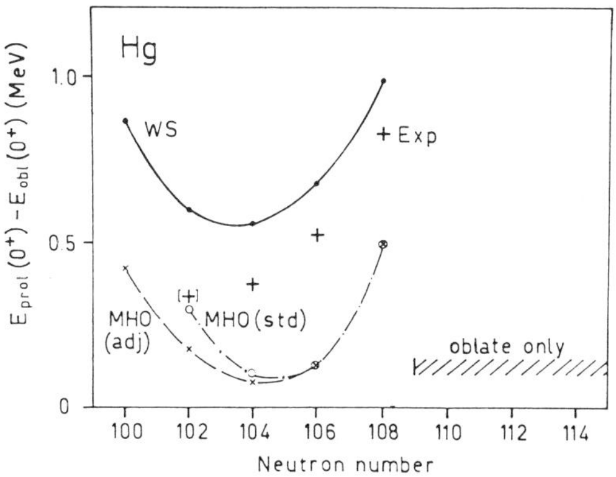


**Figure 12.** Calculated prolate-oblate energy difference obtained with the Woods-Saxon potential (WS) and with two versions of the Nilsson potential [MHO, see (Be 86b) for explanation]. The crosses show the experimental energy difference between the two lowest $0^+$ states. No corrections for a possible interaction between these states have been made. The Woods-Saxon results were obtained from two-dimensional $(\beta_2, \beta_4)$ calculations where an axial shape was assumed in order to minimize the interpolation error (expected to be of the order of 100 keV in the calculations with $\gamma \neq 0$).

$\gamma$-ray intensities are in parentheses. If correct, the prolate-oblate minimum energy separation is still dropping in $^{180}$Hg in sharp contrast to all calculations. However, it is possible that the ordering could be reversed since the $\gamma$-ray intensities have large errors. The prolate-oblate energy separation in $^{180}$Hg remains a crucial test for the theoretical calculations.

Finally, as a further example of the importance of the coexistence of states with both small (presumably oblate) and large (presumably prolate) deformations with decreasing energy separation, let us look at the differences they produce in the structures seen in the spin range of $10^+$ to $22^+$ in $^{190-182}$Hg as seen in Figure 13. In $^{190}$Hg and the heavier isotopes it is well known that the ground band is interrupted at $10^+$ by a two-quasiparticle $(i_{13/2})^2$ configuration. This same structure, $(i_{13/2})^2$ coupled to the oblate ground state, is found to coexist at similar spins in $^{188}$Hg with the prolate deformed band as seen in Figure 13. However, in $^{186,184,182}$Hg bands built on this coupling of the $(i_{13/2})^2$ two-quasiparticle to the small oblate core are not observed. As pointed out by Ma and co-workers (Ma 86), in $^{184,186}$Hg the lowest band crossings are from couplings of the $\nu i_{13/2}$ and $\pi h_{9/2}$ quasiparticles to the prolate structure, not the oblate structure. Furthermore, as the prolate band drops lower in $^{184,182}$Hg, no evidence has been found for any reasonably strong competing side bands (Ma 86, Zh 86b) as seen in $^{186}$Hg. Figure 13 clearly shows how the coexisting structures and the changes in their relative energies strongly influence the band structures in the spin range of $10^+$ to $22^+$. These results provide stringent tests for different models which include rotational alignment of pairs of particles at higher spins.

### 3.2. Odd-*A* Nuclei

Studies of the odd-*A* Tl, Hg, Au, and Pt nuclei far off stability have been offering unique insights into the structures of nuclei as seen in earlier and recent reviews, such as (Zg 80, Be 80, He 83, Zg 86). These nuclei offer significant new, and indeed presently unique, opportunities to test particle-core coupling models. The uniqueness is seen in both Tl and Au nuclei where, in Tl, the particle states can couple to both the near-spherical and well-deformed core states observed in $^{182-188}_{80}$Hg while the hole states can couple to the spherical lead ground states and, for $A \leq 196$, to coexisting more deformed excited structures. In the odd Au nuclei, as we shall see, there are identified particle states coupled to the two coexisting bands with different deformation in the Pt cores and hole states coupled to the different coexisting bands with different deformation in the Hg cores—bands built on four different coexisting structures in the same nucleus at low, overlapping energies (Zg 86). Thus, one has in the same nucleus now the unique opportunity to test how particles and holes couple to quite different deformations and structures.

An important feature of the odd-*A* Pt and Au nuclei in this region is the appearance of high spin states which come quite low in energy when the neutrons are mid-shell ($N = 104$) between 82 and 126 and which may be associated with shell model states which intrude from above the $Z = 82$ shell gap. In Tl, as shown in Figure 14 [see (Zg 80) for details], a $9/2^-$ state is observed to decrease in energy to near the ground state and then in UNISOR work (To 76, Sc 77) found to rise again near $N = 104$. The nature of the $9/2^-$ isomeric states first found in $^{193-199}_{81}$Tl come from in-beam data (Ne 70, 74) which indicated the $9/2^-$ isomers occurred when the odd proton occupies the $h_{9/2}$ orbital. This $h_{9/2}$ state is called an "intruder" state because it must cross the

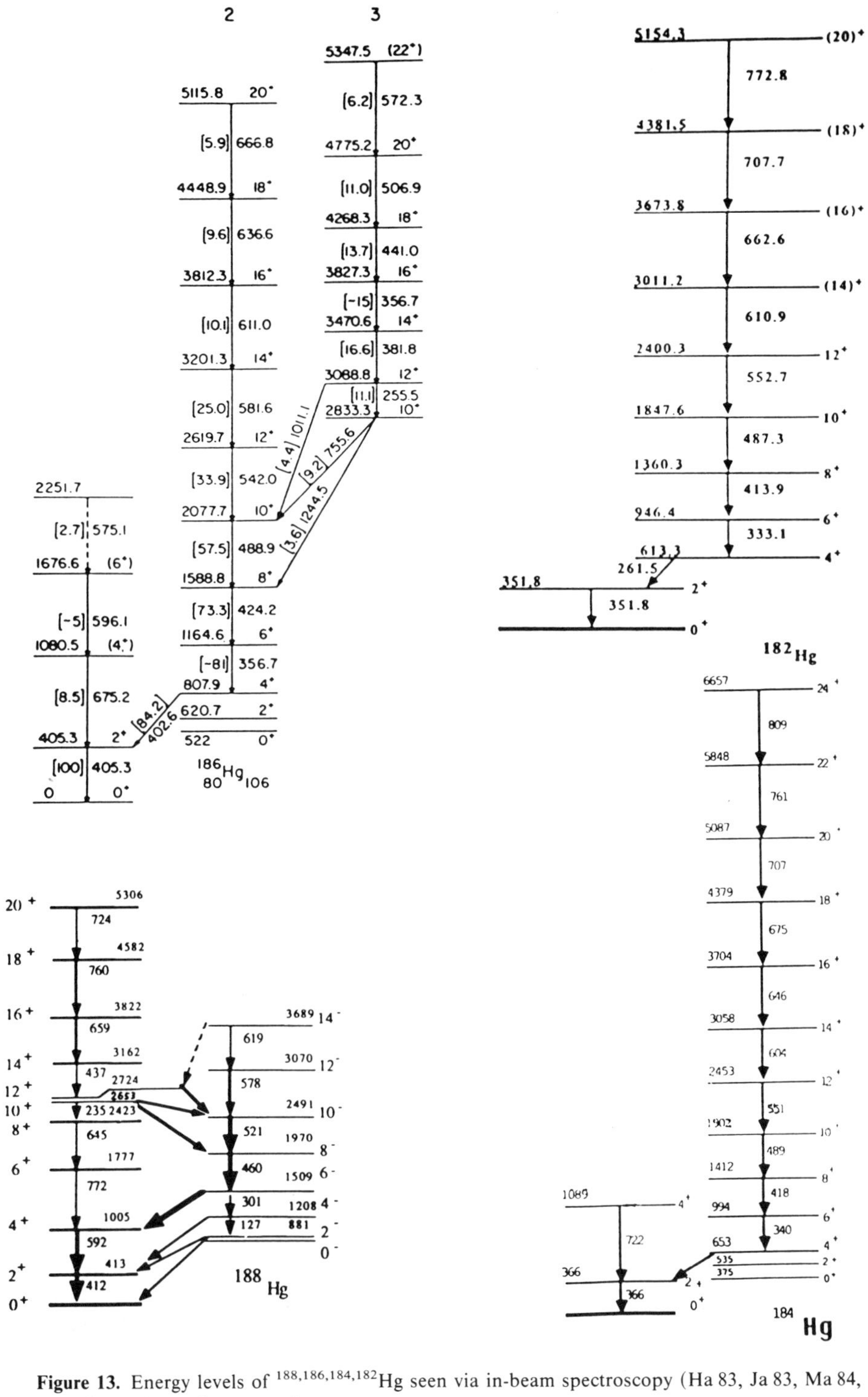


**Figure 13.** Energy levels of $^{188,186,184,182}$Hg seen via in-beam spectroscopy (Ha 83, Ja 83, Ma 84, Ma 86) are shown for comparison.

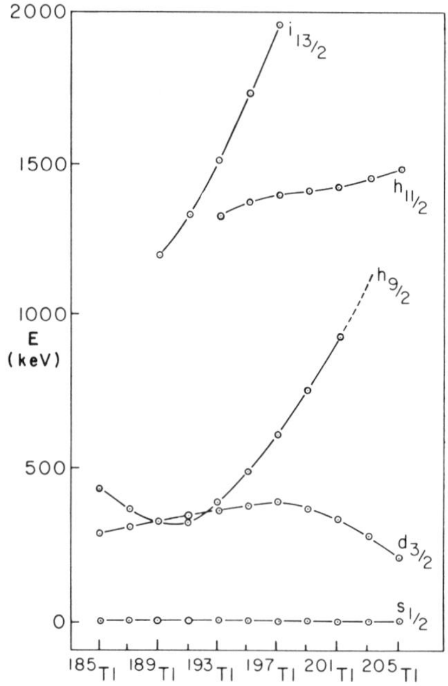


**Figure 14.** Shell model states in $^{185-205}$Tl. The $s_{1/2}$, $d_{3/2}$, and $h_{11/2}$ excitations are proton hole states coupled to the even–even Pb cores. The $h_{9/2}$ and $i_{13/2}$ excitations are proton particle states coupled to the even–even Hg cores. Note the unexpected sharp drop in energy with decreasing $A$ for the last two (Zg 80).

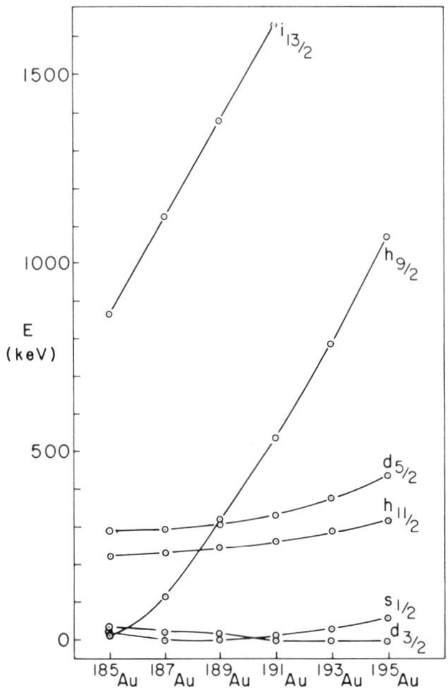


**Figure 15.** Shell model states in $^{185-195}$Au. The $s_{1/2}$, $d_{3/2}$, $d_{5/2}$, and $h_{11/2}$ excitations are proton hole states coupled to even–even Hg cores. The $h_{9/2}$ and $i_{13/2}$ excitations are proton particle states coupled to even–even Pt cores. Note the unexpected sharp drop in the energies of these latter two states with decrease in $A$ (Zg 80).

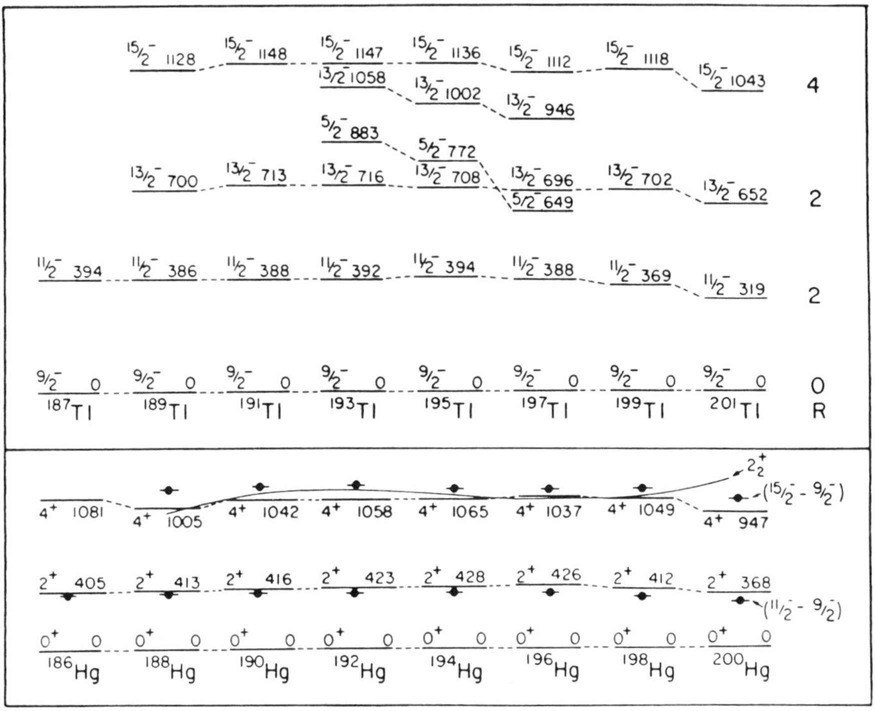

Figure 16. Systematics of the $h_{9/2}$ band structure in the odd-*A* Tl isotopes, together with excitations in the even–even Hg isotopes. A comparison of the $11/2^-$ and $15/2^-$ energies relative to the $2^+$ and $4^+$ core states is also shown in the lower part of the figure (full circles). The $2_2^+$ energy is indicated by a continuous light line. The $9/2^-$ energy was normalized to zero (Zg 80).

shell-model energy gap at $Z = 82$ to appear on the "wrong side" of the shell gap. Similar changes in energy are observed for $h_{9/2}$ and $i_{13/2}$ states in the odd-Au isotopes as shown in Figure 15 coming from on-line studies of $^{198-189}_{79}$Au at ISOLDE (Be 75), and confirmed and extended to lighter masses at UNISOR (Wo 76, Ka 78, Gr 78, Zg 80, Zg 81).

As shown in Figure 16, earlier studies at UNISOR extended the systematics of the bands built on the $h_{9/2}$ states down to $^{187}$Tl. Note in Figure 16 that there is a remarkable constancy of the moment of inertia as seen by the nearly constant $11/2^-$–$9/2^-$ energy separation. The interpretation is that the $h_{9/2}$ particle is strongly coupled to the slightly (presumably) oblate Hg cores. Also note in the lower part of Figure 16 how the $11/2^-$–$9/2^-$ and $15/2^-$–$9/2^-$ energy differences in $^{187-201}_{81}$Tl very closely track the $2^+$ and $4^+$ energies in their $^{186-200}_{80}$Hg cores where the ground-state deformations are known to be small and constant from lifetime measurements. So, while the rapid drop in the $h_{9/2}$ energy could be associated with increasing deformation as *N* decreases, the role of deformation was doubted based on the idea that there exists a simple correspondence between deformation and the rotational moment of inertia. However, recently the hyperfine structure and isotope shifts have been measured for $^{189-193}$Tl at UNISOR by means of collinear atomic bean laser spectroscopy (Bo 85). These $^{189-193}$Tl data were

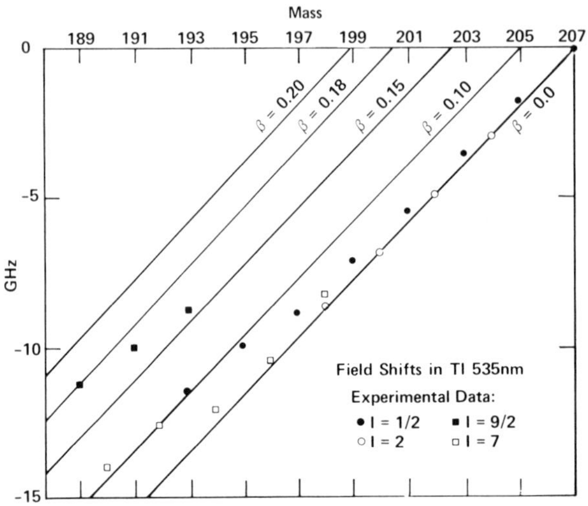


**Figure 17.** Field shifts measured for Tl isotopes. The lines show the expected field shifts for the deformations indicated from Bounds and co-workers (Bo 85).

taken with only $10^5$ separated atoms/s in the chamber. The results are shown in Figure 17, where it is clearly seen that the deformations for the $9/2^-$ states are much larger than for the $1/2^+$ ground states (shape coexistence) and that the deformation of the $9/2^-$ states increases with decreasing $N$ (Bo 85). The oblate character of the $9/2^-$, $K = 9/2$ states also is found in their work. This unique case, where the constant rotational energies first masked the increasing deformation with decreasing $N$, can be understood as a chance balance between changing deformation and neutron pairing. The theoretical calculations, see Figure 18 (Bo 85), show that as $N$ decreases, both the neutron-pairing correlations and the deformation increase but with opposite and compensating effects on the moment of inertia (Figure 18c). Note how well the calculations reproduce the constant rotational energies, increasing deformation, and the general features of the drop then rise in the $9/2^-$ energies (Figure 18).

While the $h_{11/2}$ shell model states lie quite low in the Au nuclei discussed below, they appear considerably higher in Tl and were not known prior to the UNISOR work (Co 81). The structure built on this excitation is quite different from that of the $h_{9/2}$ excitation and, as expected in the core-particle picture, indicates weak coupling to the nearly spherical Pb cores. The large differences in structure of the bands built on the $h_{9/2}$ and $h_{11/2}$ states and the rather complete separability of their structures (no interband transitions) demonstrate the importance of the core in determining the structure of these states. It is, of course, the difference in the core-particle structures of these two excitations that retards the spin-allowed transitions.

Now let us look at the odd-$A$ gold isotopes which presently provide us with the richest examples of low-lying shape coexisting structures (Zg 86, Zg 86b, La 86). Several important experimental procedures were necessary to be followed in on-line work in order to mine the new nuggets and to understand what the nuggets represented. First, it is absolutely essential that to understand an odd-$A$ nucleus one must systematically

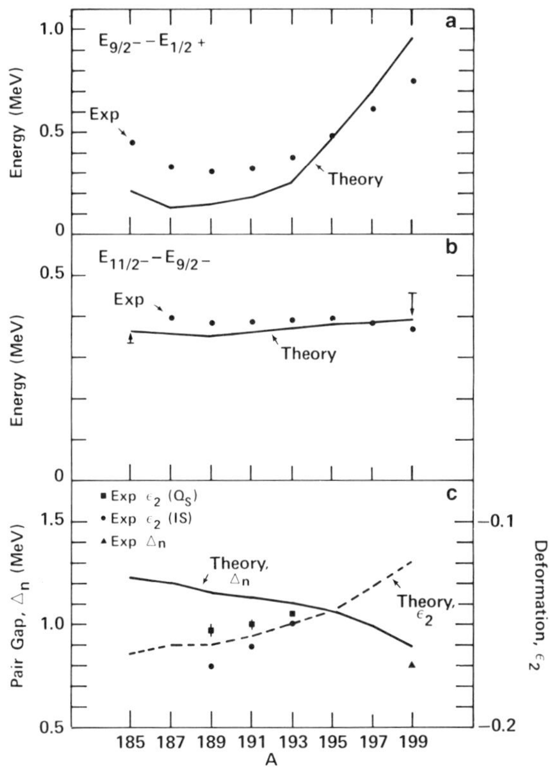


**Figure 18.** Comparisons of experiment and theory as a function of $A$ for light Tl isotopes for: (a) the $E_{9/2^-}$-$E_{1/2^+}$ energy difference in MeV; (b) the $E_{11/2^-}$-$E_{9/2^-}$ energy difference in the band; (c) the neutron pairing gap on the left and $\varepsilon_2$ deformation on the right (Bo 85).

investigate several odd-$A$ nuclei and must systematically investigate the even-even cores of higher and lower $Z$. Next, the great complexity of the nuclear level schemes with 100-300 gamma rays per isomer and the presence of high and low spin isomers for each $Z$ and more than one $Z$ present even in isotope-separated samples requires that one must obtain as many different types of data as possible from each isomer and with extremely high quality (resolution) and quantity (very good statistics). Only data of very high statistical quality with good resolution will allow one to select gates moving one channel at a time over peaks to unravel very close-lying doublets, which may hide the choice nuggets, and to obtain conversion coefficients for even weak transitions and obtain them from coincidence gates not singles spectra. Finally, separate studies of the high and low spin isomers by selective production with heavy ions are likewise essential.

The $h_{9/2}$ bands in the odd-$A$ Au isotopes exhibit behavior quite different from that of the Tl isotopes. Note again, as seen in Figure 16, the strikingly constant in energy, $\Delta I = 1$, sequences seen in the Tl isotopes where the bands are from the strong coupling of the $h_{9/2}$ particle to the slightly oblate Hg core. In contrast, as seen in Figure

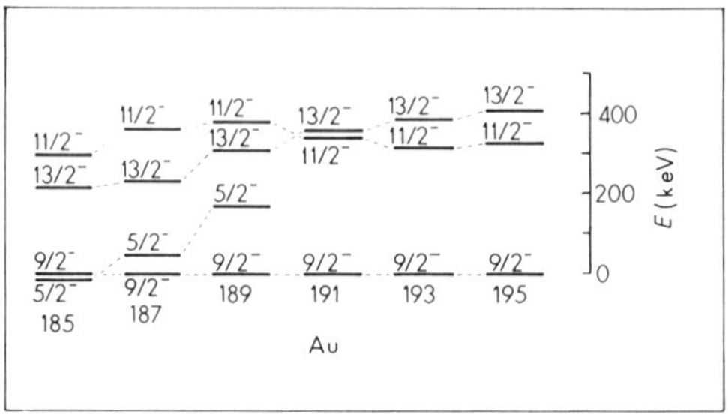


**Figure 19.** Systematics of the $h_{9/2}$ band structure in the odd-$A$ Au isotopes. The $9/2^-$ energy was normalized to zero. The critical feature is crossing of the $11/2^-$ and $13/2^-$ levels and "grounding" of the $5/2^-$ member in $^{185}$Au [(Ha 85) and modified by (Zg 86)].

19, the $h_{9/2}$ band goes from a strong coupled band in $^{195}$Au to a decoupled band in $^{185}$Au. Note the $9/2^-$, $11/2^-$, $13/2^-$ sequence for $^{191-195}$Au changes to $9/2^-$, $13/2^-$, $11/2^-$ in $^{185-189}$Au. Recall the potential-energy surfaces (Figure 10) for the Pt ground states go from oblate for $A > 190$ to prolate for $A \leqslant 186$ with $^{188,190}$Pt being $\gamma$-soft or more triaxial. Thus, the change in the Au $h_{9/2}$ band structures formed by coupling to the Pt cores reflects the strong change in their ground-state structure. However, the $h_{11/2}$ hole states are coupled to the same slightly oblate Hg cores as are the Tl $h_{9/2}$ particles, and so one has a consistency check on the Hg cores from these Au–Tl data. The data yield a consistent picture of the Hg ground-state cores. Thus one sees how indeed an odd-particle or hole can be used to probe changes in the core structures as well as test particle–core coupling models.

From systematic knowledge of the odd-$A$ neighbors and the even–even cores and data with very high statistical accuracy, Zganjar and co-workers (Zg 86, 86b) were able to establish three very important E0 + M1 + E2 transitions in $^{185}$Au and three in $^{187}$Au. Earlier, some of these transitions had been interpreted (Bo 82) as M1 + E2 with the M1 conversion electron component being anomalously large from the penetration of the electrons into the nuclear volume [see (Ha 65) for reviews of penetration effects on internal conversion]. E0 transitions also arise from the penetration of the atomic electrons into the nuclear volume (La 82, Ha 65b). The E0 transition matrix element is related to the mean-square radius, and this is very sensitive to shape changes of the nucleus. Such E0 transitions in the $\Delta I = 0$ transitions in $^{188-184}$Hg were signatures of the shape changes between the two coexisting bands in these nuclei (Ha 79). The Orsay group (Bo 82) were forced to conclude that anomalous M1 transitions were involved when they obtained conversion coefficients larger than M1 or E2 for a number of transitions between levels with different spins, and with $\Delta I \neq 0$. The UNISOR data were of sufficient statistical quality that unresolved doublets were quantitatively distinguished and the assigned anomalous M1 conversion coefficients were reinterpreted by locating another transition of nearly the same energy which connected states of the same spin ($\Delta I = 0$) and could thus contain an E0 component.

Combining all the data, Zganjar and co-workers (Zg 86) report in $^{185}$Au and $^{187}$Au four groups of states which arise from the coupling of the $\pi h_{9/2}$ particles to $0^+$ states with two different deformations in the $^{184,186}$Pt cores and the coupling of the $h_{11/2}$ hole states to $0^+$ states with two different deformations in the $^{186,188}$Hg cores which, in addition, are different from those in the Pt cores. Examples of the data for $^{185}$Au are shown in Figure 20 (Zg 86). The 213 keV transition is in cascade coincidence with both the 323 and 427 keV transitions. Comparing the e and $\gamma$ lines of the 427 keV transition

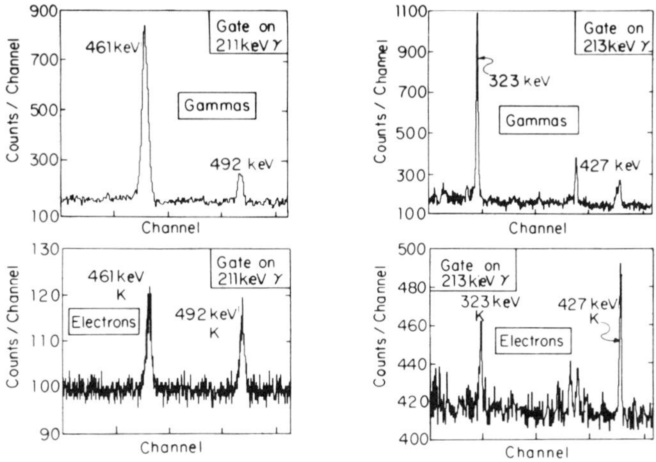


**Figure 20.** Portions of the $^{185}$Au electron and $\gamma$-spectra gated on the 213 keV $\gamma$ which is in cascade with both the 323 and 427 keV transitions. Portions of the electron and $\gamma$-spectra gated on the 211 keV $\gamma$ which is in cascade with both the 461 and 492 keV transitions (Zg 86).

to those of the E2, 323 keV transition, one sees that the 427 keV transition has a large K-conversion coefficient. Its $\alpha_K$ is $0.33 \pm 0.03$ while for an M1 $\alpha_K = 0.25$. So the 427 keV, $13/2^-$ to $13/2^-$ transition has a definite E0 admixture which indicates a shape change between the states. One cannot obtain the E0/E2 ratio for detailed comparison with theory until the M1 admixture is measured by low-temperature nuclear orientation. Similar data were found (Figure 20) for the 492 keV transition, which is also E0 + (M1) + E2, based on a measured $\alpha_K$ of 0.39(6) which is nearly three times larger than the theoretical M1 value of 0.14.

The couplings associated with the $h_{9/2}$ particle intruder states to the Pt core states are shown in Figure 21 and those with the $h_{11/2}$ hole states to the Hg core states in Figure 22. The transitions with strong E0 components feeding the $h_{9/2}$ and $h_{11/2}$ bands in $^{185,187}$Au identify the $h'_{9/2}$ and $h'_{11/2}$ bands, which result from the coupling of the

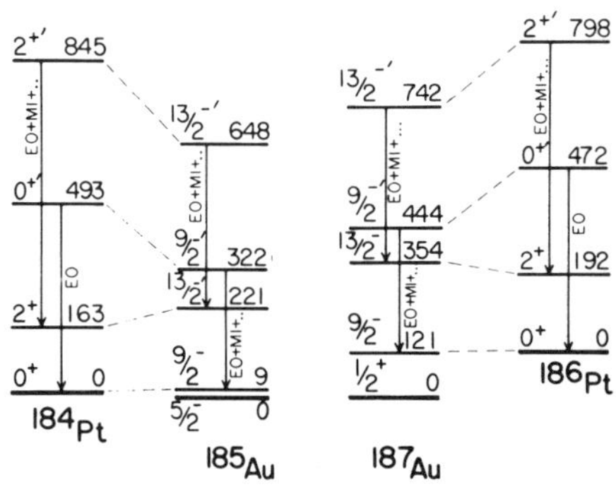


**Figure 21.** Portions of the $h_{9/2}$ and $h'_{9/2}$ bands in $^{185,187}$Au are compared to the $0^+$, $2^+$ and $0^{+\prime}$, $2^{+\prime}$ levels in $^{184,186}$Pt. The corresponding core-particle couplings are indicated by dashed lines. Only transitions with E0 components are shown (Zg 86).

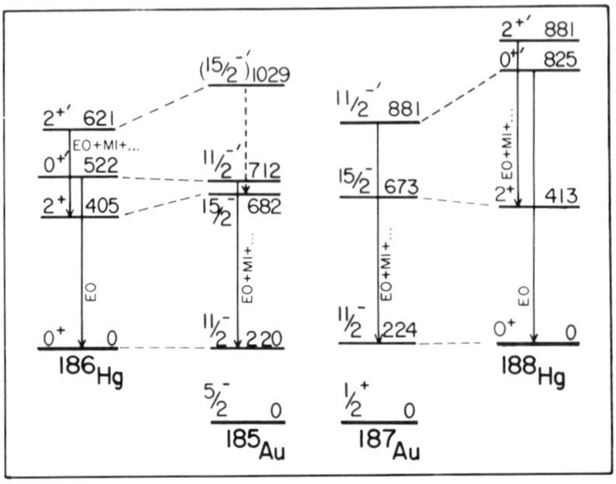


**Figure 22.** Portions of the $h_{11/2}$ and $h'_{11/2}$ bands in $^{185,187}Au$ are compared to the $0^+$, $2^+$ and $0^{+\prime}$, $2^{+\prime}$ levels in $^{186,188}Hg$. The corresponding core-particle couplings are indicated by dashed lines. Only transitions with E0 components are shown (Zg 86).

$h_{9/2}$ and $h_{11/2}$ particle and hole configurations, respectively, to the $0^{+\prime}$ and $2^{+\prime}$ states in the even–even $^{184,186}Pt$ and $^{186,188}Hg$ cores. The corresponding transitions in the even Pt and Hg cores exhibit similar significant E0 strength.

These data (Figures 21 and 22) are consistent with the interpretation that the ground bands of $^{184,186}Pt$ are well deformed and made up of a 2p–6h configuration, where a pair of particles have been promoted to the $\pi h_{9/2}$ orbital and the more near-spherical excited $0^{+\prime}$ states are 4h states (Zg 86). Of course, as the potential-energy surface calculations indicate (Be 86b) these two structures likely differ by more than just the promotion of one pair of high *j* particles. This interpretation is based on a Pauli blocking argument as well as systematics where the $8^+$ rotational state was used for normalization (Wo 81). When the odd proton is in the $h_{9/2}$ state, it blocks the 2p–6h configuration in the even–even Pt core, which is formed by the promotion of a proton pair into the $h_{9/2}$ state. The blocking would raise the core ground-state energy and give a compression of the $h_{9/2}$ and $h'_{9/2}$ separation in odd-mass Au relative to the even–even Pt cores as is seen in Figure 21. A blocking energy of approximately 170 keV was found in $^{185,187}Au$. Results from ORSAY on these odd-*A* Au isotopes were also used in the analysis (Zg 86).

The recent data on $^{189}Au$ (Ko 86b) indicates that the blocking is not occurring there to support a 4h configuration for the ground state of $^{188}Pt$ which was proposed earlier (Wo 81) and which is consistent with the trends of in-beam Pt data (Dr 86). Note also that the potential-energy surface calculations for the Pt isotopes show a shape change in going from $^{186}Pt$ to $^{188}Pt$ (Be 86b).

The levels of $^{185}Au$ also have been studied recently by in-beam spectroscopy as shown in Figure 23 (La 86). The decoupled $\Delta I = 2$ bands 1 and 4 seen there have prolate shape and are built on $\pi h_{9/2}$ and $\pi i_{13/2}$ single quasiparticles. These are also seen in radioactive-decay studies (Zg 86). The $h_{11/2}$ $\Delta I = 2$ decoupled aligned band with oblate shape begins at 220 keV as also seen in Figure 22. The 220 keV state is isomeric ($T_{1/2} = 26$ ns) (Be 83). This shape isomerism corresponds to the decay of the $\pi h_{11/2}$ oblate state to the $\pi h_{9/2}$ prolate state at 9 keV. There are two striking new features seen in Figure 23 when compared to earlier studies (Ka 78, 78b). One is the first indication of a predominantly $\pi f_{7/2}$ band with prolate shape in the deformed region below lead. The second is the observation of an evidently $\Delta I = 1$ strong coupled band which feeds into the $h_{11/2}$ decoupled, aligned band which begins at 220 keV. This

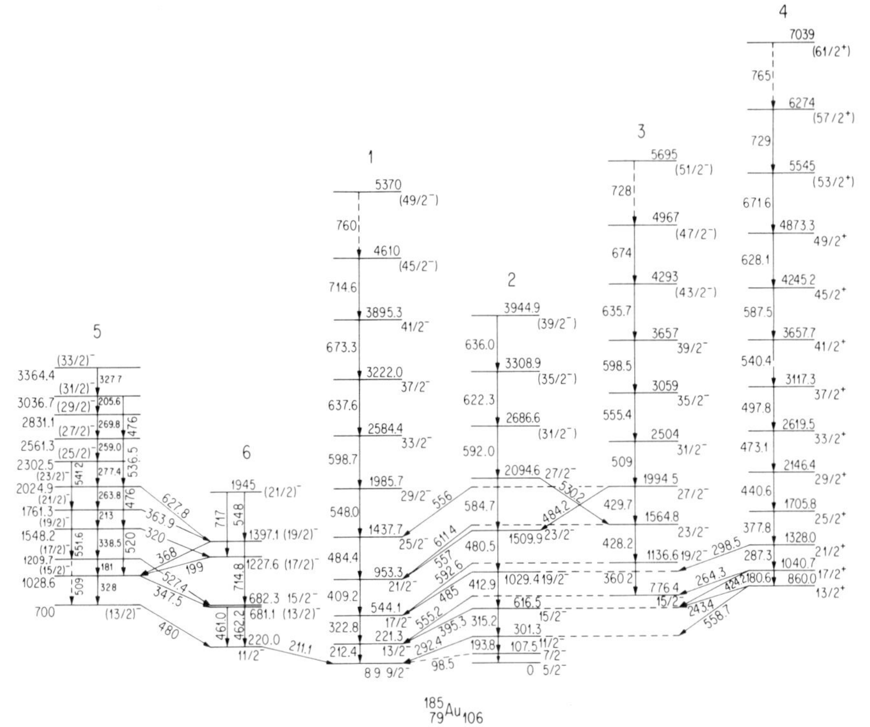


**Figure 23.** Revised level scheme of $^{185}$Au [(La 86) and private communication].

change in structure suggests that the $\Delta I = 1$ band is a $h_{11/2}$ hole state coupled to the more strongly deformed prolate shape. The $11/2^-$, 712 keV level and the 1029 keV level in $^{185}$Au, interpreted in UNISOR work to form a band based on an $h_{11/2-}$ hole coupled to the excited, deformed $^{186}$Hg core, as seen in Figure 22, are likely the first two members of this strong coupled, $\Delta I = 1$ band (Figure 23). By comparing Figures 22 and 23 the following are suggested: First, the (tentatively assigned at UNISOR) 346.4 keV transition from the 1028.6 keV level to the 682.3 keV, $15/2^-$ level was likely missed in the in-beam work. Second, in $^{186}$Hg in-beam data (Pr 73) the $2^{+\prime} \rightarrow 0^{+\prime}$ was not seen, so the analogous transition from the $15/2^{-\prime}$, 1028.6 keV level to the 712 keV $11/2^{-\prime}$ band head assigned in UNISOR work would likely not have been seen in-beam. It is not clear what the 700 keV level seen in-beam is. These questions do need to be resolved. Nevertheless, together these data firmly establish the oblate and prolate shape coexistence assigned to the lower states in the UNISOR studies (Figure 22) (Zg 86, Zg 88).

There is one problem in the picture as presented above, and that is in $^{187}$Au. If one looks at the other states in the $h'_{9/2}$ structure, as shown in Figure 24, the bands built on both the $h_{9/2}$ and $h'_{9/2}$ states have $\Delta I = 2$ sequences (Ha 85). If they do represent couplings to a prolate ground state and an excited oblate core in $^{186}$Pt, then the upper one should have a strong, coupled $\Delta I = 1$ sequence. One possible explanation of why this is not seen is that the band built with the $h_{9/2}$ structure coupled to the oblate core

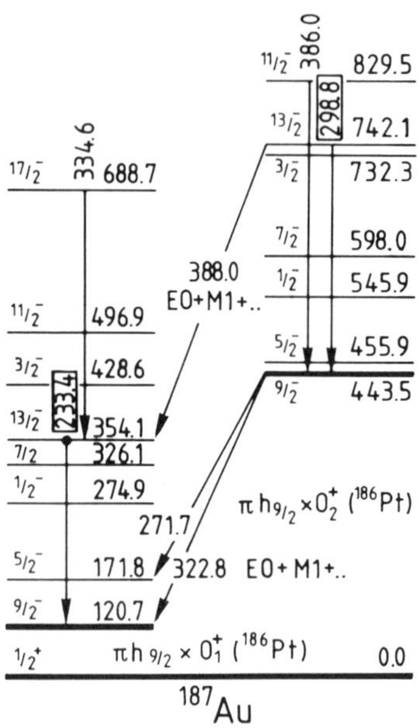


**Figure 24.** The $\pi h_{9/2} \times 0_1^+$ ($^{186}$Pt) and $\pi h_{9/2} \times 0_2^+$ ($^{186}$Pt) bands in $^{187}$Au together with several transitions which are pertinent to the assignments (Zg 81).

has its $11/2^-$ and $13/2^-$ members shifted by mixing. Also, it is possible that the softness of the oblate minima as one approaches the ground-state phase transition to $\gamma$-soft $^{188}$Pt and oblate $^{192}$Pt gives rise to perturbations in this band. Indeed, the role of triaxial shapes in these odd-$A$ and even–even cores very likely plays an important role. For example, evidence for $\gamma \neq 0$ in $^{193-197}$Tl has been presented earlier (Zg 80). These interpretations could be checked if this band could be seen to higher spins.

Deformation-energy curves were calculated for axially symmetric shapes by the Strutinsky method with a Woods–Saxon single-particle potential (La 86). These are shown in Figure 25. Note the high $j$ orbitals, $h_{9/2}$, $h_{11/2}$, and $i_{13/2}$, are seen in Figure 25 to have low-lying minima for *both* prolate and oblate shapes. Both structures are seen for the $h_{11/2}$ orbital in on-line and in-beam studies, while both are seen for the $h_{9/2}$ orbital only in on-line studies. Two approaches have been used to describe the different structures of the coexisting, more deformed prolate and less deformed, oblate bands in the Pt, Au, Hg nuclei. These are the "intruder orbitals" where the $h_{9/2}$ particle intrudes to appear below the $Z = 82$ shell gap in potential-energy surface calculations. The intruder picture is illustrated in Figure 26 where the $\pi$4h states are the oblate ground states of the heavy Pt isotopes, and the solid, down-sloping curved lines are the $\pi$(2p-6h) states formed by promoting a pair of $d_{3/2}$ particles to the $h_{9/2}$ orbital. Similarly in this picture, the even–even Hg ground states are $\pi$(2h) states, and the well-deformed bands in $^{182-188}$Hg are $\pi$(2p–4h) states with a proton pair promoted to the $h_{9/2}$ intruder. In the odd-$A$ Au isotopes, the promotion of an $h_{9/2}$ particle forms a $\pi$(1p–4h) state that couples to the $^{A-1}$Pt core while the $\pi$(3h) states couple to the $^{A+1}$Hg cores.

Earlier we discussed the results of the calculations of the potential-energy surfaces for the Pt and Hg isotopes. The competition between the prolate and oblate minima

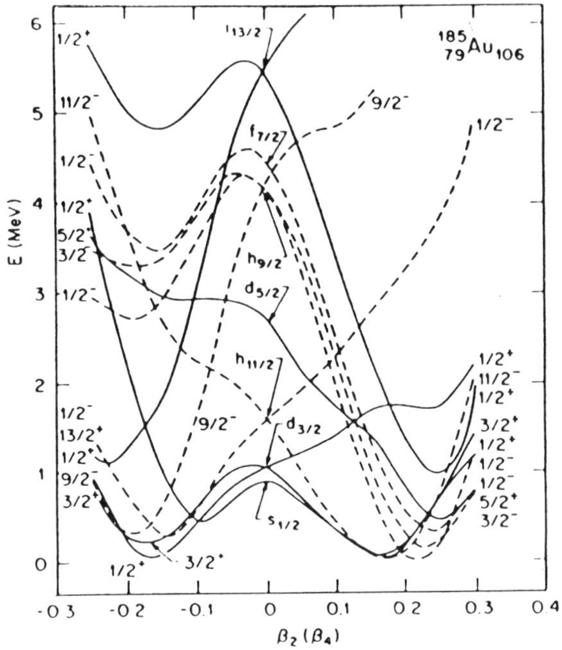


**Figure 25.** Deformation-energy curves for various bandheads in $^{185}$Au as a function of $\beta_2$. At every $\beta_2$ the minimization was carried out for the $\beta_4$ deformation parameter. Both oblate and prolate minima are shown for each bandhead (La 86).

in the even–even Pt and Hg isotopes gives rise to a ground-state phase change in Pt between $A = 192$ and 186 and to shape coexisting structures in various Pt and Hg nuclei (see Figures 9 and 10). Similar prolate-oblate competition is seen in Figure 25 for $^{185}$Au. However, these are not alternate descriptions but different ways of emphasizing the origin of the changing and competing shapes in these nuclei. Indeed when one

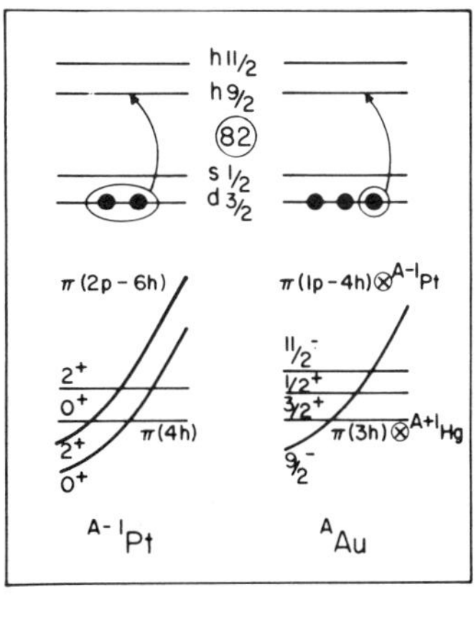


**Figure 26.** A schematic view of intruder states near $Z = 82$. The normal proton configurations for Au and Pt, for example, are $\pi(3h)$ and $\pi(4h)$, respectively. The proton intruder configurations are $\pi(1p\text{-}4h)$ and $\pi(2p\text{-}6h)$, respectively. Here p = particle and h = hole (Zg 86).

looks at the two minima shown in Figure 10 for the Pt isotopes, one finds that they correspond to structurally very different configurations: the prolate configuration corresponds to an excitation of at least 2 protons to the $h_{9/2}$ shell (the simple intruder picture), but the $h_{9/2}$ shell is not occupied at all in the oblate configuration (Be 86b). However, the prolate configuration also contains 2 or 4 (depending upon $N$) more $i_{13/2}$ neutrons than the oblate one. Thus, as could be expected, a picture of only two particles in an intruder orbital is too simplistic but, nevertheless, gives a basic picture of the physics underlying these coexisting structures. Clearly the multiple coexisting structures seen in this region, as highlighted by the four different prolate and two different oblate structures now seen in $^{185}$Au, and their competition at low spin as well as their competition in forming aligned structures at higher spins, offer an exciting wealth of new insights to test and extend our theoretical understanding.

Let us look at two final examples of shape coexistence in the odd-$A$ nuclei in this region. As already shown in Figure 6, shape coexistence has been established (Kü 77, Da 79) in $^{185}$Hg where the near-spherical $13/2^+$ isomer continues the smooth shrinking of radius with decrease in $N$ as seen in $^{187}$Hg and the heavier isotopes, while the $^{185}$Hg ground state shows a sudden large increase in mean-square radius (deformation) compared to the $A \geqslant 187$ isotope. Very recently Hannochi and co-workers (Ha 86b) studied the levels in $^{187}$Hg by in-beam spectroscopy. A $\Delta I = 2$ sequence is observed to $25/2^+$ built on the $13/2^+$ near-spherical isomeric state. In addition, the $13/2^+$ isomer is fed by a $\Delta I = 1$ band starting at $11/2^+$ and going to $45/2^+$ and becoming yrast above $25/2^+$. Their interpretation of these bands is that the $\Delta I = 2$ one built on the $13/2^+$ isomer corresponds to an oblate minima and the other to a prolate minima which is quite close in energy. Recall from Figure 6 that both the ground state and $13/2^+$ isomer in $^{187}$Hg have similar deformation. These data are the first evidence for band structures built on close-lying prolate and oblate shapes in the odd-Hg isotopes.

Finally, prolate–oblate shape coexistence has been proposed in $^{185}$Pt from studies at ISOCELE (Ro 85). The primary evidence for the coexistence of prolate and oblate structures was the observance of many levels below 840 keV, which could be interpreted in an "axial-rotor + HF quasiparticle" model for a prolate shape but also ten additional low-energy negative-parity levels which could be explained by an oblate shape. As noted above, a sign of shape coexistence is the establishment of large E0 transitions in $\Delta I = 0$ transitions between the two shapes. Two transitions which were very strongly internally converted (Very Converted Transition, VCT) were observed in the $^{185}\text{Au} \rightarrow {}^{185}\text{Pt}$ decay. These were placed as shown in Figure 27. The K-conversion coefficients of the 340 and 542 keV transitions are higher than for an M1 transition. These could be because of an E0 admixture, but could arise if the M1 conversion is anomalous because of an added contribution from penetration effects of the electrons into the nuclear volume to give an enlarged M1 conversion coefficient. However, the 542 keV transition was placed between two states, both of which were assigned prolate shape as shown in Figure 27, so that this transition was interpreted as an anomalous M1 decay. Such as assignment, as noted by Roussiere and co-workers (Ro 85), casts doubt on the 340 keV transition having any E0 component which would be a signature of the proposed oblate–prolate shape change shown in Figure 27. Thus, they concluded that these two highly converted transitions are more likely related to nuclear structure effects altering the M1 conversion process.

A correct interpretation requires measurement of the M1/E2 admixtures in the 340 and 542 keV transitions and definite spin assignments. The Leuven group has

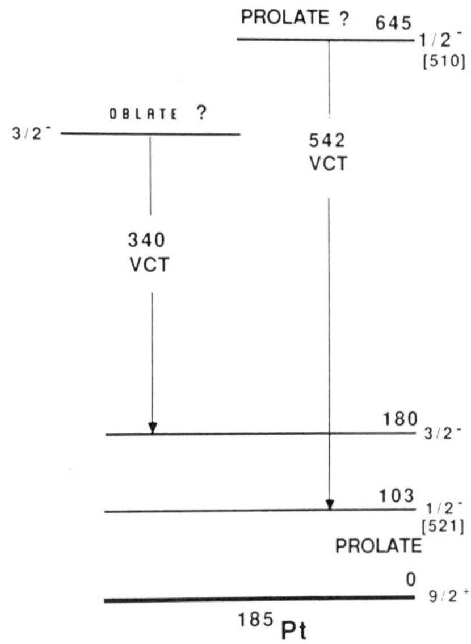


**Figure 27.** States in $^{185}$Pt depopulated by transitions with very large internal conversion coefficients (Ro 85) are shown with the assignments made there. The new assignments (Va 86) are discussed in the text.

carried out such measurements with their new on-line, low-temperature, nuclear orientation facility KOOL, which is on-line with their isotope separator on-line to their cyclotron (Va 85). In nuclear orientation the nuclei are implanted in a ferromagnetic host kept at $\leqslant 10$ mK to align the nuclear spins. Then one measures only in singles mode the angular distribution of the gamma rays from spin-oriented nuclei to obtain spins and mixing ratios. Extensive studies of Au nuclei were carried out at KOOL (va 85, va 85b, Va 86). For example, discrepancies in the magnetic moments of the $11/2^-$ state in $^{189,191}$Au were resolved to establish a striking constancy of these moments as a function of $N$ from $^{187-197}$Au. Also, measurements of $\mu$ of the low-spin Au states were extended down to $^{185,186}$Au. The resulting moments, especially of the $3^-$ ground state of $^{196}$Au, allows one to characterize the wave functions of these states. Their analysis confirms the prolate-to-oblate phase transition in these nuclei. This phase transition was the first ever predicted [see Kumar (Ku 68)]. At the Harrogate Conference, the Mainz group confirmed these results by laser spectroscopy where a sudden jump in $\delta(r^2)$ was reported. Their data (Va 86) on the $^{185,186}$Pt daughters of Au clarified the interpretations of the very converted transitions in $^{185}$Pt shown in Figure 27. The spin of the 645 keV level was found to be $3/2^-$, so this $3/2^-$–$1/2^-$ transition may have an anomalous M1 conversion coefficient from penetration effects. The $3/2^-$ spin was confirmed for the 520 keV level and the M1/E2 mixing in the 340 keV transition was measured. A definite E0 component was established in the 340 keV transition (va 85, Va 86). However, more recently, Zganjar *et al.* (Zg 88) have found that all transitions in $^{185,187}$Au and $^{185}$Pt with $K$ conversion coefficients greater than that of pure M1 transition are consistent with $\Delta I = 0$. In summary, these data firmly establish shape coexistence in $^{185}$Pt and dramatically illustrate the importance of having measured spins and measured M1/E2 mixing ratios in establishing E0 admixtures as recently emphasized (Kr 85).

# 4. Spherical and Deformed Magic Numbers and Reinforcing and Switching Shell Gaps

## 4.1. Introduction

The spherical shell model, magic numbers (Ma 50, Ha 50) which correspond to gaps in the single-particle energy spectra for protons and neutrons for spherical shapes are shown in Figure 3 as taken from Baranger and Sorenson (Ba 69). Examples of the gaps in the single-particle spectra for the spherical magic numbers 28, 40, and 50 are shown in Figure 28. One of the first important questions raised as people contemplated searches to study the properties of nuclei far from the stable and near stable nuclei in the valley of beta stability was, and still is: Are the magic numbers for the spherical shell model still valid far from stability and are there other magic numbers which are important in the unexplored regions? Double magic nuclei have unique properties and provide major tests of the spherical shell model. The earlier known double magic nuclei are ${}^{4}_{2}He_{2}$, ${}^{16}_{8}O_{8}$, ${}^{40}_{20}Ca_{20}$, ${}^{48}_{20}Ca_{28}$, ${}^{56}_{28}Ni_{28}$, and ${}^{208}_{82}Pb_{126}$. To this group some added ${}^{90}_{40}Zr_{50}$ whose energy levels are characteristic of a double closed shell. This nucleus and its

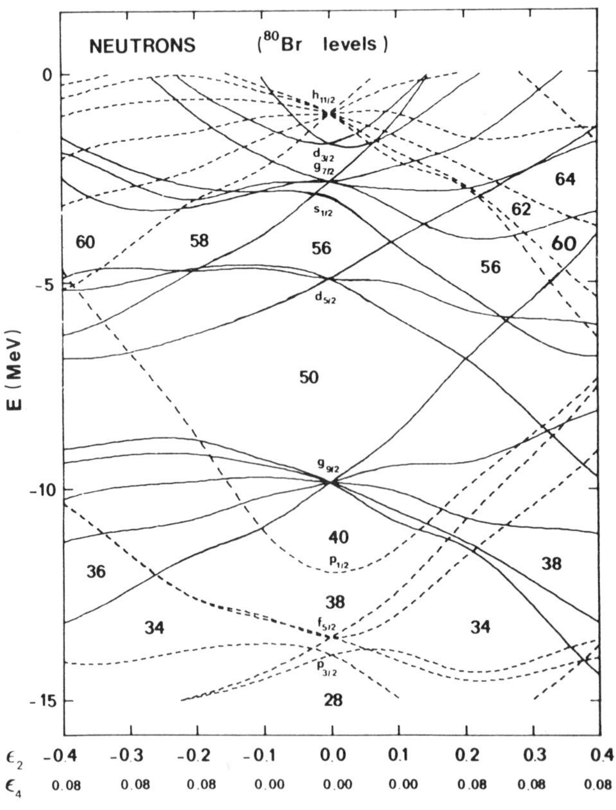


**Figure 28.** Single-particle levels calculated with a folded Yukawa single-particle potential (Be 84).

neighbors were a major reason Baranger and Sorenson (Ba 69) gave $N$ and $Z$ of 40 as magic, although for many 40 was only semimagic.

While the spherical magic numbers far from stability received essentially all the attention until the last few years, much earlier Brack and co-workers (Br 72) in their paper on shell correction effects emphasized that one can have "magic" shell gap numbers for deformed as well as spherical shapes. These "deformed magic" numbers are magic in the same sense as the spherical ones in that they give extra stability to a deformed shape. As we shall see, the neutron-rich and neutron-deficient nuclei far off stability with $Z = 34$–$40$ provide the first clear examples of "deformed magic" numbers associated with gaps in the single-particle levels at large deformation for $N$ and $Z$ of 38 and $N$ of 60, 62 as also seen in Figure 28. The importance (strength) of these new "deformed magic" numbers in determining the nuclear shape will be seen to be related to the mutual reinforcement of the protons and neutrons when both have shell gaps at the same deformation.

## 4.2. Nuclear-Shape Coexistence and New Regions of Very Strong Prolate Deformation Around $Z = 38$

Nuclear-shape coexistence was first discovered in $^{72}$Se by Hamilton and co-workers (Ha 74b, Ha 76b) and almost simultaneously in the light mercury isotopes (Ha 75) just discussed. There is a striking similarity in the moments of inertia of the yrast cascades in $^{184}$Hg and $^{72}$Se (Figure 29) which start out rising vertically, as is characteristic of a spherical vibrator, and then bend to become horizontal, as is characteristic of a deformed rotor. Their wide differences in mass should have been an early clue that nuclear-shape coexistence, far from being an isolated phenomena, in fact permeates the whole periodic table. Strongly deformed structures were not expected in the $A = 70$ region because, when both $N$ and $Z$ are less than 50, $N$ and $Z$ are never very far from a magic number, especially when 40 is considered magic as it was in 1969, as shown in Figure 3. Nuclei in the $A = 70$–$80$ region were considered more or less as rather uninteresting, near-spherical vibrators and so received little attention theoretically or even experimentally before that work (Ha 74b, Ha 76). Even with the $^{72}$Se work, the recent discovery of a new region of very strong ground-state deformation around $N = Z = 38$ by Hamilton and co-workers (Ha 81c, Pi 81, Pi 82, Zh 82) was quite unexpected. Equally surprising

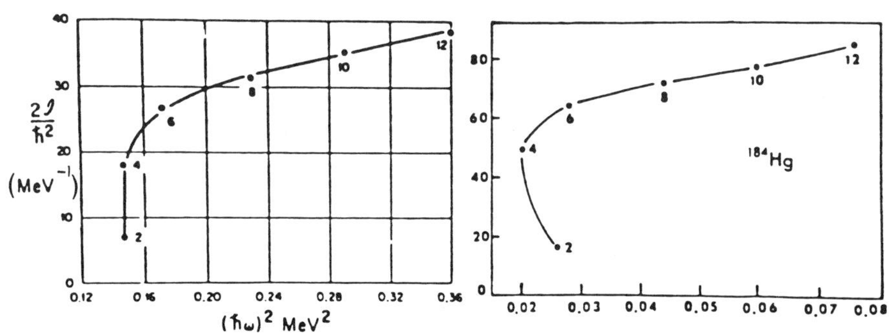


**Figure 29.** Moments of inertia of the yrast cascades in $^{72}$Se (left) and $^{184}$Hg (right) (Ha 79).

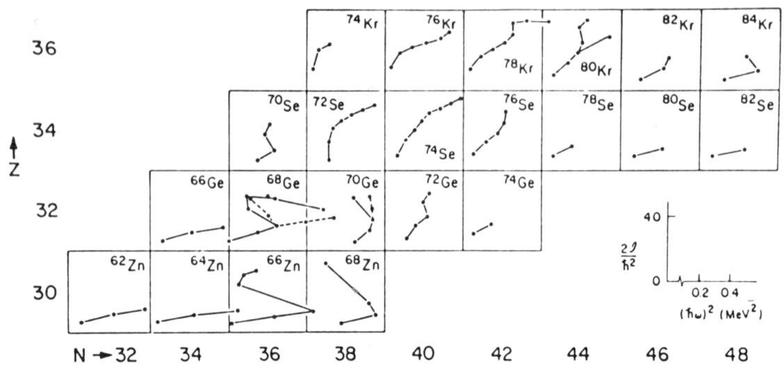


**Figure 30.** Systematics of the moments of inertia for the yrast cascades in $Z = 30$–$36$ nuclei (Ha 79c).

was the discovery of the first new region of very strong deformation in $Z = 38$, $^{100}$Sr on the other side of the $N = 50$ closed shell, by Azuma and co-workers (Az 79) as discussed in the next section. As we shall see, $Z = 38$ is important in both these new regions of very strong deformation. These discoveries, along with the richness of collective and single-particle motions uncovered in the $A = 70$ region (Ha 79c, Ha 79d) and the rapidity at which the different band structures can change with the addition of only two protons or two neutrons, have made the $A = 64$–$84$ and $A = 96$–$104$ regions for $Z$ around 38 very important new testing grounds for nuclear models.

The rapid changes with the addition of just two protons or two neutrons is illustrated by thc rapid changes in the yrast structures at low spins shown in Figure 30 (Ha 79c, Ha 79d). The richness of collective and single-particle motions now observed in the $A = 64$–$84$ region include in even–even nuclei: coexisting bands built on states with quite different deformation (including near-spherical ground states with deformed excited bands and deformed ground states with near-spherical excited bands); $\gamma$-type vibrational bands with properties like those of $\gamma$ bands in well-deformed nuclei; even-parity bands built on a pair of both proton and neutron high-$j$ $(g_{9/2})^2$ orbitals rotation aligned with the core; and odd- and even-spin negative-parity rotation-aligned bands with only one particle in $g_{9/2}$ orbital (Ha 79c, Ha 81b). Likewise, a wide variety of collective and single-particle motions are seen in odd-$A$ nuclei in this region [for example see (Fu 82)].

In-beam and on-line studies of $^{74,76}$Kr led to the discovery (Ha 81c, Pi 81, Pi 82) of strong ground-state deformation when both $N$ and $Z$ are at or near 38. This discovery illuminated the origin of the shape coexistence in this region. The levels in $^{74}$Kr and $^{76}$Kr (Figure 31) are dominated by rotational band structures which have surprisingly large B(E2) values that indicate collective rotations of a deformed nucleus. The B(E2) are 78–120 single-particle units (SPU) in the $2^+ \rightarrow 0^+$ to $10^+ \rightarrow 8^+$ transitions in $^{76}$Kr (Pi 81, Pi 82, Wi 82, Zh 82), but in $^{68,72}$Ge, for comparison, they are the order of 10–20 SPU for the $2 \rightarrow 0$ and $4 \rightarrow 2$ transitions. The moments of intertia, $\mathscr{I}$, of the yrast bands in $^{74-80}$Kr (Figure 32) become larger when going from $N = 44$ to $N = 38$. The $2^+ \rightarrow 0^+$ energies and, to a lesser degree, the $4^+ \rightarrow 2^+$ energies in $^{74-78}$Kr deviate from what one would expect by extrapolating the higher spin data with the greatest anomaly being in the $2^+ \rightarrow 0^+$ energy in $^{74}$Kr (see Figure 32), as first noted by Funke and co-workers (Fu 80). UNISOR studies of the $^{76}$Rb decay (Pi 80, Pi 81, Pi 85) show that the levels

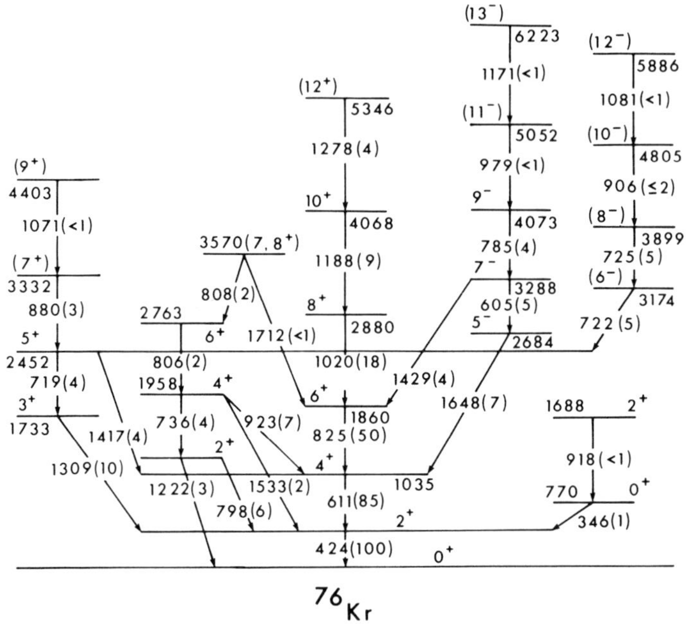


**Figure 31.** Energy levels in $^{76}Kr$ established from in-beam studies (Pi 81, Pi 82).

which feed the anomalously low-energy $0_2^+$ level have a more typical, near-spherical vibrational spacing [note the $(2^+) \rightarrow 2^+ \rightarrow 0^+$, 882–917 keV cascade in Figure 33].

Summarizing the results, the ground states of $^{74,76}Kr$ have very large prolate deformation and the $0_2^+$ level and states built on it in $^{76}Kr$ are associated with a more near-spherical, probably oblate, shape in contrast to the reverse situation in $^{72,74}Se$, where the ground states have small deformation and the $0_2^+$ states are well deformed.

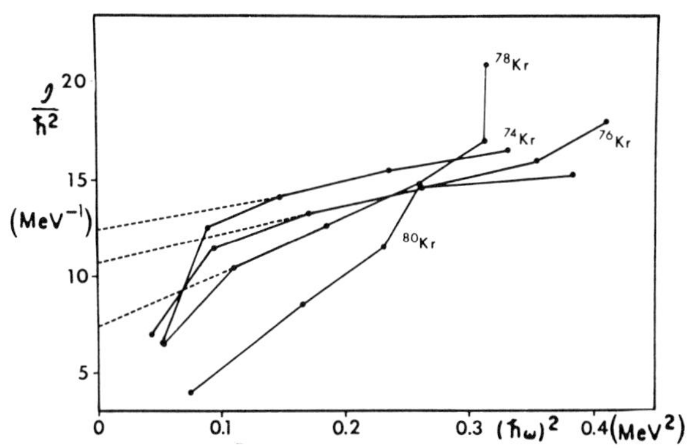


**Figure 32.** Moments of inertia of the yrast cascades in $^{74-80}Kr$ are shown. Broken lines correspond to an extrapolation back to low spin of a Harris parametrization of the states with $I = 6$–$10$ where $\mathcal{J}$ is linear (Pi 81).

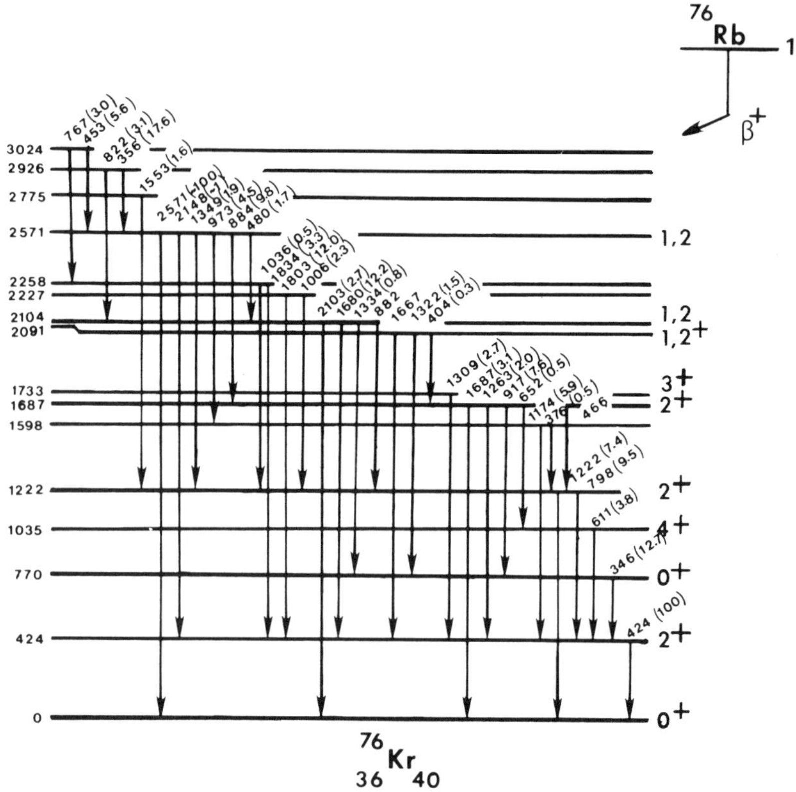


**Figure 33.** Energy levels of $^{76}$Kr from an on-line study of isotope-separated $^{76}$Br (Pi 80, Pi 85).

In $^{74,76}$Kr, the $0_1^+$ and $0_2^+$ states interact and push each other apart to enlarge the $2_1^+ \rightarrow 0_1^+$ energies and produce the anomalous behavior of $\mathscr{I}$ seen in Figure 32. These enlarged energies at first masked the ground state deformation. The very large magnitude of the deformation in $^{74}$Kr and in $^{100}$Sr (discussed later) can be seen by comparing their $2 \rightarrow 0$ energies with that of $^{240}$Pu, which has the lowest known $2_1 \rightarrow 0_1$ energy. The $2_1 \rightarrow 0_1$ energies are compared in Figure 34 where, for $^{74}$Kr and $^{100}$Sr, they are scaled by $A^{5/3}$ to compare at the same mass. The $\beta_2$ deformations in $^{74,76}$Kr and $^{100}$Sr and their neighbors are the order of 0.35 to 0.40 and thus are the largest known for nuclear ground states above $A$ of 40, "superdeformation" for ground states.

In a unique conjunction of theory and experiment, this new region of strong deformation first observed in $^{74,76}$Kr (Ha 81c, Pi 81, Ha 81b) was theoretically predicted by Möller and Nix (Mö 81, Mö 81b, Mö 82), who calculated the nuclear masses and ground-state shapes for 4023 nuclei from $^{16}$O to $^{279}$112 with a Yukawa plus exponential macroscopic model and folded Yukawa single-particle potential. Their calculations, totally independently, predicted that nuclei with both $N$ and $Z$ at or near 38 should be among the most strongly deformed ones in nature, with $\beta = 0.35$–$0.4$ as shown in Figure 35. Similar deformations were found in the potential-energy surface calculations

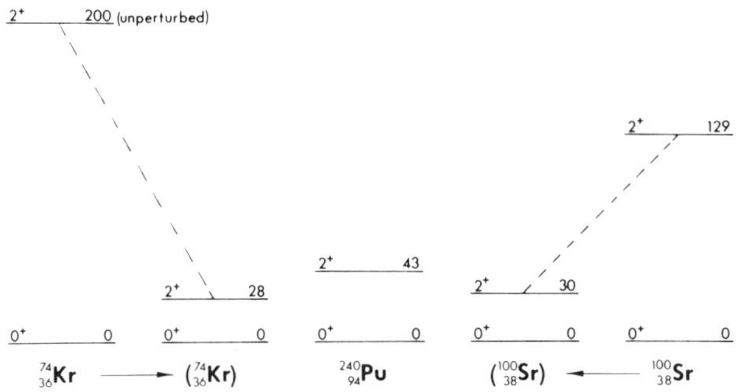


**Figure 34.** Comparison of the $2_1^+ \to 0_1^+$ energies in $^{74}$Kr and $^{100}$Sr with $^{240}$Pu where the former two energies have been scaled by $A^{5/3}$ to $A = 240$. The extrapolation in Fig. 32 was used to obtain the unperturbed $2^+$ energy for $^{74}$Kr.

in the generalized collective model for $^{74-80}$Kr (Se 84). For example, their $^{76}$Kr surface has two minima, where the ground-state wave function is centered in the deformed minimum with large $\beta$ and the $0_2^+$ wave function in the near-spherical one. For $^{78,80}$Kr these surfaces show the nuclei are soft to $\gamma$ deformation, so shape coexistence quickly disappears with increase in $N$. We will return to discussions of potential-energy surfaces later.

Subsequent studies of the levels in $^{80}$Sr (Ba 81, Ha 81c, Ha 82b, Li 82) and $^{78}$Sr (Li 82,, Ba 82, Ha 82b) confirmed the large deformation around $Z = N = 38$. The systematics of the $2_1^+ \to 0_1^+$ energies for Sr nuclei from $^{88}_{38}Sr_{50}$ at the $N = 50$ spherical gap to $^{78}_{38}Sr_{40}$ as shown in Figure 1 exhibit a smooth but rapid drop which clearly signifies the onset of large deformation, presumably going to a maximum deformation at $N = Z = 38$ in $^{76}$Sr. This was confirmed by the discovery of the levels in $^{76}$Sr (Li 88). The data on the odd-$A$ nuclei in this region, for example in $^{75}$Kr (Ra 82, He 87) and $^{77,79}$Y (Li 82), likewise confirm the large deformation ($\beta_2 = 0.35 \pm 0.4$ for $^{75}$Kr) and establish its prolate character. Similar deformation for the $g_{9/2}$ band in $^{75}$Br was independently found by Wen and co-workers (We 85) and Lühmann and co-workers (Lü 85).

Now let us look at the evidence for shape coexistence and very strong ground-state deformation in the neutron-rich nuclei around $Z = 38$, 40. Shape coexistence there was considered earlier by Sheline and co-workers (Sh 72) and Ramayya and Hamilton (Ra 74) to account for the spherical ground states and very low-energy $0_2^+$ states in $^{96,98}_{40}$Zr and ground-state deformation in $^{100}$Zr (Ch 70). However, Ramayya and Hamilton (Ra 74) already had noted that the work of Flynn and co-workers (Fl 74) indicated the low-energy $0_2^+$ state in $^{96}$Zr is not a deformed bandhead. Subsequent studies of the properties of the $0_2^+$ state in $^{98}$Zr likewise did not support this state being associated with a well-deformed structure (Si 77). Indeed, Meyer and co-workers (Me 86b) recently have found a full set of levels built on this $0_2^+$ state as shown in Figure 36, and these levels are remarkably similar to those of the near-spherical $N = 58$ isotone $^{102}_{44}$Ru. This set of $^{98}$Zr levels is interpreted (Me 86b) as arising from the promotion of a pair of $s_{1/2}$ protons to the $g_{9/2}$ intruder orbital as illustrated in Figure 36. This will be discussed more fully later.

There is a very sudden onset of large ground-state deformation at $N = 60$ in this region as discussed in more detail in the next section. The suddenness of the onset

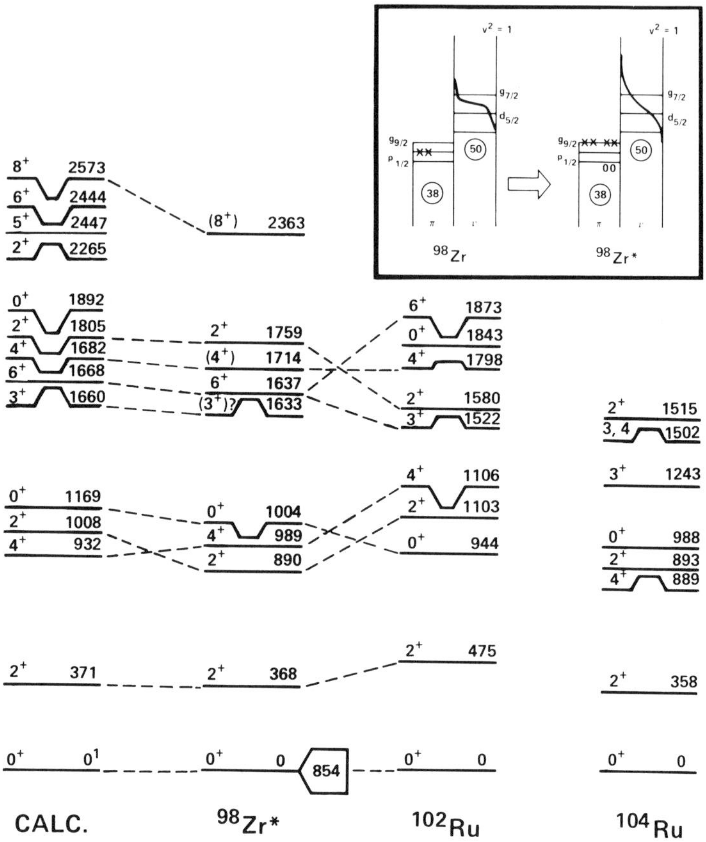


**Figure 36.** Comparison of the levels associated with the $0_2^+$, 854 keV excited state in $^{98}_{40}Zr_{58}$ with its isotone $^{102}_{44}Ru_{58}$ and with $^{104}Ru$. The insert shows the structure of the ground and $0_2^+$ states in $^{98}Zr$ (Me 86b).

and the coexisting strongly deformed and near-spherical structures in $^{100}_{40}Zr$ and $^{98}_{38}Sr$ can be seen in Figures 37 and 38. The lowest energies known for any $0_2^+$ state in even–even nuclei are at the remarkably low energies of 215.5 and 331.3 keV in $^{98}Sr$ and $^{100}Zr$, respectively (Ka 77, Sc 80). These two unusual $0_2^+$ states are interpreted (Ka 77, Sc 80) as being built on near-spherical shapes which coexist with the deformed ground states of $^{98}Sr$ and $^{100}Zr$. In $^{98}Sr$ (Figure 38) a high-energy $2_2^+$ state, characteristic of the vibrational $2^+$ state of a near-spherical nucleus, is seen on top of the $0_2^+$ state to support shape coexistence. Wohn and co-workers (Wo 86) have recently studied the decay of $^{100}Y$ to $^{100}Zr$ and observed numbers of new levels. They find evidence for bandheads of both beta ($0_\beta^+$) and gamma ($2_\gamma^+$) vibrational bands built on an axially symmetrical deformed ground state. All of the levels and their branching ratios below 1 MeV can be explained by the coexistence of a well-deformed ground state and more spherical band. If their interpretation is correct, both the rotational and vibrational characteristics

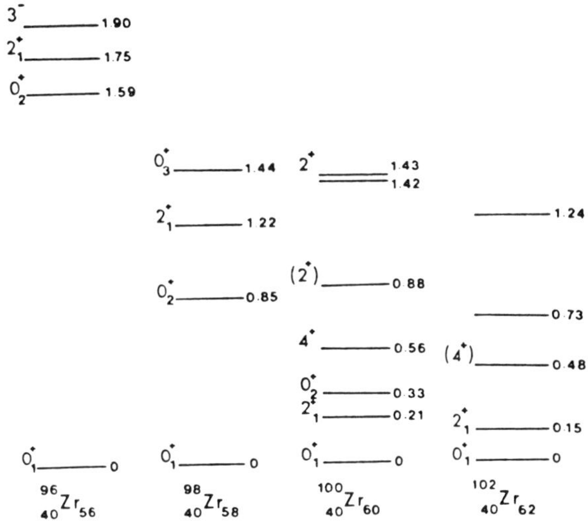


**Figure 37.** Low-lying energy levels in $^{96,98,100,102}_{40}Zr$ are compared.

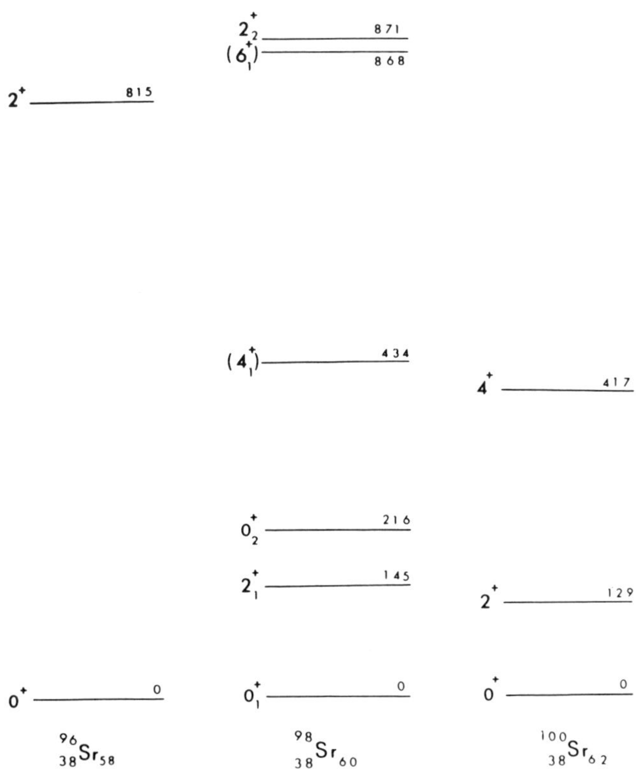


**Figure 38.** Low-lying energy levels of $^{96,98,100}_{38}Sr$ are compared.

expected for a well-deformed nucleus are now seen for the first time in $^{100}$Zr coexisting with a more spherical band.

Ragnarsson (Ra 81) has noted that their calculations (Be 81) indicate that in $^{100}$Zr with $N = 60$, shape coexistence occurs with two close-lying $0^+$ states but that for $^{98,102}$Zr with $N = 58$ and 62 one expects shape-coexisting $0^+$ states with spacings of 1.0–1.5 MeV. Studies of $^{98}$Zr have found evidence for a $0_3^+$ state at 1436 keV that has properties which indicate it has a deformation quite different from the ground state (Ka 82). These data are consistent with the expected deformed band levels of Bengtsson and co-workers (Be 81) and provide support for the shape coexistence model. However, more than two neutrons away from $N = 60$ the energy differences between the spherical and deformed shapes rise very rapidly and it becomes difficult to observe any coexistence (Be 81). A similar sharp crossing of spherical and deformed shapes with shape coexistence observed in only a narrow range of neutron numbers around $N = 60$ in the Sr nuclei also is indicated.

The level structures of odd-$A$ and odd–odd neighboring nuclei provide additional evidence for the strong ground-state deformation. Results of investigations, first of $^{99}_{39}\mathrm{Y}_{60}$ (Mo 82) and more recently of $^{99}_{38}\mathrm{Sr}_{61}$, $^{101}_{40}\mathrm{Zr}_{61}$, $^{99}_{39}\mathrm{Y}_{60}$, and $^{101}_{39}\mathrm{Y}_{62}$ with the on-line TRISTAN facility (Wo 83), are seen in Figure 39. Rotational bands are observed with large moments of inertia and deformations, $\beta \sim 0.3$–$0.4$, which are the same magnitude as those in their even–even neighbors, are found.

Peker and co-workers (Pe 86) have identified rotational bands in four odd-odd nuclei with similar large deformation, as shown in Figure 40. They (Pe 86) point out that these bands may be pairing-free bands. The first suggestion that the unusually large deformation in the $A = 100$ region may be related to the pairing-free nature of the states came from an analysis of the anomalously low energy $7/2^+$ isomeric states in the odd-$A$ $_{45}$Rh and $_{47}$Ag nuclei (Pe 86b). Normally three-particle configurations in odd-$A$ nuclei must lie above the pairing gap $2\Delta$, i.e., above 1 MeV. However, Peker and co-workers (Pe 86b) emphasize that in the $A = 100$ region the strong p–n coupling of protons in the $g_{9/2}$ and neutrons in the $g_{7/2}$ shells can lead to much reduced pairing strength as discussed by Federman and Pittel (Fe 79). The blocking effect of the unpaired

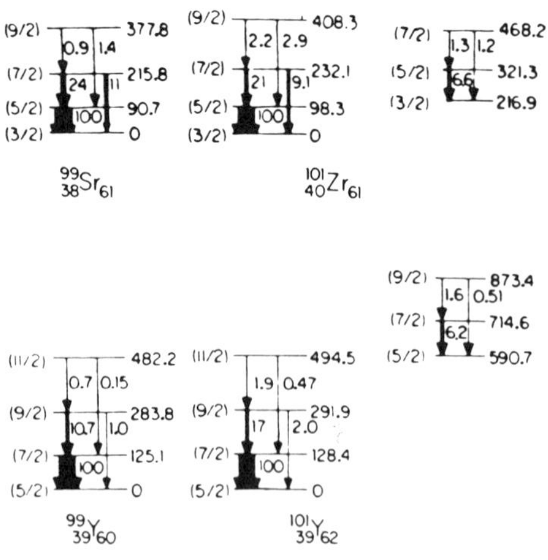


**Figure 39.** Rotational bands in deformed odd-$A$ Y isotopes and $N = 61$ isotones. Intraband $\gamma$ transitions and their relative intensities from $\beta$ decay are shown (Wo 83). The band in $^{99}$Y was reported first by Monnand and co-workers (Mo 82).

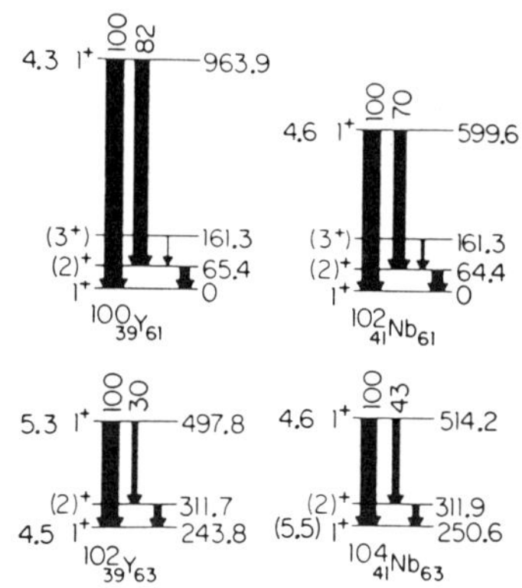


**Figure 40.** Energy levels in $^{100,102}_{\phantom{1}39}Y_{61,63}$ and $^{102,104}_{\phantom{1}41}Nb_{61,63}$ (Pe 86). Note the rotational bands with very large moments of inertia built on the $1^+$ states.

particles and the p–n coupling could suppress pairing to where $2\Delta \approx 0$, for example in a three-particle state which includes a $\pi g_{9/2}$ and $\nu g_{7/2}$ particle. It was suggested that the anomalously low energy $7/2^+$ states are $\pi(g_{9/2})^{-1}$, $\nu_1(g_{7/2})^1$, $\nu_2(d_{5/2})^1$ states which are pairing-free. This interpretation explains nicely the low log *ft* values of ≈3 for the beta decay of the $d_{5/2}$ states to the $7/2^-$ states in $^{103,105}Rh$ and $^{103,105,107,109}Ag$ and accounts for the observed gyromagnetic ratios. Heyde and Paar (He 86) have pointed out that these low log *ft* values can be explained within conventional models. However, to do so they must use the same proton–neutron configuration and so they too recognize that this can reduce pairing. Moreover, their picture would suggest that such fast beta transitions would be seen in many other regions of the periodic table, and they are not.

More convincing evidence for pairing-free states in this region come from the analysis of the odd–odd nuclei around $A = 100$ (Pe 86, Pe 86c). In even–even nuclei, the moments of inertia $\mathcal{I}^{e\text{-}e}$ are always much smaller than the rigid-body moment $\mathcal{I}_{rig}$ where, for an axially symmetric rigid rotor with deformation,

$$\mathcal{I}_{rig} = 0.4AM_nR^2[1 + (5/16\pi)^{1/2}\beta]$$

for a nucleus with $A$ nucleons of mass $M_n$ and radius $R = 1.2A^{1/3}fm$. Pairing-free deformed e–e nuclei are expected to have $X = \mathcal{I}_{exp}/\mathcal{I}_{rig} = 1$ near the ground state (Pa 75). For rare-earth and actinide nuclei with maximum deformation, $X^{e\text{-}e} \approx 0.5$ near the ground state. This reduction in $X^{e\text{-}e}$ arises from the residual pairing interaction and the $X^{e\text{-}e}$ for rare earths are well reproduced when pairing is taken into account (Na 65). Mottelson and Valatin (Mo 60) proposed that pairing-free states with very large deformation could be reached at high angular momenta. Very recently, superdeformed states have been reported in $^{152}Dy$ and $^{132}Ce$ by discrete $\gamma$-ray spectroscopy at spins of the order of 30–50$\hbar$ with $\beta_2 \sim 0.5$–$0.6$ (Tw 86, Ki 86). Indeed, the changes in nuclear shape and nuclear-shape coexistence at high spins are another major area of nuclear structure research and are offering us equally exciting, new vistas of the competition of different nuclear shapes and structures in one nucleus as discussed in the recent major review of Garrett and co-workers (Ga 86).

Federman and Pittel (Fe 79) have noted that strong proton–neutron coupling when the protons occupy the $g_{9/2}$ and the neutrons the $g_{7/2}$ orbitals in nuclei with $N \approx 60$, $Z \approx 40$ can significantly reduce pairing there. Moreover, adding an odd-proton or odd-neutron to an e–e core further reduces pairing (Na 65). In the $A \sim 100$ region,

**Table 1. Properties of $K^{\pi} = 1^{+}$ Bands Where $\mathscr{I}_{rig}$ was Calculated from $\beta = 0.3$ (Pe 86)**

| Nucleus | Bandhead energy (keV) | $X = \mathscr{I}_{exp}/\mathscr{I}_{rig}$ |
|---|---|---|
| $^{100}_{39}Y$ | 0 | 0.93 |
| $^{102}_{41}Nb$ | 0 | 0.92 |
| $^{102}_{39}Y$ | 243.8 | 0.88 |
| $^{104}_{41}Nb$ | 250.6 | 0.94 |

nuclei near the ground state have $X^{e\text{-}e} \sim 0.7$ and $X^{odd\text{-}A} \sim 0.8$. Odd–odd nuclei have a still larger quinching of pairing and they may have pairing-free states. As seen in Table 1, there are bands in $^{100,102}_{39}Y_{61,63}$ and $^{102,104}_{41}Nb_{61,63}$ which have $X$ values that approach 1 as expected for pairing-free states. The apparent collapse of pairing is taken to be related to the strong $(\pi g_{9/2}, \nu g_{7/2})$ coupling in the neutron-rich $A \sim 100$ nuclei (Pe 86b).

Final evidence comes from an analysis (Pe 86c) of the levels of an excited band in $^{98}Y$ (Lh 85) compared to $^{99}Y$ (Mo 82). In pairing-free bands, if shell effects on $\mathscr{I}$ are negligible as expected, all signature terms and higher-order terms are absent in the energy formula since $\mathscr{I} \approx \mathscr{I}_{rig}$:

$$E(I) = AI(I+1) + BI^2(I+1)^2 + CI^3(I+1)^3 + (-1)^{I+K}A_{2K}f(I, K)$$

In a $K \neq 0$ pairing-free band the signature term, which leads to oscillation in the level energies, would be absent. Figure 41 presents the four members of an excited band in $^{98}Y$ and a comparison of $\Delta E_{\gamma}(I \to I-1)$ for $^{99}Y$ and $^{98}Y$. One sees a striking difference in the $^{98}Y$ and $^{99}Y$ data with a total absence of any signature term in $^{98}Y$. Considering all these data one has strong support that we are now seeing pairing-free states at and near the ground states in neutron-rich $A \sim 100$ nuclei.

## 4.3. Role of Reinforcing Proton and Neutron Shell Gaps on Competing Nuclear Shapes, Strong Deformation, and Magic Numbers

As first pointed out by Hamilton (Ha 81b) and recently expanded by Hamilton and co-workers (Ha 84, Ha 85, Ha 85b, Ha 85c) the keys to understanding the rapid switch from near-spherical to strongly deformed ground states and vice versa for the

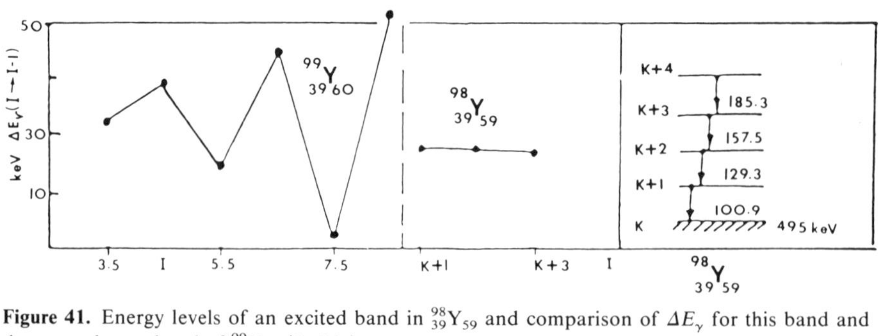


**Figure 41.** Energy levels of an excited band in $^{98}_{39}Y_{59}$ and comparison of $\Delta E_{\gamma}$ for this band and the ground-state band of $^{99}_{39}Y_{60}$ (Pe 86c).

excited bands in the Se–Kr nuclei are the importance of the new concepts of reinforcing and switching of shell gaps and what are the magic $N$ and $Z$ numbers in this region. By reinforcing shell gaps is meant the proton and neutron particle numbers are such that gaps occur in the single-particle levels for both protons and neutrons at the same deformation. Then the nuclear shape driving forces for both $Z$ and $N$ reinforce each other toward the same shape. By switching shell gaps is meant that one shell gap at one deformation may be important (magic) in stabilizing the nuclear shape for one combination of $N$ or $Z$, but there can be a switch to a nearby shell gap at a different deformation dominating at a different magic number for $N$ or $Z$ for another $N$, $Z$ combination. Schmidt and Vermeulen (Sc 80b) introduced independently for the spherical magic numbers the idea of "mutual support of magicities" from looking at atomic masses and two neutron separation energies. Zeldes (Ze 81) further emphasized their suggestion and in addition noted that the $2_1^+$ energies for $Z = 40$ nuclei indicate that the $Z = 40$ spherical magic number is strengthened by the $N = 50$ and 56 shell and subshell closures.

Shape coexistence and strong deformation for nuclei with $N \sim 38$–40 and $N \simeq 60$ are related to the number of protons which delicately controls the competition between deformed and near-spherical shapes in these regions by which shape the protons reinforce. The competition can be understood by looking at the gaps in the single-particle spectrum that stabilize the nuclear shape. For example, the spherical shell gap at 40 is magic for both $N$ and $Z$ when the other particle number for $Z$ and $N$ is at or near a strong spherical number like 28 and 50, as seen in double magic ${}^{90}_{40}Zr_{50}$ and ${}^{68}_{28}Ni_{40}$ discussed below. However, there is a switch to the deformed shell gap at 38 for both $N$ and $Z$ dominating when the other particle number corresponds to a shell gap at similar deformation as in ${}^{74,76}_{36}Kr_{38,40}$, ${}^{98}_{38}Sr_{60}$, and ${}^{100}_{40}Zr_{60}$, which are all very strongly deformed. Certainly the phenomena of reinforcing and switching shell gaps are shell-structure effects, but the idea and consequences of reinforcement of both proton and neutron shell gaps on the nuclear shape was not seen in the earlier calculations of unusual effects of Strutinski shell corrections by Brack and co-workers (Br 72). However, in their classic paper Brack and co-workers (Br 72) did emphasize that large gaps in the single-particle levels at large deformation should be equally as magic as the spherical gaps in giving the nucleus a stable shape, but now deformed. The new "deformed" shell gaps at $N = Z = 38$ and $N = 60$ disappeared before confirming their prediction.

The spherical double-magic character of the energy levels of ${}^{90}_{40}Zr_{50}$ was the first strong evidence for making $Z = 40$ a magic number, as shown in Figure 3. The levels in ${}^{68}_{28}Ni_{40}$ have been identified by Bernas and co-workers (Be 82) and likewise are characteristic of a spherical double closed-shell nucleus to support $N = 40$ as spherical magic. The levels of ${}^{96}_{40}Zr_{56}$ also are essentially those of a double closed-shell nucleus. One sees in Figure 28 the large shell gaps for a spherical shape ($\beta = 0$) for 28 and 50 and weak spherical shell gaps at 40 and 56 and somewhat weaker still at 38. In fact, ${}^{88}_{38}Sr_{50}$ is often taken as the inert double closed-shell core in shell model calculations. Note the $2_1^+$ energies in ${}^{90,96}_{40}Zr_{50,56}$ and ${}^{88}_{38}Sr_{50}$ as shown later in Figure 43 are very large, as characteristic of spherical double closed-shell nuclei. Meyer and co-workers (Me 86b) also have shown that in ${}^{98}Zr$ the $2_2^+$ level at 1590 keV is the one associated with the near-spherical ground state and not the 1222, $2_1^+$ level which is associated with the 854 keV, $0_2^+$ level. Thus, the ground state of ${}^{98}Zr$ still looks more spherical double magic like ${}^{96}Zr$ to show that the combined influence of the $Z = 40$ and $N = 56$ spherical gap reinforcement continues to $N = 58$.

Away from the $Z$ and $N = 28$ and 50 closed shells, the gaps in the Nilsson single-particle levels at $N$ and $Z \simeq 38$ for large prolate deformation, $\beta \approx 0.35$, in addition to the spherical gaps at $\beta \approx 0$ for $N = Z = 40$, were emphasized in order to explain the large ground-state deformation of ${}^{74,76}_{36}Kr_{38,40}$ (Ha 81c, Pi 81, Pi 82). However, the situation is more complex than just large prolate–spherical competition as recently emphasized by Hamilton and co-workers (Ha 84). Bengtsson and co-workers (Be 84) have presented details of the potential-energy surfaces, ground-state masses and shapes calculated for 4023 nuclei from ${}^{16}0$ to ${}^{179}112$ with a Yukawa plus exponential macroscopic model and a folded Yukawa single-particle potential. Their calculated single-particle levels for neutrons in this region (protons are similar) are shown in Figure 28. These levels differ in two respects from the modified oscillator (Nilsson) levels used earlier (Ha 81c, Pi 81, Pi 82). There are still gaps at $N$ and $Z = 38$ (Figure 28), but at larger prolate deformation $\beta_2 \sim 0.35$, which is in better agreement with the experimental values found in ${}^{74}Kr$ and ${}^{100}Sr$. However, there are also gaps at $N$ and $Z = 36$ for oblate shapes with equally large deformation ($\beta_2 \sim -0.35$). In addition, there are gaps for both prolate and oblate shapes at somewhat smaller deformation for $Z = 34$. Thus, there can be various competing shapes—large oblate (around $N \approx Z \approx 36$), large prolate ($N, Z \approx 38$), smaller oblate and prolate ($N, Z \approx 34$), and near-spherical shapes ($N, Z \approx 38, 40$)—in this region. Both the oblate and prolate deformed shapes can coexist in a delicate balance with each other and with the near-spherical shape. Very large oblate deformation is expected centered around $N = Z = 35, 36$ and very large prolate deformation centered around $N = Z = 38$. In both cases it is the combination of both $N$ and $Z$ reinforcing each other's preference for very large deformation that can drive the nucleus toward unusually large ground-state deformation, as first suggested by Hamilton (Ha 81b).

Let us briefly summarize the earlier $N = 38, 40$ data. The excited deformed band in ${}^{72}_{34}Se_{38}$ which coexists with the near-spherical ground band has large $B(E2)$ values (120 SPU) above the crossing of the two bands (Ha 76b) similar to ${}^{76}Kr$. A coexisting excited deformed band is found in $N = 40$ ${}^{74}_{34}Se_{40}$ (Ro 76) but not in $N = 42$ ${}^{76}Se$ (We 80) or the heavier-mass Se isotopes. It is possible that the near-spherical ground states of ${}^{72,74}Se$ are oblate structures associated with the shell gaps at 34 and 38 for small oblate deformation (see Figure 28). The experimental data for the observed nuclei around $N = 38$ and 40 are summarized in Table 2 (Ha 85c, Li 88). The data show that for nuclei with $N = 38$ and 40, when the protons have a strong spherical gap like 28, then 40 is like a spherical magic number. However, as $Z$ increases from 28, one begins to see low-energy deformed $0_2^+$ states as in ${}^{70,74}_{32}Ge$ and then well-developed deformed excited bands built on the $0_2^+$ states in ${}^{72,74}_{34}Se$, and then for $Z = 36$ the ground states become strongly deformed with low-energy $0_2^+$ near-spherical states and in $Z = 38$ only the strongly deformed ground states are seen. Thus, the $N = 40$ shell gap for a spherical deformation (Figure 3) is magic (important) only when it is reinforced by a proton number at a strong spherical magic shell gap like $Z = 28$. Likewise, $Z = 40$ is spherical magic only when reinforced by a strong neutron spherical magic shell gap as in ${}^{90}_{40}Zr_{50}$ and ${}^{96}_{40}Zr_{56}$.

Thus, as $Z$ increases from 28 toward 38, 40 for nuclei with $N = 38, 40$, there is a switch in importance (dominance) for the ground states from the 28, 40 spherical gaps to the $N$ and $Z = 38$ shell gaps at large deformation probably going through the small oblate deformed gap at $Z = 34$ dominating the ${}^{70,72}Ge$ and ${}^{72,74}Se$ ground states. As $N$ and $Z$ both approach 38, 40 the $N = Z = 38$ prolate deformed shell gaps reinforce

**Table 2. Summary of Experimental Data for $N = 38$ and 40 Nuclei**

| Nuclei | Observations | Heavier isotopes |
|---|---|---|
| ${}^{66,68}_{28}Ni_{38,40}$ | ${}^{68}Ni$ has spherical double magic energy levels; ${}^{66}Ni$ has spherical, large $E_{2_1^+} = 1.42$ MeV | |
| ${}^{70,72}_{32}Ge_{38,40}$ | near-spherical ground states and low-lying deformed $O_2^+$ states | $N = 42$–50 no deformed bands seen |
| ${}^{72,74}_{34}Se_{38,40}$ | near-spherical ground states and low-lying well-deformed $O_2^+$ states and deformed bands, clear shape coexistence | $N = 42$–50 no deformed bands seen |
| ${}^{74,76}_{36}Kr_{38,40}$ | strongly deformed ground states and near-spherical $O_2^+$ states | $N = 42,44$ $\gamma$-soft to $N = 50$ spherical |
| ${}^{76,78}_{38}Sr_{38,40}$ | strongly deformed ground state, but no near-spherical $O_2^+$ states observed | Smooth decrease in deformation as $N$ increases toward 50 |

each other to dominate over the $N = Z = 40$ spherical shell gaps to give very strong deformation ($\beta \sim 0.35$) to nuclei in this region (Ha 85b, Ha 85c). This switch in which shell gap dominates has been confirmed in the discovery of long-sought ${}^{80}_{40}Zr_{40}$ with the Daresbury Recoil mass separator (Li 88). The first excited $2_1^+$ energy is at 278 keV, so the ${}^{80}Zr$ ground state has large deformation though slightly less than for ${}^{78}_{38}Sr_{40}$, which has an identical $2_1^+$ energy.

Next we asked; does this picture of reinforcing and switching shell gaps illuminate any other regions of the periodic table? Indeed, this picture provides an understanding of a number of shape changes in different parts of the periodic table (Ha 81b, Ha 85, Ha 85c, Ha 85b). Let us look more closely at the very sudden onset of very large deformation one sees between ${}^{96}_{38}Sr_{58}$ and ${}^{98}_{38}Sr_{60}$ (see Figure 38) and between ${}^{98}_{40}Zr_{58}$ and ${}^{100}_{40}Zr_{60}$ (Figure 37). In $N = 60$ the (near-) spherical $0_2^+$ and deformed ground-state structures coexist in close proximity, in agreement with the theoretical calculations of Bengtsson and co-workers (Be 81). The bandhead of the deformed structure in ${}^{98}Zr$ is at 1.44 MeV (Ka 82), again in agreement with predictions. The sudden appearance and rapid drop in the deformed band coexisting with a near-spherical ground state and the switch in the ground state from near-spherical to deformed in the heavy Sr–Zr nuclei all parallel the shape coexistence and switch in ground-state shapes seen in the light Se to Sr nuclei. Further details and references to shape coexistence in these nuclei can be found elsewhere (Ha 81b, Ha 85).

The superdeformation reported in ${}^{100}_{38}Sr_{62}$ (illustrated in Figure 34) (Az 79) was attributed simply to a shell gap at $\beta \sim 0.3$–0.4 for neutrons at $N = 60$ (see Figure 28) giving rise to the potential energies shown in Figure 42. However, as pointed out (Ha 81b, Ha 85, Ha 85b) the $2_1^+$ energies (keV) in the $N = 62$ nuclei, ${}^{100}_{38}Sr_{62}$ (129.2), ${}^{102}_{40}Zr_{62}$ (151.9), and ${}^{104}_{42}Mo_{62}$ (192.2), show that the unusually strong deformation is quickly disappearing as $Z$ increases away from 38. Thus, it was interpreted (Ha 81b, Ha 85, Ha 85b) that the very large deformation seen in these heavy Sr nuclei is a result of the reinforcing of the neutron driving force toward large deformation associated with the gap at $N = 60$ for $\beta \sim 0.35$ by the proton driving force toward similarly large deformation associated with the $Z = 38$ gap for the same $\beta$, just as there is mutual

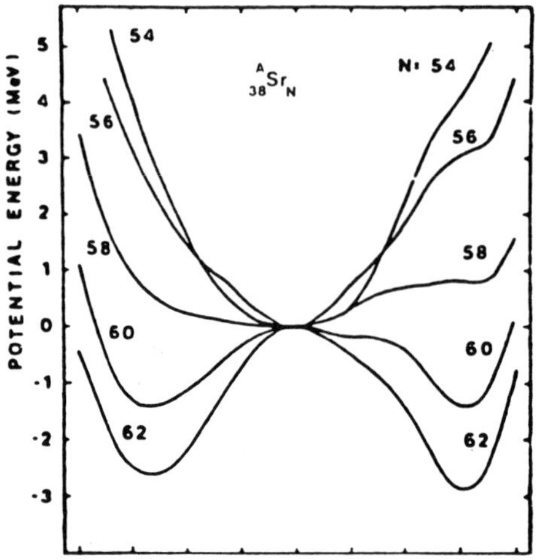


**Figure 42.** Potential-energy surfaces for $N = 54$ to 62, Sr nuclei. Note the change from one spherical minimum at $N \leq 58$ to two deformed minima at $N = 60, 62$ (Az 79).

reinforcing of the proton and neutron driving forces toward large deformation when both $N$ and $Z$ are at or near 38.

This reinforcing of $N$ and $Z$ shell gaps and switch in importance to the $Z = 38$ gap at large deformation for the heavy Sr nuclei provides an understanding of the complex and rather rapidly changing shape structures in the heavy $_{38}$Sr, $_{40}$Zr, $_{42}$Mo, $_{44}$Ru, $_{46}$Pd, and $_{48}$Cd nuclei with neutron numbers between 50 and 70, as shown in Figure 43. Starting on the left with $^{90}_{40}Zr_{50}$ and $^{88}_{38}Sr_{50}$, one sees they have unusually high

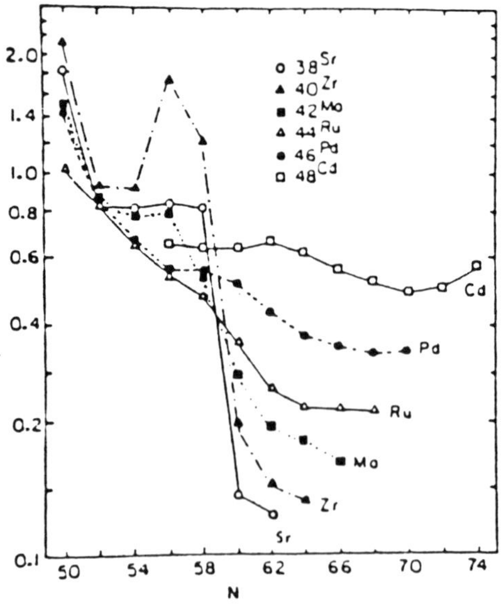


**Figure 43.** $2_1^+$ energies of heavy Sr to Cd nuclei which show the sudden onset of large deformation for $N = 60$ in Sr and Zr nuclei with $Z = 38$ and 40.

$2_1^+$ energies characteristic of double-magic nuclei. Here the driving forces of $N$ and $Z$, signified by the 40 and 50 spherical shell gaps, reinforce each other to give spherical double closed-shell nuclei. As $N$ increases, note ${}^{96}_{40}Zr_{56}$ again looks double magic. Here the spherical driving force of the $Z = 40$ gap is reinforced by the $N = 56$ weaker spherical gap (see Figure 28) and still produces a near double-magic nucleus there, too. In the Sr nuclei, while there is no peak at $N = 56$ (to signify the 38 spherical gap is not as strong as the 40 spherical one), the $N = 56$ spherical gap does lead to a high shoulder in the $2_1^+$ energies characteristic of spherical shapes out to $N = 58$. Then there is a sudden onset of deformation between $N = 58$ and 60 in both the Zr and Sr nuclei. It is sudden because the weaker $N = 56$ spherical gap reinforces the $Z = 40$ spherical gap to keep the Zr and Sr nuclei spherical as $N$ increases from 50, until suddenly the spherical driving forces are overpowered by the reinforcement of the proton and neutron driving forces toward large deformation associated with the $Z = 38$ and $N = 60$ deformed gaps. Note the Sr nuclei are considerably more deformed than the Zr nuclei at the same $N$, because now they are on the center of the proton deformed gap at 38.

Now look at the heavier $Z$ nuclei in Figure 43. For the $Z = 42$ ${}_{42}Mo$ isotopes there is a shoulder until $N = 56$, but already at $N = 58$ there is a sizable drop in the $2_1^+$ energy. So even at $Z = 42$ the reinforcing of the two weaker spherical gaps at 40 and 56 is lessening. Also, at $N = 60$ the ${}^{102}Mo$ deformation is markedly less than in ${}^{98}_{38}Sr$. By the $Z = 44$, Ru isotopes, only a relatively smooth change in deformation is seen from the somewhat less spherical ${}^{94}_{44}Ru_{50}$ to the much less deformed $N \geqslant 60$ Ru nuclei compared to the same $N$, Sr ones. There is somewhat more of a shoulder around $N = 56$, 58 in the $Z = 46$ Pd isotopes, but considerably less deformation than even in the Ru isotopes for $N \geqslant 60$, while in the $Z = 48$ Cd isotopes (2 protons away from the strong 50 shell gap) the shapes are essentially constant, near-spherical from $N = 56$ to 74. Thus in this region, too, the strong ground-state deformation is clearly related to reinforcement of the $N = 60$ gap at large deformation by the $Z = 38$ gap at similar large deformation with the magnitude of the deformation rapidly decreasing as $Z$ increases.

Federman and Pittel (Fe 79) presented what they called a unified shell model description of nuclear deformation. The basic ideas come from the work of de Shalit and Goldhaber (de 53) and Talmi (Ta 62) and Unna (Un 63) where, for example in the (2s, 1d) shell, the strong shell-model configuration mixing which was necessary to give deformation there was traced to the strong isoscalar residual interaction between neutrons and protons when they occupy the spin-orbit "partner" orbitals $1d_{5/2}$ and $1d_{3/2}$. The strong overlap between the neutron and proton orbits, especially when $n_n = n_p$ and $l_n \approx l_p$, provides the mechanism for strong interaction. This is itself a type of reinforcement between neutrons and protons in a shell model description. Federman and Pittel (Fe 79) studied the shape transitions which were known then in the heavy Zr and Mo isotopes. In the Zr nuclei, they found that the $1g_{9/2}$ proton and $1g_{7/2}$ neutron "partner" orbitals, which satisfy the above conditions, play the same role in producing the correlations required for deformation. Their calculations show that the largest number of $g_{9/2}$ protons and $g_{7/2}$ neutrons are in a low-lying $0_2^+$ collective state of ${}^{98}Zr$ as compared to the ground state, with the reverse true for the $0_1^+$ and $0_2^+$ states of ${}^{100}Zr$. They do not find any rotational band built on the $0_2^+$ state, in agreement with the recent results of Meyer and co-workers (Me 86b) (Figure 36). They suggest that the rotational band, seen in ${}^{100}Zr$, gradually develops from the excited $0_2^+$ states which are dropping

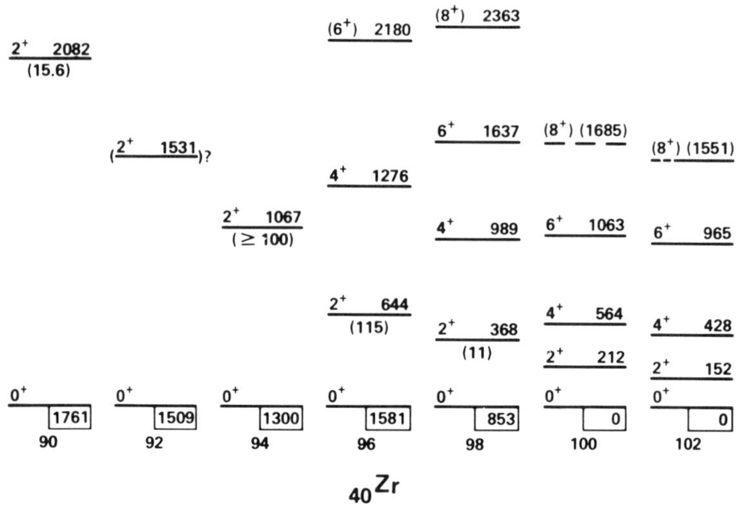


**Figure 44.** Bands built on the $0_2^+$ states with protons in the $g_{9/2}$ orbital in $^{90-98}Zr$ and deformed bands in $^{100,102}Zr$ (Me 86b).

in energy (see Figure 44) as neutrons are added, but could not calculate the members of such a band. In addition, in the Mo isotopes they find that the $1h_{11/2}$ neutrons and $1g_{9/2}$ protons also play a role in the n–p correlations in the $N > 62$ nuclei. Note these $\pi g_{9/2}$ orbitals are "intruder" orbitals in the language developed for the $h_{9/2}$ orbital in the Pt–Tl region.

Federman and Pittel (Fe 79) claim the isoscalar residual interaction between neutrons and protons provides a universal explanation of nuclear deformation throughout the periodic table. However, all of the cases they calculated involved neutrons and protons in "partner" orbitals, for example $\pi g_{9/2}$, $\nu g_{7/2}$ in the Zr–Mo nuclei. It seems clear that the n–p coupling is very important. However, it must be shown how this accounts for the behavior of the Sr through Cd nuclei shown in Figure 43 and for the $N = Z = 38$ region. The superdeformation in $^{100}Sr$ was not known to them. It is not clear how the n–p correlations from the filling of the $1g_{9/2}$ proton and $1g_{7/2}$ orbitals produce the largest deformation in the two-proton lower $Z = 38$ Sr nuclei and why the sudden deformation onset at $N = 60$ is gone by $Z = 44$. In the $N = Z = 38$ deformed region, which also was unknown to them, one does not have partner orbitals, but the condition $l_p \simeq l_n$ is fulfilled. Nevertheless, one must show how filling the $1g_{9/2}$ orbitals for both protons and neutrons would produce maximum deformation at $N = Z = 38$, large deformation in $^{74,76}_{36}Kr_{38,40}$, and quick disappearance of the deformed bands at $N \geqslant 42$. It is clear experimentally that the strongest deformations occur for $Z = 38$ when reinforced by $N = 38$ or by $N \geqslant 60$. There is also evidence in the $A = 70$ region for another shape transition to oblate shapes with smaller deformation (Ra 86) and another transiton to large oblate deformation also is expected around $N \approx Z \approx 35$, 36 as seen from Figure 28, and this is a further challenge.

Finally, in the Pt–Tl region when a pair of particles are promoted to the "intruder" $h_{9/2}$ orbital whether in the ground state or first excited state, large deformation occurs. However, large deformation is not found in the excited $0_2^+$ state which has a pair of $g_{9/2}$ particles in $^{98}$Zr and the lighter Zr nuclei. On the other had, in the Pt–Tl region one must consider the role of n–p coupling which has not been considered adequately there to date. Likewise, in the light Se, Kr, Sr nuclei, the promotion of pairs of particles to the $\pi g_{9/2}$ and $\nu g_{9/2}$ orbitals yields very large deformation for both excited states in $^{72,74}$Se and the ground states of $^{74,76}$Kr and $^{78,80}$Sr. This difference in the role of the intruder orbitals and n–p coupling in these regions needs to be understood. It would be interesting to incorporate both of these ideas, the reinforcement of the proton and neutron shell gaps and the n–p interactions in a theoretical analysis. We will return later to the importance of the n–p coupling and superdeformation in the light Se–Kr nuclei.

Another way of looking at the trends of the neutron–proton interaction systematically in nuclei may be to plot various data, such as the $2_1^+$ energies or $E_{4^+}/E_{2^+}$, against the product of the active number of neutron and proton particles or holes as proposed by Casten (Ca 85). Tabor (Ta 86) has attempted this in the light Se–Sr region. Owing to the occurrence of nuclear-shape coexistence in these nuclei, the $2_1^+$ energies show no regularity; however, the $E_{4_1^+} - E_{2_1^+}$ energies, which are characteristic of the deformed structures, do show a smooth behavior (Figure 45) in a different way emphasizing the proton–neutron reinforcement.

The potential-energy surfaces have also been calculated by the Lund–Warsaw–UNISOR collaboration (Na 86) as illustrated for the light Kr isotopes in Figure 46, and by Bonche and co-workers (Bo 85b) in the HF + BCS self-consistent approach including triaxiality and by Galeriu and co-workers (Ga 86b) who showed for the first time that a Nilsson-type potential in the Strutinsky approach can account for the shape coexisting structures. Gratifyingly, those results are basically in good agreement. The sudden appearance of clear shape coexistence at $N = 40$ as seen experimentally in Se

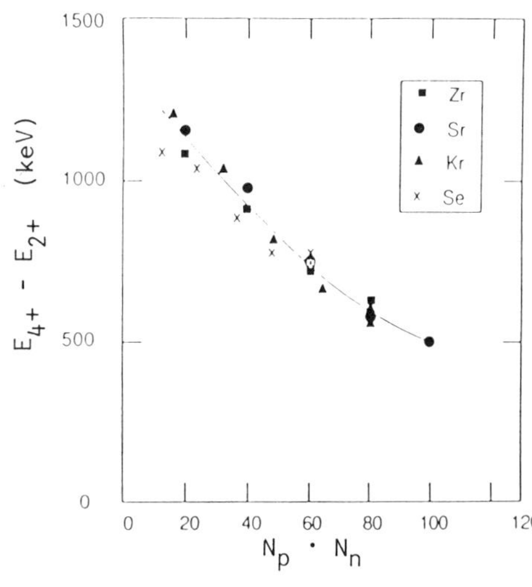


**Figure 45.** Energies $E_{4^+} - E_{2^+}$ for Se, Kr, Sr, Zr nuclei with $A$ around 70–80 vs. the product of the active number of proton and neutron particles or holes (Ta 86).

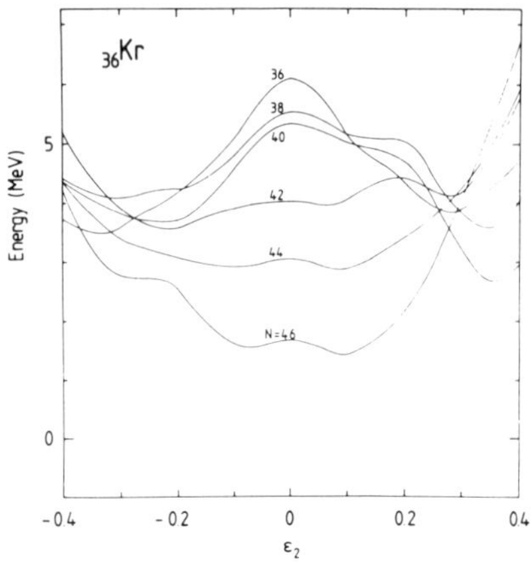


**Figure 46.** Potential energies for $N = 36$–46 Kr isotopes (Na 86). Note the shift to two minima for $N \leqslant 40$.

and Kr nuclei is nicely reproduced. More detailed examples for the $N$ = 36, 38, 40 isotones of $_{34}$Se, $_{30}$Kr, and $_{38}$Sr are shown in Figure 47 (Na 86). These PES are in very good agreement with the strong prolate deformation in $^{80,78,76}$Sr, the well-deformed prolate ground states with shape coexistence for $^{76,74,72}$Kr, and the less-deformed ground states and excited well-deformed bands in $^{74,72,70}$Se. In addition to careful consideration of triaxiality, the calculations of Bonche and co-workers (Bo 85b) suggest that the neutron and proton deformations of the heavy isotopes of Zr could be remarkably different. If this effect is verified experimentally, this would be the first example of a dissociation between shapes of the neutron and proton densities.

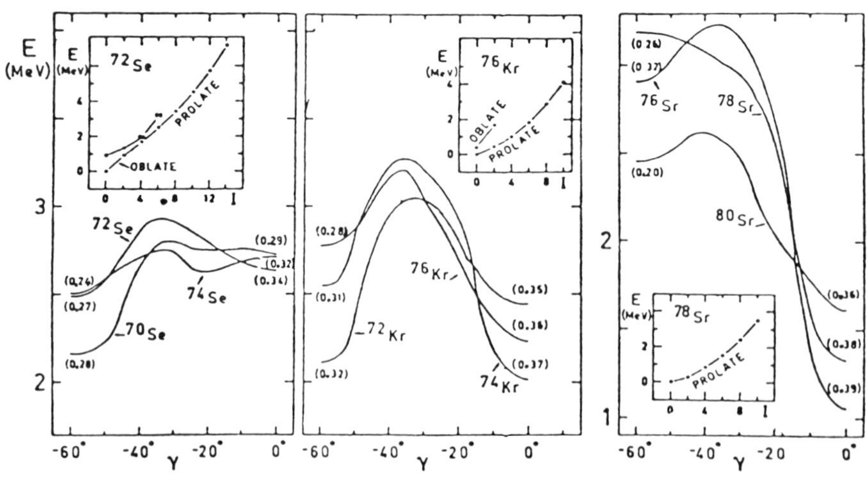


**Figure 47.** Potential energies for $^{70-74}$Se, $^{72-76}$Kr, and $^{76-80}$Sr (Na 86). The competing oblate and prolate minima are seen to be close in energy in the Se and Kr cases with a reversal as to which is lowest and a deep prolate minimum in the Sr cases.

Nazarewicz (Na 86) has emphasized that in this region, just as in the Au–Pt region, one should look at the structure of the minima being seen if one aims at a more detailed description of the low-energy excitations. The standard PES are mixtures of pieces that correspond to different shell model configurations associated with crossings between single-particle levels as a function of the deformation. Such a detailed description was made for particle number 36, in this region. So, for example, for particle number 36 the lowest shell model configuration does not contain any $g_{9/2}$ orbital for $-0.19 \leqslant \beta_2 \leqslant 0.19$. It has two occupied $g_{9/2}$ orbitals for $\beta_2 \leqslant -0.19$ and $0.19 \leqslant \beta_2 \leqslant 0.38$ and four intruder orbitals for $0.38 \leqslant \beta_2 \leqslant 0.70$. However, for $^{80}$Sr at $\beta = 0.42$, $\gamma = 15°$, the $h_{11/2}$ and $i_{13/2}$ orbitals are the "intruders."

As one looks at the PES for the Se nuclei in Figure 45, they are soft in the $\gamma$ plane so that a small barrier would be expected. However, the experimental data show very well-separated coexisting structures (Ha 74). Nazarewicz (Na 86) notes that $\gamma$ softness may be expected if one does not consider changes in the shell model configuration. In the $^{70-74}$Se nuclei, the prolate and oblate shapes differ by two to four $g_{9/2}$ intruder orbitals being filled and this leads to a much stronger separation between the prolate and oblate states than one would expect from the total PES only. Thus, here again the intruder orbital and PES pictures are not really separate ways but only different emphases in understanding the structure of these nuclei. Although it would be quite difficult, it would be most interesting to include the n–p coupling in the PES calculations. Then one would have a better microscopic understanding of the structure of the competing shapes. Ahalpara and co-workers (Ah 85) have reported preliminary results of a microscopic description of the collective states in the neutron-deficient Br to Sr nuclei in the framework of a deformed configuration mixing shell model approach based on projected Hartree–Fock Formulism starting with a $^{56}$Ni core. The results of their first calculations in the simplest possible approach are encouraging.

Before proceeding, an additional important point concerns the role of the proton and neutron $g_{9/2}$ orbitals and their possible influence in producing the large deformation and the remarkably regular band structure seen to spin $22^+$ in $^{72}$Se (see Figure 48) (Ra 86b). The anomalous behavior at low spins in Figure 48 arises from shape coexistence. Let us look at the high-spin data. Earlier it was discussed how strong n–p coupling of the $\pi g_{9/2}$ and $\nu g_{9/2}$ orbitals could lead to superdeformed, pairing-free states in the odd–odd nuclei around $A = 100$ (Pe 86, Pe 86b, Pe 86c). A strong reduction to even collapse of pairing may be occurring in $^{72}$Se and perhaps $^{70}$Se (Pe 86d). The deformed structure seen to spin $22^+$ has a remarkably high and constant moment of inertia. Its structure undoubtedly involves at least a pair of protons and a pair of neutrons in the $g_{9/2}$ orbital. Indeed, a pairing-free band would have a constant moment of inertia in agreement with the remarkably constant moment of inertia seen to high spin in $^{72}$Se (Ra 86b, Pe 86d). So strong n–p coupling of these particles could have an important influence on the disappearance of a pairing here, too.

The reinforcing shell gap picture also provides an understanding of other regions of the periodic table. First let us note, however, that the $N$ and $Z$ spherical gaps at 40 and 56 and the deformed gaps at 38 are weaker than the major spherical gaps of 2, 8, 20, 28, 50, .... The influence of the 38, 40, 56 gaps are felt only when reinforced by a similar shape-driving force from the other particle. Their effects are weaker for the addition or subtraction of even 2 particles and are considerably less important or disappear when four particles away. These examples emphasize that when a weaker shell gap is important, its importance quickly goes away with the addition of a few

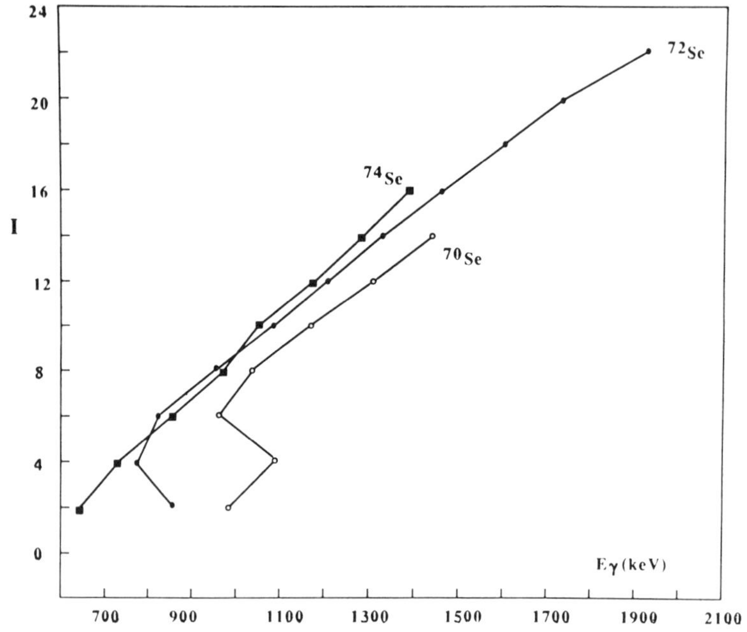


**Figure 48.** Plot of angular momentum as a function of $\gamma$-ray energies for $^{70,72,74}$Se (Ra 86b, Ra 86c).

particles. This picture gives a quite consistent interpretation of the weak $Z = 64$ spherical shell gap being reinforced by the strong $N = 82$ spherical shell gap to make $^{146}_{64}Gd_{82}$ a double-magic nucleus (Kl 81). And, as is found, its double-magic spherical character quickly disappears if one changes $Z$ or $N$ by 2 or 4 particles as discussed in the next section.

Moreover, the weak $Z = 64$ shell gap now can be seen to play the same role in the "classical" sudden onset of deformation between $N = 88$ and 90 in the rare earths as the $N = 56$ shell gap plays for the sudden onset in the Sr–Zr region. The sudden onset of deformation in the rare earths was seen 35 years ago in the isotope shift data of Brix and Kopfermann (Br 49) on $^{150,152}$Sm ($N = 88, 90$). The $2_1^+$ energies there and in the analogous Gd isotopes similarly reflect the sudden change in deformation. Recent isotope shift data (Figure 49) from laser spectroscopy studies of the $_{56}$Ba, $_{66}$Dy, $_{68}$Er, and $_{70}$Yb isotopes across the $N = 88$–$90$ region show that only for $Z = 66$ does a pronounced sudden onset of deformation occur between 88 and 90 (Ne 82, Ne 83, Mu 83). There is only a smooth gradual change to deformation as one goes away from spherical $N = 82$ structures in the $_{56}$Ba, $_{68}$Er, and $_{70}$Yb isotopes. One can see the same effect by comparing recent laser-spectroscopy measurements of the mean-square charge radii in $Z = 37$, Rb and $Z = 55$, Cs isotopes (Th 82) as shown in Figure 50a and b. There is a sudden sharp increase in the mean-square radius ($\delta\langle r^2\rangle$ is the change in $r^2$ from one nucleus to the next) at $N = 60$ for the $_{37}$Rb nuclei as the nearby $Z = 38$ and $N = 60$ gaps reinforce each other. This sudden onset of deformation at $N = 60$ in $^{97}$Rb

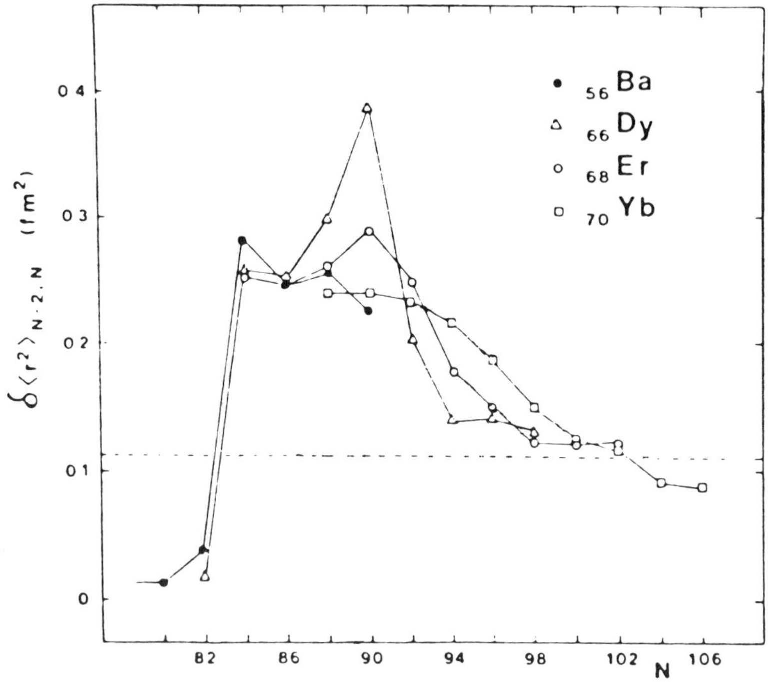


**Figure 49.** Differential increment of the nuclear mean-square charge radii for $\Delta N = 2$. Note that of the elements shown here, only Dy ($Z = 66$) exhibits the "spike" characteristic of a sudden onset of deformation between $N = 88$ and 90 (Ne 82, Ne 83).

had been seen earlier in quadrupole-moment measurements (Kl 79, Ek 81). However, in the $_{55}$Cs isotopes there is only a smooth increase in deformation with no break between 88 and 90. The suddenness of the onset of deformation in the $Z = 62$–66, Sm, Gd, Dy nuclei is the local effect of the $Z = 64$ spherical shell gap keeping these nuclei spherical further away from $N = 82$ than occurs in nuclei with $Z$ more than 2 protons removed from 64, just as $N = 56$ acts to keep the $Z = 38$, 40, Sr, Zr nuclei spherical further away from $N = 50$ than would otherwise occur (see Figure 43).

In conclusion, we see that over a sizable range of nuclei many new data on the shape and structure of nuclei far from stability can now be understood in terms of the importance of both the reinforcing and switching of proton and neutron shell gaps. New deformed magic numbers are now established and deformed double-magic nuclei associated with them. These new phenomena of reinforcing and switching shell gaps and deformed magic numbers are significant advances in the nuclear shell model and further demonstrate how studies of nuclei far from stability have opened up new understandings of the nucleus which could not have been known by studying near-stability nuclei.

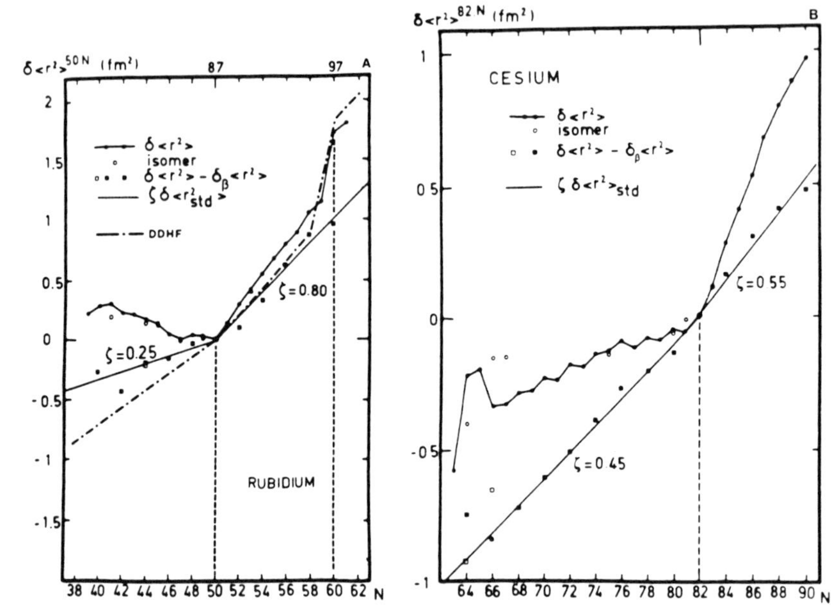


**Figure 50.** (a) Experimental variation of the mean-square charge radius of $_{37}$Rb isotopes (Th 81, Th 82). The spherical part deduced by subtraction of $\delta_\beta\langle r^2\rangle$, which reflects the deformation at constant volume, is compared to the modified standard low: $\zeta \times \delta\langle r^2\rangle_{\text{std}}$ (solid straight lines). The data are also compared to the DDHF calculated values (dot–dash). Note the shell effect at $N = 50$ and the effect of a sudden shift to large deformation between $N = 58$ and 60. (b) Experimental variation of the mean-square charge radius of $_{55}$Cs isotopes (Th 81b, Th 82). The spherical part deduced by subtraction of $\delta_\beta\langle r^2\rangle$, which reflects the deformation at constant volume, is compared to the modified standard low: $\zeta \times \delta\langle r^2\rangle_{\text{std}}$ (lower solid straight lines). Note the shell effect at $N = 82$ and the absence of any sudden change in deformation between $N = 88$ and 90.

## 5. Shell Closures and "Classical" Spherical Magic Numbers Far From Stability

An important direction in research on nuclei far from stability has been to explore the question of how magic far from stability are the "classical" magic spherical closed-shell numbers for neutrons and protons obtained from stable or near-stable nuclei as shown in Figure 3. Above, we have described the surprises in the $N$, $Z = 38$, 40 region where the new double closed-shell-type nucleus $^{68}_{28}Ni_{40}$ supports the near-stability magic numbers including 40 but well-deformed $^{74,76}Kr$ and $^{76,78}Sr$ do not. In light nuclei, a sudden onset of large deformation around $^{31}_{11}Na_{20}$ and $^{32}_{12}Mg_{20}$ has been suggested (Dé 81) but not confirmed by laser measurements (To 82). However, the

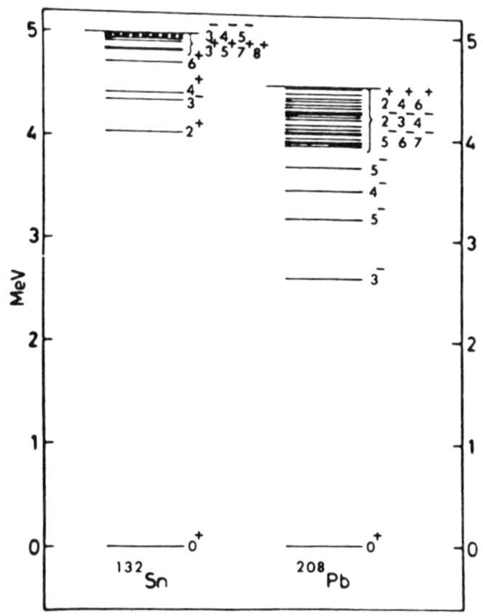


**Figure 51.** A comparison of the lowest excited states in $^{132}$Sn and $^{208}$Pb (Bl 81).

$N = 20$ spherical magic number is not as important as the $N = 14$ subshell closure far off stability (To 82).

Considerable effort has been made to probe the regions of $^{132}_{50}S_{82}$ and $^{100}_{50}Sn_{50}$ which were expected to be double closed-shell nuclei. Indeed the levels of $^{132}$Sn are characteristic of a double spherical closed-shell structure (Figure 51) (Bl 81). The $^{132}$Sn levels have been studied recently at ISOLDE to provide more detailed information for comparison with theory (Bj 86). The over 4 MeV gap to the first excited state of $^{132}$Sn is strikingly large compared to doubly magic $^{208}_{82}Pb_{126}$ with its 2.6 MeV gap. No other nucleus above $^{16}$O has a gap so large. Even allowing for the difference in size, $^{132}$Sn exhibits the strongest shell closure of any nucleus from this point of view.

Until recently, $^{208}$Pb was the only heavy spherical doubly magic nucleus for which the binding energies of the valence single-particle states were known. During the last decade, however, studies on neutron-rich fission products have provided a rather complete picture of the single-particle spectrum of spherical doubly magic $^{132}$Sn. The most extensive experimental work in the region was carried out at Studsvik several years ago by DeGreer and Holm (De 80), and more recently by Fogelberg and Blomqvist (Fo 84).

Following this latter work, where the $d_{5/2}$ level in $^{131}$Sn was observed, Leander and co-workers (Le 84) compared the observed single-particle energies with three current models and found for all three that the deviations from experiment in $^{132}$Sn are correlated with the corresponding deviations in $^{208}$Pb. The three models in question are: Hartree–Fock with the Skyrme III interaction, the folded Yukawa single-particle potential, and the Woods–Saxon single-particle potential (Le 84). The five experimental masses needed to determine single-particle energies above and below the shell gaps have all been measured; several proton states above the $Z = 50$ gap and all the neutron states between the $N = 50$ and 82 gaps are known [see references cited in (Le 84)]. In the spirit of a similar analysis by Blomqvist (Bl 81), Leander and co-workers (Le 84) demonstrated that, for the three models in question, the single-particle levels in $^{132}$Sn are similar to those in $^{208}$Pb and that a part of the discrepancy between experiment and theory is the same in $^{132}$Sn as it is in $^{208}$Pb, both for individual levels as well as the overall binding energy. Concomitant to the analysis of $^{132}$Sn in terms of its analogue $^{208}$Pb, Leander and co-workers (Le 84) estimated the proton binding energy in $^{100}$Sn ($^{56}$Ni being its analogue) and found it to be favorable for the synthesis of $^{100}$Sn (3 MeV).

While $^{100}_{50}\mathrm{Sn}_{50}$ remains unidentified, considerable progress has been made in studying this double closed-shell region through neighboring nuclei two to six nucleons removed (Ce 77, El 78, Sa 79, No 80, No 80b, Bé 81, Sc 81, Ti 81, Am 82, Ku 82, Pł 82, Tr 82, Gr 83, Ko 83, Pi 83, Wo 83b, Og 83, Ro 83). The island of $\alpha$ radioactivity just above $^{100}_{50}\mathrm{Sn}_{50}$ (Ro 83, Sc 81) shows that the reduced $\alpha$ widths there exhibit the same trend as seen just above the $N = 82$ and 126 closed shells. The GSI separator group have studied $\beta$-delayed proton emitters like $^{103,105}_{50}$Sn and $^{100,101}_{49}$In (Ti 81, Ku 82).

The systematics of the $2^+$ energies in even–even cadmium isotopes show only a small rise in $^{102}_{48}\mathrm{Cd}_{54}$ (Bé 81, Tr 82) compared to $^{104\text{-}118}_{48}$Cd (778 keV compared to 658, 633, 633, 658, and 617 keV). For odd-$A$ $^{103\text{-}109}$Cd, in-beam spectroscopy has revealed decoupled bands built on $\nu h_{11/2}$ $\nu g_{7/2}$ orbitals, which provide experimental evidence that the neutron-deficient Cd isotopes are moderately deformed and of prolate shape (Tr 82). While the rise in the $2^+$ energy between $^{104}_{48}\mathrm{Cd}_{56}$ and $^{102}_{48}\mathrm{Cd}_{54}$ indicates increased sphericity, it is not as dramatic as might be expected just four neutrons away from $N = 50$ (Tr 82). However, it must be kept in mind that these are soft transitional nuclei and not rigid rotors. The corresponding rise in the $2^+$ energy in the $Z = 46$ Pd isotopes is considerably more dramatic in going from $N = 52$ to 50, 863 to 1416 keV for $^{98}_{46}\mathrm{Pd}_{52}$ to $^{96}_{46}\mathrm{Pd}_{50}$, respectively, to indicate a strong shell closure at $N = 50$ (Ku 82, Pi 83).

The GSI separator group have explored the expected double-shell closure at $^{100}$Sn through the quenching of GT $\beta$-transitions. Beta- and gamma-spectroscopy studies of $N = 50$ $^{96}$Pd and $Z = 50$ $^{104,106,108}$Sn yielded fast $\beta$ decays with summed strengths that correspond to log $ft$ values of 3.1 to 3.5. These are interpreted as allowed GT transitions where a proton in the single-particle orbit $g_{9/2}$ goes into a spin-orbit partner $g_{7/2}$ neutron. While the experimental GT strengths agree with the theoretically expected trends as $^{100}$Sn is approached, there is a sizable difference in magnitude (see Figure 52). The experimental values are 4 to 14 times smaller than predicted by an extreme single-particle shell model. Therefore, nuclear structure and configuration mixing have to be considered. To account for the smearing out of the shell closures Towner (To 85) considered neutron and proton pairing and configuration mixing of the spin-orbit partner (core polarization). However, more recent extended pairing calculations (Do 86) in this region yield much less retardation from pairing than first found (To 85) and so do not reproduce the magnitudes of the strength functions.

A wealth of new information on nuclear radii and moments is now available in the indium and tin isotopes from experiments at GSI, ISOLDE/CERN, and at KfK Karlsruhe. From $N = 56$ to 78 in $^{105\text{-}127}$In, the ground-state magnetic dipole moment and essentially the electric quadrupole moments for $I = 9/2$ are surprisingly constant

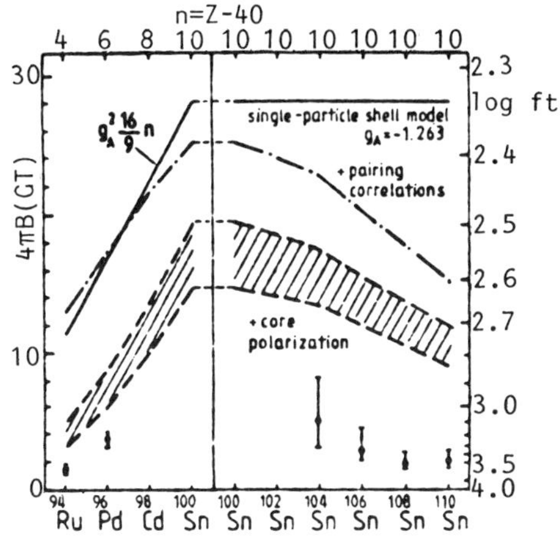


**Figure 52.** Experimentally observed total GT transition strength $B(\text{GT})$ of even-mass $N = 50$ isotones, tin isotopes, and predictions (Kl 86). Corresponding log $ft$ values are given at the right-hand scale. Full lines represent predictions from the extreme single-particle shell model as a function of the number $n$ of protons in the $g_{9/2}$ shell (left), assuming the free-neutron value of the coupling constant $g_A$. Lower transition probabilities are obtained by taking into account, at first, pairing correlations of protons and neutrons [dot-dash lines from model II (Do 86)] and then first-order core polarization from particle-hole interactions. The shaded region is obtained for different residual forces (To 85).

with magnetic moments about 80% of the Schmidt value (Lo 85, Ul 85, Eb 87). However, for $^{115\text{-}125}$In the $I = 1/2$ isomer magnetic moments drop more and more below the lower Schmidt limit with increasing mass (Eb 87). The ground-state charge radii for tin, indium, and cadmium show a parabolic behavior with small changes in deformation between the $N = 50$ and 82 shell closures as predicted by Talmi's core-polarization model (Ta 84); see Figure 53. If the deviation from a linear (droplet-like) dependence is indeed the result of deformations, the derived maximum (mid-shell) deformation parameter deduced from isotope shifts, however, is about twice as large as from quadrupole moments of $B(\text{E2})$ values in this region as seen in Figure 53. Collective effects other than a static quadrupole deformation will have to be considered. These possibly can be quantified from these data (Eb 87, Eb 86), for example by parametrizing transition-region zero-point motion (My 86).

While the ground state of all $Z = 50$ and 82 nuclei far off stability have been found to be near-spherical, nuclear-shape coexistence is seen in both tin and lead nuclei where quite unexpected low-lying deformed bands are observed. In the closed proton shell tin nuclei midway between the double closed-shell $^{100}$Sn and $^{132}$Sn deformed, excited bands coexisting with the spherical ground states were first reported by Ramayya and Hamilton (Ra 74) in $^{116}_{50}\text{Sn}_{66}$ and are now well established there (Br 79) (see Figure 54). In the tin isotopes the structure of the deformed state is assigned $\{1/2[431]^2 9/2^+[404]^{-2}\}$ associated with promoting a pair of $g_{9/2}$ protons to the $g_{7/2}$ orbital from above the $Z = 50$ closed shell. Kantele (Ka 84) has summarized their careful, detailed work on the E0 decays in the Cd and Sn nuclei, which help clarify the shape coexisting structures seen in those nuclei.

Soon after the discovery of shape coexistence in Hg nuclei, shape coexistence was predicted in closed proton shell lead nuclei very far from stability for $A \leqslant 192$ (Ma 77). Hence it was surprising when Van Duppen and co-workers (Va 84) found evidence for low-lying excited deformed structures coexisting with the spherical ground states of heavier lead isotopes up to $A = 198$. Using the LISOL facility at Leuven, they studied the decays of $^{190\text{-}198}$Bi which revealed low-lying excited $0_2^+$ states which drop in energy to become the first excited state of $^{190\text{-}194}$Pb as shown in Figure 55. In $^{196}$Pb, they

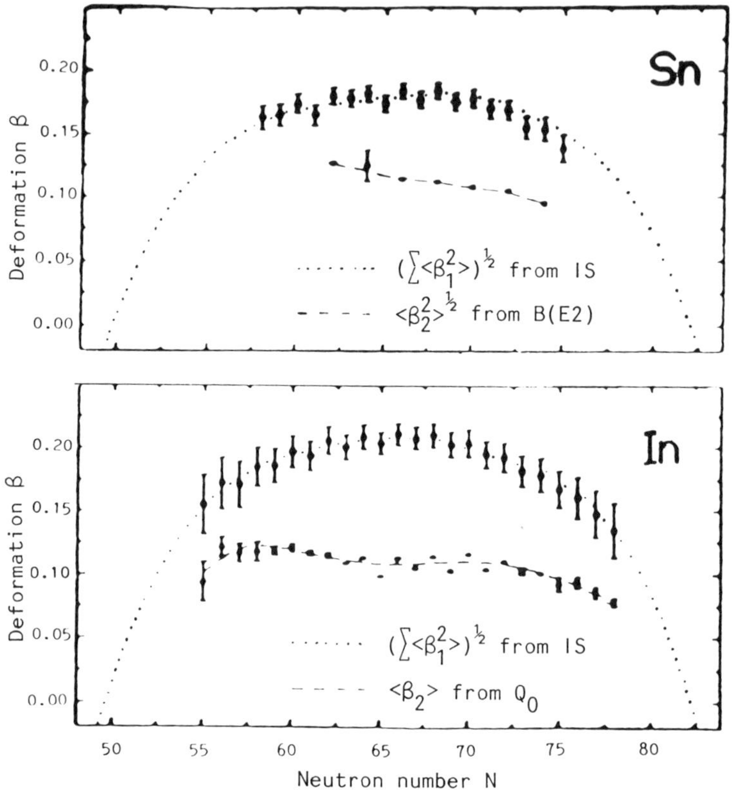


**Figure 53.** Deformation parameters $\beta$ for tin and indium isotopes. Squares indicate values obtained from isotope shift (IS) data as deviations from a linear (droplet-like) mass dependence; the dotted curve is the result from a parabolic least-squares fit. Circles mark results deduced from experimental $B$(E2) values for quadrupole moments [$^{108\text{-}111}$Sn (Eb 86), $^{112\text{-}129}$Sn (An 86), $^{106,104}$In (Lo 85), $^{107\text{-}111}$In (Ul 85), $^{104,112\text{-}127}$In (Eb 87)].

identified $2^+$ and $4^+$ states built upon this $0_2^+$ state (see Figure 56). This band has been extended to $6^+$ (Pe 86e). The difference in the $2^+ \rightarrow 0^+$ energies, 307 to 1049 keV in the excited and ground band, respectively, demonstrates the marked difference in deformation of these two bands. The $0_2^+$ states are assumed to have a 2p–2h structure based on the promotion of a pair of particles to the $h_{9/2}$ intruder orbital from above the $Z = 82$ shell. Kantele and co-workers (Ka 86, Ka 86b) extended the 2p–2h systematics to include $^{202,204}$Pb, so now this configuration is seen from $^{208}$Pb to $^{190}$Pb. They demonstrate that the picture of the proton–neutron residual interaction discussed by Heyde and co-workers (He 85) does reproduce the energies of the $0^+$ state assigned a 2p–2h configuration. However, these data present very interesting challenges because no theories have predicted low-lying collective deformed bands in any lead isotope. It will be important to establish if there are collective bands to high spin built on all these $0_2^+$ states and their degree of collectivity.

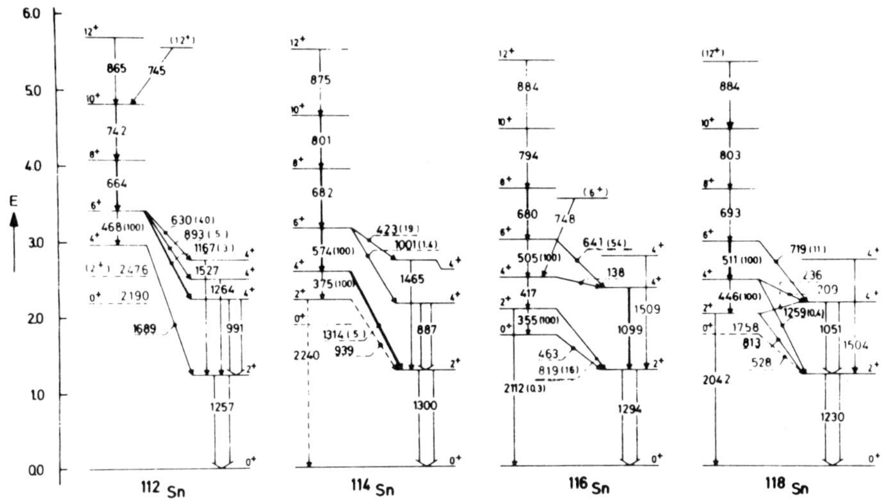


**Figure 54.** Energy levels in ${}^{112-118}_{50}Sn$ (Br 79) which exhibit near-spherical ground states and excited deformed bands.

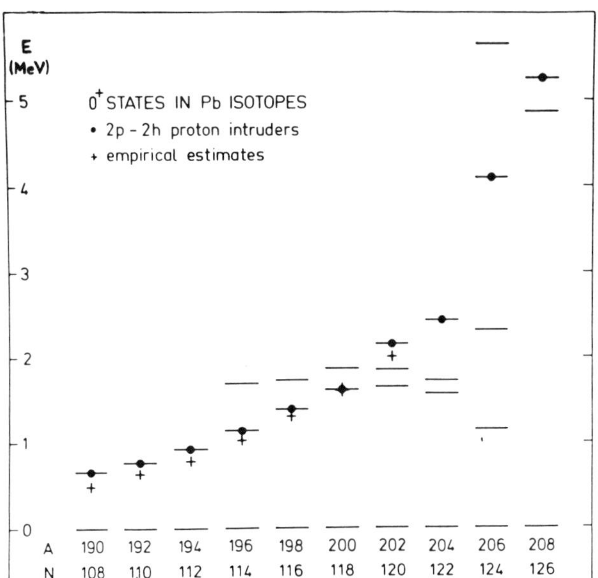


**Figure 55.** Energies of all known $0^+$ states in even-mass lead isotopes ${}^{190-208}Pb$. Candidates for the proton 2p-2h intruder states are indicated by dots. Crosses represent empirical estimates for the intruder energies according to Van Duppen and co-workers (Va 84).

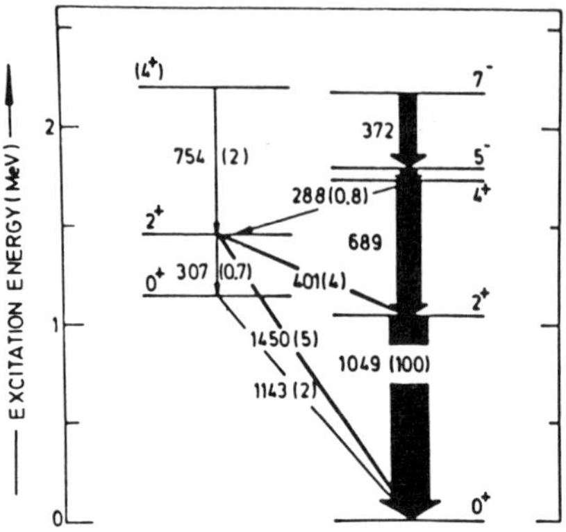


**Figure 56.** A part of the $^{196}$Pb level scheme. Numbers in parentheses are relative $\gamma$-ray intensities (Va 84).

The reduced $\alpha$ widths, which earlier exhibited a very anomalous increase between $^{186}$Pb and $^{192}$Pb (Hø 74) (see Figure 57), have been shown to behave regularly (To 84) as shown in Figure 57. The new results and the systematics of the $Z = 78$, 80, and 82 reduced $\alpha$ widths, shown in Figure 57, can be understood (To 84) as related to the disappearance of the $Z = 82$ gap in the vicinity of $N = 114$ and to the existence of slightly oblate ground states in $^{182-188}$Hg with coexisting deformed bands and more deformed prolate ground states for the $A < 190$ Pt isotopes. The $^{186-192}$Pb isotopes presumably have near-spherical or slightly oblate ground states. Then the Pb → Hg $\alpha$ decays would be between ground states with similar shape while the light Hg → Pt

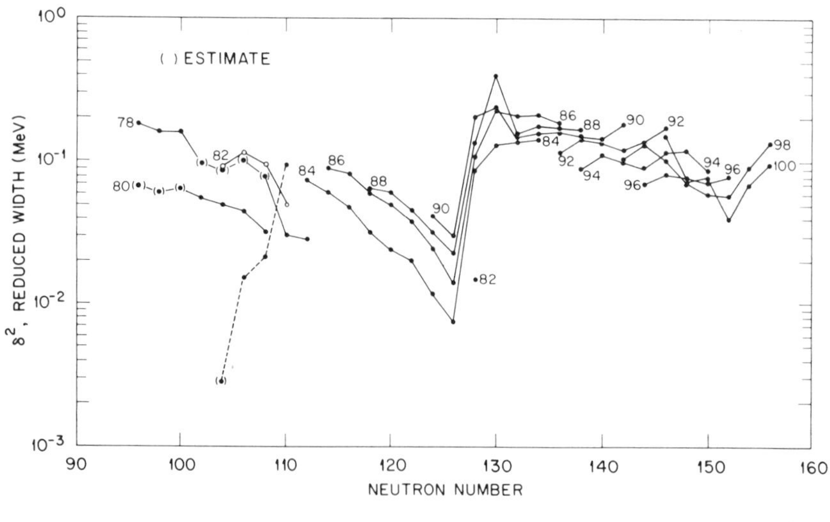


**Figure 57.** Reduced widths for s-wave $\alpha$ transitions plotted as a function of $N$ for isotopes with $Z$ from 78 to 100. The dashed line connects widths for $^{186,188,190,192}$Pb calculated from earlier data (Hø 74). Open points for $Z = 82$, connected by the full line, are widths for $^{186,188,190,192}$Pb from Toth and co-workers (To 84).

ground state $\alpha$ decays would involve a shape change. In addition, Toth and co-workers (To 84) suggest that the appreciably smaller width for $^{184}$Pb than for $^{186}$Pb and $^{188}$Pb (Figure 57) may indicate the reappearance of the $Z = 82$ gap for $N < 102$. Also note that the anomalously high, reduced $\alpha$ width for $^{218}$Ra (Figure 57) has been remeasured and found to come down to give a smooth curve for $Z = 86$ (To 86). This change weakens one piece of evidence for $\alpha$ clusters in heavy elements (To 86).

A final new discovery related to the double closed-shell structure of $^{208}_{82}\mathrm{Pb}_{126}$ is a new form of radioactivity, the decay of $^{223}_{88}$Ra to $^{209}_{82}$Pb by the emission of a $^{14}_{6}$C nucleus (Ro 84). Four years earlier, Sandulescu and co-workers (Sa 80) had predicted new types of decays of heavy nuclei intermediate between fission and $\alpha$ decay called heavy cluster radioactivities. They showed that one could expect the spontaneous emission of heavy clusters such as $^{14}$C, $^{20}$O, $^{24}$Ne and others from heavy nuclei when such a cluster would yield a daughter at or near $N = 126$, $Z = 82$ (doubly magic $^{208}_{82}\mathrm{Pb}_{126}$). The phenomenon of heavy cluster emission is a new manifestation of the strong nuclear shell structure associated with $^{208}$Pb. Now $^{14}$C emission has been observed for $^{222-226}_{88}$Ra (Ro 84, Al 84, Ga 84, Pr 85, Ku 85, Ho 85) and $^{24}$Ne emission for $^{232}_{92}$U (Ba 85), $^{233}$U (Tr 85), $^{231}_{93}$Pa (Sa 84), and $^{230}_{90}$Th (Tr 85b). These have been treated theoretically recently by Poenaru and co-workers (Po 85) and Greiner and Scheid (Gr 86). While the observed decay branch of $^{223}$Ra is very weak $(8.5 \pm 2.5) \times 10^{-10}$ $^{14}$C per $\alpha$ decay (Ro 84), this new mode offers very interesting new tests of our understanding of the structures of heavy nuclei, particularly clustering of many neutrons and protons in a nucleus as well as shell effects and nuclear shapes. (See Chapter 8 in this volume.)

It is now accepted that $^{146}_{64}\mathrm{Gd}_{82}$ has the characteristics of a spherical doubly magic nucleus and that pure $1^-$, $2^-$, and $3^-$ particle configurations are encountered systematically in its neighboring nuclei. In spite of similarities with doubly magic, $Z = 82$, $N = 126$ $^{208}$Pb and its neighbors, however, the $^{146}$Gd region is qualitatively very different and provides another excellent example that studies of far-from-stability nuclei indeed offer insights that would be difficult to obtain were our experimental knowledge limited to the near-stable nuclei. In particular, this region provides another example of the phenomena of reinforcing shell gaps, as discussed in the previous subsection. Here the strong $N = 82$ shell closure reinforces a weaker $Z = 64$ shell gap to make $^{146}$Gd doubly magic, like $N = 50$ does for $Z = 40$ in $^{90}_{40}$Zr and $Z = 28$ does for $N = 40$ in $^{68}_{28}\mathrm{Ni}_{40}$.

For the $N = 82$ isotones, the magic proton number 64 arises from a pronounced shell closure between the proton $1g_{7/2}$, $2d_{5/2}$ states and the proton $1h_{11/2}$, $2d_{3/2}$, $3s_{1/2}$ states. The experimental evidence is consistent with a gap in the single-particle energies comparable with that of $^{208}$Pb (Kl 79b). A comparison between the low-lying levels in $^{146}$Gd and $^{208}$Pb is presented in Figure 58.

The earliest indication of a single-particle energy gap at $Z = 64$ came from $\alpha$ decay (Ra 53), where it was suggested that the 65th proton may be less tightly bound than the other protons in the region. This was supported later by further $\alpha$-decay studies (To 70, Sc 76) and the systematics of spectroscopic factors measured in proton reactions (Wi 71). These indications of shell closure were solidly confirmed, for example, by the discontinuity in the $2^+$ energies of the even $Z$, $N = 82$ isotones (Og 78, Da 78) and by in-beam $\gamma$-ray and electron spectroscopy on $^{146,147}_{64}$Gd (Kl 79b). This latter work provided the first direct spectroscopic evidence for a shell closure at $Z = 64$ between the $2d_{5/2}$ and $1h_{11/2}$ shell model orbitals. On the other hand, Abbas (Ab 84) has argued (based on the similarity of the $2^+_1$ and other energy systematics) that all the $N = 82$

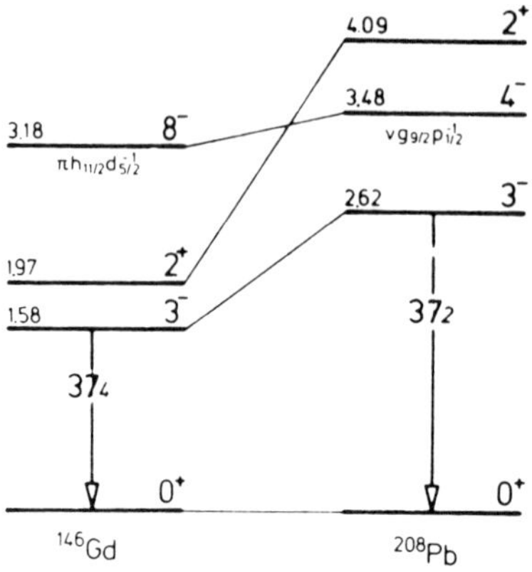


**Figure 58.** Lowest particle-hole excitations and $3^-$ and $2^+$ phonons in ${}^{146}_{64}$Gd and ${}^{208}_{82}$Pb. The $B$(E3) values are given in single-particle units (Kl 81).

isotones with $Z = 58, 60, 62, 64, 66, 68, 70$ are spherically doubly "magic" nuclei. His proposed plateau of new spherical magic numbers draws further attention to this very interesting region. Clearly the $N = 82$ closed shell has a strong spherical stabilizing influence over this region. However, one has similar striking stabilities of the $2_1^+, 4_1^+, \ldots$ and other level energies over wide ranges of neutrons numbers when the protons are near the strong closed shell $Z = 82$ (as shown in Figures 5 and 16).

Recent work (Bl 83) on masses in the $^{146}$Gd region employed a technique never before applied in mass determinations. They utilized experimental level energies and the spherical shell model to derive the ground-state masses of hitherto unknown $N = 82$ and 83 isotones of Tb, Dy, Ho, and Er which are connected by long $\alpha$-decay chains (Ho 79, Sc 82) to the known masses. The derived masses, of course, are not based on $Q$-value measurements, but they depend on experimental data and so should be more reliable than values from systematics. Further evidence for the double-magic character of $^{146}$Gd comes from two-proton separation energies as shown in Figure 59.

Figure 59 reveals a 0.5 MeV break in the nuclear mass surface at $Z = 64$. This result is supported by the elegant work of Flynn and co-workers (Fl 83) where they used radioactive targets of $^{148}$Gd ($T_{1/2} = 75$ yr) and $^{146}$Sm ($T_{1/2} = 7 \times 10^7$ yr) and the $(p, t)$ reaction to study ${}^{146}_{64}\text{Gd}_{82}$ and ${}^{144}_{62}\text{Sm}_{82}$. Their determination of the ground-state mass excess of $^{146}$Gd agrees very well with the input mass value used in the above analysis (Bl 83). Additionally, Flynn and co-workers (Fl 83) have shown that the increased excitation energy of the pairing vibration excitations, which they measured, supports the view that $^{146}$Gd definitely represents both a neutron and proton shell closure.

It is clear from Figure 59 that the irregularity in the two-proton separation energies of the $N = 82$ isotones quickly decreases when neutrons are added, and that it has vanished at $N = 86$. It should be noted that the 0.5 MeV two-proton separation energy displacement is much less than the corresponding breaks at other shell closures where pair correlations are absent. While the distinct break in the two-proton separation energies provides clear evidence for a $Z = 64$ shell gap, the pairing force is apparently just strong enough to form some correlations which are effective in smoothing out the mass surface. This is consistent with the $\alpha$-decay energies near $^{150}$Dy($^{146}$Gd + $\alpha$) which show only a small irregularity that is not at all comparable with the jump in the energies

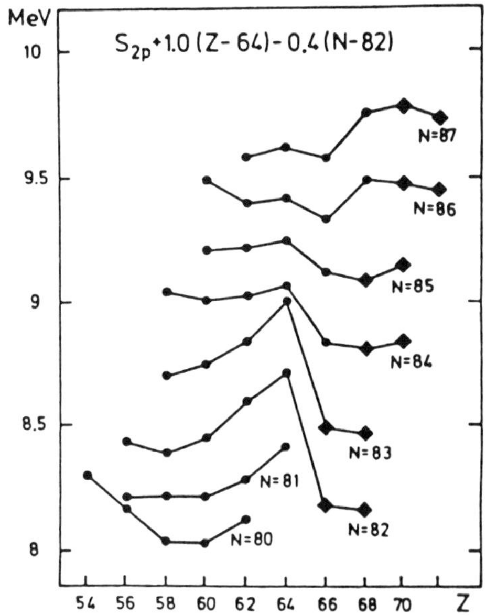


**Figure 59.** Systematics of two-proton separation energies recently discussed by Nolte (No 83) and Blomqvist and co-workers (Bl 83). Large diamonds represent new results. Note that the quantity $(Z-64)-0.4$ $(N-82)$ MeV has been added to the data points to make the presentation more clear. The conclusion is that the $Z=64$ shell is important only locally and that the corresponding discontinuity is 0.5 MeV. Spanier and co-workers (Sp 81) arrived at essentially the same conclusions (Bl 83).

observed near $^{212}$Po ($^{208}$Pb + $\alpha$) [see, for example, discussions of $\alpha$-decay energies in (To 70, Ha 79b)].

Yrast states of moderately high spin in a region of nuclei around $^{146}$Gd can be well described in terms of spherical shell model configurations based on a $^{146}$Gd core with a few valence nucleons (Kl 79b, Da 80, Wi 80, Ka 80, Pi 81b, La 81, Na 81, He 82, Pa 82, No 82, No 82b, Bl 83, Ch 83, Ho 83, St 83, St 83b, Ru 83, Wa 83, Wa 83b, Sc 84, Kl 85, To 85b, To 86). Toth and co-workers (To 85b) and Kleinheinz and co-workers (Kl 85) have recently extended the systematics of the single-particle states in odd-$Z$ $N = 82$ (see Figure 60) and even-$Z$, $N = 81$ isotones to $^{149}$Ho, $^{151}$Tm, and $^{149}$Er. There is a modest shell closure at $Z = 64$ but only when reinforced by $N = 82$ (To 85b, Kl 85). The double-magic character quickly disappears by moving a few protons or a few neutrons away. Spherical Hartree–Fock–Bogoliubov calculations with realistic interactions in a core-plus-particle description (To 85b) reproduce the trends (see Figure 60) of the single-particle energies quite well. The scale is off, probably because the effective interaction was too strong.

The interpretation of the observed spectra (in the region) as aligned multiparticle excitations is now, to a large extent, quantitatively understood. The picture which is emerging is one in which there is a delicate balance between two structures, one of the type double closed-shell core plus valence nucleons and the other more complicated with excitations involving the core and quickly leading to permanent deformation. In this regard, Chasman (Ch 83) discussed spherical, oblate, and prolate shapes in the region near $^{146}$Gd. He considered the contribution of states with large prolate deformation to the spectra of low-lying $0^+$ states which were observed in $^{146}$Gd and $^{144}$Sm (Fl 83). He showed that it was necessary to introduce strongly deformed prolate states to understand the low-lying excited $0^+$ states in nuclides near $^{146}$Gd—another example

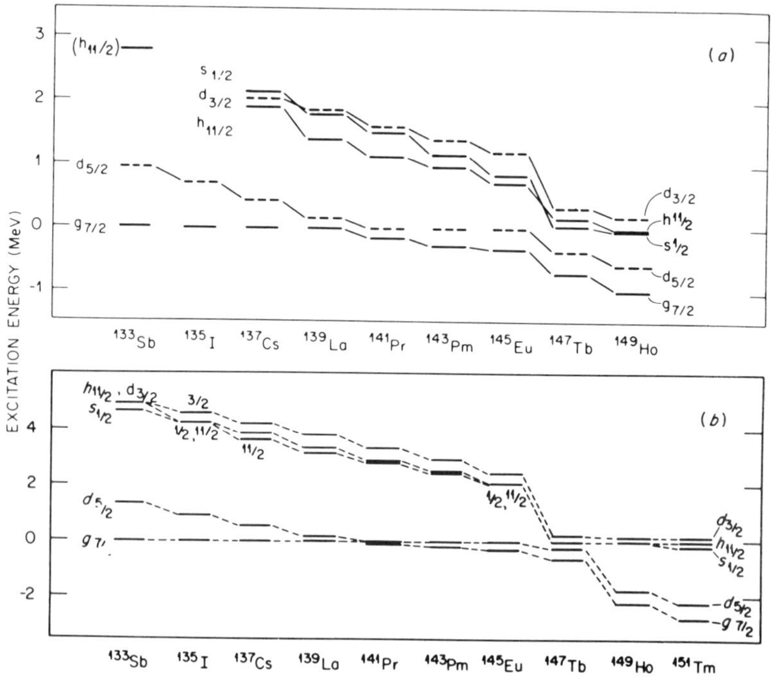


**Figure 60.** Energy systematics of single-proton states in odd-$Z$ $N = 82$ isotones: (a) experimental; (b) HFB calculations (To 85b).

of shape coexistence. The results all show that the magicity at $^{146}$Gd is strong only locally around $N = 82$. Thus, we have here too a reinforcing shell gap phenomenon where, as one moves away from the strong $N = 82$ shell closure, the magicity of $Z = 64$ rapidly disappears.

These findings, coupled with other recent data, in turn make us view nuclear structure near stability in this region in a different light. It is well known that permanent quadrupole deformation sets in very abruptly at $N = 90$ for the stable, $Z = 60, 62, 64$ nuclei, as seen from the earliest isotope shifts (Br 49) and from the sudden appearance of well-developed rotational bands. This abruptness in the onset of deformation at $N = 90$, explicable as a shell effect in the Nilsson diagram (Mo 59), may now be viewed as arising from the local reinforcing influence toward sphericity of the magic numbers $Z = 64$ and $N = 82$ [see, for example, (Ca 81, Ha 85)], which keeps the nuclei around $Z = 64$ spherical longer and so leads to a sudden onset of deformation at $N = 90$. This view is supported by laser spectroscopic studies (Ne 82, Ne 83, Mu 83, Th 81, Th 81b, Th 82) where only a gradual, smooth increase in deformation is found between $N = 88$ and 90 for Cs ($Z = 55$), Ba ($Z = 56$), Er ($Z = 68$), and Yb ($Z = 70$) (see Figures 49 and 50), while Dy ($Z = 66$) shows the sudden onset of deformation at $N = 90$ which had been observed long ago in Nd ($Z = 60$), Sm ($Z = 62$), and Gd ($Z = 64$) as discussed in more detail earlier. Thus it is the spherical $Z = 64$ shell gap which helps produce the sudden onset of deformation in the Nd to Dy nuclei. This is further supported by nuclear spectroscopic measurements on neutron-rich isotopes of $Z < 60$ and $N > 82$,

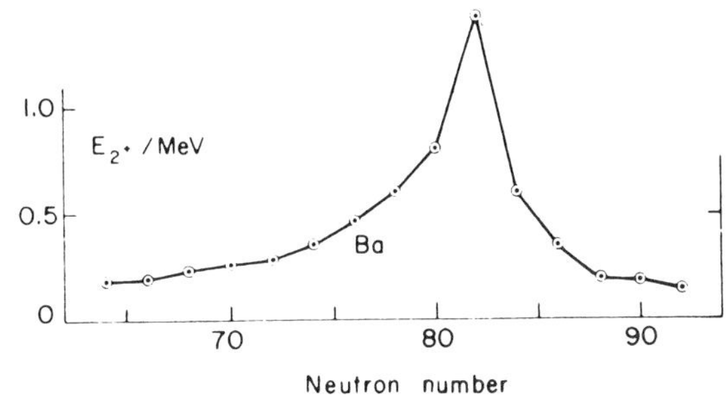


**Figure 61.** $2_1^+$ energies in barium isotopes (Ha 79b).

such as the $Z = 56$, $^{114-148}$Ba isotopes where the deformation, determined from the $2^+, 4^+, \ldots$, energies does not set in suddenly but rather follows a smooth pattern of continuously decreasing $2^+, 4^+, \ldots$, energies as $N$ increases from 82 to 92 (see Figure 61). Similar results were obtained for $^{142-150}$Ce isotopes (see Go 81).

Another highlight in the $^{146}$Gd region comes from the very strong octupole excitations which provide a wealth of new information that can be quantitatively interpreted in the framework of microscopic nuclear theory. The $3^-$ and $2^+$ one-phonon levels seen in $^{146}$Gd and $^{208}$Pb are shown in Figure 58. Although the lowest particle-hole states occur at comparable energies (thus the similarity of $^{146}$Gd and $^{208}$Pb shell model energy gaps), the energy of the vibrational levels in $^{146}$Gd is about half the energy of the corresponding states in $^{208}$Pb. This is understood as a shell structure effect because of the strong $2d_{5/2} \rightarrow 1h_{11/2}$ proton transition in the $^{146}$Gd octupole which has no counterpart in $^{208}$Pb (Kl 78, Kl 81).

After complete alignment of the $\pi h_{11/2}$, $\nu h_{9/2}$ valence particles, the yrast spectra of nuclei in the region are continued by $\Delta I = 1$ octupole excitations (Kl 81). For example, the completely aligned states in $^{148-151}$Dy ($\pi h_{11/2}\nu f_{7/2}\nu h_{9/2}$ for $^{150}$Dy) represent the final-state configuration for the $\Delta I = 1$ octupole transitions. In general, two or three units of extra angular momentum could easily be gained by lifting a valence neutron from the $i_{13/2}$ shell. That this is not observed suggests that the $\nu i_{13/2}$ excitation lies rather high, certainly more than 1.5 MeV above the $\nu f_{7/2}$ state. This situation is therefore reversed from the situation in $^{208}$Pb, and the 1 MeV $13/2^+$ state in $^{147}$Gd is mainly the octupole excitation $(\nu f_{7/2} \times 3^-)13/2^+$. The unusually low $3^-$ energies in this region made possible the identification of a two-proton octupole state in $^{147}$Gd $(\nu f_{7/2} \times 3^- \times 3^-)19/2^-$, where both stretched E3 transitions emitted in its de-excitation to the ground state were observed and their $B$(E3) values measured (Kl 82). The departures in energy and transition strength of two-phonon excitations from harmonic vibration can be quantitatively understood in terms of the microscopic composition of the states (Kl 82).

In contrast to the analogous case of the $N = 50$ isotones, the observed GT transition strengths for the $\beta$ decays of even-mass, $N = 82$ isotones above $^{146}$Gd seem to feed mainly just one $1^+$ level in the daughter nucleus, e.g., 97% in the case of $^{148}$Dy (Kl 85b). The very detailed investigation of this nucleus at ISOCELE/Orsay is presently being extended to the isotones $^{150}$Er and $^{152}$Yb at GSI. Pairing calculations for $^{148}$Dy, which were adjusted to experimental occupation numbers for $h_{11/2}$ protons in $^{144}$Sm, yield more than two $h_{11/2}$ protons in $^{148}$Dy and therefore an enhancement of the GT transition strength and, together with corrections for ground-state correlations in $^{148}$Dy, a surprisingly low quenching factor of 0.15 (Kl 85b). More clearly than in the cases of $^{106,108,110}$Sn, GT quenching for $^{148}$Dy seems stronger than in the sd and fp shells. In the magic nuclei around $^{100}$Sn and $^{146}$Gd, the retardation of the GT strength related to pairing correlations and first-order core polarization is much smaller than the other nuclei and of comparable size to that from the "higher-order terms," so that here the effect of GT quenching in medium mass nuclei can be studied more effectively. Thus, investigations are in progress at GSI and partially in collaboration with ISOLDE at CERN to search for and study the $\beta$ decays of $^{98,100}$Cd, $^{102}$Sn, $^{150}$Er, and $^{152}$Yb. Recently the $\beta^+$ decays of $^{152}$Yb (To 87) and $^{150}$Er (To 87b) have been investigated. In each case several states were found in addition to the single, previously known $1^+$ levels in the odd–odd daughters. However, there was no indication of any sizable missing GT strength to account for the retardation factors noted earlier (No 82b) for both beta decays.

Finally, in connection with the discovery of proton radioactivity at mass 147, two new $\beta$-delayed proton emitters, $^{147}_{68}$Er and $^{147}_{66}$Dy, were observed (Kl 83, Ki 83, To 84b). It is striking that, while $^{147}_{68}$Er has no observable fine structure, the isotope $^{147}_{66}$Dy shows a very pronounced fine structure corresponding to a level density considerably lower than that predicted by the semiempirical systematics, while $^{149}$Er lies between the two extremes (see Figure 62). This at first (To 84b) was interpreted as an effect of the nearby doubly magic $^{146}$Gd. Very recently Toth and co-workers (To 86b) identified the next $N = 81$ isotone $^{151}_{70}$Yb. Its $\beta$-delayed proton spectra is shown in Figure 63. An analysis of those and p–$\gamma$ spectra led to an interpretation that the higher-energy $\beta$-delayed protons without structure are from the $11/2^-$ isomer of $^{151}$Yb while the lower-energy proton spectrum with the structure is from the decay of the $1/2^+$ ground state with the protons feeding the ground state. Toth and co-workers (To 86b) concluded that rather than being a departure from the $Z = 64$ closure, the dramatic increase in the number of statistical protons in the $^{149}$Er spectrum can now be understood in terms of an enlarged $\beta$-decay Q window and a decreased isomeric decay branch. The persistence of the intense peaks in $^{151}$Yb decay, even though $^{151}_{69}$Tm$_{82}$ has 5 protons beyond $Z = 64$, clearly shows that the low-level density of states at ~4.5 MeV is a consequence of the $N = 82$ shell, not the $Z = 64$ one. So, in contrast to earlier speculation (To 84b), the effect of the $Z = 64$ closure is minimal (To 86b). This conclusion is reinforced by the fact that the $^{145}_{66}$Dy and $^{147}_{68}$Er proton spectra, which sample levels in the $N = 80$ isotones, $^{145}$Tb ($Z = 65$) and $^{147}$Ho ($Z = 67$) are essentially structureless (Sc 84, To 84c).

Another very interesting region is that just below $^{146}$Gd, in particular region 1 (see Figure 3). As noted earlier, nuclei with $N$ and $Z$ far removed from any spherical magic numbers are expected to be deformed. Current research is pushing more and more into the far-from-stability region 1 in Figure 3. The deformation in this region has been reviewed by Leander and Möller (Le 82). Of particular interest is their prediction of an abrupt transition to large deformation for Pm and Sm nuclei at $N \approx 73$. The nearest-to-stability phase transition is predicted (Le 82) to take place between $^{136}$Pm ($\varepsilon_2 = 0.20$) and $^{135}$Pm ($\varepsilon_2 = 0.29$). The available experimental data clearly show that this region will be a rich source of nuclear structure information, regardless of whether or not the exact predictions of Leander and Möller (Le 82) are borne out. Particular topics of interest in the region include the following: the experimental determination of the Nilsson orbitals; a comparison of the rotational properties with those on the neutron-rich side of magic number $N = 82$ since the two regions could be inherently different because of the lower particle numbers for $N < 82$ or because of the absence

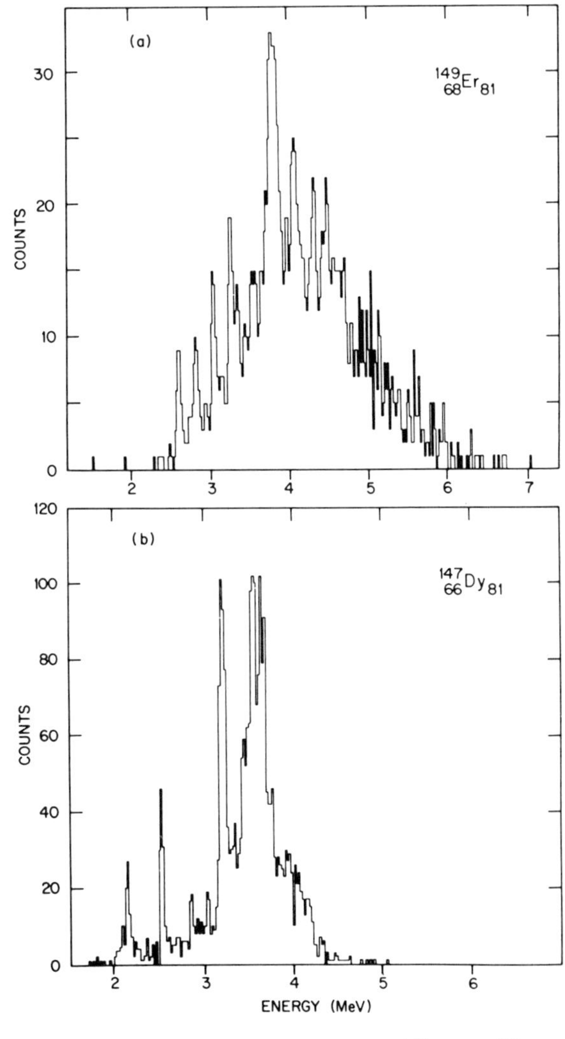


**Figure 62.** $\beta$-delayed proton spectra for $^{149}_{68}$Er and $^{147}_{66}$Dy.

in the valence space of the $i_{13/2}$ neutron-intruder shell which is known to play a fundamental role in the rare-earth $N > 82$ nuclei; and, of course, to search for the predicted phase transition which might be analogous, in some respects, to the abrupt phase transition between the $N = 88$ and 90 isotopes on the other side of the $N = 82$ shell closure. If such a sudden phase transition in the neutron-deficient Sm and Pm isotopes is observed, it could be indicative of the existence of another semimagic proton number which may be operative in this region and showing its effect just as the $Z = 64$ spherical gap exhibited its effect in the sudden onset of deformation in the rare earths between $N = 88$–90, as discussed above.

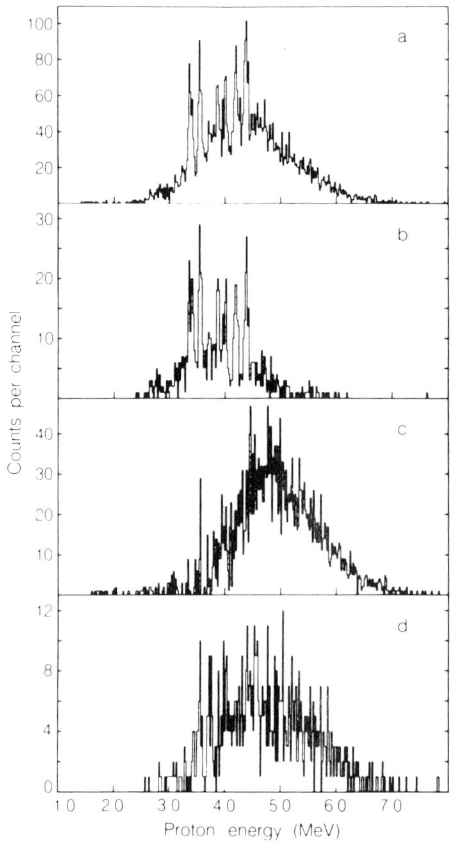


**Figure 63.** Delayed protons from $^{151}Yb$ $\beta$ decay: (a) singles spectrum, (b) spectrum in coincidence with positrons, (c) spectrum generated by subtracting part (b) from part (a), and (d) spectrum in coincidence with Tm K–X rays (To 86b).

# 6. New Directions in Studying Nuclei Far From Stability

## 6.1. Introduction

Isotope separators on-line with accelerators and with nuclear reactors have been the prime source of the new insights into nuclear structures at low spins and the decay modes of nuclei far from stability. Two recent developments are opening up many exciting, new, research opportunities which were previously very, very difficult or impossible with far-off-stability nuclei. One is the addition of on-line low-temperature, nuclear orientation (NO) refrigerators to isotope separators on-line (ISOL) with accelerators, and the other is recoil mass spectrometers (RMS).

The NO devices open up measurements of spins of essentially all the excited states and, with a Compton polarimeter, measurements of their parities and multipole mixing ratios of all the transitions from states populated in even very complex, far-off-stability radioactive decays with half-lives down to about 1 m and low production rates. One can also measure $\alpha$ and $\beta$ angular correlations from oriented nuclei which give insight into the nuclear shape and decay properties. With the addition of nuclear magnetic resonance (NMR) measurements to NO, the spins and magnetic moments of ground and isomeric states can be uniquely measured with high precision.

An RMS opens up both in-beam spectroscopy studies of far-off-stability nuclei with very low production cross-sections and studies of the radioactive decays of (1) elements which are difficult to extract from conventional ISOL ion sources such as Zr, Mo, and Ta and (2) those with very short half-lives down to ms and even $\mu$s.

Some of the opportunities being opened up are illustrated by recent examples discussed below, such as: the discovery of nuclear-shape coexistence in $^{118,120}I^{m,g}$ nuclei (Sh 85, Gr 86b), $\alpha$-directional correlations and their dependence on the nuclear deformation (Wo 86b), a sudden phase transition from very large prolate to oblate shapes between $N = Z = 38$ and $N = Z = 36$ (Ra 86), and large deformation in $^{124}$Ce (Yi 86).

## 6.2. On-Line Nuclear Orientation

Low-temperature helium refrigerators have been used for low-temperature nuclear orientation for numbers of years in a wide variety of basic and applied research areas; for examples, see the Proceedings of the Leuven Conference on Nuclear Orientation and Nuclei Far From Stability (De 85). Recently, low-temperature helium refrigerators have been coupled directly on-line to isotope separators; first as a stand-alone device for long-lived isotopes in Bonn (He 78) and now on-line to separators on-line to accelerators at Leuven (Va 81, Va 85) and Daresbury (St 81, Gr 83b, Wa 85). A new NO facility on-line to an ISOL coupled to two accelerators was installed (Fall–Winter 1986–87) at UNISOR (Ha 86c, Gi 86, Gi 88) and another is underway at Dubna (Fi 85).

To illustrate the technique, a schematic diagram of the UNISOR on-line nuclear orientation (OLNO) facility is shown in Figure 64. A particle beam from the accelerator continuously strikes a target from which the various far-from-stability nuclei being produced are extracted and pass through an isotope separator from which beams of radioactive nuclei separated by mass ($A = N + Z$) emerge. A beam of one separated

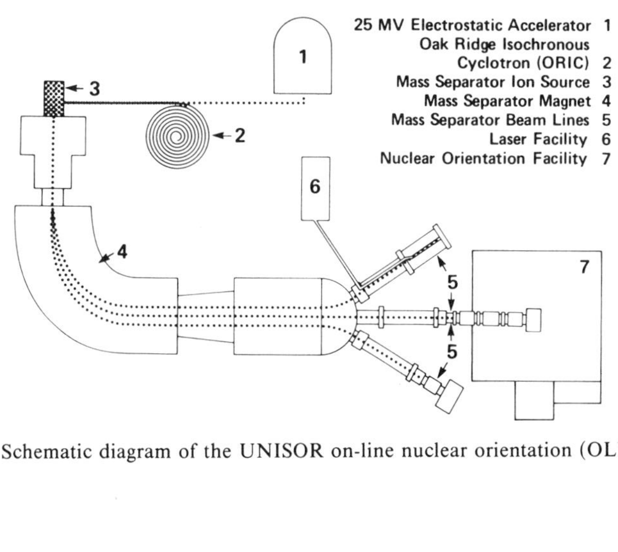


**Figure 64.** Schematic diagram of the UNISOR on-line nuclear orientation (OLNO) facility.

mass at a time can be directed from the separator down an extended beam line, which is cooled to the order of 4 K, and implanted into a foil, which can be cooled to 10 mK and which has an external magnetic field applied to align the nuclei through interaction of the magnetic field and magnetic moment. Details of the principles behind NO measurements may be found in various reviews [for example, see (St 85)]. Suffice it to note that measurements of the temperature dependence of the angular anisotropy of radiation emitted from oriented nuclei yield information on the nuclear moments and, with NMR, the spins of the parent state, spectroscopic information on the levels in the daughter nucleus and, for $\beta$ and $\alpha$ decay, information on the decay process and nuclear shape.

In addition to the many individual research papers in the Leuven Symposium Proceedings (De 85), Krane and Hamilton (Kr 85) summarized many of the recent highlights and described the exciting opportunities for future research. Some of the advantages and new research opportunities opened up by on-line nuclear orientation facilities are reviewed briefly here and elsewhere (Ha 87). Recent, significant results which demonstrate the power and importance of such studies are described. The emphasis of this section is on nuclear structure information which can be gained from OLNO measurements.

From NO studies, one can extract the spins of excited nuclear states and the multipole mixing ratios of $\gamma$-ray transitions depopulating these states. The big advantage of a NO device is that $\gamma$-ray data are accumulated in the singles mode from the oriented nuclei. One can accumulate sufficient data in much shorter lengths of time (less beam time required) and measure the properties of very weak ($\leqslant 1\%$) transitions and weakly ($\leqslant 1\%$) populated states in even very complex decays. Thus, one can obtain detailed information on essentially all the excited states in a complex decay. In addition, since the cross-sections for the production of very-far-off-stability neutron-deficient nuclei with heavy ions decrease rapidly as $N$ decreases, one reaches nuclei where $\gamma$-$\gamma$ directional correlations require prohibitive beam time for even the stronger transitions in a decay. Thus, NO studies can extend our knowledge to much farther off stability nuclei. Since the nuclei are oriented, measurement of the polarization of the $\gamma$ rays from these oriented nuclei can be used to determine the parity change and the parity of a level if the other parity is known. In addition, the polarization data may be used to distinguish between ambiguities for spin and $\delta$ in the NO data, as has been shown in polarization studies of nuclei oriented in a heavy-ion reaction (Ro 80).

Rikovska and co-workers (Ri 85b) have used three individual Ge detectors (two horizontal and one vertical) and two detectors on a turntable under the sample to make polarization measurements from oriented nuclei at Daresbury. They studied the decay of $^{192}$Ir to $^{192}$Pt and $^{192}$Os and found good agreement with earlier data. A Compton polarimeter, with four intrinsic Ge detectors in one housing to allow simultaneous acquisition of parallel and perpendicular scattering events, was designed to significantly reduce the data acquisition time for in-beam studies (Ra 84). It can likewise be used with a NO facility.

One might think that in a decay with a few hundred $\gamma$-rays, the ability to study 1% transitions is not really important. However, such transitions may be crucial to testing new theoretical ideas. For example, one of the important developments in studies of collective nuclear structure phenomena is the search for dynamical supersymmetry. The concept of supersymmetry (SUSY) was proposed by high-energy physicists about a decade ago in an attempt to unify all forces in nature. While it may be intellectually satisfying to have a unifying theoretical scheme, there was little empirical evidence for SUSY to support its validity until recently. Iachello (Ia 80) first proposed that SUSY may be applicable to collective nuclear structures in nuclei where a wealth of empirical data can be obtained for verification. A brief review of searches for SUSY in nuclear physics is found in (Ha 85). As emphasized there (Ha 85), the important point is not simply that one can extend SUSY searches to nuclei far from stability, but crucial tests of SUSY are often found in the E2 branching ratios involving transitions which are forbidden. Thus, important tests arose in knowing the properties of weakly populated states and E2/M1 mixing ratios of highly hindered transitions. A NO facility can extract from singles data such information on these important weak transitions. Very recently an exciting, new approach to nuclear collective phenomena, based on a fermion dynamical symmetry model, has been proposed (Wu 86). Nuclear collective phenomena continue to be a major testing ground for our theoretical understanding, and the detailed nuclear structure information from NO studies will continue to be essential in testing our ideas, including symmetries in physics.

Of course, E2/M1 mixing ratios have long provided important tests of nuclear models [see the review of Lange and co-workers (La 82)]. Such data are needed in all structure studies. In studies of shape coexistence in the Tl–Hg–Au–Pt region discussed earlier, little, if any, information is known about E2/M1 mixing ratios. Thus, the E0/E2 mixing ratios in the $\Delta I = 0$ transitions, which provide important data on nuclear-shape changes, are uncertain because the E2/M1 ratios of the $\gamma$ transitions are not known. Indeed, in some cases it is not possible to distinguish between an E0/E2 and an E2/M1 admixture from conversion coefficient data alone. Hence, NO data can provide essential data to establish the magnitude of the E0 admixtures in $\Delta I = 0$ transitions. Examples of important cases are found in the odd-$A$ Au and Tl isotopes, where E0 transitions can be signatures of transitions between states coupled to states with different deformations in the Pt or Hg cores (Ha 85, Zg 81, Wo 85).

Another important area opened up by an OLNO facility is measurement of the spins and magnetic moments of ground and isomeric states. By combining data from NMR measurements on oriented nuclei and NO measurements, unique values of the nuclear spin and very precise values of the magnetic moment can be obtained. There are many cases of nuclei far off stability where there exist extensive in-beam data which establish numerous band structures, but the ground-state spin is not known; examples are $^{73,75}$Br (We 85). Knowledge of the ground-state spins would lead to unique assignments of the spins of the observed bands. The properties of these nuclei are important in establishing the predicted, very large prolate to large oblate shape change, as discussed in the next section. In addition, a precise measurement of the $g$-factor can often yield a unique assignment of a Nilsson orbital. The NMR data also can provide crucial data on whether the implanted atoms are in only one or more sites in the foil.

Now let us look at some highlights of the first on-line NO and NMR/NO studies. One of the challenges of on-line NO work was whether one could measure $\alpha$ anisotropies from oriented nuclei. Such studies required solving problems associated with placing the solid-state detectors inside the cryostat at 4 K. The interest in $\alpha$ correlations is related to the fact that the direction of the $\alpha$ emission is correlated with the shape of the emitting nucleus. Hill and Wheeler (Hi 53) predicted enhanced $\alpha$ emission from the ends of a prolate-deformed nucleus, and this is confirmed. The Leuven group (Wo 85b) have pointed out that with on-line NO one could study $\alpha$ decay of a chain of isotopes to very far off stability to search for sudden changes in the nuclear shape of a particular element as the neutron number decreased. The Leuven group have solved the problems associated with low-temperature on-line $\alpha$ studies (Wo 86b, Wo 85b). In their first studies, they found clear differences in the $\alpha$ anisotropies as a function of $1/T_{exp}$ ($T$ is the temperature) for $^{199m}$Po and $^{201}$At (Wo 85b). More recently they had the first measurements of a series of isotopes, $^{199,201,203}$At (Wo 86b, Wo 85b), and more recently $^{205}$At (Va 86), as shown in Figure 65. Strong increases in the anisotropies are observed as the neutron number goes away from $N = 126$, as shown in Figure 65. The $^{199-205}$At $\alpha$ decays involve $\pi h_{9/2} \rightarrow \pi h_{9/2}$ transitions. Since all other experimental conditions are identical for the first three isotopes studied (Wo 86b), the difference in the anisotropies was most easily explained by changes in the nuclear deformation. The mixing ratio, $\delta$, of the $\alpha$ probability with $L = 2$ and $L = 0$ angular momenta is responsible for the variation. The $\alpha$-emission probability is very sensitive to the nuclear radius and is much enhanced in the region of the nuclear surface furthest from the center of the nucleus. In a first qualitative way, Wouters and co-workers (Wo 86b, Wo 85b) note that a change in $\delta = a_2/a_0$ correlates with a change in deformation, so that the larger the $P_2$ deformation, the more $L = 2$, $\alpha$ particles are emitted compared to the $L = 0$, isotropic ones. However, the recently observed change in sign for $^{205}$At (Figure 65) does not agree with the main picture (Va 86). Several theorists are seeking an explanation of this puzzle. This work also represents not only a breakthrough in techniques to do $\alpha$-NO studies, but the half-life of $^{199}$At is only 7.0 s. This is a record short half-life for OLNO work.

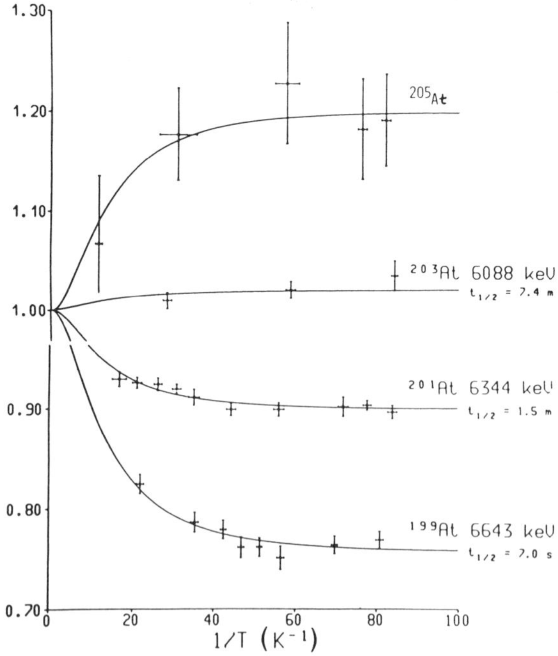


**Figure 65.** Anisotropies of the $\alpha$ emission as a function of $1/T$ in the decay of $^{199,201,203,205}$At (Wo 86b, Va 86).

Very recently Vanneste and co-workers (Va 86b) have shown that one can measure $\beta$ anisotropies in OLNO to extract information on the weak interaction, including the Cabibbo angle and the quenching of $C_A$ in nuclear matter. This opens up the study of the weak interaction in numbers of more mirror nuclei. Previously the $\beta$ decay of only two mirror nuclei had been studied. Indeed almost ideal $\beta$ sources can be prepared in on-line studies.

Let us look at examples from Leuven and Daresbury which clearly demonstrate the power of OLNO techniques in nuclear structure studies. Vandeplassche and co-workers (Va 85, Va 86c) have shown in OLNO studies of $^{106,108,110}$Cd that one can measure anisotropies for all transitions with intensities >1% and extract spins and multipole mixing ratios to provide important clarification of the level structures of these nuclei. Similar studies were done by van Walle and co-workers (va 86) of the decays of $^{101-104}$Ag to $^{101-104}$Pd. The systematics of the odd-$A$ Pd nuclei were extended to $^{101}$Pd where spin assignments were made to 10 low-lying levels. An interesting result was the observance of a spin change in the $\delta$ values in the even-Pd isotopes between $^{102}_{46}\mathrm{Pd}_{56}$ and $^{104,106}_{96}\mathrm{Pd}_{56,60}$. A similar behavior was detected in studying the Pt daughters from oriented Au nuclei. This change has not been reproduced by any theory. They suggest that the reversal in sign of $\delta$ could be related to a subshell closure.

This group also demonstrated the power of combining NMR/NO to measure the magnetic moments of $^{106,108}$In and $^{104}$Ag$^g$ (va 86). The NMR resonance curves are shown in Figure 66. The magnetic moment of $^{104}$Ag ($5^+$) was measured with high precision to be $\mu = 3.914(8)\mu_N$. The high precision of the NMR/NO results is seen by comparing with the previous atomic beam measurement of $4.0^{+0.2}_{-0.1}\mu_N$. This value enabled them to calibrate their magnetic moment data on $^{101-104}$Ag. These data indicate a transition from a $(\pi g_{9/2})^{-3}$ $7/2^+$ to $(\pi g_{9/2})^{-3}$ $9/2$ configuration in going from $^{103}$Ag to $^{101}$Ag and in going from $^{106}$Ag to $^{102}$Ag with $^{104}$Ag a mixture of the two proton configurations. These data demonstrate how precision measurements of $\mu$ with NMR/NO can be used to extract detailed information on the configuration of a nuclear state.

Vandeplassche and co-workers (Va 86d) in studies of $^{106,108}$In have also shown that one can do on-line NMR/NO studies to measure the spin and accurate $\mu$ of the parent state. The earlier spin assignments for $^{106}$In, for example, oscillate between 6 and 7 even in the most recent measurement (Wi 80b, Ro 84b). Their NMR results yield $g(^{106}\mathrm{In}) = 0.7030(19)$, which gives for spins 6 or 7 magnetic moments of 4.218(11) and 4.921(13), respectively (Va 86d). These data were combined with an integral measurement on the $6^+ \rightarrow 4^+$, 998 keV E2 transition. The fit to these data for spins 6 and 7, as shown in Figure 67, clearly exclude spin 6. Similar data for $^{108}$In yield a spin of 7 to remove similar differences in recent spin assignments there. An unexpected and unexplained drop in $\mu$ was found for $^{108}$In compared to $^{106,110}$In. This drop suggests that configuration mixing has not been properly identified there.

Green and co-workers (Gr 86c) at Daresbury have measured, by the NMR/NO technique, the magnetic moment of $^{116}$Sb$^g$ ($T_{1/2} = 16$ m, $I^\pi = 3^+$). One of the advantages to be noted about the NMR measurement is that, since one is only interested in the frequency at which the anisotropy is destroyed, NaI detectors may be used. They found $|\mu| = 2.715(9)\mu_N$ for the ground state of $^{116}$Sb where the sign was taken as positive from the known moments in $^{118,120}$Sb. By using the known information on the neighboring odd-$A$, odd-proton (antimony) and odd-neutron (tin) isotopes, they showed that the differences in the magnetic moments of the $3^+$ ground states of $^{116}$Sb and $^{120}$Sb

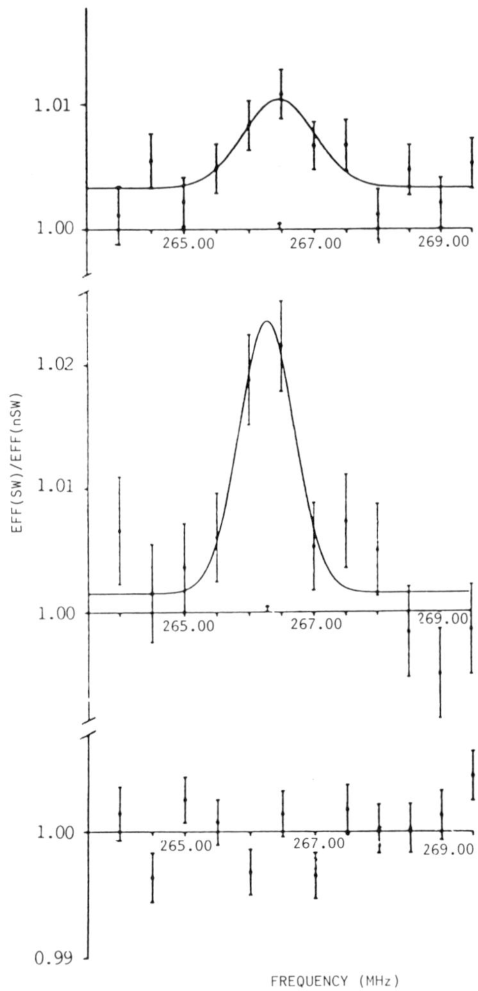


**Figure 66.** NMR/ON curves for $^{104}Ag^{g}$ (va 86).

arise from a variation in the moment of the associated odd-neutron without any contribution from the odd-proton, which plays only a spectator role in these states.

A final example of the power of NO studies is the beautiful systematic measurements of the magnetic moments of $N = 64$–69, $^{117}_{53}I$, $^{118}I^{g}$, $^{118}I^{m}$, $^{119}I$, $^{120}I^{g}$, $^{120}I^{m}$, $^{121}I$, and $^{122}I$ by Green and co-workers (Gr 86b) and Shaw and co-workers (Sh 85). These data provide new examples of nuclear-shape coexistence and the influence of the $N = 64$ subshell gap. Figure 68 shows the measured anisotropies, and Table 3 gives the measured moments. They had already reported (Sh 85) that the magnetic moment $|\mu| = 2.0 \pm 0.2$ nm [determined with improved accuracy in (Gr 86c)] of $^{118}I^{g}$ with $I^{\pi} = 2^{-}$ (see Figure 69) can be understood as arising from the spherical $(\pi g_{7/2}, \nu h_{11/2})_{2^-}$

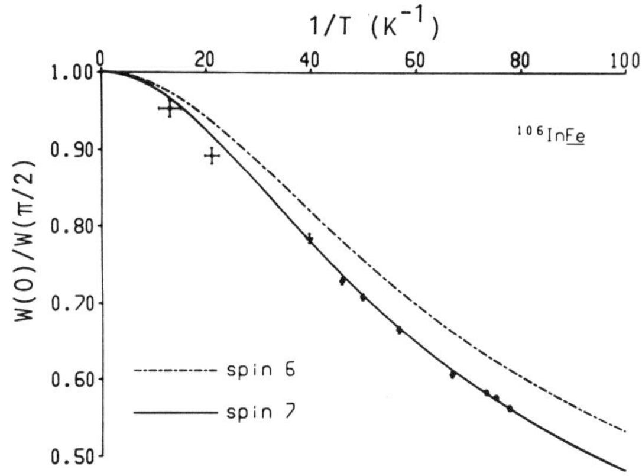


**Figure 67.** Anisotropy of the 998 keV transition in $^{106}$Cd (Va 86d).

**Table 3. Measured Magnetic Moments of Iodine Isotopes (Gr 86b)**

| Isotope | $T_{1/2}$ | $I^\pi$ | $\lvert\mu\rvert$(nm) |
|---|---|---|---|
| $^{117}$I | 2.2 m | $(\frac{5}{2}^+)$ | 3.1 (2) |
| $^{118}$I$^g$ | 13 m | $2^-$ | 2.0 (2) |
| $^{118}$I$^m$ | 8 m | $(7^-)$ | 4.2 (2) |
| $^{119}$I | 19 m | $(\frac{5}{2}^+)$ | 2.9 (1) |
| $^{120}$I$^g$ | 1.4 h | $2^-$ | 1.23 (3) |
| $^{120}$I$^m$ | 53 m | $(7^-)$ | 4.2 (2) |
| $^{121}$I | 2.1 h | $\frac{5}{2}^+$ | 2.3 (1) |
| $^{122}$I | 3.6 m | $1^+$ | 0.94 (3) |

configuration (the same as for the $^{122}$Sb $2^-$ ground state). However, the large moment $|\mu| = 4.2 \pm 0.2$ nm for $^{118}$I$^m$ with $I^\pi = (7^-)$ can only be understood if there is considerable quadrupole deformation. With a deformation $\beta \gtrsim 0.2$, the $\{\pi 9/2^+[404], \nu 5/2^-[532]\}K^\pi = 7^-$ configuration is close to the Fermi surface with a calculated moment of $\mu = 4.4$ nm (using $g_s = 0.6 g_s^{\text{free}}$). This value is in good agrement with the experimental results. Other configurations close to the Fermi surface fail to yield consistent moments and $I^\pi$. Their new moment for $^{120}$I$^m$ of $4.2 \pm 0.2$ nm strongly suggests a similar deformation and shape coexistence there, too. They (Gr 86b) used these results and the angular distributions of the rotational band above the isomer (Qu 84) to extract a $\beta = 0.21 \pm 0.04$ to support the rotational model assumption (Qu 84) for the band and in agreement with the moment data. In $^{118}$I, the isomer and ground state are only 100 keV apart and have similar half-lives for $\beta$ decay. These two states provide a dramatic new example of nuclear-shape coexistence.

Finally, Green and co-workers (Gr 86b) note that the ground-state moment of $^{120}$I$^g$ is almost a factor of two smaller than that of $^{118}$I$^g$. This difference can be explained if $^{120}_{53}$I$^g_{67}$ is weakly deformed and $^{118}_{53}$I$^g_{65}$ is not. They suggest this is evidence for an $N = 64$

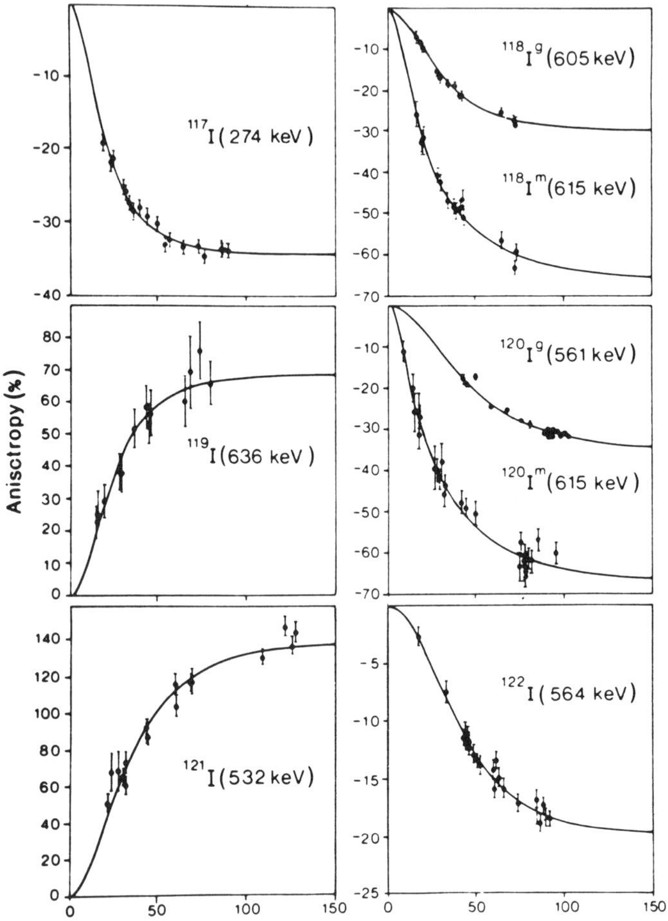


**Figure 68.** Examples of gamma-ray anisotropies as a function of inverse temperature for iodine isotopes (Gr 86b).

spherical subshell closure stabilizing the spherical shape more for $N = 65$, $^{118}I^{g}$ than for $N = 67$, $^{120}I^{g}$. They also considered the effects of the $N = 64$ subshell closure on the $^{118,120}I^{m}$ excitation energies which should also be influenced. However, the excitation energy of $^{118}I^{m}$ appears to be lower than in $^{120}I^{m}$, contrary to this expectation. They suggest that the extra active protons which were created in the formation of the $g_{9/2}$ proton hole reduce significantly the effect of the $N = 64$ subshell closure. The systematics of the region suggest that when the active number of protons is four or more, then $N = 64$ has no significant closure effect. Therefore, with the three active protons in $^{118,120}I^{g}$ the $N = 64$ subshell closure affects the magnetic moment but does not affect $^{118,120}I^{m}$ with their five active protons.

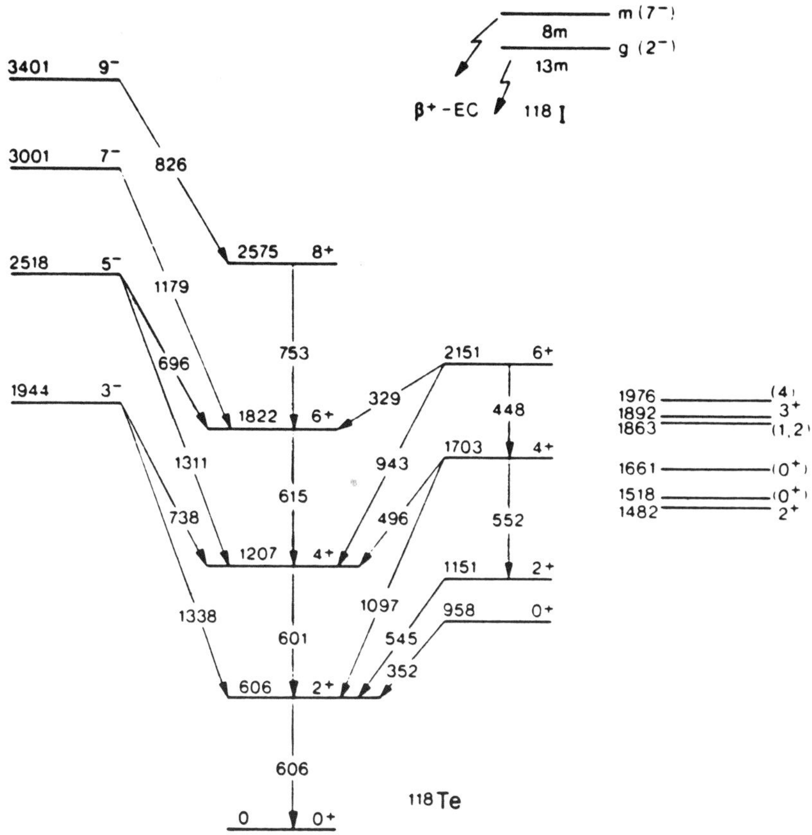


**Figure 69.** Decay schemes of $^{118}I^{g,m}$. The excited levels of $^{118}$Te show both near-spherical and deformed structures (Sh 85).

Part of the extensive studies of nuclei in the $Z = 82$ region, especially the Au nuclei (va 85) by OLNO, was described in Section 3.2.

## 6.3. Recoil Mass Spectrometers

The first use of a recoil mass spectrometer (RMS) to probe nuclei far off stability by in-beam $\gamma$-ray spectrometer was in studies in the $A = 70$ region in a Vanderbilt-Rochester-LSU-Idaho-Florida-Köln collaboration (We 85, Ha 84b, Ra 85, Zh 86c). The detector system used at the University of Rochester RMS is illustrated in Figure 70. With heavy ions, the cross-sections for the production of nuclei very far off stability can decrease to 1% and less of the total cross-section. Studies via traditional in-beam spectroscopy are impossible in such cases. An RMS facility opens up in-beam studies of such low cross-section products. Even so, only an RMS tag is not sufficient to unscramble the $\gamma$ rays from weak channels, from strong channels with the same mass (e.g., the 3p $^{71}$As, 2pn $^{71}$Se, and p2n $^{21}$Br channels with relative cross-sections of 180,

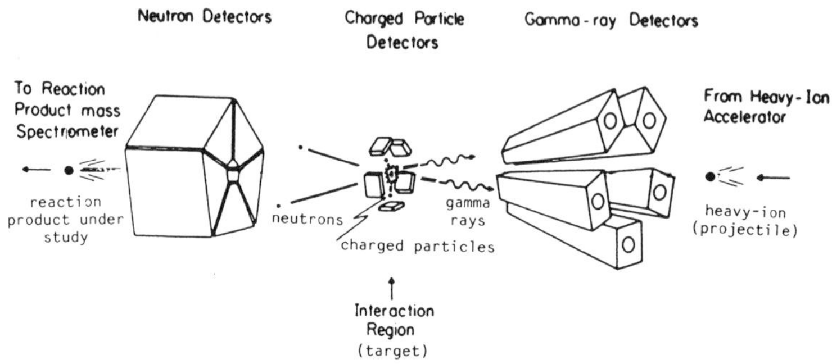


**Figure 70.** Schematic diagram of the system developed jointly by Vanderbilt-LSU-Florida-Idaho-Rochester groups for in-beam $\gamma$-ray spectroscopy with the Rochester RMS.

160, and 2.5) (Ra 86, Ra 85, Zh 86c). Two different approaches have been used. For mass 70 work, a five-sector neutron multiplicity filter covering nearly a $2\pi$ solid angle was placed in the forward direction and four NaI detectors especially built for detecting fast protons and alphas were placed at 90° as shown in Figure 70. The following types of data were collected: RMS-$\gamma$, RMS-n-$\gamma$, RMS-p-$\gamma$, as well as n-$\gamma$ and n-n-$\gamma$. At Daresbury a $\Delta E$-$E$ detector at the focal plane of the RMS enables one to measure $Z$ when an inverse reaction is used to give the recoils sufficient energy to discriminate the $\Delta E$ signals from different $Z$ recoils (Yi 86, Li 88).

The complete separation of the different $\gamma$ rays in coincidence with different masses for reactions like $^{16}$O on $^{58}$Ni and $^{28}$Si on $^{54}$Fe is shown in Figure 71 (Ra 86). In Figure 72, one can see how the $\gamma$ rays associated with the different reaction channels are separated by gating on protons and neutrons in addition to gating on mass 71. Mass gates at two different energies are shown in Figure 73. The intensity of the mass 71 recoils increases at the higher energy because of the rising three particles out cross-sections. The 894 keV transition was assigned to $^{71}$Se for the first time from its presence in both the neutron-gated and proton-gated spectra (Figure 72) (Ra 86). As shown in a later decay scheme, this is placed in the $^{71}$Se levels as a $11/2^+ \rightarrow 9/2^+$ transition, which is important in understanding the structure of this nucleus. In more recent work, $^{72}_{37}$Kr was observed in the reaction $^{58}$Ni on an $^{16}$O target with a cross-section of $60 \pm 15\ \mu$b (Li 88).

Results of the channel separation technique used in Daresbury are illustrated in Figure 74. The complete separation of the $\gamma$ rays in coincidence with the different $Z(\Delta E$ signals) and in coincidence with a recoil mass of 124 is shown in Figure 74. The data are particularly impressive when one realizes that $^{124}$Ce is only about 2% of the total cross-section. These $\gamma$ rays and excited levels in $^{124}$Ce are identified for the first time from this spectrum (Yi 86).

The coexistence of bands of states built on large prolate and small oblate (or near-spherical) nuclear shapes in $^{72}$Se and a new region of very strong ($\beta \sim 0.35$) prolate deformed ground states centered around $N = Z = 38$ led to the discovery of new shell gaps at $Z = 38$ and $N = 38$, 60, which stabilize the nucleus for a deformed shape (large prolate $\beta \sim 0.35$), and to the importance of the reinforcing of proton and neutron shell gaps on the shape of a nucleus (Ha 81b, Ha 85b). The $N = Z = 38$ shell

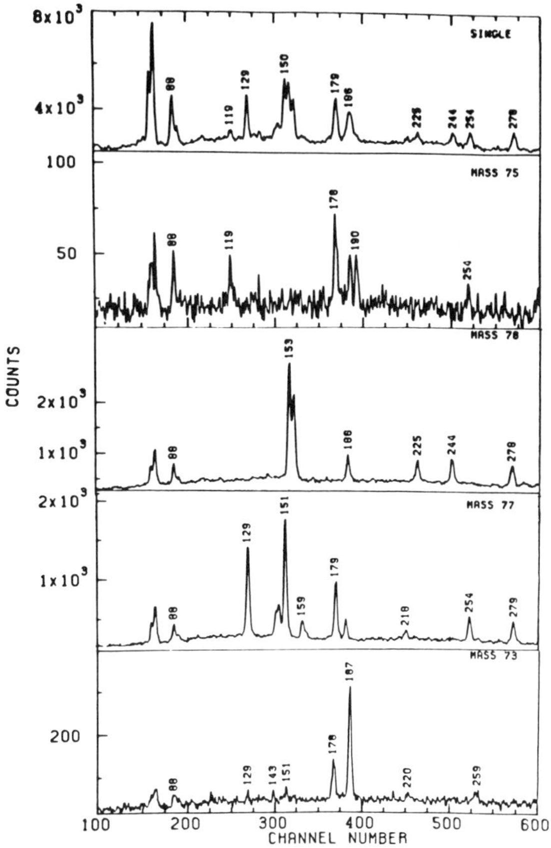


**Figure 71.** Gamma rays in coincidence with different recoil masses (Ha 84b, Ra 86). In this reaction, the recoils at mass 73 are about 5% of those at mass 78.

gaps at large prolate deformation are seen in the single-particle levels in Figure 28, from Bengtsson and co-workers (Be 84).

Also seen in Figure 28 are shell gaps at modest oblate deformation ($\beta \sim -0.2$) for $N = Z = 34$ and at large oblate deformation ($\beta \sim -0.35$) for $N = Z = 36$. Based on these levels, the small deformed ground states, which coexist with large ($\beta \gtrsim 0.25$) prolate deformed states in $^{72,74}_{34}Se_{38,40}$, could be of small oblate shape associated with the $N = Z = 34$ gaps. Theoretically, the large oblate ($\beta \sim -0.35$) shell gaps at 36 should be as effective as the established (Ha 85b, Ha 85c) $N = Z = 38$ shell gaps at large prolate deformation. It is important for the development of our theoretical understanding of this region to establish the presence or absence of the expected large and small oblate deformations which would be centered around $N \approx Z \approx 36$ or 35 and $N \approx Z \approx 34$.

The odd-$A$ nuclei in this region offer more easily observable differences in their level structures to distinguish between large prolate ($b \sim 0.35$) and large oblate ($\beta \sim -0.35$) deformations. Studies of $^{75}_{36}Kr_{39}$ (Ra 82, He 87) and $^{75}_{35}Br_{40}$ (We 85) revealed $g_{9/2}$

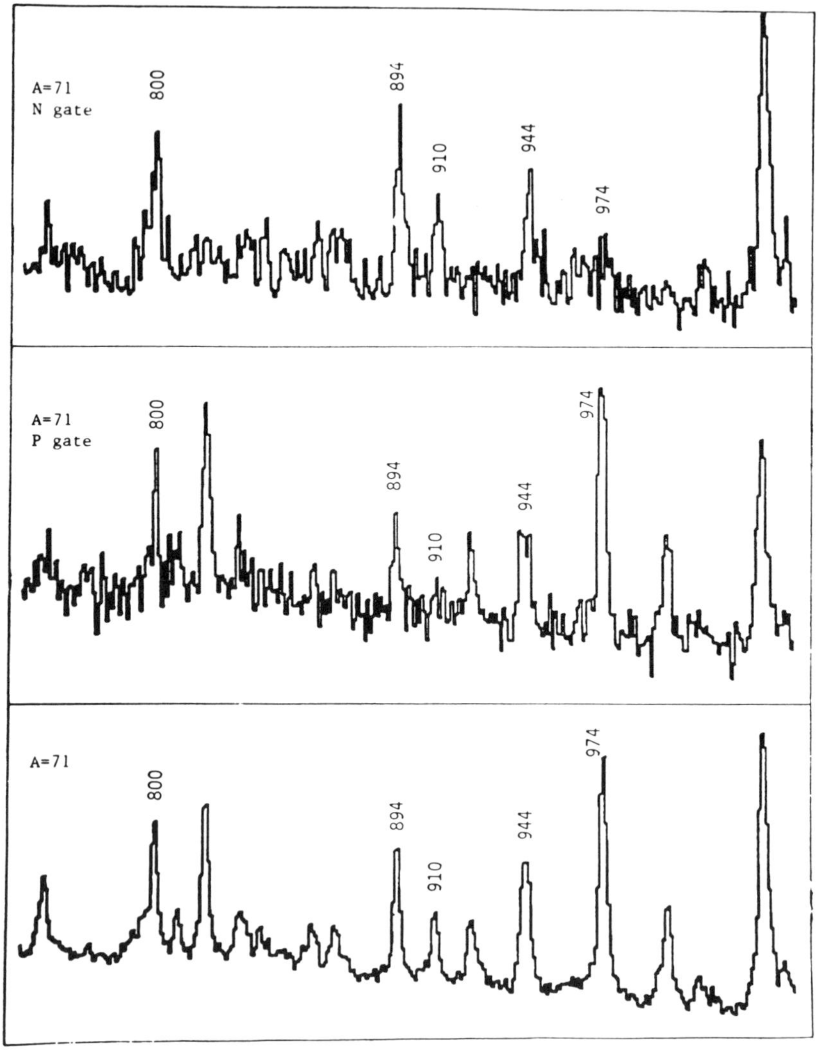


**Figure 72.** Partial $\gamma$-ray spectrum from the reaction ($^{16}O+^{58}Ni$), using the RMS facility at the University of Rochester for RMS-$\gamma$ (bottom), RMS-p-$\gamma$ and RMS-n-$\gamma$ (top) coincidences (Ra 86).

rotational bands characteristic of a large prolate shape, $\beta \gtrsim 0.3$. However, $^{73}_{35}Br_{38}$ or at least $^{71}_{35}Br_{36}$ could be expected to exhibit a sudden shift from a strong prolate to strong oblate ground-state shape. Thus, searches were initiated to identify the levels of these isotopes. Unfortunately, the cross-sections for the production of far-off-stability isotopes in this region decrease rapidly as $N$ decreases. Thus, new techniques were required to carry out such studies.

For the first time, a recoil mass spectrometer was used in coincidence with $\gamma$ rays and separately with protons and neutrons to identify the levels in $^{73}Br$. The Rochester

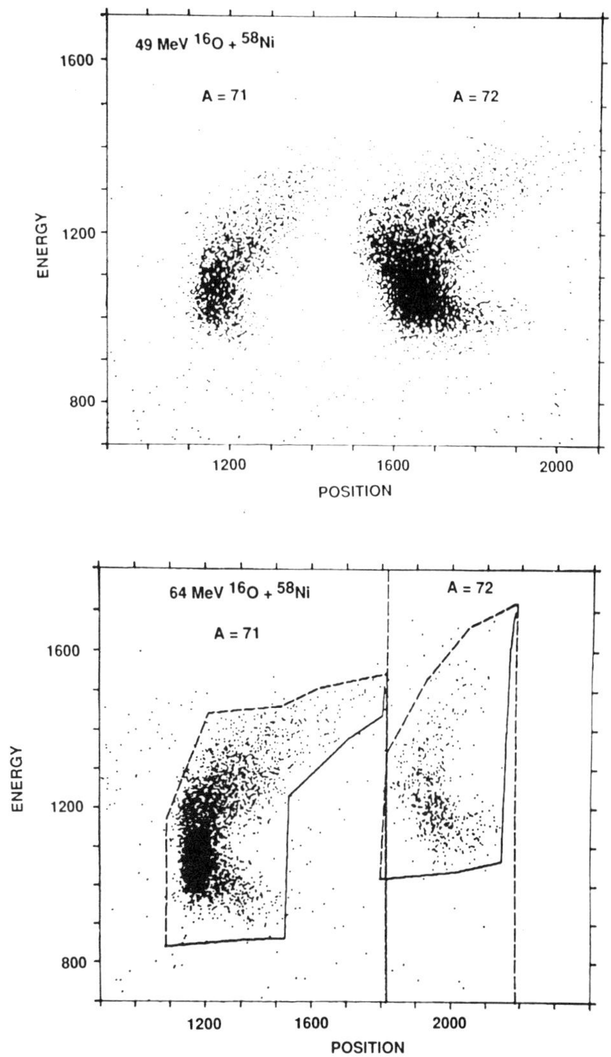


**Figure 73.** Two-dimensional recoil mass spectra for mass 71 and 72 in the focal plane of the recoil mass spectrometer for two different beam energies, 49 MeV and 64 MeV (Ra 86, Zh 86c).

recoil mass spectrometer (Co 81b) was used in these studies. First, a five-sector neutron detector (Pi 84) was used to do RMS-n-$\gamma$ coincidences. In the most recent experiment (Ra 86, Zh 86c) (see Figure 70), four large solid-angle NaI charged particle detectors were added in addition to measure RMS-p-$\gamma$ coincidences. For this last run an improved version of an earlier detector (Pi 84) was used. The new detector has five separate sectors and has the photomultipliers perpendicular to the beam to conserve space along the beam line. Identification of the $^{73}$Br levels was aided by n-$\gamma$ studies with the five-sector neutron detector at the Holifield Heavy Ion Research Facility and separate

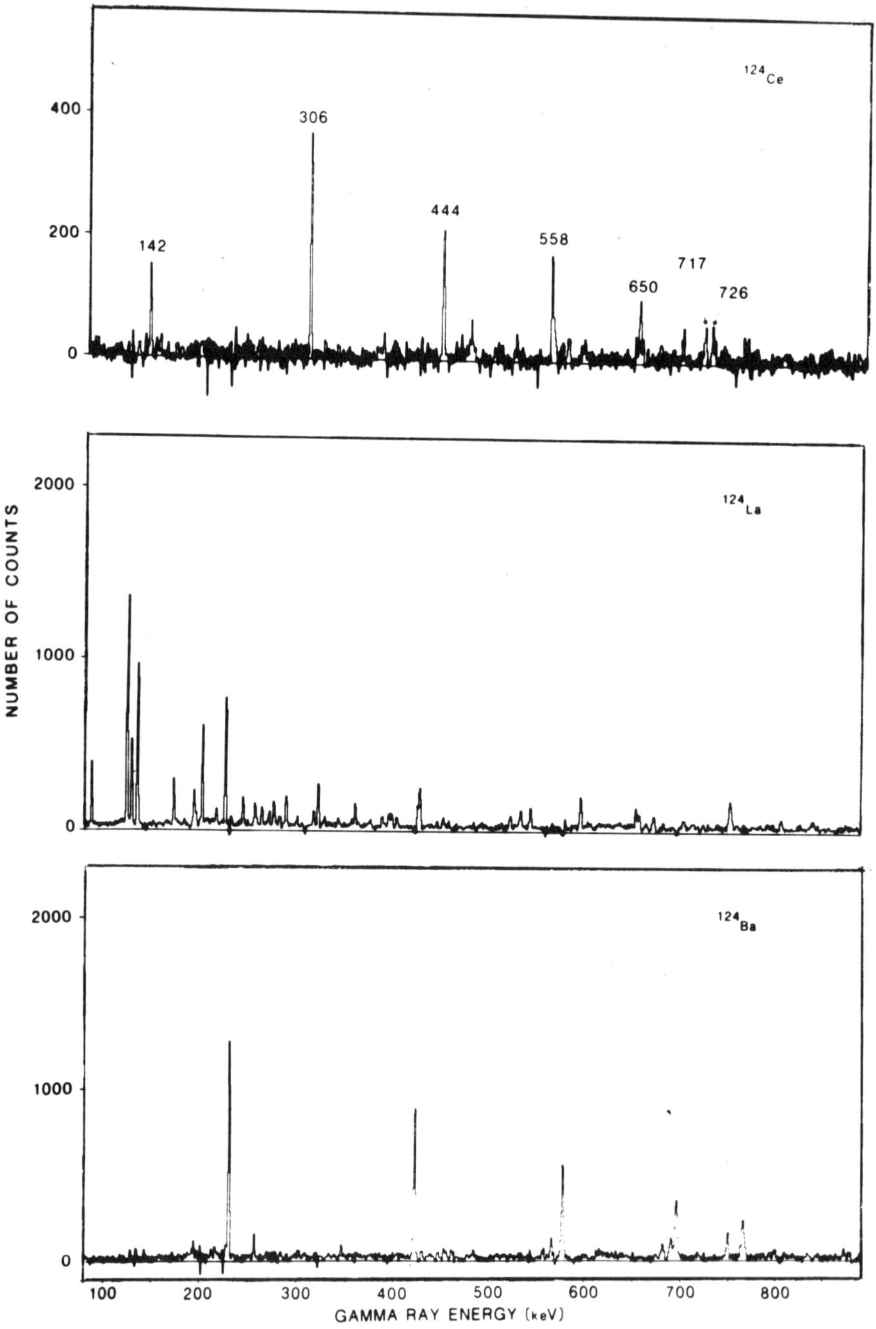


**Figure 74.** $\gamma$ rays in coincidence with mass 124 and $\Delta E(Z)$ signals from $^{124}$Ce, $^{124}$Ba, and $^{124}$La recoils (Yi 86).

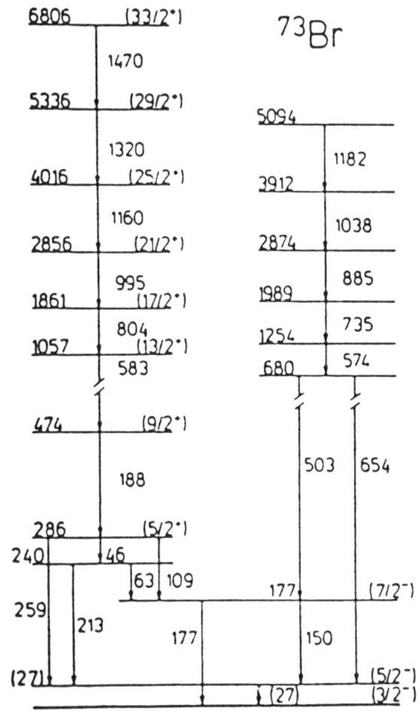


**Figure 75.** Levels in $^{73}$Br (We 85 and subsequent analysis, Wö 85).

$\gamma$-$\gamma$ coincidence runs. The band structures identified for the first time in $^{73}$Br (We 85) are shown in Figure 75. Of interest is the band built on the $9/2^+$ level. Subsequent work of Wörmann and co-workers (Wö 85) has added to the low spin structure below this band as shown. The $g_{9/2^+}$ band in $^{73}$Br is essentially identical to that in $^{75}$Br.

Potential-energy surface calculations (Na 85, Na 86, Lü 85, Du 85) indicate that both the $^{72}_{34}Se_{38}$ and $^{74}_{34}Se_{40}$ ground states are very flat in the $\gamma$ plane with prolate-oblate energy differences, $\delta E$, of the order of 160 keV while the $^{74}_{36}Kr_{38}$ and $^{76}_{36}Kr_{40}$ ground states have large prolate deformations, $\beta_2 \approx 0.35$, and large prolate-oblate energy differences, $\delta E \approx 600$ keV. Very collective rotational patterns characteristic of prolate shapes are clearly visible in the low-lying $p_{3/2}$, $f_{5/2}$, $g_{9/2}$ bands in $^{73}_{35}Br_{38}$ and $^{75}_{35}Br_{40}$, so these nuclei are clearly prolate well-deformed like $^{74,76}$Kr.

In seeking to identify levels in$^{71}$Br, RMS-n-$\gamma$ and RMS-p-$\gamma$ coincidence measurements were made at the University of Rochester tandem and $\gamma$-$\gamma$, n-$\gamma$, n-n-$\gamma$, and n-$\gamma$-$\gamma$ coincidences were made at the University of Notre Dame tandem on the $^{58}$Ni + $^{16}$O $\rightarrow$ ($^{74}$Kr)* reaction. The mass 71 products in the RMS are $^{71}$As(3p), $^{71}$Se(2pn), and $^{71}$Br(p, 2n). New lines in $^{71}$As, $^{71}$Se, and tentatively $^{71}$Br were identified by comparing RMS-n-$\gamma$ and RMS-p-$\gamma$ spectra with each other and with the singles spectra (Ra 86). From Figure 72 one can see that the 894 keV transition can be assigned to $^{71}$Se by its presence in both the proton- and neutron-gated spectra. The $\gamma$-$\gamma$ and $\gamma(\theta)$ data at Köln were reanalyzed (Ra 86), and the 894 keV transition in $^{71}$Se was shown to be an $11/2^+ \rightarrow 9/2^+$ transition with an E2/M1 mixing ratio of +0.5. A new sequence of levels built on it was found, as shown in Figure 76 (Eb 88). Based on a preliminary analysis (Le 86) of all the members of the $9/2^+$ band in $^{71}_{34}Se_{37}$, this band is more predominately an oblate structure with $|\beta_2| \sim 0.24$ and $\gamma \sim 32°$. A fit to $\delta$ yields a larger $\gamma$ near 50°.

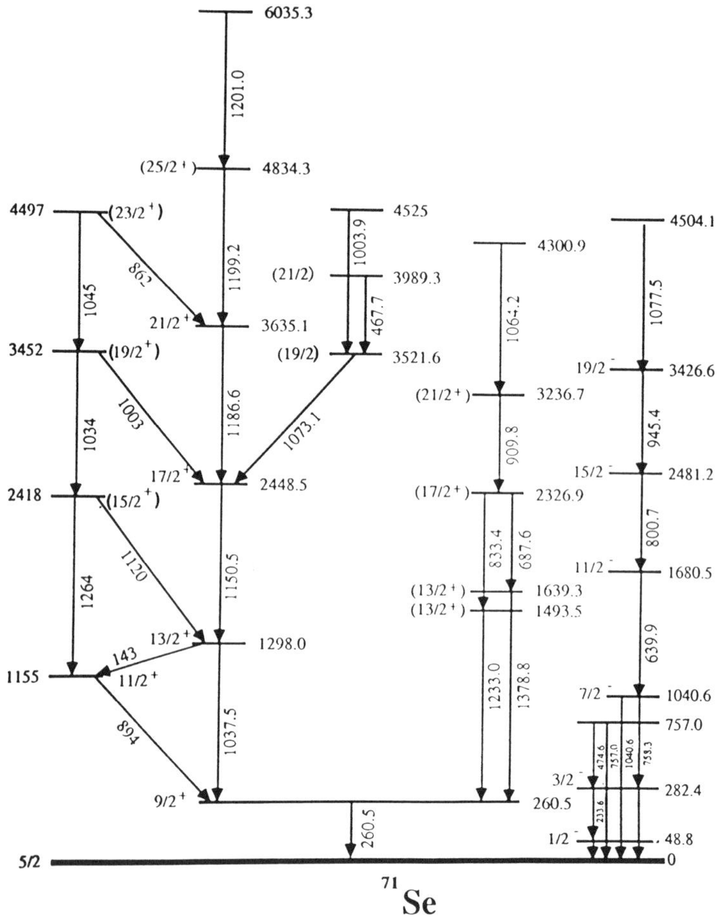


**Figure 76.** Level structure of $^{71}$Se (Ra 86, Eb 88).

This is the first direct evidence in this mass region for oblate structures, which have been predicted in various theoretical calculations for many years. This band is presumably associated with the oblate shell gap at $\beta \sim -0.2$ in Figure 28. The other bands in $^{71}$Se have larger deformation and are presumably associated with the large prolate deformation associated with the $N = 38$ shell gap (Ha 85b, Ha 85c). Candidates for transitions in $^{71}$Br have been identified but no level structure established. New studies with the Rochester RMS are planned.

Earlier we discussed how the large and very constant moment of inertia of $^{72}$Se and $^{70}$Se may be related to the collapse of pairing in these nuclei. Recently, Heese and co-workers (He 86b) measured the $B$(E2) values for $^{70}$Se and compared them with $^{72}$Se and $^{68}$Ge as shown in Figure 77. Note $^{70}$Se shows a quite different change in $I$ with $\hbar\omega$ at low spin compared to $^{72}$Se but is very similar above $8^+$. The $4^+ \rightarrow 2^+$ and $6^+ \rightarrow 4^+$ $B$(E2) values drop rapidly in $^{70}$Se compared to their sharp rise in $^{72}$Se. For $^{70}$Se,

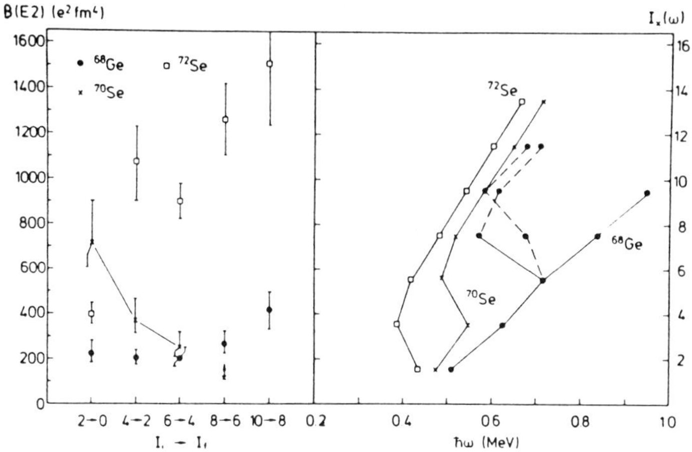


**Figure 77.** $B$(E2) and $I_x$ values for yrast transitions in $^{70}$Se, $^{72}$Se, and $^{68}$Ge (Ah 81, Ra 86c, He 86b, Ha 76b, Li 77, Pa 83, de 81) are compared (He 86b).

$\beta_2 = 0.30(3)$ for the ground state while the presumably prolate excited band in $^{72}$Se has $\beta_2 = 0.31(2)$. The difference in $^{70}$Se and $^{72}$Se may reflect a shape transition from a modest oblate ground state to a large prolate excited band in $^{72}$Se, but large prolate to large oblate shape in $^{70}$Se. For the coexisting structures, clearly further studies are needed to resolve this puzzle. Other unusual effects on transition rates have been observed in this region related to shape changes between bands, such as the fast M1 transition seen in $^{81}$Kr (Figure 78) (Fu 83). The origin of the fast M1 transition is related to how the three quasiparticles couple in the $21/2^+$ state (Dö 82, Fu 83).

Nuclei with $Z > 50$ and $N < 82$ are expected to have sizable ground-state deformation with the nuclei furthest from stability having the largest deformation, $\beta_2 \sim 0.25$–$0.30$ [see for example the calculations in (Mö 81, Le 82, Åb 82)]. The Daresbury Recoil Separator was used to identify, for the first time, $\gamma$ transitions and levels in $^{124}$Ce (Yi 86). If one gates the prompt, in-beam $\gamma$ rays at the target with only the recoils at mass 124, transitions in $^{124}$Ba and $^{124}$La would dominate the spectrum since $^{124}$Ce produced in the reaction $^{54}$Fe($^{74}$Se, 2p2n)$^{124}$Ce is only about 2% of the total cross-section. Thus, the recoil-mass-gated $\gamma$ spectra are too complex to identify $^{124}$Ce. To identify $^{124}$Ce, a $\Delta E$–$E$ recoil detector was added at the focal plane of the RMS. By using an inverse reaction, $^{74}$Se on a $^{54}$Fe target, the recoils has sufficient velocity ($V/C \sim 5.3\%$) that the $\Delta E$ signals from the different $Z$ recoils were sufficiently well-separated that they would be used as an additional gate on the $\gamma$ spectra. The results of the RMS-$\Delta E(Z)$-$\gamma$ coincidence are shown in Figure 74. The $\gamma$ rays were detected in an array of six bismuth germanate-shielded germanium detectors. One sees that there is excellent separation and unique determination of the 2% cross-section $^{124}$Ce lines from the stronger ones. The BGO Compton suppression was also crucial in reducing the background in the spectrum. The transitions in $^{124}$Ce were assumed to be a $2^+$, $4^+$, $6^+$ sequence. The energy of the $2^+$ state was used to determine the deformation.

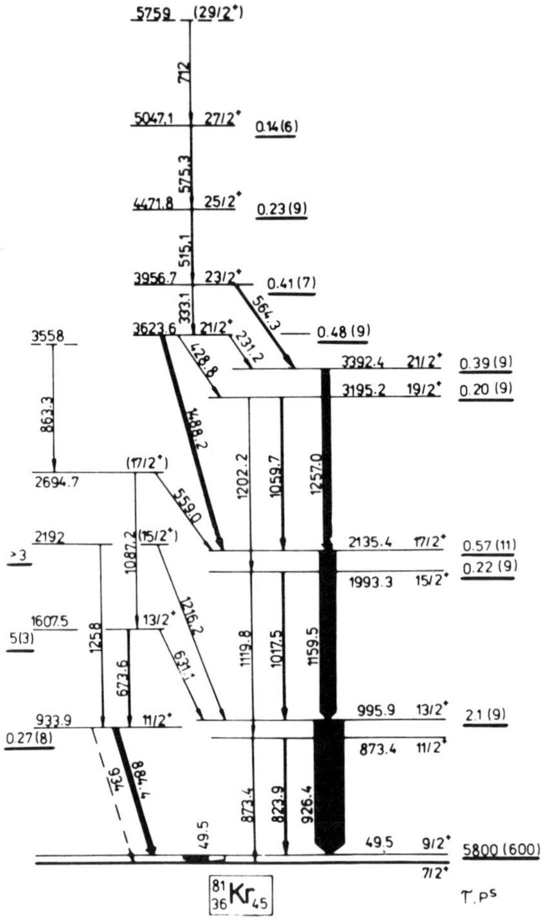


**Figure 78.** Positive-parity levels in $^{81}$Kr observed in the $(\alpha, 3n)$ reaction (Fu 83).

Their results are given in Table 4 (Yi 86) along with earlier results for $^{126-132}$Ce and are compared there with theoretical calculations. They also carried out potential-energy surface calculations similar to those of Åberg (Åb 82) but with a hexadecapole rather than a triaxial deformation. The $^{124}$Ce results (Yi 86) confirm the predictions that the cerium isotopes become more deformed as the neutron number decreases. However, their data indicate a larger deformation than any of the published calculations but in line with their own calculations. Note also the $E_{4^+}/E_{2^+}$ ratio increases regularly toward the rigid rotor value with decrease in $N$. As additional examples, very recently the use of inverse reactions in the Daresbury RMS was crucial to the $Z$ identification of the new $N = Z$ nuclei $^{72}_{36}$Kr, $^{76}_{38}$Sr, and $^{80}_{40}$Zr (Li 88).

Finally, future use of RMS facilities will include separation of elements such as Zr, Mo, Ta which are hard to extract from present ISOL systems and of elements with half-lives down to ms and even $\mu$s. The extraction of these presently inaccessible

**Table 4. Deformations (Experimental and Calculated) and Ratios of the Excitation Energies of the First $4^+$ and $2^+$ States in $^{124-132}$Ce**

| | Experiment[a] | | Calculations (Le 82) | | (Yi 86) | | (Ab 82)[b] |
|---|---|---|---|---|---|---|---|
| | $\varepsilon_2$ | $\frac{E(4^+)}{E(2^+)}$ | $\varepsilon_2$ | $\varepsilon_4$ | $\varepsilon_2$ | $\varepsilon_4$ | $\varepsilon_2$ |
| $^{124}$Ce | 0.31 | 3.15 | 0.28 | 0.005 | 0.31 | 0.020 | |
| $^{126}$Ce | 0.28 | 3.06 | 0.27 | 0.020 | 0.30 | 0.027 | 0.27 |
| $^{128}$Ce | 0.25 | 2.93 | 0.26 | 0.035 | 0.27 | 0.028 | 0.24 |
| $^{130}$Ce | 0.22 | 2.79 | 0.22 | 0.200 | 0.23 | 0.021 | 0.21 |
| $^{132}$Ce | 0.19 | 2.64 | 0.19 | 0.025 | 0.18 | 0.013 | 0.18 |

[a] The data are taken from (Yi 86, Li 85, Wa 68).
[b] The calculations of Åberg were in the $\varepsilon_2$-$\gamma$ plane and did not include $^{124}$Ce.

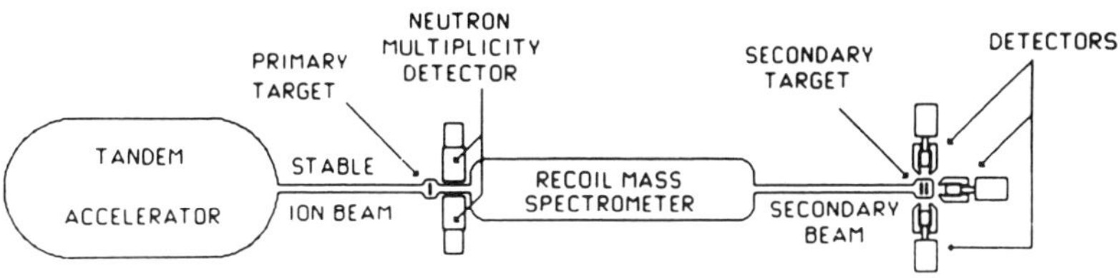


**Figure 79.** A schematic diagram of an experiment which uses a radioactive beam separated by a recoil mass spectrometer (Pi 86b).

isotopes with a RMS will open up studies of nuclear structure and decay processes of new elements and new short-lived isotopes very far off stability (Pi 86, Ki 85, Ki 86b). They will also be important in giving beams of radioactive isotopes for many new types of experiments (Pi 86, Ki 86b). A simple arrangement is illustrated in Figure 79 (Pi 86b). For example, these separated radioactive beams can be Coulomb excited. This should be possible with strongly deformed far-off-stability nuclei like $^{76}$Sr because of their large $B$(E2) values. Finally, the radioactive beams produced in inverse reactions can be used to strike a light target and produce new nuclei very far off stability, which are inaccessible by present beams.

## 7. Other New Directions

There are other new directions underway that will have significant impacts. Already at GSI, they are using multinucleon transfer reactions to produce neutron-rich nuclei inaccessible by fission from the Cr to Ni region (Bo 85c) to the Ra–Ac region (Gi 86b). Such experiments are already providing significant data to test different theories of nucleosynthesis. The new measured half-lives are systematically shorter in the Cr–Ni region than predicted. At Berkeley, heavy-ion multinucleon transfer reactions are being used to selectively produce actinide and transactinide nuclei for study [for example, see (We 87)]. The use of fragmentation reactions to probe very-far-off-stability neutron-rich nuclei will be a major research area as shown by the first Berkeley work (Sy 79, We 79) and recent GANIL work (Po 85b). In addition, there are new systems under construction or just completed, like ISOLDE III with a two-stage separator to reduce cross-contamination, the SPIN-2-YASNAPP-2 facility at Dubna (Fi 85), and the Los Alamos Time-Of-Flight-Isochronous (TOFI) recoil spectrometer which has a time dispersion that is a function of mass/charge. To illustrate its power, the TOFI system was used recently to make direct mass measurements of 21 neutron-rich nuclei (Vi 86). Improvements in laser techniques will continue to open up new areas. The new 18 Tm synchrotron (SIS), with a 10m storage ring and a recoil mass spectrometer which can provide radioactive beams to the storage ring and back to the accelerator, will also be a powerful new facility to study nuclei still farther off stability (Ki 85, Ki 87, Ro 86).

All these and other developments offer much promise for the future. Studies of nuclei far from stability will continue to be a major topic at the forefront of nuclear science with an ever-widening scope and with increasing opportunities in applied science, because of the unique vistas they offer.

## Acknowledgments

Many people are thanked for their help in the preparation of this manuscript, in particular A. V. Ramayya, E. F. Zganjar, L. Chaturvedi, J. Kormicki, L. K. Peker, H. K. Carter, and K. S. Toth, along with numerous colleagues who supplied reprints, preprints, and figures from their work.

This work was supported in part by the U.S. Department of Energy under Basic Energy Science contracts, No. DE-AS05-76ER05034 with Vanderbilt University and No. DE-AC05-760R00033 with UNISOR.

## References

(Ab 84) Afsar Abbas, *Phys. Rev. C* **29**:1033 (1984).

(Åb 82) S. Åberg, *Phys. Scr.* **25**:23 (1982).

(Ah 81) A. Ahmed, A. V. Ramayya, D. L. Sastry, J. H. Hamilton, R. B. Piercey, H. Kawakami, A. P. de Lima, C. F. Maguire, R. L. Robinson, H. J. Kim, J. C. Wells, and A. C. Rester, *Phys. Rev. C* **24**:1486 (1981).

(Ah 85) D. P. Ahalpara, A. Abzouzi, and K. H. Bhatt, in *Nucleus-Nucleus Collisions from the Coulomb Barrier Up To The Quark-Gluon Plasma*, in *Progress in Particle and Nuclear Physics* (A. Faessler, ed.), Pergamon Press, New York (1985), p. 135.

(Al 84) D. V. Alexandrov, A. F. Beliatsky, Yu. A. Gluhov, E. Yu Nikolsky, B. G. Novatsky, A. A. Ogloblin, and D. N. Stepanov, *Pis'ma Zh. Eksp. Teor. Fiz.* **40**:152 (1984).

(Am 82) A. Amusa and R. D. Lawson, *Z. Phys. A* **307**:333 (1982).

(An 86) M. Anselment, K. Bekk, A. Hanser, H. Hoeffgen, G. Meisel, S. Göring, H. Rebel, and G. Schatz, *Phys. Rev. C* **34**:1052 (1986).

(Ar 74) A. Arima and F. Iachello, *Phys. Rev. Lett.* **25**:1069 (1974).

(Ar 81) A. Arima and F. Iachello, *Ann. Rev. Nucl. Sci.* **31**:75 (1981).

(Az 79) R. E. Azuma, G. L. Borchert, L.-C. Carraz, P. G. Hansen, B. Jonson, S. Mattsson, O. B. Nielsen, G. Nyman, I. Ragnarsson, and H. L. Ravn, *Phys. Lett.* **86B**:5 (1979).

(Ba 69) M. Baranger and R. A. Sorenson, *Sci. Am.*, p. 58 (Aug. 1969).

(Ba 81) M. Barclay, L. Cleemann, A. V. Ramayya, J. H. Hamilton, J. D. Cole, A. de Lima, W.-C. Ma, S. C. Pancholi, R. Soundranayagam, H. Yamada, W. Ellerby, R. L. Robinson, and W. Milner, private communication (1981), reported in (Ha 81c).

(Ba 82) M. Barclay, L. Cleemann, J. H. Hamilton, A. V. Ramayya, J. D. Cole, and W.-C. Ma, private communication (1982), reported in (Ha 82b).

(Ba 85) S. W. Barwick, R. B. Price, and J. D. Stevenson, *Phys. Rev. C* **31**:1984 (1985).

(Be 66) I. Bergström, *Nucl. Instrum. Methods* **43**:116 (1966).

(Be 66b) I. Bergström, *Nucl. Instrum. Methods* **43**:129 (1966).

(Be 75) V. Berg, C. Bourgeois, and R. Foucher, *J. Phys. (Paris)* **36**:613 (1975).

(Bé 77) R. Béraud, M. Meyer, M. G. Desthuilliers, C. Bourgeois, P. Kilcher, and Jocelyne Letessier, *Nucl. Phys. A* **284**:221 (1977).

(Be 79) R. Bengtsson and S. Frauendorf, *Nucl. Phys. A* **314**:27 (1979).

(Be 79b) R. Bengtsson and S. Frauendorf, *Nucl. Phys. A* **327**:139 (1979).

(Be 80) V. Berg, in *Future Directions in Studies of Nuclei Far From Stability* (J. H. Hamilton, E. H. Spejewski, C. R. Bingham, and E. F. Zganjar, eds.), North-Holland, Amsterdam (1980), p. 353.

(Be 81) R. Bengtsson, I. Ragnarsson, J.-Y. Zhang, and S. Åberg, in *Proceedings of the Fourth International Conference on Nuclei Far From Stability*, CERN 80-09 (1981), p. 509.

(Bé 81) R. Béraud, J. Tréherne, A. Charvet, R. Duffait, J. Genevey, A. Gizon, J. Gizon, and M. Meyer, *Z. Phys. A* **299**:279 (1981).

(Be 82) M. Bernas, Ph. Dessagne, M. Lanagevin, J. Payet, F. Pougheon, and P. Roussel, *Phys. Lett.* **113B**:279 (1982).

(Be 83) V. Berg, Z. Hu, J. Oms, and C. Ekström, *Nucl Phys. A* **410**:445 (1983).

(Be 84) R. Bengtsson, P. Möller, J. R. Nix, and J.-Y. Zhang, *Phys. Scr.* **29**:402 (1984).

(Be 85) T. Bengtsson and I. Ragnarsson, *Nucl. Phys. A* **436**:14 (1985).

(Be 86) R. Bengtsson, J.-Y. Zhang, J. H. Hamilton, and L. K. Peker, *J. Phys. G Lett.* **12**:L223 (1986).

(Be 86b) R. Bengtsson, T. Bengtsson, J. Dudek, G. A. Leander, W. Nazarewicz, and J.-Y. Zhang, *Phys. Lett.* **183B** (1987).

(Bj 86) T. Björnstad, M. J. G. Borge, J. Blomqvist, R. D. von Dincklage, G. T. Ewan, P. Hoff, B. Jonson, K. Kawade, A. Kerek, O. Klepper, G. Lövhöiden, S. Mattsson, G. Nyman, and H. L. Ravn, *Nucl. Phys. A* **453**:463 (1986).

(Bl 81) J. Blomqvist, in *Proceedings of the Fourth International Conference on Nuclei Far From Stability*, CERN 81-09 (1981), p. 536.

(Bl 83) J. Blomqvist, P. Kleinheinz, and P. J. Daly, *Z. Phys. A* **312**:27 (1983).

(Bo 53) A. Bohr and B. R. Mottelson, *Dan. Mat. Fys. Medd.* **27**:No. 16 (1953).

(Bo 72) J. Bonn, G. Huber, H. Kluge, and E. W. Otten, *Phys. Lett.* **38B**:308 (1972).

(Bo 76) C. Bourgeois, M. Bouet, A. Caruette, A. Ferro, R. Foucher, J. Fournet, A. Höglund, L. Kotfila, G. Landois, C. F. Liang, B. Merlant, J. Obert, A. Peghaire, J. C. Putaux, J. L. Sarrouy, W. Watzig, A. Wojtasiewicz, and V. Berg, *J. Phys. (Paris)* **37**:49 (1976).

(Bo 82) C. Bourgeois, P. Kilcher, B. Roussiere, J. Sauvage-Letessier, M. G. Porquet, and the ISOCELE Collaboration, *Nucl. Phys. A* **386**:308 (1982).

(Bo 85) J. A. Bounds, C. R. Bingham, P. Juncar, H. K. Carter, G. A. Leander, R. L. Mlekodaj, E. H. Spejewski, and W. M. Fairbank, Jr., *Phys. Rev. Lett.* **55**:2269 (1985).

(Bo 85b) P. Bonche, H. Flocard, P. H. Heenen, S. J. Krieger, and M. S. Weiss, *Nucl. Phys. A* **443**:39 (1985).

(Bo 85c) U. Bosch, W.-D. Schmidt-Ott, P. Tidemand-Petersson, E. Runte, W. Hillebrandt, M. Lechle, F.-K. Thielmann, R. Kirchner, O. Klepper, E. Roeckl, K. Rykaczewski, D. Schardt, N. Kaffre, M. Bernas, Ph. Dessagne, and W. Kurcewicz, *Phys. Lett.* **164B**:22 (1985).

(Br 49) P. Brix and H. Kopfermann, *Z. Phys.* **126**:344 (1949).

(Br 72) M. Brack, J. Damgaard, A. S. Jensen, H. C. Pauli, V. M. Strutinsky, and C. Y. Wong, *Rev. Mod. Phys.* **44**:320 (1972).

(Br 79) J. Bron, W. H. A. Hesselink, A. Van Poelgeest, J. J. A. Zalmstra, M. J. Uitzinger, and H. Verheul, *Nucl. Phys. A* **318**:335 (1979).

(Ca 81) R. F. Casten, D. D. Warner, D. S. Brenner, and R. L. Gill, *Phys. Rev. Lett.* **47**:1433 (1981).

(Ca 85) R. F. Casten, *Phys. Lett.* **152B**:145 (1985).

(Ce 77) J. Cerny and J. C. Hardy, *Ann. Rev. Nucl. Sci.* **27**:33 (1977).

(Ch 70) E. Cheifetz, R. C. Jarod, S. G. Thompson, and J. B. Wilhelmy, *Phys. Rev. Lett.* **25**:38 (1970).

(Ch 83) R. R. Chasman, *Phys. Rev. C* **28**:1374 (1983).

(Co 76) J. D. Cole, J. H. Hamilton, A. V. Ramayya, W. G. Nettles, H. Kawakami, E. H. Spejewski, M. A. Ijaz, K. S. Toth, E. L. Robinson, K. S. R. Sastry, J. Lin, F. T. Avignone, W. H. Brantley, and P. V. G. Rao, *Phys. Rev. Lett.* **37**:1185 (1976).

(Co 81) L. L. Collims, L. L. Riedinger, A. C. Kahler, C. R. Bingham, G. D. O'Kelley, J. L. Wood, R. W. Fink, A. G. Schmidt, E. H. Spejewski, H. K. Carter, R. L. Mlekodaj, E. F. Zganjar, and J. H. Hamilton, private communication (1981), reported in (Ha 82).

(Co 81b) T. Cormier and P. M. Stwertka, *Nucl. Instrum. Methods* **184**:423 (1981).

(Da 78) P. J. Daly, P. Kleinheinz, R. Braoda, A. M. Stefanini, S. Lunardi, H. Backe, L. Richter, R. Willwater, and F. Weik, *Z. Phys. A* **288**:103 (1978).

(Da 79) P. Dabkiewicz, F. Buchinger, H. Fischer, H. J. Kluge, H. Kremmling, T. Kühl, A. C. Müller, and H. A. Schuessler, *Phys. Lett.* **82B**:199 (1979).

(Da 80) P. J. Daly, P. Kleinheinz, R. Broda, S. Lunardi, H. Backe, and J. Blomqvist, *Z. Phys. A* **298**:173 (1980).

(de 53) A. de Shalit and M. Goldhaber, *Phys. Rev.* **92**:1211 (1953).

(De 80) L. E. DeGreer and G. B. Holm, *Phys. Rev. C* **22**:2163 (1980).

(de 81) A. P. de Lima, A. V. Ramayya, J. H. Hamilton, B. van Nooijen, R. M. Ronningen, H. Kawakami, R. B. Piercey, E. de Lima, R. L. Robinson, H. J. Kim, L. K. Peker, F. A. Rickey, R. Popli, A. J. Caffrey, and J. C. Wells, *Phys. Rev. C* **23**:213 (1981).

(Dé 81) C. Détraz, in *Proceedings of the Fourth International Conference on Nuclei Far From Stability*, CERN 81-09 (1981), p. 311.

(De 85) B. I. Deutch and L. Vanneste, eds., *Proceedings of the International Symposium on Nuclear Orientation and Nuclei Far From Stability*, Scientific Publ. Co., Basel-Switzerland (1985).

(Dö 82) F. Dönau and S. Frauendorf, in *Proceedings of the International Summer School*, Poiana Brasov, Romania (1982).

(Do 86) J. Dobaczewski and co-workers, in *Proceedings of the International Symposium on Weak and Electromagnetic Interactions in Nuclei*, Springer-Verlag, Heidelberg (1986).

(Dr 78) G. D. Dracoulis, P. M. Walker, and A. Johnson, *J. Phys. G* **4**:714 (1978).

(Dr 86) G. D. Dracoulis, A. E. Stuchbery, A. P. Byrne, A. R. Poletti, S. J. Poletti, J. Gerl, and R. A. Bark, *J. Phys. G* **12**:L97 (1986).

(Du 82) J. L. Durell, G. D. Dracoulis, C. Fahlander, and A. D. Byrne, *Phys. Lett.* **115B**:367 (1982).

(Du 85) J. Dudek and W. Nazarewicz, private communication (1985).

(Eb 86) J. Eberz, U. Dinger, G. Huber, H. Lochmann, R. Menges, R. Neugart, R. Kirchner, O. Klepper, T. Kühl, D. Marx, G. Ulm, and K. Wendt, *Nucl. Phys.* **A464**:9 (1987).

(Eb 86b) J. Eberz, U. Dinger, T. Horiguchi, G. Huber, H. Lochmann, R. Menges, R. Kirchner, O. Klepper, T. Kühl, D. Marx, E. Roeckl, D. Schardt, and G. Ulm, *Z. Phys. A* **323**:119 (1986) and *GSI Scientific Report* 86-1 (1985), p. 49.

(Eb 88) J. Eberth, M. Wiosna, T. Mylaeus, N. Schmal, S. Skoda, J. Busch, W. Teichert, P. v. Brentano, J. H. Hamilton, A. V. Ramayya, X. Zhao, C. F. Maguire, W.-C. Ma,

J. Kormicki, S. Wen, L. Chaturvedi, Z.-M. Chen, T. M. Cormier, M. Satteson, J. D. Cole, R. B. Piercey, M. A. Herath-Banda, E. F. Zganjar, M. O. Kortelahti, G. A. Leander, and W. Nazarewicz, in *Nuclei Far From Stability*, Fifth International Conference (Ian S. Towner, ed.), American Institute of Physics Conference Proceeding 164 (1988), p. 268.

(Ek 81) C. Ekström, in *Proceedings of the Fourth International Conference on Nuclei Far From Stability*, CERN 81-09 (1981), p. 12.

(El 78) T. Elmroth, E. Hagberg, P. G. Hansen, J. C. Hardy, B. Jonson, H. L. Ravn, and P. Petersson-Tidemand, *Nucl. Phys. A* **304**:493 (1978).

(Fe 79) P. Federman and S. Pittel, *Phys. Rev. C* **20**:820 (1979).

(Fi 85) M. Finger, T. I. Kracikova, N. A. Lebedev, V. N. Pavlov, I. Prochazka, V. M. Tsupko-Sitnikov, Z. Janout, J. Konicek, M. Rotter, B. Sedlak, and J. Dubak, in *Proceedings of the International Symposium on Nuclear Orientation and Nuclei Far From Stability* (B. I. Deutch and L. Vanneste, eds.), Scientific Publ. Co., Basel-Switzerland (1985), p. 461.

(Fl 74) E. R. Flynn, J. G. Beery, and H. G. Blair, *Nucl. Phys. A* **218**:285 (1974).

(Fl 76) S. Flanagan, R. Chapman, J. L. Durell, W. Gelletly, and J. N. Mo, *J. Phys. G* **2**:589 (1976).

(Fl 83) E. R. Flynn, J. van der Plicht, J. B. Wilhelmy, L. G. Mann, G. L. Struble, and R. G. Lanier, *Phys. Rev. C* **28**:97 (1983).

(Fo 66) W. Forsling, C. J. Herrlander, and H. Ryde, eds., *Proceedings of the International Symposium on Why and How Should We Investigate Nucleides Far From Stability*, Almqvist and Wiksell, Stockholm (1966).

(Fo 84) B. Fogelberg and J. Blomqvist, *Phys. Lett.* **137B**:20 (1984).

(Fu 80) L. Funke, G. Winter, J. Döring, F. Dubbers, P. Kemnitz, and E. Will, in *Future Directions in Studies of Nuclei Far From Stability* (J. H. Hamilton, E. H. Spejewski, C. R. Bingham, and E. F. Zganjar, eds.), North-Holland, Amsterdam (1980), p. 231.

(Fu 82) L. Funke, in *Dynamics of Nuclear Collective Motion* (K. Ogawa and K. Tanabe, eds.), Bando Printing Co., Osaka (1982), p. 495.

(Fu 83) L. Funke, F. Dönau, J. Döring, P. Kemnitz, E. Will, G. Winter, L. Hildingsson, A. Johnson, and Th. Lindblad, *Phys. Lett.* **120B**:301 (1983).

(Ga 84) S. Gales, E. Hourani, M. Hussonnois, J. P. Shapira, L. Stab, and M. Vergnes, *Phys. Rev. Lett.* **53**:759 (1984).

(Ga 86) J. D. Garrett, G. B. Hagemann, and B. Herskind, in *Annual Review of Nuclear and Particle Science, Vol. 36* (J. D. Jackson, ed.), Annual Reviews Inc., Palo Alto (1986), p. 419.

(Ga 86b) D. Galeriu, D. Bucurescu, and M. Ivascu, *J. Phys. G* **12**:329 (1986).

(Gi 86) C. Girit, D. Olive, J. H. Hamilton, and H. K. Carter, ORNL Annual Report (1986).

(Gi 86b) K.-L. Gippert, E. Runte, W.-D. Schmidt-Ott, P. Tidemand-Petersson, N. Kaffrell, P. Peuser, R. Kirchner, O. Klepper, W. Kurcewicz, P. O. Larsson, E. Roeckl, D. Schardt, and K. Rykaczewski, *Nucl. Phys. A* **453**:1 (1986).

(Gi 88) C. Girit, G. D. Alton, C. R. Bingham, H. K. Carter, M. L. Simpson, J. D. Cole, J. H. Hamilton, B. D. Kern, K. S. Krane, and E. F. Zganjar, in *Nuclei Far From Stability*, Fifth International Conference (Ian S. Towner, ed.), American Institute of Physics Conference Proceedings 164 (1988), p. 849.

(Go 81) G. M. Gowdy, R. E. Chrien, Y. Y. Chu, R. L. Gill, H. I. Liou, M. Schmid, M. L. Stelts, K. Sistemich, F. K. Wohn, H. Yamamoto, D. S. Brenner, T. R. Yeh, R. A. Meyer, C. Chung, W. B. Walters, and R. F. Petry, in *Proceedings of the Fourth International Conference on Nuclei Far From Stability*, CERN 81-09 (1981), p. 562.

(Gr 65) W. Greiner, *Proceedings of the International Conference on Neutron Capture*, Jülich, W. Germany (1965).

(Gr 78) M. A. Grimm, Ph.D. Thesis, Georgia Institute of Technology (1978).

(Gr 83) H. Grawe and H. Haas, *Phys. Lett.* **120**:63 (1983).

(Gr 83b) V. R. Green, W. D. Hamilton, S. J. Robinson, N. J. Stone, and P. M. Walker, *Hyperfine Interalt.* **15/16**:979 (1983).

(Gr 86) M. Greiner and W. Scheid, *J. Phys. G* **12**:L229 (1986).

(Gr 86b) V. R. Green, N. J. Stone, T. L. Shaw, J. Rikovska, K. S. Krane, P. M. Walker, and I. S. Grant, *Phys. Lett. B* **173**:115 (1986).

(Gr 86c) V. R. Green, C. J. Ashworth, J. Rikovska, T. L. Shaw, N. J. Stone, P. M. Walker, and I. S. Grant, *Phys. Lett. B* **177**:159 (1986).

(Ha 50) O. Haxel, J. H. D. Jensen, and H. E. Suess, *Z. Phys.* **128**:295 (1950).

(Ha 65) J. H. Hamilton, ed., *Internal Conversion Processes*, Academic Press, New York (1965).

(Ha 65b) J. H. Hamilton, in *Internal Conversion Processes* (J. H. Hamilton, ed.), Academic Press, New York (1965), p. 1.

(Ha 74) J. H. Hamilton, E. H. Spejewski, R. L. Mlekodaj, W. D. Schmidt-Ott, R. W. Fink, Z. Xenoulis, K. R. Baker, J. L. Wood, G. Gowdy, H. K. Carter, B. D. Kern, K. J. Hofstetter, J. L. Weil, E. F. Zganjar, K. S. Sastry, F. T. Avignone, C. R. Bingham, L. L. Riedinger, L. Harwood, F. Turner, I. A. Sellin, D. J. Pegg, J. Lin, A. V. Ramayya, R. S. Lee, G. Garcia-Bermudez, E. Bosworth, K. S. Toth, and N. R. Johnson, *Izv. Akad., Nauk USSR, Ser. Fiz.* **38**:22 (1974) (English translation).

(Ha 74b) J. H. Hamilton, A. V. Ramayya, W. T. Pinkston, R. M. Ronningen, G. Garcia-Bermudez, R. L. Robinson, H. J. Kim, R. O. Sayer, and H. K. Carter, *Phys. Rev. Lett.* **32**:239 (1974).

(Ha 75) J. H. Hamilton, A. V. Ramayya, E. L. Bosworth, W. Lourens, J. D. Cole, B. van Nooijen, G. Garcia-Bermudez, B. Martin, B. M. Subba Rao, H. Kawakami, L. L. Riedinger, C. R. Bingham, F. Turner, E. F. Zganjar, E. H. Spejewski, H. K. Carter, R. L. Mlekodaj, W. D. Schmidt-Ott, K. R. Baker, R. W. Fink, G. W. Gowdy, J. L. Wood, A. Xenoulis, B. D. Kern, K. J. Hofstetter, J. L. Weil, K. S. Toth, M. A. Ijaz, and K. S. R. Sastry, *Phys. Rev. Lett.* **35**:562 (1975).

(Ha 76) J. H. Hamilton, in *Proceedings of the International Conference on Selected Topics in Nuclear Structure, Vol. I* (V. G. Soloviev *et al.*, eds.), JINR, Dubna (1976), p. 303.

(Ha 76b) J. H. Hamilton, H. L. Crowell, R. L. Robinson, A. V. Ramayya, W. E. Collins, R. M. Ronningen, V. Maruhn-Rezwani, J. Maruhn, N. C. Singhal, H. J. Kim, R. O. Sayer, T. Magee, and L. C. Whitlock, *Phys. Rev. Lett.* **36**:340 (1976).

(Ha 78) E. Hagberg, P. G. Hansen, P. Hornshøj, B. Jonson, S. Mattsson, and P. Tidemand-Petersson, *Phys. Lett.* **78B**:44 (1978).

(Ha 79) J. H. Hamilton, *Nucleonika* **24**:561 (1979).

(Ha 79b) P. G. Hansen, *Ann. Rev. Nucl. Particle Sci.* **29**:69 (1979).

(Ha 79c) J. H. Hamilton, R. L. Robinson, and A. V. Ramayya, in *Nuclear Interactions* (B. A. Robson, ed.), Springer-Verlag, Berlin (1979), p. 253.

(Ha 79d) J. H. Hamilton, R. L. Robinson, and A. V. Ramayya, *Physica Energiae Fortis et Physica Nuclearis* **3**:355 (1979) [in Chinese].

(Ha 80) J. H. Hamilton, E. H. Spejewski, C. R. Bingham, and E. F. Zganjar, eds., *Future Directions in Studies on Nuclei Far From Stability*, North-Holland, Amsterdam (1980).

(Ha 81) P. G. Hansen and O. B. Nielson, eds., *Proceedings of the Fourth International Conference on Nuclei Far From Stability*, CERN 81-09 (1981).

(Ha 81b) J. H. Hamilton, in *Proceedings of the International Symposium on Nuclear Collectivity* (A. Arima and T. Marumori, eds.), Institute for Nuclear Studies, Tokyo (1981), p. 87.

(Ha 81c) J. H. Hamilton, R. B. Piercey, R. Soundranayagam, A. V. Ramayya, C. F. Maguire, X.-J. Sun, Z. Z. Zhao, J. Roth, L. Cleemann, J. Eberth, T. Heck, W. Neumann, M. Nolte, R. L. Robinson, H. J. Kim, S. Frauendorf, J. Döring, L. Funke, G. Winter, J. C. Wells, J. Lin, A. C. Rester, and H. K. Carter, in *Proceedings of the Fourth International Conference on Nuclei Far From Statiliby*, CERN 81-09 (1986), p. 391.

(Ha 82) J. H. Hamilton, in *Heavy Ion Collisions, Vol. 3* (R. Bock, ed.), North-Holland, Amsterdam (1982), p. 574.

(Ha 82b) J. H. Hamilton, in *Proceedings of the La Rabida International Summer School on Heavy Ion Collisions*, in *Lecture Notes in Physics, Vol. 168* (G. Madurga and M. Lozano, eds.), Springer-Verlag, Berlin (1982), p. 287.

(Ha 83) K. Hardt, Y. K. Agarwal, C. Günther, M. Guttormsen, R. Kroth, J. Recht, F. A. Beck, T. Byrski, J. C. Merdinger, A. Nourredine, D. C. Radford, J. P. Vivien, and C. Bourgeois, *Z. Phys. A* **312**:251 (1983).

(Ha 84) J. H. Hamilton, A. V. Ramayya, C. F. Maguire, R. B. Piercey, R. Bengtsson, P. Möller, J. R. Nix, J.-Y. Zhang, R. L. Robinson, and S. Frauendorf, *J. Phys. G* **10**:L87 (1984).

(Ha 84b) J. H. Hamilton, in *Proceedings of the Fifth Adriatic International Conference on Nuclear Physics*, in *Fundamental Problems in Heavy Ion Collisions* (N. Cindro *et al.*, eds.), World Scientific Publ., Singapore (1984), p. 111.

(Ha 85) J. H. Hamilton, P. G. Hansen, and E. F. Zganjar, *Rep. Prog. Phys.* **48**:631 (1985).

(Ha 85b) J. H. Hamilton, in *Nucleus-Nucleus Collisions from the Coulomb Barrier Up to the Quark-Gluon Plasma*, in *Progress in Particle and Nuclear Physics* (A. Faessler, ed.), Pergamon Press, New York (1985), p. 107.

(Ha 85c) J. H. Hamilton, in *Proceedings of the International Symposium on Nuclear Shell Models* (M. Vallieres and B. H. Wildenthal, eds.), World Scientific Publ., Singapore (1985), p. 31.

(Ha 86) J. H. Hamilton and J. Maruhn, *Sci. Am.*, p. 80 (July 1986).

(Ha 86b) F. Hannachi, G. Bastin, M. G. Porquet, C. Schück, J. P. Thibaud, C. Bourgeois, L. Hildingsson, D. Jerrestam, N. Perrin, H. Sergolle, F. A. Beck, T. Byrski, and J. C. Merdinger, *Z. Phys.* **325**:371 (1986).

(Ha 86c) J. H. Hamilton, in *Nuclei Off The Line Of Stability* (R. A. Meyer and D. Brenner, eds.), ACS Symposium 324, American Chemical Society Press, Washington (1986), p. 510.

(Ha 87) J. H. Hamilton, in *Proceedings of the International School-Seminar on Heavy Ion Physics* (Yu. T. Oganessian, ed.), Dubna, Russia (1987), p. 275.

(He 78) P. Herzog, H.-R. Folle, K. Freitag, A. Kluge, M. Reuschenbach, and E. Bodenstedt, *Nucl. Instrum. Methods* **155**:421 (1978).

(He 80) P. O. Hess, J. A. Maruhn, and W. Greiner, in *Future Directions in Studies of Nuclei Far From Stability* (J. H. Hamilton, E. H. Spejewski, C. R. Bingham, and E. F. Zganjar, eds.), North-Holland, Amsterdam (1980), p. 151.

(He 82) E. Helppi, Y. H. Chung, P. J. Daly, S. R. Faber, A. Pakkaneu, I. Ahmad, P. Chowdhury, Z. W. Grabowski, T. L. Khoo, R. D. Lawson, and J. Blomqvist, *Phys. Lett.* **115B**:11 (1982).

(He 83) K. Heyde, P. van Isacker, M. Waroquier, J. L. Wood, and R. A. Meyer, *Phys. Rep.* **102**:291 (1983).

(He 85) K. Heyde, P. van Isacker, R. F. Casten, and J. L. Wood, *Phys. Lett.* **155B**:303 (1985).

(He 86) K. Heyde and V. Paar, *Phys. Lett. B* **179**:1 (1986).

(He 86b) J. Heese, K. P. Lieb, L. Lühman, F. Raether, B. Wörmann, D. Alber, H. Grawe, J. Elberth, and T. Myläus, in *Nuclear Structure, Reactions and Symmetries* (R. A. Meyer and V. Paar, eds.), World Scientific Publ., Singapore (1986).

(He 87) M. A. Herath-Banda, A. V. Ramayya, L. Cleemann, J. Eberth, J. Roth, T. Heck, N. Schmal, T. Mylaeus, W. Koenig, B. Martin, K. Bethge, and G. A. Leander, *J. Phys. G* **13**:43 (1986).

(Hi 53) D. L. Hill and J. A. Wheeler, *Phys. Rev.* **89**:1102 (1953).

(Hø 74) P. Hørnshøj, P. G. Hansen, B. Jonson, H. L. Ravn, L. Westgaarad, and O. B. Nielsen, *Nucl. Phys. A* **230**:365 (1974).

(Ho 79) S. Hofmann, W. Faust, W. Münzenberg, W. Reisdorf, P. Armbruster, K. Güttner, and H. Ewald, *Z. Phys. A* **291**:53 (1979).

(Ho 83) D. Horn, O. Häusser, I. S. Towner, H. R. Andrews, M. A. Lone, and P. Taras, *Phys. Rev. Lett.* **50**:1447 (1983).

(Ho 85) E. Hourani, M. Houssonnois, L. Stab, L. Brillard, S. Gales, and J. P. Shapira, *Phys. Lett.* **160B**:375 (1985).

(Ia 80) F. Iachello, *Phys. Rev. Lett.* **44**:772 (1980).

(Ja 83) R. V. F. Janssens, P. Chowdhury, H. Emling, D. Frekers, T. L. Khoo, W. Kühn, Y. H. Chung, P. J. Daly, Z. W. Grabowski, M. Kortelahti, S. Frauendorf, and J. Y. Zhang, *Phys. Lett.* **131B**:35 (1983).

(Ka 77) T. A. Kahn, W. D. Lauppe, K. Sistemich, H. Lawin, G. Sadler, and H. A. Selic, *Z. Phys. A* **283**:105 (1977).

(Ka 78) A. C. Kahler, Ph.D. Thesis, University of Tennessee (1978).

(Ka 78b) A. C. Kahler, L. L. Riedinger, N. R. Johnson, R. L. Robinson, E. F. Zganjar, A. Viskanathan, D. R. Zolnowski, M. B. Hughers, and T. T. Sugihara, *Phys. Lett.* **72B**:443 (1978).

(Ka 80) T. A. Kahn and M. W. Johns, *Can. J. Phys.* **58**:1617 (1980).

(Ka 82) K. Kawade, G. Battistuzzi, H. Lawin, H. A. Selic, K. Sistemic, F. Schussler, E. Monnand, J. A. Pinston, B. Pfeiffer, and G. Jung, *Z. Phys. A* **304**:293 (1982).

(Ka 84) J. Kantele, in *Heavy Ions and Nuclear Structure* (B. Sikora and Z. Wilhelmi, eds.), Harwood Academic, New York (1984), p. 390.

(Ka 86) J. Kantele, R. Julin, M. Luontma, and A. Passoja, in *Nuclei Off The Line Of Stability* (R. A. Meyer and D. Brenner, eds.), ACS Symposium 324, American Chemical Society Press, Washington (1986), p. 221.

(Ka 86b) J. Kantele, M. Luontama, W. Trzaska, R. Julin, A. Passoja, and K. Heyde, *Phys. Lett. B* **171**:151 (1986).

(Ki 83) R. Kirchner, O. Klepper, W. Kurcewicz, P. O. Larsson, E. Roeckl, D. Schardt, G. Nyman, and P. Tidemand-Petersson, in *Proceedings of the International Workshop on Gross Properties of Nuclei and Nuclear Excitations XI* (H. Feldmeier, ed)., Institut für Kernphysik, Darmstadt (1983), p. 60.

(Ki 85) P. Kienle, "The SIS-ESR Project of GSI", in *GSI-85-16* (1985).

(Ki 86) A. J. Kirwan, G. C. Ball, P. J. Bishop, M. J. Godfrey, P. J. Nolan, D. J. Thornley, D. J. G. Love, and A. H. Nelson, *Phys. Rev. Lett.* **58**:467 (1986).

(Ki 87) P. Kienle, in *Proceedings of the International School-Seminar on Heavy Ion Physics*, Dubna, Russia (1987).

(Kl 78) P. Kleinheinz, M. Ogawa, R. Broda, P. J. Daly, D. Haenni, H. Beuscher, and A. Kleinrahm, *Z. Phys. A* **286**:27 (1978).

(Kl 79) P. Kleinheinz, R. Broda, P. J. Daly, S. Lundardi, M. Ogawa, and J. Blomqvist, *Z. Phys. A* **290**:279 (1979).

(Kl 79b) W. Klempt, J. Bonn, and R. Neugart, *Phys. Lett.* **82B**:47 (1979).

(Kl 81) P. Kleinheinz, *Phys. Scr.* **24**:236 (1981).

(Kl 82) P. Kleinheinz, J. Styczen, M. Piiparinen, J. Blomqvist, and M. Kortelahti, *Phys. Rev. Lett.* **48**:1457 (1982).

(Kl 83) O. Klepper, in *Lectures at International School-Seminar on Heavy-Ion Physics* (Alushta, USSR) JINIR, Dubna (1983), p. 86.

(Kl 85) P. Kleinheinz, B. Rubio, M. Ogawa, M. Piiparinen, A. Plochocki, D. Schardt, R. Barden, O. Klepper, R. Kirchner, and E. Roeckl, *Z. Phys. A* **323**:705 (1985).

(Kl 85b) P. Kleinheinz, K. Zuber, C. Conci, C. Protop, J. Zuber, C. F. Liang, P. Paris, and J. Blomqvist, *Phys. Rev. Lett.* **55**:2664 (1985).

(Kl 86) O. Klepper and E. Roeckl, GSI Scientific Report Nachrichten 9-86 (1986), p. 3.

(Ko 83) P. Komninos and E. Nolte, *Z. Phys. A* **310**:137 (1983).

(Ko 86) J. Kormicki, L. Chaturvedi, X. Zhao, W.-C. Ma, A. V. Ramayya, J. H. Hamilton, C. F. Maguire, Z. Chen, C. Girit, N. R. Johnson, C. Baktash, I. Y. Lee, M. L. Halbert, F. K. McGowan, D. Henry, M. Kortelahti, E. F. Zganjar, private communication (1986).

(Ko 86b) M. Kortelahti, E. F. Zganjar, and J. L. Wood, ORNL Progress Report (1986), p. 150.

(Kr 85) K. S. Krane and J. H. Hamilton, in *Proceedings of the International Symposium on Nuclear Orientation and Nuclei Far From Stability* (B. E. Deutch and L. Vanneste, eds.), Scientific Publ., Basel-Switzerland (1985), p. 599.

(Ku 68) K. Kumar, in *Proceedings of the International Symposium on Nuclear Structure*, Dubna Symposium, I.A.E.A./Vienna (1968), p. 149.

(Ku 75) K. Kumar, in *The Electromagnetic Interaction in Nuclear Spectroscopy* (W. D. Hamilton, ed.), North-Holland, Amsterdam (1975), p. 119.

(Kü 77) T. Kühl, P. Dabkiewicz, C. Duke, H. Fischer, H. J. Klüge, H. Kremmling, and E. W. Otten, *Phys. Rev. Lett.* **39**:180 (1977).

(Ku 82) W. Kurcewicz, E. F. Zganjar, R. Kirchner, O. Klepper, E. Roeckl, P. Komninos, E. Nolte, D. Schardt, and P. Tidemand-Petersson, *Z. Phys. A* **308**:21 (1982).

(Ku 85) W. Kutschera, J. Ahmad, S. G. Armato, III, A. M. Friedman, J. E. Gindler, W. Henning, T. Ishii, M. Paul, and K. E. Rehm, *Phys. Rev. C* **32**:2036 (1985).

(La 81) M. Langevin, C. Détraz, M. Guillemaud, F. Naulin, M. Epherre, R. Klapisch, S. K. T. Mark, Simon de Saint, C. Thibault, and F. Touchard, in *Proceedings of the Fourth International Conference on Nuclei Far From Stability*, CERN 81-09 (1981), p. 297.

(La 82) J. Lange, K. Kumar, and J. H. Hamilton, *Rev. Mod. Phys.* **54**:119 (1982).

(La 86) A. J. Larabee, M. P. Carpenter, L. L. Riedinger, L. H. Courtney, J. C. Waddington, V. P. Janzen, W. Nazarewicz, J.-Y. Zhang, R. Bengtsson, and G. A. Leander, *Phys. Lett.* **169B**:21 (1986).

(Le 82) G. A. Leander and P. Möller, *Phys. Lett.* **110B**:17 (1982).

(Le 84) G. A. Leander, J. Dudek, W. Nazarewicz, J. R. Nix, and P. H. Quentin, *Phys. Rev. C* **30**:416 (1984).

(Le 86) G. A. Leander, J. H. Hamilton, and A. V. Ramayya, private communication (1986).

(Lh 85) G. Lhersonneau and co-workers, Annual Report (Jülich, 1985), p. 91.

(Li 77) K. P. Lieb and J. J. Kolata, *Phys. Rev. C* **15**:939 (1977).

(Li 82) C. J. Lister, B. J. Varley, H. G. Price, and J. W. Olness, *Phys. Rev. Lett.* **49**:308 (1982).

(Li 85) C. J. Lister, B. J. Varley, R. Moscrop, W. Gelletly, P. J. Nolan, D. J. G. Love, P. J. Bishop, A. Kirwan, D. J. Thornley, L. Ying, R. Wadsworth, J. M. O'Donnell, H. G. Price, and A. H. Nelson, *Phys. Rev. Lett.* **55**:810 (1985).

(Li 88) C. J. Lister, B. J. Varley, W. Gelletly, A. A. Chishti, A. N. James, T. Morrison, H. G. Price, J. Simpson, and O. Skeppstedt, in *Nuclei Far From Stability*, Fifth International Conference (Ian S. Towner, ed.), American Institute of Physics Conference Proceeding 164 (1988), p. 354.

(Lo 85) H. Lochmann, U. Dinger, J. Eberz, G. Huber, R. Menges, G. Ulm, R. Kirchner, O. Klepper, T. U. Kühl, D. Marx, and D. Schardt, *Z. Phys. A* **322**:703 (1985).

(Lü 85) L. Lühmann, M. Debray, K. P. Lieb, W. Nazarewicz, B. Wörmann, J. Eberth, and K. Heck, *Phys. Rev. C* **31**:828 (1985).

(Ma 50) M. G. Mayer, *Phys. Rev.* **78**:16 (1950).

(Ma 77) F. R. May, V. V. Pashkevich, S. Frauendorf, *Phys. Lett. B* **68B**:113 (1977).

(Ma 84) W.-C. Ma, A. V. Ramayya, J. H. Hamilton, S. J. Robinson, M. E. Barclay, K. Zhao, J. D. Cole, E. F. Zganjar, and E. H. Spejewski, *Phys. Lett.* **139B**:276 (1984).

(Ma 86) W.-C. Ma, A. V. Ramayya, J. H. Hamilton, S. J. Robinson, J. D. Cole, E. F. Zganjar, E. H. Spejewski, R. Bengtsson, W. Nazarewicz, and J.-Y. Zhang, *Phys. Lett.* **167B**:277 (1986).

(Me 86) R. A. Meyer and D. Brenner, eds., *Nuclei Off The Line of Stability*, ACS Symposium 324, American Chemical Society Press, Washington (1986).

(Me 86b) R. A. Meyer, E. A. Henry, L. G. Mann, and K. Heyde, *Phys. Lett. B* **177**:271 (1986).

(Mo 56) H. Morinaga, *Phys. Rev.* **101**:254 (1956).

(Mo 59) B. R. Mottelson and S. G. Nilsson, *Mat. Fys. Meddr. Dansk. Vidensk. Selsk* **1**: No. 8 (1959).

(Mo 60) R. B. Mottelson and J. G. Valatin, *Phys. Rev. Lett.* **5**:511 (1960).

(Mö 81) P. Möller, and J. R. Nix, *Nucl. Phys. A* **301**:199 (1981).

(Mö 81b) P. Möller and J. R. Nix, *At. Data Nucl. Data Tables* **26**:165 (1981).

(Mö 82) P. Möller and J. R. Nix, *Am. Chem. Soc. 183rd Nat. Meeting, Abs. Nucl.* 16 (1982).

(Mo 82) E. Monnand, J. A. Pinston, F. Schussler, B. Pfeiffer, H. Lawin, G. Battistuzzi, K. Shizuma, and K. Sistemich, *Z. Phys. A* **306**:183 (1982).

(Mu 83) A. C. Mueller, F. Buchinger, W. Klempt, E. W. Otten, R. Neugart, C. Ekström, and J. Heinemeier, *Nucl. Phys. A* **403**:234 (1983).

(My 86) W. D. Myers, private communication (1986).

(Na 65) O. Nathan and S. G. Nilsson, in *Alpha-, Beta-, and Gamma-Ray Spectroscopy, Vol. I* (K. Siegbahn, ed.), North-Holland, Amsterdam (1965), p. 601.

(Na 81) Y. Nagai, J. Styczen, M. Piiparinen, P. Kleinheinz, D. Bazzacco, P. v. Brentano, K. O. Zell, and J. Blomqvist, *Phys. Rev. Lett.* **47**:1259 (1981).

(Na 85) W. Nazarewicz, J. Dudek, R. Bengtsson, T. Bengtsson, and T. Ragnarsson, *Nucl. Phys. A* **435**:397 (1985).

(Na 86) W. Nazarewicz, in *Nuclear Structure, Reactions and Symmetries* (R. A. Meyer and V. Paar, eds.), Vol. 2, World Scientific, Singapore (1986), p. 1079.

(Ne 70) J. O. Newton, S. D. Cirilov, F. S. Stephens, and R. M. Diamond, *Nucl. Phys. A* **148**:593 (1970).

(Ne 74) J. O. Newton, F. S. Stephens, and R. M. Diamond, *Nucl. Phys. A* **236**:225 (1974).

(Ne 82) R. Neugart, in *Lasers in Nuclear Physics* (C. E. Bemis and H. K. Carter, eds.), Harwood, New York (1982), p. 231.

(Ne 83) R. Neugart, W. Klempt, F. Buchinger, A. C. Mueller, K. Wendt, and C. Ekström, in *Proceedings of International Workshop on Gross Properties of Nuclei and Nuclear Excitations XI* (H. Feldmeier, ed.), Institut für Kernphysik, Darmstadt (1983), p. 36.

(No 80) E. Nolte and H. Hick, *Phys. Lett.* **97B**:55 (1980).

(No 80b) E. Nolte, G. Korschinek, and U. Heim, *Z. Phys. A* **298**:191 (1980).

(No 82) E. Nolte, G. Colombo, S. Z. Gui, G. Korschinek, W. Schollmeier, P. Kubik, S. Gustavsson, R. Geier, and H. Morinaga, *Z. Phys. A* **306**:211 (1982).

(No 82b) E. Nolte, G. Korschinek, and Ch. Setzensack, *Z. Phys. A* **309**:33 (1982).

(No 83) E. Nolte, in *Proceedings of International Workshop on Gross Properties of Nuclei and Nuclear Excitations XI* (H. Feldmeier, ed.), Institut für Kernphysik, Darmstadt (1983), p. 42.

(Og 78) M. Ogawa, R. Broda, K. Zell, P. J. Daly, and P. Kleinheinz, *Phys. Rev. Lett.* **41**:289 (1978).

(Og 83) K. Ogawa, *Phys. Rev. C* **28**:958 (1983).

(Pa 75) V. V. Pashkevich and S. Frauendorf, *Yad. Fiz.* **20**:1122 (1975) [*Sov. J. Nucl. Phys.* **20**:588 (1975)].

(Pa 82) A. Pakkanen, J. Muhonen, and M. Piiparinen, *Nucl. Phys. A* **373**:237 (1982).

(Pa 83) T. Paradellis and C. A. Kalfas, *Phys. Rev. C* **25**:350 (1983) and references cited therein.

(Pe 86) L. K. Peker, F. K. Wohn, John C. Hill, and R. F. Petry, *Phys. Lett.* **169B**:323 (1986).

(Pe 86b) L. K. Peker, J. H. Hamilton, and P. G. Hansen, *Phys. Lett.* **167B**:283 (1986).

(Pe 86c) L. K. Peker and J. H. Hamilton, *Bull. Am. Phys. Soc.* **31**:1765 (1986).

(Pe 86d) L. K. Peker, J. H. Hamilton, and A. V. Ramayya, private communication (1986).

(Pe 86e) J. Penninga, W. H. A. Hasselink, A. Stolk, H. Verheul, J. K. Ho, J. van Klinken, H. J. Riezeonbos, M. J. A. DeVoigt, and A. Zemel, in *Nuclei Off The Line of Stability* (R. A. Meyer and D. Brenner, eds.), ACS Symposium 324, Am. Chem. Soc. Press, Washington (1986), p. 252.

(Pi 80) R. B. Piercey and J. H. Hamilton, *Proceedings of the XVIII Winter School,* in "Selected Topics in Nuclear Research" (A. Balanda and Z. Stachura, eds.), Jagiellonian University, Krakow (1980), p. 332.

(Pi 81) R. B. Piercey, J. H. Hamilton, R. Soundranayagam, A. V. Ramayya, C. F. Maguire, X. J. Sun, Z. Z. Zhao, R. L. Robinson, H. J. Kim, S. Frauendorf, J. Döring, L. Funke, G. Winter, J. Roth, L. Cleemann, J. Eberth, W. Neumann, J. C. Wells, J. Lin, A. C. Rester, and H. K. Carter, *Phys. Rev. Lett.* **47**:1514 (1981).

(Pi 81b) M. Piiparinen, R. Pengo, Y. Nagai, E. Hammaren, and P. Kleinheinz, *Z. Phys. A* **300**:133 (1981).

(Pi 82) R. B. Piercey, A. V. Ramayya, J. H. Hamilton, X.-J. Sun, Z. Z. Zhao, R. L. Robinson, H. J. Kim, and J. C. Wells, *Phys. Rev. C* **25**:1914 (1982).

(Pi 83) W. F. Piel, Jr., G. Scharff-Goldhaber, C. J. Lister, and B. J. Varley, *Phys. Rev. C* **28**:209 (1983).

(Pi 84) R. B. Piercey, F. E. Dunnam, M. L. Muga, A. C. Rester, A. V. Ramayya, J. H. Hamilton, J. Eberth, and E. F. Zganjar, *Proceedings of the International Conference on Instrumentation for Heavy-Ion Nuclear Research, CONF-841008* (1984), p. 27.

(Pi 85) R. B. Piercey, A. C. Rester, J. H. Hamilton, A. V. Ramayya, H. K. Carter, R. L. Robinson, H. J. Kim, J. C. Wells, and J. Lin, *Phys. Rev. C* **32**:625 (1985).

(Pi 86) R. B. Piercey, M. A. Herath-Banda, and T. Cormier, *Proceedings of the Ninth International Conference on the Application of Accelerators in Research and Industry* (J. L. Duggan and I. L. Morgan, eds.), North Texas State University, Denton, TX (1986), pp. 464–467.

(Pi 86b) R. B. Piercey, Research Proposal, University of Florida (1986).

(Pł 82) A. Płochocki, J. Zylicz, R. Kirchner, O. Klepper, E. Roeckl, P. Tidemand-Petersson, I. S. Grant, and P. Misaelides, *Nucl. Phys. A* **388**:91 (1982).

(Po 85) D. N. Poenaru, M. Ivascu, A. Sandulescu, and W. Greiner, *Phys. Rev. C* **32**:572 (1985).

(Po 85b) F. Pougheon, R. Anne, M. Bernas, C. Détraz, J. Galin, D. Geurreau, S. Hoath, D. Guillemaud-Mueller, J. C. Jacmart, M. Langevin, A. C. Mueller, F. Naulin, and E. Quiniou, in *Proceedings of the Workshop on Intermediate Energy Heavy Ion Physics, CONF-8509176*, Oak Ridge National Lab. (1985), p. L-3.

(Pr 73) D. Proetel, R. M. Diamond, T. Kienle, J. R. Leigh, K. H. Maier, and F. S. Stephens, *Phys. Rev. Lett.* **31**:896 (1973).

(Pr 85) P. B. Price, J. D. Stevenson, and S. W. Barwick, *Phys. Rev. Lett.* **54**:297 (1985).

(Qu 84) M. A. Quader, W. F. Piel, Jr., S. Vajda, W. A. Watson, III, F. C. Yang, and D. B. Fossan, *Phys. Rev.* **30**:1772 (1984).

(Ra 53) J. O. Rasmussen, S. G. Thompson, and A. Ghiorso, *Phys. Rev.* **89**:33 (1953).

(Ra 74) A. V. Ramayya and J. H. Hamilton, in *Gamma Ray Transition Probabilities* (S. C. Pancholi and S. L. Gupta, eds.), University of Delhi Press, Delhi (1974), p. 125.

(Ra 81) I. Ragnarsson, in *Proceedings of the Fourth International Conference on Nuclei Far From Stability*, CERN 81-09 (1981), p. 434.

(Ra 82) A. V. Ramayya and J. Eberth, in *Lecture Notes in Physics, Vol. 168* (G. Madurga and M. Lozano, eds.), Springer-Verlag, Berlin (1982), p. 317.

(Ra 84) A. V. Ramayya, S. Robinson, M. Herath-Banda, W.-C. Ma, J. H. Hamilton, P. Haskins, F. E. Dunnam, L. Muga, R. B. Piercey, and A. C. Rester, in *International Conference on Instrumentation for Heavy-Ion Nuclear Research, CONF-841005*, Oak Ridge National Laboratory (1984), p. 23.

(Ra 85) A. V. Ramayya, W.-C. Ma, C. F. Maguire, J. H. Hamilton, S. Wen, X. Zhao, T. M. Cormier, P. M. Stwertka, J. D. Cole, E. F. Zganjar, R. B. Piercey, M. A. Herath-Banda, and J. Eberth, *Bull. Am. Phys. Soc.*, **30**:1257 (1985).

(Ra 86) A. V. Ramayya, X. Zhao, J. H. Hamilton, C. F. Maguire, W. C. Ma, J. Kormicki, S. Wen, L. Chaturvedi, Z.-M. Chen, T. M. Cormier, M. D. Kortelahti, J. D. Cole, R. B. Piercey, M. A. Herath-Banda, G. Leander, J. Eberth, N. Schmal, T. Heck, L. Cleemann, J. Eberth, T. Mylaeus, W. Neumann, M. Tolte, and M. Wiosna, to be published (1986).

(Ra 86b) A. V. Ramayya, T. Mylaeus, J. Busch, J. Eberth, N. Schmal, M. Wiosna, P. V. Brentano, K. O. Zell, K. H. Maier, and H. Kluge, *Bull. Am. Phys. Soc.* **31**:772 (1986).

(Ra 86c) A. V. Ramayya, T. Mylaeus, J. Busch, J. Eberth, N. Schmal, M. Wiosna, P. V. Brentano, K. O. Zell, K. H. Maier, H. Kluge, S. K. Moore, W. G. Nettles, E. F. Jones, and J. H. Hamilton, *Bull. Am. Phys. Soc.*, **31**:1764 (1986).

(Ra 88) H. L. Ravn and B. W. Allardyce, in *Treatise on Heavy-Ion Science, Vol. 8* (D. Allan Bromley, ed.), Plenum Press, New York (1988), this volume.

(Ri 85) H. J. Riezebos, J. F. W. Jansen, R. Kaczarowski, and M. J. A. de Voigt, in *Symposium on Electromagnetic Properties of High Spin States*, Stockholm (1985).

(Ri 85b) J. Rikovska, N. J. Stone, and V. R. Green, in *Proceedings of the International Symposium on Nuclear Orientation and Nuclei Far From Stability* (B. I. Deutch and L. Vanneste, eds.), Scientific Publ. Co., Basel—Switzerland (1985), p. 355.

(Ro 76) R. M. Ronningen, A. V. Ramayya, J. H. Hamilton, W. Lourens, J. Lange, H. K. Carter, and R. O. Sayer, *Nucl. Phys. A* **261**:439 (1976).

(Ro 83) E. Roeckl, *Nucl. Phys. A* **400**:131C (1983).

(Ro 80) R. L. Robinson, H. J. Kim, R. O. Sayer, W. T. Milner, R. B. Piercey, J. H. Hamilton, A. V. Ramayya, J. C. Wells, and A. J. Caffrey, *Phys. Rev.* **21**:603 (1980).

(Ro 84) H. J. Rose and G. A. Jones, *Nature* **307**:245 (1984).

(Ro 84b) B. Roussiere, P. Kilcher, J. Sauvage-Letessier, C. Bourgeois, R. Béraud, R. Duffait, M. Meyer, J. Genevey-Rivier, and J. Treherne, *Nucl. Phys. A* **419**:61 (1984).

(Ro 85) B. Roussiere, C. Bourgeois, P. Kilcher, J. Sauvage, and M. G. Porquet, *Nucl. Phys. A* **438**:93 (1985).

(Ro 86) E. Roeckl, in *Nuclei Off The Line Of Stability* (R. A. Meyer and D. Brenner, eds.), ACS Symposium 324, American Chemical Society Press, Washington (1986), p. 439.

(Ru 73) N. Rud, D. Ward, H. R. Andrews, R. L. Graham, and J. S. Geiger, *Phys. Rev. Lett.* **31**:1421 (1973).

(Ru 83) N. Rud, D. Ward, H. R. Andrews, P. Taras, and J. Keinonen, *Phys. Lett.* **100B**:17 (1983).

(Sa 79) L. E. Samuelson, J. A. Grau, S. I. Popik, F. A. Rickey, and P. C. Simms, *Phys. Rev. C* **19**:73 (1979).

(Sa 80) A. Sandulescu, D. N. Poenaru, and W. Greiner, *Sov. J. Part. Nucl.* **11**:528 (1980).

(Sa 84) A. Sandulescu, Yu. S. Zamyatnin, J. A. Lebedev, B. F. Myasoedov, S. P. Tretyakova, and D. Hasegan, *JINIR Rapid Communications* **5**:5 (1984).

(Sc 76) W. D. Schmidt-Ott and K. S. Toth, *Phys. Rev. C* **13**:2574 (1976).

(Sc 77) A. G. Schmidt, R. L. Mlekodaj, E. L. Robinson, F. T. Avignone, J. Lin, G. M. Gowdy, J. L. Wood, and R. W. Fink, *Phys. Lett.* **66B**:133 (1977).

(Sc 80) F. Schussler, J. A. Pinston, E. Monnand, A. Moussa, G. Jung, E. Koglin, B. Pfeiffer, R. V. F. Janssens, and J. Van Klinkin, *Nucl. Phys. A* **339**:415 (1980).

(Sc 80b) K. H. Schmidt and D. Vermeulen, in *Atomic Masses and Fundamental Constants 6* (J. Nolan and W. Benenson, eds.), Plenum Press, New York (1980), p. 119 and references cited therein.

(Sc 81) D. Schardt, T. Batsch, R. Kirschner, O. Klepper, W. Kurcewicz, E. Roeckl, and P. Tidemand-Petersson, *Nucl. Phys. A* **368**:153 (1981).

(Sc 82) U. J. Schrewe, E. Hagberg, H. Schmeing, J. C. Hardy, V. T. Koslowsky, K. S. Sharma, and E. T. H. Clifford, *Phys. Rev. C* **25**:3091 (1982).

(Sc 84) D. Schardt and co-workers, in *Proceedings of the Seventh Conference on Atomic Masses and Fundamental Constants* (AMCO-7) (O. Klepper, ed.), Darmstadt-Seeheim (1984), p. 222.

(Sc 86) K. W. Schmid, F. Grümmer, M. Kyotoku, and Amand Faessler, *Nucl. Phys. A* **452**:493 (1986).

(Se 82) M. Seiwert, J. A. Maruhn, and W. Greiner, in *High Angular Momentum Properties of Nuclei* (N. R. Johnson, ed.), Harwood Academic, New York (1982), p. 249.

(Se 84) M. Seiwert, A. V. Ramayya, and J. Maruhn, *Phys. Rev. C* **29**:284 (1984).

(Sh 72) R. K. Sheline, I. Ragnarsson, and S. G. Nilsson, *Phys. Lett.* **41B**:115 (1972).

(Sh 85) T. L. Shaw, V. R. Green, N. J. Stone, J. Rikovska, P. M. Walker, S. Collins, S. A. Hamada, W. D. Hamilton, and I. S. Grant, *Phys. Lett.* **153B**:221 (1985).

(Si 77) K. Sistemich, G. Sadler, T. A. Khan, H. Lawin, W. D. Lauppe, H. A. Selic, F. Schussler, J. Blachot, E. Monnand, J. P. Bocquet, and B. Pfeiffer, *Z. Phys. A* **281**:169 (1977).

(Si 86) R. S. Simon, K. H. Schmidt, F. P. Heßberger, S. Hlavac, M. Honusek, G. Münzenberg, H. G. Clerc, U. Gollerthan, and W. Schwab, *Z. Phys. A* **325**:197 (1986).

(So 66) V. G. Soloviev, *Phys. Lett.* **21**:311 (1966).

(Sp 81) L. Spanier, S. Z. Gui, H. Hick, and E. Nolte, *Z. Phys. A* **299**:113 (1981).

(St 81) N. J. Stone and W. D. Hamilton, *Hyperfine Interact.* **10**:1219 (1981).

(St 83) J. Styczen, Y. Nagai, M. Piiparinen, A. Ercan, and P. Kleinheinz, *Phys. Rev. Lett.* **50**:1752 (1983).

(St 83b) J. Styczen, M. Piiparinen, Y. Nagai, P. Kleinheinz, D. Bazzacco, J. Eberth, and J. Blomqvist, *Z. Phys. A* **312**:149 (1983).

(St 85) N. J. Stone, in *Proceedings of the International Symposium on Nuclear Orientation and Nuclei Far From Stability* (B. I. Deutch and L. Vanneste, eds.), Scientific Publ. Co., Basel—Switzerland (1985), p. 1.

(Sy 79) T. J. M. Symons, Y. P. Viyogi, G. D. Westgall, P. Doll, D. E. Greiner, H. Faraggi, P. J. Linstrom, D. K. Scott, H. J. Crawford, and C. McParland, *Phys. Rev. Lett.* **42**:40 (1979).

(Ta 62) I. Talmi, *Rev. Mod. Phys.* **34**:704 (1962).

(Ta 84) I. Talmi, *Nucl. Phys. A* **423**:189 (1984).

(Ta 86) S. L. Tabor, *Phys. Rev.* **34C**:311 (1986).

(Th 81) C. Thibault, F. Touchard, S. Büttgenbach, R. Klapisch, M. de Saint Simon, H. T. Duong, P. Jacquinot, P. Juncar, S. Liberman, P. Pillet, J. Pinard, J. L. Vialle, A. Pesnelle, and G. Huber, *Phys. Rev. C* **23**:2720 (1981).

(Th 81b) C. Thibault, F. Touchard, S. Büttgenbach, R. Klapisch, M. de Saint Simon, H. T. Duong, P. Jacquinot, P. Juncar, S. Liberman, P. Pillet, J. Pinard, J. L. Vialle, A. Pasnelle, and G. Huber, *Nucl. Phys. A* **367**:1 (1981).

(Th 82) C. Thibault and F. Touchard, in *Lasers in Nuclear Physics* (C. E. Bemis and H. K. Carter, eds.), Harwood, New York (1982), p. 113.

(Ti 81) P. Tidemand-Petersson, R. Kirchner, O. Klepper, W. Kurcewicz, E. Roeckl, and E. F. Zganjar, *Z. Phys. A* **302**:343 (1981).

(To 70) K. S. Toth, R. L. Hahn, M. A. Ijaz, and W. M. Sample, *Phys. Rev. C* **2**:1480 (1970).

(To 76) K. S. Toth, M. A. Ijaz, J. Lin, E. L. Robinson, B. O. Hannah, E. H. Spejewski, J. D. Cole, J. H. Hamilton, and A. V. Ramayya, *Phys. Lett.* **63B**:150 (1976).

(To 82) F. Touchard, J. M. Serre, S. Büttgenbach, P. Guimbal, R. Klapisch, M. de Saint Simon, C. Thibault, H. T. Duong, P. Juncar, S. Liberman, J. Pinard, and J. L. Vialle, *Phys. Rev. C* **25**:2756 (1982).

(To 84) K. S. Toth, Y. A. Ellis-Akovali, C. R. Bingham, D. M. Moltz, D. C. Sousa, H. K. Carter, R. L. Mlekodaj, and E. H. Spejewski, *Phys. Rev. Lett.* **53**:1623 (1984).

(To 84b) K. S. Toth, D. M. Moltz, E. C. Schloomer, M. D. Cable, F. T. Avignone, III, and Y. A. Ellis-Akovali, *Phys. Rev. C* **30**:712 (1984).

(To 84c) K. S. Toth and co-workers, in *Proceedings of the Seventh Conference on Atomic Masses and Fundamental Constants* (AMCO-7) (O. Klepper, ed.), Darmstadt-Seeheim (1984), p. 237.

(To 85) I. S. Towner, *Nucl. Phys. A* **444**:402 (1985) and private communication.

(To 85b) K. S. Toth, Y. A. Ellis-Akovali, F. T. Avignone, III, R. S. Moore, D. M. Moltz, J. M. Nitschke, P. A. Wilmarth, P. K. Lemmertz, D. C. Sousa, and A. L. Goodman, *Phys. Rev. C* **32**:342 (1985).

(To 86) K. S. Toth, H. J. Kim, M. N. Rao, and R. L. Mlekodaj, *Phys. Rev. Lett.* **56**:2360 (1986).

(To 86b) K. S. Toth, Y. A. Ellis-Akovali, J. M. Nitschke, P. A. Wilmarth, P. K. Lemmertz, D. M. Moltz, and F. T. Avignone, III, *Phys. Lett. B* **178**:150 (1986).

(To 87) K. S. Toth, D. C. Sousa, J. M. Nitschke, and P. A. Wilmarth, *Phys. Rev. C* **35**:310 (1987).

(To 87b) K. S. Toth, D. C. Sousa, J. M. Nitschke, and P. A. Wilmarth, *Phys. Rev. C* **35**:620 (1987).

(To 88) I. S. Towner, (ed.), *Nuclei Far From Stability*, Fifth International Conference, American Institute of Physics Conference Proceeding 164 (1988).

(Tr 82) J. Tréherne, J. Genevey, A. Gizon, J. Gizon, R. Beraud, A. Charvet, R. Duffait, A. Emsallem, and M. Meyer, *Z. Phys. A* **309**:135 (1982).

(Tr 85) S. P. Tretyakova, A. Sandulescu, Yu S. Zamyatnin, Yu S. Korotkin, and V. L. Micheev, *JINIR Rapid Communications* **7**:23 (1985).

(Tr 85b) S. P. Tretyakova, A. Sandulescu, V. L. Micheev, D. Hasegan, J. A. Lebedev, Yu S. Zamyatkin, Yu S. Korotkin, and B. F. Myasoedov, *JINIR Rapid Communications* **13**:34 (1985).

(Tw 86) P. J. Twin, B. M. Nyakó, A. H. Nelson, J. Simpson, M. A. Bentley, H. W. Cranmer-Gordon, P. D. Forsyth, D. Howe, A. R. Mokhtar, J. D. Morrison, J. F. Sharpey-Schafer, and G. Sletten, *Phys. Rev. Lett.* **57**:811 (1986).

(Ul 85) G. Ulm, J. Eberz, G. Huber, H. Lochmann, R. Menges, R. Kirchner, O. Klepper, T. Kühl, P. O. Larsson, D. Marx, D. Murnick, and D. Schardt, *Z. Phys. A* **321**:395 (1985).

(Un 63) I. Unna, *Phys. Rev.* **132**:2225 (1963).

(Va 81) D. Vandeplassche, L. Vanneste, H. Pattyn, J. Geenen, C. Nuytten, and E. van Walle, *Nucl. Instrum. Methods* **186**:211 (1981).

(Va 84) P. Van Duppen, E. Coenen, K. Deneffe, M. Huyse, K. Heyde, and P. van Isacker, *Phys. Rev. Lett.* **52**:1974 (1984).

(Va 85) D. Vandeplassche, E. van Walle, J. Wouters, N. Severijns, and L. Vanneste, in *Proceedings of the International Symposium on Nuclear Orientation and Nuclei Far From Stability* (B. I. Deutch and L. Vanneste, eds.), Scientific Publ. Co., Basel—Switzerland (1985), p. 493.

(va 85b) E. van Walle, Ph.D. Thesis, Leuven (1985), to be published.

(va 85b) E. van Walle, J. Wouters, D. Vandeplassche, N. Severijns, and L. Vanneste, *Hyperfine Interact.* **22**:507 (1985).

(Va 86) L. Vanneste, private communication.

(Va 86b) L. Vanneste, N. Severijns, J. Wouters, D. Vandeplassche, and E. van Walle, private communication.

(Va 86c) D. Vandeplassche, E. van Walle, J. Wouters, N. Severijns, and L. Vanneste, *1984/85 Progress Report* (M. Rots, ed.), Inst. voor Kern-en Stralingsfysika, Katholieke Univ., Leuven, Belgium (1986), p. 12.

(Va 86d) D. Vandeplassche, E. van Walle, J. Wouters, N. Severijns, and L. Vanneste, *Phys. Rev. Lett.* **57**:2641 (1986).

(va 86) E. van Walle, D. Vandeplassche, J. Wouters, N. Severijns, and L. Vanneste, in *1984/1985 Progress Report* (M. Rots, ed.), Inst. voor Kern-en Stralingsfysika, Katholieke Univ., Leuven, Belgium (1986), p. 12.

(Vi 86) D. J. Vieira, J. M. Wouters, K. Vaziri, R. H. Kraus, Jr., J. Wollnik, G. W. Butler, F. K. Wohn, and A. H. Wapstra, *Phys. Rev. Lett.* **57**:3253 (1986).

(Wa 68) D. Ward, R. M. Diamond, and F. S. Stephens, *Nucl. Phys. A* **117**:309 (1968).

(Wa 83) D. Ward, H. R. Andrews, B. Haas, P. Teras, and N. Rud, *Nucl. Phys. A* **397**:161 (1983).

(Wa 83b) D. Ward, G. D. Dracoulis, J. R. Leigh, R. J. Charity, D. J. Hinde, and J. O. Newton, *Nucl. Phys. A* **406**:591 (1983).

(Wa 85) P. M. Walker, in *Proceedings of the International Symposium on Nuclear Orientation and Nuclei Far From Stability* (B. I. Deutch and L. Vanneste, eds.), Scientific Publ. Co., Basel—Switzerland (1985), p. 475.

(We 79) G. D. Westfall, T. J. M. Symons, D. E. Greiner, H. H. Heckman, P. J. Linstrom, J. Mahoney, A. C. Shotter, D. K. Scott, H. J. Crawford, C. McParland, T. C. Awes, C. K. Gelbke, and J. M. Kidd, *Phys. Rev. Lett.* **43**:1859 (1979).

(We 80) J. C. Wells, Jr., R. L. Robinson, H. J. Kim, R. O. Sayer, R. B. Piercey, A. V. Ramayya, J. H. Hamilton, and C. F. Maguire, *Phys. Rev. C* **22**:1126 (1980).

(We 85) S. Wen, A. V. Ramayya, S. J. Robinson, C. F. Maguire, W.-C. Ma, X. Zhao, T. M. Cormier, P. M. Stwertka, J. D. Cole, E. F. Zganjar, R. B. Piercey, M. A. Herath-Banda, J. Eberth, and M. Wiosna, *J. Phys. G* **11**:L173 (1985).

(We 87) Robert B. Welch, Kenton J. Moody, Kenneth E. Gregorich, Diana Lee, and Glenn T. Seaborg, *Phys. Rev. C* **35**:204 (1987).

(Wi 71) B. H. Wildenthal, E. Newman, and R. L. Auble, *Phys. Rev. C* **3**:1199 (1971).

(Wi 80) J. Wilson, S. R. Faber, P. J. Daly, I. Ahmad, J. Borggreen, P. Chowdhury, T. L. Khoo, R. D. Lawson, R. K. Smith, and J. Blomqvist, *Z. Phys. A* **296**:185 (1980).

(Wi 80b) I. N. Wischnewski, H. V. Klapdor, P. Herges, H. Fromm, and W. A. Zheldonozhski, *Z. Phys. A* **298**:21 (1980).

(Wi 82) G. Winter, J. Döring, L. Funke, P. Kemnitz, and W. Will, in *International Symposium on Dynamics of Nuclear Collective Motion*, Osaka University, Osaka, Contributed papers (1982), p. 21.

(Wo 76) J. L. Wood, R. W. Fink, E. F. Zganjar, and J. Meyer-ter-Vehn, *Phys. Rev. C* **14**:682 (1976).

(Wo 81) J. L. Wood, in *Proceedings of the Fourth International Conference on Nuclei Far From Stability*, CERN 81-09 (1981), p. 612.

(Wo 83) F. K. Wohn, J. C. Hill, R. F. Petry, H. Dejbakhsh, Z. Berant, and R. L. Gill, *Phys. Rev. Lett.* **51**:873 (1983).

(Wo 83b) J. M. Wouters, H. M. Thierens, J. Äystö, M. D. Cable, P. E. Haustein, R. F. Parry, and J. Cerny, *Phys. Rev. C* **27**:1745 (1983).

(Wo 85) John L. Wood, in *Proceedings of the International Symposium on Nuclear Orientation and Nuclei Far From Stability* (B. I. Deutch and L. Vanneste, eds.), Scientific Publ. Co., Basel—Switzerland (1985), p. 379.

(Wo 85b) J. Wouters, D. Vandeplassche, E. van Walle, N. Severijns, and L. Vanneste, in *Proceedings of the International Symposium on Nuclear Orientation and Nuclei Far From Stability* (B. I. Deutch and L. Vanneste, eds.), Scientific Publ. Co., Basel—Switzerland (1985), p. 527.

(Wö 85) B. Wörmann, J. Heese, K. P. Lieb, L. Lühmann, F. Raether, D. Alber, H. Grawe, and B. Spellmeyer, *Z. Phys. A* **322**:171 (1985).

(Wo 86) F. K. Wohn, John C. Hill, C. B. Howard, K. Sistemich, R. F. Petry, R. L. Gill, H. Mach, and A. Piotrowski, *Phys. Rev. C* **33**:677 (1986).

(Wo 86b) J. Wouters, D. Vandeplassche, E. van Walle, N. Severijns, and L. Vanneste, *Phys. Rev. Lett.* **56**:1901 (1986).

(Wu 86) Cheng-Li Wu, Da Hsuan Feng, Xuan-Gen Chen, Jin-Quan Chen, and Michael W. Guidry, *Phys. Lett.* **168B**:313 (1986).

(Wu 87) C.-L. Wu, D. H. Feng, X.-G. Chen, J.-Q. Chen, and M. W. Guidry, *Phys. Rev. C* **36**:1157 (1987).

(Yi 86) K. L. Ying, P. J. Bishop, A. N. James, A. J. Kirwan, D. J. G. Love, T. P. Morrison, P. J. Nolan, D. C. B. Watson, K. A. Connell, A. H. Nelson, and J. Simpson, *J. Phys. G* **12**:L211 (1986).

(Ze 81) N. Zeldes, in *Proceedings of the Fourth International Conference on Nuclei Far From Stability*, CERN 81-09 (1981), p. 93.

(Zg 80) E. F. Zganjar, in *Future Directions in Studies of Nuclei Far From Stability* (J. H. Hamilton, E. H. Spejwski, C. R. Bingham, and E. F. Zganjar, eds.), North-Holland, Amsterdam (1980), p. 49.

(Zg 81) E. F. Zganjar, J. D. Cole, J. L. Wood, and M. A. Grimm, in *Proceedings of the Fourth International Conference on Nuclei Far From Stability*, CERN 81-09 (1981), p. 630.

(Zg 86) E. F. Zganjar, J. L. Wood, C. D. Papnicolopoulos, and M. Kortelahti, in *Nuclear Structure, Reactions and Symmetries* (R. A. Meyer and V. Paar, eds.), Vol. 2, World Scientific, Singapore (1986), p. 716.

(Zg 86b) E. F. Zganjar, C. D. Papanicolopoulos, J. L. Wood, R. A. Braga, R. W. Fink, A. J. Larabee, M. Carpenter, D. Love, C. R. Bingham, L. L. Riedinger, and J. C. Waddington, in *Nuclei Off the Line of Stability* (R. A. Meyer and D. Brenner, eds.), ACS Symposium 324, American Chemical Society Press, Washington (1986), p. 245.

(Zg 88) E. F. Zganjar, M. O. Kortelahti, J. L. Wood, and C. D. Papanicolopoulos, in *Nuclei Far From Stability*, Fifth International Conference (Ian S. Towner, ed.), American Institute of Physics Conference Proceeding 164 (1988), p. 313.

(Zh 82) Z. Z. Zhao, X.-J. Sun, R. B. Piercey, J. H. Hamilton, C. F. Maguire, A. V. Ramayya, R. L. Robinson, H. J. Kim, J. C. Wells, *Phys. Energiae Fortis er Phys. Nucl.* **6**:255 (1982) [in Chinese].

(Zh 86) J.-Y. Zhang and co-workers, University of Tennessee Preprint (1986).

(Zh 86b) X. Zhao, A. V. Ramayya, J. H. Hamilton, C. F. Maguire, W.-C. Ma, L. Chaturvedi, J. Kormicki, Z. Chen, C. Girit, N. R. Johnson, C. Baktash, I. Y. Lee, M. L. Halbert, F. K. McGowan, D. Hendry, M. Kortelahti, and E. F. Zganjar, private communication (1986).

(Zh 86c) X. Zhao, A. V. Ramayya, C. F. Maguire, J. Kormicki, W.-C. Ma, S. Wen, J. H. Hamilton Z.-M. Chen, L. Chaturvedi, T. C. Cormier, M. Statteson, E. F. Zganjar, M. Kortelahti, J. D. Cole, J. Eberth, N. Schmal, R. B. Piercey, and M. A. Herath-Banda, *Bull. Am. Phys. Soc.* **31**:772 (1986) and private communication.

# 2

VADIM V. VOLKOV is a senior scientist and group leader in the Laboratory for Nuclear Reactions in the Joint Institute for Nuclear Research at Dubna, USSR. He is a 1950 graduate from the M. V. Lomonosor Moscow State University in physics. Prior to joining the Dubna laboratory staff in 1960, he worked at the S. I. Vavilov Institute of Physical Problems and at the I. V. Kurchatov Institute of Atomic Energy in Moscow. He was awarded his Doctorate in Physico-Mathematical Sciences in 1972 and in 1975 was the winner of a USSR State Prize for his work on nuclei far from stability.

2

# Production of Nuclei Far From Stability

VADIM V. VOLKOV

## 1. Introduction

### 1.1. Some History

In nuclear physics, as in other research activities, investigators began exploring objects which Nature had provided, namely, stable and naturally radioactive nuclei. The first step beyond the boundaries of the world of existing nuclei was taken by Irene and Frederic Joliot-Curie. In 1934 they produced the first artificially radioactive nuclei. The discovery of the fission of uranium nuclei by Hahn and Strassmann and the subsequent construction of nuclear reactors provided physicists with a wide range of neutron-rich isotopes of medium and heavy elements. In the late 1940s the first high-energy particle accelerators came into operation bringing with them spallation reactions as a source of nuclei far from the region of stability.

Theoretical estimates showed that up to 6000 nuclides might lie within the boundaries of the nucleon stability of nuclei. Physicists now had an extensive field for research, with a multitude of exciting scientific problems to be tackled, such as where the boundaries of the nucleon stability of nuclei lie, whether or not new modes of radioactive decay are feasible, and how the nuclear properties change as the neutron or proton excess in the nucleus increases. The synthesis of new transuranium elements and, somewhat later, of superheavy nuclei was a challenging task. Gradually it became more and more obvious that our understanding of atomic nuclei cannot be complete if it was only based on studies of the properties of stable and naturally radioactive isotopes. The production and investigation of nuclei far from the region of stability gave rise to a separate trend of research.

The first International Conference on Nuclei Far From Stability was held in August 1966 in the small Swedish town of Lysekil and could be considered the formal recognition of this trend. The conference was concerned with two principal questions: "Why and how should we investigate nuclides far off the stability line?" Will science and technology benefit from these investigations? What are the ways of producing these nuclei, the choice of nuclear reactions, and the methods of separation,

VADIM V. VOLKOV • Laboratory of Nuclear Reactions, Joint Institute for Nuclear Research, 101000 Moscow, USSR.

identification, and study of the new nuclides? In the review talks presented at the Lysekil Conference attempts were made to tackle these questions, naturally, at the level of current knowledge.

Since the Lysekil Conference nuclei far from the region of stability have been studied as part of a rapidly developing field of science. It involves several hundred experimentalists and theoreticians working in many countries. The Lysekil Conference was followed by three specialized international conferences, which took place in Leysin (Switzerland), in Cargese, Corsica (France), and in Helsingør (Denmark). The proceedings of those conferences show how this trend has continued to expand and document its impressive achievements.

## 1.2. Prerequisites for the Study of Nuclei Far From Stability

In order for nuclei far from stability to be studied, there must first be a source, such as a particle accelerator or nuclear reactor. Then an effective system is needed for selecting the nuclei of interest. This is because a great many different nuclides are produced in the nuclear reactions induced by accelerated particles and in the thermal-neutron fission of uranium-235. The quantities in which nuclei far from stability can be obtained are orders of magnitude smaller than are possible for nuclei close to the stability region. The reaction products can only be separated using special devices, the so-called selectors. Finally, detector systems are needed to identify the new isotopes and to investigate their nuclear properties, such as the mode of radioactive decay, mass, spin, and shape of the nuclei, and the spectra of the excited states.

The studies of nuclei far from stability have progressed due to the development of acceleration techniques, on the one hand, and to the improvement of isotope separation and detection methods, on the other hand. As higher energies, more intense beams of accelerated particles, and new kinds of bombarding particles became available, the studies of nuclei far from stability were opened up still further. The new possibilities enabled scientists to push toward the nucleon stability boundary and to make their investigations of the properties of the new nuclides more detailed.

Stable and naturally radioactive isotopes are the usual original material for producing nuclei far from the limit of stability. As a rule, the production of new nuclei requires a great deal of energy since the reactions used to obtain them are endoenergetic. The $Q$ value of the reaction may be several tens of MeV, with $Q$ rising the farther the nuclide lies from the stability region. Therefore, the projectile energy is important and often crucial in the production of nuclei far from stability. For example, the cross-sections of spallation reactions increase by three orders of magnitude when the energy is increased from 0.1 to 2 GeV. The exotic reaction ($^4$He, $^8$He) used to produce neutron-deficient isotopes of light nuclei may only have a noticeable cross-section for $^4$He ion energies of the order of 100 MeV. An exception to this rule is the thermal-neutron fission of uranium-235. The fission cross-section exceeds 500 barn and, as a result of fission, very many neutron-rich isotopes are produced with considerable yields. We cannot but be grateful to Nature for this reaction; it is a gift for those involved in the studies of nucleon far from stability.

Beam intensity is also of great importance in the advance toward the nucleon stability limit. In a nuclear reaction the nuclear yields decrease rapidly as the distance of the nucleus from the stability region increases, the most exotic nuclides having very low production cross-sections.

## 1.3. Heavy Ions in the Studies of Nuclei Far From Stability

The advent of heavy-ion accelerators opened up new vistas for research on nuclei far from stability. From the beginning it was clear that heavy ions were unique projectiles for synthesizing new transuranium elements. Indeed, it was this objective that stimulated the development of heavy-ion accelerators. All the transuranium elements with atomic numbers from 102 to 109 were synthesized in heavy-ion induced reactions. The complete-fusion reactions between two complex nuclei were quickly recognized as being an efficient means of producing neutron-deficient isotopes of many elements, thus opening up the possibility of an advance toward the limit of the proton stability of nuclei. Several new modes of radioactive decay, viz., delayed proton emission, delayed fission, and proton decay from the ground state, were discovered as a result of complete fusion studies.

Somewhat later multinucleon transfer reactions, which occur during grazing collisions of two complex nuclei, proved to be a rich source of nuclei far off stability. This reaction type was especially successful in producing light-element nuclei with large neutron excesses and several dozen new isotopes have been obtained.

The construction of heavy-ion accelerators for intermediate (tens of MeV per nucleon) and relativistic (hundreds of MeV per nucleon) energies gave a new impetus to studies with heavy ions. The typical reaction for these energies is projectile fragmentation in a peripheral collision with the target nucleus. The pronounced forward peaking of the products for projectile fragmentation and the fact that the fragment and projectile velocities are close provide favorable conditions for the separation of nuclei far from stability and for their efficient collection in the detector system. At intermediate energies the target may be several dozens of milligrams per $cm^2$ thick, while for relativistic energies thicknesses of hundreds and even thousands of milligrams per $cm^2$ are common. In recent years fragmentation reactions have yielded very exotic light-element isotopes and provided a further advance toward the limit of the nucleon stability of nuclei.

The concentration of the yield at small angles and the comparatively small velocity spread of the projectile fragments at relativistic energies opened up the possibility of using secondary particle beams in the studies of nuclei far from stability. Experiments with the nuclear scattering of projectile fragments have given valuable information about the size of the neutron-rich isotopes of the lightest elements.

## 1.4. Outline of this Chapter

The present chapter deals with the production of nuclei far from the stability region in reactions induced by heavy ions at low, intermediate, and relativistic energies. The methods for selecting and identifying heavy-ion reaction products are described, and the important scientific results are presented.

The chapter consists of an introduction, three sections, and a conclusion. Section 2 deals with complete-fusion reactions between two complex nuclei. The methods for selecting and identifying compound-nucleus decay products are described and the important data are presented.

The third section describes multinucleon transfer reactions with heavy ions. We only discuss those aspects of the reaction that pertain to the main topic, together with the methods for selecting and identifying the multinucleon transfer reaction products. The results so far achieved are considered.

The fourth section concerns fragmentation reactions as efficient methods for producing nuclei far from stability and for studying their properties. The methods for separating and identifying projectile fragments and the relevant scientific results are considered.

To conclude we summarize the investigations aimed at producing nuclei far from stability in heavy-ion reactions and discuss their future prospects.

## 2. Production of Nuclei Far From Stability by Complete-Fusion Reactions

### 2.1. Neutron-Deficient Nuclei

#### *2.1.1. Neutron Deficit as a Result of the Fusion of Two Nuclei*

In stable and naturally radioactive isotopes, the neutron-to-proton ratio $N/Z$ increases with increasing atomic and mass numbers. In the most abundant light-element isotopes the $N/Z$ ratio is unity, while for the isotopes of heavy elements it exceeds 1.5 and approaches 1.6. This feature of the nucleon composition of naturally occurring nuclei makes is possible, in principle, to produce the neutron-deficient isotopes of medium and heavy elements by the complete fusion of lighter nuclei.

Indeed, the fusion of two $^{58}$Ni nuclei leads to $^{116}$Ba which has a neutron deficit of 14. The fusion of two $^{90}$Zr nuclei results in the formation of the compound nucleus $^{180}$Hg, which contains 16 fewer neutrons than the lightest natural mercury isotope $^{196}$Hg. The dashed line in Figure 1 indicates the lightest compound nuclei (LCN) for elements from tin to lead that can be produced by the complete fusion of two stable nuclei. The greatest deficit is achieved by fusing nuclei with close $Z$ and $A$ values and using the lightest stable isotopes as the target and projectile.

Modern heavy-ion accelerators can generate ion beams of practically all the elements. Experimentalists can now choose any combination of projectile and target nucleus and use complete-fusion reactions to produce nuclei with large neutron deficits.

#### *2.1.2. Limiting Factors for the Complete Fusion of Nuclei*

The wide choice of target-projectile combinations naturally entails questions as to whether complete nuclear fusion is always possible or whether there are limitations to the formation of compound nuclei, assuming that the heavy-ion energy is enough to overcome the entrance Coulomb barrier. In fact, limitations do exist. Some of the limits were discovered in the infancy of nuclear physics with heavy ions and others were only established recently.

One restriction on the complete fusion of nuclei is associated with the large angular momenta occurring in heavy-ion reactions. Experimental data indicate that for each target-projectile combination and for a chosen collision energy, there exists a critical angular momentum $l_{cr}$ above which the projectile nucleus cannot be trapped by the target nuclei, this being a necessary condition for compound-nucleus formation.

A theoretical analysis of the stability of a rapidly rotating nuclear drop (Co 74) has led to the determination of limiting spins $I_{cr}$ which the nucleus can endure without

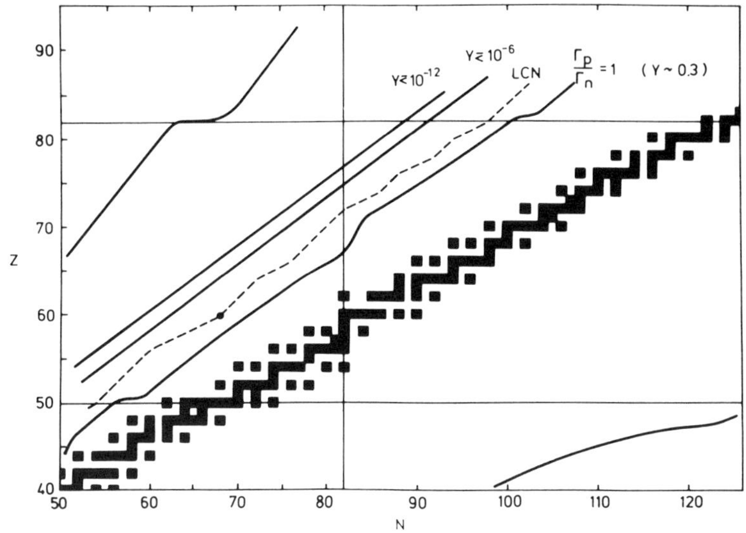


**Figure 1.** Part of the Chart of the Nuclides with $40 \leqslant Z \leqslant 82$. Filled squares indicate stable isotopes, the dashed line labeled as LCN denotes the most neutron-deficient isotopes produced as a result of the complete fusion of stable nuclei, the thick line shows isotopes with equal probabilities for neutron and proton emission, thin lines indicate the isotopes having yields of the $\sim 10^{-6}$th and $\sim 10^{-12}$th fractions of the compound nucleus cross-section. Limits of stability of the nuclides with respect to neutron and proton emission are also indicated (St 71).

being destroyed instantly. The relevant evidence is presented in Figure 2, where $I_{cr}$ is given as a function of the mass number of the nucleus.

The critical angular momentum $l_{cr}$ may either be both larger or smaller than the limiting spin $I_{cr}$. If larger, then the limitation for complete fusion depends on the entrance reaction channel, $l < l_{cr}$. If smaller, it depends on the limiting spin and the only contributions to the compound-nucleus formation are those partial waves for which the fission barrier of the compound nucleus does not vanish, i.e., $l < l_{Bf}$.

Now we shall turn to head-on collisions of nuclei in which centrifugal forces no longer play a crucial role in the trapping of the projectile by the target nucleus. Until recently it was thought that any two nuclei, even uranium, can completely fuse if they collide at sufficiently high kinetic energies. With the advent of accelerators capable of producing very heavy ions, we can test this assumption experimentally.

Figure 3 illustrates several observations of complete fusion between nuclei. The axes give the atomic numbers of the colliding nuclei, $Z_1$ and $Z_2$. The closed circles indicate the target-projectile systems in which compound-nucleus formation has been observed, and the crosses show the target-projectile combinations in which it has not been observed. The figure shows that the nuclear fusion materializes best when the target-projectile system is asymmetric. Reactions involving asymmetric systems have yielded products up to element 109, while symmetric combinations have only generated products up to uranium.

Swiatecki's analysis of the nuclear interaction using the macroscopic dynamical approach (Sw 81) revealed a dynamical barrier to the complete fusion of a heavy target

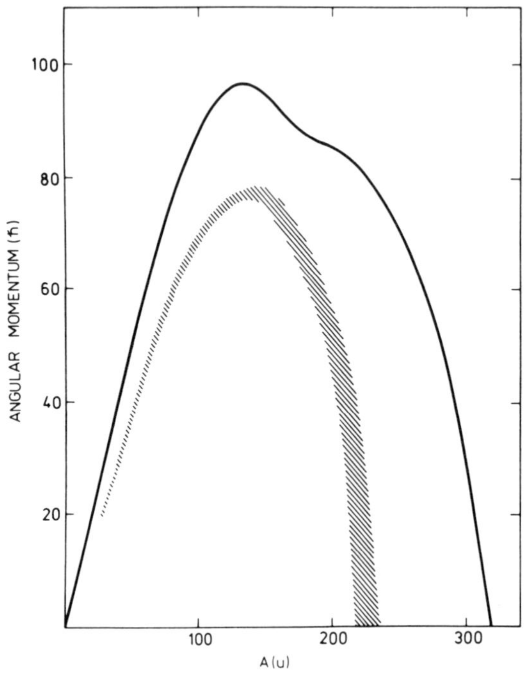


**Figure 2.** Liquid-drop model prediction of the angular momentum at which the fission barrier of beta-stable nuclei with mass number $A$ vanishes (solid curve). The hatched area indicates the angular momentum at which the fission barrier is equal to the neutron separation energy (Pl 74).

nuclei and a heavy projectile. Additional energy over the entrance potential barrier—an extra push—is needed for the fusion. In order to form a compound nucleus even more energy is required, and this has been termed an extra-extra push, by Bjornholm and Swiatecki (Bj 82).

The extra-extra push values calculated as a function of the atomic number of the colliding nuclei are given in Figure 4. It can be seen that the extra-extra push reaches its maximum value for $Z_1 = Z_2$ when the $Z$ value of the compound nucleus is fixed. But symmetric projectile-target systems yield compound nuclei with the greatest neutron deficits.

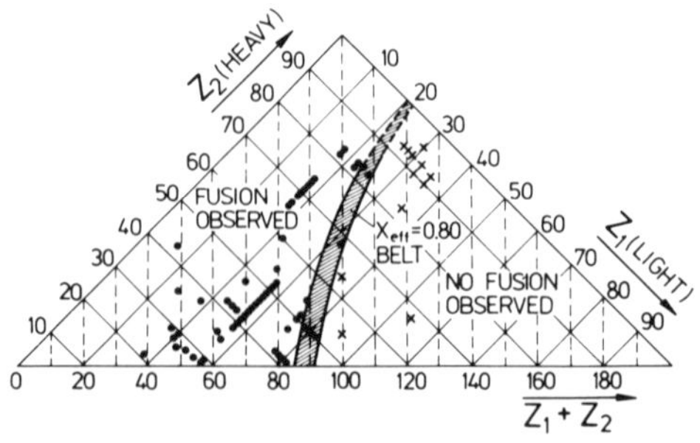


**Figure 3.** Compilation of the fusion-reaction studies: Full points mark cases where evaporation residues were detected (at GSI), crosses stand for nonobservation of fusion (in different laboratories). The hatched area corresponds to an $X_{\text{eff}}$ value of 0.80, $X_{\text{eff}}$ being the ratio of attractive nuclear forces and repulsive Coulomb forces in a configuration of two touching spheres (Ar 83).

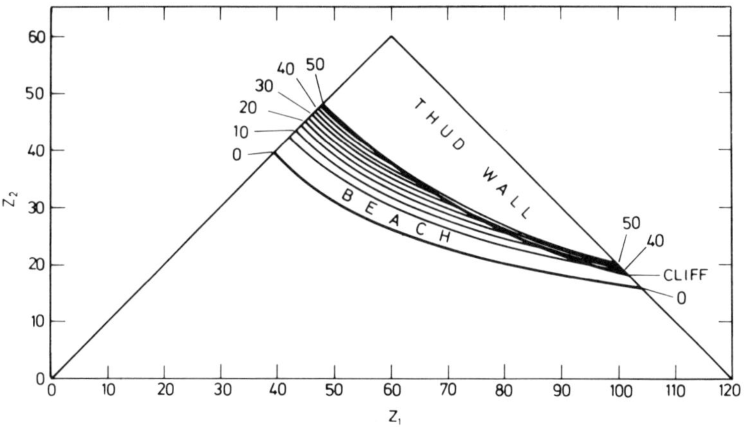


**Figure 4.** Contour lines (in MeV) of the estimated extra-extra push over the interaction barrier needed to make compound nucleus (or, at least, to form a spherical composite nucleus) out of two nuclei with atomic numbers $Z_1$ and $Z_2$ (schematic model, not to be compared with experiment). In the region "thud wall," the extra-extra push rises to very high values (Bj 82).

Figure 4 also shows that the extra-extra push increases drastically for $40 \leqslant Z_1$, $Z_2 \leqslant 50$. This is the reason why element 100—fermium—could not be produced by a symmetric combination of colliding nuclei.

The existence of the dynamical barrier preventing heavy nuclei from fusing completely is in principle important in the synthesis of new transuranium elements and superheavy nuclei. However, if we restrict ourselves to nuclei far from stability with $Z \leqslant 92$, any target–projectile combination will allow complete fusion and compound-nucleus formation.

### *2.1.3. Yields of Neutron-Deficient Nuclei*

The yield of a neutron-deficient nucleus produced as a result of compound-nucleus decay depends on two factors: the production cross-section of the compound nucleus and the competition between different channels for its decay. We shall briefly consider each factor separately.

*2.1.3a. The Production Cross-Section of a Compound Nucleus* is given by equation (2.1), which expresses the overall contribution from partial waves with different angular momenta from $l = 0$ to $l = l_{cr}$:

$$\sigma_{CN} = \pi \lambda\!\!\!^{-2} l_{cr}(l_{cr} + 1) \tag{2.1}$$

If $l_{Bf} < l_{cr}$, then $l_{cr}$ in equation (2.1) is replaced by $l_{Bf}$, the angular momentum at which the fission barrier of the compound nucleus vanishes.

A typical form of the dependence of $\sigma_{CN}$ on the collision energy is presented in Figure 5. The cross-section increases rapidly in the vicinity of the Coulomb barrier, reaches a maximum, and then begins to decrease. The rise in the cross-section is associated with increasingly more partial waves contributing to $\sigma_{CN}$, and it falls because $l_{cr}$ increases more slowly than $\lambda\!\!\!^{-}$ decreases.

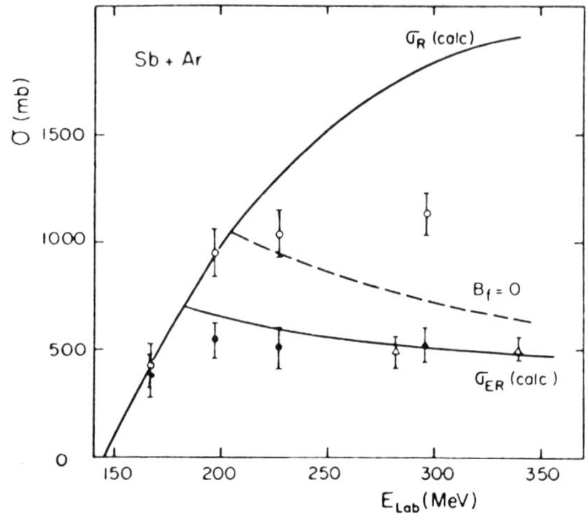


**Figure 5.** Fusion excitation functions for the system $^{40}Ar + {}^{121}Sb$. The lower points are cross-sections for evaporation-residue formation [full circles from (Ga 75), open triangles from (Br 76)] and the upper points are obtained by adding the fission cross-sections from (Ta 75). The full curves give theoretical predictions of the total reaction cross-section (upper) and the cross-section for evaporation-residue formation (lower). The dashed curve results from the assumption that all angular momenta contribute to the cross-section for which the calculated fission barrier of the compound nucleus ($B_f$) is larger than zero (Ga 75).

As the mass of the projectile increases, $l_{cr}$ initially increases up to mass approximately 50 and then begins to decrease slowly. The $\mathcal{X}$ value is inversely proportional to the reduced mass of the projectile. At last, if we compare the complete-fusion cross-sections for different heavy ions at the same energy per nucleon, they are smaller for the heavier ions. Figure 6 illustrates this dependence.

Heavier ions are advantageous for producing compound nuclei with very large neutron deficits. We note however that the decrease in the fusion cross-section with increasing projectile mass is not dramatic in character.

*2.1.3b. Competition Between the Channels of Compound-Nucleus Decay.* The excitation energy of the compound nucleus in heavy-ion reactions may reach many tens of MeV. The deexcitation of the compound nucleus occurs by the emission of neutrons, light charged particles, and $\gamma$ rays. Competition between the numerous decay channels leads to the formation of a considerable number of differing daughter nuclei.

Neutron emission leads to an increase in the neutron deficit, thus pushing the nucleus farther from the region of stability. The emission of protons and $\alpha$ particles acts in the opposite direction. Light charged particles have to overcome the Coulomb barrier of the compound nucleus, while no barrier exists for neutrons. Generally speaking, therefore, neutron emission is preferable. By increasing the energy of heavy ions and, at the same time, the excitation energy of the compound nucleus, the emission of a fairly large number of neutrons can be achieved. Figure 7 illustrates the shape of the excitation functions for the (HI, $x$n) reactions, measured during the $^{12}C$-ion bombardment of $^{130}Te$. Three to twelve neutrons are observed to have been emitted in the ion energy range between 30 and 160 MeV; see Oganessian *et al.* (Og 75). Hence it would seem that the production of neutron-deficient nuclei requires an increased projectile energy in order to get highly excited compound nuclei which will then deexcite mainly by neutron emission.

It should, however, be noted that in this study (Og 75) a very asymmetric combination of projectile and target nuclei was used, the target being a heavy tellurium isotope.

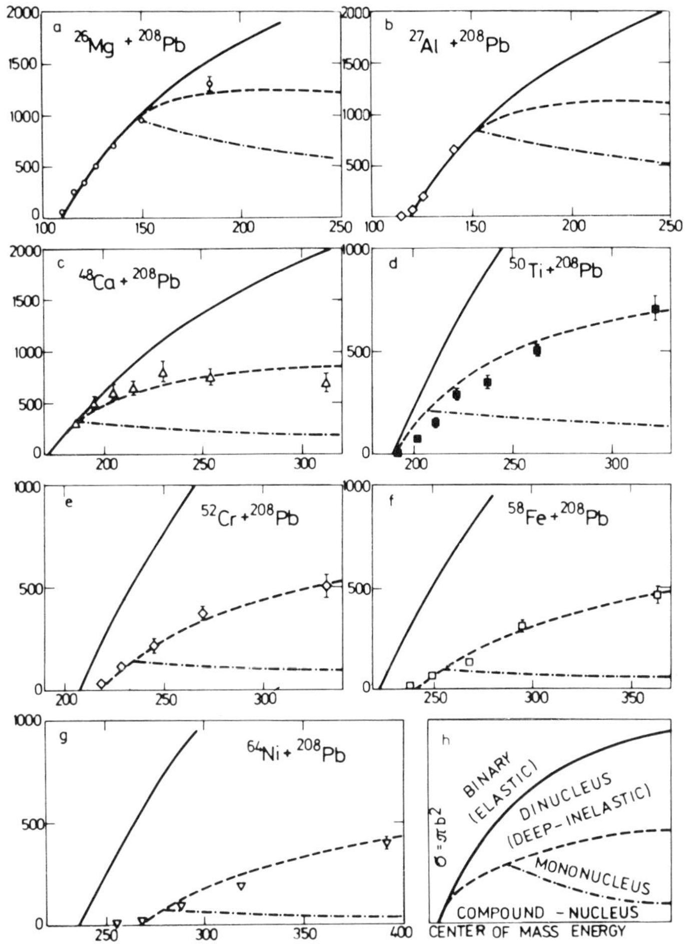


**Figure 6.** Excitation functions for various types of reactions. Figure (h) illustrates the qualitative distinction between binary (elastic and quasi-elastic) reactions, dinucleus (deep-inelastic) reactions, mononucleus (fast-fission) reactions, and compound-nucleus reactions (Bj 82).

But how do the different channels of compound-nucleus decay compete when symmetric combinations of natural isotopes collide?

In the early seventies, some studies were carried out by Stephens *et al.* and Winn *et al.* (St 71, Wi 72) to investigate the possibility of achieving the greatest neutron deficit in complete-fusion reactions. The investigation concerned compound nuclei lying between zirconium and astatine. Target–projectile systems were used (St 71) which produced a maximum neutron deficit in the compound nucleus by the initial state. They were symmetric or nearly symmetric combinations of projectile and target nuclei with the lightest natural isotopes.

In the study (Wi 72) the charge asymmetry of the entrance channel was varied in order to reveal its influence on compound-nucleus decay. The excitation energy of the

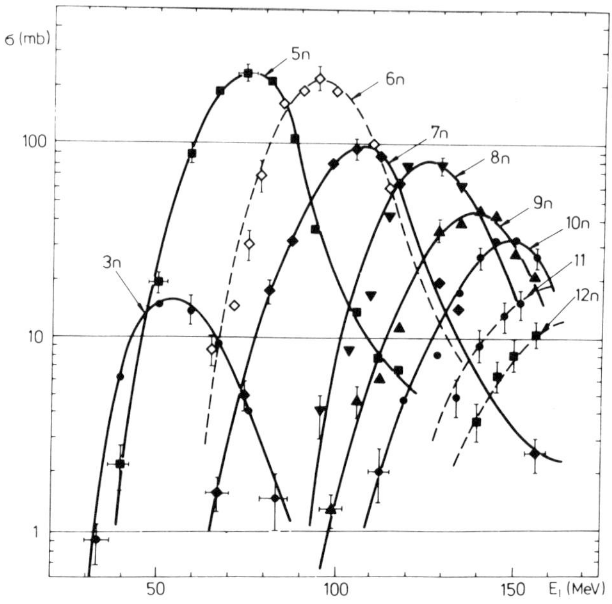


**Figure 7.** Excitation functions for the ($^{12}C$, $xn$) reactions with $x$ between 3 and 12. These have been obtained by bombarding $^{130}Te$ with $^{12}C$ ions at energies from 30 to 160 MeV (Og 75).

compound nucleus varies widely and the values at which a given isotope has a maximum yield have been found.

In both papers the compound-nucleus cross-section was identified with the cross-section of the reaction, $\sigma_R$. The decay of an excited compound nucleus was calculated within the framework of the statistical model. The competing decay channels were the emission of neutrons, protons, and $\alpha$ particles. Neither the fission contribution nor competition with $\gamma$-ray emission were taken into account. The fraction of excitation energy spent on nuclear rotation was taken into account in (Wi 72). The compound nucleus was assumed to have a rigid-body moment of inertia. The compound-nucleus deformation due to rotation was ignored.

In the study (St 71) it was assumed that the compound-nucleus temperature would remain constant, equal to 1.5 MeV, during the entire cascade of light particle emission. Despite all these simplifying assumptions, the main conclusions in these works concerning the specific features of compound-nucleus deexcitation with increasing neutron deficit in it are beyond doubt.

Figures 8a, 8b, and 8c demonstrate how the decay routes of an excited compound nucleus change as the charge (mass) asymmetry of the reaction entrance channel varies.

The fusion of $^{128}Te$ and $^{12}C$ produces the compound nucleus $^{140}Ce$, the most abundant natural isotope of cerium. In this reaction the compound nucleus happens to lie on the very bottom of the valley of $\beta$-stability. This is not causal as $^{128}Te$ is one of the two heavy stable isotopes of tellurium. It can be seen from Figure 8a that the largest yield exceeded a hundred millibarns and was for isotopes produced by the emission of 3 to 6 neutrons, or a proton and four or five neutrons, or an $\alpha$ particle and four neutrons.

In the reaction $^{92}Mo + ^{58}Ni \rightarrow ^{150}Yb$, the charge and mass asymmetries are not considerable. The use of the lightest isotopes of nickel and molybdenum results in a deficit of 18 neutrons on the $^{150}Yb$ compound nucleus. Of the stable isotopes of ytterbium, the lightest has mass number 168. The most likely channels of compound-nucleus decay are quite different (Figure 8b). The isotopes corresponding to the emission of 3 to 5 protons are produced with cross-sections larger than a hundred millibarns.

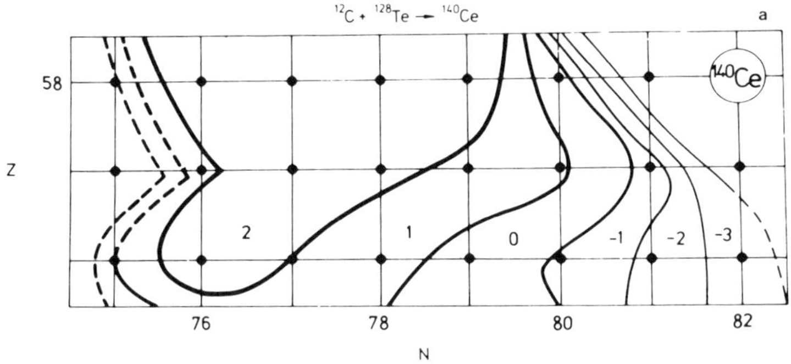


**Figure 8a.** Contour diagram for maximum yields following decay of $^{140}Ce$ produced by $^{12}C + ^{128}Te$. The numbers are the logarithms (base 10) for the maximum yields indicated by the contour lines in units of mb, e.g., 2 means hundreds of mb (Wi 72).

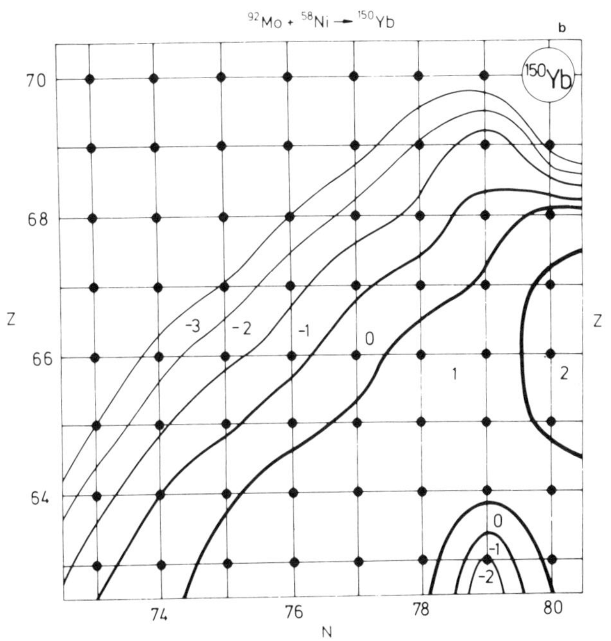


**Figure 8b.** The same as in Figure 8a, for the reaction $^{92}Mo + ^{58}Ni \rightarrow ^{150}Yb$ (Wi 72).

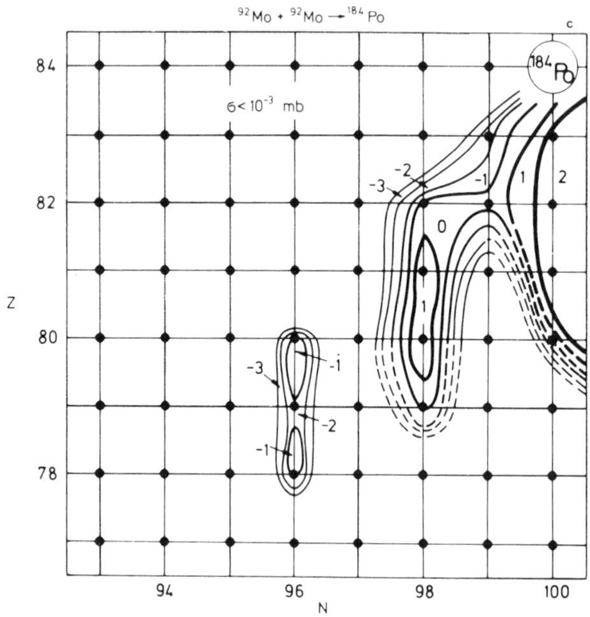


**Figure 8c.** The same as in Figure 8a, for the reaction $^{92}Mo + ^{92}Mo \rightarrow ^{184}Po$ (Wi 72).

The cross-section for the emission of one and two neutrons is smaller than one microbarn and the cross-section for the emission of an even larger number of neutrons is vanishingly small.

In the symmetric target–projectile system $^{92}Mo + ^{92}Mo \rightarrow ^{184}Po$ the initial compound nucleus $^{184}Po$ contains 26 fewer neutrons than the most stable isotope of polonium, $^{210}Po$. It can be seen from Figure 8c that here again the emission of one to three protons, or an $\alpha$ particle, and one or two protons are the most intensive channels for $^{184}Po$ nucleus decay. The cross-section of the reaction channels involving the emission of one or two neutrons is several orders of magnitude smaller than one microbarn.

Figure 9 shows the thresholds of neutron, proton, and alpha emission for various isotopes of osmium and cerium, giving some idea of the reasons for the suppression of neutron emission from compound nuclei with large neutron deficits. The neutron-emission threshold coincides with the neutron binding energy $B_n$, and for protons and $\alpha$ particles it also includes the exit Coulomb barriers. The Coulomb barriers for cerium are respectively 8 and 15 MeV, and 9.7 and 18.3 MeV for osmium. It can be seen that the proton-emission threshold decreases with increasing neutron deficit in the nucleus, while the neutron-emission threshold increases. In the expression for the ratio between the neutron- and proton-emission probabilities $P_n/P_p$ the thresholds are part of the exponent, i.e.,

$$P_n/P_p \sim \exp[-(B_n - B_p^*)/T] \tag{2.2}$$

where $T$ is the temperature of the compound nucleus. A difference of several MeV

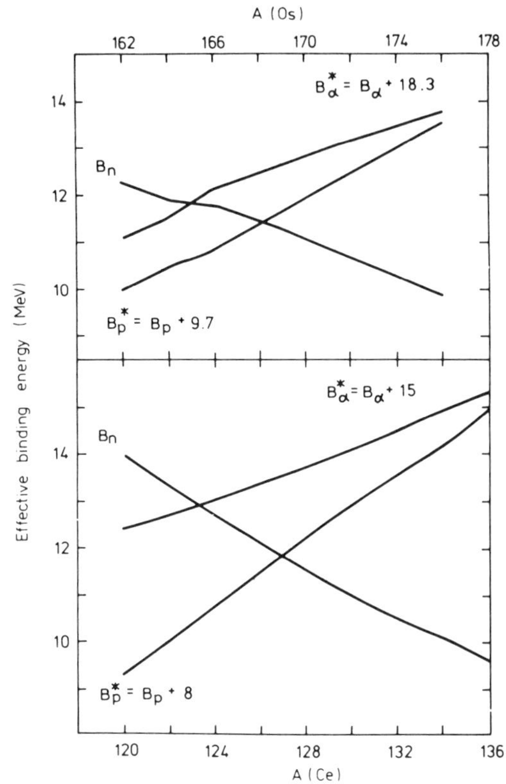


**Figure 9.** Effective binding energies for a neutron $B_n$, a proton $B_p^*$, and an alpha particle $B_\alpha^*$, as functions of the mass number of Os and Ce isotopes (St 71).

between the thresholds is enough to suppress compound-nucleus decay channels involving the emission of neutrons alone.

Figure 1, which is taken from (St 71), demonstrates the possible advance in the direction of the limit of nucleon stability in nuclei with a large neutron deficit. The solid line on the right indicates the nuclei for which the probabilities of neutron and proton emission are equal, $\Gamma_p/\Gamma_n = 1$. For these nuclei, the production cross-section for the neutron-deficient nucleus is about 30% of the compound-nucleus cross-section. Reaction products with larger neutron deficit than on the LCN line and resulting from neutron emission have much smaller yields. In the case of the emission of 4 to 6 neutrons, the cross-section is one million or more times smaller than those for the nuclides on the LCN line, and another million times smaller if another 3–2 neutrons are ejected. The two lines on the left of the dashed line indicate nuclei with yields of, respectively, about one $10^{-6}$th and one $10^{-12}$th of the yields of the compound nuclei on the LCN line. The nucleon stability boundaries from this $Z$ region are also shown in Figure 1.

From the analysis of (St 71) and (Wi 72) one can draw the following conclusions:

1. The most efficient way of producing nuclei with large neutron deficits by complete-fusion reactions is to use symmetric target-projectile combinations and the lightest stable isotopes.

2. Increasing the neutron deficit further by neutron emission from the compound nucleus has limitations.
3. The use of reactions involving the emission of protons and $\alpha$ particles is also possible. In this case the target–projectile combinations should be chosen so as to produce a compound nucleus with a larger $Z$ than that of the isotopes of interest, and the heavy-ion energy should lie in the vicinity of the Coulomb barrier.
4. Complete-fusion reactions offer the possibility of studying the nuclear properties of a large number of as yet unknown neutron-deficient isotopes.

## 2.2. Separation and Identification of Complete-Fusion Reaction Products

### *2.2.1. The Recoil Method*

In complete-fusion reactions the compound nucleus receives all the momentum of the projectile and its kinetic energy $E_{CN}$ is defined by

$$E_{CN} = [M_p/(M_p + M_t)] \cdot E_p \tag{2.3}$$

where $M_p$ and $M_t$ are the masses of the projectile and the target nucleus, and $E_p$ is the projectile kinetic energy. In heavy-ion reactions $E_{CN}$ may reach many tens and even hundreds of MeV. The nucleons evaporated from an excited compound nucleus only carry off a small part of the momentum and kinetic energy. The evaporation residues, i.e., the nuclei which the experimentalists deal with, also possess high kinetic energy, their average momentum being close to that of the compound nucleus.

This feature of the kinematics of heavy-ion induced complete-fusion reactions opens up the possibility of separating the evaporation residues from the target material rapidly and efficiently. With a 1–2 mg/cm$^2$ target, all the evaporation residues can be knocked out of the target. This method for separating evaporation residues is called the recoil method. It was first used in experiments aimed at synthesizing new transuranium elements and it was subsequently widely used in experiments to produce nuclei far from stability.

The shape of the excitation functions of the reactions used to produce isotopes with a neutron deficit is also a favorable factor for the application of the recoil method. As can be seen from Figure 7, the excitation functions of reactions involving the evaporation of different numbers of neutrons cover comparatively narrow energy ranges. The excitation function half-widths (FWHM) lie between 15 and 20 MeV. A similar shape characterizes the excitation functions of the reactions in which protons and neutrons are evaporated (HI, $x$p, $y$n). We can also state that different channels of excited compound-nucleus decay possess energy selectivity. The only contributor to the production of a certain isotope is the target layer in which the heavy-ion energy remains within certain limits. For sufficiently heavy ions the effective thickness of the target is comparable with recoil ranges. The correct choice of projectile energy ensures that all or most of the recoil nuclei of interest are knocked out from the target.

The correlation between the recoil range and the effective thickness of the target is due to the mass and energy of the bombarding nucleus. As the mass and energy of the heavy ion increase, the recoil ranges become longer while the effective target thickness decreases as a result of the growing specific ionization of the heavy ion.

In the preceding section we noted that symmetric target–projectile systems lead to the greatest neutron deficits. These combinations are also best when using the recoil method since the compound nucleus receives half of the heavy-ion kinetic energy.

Thus, the extremely difficult problem of separating nuclear reaction products from the target material, which experimentalists face when producing nuclei far from stability, is solved successfully by using the recoil method in complete-fusion reactions.

### *2.2.2. Separating Recoil Nuclei from the Bombarding Heavy Ions*

There are no particular problems when separating reaction products from the projectiles after proton-induced spallation reactions or thermal-neutron fission. The fast protons that have struck the target are retarded by a stopper placed some distance from the target. Heavy nuclear shielding protects detectors for recording uranium fission products from direct neutron beams. The situation is considerably more complicated when detecting compound-nucleus decay products after heavy-ion reactions. Evaporation residues with momenta close to those of the ions move together with the intense heavy-ion flux.

Separating recoil nuclei from the projectiles is the principal technical difficulty during experiments with heavy ions.

There are two main approaches to the problem. In one approach the recoil nuclei are retarded and stopped in a gaseous or solid medium. Then they are extracted somehow from the stopping zone and transported to a detector system. Once they have lost their kinetic energy the heavy ions move due to thermal energy, and so are unable to affect the radioactive radiation detectors. The recoil nuclei can be transported to the detectors in a gas jet, or they can be re-ionized, accelerated by an electric field, and then transported as ions to the radioactive radiation detectors.

In the second approach, the recoil nuclei are separated from the heavy-ion beam using a combination of magnetic and electric fields, without the first stopping the recoil nuclei. In this case the considerable difference between the velocities of the heavy ions and recoil nuclei is exploited. This method is called kinematical separation. The recoil nuclei, once separated from the ion beam, are knocked into a solid-state detector which is used to identify them and to study their radioactive decay properties.

### *2.2.3. The Gas-Jet Technique*

In the early sixties a simple and effective method was proposed by Friedman and Macfarlane (Fr 62, Ma 63) for separating recoil nuclei and transporting them to radioactive radiation detectors, namely, a gas jet.

Figure 10 is a schematic diagram of one of the first facilities using the gas-jet technique. The facility was used to produce neutron-deficient isotopes that emit delayed protons after complete-fusion reactions induced by heavy ions.

The recoil nuclei were knocked out of a thin target and stopped in a chamber containing helium at a pressure of about one atmosphere. Heavy ions shoot through the gas volume and reach a projectile beam collector.

The stopping chamber has a small aperture 0.4 mm in diameter, through which the helium jet passes into a second, continually evacuated chamber. The reaction products, which are stopped in the helium, are carried out with the helium jet. Having reached the activity collector, the reaction products adhere to its surface and the

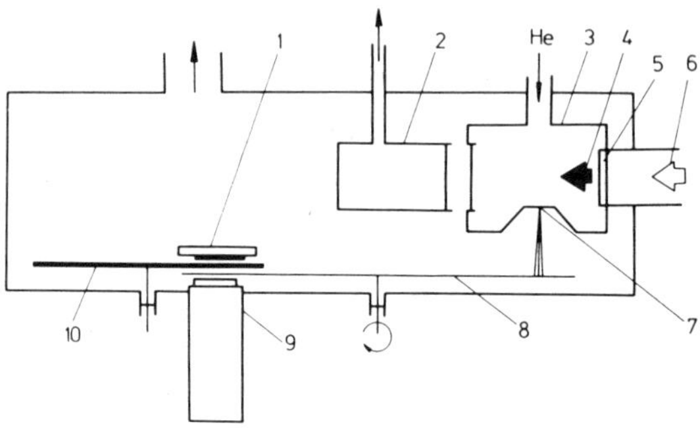


**Figure 10.** Schematic drawing of the experimental setup designed to collect activity using the gas-jet method. The gas jet flows from a small-diameter hole: (1) surface-barrier detector; (2) ion-beam collector; (3) stopping chamber; (4) recoil nuclei; (5) target; (6) beam; (7) gas jet; (8) activity collector; (9) scintillation $\beta$-counter; (10) calibration $\alpha$-source holder (Ka 81).

scattered helium is pumped out. The collector is turned from time to time to place the spot with the reaction products in front of the detector.

In the gas-jet technique the gas fulfils two functions: it stops the recoil nuclei and transports them to the detectors. Subsequently, the transport function was extended by using a capillary tube instead of an aperture in the stopping chamber. It turned out that a tube can carry recoil nuclei considerable distances (several tens of meters) together with the helium jet without any substantial loss in efficiency and without noticeable delay in the transport of the products to the detectors. Tubes made from stainless steel, teflon, and polyethylene and with inside diameters of 0.5–1 mm have been used. Capillary tubes to carry the helium jet are a successful solution to the problem of transporting recoil nuclei into zones with low background radiation.

The efficiency with which recoil nuclei are transported from the stopping zone via capillaries is closely related to the physicochemical processes occurring in the stopping zone. These processes are caused by the strong ionization produced by the heavy-ion beam in the stopping volume of the gas. For example, it has been found that very pure helium is less efficient than helium containing impurities. The addition of water, benzene, ethylene, or other hydrocarbon vapors to the helium increases the activity yield. Some studies have shown that the formation in the stopping zone of macromolecules with atomic masses of $10^6$–$10^8$ amu is crucial for the efficiency of the gas-jet technique (Ju 71). The macromolecules act as carriers of the recoil nuclei, which themselves easily stick to the walls of the stopping chamber or the inside walls of the capillaries. The average efficiency of the gas-jet method is between 30% and 50%. The capillary length has no essential influence on the efficiency.

The transport time in the gas-jet technique is mainly determined by the time required for the recoil nuclei to leave the stopping chamber through the exit aperture or to enter the capillary. Transport along the capillary itself takes a very small fraction of the whole transport time. The gas-jet method can be used under favorable conditions to separate recoil nuclei and measure their radioactive decay half-lives for periods of the order of several milliseconds.

The gas-jet technique was essential at the start of the studies of the neutron-deficient isotopes formed by the complete fusion of two complex nuclei. The delayed proton emitters were separated and investigated and the lightest $\alpha$ emitters—tellurium isotopes—were observed for the first time (Ma 65) using this method.

While paying tribute to the simplicity and effectiveness of the gas-jet technique, we must mention its limitations. These are primarily associated with the absence of selectivity in the separation of recoil nuclei. The helium jet carries all the products of compound-nucleus decay out of the stopping volume. They produce a dense background in the detectors from which it is difficult to isolate the activity of interest. The difficulties get worse the farther the nuclides of interest are from the stability region. An increase in the neutron deficit in the nucleus produced as a result of complete fusion is inevitably associated with a decrease in its yield. There is a need for a facility which can separate out from the variety of compound-nucleus decay products only those which are of interest to us, viz., nuclei with very large neutron deficits.

An on-line isotope separator operated at a heavy-ion accelerator is one such facility.

### 2.2.4. An On-Line Isotope Separator

The on-line isotope separator ISOL operated in combination with an accelerator or a reactor is one of the main experimental techniques used to separate and study nuclei far from stability. An on-line isotope separator was first operated with a heavy-ion accelerator by Tarantin and co-workers (Ta 65, Ta 66, Ta 70) at the Laboratory of Nuclear Reactions at Dubna. It was used to separate the neutron-rich isotopes of light elements produced in transfer reactions and to study their properties (Ka 71, Gu 73, Gu 74).

The ISOL technique is described in detail by H. L. Ravn and B. W. Allardyce in Chapter 5. We shall restrict ourselves to a brief description of the ISOL method to give the reader an idea of how the method compares to the other methods for separating and studying nuclei far from stability produced by complete-fusion reactions.

The main units of an on-line isotope separator operated in combination with a heavy-ion accelerator are shown in Figures 11a and 11b. They are a joint target-ion source unit, and electrostatic system for accelerating the ions and forming the ion beam, a separating magnet, an electrostatic system for separating ion beams with different $A$ values, a system for transporting the separated reaction products to the detectors, and the detectors.

A thin target placed very near the ion source is continuously bombarded by heavy ions (Figure 11b). The nuclei recoiling from the target enter the ion source to be retarded in a catcher. A thin tungsten foil serves as an entrance window for the reaction products. A thin carbon foil is placed in front of the window and acts as a heat screen for the target. The high temperature inside the ion source causes the reaction products to be knocked into the catcher, diffuse to its surface, and then evaporate and reach the ion-source chamber. Here they are either ionized in gas-discharge plasma or on the hot surface of the chamber.

The ions leave the ion source through a small orifice and reach the electric field produced by an extracting electrode. The ions are accelerated to an energy of several tens of keV and directed into a magnetic separator. Since the energy of the ion thermal motion may be as low as a few tenths of an eV even at 2500–3000 K, the ion beam is very monochromatic. The small exit slit of the ion source and the geometry of the accelerating electric field produce a good emittance of the ion beam.

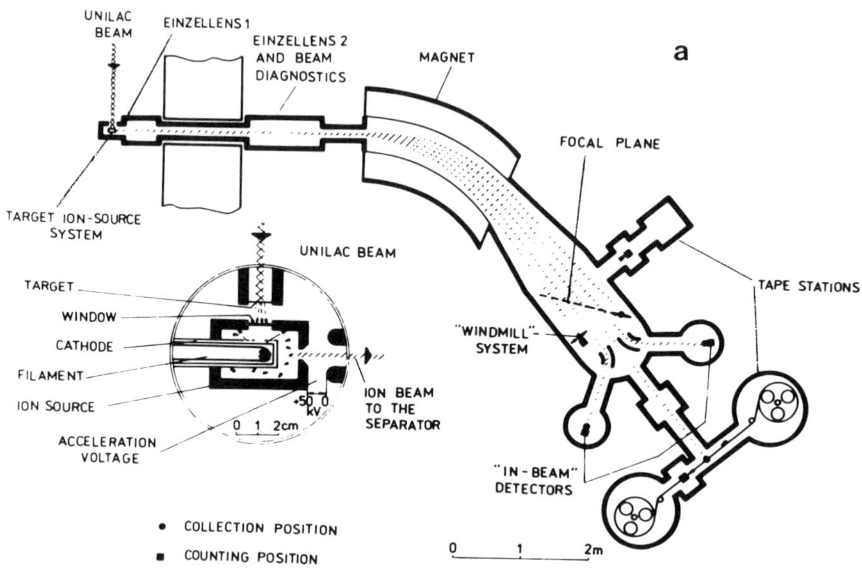


**Figure 11a.** Layout of the GSI mass-separator facility (Br 81).

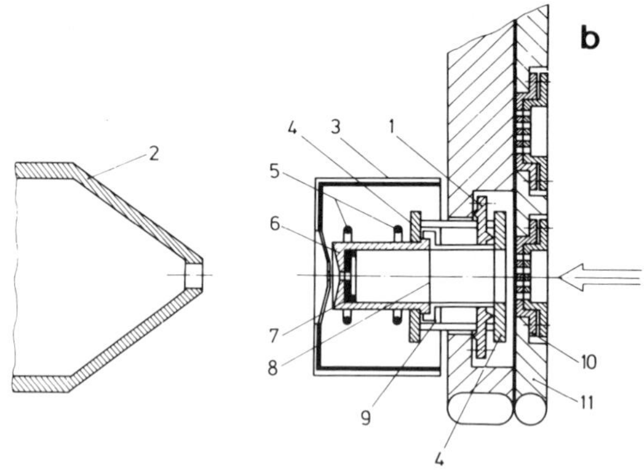


**Figure 11b.** Schematic drawing of the surface-ionization ion source BEMS-2: (1) titanium insert; (2) extraction electrode; (3) multilayer thermal screen; (4) supporting tantalum rings; (5) tungsten spiral heater of the ionizer; (6) tungsten ionizer; (7) tantalum recoil catcher; (8) thin tungsten foil; (9) thin-sall tantalum cylinder for foil clamping; (10) cassette containing two targets; (11) cooled copper holder (Ka 81).

The mass-separator magnetic field separates ions with respect to $q/A$, where $q$ is the ion charge and $A$ is its mass number. Since most of the ions are singly charged, the ions are separated according to mass number $A$. The configuration of the magnet boundaries and special magnetic screens focus the beam both horizontally and vertically and can be used to obtain small images. Electrostatic lenses placed in front of the separating magnet are also used to focus the beam.

Electrostatic deflectors are placed at the exit of the magnet. They are designed for increasing the angle between the central trajectory and the trajectories of adjacent isotopes with larger and smaller $A$ values. This set-up can be used to study several ion beams simultaneously.

Separated reaction products with a certain A value reach the catchers, which transport them to the radioactive radiation detectors. Tape-recorder-type mechanisms operated in a pulsed mode are often used to transport the products. It is then possible to avoid any accumulation on the catcher of long-lived radioactivity due to the decay of isobars closer to the stability region and, therefore, produced with large yields.

The radioactive decay of the separated reaction products is studied using detectors for $\alpha$ particles, protons, electrons, positrons, delayed neutrons, delayed fission fragments, $\gamma$ rays, and X rays. Correlation measurements are important because radioactive decay chains leading to known radioactive isotopes are typical for nuclei far from stability.

The principal parameters of an on-line separator designed to work with a heavy-ion accelerator are its resolving power, efficiency, speed of operation, and chemical selectivity.

The mass resolution $M/\Delta M$ may reach a value of several thousand, so that the spatial separation of neighboring isotopes, including those of heavy elements, presents no problem.

An ISOL separator used on a heavy-ion beam may have a widely varying efficiency, from several tens to a hundredth of one percent. The efficiency depends on the physicochemical properties of the element being separated, namely, on its melting point, its volatility at the operating temperature of the ion source, and on its ionization potential. For example, the alkali metal isotopes are separated very efficiently. These elements readily lose an electron and take on the electron configuration of noble gases, which are very volatile. By contrast, isotopes of elements with high melting points and low volatilities (e.g., W, Ta and Re) are difficult to separate using a mass separator.

The speed of ISOL operation is determined by the average time required for separating the isotopes of a given element. It takes fractions of a millisecond for an ion to travel from the exit orifice of the ion source to the collector. Therefore, the operating speed mainly depends on the ion holdup time in the ion source. The holdup time includes the time it takes for the atom to diffuse to the catcher surface, its residence time on the catcher surface, the time needed to ionize it and, finally, the time taken for the ion to travel inside the ion-source chamber to the exit orifice. Once it has collided with the walls of the ion-source chamber, an atom or an ion may be retained on the wall surface for some time. The melting point, volatility, and ionization potential are the main factors that determine how long the atom is retained in the ion source. Figure 12 illustrates the temperature dependence of the retention time in the ion source for a number of elements (Ka 81). There are now some ion sources designed to operate at temperatures of 2700–3000 K (Ki 81).

The isotope separator only separates reaction products according to their mass number. Isotopes of several elements may reach the same place on the focal plane, thus creating difficulties when studying nuclei farthest from stability. The yield of these nuclei is small and it is very difficult to detect their decay against the background from the radioactive radiation of the more stable isobars.

The way the efficiency and the operating speed of the ion source depend on the chemical properties of an element can be exploited to separate the isotopes of certain

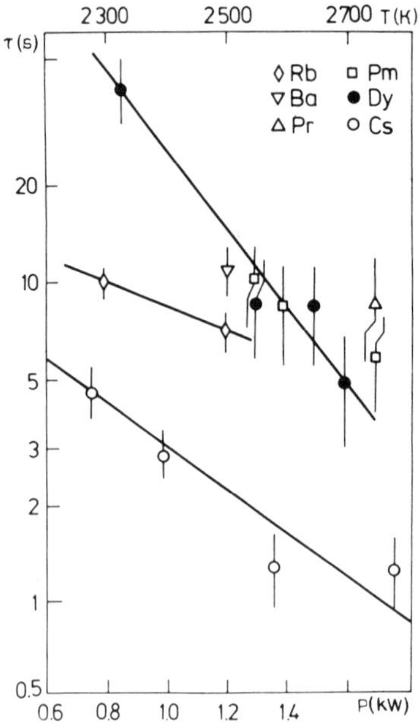


**Figure 12.** Average holdup time for several elements in the ion source BEMS-2, as a function of the ionizer heating power (temperature) (Ka 81).

elements preferentially. At low temperatures an ion source operated in a gas-discharge mode preferentially generated more volatile elements, such as the noble gases, the alkali metals, mercury, thallium, and bismuth. Yields in which the alkali and alkali-earth elements predominate are obtained by using thermal ionization on the inner surface of the ion source rather than ionization in a gas discharge.

An ISOL system has been successfully used to separate and identify a large number of new isotopes. It has also generated valuable information about the radioactive properties of nuclei far from the region of stability.

At the same time the technique has some operating limitations. The necessity of extracting the reaction products from a solid-state catcher by diffusion, the sorption and desorption of atoms on the inside surface of the ion source, and the necessity of ionizing them result in a noticeable delay between the moment a recoil nucleus is emitted from the target and the moment it is delivered to the radioactive radiation detectors. The farther a nuclide is from the region of stability, the shorter its lifetime, and hence the time delay is important.

Because of the low yields and long retention times in the ion source of elements with high melting points and low volatilities, the ISOL technique is unsuitable for separating them.

### *2.2.5. Combination of Gas Jet and Ion Source in an ISOL System*

One way of decreasing the time delay and extending the range of elements to be separated is to replace the solid-state catcher by a gas one, thus enabling the reaction products to be transported from the stopping volume by a gas jet. The idea of using the gas-jet technique in combination with the ISOL one was suggested by Nitschke

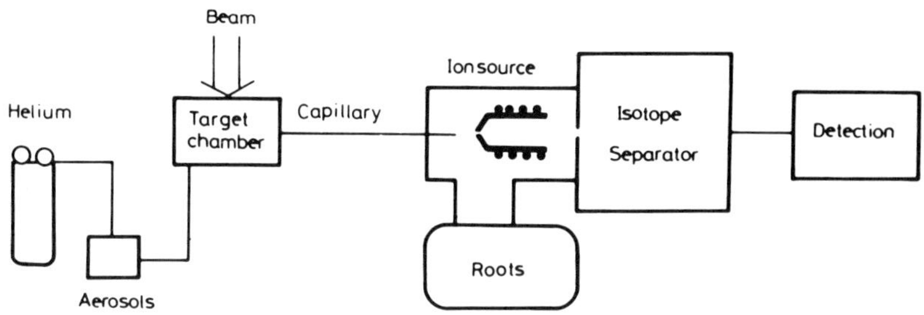


**Figure 13.** Schematic drawing of the He jet coupled with ISOL (Är 87).

and first implemented on a heavy-ion beam in the RAMA facility (Recoil Atom Mass Analyzer) (Ni 70).

A schematic diagram of the principal units of an ISOL system together with a gas catcher and a helium jet is presented in Figure 13. Recoil nuclei are knocked out of the target and stopped in helium, which is fed into the target chamber continuously. The helium at pressure completely stops and thermalizes the recoil nuclei. Special additives—aerosols—are introduced into the helium, which form macromolecules in the ionization zone, and these molecules absorb the reaction products and transport them, together with the helium jet, through the capillary. The helium atoms and macromolecules are separated on a skimmer. Having escaped from the capillary, the helium atoms form a widely dispersing flux. Most hit the skimmer walls, recoil, and are then evacuated by a powerful forevacuum pump. When they leave the capillary, the macromolecules containing reaction products have the same velocity as the helium atoms. However, their cone of dispersion is very narrow, within 1-2°, because of their large masses. The macromolecules pass the skimmer hole and enter the ion-source chamber. The gap between the ion source and the skimmer is evacuated by a diffusion pump.

In this system the ion source fulfils two functions. It destroys the macromolecules to release the atoms of the reaction products and ionizes them. A high-voltage electrode extracts the ions from the source and gives them the velocity required for the separation.

The operating speed of this combined ISOL system depends on the time required for the aerosols containing the reaction products to be extracted from the gas volume, the length of time the aerosols spend in the capillary tube, the holdup time in the ion source and, finally, the time required for them to pass through the ion-optical system of the isotope separator. The main delay is associated with the time required to transport the macromolecules from the stopping volume and with the holdup time of reaction products in the ion source. Several attempts have been made to reduce these times. For instance, several extracting capillaries have been placed along the length of the gas volume in which recoil nuclei are stopped, and the size of the ion-source chamber has been decreased (Mo 80). As a result, isotopes with half-lives of up to 0.1 s (Äy 81) have been separated and identified.

Table 1 (Äy 81) gives some idea of the efficiency of this combined system, which is of the order of 0.01-0.02%. An increase in the ion-source temperature enhances the efficiency, especially for the separation of the isotopes of rare-earth elements.

The combination of a gas jet with an ion source in an ISOL system considerably reduces the time required for isotope analysis and extends the range of the elements

**Table 1. RAMA Efficiencies (Mo 80)**

| | $^{20}$Na (%) | $^{111}$Te (%) | $^{150}$Dy (%) | $^{211}$At (%) |
|---|---|---|---|---|
| He jet | 20[a] | 60 | 10[a] | 15[a] |
| Skimmer | 70 | 60 | 60 | 70 |
| Ion source | 0.23 | 0.12 | ~0.2 | 0.2 |
| Magnetic analysis[b] | 50 | 50 | 50 | 50 |
| Overall | 0.016 | 0.02 | 0.01 | 0.01 |

[a] Not optimized.
[b] Calculated based on measured ion-source emittance.

that can be analyzed, first of all, in the rare-earth region. At the same time, we cannot help but see that only one of the elements of an ISOL system, namely, the solid-state catcher which limits the operating speed of the system and its versatility, has been eliminated in a combined system. The second limiting element, the ion source, remains in the combination practically unchanged.

The purpose of the ion source is to re-ionize the reaction products, which during the thermalization are transformed from positively charged ions to neutral atoms. Whether it is still possible to use the ability of recoil nuclei to escape from a target in the form of positive ions so that they can be accelerated in the electric field of the isotope separator remains an open question.

### *2.2.6. An Isotope Separator in Conjunction With an Ion Guide*

The fascinating idea of extracting thermalized recoil nuclei as positive ions with a gas jet for a subsequent isotopic analysis was realized by Ärje *et al.* at the University of Jyväskylä, initially only for the $\alpha$-decay products of heavy elements (Är 81). The method was employed rather successfully to separate the products of reactions induced by light projectiles, protons, and $\alpha$ particles (Är 87). Recently work has been underway to extend the method to heavy-ion reactions. Facilities based on this technique are called Ion Guide Isotope Separation On-Line abbreviated as IGISOL (Är 87).

The initial part of an IGISOL facility for light projectiles is presented in Figure 14.

A thin target is placed inside a small chamber, into which helium is fed under pressure. The helium completely stops the recoil nuclei and thermalizes them. The recoil nuclei emerge from the target as positive ions with charges depending on their velocity. Once stopped and thermalized they collide repeatedly with helium atoms, most of which are not ionized. Since the ionization potential of the helium atoms is high (26 eV), the charge state of the recoil nuclei ceases to change once they reach the single-charge state. The capture of an elelctron from a helium atom by a singly charged ion would require a considerable amount of energy in terms of thermal motion.

The target chamber has a small hole through which the helium flows out into a low-pressure region, carrying with it the singly charged ions of the recoil nuclei. A skimmer, to which a small potential is applied to provide the extraction of positive ions, is placed in front of the hole to the target chamber. When neutral helium atoms hit the skimmer, they recoil away and are exhausted by a powerful forevacuum pump.

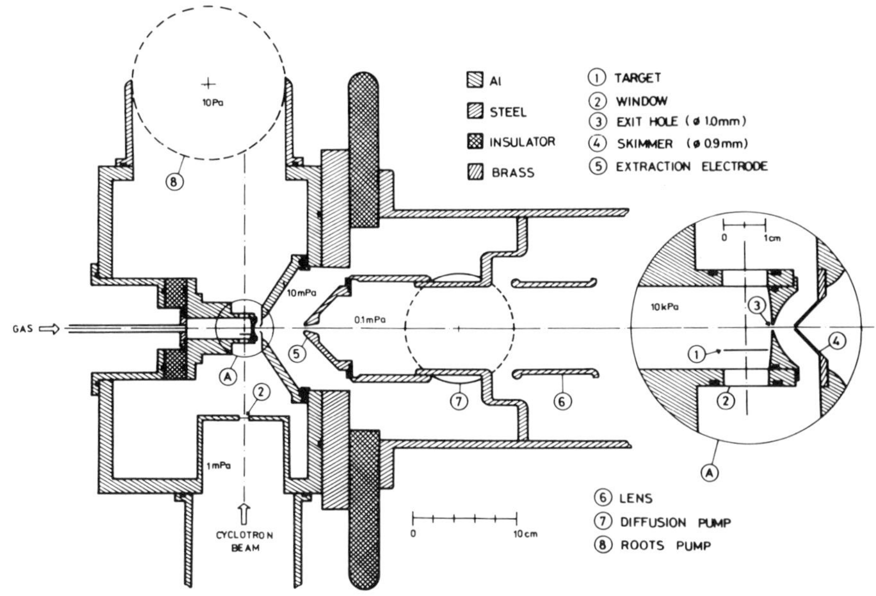


**Figure 14.** Sectional view of the initial part of the IGISOL system (Är 87).

The ions of reaction products are directed by the electric field to the skimmer hole and enter the accelerating electric field of a high-voltage electrode. The skimmer and the high-voltage electrode together form an electrostatic lens along the axis of which lie the centers of the holes in the source chamber, in the skimmer, and the high-voltage electrode. The accelerated ions are sent into a magnetic separator where they are separated according to their masses.

The rapid operation of the IGISOL system and its efficiency in separating the products of light-projectile reactions appear to be rather promising.

Figure 15 shows some data on the overall efficiency of an IGISOL system used to separate a number of nuclides as they decay from short-lived isomeric states. The

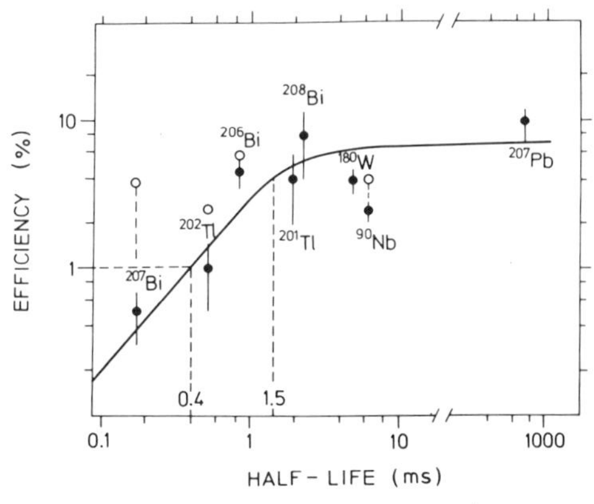


**Figure 15.** Overall transmission efficiency for some heavy nuclides in their isomeric states as a function of half-life. Error bars are due to statistical uncertainties in the measurements. Open circles correspond to efficiencies corrected with respect to the evacuation time of the target chamber for $^{207}$Bi, $^{202}$Tl, and $^{206}$Bi, and to stopping efficiency for $^{90}$Nb. The reactions were proton induced at energies of 16–20 MeV (Är 87).

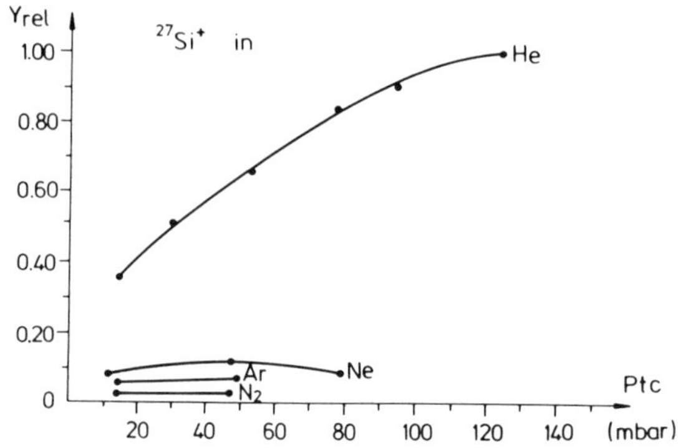


**Figure 16.** Relative yields of the $^{27}$Si recoil ions thermalized in helium, neon, argon, and nitrogen buffer gases. The reaction was p + $^{27}$Al at 16 MeV (Är 87).

efficiency is given as a function of the half-life of the isomer. The isomeric states were excited by bombarding targets with 16–18 MeV protons (Är 87). For half-lives exceeding 2 ms the efficiency approaches 10%. The total transport time in the IGISOL facility can be estimated from the fall in activity of the shorter-lived isomers. It amounts to about 1.5 ms, that is, close to the calculated time for the evacuation of an ion from the target chamber. Note that for a high-melting element, tungsten, the efficiency was almost the same as it was for the more volatile elements Tl, Pb, and Bi.

Figure 16 shows how the system efficiency was affected by the helium pressure and the contents of other gases. The (p, n) reaction product $^{27}$Si was separated by bombarding $^{27}$Al with 16 MeV protons (Är 87). If gases with lower ionization potentials than that of helium are used, the efficiency falls, even though these gases have better stopping powers.

Heavy ions introduce two unfavorable factors for this method. First, the recoil nuclear ranges for heavy ions are considerably longer than they are for light particles. This necessitates significantly bigger target chambers, inevitably reducing the operating speed and efficiency of the system. Second, the higher ionization density produced by heavy ions in the target chamber may neutralize the thermalized ions of the recoil nuclei.

However, the advantages of the IGISOL technique, i.e., its rapidity of operation and the ability to separate isotopes of elements with high melting points and low volatilities, are so attractive that IGISOL facilities are being built for heavy-ion beams in several laboratories. Experience with these facilities will indicate directly whether the IGISOL technique will be widely used to separate the nuclei far from stability produced by heavy-ion reactions.

### 2.2.7. *Kinematic Separators*

Now we shall consider the system called kinematic separators. In these systems the recoil nuclei are isolated due to differences between the kinematic characteristics of the reaction products and the projectiles.

Kinematic separators employ the ability of complete-fusion products to emerge at 0° in a very narrow solid angle. This means they can be separated using magnetic and electric fields and efficiently collected on detectors. The flight time in the analyzing system, even if its length is several meters, is of the order of one microsecond. If the recoil nuclei arrive directly at the detectors, the chemical properties of the element do not affect the efficiency and speed of its separation.

At the same time, there is a serious technical problem in separating the reaction products from the bombarding particles. In many of the experiments for producing nuclei far from stability the heavy-ion beam intensity exceeds $10^{12}$ particles per second. Detector loads at the level of $10^8$ p/s are close to the maximum permissible. When separating products with small formation cross-sections, which are typical for nuclei far from stability, the lowest possible loads are desirable.

The most efficient method for isolating complete-fusion products from the heavy ions is to use crossed electric and magnetic fields in a Wein velocity filter. This can spatially separate the trajectories of the recoil nuclei and projectiles. This principle is employed in the well-known SHIP facility (Selector Heavy-Ion Reaction Products), which is installed on a heavy-ion beam from the UNILAC at GSI, Darmstadt (Mü 81). A schematic diagram of the facility is given in Figure 17.

The complete-fusion products knocked out of the target pass through a thin carbon foil and a limiting aperture. The foil unifies the charge distribution of the reaction products and the aperture limits the angular emittance of the particles. A triplet of quadrupole lenses focuses the particles.

At the first stage the reaction products and projectiles are separated in the electric field. Since the reaction products have lower electric rigidities, they are deflected through larger angles than are the heavy ions. Two magnetic dipoles then increase the spatial separation of the projectiles and the recoil nuclei. The dispersion reaches a maximum value between the first and second pairs of the magnetic dipoles, where a variable-aperture diaphragm is placed to separate the recoil nuclei from the heavy ions. The deflected heavy ions hit a beam collector while the reaction products reach the second part of the facility. This part, too, contains magnetic and electric elements, but with the directions of the magnetic and electric fields reversed. In this way the reaction products are focused onto the detectors and are additionally decontaminated from the heavy ions and scattered particles.

The decontamination factor depends on the mass ratio between the projectile and the target nucleus and is between $10^{12}$ and $10^7$. The first figure relates to Ar-induced reactions and the second to reactions induced by Xe ions.

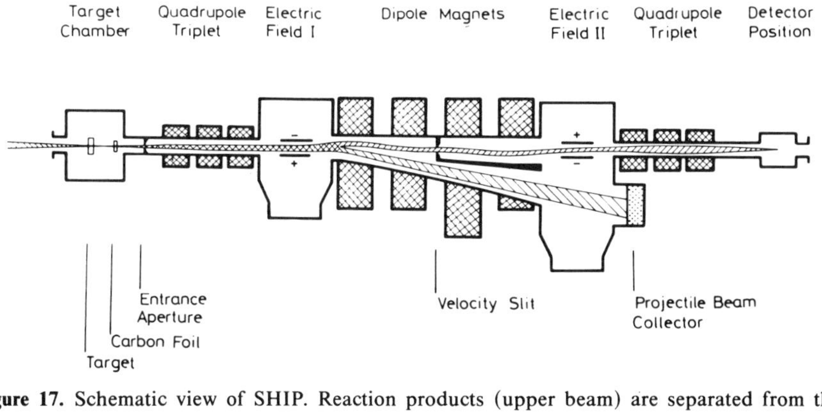


**Figure 17.** Schematic view of SHIP. Reaction products (upper beam) are separated from the projectile (lower beam) and focused in the detector position (Mü 79).

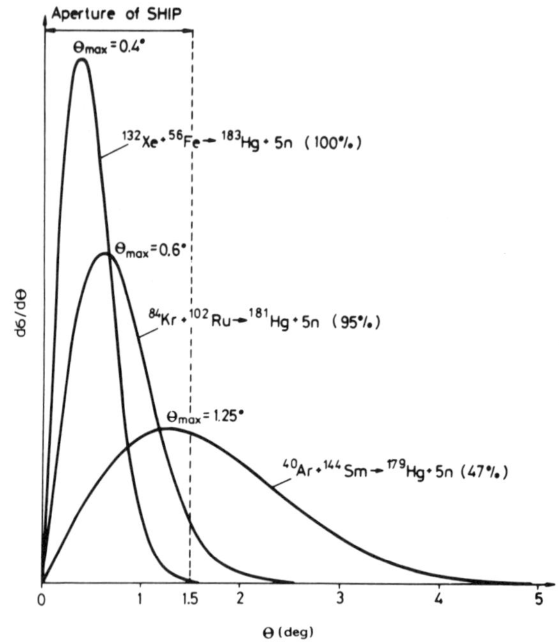


**Figure 18.** Angular distribution of the evaporation residues for different reactions. The fraction within the aperture of SHIP is given in parentheses. $\theta_{max}$ is the position of the maximum of $d\sigma/d\theta$. Target thickness is 0.1 mg/cm$^2$ (Mü 79).

The efficiency with which the recoil nuclei are collected on the detectors also depends on the ratio between the projectile and target nucleus masses (Figure 18). It is also affected by the nature and number of the light particles emitted by an excited compound nucleus. For ions with masses larger than that of the target nucleus, the recoil collection efficiency reaches 100% if neutrons alone are evaporated. If $\alpha$ particles are evaporated, the efficiency falls severalfold because the angular divergence of the recoil nuclei trajectories is increased. The path length in SHIP is 11 m, which yields times of flight between 0.5 and 3 $\mu$s depending on the masses of the target nucleus and the projectile. The principal characteristics of the SHIP facility are given in Table 2.

In order to identify a reaction product with $A < 100$ the time of flight (TOF), specific ionization $\Delta E$, and total energy $E$ are measured. The 11 m between the target and detector is the path length. One of the time signals is taken from the high-frequency micropulses of the UNILAC. The second time signal is produced by the secondary electrons which are knocked out of a thin carbon foil when a reaction product passes through it. Quantity $\Delta E$ is measured using a thin detector, either gas-filled or semiconductor, and $E$ is measured by a total absorption semiconductor detector. The $Z$ and $A$ resolutions are $Z/\Delta Z \sim 50$ and $A/\Delta A \sim 130$, respectively.

Heavy neutron-deficient nuclei produced in complete-fusion reactions can be identified by detecting the $\alpha$ decay of nuclides with $Z > 70$ and $A > 150$. Recoil nuclei are implanted into position-sensitive semiconductor detectors. The place where a recoil nucleus hits a detector is established and the energy release is the signal for $\alpha$-particle

**Table 2. Specification of SHIP (Mü 79)**

| | |
|---|---|
| *Target* | |
| Diameter | 15 mm |
| Beamspot diameter | (3–10) mm |
| Thickness | (0.1–1) $mg/cm^2$ |
| *Acceptances* | |
| Entrance aperture | $52 \times 52\ mrad^2$ |
| Solid angle | 2.7 msr |
| Relative charge width | 20% |
| Relative velocity width | 10% |
| Efficiency of the filter for these acceptances generally | >30% |
| *Quadrupoles* | |
| Aperture radius | 7.5 cm |
| Ion length | 25 cm |
| Maximum field gradient | 9.5 T/m |
| $B\rho_{max}$ | 1.2 Tm |
| *Electric field* | |
| Maximum deflection voltage | 600 kV |
| Gap | 15 cm |
| $F\rho_{max}$ | 25 MV |
| *Dipolemagnets (C-Type)* | |
| Maximum field | 0.75 T |
| *Other parameters* | |
| Overall background rejection | $10^{12}$–$10^{7}$ |
| Rejection background outside allowed velocity | $>10^{16}$–$10^{7}$ |
| Separation time | (0.5–3) $\mu s$ |
| fwhm of beamspot | $10 \times 30\ mm^2$ |
| Active detector area | (450–1600) $mm^2$ |
| Velocity dispersion | 2.2 mm/% |
| Velocity resolution | <2% |

observation. The spatial and time correlation of $\alpha$ particles are used to re-establish the chains of $\alpha$ decays.

These chains begin with the $\alpha$ decay of a nucleus far from stability and are completed by the $\alpha$ decay of a known isotope. By ascending the $\alpha$-decay chain the $Z$ value of the initial nucleus can be determined and the $\alpha$-particle energies involved in the $\alpha$-decay chain permits its mass estimate (Ho 81).

### *2.2.8. Gas-Filled Mass Separators*

Another system which separates recoil nuclei and projectiles without stopping them is a gas-filled mass separator.

The idea was first suggested in 1958 by Cohen and Fulmer (Co 58), who used it to study the charge distribution of fission fragments. Later, Armbruster *et al.* (Ar 61,

Ar 71d, Si 75) carried out a detailed investigation of the technique and built a major facility for the Julich reactor (FRG) in order to study the charge distribution of fission fragments and the nuclear spectra of the short-lived neutron-rich isotopes produced by fission.

The first on-line gas-filled mass separator for a heavy-ion beam was constructed by Karnaukhov and co-workers at the Laboratory of Nuclear Reactions at Dubna (Ka 69, Ba 70). It was used to study the radioactive decay of the neutron-deficient isotopes produced by complete-fusion reactions.

The principle of operating a gas-filled mass separator can be described in the following way. Imagine that a target to be bombarded by a heavy-ion beam is placed at the entrance focus of a magnetic analyzer. The recoil nuclei—complete-fusion products—have a rather broad (several tens of percent) ionic-charge distribution and a considerable spread of velocities. The latter is due to the energy losses in a target with a finite thickness. After passing through a conventional magnetic analyzer with a vacuum chamber, the reaction products will hit different points of the focal plane in accordance with their magnetic rigidities. Because of the spread in the ionic charge and velocity, the recoil nuclei will be scattered over the focal plane.

The situation is very different if the vacuum chamber of the magnetic analyzer is filled with gas at a low pressure. Due to collisions with the atoms or molecules of the gas, the ionic charge of the reaction product will change continuously. If the energy losses in the gas are only a small fraction of the kinetic energy of the reaction product, its velocity will remain practically constant along the whole trajectory. The ionic charge of the reaction product $q$ will therefore fluctuate around some average value $\bar{q}$, the magnitude of which for a given gas filling is determined by the product velocity $v$ and atomic number $Z$. To a first approximation we have

$$\bar{q} = (v/v_0)Z^{1/3} \tag{2.4}$$

where $v_0$ is the velocity of an electron in a hydrogen atom, $v_0 = c/137$.

The magnetic rigidity of the reaction product $B\rho$ can then be written as

$$B\rho(T\cdot m) = 0.0227A(v/v_0)\bar{q}^{-1} \sim AZ^{-1/3} \tag{2.5}$$

It can be seen from this equation that the magnetic rigidity of a complete-fusion product does not depend on either its initial ionic charge or its initial velocity. In other words, a gas-filled mass separator has the rather valuable property of being able to focus recoil nuclei according to their charge and mass number.

The mass resolution of a gas-filled mass separator is mainly determined by ionic-charge fluctuations during the charge-exchange process and by the multiple angular scattering of the ions on the atoms (molecules) of the filler gas. In reality, the relationship between $\bar{q}$ and $v$ is more complicated (Gi 86),

$$\bar{q}/Z = 1 - C_1 \exp[-C_2(v/v_0)Z^{-2/3}] \tag{2.6}$$

where $C_1$ and $C_2$ are parameters deduced by fitting the experimental data to equation (2.6). For $C_1 = C_2 = 1$ a first-order expansion of the exponential gives the expression of equation (2.4). In approximation (2.6) the velocity spread of the recoil nuclei contributes to the resolving power of the separator.

Figure 19 shows calculated and experimental data on the resolving power of a gas-filled separator, the SASSY (Small Angle Separating System) and the contributions to it from various components (Gh 86). We can see that the smallest image of the ion

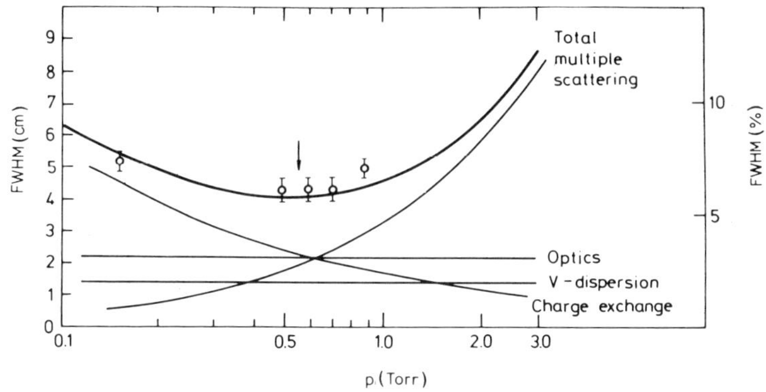


**Figure 19.** Calculated and experimental widths of the distribution, at the focal plane, for the evaporation residues with $Z = 84$ and $v/v_0 = 2.6$ from the reaction $^{40}Ar + ^{164}Dy$. The arrow denotes the minimum values for the width (Gh 86).

source is obtained at a pressure of 0.5 Torr, which corresponds to mass resolution $A/\Delta A \sim 15$.

Gas-filled mass separators are very rapid in operation: the time of flight of the recoil nuclei from the target to the detector is of the order of one microsecond. Its efficiency reaches 20%, and the decontamination factor from the beam is about $10^{15}$. However, the mass resolution of a gas-filled mass separator is too small for the unambiguous identification of reaction products, especially for nuclei of heavy elements. Therefore, the time-of-flight technique is used to determine the mass number of the reaction product. The atomic number of a fairly heavy reaction product is established by detecting the $\alpha$-decay chain of related isotopes. Figure 20, which shows a schematic diagram of the SASSY, gives some idea of the main elements of a gas-filled mass separator (Gh 86).

Interest in the gas-filled mass separator has grown recently because of its high efficiency, the independence of the efficiency from the chemical properties of the reaction products, its high operating speed, the high decontamination factor, and its relative simplicity.

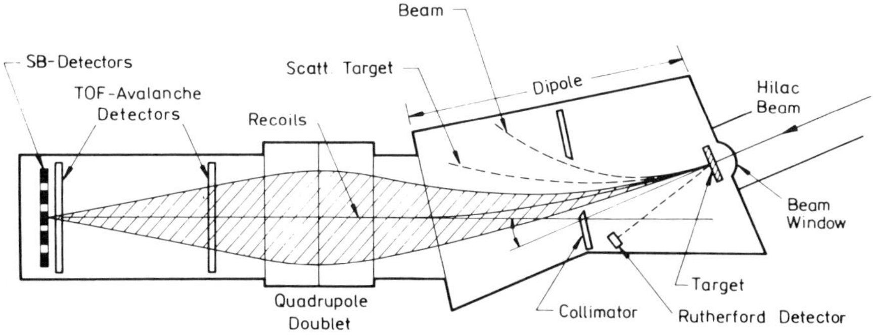


**Figure 20.** Schematic representation of SASSY. Evaporation residues move in the hatched area. Target and beam trajectories are shown for heavy target recoils and light beam (Gh 86).

## 2.3. New Processes and Nuclides Observed in Complete-Fusion Reactions

### *2.3.1. New Modes of Radioactive Decay*

*2.3.1a. Delayed Proton Emission.* As the neutron deficit grows, the difference between the neutron and proton binding energies, $B_n$ and $B_p$, in the nucleus increases. Correspondingly, the $\beta^+$ decay energy also increases. The $Q_\beta$ value may exceed $B_p$ in nuclei far from the region of stability, and as a consequence a daughter nucleus after $\beta^+$ decay may have a higher excitation energy than the proton binding energy. Such a nucleus will emit a proton with a high probability, the proton emission taking very little time. However, the rate of proton activity falls exponentially with the half-life of the preceding $\beta^+$ emitter. This phenomenon has been called delayed proton emission.

The first delayed proton emitter, the $^{17}$Ne nucleus, was discovered in 1962 by Karnaukhov, Ter-Akopian, and Subbotin at the JINR Laboratory of Nuclear Reactions, after bombarding a natural nickel foil with 130 MeV $^{20}$Ne ions (Ka 62, Ka 63, Ka 63b, Fl 64). Subsequently, Karnaukhov with co-workers discovered and investigated a considerable number of delayed proton emitters produced in complete-fusion reactions (Bo 76). The projectiles were $^{32}$S ions at energies of 180–190 MeV and enriched isotopes $^{92}$Mo, $^{96}$Ru, $^{102}$Pd, $^{106}$Cd, and $^{112}$Sn were the targets. An ISOL system with an ion source operated in the surface thermal-ionization mode was used to separate the isotopes (Ka 74).

Delayed proton emission is typical for neutron-deficient isotopes that lie far from stability. Studies of this phenomenon and of the energy spectra of delayed protons in particular can be used to obtain information about the level densities of excited nuclei, the $\beta$-decay strength functions, and the masses of nuclei far from stability.

*2.3.1b. Delayed Fission.* A heavy neutron-deficient nucleus, having undergone $\beta^+$ decay or K capture, may obtain an excitation energy comparable to the fission barrier and then divide. This process is termed delayed fission, by analogy with delayed proton emission. Delayed fission was discovered in 1966 by Kuznetsov, Skobelev, and Flerov (Ku 66, Ku 66b), at the JINR Laboratory of Nuclear Reactions. In complete-fusion reactions accompanied by neutron emission, namely $^{209}$Bi($^{22}$Ne, 3n), $^{230}$Th($^{10}$B, 8n), and $^{230}$Th($^{10}$B, 6n), the neptunium and americium neutron-deficient isotopes—$^{228}$Np, $^{232}$Am, and $^{234}$Am—were produced and underwent fission after electron capture. The half-lives of delayed fission were determined by the electron-capture periods of the parent nuclei, and for the above daughter isotopes they were 1.0, 1.0, and 2.6 min, respectively (Ku 66, Ku 66b, Ku 67).

Several emitters of delayed fragments were later produced in complete-fusion reactions involving neutron evaporation (Og 85, So 77, Ga 80).

We can conclude from the systematics of $\beta^+$-decay energies (K capture) and fission barriers that delayed fission must be quite a common phenomenon for heavy neutron-deficient nuclei (Be 69). However, the production of delayed fragment emitters presents a complicated problem. The predominance of the fission channel in the deexcitation of a heavy compound nucleus decreases the yield of the emitters of delayed fragments, especially if a considerable number of neutrons is emitted. The probability of fission following K capture is some $10^{-7}$–$10^{-4}$. In general, the formation cross-sections for delayed fragment emitters are about $10^{-33}$ cm$^2$ in magnitude and this naturally makes the detailed study of this process difficult. At the same time, delayed fission can provide unique information about the fission barriers of heavy nuclei far from the region of stability, information inaccessible to other methods of research (Og 85).

*2.3.1c. Proton Emission from the Ground State.* In the presence of a very great neutron deficit the proton binding energy in the nucleus becomes negative, thus making proton emission from the ground state possible. The proton in such a nucleus is held by the Coulomb barrier alone, the situation being analogous to $\alpha$ decay.

In the 1950s and 1960s Djelepov (Dj 51), Karnaukhov and Tarantin (Ka 60), and Goldansky (Go 60) made the first theoretical estimates of the possibility of proton decay from the nuclear ground state. However, about two decades passed before the new mode of radioactive decay was reliably detected.

The production of sufficiently intense beams of heavy ions, such as $^{58}$Ni, and the construction of rapid sensitive facilities that can reliably separate and identify reaction products with small production cross-sections were crucial.

Proton decay from the nuclear ground state was discovered in 1981 at GSI, Darmstadt, simultaneously by the groups of Hofmann (Ho 81, Ho 82) and Roeckl (Kl 82b). Both groups used complete-fusion reactions and $^{58}$Ni ions. They chose the lightest stable isotopes of ruthenium and molybdenum, i.e., $^{96}$Ru and $^{92}$Mo, for the targets. The proton emitters were separated and identified using the SHIP facility and an on-line isotope separator operated at the UNILAC accelerator.

The fusion of the $^{96}$Ru and $^{58}$Ni nuclei led to the formation of the compound nucleus $^{154}$Hf with an excitation energy of 49 MeV. This nucleus contains 20 fewer neutrons than the lightest hafnium isotope $^{174}$Hf, and 26 fewer neutrons than the most abundant hafnium isotope $^{180}$Hf. The evaporation of one proton and two neutrons from $^{154}$Hf leads to the formation of the $^{151}$Lu nucleus, which emits protons from the ground state. The proton energy was $1231 \pm 3$ keV and the half-life of the emitter was $85 \pm 10$ ms (Ho 81, Ho 82) (see Figure 21a).

The $^{58}\text{Ni} + {}^{92}\text{Mo}$ reaction produced the compound nucleus $^{150}$Yb with an excitation energy of 51 MeV. This nucleus contains 18 fewer neutrons than the lightest stable isotope $^{168}$Yb, and 24 fewer neutrons than the most abundant isotope $^{174}$Yb. After the emission of a proton and two neutrons the $^{147}$Tm nucleus is left, and this decays from the ground state with a half-life of $0.42 \pm 0.10$ s by emitting protons with an energy of $1117 \pm 6$ keV (Kl 82b). Later, more accurate measurements of the half-life yielded the value $560 \pm 40$ ms (Figure 21b).

In both reactions the protons were identified using a $\Delta E$-$E$ telescope. The mass of the proton emitter $^{147}$Tm was established using a mass separator. The proton emitter $^{151}$Lu was identified in the velocity selector SHIP by measuring the excitation functions and angular distributions of the reaction products, and by using cross reactions. The delayed protons were eliminated by detecting the proton-positron and proton-X ray coincidences.

A comparison of the calculated and experimentally observed half-lives indicated a hindrance factor of about 100 for proton emission. This was interpreted as being due to the considerable change in the angular momentum involved in the proton decays: $^{151}\text{Lu}(11/2^-) \to {}^{150}\text{Yb}(0^+)$ and $^{147}\text{Tm}(11/2^-) \to {}^{146}\text{Er}(0^+)$ (Ho 87).

Proton emission from the ground state was later observed in the neutron-deficient isotopes of cesium and iodine (Ho 87): $^{113}\text{Cs}(5/2^+) \to {}^{108}\text{Te}(0^+)$, $E_p = 958 \pm 4$ keV, $T_{1/2} = 33 \pm 7\ \mu$s; and $^{109}\text{I}(5/2^+) \to {}^{108}\text{Te}(0^+)$, $E_p = 811 \pm 5$ keV, $T_{1/2} = 109 \pm 17\ \mu$s. Of the heavy nuclides, the neutron-deficient isotope of bismuth, $^{185}$Bi, is expected to be a

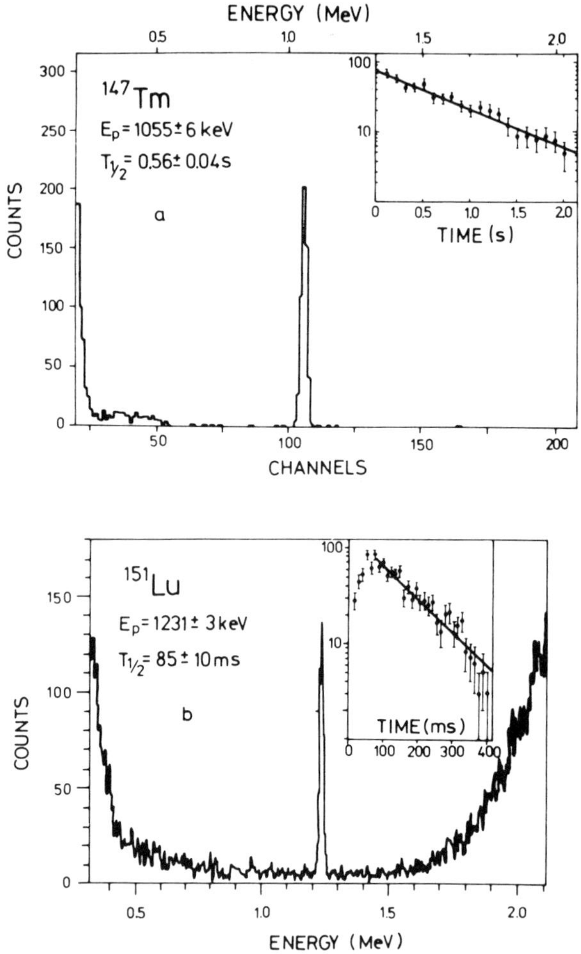


**Figure 21.** (a) Energy spectrum obtained at mass 147 from investigating $^{56}Ni + ^{92}Mo$ reactions at the GSI on-line mass separator; the 1055-keV proton line is assigned to the direct decay of 0.56-s $^{147}Tm$. (b) Energy spectrum obtained from $^{58}Ni + ^{96}Ru$ reactions at the velocity filter SHIP; the 1231-keV proton line is assigned to the direct decay of 85-ms $^{151}Lu$ (Ro 83).

proton emitter with a detectable half-life (Ho 87). The studies of proton decay from the ground state produce valuable information about the nucleon states of nuclei very far from the region of stability.

### 2.3.2. *Alpha Decay of Neutron-Deficient Isotopes of Elements up to Lead*

A large number of the new neutron-deficient isotopes of elements below lead and which decay by $\alpha$-particle emission were produced by complete-fusion reactions induced by $^{58}Ni$, $^{92}Mo$, and $^{107}Ag$ ions. These studies were conducted by Armbruster,

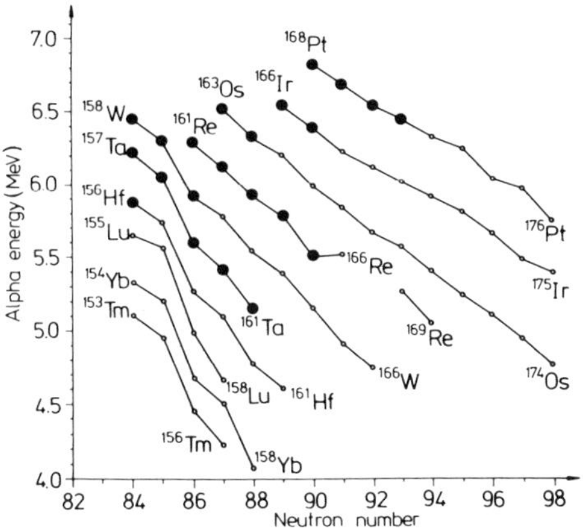


**Figure 22.** Alpha systematics of neutron-deficient isotopes of elements from Tm to Pt. The new isotopes are marked as bigger dots (Ho 81).

Hofmann, and Münzenberg *et al.* (Ho 79, 81, 81b). Elements between Zr and Sn, Fe and Mo, and V and Ni were used as target materials. The experiments were carried out at the UNILAC at GSI, Darmstadt. The SHIP velocity filter was used to separate the products of complete-fusion reactions. The new isotopes were identified by measuring the spatial and time correlations of related $\alpha$ decays in the decay chains (Mü 81).

In Figure 22 the new neutron-deficient isotopes which undergo $\alpha$ decay are marked by filled circles. There is a good correlation between the $\alpha$-decay energies of the new and previously known isotopes.

Six new $\beta$ emitters and two high-spin isomers decaying by $\alpha$ emission have also been identified (Ho 81, 81b).

The observation of $\alpha$ decay in the neutron-deficient nuclei of elements below lead can be used to obtain fairly accurate estimates of the masses of nuclei very far from stability. This is done by measuring the energies of the $\alpha$ particles involved in the $\alpha$-decay chains leading to an isotope with known mass.

The reduced $\alpha$-decay widths are closely related to the shell structure of the nuclei and carry information about the effects of closed shells and subshells in different regions of the $Z$ and $N$ of the nucleus (Figure 23).

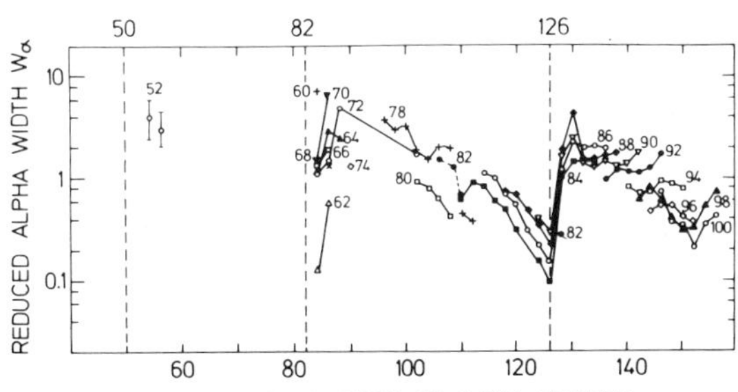


**Figure 23.** Reduced width $W$ for s-wave $\alpha$ decay of even nuclei as a function of the neutron number of the emitter (Ro 83).

#### *2.3.3. En Route to the Doubly Magic Nucleus $^{100}Sn$*

It is known that stable isotopes include five doubly magic nuclei, viz. $^{4}$He, $^{16}$O, $^{40}$Ca, $^{48}$Ca, and $^{208}$Pb, in which the proton and neutron shells are filled completely. New doubly magic nuclides can only be produced by going outside the stability region.

Light-projectile induced reactions have resulted in the doubly magic nucleus $^{56}$Ni (28 protons and 28 neutrons), which lies near the region of stability. It was produced in the reaction $^{54}$Fe ($\alpha$, 2n) in the early 1950s.

The doubly magic nucleus of tin $^{132}$Sn, which contains 50 protons and 82 neutrons, is ten neutrons away from the most stable isotope of tin, i.e., $^{122}$Sn. Nevertheless, this nucleus was produced by the thermal-neutron fission of uranium in the mid-50s. Then followed a long interval.

Of the other doubly magic nuclei, $^{100}$Sn is of special interest. It is the heaviest doubly magic nucleus with equal numbers of protons and neutrons that might reasonably be produced experimentally. The $^{164}$Pb nucleus lies beyond the limits of nucleon stability. The only possible way of producing $^{100}$Sn is the use of the complete-fusion reactions induced by heavy ions. It is, however, a difficult experimental problem. The production cross-section of this exotic nucleus, which has 20 fewer neutrons than the most abundant nucleus of tin, i.e. $^{120}$Sn, is rather small.

Roeckl with co-workers (Ro 83) conducted experiments to produce $^{100}$Sn using a mass separator at GSI, Darmstadt. So far, the attempts to produce this attractive nucleus have been unsuccessful. However, the experimentalists have approached it rather closely by producing the adjacent nuclides by complete-fusion reactions (Figure 24).

The study of the properties of these nuclides provides positive evidence for the strong effects of the closing of the proton and neutron shells at $Z = N = 50$. This can be seen in the $\alpha$ decay of tellurium isotopes with neutron numbers 54, 55, and 56 (see Figure 24), in the decrease in the value of the level density parameter $a$ for the nucleus $^{99}$Ag, and in the positive-parity level scheme of $^{96}_{46}Pd_{50}$. We hope that in the near future the doubly magic nucleus $^{100}$Sn will be produced in complete-fusion reactions and experimentalists will be able to begin their direct investigation.

## 3. Production of Nuclei Far From Stability in Multinucleon Transfer Reactions

### 3.1. First Steps

It was clear right from the start of investigations with heavy ions that complete-fusion reactions could be used to obtain nuclei far from stability. The situation as regards reactions involving the exchange of nucleons was more complicated. In the 1960s both experimentalists and theoreticians began paying attention to few-nucleon transfer reactions and particularly those involving the transfer of a single neutron. It was hoped that these reactions could be used to obtain information on the reduced neutron widths in the nucleus.

The first experiment in which many nucleons were observed to have been transferred was carried out at the JINR Laboratory of Nuclear Reactions in 1963 by Volkov and his co-workers (Gr 63). A large number of rare-earth isotopes were separated

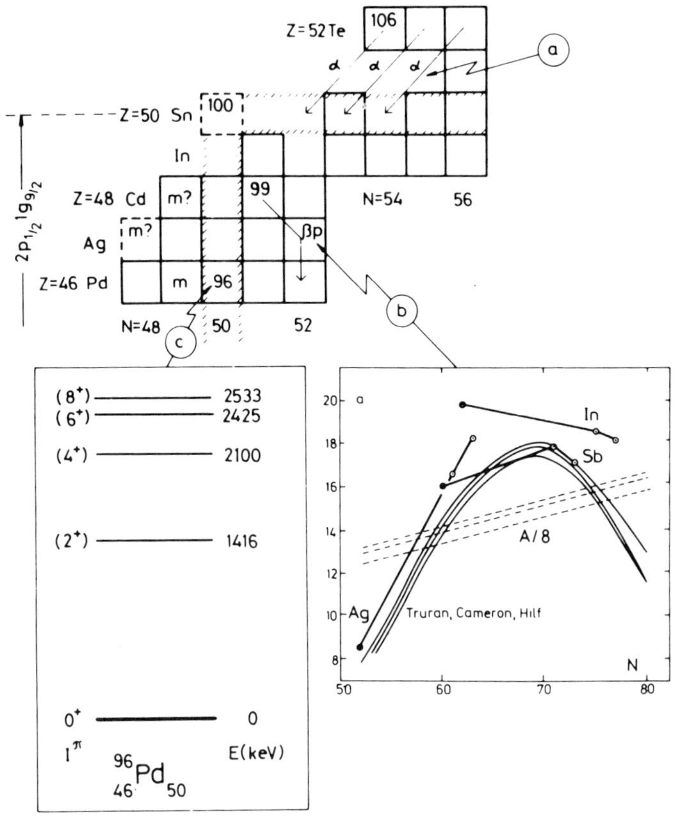


**Figure 24.** Approaches toward the structure of the doubly closed shell nucleus $^{100}_{50}Sn$ using (a) $\alpha$-decay data, (b) fluctuation analysis of $\beta$-delayed proton spectra, and (c) interpretation of positive-parity states in even $N = 50$ nuclei within a shell model. In the plot (b), the parameter $a$ of the Fermi-gas level density formula is shown as deduced from fluctuation analyses (full points), from neutron-resonance data (open points), and from semiempirical models (Ro 83).

radiochemically after the $^{181}Ta + ^{20}Ne$ (190 MeV) and $^{159}Tb + ^{22}Ne$ (160 MeV) reactions. These could only have been formed by the transfer from one nucleus to another of up to twenty nucleons. This showed the way for multinucleon transfer reactions for obtaining nuclei far from stability.

Volkov and his co-workers also studied reactions involving the pickup of one, two, and three neutrons, such as ($^{15}N, ^{16}N$), ($^{14}N, ^{16}N$), ($^{15}N, ^{17}N$), and ($^{14}N, ^{17}N$) (Fl 61, Vo 62, 62b, 63, Ar 66, 70, Po 66). Delayed neutrons from the decay of $^{17}N$ and energetic $\gamma$ rays from the decay of $^{16}N$ were recorded. It was established (see Figure 25) that the decrease in the reaction cross-section with the increasing number of neutrons picked up is moderate. The neutron-pickup cross-section increases with atomic and mass number of the target nucleus, which reflects a decrease in the neutron binding energy in the donor nucleus. This relationship between the cross-section and the neutron

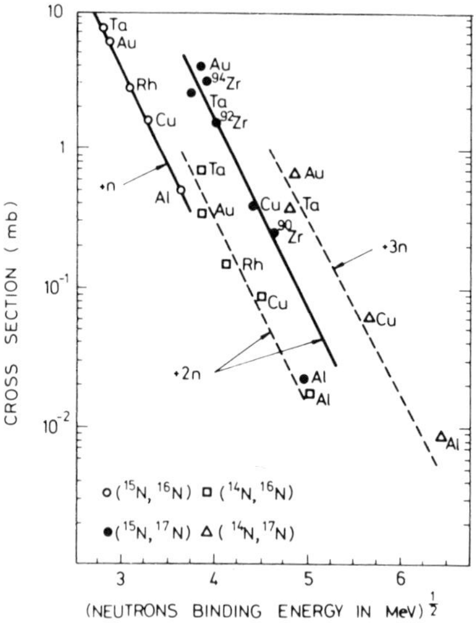


**Figure 25.** Systematics of the cross-sections for neutron-pickup reactions. The abscissa gives the square root of the binding energy of the neutrons transferred in the donor nucleus. Comparison of the cross-sections is made at an energy 30 MeV above the Coulomb barrier (Ar 70).

binding energy in the donor nucleus is even clearly apparent in the ($^{15}$N, $^{17}$N) reaction on separated isotopes of zirconium (Figure 26).

The excitation functions for reactions involving the pickup of neutrons have monotonically increasing cross-sections with ion energy (Figure 27). The neutron-rich isotope of nitrogen $^{17}$N is formed both by neutron-pickup reactions and by proton-stripping reactions ($^{20}$Ne, $^{17}$N), as well as by the exchange of neutrons and protons ($^{16}$O, $^{17}$N). The excitation functions for these reactions are similar in shape to those for neutron pickup and have close cross-section values (Figure 28).

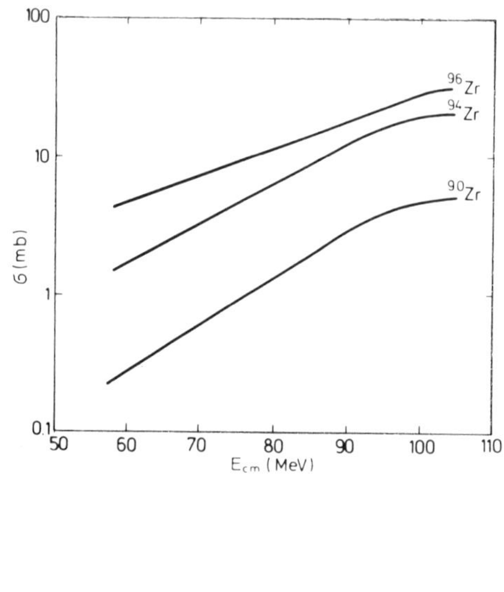


**Figure 26.** Cross-sections for a two-neutron pickup in the reaction ($^{15}$N, $^{17}$N) on $^{90}$Zr, $^{94}$Zr, and $^{96}$Zr targets (Vo 63).

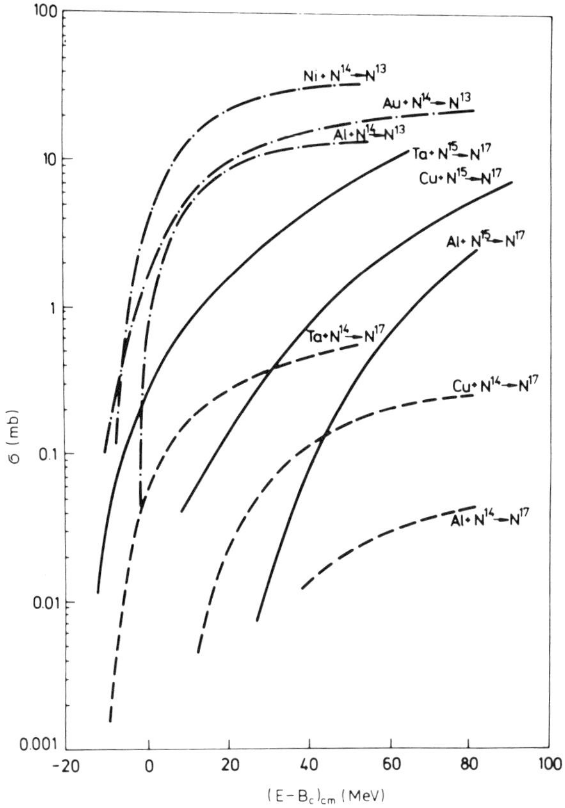


**Figure 27.** Excitation functions for reactions involving neutron stripping and the pickup of two and three neutrons. The ion energy is set above the Coulomb barrier (Vo 62b).

The angular distributions of neutron-pickup reactions are similar to those for quasi-elastic processes, the maximum of the differential cross-section lying in the region of the scattering angle for grazing nuclear collisions (see Figure 29).

Experimental data from neutron-pickup reactions strongly suggest their use in obtaining new light-element neutron-rich isotopes. The data also defined the conditions for the first experiment designed to obtain neutron-rich light-element isotopes. The experiment was carried out in 1969 by Artukh, Volkov, Gridnev, and Mikheev (Ar 69). The $^{232}$Th nucleus was chosen as the target and $^{18}$O as the projectile. The $^{18}$O(+3) ions were accelerated in the U-300 cyclotron to energies of 122 MeV. The products of the transfer reaction were detected at an angle of 60° to the beam, this being close to the Rutherford scattering angle for grazing collisions between thorium and oxygen. The new isotopes were identified using a magnetic spectrometer, in the focal plane of which was placed a telescope of $\Delta E$–$E$ semiconductor detectors.

The results of this first experiment are given in Figure 30. In addition to known isotopes, the following new neutron-rich nuclei were detected: $^{18}$C, $^{20}$N, and $^{22}$O (Ar 69).

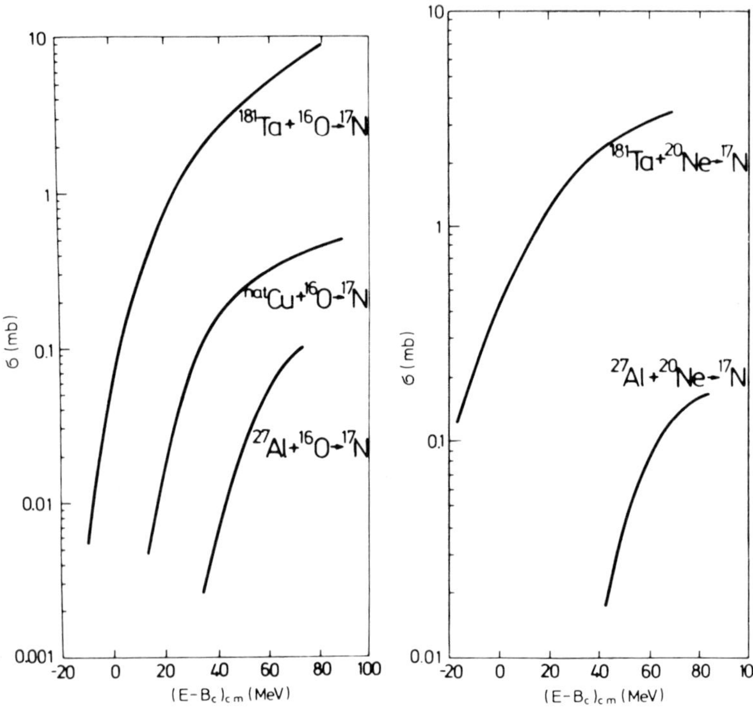


**Figure 28.** Excitation functions for the reactions ($^{20}$Ne, $^{17}$N) and ($^{16}$O, $^{17}$N). The ion energy is set above the Coulomb barrier (Vo 62b, 63).

## 3.2. Multinucleon Transfer Reactions and the Production of Nuclei Far From Stability

Multinucleon transfers are considered in more detail by Schröder and Huizenga in the chapter on "Damped Nuclear Reactions" (Sc 84). We shall therefore confine ourselves to those aspects which are significant for the production of nuclei far from stability.

### *3.2.1. Interactions between Nuclei in Multinucleon Transfers*

The reactions that occur between nuclei at energies above the Coulomb barrier are classified in Figure 31. We are interested in collisions with angular momenta above a critical value, $l > l_{cr}$, since these lead to multinucleon transfers.

There is a qualitative change in the nature of the interactions between the nuclei as the angular momentum is decreased from $l_s$ to $l_{cr}$. When $l \approx l_s$ the nuclei only briefly touch each other's surfaces, the kinetic-energy loss is small, and the contact time is of the order of $10^{-22}$ s. The interaction has the form of quasi-elastic scattering and is mainly governed by the Coulomb forces acting between the nuclei. During collisions with $l = l_{cr}$ the surfaces of the nuclei greatly overlap and the nuclei strongly interact.

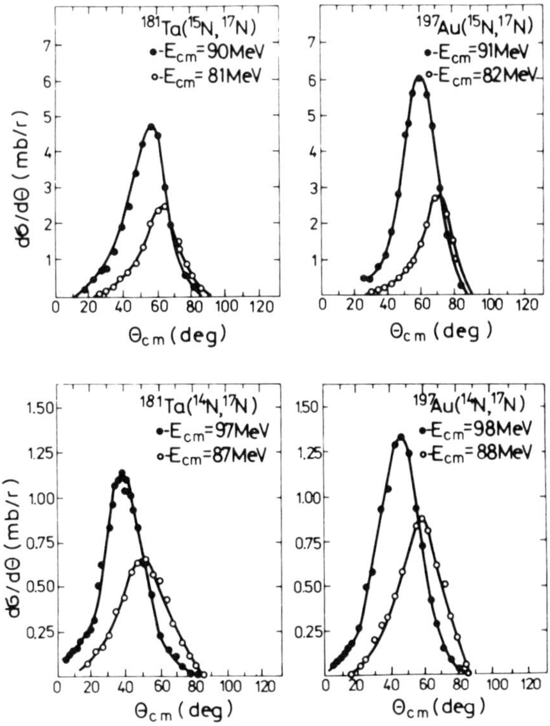


**Figure 29.** Differential cross-sections for reactions involving the pickup of two and three neutrons (Ar 70).

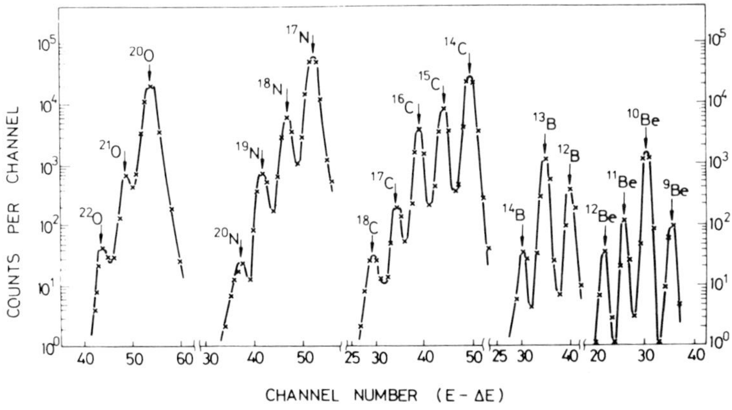


**Figure 30.** Yields of beryllium, boron, carbon, nitrogen, and oxygen isotopes from bombardment of a $^{232}Th$ target with 122 MeV $^{18}O$ ions (Ar 69).

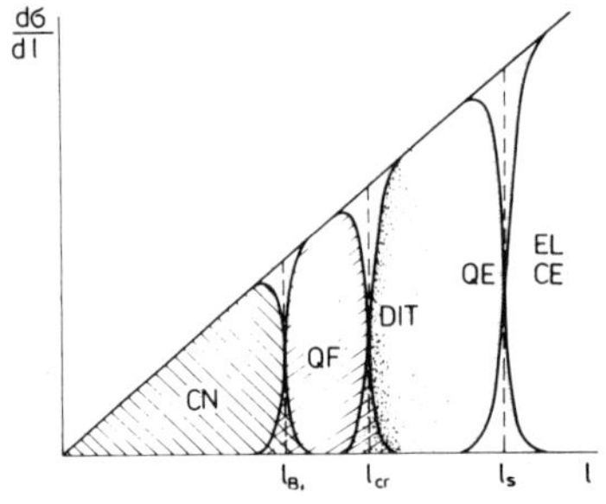


**Figure 31.** Schematic illustration of the $l$ dependence of the partial cross-section for different reactions: (CN)—compound nucleus, (QF)—quasi-fission, (DIT)—deep inelastic transfers, (QE)—quasi-elastic transfers, (EL)—elastic scattering, (CE)—Coulomb excitation.

The kinetic energy of the collision is completely dissipated and a dinuclear system is formed (Vo 74, 78). This sort of collision is called deep inelastic (Vo 75).

Two very different reaction mechanisms take place during quasi-elastic and deep-inelastic collisions. At intermediate angular momenta, the reaction mechanism may combine separate properties of both quasi-elastic and deep-inelastic collisions. As the angular momentum decreases and the kinetic-energy dissipation increases, the reaction mechanism gradually becomes less quasi-elastic and more deep-inelastic. Sometimes deep-inelastic and intermediate collisions combine in a single group of dissipative collisions (Go 80).

The equilibrium distributions of the thermal excitation energy and the neutron-to-proton ratio $N/Z$ are established fairly quickly during dissipative collisions between nuclei, of the order of $10^{-22}$ s. This property promotes the formation of very neutron-rich light-element nuclei during multinucleon transfer reactions in which a heavy-element target is bombarded by moderately heavy ions.

The mass-distribution equilibrium does not have enough time to become established during a dissipative collision. On the other hand, there is time for a considerable number of nucleons to be transferred from one nucleus to the other. A radical rearrangement is possible during a deep-inelastic collision.

The direction in which a system evolves is determined by the potential energy, which is a function of the system's charge (or mass) asymmetry and the angular momentum of the collision. Figure 32 gives some idea of the shape of the system's potential energy for several combinations of colliding nuclei, as given by Moretto and Schmidt (Mo 76).

The statistical nature of the nucleon exchange between nuclei means that there is a whole set of evolutionary pathways along which a system may move in the $Z,A$ space of the interacting nuclei. Any system configuration may be realized with some probability, as long as the energy, nucleon number, and charge are conserved.

In dissipative collisions the Coulomb and centrifugal forces acting between the nuclei exceed the nuclear-attraction forces and so the system disintegrates before an equilibrium configuration has been attained. The lifetime of the system depends on the angular momentum of the collision, falling with increasing $l$.

The system may decay at various stages of its evolution and the statistical character of nucleon exchange gives rise to many decay channels, which may yield dozens of different isotopes. This means multinucleon transfer reactions can be used to produce nuclei far from stability.

Differential production cross-sections for the isotopes from chlorine to helium are shown in Figure 33. These isotopes were formed as the products of multinucleon

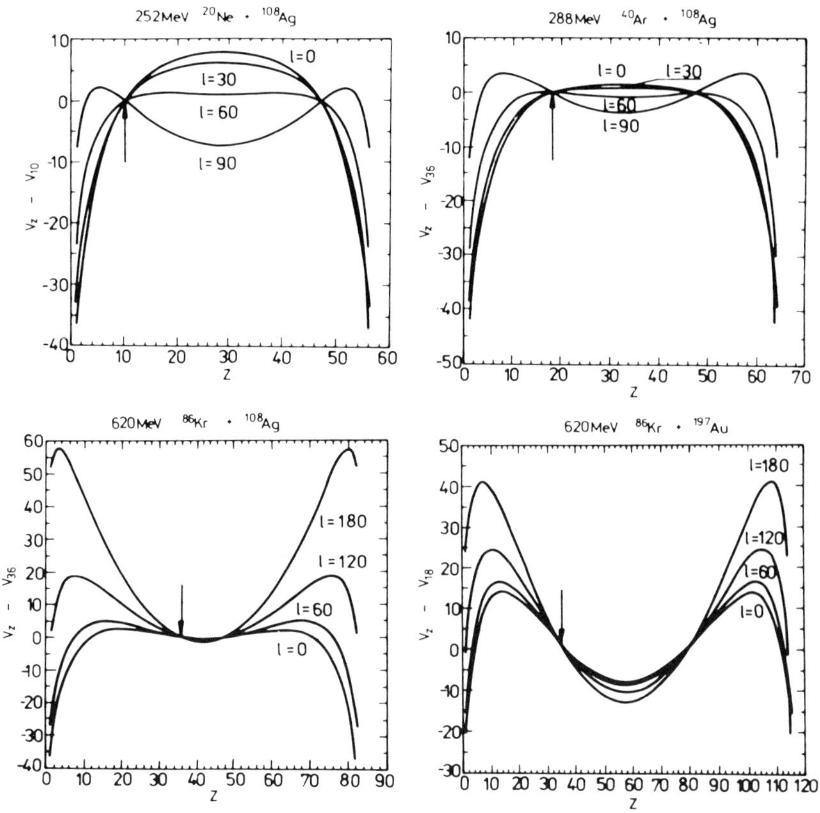


**Figure 32.** Potential energies of the dinuclear systems formed in various target-projectile combinations, as functions of the atomic number of a fragment and of the angular momentum of the collision. The potential energy of the initial system is taken as a reference point. The initial charge asymmetry is indicated by an arrow. The calculation is made for the liquid-drop mass (Mo 76).

transfers by bombardng a target of natural silver with $^{40}$Ar ions at an energy of 285 MeV (Ar 81). The spectra were recorded at 40°, the Rutherford scattering angle for grazing collisions being 27°. The energy spectra of the isotopes and the angular distributions of the elements measured in this reaction indicate that the products were formed during deep-inelastic collisions between the argon and the silver. Figure 33 gives some idea of the variety of multinucleon transfer products formed during deep-inelastic collisions.

Figure 34 shows how the charge dispersion of multinucleon transfer reaction products depends on the dissipation of the collision kinetic energy (Ho 82b). Hoover *et al.* investigated the reaction between $^{165}$Ho and $^{56}$Fe (462 MeV) and detected the reaction products similar to the projectile nucleus, the products only being separated by atomic number. As can be seen from the figure, the nuclides formed in quasi-elastic collisions were close in $Z$ to the projectile, while those formed in deep-inelastic collisions had widely varying atomic numbers, although the production cross-section decreased with increasing distance of the atomic number from that of the projectile.

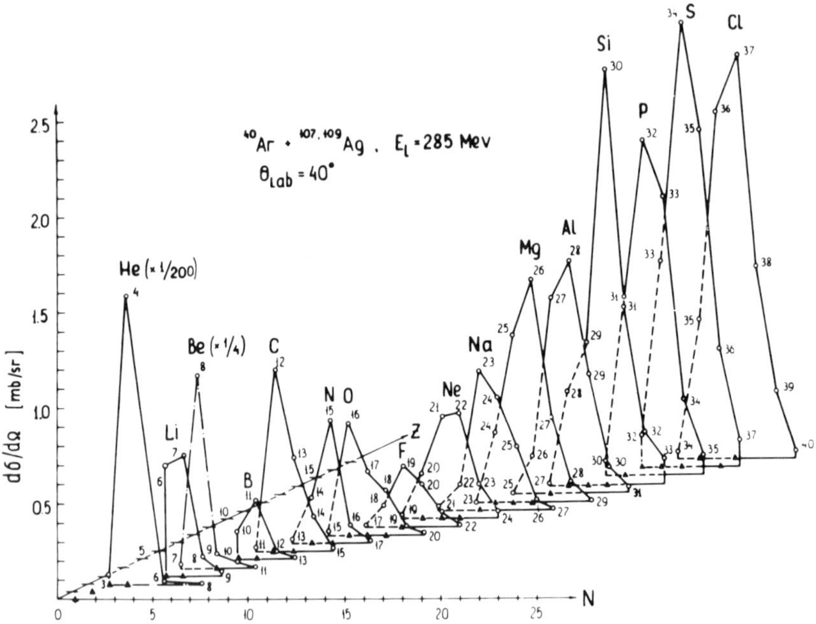

**Figure 33.** Differential production cross-sections for the isotopes of the elements from He to Cl as multinucleon transfer products from the reaction $^{nat}Ag + ^{40}Ar$ (285 MeV); the emission angle is 40°. Horizontal axes give the numbers of neutrons and protons in the isotope (Ar 81).

The charge dispersion of the products of a multinucleon transfer as a function of the dissipation of the collision's kinetic energy is shown explicitly in Figure 35 (Mi 84). These data refer to the $^{144}Sm + ^{84}Kr$ reaction for three energies of the projectile. The jump in the charge dispersion can be seen for the transition from the quasi-elastic to dissipative mechanism. The charge dispersion maximum accessible in deep-inelastic collisions grows with increasing ion energy.

### *3.2.2.* $Q_{gg}$ *Systematics*

An important piece of information for the production of nuclei far from stability is the way in which the production cross-section decreases with distance from the stability line. This relation is described by the $Q_{gg}$ systematics for multinucleon transfer reactions.

The $Q_{gg}$ systematics were established in 1971 at the JINR Laboratory of Nuclear Reactions (Ar 71). Later, experiments carried out by Volkov and his co-workers led him to introduce a pairing correction which cleared up the physical sense of the slope and interval between the elemental lines, arguing from ideas concerning the formation of a dinuclear system (Vo 74).

The $Q_{gg}$ systematics for the $^{232}Th + ^{22}Ne$ (174 MeV) reaction are shown in Figure 36 (Ar 73, Vo 78). The differential production cross-section for the reaction products emitted at 40° is plotted along the ordinate, and the $Q_{gg}$ values with the pairing

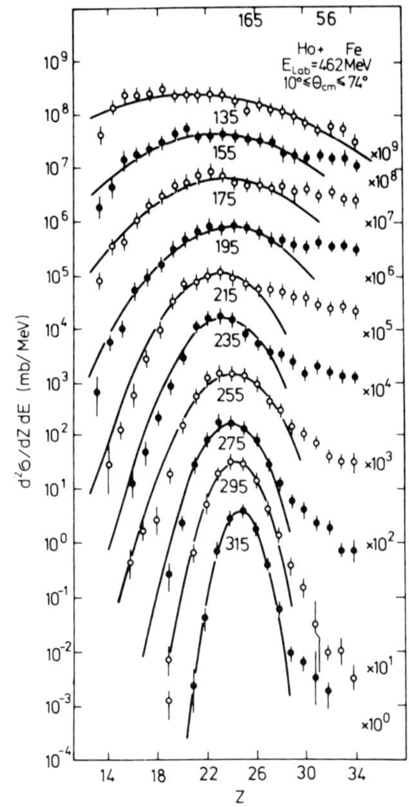


**Figure 34.** $Z$ distribution of fragments formed in the $^{165}$Ho + $^{56}$Fe reaction, plotted as a function of the total kinetic energy. Solid curves drawn through the data points are the results of Gaussian fits (Ho 82b).

corrections $\delta(\mathrm{p})$ and $\delta(\mathrm{n})$ are plotted along the abscissa. Here $Q_{gg} = (M_1 + M_2) - (M_3 + M_4)$ is the difference between the masses of the initial and final nuclei in energy units.

The remarkable feature of $Q_{gg}$ systematics is that the production cross-sections for the isotopes of a given element (as products of multinucleon transfers) lie along a straight line the slope of which is the same from element to element. The $Q_{gg}$ systematics can therefore be used to estimate the production cross-section for an unknown isotope, although a semiempirical extrapolation of the new nuclear mass is still needed. It should be emphasized tht the production cross-sections of nuclides with small binding energies for one or two neutrons do not deviate from $Q_{gg}$ systematics.

It has been verified that the $Q_{gg}$ systematics are satisfied for many combinations of projectile and target nuclei. The $Q_{gg}$ systematics for the products of deep-inelastic transfers in the $^{nat}$Ag + $^{40}$Ar reaction (285 MeV) are shown in Figure 37 (Ar 81). As can be seen, the isotopes of helium to oxygen, inclusively, satisfy $Q_{gg}$ systematics very well. The deviations for the heavier-element isotopes are associated with secondary nuclear processes, the main one being the evaporation of neutrons from the excited products of the multinucleon transfer. Once these have been taken into account, the production cross-sections for the heavier products are well described by $Q_{gg}$ systematics.

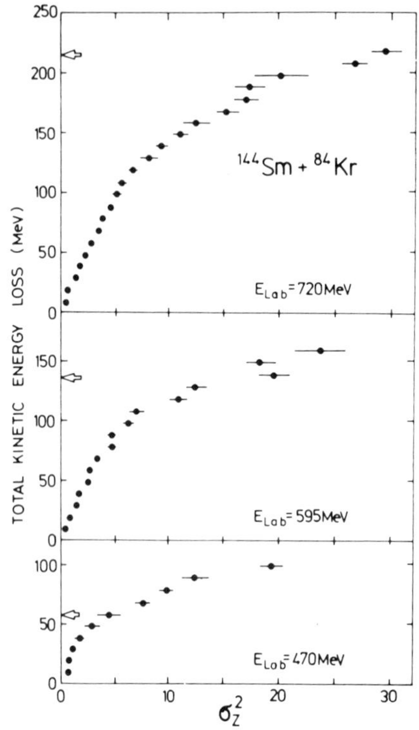


**Figure 35.** Total kinetic energy loss $E_{\text{loss}}$ plotted vs. variance $\sigma_z^2$ of the charge distribution of projectile-like fragments from the reaction $^{144}$Sm + $^{84}$Kr measured at three bombarding energies. Arrows indicate the initially available total kinetic energy above the barrier in each case (Mi 84).

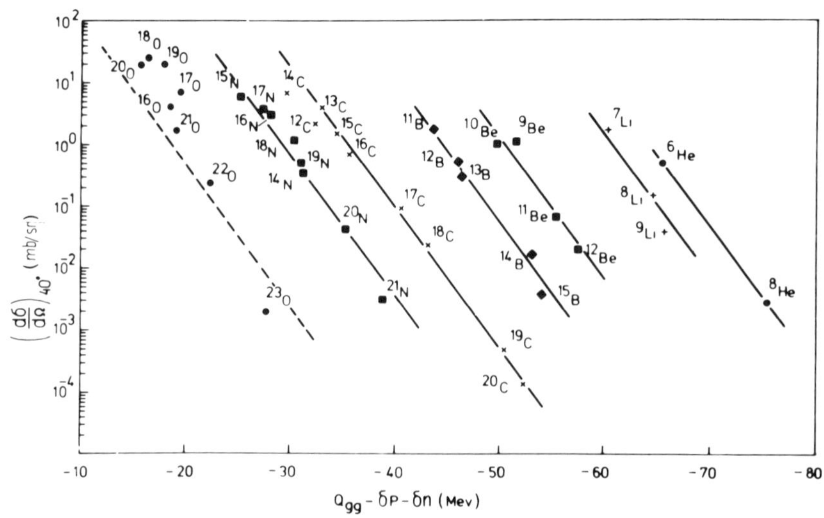


**Figure 36.** $Q_{gg}$ systematics of production cross-sections for isotopes formed as multinucleon transfer products in the $^{232}$Th + $^{22}$Ne (174 MeV) reaction; the detection angle is 40° (Ar 73, Vo 78).

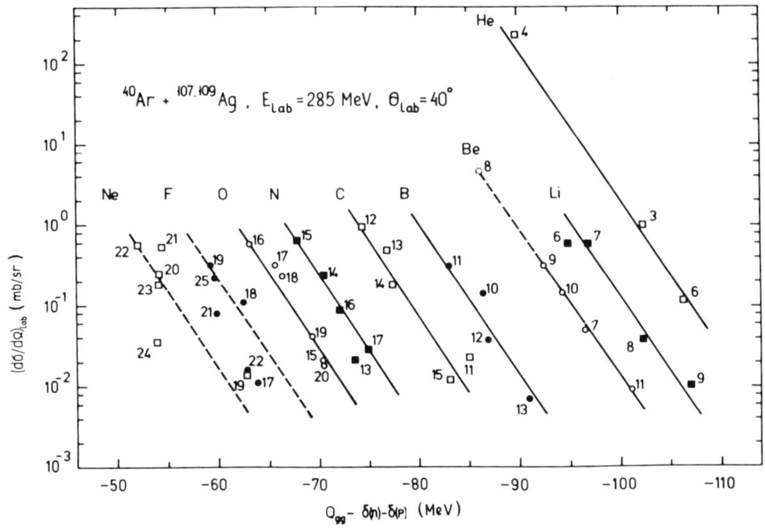


**Figure 37.** $Q_{gg}$ systematics of production cross-sections for isotopes formed as multinucleon transfer products in the $^{nat}$Ag + $^{40}$Ar (285 MeV) reaction; the detection angle is 40° (Ar 81).

This can be seen from Figure 38 where Organessian's experimental data on the production cross-sections of the neutron-rich isotopes of gold formed by bombarding $^{181}$Ta with $^{40}$Ar, $^{86}$Kr, and $^{136}$Xe (Og 73) are compared with those calculated using $Q_{gg}$ systematics after allowing for the evaporation of neutrons from the multinucleon transfer products (Ch 78).

The reason for the applicability of $Q_{gg}$ systematics to different nuclear systems is the establishment of partial statistical equilibrium in a dinuclear system.

There is no potential barrier in a deep-inelastic collision with $l > l_{cr}$ preventing the dinuclear system from disintegrating because the Coulomb and centrifugal forces prevail over the attractive ones. The disintegration of the dinuclear system along one of the exit channels of the reaction is determined by statistical factors and, above all, by the attainable volume in phase space for the disintegration.

Let us suppose that $U_i$ is the thermal excitation energy of an initial dinuclear system that has been formed during a deep-inelastic collision with angular momentum $l_i > l_{cr}$. As the dinuclear system evolves toward the equilibrium configuration, the two nuclei retain to a large extent their individualities (Vo 78). The thermal excitation energy $U_f$ of an intermediate dinuclear configuration before it disintegrates will therefore be

$$U_f = U_i + Q_{gg} + \Delta E_c + \Delta E_{rot} - \delta(p) - \delta(n) \tag{3.1}$$

where $Q_{gg}$ is the energy spent on rearranging the nuclei, $\Delta E_c$ is the change in the system Coulomb energy due mainly to the redistribution of protons between the nuclei and to the deformation of the system, $\Delta E_{rot}$ is the change in the system rotational energy due to the change in its moment of inertia, and $\delta(p)$ and $\delta(n)$ are the pairing corrections. The corrections are to take into account the fact that, during deep-inelastic transfers, nucleons from the donor nucleus are transferred into the excited levels of the acceptor

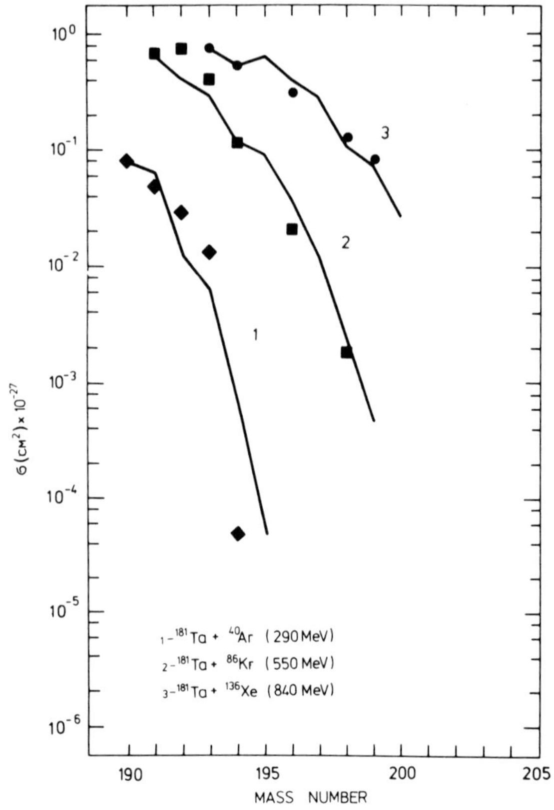


**Figure 38.** Comparison of experimental cross-sections for forming gold isotopes by bombarding $^{181}$Ta with $^{40}$Ar, $^{86}$Kr, and $^{136}$Xe ions, from (Og 73), with the calculations of the relative yields of these isotopes according to the $Q_{gg}$ systematics. Neutron emission from excited reaction products is taken into account. Calculated data are normalized to experimental results for the lightest isotopes (Ch 78).

nucleus and, as a rule, are not paired (Vo 78). Quantity $Q_{gg}$ characterizes the loss of energy due to the rearrangement, given that the nucleons are exchanged from the ground state in the donor to the ground state in the acceptor nucleus.

If we use a constant temperature expression for the density of the states of the dinuclear system, i.e.,

$$\rho \approx \exp(U_f/T) \tag{3.2}$$

then the probability of disintegration of the dinuclear system along a given reaction channel $W_i$ will be

$$W_i \approx \exp\left[\frac{Q_{gg} + \Delta E_c + \Delta E_{rot} - \delta(p) - \delta(n)}{T}\right] \tag{3.3}$$

By plotting $Q_{gg} - \delta(p) - \delta(n)$ along the abscissa and the logarithm of the cross-section along the ordinate, we obtain the regularity that is characteristic of $Q_{gg}$ systematics, i.e., the isotopic cross-sections lie along straight lines, one for each element.

The slope of this line characterizes the temperature of the dinuclear system under the conditions of partial statistical equilibrium. The displacement of the lines for the elements along the abscissa reflects the changes in the Coulomb energy of the dinuclear system during a transfer of protons and the system's deformation. The scale of the change on the rotational energy during a transfer of nucleons to a heavy target nucleus is much smaller; it can be neglected in a first approximation.

The $Q_{gg}$ systematics is the result of the establishment of partial statistical equilibrium in the system of two interacting nuclei, and so $Q_{gg}$ systematics are realized more completely during deep-inelastic transfers. Experiments have also shown that $Q_{gg}$ systematics retains its validity within known limits even during dissipative collisions, and this indicates that the equilibrium distribution of the thermal-excitation energy between the interacting nuclei is established very quickly (Ar 73).

### *3.2.3. Angular Distributions of Multinucleon Transfer Products*

When nuclei far from stable are detected "in beam," it is important for the experimentalist to know the type of the angular distribution of the multinucleon transfer reaction products.

These angular distributions are very different. They vary in shape from isotropic ones, $d\sigma/d\theta$ = const, to concentrated in a narrow interval of angles. The shape of the angular distribution depends on the masses of the colliding nuclei, on the collision kinetic energy and the degree of its dissipation, and on the number of nucleons transferred [see Schröder and Huizenga's review (Sc 84)].

Partial angular distributions reflecting different kinetic-energy losses are shown in Figure 39 for the products of reactions involving the transfer of few nucleons (Cl, K) and many nucleons (Al). The data were obtained by Artukh *et al.* (Ar 73) for the bombardment of $^{232}$Th by $^{40}$Ar ions at 297 MeV.

Deep-inelastic transfers, irrespective of the number of transferred nucleons, are characterized by an exponentially decreasing differential cross-section with increasing angle of emission of the reaction products. The angular distribution of quasi-elastic few-nucleon transfers has a sharp maximum at the angle corresponding to Rutherford scattering for grazing collisions. The maximum widens as the inelasticity rises, while the maximum itself moves to smaller angles.

When ions heavier than $^{40}$Ar are used, the maximum at the grazing-collision angle becomes sharper in the angular distribution of the deep-inelastic transfers, even though a substantial number of nucleons are transferred; see Figure 40 (Ru 77).

The variation in the angular distribution of the various isotopes resulting from multinucleon transfers is of great interest. The production cross-sections of the isotopes of elements from beryllium to fluorine are compared in Figure 41 for emission angles of 12° and 40°. The cross-sections were measured by Artukh *et al.* for the $^{232}$Th + $^{22}$Ne (174 MeV) reaction (Ar 77). Note that the production cross-sections of the neutron-rich isotopes from boron to fluorine increase from 12° to 40°. This angle is, for this reaction, close to that for Rutherford scattering for a grazing collision. The cross-sections at both angles are both nearly the same only for beryllium. The width of the energy distribution is narrower for neutron-rich isotopes (Figure 42), which also points to the peripheral nature of the collisions which form neutron-rich isotopes.

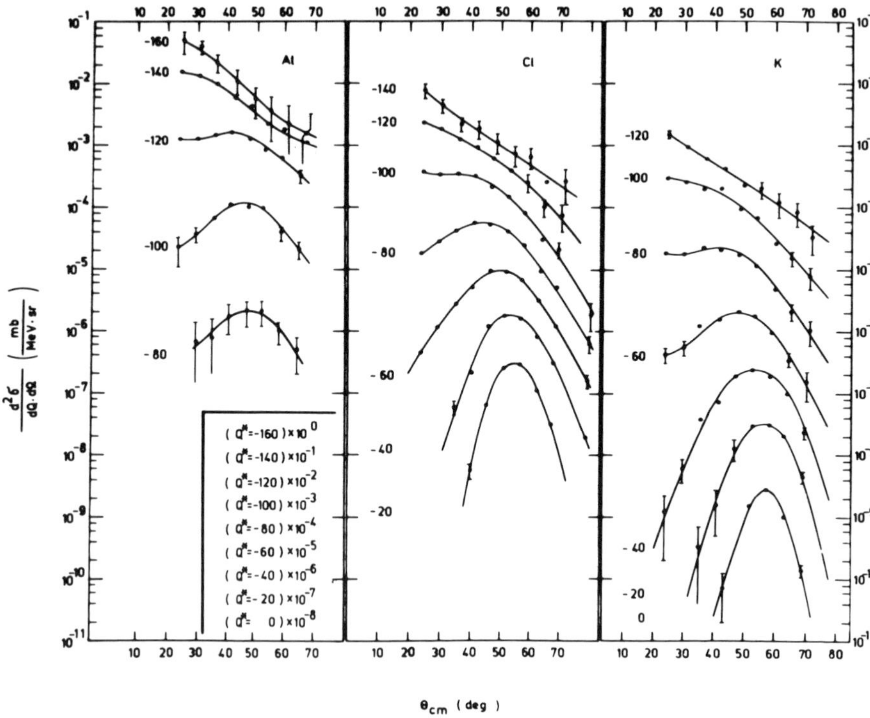


**Figure 39.** Partial angular distribution of Al, Cl, and K with the fixed value $Q = Q^*$ formed as multinucleon transfer products in the $^{232}$Th + 40Ar (288 MeV) reaction. The reaction $Q$ values are given on the left of the curves. All the points and curves for the given value of $Q^*$ have been multiplied by the coefficient given in the left-hand side of the figure (Ar 73).

## 3.3. Separation and Identification of Nuclei Far From Stability Resulting from Multinucleon Transfers

Two types of experiment are conducted to investigate the nuclei far from stability that result from multinucleon transfers. The first type of experiment is aimed to obtain new nuclides in the direction of the nucleon-stability limit. The technique must ensure that a new nucleus is reliably identified. That a nucleus is detected is in itself proof that it is stable against the emission of a nucleon. This sort of experiment is conducted "in beam" without first slowing the reaction products.

The second kind of experiment is to study the properties of nuclides far from stability, e.g., their radioactive decay modes, half-lives, the energies of the particles and gamma rays emitted. Usually these experiments are conducted using an ISOL system.

When identifying a new nuclide, it is sufficient to have a detection rate of a few nuclei per hour. A study of its nuclear properties requires a yield of hundreds or thousands of nuclei per second.

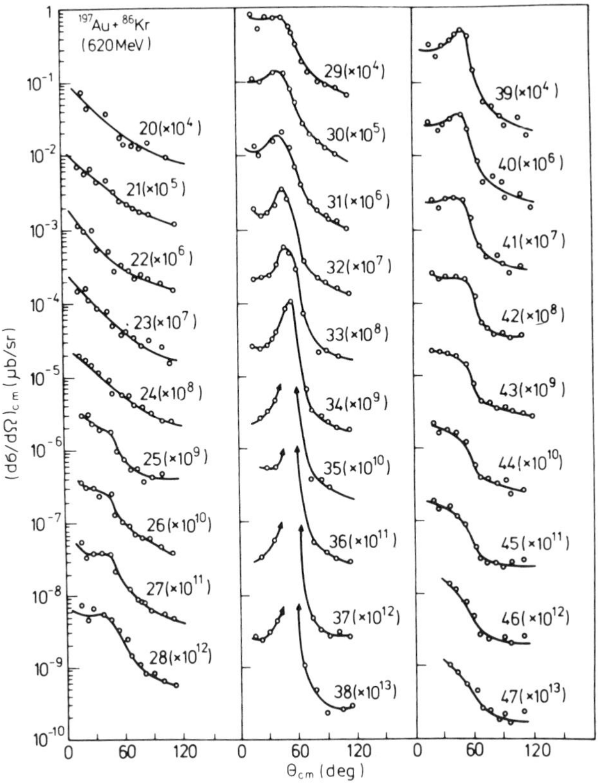


**Figure 40.** Dependence of the angular distributions of deep-inelastic transfer products upon the atomic number of the light fragment produced in the reaction $^{197}Au + ^{86}Kr$ (720 MeV). The light fragment atomic number and the multiplying factors to be used to obtain the absolute values of differential cross-sections are indicated (Ru 77)

### *3.3.1. In-Beam Identification of Multinucleon Transfer Products*

*3.3.1a. Magnetic Analysis Combined with the $\Delta E-E$ Technique.* Experience of the in-beam identification of new neutron-rich light-element isotopes had been accumulated for the spallation reactions induced by high-energy protons before any experiments on the production of nuclei far from stability had been carried out using multinucleon transfer reactions. Poskantzer, Hyde, Cerny, Thomas, and various other investigators used specific ionization $\Delta E$, energy $E$, and time-of-flight $t$ measurements in their experiments (Po 66b, Po 68, Th 68, Ra 70, Bu 70). At nonrelativistic velocities and for rather light nuclei, the following are valid:

$$\Delta E \cdot E \approx mZ^2 \tag{3.4}$$

$$m = 2E/v^2, \qquad v = d/t \tag{3.5}$$

where $d$ is the length of the flight path, while $v$, $m$, and $Z$ are the velocity, mass, and atomic number, respectively, of the reaction products. Equations (3.4) and (3.5) can

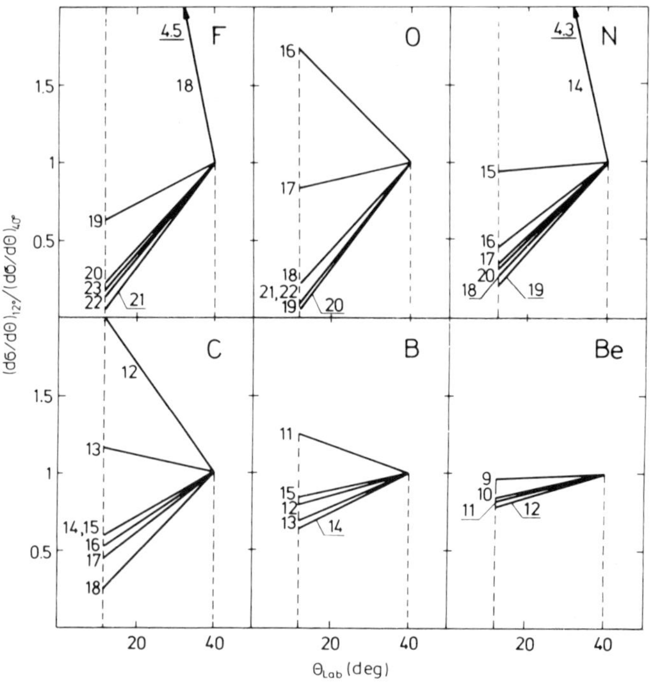

**Figure 41.** Ratio of differential cross-sections $d\sigma/d\theta$ for producing isotopes of the elements from beryllium to fluorine in the reaction $^{232}Th + {}^{22}Ne$ (174 MeV) at emission angles of 12° and 40° (Ar 77).

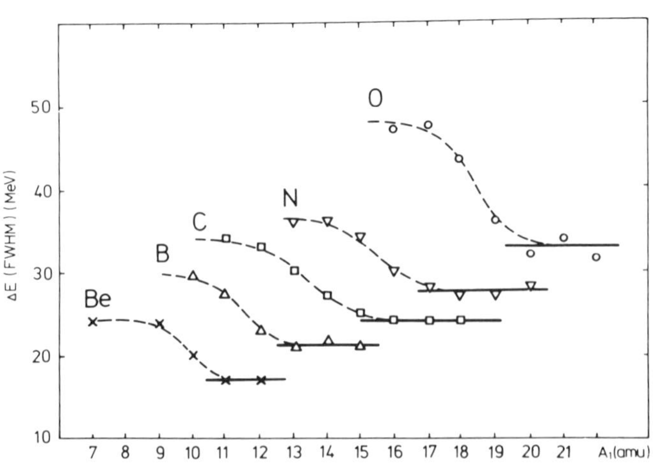

**Figure 42.** Widths of the energy spectra (FWHM) of the isotopes of the elements from beryllium to oxygen produced in the reaction $^{232}Th + {}^{22}Ne$ (174 MeV) at an emission angle of 12° (Ar 77).

be used to identify a new isotope. Many neutron-rich light-element isotopes were first detected using this technique.

However, the limitations of the $(\Delta E-E)$-TOF method were already apparent in spallation reactions. The statistical spread in the $\Delta E$ detector and the impossibility of increasing the length of the flight path without a considerable loss of sensitivity limited the method to elements up to oxygen and isotopes with mass numbers up to 20.

Several extra serious difficulties, associated with elastic and inelastic scattering of the projectiles, emerged when the $(\Delta E-E)$-TOF method was used to identify products from multinucleon transfers involving heavy ions. The cross-sections of elastic and inelastic scattering are many orders of magnitude larger than the cross-sections of multinucleon transfers which lead to nuclei far from stability. In order to reduce the loading on the $\Delta E$ and $E$ detectors by scattering projectile nuclei, the intensity of the heavy ions has to be reduced, which in turn decreases the yield of nuclei very far from the stability line, thus increasing experimental time to unrealistic values.

The situation was resolved by introducing a magnetic analyzer to separate the products of multinucleon transfer reactions from the heavy-ion scattering. This was first done by Jacmart and his co-workers at the Institut de Physique Nucleaire at Orsay (Ja 66). A moderately sized semiconductor detector was placed at the focal plane of a magnetic analyzer and measured the energy $E$ of the reaction products. The equation for the magnetic analyzer has the form

$$E = \text{const}(B\rho)^2 q^2/m \tag{3.6}$$

where $B$ is the magnetic induction, $\rho$ is the radius of the product's trajectory, $q$ is its ionic charge, and $m$ is its mass.

The magnetic rigidities of elastically and inelastically scattered projectile nuclei differ considerably from those of very neutron-rich light-element nuclei formed in multinucleon transfer reactions. It is possible to adjust a system to record only nuclei far from stability by choosing an appropriate magnetic field.

Semiconductor detectors can measure energies to within 1%, which means that a reaction product's $q^2/m$ parameter can be measured very accurately.

However, a magnetic analyzer with an $E$ detector in the focal plane cannot identify new neutron-rich light-element isotopes. This is because multinucleon transfer products leave the target in various charge states. A magnetic analyzer-$E$ detector system cannot in principle be used to distinguish isobars with the same ionic charge, e.g., $^{22}O^{+8}$, $^{22}F^{+8}$, $^{22}Ne^{+8}$. Isobars close to the stability region contribute most to the yield of transfer reaction products with a given mass number. Furthermore, it is always possible to find among the many products of a multinucleon transfer reaction pairs of nuclides with similar identification parameters, e.g., $^{4}He^{+2}$ and $^{9}Li^{+3}$, and $^{14}B^{+5}$, and $^{9}Be^{+4}$. An MA-$E$ detector system cannot separate them.

A way out was found in Dubna at the JINR Laboratory of Nuclear Reactions. Volkov and his co-workers combined a magnetic analyzer with the $\Delta E-E$ technique (Ar 70b) and this overcame the difficulties encountered by the MA-$E$ method. A telescope consisting of a thin semiconductor $\Delta E$ detector and a total-absorption $E$ detector is placed in the focal plane of a magnetic analyzer. The signals from the $\Delta E$ and $E$ detectors are fed into a multichannel analyzer working in a two-dimensional mode. The degeneration due to isobars with the same ionic charge is resolved because the specific ionization in the $\Delta E$ detector is determined by the atomic number $Z$ of the reaction product.

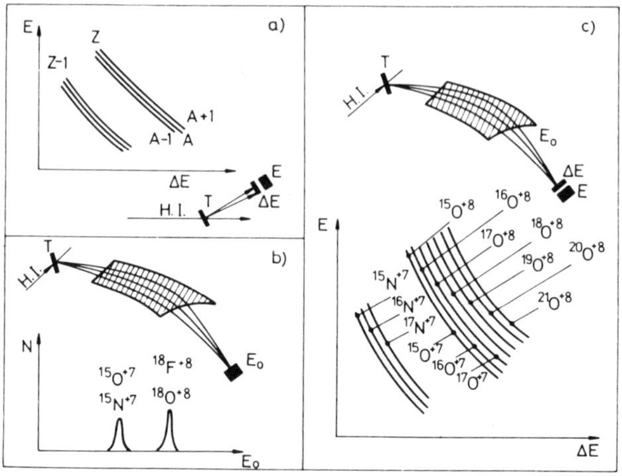


**Figure 43.** Comparison of experimental data obtained by different methods: (a) the $\Delta E$-$E$ technique; (b) the combination of magnetic analysis and $E_0$ measurement; (c) the combination of magnetic analysis and the $\Delta E$-$E$ technique (Ar 70b).

The forms of recorded information are compared in Figure 43 for the $\Delta E$-$E$, MA-$E$, and MA-($\Delta E$-$E$) methods. The continuous hyperbola characteristic of the $\Delta E$-$E$ method is broken into separate points, each of which corresponds to a particular isotope with a particular charge state, when it is combined with a magnetic analyzer (see Figure 44).

When $\Delta E \ll E$ and $q = Z$, we obtain from equations (3.4) and (3.5)

$$\Delta E \approx \text{const} \cdot A^2 \tag{3.7}$$

A comparison of relations (3.4) and (3.7) shows that the specific ionization difference between two neighboring isotopes of an element $Z$ in the $\Delta E$ detector is twice as large when it is used in a MA-($\Delta E$-$E$) system than when it is used in a $\Delta E$-$E$ technique. The specific ionization difference may be four times larger in a MA-($\Delta E$-$E$) system if the isotopes being separated in the $\Delta E$ and $E$ detectors have comparable energy losses (Ar 70b). Figure 45 shows the isotopic resolution along the $E$-$\Delta E$ and $\Delta E$ axes.

The MA-($\Delta E$-$E$) combination turned out to be a very efficient method for identifying nuclei far from stability. Volkov and his co-workers were able with this method to identify 28 new neutron-rich isotopes for elements from carbon to chlorine, and to establish the $Q_{gg}$ systematics for the production cross-sections of isotopes formed in multinucleon transfer reactions (Ar 70f, Vo 71, Vo 73, Vo 78).

*3.3.1b. Magnetic Analysis Combined with $\Delta E$-$E$ and TOF.* Although the MA-($\Delta E$-$E$) detector system reliably separates and identifies the products of transfer reactions in most cases, it does not completely avoid confusing signals from the $\Delta E$ and $E$ detectors for two different nuclides. The danger comes when neutron-rich light-element isotopes with very small production cross-sections are being detected. Sometimes, given the large variety of reaction products from multinucleon transfers and the width of their energy spectra, it happens that the $Z_{\text{eff}}^2 m$ product ($Z_{\text{eff}}$ is the effective $Z$ of a

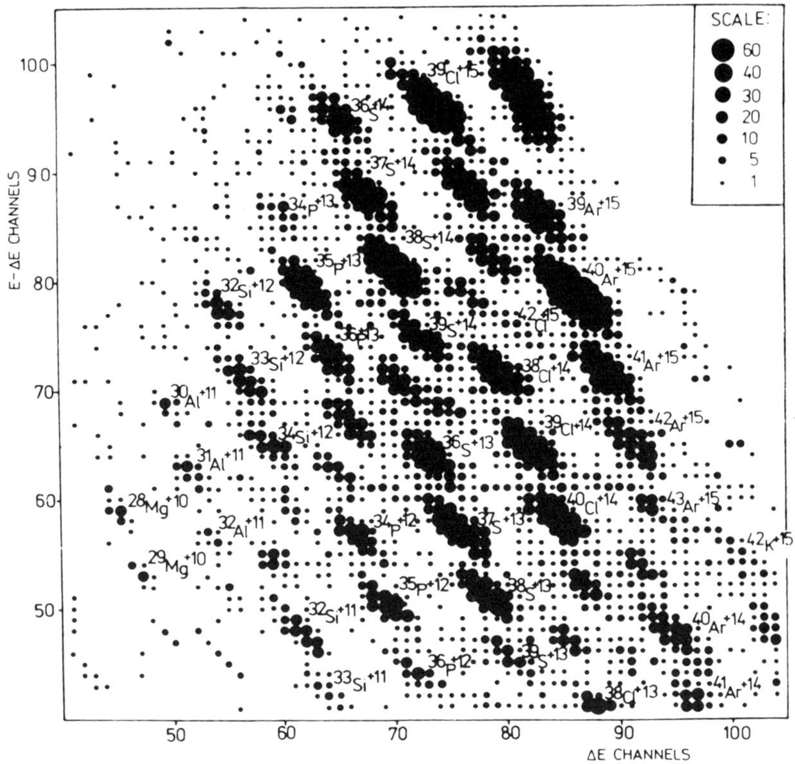


**Figure 44.** Two-dimensional $\Delta E$, $E$-$\Delta E$ spectrum obtained in bombardment of a $^{232}$Th target with 290 MeV $^{40}$Ar ions. Combination of the magnetic analysis and the $\Delta E$-$E$ technique was used as the detection system (Ar 71c).

product when it passes through the $\Delta E$ detector) of a neutron-rich isotope of an element with atomic number $Z$ is the same as that of a neutron-deficient isotope of the element with atomic number $Z + 1$, and their energy spectra partially overlap. The neutron-deficient isotope will nearly always have charge states one or two units lower than the charge state in which the neutron-rich isotope is recorded. In this case the parameter $q^2/m$ for both isotopes may be similar and both isotopes will arrive at the $\Delta E$-$E$ telescope for the same magnetic field. For example, it is possible that $^{13}$N and $^{14}$N will be recorded instead of $^{19}$C and $^{20}$C, or $^{8}$B instead of $^{14}$Be. The interfering isotopes are closer to the stability region and have larger yields. It is impossible under these conditions to separate a neutron-rich isotope with a small production cross-section using an MA-($\Delta E$-$E$) system.

The large difference between the masses of the neutron-rich and the interfering isotopes indicates that they can be separated using a time-of-flight measurement for the distance from the target to the $\Delta E$-$E$ telescope. Since the particles are focused by the magnetic field, the flight path may be several meters long. This would yield a tolerable accuracy for the mass-number measurement, even at time resolutions of a few nanoseconds.

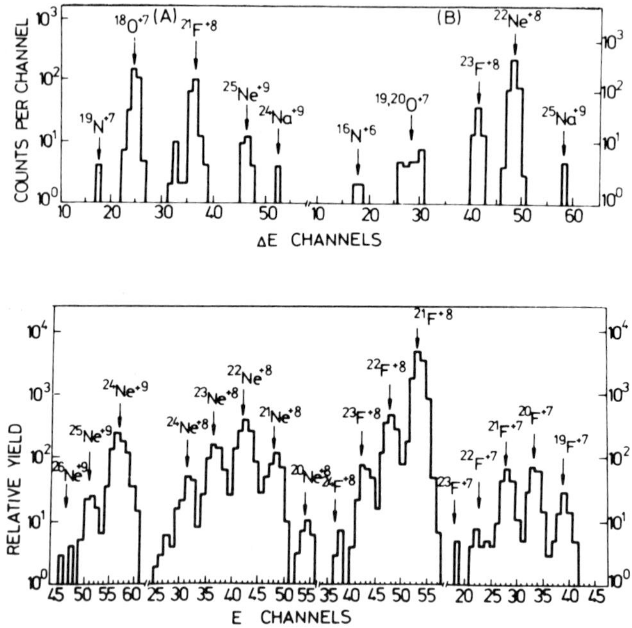


**Figure 45.** Spectra from the $\Delta E$ detector corresponding to the cutoffs at the two-dimensional spectrum obtained in the $^{232}Th + ^{22}Ne$ (174 MeV) reaction: in the maxima of $^{25}Ne^{9+}$ (A) and in the maxima of $^{23}F^{8+}$ (B) yields (at the top). Ne and F isotope yields are obtained from the two-dimensional spectrum (at the bottom) (Ar 70b).

The first detection system including magnetic analysis, the $\Delta E$–$E$ technique, and time-of-flight measurements was developed by Artukh and co-workers at the JINR Laboratory of Nuclear Reactions and applied to identify light-element neutron-rich isotopes (Ar 74). It identified the exotic isotopes $^{14}Be$, $^{19}C$, and $^{20}C$, which were formed by the bombardment of $^{232}Th$ with $^{15}N$ and $^{22}Ne$ (Ar 74, Ar 75).

The high-frequency voltage across the U-300 cyclotron dees was used to form the start signal. The time width of a cluster of accelerated ions was about 2–3 ns with 200 ns between clusters. The stop signal was taken from the $\Delta E$ detector. The average flight time of a reaction product from the target to the detector is around 150 ns. The yields of the neutron-rich isotopes of helium, lithium, and beryllium from the $^{232}Th + ^{15}N$ (145 MeV) reaction are shown in Figure 46. The decisive role of the separation of the reaction products in terms of their flight times can be seen for the reliable identification of $^{14}Be$.

A similar sort of detector system, MA-($\Delta E$–$E$)-TOF, was set up at the Institut de Physique Nucleaire at Orsay by Stephan *et al.* (St 76). A thin plastic scintillator was used to provide the start signal, and the stop signal was taken from a semiconductor $E$ detector. The plastic scintillator was placed at an angle to the average trajectory, thus compensating for the scatter of the trajectory lengths in the magnetic analyzer. For a flight path of 309 cm, and an input solid angle of $1.8 \times 10^{-3}$ sr, the time resolution was 0.7 ns. The separation of the isotopes of iron using this system is shown in Figure

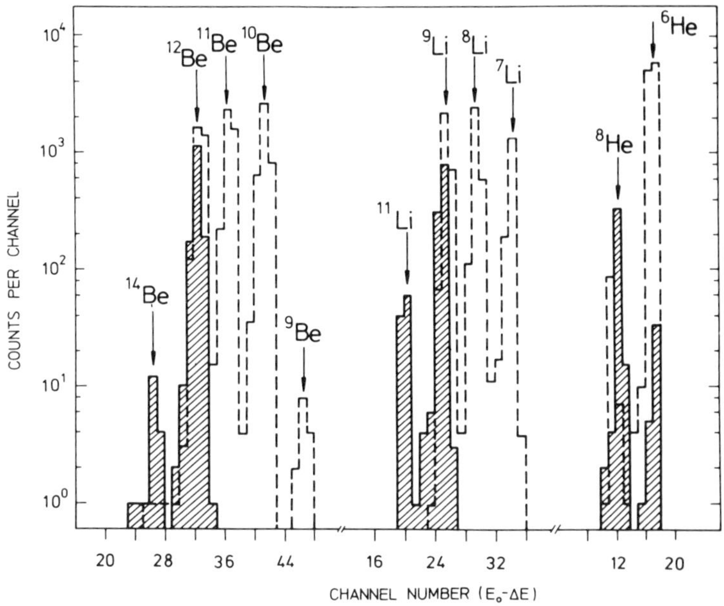


**Figure 46.** Yields of Be, Li, and He isotopes from the bombardment of $^{232}$Th with 145 MeV $^{15}$N ions. The dashed line corresponds to calibration measurements without time-of-flight discrimination. Shaded peaks are obtained in the main measurements with time-of-flight discrimination. The ion flux passing through the target in the main measurements was 30 times as large as that in calibration measurements (Ar 74).

47. For more detail about the detector systems used to separate and identify multinucleon transfer products, see Volkov's review (Vo 79).

*3.3.1c. Unambiguity of the Identification of Multinucleon Transfer Products Using a MA-($\Delta E$-$E$)-TOF Detector System.* A product of a multinucleon transfer reaction leaving the target at an angle $\theta$ possesses certain $Z$, $A$, $q$, and $E$ values. These properties determine its trajectory in the magnetic field of the analyzer, the amount of energy released in a thin $\Delta E$ detector and a total absorption $E$ detector, and the time of flight along a given flight path. The measured $E$, $\Delta E$, $\rho$, and $t$ are related to the properties we want, viz., $Z$, $A$, $q$, and $E$, via three equations (3.4), (3.5), and (3.6).

The mass of a reaction product is determined by measuring its kinetic energy and flight time along a path length $d$. The atomic number $Z$ is more difficult to ascertain because $Z_{\rm eff}$ enters in equations (3.4).

The effective atomic number $Z_{\rm eff}$, which appears when the product is passing through the $\Delta E$ detector, may differ significantly from the real atomic number $Z$, and this difference increases with the $Z$ and $A$ values of the product. Another equation is therefore needed in which the atomic number is explicit. Experimental and calculated data on the ranges of particles in substances is therefore used. The range $R$ depends on the atomic number, mass, and kinetic energy of the reaction product, $R = R(A, Z, E)$. It can either be tabulated or plotted on a graph. It is always possible to

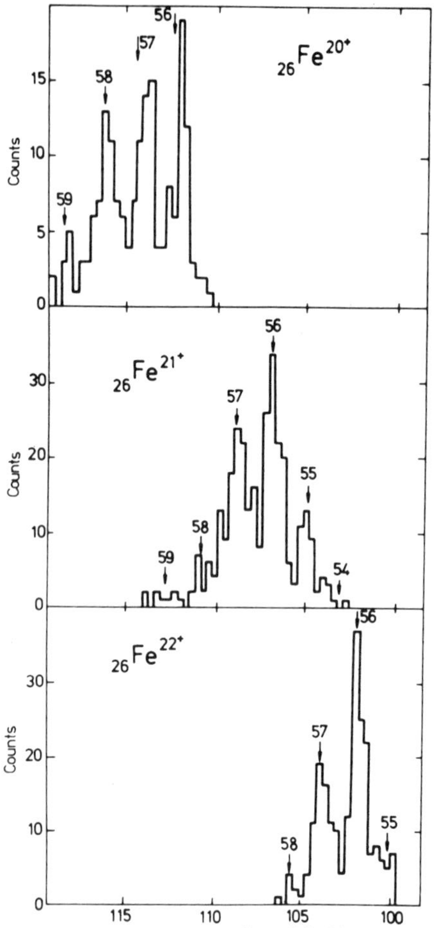


**Figure 47.** Identification of iron isotopes using the MA-($\Delta E$-$E$)-TOF technique combination in the reaction $^{65}Cu + ^{124}Sn$. Mass numbers and charge states of the isotopes are indicated (St 76).

calculate the energy loss in a thin $\Delta E$ detector using data on $R$ for any isotope and given energy. Thus the ionization loss in the detector can be found as a function of the reaction product's properties, i.e.,

$$\Delta E = f(Z, A, E) \tag{3.8}$$

This can be considered a fourth identification equation for the MA-($\Delta E$-$E$)-TOF detector. A product's mass is determined from equation (3.5) independently of its atomic number. By contrast, the mass is needed in order to obtain $Z$.

### 3.3.2. *The ISOL System for Separating Multinucleon Transfer Products*

The construction and specifications of an ISOL system to separate the products resulting from the complete fusion of two nuclei were considered in Section 2.2.4. The systems can also be used for multinucleon transfer reactions. The main change is to the target unit of the ion source. This is needed because multinucleon transfer products have much longer ranges than complete-fusion products. The excitation functions for the complete-fusion and multinucleon-transfer reactions are also very different. The former have the shape of curves with maxima around 20 MeV in width, while the second are characterized by cross-sections that are monotonically increasing with heavy-ion energy. The recoiling complete-fusion products escape from the target within a very small solid angle around 0°. The angular distribution of multinucleon transfers which lead to neutron-rich isotopes has a maximum near the Rutherford angle for grazing collisions. This angle grows with decreasing ion energy.

In order to obtain nuclei far from stability, the target is increased in thickness by an order of magnitude, and the target function is extended to the catcher. In experiments on the GSI mass separator at Darmstadt the thickness of the tungsten target was increased to 46.5 mg/cm$^2$ (Ru 85). When this target was bombarded with $^{82}$Se ions at 11.5 MeV/nucleon, neutron-rich isotopes of Mn, Co, Fe, Ni, and Cu were produced. The tantalum catcher inside the ion source was also used as a target, because the $^{82}$Se ions still had enough energy after passing through the tungsten target to induce multinucleon transfer reactions.

The considerable ranges of multinucleon transfer products may make it difficult to use IGISOL-type systems to separate and identify them.

## 3.4. New Nuclides from Multinucleon Transfer Reactions and their Properties

### 3.4.1. *To the Limit of Nucleon Stability*

Very neutron-rich light-element isotopes are the most interesting nuclei far from stability. They have relatively large numbers of neutrons in comparison to protons, the $^{8}$He nucleus, for example, containing three times as many neutrons as it does protons. These nuclei could be used to give experimental answers to questions about the nucleon-stability limits, and studying their nucleon properties can verify experimentally the validity of various theoretical models under extreme conditions.

The main way of obtaining very neutron-rich light-element isotopes until the end of the 1960s was from spallation reactions induced by energetic protons. Poskantzer, Hyde, Cerny, and others (Po 66b, Po 68) first identified $^{11}$Li, $^{14}$B, $^{15}$B, and $^{17}$C, while Thomas *et al.* (Th 68) first observed $^{19}$N and $^{21}$O, and Reisbeck *et al.* (Ra 70) discovered $^{19}$C.

A new series of experiments was started at the JINR Laboratory of Nuclear Reactions in Dubna at the end of the 1960s. The efficiency of multinucleon transfer at producing very neutron-rich light-element isotopes was demonstrated. All the known neutron-rich isotopes were observed by bombarding $^{232}$Th with $^{18}$O at 122 MeV for all the elements from beryllium to oxygen, and the nuclides $^{18}$C, $^{20}$N, and $^{22}$O were detected for the first time. A six-hour exposure produced yields of $^{17}$C, $^{19}$N, and $^{21}$O hundreds of times larger than the yields obtained from high-energy proton experiments (Ar 69).

Later Volkov and co-workers (Ar 70c, Ar 70d, Ar 70e, Ar 70f, Ar 71b, Ar 72, Ar 74, Ar 75) carried out experiments using $^{232}$Th targets that were several milligrams per square centimeter thick and irradiated with $^{11}$B, $^{15}$N, $^{16}$O, $^{22}$Ne, and $^{40}$Ar ions at energies of 7–9.7 MeV/nucleon. Very neutron-rich light-element nuclei resulting from multinucleon transfers were detected using an MA-($\Delta E - E$) detector system. The time-of-flight

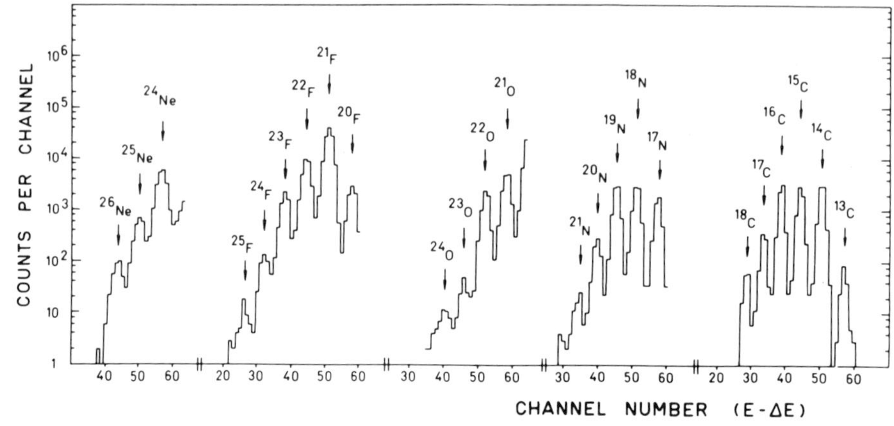


**Figure 48.** Yields of neon, fluorine, oxygen, nitrogen, and carbon isotopes from the $^{232}$Th + $^{22}$Ne (174 MeV) reaction; the $^{22}$Ne flux through the target amounts to $5.1 \times 10^{16}$ particles (Ar 70d).

technique was added and the $^{14}$Be, $^{19}$C, and $^{20}$C nuclei identified. The results of some of these experiments are presented in Figures 48–50.

All of the heaviest isotopes that were then known for the elements from He to Ne and from Mg to Ar were obtained, and 29 new isotopes for the elements C, N, O, F, Ne, Mg, Al, Si, P, S, and Cl (see Table 3) were identified for the first time. A three-to-four

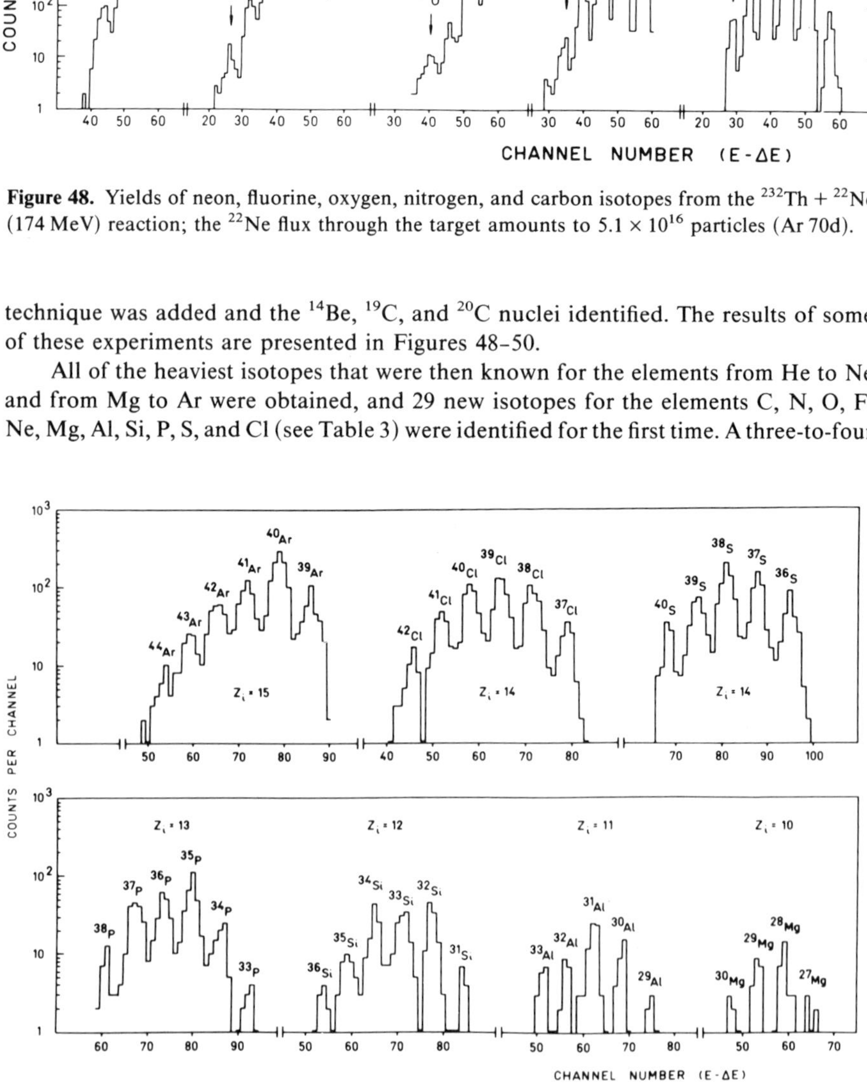


**Figure 49.** Yields of different isotopes of Mg, Al, Si, P, S, Cl, and Ar produced in the $^{232}$Th + $^{40}$Ar (290 MeV) reaction. The charge states are indicated (Ar 71c).

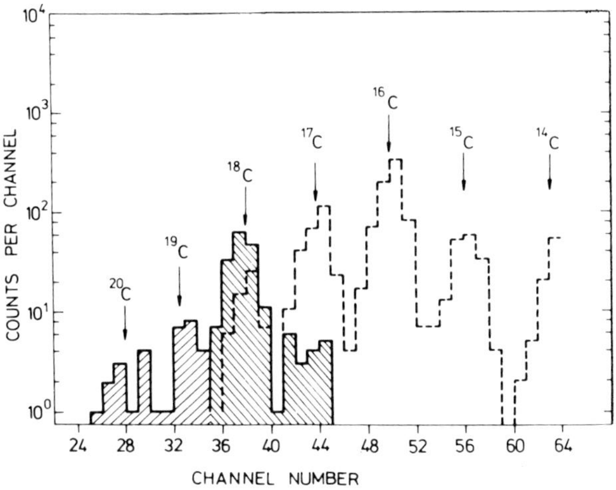


**Figure 50.** Yields of carbon isotopes from the reaction $^{232}Th + {}^{22}Ne$ (174 MeV). The detection angle is 40°, magnetic rigidity $B\rho = 9.26$ kG · m. The results of calibration measurements without TOF discrimination are shown by a dashed line. The shaded peaks are obtained in measurements with TOF discrimination. The ion flux passing through the target in the principal measurements is a factor of 60 higher than in calibration measurements (Ar 75).

**Table 3. Isotopes of Light Elements with Large Neutron Excess Produced in Multinucleon Transfer Reactions with Heavy Ions (Vo 73)**

| | | | | | | | | | | | | | | | |
|---|---|---|---|---|---|---|---|---|---|---|---|---|---|---|---|
| (a) Known isotopes | $^{8}$He, | $^{11}$Li, | $^{14}$Be, | $^{15}$B, | $^{17}$C, | $^{19}$N, | $^{21}$O, | $^{22}$F, | $^{24}$Ne, | $^{28}$Mg, | $^{30}$Al, | $^{32}$Si, | $^{34}$P, | $^{38}$S, | $^{40}$Cl |
| (b) New isotopes | | | | | $^{18}$C, | $^{20}$N, | $^{22}$O, | $^{23}$F, | $^{25}$Ne, | $^{29}$Mg, | $^{31}$Al, | $^{33}$Si, | $^{35}$P, | $^{39}$S, | $^{41}$Cl |
| | | | | | $^{20}$C, | $^{21}$N, | $^{23}$O, | $^{24}$F, | $^{26}$Ne, | $^{30}$Mg, | $^{32}$Al, | $^{34}$Si, | $^{36}$P, | $^{40}$S, | $^{42}$Cl |
| | | | | | | | $^{24}$O, | $^{25}$F | | | $^{33}$Al, | $^{35}$Si, | $^{37}$P | | |
| | | | | | | | | | | | | $^{36}$Si, | $^{38}$P | | |

neutron advance had been made in the direction of larger neutron excesses for some of these elements. Reviews of this work can be found in (Ar 70f, Vo 71, Vo 73).

The nucleon transfer reactions that contribute most to the formation of neutron-rich light-element isotopes are given in Table 4. They are mainly reactions involving neutron pickup by the projectile nucleus, proton stripping from the projectile, and nucleon exchange. In the last case the projectile nucleus gives up protons to the target nucleus and receives neutrons in exchange.

Multinucleon transfers were also used at the IPN, Orsay to obtain neutron-rich light-element isotopes. Auger *et al.* (Au 79) identified the new isotopes $^{37}Si$, $^{40}P$, $^{41}S$, and $^{42}S$ after the $^{238}U + {}^{40}Ar$ (263 MeV) reaction. They noted that at energies just over that of the Coulomb barrier, most of the neutron-rich isotopes are produced for elements with atomic numbers close to that of the projectile nucleus.

By increasing the energy of the $^{40}Ar$ ions to 340 MeV, Guerreau *et al.* (Gu 80) succeeded in producing seven new neutron-rich isotopes for elements with atomic

**Table 4. Types of Reactions Resulting in the Production of Neutron-Rich Isotopes of Light Elements (Vo 73)**

| | |
|---|---|
| 1. Neutron pickup | |
| $^{18}O(+1n) \rightarrow {}^{19}O$ | $^{22}Ne(+1n) \rightarrow {}^{23}Ne$ |
| $^{18}O(+2n) \rightarrow {}^{20}O$ | $^{22}Ne(+2n) \rightarrow {}^{24}Ne$ |
| $^{18}O(+3n) \rightarrow {}^{21}O$ | $^{22}Ne(+3n) \rightarrow {}^{25}Ne$ |
| $^{18}O(+4n) \rightarrow {}^{22}O$ | $^{22}Ne(+4n) \rightarrow {}^{26}Ne$ |
| 2. Proton stripping | |
| $^{11}B(-3p) \rightarrow {}^{8}He$ | $^{15}N(-3p) \rightarrow {}^{12}Be$ |
| $^{15}N(-4p) \rightarrow {}^{11}Li$ | $^{16}O(-4p) \rightarrow {}^{12}Be$ |
| | $^{16}O(-5p) \rightarrow {}^{11}Li$ |
| 3. Nucleon exchange | |
| $^{15}N(-2p, +2n) \rightarrow {}^{15}B$ | $^{22}Ne(-4p, +2n) \rightarrow {}^{20}C$ |
| $^{16}O(-3p, +2n) \rightarrow {}^{15}B$ | $^{22}Ne(-3p, +2n) \rightarrow {}^{21}N$ |
| $^{18}O(-2p, +2n) \rightarrow {}^{18}C$ | $^{22}Ne(-2p, +3n) \rightarrow {}^{23}O$ |
| $^{18}O(-1p, +3n) \rightarrow {}^{20}N$ | $^{22}Ne(-2p, +4n) \rightarrow {}^{24}O$ |
| | $^{22}Ne(-1p, +4n) \rightarrow {}^{25}F$ |

numbers markedly higher than that of the projectile, i.e., $^{54}Ti$, $^{56}V$, $^{58}Cr$, $^{59}Cr$, $^{61}Mn$, $^{63}Fe$, and $^{64}Fe$ (Figure 51). Breuer *et al.* (Br 80) at LBL in Berkeley used $^{56}Fe$ ions at 8.3 MeV/nucleon to bombard $^{238}U$ and detected several new isotopes, viz., $^{52}Sc$, $^{53}Sc$, $^{54}Ti$, $^{55}Ti$, $^{56}V$, $^{58}Cr$, and $^{59}Cr$. Both groups used the ($\Delta E - E$)-TOF technique to identify the new isotopes. The large yields from multinucleon-transfer reactions enabled them to use flight paths of up to 100 cm.

Thus we see that multinucleon transfer is an efficient way of obtaining new neutron-rich isotopes for the light and medium-weight elements, and at the same time enables us to get closer to the nucleon-stability limit.

### *3.4.2. Beyond the Nucleon-Stability Limit*

Establishing the position of the nucleon-stability limit for very neutron-rich light nuclei is very difficult. Coulomb repulsions rise fairly quickly and so impose a limit in the direction of large excesses of protons. The existence of a nucleon-stability limit for the neutron-deficient isotopes of light elements is not therefore in doubt.

The situation is not so clear as to the existence of a nucleon-stability limit for the neutron-rich isotopes of light elements, and an experimental verification is needed. The nucleon stabilities of nuclei such as $^{8}He$ and $^{11}Li$ demonstrate that the excess of neutrons over protons in light elements can be large. How far can this be extended in the $N$ direction, can more and more neutrons be added to a nucleus with a fixed $Z$? Is it possible to obtain nuclei consisting almost entirely of neutrons, a sort of neutron drop?

The shell model of the nucleus rules out this possibility. Pauli's exclusion principle for the nucleus requires far more levels in a neutron-only nucleus than in a nucleus containing the same number of neutrons and protons. It is possible to accommodate

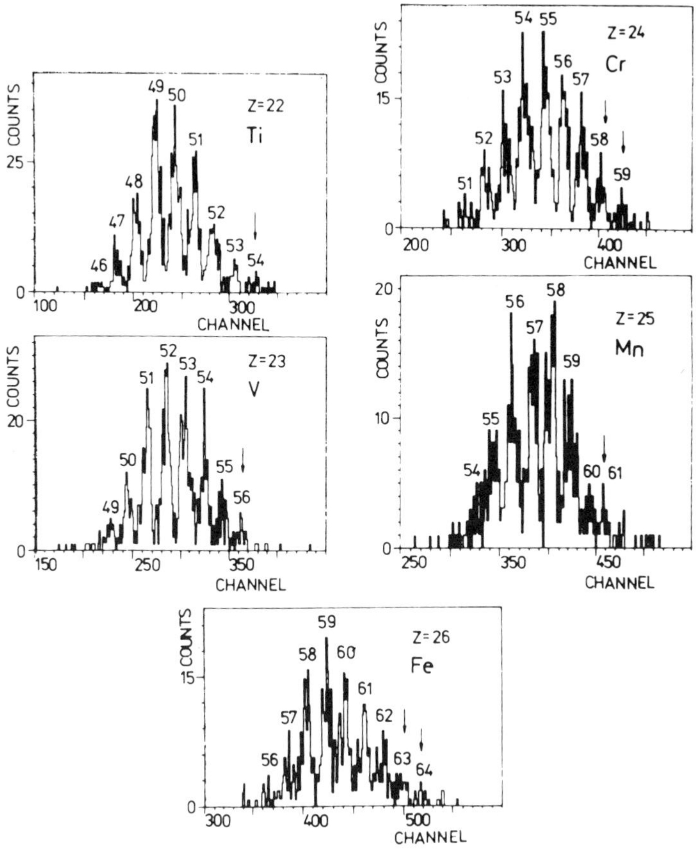


**Figure 51.** Ti, V, Cr, Mn, and Fe isotopes detected in the reaction $^{238}U + {}^{40}Ar$ (340 MeV). Isotopes observed for the first time are indicated by arrows (Gu 80).

two pairs of different nucleons but only one pair of similar nucleons at each level. A theoretical investigation of the positions of the energy levels for a nucleus with a fixed nucleon number using the shell model (Vi 69) showed that the upper energy levels for a light-element nucleus with a large number of neutrons would, if filled solely by neutrons, lie outside the potential well, i.e., the neutrons would not be bound to the nucleus. It is entirely clear that such nuclei would be unstable against the emission of a single or pair of neutrons.

However, several qualifications must be added to conclusions from the shell model, because we are dealing with nuclei under extreme conditions. A warning signal was the nucleon stability of the $^{11}Li$ nucleus. Three different shell-model calculations completed at the end of the 1960s predicted that the Li nucleus would be unstable (Co 65, Vi 69, Ga 66). However, three experimental groups, using different reactions and techniques, identified (Po 66b, Kl 69, Ar 70e) this nuclide, measured its mass, determined its radioactive decay mode, and established its half-life (Kl 69).

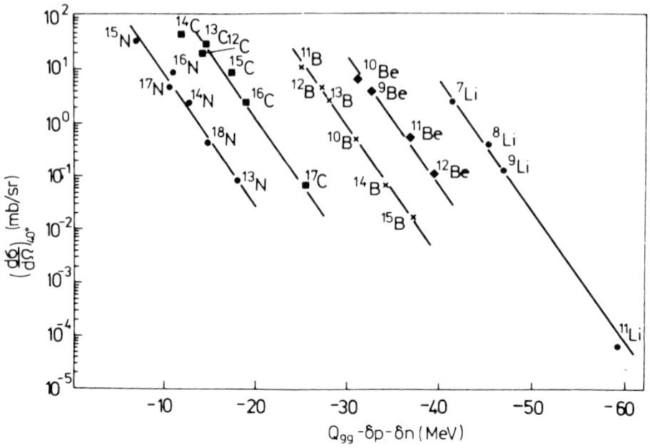


**Figure 52.** $Q_{gg}$ systematics of differential cross-sections for forming multinucleon-transfer products in the reaction $^{232}Th + {}^{16}O$ (137 MeV). The detection angle is 40° (Vo 75b).

In 1971 an attempt was made at the JINR Laboratory of Nuclear Reactions to go beyond the nucleon-stability limit by trying to produce $^{10}He$ using a multinucleon-transfer reaction. The $^{10}He$ nucleus is of especial interest in that it should be doubly magic. Its protons would fill the 1s shell, and the neutrons the 1s and 1p shells. The shell model predicted that $^{10}He$ would be unstable (Vi 69), although preliminary calculations using the K-harmonic method did not rule out its existence (Ba 69).

The attempt to get $^{10}He$ was made by Volkov and his co-workers (Ar 71b). A reaction involving the stripping of five protons from a $^{15}N$ nucleus as it interacted with $^{232}Th$ was selected. The $^{15}N$ ions were accelerated in the U-300 cyclotron to energies of 145 MeV, the high energy being needed to overcome the larger negative $Q$ for the reactions in which $^{10}He$ is produced. It had been established in previous experiments that the projectile nucleus readily transferred protons to the heavy target nucleus. The $^{232}Th + {}^{15}N$ (145 MeV) reaction yielded products corresponding to the stripping of between one and four protons from the projectile. The $^{232}Th + {}^{16}O$ (137 MeV) reaction had yielded $^{11}Li$, which involved the stripping of five protons from the projectile. The production cross-section of $^{11}Li$ satisfies $Q_{gg}$ systematics (Figure 52).

The $^{10}He$ and several other transfer products should be recorded at 40° in a MA-($\Delta E-E$) detector system. The optimum magnetic rigidity was chosen for the recording of $^{10}He$ on the basis of the energy spectra for the reactions involving the stripping of one to four protons and for reactions involving the loss from the projectile of two neutrons as well as the protons (Figure 53).

The target was bombarded for 20 h at an average beam intensity of 2 $\mu$A. If $^{10}He$ were nucleon stable, then according to the $Q_{gg}$ systematics (see Figure 54) there should have been some 80 events at the appropriate place in the two-dimensional $\Delta E$, $E-\Delta E$ matrix. However, not even one event was recorded in that region. The detection of the event would have corresponded to a cross-section of $3.5 \times 10^{-33}$ cm$^2$. Later, Oganessian *et al.* pushed the cross-section threshold for $^{10}He$ down another order of magnitude using the $^{232}Th$ ($^{11}B$, $^{10}He$) reaction (Og 82).

Since the production cross-sections of nuclides with low binding energies for a neutron or pair of neutrons, e.g., $^{11}Li$, $^{11}Be$, $^{14}B$, or $^{19}C$, all obey $Q_{gg}$ systematics, the

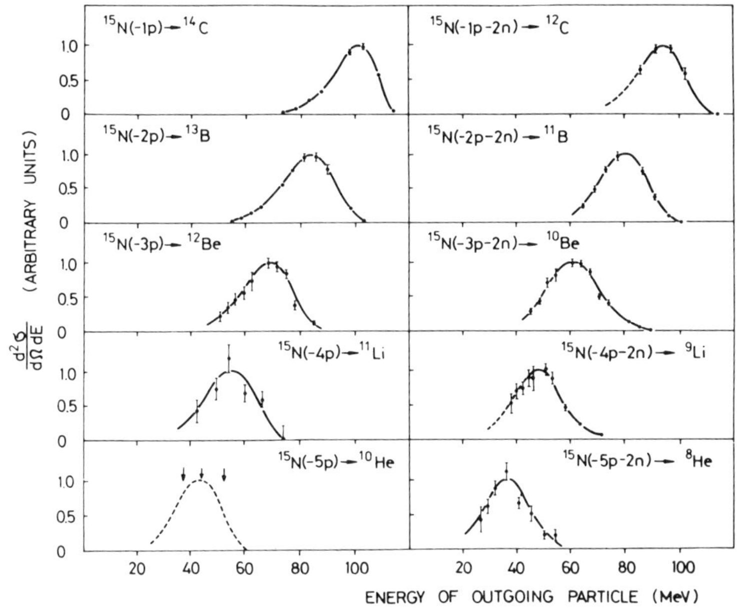


**Figure 53.** Energy spectra of the products of stripping from the projectile nucleus of a few protons ($-x_p$) and a few protons and two neutrons ($-xp$, $-2n$) in the reaction $^{232}Th + ^{15}N$ (145 MeV). The detection angle is 40°. The expected energy spectrum of the ($-5p$) reaction is shown by a dashed line. The energies at which $^{10}He$ was sought (Ar 71b) are indicated by arrows (Ar 71b).

absence of $^{10}He$ should be taken as a definite indication that it is not nucleon-stable. We note too that $^{10}He$ has not been detected either in spallation reactions with high-energy protons (Po 66b), or the fission of $^{235}U$ with thermal neutrons (Vo 69), or after the spontaneous fission of $^{252}Cf$ (Co 67). We can therefore assert that $^{8}He$ is the last nucleon-stable neutron-rich nucleus for $Z = 2$.

### *3.4.3. Investigating the Properties of Very Neutron-Rich Nuclei*

The nucleon-stability limits for medium-mass nuclei against the emission of a neutron is far from the stability line. It does not seem that it will be possible to reach it in the near future using the nuclear reactions known at present. However, the nuclear properties of the neutron-rich isotopes in this $Z$ region are of considerable interest.

The half-lives of the neutron-rich isotopes of elements in the neighborhood of iron are used in nucleosynthesis theories to determine the abundances of the elements in the Universe. The half-magic nucleus $^{68}Ni$ and the $^{67}Fe$ nucleus, which is predicted to emit neutrons from its isomeric state, both lie in this region.

The region is practically inaccessible for the traditional source of neutron-rich isotopes, viz., the fission of $^{235}U$ by thermal neutrons, because the yields of the isotopes of interest are so small. An effective method, however, is the multinucleon-transfer reaction using ions of the heavier nucleus argon.

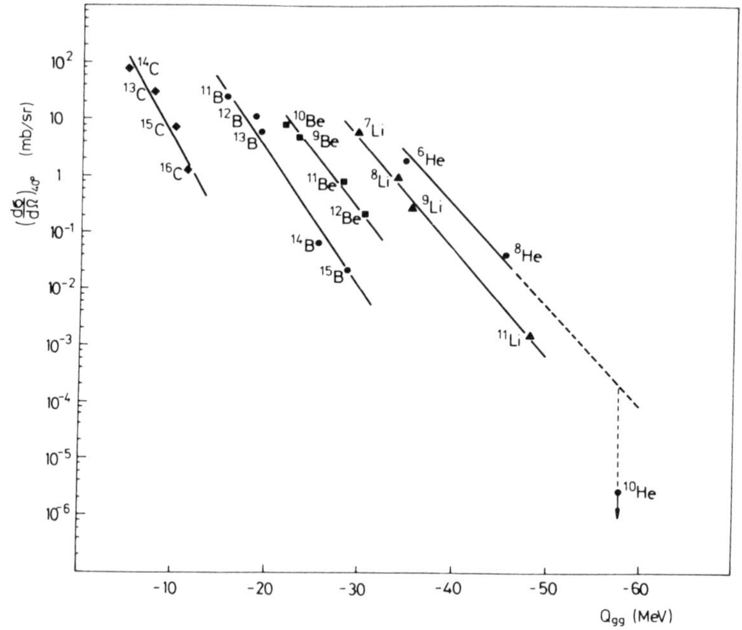


**Figure 54.** Search for $^{10}$He. The $Q_{gg}$ systematics of production cross-sections for the He, Be, Li, B, and C isotopes formed in the reaction $^{232}$Th + $^{15}$N (145 MeV); the detection angle is 40°. For $^{10}$He $Q_{gg}$ corresponds to the stability threshold $E(2n) = 0$. The point with an arrow indicates the value of the $^{10}$He production cross-section in the case of detection of one event. No one event has been detected in the experiment (Ar 71b).

The technique used to separate and study the isotopes—the ISOL method—generated information about the properties of these nuclides, such as radioactive-decay mode, half-life, decay energy, and lifetimes in the second range.

Most of the experimental information was obtained at GSI in Darmstadt by Roeckl and his co-workers using an on-line mass separator with the UNILAC accelerator (Ro 83).

A tungsten-foil incorporated into the ion source was bombarded with $^{76}$Ge ions at 8 MeV/nucleon or with $^{82}$Se ions at 11.5 MeV/nucleon. The ion source was operated at 2500 °C with thermal-surface ionization. Tantalum foil was used as a catcher. The detection module of the mass separator also ensured the registration of electrons, $\gamma$ rays, and X rays.

The first isotopes to be identified in these experiments (Ru 83, 85) were $^{62}$Mn, $^{63}$Fe, $^{71}$Cu, $^{72}$Cu, and $^{73}$Cu. The $\beta$-decay half-lives, $\beta$-transition energies, and the $\gamma$-ray energies accompanying the $\beta$ decays were all measured, and the half-lives of the $^{61}$Mn, $^{64}$Fe, $^{65}$Co, $^{67}$Ni, and $^{69}$Ni isotopes were measured for the first time (Ru 85). The yields of the neutron-rich isotopes of the elements from chromium to astatine are shown in Figure 55 as products of multinucleon-transfer reactions. The yields were measured at the focal plane of the GSI mass separator and were obtained by bombarding natural tungsten by $^{82}$Se ions at 11.5 MeV/nucleon. The yields were normalized by the intensity of the $^{82}$Se beam, which was 10 particles · nA.

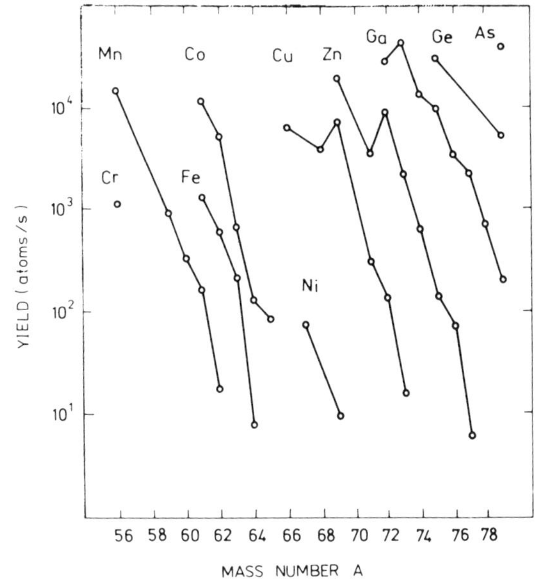


**Figure 55.** Yields of observed reaction products, measured in the focal plane of the mass separator and normalized to a $^{82}$Se beam intensity of 10 particles · nA. Plotted points for isotopes of the same element are connected by a line (Ru 85).

Multinucleon transfer using ions such as xenon has turned out to be an effective method of obtaining the neutron-rich isotopes of heavy rare-earth elements. This series of nuclides was also barely accessible by the thermal-neutron fission of uranium-235. Roeckl and his colleagues at the GSI bombarded a natural tungsten target with $^{136}$Xe ions at 9 MeV/nucleon. The ion source was operated in the thermal-ionization mode. Multinucleon-transfer reactions yielded the neutron-rich isotopes of ytterbium and lutecium, $^{179}$Yb, $^{181}$Lu, and $^{182}$Lu (Ki 82), which were identified for the first time. Their half-lives, $\beta$-transition energies, and the energies of the corresponding $\gamma$ quanta were also measured. This research group also used $^{136}$Xe at the higher energy of 11.7 MeV/nucleon and identified another isotope of lutecium, $^{183}$Lu, and measured its half-life (Ry 83).

Figure 56 contains the isotopic yields for the elements from francium to neptunium. These data were obtained by Gippert *et al.* (Gi 86) by bombarding tungsten and tantalum targets with $^{238}$U ions accelerated to 11.4 MeV/nucleon on the GSI's UNILAC. The multinucleon-transfer products included $^{232}$Ra, $^{232}$Ac, and $^{234}$Ac, which were identified for the first time. The figure also shows the potential-energy curves for a dinuclear system, which gives some idea of the way changes in the systems's potential energy affects the direction of the nucleon transfer. As before, the $\beta$-decay half-lives, the $\beta$-transition energies, and the accompanying $\gamma$-quanta energies were measured.

The experimental $\beta$-decay half-lives for neutron-rich isotopes are compared in Figure 57 (Ro 87) over a wide range of $Z$ and $A$ with results calculated using the microscopic (Kl 84) and gross (Ta 73) theories of $\beta$ decay. It can be seen that experiment and theory are in good agreement up to copper. However, the experimental half-lives

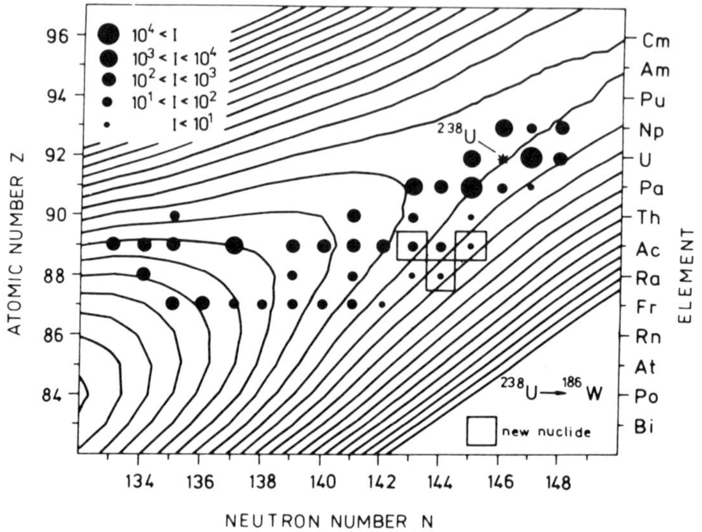


**Figure 56.** Part of the chart of the nuclides with the observed nuclei. The size of the dots corresponds to the yield $I$, given in atoms/s at 10 particles · nA beam current. Assuming a separator efficiency of 10%, 1000 atoms/s represent a cross-section of 2 mb. For comparison, contour lines of the potential-energy surface for 6.5 MeV/u $^{238}U$ and $^{186}W$ are displayed. The line spacing is 2 MeV (Gi 86).

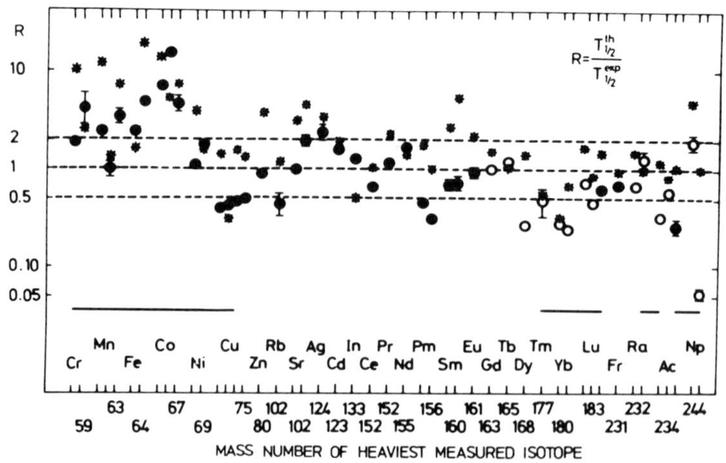


**Figure 57.** Ratios $R$ of predicted ($T^{th}_{1/2}$) to measured ($T^{exp}_{1/2}$) new $\beta$-decay half-lives. Circles indicate calculations from a microscopic model (Kl 84) and stars those from the gross theory of $\beta$ decay (Ta 73) using decay energies from the mass formula of (Mo 81) in the latter case. Full and open circles correspond to half-lives shorter or longer than 60 s, respectively. Horizontal bars indicate regions where the new isotopes were produced in multinucleon-transfer reactions (Ro 87).

of the Cr, Mn, Fe, and Co isotopes are systematically shorter than those predicted by theory. This divergence should be reflected in the calculations of the fast neutron capture processes (r processes) that are being studied in astrophysics (Bo 85).

# 4. Projectile Fragmentation as a Method for Producing Nuclei Far From Stability

## 4.1. Relativistic Energies

### *4.1.1. From Low to Relativistic Energies*

The development of heavy-ion studies proceeded so that the breakthrough into the region of high energies had first taken place at relativistic energies. Only a decade later did intermediate-energy accelerators appear, at which heavy ions were accelerated to energies of several tens of MeV per nucleon. The leap by nearly three orders in ion energy became possible due to the successful use of the then available 1–10 GeV proton accelerators for heavy-ion acceleration.

During the 1960s proton accelerators for energies of several tens of GeV were put into operation in Brookhaven and Argonne (USA), at CERN, and in Serpukhov (USSR), and elementary particle physics studies centered around these facilities. Accelerators for energies below 10 GeV seemed to have exhausted themselves. Some accelerators, such as the PPA in the USA and Nimrod in Great Britain, were shut down. However, scientists at Berkeley, Dubna, and Saclay found a new application for the old accelerators by converting them to accelerate light-element nuclei.

Theoretical studies of relativistic collisions between two complex nuclei predicted the existence of new, previously unstudied, physical phenomena which could not occur in nucleon–nucleon or nucleon–nucleus collisions (Ba 71, Ba 71b, Ba 75, Ba 80).

Very good conditions for producing heavy ions with relativistic energies were created at the Lawrence Berkeley Laboratory, University of California, Berkeley. Two facilities existed at Berkeley in close proximity to each other, namely, the linear heavy-ion accelerator (SUPERHILAC) and the 8.4 GeV proton accelerator (the Bevatron). The two accelerators were combined into a single acceleration system. The SUPERHILAC was operated as an injector of fully ionized light nuclei into the Bevatron. The system was called the Bevalac. By 1971 the Bevalac has accelerated carbon, nitrogen, and oxygen nuclei to energies of 2.1 GeV per nucleon. Somewhat later a system for producing a high vacuum ($10^{-9}$–$10^{-10}$ torr) was installed in the Bevatron chamber, and the ion sources were improved. This enabled research teams to accelerate the ions of many elements, including uranium, to relativistic energies (Gr 79).

At the JINR High Energy Laboratory, Dubna, new types of ion sources were designed, which could produce fully stripped light-element ions that could subsequently be accelerated in the Synchrophasotron. In 1977 carbon–neon ion beams with energies of 4.2 GeV/nucleon were produced, record-breaking energies for relativistic heavy ions (Ku 83). These high energies opened up a possibility for intensive studies of manifestations of the quark structure of nucleons in nuclei (Ba 80). At present the Synchrophasotron accelerates fully stripped ions of elements up to $^{28}$Si with energies of 4.2 GeV/nucleon.

At the Saturn accelerator at Saclay, $^{12}C$, $^{14}N$, and $^{20}Ne$ ions have been accelerated to energies of 1.1 GeV/nucleon since 1984 (Ra 86).

After having been converted to heavy-ion accelerators, these old facilities have been given a second lease of life.

### *4.1.2. Projectile Fragmentation and the Production of Nuclei Far From Stability*

Even in the first inclusive experiments involving the bombardment of targets with $^{12}C$, $^{14}N$, $^{16}O$, and $^{20}Ne$ ions at energies of 1–2 GeV/nucleon showed that relativistic collisions between two nuclei led to isotopes far from the stability line (Gr 75, Li 75). Figure 58 shows the spectrum of the magnetic rigidities of various carbon isotopes at 0°. It was obtained by bombarding a $CH_2$ target with 2.1 GeV/nucleon $^{16}O$ ions. The range of the isotopes detected extends from the lightest nucleon-stable carbon isotope, $^{9}C$, to the neutron-rich isotope $^{15}C$.

These spectra have two peculiar features. First, the velocities of the isotopes, for the most probable momentum values, practically coincide with that of the oxygen projectile. The arrows in the figure indicate the corresponding values of the momenta. The second feature is that the momentum distributions of all the isotopes have close half-widths which are substantially smaller in magnitude than the momentum itself. This type of the momentum distribution suggests that all the isotopes were produced in peripheral collisions between the projectile and the target nucleus. The energies of

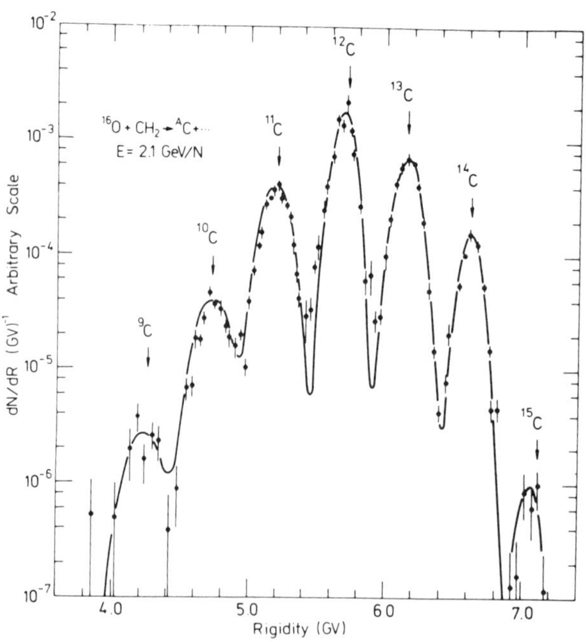


**Figure 58.** Rigidity spectrum of the carbon isotopes produced by the fragmentation of 2.1-A GeV $^{16}O$ projectiles. Arrows indicate the rigidities of each isotope at beam velocity (Gr 75).

the reaction products, in the frame of the bombarding nucleus, turned out to be small. This kind of interaction between two nuclei colliding at relativistic energies has been termed projectile fragmentation.

Projectile fragmentation in relativistic collisions has been investigated by many experimentalists with various target materials, projectiles, and kinetic energies. A number of theoretical models have been suggested to describe quantitatively the fragmentation process. The relevant experimental studies are covered in detail by Friedlander and Heckman (Fr 85) and the theoretical approaches to projectile fragmentation are discussed by Maruhn and Greiner (Ma 85) in the fourth volume of the Treatise on Heavy-Ion Science.

In this chapter we shall be considering those aspects of projectile fragmentation which are essential in order to produce nuclei far from stability.

The conservation of the velocity of the projectile fragments and of the direction of projectile motion is of fundamental importance in this case. Projectile fragmentation occurs in peripheral collisions involving low energy losses for nuclear friction and in which relatively low excitation energies are imparted to the fragments. The momentum dispersions along and perpendicular to the beam are close in the frame of the projectile $\sigma_{\parallel} \approx \sigma_{\perp}$, indicating the isotropy of the angular distribution of the projectile fragments in the frame of the bombarding nucleus. The absolute values of $\sigma_{\parallel}$ and $\sigma_{\perp}$ are considerably smaller than the average momenta of the projectile fragments.

These features mean that projectile fragmentation involves strong kinematic focusing for the reaction products in the laboratory system of coordinates. The overwhelming majority of the products are emitted at an angle of 0° in a narrow cone with a half-width of about 1°. The product velocities also group round an average value close to the projectile velocity.

Kinematic focusing ensures that the collection of the fragments on the detector of the analyzing system is efficient and that the $Z$ and $A$ resolutions may be high. A considerable portion (several tens of percent) of the total number of fragments will enter a magnetic analyzer with an acceptance angle as small as 1 mrad.

The close velocities of the projectile fragments facilitate the spatial separation of the products according to the $A/Z$ ratio with the help of a magnetic analyzer. The high velocities lead to the loss of all electrons by the projectile fragments during their interaction with the target. The presence of only one charge state and long ranges in the detectors make it easier to identify separate isotopes. The high kinetic energies of the projectile fragments allow targets with thicknesses of up to several grams per square centimeter to be used. The efficiency with which the projectile fragments are collected and detected can therefore be made much higher than that for reaction products at low bombarding energies. This factor alone makes relativistic heavy-ion beams a competitive method for the production of nuclei far from stability even though the intensities of the beams are many orders of magnitude lower than those of low-energy heavy-ion beams.

The rate at which the production cross-section falls with increasing distance of the isotopes from the region of stability is an important characteristic of the nuclear reactions leading to exotic nuclei. The theoretical calculation of this characteristic of projectile fragmentation reactions is rather difficult. The rate of fall of the cross-sections reflects not only the regularity of forming projectile fragments, but also the effects of the subsequent nucleon evaporation from excited states. Weak nucleon binding in exotic nuclei makes them especially sensitive to the excitation energy, which is a difficult quantity to calculate. Therefore experiment plays a crucial role in disclosing this characteristic of the projectile fragmentation reactions. We shall revert to this topic later when we discuss experimental results on the production of nuclei far from stability.

### 4.1.3. *The Choice of Projectile and Target and the Role of Bombarding Energy*

When attempting to produce nuclei far from stability the experimentalist must choose the heavy ion and the target nucleus. Projectile fragmentation reactions lead to a great variety of products with different values of $Z$ and $A$. Because the difference between the nucleon momenta in the target nucleus and in the projectile is large, projectile nuclei do not pick up nucleons. At the same time it should be possible to produce nuclei with atomic numbers one and two greater than the $Z$ of the projectile due to single or double charge exchange. However, experiments have shown that the cross-sections of nucleon charge exchange processes are small. Therefore, projectiles with a noticeably larger value of $Z$ are used to produce exotic isotopes of a given element.

The nucleon-stability limit for neutron-deficient nuclei is closer to the stability region than is the limit for neutron-rich nuclei. By the time experiments aimed at producing exotic nuclei via projectile fragmentation reactions were begun, the position of the nucleon-stability limit had been established for many of the neutron-deficient isotopes of light elements. In the case of neutron-rich isotopes the limit had been reached for $Z \leq 4$. That was why projectile fragmentation experiments to produce exotic nuclei were initially aimed at obtaining the neutron-rich isotopes of light elements.

Both experimental evidence and the theoretical models of projectile fragmentation suggested that the fragmentation of a nucleus with atomic number $Z$ and mass number $A$ might lead to a nucleus whose atomic number $Z_f$ and mass number $A_f$ met the conditions $Z_f \leq Z$ and $A_f \leq A$. Naturally, a projectile having a large neutron excess would therefore be expected to fragment into a larger number of neutron-rich isotopes. Indeed, the yield of neutron-rich sodium isotopes produced by bombarding a beryllium target with 215 MeV/nucleon $^{48}Ca$ ions turned out to be an order of magnitude higher than that observed when a similar target was bombarded with 205 MeV/nucleon $^{40}Ar$ ions (We 79).

The neutron excess in stable nuclei increases with $Z$. In the $^{76}_{32}Ge$ nucleus it amounts to 12 neutrons, i.e., 4 neutrons more than in the $^{48}_{20}Ca$ nucleus. We might then ask whether the fragmentation of a nucleus such as $^{76}Ge$ get us closer to the nucleon-stability limit. An analysis by Harvey (Ha 85) showed that the degree to which the nuclear surfaces overlap during a projectile-fragmentation reaction grows as the number of nucleons lost by the projectile increases. The more the nuclear surfaces overlap, the greater the excitation energies become in the projectile fragments, and hence the lower the survival of the exotic nuclei. Thus, the experimentalists have the last word concerning the use of heavier ions than calcium for the production of new neutron-rich isotopes of light elements.

An analysis of the energy dependence of the cross-section for forming projectile fragments has demonstrated that after an energy of about 1 GeV/nucleon the cross-sections can be factorized, i.e.,

$$\sigma_F = \sigma_T \cdot \gamma_{pf} \tag{4.1}$$

Here $\sigma_T$ is a factor that depends on the target alone, neither the projectile characteristics nor the collision energy affecting it. It turns out that $\sigma_T \sim A^{1/4}$. This reflects the surface character of projectile-fragmentation reactions. The $\gamma_{pf}$ factor is determined by the properties of the projectile and the fragment, but does not depend on the nuclear properties of the target.

Equation (4.1) describes properties of projectile-fragmentation reactions at fairly high energies. The target nucleus, owing to the geometry of its surface, initiates projectile fragmentation by inducing its excitation and decay. However, the relationship between the different fragmentation channels of the projectile is determined by the properties of the nucleus itself, namely, the "reduced widths" of its virtual fragmentation into different pairs of nuclei (Fr 83).

The rather weak dependence of $\sigma_T$ on $A$ determines the choice of target needed to produce nuclei far from stability. The energy losses in the target due to ionization grow with target $Z$. The average angle of particle deflection also increases with $Z$ in cases of single and multiple Coulomb scattering. A beryllium target has proved convenient for most experiments. In fact, almost all the experiments aimed at producing nuclei far from stability by projectile fragmentation at relativistic energies have been carried out using beryllium targets.

The relativistic energies that can be imparted to a projectile are determined by the ability of the accelerator itself. Studies of the energy dependence of the fragmentation of nuclei ranging from $^{12}C$ to $^{40}Ar$ have demonstrated that the production cross-sections of various isotopes reach a plateau at an energy of about 1 GeV/nucleon (Li 75). This phenomenon has been called limiting fragmentation. Although increasing the energy beyond this limit does not lead to larger cross-sections for the formation of projectile fragments, it makes possible the use of thicker targets. Furthermore, the momentum dispersions $\sigma_{\parallel}$ and $\sigma_{\perp}$ of projectile fragments, in the frame of the bombarding nucleus, is independent of the heavy-ion energy under the conditions of limiting fragmentation. Increasing heavy-ion energy decreases the angular spread of the projectile fragments. The beam of projectile fragments becomes similar to the bombarding beam, thus helping the identification of the exotic nuclei and offering the possibility of secondary beam experiments.

### 4.1.4. *Identification of Projectile Fragments*

Figure 59 shows a schematic view of the experimental setup used to identify nuclei produces as a result of projectile fragmentation (Sy 81). The principal unit is a magnetic analyzer designed for the spatial separation of isotopic beams with different $A/Z$ ratios. A pair of quadrupole lenses focus the projectile fragment beams into the detector system. The latter is located in air and outside a $7 \times 3\ m^2$ vacuum tank, provided with windows for the exit of the reaction products.

The detector system consists of two position-sensitive detectors to measure the horizontal and vertical deflections of the beam, 12 $\Delta E$ detectors to measure the specific ionization energy and range of each particle, and a veto detector to reject particles which have passed through all the 12 $\Delta E$ detectors. The position-sensitive Si(Li) detectors are 6 cm in diameter and 500 $\mu$m thick, and the Si(Li) $\Delta E$ detectors are 5 cm in diameter and 5 mm thick. The veto detector is made from a plastic scintillator.

For each $\Delta E$ detector, the first product identification signal is determined according to the formula

$$PI_i = [(E_i + \Delta E_i)^n - E_i^n]/S_i \propto M^{n-1}Z^2 \tag{4.2}$$

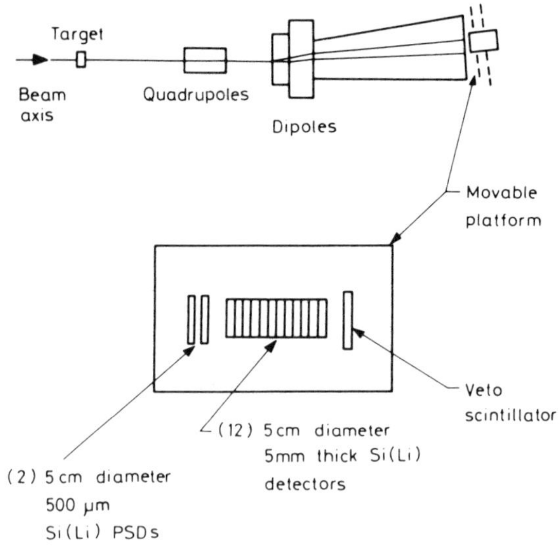


**Figure 59.** Experimental layout for detection of fragments of 212 MeV/nucleon $^{48}$Ca (Sy 81).

where $\Delta E_i$ is the energy loss in the $i$th $\Delta E$ detector, $E_i$ is the total energy released by the product in the stack of $\Delta E$ detectors including the detector in which it is stopped, $S_i$ is the thickness of the $i$th detector, $n$ is a parameter whose value varies somewhat from element to element ($n \sim 1.78$), and $M$ and $Z$ are the mass and charge of the product. The $I_i$ signals arriving from all $\Delta E$ detectors are combined to obtain the weighted mean $\bar{I}$ and the function $\chi^2$ defined by the relation

$$\chi^2 = \sum_{i=1}^{s-1} \left( \frac{I_i - \bar{I}}{\varepsilon_i^2} \right) \tag{4.3}$$

where $\varepsilon_i$ is the area of each $I_i$ signal. The mass separation is substantially improved by rejecting particles with large $\chi^2$ values.

The second identification signal is formed by measuring the particle kinetic energy $T$ and the linear deflection $D$ in the detector plane due to the magnetic field, and is given by

$$PI = k/TD^2 - T/2Z^2 \propto M/Z^2 \tag{4.4}$$

where $k$ is a spectrometer constant determined in a calibration experiment using particles with known $Z$, $A$, and $T$ values. The $Z$ and $A$ resolutions are 0.2 e unit and 0.2–0.3 amu, respectively.

Other types of detector were also used in the same experimental setup. For example, projectile fragments were stopped by a stack of Lexan plastic foils (St 81). The stack was thick enough to stop all products with $Z > 5$. The plastic foil was treated chemically to reveal the tracks left by the products and then the track lengths and diameters near the end of the paths were measured. These data can be used to identify the atomic number and kinetic energy of each product, from which the product mass can be determined according to the linear deflection of the trajectory in the magnetic field. With this type of detector, the charge and mass resolutions are $\sigma_Z \approx 0.2$ e unit and $\sigma_A \approx 0.2$ amu, respectively.

Musser and Stevenson (St 83, Mu 84b) used a detector telescope consisting of a wire-chamber hodoscope, a front scintillator paddle, a set of threshold Cherenkov counters, and a back scintillator paddle. The product charge was determined from the scintillator and Cherenkov-counter signals. The product mass could be found from the particle rigidity in combination with the Cherenkov signal measurements. When detecting fragments from 670 MeV/nucleon $^{56}$Fe nuclei the resolution was $\sigma_Z = 0.1$ e unit and $\sigma_A = 0.25$ amu in charge and mass, respectively.

### *4.1.5. New Neutron-Rich Isotopes of Light Elements*

The pioneering work on new neutron-rich light-element isotopes using projectile-fragmentation reactions at relativistic energies was done by Symons *et al.* (Sy 79). A 860 mg/cm$^2$ Be target was bombarded by $^{40}$Ar ions with energies of 205 MeV/nucleon. The projectile fragments were detected and the separate isotopes identified using the experimental setup described in the previous section. However, a 3-plane multiwire proportional chamber placed in front of the $\Delta E$ detectors was used to measure the magnetic-field deflection of the products. The intensity of the $^{40}$Ar ion beam incident upon the target was equal to about $5 \times 10^6$ pps. The target exposure ranged from 20 min to 8 h. Figure 60 shows the yields of the various isotopes of neon and aluminum. Together with the known isotopes, two new neutron-rich isotopes, $^{28}$Ne and $^{35}$Al, were detected for the first time. The ratio of isotopic yields given in Figure 60 does not correspond exactly to the cross-sections of their production as the exposure duration was different for different magnetic rigidities.

An especially "fine catch" of new neutron-rich light-element isotopes was made by Westfall *et al.* (We 79). Instead of $^{40}$Ar projectiles they used $^{48}$Ca, which has twice the neutron excess. Nuclei of $^{48}$Ca with energies of 212 MeV/nucleon bombarded a 890 mg/cm$^2$ beryllium target. The layout of the experiment is shown in Figure 59. A pair of position-sensitive Si(Li) detectors was used to measure the horizontal and vertical deflections of the projectile-fragment beam. Westfall *et al.* succeeded for the first time in observing 14 new neutron-rich isotopes: $^{22}$N, $^{26}$F, $^{33}$Mg, $^{34}$Mg, $^{36}$Al, $^{37}$Al,

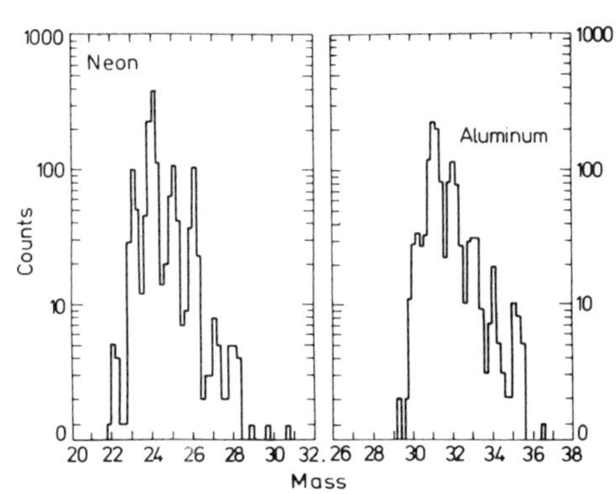


**Figure 60.** Neon and aluminum isotopes detected in $^{40}$Ar fragmentation at an energy of 205 MeV/nucleon (Sy 79).

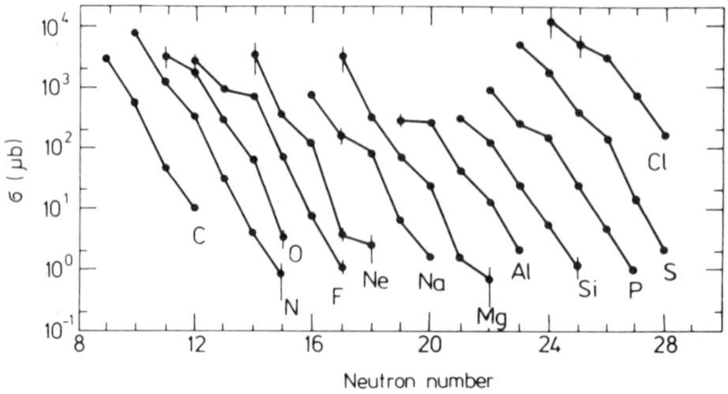


**Figure 61.** Production cross-sections for the elements observed in the fragmentation of 212 MeV/amu $^{48}$Ca by a beryllium target. Lines are to guide the eye (We 79).

$^{38}$Si, $^{39}$Si, $^{41}$P, $^{42}$P, $^{43}$S, $^{44}$S, and $^{44}$Cl, $^{45}$Cl. The existing theoretical models predict that all these isotopes should be stable against the emission of one or two neutrons.

The measurements of Westfall *et al.* of the production cross-sections for this large number of light-element isotopes allow one to plot production cross-section as a function of isotope neutron number for elements ranging from C to Cl (Figure 61). On the average, the isotopic production cross-section decreases by one order of magnitude as the number of neutrons in the isotope increases by one. Westfall *et al.* noted the effect of nuclear structure on the isotopic yield. Isotopes with an even number of neutrons have somewhat larger yields. The production cross-section for oxygen isotopes decreases with increasing neutron excess more rapidly than the corresponding curve for isotopes of the adjacent odd-$Z$ elements. Westfall *et al.* account for this effect as being due to the smaller number of bound excited levels in oxygen neutron-rich isotopes.

The use of a plastic detector stack for recording projectile fragments by Stevenson and Price (St 81) allowed them to improve the sensitivity of their experimental setup and to get closer to the nucleon-stability limit for light-element nuclei with large neutron excesses. By bombarding a 890 mg/cm$^2$ beryllium target by $^{48}$Ca ions with energies of 213 MeV/nucleon, they succeeded in detecting $^{20}$C and $^{27}$F nuclei (Figure 62). The $^{27}$F nucleus was detected for the first time, while the $^{20}$C isotope had been observed at Dubna earlier (Ar 75).

The heaviest boron isotope $^{19}$B predicted as being stable against nucleon emission was first produced by Musser and Stevenson (Mu 84b) (Figure 63) as a projectile fragment by bombarding 670 MeV/nucleon $^{56}$Fe nuclei against a beryllium target. They used larger energies than had been possible in previous experiments and this allowed them to increase the beryllium target thickness by nearly an order to magnitude to 7.9 g/cm$^2$. In that experiment a very deep fragmentation of the projectile had been realized. In order to produce a $^{19}$B nucleus, 21 protons and 16 neutrons had to be "chopped" off the $^{56}$Fe nucleus. This result was indicative of the possibility of using fairly heavy projectiles with a considerable neutron excess for an advance toward the boundary of nucleon stability of nuclei.

Projectile fragmentation at relativistic energies has proved to be an efficient method for producing nuclei far from stability. During a relatively short period of time, 19 new neutron-rich light-element isotopes were produced, the nucleon-stability limit was

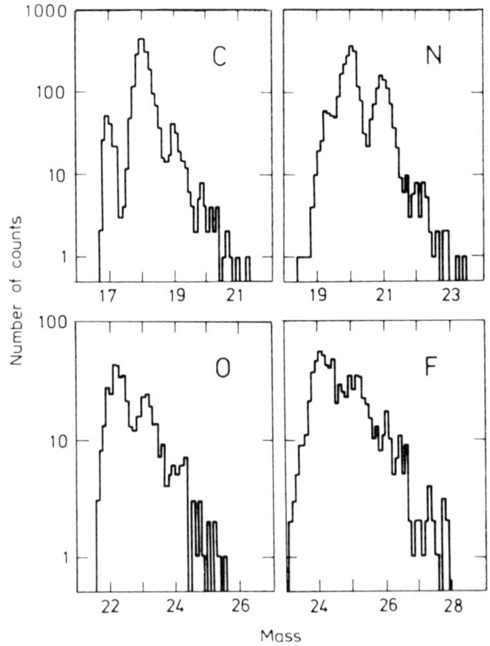


**Figure 62.** Mass histograms of neutron-rich isotopes of C, N, O, and F observed in experiment. Peak heights do not directly reflect the relative abundances of the isotopes (St 81).

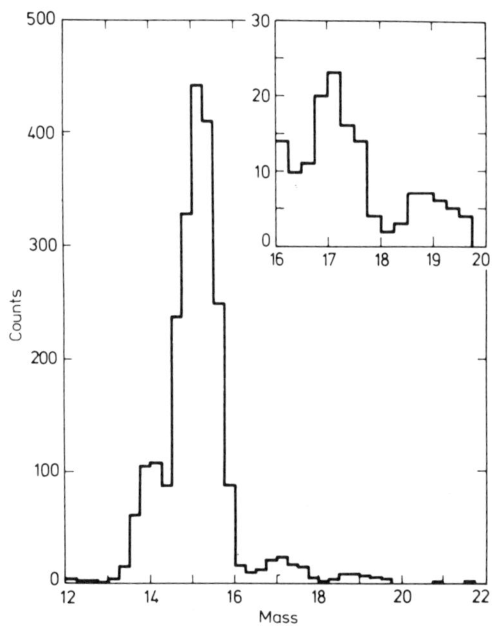


**Figure 63.** Histogram of boron isotopes $^{15}B$ to $^{19}B$. Inset: $^{17}B$ and $^{19}B$ counts (Mu 84b).

reached for $Z = 5$, and experiments began on the study of the radioactive decay of exotic nuclei (Sy 81).

### 4.1.6. *Secondary Beams*

As the bombarding energy is increased, the beams of projectile fragments become more and more like their primary ion beams. This opens the way to the use of the projectile fragments as secondary beams. The first such studies were performed at the Bevalac (Berkeley) by Tanihata *et al.* (Ta 85). A 4.7 g/cm$^2$ beryllium target was bombarded with $^{11}B$ and $^{20}Ne$ ion beams at energies of 800 MeV/nucleon. The projectile fragmentation products, i.e., the helium isotopes $^3He$, $^4He$, $^6He$, and $^8He$, and the lithium isotopes $^6Li$, $^7Li$, $^8Li$, $^9Li$, and $^{11}Li$, were separated by a magnetic analyzer according to magnetic rigidity. The beam of one of the isotopes was redirected onto the target, Be, C and Al being used as target materials. By using the method of particle knockout from the beam, the interaction cross-section between the projectile fragments and the target nuclei, $\sigma_I$, was measured:

$$\sigma_I = \frac{1}{N_t}\log(\gamma_0/\gamma) \tag{4.5}$$

Here $\gamma$ is the ratio of the number of noninteracting projectiles to the total number of projectiles that have struck the target; $\gamma_0$ is the same ratio but in the absence of a target. In all measurements $\gamma_0$ exceeded 0.95. The target thicknesses for the secondary beams were chosen so that the $\gamma$ value lay between 0.4 and 0.8. The interaction cross-sections were determined for all the nucleon-stable isotopes of helium and lithium. On the basis of the data obtained, the isotopic radii were calculated by the formula

$$\sigma_I(p, t) = \pi[R(p) + R(t)]^2 \tag{4.6}$$

where $R(t)$ and $R(p)$ are the radii of the target and projectile nuclei, respectively. The validity of equation (4.6) was verified by comparing data obtained with different targets.

The results of (Ta 85) are displayed in Figure 64. Note the large radius of the $^{11}Li$ nucleus. It is substantially bigger than the radius of $^{12}C$, thus reflecting the weakly bound nature of $^{11}Li$. It can also be seen that the $^6He$ and $^8He$ nuclei are larger than the corresponding isotopes of lithium. The presence in the nucleus of one or two loosely bound neutrons causes the nucleus "swell."

## 4.2. Intermediate Energies

### 4.2.1. *Heavy-Ion Acceleration to Intermediate Energies*

Heavy ions with intermediate energies were first produced at the existing light-particle accelerators. In 1978 $^{12}C^{4+}$ and $^{18}O^{6+}$ ions were accelerated to energies of 86 MeV/nucleon at the CERN 600 MeV Proton Synchrotron (Al 81). The $^{12}C$ beam intensity reached several times $10^{11}$ pps. The ion beams were used in experiments aimed at studying nuclear interactions in intermediate-energy collisions (Gu 83, Ly 82). However, by the early 1970s it had become apparent that special-purpose heavy-ion accelerators had to be designed to study intermediate-energy processes.

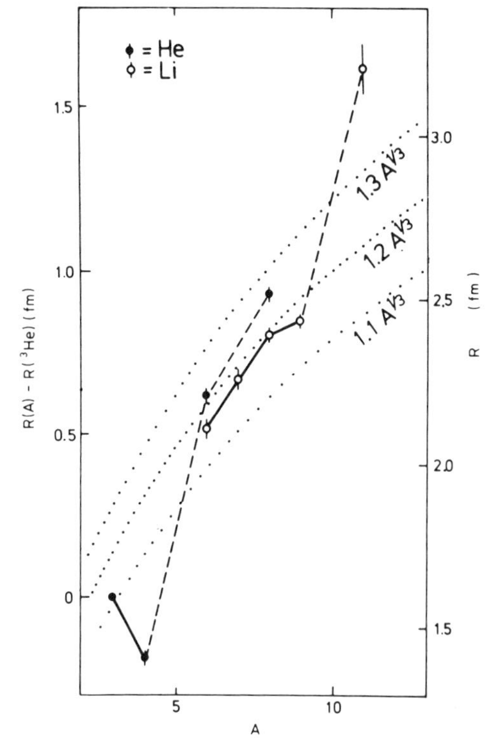


**Figure 64.** Nuclear radii of the helium and lithium isotopes obtained by the method of secondary beam scattering (Ta 85).

Acceleration to energies of several tens of MeV/nucleon is only possible for highly charged heavy ions, because otherwise the size and cost of the accelerator would be unacceptably large. The importance of a high charge state for a heavy ion can be clearly seen from the following equation, which gives the energy per nucleon for a heavy ion accelerated in a cyclotron:

$$E/A = Kq^2/A^2 \tag{4.7}$$

where $q$ is the ion charge in units of the charge on an electron e; $A$ is the ion mass number; and $K$ is the energy parameter of the cyclotron, $K \approx 0.5 B_f^2 R_f^2$, with $B_f$ being the induction of the average magnetic field in kG, and $R_f$ the final acceleration radius in meters. The value of $K$ varies from several tens to several hundreds, depending on the size of the cyclotron and the magnitude of the magnetic field.

The production of high-charge-state ions directly in the ion source of the accelerator is quite difficult. The ion yield drops quickly with increasing charge. The difficulty can be overcome using a two-stage acceleration combining two accelerators into a tandem system. In the first accelerator, ions with low charges are accelerated to 0.1–1 MeV/nucleon and then sent through a thin carbon foil. The ions are stripped of all or most of their electrons and then accelerated in their high charge states by the second accelerator.

The two-stage method for accelerating heavy ions has been extensively utilized in low-energy facilities. Various combinations of accelerators for energies of up to

10 MeV/nucleon have been tried, for example, two cyclotrons, two linacs, linacs and a cyclotron, a tandem electrostatic generator and a cyclotron.

A tandem system of two cyclotrons—the classical 3-meter and 2-meter isochronous cyclotrons—was first built at Dubna in 1971 (Sh 72). The system can accelerate xenon ions to energies of 6.5 MeV/nucleon, which were used to synthesize superheavy elements (Fl 74). Experience in operating the system of two cyclotrons in tandem has shown that it can accelerate a wide range of heavy ions to high beam intensities. This system has been used in a number of intermediate-energy, heavy-ion acceleration facilities.

The first tandem system of two cyclotrons designed to produce intermediate-energy heavy ions, the SARA (Systeme Accelerateur Rhone/Alpes), was put into operation in early 1982 in Grenoble, France (Lo 84). The system consists of a classical cyclotron (magnet pole diameter $D = 2.12$ m, $K = 88$) and an isochronous four-sector cyclotron ($D = 2.28$ m, $K = 160$). In the SARA system $^{12}C$, $^{20}Ne$, and $^{40}Ar$ ions are accelerated to energies of 30–40 MeV/nucleon at intensities of about $2 \times 10^{11}$ pps and about $10^9$ pps, respectively.

Figure 65 shows the layout of the major accelerator facility for intermediate-energy heavy ions (GANIL, Grand Accelerateur National d'Ions Lourds) that has been operated at Caen in France since early 1983 (Fe 81). GANIL is a tandem system of two identical four-sector cyclotrons. The external diameter of each cyclotron magnet is 8 m, the beam extraction radius is 300 cm, and the magnet weighs 1700 tonnes. The $K$ value for both cyclotrons is 400. Two small classical cyclotrons ($K = 30$) are used as heavy-ion injectors: the first being in combination with a PIG ion source and the other with an ECR-type ion source. When accelerated in the first cyclotron $^{40}Ar$ ions have a charge state of $4^+$ and $^{16}O$ ions a charge state of $3^+$. The external beam intensity is $6 \times 10^{12}$–$10^{13}$ pps, with energies of about 0.5–0.7 MeV/nucleon. After passing through the carbon foil the $^{16}O$ ions are completely stripped of electrons, $^{40}Ar$ ions have the most probable charge state of $16^+$, and $^{86}Kr$ one of $26^+$. The energies of the $^{16}O$, $^{40}Ar$, and $^{86}Kr$ ions extracted from the second cyclotron are 94, 44, and 35 MeV/nucleon at intensities of $\sim 10^{12}$, $\sim 4 \times 10^{11}$, and $\sim 2 \times 10^{10}$ pps, respectively.

In recent years, the heavy-ion beams produced by the GANIL accelerator have yielded valuable information about nuclear interactions in intermediate-energy collisions and twenty new light-element isotopes, both very neutron-rich and very neutron-deficient, have been detected.

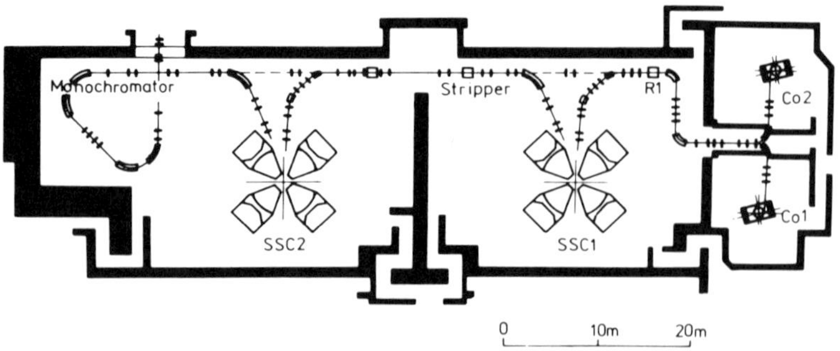


**Figure 65.** General layout of the GANIL accelerator (Fe 81).

It was planned to construct a tandem system of two superconducting cyclotrons with $K_1 = 500$ and $K_2 = 800$ at Michigan State University, USA (Mi 84b). The first superconducting cyclotron with $K_1 = 500$ was put into operation in late 1982. It has accelerated light-element ions to an energy of 35 MeV/nucleon.

At the Laboratory of Nuclear Reactions at Dubna, a tandem is being built consisting of two 4-meter isochronous cyclotrons (U400 and U400M) with an increased magnetic field (Fl 84). The first cyclotron, the U400, has an isochronous magnetic field in the range from 19 to 21.4 kG ($K = 700$) and was put into operation in 1978. It can accelerate ions with masses ranging between 4 and 250 amu to energies from 35 to 1.7 MeV/nucleon, respectively. In the class of accelerators designed for energies of up to 20 MeV/nucleon it has produced the most intense beams for heavy ions with $A < 90$ (Og 83). The U400–U400M cyclotron tandem will accelerate elements ranging from oxygen to uranium to energies of 120–20 MeV/nucleon, respectively, with very high intensities of the ion beams.

A tandem cyclotron system consisting of a classical cyclotron ($K = 69$) and a separated-sector cyclotron ($K = 450$) is being constructed at Lanzhou, China (We 81).

Various other tandem accelerator systems for producing ions of intermediate energies are under construction. At the RIKEN Cyclotron Laboratory of the Institute of Physical and Chemical Research, at Saitama in Japan, a system consisting of a linac and a four-sector cyclotron ($D = 12.6$ m, $K = 540$) is under construction (Ka 84). An accelerator system consisting of a three-sector superconducting cyclotron with $K = 800$ and a 15 MV tandem is being built at the National Institute of Nuclear Physics in Milan, Italy (Ac 84). A similar combination of a four-sector superconducting cyclotron ($K = 520$) and a 13 MV tandem has been put into operation at Chalk River, Canada (Or 84).

### 4.2.2. Nuclear Interaction Mechanisms at Intermediate Energies

*4.2.2a. From Low and Relativistic to Intermediate Energies.* The mechanisms governing nuclear interactions at low and at relativistic energies differ in principle. Collisions between nuclei at low energies do not lead to their immediate destruction. The nuclei interact with each other as tightly bound systems of nucleons without losing their individuality during the collision.

In quasi-elastic collisions the nuclei exchange nucleons get slightly excited and continue to move along the trajectories determined by the average field produced by the joint action of all the nucleons in each nucleus.

In deep-inelastic collisions, the dissipation of significant amounts of kinetic energy results in the formation of a dinuclear system, within which the nuclei still retain their individual features. An important consequence of deep-inelastic collisions is the evolution of the dinuclear system, which proceeds much more slowly than the dissipation of the kinetic energy of the collision. The process is completed either by the formation of a compound nucleus or by the decay of the system into two fragments.

In contrast, collisions at relativistic energies are characterized by the destruction, whether partial or total, of the nuclei at the moment of their collision. The result of the collision depends on the collision geometry, e.g., the degree of an overlap between the nuclear volumes. Only the nucleons in the overlap region interact, the rest of the nucleons being spectators as it were. The collision proceeds vary rapidly, lasting about

$10^{-22}$ s. In grazing collisions, the projectile nucleons in the overlap region are "cut" off and the remainder of the projectile continues to move in the same direction with the initial velocity. The number of neutrons and protons in the overlap region vary randomly for each nucleus. Consequently, the nucleon composition of the residual nuclei will also change. In knock-on collisions, the interaction involves most of the nucleons in the two nuclei, a considerable amount of energy is released, and the nuclei literally explode. The products of the microexplosion include nucleons, $\alpha$ particles, and light-element nuclei.

In order to produce nuclei far from stability at low energies almost every kind of nuclear reaction can be used, namely, fusion, deep-inelastic transfer, and quasi-elastic transfer. At relativistic energies, only projectile-fragmentation reactions can be used since kinematical focusing occurs only for the products of these reactions.

The sorts of questions that can be answered by experimentalists using heavy-ion intermediate energy include: How do nuclear interactions proceed in intermediate-energy collisions? Which mechanisms typical at low and relativistic energies remain in this intermediate region? Do new nuclear processes occur which are peculiar to intermediate energies? What nuclear processes can be used to advance toward the nucleon-stability limit of nuclei?

A considerable quantity of experimental information has now been accumulated, and this has been reviewed at recent international conferences. Several theoretical models are being actively developed to enable the interpretation and quantitative description of the patterns that have been observed. As a whole, however, this field of nuclear physics research using heavy ions is still in its infancy. Considerable efforts should be made by both experimentalists and theoreticians to obtain a unified picture of the nuclear interactions at intermediate energies. The scope of this review permits only a very general description of the nuclear reactions induced by heavy ions at intermediate energies. They will be of interest to us in the context of the production of nuclei far from stability.

*4.2.2b. Extinction of the Compound Nucleus.* The energy released in a central collision at intermediate energies can approach or even exceed the total nucleon binding energy of both nuclei. The behavior of the nucleus with increasing excitation energy has been theoretically analyzed by Bondorf, Mishutin *et al.* (Bo 86, Mi 85) under the assumption of the establishment of statistical equilibrium. It turns out that at excitation energies of 5–6 MeV/nucleon a phase transition in the nucleus takes place from the liquid-drop state to a state corresponding to a mixture of nucleons and all the possible clusters—nuclei of light elements. This state is extremely unstable. The nucleus explodes by decaying into nucleons and light fragments. The process has been called multifragmentation. The nuclear stars with multiple rays, which have been observed in photo-emulsions irradiated by intermediate-energy heavy ions, seem to be evidence for this multifragmentation process.

The compound nucleus, if formed, possesses all the momentum of the projectile. This results in a laboratory angle between the fission fragments being smaller than $\pi$. The larger the momentum the compound nucleus receives, the further the angle is from $\pi$. Measurements of the angular correlations of fission fragments can be used to establish whether we are dealing with the fission of a compound nucleus or a nucleus produced by incomplete fusion. Some data on the angular correlation of the fission fragments produced by bombarding $^{232}$Th by $^{40}$Ar ions at different energies (Co 85) are presented

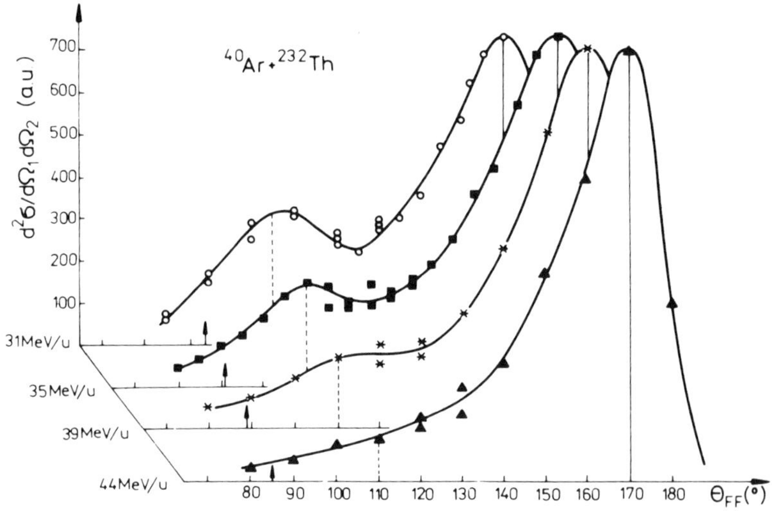


**Figure 66.** In-plane angular correlations of fission fragments for the system $^{40}Ar + ^{232}Th$ at 31, 35, 39, and 44 MeV/u. Curves are drawn to guide the eye. The vertical lines at each energy correspond to $\theta_{ff} = 170°$ and 110° (about 0.8 and 7 GeV/c, respectively); the arrows indicate the locations of the full momentum transfers (Co 85).

in Figure 66. The maxima at angles close to $\pi$ correspond to fission in peripheral collisions involving the transfer of a fraction of the projectile momentum to the target nucleus while the maximum at the smaller angle corresponds to fission in central collisions. The arrows indicate the separation angle between the fission fragments when all the momentum is transferred to the compound nucleus. These data indicate that the complete-fusion process would cease if the $^{40}Ar$ ion energy increases to 40–50 MeV/nucleon. The noncoincidence between the positions of the maxima and the arrows occurs because the fission of the highly excited compound nucleus is preceded by the emission of the considerable number of nucleons, which carry off part of the initial momentum. The energy per nucleon at which compound-nucleus formation ceases depends on the mass of the bombarding nucleus. It has been shown experimentally that it decreases with increasing ion mass (Le 86).

*4.2.2c. Extinction of Deep-Inelastic Transfer Reactions.* And what happens to deep-inelastic transfer reactions in nuclear collisions at intermediate energies? Figure 67 shows how the energy spectrum of the $^{34}S$ nuclei produced by bombarding $^{58}Ni$ and $^{12}C$ by $^{40}Ar$ ions changes with increasing collision energy (Gu 85). At 7 MeV/nucleon the energy spectrum of $^{34}S$ has two pronounced maxima. One corresponds to nucleon transfer in quasi-elastic collisions and the other the transfer in deep-inelastic collisions. At 26.5 MeV/nucleon, the contribution from deep-inelastic collisions falls rapidly, vanishing completely at an energy of 44 MeV/nucleon. The shape of the energy spectrum at 44 MeV/nucleon is close to that characteristic for projectile fragmentation. The asymmetry of the energy spectrum indicates the formation of a small amount of $^{34}S$ in processes associated with the partial dissipation of the kinetic energy. However, the

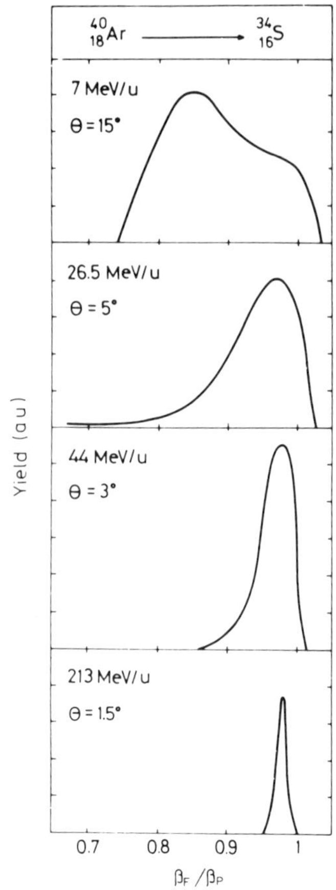


**Figure 67.** Evolution with energy of the $^{34}S$ velocity spectrum measured around the grazing angle (Gu 85).

main contribution to the $^{34}S$ yield comes from the fragmentation of the projectile. Thus, a deep-inelastic transfer contribution to the cross-section for forming the corresponding reaction products decreases with increasing collision energy. At an energy of about 40 MeV/nucleon this nuclear process also dies away. The energy at which deep-inelastic transfer reactions cease depends on the heavy-ion mass (Da 86).

*4.2.2d. Quasi-Elastic Transfer Reactions.* In contrast to relativistic energies, reaction products heavier than the projectile can form at intermediate energies. For example, the nuclei of $^{41}K$, $^{41}Ca$, and $^{40}Ca$ have been detected after the $^{40}Ar$ bombardment of Al and Ta targets at an energy of 27 MeV/nucleon (Me 85). The $^{41}K$ formation is associated with a proton pickup, and the production of $^{41}Ca$ and $^{40}Ca$ results from the pickup of two protons and the transfer of, respectively, one and two neutrons to the target nucleus. The angular distributions of these products concentrate around 0° angle and their velocities are close to that of the projectile. Fragmentation cannot lead to products that are heavier than the projectile. A theoretical analysis of the angular distributions and energy spectra of these products in terms of the theory of direct-surface quasi-elastic-type reactions yielded results in good agreement with the experimental data (Me 85).

*4.2.2e. The Dominant Role of Projectile Fragmentation.* A characteristic feature of projectile fragmentation is the value of the $\sigma_0$ parameter in the expression for the variance of the energy distributions of the fragmentation products. Figure 68 presents the $\sigma_0$ data for a large number of products from two $^{40}$Ar-induced reactions at relativistic and intermediate energies (Gu 85). At relativistic energies (Figure 68b) the $\sigma_0$ value is practically unchanged for all the reaction products. Its average value, $\langle\sigma_0\rangle = 94 \pm 5$ MeV/s, is close to the $\sigma_0$ value given by the fragmentation models (Go 74, Be 81, Mu 84). At an intermediate energy (Figure 68a), all the reaction products can be divided into two groups according to the $\sigma_0$ values. The products in the first group are formed as a result of the loss of a considerable number of nucleons from the projectile. These products have the $\sigma_0$ value close to that of the projectile fragments. The second group contains the products which are close in mass to the bombarding nucleus and have a lower value of $\sigma_0$. The two groups also differ in their angular distributions. The products

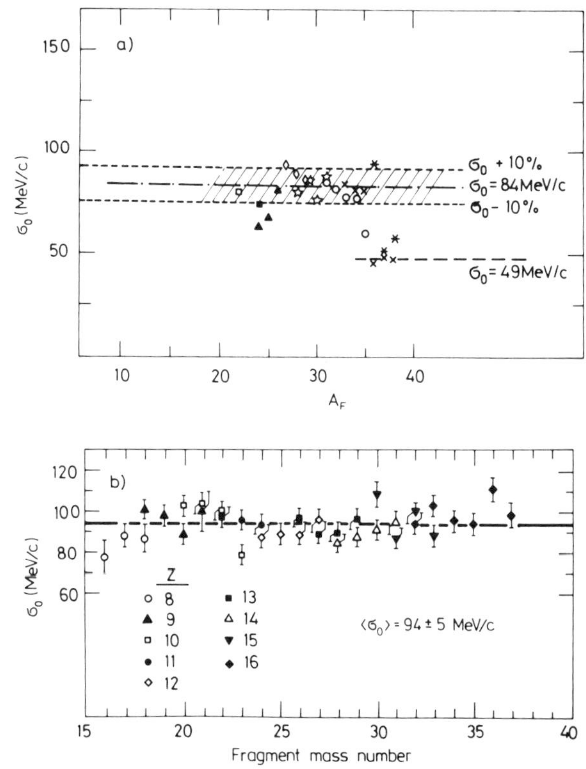


**Figure 68.** Evolution with energy of the experimental reduced width: (a) 27 MeV/u Ar + $^{64}$Zn, (b) 213 MeV/u Ar + $^{12}$C (Gu 85).

with smaller $\sigma_0$ are concentrated in a narrower angle about 0° and their velocities are closer to the velocity of the bombarding nucleus. These kinematic features of the second-group products indicate that they are formed as a result of few-nucleon transfer reactions occurring in surface quasi-elastic collisions. Naturally, the products with the larger $\sigma_0$ values should be attributed to the projectile-fragmentation process.

Thus, there are two nuclear processes that occur during collisions between nuclei with intermediate energies and which yield products focused kinematically. These are projectile fragmentation and quasi-elastic few-nucleon transfer reactions. At energies of 30 MeV/nucleon the projectile-fragmentation process makes the main contribution to the yield of isotopes with considerably smaller $Z$ and $A$ values than those of the bombarding nuclei.

The dispersion of the angular distributions and the relative spread in the velocities of the projectile fragments are significantly larger at intermediate energies than they are at relativistic energies. On the contrary, the admissible thickness of the targets is approximately an order of magnitude smaller. Although these factors reduce the detectable yield of projectile fragments, they are compensated for by the higher intensities of heavy-ion beams with intermediate energies. In recent years, an advance toward the nucleon-stability limit was possible primarily owing to the use of projectile-fragmentation reactions involved in intermediate-energy collisions (Gu 86).

*4.2.2f. Quasi-Elastic Transfer Reactions and Secondary Beams.* The kinematic features of quasi-elastic transfer products, such as the yield concentration around 0° and the smaller energy dispersion than in fragmentation, make these reactions extremely convenient for the production of secondary particle beams. Experiments carried out using the LISE facility at the GANIL have shown that by bombarding a 567 mg/cm$^2$ beryllium target with $^{18}$O nuclei at 65 MeV/nucleon secondary beams of carbon and nitrogen neutron-rich isotopes ($^{14}$C, $^{16}$C, $^{17}$N, and $^{18}$N) can be produced with intensities of, respectively, $10^{-4}$, $2 \times 10^{-6}$, $5 \times 10^{-5}$ and $2 \times 10^{-6}$ of the intensity of the $^{18}$O beam and with energies of about 50 MeV/nucleon (Bi 86). With an $^{18}$O beam intensity of about $10^{12}$ pps at GANIL, the intensities of these secondary beams are high enough for a number of experiments to be performed. Sufficiently pure secondary beams of $^{38}$S, $^{41}$K, $^{38}$Ar, and $^{39}$Ar with intensities ranging from $0.6 \times 10^{-5}$ ($^{38}$S) to $3 \times 10^{-4}$ pps ($^{39}$Ar) and with energies of 32–35 MeV/nucleon have been obtained by bombarding a 100 mg/cm$^2$ beryllium target with $^{40}$Ar at an energy of 44 MeV/nucleon (Bi 86). However, the possibilities of quasi-elastic transfer reactions for producing exotic isotopes are limited because the cross-sections of these reaction products decrease rapidly as the number of nucleons transferred is increased (Figure 69).

*4.2.2g. New Kind of Reaction?* Products whose properties cannot be interpreted in terms of the reactions known for relativistic and low energies (Da 86) have been observed in the $^{84}$Kr bombardment of $^{197}$Au and $^{93}$Nb targets at energies of 35 MeV/nucleon. We mean here the light fragments with atomic numbers below 30 having their maximum yields at 0°. However, the differential production cross-sections decreased more slowly with increasing emission angle than has been observed in projectile fragmentation and the velocities of the products were much lower than the velocities of the bombarding nucleus. This indicates that they are formed during collisions involving the dissipation of a great deal of kinetic energy and governed by mechanisms which have not yet been completely understood. It is possible that we are

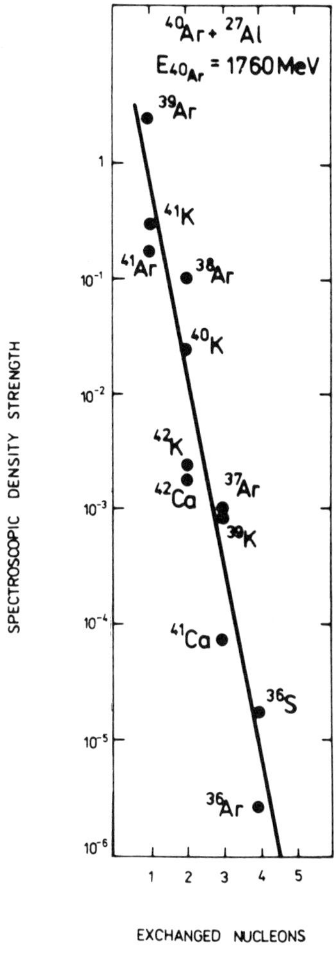


**Figure 69.** Spectroscopic level-density strength for direct transfer determined at 2.5° laboratory angle for the reaction $^{40}Ar + ^{27}Al$. The solid line is just to guide the eye (Me 85).

dealing with a new nuclear process that is peculiar to collisions involving heavy nuclei. Since further experimental information about the yields of the various isotopes from these reactions is absent, we cannot yet say anything positive about whether they may be used to produce nuclei far from the region of stability.

### *4.2.3. Experimental Facilities*

The kinematic characteristics of the products of projectile fragmentation and quasi-elastic transfer at intermediate energies make magneto-optical systems look very suitable to analyze, collect, and identify them. These systems incorporate dispersive and focusing units, first to separate the products spatially according to their magnetic rigidities, and then to focus the products of interest onto the detector. The reaction products are identified, viz. their atomic ($Z$) and mass ($A$) numbers determined, using $\Delta E$, $E$ detectors and by measuring their times of flight for a certain path length.

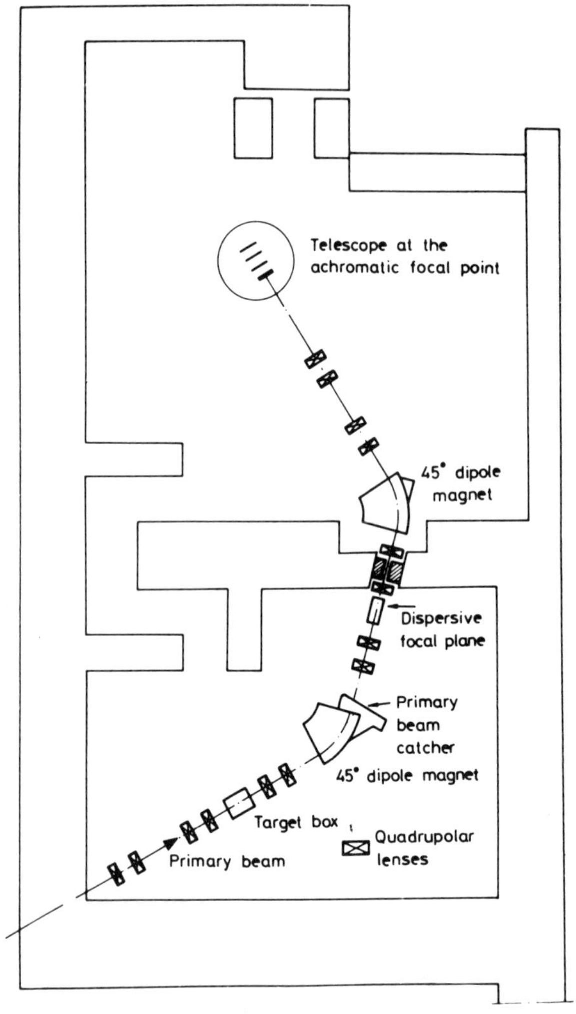


**Figure 70.** General layout of the triple-focusing LISE spectrometer (Gu 85c).

*4.2.3a The LISE Facility.* Figure 70 shows a schematical view of the LISE [Ligne d'Ions Super Epluches (Line of Super Stripped Ions)] experimental facility. The facility is installed on a beam from GANIL (Gu 85c). The LISE magneto-optical system consists of two 45° dipole magnets and five pairs of quadrupole lenses. The first dipole magnet serves as a dispersive unit separating reaction products spatially according to their magnetic rigidities. The products of projectile fragmentation and quasi-elastic transfer reactions will be separated according to the $q/A$ ratio. The second dipole magnet compensates for the dispersion of the first one and focuses the product beam onto the detector. A variable-width vertical slit in the focal plane of the first dipole is used to isolate the reaction products of interest. A two-dipole system provides double achromatic focusing both in angle and energy. The angular acceptance is 1 msr, and the maximum acceptance in magnetic rigidity is $\Delta H\rho/H\rho = \pm 2.5\%$.

The magneto-optical characteristics of the LISE facility are not changed if a degrader foil is placed between the dipoles. Reaction products with the same magnetic rigidities and $q/A$ ratios but different $Z$ will have different magnetic rigidities after they have passed through the degrader, and this naturally facilitates their separation. This feature is employed to identify exotic nuclei and in the production of secondary heavy-ion beams.

A telescope of two $\Delta E$ detectors and two $E$ detectors is placed at the achromatic focus of the facility. The surface-barrier silicon $\Delta E$ detectors are between 300 and 1000 $\mu$m thick. The thickness is chosen depending on the energy, $Z$, and $A$ of the reaction products. The Si(Li) $E$ detector is from 4000 to 5000 $\mu$m thick. The veto detector is the same thickness.

The time of flight of the reaction products is measured over an 18-meter distance between the target and the $E$ detector. The LISE magneto-optical properties lead to a $\pm 1$ mm spread in the trajectory lengths. The start and stop signals arrive from the $E$ detector and from a high-frequency phase of the second cyclotron. A neutron detector controls the time correlation between the high frequency and the time when the accelerated ion clusters hit the target. This detector is placed near the heavy-ion beam stopper at the exit from the first dipole magnet. The time resolution is 800 ps, which provides for time-of-flight measurements with accuracies of about 1%. At the LISE facility, resolutions in $Z$ and $A$ were obtained better than 1.5% and 1%, respectively.

*4.2.3b. The Magnetic Spectrometer SPEG.* The high-resolution magnetic spectrometer SPEG ($\Delta E/E = 10^{-4}$) has also been installed on a beam from the GANIL (Mi 86). It not only can determine the $Z$ and $A$ of a reaction product, but also distinguish transitions between individual states in quasi-elastic few-nucleon transfer reactions.

The use of the SPEG facility for the direct measurement of the masses of nuclei far from stability—projectile fragments—is of interest. The long path chosen for the measurement of the times of flight of the fragmentation products was a good decision. The target is placed on an external beam near the second cyclotron. Thus, the path length includes the line along which the ion beam is monochromatized and the line along which the beam is transported to the SPEG. The full path length reached 116 m, thus making it possible to measure the reaction-product velocity, $\Delta V/V \sim 5 \times 10^{-4}$, very accurately. The masses of the nuclei were determined from

$$B\rho/V = m/Ze \tag{4.8}$$

The mass excesses for a number of neutron-rich isotopes with $7 \leqslant Z \leqslant 9$ were determined to an accuracy of about 500 keV (Mi 86).

*4.2.3c. The Recoil-Mass Spectrometer at Michigan State University (MSU).* The MSU mass spectrometer was designed to operate with the first stage of a superconducting cyclotron system. It can separate and identify the products of projectile fragmentation and transfer reactions at energies of up to 30 MeV/nucleon (En 85). A dipole magnet and a Wien velocity filter serve as dispersing units in the facility. The velocity filter consists of crossed electric and magnetic fields. It can separate the reaction products and the bombarding ions. Two quadrupole-lens doublets placed in front of the Wien filter and behind the dipole magnet focus the reaction-product beam onto the detector system. The spectrometer can reliably separate reaction products with the small production cross-sections typical for isotopes far from stability. The focal-plane detector consists of a two-dimensional position-sensitive proportional gas-filled counter and a semiconductor detector $\Delta E-E$ telescope. The telescope, in turn, consists of a 300 mm$^2$ × 100 $\mu$m surface-barrier Si $\Delta E$ detector and a 500 mm$^2$ × 5 mm Si(Li) $E$ detector. It identifies the isotope and at the same time serves as a $\beta$ detector for the reaction products. The half-lives of $^{14}$Be ($T_{1/2} = 4.2 \pm 0.7$ ms) and $^{17}$C ($T_{1/2} = 202 \pm 17$ ms) were first obtained on this facility (Sh 86).

### 4.2.4. New Isotopes of Light Elements

*4.2.4a. Neutron-Rich Isotopes.* $^{40}$Ar ions with energies of 44 MeV/nucleon and $^{86}$Kr ions at 33 and 44 MeV/nucleon were used at the GANIL to produce light-element isotopes with large neutron excesses. The isotopes were separated and identified using the LISE facility. A comparison of results obtained by bombarding Ta and Be targets has shown that the yield of neutron-rich light-element isotopes is much larger on Ta than on the Be targets (Gu 85). In contrast to relativistic energies, at which the yield of projectile fragments was determined by the geometry of the collision, the structure of the target nucleus plays an essential role at intermediate energies.

Figure 71 presents some data reported by Langevin *et al.* (La 85b) on the yields of the neutron-rich isotopes of elements between B and Ne. The data were obtained by bombarding a tantalum target by $^{40}$Ar ions with energies of 44 MeV/nucleon. This pioneering work demonstrated the potential possibilities of using intermediate-energy heavy ions for an advance toward the nucleon-stability limit. Four new isotopes of carbon, nitrogen, and neon, namely $^{22}$C, $^{23}$N, $^{29}$Ne, and $^{30}$Ne, were produced for the first time. In addition, the nucleon stability of the $^{19}$B isotope, which had been observed earlier, was confirmed and the nucleon instability of $^{18}$B, $^{21}$C, and $^{25}$O was demonstrated. There had already been some indication for the existence of the $^{21}$C and $^{25}$O nuclei (St 81).

The existing theoretical models and extrapolation relationships all predict that $^{23}$N is the last nucleon-stable neutron-rich nucleus for $Z = 7$. Experimentalists have thus reached the nucleon-stability limit for neutron-rich nuclei for $Z = 7$ as well as for $Z = 5$.

Most theoretical estimates, except for the work by Uno and Yamada (Un 82), predict that $^{29}$Ne will be nucleon unstable. However, this nucleus has been successfully detected by Langevin *et al.* (La 85b).

Thus, that an exotic nucleus has been detected in the vicinity of the nucleon-stability limit is in itself valuable information. It helps us select theoretical models which most exactly describe the properties of nuclei far from stability.

Guillemaud-Mueller *et al.* (Gu 85c) used $^{86}$Kr ions with energies of 33 MeV/nucleon to produce light-element neutron-rich isotopes. These ions are the heaviest intermediate-energy ions so far used for the purpose. As before, the experiment was carried out on the LISE facility. A 63 mg/cm$^2$ titanium and a 73 mg/cm$^2$ tantalum target were used. The average ion flux incident on the target was $2.5 \times 10^{10}$ pps. The

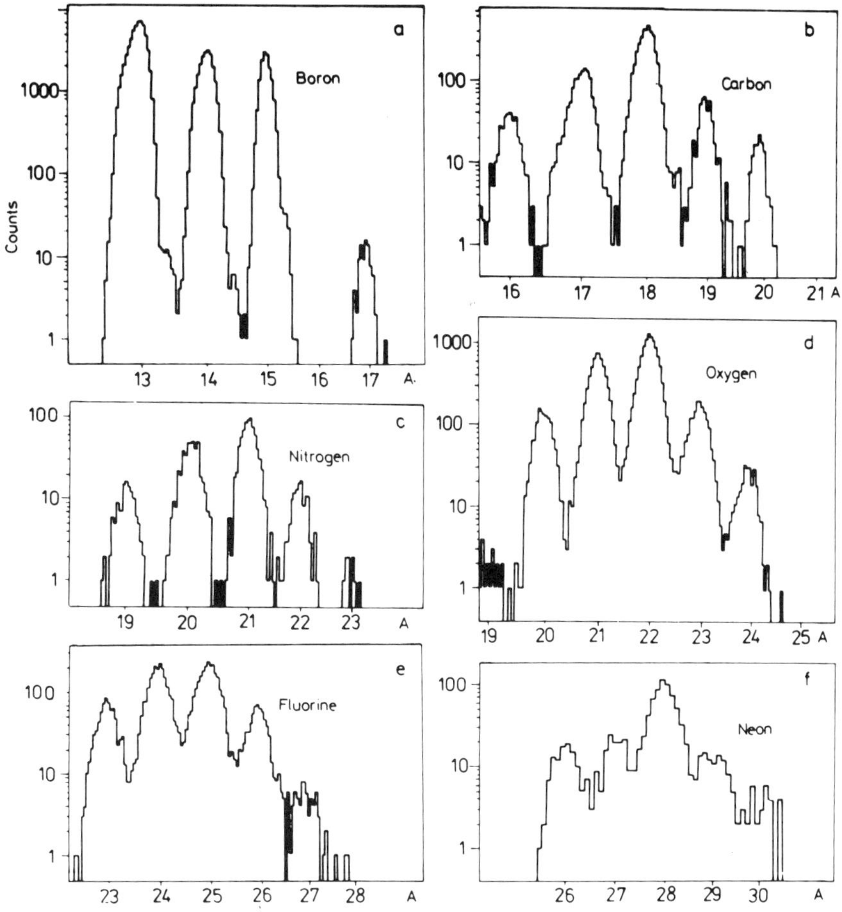


**Figure 71.** Mass histograms of neutron-rich isotopes for boron (a), carbon (b), nigrogen (c), oxygen (d), fluorine (e), and neon (f). Histograms (a), (d), (c), and (f) were obtained at $B\rho = 2.66$ T · m (optimal setting for $^{15,16}B$, $^{23\text{-}25}O$, $^{26\text{-}28}F$, and $^{29\text{-}31}Ne$, respectively), (b) and (c) at $B\rho = 2.92$ T · m (optimal setting for $^{19\text{-}21}C$ and $^{21\text{-}23}N$, respectively) (La 85).

background from scattering ions in lower charge states and from reaction products with large production cross-sections was considerably reduced by placing a 5.3 mg/cm$^2$ Al absorber in the focal plane between the two dipole magnets. Fourteen new neutron-rich isotopes of elements from argon to cobalt—$^{47}Ar$, $^{57}Ti$, $^{59}V$, $^{60}V$, $^{61}Cr$, $^{62}Cr$, $^{64}Mn$, $^{65}Mn$, $^{66}Fe$, $^{67}Fe$, $^{68}Fe$, $^{68}Co$, $^{69}Co$, and $^{70}Co$—were detected in the experiment for the first time (Figure 72).

The nucleon stability of these neutron-rich isotopes is beyond doubt since they are far off the nucleon-stability limit. At the same time, the isotopes are of great interest for the spectroscopic studies of nuclear properties in the new region of $Z$ and $A$. Such

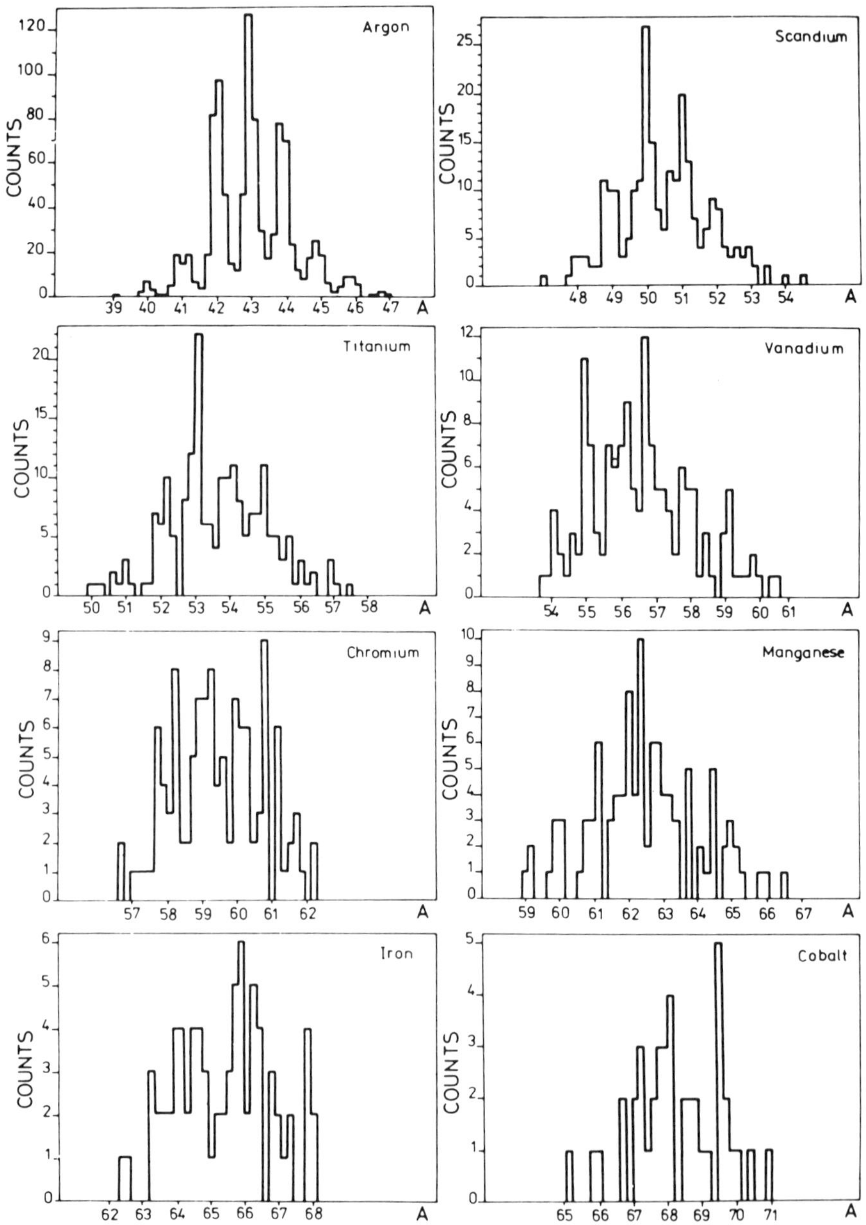


**Figure 72.** Mass histograms of neutron-rich isotopes of scandium, titanium, vanadium, chromium, manganese, iron, and cobalt, from the irradiation of the titanium target by 35 MeV/u krypton projectiles. The argon spectrum is the sum of the spectra obtained with the tantalum and titanium targets. The observed yield of an isotope does not represent a relative production cross-section (Gu 85c).

studies will become more wide-ranging as the beam intensity and the energy of krypton ions are increased, as is planned at GANIL in the near future.

*4.2.4b. Neutron-Deficient Isotopes.* Projectile fragmentation has also proved to be an efficient means of producing neutron-deficient isotopes of light nuclei and of getting close to the proton-stability limit. Langevin *et al.* (La 86) detected a number of light-nuclei isotopes with $T_z = -5/2$, namely, $^{23}$Si, $^{27}$S, $^{31}$Ar, and $^{35}$Ca. Except for $^{23}$Si, all of these nuclei were recorded for the first time. According to theoretical predictions, these isotopes lie on the boundary of nuclear stability against the emission of one or two protons.

A 92 mg/cm$^2$ natural-nickel target was bombarded in the experiment by 77 MeV/nucleon $^{40}$Ca ions. Naturally occurring nickel contains 67.7% $^{58}$Ni, the lightest of its stable isotopes. This is a favorable factor in the production of neutron-deficient isotopes.

Experiments have been carried out on the LISE facility with a $^{40}$Ca ion beam at an intensity of $1.5 \times 10^{10}$ pps and with a 14-hour exposure. The magnetic rigidity acceptance was reduced to 0.5%, since neutron-deficient isotopes are closer in magnetic rigidity to the heavy-ion beam than are neutron-rich isotopes. The background was decreased by placing a 5.3 mg/cm$^2$ Al absorber in the intermediate focus of the first dipole magnet.

The results are presented in Figure 73. We can see that the yields of isotopes with $T_z = -3$ decrease rapidly after isotopes with $T_z = -5/2$. Theoretical extrapolations for

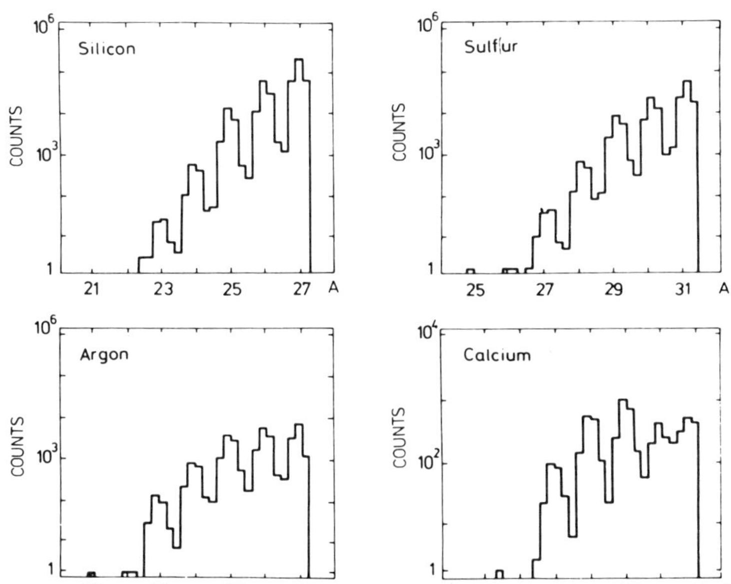


**Figure 73.** Mass distributions for the elements Si, S, Ar, and Ca. They have been obtained by summing up the data from 14 hours integration time at a magnetic rigidity setting of $B\rho = 2.10$ T · m and 6 hours at $B\rho = 2.15$ T · m. Thus 74, 112, 315, and 246 events are observed for $^{23}$Si, $^{27}$Si, $^{31}$Ar, and $^{35}$Ca, respectively (La 86).

the mass excess and the binding energy of one and two protons suggest that the $^{30}Ar$ and $^{34}Ca$ isotopes, which follow $^{31}Ar$ and $^{35}Ca$, are undoubtedly unstable with respect to decay involving the emission of two protons.

The two-proton binding energy in $^{22}Si$, which follows $^{23}Si$, is estimated as $16 \pm 95$ keV. This indicates that $^{23}Si$ may also be the last nucleon-stable nucleus.

For $^{31}Ar$, theoretical estimates give a value of $189 \pm 85$ keV (La 86) for the two-proton binding energy. In other words, $^{31}Ar$ is expected to be unstable against the emission of two protons. That it has been detected shows how difficult it is to make reliable theoretical estimates for nuclear masses in the vicinity of the nucleon-stability limit and that direct experiments are needed to locate its position. The low two-proton binding energy in the nucleus $^{31}Ar$ make it promising for the experimental observation of two-proton decay from the ground state.

Thus, we see that already the initial experiments with intermediate-energy ions have allowed us to go beyond the nucleon-stability limit for neutron-deficient nuclei with $Z$ values between 14 and 20. When $^{58}Ni$ ion beams become available, we should see the proton-stability limit reached for nuclei up to $Z = 28$ (La 86).

## 5. Conclusion

Over two decades have elapsed since heavy-ion reactions were first used to produce nuclei far from stability. During this period of time delayed protons, delayed fission, and proton emission from ground-state nuclei have been discovered, and many dozens of new neutron-deficient and neutron-rich isotopes have been produced. The reactions induced by heavy ions have enabled experimentalists to reach the nucleon-stability limits for light nuclei with respect to the emission of a proton or a neutron. A great deal of information has been obtained about the properties of nuclei far from stability. It has led to a better knowledge of nuclear structure and to a deeper understanding of the processes of nucleonsynthesis in the Universe. The first experiments with secondary beams have been carried out. We can therefore state that heavy-ion reactions have proved to be a principal method for producing nuclei far from stability.

In contrast to the thermal neutron-induced fission of $^{235}U$ and to the deep-spallation reactions induced by high-energy protons, heavy-ion reactions offer a much wider choice of projectiles and target nuclei. Modern heavy-ion accelerators are capable of accelerating the ions of almost all the elements in the periodic table. The experimentalist may choose the optimal target–projectile combination for obtaining the maximum yield of the isotope of interest in the presence of the lowest background from interfering reaction products.

Further progress in the study of nuclei far from stability using heavy ions was closely connected with an increase in the intensities and energies of the beams of intermediate- and relativistic-energy ions. It is just heavy ions that have permitted the impressive progress toward the limits of the nucleon stability of nuclei we have seen in recent years.

Calculations and experimental data indicate that $^{48}Ca$ ions at intermediate and relativistic energies are especially promising for the production of neutron-rich isotopes of light elements. It is conceivable that the nucleon-stability drip-line for the neutron-rich isotopes of carbon, nitrogen, and possibly oxygen will be crossed in $^{48}Ca$ fragmentation reactions alone.

In stable nuclei the neutron excess increases with increasing atomic number. This, in principle, opens up the possibility of producing new neutron-rich isotopes by their fragmentation in the collisions at intermediate and relativistic energies. The heaviest isotope of boron $^{19}B$ was produced by the fragmentation of $^{56}Fe$. In this respect the fragmentation at intermediate energies of $^{238}U$ nuclei, which each possess an excess of 54 neutrons, looks particularly fascinating. We should remember that this property of uranium nuclei is exploited in two other effective methods for producing neutron-rich isotopes, namely, the thermal neutron-induced fission of $^{235}U$ and the deep spallation of uranium by high-energy protons.

An extremely great neutron excess occurs in a system of two colliding nuclei of uranium. How such a system will behave in collisions at intermediate and, all the more, at relativistic energies is an interesting question. It is possible that specific and as yet unknown reaction channels may arise in interactions between massive nuclear systems and these channels may turn out to be efficient at producing neutron-rich isotopes.

The production of intense heavy-ion beams with intermediate and relativistic energies will greatly stimulate the development of a new trend in the studies of nuclei far from stability, namely, the use of secondary heavy-ion beams. In this case, the nuclei far from stability will themselves be the projectiles. The first experiments carried out along this line of research have proved encouraging. They have generated unique information about the nuclear sizes of the neutron-rich isotopes of helium and lithium.

When speaking of the use of intermediate- and relativistic-energy heavy ions for the production of nuclei far from stability we should not forget about the great possibilities offered by heavy ions at lower energies. The fusion of nuclei, especially those with close masses, remains the most efficient way of producing the very neutron-deficient nuclei of medium and heavy elements. This seems to be the route to the doubly magic tin nucleus $^{100}Sn$. The multinucleon transfer reactions induced by sufficiently heavy ions make it possible to reach those regions of the nuclidic chart which are inaccessible or hard to reach from the deep-spallation reactions induced by high-energy protons. The extensive region of the transmendelevium elements which also belong to nuclei far from stability can only be investigated with the help of low-energy heavy ions.

The difficulties encountered in separating and identifying nuclei far from stability have led to complex detector systems, kinematic separators, and ISOL systems. In moving farther from the region of stability, the requirements imposed on the rapidity, selectivity, and sensitivity of the experimental apparatus become more severe. An increase in the heavy-ion beam intensity will inevitably be accompanied by a larger background from the numerous reaction products.

Kinematic separators and in-beam detector systems, which have rapid operating speeds, must be developed to increase their selectivity and sensitivity. One solution to this problem is the gas-filled mass separator at the exit of the well-known SHIP facility.

In ISOL-type facilities the rapidity and chemical selectivity of the mass-separator ion source must be enhanced. The last few years have seen impressive progress in this field. A considerable reduction in the time required for isotopic separation can be achieved in IGISOL systems. However, there is a need for detailed studies of the efficiency of their operation on heavy-ion beams and of the chemical selectivity of these systems.

In order to perform investigations using intermediate- and relativistic-energy ions some major experimental facilities have been built which are nearly as large as the experimental facilities used in elementary particle physics. As much information as possible about the net results of collisions between two nuclei is needed to study the fragmentation processes which lead to the formation of nuclei far from stability. This can be done by using "global" systems that can detect the majority of the reaction products and radiations resulting from the collision of two nuclei. On the other hand, the isolation of nuclides far from stability against the background of the numerous reaction products formed during projectile and target nucleus fragmentation required a system with a narrow transmission window. Such a system would be efficient and selective in detecting the isotopes of interest in the presence of a variety of other reaction products. The development of experimental facilities for use with heavy-ion beams at intermediate and relativistic energies proceeds in this direction.

To conclude this chapter the following is emphasized. Even though nuclei far from stability have been studied for more than two decades, the researchers still face a wide field and a diversity of exciting scientific problems to tackle. There can be no doubt that studies carried out with heavy ions at low, intermediate, and relativistic energies will make an increasingly great contribution to the solution of these problems.

## Acknowledgments

The author expresses his deep gratitude to Academician Georgy N. Flerov, who pioneered this trend of heavy-ion physics and constantly stimulated its development. His lively interest in the author's work on this chapter and his helpful comments have to a great extent determined its content.

As a veteran of heavy-ion physics the author highly values the tremendous amount of work by Professor D. A. Allan Bromley involved in creating the multivolume *Treatise on Heavy-Ion Science* and appreciates his invitation to prepare this chapter.

The author is much indebted to his colleagues A. G. Artukh, C. F. Gridnev, V. L. Mikheev, A. N. Mezentsev, and J. Wilczynski for their fruitful cooperation in reviewing the research covered in this chapter. He is also grateful to Yu. P. Gangrsky, who read the manuscript, made some useful comments, and gave advice.

The author cordially thanks L. V. Pashkevich, who translated the manuscript into English within a short period of time, and also R. Hainsworth, who was very helpful in the preparation of the English version of the chapter.

The assistance of all JINR staff members in the preparation of the manuscript for publication is gratefully acknowledged. Special thanks are expressed to V. R. Sarantseva whose support has facilitated the preparation of the chapter.

The author thanks Z. P. Kopylova for her careful drafting of the numerous figures.

The author is especially indebted to his wife, Anna, for her support in work on the text and for her great help in preparing the chapter for publication.

The author acknowledges the courtesy of the American Physical Society, North-Holland Publ. Co., and Springer-Verlag who have granted permission to reproduce some figures from their publications in journals and proceedings of international conferences, and expresses his deep gratitude to those authors whose figures have been reproduced in this chapter.

## References


(Ac 84) E. Acerbi, F. Aghion, F. Alessandria, G. Baccaglioni, G. Bellomo, C. Birattari, C. De Martinis, E. Fabrici, D. Giove, W. Giussani, G. Gualeni, P. Michelato, L. Milinkovic, C. Pagani, E. Panzeri, F. Resmini, L. Rossi, A. Salomone, L. Serafini, and G. Varisco, *Tenth International Conference on Cyclotrons and Their Applications*, April 30–May 3 1984, East Lansing, Michigan, U.S.A. (F. Marti, ed.), IEEE, New York (1984), p. 251.

(Al 81) B. W. Allardyce, *Ninth International Conference on Cyclotrons and Their Applications*, September 1981, Caen, France (G. Gendreau, ed.), Les Editions de Physique, Les Ulis Cedex, France (1981), p. 55.

(Ar 61) P. Armbruster, *Nucleonica* **3**:188 (1961).

(Ar 66) A. G. Artukh, V. V. Volkov, and T. Kwiecinska, *Yad. Fiz.* **4**:1165 (1966).

(Ar 69) A. G. Artukh, G. F. Gridnev, V. L. Mikheev, and V. V. Volkov, *Nucl. Phys. A* **137**:348 (1969).

(Ar 70) A. G. Artukh, V. V. Volkov, L. Pomorski, and J. Tys, Communication P7-5494, JINR, Dubna (1970).

(Ar 70b) A. G. Artukh, V. V. Avdeichikov, J. Erö, G. F. Gridnev, V. L. Mikheev, and V. V. Volkov, *Nucl. Instrum. Methods* **83**:72 (1970).

(Ar 70c) A. G. Artukh, V. V. Avdeichikov, G. F. Gridnev, V. L. Mikheev, V. V. Volkov, and J. Wilczynski, *Phys. Lett.* **31B**:129 (1970).

(Ar 70d) A. G. Artukh, V. V. Avdeichikov, L. P. Chelnokov, G. F. Gridnev, V. L. Mikheev, V. I. Vakatov, V. V. Volkov, and J. Wilczynski, *Phys. Lett.* **32B**:43 (1970).

(Ar 70e) A. G. Artukh, V. V. Avdeichikov, J. Erö, G. F. Gridnev, V. L. Mikheev, V. V. Volkov, and J. Wilczynski, *Phys. Lett.* **33B**:407 (1970).

(Ar 70f) A. G. Artukh, V. V. Avdeichikov, J. Erö, G. F. Gridnev, V. L. Mikheev, V. V. Volkov, and J. Wilczynski, Proceedings of the International Conference on the Properties of Nuclei Far From the Region of Beta-Stability, Leysin, Switzerland, August 31–September 4, 1970, report CERN 70-30, Vol. 1, Geneva (1970), p. 47.

(Ar 71) A. G. Artukh, V. V. Avdeichikov, J. Erö, C. F. Gridnev, V. L. Mikheev, V. V. Volkov, and J. Wilczynski, *Nucl. Phys. A* **160**:511 (1971).

(Ar 71b) A. G. Artukh, V. V. Avdeichikov, G. F. Gridnev, V. L. Mikheev, V. V. Volkov, and J. Wilczynski, *Nucl. Phys. A* **168**:321 (1971).

(Ar 71c) A. G. Artukh, V. V. Avdeichikov, G. F. Gridnev, V. L. Mikheev, V. V. Volkov, and J. Wilczynski, *Nucl. Phys. A* **176**:284 (1971).

(Ar 71d) P. Armbruster, J. Eidens, J. W. Grüter, H. Lawin, E. Roeckl, and K. Sistemich, *Nucl. Instrum. Methods* **91**:499 (1971).

(Ar 72) A. G. Artukh, G. F. Gridnev, V. L. Mikheev, V. V. Volkov, and J. Wilczynski, *Nucl. Phys. A* **192**:170 (1972).

(Ar 73) A. G. Artukh, G. F. Gridnev, V. L. Mikheev, V. V. Volkov, and J. Wilczynski, *Nucl. Phys. A* **211**:299 (1973).

(Ar 73b) A. G. Artukh, G. F. Gridnev, V. L. Mikheev, V. V. Volkov, and J. Wilczynski, *Nucl. Phys. A* **215**:91 (1973).

(Ar 74) A. G. Artukh, V. V. Avdeichikov, G. F. Gridnev, A. N. Mezentsev, V. L. Mikheev, and V. V. Volkov, preprint E7-8298, JINR, Dubna (1974).

(Ar 75) A. C. Artukh, V. V. Avdeichikov, V. V. Volkov, G. F. Gridnev, A. N. Mezentsev, V. L. Mikheev, Tezisi dokladov XXV Soveschania po jadernoi spectroskopii i structure atomnogo jadra, janvar 1975, Leningrad, Nauka, Leningrad (1975), p. 223.

(Ar 77) A. G. Artukh, G. F. Gridnev, V. L. Mikheev, and V. V. Volkov, *Nucl. Phys. A* **283**:350 (1977).

(Ar 81) A. G. Artukh, G. F. Gridnev, M. Gruszecki, W. Karcz, A. N. Mezentsev, V. L. Mikheev, L. Pomorski, A. Popescu, D. G. Popescu, and V. V. Volkov, *Z. Phys. A* **303**:41 (1981).

(Är 81) J. Ärje, J. Äystö, J. Honkanen, K. Valli, and A. Hautojärvi, *Nucl. Instrum. Methods* **186**:149 (1981).

(Ar 83) P. Armbruster, *The Neutron and its Applications 1982*, Institute of Physics Conference, Series number 64 (P. Schofield ed.), Bristol and London (1983), p. 105.

(Är 86) J. Ärje, Ion Guide Method for Isotope Separator, Research Report No. 3/86 (Doctor Thesis), Department of Physics, University of Jyväskylä, Jyväskylä, Finland (1986).

(Är 87) J. Ärje, Proceedings of the International School—Seminar on Heavy Ion Physics (Dubna, September 1986), report D7-87-68, JINR, Dubna (1987), p. 649.

(Au 79) P. Auger, T. H. Chiang, J. Galin, B. Gatty, D. Guerreau, E. Nolte, J. Pouthas, X. Tarrago, and J. Girard, *Z. Phys. A* **289**:255 (1979).

(Äy 81) J. Äystö, M. D. Cable, R. F. Parry, J. M. Wouters, D. M. Moltz, and J. Cerny, *Phys. Rev. C* **23**:879 (1981).

(Ba 69) A. I. Baz, V. F. Demin, and M. V. Zhukov, *Yad. Fiz.* **9**:1184 (1969).

(Ba 70) I. Bacho, D. D. Bogdanov, Sh. Darotsci, V. A. Karnaukhov, and L. A. Petrov, *Prib. Tekh. Eksp.* **2**:43 (1970).

(Ba 71) A. M. Baldin, preprint P7-5808, JINR, Dubna (1971).

(Ba 71b) A. M. Baldin, International Conference on Heavy Ion Physics (Dubna, February 1971), report D7-5769, JINR, Dubna (1971), p. 607.

(Ba 75) A. M. Baldin, preprint E2-9138, JINR, Dubna (1975).

(Ba 80) A. M. Baldin, preprint E1-80-174, JINR, Dubna (1980).

(Ba 85) A. M. Baldin, *Nucleus-Nucleus Collisions II*, Proceedings of the Second International Conference on Nucleus-Nucleus Collisions, Visby, Sweden, June 1985 (H. A. Gustafsson, B. Jakobsson, I. Otterlund, and K. Aleklett, eds.), North-Holland, Amsterdam (1986), p. 203c.

(Be 69) E. E. Berlovich and Yu. N. Novikov, *Dokl. Akad. Nauk USSR* **185**:1025 (1969).

(Be 81) G. Bertsch, *Phys. Rev. Lett.* **46:**472 (1981).

(Bi 86) R. Bimbot, S. Della-Negra, M. Manasijevic, P. Aguer, G. Bastin, R. Anne, H. Delagrange, Y. Schutz, F. Hubert, Y. Gono, and K. Natanaka, Proceedings of the International Conference on Heavy Ion Nuclear Collisions in the Fermi Energy Domain (Caen, France, May 1986), *J. Phys.* (*Paris*), *Colloq. C4*, **47**:241 (1986).

(Bj 82) S. Bjørnholm and W. J. Swiatecki, *Nucl. Phys. A* **391**:471 (1982).

(Bo 76) D. D. Bogdanov, A. V. Demyanov, V. A. Karnaukhov, L. A. Petrov, A. Plohocki, V. G. Subbotin, and J. Voboril, Proceedings of the 3rd International Conference on Nuclei Far From Stability, Cargèse, Corsica (France), May 1976, report CERN 76-13, Geneva (1976), p. 299.

(Bo 85) U. Bosh, W. D. Schmidt-Ott, P. Tidemand-Petersson, E. Runte, W. Hillebrandt, M. Lechle, F. K. Thielemann, R. Kirchner, O. Klepper, E. Roeckl, K. Rykaczewski, D. Schardt, N. Kaffrell, M. Bernas, P. Dessagne, and W. Kurcewicz, *Phys. Lett.* **164B**:22 (1985).

(Bo 86) J. P. Bondorf, Proceedings of the International Conference on Heavy Ion Nuclear Collisions in the Fermi Energy Domain (Caen, France, May 1986), *J. Phys.* (*Paris*) *Colloq. C4*, **47**:263 (1986).

(Br 76) H. C. Britt, B. H. Erkkila, R. H. Stokes, H. H. Gutbrod, F. Plasil, R. L. Ferguson, and M. Blann, *Phys. Rev. C* **13**:1483 (1976).

(Br 80) H. Breuer, K. L. Wolf, B. G. Glagola, K. K. Kwiatkowski, A. C. Mignerey, V. E. Viola, W. W. Wilcke, W. V. Schröder, J. R. Huizenga, D. Hilsher, and J. Birkelund, *Phys. Rev. C* **22**:2454 (1980).

(Br 81) C. Bruske, K. H. Burkard, W. Hüller, R. Kirchner, O. Klepper, and E. Roeckl, *Nucl. Instrum. Methods* **186**:61 (1981).

(Bu 70) G. W. Butler, A. M. Poskanzer, and D. A. Landis, *Nucl. Instrum. Methods* **89**:189 (1970).

(Ch 78) E. A. Cherepanov, Lee Gen Soo, A. N. Mezentsev, and V. V. Volkov, Communication E7-11364, JINR, Dubna (1978).

(Co 58) L. Cohen and C. B. Fulmer, *Nucl. Phys.* **6**:547 (1958).

(Co 65) S. Cohen and D. Kurath, *Nucl. Phys.* **73**:1 (1965).

(Co 67) S. W. Cosper, J. Cerny, and R. G. Gatti, *Phys. Rev.* **154**:1193 (1967).

(Co 74) S. Cohen, F. Plasil, and W. J. Swiatecki, *Ann. Phys.* (*N.Y.*) **82**:577 (1974).

(Co 85) M. Conjeaud, S. Harar, M. Mostefal, E. C. Pollacco, C. Volant, Y. Cassagnou, R. Dayras, R. Legrain, H. Oeschler, and F. Saint-Laurent, *Phys. Lett.* **159B**:244 (1985).

(Da 86) D. Dalili, R. Lucas, C. Ngô, C. Cerruti, S. Leray, C. Mazur, M. Ribrad, T. Suomijärvi, M. Berlanger, S. Chiodelli, A. Demeyer, and D. Guinet, *Nucl. Phys. A* **454**:163 (1986).

(Da 86b) R. Dayras, Proceedings of the International Conference on Heavy Ion Nuclear Collisions in the Fermi Energy Domain (Caen, France, May 1986), *J. Phys.* (*Paris*), *Colloq. C4*, **47**:13 (1986).

(De 66) A. V. Demyanov, N. I. Tarantin, Yu. A. Dyachikhin, and A. P. Kabachenko, *Prib. Tekh. Eksp.* No. **2**:28 (1966).

(Dj 51) B. S. Djelepov, *Izv. Akad. Nauk USSR, Ser. Fys.* **15**:498 (1951).

(Do 77) E. D. Donez, *IEEE Trans. Nucl. Sci.* **NS-23**:897 (1977).

(En 85) H. A. Enge, in *Treatise on Heavy-Ion Science*, Vol. 7 (D. Allan Bromley, ed.), Plenum Press, New York (1985), p. 403.

(Fe 81) J. Fermé, M. Gouttefangeas, and the GANIL group, *Ninth International Conference on Cyclotrons and Their Applications*, September 1981, Caen, France (G. Gendreau, ed.), Les Editions de Physique, Les Ulis Cedex (France) (1981), p. 3.

(Fl 61) G. N. Flerov, V. V. Volkov, L. Pomorski, and J. Tys, *Zh. Eksp. Teor. Fiz.* **41**:1365 (1961).

(Fl 64) G. N. Flerov, V. A. Karnaukhov, G. M. Ter-Akopian, L. A. Petrov, and V. G. Subbotin, *Zh. Eksp. Teor. Fiz.* **47**:419 (1964).

(Fl 74) G. N. Flerov, *Proceedings of the International Conference on Reactions Between Complex Nuclei*, Nashville, June 1974, Vol. II (R. L. Robinson, F. K. McGowan, J. B. Ball, and J. H. Hamilton, eds.), North-Holland, Amsterdam (1974), p. 459.

(Fl 84) G. N. Flerov, Yu. Ts. Oganessian, S. L. Bogomolov, B. N. Gikal, G. G. Gulbekyan, A. I. Ivanenko, V. V. Kamanin, B. A. Klenin, I. V. Kolesov, V. B. Kutner, V. N. Melnikov, E. A. Minin, A. M. Morduev, R. Ts. Oganessian, A. S. Pasyuk, Yu. E. Penionzhkevich, K. I. Semin, B. V. Fefilov, and V. A. Chugreev, *Heavy Ions Cyclotron Facility, Laboratory of Nuclear Reactions JINR* (in Russian), report 9-84-555, JINR, Dubna (1984).

(Fr 62) A. M. Friedman and W. C. Mohr, *Nucl. Instrum. Methods* **17**:78 (1962).

(Fr 83) W. A. Friedman, *Phys. Rev. C* **27**:569 (1983).

(Fr 85) E. M. Friedlander and H. H. Heckman, in *Treatise on Heavy-Ion Science*, Vol. 4 (D. Allan Bromley, ed.), Plenum Press, New York (1985), p. 403.

(Ga 66) G. T. Garvey and I. Kelson, *Phys. Rev. Lett.* **16**:197 (1966).

(Ga 75) H. Gauvin, D. Guerreau, Y. Le Beyec, M. Lefort, F. Plasil, and X. Tarrago, *Phys. Lett.* **58B**:163 (1975).

(Ga 80) Yu. P. Gangrsky, M. B. Miller, L. V. Mikhailov, and I. F. Kharisov, *Yad. Fiz.* **31**:306 (1980).

(Gh 86) A. Ghiorso, S. Yashita, M. E. Leino, L. Frank, J. Kalnins, P. Armbruster, J. P. Dufour, and P. K. Lemmertz, Lawrence Berkeley Laboratory Report, LBL-21975, Berkeley (1986).

(Gi 86) K. L. Gippert, E. Runte, W. D. Schmidt-Ott, P. Tidemand-Petersson, N. Kaffrel, P. Peuser, R. Kirchner, O. Klepper, W. Kurcewicz, P. O. Larsson, E. Roeckl, D. Schardt, and K. Rykaczewski, *Nucl. Phys. A* **453**:1 (1986).

(Go 60) V. I. Goldansky, *Zh. Eksp. Teor. Fiz.* **39**:497 (1960).

(Go 74) A. S. Goldhaber, *Phys. Lett.* **53B**:306 (1974).

(Go 80) A. Gobbi and W. Nörenberg, in *Heavy Ion Collisions*, Vol. 2 (R. Rock, ed.), North-Holland, Amsterdam (1980), p. 128.

(Gr 63) W. Grochulski, T. Kwiecińska, Lian Go-chan, E. Lozyński, J. Malý, L. K. Tarasov, and V. V. Volkov, *Proceedings of the Third Conference on Reactions between Complex Nuclei, Asilomar (Pacific Grove, California), April 1963* (A. Ghiorso, R. M. Diamond, and H. E. Conzett, eds.), University of California Press, Berkeley and Los Angeles (1963), p. 120.

Gr 75) D. E. Greiner, P. J. Lindstrom, H. H. Heckman, B. Cork, and F. S. Biesser, *Phys. Rev. Lett.* **35:**152 (1975).

(Gr 79) H. Grunder, Lawrence Berkeley Laboratory Report LBL-PUB 5014 (1979).

(Gu 73) H. Guratzsch, A. P. Kabachenko, I. V. Kusnetsov, K. Siwek-Wilczynska, and N. I. Tarantin, *Nucl. Phys. A* **205**:574 (1973).

(Gu 74) H. Guratzsch, A. P. Kabachenko, I. V. Kusnetsov and N. I. Tarantin, *Yad. Fiz.* **19**:1167 (1974).

(Gu 80) D. Guerreau, J. Galin, B. Gatty, X. Tarrago, J. Girard, R. Lucas, and C. Ngô, *Z. Phys. A* **295**:105 (1980).

(Gu 83) C. Guet, *Nucl. Phys. A* **400**:191c (1983).

(Gu 85) D. Guerreau, Proceedings of the Second International Conference on Nucleus-Nucleus Collisions, Visby, Sweden, June 1985, *Nucl. Phys. A* **447**:37c (1985).

(Gu 85b) D. Guerreau, Proceedings of XXIV International Winter Meeting on Nuclear Physics, Bormio, 1985; preprint CENBG R 8601 (1986).

(Gu 85c) D. Guillemaud-Mueller, A. C. Mueller, D. Guerreau, F. Pougheon, R. Anne, M. Bernas, J. Galin, J. C. Jacmart, M. Langevin, F. Naulin, E. Quiniou, and C. Détraz, *Z. Phys. A* **322**:415 (1985).

(Gu 86) D. Guerreau, Proceedings of the International Conference on Heavy Ion Nuclear Collisions in the Fermi Energy Domain (Caen, France, May 1986), *J. Phys. (Paris), Colloq. C4*, **47**:207 (1986).

(Ha 85) B. G. Harvey, *Nucl. Phys. A* **444**:498 (1985).

(Ho 79) S. Hofmann, W. Faust, G. Münzenberg, W. Reisdorf, P. Armbruster, K. Güttner, and H. Ewald, *Z. Phys. A* **291:**53 (1979).

(Ho 81) S. Hofmann, G. Münzenberg, W. Faust, F. Heßberger, W. Reisdorf, J. R. H. Schneider, P. Armbruster, K. Güttner, and B. Thuma, Proceedings of the 4th International Conference on Nuclei Far From Stability, Helsingør (Denmark), June 1981, report CERN 81-09, Geneva (1981), p. 190.

(Ho 81b) S. Hofmann, G. Münzenberg, F. P. Heßberger, W. Reisdorf, P. Armbruster, and B. Thuma, *Z. Phys. A* **299**: 281 (1981).

(Ho 82) S. Hofmann, W. Reisdorf, G. Münzenberg, F. P. Heßberger, J. R. H. Schneider, and P. Armbruster, *Z. Phys. A* **305**:111 (1982).

(Ho 82b) A. D. Hoover, J. R. Birkelund, D. Hilscher, W. V. Schröder, W. W. Wilcke, J. R. Huizenga, H. Breuer, A. C. Mignerey, V. E. Viola, Jr., and K. L. Wolf, *Phys. Rev. C* **25**:256 (1982).

(Ho 87) S. Hofmann, P. Armbruster, F. P. Hessberger, G. Münzenberg, H. J. Schött, and I. Zychor, Proceedings of the International School-Seminar on Heavy Ion Physics (Dubna, September 1986), report JINR D7-87-68, Dubna (1987), p. 378.

(Ja 66) J. C. Jackmart, M. Liu, F. Mazloum, and M. Riou, Proceedings of the International Conference on Heavy Ion Physics, Dubna, October 1966, section 3, Nuclear Reactions, report D7-5342, JINR, Dubna (1970), p. 137.

(Ju 71) H. Junglas, R. D. Macfarlane, and Y. Fares, *Phys. Rev. Lett.* **27**:556 (1971).

(Ka 60) V. A. Karnaukhov and N. I. Tarantin, *Zh. Eksp. Teor. Fiz.* **39**:1106 (1960).

(Ka 62) V. A. Karnaukhov, G. M. Ter-Akopian, and V. G. Subbotin, preprint P-1022, JINR, Dubna (1962).

(Ka 63) V. A. Karnaukhov, G. M. Ter-Akopian, and V. G. Subbotin, *Proceedings of the Third Conference on Reactions between Complex Nuclei, Asilomar (Pacific Grove, California), April 14–18, 1963* (A. Ghiorso, R. M. Diamond, and H. E. Conzett, eds.), University of California Press, Berkeley and Los Angeles (1963), p. 434.

(Ka 63b) V. A. Karnaukhov, G. M. Ter-Akopian, L. A. Petrov, and V. G. Subbotin, *Zh. Eksp. Teor. Fiz.* **45**:1280 (1963).

(Ka 69) V. A. Karnaukhov, L. A. Rubinskaya, G. M. Ter-Akopian, V. N. Titov, and V. A. Chugreev, Communication P13-4454, JINR, Dubna (1969).

(Ka 71) A. P. Kabachenko, I. V. Kuznetsov, K. Siwek-Wilczynska, E. A. Skakun, and N. I. Tarantin, Proc. Int. Conf. on Heavy Ion Physics, Dubna, February 1971, Report D7-5769, JINR, p. 204.

(Ka 74) V. A. Karnaukhov, D. D. Bogdanov, A. V. Demyanov, G. I. Koval, and L. A. Petrov, *Nucl. Instrum. Methods* **120**:69 (1974).

(Ka 81) V. A. Karnaukhov and L. A. Petrov, *Nuclei Far From the Beta-Stability Line* (in Russian), Energoizdat, Moscow (1981), p. 23.

(Ka 84) H. Kamitsubo, Tenth International Conference on Cyclotrons and Their Applications, April 29–May 3 1984, East Lansing, Michigan, U.S.A. (F. Marti, ed.), IEEE, New York (1984), p. 257.

(Ki 81) R. Kirchner, *Nucl. Instrum. Methods* **186**:275 (1981).

(Ki 82) R. Kirchner, O. Klepper, W. Kurcewicz, E. Roeckl, E. F. Zganiar, E. Runte, W. D. Schmidt-Ott, P. Tidemand-Petersson, N. Kaffrell, P. Peuser, and K. Rykaczewski, *Nucl. Phys. A* **378**:549 (1982).

(Kl 69) R. Klapisch, C. Thibault-Philippe, C. Detraz, J. Chaumont, R. Bernas, and E. Beck, *Phys. Rev. Lett.* **23**:625 (1969).

(Kl 82) H. V. Klapdor, J. Metzinger, and T. Oda, *Z. Phys. A* **309**:91 (1982).

(Kl 82b) O. Klepper, T. Batsch, S. Hofmann, R. Kirchner, W. Kurcewicz, W. Reisdorf, E. Roeckl, D. Schardt, and G. Nyman, *Z. Phys. A* **305**:125 (1982).

(Kl 84) H. V. Klapdor, J. Metzinger, and T. Oda, *At. Data Nucl. Data Tables* **31**:81 (1984).

(Ku 66) V. I. Kusnetsov, N. K. Skobelev, and G. N. Flerov, *Yad. Fiz.* **4**:99 (1966).

(Ku 66b) V. I. Kusnetsov, N. K. Skobelev, and G. N. Flerov, *Yad. Fiz.* **4**:279 (1966).

(Ku 67) V. I. Kusnetsov, N. K. Skobelev, and G. N. Flerov, *Yad. Fiz.* **5**:271 (1967).

(Ku 83) A. A. Kuznetsov, in *Nucleus-Nucleus Collisions*, Proceedings of the International Conference on Nucleus-Nucleus Collisions, Michigan State University, September 26–October 1, 1982, Invited Papers (G. F. Bertsch, C. K. Gelbke, and D. K. Scott, eds.), North-Holland, Amsterdam (1983), p. 493c.

(La 85) M. Langevin, E. Quiniou, M. Bernas, J. Galin, J. C. Jacmart, F. Naulin, F. Pougheon, R. Anne, C. Détraz, D. Guerreau, D. Guillemaud-Mueller, and A. C. Mueller, *Phys. Lett.* **150B**:71 (1985).

(La 86) M. Langevin, A. C. Mueller, D. Guillemaud-Mueller, M. G. Saint-Laurent, R. Anne, M. Bernas, J. Galin, D. Guerreau, J. C. Jacmart, S. D. Hoath, F. Naulin, F. Pougheon, E. Quiniou, and C. Détraz, *Nucl. Phys. A* **455**:149 (1986).

(Le 86) S. Leray, Proceedings of the International Conference on Heavy Ion Nuclear Collisions in the Fermi Energy Domain (Caen, France, May 1986), *J. Phys. (Paris), Colloq. C4*, **47**:275 (1986).

(Li 75) P. J. Lindstrom, D. E. Greiner, H. H. Heckman, B. Cork, and F. S. Bieser, Lawrence Berkeley Laboratory report LBL-3650 (1975).

(Lo 84) J. M. Loiseaux and M. Fruneau, *Tenth International Conference on Cyclotrons and Their Applications*, April 30–May 3 1984, East Lansing, Michigan, U.S.A. (F. Marti, ed.), IEEE, New York (1984).

(Ly 82) U. Lynen, H. Ho, W. Kühn, D. Pelte, U. Winkler, W. F. J. Müller, Y. T. Chu, P. Doll, A. Gobbi, K Hildenbrand, A. Olmi, H. Sann, H. Stelzer, R. Bock, H. Löhner, R. Glasow, and R. Santo, *Nucl. Phys. A* **387**:129C (1982).

(Ma 63) R. D. Macfarlane and R. D. Griffioen, *Nucl. Instrum. Methods* **24**:461 (1963).

(Ma 65) R. D. Macfarlane, D. Ronald, and A. Siivola, *Phys. Rev. Lett.* **14**:114 (1965).

(Ma 85) J. A. Maruhn and W. Greiner, in *Treatise on Heavy-Ion Science*, Vol. 4 (D. Allan Bromley, ed.), Plenum Press, New York (1985), p. 465.

(Me 85) M. C. Mermaz, R. Dayras, J. Barrette, B. Berthier, D. M. De Castro Rizzo, O. Cisse, R. Legrain, A. Pagano, E. Pollacco, H. Delagrange, W. Mittag, B. Heusch, G. Lanzano, and A. Palmeri, *Nucl. Phys. A* **441**:129 (1985).

(Mi 84) A. C. Mignerey, K. L. Wolf, D. G. Raich, V. E. Viola, Jr., J. R. Birkleund, W. U. Schröder, and J. R. Huizenga, *Phys. Rev. C* **29**:158 (1984).

(Mi 84b) P. S. Miller, *Tenth International Conference on Cyclotrons and Their Applications*, April 30–May 3 1984, East Lansing, Michigan, U.S.A. (F. Marti, ed.), IEEE, New York (1984), p. 11.

(Mi 85) I. N. Mishustin, Proceedings of the Second International Conference on Nucleus-Nucleus Collisions (Visby, Sweden, June 1985), *Nucl. Phys. A* **447**:67c (1985).

(Mi 86) W. Mittag, L. Biachi, A. Cunsolo, B. Fernandez, A. Foti, J. Gastebois, A. Gilibert, C. Grégorie, Y. Schutz, C. Stephan, and A. Péghaire, XXIV International Winter Meeting on Nuclear Physics, Bormio, 1985; preprint CENBGR8601 (1986).

(Mo 76) L. G. Moretto and R. Schmitt, *J. Phys. (Paris), Colloq. C5*, **37**:109 (1976).

(Mo 80) D. M. Moltz, R. A. Gough, M. S. Zisman, D. J. Vieira, H. C. Evans, and J. Cerny, *Nucl. Instrum. Methods* **172:**507 (1980).

(Mo 81) P. Möller, and J. R. Nix, *Nucl. Phys. A* **361**:117 (1981).

(Mu 84) M. J. Murphy, *Phys. Lett.* **135B**:25 (1984).

(Mu 84b) J. A. Musser and J. D. Stevenson, *Phys. Rev. Lett.* **53**:2544 (1984).

(Mü 79), G. Münzenberg, W. Faust, S. Hofmann, P. Armbruster, K. Güttner, and H. Ewald, *Nucl. Instrum. Methods* **161**:65 (1979).

(Mü 81) G. Münzenberg, W. Faust, F. P. Hessberger, S. Hofman, W. Reisdorf, K. H. Schmidt, W. F. W. Schneider, H. Schött, P. Armbruster, K. Güttner, B. Thuma, H. Ewald, and D. Vermeulen, *Nucl. Instrum. Methods* **186**:423 (1981).

(Ni 70) J. M. Nitschke, Proceedings of the International Conference on Properties of Nuclei Far From the Region of Beta-Stability, Leysin, Switzerland, August 31–September 4, 1970, report CERN 70-30, Vol. 1 (1970), p. 153.

(Og 73) Yu. Ts. Oganessian, Yu. E. Penionzhkevich, Nguyen-Tak-Anh, D. M. Nadkarnee, K. A. Gavrilov, Kim de En, and M. Yussonua, *Yad. Fiz.* **18**:734 (1973).

(Og 75) Yu. Ts. Oganessian, Yu. E. Penionzhkevich, and A. O. Shamsutdinov, *Acta Phys. Pol.* **B6**:323 (1975).

(Og 76) Yu. Ts. Oganessian, A. G. Demin, N. A. Danilov, G. N. Flerov, M. P. Ivanov, A. S. Iljinov, N. N. Kolesnikov, B. N. Markov, V. M. Plotko, and S. P. Tretyakova, *Nucl. Phys. A* **273**:505 (1976).

(Og 82) Yu. Ts. Oganessian, Yu. E. Penionzhkevich, R. Kalpakchieva, F. Pawlat, C. Borcea, A. V. Belozyorov, Yu. P. Kharitonov, S. P. Tretyakova, V. G. Subbotin, S. M. Lukyanov, N. V. Pronin, and A. A. Bykov, *Zh. Eksp. Teor. Fiz. Pis'ma* **36**:104 (1982).

(Og 83) Yu. Ts. Oganessian, G. G. Gulbekyan, B. N. Gikal, A. I. Ivanenko, B. N. Klenin, S. I. Kozlov, I. V. Kolesov, V. B. Kutner, V. N. Melnikov, R. Ts. Oganessian, A. S. Pasyuk, and V. A. Chugreev, Proceedings of the International School-Seminar on Heavy Ion Physics, Alushta, USSR, April 1983, report D7-83-644, JINR, Dubna (1983), p. 55.

(Og 85) Yu. Ts. Oganessian and Yu. Lazarev, in *Treatise on Heavy-Ion Science*, Vol. 4 (D. Allan Bromley, ed.), Plenum Press, New York (1985), p. 40.

(Or 84) J. H. Ormrod, C. B. Bigham, E. A. Heighway, J. D. Hepburn, C. R. Hoffmann, J. A. Hilbert, and H. R. Schneider, *Tenth International Conference on Cyclotrons and Their Applications*, April 29–May 3 1984, East Lansing, Michigan, U.S.A. (F. Marti, ed.), IEEE, New York (1984), p. 245.

(Pl 74) F. Plasil, in *Proceedings of the International Conference on Reactions between Complex Nuclei* (R. L. Robinson, F. K. McGowan, J. B. Ball, and J. H. Hamilton, eds.) North-Holland, Amsterdam (1974), Vol. 2, p. 107.

(Po 66) L. Pomorski, J. Tys, and V. V. Volkov, *Phys. Lett.* **23**:369 (1966).

(Po 66b) A. M. Poskanzer, S. W. Cosper, E. K. Hyde, and J. Cerny, *Phys. Rev. Lett.* **17**:1271 (1966).

(Po 68) A. M. Poskanzer, G. W. Butler, E. K. Hyde, J. Cerny, D. A. Landis, and F. S. Golding, *Phys. Lett.* **27B**:414 (1968).

(Po 69) L. Pomorski, J. Tys, and V. V. Volkov, preprint JINR, E7-4877, Dubna (1969).

(Po 86) F. Pougheon, D. Guillemaud-Mueller, E. Quintiou, M. G. St. Laurent, R. Anne, D. Bazin, M. Bernas, D. Guerreau, J. C. Jacmart, S. D. Hoath, A. C. Mueller, and C. Detraz, *Europhys. Lett.* **2**:505 (1986).

(Ra 70) G. M. Raisbeck, P. Boerstling, P. Riesenfeldt, T. D. Thomas, R. Klapisch, and G. T. Garvey, *International Conference on High Energy Physics and Nuclear Structure, 1969*, Plenum Press, New York-London (1970), p. 341.

(Ra 86) P. Radvani, *Nucleus-Nucleus Collisions II*, Proceedings of the Second International Conference on Nucleus-Nucleus Collisions, Visby, Sweden, June 1985 (H. A. Gustafsson, B. Jakobsson, I. Otterlund, and K. Aleklett, eds.), North-Holland, Amsterdam (1986), p. 435c.

(Ro 83) E. Roeckl, *Proceedings of the International Conference on Nucleus-Nucleus Collisions, Michigan State University, September 26–October 1 1982, Invited Papers* (G. F. Bertsch, C. K. Gelbke, and D. K. Scott, eds.), North-Holland, Amsterdam (1983), p. 131c.

(Ro 87) E. Roeckl, Proceedings of the International School-Seminar on Heavy Ion Physics (Dubna, September, 1986), report D7-87-68, JINR, Dubna (1987), p. 636.

(Ru 77) P. Russo, R. P. Schmitt, G. J. Wozniak, R. C. Jared, P. Glassel, B. Cauvin, J. S. Sventek, and L. G. Moretto, *Nucl. Phys. A* **281:**509 (1977).

(Ru 83) E. Runte, W. D. Schmidt-Ott, P. Tidemand-Petersson, R. Kirchner, O. Klepper, W. Kurcewicz, E. Roeckl, N. Kaffrell, P. Peuser, K. Rykaczewski, M. Bernas, P. Dessagne, and M. Langevin, *Nucl. Phys. A* **399**:163 (1983).

(Ru 85) E. Runte, K. L. Gippert, W. D. Schmidt-Ott, P. Tidemand-Petersson, L. Ziegeler, R. Kirchner, O. Klepper, P. O. Larsson, E. Roeckl, D. Schardt, N. Kaffrel, P. Peuser, M. Bernas, P. Dessagne, M. Langevin, and K. Rykaczewski, *Nucl. Phys. A* **441**:237 (1985).

(Ry 83) K. Rykaczewski, R. Kirchner, W. Kurcewicz, D. Schardt, N. Kaffrell, P. Peuser, E. Runte, W. D. Schmidt-Ott, P. Tidemand-Petersson, and K. L. Gippert, *Z. Phys. A* **309**:273 (1983).

(Sc 84) W. U. Schröder and J. R. Huizenga, in *Treatise on Heavy-Ion Science*, Vol. 2 (D. A. Bromley, ed.), Plenum Press, New York (1984), p. 113.

(Sh 72) I. A. Shelaev *et al.*, Proceedings of the Sixth International Cyclotron Conference, New York (1972), p. 232.

(Sh 86) B. Sherrill, W. Benenson, B. A. Brown, M. S. Curtin, L. Harwood, E. Kashy, D. Mikolas, J. A., Nolen, Jr., J. Stevenson, J. S. Winfield, Q. Xie, and R. Sherr, Proceedings of International Conference on Heavy Ion Nuclear Collisions in the Fermi Energy Domain (Caen, France, May 1986), *J. Phys. (Paris), Colloq. C4*, **47**:245 (1986).

(Si 75) K. Sistemich, J. W. Grüter, H. Lawin, J., Eidens, R. Fabbri, T. A. Khan, W. D. Lauppe, G. Sadler, H. A. Seilé, and M. Shaanan, *Nucl. Instrum. Methods* **130**:491 (1975).

(So 77) L. P. Somerville, A. Ghiorso, M. J. Nurmia, and G. T. Seaborg, Berkeley Annual Report (Nuclear Science), LBL-6575, Berkeley (1977), p. 39.

(St 71) F. S. Stephens, J. R. Leigh, and R. M. Diamond, *Nucl. Phys. A* **170**:321 (1971).

(St 76) C. Stephan, N. Frascaria, J. P. Garron, J. C. Jacmart, N. Poffé, and L. Tassan-Got, Proceedings of the 3rd International Conference on Nuclei Far From Stability, Cargése, Corsica, May 1976, report CERN 76-13, Geneva (1976), p. 46.

(St 81) J. D. Stevenson and P. B. Price, *Phys. Rev. C* **24**:2102 (1981).

(St 83) J. D. Stevenson and J. A. Musser, *Nucl. Instrum. Methods* **213**:285 (1983).

(Sw 81) W. J. Swiatecki, *Phys. Scr.* **24**:113 (1981).

(Sy 79) T. J. M. Symons, Y. P. Viyogi, G. D. Westfall, P. Doll, D. E. Greiner, H. Faraggi, P. J. Lindstrom, D. K. Scott, H. J. Crawford, and C. McParland, *Phys. Rev. Lett.* **42**:40 (1979).

(Sy 81) T. J. M. Symons, Proceedings of the 4th International Conference on Nuclei Far From Stability, Helsingør (Denmark), June 1981, report CERN 81-09, Geneva (1981), p. 668.

(Ta 65) N. I. Tarantin, A. V. Demyanov, Yu. A. Dyachikhin, and A. P. Kabachenko, *Nucl. Instrum. Methods* **38**:103 (1965).

(Ta 66) N. I. Tarantin, A. V. Demyanov, *Prib. Tekh. Eksp.* No. **1**:41 (1966).

(Ta 70) N. I. Tarantin, A. P. Kabachenko, I. V. Kusnetsov, Yu. A. Dyachikhin, N. S. Ivanov, and Om Zai Khun, Proceedings of the International Conference on Electromagnetic Isotope Separators and the Techniques of Their Applications, Marburg, September 7–10, 1970 (H. Wagner and W. Walcher, eds.). Phys. Inst. Univer. Marburg, Forschunbericht BMBW-FBK70-28 (1970), p. 59.

(Ta 73) K. Takahashi, M. Yamada, and T. Kondoh, *At. Data Nucl. Data Tables* **12**:101 (1973).

(Ta 75) B. Tamain, C. Ngô, J. Péter, and F. Hanappe, *Nucl. Phys. A* **252**:187 (1975).

(Ta 85) I. Tanihata, O. Yamakawa, H. Hamagaki, O. Hashimoto, S. Nagamiya, Y. Shida, N. Yoshikawa, K. Sugimoto, T. Kobayashi, D. E. Greiner, N. Takahashi, and Y. Nojiri, *Nucleus–Nucleus Collisions II* (Second International Conference on Nucleus–Nucleus Collisions, Visby, Sweden, 10–14 June, 1985), Vol. I: Contributed Papers (B. Jakobsson and K. Aleklett, eds.), Lund and Studsvik (1985), p. 33.

(Th 68) T. D. Thomas, G. M. Raisbeck, P. Boerstling, G. T. Garvey, and R. P. Lynch, *Phys. Lett.* **27B**:504 (1968).

(Un 82) M. Uno and M. Yamada, INS report NUMA 40 (1982).

(Vi 69) B. N. Vinogradov and P. E. Nemirovsky, *Yad. Fiz.* **10**:505 (1969).

(Vo 62) V. V. Volkov, L. Pomorski, J. Tys, and G. N. Flerov, *Zh. Eksp. Teor. Fiz.* **42**:635 (1962).

(Vo 62b) V. V. Volkov, L. Pomorski, J. Tys, and G. N. Flerov, *Zh. Eksp. Teor. Fiz.* **43**:865 (1962).

(Vo 63) V. V. Volkov, L. Pomorski, J. Tys, and J. Wilczynski, *Zh. Eksp. Teor. Fiz.* **45**:897 (1963).

(Vo 69) A. A. Vorobjev, G. T. Grachev, Yu. K. Zalite, I. A. Kondurov, A. M. Nikitin, and D. M. Seliverstov, Communication Fiz. Technic. Inst. FTI-232, Leningrad (1969).

(Vo 71) V. V. Volkov, *Part. Nucl.* **2**:285 (1971).

(Vo 73) V. V. Volkov, *Proceedings of the International Conference on Nuclear Physics, Munich, August 27–September 1, 1973*, Vol. 2, North-Holland, Amsterdam (1973), p. 279.

(Vo 74) V. V. Volkov, *Proceedings of the International Conference on Reactions between Complex Nuclei, Nashville, Tennessee, June 1974*, Vol. 2 (R. L. Robinson, F. K. McGowan, J. B. Ball, and J. H. Hamilton, eds.), North-Holland, Amsterdam (1974), p. 363.

(Vo 75) V. V. Volkov, in *Lecture Notes in Physics 33*, Classical and Quantum Mechanical Aspects of Heavy Ion Collisions (H. L. Harney, P. Braun Munzinger, and C. K. Gelbke, eds.), Springer-Verlag, Berlin, Heidelberg (1975), p. 253.

(Vo 75b) V. V. Volkob, *Part. Nucl.* **6**:1040 (1975).

(Vo 78) V. V. Volkov, *Phys. Rep.* **44**:93 (1978).

(Vo 79) V. V. Volkov, in *Detectors in Nuclear Science* (D. Allan Bromley, ed.), North-Holland, Amsterdam (1979), p. 623.

(Vo 80), V. V. Volkov, Proceedings of the International Conference on Extreme States in Nuclear Systems, Dresdan, GDR, February, 1980, Vol. 1 (H. Prade and S. Tesch, eds.), report ZfK-430, Rossendorf (1980), p. 1.

(We 79) G. D. Westfall, T. J. M. Symons, D. E. Greiner, H. H. Heckman, P. J. Lindstrom, J. Mahoney, A. C. Shotter, D. K. Scott, H. J. Crawford, C. McParland, T. C. Awes, C. K. Gelbke, and J. M. Kidd, *Phys. Rev. Lett.* **43**:1859 (1979).

(We 81) B. W. Wei, *Ninth International Conference on Cyclotrons and Their Applications*, September 1981, Caen, France (G. Gendreau, ed.), Les Editions de Physique, Les Ulis Cedex, France (1981), p. 23.

(Wi 72) W. G. Winn, H. H. Gutbrod, and M. Blann, *Nucl. Phys.* A **188**:423 (1972).

# 3

JUHA ÄYSTÖ (left) is a Senior Researcher of the Academy of Finland and is presently carrying out research on nuclides far from stability at the Department of Physics of the University of Jyväskylä in Finland. He received his Ph.D. degree in Physics from the University of Jyväskylä in 1977. He spent two years as a postdoctoral fellow in Lawrence Berkeley Laboratory in 1978–1979. Since 1980 he has held various positions in the University of Jyväskylä and became a researcher of the Academy of Finland in 1983. He spent one year as a visiting staff scientist in Lawrence Berkeley Laboratory in 1985 carrying out research on exotic proton-rich nuclei.

JOSEPH CERNY (right) is a Professor of Chemistry, Provost for Research, and Dean of the Graduate Division at the University of California, Berkeley, California. He received his Ph.D in Nuclear Chemistry from the University of California at Berkeley in 1961. He received the E. O. Lawrence Award of the Atomic Energy Commission in 1974 and the American Chemical Society's Award for Nuclear Chemistry in 1984, both for his studies of exotic nuclei and new radioactivities. He has served as Chairman of the Chemistry Department at the University of California at Berkeley and as an Associate Director of the Lawrence Berkeley Laboratory, as well as on numerous advisory and visiting committees to government and national laboratories. Recipient of a Guggenheim Fellowship in 1969–70 (Oxford) and a Humboldt Award in 1985 (GSI, Darmstadt), Cerny is a fellow of the American Physical Society and of the American Association for the Advancement of Science.

# Proton-Rich Light Nuclei

JUHA ÄYSTÖ AND JOSEPH CERNY

## 1. Introduction

A basic question for a given element of how few or how many neutrons can be contained in its nucleus and still result in a "bound" system, and the resulting consequences on our present concept of nuclear structure, have been the focus of much recent research on highly unstable nuclei. It is only during the last several years that the limit of nuclear stability has been reached in a few areas of the nuclear chart. Except among the lightest elements, knowledge of the limit on the neutron-excess side is still beyond our experimental capability. For the heavier elements, the gap between the most neutron-rich isotope known and the isotope which is predicted to define the neutron drip-line more or less increases monotonically with element number. The situation is quite different for neutron-deficient nuclides. Here, the boundary region, in which the proton binding energy goes to zero, has been essentially delineated through the sd shell and established to some extent in the $A = 100$–$160$ region. This review gives the present status of research on light ($A \leqslant 60$) proton-rich nuclides with an emphasis on nuclei with a proton number $Z$ substantially larger than the neutron number $N$.

Due to the strong Coulomb repulsion between protons in the nucleus, the proton drip-line is much closer to the valley of stable nuclei than the neutron drip-line. Also, this repulsion increases the decay energies very rapidly when moving out toward the drip-line. These exotic, proton-rich nuclear species can offer unique opportunities to study new radioactive-decay phenomena. Studies of the decay of light nuclei with $N = Z - 3$ produced in compound-nucleus reactions induced primarily by light-ion beams was pioneered by detecting their $\beta$-delayed proton emission. Similar experiments with heavy-ion beams extended these measurements and lead also to the first observation of direct proton radioactivity (Ja 70, Ce 70). Further, the novel application of multineutron and certain multinucleon transfer reactions provided mass measurements for a number of bound and unbound light, proton-rich nuclei. In general, these experiments, together with progress in studies of some very neutron-rich light nuclides, provided a

JUHA ÄYSTÖ • Department of Physics, University of Jyväskylä, Jyväskylä, Finland. JOSEPH CERNY • Department of Chemistry and Lawrence Berkeley Laboratory, University of California at Berkeley, Berkeley, California 94720.

firm basis for more reliable mass predictions of nuclides near the proton drip-line, particularly when the Kelson–Garvey relation (Ke 66) based on charge symmetry can be employed.

If one takes the 1976 Conference on Nuclei Far From Stability (Cargese) as a reference (CERN report 76-13), much progress has been made in exploring the decay properties of highly proton-rich nuclides through the titanium isotopes, so that the experimental frontier in this region generally lies among the nuclides with isobaric spin projection $T_z = (N - Z)/2 = -2$ and $-\frac{5}{2}$. By contrast, in the region between the vanadium and the zinc isotopes, relatively little is known about the decay properties of nuclei that are more proton rich than those with $T_z = -\frac{1}{2}$, though quite recently the existence of a number of new, proton-rich nuclides in this region has been demonstrated (La 86, Po 86). These features are clearly observable in Table 1, which summarizes the

**Table 1. Discoveries of Bound Proton-Rich Nuclides with $T_z \leqslant -\frac{3}{2}$**

| Nuclide | $T_z$ | Reaction | Observables | Year | Reference |
|---|---|---|---|---|---|
| $^{9}$C | $-\frac{3}{2}$ | ($^3$He, $^6$He) | $Q$-value | 64 | (Ce 64) |
| $^{13}$O | $-\frac{3}{2}$ | (p, 2n) | $\beta$p-decay; $T_{1/2}$ | 65 | (Mc 65) |
| $^{17}$Ne | $-\frac{3}{2}$ | (p, 3n) | $\beta$p-decay; $T_{1/2}$ | 63 | (Ba 63) |
| $^{21}$Mg | $-\frac{3}{2}$ | (p, 3n) | $\beta$p-decay; $T_{1/2}$ | 63 | (Ba 63) |
| $^{20}$Mg | $-2$ | ($\alpha$, $^8$He) | $Q$-value | 74 | (Ro 74) |
| $^{23}$Al | $-\frac{3}{2}$ | (p, $^6$He) | $Q$-value | 69 | (Ce 69) |
| $^{22}$Al | $-2$ | ($^3$He, p4n) | $\beta$p-decay; $T_{1/2}$ | 82 | (Ca 82) |
| $^{25}$Si | $-\frac{3}{2}$ | (p, 3n) | $\beta$p-decay; $T_{1/2}$ | 63 | (Ba 63) |
| $^{24}$Si | $-2$ | ($^3$He, 3n) | $\beta$p-decay; $T_{1/2}$ | 79 | (Äy 79) |
| $^{23}$Si | $-\frac{5}{2}$ | $^{40}$Ca-fragm. | identified | 86 | (La 86) |
| $^{27}$P | $-\frac{3}{2}$ | ($^3$He, $^8$Li) | $Q$-value | 77 | (Be 77) |
| $^{26}$P | $-2$ | ($^3$He, p4n) | $\beta$p-decay; $T_{1/2}$ | 83 | (Ca 83b) |
| $^{29}$S | $-\frac{3}{2}$ | (p, 3n) | $\beta$p-decay; $T_{1/2}$ | 64 | (Ha 64) |
| $^{28}$S | $-2$ | ($\pi^+$, $\pi^-$) | $Q$-value | 82 | (Mo 82) |
| $^{27}$S | $-\frac{5}{2}$ | $^{40}$Ca-fragm. | identified | 86 | (La 86) |
| $^{31}$Cl | $-\frac{3}{2}$ | ($^3$He, $^8$Li) | $Q$-value | 77 | (Be 77) |
| $^{33}$Ar | $-\frac{3}{2}$ | ($^3$He, 2n) | $\beta$p-decay; $T_{1/2}$ | 64 | (Re 64) |
| $^{32}$Ar | $-2$ | spallation | $\beta$p-decay; $T_{1/2}$ | 77 | (Ha 77) |
| $^{31}$Ar | $-\frac{5}{2}$ | $^{40}$Ca-fragm. | identified | 86 | (La 86) |
| $^{35}$K | $-\frac{3}{2}$ | ($^3$He, $^8$Li) | $Q$-value | 76 | (Be 76) |
| $^{37}$Ca | $-\frac{3}{2}$ | (p, 3n) | $\beta$p-decay; $T_{1/2}$ | 64 | (Ha 64b) |
| | | ($^3$He, 2n) | $\beta$p-decay; $T_{1/2}$ | | (Re 64) |
| $^{36}$Ca | $-2$ | ($\alpha$, $^8$He) | $Q$-value | 77 | (Tr 77) |
| $^{35}$Ca | $-\frac{5}{2}$ | ($^3$He, $\alpha$4n) | $\beta$2p-decay; $T_{1/2}$ | 85 | (Äy 85) |
| $^{41}$Ti | $-\frac{3}{2}$ | ($^3$He, 2n) | $\beta$p-decay; $T_{1/2}$ | 64 | (Re 64) |
| $^{40}$Ti | $-2$ | ($\pi^+$, $\pi^-$) | $Q$-value | 82 | (Mo 82) |
| $^{43}$V | $-\frac{3}{2}$ | $^{58}$Ni-fragm. | identified | 86 | (Po 86)<br>(De 86) |

**Table 1.** *(Continued)*

| Nuclide | $T_z$ | Reaction | Observables | Year | Reference |
|---|---|---|---|---|---|
| $^{45}$Cr | $-\frac{3}{2}$ | ($^{16}$O, 3n) | $\beta$p-decay; $T_{1/2}$ | 74 | (Ja 74) |
| $^{44}$Cr | $-2$ | $^{58}$Ni-fragm. | identified | 86 | (Po 86) |
| $^{47}$Mn | $-\frac{3}{2}$ | $^{58}$Ni-fragm. | identified | 86 | (Po 86) |
| $^{46}$Mn | $-2$ | $^{58}$Ni-fragm. | identified | 86 | (Po 86) |
| $^{49}$Fe | $-\frac{3}{2}$ | ($^{12}$C, 3n) | $\beta$p-decay; $T_{1/2}$ | 70 | (Ce 70b) |
| $^{48}$Fe | $-2$ | $^{58}$Ni-fragm. | identified | 86 | (Po 86) |
| $^{51}$Co | $-\frac{3}{2}$ | $^{58}$Ni-fragm. | identified | 86 | (Po 86) |
| $^{50}$Co | $-2$ | $^{58}$Ni-fragm. | identified | 86 | (Po 86) |
| $^{53}$Ni | $-\frac{3}{2}$ | ($^{16}$O, 3n) | $\beta$p-decay; $T_{1/2}$ | 76 | (Vi 76) |
| $^{52}$Ni | $-2$ | $^{58}$Ni-fragm. | identified | 86 | (Po 86) |
| $^{51}$Ni | $-\frac{5}{2}$ | $^{58}$Ni-fragm. | identified | 86 | (Po 86) |
| $^{55}$Cu | $-\frac{3}{2}$ | $^{58}$Ni-fragm. | identified | 86 | (Po 86) |
| $^{57}$Zn | $-\frac{3}{2}$ | ($^{20}$Ne, 3n) | $\beta$p-decay; $T_{1/2}$ | 76 | (Vi 76) |
| $^{61}$Ge | $-\frac{3}{2}$ | ($^{24}$Mg, 3n) | $\beta$p-decay; $T_{1/2}$ | 76(86) | (Ce 76, Ho 86) |

discoveries of proton-rich nuclides with $T_z \leqslant -\frac{3}{2}$. Rapid progress since the 1976 Cargese conference is demonstrated by the fact that at least half of the isotopes noted were initially observed during this period. Illustrating the most recent advances in this field are:

1. The discovery of the $\beta$-delayed two-proton radioactivity of $^{22}$Al (Ca 83) and $^{26}$P (Ho 83).
2. The discovery of a series of $T_z = -\frac{5}{2}$ nuclei in the sd shell via $\beta$-delayed two-proton decay (Äy 85) and heavy-ion fragmentation (La 86).
3. The production of an intense source of a $T_z = -2$ activity, $^{32}$Ar, permitting detailed spectroscopic investigations (Bj 85).
4. The establishment of the existence of ten new proton-rich nuclides in the fp shell (Po 86).

A number of interesting problems in nuclear physics can be explored in the study of light nuclei up to the proton drip-line. Among these are the effects of charge symmetry and charge independence, especially in complete isobaric multiplets; the question of the proton-stability limit of nuclei; the existence of new radioactive-decay modes; the study of the weak interaction in nuclei via high-energy Fermi and Gamow–Teller $\beta$ decays; and tests of large-basis shell-model calculations in a variety of applications. Also, the role of nuclear processes associated with quite proton-rich nuclei are increasingly interesting from the standpoint of nuclear astrophysics.

## 2. Techniques for the Study of Light Proton-Rich Nuclei

### 2.1. General

As nuclei move further from the valley of $\beta$ stability on the proton-rich side, their total binding energy rapidly decreases, due to the increasing Coulombic repulsion, which results in two major difficulties in their studies. One is related to the reaction mechanism for their production and the other to their nuclear properties. The relatively large reaction $Q$-values which are required cause high excitations in the nuclear systems involved and open up many competing reaction channels favoring nuclei closer to stability. Damping this excitation can be very crucial for nuclei produced near the limit of proton stability, because their ground states are often the only bound ones.

Another difficulty is related to the decay properties, since the steep mass surface causes $\beta$-decay energies to increase strongly away from the valley of $\beta$ stability. Assuming that $\beta$ decay is the only available decay mode, the half-lives of the least-bound proton-rich nuclei are in the 5–50 ms range. At the drip-line, other decay modes may also be present and proton or two-proton decay may occur with half-lives considerably shorter than those predicted for the $\beta$-decay process. An example of the curvature of the mass parabola is given in Figure 1, which displays the mass excesses of the $A = 35$ nuclides as a function of their isospin projection $T_z$. The $T_z = -\frac{5}{2}$ nucleus, $^{35}$Ca,

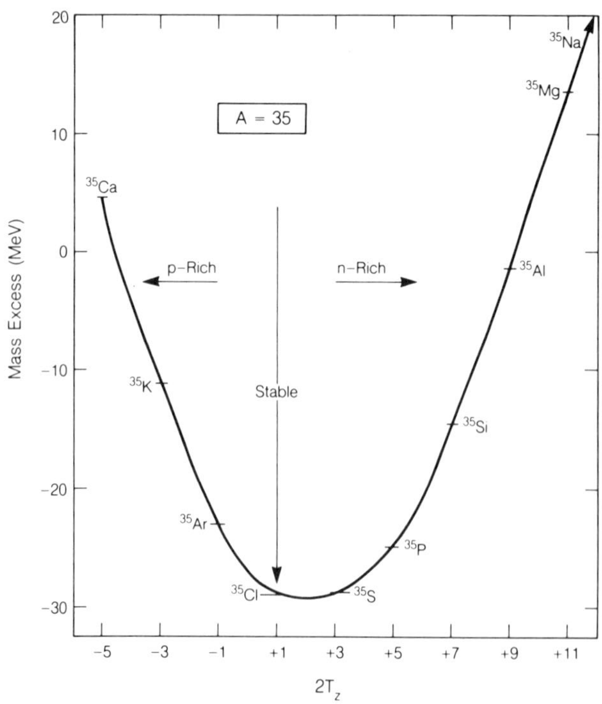


**Figure 1.** Mass excess parabola for the $A = 35$ isobar. Masses are from the latest mass evaluation (Wa 85), Kelson-Garvey predictions, and reference (Äy 85).

is the most proton-rich bound member of this multiplet, while the most neutron-rich bound isotope is expected to be $^{35}$Na, with a $T_z$ of $+\frac{13}{2}$.

In the following sections, the reactions and techniques used to produce and study highly proton-rich nuclei are reviewed. Depending on the experimental approach, the objective may be the determination of a nuclide's "existence" or particle stability, its ground-state mass, its decay modes and half-life, or its excited states and their decay.

## 2.2. Light-Ion-Induced Reactions

Traditionally, light-ion-induced fusion evaporation reactions have been used to produce proton-rich nuclei through the titanium isotopes. This has arisen from both the availability of high-quality proton and $^3$He beams as well as from the favorable combination of protons and neutrons in the fused system. Similarly, proton-induced spallation reactions can be used to produce very proton-rich light nuclei (Ra 79). Transfer reactions induced by light ions, such as (p, $^6$He) and ($\alpha$, $^8$He), have had a very important impact on the discoveries and especially the mass measurements of highly proton-rich nuclei (Ce 68, Ro 74). Reactions involving the use of pions as projectiles (Mo 82) or as reaction products (Sh 84) have recently also been exploited successfully.

The families of the $T_z \geqslant -\frac{3}{2}$ nuclides in the sd shell and the $T_z \geqslant -\frac{1}{2}$ nuclides in the $f_{7/2}$ shell can be relatively easily produced via low-energy proton and $^3$He-induced reactions. Detailed spectroscopy of the radioactive decays of these nuclei can, in general, be performed with rather simple experimental setups. Typical examples of such techniques are: pneumatic rabbit transport systems (Pa 77), gas transport techniques (Mo 80), and direct target (Mi 86) or recoil catcher (Ho 86) counting employing pulsed beams. With increasing bombarding energy to make nuclei further from stability, additional reaction channels open up and more sophisticated experimental techniques become necessary. In the sd shell, light-ion-induced reactions, such as ($^3$He, ypxn), can, in principle, be used to reach the limit of proton stability. Figure 2 shows an example of a $\beta$-delayed proton spectrum recorded during the bombardment of a calcium target with a 135 MeV $^3$He beam, using the He-jet technique. We note that the experimental cross-sections to produce $^{37}$Ca, $^{36}$Ca, and $^{35}$Ca isotopes via $^3$He + $^{40}$Ca reactions at the optimized energies for each are of the order of 100 $\mu$b, 1 $\mu$b, and 10 nb, respectively; see the literature (Äy 81, Äy 85). Due to the lack of $Z = N$ targets above $^{40}$Ca, these reactions have been little used for the production of quite proton-rich nuclides in and above the $f_{7/2}$ shell.

Light-ion-induced transfer reactions have played a crucial role in identifying and measuring the masses of a large number of proton-rich nuclei. These experiments are based on the two-body character of the reactions employed. When three of the four involved masses are known, the determination of the reaction $Q$-value establishes the mass for the fourth nucleus. These experiments are frequently carried out using magnetic spectrometers and/or time-of-flight techniques. A list of commonly used two-body reactions, involving transfers of varying numbers of nucleons, is given in Table 2. A special feature in these transfer reactions is that they can be used to determine the masses of particle unbound exotic nuclei, such as $^8$C, $^{12}$O, $^{16}$Ne, and $^{15}$F (Ro 74, Ke 78).

One of the most successful techniques, high-energy proton spallation, has had a strong impact on studies of neutron-rich and neutron-deficient nuclei (Ra 79). Because of the high production rates for a large number of nuclides, covering a wide range in

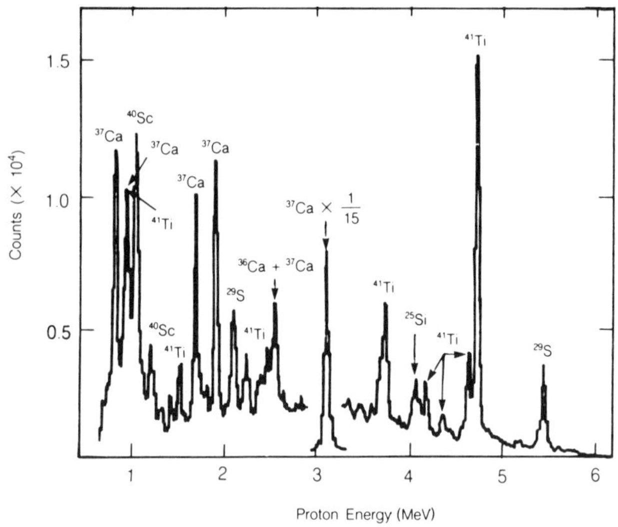


**Figure 2.** Spectrum of $\beta$-delayed proton emitters produced in the bombardment of $^{40}Ca$ by 135 MeV $^{3}He$. See text.

**Table 2. Transfer Reactions Employed in the Discoveries and Mass Measurements of Proton-Rich Nuclei**

| Reaction | Typical $Q$-value (MeV) | Reference |
|---|---|---|
| (p, $^{6}$He) | −30 | (Ce 69, Mu 75) |
| ($^{3}$He, $^{6}$He) | −20 | (Ce 64, Mu 77) |
| ($^{3}$He, $^{8}$Li) | −30 | (Be 75, Be 77) |
| ($^{3}$He, $^{9}$Li) | −40 | (Ro 76) |
| ($^{4}$He, $^{8}$He) | −60 | (Ce 66, Ro 74) |
| ($^{7}$Li, $^{8}$He) | −30 | (Sh 84, Sh 85) |

mass and charge, efficient and chemically selective mass-separation techniques become a necessity. Such reactions and techniques, as described by Ravn and Allardyce in Chapter 5, have not been widely applicable to light proton-rich nuclides. However, as a special case, one should note the discovery of the first $T_z = -2$ radioactivity, $^{32}Ar$, by the ISOLDE group (Ha 77). It is obvious that when some technological problems are solved, this technique may also become an invaluable approach to studies of the most proton-rich light nuclei.

A new tool in the spectroscopy of proton-rich nuclei has been the use of pions as projectiles or reaction products. A nice example of such a reaction in which pions are observed in the final state is $^{58}Ni(p, \pi^{-})^{59}Zn$ (Sh 84). Due to a negligible background, this quite low-yield reaction (0.08 nb/sr) proved to be very useful in obtaining a precise mass measurement ($\sigma = 40$ keV). For some years, the pion double-charge exchange reaction ($\pi^+$, $\pi^-$) has been used in mass measurements of several proton-rich exotic nuclei (Mo 82). As examples, we note the mass determinations of the $^{40}$Ti and $^{58}$Zn (Se 86) nuclides, which could not be measured by other established techniques. Typically, the precision associated with the ($\pi^+$, $\pi^-$) mass measurements is 50 keV, a value which is quite comparable in order of magnitude to those observed in transfer reactions.

## 2.3. Heavy-Ion-Induced Reactions

Heavy-ion-induced fusion reactions are normally not competitive with light-ion-induced reactions in producing proton-rich nuclei in the sd shell. Their power, however, becomes obvious when the production of isotopes of elements above titanium is desired, since $Z = N$ targets are no longer available. Our rather limited knowledge of the decays of proton-rich nuclei in the $f_{7/2}$ shell beyond the $T_z = -\frac{1}{2}$ series comes exclusively from studies that employed heavy-ion-induced fusion reactions. This is well demonstrated by the series of discoveries of the $T_z = -\frac{3}{2}$, $A = 4n + 1$ nuclides from $^{45}$Cr to $^{61}$Ge (see Table 1). However, a general problem associated with the application of these fusion reactions for producing highly proton-rich isotopes is their small cross-section due to the high proton-emission probabilities during the evaporation cascade. Recently, kinematically inverse reactions, such as $^{40}$Ca + $^{12}$C used in the production of the $T_z = -1$ nucleus $^{48}$Mn, have provided a new variation in the production of heavier proton-rich nuclei (Se 86b).

Heavy-ion-induced deep-inelastic and fragmentation reactions have already been used very successfully in studies of the existence (and some half-life measurements) of quite neutron-rich nuclei. Pioneering work at the Bevalac accelerator using the 200 MeV/nucleon $^{48}$Ca beam (We 79) and subsequent experiments with the 44 MeV/nucleon $^{40}$Ar and 33 MeV/nucleon $^{86}$Kr beams at GANIL (De 86) have established projectile fragmentation as an effective mechanism for producing very neutron-rich isotopes of light and medium heavy elements.

The use of heavy-ion-induced transfer or fragmentation reactions to produce proton-rich nuclei has just begun to be dramatically exploited at GANIL. In the first experiment, using 77 MeV/nucleon $^{40}$Ca projectiles on a 92 mg/cm$^2$ nickel target, a high production rate for very proton-rich nuclei was observed (La 86) and led to the discovery of several $T_z = -\frac{5}{2}$ nuclides, such as $^{31}$Ar. The high beam intensity available at GANIL, the kinematic properties of the projectile fragmentation process, and the excellent performance of the doubly achromatic magnetic spectrometer LISE (discussed below) permitted the identification of isotopes in the vicinity of the predicted proton drip-line as seen in Figure 3. A subsequent experiment using 55 MeV/nucleon $^{58}$Ni resulted in the discovery of twelve new proton-rich isotopes in the fp shell, as seen in Figure 4 (Po 86).

It has turned out that the projectile-like fragmentation reaction has much higher production cross-sections (10–100 times) for proton-rich isotopes than those arising in more conventional fusion and spallation reactions. So far, the applicability of the fragmentation reaction has been limited in these early experiments to the identification of nuclides. The "collection" of nuclides from such reactions for decay studies presents a further experimental challenge. However, the successful decay-time measurements for some neutron-rich nuclides both at GANIL (Du 86) and at the MSU superconducting cyclotron laboratory (Cu 86) clearly show promising prospects for the future.

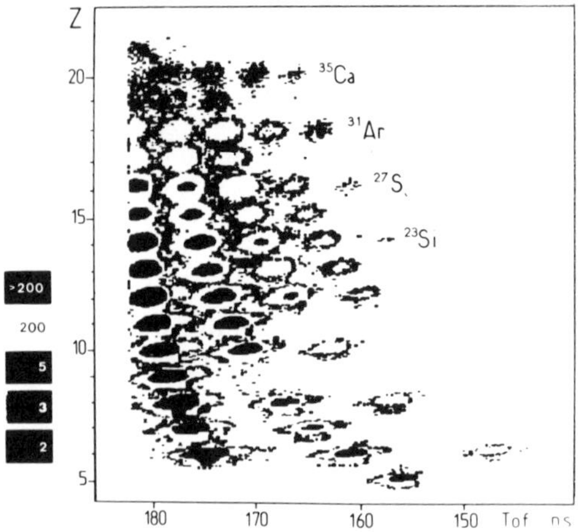


**Figure 3.** Two-dimensional plot of the events corresponding to proton-rich products in the fragmentation of 77 MeV/nucleon $^{40}$Ca on a nickel target. Four new $T_z = -\frac{5}{2}$ nuclides are clearly seen in this plot.

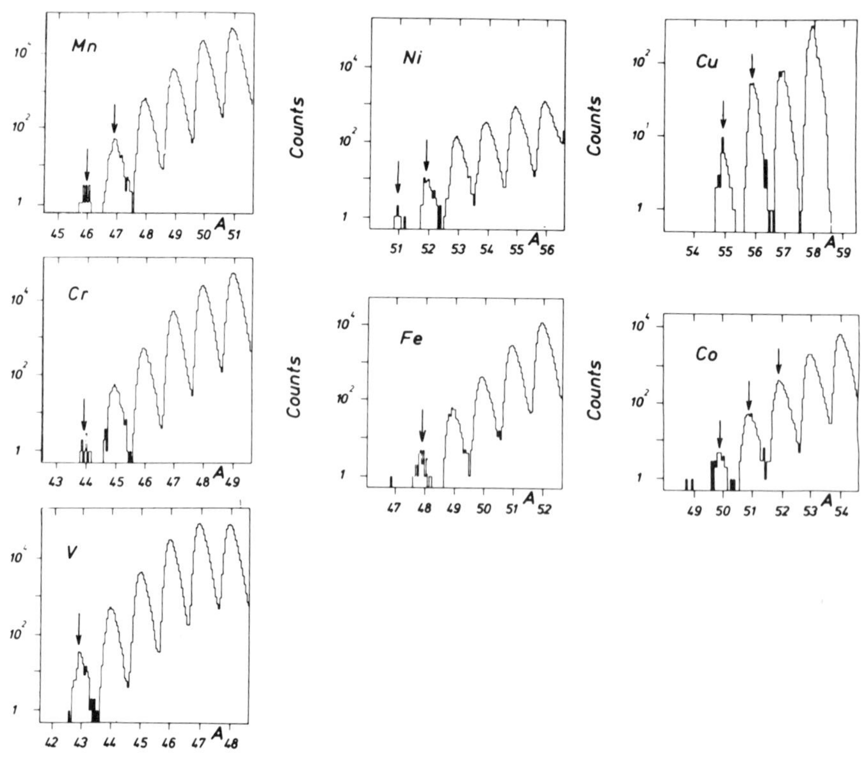


**Figure 4.** Yield distributions for isotopes of a given element obtained with the LISE spectrometer. New isotopes are marked by arrows. See text.

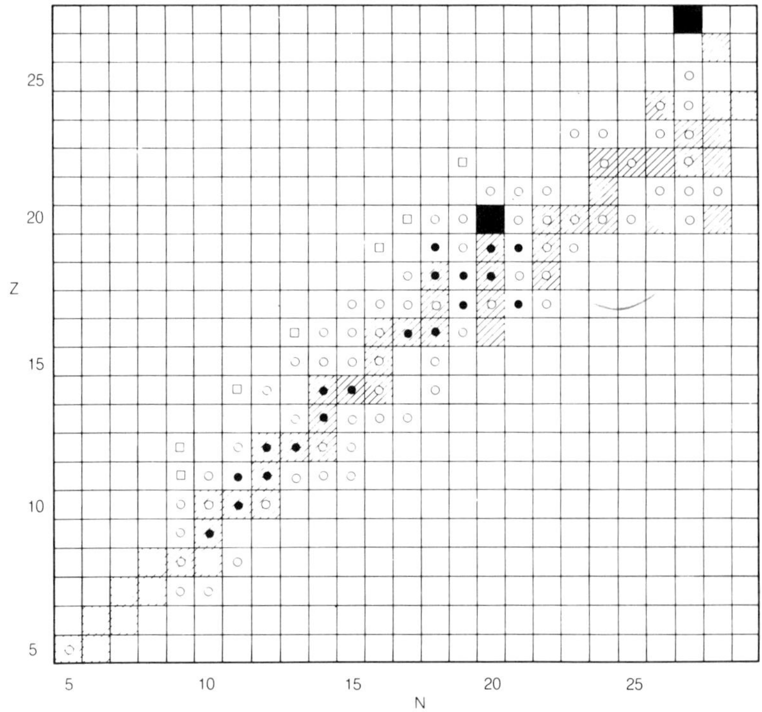


**Figure 5.** Distribution of observed fragments in the $Z-N$ plane in the reaction of 30 MeV/nucleon $^{14}N$ + calcium. Black squares represent the target and the compound system, respectively. Filled circles represent cross-sections >15 mb; open circles, cross-sections of 1–15 mb; and open squares, nuclei detected by $\beta$-delayed particle emission. The stable nuclei are cross-hatched.

A survey has recently been done in which a heavier target ($A \approx 200$) was bombarded with intermediate-energy (10–100 MeV/nucleon) light heavy ions. The results showed that the target-like nuclides were produced in reasonable quantities (Du 82). Consequently, the same idea was later tested with a lighter system by bombarding a calcium target with 15 and 30 MeV/nucleon $^{12}C$ and $^{14}N$ beams. In these experiments, in-beam and on-line radioactivity techniques were used to detect low recoil energy residues (Ol 84). The distribution of observed fragments in the $Z-N$ plane is shown in Figure 5. The charge dispersion is centered on the stable isotopes and has a half-width of about 2.5 atomic mass units; the most proton-rich nuclides observed were those with $T_z = -\frac{3}{2}$.

## 2.4. Experimental Techniques

A number of methods have been developed over the last several years to permit the study of light nuclei far from stability. The feature common to these experiments is the very small yield of the nuclei of interest, which must be detected in the presence of the primary beam and/or in the presence of the bulk of the other simultaneously produced nuclides. The objectives of these experimental studies are:

1. To establish and identify the particle stable nuclei.
2. To measure masses.
3. To investigate decay modes and measure half-lives.
4. To study excited nuclear states.

In principle, the techniques can be divided into two categories, one based on the prompt detection of the reaction products and the other based on on-line radioactivity measurements. Many of the commonly used techniques will be discussed here only briefly, since they have already been discussed in Volume 7 of *Treatise on Heavy-Ion Science* and in other chapters of the present volume. A special emphasis here will be on certain on-line radioactivity techniques, such as the helium-jet transport method and its application in on-line isotope separation. Also, the emphasis will be on those established techniques that are most relevant for current studies of light, proton-rich isotopes.

*Magnetic Spectrometers.* In studies of proton-rich light nuclei, several types of magnetic spectrometers have been used. Since these devices are used to detect energetic, promptly emitted particles, the associated transit times are very short, being typically of the order of 100 ns. With the selection of the magnetic field and emission angle, one can choose final-state particles that have resulted either from exotic transfer or fragmentation processes. These spectrometers can be used in the energy dispersive or in the charge-mass dispersive modes. The former has been used very successfully in $Q$-value measurements induced by light projectiles, such as in the $(\alpha, {}^8He)$ reaction. Normally, a light ejectile is observed and the mass of the nucleus of interest is derived from the measured $Q$-value. Good examples of the spectrometers that have been used for this type of experiments are the Enge split-pole spectrographs at Michigan State University and Texas A&M University as well as the QSD spectrometer at Lawrence Berkeley Laboratory (He 74, En 79). An essential part of all these spectrometers is a focal plane detector system, which is often a combination of several gas, scintillation, and solid-state counters (Fu 85). Particle identification is based on differential energy loss and total energy measured at the focal plane, the time of flight through the spectrometer, and the precise knowledge of the magnetic rigidity $B\rho$ of the particle. The accuracy of the $Q$-value measurement can typically be on the order of 50 keV. Figure 6 shows an application of the energy dispersive use of a heavy-ion spectrometer (Tr 80) to establish the mass of ${}^{24}Si$.

In recent GANIL experiments with the LISE spectrometer, many new proton- and neutron-rich nuclides have been detected (De 86). This spectrometer is a doubly achromatic system, consisting essentially of two identical magnetic dipoles, set at 0° as seen in Figure 7. With such a combination, it is possible to focus the fragments at the achromatic focal point. This improves and simplifies the focal-point detector considerably. Very good time-of-flight resolution and high precision in the energy-loss measurement with solid-state detectors result in an excellent $Z$ resolution of $\leq 1\%$. The mass resolution of the device is of the order of 0.5%. Use of a multiple differential energy-loss detector array and a parallel-plate avalanche counter for precise $B\rho$ determination ensures very low background conditions. This is beautifully demonstrated

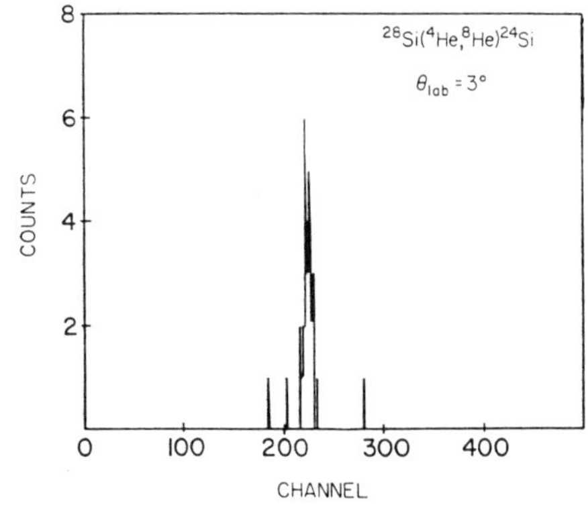


**Figure 6.** Energy spectrum of $^{8}He$ from the bombardment of $^{28}Si$ by 128.8 MeV $^{4}He$.

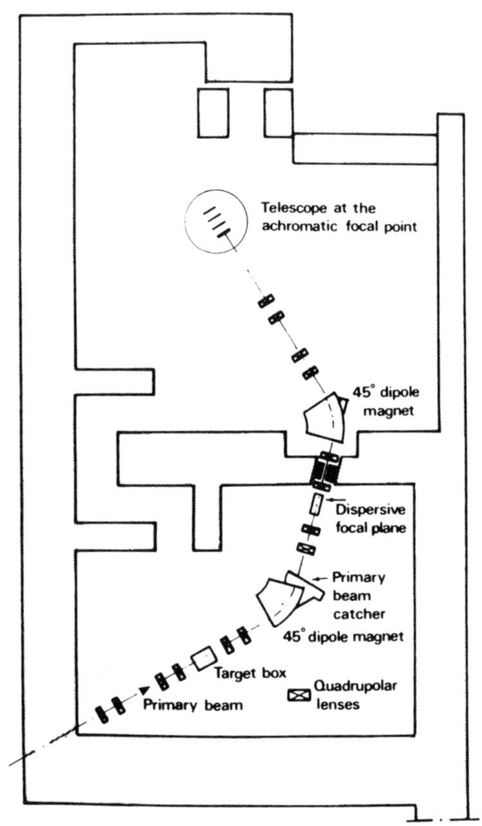


**Figure 7.** General layout of the doubly achromatic spectrometer LISE at the heavy-ion accelerator GANIL in Caen, France.

by the observation of many new proton-rich nuclides, as already shown in Figures 3 and 4.

*Recoil Separators.* The basic idea behind the recoil separator is the separation of the reaction products from the primary beam at forward angles. A well-known separator is SHIP, which is located at the UNILAC heavy-ion accelerator in Darmstadt (Mu 81). Some of the newer recoil separators are also capable of providing mass information as well. An example of a recoil separator employed for studies of light nuclei far from stability is the 14.3-meter-long reaction product mass separator at the National Superconducting Cyclotron Laboratory at MSU (Ha 81, Mi 86b). It is designed for heavy-ion energies above 5 MeV/nucleon, the present upper limit being 30 MeV/nucleon. It is a triply focusing device and is not fixed in angle, but can be varied between 0° and 30°. The mass resolving power of the separator is 100 with a 1 msr solid angle and 16% energy range.

*Helium-Jet Transport and Recoil Catching Techniques.* Techniques based on stopping and transporting reaction product recoils employing a gaseous or a solid medium have had a very central role in studies of $\beta$-decay properties of light proton-rich nuclides. Much of the pioneering work on proton-rich nuclei was done using direct recoil stopping or catching techniques. However, recent studies on the decay properties of nuclei with $T_z \leqslant -1$ have utilized the He-jet technique (Ma 74), which is illustrated in Figure 8 (Mo 80). The accelerator beam passes through a target chamber via a few mg/cm$^2$ thick metallic entrance and exit window foils. Reaction recoils ejected from

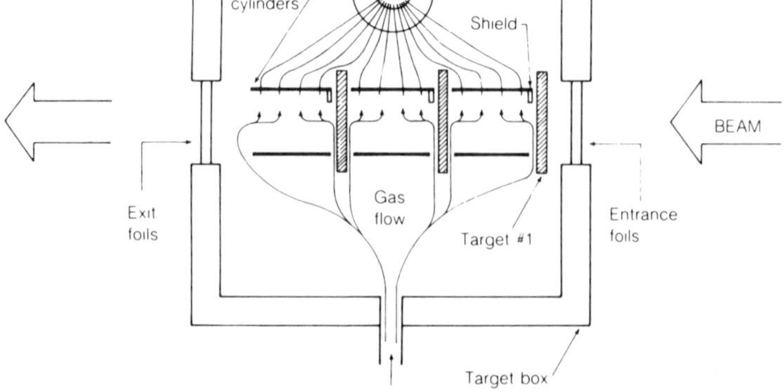


**Figure 8.** Principle of the helium-jet recoil transport technique; in this case, a system with three targets is shown.

the target(s) are stopped in 1–2 atm of helium. A continuous flow of helium transports the thermalized activity from the target chamber, via a capillary, to a low-background collection and detection area. In different experiments, capillaries vary in length from a few centimeters to several meters with diameters in the 1–5 mm range. Helium flow rates can be as high as 200 $cm^3/s$ (STP).

A high transport efficiency through the capillary requires the addition of cluster-forming material to the helium. Various substances have been used including organic vapors, water vapor, and inorganic vapors such as NaCl and $PbCl_2$. The clusters are either preformed or produced by the interaction of the beam with these additives. The optimum size of a cluster for high transmission is of the order of 0.1 $\mu$m (Wi 75). Typically, efficiencies of several tens of percent with total delay times as short as 20–100 ms have been observed.

After leaving the capillary, the activity is normally collected on a foil, on a catcher wheel, or on a movable tape for subsequent spectroscopy. It can also be transported into the ion source of an isotope separator for mass analysis. Even without mass separation, the He-jet technique in combination with light-ion-induced fusion reactions has proven very productive in searches for new activities as well as in the $\beta$-$\gamma$ and $\beta$-delayed particle spectroscopy of proton-rich nuclei. This technique produces extremely thin samples for $\beta$-delayed proton or $\alpha$-particle measurements, permitting high-resolution studies, as is illustrated in the spectrum of $^{40}Sc$ decay shown in Figure 9 (Ho 82).

The detection instrumentation consists of a specially tailored detector unit and a device to transport the activity to and/or away from the detectors. Several automatic catcher-wheel and tape-transport systems have been built for various purposes, the fastest of which can move the activity with speeds up to nearly 2 m/s. An example of a special setup is the multi-element charged particle telescope system

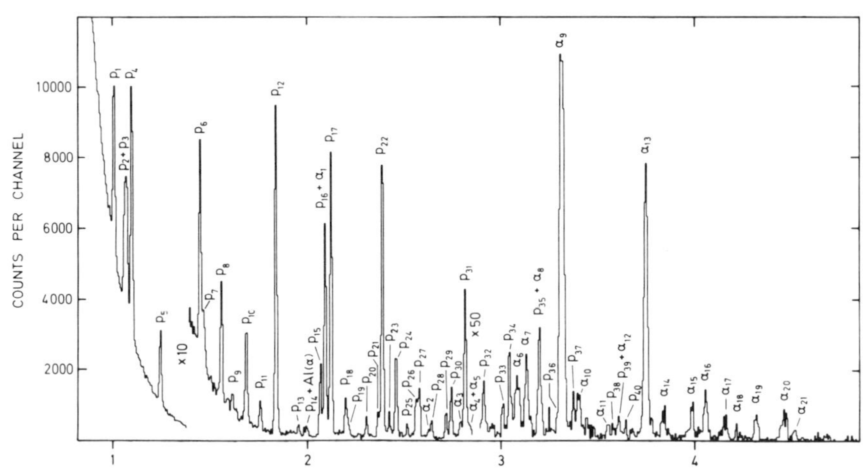


**Figure 9.** High-resolution $\beta$-delayed proton and $\alpha$ spectrum of $^{40}Sc$. The activity was produced via the 20 MeV proton + $^{40}Ca$ reaction and detected in a single 100-$\mu$m-thick surface-barrier detector.

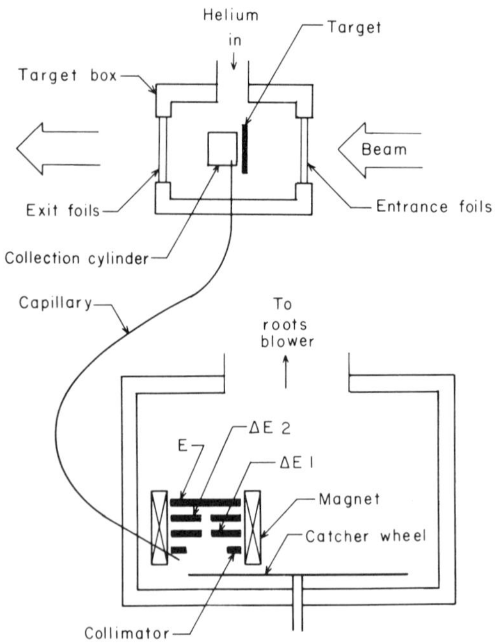


**Figure 10.** Schematic diagram of a multielement detector telescope system.

used in studying the $\beta$-delayed two-proton radioactivity of $^{22}$Al, as shown in Figure 10 (Ca 84).

For activities having half-lives below 50 ms, direct target or catcher counting methods become competitive with He-jet transport. An example of such an approach is the setup shown in Figure 11, which was used to study the $\beta$-delayed proton emission of heavy $T_z = -\frac{3}{2}$, $A = 4n + 1$ nuclides up to $^{61}$Ge (Ho 86). In these experiments, $\beta$-delayed protons emitted from precursors *retained in the target* were detected by semiconductor counter telescopes. Pulsed heavy-ion beams of $^{16}$O, $^{20}$Ne, and $^{24}$Mg were used to produce the precursors in the irradiations of natural calcium targets rotated to an angle of 70°. A slotted, rotating wheel system controlled the pulsing of the beam and protected the detectors while the target was irradiated (Vi 78). A similar setup (Mi 86) was also recently used by the Sendai group in studies of mirror $\beta$-decays in the fp shell and in the discovery of $^{57}$Cu (Sh 85b).

*On-Line Isotope Separators.* On-line isotope separators equipped with fast and efficient ion sources have played a central role in the study of many nuclei far from stability. Because on-line separators are dealt with in a separate chapter of this volume, only the approaches based on the He jet and used in experiments to study proton-rich light nuclei are described here. With this technique, product nuclei recoiling out of the target are stopped in helium and are transported through a capillary into the ion source of an isotope separator. This permits the transport of isotopes of all elements into the ion source so that a vast majority of the elements should be available for study. After the ion source, the singly charged radioactive nuclides are accelerated and separated by mass in a dipole magnet.

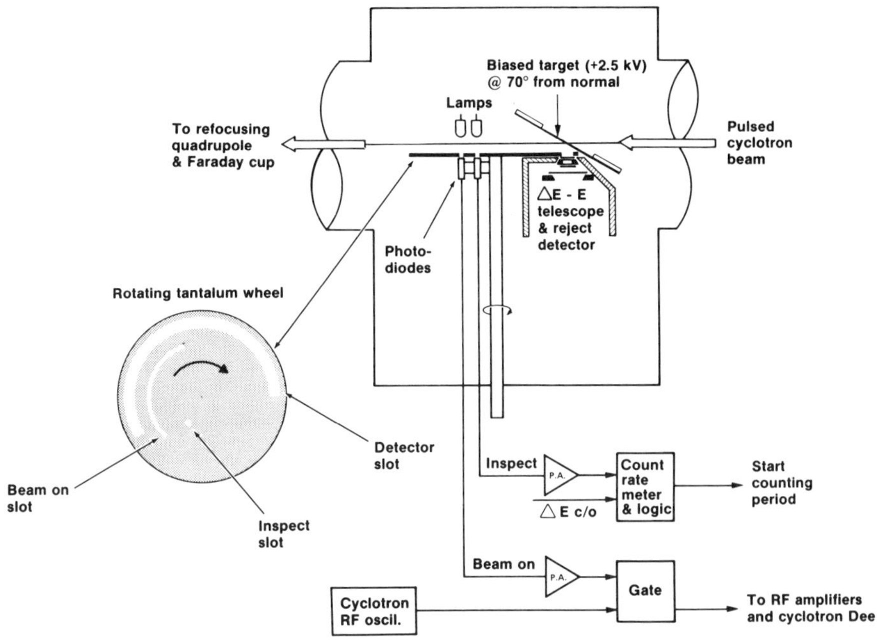


**Figure 11.** A setup used for the direct detection of $\beta$-delayed protons.

Figure 12 shows one of the He-jet separator systems, RAMA, or Recoil Atom Mass Analyzer, that was used in studies of the $T_z = -2$ nuclides $^{20}$Mg, $^{24}$Si, and $^{36}$Ca (Mo 80, Äy 80, Äy 81). This He-jet setup employs 1.3–1.5 atm pressure and a multiple-target arrangement. A single capillary is used to transport the activity into the hollow-cathode ion source of the mass separator. By employing a large Roots blower in conjunction with a skimmer, most of the helium is removed. Ionized recoils are extracted from the ion source using an 18 kV potential and are subsequently passed through a series of optical elements to focus the extracted beam before it is mass-analyzed. A velocity filter is used ahead of the analyzing magnet to divert the strong helium beam produced by the arc support gas and hence to reduce any space-charge problems. The total transit time for recoils to pass through this system is as low as 180 ms and has permitted measurements of isotopes with half-lives less than 100 ms. Recently, two other He-jet coupled high-current separators have become operational, one at the SARA accelerator in Grenoble (Pl 86) and the other at Chalk River Nuclear Laboratories (Sc 86).

Basically, two different types of detection techniques are used at the RAMA system. One was designed for detecting short-lived ($T_{1/2} \leq 0.5$ s) $\beta$-delayed particle radioactivities and incorporates a 50 $\mu$g/cm$^2$ carbon foil, on which the mass-separated beam of interest is implanted in front of solid-state telescopes. For $\beta$, $\gamma$, X-ray, and conversion electron measurements of activities with half-lives >0.5 s, a tape-drive system for transportation of the implanted activity is used. This approach (Mo 81) incorporates a two-stage differentially pumped entrance and exit system with the primary vacuum in the detector box $\sim 5 \times 10^{-6}$ torr. Common computer tape is pulled over a series of

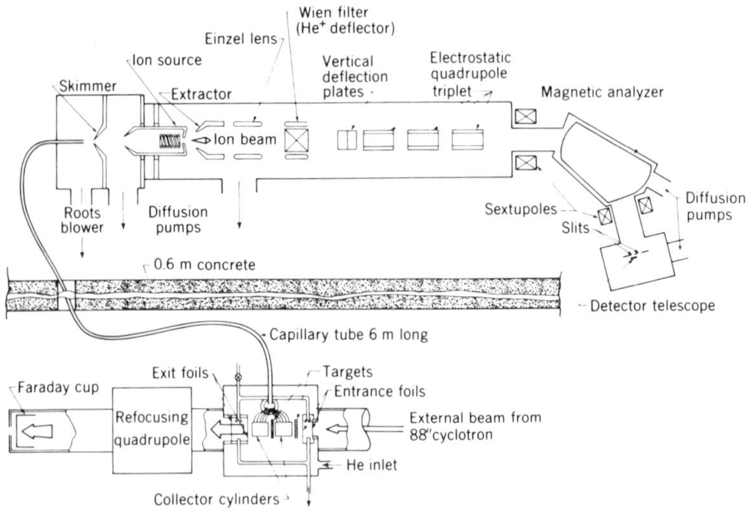


**Figure 12.** Overall schematic diagram of the on-line mass separator system RAMA.

rollers at 75 cm/s, moving a collected sample of activity from the focal plane to the appropriate detector station.

General difficulties associated with rapid separations of short-lived nuclei of refractory elements continue to require the development of new separation techniques. A novel approach is the recently constructed ion-guide isotope separator on-line (IGISOL) system (Är 86) which is shown in Figure 13. It operates by thermalizing the primary recoil ions in helium and transporting them in ionic form by helium flow through a differential pumping system into the acceleration chamber of an isotope separator. With this approach, separation times several orders of magnitude shorter that those typical for the ion source based systems can be achieved (Är 85). This is well demonstrated by the observation of several activities with half-lives less than 1 ms, including the 440 $\mu$s $\frac{1}{2}^+$ isomeric state of $^{43}$Sc. This technique has already been successfully used in studies of the decay properties of the $T_z = -\frac{1}{2}$ mirror nuclides $^{43}$Ti, $^{51}$Fe, $^{55}$Ni, and $^{57}$Cu, which all have half-lives in the 100 ms regime (Ho 86b, Äy 84, Yo 86). The overall efficiency of IGISOL is on the order of 1% with a mass resolving power of 375. This technique is well suited for use with light-ion-induced reactions, and active research on the technique and its potential for heavy-ion-induced reactions and fission is being conducted in several laboratories.

## 3. Nuclear Masses and Stability

### 3.1. General

Since the mass of a nucleus is determined by its binding energy, the extension of the limits of known nuclei is important toward a better understanding of nuclear stability. Moreover, the rapid decrease of the nuclear binding energy when moving outward from the valley of stable nuclei permits the observation of new radioactivities.

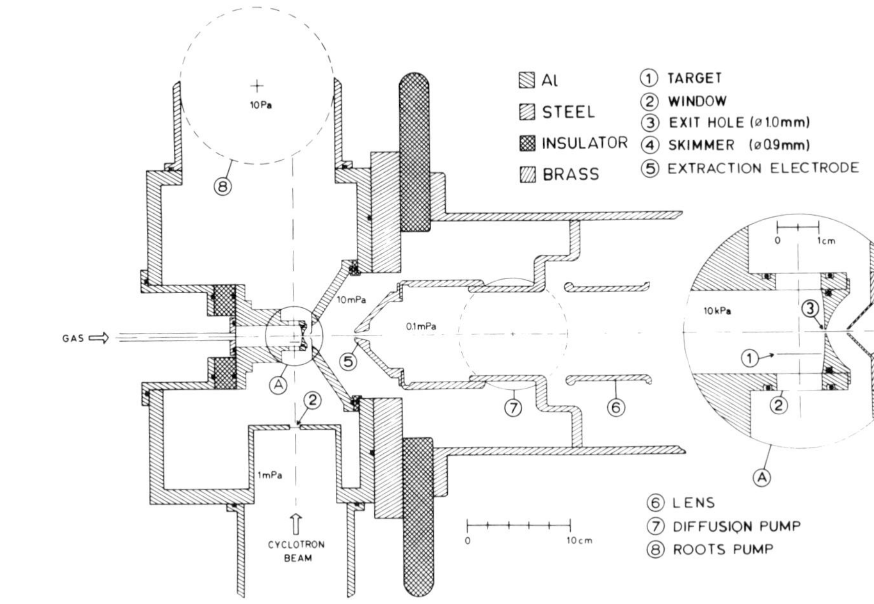


**Figure 13.** Schematic diagram of the ion guide and the accelerating stage of the isotope separator IGISOL.

Analysis of mass or binding energy systematics can be used to evaluate changes in nuclear shapes as well as to study shell effects and the onset of deformation.

Figure 14 summarizes our present experimental knowledge of the existence of proton-rich light nuclei and incorporates predictions based on the charge-symmetric Kelson–Garvey relation (Ke 66) (see Section 3.3). This figure clearly shows that the proton drip-line has practically been reached through the sd shell. Future studies will concentrate on the drip-line and will deal with topics such as completing high isospin multiplets, new modes of radioactive decay, and general tests of current mass theories for very proton-rich nuclear systems. Above $Z = 20$, the nuclear properties and, until very lately, even the existence of quite proton-rich nuclides have been very poorly known. Intermediate-energy heavy-ion reactions coupled to fast on-line analyzing systems are now opening up exciting studies of proton-rich nuclides in this region of the nuclear chart. An example is again the results from GANIL, shown in Figure 4, where the bound character of the new proton-rich isotopes $^{43}V$, $^{44}Cr$, $^{46,47}Mn$, $^{48}Fe$, $^{50,51,52}Co$, $^{51,52}Ni$, and $^{55,56}Cu$ was established in the reactions of 55 MeV/nucleon $^{58}Ni + {}^{58}Ni$ (Po 86).

## 3.2. Experimental Masses of Proton-Rich Light Nuclei

Based on our level of knowledge of their masses, these proton-rich nuclides can be divided into three subgroups. The masses of 28 $Z > N$ nuclides up to sodium ($Z = 11$) are very well known and proton binding energies across the proton drip-line have been measured. From magnesium up to titanium ($Z = 22$), nuclei establishing

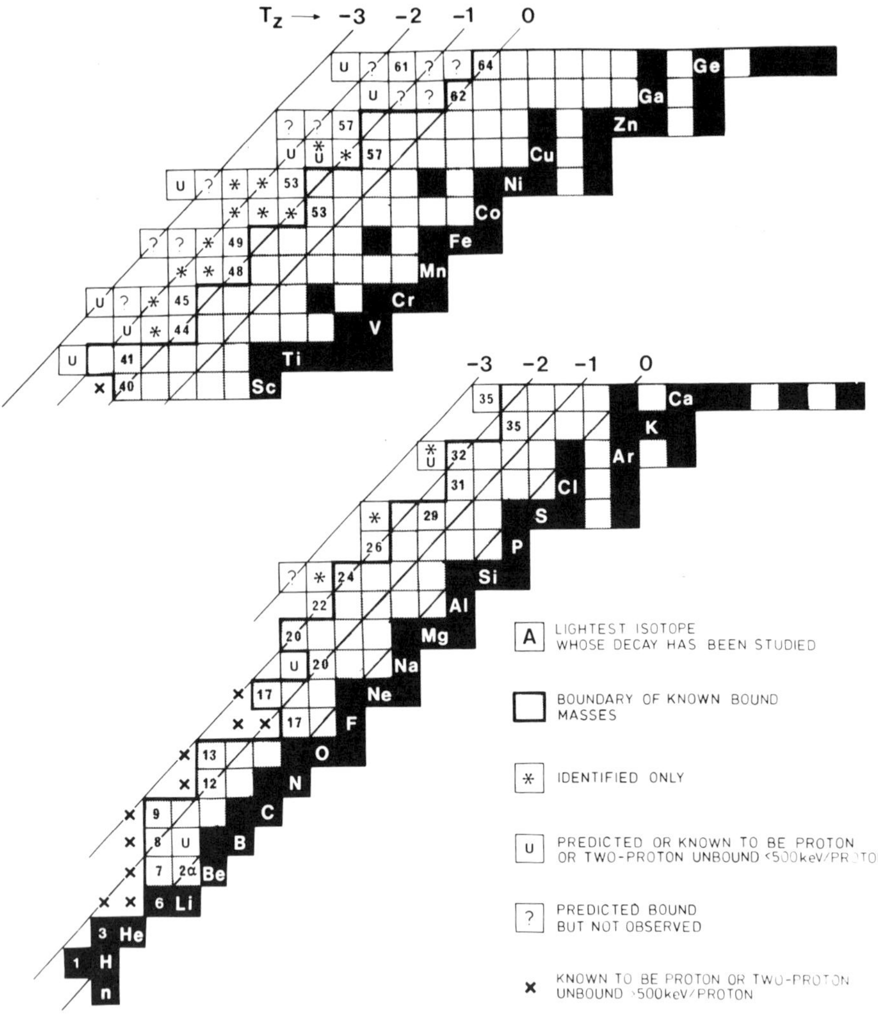


**Figure 14.** A chart of the nuclides for light, proton-rich isotopes.

the drip-line are the focus of current research and the location of the drip-line has already been determined from the odd-$Z$ elements Al, P, Cl, K, and Sc. The experimentally known $Z > N$ masses in this group total some 39. Above titanium, however, experimental knowledge of the masses of proton-rich nuclei is very scarce. Only 12 masses of $Z > N$ nuclei above vanadium ($Z = 23$) are known, a very small fraction of the total number of nuclei which extend to and through the proton drip-line in this mass region. A summary of the experimental mass for the most proton-rich bound nuclide for each element or an IMME prediction (see below) if available is given in Table 3. Table 4 gives the experimental one- and two-proton separation energies of

**Table 3. Mass Excesses of the Most Proton-Rich Experimentally Measured Bound Isotopes for Each Element**

| Nuclide | $T_z$ | Mass excess (keV) | Method | Reference (Wa 85) (unless noted) |
|---|---|---|---|---|
| $^7$Be | $-\frac{1}{2}$ | 15768.7(.8) | (p, n) | |
| $^8$B | $-1$ | 22920.3(1.2) | ($^3$He, n) | |
| $^9$C | $-\frac{3}{2}$ | 28913.2(2.2) | ($^3$He, n), ($^3$He, $^6$He) | |
| $^{12}$N | $-1$ | 17338.1(1.0) | (p, t) | |
| $^{13}$O | $-\frac{3}{2}$ | 23111(10) | ($^3$He, $^6$He) | |
| $^{17}$F | $-\frac{1}{2}$ | 1951.54(.24) | (p, $\gamma$), (d, n) | |
| $^{17}$Ne | $-\frac{3}{2}$ | 16480(50) | ($^3$He, $^6$He) | |
| $^{20}$Na | $-1$ | 6841(7) | (p, n) | |
| $^{20}$Mg | $-2$ | 17572(27) | ($\alpha$, $^8$He) | |
| $^{22}$Al | $-2$ | 18040(90) | IMME[a] | |
| $^{23}$Al | $-\frac{3}{2}$ | 6767(25) | (p, $^6$He), ($^3$He, $^8$Li) | |
| $^{24}$Si | $-2$ | 10755(19) | ($\alpha$, $^8$He), ($\pi^+$, $\pi^-$) | |
| $^{27}$P | $-\frac{3}{2}$ | −750(40) | ($^3$He, $^8$Li) | |
| $^{28}$S | $-2$ | 4130(160) | ($\pi^+$, $\pi^-$) | |
| $^{31}$Cl | $-\frac{3}{2}$ | −7070(50) | ($^3$He, $^8$Li) | |
| $^{32}$Ar | $-2$ | −2180(50) | ($\pi^+$, $\pi^-$) | |
| $^{35}$K | $-\frac{3}{2}$ | −11168(20) | ($^3$He, $^8$Li) | |
| $^{35}$Ca | $-\frac{5}{2}$ | 4453(60) | IMME[a] | (Äy 85) |
| $^{36}$Ca | $-2$ | −6440(40) | ($\alpha$, $^8$He) | |
| $^{40}$Sc | $-1$ | −20527(4) | (p, n) | |
| $^{40}$Ti | $-2$ | −8790(160) | ($\pi^+$, $\pi^-$) | (Mo 82) |
| $^{45}$V | $-\frac{1}{2}$ | −31875(17) | (p, $^6$He) | |
| $^{45}$Cr | $-\frac{3}{2}$ | −19460(150) | IMME[a] | |
| $^{46}$Cr | $-1$ | −29472(20) | ($^{16}$O, 2n) | |
| $^{49}$Mn | $-\frac{1}{2}$ | −37611(24) | (p, $^6$He) | |
| $^{49}$Fe | $-\frac{3}{2}$ | −24470(160) | IMME[a] | |
| $^{50}$Fe | $-1$ | −34470(60) | ($\alpha$, $^8$He) | |
| $^{53}$Co | $-\frac{1}{2}$ | −42640(18) | (p, $^6$He) | |
| $^{53}$Ni | $-\frac{3}{2}$ | −29410(180) | IMME[a] | |
| $^{54}$Ni | $-1$ | −39210(50) | ($\alpha$, $^8$He) | |
| $^{57}$Cu | $-\frac{1}{2}$ | −47350(50) | ($^7$Li, $^8$He) | (Sh 85) |
| | | −47303(15) | ($^7$Li, $^8$He) | (Ga 86) |
| $^{57}$Zn | $-\frac{3}{2}$ | −32610(130) | IMME[a] | |
| $^{58}$Zn | $-1$ | −42295(50) | ($\pi^+$, $\pi^-$) | (Se 86) |

[a] IMME denotes mass derived by using the quadratic isobaric multiplet mass equation, if three or more members of the multiplet are known.

the unbound isotopes. The masses of these proton-rich nuclei have predominantly been obtained via reaction $Q$-value measurements as indicated in these tables.

Future studies will require even more exotic reactions than have been used to date, some of which—such as ($^3$He, $^8$He)—are known (Ko 77) to have extremely low cross-sections. Such experiments will be very difficult and other techniques, such as direct

**Table 4. Experimental One- and Two-Proton Separation Energies of Unbound Proton-Rich Isotopes**

| Nuclide | $T_z$ | $S_p$ (keV) | $S_{2p}$ (keV) | Method | Reference (Wa 85) (unless noted) |
|---|---|---|---|---|---|
| $^4$Li | −1 | −2900(300) | 2590(300) | (p, t) | |
| $^5$Li | $-\frac{1}{2}$ | −1970(50) | 17850(50) | (p, γ) | |
| $^6$Be | −1 | 590(50) | −1371(5) | (p, n), ($^3$He, t) | |
| $^7$B | $-\frac{3}{2}$ | −2210(70) | −1610(90) | ($^3$He, $^6$He), ($\pi^+$, $\pi^-$) | |
| $^9$B | $-\frac{1}{2}$ | −185.1(1.0) | 17069.1(1.2) | (p, n) | |
| $^8$C | −2 | 60(70) | −2143(25) | (α, $^8$He) | |
| $^{11}$N | $-\frac{3}{2}$ | −1920(150) | 2080(150) | ($^3$He, $^6$He) | |
| $^{12}$O | −2 | 140(160) | −1780(40) | (α, $^8$He) | |
| $^{15}$F | $-\frac{3}{2}$ | −1480(130) | 3150(130) | ($^3$He, $^8$Li) | |
| $^{16}$F | −1 | −536(8) | 6761(8) | ($^8$He, n), ($^3$He, t) | |
| $^{16}$Ne | −2 | 70(140) | −1404(20) | (α, $^8$He), ($\pi^+$, $\pi^-$) | |
| $^{19}$Na | $-\frac{3}{2}$ | −320(13) | 3601(12) | (p, $^6$He), ($^3$He, $^8$Li) | |
| $^{39}$Sc | $-\frac{3}{2}$ | −630(40) | 3920(40) | ($^7$Li, $^8$He) | (Sh 84b) |

mass measurements and a return to decay-energy measurements, will have to be employed.

## 3.3. Mass Predictions

Most of the current mass formulas are primarily intended for mass predictions and calculations over large mass regions. These mass calculations can be broadly divided into three groups that are based (1) on a liquid-drop model with various methods to take account of shell corrections, (2) on a shell-model approach often in a semiempirical way, or (3) on mass relations linking specific known masses of neighboring nuclei to predict an unknown mass. An extensive survey of these different formulas was given by Maripuu (Ma 76).

For light nuclei, independent particle-mass relations introduced by Garvey and Kelson (Ga 66, Ke 66, Ga 69) many years ago are still considered as some of the best predictive approaches, especially for proton-rich nuclei. A detailed discussion of mass relations and their applications is given in a recent article by Jänecke and Comay (Jä 85b). Of particular interest to us are methods of predicting masses of proton-rich nuclides as discussed below.

### *3.3.1. Kelson-Garvey Mass Relations*

Due to the charge symmetry of the nuclear force, a set of mass relations can be generated for proton-rich nuclei from the approach of Kelson and Garvey (Ke 66) which connects the masses across the $Z = N$ line in a symmetric way. One generalized version for the mass difference between mirror nuclides with $T_z = -T$ and $+T$ can be written as

$$M(A, T_z = -T) - M(A, T_z = +T) = \sum_{A^1=A-(2T-1)}^{A+(2T-1)} [M(A^1, -\tfrac{1}{2}) - M(A^1, +\tfrac{1}{2})] \quad (1)$$

**Table 5. Kelson–Garvey Mass Predictions Compared to Experiment for Some $T_z = -2$ Isotopes**

| Nuclide | Difference (TH-EXP) (keV) | Exp. precision (keV) |
|---|---|---|
| $^{20}Mg$ | −172 | 27 |
| $^{24}Si$ | −11 | 19 |
| $^{28}S$ | 280 | 160 |
| $^{28}S^{a}$ | 205 | 20 |
| $^{32}Ar$ | −13 | 50 |
| $^{36}Ca$ | 2 | 40 |
| $^{40}Ti$ | −208 | 160 |
| $^{40}Ti^{a}$ | 66 | 11 |

[a] Experimental mass derived from the quadratic IMME; see equation (2).

Relation (1) can be used to predict the mass of a proton-rich nuclide with $T_z = -T$ from the known masses of the appropriate $T = \frac{1}{2}$ mirror pairs as well as the mass of its $T_z = +T$ mirror nucleus. These $T = \frac{1}{2}$ masses are known up to zinc. An example of the utility of this relation is given in Table 5, which shows a comparison between the calculated and experimental masses of $T_z = -2$, $A = 4n$ nuclei from magnesium to titanium. All input masses used in the predictions are known experimentally to better than a few keV. In general, the predictive power of the relation is good to excellent.

### *3.3.2. Isobaric Multiplet Mass Equation*

Isobaric multiplets play a significant role both in predicting masses of highly proton-rich light nuclides and in contributing to our basic understanding of charge-dependent effects in nuclear forces. The masses of analog states in an isospin multiplet can be given by a quadratic relationship, introduced by Wigner in 1957, as follows (Wi 57):

$$M(A, T, T_z) = a(A, T) + b(A, T)T_z + c(A, T)T_z^2 \tag{2}$$

This isobaric multiplet mass equation (IMME) arises via first-order perturbation theory from the assumptions that the wave functions of the members of an isospin multiplet are identical and that only two-body forces are responsible for charge-dependent effects in nuclei. Possible deviations from this quadratic form can be due to higher-order Coulomb or other charge-dependent effects as well as to various isospin-mixing effects.

Deviations from the quadratic IMME are generally represented by the additional terms $d(A, T)T_z^3$ and $e(A, T)T_z^4$ in which the coefficients can be derived from second-order perturbation theory. These deviations can only be studied with isospin multiplets which have $T \geqslant \frac{3}{2}$. There are fifteen isospin quartets from mass number 7 to mass number 41, in which all masses of the multiplet are known, permitting an exact test of the IMME. Excellent agreement with the quadratic form has generally been observed

(Be 79, An 86). However, a persistent deviation has been found for the accurately known mass-9 quartet in the form of a $d$ coefficient of 5.2 ± 1.7 keV. This significant result has partly been explained by isospin-mixing effects in the $T_z = -\frac{1}{2}$ and $+\frac{1}{2}$ members, by the expansion of the least bound proton orbit in $^9C$, as well as by non-Coulomb charge-dependent forces (Be 79, Be 70); however, a fully satisfactory explanation has not been given so far.

Until fairly recently, experimental information on the masses of certain members of the isospin quintets has been rather limited. During the last few years, however, new data have been obtained, especially for the $T_z = -1$ and $-2$ members of the multiplets [see review by Antony *et al.* (An 86)]. In addition to the $d$ coefficients, isospin quintets also permit the study of possible nonzero $e$ coefficients. For example, any shift of a $T = 2$ state in a $T_z = 0$ nucleus due to isospin mixing by $T = 0$ and $T = 1$ states would produce an $e$ coefficient, while mixing in the $T_z = -1$ and $+1$ members affects both $d$ and $e$ coefficients. Experimental data on quintets now include nine multiplets with at least four known masses and five cases with all five members known.

Figure 15 presents a summary [based primarily on data in Antony *et al.* (An 86)] of the experimental status of the $d$ and $e$ coefficients in isospin quintets. According to this figure and evaluations of these high isospin multiplets (Äy 81, An 86, En 78), no strong deviations from the quadratic IMME have been observed, not even among the quintets with $A = 8$, 12, and 16, in which the $T_z = -2$ nuclei $^8C$, $^{12}O$, and $^{16}Ne$ are unbound to prompt nucleon emission (Ke 78). [The lightest quintet with all of its members bound to isospin particle decay modes is $A = 20$ (Mo 79).] This is a striking feature of the IMME; it is valid for unbound nuclei as well as for bound ones. Generally, the currently available data on isospin quintets support the validity of the simple

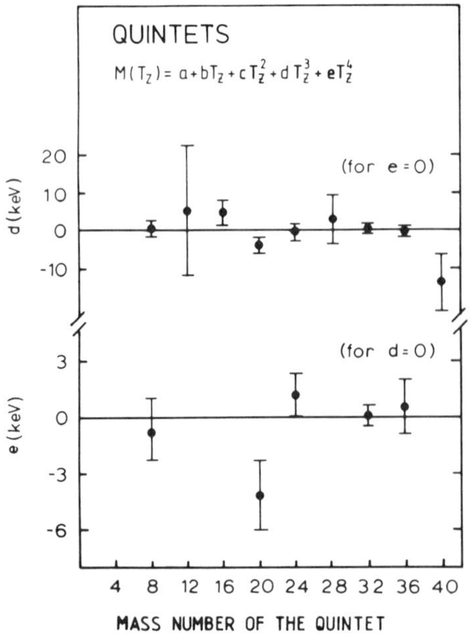


**Figure 15.** The $d$ and $e$ coefficients of the IMME plotted as a function of the mass number of the quintet.

quadratic mass equation and provide no evidence for substantial higher-order charge-dependent effects in the nuclear interaction. The only significant deviations from the quadratic form are observed in masses 20 and 40, where the effect could be statistical.

The isobaric multiplet mass equation can be used to predict the ground-state mass of a very proton-rich nucleus via the mass of its mirror. Deriving the unknown $T_z = -T$ mass requires a knowledge of the $b$ coefficient in the equation

$$M(T_z = -T) = M(T_z = +T) - 2bT_z \tag{3}$$

This has been done in two different ways. The most reliable case is that in which three of more members of the multiplet are known; this permits direct calculation of all three coefficients in the quadratic IMME. Masses calculated via this approach can normally be considered very reliable, with quite low predicted errors (see Table 5). Unfortunately, inadequate experimental information on high isospin multiplets limits the *general* use of this approach to nuclei with $A \leqslant 40$ and $T_z \geqslant -2$. Above $A = 40$, only for a few isobaric quartets can predictions be made for the masses of some $T_z = -\frac{3}{2}$ nuclides.

A second approach can be illustrated by recent calculations of Antony *et al.* (An 86). They calculate the $b$ coefficient of equation (3) from the following formula:

$$b = \Delta_{\mathrm{nH}} - E_{\mathrm{c}}^{(1)} = -720.4\,A^{2/3} + 1808.6 \qquad (\mathrm{keV}) \tag{4}$$

where $\Delta_{\mathrm{nH}}$ is the neutron-hydrogen atom mass difference and the vector part of the nuclear Coulomb energy, $E_{\mathrm{c}}^{(1)}$, is derived from a fit to the experimental Coulomb displacement energies for $A = 9$–60. This relation for $b$ assumes that its value is a constant for all multiplets of a given mass; it then provides a simple way of predicting masses of very proton-rich nuclei up to $A = 60$. The uncertainty of the predicted masses resulting from this approach is typically 150, 200, 300, and 400 keV for $T_z = -\frac{3}{2}$, $-2$, $-\frac{5}{2}$, and $-3$ nuclides, respectively. An overall accuracy of better than 200 keV was obtained for 70% of the cases tested against the 1985 Mass Tables (Wa 85). However, systematic deviations show up clearly, especially just above $A = 40$.

### 3.3.3. *Mass Predictions for* $^{35}$*Ca*

An interesting comparison among the values given by the above mass relations can be made in the case of the most proton-rich nucleus whose decay is known, $T_z = -\frac{5}{2}$ $^{35}$Ca. This nucleus was discovered via its $\beta$-delayed two-proton decay (Äy 85) and its existence was later confirmed in a projectile-fragmentation reaction (La 86). The precise determination of the energy of two-proton emission from the $T = \frac{5}{2}$ isobaric analog state in the $^{35}$K daughter, populated via superallowed $\beta$ decay, resulted in a mass excess of $-2115(45)$ keV and an excitation energy of 9053(45) keV for the $T = \frac{5}{2}$ state. Since this measurement provided the third member of the $A = 35$, $T = \frac{5}{2}$ isospin *sextuplet*, it was possible to use the isobaric multiplet mass equation to predict the mass of the $^{35}$Ca ground state (Äy 85). The mass excess of the lowest $T = \frac{5}{2}$ state in $^{35}$S had been determined by the (p, $^3$He) reaction (Gu 75), and the third known member is the ground state of $^{35}$P, obtained as a weighted average of four measurements. The quadratic relation could then be used to predict a value of 4453(60) keV for the ground state of $^{35}$Ca.

In the second approach, using the empirical $b$ coefficient from equation (4), we get a value of 4558(360) keV in agreement with the value from the IMME. Finally, the

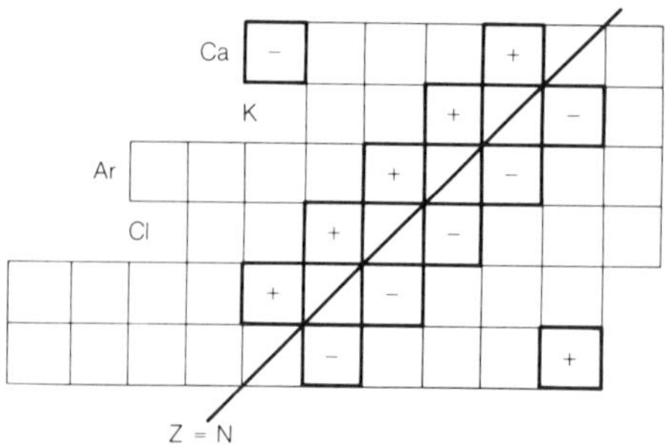


**Figure 16.** Masses involved in the prediction of the mass of $^{35}Ca$ using the Kelson–Garvey relation.

corresponding prediction for the $^{35}Ca$ mass excess from the Kelson–Garvey relation (illustrated in Figure 16) gives a mass excess of 4777 keV, which is significantly different from the above results, even though the input masses are all known to a few keV or less. This deviation indicates that $^{35}Ca$ is probably ~300 keV better bound than predicted by the Kelson–Garvey relation which agrees with what appears to be a common result that this relation tends to underpredict the binding of quite proton-rich nuclei.

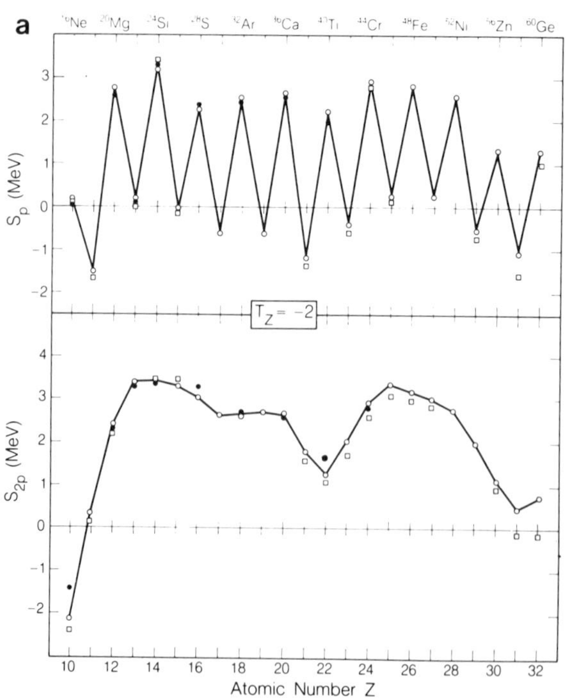


**Figure 17.** One- and two-proton separation energies for (a) $T_z = -2$; (b) $T_z = -\frac{5}{2}$; and (c) $T_z = -3$ nuclides. See text.

### 3.4. Proton Drip-Line Below $A = 60$

Predictions of masses and binding energies form an important guide for experimental searches for unknown new nuclides, new radioactive-decay phenomena, and for mass measurements themselves. Figure 17a,b,c summarize our experimental knowledge and predictions of one- and two-proton binding energies for proton-rich nuclides from neon to germanium. Only the most proton-rich nuclides with $T_z = -2, -\frac{5}{2}$ and $-3$ are shown. [The series with $T_z = -\frac{3}{2}$ is well known (Wa 85) and is not shown here. There are only two nuclei belonging to this series that are *known* to be proton unbound, $^{19}$Na and $^{39}$Sc, and two more that are predicted to be slightly unbound against proton emission, both by about 300 keV, $^{55}$Cu and $^{59}$Ga. However, $^{55}$Cu has recently been seen among the fragmentation products at GANIL, suggesting a bound character or a finite lifetime against proton decay.] The symbols in the binding-energy plots are as follows: the full black circles mean that the binding energy value is taken from the 1985 Mass Tables (Wa 85), the open circles mean that the value has been calculated with the Kelson–Garvey relation, and the open squares mean that the value is based on the empirical IMME as applied in Antony *et al.* (An 86). If the two predicted values are within 100 keV, only an open circle is shown.

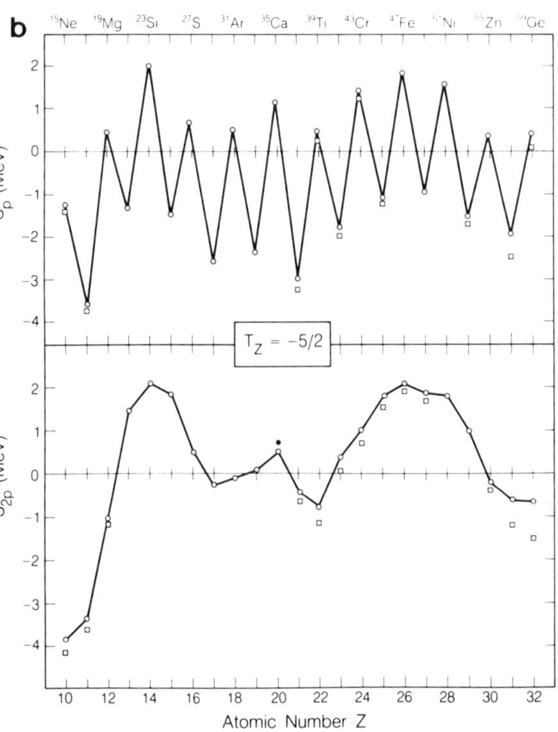


**Figure 17b**

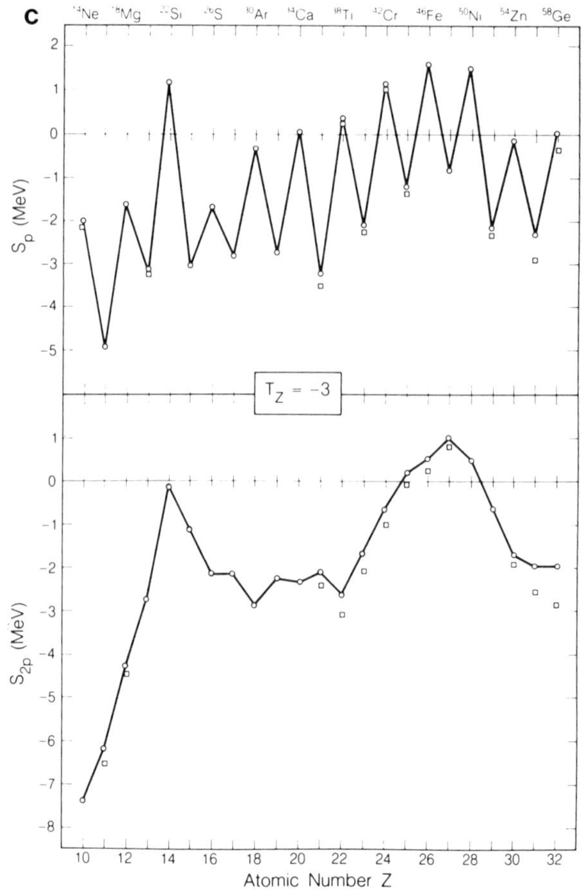


**Figure 17c**

With regard to $T_z = -2$ nuclei, all even-$Z$ ones except $^{16}$Ne are known or predicted to be bound. Of the odd-$Z$ isotopes, in addition to the sd-shell ones, $^{18}$Na, $^{30}$Cl, and $^{34}$K, in the fp shell $^{38}$Sc, $^{42}$V, $^{54}$Cu, and $^{58}$Ga are expected to be unbound. A recent experiment at GANIL established the existence of all the $T_z = -2$ isotopes from $^{44}$Cr to $^{52}$Ni, in accordance with these predictions (Po 86). An interesting case just at the drip-line is $^{26}$P, which has been observed to $\beta$-decay with a 20 ms half-life. According to barrier-penetrability calculations, this nucleus can be unbound by about 50 keV without significantly affecting its $\beta$ half-life (Ca 83b).

There are eight $T_z = -\frac{5}{2}$ nuclides that are predicted to be bound or nearly bound. Four of these in the sd shell have been confirmed experimentally (La 86, De 86) and one of these, $^{35}$Ca, has a measured $\beta$-decay half-life of 50 ms (Äy 85). Two $T_z = -\frac{5}{2}$ nuclides $^{31}$Ar and $^{39}$Ti are both predicted bound to proton but unbound to two-proton emission and are candidates for two-proton radioactivity. The predicted 2p separation energies are −191 and −76 keV for $^{31}$Ar with the Kelson–Garvey and with the IMME, respectively. The same predictions for $^{39}$Ti are −785 keV (K–G) and −1189 keV (IMME).

Finally, there are only four $T_z = -3$ nuclides, three of which are in the $f_{7/2}$ shell, which are expected to be bound or nearly bound. Again, two of these are candidates for two-proton radioactivity, $^{22}$Si and $^{42}$Cr. No nuclides belonging to this group have so far been observed. There is only one $T_z = -\frac{7}{2}$ nucleus, $^{49}$Ni, that could in principle be observable. It is predicted to be bound by about 700 keV to proton emission and unbound to two-proton emission by only 200 keV in both the Kelson–Garvey and the IMME predictions.

*Note added in proof.* The first, and up to now the only $T_z = -3$ nuclide $^{22}$Si was discovered at GANIL using the LISE spectrometer and the fragmentation of an $^{36}$Ar beam. Only the existence of $^{22}$Si has so far been shown. [See M. G. Saint-Laurent *et al.*, *Phys. Rev. Lett.* **59**:33 (1987).] In addition to $^{35}$Ca, the decays of two other $T_z = -\frac{5}{2}$ nuclides have been observed. $\beta$-delayed proton groups have been seen at GANIL from 15 ms $^{31}$Ar [V. Borrel *et al.*, *Nucl. Phys.* **A473**:331 (1987)] and 16 ms $^{27}$S [J. C. Jacmart *et al.*, *Nuclei Far From Stability*, Fifth International Conference (I. S. Towner, ed.), American Institute of Physics Conference Proceedings 164 (1988), p. 690]. This last reference volume contains several other articles on proton-rich nuclei.

## 4. Decay Modes of Proton-Rich Nuclei

Proton-rich nuclides far from the valley of $\beta$ stability provide a variety of radioactive decay paths. Their large $\beta$-decay energies permit the population of highly excited states in the daughter nuclides so that delayed proton and $\alpha$ emission are common phenomena. More exotic decays are also possible, which the recent observation (Ca 83) of $\beta$-delayed two-proton radioactivity demonstrates.

Direct proton radioactivity was first observed from the proton-rich isomer $^{53m}$Co (Ja 70, Ce 70). This many-particle high-spin state emits 1.59 MeV protons with a 247 ms half-life (Ce 72). Later, direct proton emission from the ground state was discovered for the first time for the neutron-deficient nuclides $^{151}$Lu and $^{147}$Tm (Ho 81, Ho 82b, Kl 82). This decay mode is also being sought among the proton-rich nuclides below $A = 60$. Another direct decay, the emission of a proton pair, has also been predicted, but has not yet been observed.

### 4.1. $\beta$ Decay Far From Stability

Allowed $\beta$ decay of a proton-rich nucleus is mediated via the Fermi and Gamow–Teller decays. Figure 18 shows schematically the decay pattern of a proton-rich nucleus. In addition to a fast Fermi component to the isobaric analog state, the decay consists of several allowed Gamow–Teller decays to low-lying states and to a cluster of states around the analog state. In general, the role of forbidden transitions is not essential in the following discussion. Also, it should be emphasized that the large $\beta$-decay

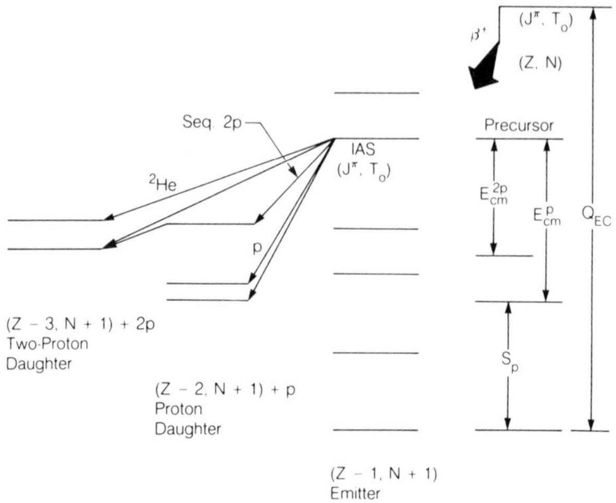


**Figure 18.** Decay pattern of a proton-rich nucleus. See text.

energies of proton-rich nuclides make possible the study of the $\beta$ strength function over a considerably wider energy range than is possible for the decays of nuclei nearer stability.

Using the approach of Brown and Wildenthal (Br 85), the partial half-life $t$, associated with allowed $\beta$ decay to a specific final state, can be related to the corresponding nuclear matrix elements via the expression

$$f_\beta t = \frac{K}{G_V^2\langle\tau\rangle^2 + G_A^2\langle\sigma\tau\rangle^2} \tag{5}$$

where $f_\beta$ is the statistical rate function or phase-space factor containing the energy dependence of the transition. The constant $K$ is given by

$$\begin{aligned} K &= 2\pi^3 \ln 2\, \hbar^7/(m_e^5 c^4) \\ &= 1.230618\ 10^{-94}\ \text{erg}^2\ \text{cm}^6\ \text{s} \end{aligned}$$

and the weak interaction vector and axial vector coupling constants are denoted by $G_V$ and $G_A$, respectively.

The Fermi transition probability $B(\text{F})$, $\langle\tau\rangle^2$, can be written as

$$B(\text{F}) = \langle\tau\rangle^2 = [T(T+1) - T_{zi}T_{zf}]\delta_{if}(1-\delta_c) \tag{6}$$

where $T$ is the isospin of the initial and the final state while $T_{zi}$ and $T_{zf}$ are their corresponding $z$ projections. The correction $(1 - \delta_c)$ arises from the incomplete overlap between the initial and final nuclear wave functions due to isospin mixing or charge-dependent effects. The Fermi transition probability can be shown to be nonzero only for the $\Delta T = 0$, $\Delta J = 0$, and $\Delta\pi = 0$ transitions, i.e., for transitions between states of the same isospin multiplet.

The Gamow–Teller transition probability $B(\mathrm{GT})$ is related to the corresponding matrix element $\langle\sigma\tau\rangle$ via the expression

$$B(\mathrm{GT}) = (G_\mathrm{A}/G_\mathrm{V})^2\langle\sigma\tau\rangle^2 \tag{7}$$

The matrix element $\langle\sigma\tau\rangle$ is composed of a sum over all the nucleons in the nucleus

$$\langle\sigma\tau\rangle = \left\langle \mathrm{f} \left| \sum_K^A \sigma(k)\tau(k) \right| \mathrm{i} \right\rangle \tag{8}$$

and its magnitude is determined by the explicit details of the nuclear wave functions involved. The reduction of this matrix element in spin and isospin space and the conservation of parity results in the selection rules $\Delta J = 0, 1$; no $J_\mathrm{i} = 0 \to J_\mathrm{f} = 0$; $\Delta T = 0, 1$ and $\Delta\pi =$ no for allowed Gamow–Teller transitions. The matrix element is readily calculable within the framework of the nuclear-shell model. Examples of such calculations are the large-basis sd-shell calculations by Wildenthal and Brown (Wi 84, Br 85) and fp-shell calculations by Ogawa *et al.* (Og 85) and Sekine *et al.* (Se 86b).

Values for the coupling constants $G_\mathrm{V}$ and $G_\mathrm{A}$ are obtained from experiment. The constant $C = K/G_\mathrm{V}^2$ is obtained from the pure $0^+ \to 0^+$ Fermi transitions and the result deduced from a large set of data is $C = 6170 \pm 4$ s (Wi 78). The ratio of the coupling constants is obtained from the decay of the free neutron and the latest analysis by Wilkinson (Wi 82) resulted in the value $|G_\mathrm{A}/G_\mathrm{V}| = 1.2606 \pm 0.0075$. The phase-space factor $f_\beta$ in expression (5) contains the energy dependence of the transition and is, for example, calculable using the procedure of Wilkinson (Wi 73). In a precise scheme, slightly different phase factors are associated with the Fermi and Gamow–Teller decays (Wi 73, Wi 74). In general, it should be noted that the energy dependence of $f_\beta$ is very strong, proportional to $E_\beta^5$.

In order to evaluate and compare the contributions of the Fermi and Gamow–Teller strengths, it is useful to consider their sum rules. The total Fermi strength depends on the isospin of the analog states involved, according to equation (6), and is concentrated on this one transition. For transitions between mirror nuclei, involving the small value of isospin $T = \frac{1}{2}$, this strength is $B(\mathrm{F}) = 1$, while for the highest isospin transition observed so far, $T = \frac{5}{2}$, this strength is $B(\mathrm{F}) = 5$.

The Gamow–Teller sum rule relates the sums of the strengths of the $\beta^+$ and $\beta^-$ decays of a single initial state to all possible final states to the neutron and proton numbers of the parent state:

$$\sum_\mathrm{f} [B(\mathrm{GT}+)_{\mathrm{i}\to\mathrm{f}} - B(\mathrm{GT}-)_{\mathrm{i}\to\mathrm{f}}] = (G_\mathrm{A}/G_\mathrm{V})^2 3(Z_\mathrm{i} - N_\mathrm{i}) \tag{9}$$

For the $\beta$ decay of proton-rich nuclei, this relation provides the lower limit for the sum over $B(\mathrm{GT}+)$. For nuclei with a proton excess, the theoretical results (Br 85) suggest that the sum over $B(\mathrm{GT}-)$ is usually much smaller than the sum over $B(\mathrm{GT}+)$. The strength included in the above sum rule is distributed over a wide range of final-state excitation energies. Its experimental study has drawn considerable interest after it was systemically found via (p, n) reactions that only about half of the sum-rule strength was found for $B(\mathrm{GT}-)$ (Go 82). $\beta$ decay of proton-rich nuclei offers a direct approach for the study of the GT strength and its distribution; it does not suffer from the evaluation difficulties encountered with the (p, n) data and additionally provides much better energy resolution. However, a major drawback is a limited energy window, a problem not present in the (p, n) studies.

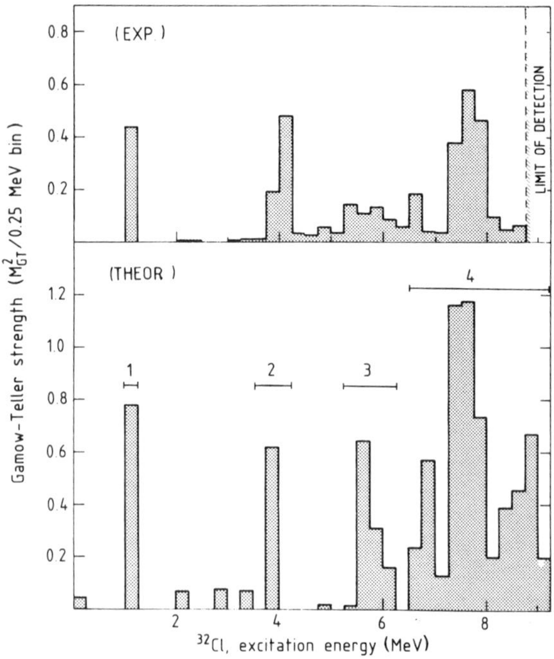


**Figure 19.** Experimental and theoretical Gamow–Teller strength distributions of $^{32}$Ar. The Gamow–Teller giant resonance is observed at 7.5 MeV. See text.

There are several experimental studies of the $\beta$ decay of $T_z = -\frac{1}{2}$, $-1$ and $-\frac{3}{2}$ nuclei in which a quenching of the Gamow–Teller strength, similar to that observed in the (p, n) results, has been observed (see Sections 4.2 and 5.2). Recent work on the decay of $^{32}$Ar has provided the first test with a $T_z = -2$ nucleus (Bj 85). This exotic isotope is now available at ISOLDE as a radioactive beam with an intensity of 500 atoms/s. Figure 19 shows the experimental Gamow–Teller strength distribution as determined from $\beta$-delayed $\gamma$, proton, and $\alpha$ emission. The lower part of the figure shows the theoretical strength calculated by the large-basis shell-model approach of Muller *et al.* (Mu 84) using the Chung–Wildenthal interaction. Good agreement in energy is observed; the concentration of strength around 7.5 MeV can be interpreted as a spin-isospin flip resonance on the $0^+$, $T = 2$ isobaric analog state. This resonance has also been seen in other proton-rich nuclei in the sd shell. Comparison of the observed strength with the shell-model prediction results in an overall missing factor of 0.49. It should be noted that in the case of $^{32}$Ar, 62% of the strength from the shell-model prediction lies inside the $\beta$-decay energy window. This observed quenching is in good agreement with the systematic study of Wildenthal *et al.* (Wi 84, Br 85) for neutron-rich light nuclei. A quenching of the Gamow–Teller transition probability $B$(GT) can be explained both by collective effects in nuclear structure as well as by internal degrees of freedom of decaying nucleons. Hence, it is customary to associate this quenching either with the Gamow–Teller matrix element itself or with the axial vector to vector coupling constant ratio ($G_A/G_V$); see equation (7).

The total Gamow–Teller and Fermi strengths are also closely related to the $\beta$-decay half-life $T_{1/2}$:

$$\frac{1}{T_{1/2}} = \sum_{\mathrm{f}} \left[ \left( \frac{1}{f_\beta} \frac{C}{B(\mathrm{F})\delta_{\mathrm{if}} + B(\mathrm{GT})} \right)_{\mathrm{f}} \right]^{-1} \tag{10}$$

For example, shell-model calculations such as those by Brown and Wildenthal (Br 85), with a quenching factor folded in, can also be used for half-life predictions of unknown isotopes. Such an approach was recently applied to the $T_z = -\frac{3}{2}$ nuclides $^{27}$P and $^{31}$Cl. The half-lives predicted by the shell model were 215 and 159 ms, respectively; the observed half-lifes of $260 \pm 80$ and $150 \pm 25$ ms are in excellent agreement with this prediction (Äy 85b).

## 4.2. $\beta$-Delayed Particle Decay

Experimental information on the $\beta$ strength is normally derived from measurements of delayed radiations from excited states of the daughter nucleus. For proton-rich nuclides, the observed delayed particle emissions include protons and $\alpha$ particles, which are the most characteristic decay modes for nuclei with $T_z \leqslant -\frac{3}{2}$. The more common mode of $\beta$-delayed proton emission has had a very important role in successful searches for proton-rich nuclei up to $A = 60$. Substantial growth in our knowledge of this phenomenon has occurred since an earlier review (Ce 77) and the study of this decay mode has become a rich source of information for $\beta$-strength studies as well as for the spectroscopy of highly excited states.

As nuclei approach the limit of proton stability, $\beta$-decay energies increase rapidly with respect to proton and $\alpha$-particle separation energies, resulting in the emission of charged particles from excited states. Also, for a given element the more proton-rich isotopes show rapidly increasing isobaric mass differences and the analog states of the $\beta$ parents also become particle unbound. Among light nuclei, where states in the emitter nuclei are well separated, individual transitions are resolved and can be related to the preceding $\beta$ decay.

The intensity distribution of the delayed particle spectrum is determined by two factors: (1) the $\beta$ transition probability distribution and (2) the partial width for particle decay from the initial state i in the emitter to all possible final states f in the daughter. It can be expressed as

$$I_\nu(E_\nu) = \sum_{\mathrm{if}} I_\beta^{\mathrm{i}} \frac{\Gamma_\nu^{\mathrm{if}}}{\Gamma^{\mathrm{i}}} \tag{11}$$

where $I_\beta^{\mathrm{i}}$ is the intensity of $\beta$ decay to the excited state i, and $\Gamma_\nu^{\mathrm{if}}$ is the partial width for emission of particle $\nu$ with an energy $E_\nu$ from the initial state i to the final state f; $\Gamma^{\mathrm{i}}$ is the total decay width of the state i and can be expressed in the form

$$\Gamma^{\mathrm{i}} = \Gamma_\gamma^{\mathrm{i}} + \sum_\nu \left[ \sum_{\mathrm{f}} \Gamma_\nu^{\mathrm{if}}(E_\nu) \right] \tag{12}$$

In cases where the level spacings are large compared with their widths and the experimental energy resolution, the energy spectrum consists of individual peaks and detailed spectroscopic information can be obtained for individual levels. When the level density increases so that experiment cannot resolve individual transitions, a statistical approach has to be employed (Jo 76). Most light $\beta$-delayed particle emitters belong to the former category, though some of the heavier ones and those with higher decay energies, for example $^{40}$Sc and $^{59}$Zn, are transitional cases between the two approaches.

The particle decay of an unbound state depends on the penetrability $P_l$ through the Coulomb and centrifugal barriers and on the reduced width $\gamma_\nu$, as follows:

$$\Gamma_\nu = 2\gamma_\nu^2 P_l \tag{13}$$

The penetrability is often calculated employing the well-known regular and irregular Coulomb wave functions $F_l$ and $G_l$, respectively. When the nuclear mass is below 60, the lower energy cutoff in the observable particle spectra is determined by the penetrability and is typically of the order of 0.5 MeV for protons and 1.5–2 MeV for $\alpha$ particles. The upper limit in the energy spectra is determined by the $\beta$-decay energy window and especially by the strong energy dependence of the $\beta$-decay phase-space factor $f_\beta$. However, observed energies of $\beta$-delayed particles (protons) can be as high as 10 MeV.

### 4.2.1. $T_z = -\frac{1}{2}$ Nuclei

The $\beta$ decay of these nuclei is characterized by a strong combined Fermi and Gamow–Teller decay to the mirror state, the ground state of the $T_z = +\frac{1}{2}$ daughter nucleus. Only among heavier nuclides is it possible to observe particle emission from excited states of the daughter. The only known case is $^{59}$Zn (Ho 81b), the delayed proton spectrum of which is shown in Figure 20. This nucleus was actually discovered via this decay mode. Due to the effect of the phase-space factor, only 0.23% of the $\beta$

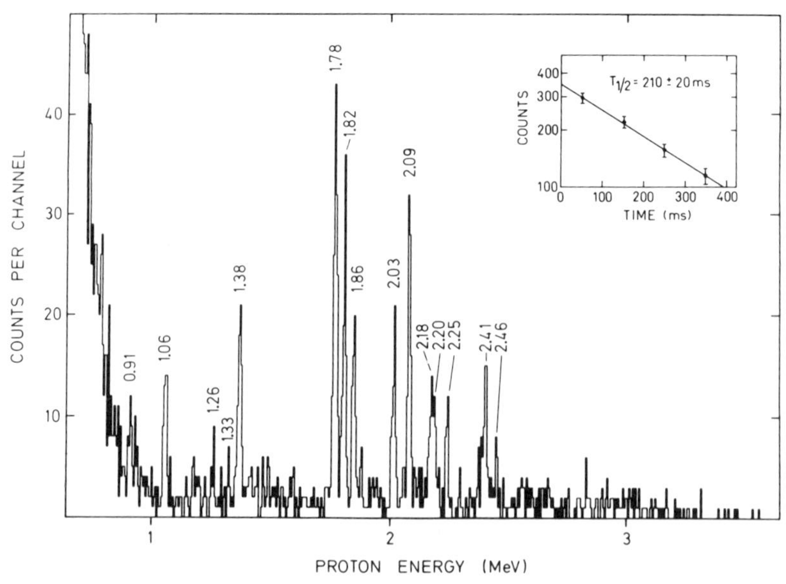


**Figure 20.** The $\beta$-delayed proton spectrum of the $T_z = -\frac{1}{2}$ nucleus $^{59}$Zn.

decays populate proton-emitting states, although the major part of the Gamow–Teller strength is in these transitions.

### 4.2.2. $T_z = 1$ Nuclei

Several odd–odd $\beta$-delayed proton and $\alpha$ emitters are known ranging from $^8B$ to $^{48}Mn$. The latest and heaviest one is $^{48}Mn$, which has been discovered at the GSI on-line mass separator (Se 86b). These precursors generally are weak proton (and/or $\alpha$) emitters, since the isobaric analog state lies below the proton-separation energy. Considerable information on $T_z = -1$ nuclei was obtained in He-jet experiments at Jyväskylä (Ho 80, Es 80, Ho 86c), where two new precursors, $^{28}P$ and $^{36}K$, were discovered.

An example of these experiments is shown in Figure 9 (Section 2.4), which displays the particle spectrum of $^{40}Sc$ with a total number of proton and $\alpha$ lines of 40 and 21, respectively (Ho 82). Most of these lines originate from known resonances. An anomalously high $\alpha$-to-proton intensity ratio was observed, which was explained by the selective population in the $\beta$ decay of $\alpha$-cluster states at excitation energies near the $f_{7/2}\, d_{5/2}$ and $f_{5/2}\, d_{3/2}$ doorway states (Ho 86c). These $T_z = -1$ nuclides provide good opportunities for studying Gamow–Teller strength, because they have high decay energies and high production cross-sections.

### 4.2.3. $T_z = -\frac{3}{2}$ Nuclei

The series of even-$Z$, $T_z = -\frac{3}{2}$ nuclides contains the oldest-known delayed-proton emitters. They are particularly favored by strong proton branches, ranging from 12 to 100 percent per disintegration. All the currently known members of this series were known at the time of the review by Cerny and Hardy (Ce 77). The proton binding energies in the emitter nuclides are generally low and $\beta$-decay branchings to proton-emitting states, including the isobaric analog state, consequently very high.

A typical example of a proton spectrum and the decay scheme of an sd-shell precursor, $^{29}S$, is shown in Figure 21. The activity was produced via the ($^3He$, 2n) reaction and transported using the He-jet technique; protons were detected with a surface barrier detector telescope (Vi 79). Since in general the isobaric analog state decays via isospin-forbidden proton decay, it is possible to determine the extent of isospin mixing in this special state [see equation (6)]. For example, in the experiment on $^{29}S$, the isospin purity of the $T = \frac{3}{2}$ analog state could be deduced to be $95 \pm 9/5\%$.

Proton emission from the strongly populated $T = \frac{3}{2}$ analog state has also been used as a tool to search for new proton-rich isotopes of heavier elements such as $^{45}Cr$ (Ja 74), $^{49}Fe$ (Ce 70b), $^{53}Ni$ (Vi 76), $^{57}Zn$ (Vi 76), and $^{61}Ge$ (Ce 76, Ho 86). An example from such an experiment is shown in Figure 22, which displays a proton spectrum following the $\beta$ decay of $^{61}Ge$, the heaviest known $T_z = -\frac{3}{2}$ nucleus (Ho 86). [So far, searches for the next higher member, $^{65}Se$, have been unsuccessful (Ho 86).] The activity was produced via the $^{24}Mg + {}^{40}Ca$ reaction. In these experiments, both the He-jet and the direct recoil counting techniques, as described in Section 2.4, were used. Based on an analysis of the Coulomb displacement energies in this multiplet, the peak was assigned to the decay of the analog state.

The series of odd-$Z$, $T_z = -\frac{3}{2}$ delayed proton precursors are associated with low branching for this mode, due to the high proton binding in their even-$Z$ daughter

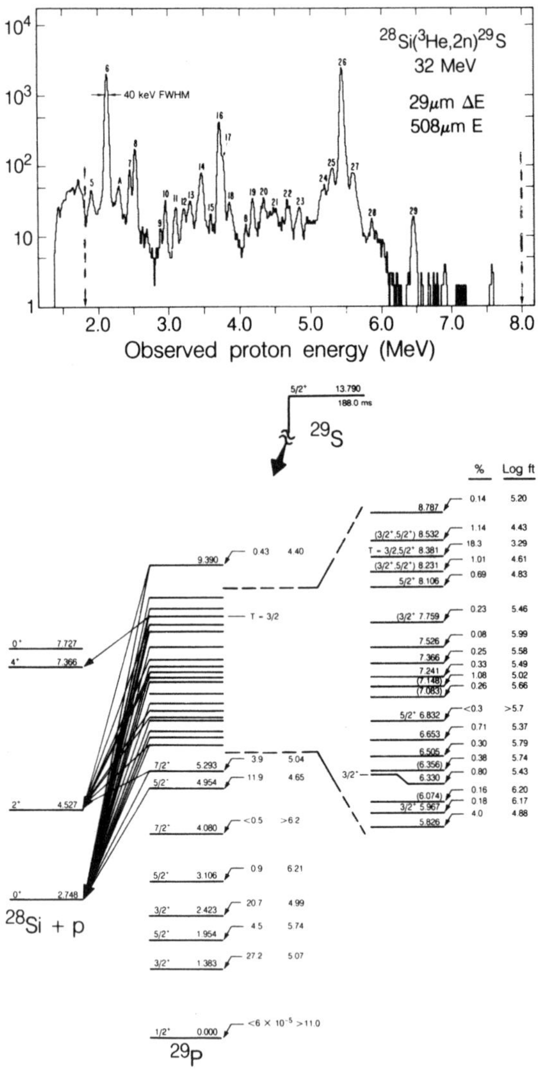


**Figure 21.** Top: The β-delayed proton spectrum from the decay of $^{29}S$. Dashed arrows define the energy limits over which protons could be observed. Bottom: The decay scheme of $^{29}S$.

nuclei. Only one such nucleus, $^{23}Al$, was known (Go 72) at the time of the earlier review (Ce 77). Three more cases have since been discovered: $^{27}P$ (Äy 85b), $^{31}Cl$ (Äy 82), and $^{35}K$ (Ew 80). The latter two nuclides are the most proton-rich, bound isotopes of these elements. No decays of precursors belonging to this series have been observed in the $f_{7/2}$ shell and above. In general, the β-delayed proton measurements of $T_z = -\frac{3}{2}$ nuclides have also provided valuable information for testing the quadratic isobaric multiplet mass equation (Be 79).

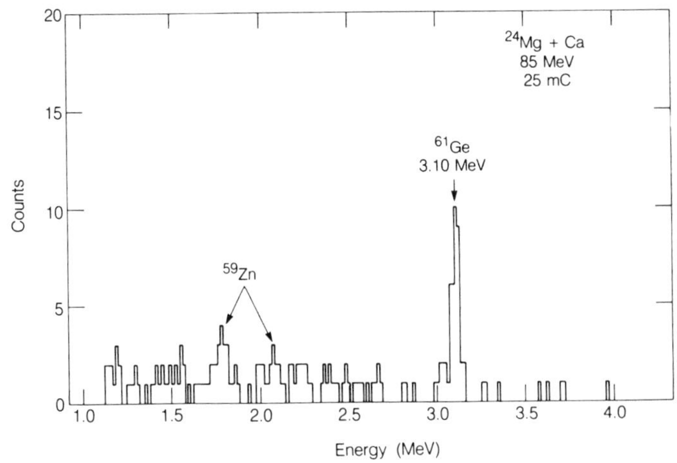


**Figure 22.** The $\beta$-delayed proton spectrum of the $T_z = -\frac{3}{2}$ nuclide $^{61}$Ge.

### 4.2.4. $T_z = -2$ Nuclei

These nuclides form a very interesting group of $\beta$-delayed particle emitters. Their $\beta$ decay is dominated by a strong Fermi decay, with a strength $B(\mathrm{F}) = 4$, to the proton unbound isobaric analog state. Measurement of the energy of this isospin-forbidden proton decay can provide the mass for the $T_z = -1$ member of the isospin quintet. Considerable progress in the spectroscopy of these nuclei has been made only recently. The more stable series of $T_z = -2$ nuclei with even $Z$ has four known members; they are, in the order of their discovery, $^{32}$Ar (Ha 77), $^{20}$Mg (Mo 79), $^{24}$Si (Äy 79), and $^{36}$Ca (Äy 81).

The above nuclides were found using on-line mass-separation techniques, which was a necessity due to the very high background of strongly produced $T_z = -\frac{3}{2}$ precursors closer to stability. In general, low production cross-sections have limited spectroscopic measurements to only the observation of delayed protons emitted from the analog state. However, in a recent experiment at ISOLDE, a quite high production rate (500 atoms/s) of mass-separated $^{32}$Ar was obtained, permitting a detailed study of the $\beta$-strength function; see Section 4.1 and Figure 19. Several other even-$Z$ precursors are known or predicted to exist, and only difficulties in mass-separation techniques have prevented the observation of $^{28}$S and $^{40}$Ti decays. However, recently a mass-separated beam of $^{41}$Ti has become available via the ion-guide on-line mass-separation technique (Sh 85c), providing promise for the future. Further studies of even-$Z$, $T_z = -2$ nuclei will obviously concentrate on increasing the yields of the sd-shell precursors and on searches for new isotopes that are predicted to be bound and to extend beyond $^{56}$Zn.

The odd-$Z$ members of the $T_z = -2$ nuclides are expected to be less bound than their even-$Z$ partners; their half-lives are shorter and production cross-sections tend to be smaller. Two such nuclei, $^{22}$Al (Ca 82) and $^{26}$P (Ca 83b), were recently discovered via their $\beta$-delayed proton decay by utilizing a very fast He-jet technique and 110 MeV

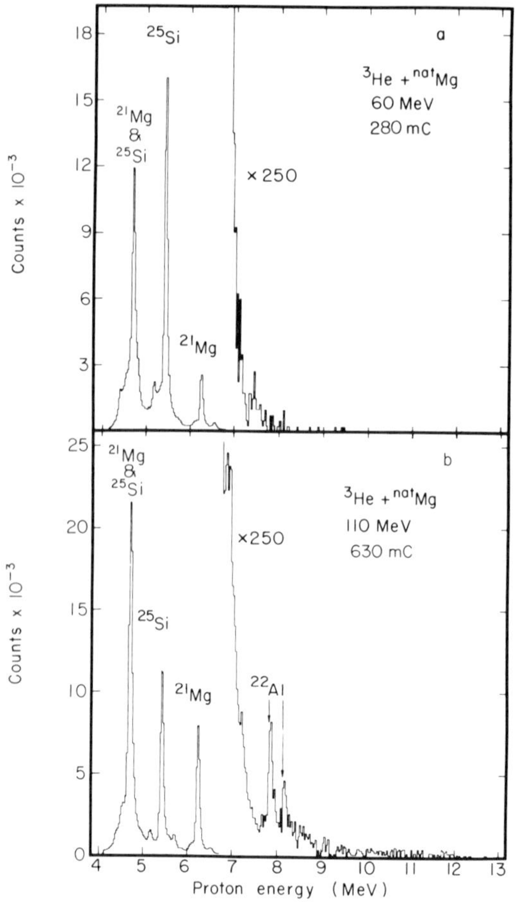


**Figure 23.** $\beta$-delayed proton spectra: (a) below threshold for $^{22}$Al production; (b) at an optimal energy for $^{22}$Al production. Note the strong background of $T_z = -\frac{3}{2}$ nuclides.

$^3$He induced reactions. An example of such a spectrum from the decay of $^{22}$Al is shown in Figure 23b. The identification of the proton groups labeled "$^{22}$Al" are based on predicted energies and the known difference of the energies of the final states in $^{21}$Na. Also, the below-threshold bombardment shown in Figure 23a did not show any sign of these groups. The decay scheme (shown in Figure 25 below) features one of the highest observed $\beta$-decay energies for proton-rich nuclei, being nearly twice that of the even-$Z$, $T_z = -2$ nuclei.

### 4.2.5. *Nuclei with* $T_z \leq -\frac{5}{2}$

Although four $T_z = -\frac{5}{2}$ nuclides have very recently been shown to exist in the sd shell, $^{23}$Si, $^{27}$S, $^{31}$Ar, and $^{35}$Ca (La 86), the decay of only the last has so far been observed. $^{35}$Ca was discovered via its $\beta$-delayed two-proton emission (see below), but no single protons could be observed due to background in this He-jet experiment (Äy 85, Mo 85). Further $\beta$-delayed proton emitters with $T_z \leq -\frac{5}{2}$ are predicted to exist

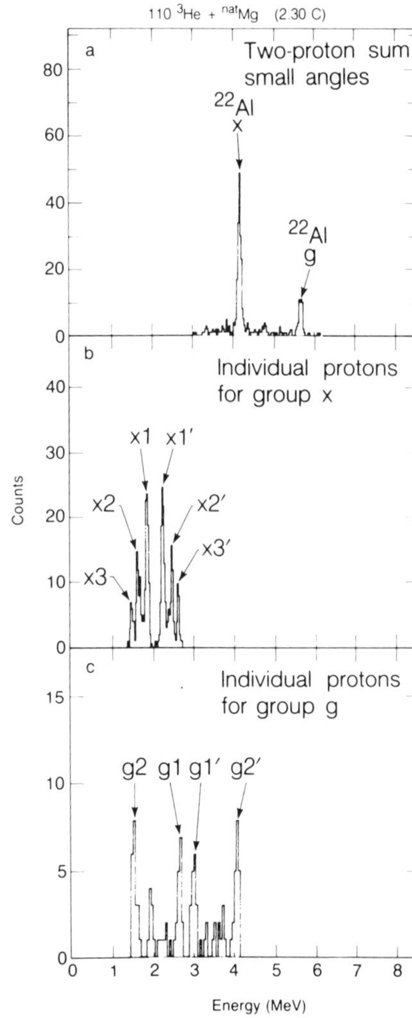


**Figure 24.** Proton–proton coincidence spectra following the decay of $^{22}$Al: (a) two-proton summed-energy spectrum; groups x and g correspond to transitions to the $^{20}$Ne first excited state and ground state, respectively; (b) and (c) show the individual proton spectra corresponding to these two transitions.

in the fp shell and will be the subjects of future experiments; indeed, the existence of $^{51}$Ni has just been established (Po 86).

## 4.3. β-Delayed Two-Proton Emission

Two-proton radioactivity, the emission of a pair of protons from the nuclear ground state, is a result of nuclear pairing and has long been proposed for some nuclei far from the valley of stability (Go 60, Go 61). It has been only relatively recently, however, that $\beta$-delayed two-proton radioactivity has been predicted as a potential decay mode for some proton-rich nuclei, especially for those in the odd–odd, $T_z = -2$ series (Go 80). The discoveries of two such nuclei, $^{22}$Al and $^{26}$P, via their $\beta$-delayed single-proton emission, made them prime candidates for a search for this new mode of decay. These initial experiments established that the $T = 2$ analog states in the daughter nuclei were unbound by 6.1 and 5.3 MeV, respectively, to the emission of two protons. Subsequent proton–proton coincidence experiments (Ca 83, Ho 83) then detected the first cases of $\beta$-delayed two-proton radioactivity. In principle, these two protons can be emitted in a sequential or in a simultaneous process and since their decay will be isospin forbidden, the analog state will be narrow. The question of the emission of $^2$He, a correlated pair of protons in an $^1S_0$ configuration, deserves special attention in connection with this decay mode.

$\beta$-delayed two-proton spectra from both $^{22}$Al and $^{26}$P were observed in the bombardment of Mg and Si targets with 110 MeV $^3$He beams; the spectrum from $^{22}$Al decay is shown in Figure 24. Again, as in the case of $\beta$-delayed single-proton experiments on these nuclei (Ca 82, Ca 83b), the activities were transported to the detection site with a fast He-jet apparatus. Coincident protons shown in Figure 24 were detected with the three-element surface barrier telescope system shown schematically in Figure 10. The telescope consisted of thin $\Delta E1$, and $\Delta E2$, and thick $E$ detectors, with each circular $\Delta E$ detector fabricated such that the surface contact on one side was divided along the center line (the coincident protons stop in the $\Delta E2$ detectors). This system covered an angular range from 0° to 70° between the protons. For $^{22}$Al, two, two-proton groups are observed and can be shown to correspond to transitions from the $T = 2$ analog state in $^{22}$Mg to the first excited state and the ground state of $^{20}$Ne. In a similar manner, for $^{26}$P, a single, two-proton peak was observed which could be associated with the decay of the $T = 2$ analog state in $^{26}$Si to the ground state of $^{24}$Mg. Individual coincident proton spectra of the two telescopes are composed of several discrete proton groups—indicating that protons are emitted sequentially via excited states of the intermediate nucleus (Ca 84). Although these experiments established this new radioactive-decay mode, its mechanism is to some extent uncertain.

### 4.3.1. *Mechanism for Two-Proton Decay*

In principle, the two protons can result from several decay mechanisms, which include decay involving a final-state interaction between the two protons ($^2$He emission), uncoupled simultaneous emission, or sequential emission (Ca 84). In the case of $^2$He emission, a strong angular correlation at small angles between the emitted protons is expected. A most probable angle of emission of 30° can be deduced from the $^2$He breakup energy distribution observed in direct reaction studies (Ja 78, St 79). Individual proton spectra should follow a distribution symmetric about $E_1^{\mathrm{LAB}} = E_2^{\mathrm{LAB}}$ with its shape determined by the final-state interaction energy $\varepsilon$. In an uncoupled simultaneous emission, where no intermediate states are formed, a continuum of individual proton energies, such as in the case of $^2$He, is expected. Unlike $^2$He emission though, a nearly isotropic angular correlation between the protons is expected.

Sequential emission is analogous to $\gamma$-ray cascades; the angular correlation is determined by the angular momenta of the initial, intermediate, and final states. However, for protons from $^{22}$Al and $^{26}$P decay, these correlations are not expected to be large and the maximum anisotropy should be less than 10–20%. The energies of the detected individual protons are determined by the energy of the intermediate state, the order of emission, and by the emission angle between the protons.

Based on energy correlation measurements both with a small (0°–70°) and large (70°–170°) angle setup, the existence of a sequential mechanism was established for

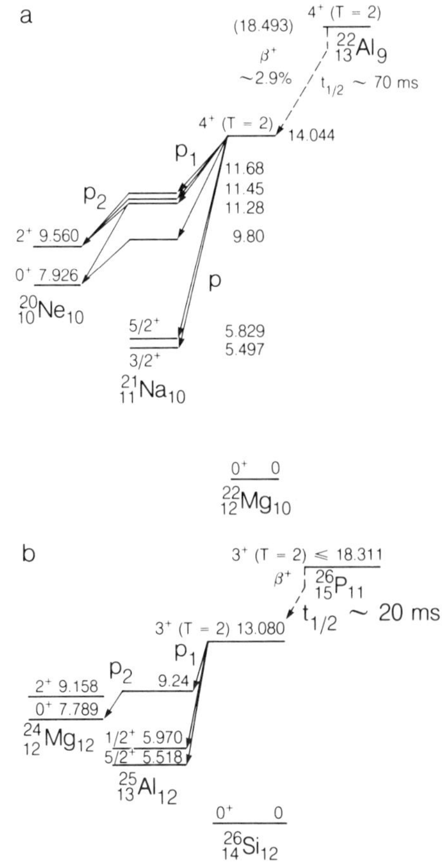


**Figure 25.** β-delayed proton and two-proton decay schemes of (a) $^{22}Al$ and (b) $^{26}P$.

both $^{22}Al$ and $^{26}P$ (Ca 84). In addition, due to spin-parity conservation, $^{2}He$ emission is excluded for the observed $^{26}P$ decay. The order of emission of the protons could also be deduced from these experiments at two different angles, and the decay schemes are shown in Figure 25.

Due to some puzzling features associated with the overall shapes of the individual proton spectra and the 2p/1p ratio observed for $^{22}Al$ at small and large angles, further experiments were conducted. A small component of two-proton decay proceeding by another mechanism such as $^{2}He$ emission remained a possibility and a more detailed angular correlation was obtained by utilizing position-sensitive detectors. Due to limited statistics, only the transition feeding the first excited state of $^{20}Ne$ could be analyzed. The result shown in Figure 26 (Ja 85) is essentially isotropic, but cannot exclude a small component of diproton ($^{2}He$) emission at small relative angles; large errors arise from the poor statistics of this low-yield experiment. The dotted line in Figure 26 has been calculated using the breakup properties of $^{2}He$ observed in reaction studies.

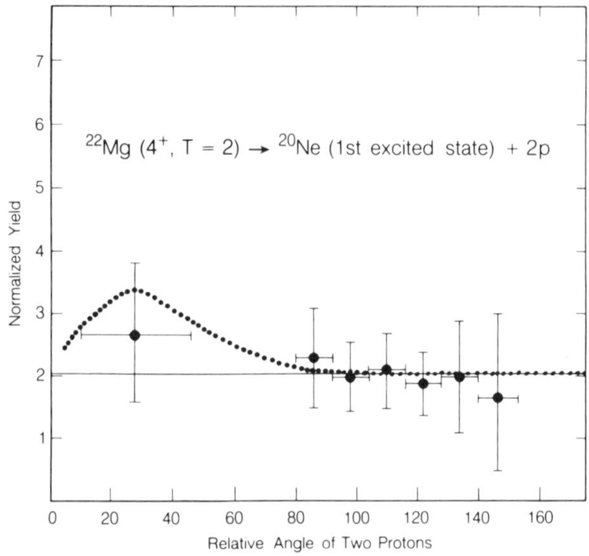


**Figure 26.** Normalized angular correlation for the two protons from the $\beta$-delayed two-proton emission of $^{22}$Al. The dotted line corresponds to a 15% admixture of $^2$He emission to an otherwise isotropic distribution.

## 4.4. Search for New Nuclides via $\beta$-Delayed Two-Proton Decay

In addition to being a physically interesting decay mode, $\beta$-delayed two-proton emission offers an excellent tool with which to search for new nuclides in the vicinity of the proton drip-line, where this mode of radioactivity is expected to form a major decay channel for the isobaric analog states. Figure 27 displays all potential, strong $\beta$-delayed single- and two-proton emitters. A decay mode is labeled "strong," if the analog state is unbound against these decay modes. This does not actually guarantee a large branch, since the analog state may decay predominantly by another mode, but it is a good first criterion. Also shown are probable direct proton and two-proton emitters.

It can be seen from Figure 27 that the odd-odd, $T_z = -2$ series in the fp shell forms a potential group of $\beta$-delayed two-proton emitters, the prime candidates being $^{46}$Mn and $^{50}$Co. The even-$Z$, $T_z = -\frac{5}{2}$ nuclei throughout the sd and fp shells can be predicted to have a rather strong branch of 2p decay from their analog states in the $T_z = -\frac{3}{2}$ daughter nuclei. Similarly, the $\beta$-decays of all other more proton-rich nuclei are expected to be associated with the emission of two protons. This group includes nuclides such as $^{22}$Si ($T_z = -3$), $^{50}$Ni ($T_z = -3$), and $^{49}$Ni ($T_z = -\frac{7}{2}$).

The systematic search for new $\beta$-2p emitters led recently to the discovery of the first $T_z = -\frac{5}{2}$ nucleus $^{35}$Ca (Äy 85), produced by bombarding a natural-calcium target with a 135 MeV $^3$He beam. Recoil nuclides were transported with a fast He jet and two-proton decay was observed by a detector setup similar to that shown in Figure 10. The observation of this nucleus via its $\beta$-delayed single-proton emission was prevented by strong production of the $T_z = -\frac{3}{2}$ $\beta$-delayed proton precursors $^{41}$Ti, $^{37}$Ca, $^{29}$S, and

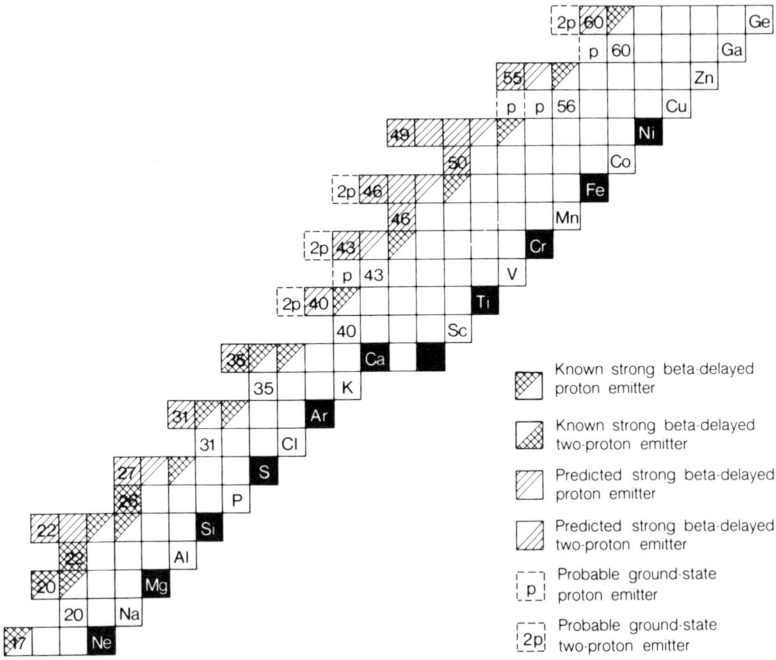


**Figure 27.** Chart of the nuclides showing potential $\beta$-delayed (and direct) proton and two-proton emitters.

$^{25}$Si. Background-free conditions for observing $\beta - 2p$ decay were provided by the unbound character of the possible "background" nuclides $^{30}$Cl, $^{34}$K, and $^{38}$Sc. The $T_z = -\frac{5}{2}$ nuclides $^{31}$Ar and $^{39}$Ti, possible products in the reaction, could be excluded, because Ar activities are not detected in a He-jet experiment and $^{39}$Ti is almost certainly unbound to ground-state two-proton emission and does not possess similar decay energies to those predicted for $^{35}$Ca.

The observed $\beta$-delayed two-proton spectrum of $^{35}$Ca is shown in Figure 28. A half-life of $50 \pm 30$ ms was measured for the groups labeled G and X. Good agreement was observed between the measured ground-state energy and that predicted from Coulomb displacement energy calculations; moreover, the energy difference between the two groups agrees with the known energy difference for decays to the ground state (G) and the excited state (X) at 811 keV in $^{33}$Cl. The distribution of individual proton energies suggests a sequential decay via intermediate states in $^{34}$Ar. Based on this assumption, a partial decay scheme for $^{35}$Ca is shown in Figure 29.

## 4.5. Proton Radioactivity

Proton radioactivity is the single-step emission of protons from a narrow nuclear state, be it the ground or an isomeric state, with a rate determined by penetration through the Coulomb and centrifugal barriers. The experimental study of direct proton radioactivity provides, via the observed half-life and decay energy, information about

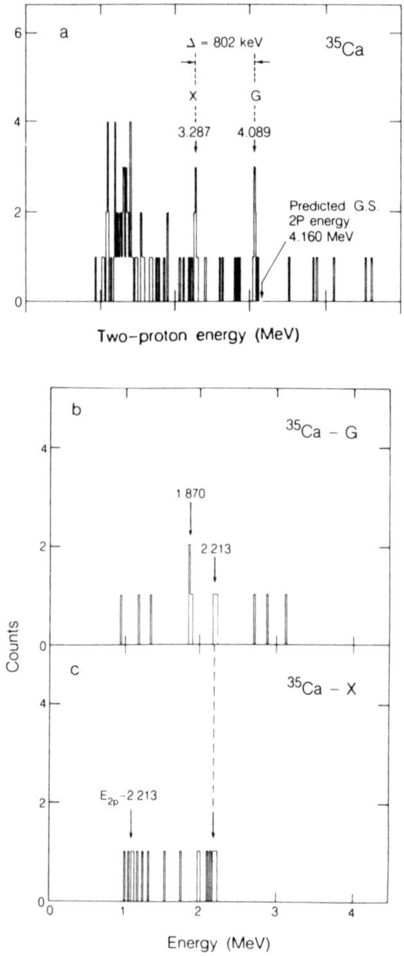


**Figure 28.** (a) $\beta$-delayed two-proton sum spectrum of $^{35}$Ca and the corresponding individual proton spectra (b) and (c). See text.

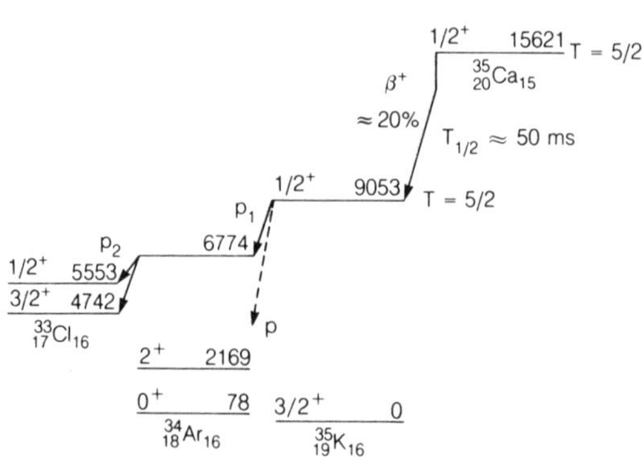


**Figure 29.** Proposed partial decay scheme for the $\beta$-delayed two-proton emission of $^{35}$Ca.

the reduced proton width for comparison with model calculations. In general, the phenomenon is expected to occur for highly unstable nuclei, just beyond the proton drip-line. However, the first proton radioactive nucleus discovered, an isomer of $^{53}$Co, lies rather close to the valley of $\beta$-stability (Ja 70, Ce 70). Subsequent searches for this phenomenon among heavier nuclei have established the existence of the proton radioactive nuclei $^{109}$I (Fa 84), $^{113}$Cs (Fa 84), $^{147}$Tm (Ho 82b, Kl 82), and $^{151}$Lu (Ho 81, Ho 82b), whose proton emission is believed to take place from their ground states [also see (Ho 84)].

A proton energy spectrum observed from the decay of $^{53m}$Co (Ce 72) is shown in Figure 30, together with the decay scheme of this spin-gap isomer. The partial half-life of 17 s is surprisingly long for direct emission of a 1.59 MeV proton: transmission through the Coulomb and $l = 9$ centrifugal barriers leads to an expected half-life of 60 ns. The additional strong hindrance can be explained as due to the high sensitivity of the decay rate to the residual nuclear interaction (Ce 77).

Although ground-state proton radioactivity should be observable among the very proton-rich light nuclei, its experimental detection will be difficult. In contrast to the proton radioactivities observed in heavier nuclei, much smaller decay energies and production cross-sections are involved. Also, the effective half-life window set by the $\beta$-decay probability as the upper limit, together with the proton-barrier penetrability, strongly restrict the number of possible candidates. It can be estimated that a proton-radioactive light nucleus should be unbound by more than 50 keV but less than of the order of 500 keV.

The most promising candidates below $A = 60$ for measurable ground-state proton emitters are, according to their predicted decay energies, $^{42}$V ($T_z = -2$), $^{54}$Cu ($T_z = -2$), and $^{55}$Cu ($T_z = -\frac{3}{2}$); $^{39}$Sc has recently been shown (Sh 84b) to be unbound by 630 ± 40 keV. In fact, $^{55}$Cu has been observed via fragmentation reactions in a spectrometer

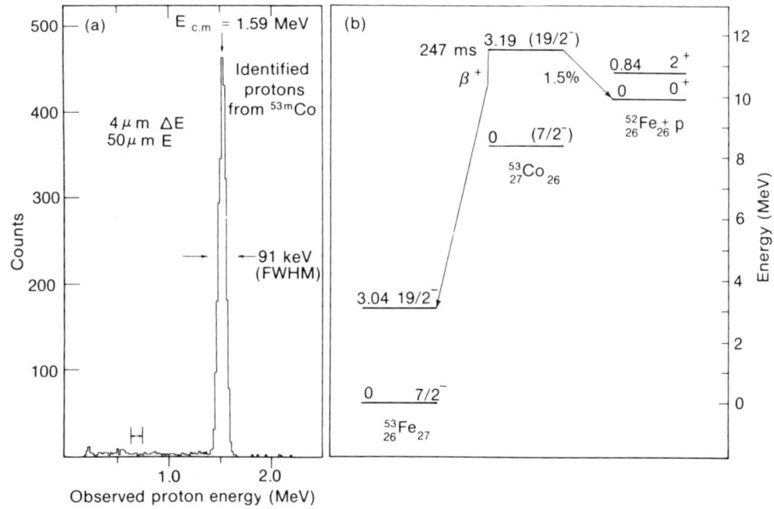


**Figure 30.** (a) Proton energy spectrum from the decay of $^{53m}$Co produced by the $^{54}$Fe(p, 2n) reaction. (b) Decay scheme of $^{53m}$Co.

experiment (Po 86) indicating a finite lifetime for this nucleus. All these nuclides are expected to be unbound by 200–300 keV from Kelson–Garvey predictions. A particular interest in characterizing $^{42}$V, $^{54}$Cu, and $^{55}$Cu as ground-state proton emitters, in addition to confirming existing approaches to calculating the proton drip-line in this mass region, is that the spectroscopic information from their decay should be directly comparable with theory, since one anticipates simple $f_{7/2}$ or $p_{3/2}$ proton configurations.

### 4.6. Two-Proton Radioactivity

Direct two-proton radioactivity was originally proposed by Goldanskii (Go 60). No experimental observations of this radioactive-decay mode have as yet been reported. Paralleling direct single-proton decay, it should occur near the proton drip-line throughout the nuclidic chart. Such decays could also occur from many-particle isomeric states (Go 71). Several light proton-rich nuclei are known not to exist due to their instability against two-proton emission. These nuclides, as shown in Table 4, are $^{6}$Be, $^{8}$C, $^{12}$O, and $^{16}$Ne. The most promising candidates for diproton emitters among the proton-rich nuclides, with their two-proton separation energies predicted by the Kelson–Garvey method, are $^{31}$Ar ($S_{2p} = -191$ keV), $^{39}$Ti ($-785$ keV), and $^{42}$Cr ($-691$ keV). Should this decay be observed, its mechanism is of considerable interest. The emission probability of the two protons and their energy and angular correlations will be closely related to the formation probability and to the mutual interaction between the protons as this cluster penetrates the Coulomb and centrifugal barriers. As noted by Goldanskii (Go 66), this phenomenon can be compared with the tunneling of electrons between metals in superconducting and normal states, providing an analog of the Josephson effect to be observed in nuclear transitions.

## 5. Recent Spectroscopy Above $Z = 22$

### 5.1. The Spectroscopy of $T_z = -\frac{1}{2}$ Nuclei

Progress in the nuclear spectroscopy of proton-rich nuclei above titanium has not been as rapid as that of lighter nuclei. Apart from our knowledge of the $T_z = -\frac{3}{2}$, $A = 4n + 1$ series of strong $\beta$-delayed proton emitters discussed earlier, most of the investigations involve $T_z = -\frac{1}{2}$ nuclei.

One of the main motivations in studies of the $\beta$ decays of these nuclides is the determination of the Gamow–Teller strength and especially its quenching, as already observed in the sd-shell region (Br 85) and in (p, n) reactions (Go 82). Since large-scale shell-model calculations are now availabe for fp-shell nuclei, a similar model-dependent study of this quenching is becoming possible for heavier nuclei. Recently, data on $\beta$ transitions from parent $T_z = -\frac{1}{2}$ nuclei up to $^{59}$Zn have become available, populating both ground and excited states in the daughter nuclei. [Although the $\beta$ decay of $^{71}$Kr (Ew 81) has been measured, neither its mass nor any decay to excited states has been determined.] These "exotic" nuclei have a large decay energy window, which allows the study of a good part of the Gamow–Teller strength distribution.

Experimental measurements of mirror $\beta$ decays in this mass region are difficult due to background from other simultaneously produced short-lived $\beta$ activities. For this reason, progress has only been possible through the use of on-line mass-separation

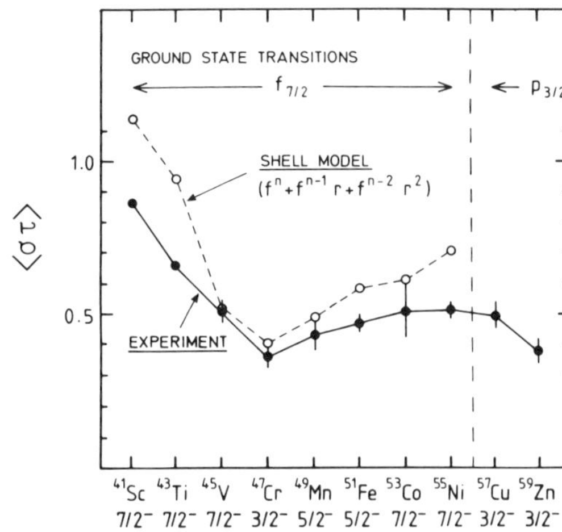


**Figure 31.** Experimental and theoretical matrix elements for $\beta$ transitions between the ground states of $T = \frac{1}{2}$ mirror nuclei. In the shell-model calculations f means the $f_{7/2}$ orbit while r denotes the $1p_{3/2}$, $0f_{5/2}$, and $1p_{1/2}$ orbits.

techniques. For example, the mirror transitions from $^{51}$Fe and $^{55}$Ni were measured using the new Ion Guide Isotope Separator On-Line technique (see Section 2.4).

Ground-state transitions between mirror states normally possess large Gamow–Teller transition strengths. These special transitions should be less sensitive to spurious effects from configuration mixing and they should provide a good test of the major components of the calculated wave functions. Figure 31 summarizes all of the present experimental data on mirror transitions above calcium. The transition matrix element $\langle\sigma\tau\rangle$ has been extracted from experiment by using the relations of Section 4.1 and the free nucleon value for the ratio of the axial vector to vector coupling constant $G_A/G_V$. Experimental values are taken from (Bu 85), except for the values for $^{43}$Ti, $^{57}$Cu, and $^{59}$Zn, which are from (Ho 86d), (Mi 85, Sh 86), and (Ho 81b, Ar 81), respectively. The value for $^{45}$Vi is an estimate, deduced from the experimental half-life and a theoretical value for the branching (Ho 82c). Theoretical values for the $f_{7/2}$ nuclei have been calculated with a shell-model code based on a proton–neutron coupling formalism, using $^{40}$Ca as an inert core, and taking into account particle excitation from the $0f_{7/2}$ orbital to the upper fp-shell, containing the $1p_{3/2}$, $0f_{5/2}$, and $1p_{1/2}$ orbits. The maximum number of particles raised to upper orbitals is two in this calculation, which is the current practical limit and has been shown to give good results throughout the $f_{7/2}$ shell (Mi 86c). This calculation used single-particle matrix elements from free nucleon decay.

The results in Figure 31 show that, in the lower and upper part of the $f_{7/2}$ shell, a systematic quenching of the experimental strength with respect to the shell-model calculation is observed in accordance with earlier observations both in the sd shell (Br 85) and in the $f_{7/2}$ shell (Og 86). In the middle of the shell, a less significant difference is observed, which deserves special attention in future shell-model analyses. It is also quite clear from the data that the behavior of the values of the matrix element across the $Z = 28$ proton number does not strongly indicate the single-particle character of $^{57}$Cu, but instead reflects the importance of the 2p–2h and/or 4p–4h excitations from the $f_{7/2}$ shell.

In one of the cases of Figure 31, $^{43}$Ti, a more detailed analysis has recently been provided (Ho 86d). Altogether 10 $\beta$ transitions to excited states in $^{43}$Sc were observed, corresponding to an integrated branching of 9.8(8)%. The total experimental strength $\sum\langle\sigma\tau\rangle^2$ to states below 5 MeV excitation is 1.35 in absolute units, which again is reduced when compared with the theoretical strength of about two over the same energy interval. It should be noted, however, that the shell-model result for the $\beta^+$ strength below 5 MeV is less than half of the total calculated strength. Thus, only a limited experimental window on the quenching of the total $\beta$ strength can be observed. Also, due to the strong energy dependence in $\beta$ decay, even strong Gamow–Teller transitions to highly excited states may remain undetected, which can lead to a severe underestimation of the total experimental strength, as shown in the recent study of the decay of $^{33}$Ar (Bo 86).

## 5.2. Toward the Spectroscopy of $T_z \leq -1$ Nuclei

The only reported $T_z = -1$ nuclei with $Z \geq 22$, whose decay has been characterized to any extent, are $^{44}$V (Ce 71), $^{46}$Cr (Zi 72), and $^{48}$Mn (Se 86b). The last has been produced via the $^{40}$Ca + $^{12}$C reaction and analyzed with the GSI on-line isotope separator. The good yield of this 150 ms activity permitted detailed spectroscopic studies via its $\beta$-delayed particle and $\gamma$ decay, making possible a detailed test of shell-model calculations in the middle of the $f_{7/2}$ shell. The decay scheme is shown in Figure 32

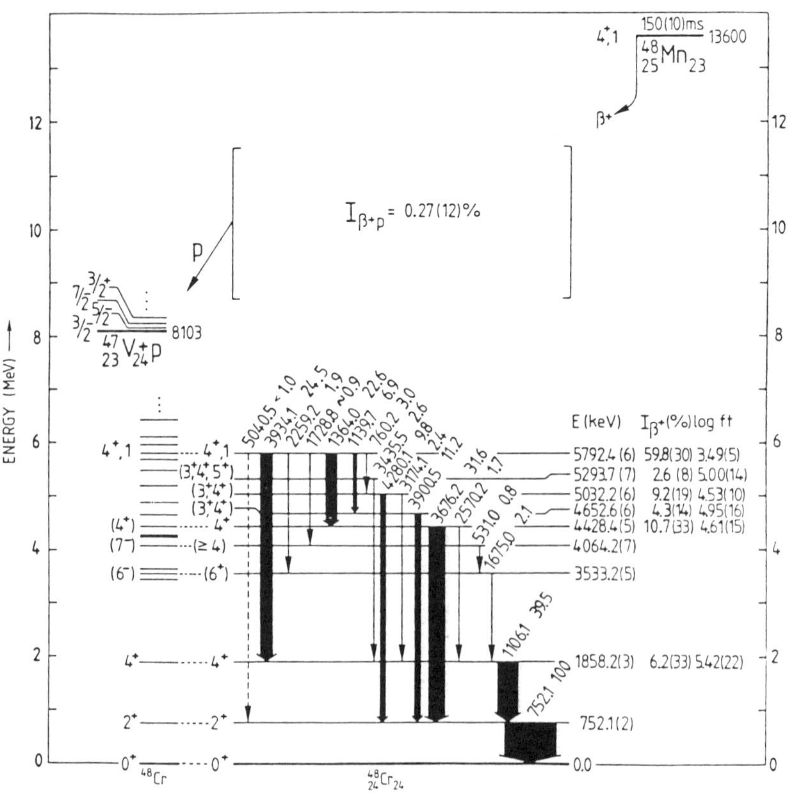


**Figure 32.** The decay scheme of $^{48}$Mn.

(Se 86b). The experimental Gamow–Teller strength and its distribution in the energy interval 0–5.8 MeV yields a quenching factor of 0.6(1) when compared with a shell-model calculation including one-particle promotion from the $f_{7/2}$ orbital to the $1p_{3/2}$, $0f_{5/2}$, and $1p_{1/2}$ orbitals (Se 86b).

The detailed study of $^{48}$Mn decay, together with new spectroscopy on the $T_z = -\frac{3}{2}$ series of nuclei, mark an important step in the investigations of more proton-rich nuclei above titanium. They have been produced in heavy-ion-induced reactions and have been analyzed by increasingly more sophisticated experimental techniques. It is clear that future studies will not only provide important new information on the $\beta$-strength functions, but also additional information on nuclear structure in general in the $(Z, N)$ plane of these very proton-rich nuclei.

## References


(An 86) M. S. Antony, J. Britz, and A. Pape, *At. Data Nucl. Data Tables* **34**:279 (1986).

(Ar 81) Y. Arai, M. Fujioka, E. Tanaka, T. Shinozuka, H. Miyatake, M. Yoshii, and T. Ishimatsu, *Phys. Lett.* **104B**:186 (1981).

(Är 85) J. Ärje, J. Äystö, H. Hyvönen, P. Taskinen, V. Koponen, and J. Honkanen, *Phys. Rev. Lett.* **54**:99 (1985).

(Är 86) J. Ärje, J. Äystö, H. Hyvönen, P. Taskinen, V. Koponen, J. Honkanen, and K. Valli, *Nucl. Instrum. Methods* **247**:431 (1986).

(Äy 79) J. Äystö, D. M. Moltz, M. D. Cable, R. D. von Dincklage, R. F. Parry, J. M. Wouters, and J. Cerny, *Phys. Lett.* **82B**:43 (1979).

(Äy 80) J. Äystö and J. Cerny, in *Future Directions in Studies of Nuclei Far From Stability* (J. H. Hamilton *et al.*, eds.), North-Holland, Amsterdam, 1980), p. 257.

(Äy 81) J. Äystö, M. D. Cable, R. F. Parry, J. M. Wouters, D. M. Moltz, and J. Cerny, *Phys. Rev. C* **23**:879 (1981).

(Äy 82) J. Äystö, J. Honkanen, K. Vierinen, A. Hautojärvi, K. Eskola, and S. Messelt, *Phys. Lett.* **110B**:437 (1982).

(Äy 84) J. Äystö, J. Ärje, V. Koponen, P. Taskinen, H. Hyvönen, A. Hautojärvi, and K. Vierinen, *Phys. Lett.* **138B**:369 (1984).

(Äy 85) J. Äystö, D. M. Moltz, X. J. Xu, J. E. Reiff, and J. Cerny, *Phys. Rev. Lett.* **55**:1384 (1985).

(Äy 85b) J. Äystö, X. J. Xu, D. M. Moltz, J. E. Reiff, J. Cerny, and B. H. Wildenthal, *Phys. Rev. C* **32**:1700 (1985).

(Ba 63) R. Barton, R. McPherson, R. E. Bell, W. R. Frisken, W. T. Link, and R. B. Moore, *Can. J. Phys.* **41**:2007 (1963).

(Be 70) G. Bertsch and S. Kahana, *Phys. Lett.* **33B**:193 (1970).

(Be 75) W. Benenson, A. Guichard, E. Kashy, D. Mueller, H. Nann, and L. W. Robinson, *Phys. Lett.* **58**:46 (1975).

(Be 76) W. Benenson, E. Kashy, D. Mueller, and H. Nann, Proc. Int. Conf. on Nuclei Far From Stability, 3rd., Cargese, Corsica, 1976; report CERN 76-13, p. 235.

(Be 77) W. Benenson, D. Mueller, E. Kashy, H. Nann, and L. W. Robinson, *Phys. Rev. C* **15**:1187 (1977).

(Be 79) W. Benenson and E. Kashy, *Rev. Mod. Phys.* **51**:527 (1979).

(Bj 85) T. Björnstadt, M. J. G. Borge, P. Dessagne, R. D. von Dincklage, G. T. Ewan, P. G. Hansen, A. Huck, B. Jonson, G. Klotz, A. Knipper, P. O. Larsson, G. Nyman, H. L. Ravn, C. Richard-Serre, K. Riisager, D. Schardt, and G. Walter, *Nucl. Phys. A* **443**:283 (1985).

(Bo 86) M. J. G. Borge, P. Dessagne, G. T. Ewan, P. G. Hansen, A. Huck, B. Jonson, G. Klotz, A. Knipper, S. Mattsson, G. Nyman, C. Richard-Serre, K. Riisager, G. Walter, and the ISOLDE Coll., *Phys. Scr.* **36**:218 (1987).

(Br 85) B. A. Brown and B. H. Wildenthal, *At. Data Nucl. Data Tables* **33**:347 (1985).

(Bu 85) T. W. Burrows, J. W. Olness, and D. E. Alburger, *Phys. Rev. C* **31**:1490 (1985).

(Ca 82) M. D. Cable, J. Honkanen, R. F. Parry, H. M. Thierens, J. M. Wouters, Z. Y. Zhou, and J. Cerny, *Phys. Rev. C* **26**:1778 (1982).

(Ca 83) M. D. Cable, J. Honkanen, R. F. Parry, S. H. Zhou, Z. Y. Zhou, and J. Cerny, *Phys. Rev. Lett.* **50**:404 (1983).

(Ca 83b) M. D. Cable, J. Honkanen, R. F. Parry, S. H. Zhou, Z. Y. Zhou, and J. Cerny, *Phys. Lett.* **123B**:25 (1983).

(Ca 84) M. D. Cable, J. Honkanen, E. C. Schloemer, M. Ahmed, J. E. Reiff, Z. Y. Zhou, and J. Cerny, *Phys. Rev. C* **30**:1276 (1984).

(Ce 64) J. Cerny, R. H. Pehl, F. S. Goulding, and D. A. Landis, *Phys. Rev. Lett.* **13**:726 (1964).

(Ce 66) J. Cerny, S. W. Cosper, G. W. Butler, R. H. Pehl, F. S. Goulding, D. A. Landis, and C. Detraz, *Phys. Rev. Lett.* **16**:469 (1966).

(Ce 68) J. Cerny, *Ann. Rev. Nucl. Sci.* **18**:27 (1968).

(Ce 69) J. Cerny, R. A. Mendelson, G. J. Wozniak, J. E. Esterl, and J. C. Hardy, *Phys. Rev. Lett.* **22**:612 (1969).

(Ce 70) J. Cerny, J. E. Esterl, R. A. Gough, and R. G. Sextro, *Phys. Lett.* **33B**:284 (1970).

(Ce 70b) J. Cerny, C. U. Cardinal, H. C. Evans, K. P. Jackson, and N. A. Jelley, *Phys. Rev. Lett.* **24**:1128 (1970).

(Ce 71) J. Cerny, D. R. Goosman, and D. E. Alburger, *Phys. Lett.* **37B**:380 (1971).

(Ce 72) J. Cerny, R. A. Gough, R. G. Sextro, and J. E. Esterl, *Nucl. Phys. A* **188**:666 (1972).

(Ce 76) J. Cerny, Proc. Int. Conf. on Nuclei Far From Stability, 3rd., Cargese, Corsica, 1976; report CERN 76-13, p. 225.

(Ce 77) J. Cerny and J. C. Hardy, *Ann. Rev. Nucl. Sci.* **27**:333 (1977).

(Cu 86) M. S. Curtin, L. H. Harwood, J. A. Nolen, B. Sherrill, Z. Q. Xie, and B. A. Brown, *Phys. Rev. Lett.* **56**:34 (1986).

(De 86) C. Detraz, *Proc. Int. Conf. Nucl. Phys.*, Harrogate, U.K. (J. L. Durell, J. M. Irvine, and G. C. Morrison, eds.), IOP, Bristol (1986), Vol. 2, p. 495.

(Du 82) J. P. Dufour, H. Delagrange, R. Del Moral, A. Fleury, F. Hubert, Y. Llabador, M. B. Mauhourat, K. H. Schmidt, and A. Lleres, *Nucl. Phys. A* **387**:157 (1982).

(Du 86) J. P. Dufour, R. Del Moral, A. Fleury, F. Hubert, D. Jean, M. S. Pravikoff, H. Delagrange, H. Geissel, and K. H. Schmidt, *Z. Phys. A* **324**:487 (1986).

(En 78) P. M. Endt and C. van der Leun, *Nucl. Phys. A* **310**:1 (1978).

(En 79) H. A. Enge, *Nucl. Instrum. Methods* **162**:161 (1979).

(Es 80) K. Eskola, M. Riihonen, K. Vierinen, J. Honkanen, M. Kortelahti, and K. Valli, *Nucl. Phys. A* **341**:365 (1980).

(Ew 80) G. T. Ewan, E. Hagberg, J. C. Hardy, B. Jonson, S. Mattsson, P. Tidemand-Petersson, and I. S. Towner, *Nucl. Phys. A* **343**:109 (1980).

(Ew 81) G. T. Ewan, E. Hagberg, P. G. Hansen, B. Jonson, S. Mattsson, H. L. Ravn, and P. Tidemand-Petersson, *Nucl. Phys. A* **352**:13 (1981).

(Fa 84) T. Faesterman, A. Gillitzer, K. Hartel, P. Kienle, and E. Nolte, *Phys. Lett.* **137B**:23 (1984).

(Fu 85) H. W. Fulbright, in *Treatise on Heavy-Ion Science* (D. Allan Bromley, ed.), Vol. 7, Plenum Press, New York (1985), p. 179.

(Ga 66) G. T. Garvey and I. Kelson, *Phys. Rev. Lett.* **16**:197 (1966).

(Ga 69) G. T. Garvey, W. J. Gerace, R. L. Jaffe, I. Talmi, and I. Kelson, *Rev. Mod. Phys.* **41**:S1 (1969).

(Ga 86) C. A. Gagliardi, D. R. Semon, R. E. Tribble, and L. A. Van Ausdeln, *Phys. Rev. C* **34**:1663 (1986).

(Go 60) V. I. Gol'danskii, *Nucl. Phys.* **19**:482 (1960).

(Go 61) V. I. Gol'danskii, *Nucl. Phys.* **27**:648 (1961).

(Go 66) V. I. Gol'danskii, Usp. Fiz. Nauk **87**:255 (1966) [*Sov. Phys. Usp.* **8**:770 (1966)].

(Go 71) V. I. Gol'danskii and L. K. Peker, *Pis'ma Zh. Eksp. Teor. Fiz.* **13**:577 (1971) [*Sov. Phys.—JETP Lett.* **13**:412 (1971)].

(Go 72) R. A. Gough, R. G. Sextro, and J. Cerny, *Phys. Rev. Lett.* **28**:510 (1972).

(Go 80) V. I. Gol'danskii, *Pis'ma Zh. Eksp. Theor. Fiz.* **32**:572 (1980) [*Sov. Phys.—JETP Lett.* **32**:554 (1980)].

(Go 82) C. Goodman, *Nucl. Phys. A* **374**:241c (1982).

(Gu 75) A. Guichard, H. Nann, and B. H. Wildenthal, *Phys. Rev. C* **12**:1109 (1975).

(Ha 64) J. C. Hardy and R. I. Verrall, *Phys. Lett.* **13**:148 (1964).

(Ha 64b) J. C. Hardy and R. I. Verrall, *Phys. Rev. Lett.* **13**:764 (1964).

(Ha 77) E. Hagberg, P. G. Hansen, J. C. Hardy, A. Huck, B. Jonson, S. Mattsson, H. L. Ravn, P. Tidemand-Petersson, and G. Walter, *Phys. Rev. Lett.* **39**:792 (1977).

(Ha 81) L. H. Harwood and J. A. Nolen, *Nucl. Instrum. Methods* **186**:435 (1981).

(He 74) D. L. Hendrie, in *Nuclear Spectroscopy and Reactions* (J. Cerny, ed.), Part A, Academic Press, New York (1974), p. 365.

(Ho 80) J. Honkanen, M. Kortelahti, K. Valli, A. Hautojärvi, and K. Vierinen, *Nucl. Phys. A* **330**:429 (1980).

(Ho 81) S. Hofmann, G. Münzenberg, W. Faust, F. Hessberger, W. Reisdorf, J. R. H. Schneider, P. Armbruster, K. Güttner, and B. Thuma, Proc. Int. Conf. Nuclei Far From Stability, 4th, Helsingor, Denmark, 1981; CERN-Report 81-09, p. 190.

(Ho 81b) J. Honkanen, M. Kortelahti, K. Eskola, and K. Vierinen, *Nucl. Phys. A* **366**:109 (1981).

(Ho 82) J. Honkanen, J. Äystö, M. Kortelahti, K. Eskola, K. Vierinen, and A. Hautojärvi, *Nucl. Phys. A* **380**:410 (1982).

(Ho 82b) S. Hofmann, G. Münzenberg, F. P. Hessberger, W. Reisdorf, J. R. H. Schneider, and P. Armbruster, *Z. Phys. A* **305**:111 (1982).

(Ho 82c) P. Hornshøj, J. Kolind, and N. Rud, *Phys. Lett.* **116B**:4 (1982).

(Ho 83) J. Honkanen, M. D. Cable, R. F. Parry, S. H. Zhou, Z. Y. Zhou, and J. Cerny, *Phys. Lett.* **133B**:146 (1983).

(Ho 84) S. Hofmann, Y. K. Agarwal, P. Armbruster, F. P. Hessberger, P. O. Larsson, G. Münzenberg, K. Poppensieker, W. Reisdorf, J. R. H. Schneider, and H. J. Schött, in Proc. 7th Int. Conf. on Atomic Masses and Fundamental Constants, Darmstadt-Seeheim, W. Germany, 1984, p. 184.

(Ho 86) M. A. C. Hotchkis, J. E. Reiff, D. J. Vieira, F. Blonningen, T. F. Lang, D. M. Moltz, X. J. Xu, and J. Cerny, *Phys. Rev. C* **35**:315 (1987).

(Ho 86b) J. Honkanen, V. Koponen, H. Hyvönen, P. Taskinen, J. Äystö, K. Ogawa, and K. Eskola, in *Proc. Int. Symp. on Weak and Electromagnetic Interactions in Nuclei, Heidelberg* (H. V. Klapdor, ed.), Springer-Verlag, Berlin (1986), p. 253.

(Ho 86c) J. Honkanen, J. Äystö, and K. Eskola, *Phys. Scr.* **34**:608 (1986).

(Ho 86d) J. Honkanen, V. Koponen, H. Hyvönen, P. Taskinen, J. Äystö, and K. Ogawa, *Nucl. Phys.* **A470**:489 (1987).

(Ja 70) K. P. Jackson, C. U. Cardinal, H. C. Evans, N. A. Jelley, and J. Cerny, *Phys. Lett.* **33B**:281 (1970).

(Ja 74) K. P. Jackson, J. C. Hardy, H. Schmeing, R. L. Graham, J. S. Geiger, and K. W. Allen, *Phys. Lett.* **B49**:341 (1974).

(Ja 78) R. Jahn, D. P. Stahel, G. J. Wozniak, R. J. de Meijer, and J. Cerny, *Phys. Rev. C* **18**:9 (1978).

(Ja 85) R. Jahn, R. L. McGrath, D. M. Moltz, J. E. Reiff, X. J. Xu, J. Äystö, and J. Cerny, *Phys. Rev. C* **31**:1576 (1985).

(Jä 85b) J. Jänecke and E. Comay, *Nucl. Phys. A* **436**:108 (1985).

(Jo 76) B. Jonson, E. Hagberg, P. G. Hansen, P. Hornshøj, and P. Tidemand-Petersson, Proc. Int. Conf. on Nuclei Far From Stability, 3rd., Cargese, Corsica, 1976; report CERN 76-13, p. 277.

(Ke 66) I. Kelson and G. T. Garvey, *Phys. Lett.* **23**:689 (1966).

(Ke 78) G. J. KeKelis, M. S. Zisman, D. K. Scott, R. Jahn, D. J. Vieira, J. Cerny, and F. Ajzenberg-Selove, *Phys. Rev. C* **17**:1929 (1978).

(Kl 82) O. Klepper, T. Batsch, S. Hofmann, R. Kirchner, W. Kurzewicz, W. Reisdorf, E. Roeckl, D. Schardt, and G. Nyman, *Z. Phys. A* **305**:125 (1982).

(Ko 77) R. Kouzes, J. Lind, W. H. Moore, R. G. H. Robertson, and R. Sherr, *Nucl. Phys. A* **286**:253 (1977).

(La 86) M. Langevin, A. C. Mueller, D. Guillemaud-Mueller, M. G. Saint-Laurent, R. Anne, M. Bernas, J. Galin, D. Guerreau, J. C. Jacmart, S. D. Hoath, F. Naulin, F. Pougheon, E. Guiniou, and C. Detraz, *Nucl. Phys. A* **455**:149 (1986).

(Ma 74) R. D. Macfarlane and Wm. C. McHarris, in *Nuclear Spectroscopy and Reactions* (J. Cerny, ed.), Part A, Academic Press, New York (1974), p. 243.

(Ma 76) S. Maripuu, *At. Data Nucl. Data Tables* **17**:411 (1976).

(Mc 65) R. McPherson, R. A. Esterlund, A. M. Poskanzer, and P. L. Reeder, *Phys. Rev.* **140**:B1513 (1965).

(Mi 85) H. Miyatake, M. Yoshii, H. Hama, T. Kamiya, T. Ishimatsu, T. Shinozuka, M. Fujioka, and K. Ogawa, *Phys. Rev. C* **30**:2111 (1985).

(Mi 86) H. Miyatake, H. Hama, T. Kamiya, M. Yoshii, T. Shinozuka, and M. Fujioka, *Nucl. Instrum. Methods* **A245**:271 (1986).

(Mi 86b) D. M. Mikolas, B. A. Brown, W. Benenson, L. H. Harwood, E. Kashy, J. A. Nolen Jr., B. Sherrill, J. Stevenson, J. Winfield, Z. Q. Xie, and R. Sherr, *Phys. Rev. C* **37**:766 (1988).

(Mi 86c) H. Miyatake, K. Ogawa, T. Shinozuka, and M. Fujioka, *Nucl. Phys.* **A470**:328 (1987).

(Mo 79) D. M. Moltz, J. Äystö, M. D. Cable, R. D. von Dincklage, R. F. Parry, J. M. Wouters, and J. Cerny, *Phys. Rev. Lett.* **42**:43 (1979).

(Mo 80) D. M. Moltz, J. M. Wouters, J. Äystö, M. D. Cable, R. D. von Dincklage, and J. Cerny, *Nucl. Instrum. Methods* **172**:519 (1980).

(Mo 81) D. M. Moltz, J. Äystö, M. D. Cable, R. F. Parry, P. E. Haustein, J. M. Wouters, and J. Cerny, *Nucl. Instrum. Methods* **186**:141 (1981).

(Mo 82) C. L. Morris, H. T. Fortune, L. C. Bland, R. Gilman, S. J. Greene, W. B. Cottingame, D. B. Holtkamp, G. R. Burleson, and C. F. Moore, *Phys. Rev. C* **25**:3218 (1982).

(Mo 85) D. M. Moltz, J. Äystö, M. A. C. Hotchkis, and J. Cerny, *Proc. ACS Symp. Nuclei Off the Line of Stability* (R. A. Meyer and D. S. Brenner, eds.), ACS Symposium Series (1986), p. 448.

(Mu 75) D. Mueller, E. Kashy, W. Benenson, and H. Nann, *Phys. Rev. C* **12**:51 (1975).

(Mu 77) D. Mueller, E. Kashy, and W. Benenson, *Phys. Rev. C* **15**:1282 (1977).

(Mü 81) G. Münzenberg, W. Faust, F. P. Hessberger, S. Hofmann, W. Reisdorf, K.-H. Schmidt, W. F. W. Schneider, H. Schött, P. Armbruster, K. Güttner, B. Thuma, H. Ewald, and D. Vermeulen, *Nucl. Instrum. Methods* **186**:423 (1981).

(Mü 84) W. Müller, B. C. Metsch, W. Knupfer, and A. Richter, *Nucl. Phys. A* **430**:61 (1984).

(Og 85) K. Ogawa, private communication.

(Ol 84) T. Ollivier, S. Andre, J. Äystö, D. Barneoud, R. Beraud, A. Charvet, K. Deneffe, R. Duffait, A. Emsallem, C. Foin, J. Genevey, A. Gizon, N. Idrissi, R. Kossakowski, M. Meyer, and J. Treherne, Proc. 7th Int. Conf. on Atomic Masses and Fundamental Constants, Darmstadt-Seeheim, W. Germany, 1984, p. 82.

(Pa 77) L. A. Parks, C. N. Davids, B. G. Nardi, and J. N. Worthington, *Nucl. Instrum. Methods* **143**:93 (1977).

(Pl 86) A. Plantier, R. Beraud, T. Ollivier, A. Charvet, R. Duffait, A. Emsallem, N. Redon, V. Boninchi, S. Vanzetto, A. Gizon, J. Treherne, N. Idrissi, J. Genevey, and J. L. Vieux-Rochaz, *Nucl. Instrum. Methods* **B26**:314 (1987).

(Po 86) F. Pougheon, J. C. Jacmart, E. Quiniou, R. Anne, D. Bazin, V. Borrel, J. Galin, D. Guerreau, D. Guillemaud-Mueller, M. G. Saint-Laurent, A. C. Mueller, E. Roeckl, and C. Detraz, *Z. Phys. A* **327**:17 (1987).

(Ra 79) H. L. Ravn, *Phys. Rep.* **54**:201 (1979).

(Re 64) P. L. Reeder, A. M. Poskanzer, and R. A. Esterlund, *Phys. Rev. Lett.* **13**:767 (1964).

(Ro 74) R. G. H. Robertson, S. Martin, W. R. Falk, D. Ingham, and A. Djaloeis, *Phys. Rev. Lett.* **32**:1207 (1974).

(Ro 76) R. G. H. Robertson, W. Benenson, E. Kashy, and D. Mueller, *Phys. Rev. C* **13**:1018 (1976).

(Sc 86) H. Schmeing, J. S. Wills, E. Hagberg, J. C. Hardy, V. T. Koslowsky, and W. L. Perry, *Nucl. Instrum. Methods* **B26**:321 (1987).

(Se 86) K. K. Seth, S. Iversen, M. Kaletka, D. Barlow, A. Saha, and R. Soundrarayagam, *Phys. Lett.* **B173**:397 (1986).

(Se 86b) T. Sekine, J. Cerny, R. Kirchner, O. Klepper, V. T. Koslowsky, A. Plochocki, E. Roeckl, D. Schardt, B. Sherrill, and B. A. Brown, *Nucl. Phys. A* **467**:93 (1987).

(Sh 84) B. Sherrill, K. Beard, W. Benenson, C. Bloch, B. A. Brown, E. Kashy, J. A. Nolen Jr., A. D. Panagiotou, J. van der Plicht, and J. S. Winfield, Proc. 7th Int. Conf. on Atomic Masses and Fundamental Constants, Darmstadt-Seeheim, W. Germany, 1984, p. 82.

(Sh 84b) B. Sherrill, private communication and M. F. Mohar, E. Adamides, W. Benenson, C. Bloch, B. A. Brown,, J. Clayton, E. Kashy, M. Lowe, J. A. Nolen, Jr., W. E. Ormand, J. van der Plicht, B. Sherrill, J. Stevenson, and J. S. Winfield, *Phys. Rev. C* (to be published).

(Sh 85) B. Sherrill, K. Beard, W. Benenson, C. Bloch, B. A. Brown, E. Kashy, J. A. Nolen Jr., A. D. Panagiotou, J. van der Plicht, and J. S. Winfield, *Phys. Rev. C* **31**:875 (1985).

(Sh 85b) T. Shinozuka, M. Fujioka, H. Miyatake, M. Yoshii, H. Hama, and T. Kamiya, *Phys. Rev. C* **30**:2111 (1985).

(Sh 85c) T. Shinozuka, H. Hama, T. Kaguchi, M. Yoshii, J. Äystö, and M. Fujioka, unpublished; see also (Yo 86).

(Sh 86) T. Shinozuka, private communication, to be published.

(St 79) D. P. Stahel, R. Jahn, G. J. Wozniak, and J. Cerny, *Phys. Rev. C* **20**:1680 (1979).

(Tr 77) R. E. Tribble, J. D. Cossairt, and R. A. Kenefick, *Phys. Rev. C* **15**:2028 (1977).

(Tr 80) R. E. Tribble, D. M. Tanner, and A. F. Zeller, *Phys. Rev. C* **22**:17 (1980).

(Vi 76) D. J. Vieira, D. F. Sherman, M. S. Zisman, R. A. Gough, and J. Cerny, *Phys. Lett.* **B60**:261 (1976).

(Vi 78) D. J. Vieira, Ph.D. Thesis, Report No. LBL-7161, unpublished (1978).

(Vi 79) D. J. Vieira, R. A. Gough, and J. Cerny, *Phys. Rev. C* **19**:177 (1979).

(Wa 85) A. H. Wapstra and G. Audi, *Nucl. Phys. A* **432**:1 (1985).

(We 79) G. D. Westfall, T. J. M. Symons, H. J. Crawford, D. E. Greiner, H. H. Heckman, P. J. Lindstrom, J. Mahoney, D. K. Scott, A. C. Shotter, T. C. Awes, C. K. Gelbke, and J. M. Kidd, *Phys. Rev. Lett.* **43**:1859 (1979).

(Wi 57) E. P. Wigner, in Proc. of the Robert A. Welch Foundation Conference on Chemical Research, Houston, 1957 (ed. W. O. Millikan), p. 67, unpublished.

(Wi 73) D. H. Wilkinson, *Nucl. Phys. A* **209**:470 (1973).

(Wi 74) D. H. Wilkinson, *Nucl. Phys. A* **238**:58 (1974).

(Wi 75) W. J. Wiesehahn, G. Bischoff, and J. M. D'Auria, *Nucl. Instrum. Methods* **124**:221 (1975).

(Wi 78) D. H. Wilkinson, A. Gallman, and D. E. Alburger, *Phys. Rev. C* **18**:401 (1978).

(Wi 82) D. H. Wilkinson, *Nucl. Phys. A* **377**:474 (1982).

(Wi 84) B. H. Wildenthal, *Prog. Part. Nucl. Phys.* **11**:5 (1984).

(Yo 86) M. Yoshii, H. Hama, K. Taguchi, T. Ishimatsu, T. Shinozuka, M. Fujioka, and J. Ärje, Proc. 11th Int. Conf. on Electromagnetic Isotope Separators and Techniques Related to their Applications, Los Alamos, NM, *Nucl. Instrum. Methods* **B26**:410 (1987).

(Zi 72) J. Zioni, A. A. Jaffe, E. Friedman, N. Haik, R. Schectman, and D. Nir, *Nucl. Phys. A* **181**:465 (1972).

# 4

ALEXEI A. OGLOBLIN (left) is Head of the Department of Nuclear Interaction at the Kurchatov Institute of Atomic Energy at Moscow, USSR. He received his advanced education at the Moscow University, graduating in 1953. In 1962 he received his Candidat of Physicomatematical Sciences degree and 1970 he was awarded the Doctor of Physicomathematical Sciences degree. He is Professor in Nuclear and Particle Physics. His main works concern the study of nuclear reactions with light heavy ions.

YURI E. PENIONZHKEVICH (right) is Head of the Section of Nuclear Reactions Mechanisms, Laboratory of Nuclear Reactions, the Joint Institute for Nuclear Research at Dubna, USSR. He graduated from the Voronezh University in 1964; since then he has been on the research staff in Dubna. In 1970 he received his Kandidat of Physico-Mathematical Sciences degree and in 1985 he received his Doctor of Physico-Mathematical Sciences degree.

# Very Neutron-Rich Very Light Nuclei

ALEXEI A. OGLOBLIN AND YURI E. PENIONZHKEVICH

## Introduction

The present review deals with very light neutron-rich nuclei. Light nuclei have always attracted attention due to their varied properties. If structural characteristics of medium and heavy nuclei change rather smoothly while undergoing only isolated jumps, the structure of light nuclei is clearly individual in its character: two neighboring light nuclei are often very much unlike one another. This makes their study very interesting, on the one hand, but on the other hand, it is difficult to reveal general tendencies.

Our interest in light nuclei with large neutron excess is connected precisely with our aim of finding characteristic properties of nuclei enriched by neutrons. It will be seen from what follows that at present one can obtain information on nuclei with large neutron excess $[(N - Z)/A]$ only by studying light nuclei. It is possible to reach the neutron drip-line and to observe nuclei which decay by neutron emission only in the region of very light nuclei. To what extent these data are general is a separate problem.

Interest in nuclei enriched by neutrons is determined by many factors. The most important of them is that the position of the neutron drip-line, even the very fact of its existence, is in essence an open question. Though there exist a great number of nuclear mass calculations, none of them is reliable enough far away from the stability valley. It has been discussed more than once in the literature whether nuclei having a neutron excess much larger than predicted could exist, and whether nuclei consisting only of neutrons could be observed.

If it were demonstrated that nuclei with an abnormally large neutron excess (we shall call them neutron nuclei) could exist, it would require a qualitative change of our present concepts of nuclear structure. This would undoubtedly affect other fields of science, such as astrophysics. The existence of, and possibility of obtaining, long-lived neutron nuclei could also have practical applications, since man would then have "canned neutrons."

---

ALEXEI A. OGLOBLIN • Department of Nuclear Interaction, Kurchatov Institute of Atomic Energy, 123436 Moscow, USSR. YURI E. PENIONZHKEVICH • Laboratory of Nuclear Reactions, Joint Institute for Nuclear Research, 101000 Moscow, USSR.

Modern theories of nuclear matter and calculations of nuclear masses deny such a possibility. However, this conclusion cannot be treated as absolute; only experiment can give the answer to questions regarding the position of the neutron drip-line at our present level of development of theoretical concepts.

Progress made in experimental techniques and, most important, in creating a new generation of heavy-ion accelerators has allowed many tens of new neutron-rich nuclides to be synthesized recently. In studying these nuclei new interesting phenomena and data on their structure have been obtained. However, progress has been slow in terms of the solution to the main problem, i.e., the establishment of the neutron drip-line, which reflects extremely great experimental difficulties.

One can speak about reaching the stability line only if a nucleus unstable against neutron emission is detected. Certainly, this condition is not sufficient. Nevertheless, the appearance of a nucleus emitting a neutron when the neutron excess is increasing suggests the onset of a new situation. Presently, only eight such nuclides are known, the lightest of them being the dineutron and the heaviest, $^{13}$Be. We have decided, therefore, to limit this chapter to elements up to boron while considering both neutron-stable nuclei lying near the boundary and neutron-decaying ones; proceeding from modern experimental data, our main aim is to find out what can be established about neutron instability of light nuclei.

The works of Soviet authors are not always known to a broader foreign audience, so we have tried to correct this situation by giving a brief account of these studies where necessary, instead of just referring to them as is generally the case.

# 1. The Neutron Drip-Line

## 1.1. The $N$–$Z$ Diagram

The question of the stability boundaries of atomic nuclei is among the fundamental ones in nuclear physics and has always attracted much attention. By studying the properties of nuclides lying near the nuclear-stability line one can investigate nuclear matter under extreme conditions where new properties are manifested, including new types of radioactive decay, and obtain experimental information for checking various model concepts of nucleus structure, on the basis of which nuclear properties are predicted in various nuclear-mass regions.

The present-day situation is shown in Figure 1. About 300 stable nuclei are known (to be more exact, nuclei whose lifetime is great compared with that of the solar system). These nuclides occupy the so-called stability valley. For light nuclei this valley lies along the $N = Z$ line; for heavy nuclei it corresponds to a neutron excess $(N - Z)/A =$ 0.21–0.23.

At the present time about 1600 radioactive nuclides undergoing $\alpha$ and $\beta$ decay and spontaneous fission with shorter lifetimes have been observed. It follows from general nuclear-matter considerations that at a certain ratio of protons to neutrons, neutron instability occurs in a nucleus. Instability against proton emission has been detected experimentally in the region of the lightest nuclei and for those with $A \simeq$ 100–150. Figure 1 shows the predicted proton-stability line corresponding to a lifetime of $10^{-10}$ s. The proton instability of nuclei is caused, first of all, by increasing Coulomb

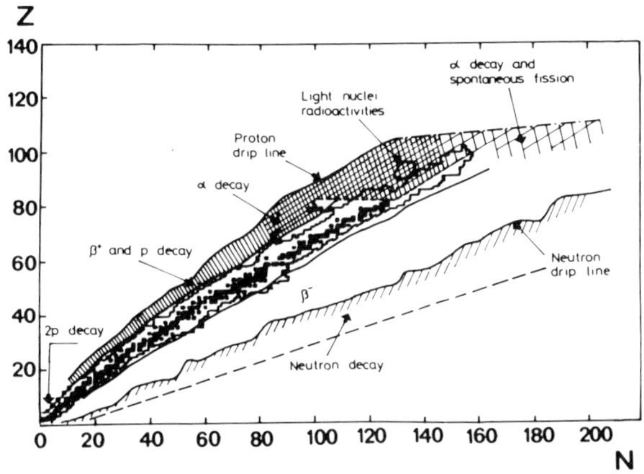


**Figure 1.** Chart of nuclei, based on the scheme given by Flerov (Fl 68). Black squares represent stable nuclei; they are displayed between two lines that separate the domain of studied nuclei from that of as yet undiscovered nuclei. The outer two lines represent the calculated proton and neutron drip-lines, respectively. The region below the neutron drip-line may still contain quasi-stationary states; the dashed line represents the hypothetical border beyond which such states disappear.

forces with increase in the number of protons and the lifetime is determined by penetrability of the Coulomb and centrifugal barriers.

The very fact of the existence of the proton-stability line is not surprising since the Coulomb interaction must, evidently, disintegrate the nucleus sooner or later. Data on the lifetime of proton-decaying nuclei, scarce as they are, show that nothing unexpected takes place here and current calculations are in rather good agreement with experiment. The most interesting thing in studying proton-rich nuclei is that it is sometimes possible to detect quite unexpected correlations in the nucleon motion in this region. The simplest well-known example is the competition between $\alpha$ decay and proton decay. Observation of a comparatively high probability of spontaneous emission of light nuclei from heavy nuclei turned out to be quite unexpected (Ro 84, Al 84b). This phenomenon manifested by heavy nuclei with rather moderate proton excess establishes a connection between such seemingly different features of unstable nuclei as $\alpha$ decay and spontaneous fission (Og 85). Finally, two-proton radioactivity which may compete with single-proton decay has been predicted (Go 60) but has not been discovered as yet.

The instability of nuclei against the emission of neutrons appears as a consequence of the symmetry energy increase with the number of neutrons which, according to the semiempirical formula for nucleus masses (Be 71), is given by

$$E_{\text{sym}} = 28.1(N - Z)^2 A^{-1}(1 - 1.18A^{-1/3}) \tag{1.1}$$

Knowing the nuclear energy $E(A, Z)$, one can obtain the chemical potential of a neutron in the nucleus:

$$\mu_n(A, Z) = \left.\frac{\partial E(A, Z)}{\partial A}\right|_{Z=\text{const}} \tag{1.2}$$

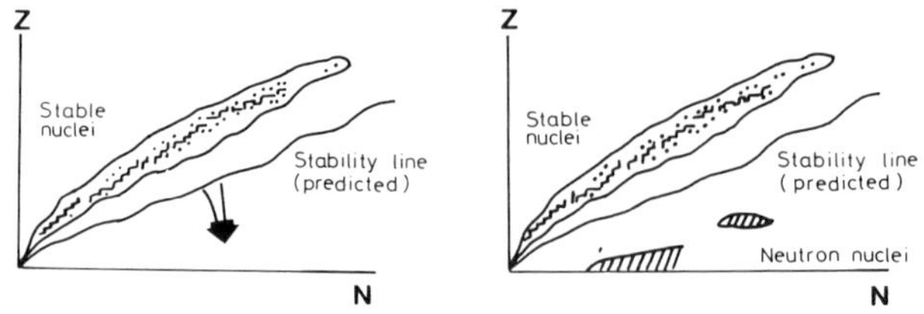


**Figure 2.** Neutron nuclei. The borders of stabiity are obtained as extreme extrapolations, so two possibilities appear: first (left side), this border may turn out to lie sensibly lower or not exist at all; second (right side), some islands may show up beyond the stability border. In both cases nuclei with high neutron excess are expected.

If we proceed from the semiempirical formula, $\mu_n$ becomes zero at a relative neutron excess $(N - Z)/A = 0.36$. Consequently, no stable (against neutron emission) nuclei with large neutron excess should exist and this value of $(N - Z)/A$ then defines the neutron-stability line.

Evidently, this result is qualitative in its character as the semiempirical formula parameters have been obtained by fitting the masses of all known nuclei and stability-line calculations represent a major extrapolation. For example, for isotopes of lead the transition from $(N - Z)/A = 0.21(^{208}Pb)$ *to* $(N - Z)/A = 0.36$ corresponds to the $^{256}$Pb nucleus. The heaviest known isotope of lead is $^{214}$Pb. Thus, it is clear how far the predicted stability line is from the region of investigated nuclei: 42 more neutron-rich isotopes of lead should be discovered in addition to the 6 presently known.

There are a great number of formulas available for the calculation of the mass of nuclei, which take into account shell corrections, etc. [see, for example, (Ne 59)]. We shall not discuss them here, since they all are applicable only near the stability valley. Any calculation of the limits of nuclear existence is an extrapolation of known properties to a region with a very different $N/Z$ ratio. The structure of such nuclei far from stability may turn out to be absolutely different (Bo 69) and, in principle, it is not excluded that there exists a situation in which there is no neutron drip-line at all or in which the existence of a second stability region or of islands with abnormally large $N/Z$ could turn out to be possible (Figure 2).

## 1.2. Physical Grounds for the Possible Existence of Neutron Nuclei

We stress once more that finite nuclear mass calculations, as well as those of nuclear matter, yield negative predictions for the possible existence of nuclei with abnormally large neutron excess. Nevertheless, qualitative considerations can be presented suggesting that this possibility cannot be excluded on a purely theoretical basis.

The analogy between neutron matter and liquid helium is well known. Two neutrons as well as two atoms of helium form no bound system. Their interaction potentials are very similar (Figure 3); they are attractive but not strong enough to form a stable nucleus or molecule, respectively. At the same time a very great number of helium atoms obviously can form a liquid drop.

There arises a question as to whether an analogous situation can occur with neutrons as well. The dineutron is, indeed, unbound—but just by 70 keV. A small

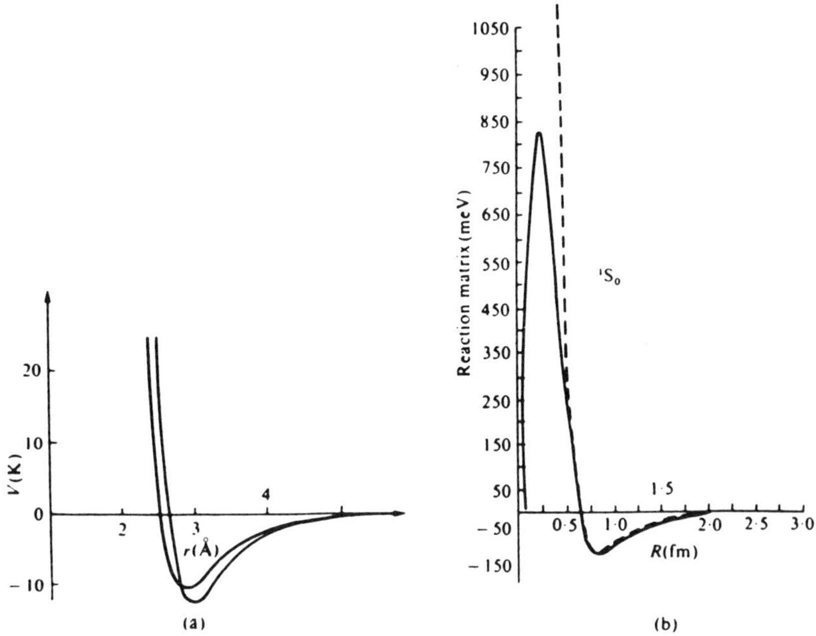


**Figure 3.** Comparison between the $^3$He interatomic potential and the neutron–neutron potential in the singlet state (Ir 75). (a) Two versions of the $^3$He potential. These should be compared with the dashed curve in (b) which represents the neutron–neutron $^1S_0$ interaction. The solid line is the effective Brueckner-Goldstone potential.

deepening of the interaction potential would thus make the dineutron stable. If the required additional attraction can be achieved by adding neutrons to the system, a drop of neutron matter might appear beginning with some critical neutron number.

In the seventies Baz and his collaborators tried to establish this number of neutrons beginning from which one could expect the appearance of a bound system at a microscopic level. The main conclusion was that a neutron nucleus could, indeed, be stabilized by slight variation of the potentials which, practically, did not change the calculated nucleon–nucleon scattering sphases at small energies, if the number of neutrons was no less than some tens.

The basic reasoning was approximately as follows (Ba 72). The main contribution to the binding energy of nuclei comes from central forces between nucleons. These forces depend on the spin-isospin state of the nucleon pair. There are four such potentials $U_{2S+1,2T+1}$: $U_{31}(r)$, $U_{13}(r)$, $U_{33}(r)$, and $U_{11}(r)$. The first two act in even states of relative motion of the nucleon pair and the last two in odd states. For example, for the dineutron ($S = 1$, $T = 0$) the potential $U_{31}$ acts; in the singlet state ($S = 0$, $T = 1$) of two nucleons, $U_{13}$ is the appropriate potential.

In the lightest nuclei ($A \leqslant 4$) all nucleons occupy states with $l = 0$ relative to one another and only the "even" potentials $U_{31}$ and $U_{13}$ contribute to the energy. In nuclei heavier than $^4$He, the Pauli principle prohibits the outer nucleons from being in the $1S$ state and some of the nucleon pairs are necessarily in relative motion states with

$l > 0$. Correspondingly, all four potentials contribute to the binding energy. The probability for a pair of nucleons to be in a nucleus $A$ in states with some definite values of $S$ and $T$ depends on the actual $A$, $Z$, and $N$ values.

It turns out that for all nuclei, except for the lightest ones ($A < 10$), most of the nucleon pairs are in the state with $S = T = 1$. Therefore, the binding energy of nuclei with $A > 10$ should depend strongly on the nucleon–nucleon potential component $U_{33}(r)$. It is important that this component, effective in states with $l \geqslant 1$, is not significant in problems relating to either the deuteron or low-energy scattering.

As an example, note that in work by Baz and Jukov (Ba 71) the nucleon–nucleon potentials were selected on the basis of agreement between experimental and calculated binding energies and radii of the nuclei $^{16}O$ and $^{40}Ca$, as well as the values of singlet and triplet nucleon–nucleon scattering lengths and effective radii. For the potentials $U_{33}$ and $U_{11}$ several sets of values were obtained, each describing the totality of experimental data equally well.

There are only two central potentials effective in a purely neutron nucleus: $U_{13}(r)$, given in Fig. 3, and $U_{33}(r)$.

The neutron-system energy has been calculated by the $K$-harmonic technique (Si 69, Ba 70). To simplify the calculations it was assumed that the radial dependence of the potentials $U_{33}$ and $U_{13}$ had a Gaussian form. It has turned out that if

$$U_{13} + 3U_{33} < 0 \tag{1.3}$$

there exists some $N_{\min}$ above which a multineutron system is predicted to become stable.

For a rough estimate, it is assumed that $U_{13} = U_{33} < 0$. Then for $N_{\min}$ we obtain the condition

$$\frac{ma^2}{n^2}|U_{13}| > \frac{0.13}{\varepsilon(N)} N_{\min}^{-1/6} \tag{1.4}$$

where $\varepsilon(N)$ has a value of the order of 0.1 depending weakly on $N$. This inequality can be compared with the condition of the bound dineutron existence:

$$S \equiv \frac{m}{h^2}|U_{13}|a^2 > 1 \tag{1.5}$$

Thus, even if the value of $U_{13}$ is too small for the dineutron to exist, it can turn out to be sufficient for formation of a bound system of many neutrons. If, for example, we take $S = 0.5$, then $N_{\min} = 64$ is obtained for the minimum number of neutrons forming such a bound system.

Within the framework of the $K$-harmonic technique some exact calcuations have been carried out for five different sets of potentials. The result is as follows. If the number of neutrons is less than 100, none of the sets leads to the existence of a bound multineutron state. At $N > 112$ such a bound state appeared for three out of five of the potential sets. For the remaining two sets the system failed to become stable at any number of neutrons. All these potentials were not arbitrary but were appropriate for calculations of the binding energies and radii of many nuclei, lying along the stability valley.

We have purposely treated these calculations of Baz in considerable detail in order to show that there exists an alternative approach, as compared with the generally accepted one, to the problem of neutron nuclei, and to illustrate the above-mentioned statement that at present there are no final theoretical arguments either in favor of or against the existence of stable neutron nuclei.

There is another principal possibility of obtaining stable neutron systems. It is connected with the existence of a hypothetical phase of nuclear matter, i.e., a pion condensate (Mi 72).

The phenomenon of pion condensation, if it exists, implies that at some critical density of nuclear matter there appears an instability of the vacuum against the spontaneous production of pions.

The energy density of nuclear matter taking this into accont can be written as

$$\varepsilon(\rho) = \varepsilon_0(\rho) - (\gamma/2)(\rho - \rho^*)^2 \tag{1.6}$$

where $\varepsilon_0(\rho)$ is the energy density of the conventional state of nuclear matter with a minimum at the equilibrium nuclear density $\rho_0$ and the second term is the condensation energy ($\rho^*$ is the critical density). The constant $\gamma$ equals the pion rest mass $m_\pi c^2$ in order of magnitude.

For conventional nuclei the effect of pion condensation (if $\rho^* > \rho_0$) on the equation of state would be manifested in a slowing down of the energy increase with the nuclear matter density and for a sufficiently developed condensate, by the appearance of a second minimum (Figure 4). The presence of this second minimum would correspond to the formation of a superdense nuclear state—a metastable or even a stable one.

For neutron matter having no minimum in the curve of $\varepsilon_0(\rho)$, the effect of a pion condensation would manifest itself (bottom part of Figure 4) in the appearance of a maximum in the equation of state $\varepsilon(\rho)$ at the density

$$\rho_{\max} = \rho^* + \frac{1}{\gamma}\left(\frac{d\varepsilon_0}{d\rho}\right)_{\rho_{\max}} \tag{1.7}$$

As has been noted by Migdal (Mi 78), it follows from the very fact that the curve $\varepsilon(\rho)$ has a maximum that it should have a minimum at high densities, since repulsion

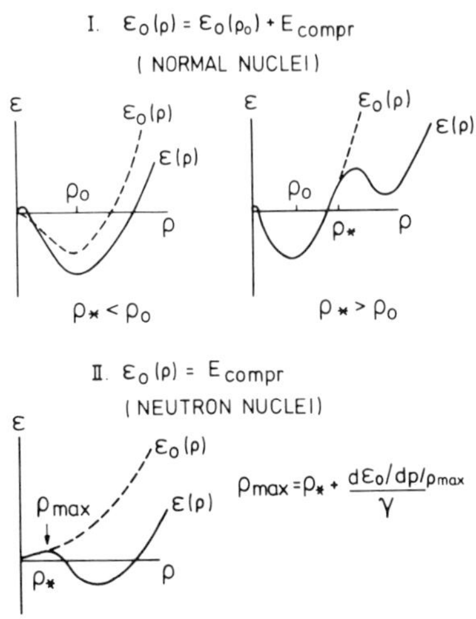


**Figure 4.** Nuclear energy density $\varepsilon$ dependence on nuclear density ($\rho$) for normal nuclei (top) and for neutron nuclei (bottom); $\rho_0$ is the equilibrium density, $\rho^*$ is the critical density corresponding to phase transition to the pion condensate. Dashed curves denote without pion condensate, full curves with pion condensate.

between nucleons at small distances increases with density and nuclear compressibility decreases. It is seen from equation (1.6) that the effect is independent of the value of the critical density $\rho^*$ and the necessary condition for the existence of neutron nuclei (stable, or metastable decaying into neutrons) is just the existence of the pion condensation in neutron matter ($\pi^+\pi^-$ pairs or $\pi^0$ mesons).

Thus, even if there exists no conventional neutron matter, a pion condensation would make it stable.

At present, it is more or less clear that the pion condensation does not develop in conventional finite nuclei. The question of whether a phase transition can occur in nucleus–nucleus collisions involving nuclear matter compression remains open. The existing experimental data do not, however, suggest such a possibility. The expected appearance of a quark-gluon plasma strongly limits the density and temperature range at which the pion condensate could develop. Nevertheless, it is too early to drop the beautiful idea of the possible appearance of neutron nuclei as a consequence of the stabilizing effect of a pion condensation.

## 1.3. Structure of the Neutron–Stability Line

It follows from the above that the existence of neutron nuclei should be considered only as a possibility to be checked experimentally. There are two aspects to be dealt with in this way. First, one should directly advance toward the expected neutron-stability line through synthesis of new nuclei with ever-larger neutron excess until neutron-decaying nuclei are detected. Second, the properties of such nuclei (their dimensions, types of decay, energy spectra, etc.) should be investigated.

Let us try, on the basis of our knowledge of nuclei near the stability valley, to derive some conclusions regarding the properties of nuclei in the neutron drip-line region while naturally taking into consideration the above-mentioned reservations on the accuracy of the extrapolation involved.

It is obvious that nucleon-stable, neutron-rich nuclei will be $\beta^-$ radioactive and it is extremely probable that the $\beta$ decay to excited states of the daughter nucleus will be accompanied by delayed neutron emission. With the increase in neutron excess the binding energy of the last neutron will decrease and its value will become extremely small as the stability line is approached. It is clear that the discovery of any type of binding-energy anomalous behavior with increasing neutron excess would arouse particular interest, since it might serve as an indication of qualitatively new properties.

It is very difficult to predict the structure of neutron-rich nuclei. Owing to their low binding energy, the wave-function tails of the last neutrons should extent to large radius. As a result, the nuclear border will be smeared, while the root-mean-square radius will increase. There are grounds to assume that the neutron radius will be larger than the proton radius although, as shown by comparison of the radii of the calcium isotopes $^{40}Ca$–$^{48}Ca$ (Og 78), this difference can be small.

It seems preferable to describe the structure of such nuclei proceeding from a shell model than, for example, on the basis of a cluster model (for light nuclei). In any case, the excited-state spectroscopy is particularly interesting.

The pair interaction should play an outstanding role near the neutron-stability line. In many cases the pairing energy (2–3 MeV) will turn out to be higher than the binding energy of the last neutron and should result in clearly defined even–odd effects.

The addition of one more neutron to an unstable nucleus with an odd number of neutrons will cause a considerable increase in the stability and might even change the sign of the binding energy. It should be expected that odd (with respect to neutrons) nucleon-unstable nuclei will alternate with even nucleon-stable ones. The boron isotopes provide an example of such a chain: nuclei $^{15}B$–$^{17}B$–$^{19}B$ having 10, 12, and 14 neutrons, respectively, are $\beta$ radioactive and $^{16}B$–$^{18}B$ with 11 and 13 neutrons are unstable against neutron emission.

Thus, the neutron-stability boundary should be presented in an $N$–$Z$ diagram not as a line, but as a band of greater or lesser width. The dimensions of this band in $N$ are determined by the competition between the general decrease of stability with increasing number of neutrons and the stabilizing effect of the pairing forces.

Now, let us briefly consider the expected properties of neutron-decaying nuclei. For them there is no Coulomb barrier ensuring long lifetimes as in the case of proton-rich nuclei and their decay can be inhibited for two reasons: because of the necessity to overcome the centrifugal barrier (for neutrons with $l > 0$) and if, as a result of the decay, the initial configuration of nucleons must change appreciably.

Centrifugal barrier penetrability estimates show that at $l = 6$ and a decay energy, $E = 0.5$ MeV, the lifetime can reach $10^{-16}$ s. The structural hindrances can further increase this time.

Though the lifetimes of neutron-decaying nuclei are considerably smaller than, say, the proton-decaying ones in the region of $A \simeq 100$ and cannot be measured directly, the physical reason for their existence is one and the same, i.e., the passage through the barrier. Therefore, it would be fully appropriate to speak of neutron radioactivity and we shall sometimes use this term later on.

For neutron radioactivity any competition with other types of decay could hardly be expected. In this respect the neutron-rich nuclei differ from neutron-deficient ones for which, as has been noted above, a large variety of decay types is observed.

The decay energy of the neutron-radioactive nuclei should increase with growing neutron excess and the lifetime should decrease gradually reaching the free emission time. Otherwise, if no qualitatively new effects stabilizing nuclei enriched by neutrons are manifested, there would appear gradually the so-called neutron sea of instability, in which there exist no quantum systems.

## 1.4. Light Neutron-Rich Nuclei at the Stability Line

It has already been noted above that with up-to-date facilities for nuclear-physics experiments one may hope to reach the neutron-stability line only for light nuclei. Figure 5 presents a section of the $N$–$Z$ diagram in this region.

Stable, as well as neutron-decaying, nuclides have been discovered for nuclei with $Z \leqslant 4$. The relative neutron excess $(N - Z)/A$ of these nuclei reaches, and exceeds, the above-mentioned characteristic value of 0.36. For these nuclei the alternation of even stable and odd decaying nuclei is observed as expected near the neutron-stability line. As far as some of the lightest nuclei are concerned (such as 4n, $^{5}H$), there are data indicating that they do not exist even in the form of short-lived quantum systems.

For larger $Z$, only $\beta^-$-radioactive nuclei are observed. For boron and carbon isotopes the neutron-even nuclides apparently alternate with the odd neutron-decaying ones, though there is no direct evidence that, say, $^{16}B$ or $^{18}B$ are nucleon-unstable.

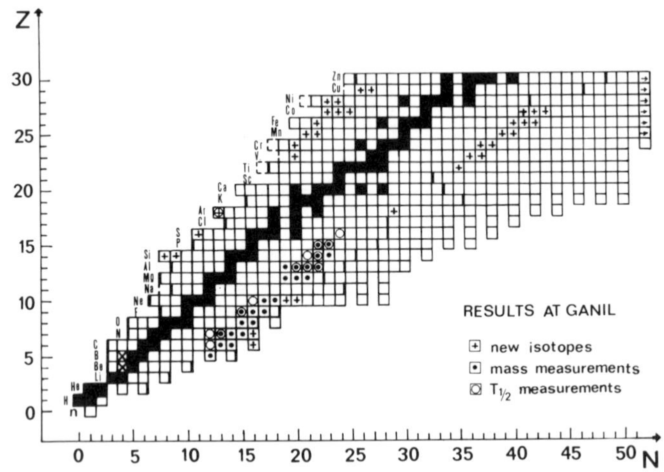


**Figure 5.** Chart of the light nuclei showing the last results obtained at GANIL (De 86). Thick lines indicate the limits of previously known isotopes; crosses, the last new isotopes obtained at GANIL; black dots, isotopes for which the mass is measured; open circles, isotopes for which the half life is measured.

For heavier nuclei the predicted neutron-stability boundary has not been reached experimentally as yet, though recently considerable progress has been made in this field: several tens of new isotopes have been synthesized.

Thus, at present, all conclusions that can be drawn on the position of the neutron-stability line, its structure, and the special features of nuclei with neutron excess $(N - Z)/A \geqslant 0.36$ on the basis of experimental data concern only the group of the lightest nuclides $(Z \leqslant 5)$. Their systematics will be treated in Section 4.

## 2. Nuclear Reactions with Formation of Neutron-Rich Isotopes of the Lightest Elements

In principle, neutron-rich nuclei can be obtained as a result of any nuclear collision, if the number of neutrons and protons, as well as the initial energy and the $Q$ value are sufficient to obtain the desired nuclear system. This accounts for the great diversity of processes used for these purposes and with good results. The reactions for synthesis of neutron-rich nuclei can be divided into two types: first, reactions used to obtain nucleon-stable, i.e. radioactive, nuclei; second, reactions used to obtain nuclei decaying by emission of neutrons and having lifetimes of the order of the characteristic nuclear ones. In the first case the neutron-rich nuclei being formed can be detected in terms of the products of their radioactive decay or directly as particles. In the second case the resulting nuclei are identified by the measured characteristics of so-called associated particles.

In Sections 2.1–2.5 we consider the most popular nuclear reactions used to obtain nucleon-stable nuclei; the reactions for production of neutron-decaying nuclei are treated in Sections 2.6–2.8 and Section 2.9 deals with some hypothetical processes.

## 2.1. Deep-Inelastic Transfer Reactions

The problem of obtaining nuclei in reactions with excitation energies not exceeding the nucleon binding energy in these nuclei is fundamental for the synthesis of nuclei near the nuclear-stability boundary. From this point of view it seems to us that deep-inelastic transfer reactions, the reactions accompanied by emission of fast charged particles, charge exchange reactions induced by heavy ions, and fragmentation reactions at energies $E_{lab} \geqslant 40$ MeV/nucleon are particularly effective.

The first evidence that nucleon-transfer reactions in the collision of two compound nuclei can be not only quasi-elastic but also accompanied by considerable loss of kinetic energy was obtained in Dubna at the end of the 1960s (Vi 67). Further intensive development of theoretical and experimental work on the characteristics of these reactions has allowed a number of important regularities of the deep-inelastic transfer reaction product formation to be obtained, on the basis of which it has been concluded that these reactions are promising for the synthesis of new nuclei—neutron-rich nuclei of light elements included (Vo 71). Oganessian *et al.* (Og 70a, Og 70b) have shown that the distribution of multinucleon transfer reaction product formation cross-sections, at some fixed number of protons, was described with good accuracy by a Gaussian distribution which allowed a conclusion to be drawn on the statistical character of this process. As for the isotopes to be obtained far from the stability line, the ratio between the number of neutrons and protons in the products ($\bar{N}/Z$) is of particular interest. It follows from the published data (Og 70a) that $\bar{N}/Z$ in the reaction products is close to that ratio in the compound system. A slight decrease of the ratio $\bar{N}/Z$, as compared with the compound system, is explained by evaporation of neutrons from the excited reaction products. Karamyan *et al.* (Ka 69) attempted to describe the isotopic distributions of the products of deep-inelastic transfer reactions with the aid of Gaussians whose parameters were determined by the standard rules used for the construction of fission fragment isotope distributions. According to the latter, the dependence of the most probable charge for a given mass, $Z_p(A_f)$, was determined from the condition that the potential-energy minimum for fragments being formed at the moment of fission, and the isotope distribution width calculated on a statistical-model basis, depended on the thermal energy of the compound system and the fission asymmetry. In the case of isotopic distributions of gold from the reactions $^{181}Ta + {}^{136}Xe$ and $^{238}U + {}^{129,136}Xe$, a satisfactory reproduction of the data was obtained in the calculations suggesting a statistical formation of the deep-inelastic transfer reaction products in much the same way as in the nuclear fission case. Volkov and collaborators (Vo 75) proposed describing the dependence of the formation cross-section of various isotopes on neutron number with the air of the so-called $Q_{gg}$ systematics. According to these the isotopic formation cross-section is proportional to the energy which would be spent to obtain the given isotope where the final nuclear products were produced in their ground states. In this case $Q_{gg}$ is determined b the mass difference of the nuclei in the initial and final states $[Q_{gg} = (M_1 + M_2) - (M_3 + M_4)]$. A more exact expression for the isotopic production cross-sections, with account taken of the statistical character of the deep-inelastic transfer process and with corrections for nucleon unpairing (Vo 82), is written in the form

$$\sigma \sim \exp\{[Q_{gg} + \Delta E_c + \Delta E_{rot} - \delta(p) - \delta(n)]/T\} \tag{2.1}$$

where $E_c$ and $E_{rot}$ are the Coulomb and rotational energy variations of the system, $\delta(n)$ and $\delta(p)$ are the corrections for neutron and proton unpairing, and $T$ is the double nuclear system temperature. This dependence satisfactorily reproduced experimental regularities of isotopic production in transfer reactions.

Thus, such characteristics of the transfer reactions as relatively high cross-sections for isotopic production, and broad isotopic distributions of products, make them effective for synthesis of neutron-rich isotopes of light elements. To this end, use is made of nucleon-exchange reactions between bombarding and target nuclei. At the Laboratory of Nuclear Reactions in Dubna researchers succeeded in synthesizing 29 new isotopes of carbon, nitrogen, fluorine, neon, aluminum, silicon, phosphorus, sulfur, and chlorine in transfer reactions with $^{11}$B, $^{15}$N, $_{18}$O, $^{22}$Ne, and $^{40}$Ar ions (Vo 71). By irradiating $^{238}$U with $^{40}$Ar ions at energies of 263 MeV (Au 79) and 340 MeV, 11 new isotopes of elements from silicon to iron were obtained (Gu 80).

Meanwhile, attempts to synthesize neutron-rich isotopes of the lightest elements with a low neutron binding energy [$^{11}$Li ($B_{2n}$ = 0.17 MeV), $^{14}$Be ($B_{2n}$ = 0.2 MeV)] have shown that their formation cross-sections in transfer reactions are considerably smaller than those expected by extrapolation from yields of the rest of the lithium and beryllium isotopes (see Figure 6). Possibly, this reflects weak stability of these nuclei. As a result, their formation cross-section is strongly suppressed since a part of the energy realized in the transfer reaction goes over into the internal excitation energy of the reaction products. It follows from the above that prediction of the weakly bound nuclear formation cross-sections in transfer reactions by means of extrapolation from those for neighboring isotopes is not reliable. In this connection, one may assume that the conclusion regarding the nonexistence of $^{10}$He, made elsewhere (Og 82), from its lack

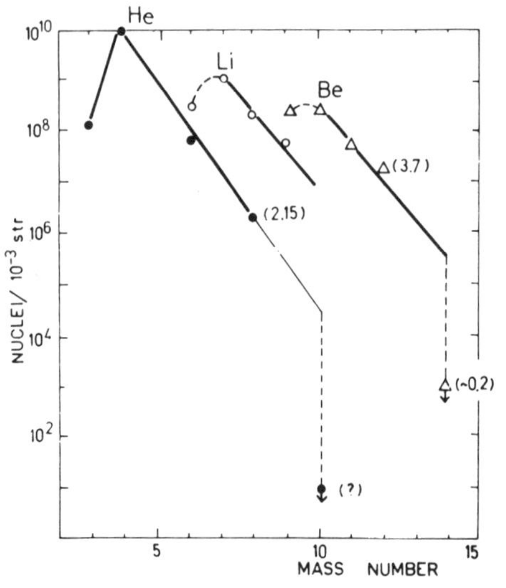


**Figure 6.** Yield of He, Li, and Be isotopes from the reaction $^{232}$Th + $^{11}$B (89 MeV); the neutron binding energy (in MeV) is given in parentheses.

of observation in a transfer reaction with an expected cross-section, is not absolute. To establish the degree of applicability of deep-inelastic transfers to the synthesis of light nuclei with a low binding energy requires further investigation.

## 2.2. Reactions with Heavy Ions Accompanied by Emission of Fast Particles and Nuclei

Reactions with emission of fast particles and nuclei are of great interest for the synthesis of neutron-rich nuclei of the lightest elements. Presently, much attention is being paid to their study stimulated both by the unusual mechanisms of particle formation and by the possibilities opened up by these little-studied processes for obtaining nuclei with various new properties, including exotic nuclei. More than 20 years ago Britt and Quinton (Br 61) observed two components in the $\alpha$-particle distributions from heavy-ion reactions while measuring the energy spectra of protons and $\alpha$ particles. One of the components comes from the evaporation of $\alpha$ particles from the compound nucleus and the other, the high-energy component, has a forward-directed angular distribution and a velocity spectrum of the form $P(V) \sim V^2 \exp(-V^2/V_0^2)$ with a maximum near the bombarding-ion velocity. The $\alpha$-particle formation cross-section, measured at forward-scattering angles in that work, turned out to be considerably greater than the proton formation one, while at backward-scattering angles the relation of the proton and $\alpha$-particle cross-sections could be reproduced by the standard evaporation model. Subsequent experiments with ions heavier than $^{12}C$ have shown that other, heavier charged particles can be emitted in reactions with heavy ions at forward-scattering angles with a relatively high cross-section. Table 1 presents cross-section values for isotopes of elements from H to Be as measured (Bo 82) at an angle of 0° with respect to a projectile beam for the reaction $^{181}Ta + ^{22}Ne$. Also given are calculated cross-section values obtained using a phenomenological "sum rule" model (Wi 80). This model is based on the assumption of transfer of a projectile fragment to the target nucleus and satisfactorily describes experimental cross-sections for light charged particle formation. At the same time, it is not presently possible to reproduce other important characteristics of the process, such as the energy spectra of particles and nuclei or angular distributions, within the framework of any single model (Pe 86). Some experimental results on these characteristics are presented with a view to possibly using reactions of this type for synthesizing exotic nuclei.

Figure 7 presents the energy spectra of hydrogen and helium isotopes measured in the $^{232}Th + ^{22}Ne$ reaction (Bo 84) at 0°. All the energy spectra of light charged

**Table 1. Formation Cross-Sections for Charged Particles from the $^{181}Ta + ^{22}Ne$ (178 MeV) Reaction[a]**

| Cross-section ($10^{-27}$ cm$^2$) | p | d | t | $^3He$ | $^4He$ | $^6He$ | $^8He$ | $^6Li$ | $^7Li$ | $^8Li$ | $^9Li$ | $^9Be$ | $^{10}Be$ |
|---|---|---|---|---|---|---|---|---|---|---|---|---|---|
| $\sigma_{exp}$ | 12.2 | 16.8 | 11.0 | 0.65 | 460 | 1.1 | 0.002 | 1.3 | 3.5 | 0.55 | 0.066 | 9.0 | 6.5 |
| $\sigma_{calc}$ | 22.5 | 3.6 | 3.02 | 0.57 | 270 | 0.8 | 0.0006 | 2.3 | 5.2 | 0.22 | 0.47 | 9.8 | 5.5 |

[a] Experimental data are from the literature (Pe 86). Calculated cross-sections were obtained using a "sum rule" model (Wi 80).

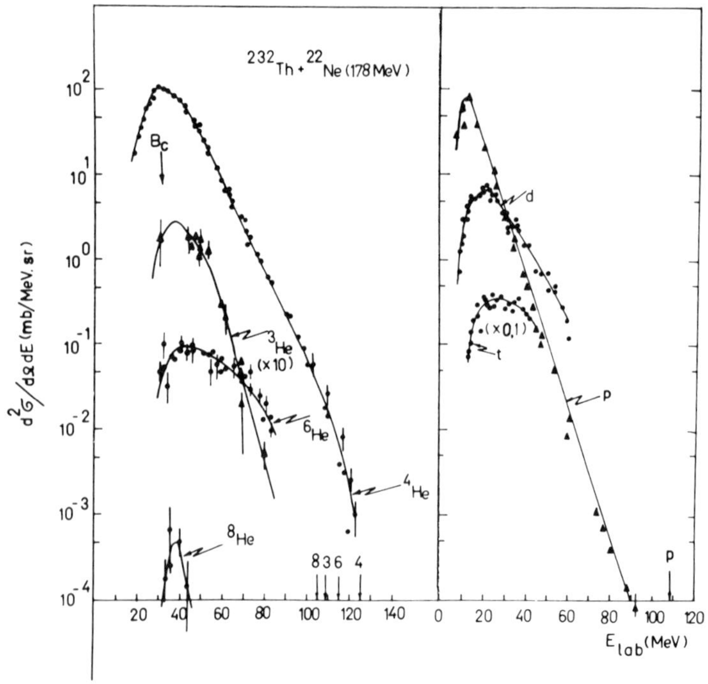


**Figure 7.** Energy spectra of H and He from the reaction $^{232}$Th + $^{22}$Ne (178 MeV) measured at 0° with respect to the beam direction (Bo 84). Full curves are drawn to guide the eye. The arrow in the upper part of the figure is the exit-channel Coulomb barrier *B*. The arrow on the $E_{\rm lab}$ axis indicates the two-body kinematic limit (Bo 84).

particles and nuclei decrease exponentially with energy. If the spectrum slope is characterized by a fitting parameter *T* which, in the case of a statistical process, would have the meaning of a nuclear, effective temperature, a thermodynamic approach can be used to describe the spectra. In accord with this approach, the energy distribution of emitted particles has the form

$$N_x = N_0 E_x \sigma_c(E_x) \exp(-E_x/T) \tag{2.2}$$

where $E_x$ and $\sigma_c$ are emitted particle energy and production cross-section.

Values of parameter *T* were obtained (Bo 84) and it was shown that in all cases those *T*-values exceed twofold, threefold, or more the temperature value calculated for the compound nucleus formed in the same reaction. For a given element the effective temperature increases with growth of the target mass and the emitted particle mass (except for $^8$He nuclei). The energy spectrum exponential approaches a limit determined by the reaction energetics, the so-called kinematic limit of the reaction (in Figure 7 this kinematic limit is marked by an arrow). By studying the energy spectra near the maximum energies possible for given reactions, it was shown that for all investigated reactions with heavy ions (Bo 82, Bo 84) the maximum energy of light nuclear products was only a few MeV lower than the maximum possible ($E_{max}$) which the nucleus can carry in a binary process.

The cross-section for light nuclear emission at forward angles depends strongly on the combination of colliding nuclei. The analysis of absolute cross-sections for formation of high-energy light particles obtained in reactions with various targets allowed a correlation to be established between the cross-section and the binding energy of these particles in the target nucleus (Pe 86). The yield of fast $\alpha$ particles rises with increasing energy release in its separation from the target nucleus.

The reaction product angular distribution is an important characteristic of the process. For reactions with fast particle emission the angular distributions of particles with various energies differ greatly. Experimental investigation of energy spectra of nuclei with $Z = 1$-$4$ over a wide angular range showed that angular distributions for different sections of the energy spectrum behaved differently: for high-energy nuclei they are directed markedly forward, while the soft part of the spectrum only varies slightly with angle. The cross-sections for heavier particles (lithium and beryllium isotopes) drop much more sharply with increasing angle. It has been shown (Bo 84) that in reactions with heavy ions at 10 MeV/nucleon energy the nuclei with $Z > 3$ emitted at angles greater than 40° were not observed within the experimental sensitivity limits ($10^{-33}$ cm$^2$/MeV · sr). This suggests that in these reactions such nuclei are formed only via a direct process.

It follows from the experimental results obtained in heavy-ion reactions accompanied by high-energy charged particle and nuclear emission (from hydrogen isotopes to those of beryllium) after the emission of the particle—preferably in the forward direction—that residual nuclei are formed with excitation energy depending on that carried by the particle. Thus, reactions accompanied by emission of a particle with energy exactly at the kinematic limit result in formation of residual nuclei with zero excitation energy. A slight difference in the particle energy spectrum boundary from the kinematic limit can be determined by residual nucleus angular momentum. This momentum depends on the reaction type and emitted particle, as well as on the bombarding energy. Thus, in the case of Be nuclear emission, their spectrum boundary energy reaches the kinematic limit practically at relatively high cross-sections ($10^{-30}$ cm$^2$/MeV · sr) and such reactions may turn out to be most favorable in obtaining nuclei in their ground state, which is particularly important for the synthesis of weakly bound nuclei lying near the nuclear-stability line, such as $^{10}$He or $^{14}$Be.

The authors of this work (Og 82) carried out experiments searching for $^{10}$He in reactions with fast nucleus emission. To this end, the $^{232}$Th + $^{11}$B reaction was chosen in which the highest yield of helium isotopes was observed at forward-scattering angles. The energy spectra of helium isotopes obtained in that work are given in Figure 8. The assumed cross-section for $^{10}$He nucleus formation, obtained by extrapolation from helium light isotope cross-sections, should be $\sim 5 \times 10^{-30}$ cm$^2$/sr in this reaction. However, the experimentally obtained upper limit for $^{10}$He nuclei formation was $5 \times 10^{-34}$ cm$^2$/sr. Thus, the $^{10}$He yield turned out to be much less than expected by extrapolation from other He light isotope formation cross-sections. This fact apparently indicates the instability of $^{10}$He. However, this statement cannot be considered absolute. One cannot completely exclude the possibility that, because of its weak nucleon stability, the formation cross-section for $^{10}$He is strongly suppressed by virtue of the fact that part of the energy realized in the reaction, no matter how insignificant it might be, can go over into internal excitation of the reaction products. To answer the question of the

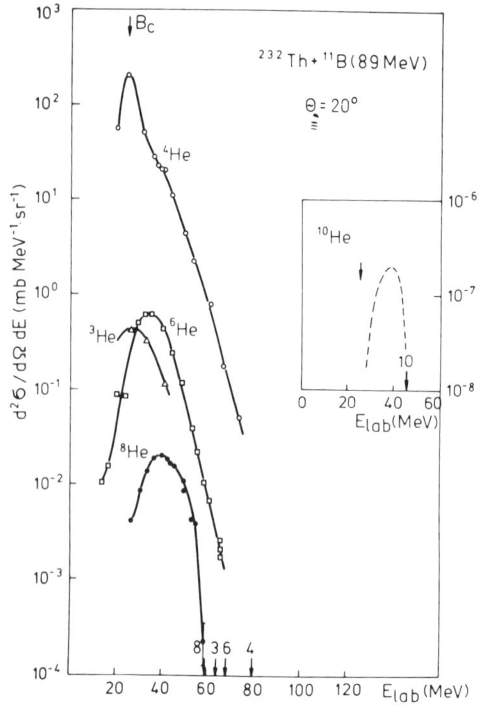


**Figure 8.** Energy spectra of He isotopes from the reaction $^{232}Th + ^{11}B$ (89 MeV) measured at 0° with respect to the beam direction (Og 82). Full curves are drawn to guide the eye. The arrow in the upper part of the figure is the exit-channel Coulomb barrier $B_c$. The arrow on the $E_{lab}$ axis indicates the two-body kinematic limit. The inset shows the presumable $^{10}He$ spectrum from this reaction.

stability of such weakly bound nuclei unequivocally, one should obtain them in their ground states in some nuclear reaction. We hold that charge exchange reactions with heavy ions can be used effectively for this purpose. They will be treated below.

## 2.3. Ternary Nuclear Fission

Spontaneous fission of heavy nuclei may be accompanied, with a certain probability, by emission of light charged particles (so-called ternary fission). In most cases they are emitted perpendicular to the fission axis. By investigating the ternary fission mechanism, it was shown that light particles appeared the moment the nucleus broke into heavy fragments and, with high probability, the light particles could be neutron-rich. Therefore, spontaneous fission of heavy nuclei is used to obtain neutron-rich isotopes of hydrogen, helium, lithium and beryllium (Wh 67) and, recently, even for heavier nuclei (Mi 84). Figure 9 presents experimental yields of hydrogen and helium isotopes measured in the spontaneous fission of $^{252}Cf$ (Wh 67). The ratio of $^4He$ to $^8He$ amounts to ~1000, which is considerably higher than that found in transfer reactions with heavy ions. But the limited quantity of fissile matter used in obtaining these experimentally means that the total yield of the lightest element isotopes is negligible. This yield is considerably higher in the case of induced fission. The ratio of ternary fission (fission into two fragments and the $^4He$ nucleus) to binary fission, observed at low excitation energies, increases with excitation energy from $0.0023 \pm 0.0001$ for thermal neutron fission to $0.0027 \pm 0.0004$ in the fission-inducing reaction, p(42 MeV) + $^{238}U$ (Ra 72).

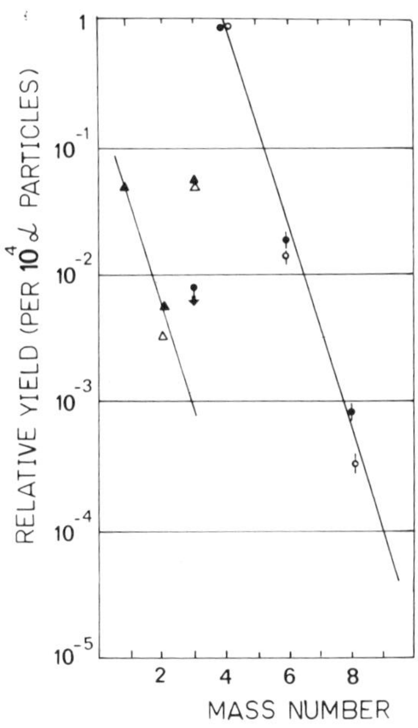


**Figure 9.** Yield of He (circles) and H (triangles) isotopes from spontaneous fission of $^{252}$Cf (full circles and triangles) (Wh 67) and from neutron-induced fission of $^{235}$U (open circles and triangles) (Yo 69).

As shown elsewhere (Du 84, So 85), the probability of ternary fission with the emission of light particles and nuclei increases in reactions with heavy ions, in which considerably higher excitation energy and angular momentum of the fissile compound nucleus are realized. Thus, for the reaction $^{16}$O(144 MeV) + $^{232}$Th the ratio of ternary- to-binary fission is $0.005 \pm 0.002$. The relatively high excitation energy of the compound nucleus is redistributed between the three fission fragments which may result in breakup of any weakly bound isotopes of the lightest elements formed in the ternary fission.

Experiments with $^{252}$Cf may provide examples of the search for neutron-rich nuclei in ternary fission. The fission of $^{252}$Cf was used to search for $^{10}$He and $^{7}$H isotopes emitted as the third fragment. It was shown (Co 67) that no emission of $^{10}$He was observed. However, it was reported (Ba 80) that, in the decay of $^{252}$Cf, several events were observed which could be explained by the emission of the $^{10}$He nuclei. As compared with (Co 67) the $^{10}$He registration threshold was lowered to 14.5 MeV. In this connection the search for ternary fission of $^{252}$Cf with emission of $^{10}$He was carried out (Al 82a) with a still lower threshold.

The measurements were made with a source of $^{252}$Cf weighing 10 $\mu$g; the emitted particles were detected by a $\Delta E-E$ telescope. No emission of $^{10}$He nuclei was observed. The upper limit of the $^{10}$He yield with respect to that of $^{8}$He was $Y(^{10}\text{He})/Y(^{8}\text{He}) \leqslant 4 \times 10^{-3}$, which was one fourth of the value obtained in (Ba 80) (altogether, $6.2 \times 10^4$ nuclei of $^{6}$He and $2.3 \times 10^3$ nuclei of $^{8}$He were detected).

Alexandrov *et al.* (A 182) also carried out a search for nucleon-stable $^7$H. If we assume the binding energies of $^7$H and $^8$He to be close in magnitude and their structure to be similar (4 neutrons in the p shell together with the core), the following estimate of the $^7$H nuclide yield can be obtained on the basis of the known relation between the isotopic yields of helium $Y(^8\text{He})$, $Y(^4\text{He})$ and those of tritons $Y(^3\text{H})$ (Co 67):

$$Y(^7\text{H}) \simeq Y(^3\text{H}) \cdot Y(^8\text{He})/Y(^4\text{He}) = Y(^3\text{H}) \cdot 10^{-3} \qquad (2.3)$$

The experiment gave $Y(^7\text{H})/Y(^3\text{H}) \leqslant 1 \times 10^{-4}$. During the same measurements the emission of $^{12}$He and $^{14}$He nuclei could be observed. Thus, the investigation of spontaneous fission of $^{252}$Cf indicates the nuclear instability of both the $^7$H and $^{10}$He isotopes.

## 2.4. Target Fragmentation Reactions

Target nuclei disintegration is most effective if one uses protons with an energy of a few GeV and targets with $Z > 80$ (lead, uranium). This method is based, first, on the fact that the target nucleus contains sufficiently many neutrons to form a very neutron-rich nucleus and, second, even after the completion of direct interactions the energy left by the incident proton in the residual nucleus is quite enough for a decay with a large negative $Q$ value to take place.

Isotopic distributions of light products of the fragmentation reactions have an asymmetric shape. In the neutron-excess region $N/Z$ decreases almost linearly with $A$, which makes these reactions highly effective for obtaining neutron-rich nuclei. The value of the isotopic distribution maximum ($A_p$) depends on the target mass number. With increase in the $N/Z$ of the target, the $N/Z$ value for the fragmentation reaction products rises, though this ratio increases much slower than does $N/Z$ of the target. The isotopic distribution width also rises with increasing $A$ of the target. Poskanzer *et al.* (Po 71) have shown that the increase of the bombarding proton energy from 3 to 25 GeV changes only weakly the shape of the proton-induced distributions of lithium isotopes formed in uranium target fragmentation. However, the effective cross-section of isotope formation increases sharply with rising energy up to 10 GeV and remains practically constant up to $E_p = 30$ GeV.

Bogatin *et al.* (Bo 80) analyzed data systematics on fragment formation cross-sections versus their neutron excess. Figure 10 presents the fragment formation cross-section ratios $R = \sigma(A_f + 2)/\sigma(A_f)$ which for fragments differ by two neutrons from their average neutron excess $\bar{\eta} = \langle (N_f - Z_f)/A_f \rangle$. All the known data on the fragments from He to Na fall within a narrow band, which demonstrated the striking fall off of the fragment formation cross-sections (from hundreds of mb for the $^6$He and $^7$Li isotopes to fractions of mb for $^{32}$Na and $^{33}$Na). The spread of points (within a factor of 2) is partially connected with the use of experimental data over a wide range of nuclear targets, from Mo to U.

It is clear from Figure 10 that a three-order-of-magnitude decrease in the ratio $R$ is observed in the range $\bar{\eta} = 0$–$0.15$. Within the 0.15–0.3 interval the drop is slowed down (~approximately a factor of 10 decrease). Naturally, the problem of the $R$ behavior for still higher $\bar{\eta}$ is of particular interest. It is possible that for $\bar{\eta} = 0.3$ the decrease stops and, possibly, even some increase may take place. If this is

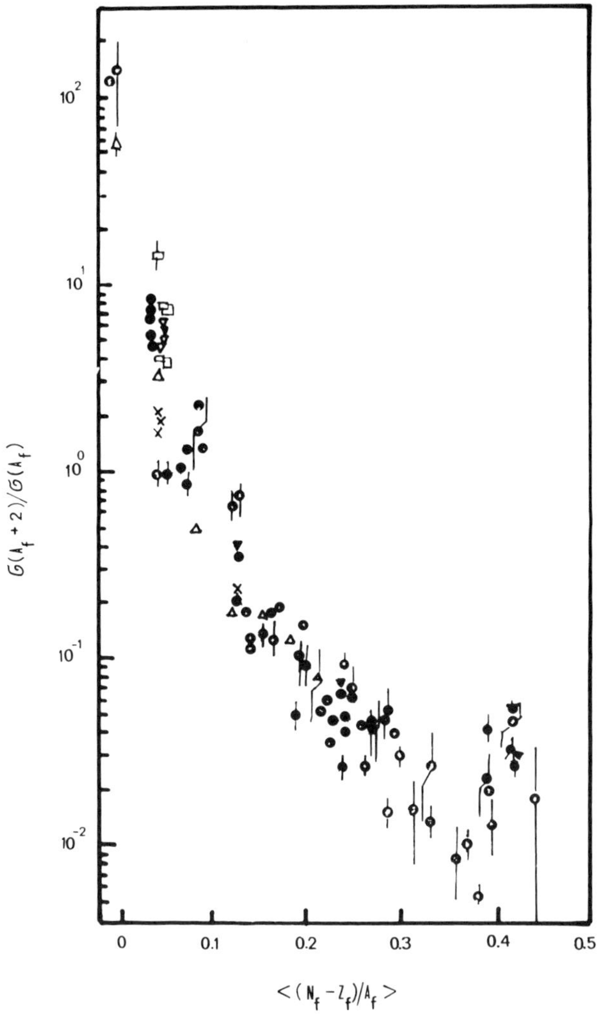


**Figure 10.** Dependence of the cross-section ratio $\sigma(A_f+2)/\sigma(A_f)$ for fragment production on the $\eta$ parameter defined as $\eta = \langle (N_f - Z_f)/A_f \rangle$ in high-energy reactions (Bo 80). Use was made of fragmentation data: $^{112,124}Sn$ + p (6.7 GeV), ●; $^{238}U$ + p (4.8 and 5.5 GeV), ○; Ag + p (5.3 GeV), ×; $^{100}Mo$ + p (24 GeV), △; $^{124}Sn$, $^{238}U$ + d (3.1 GeV), ▼; $^{181}Ta$ + p (10 and 30 GeV), □; Pb + p (10 and 30 GeV), ▽.

so, the probability of detecting nuclei with an extremely high neutron excess rises considerably.

The estimates of the $^8n$, $^{10}He$, $^{13}Li$ formation cross-sections made elsewhere (Lo 83) on the basis of the above-mentioned systematics, and some additional considerations, yield the following values (for a uranium target): $^8n \simeq 10^{-30}$ cm$^2$, $^{10}He \simeq 10^{-29}$ cm$^2$, and $^{13}Li \simeq 10^{-30}$ cm$^2$.

## 2.5. Projectile Fragmentation Reactions

With the appearance of heavy-ion beams possessing an energy of $E >$ 40 MeV/nucleon, bombarding ion fragmentation began to be used for the synthesis of exotic nuclei. Thus, in experiments performed at Berkeley (BEVALAC) with ions of $^{40}$Ar and at an energy of ~200 MeV/nucleon (Ta 85a) it was shown that fragmentation reactions allowed one to approach the stability line in the neutron-rich nuclei region of light elements. In irradiating the target with 0.8 GeV $^{12}$C ions a secondary beam of radioactive nuclei of $^{6}$He, $^{8}$He, and $^{11}$Li was obtained. With their use experiments on the effective radii of these nuclei were carried out (Ta 85b). Many new neutron-rich isotopes of elements in the region $5 \leq Z \leq 27$ were obtained in projectile fragmentation reactions at 40–100 MeV/nucleon intermediate energies at the GANIL accelerator in France.

The mechanism of fragmentation reactions at high and intermediate energies was treated in detail elsewhere (Gr 86). We mention just some properties of these reactions important for the synthesis of exotic nuclei. Products formed in fragmentation reactions have a high dispersion over mass and charge. Comparison of isotopic distributions for light elements obtained in various fragmentation reactions has shown the following: Isotopic distributions at intermediate and high energies for one and the same reaction differ only slightly and their half-width amounts to 8–10 units (Bo 83). For high energies the isotopic distribution width and isotope formation cross-section depend weakly on the bombarding ion and target nucleus (Sy 79). On the other hand, at intermediate energies the ratio $N/Z$ in the products being formed is highly sensitive to the ratio $N/Z$ in the projectile; this makes ions with a higher ratio $N/Z$ more preferable for the synthesis of neutron-rich nuclei of the lightest elements. The calculations of Guerreau (Gu 86) showed that from this point of view, reactions with $^{48}$Ca ions accelerated to intermediate energies may turn out to be highly promising for the synthesis of neutron-rich nuclei of the lightest elements.

An important feature of fragmentation reactions is that at high energies their products have a very narrow angular distribution peaked near 0° and fragment velocities close to that of the projectile (Gr 86). At intermediate energies the half-width of this angular distribution is also relatively small (~5°) but depends somewhat on the mass of the fragments (Bo 83) whose velocities correspond to that of the projectile.

The above-mentioned properties of projectile fragmentation reactions at intermediate and high energies make them a highly effective means of obtaining neutron-rich nuclei of the lightest elements and allow one to use them to obtain secondary beams of radioactive nuclei.

## 2.6. Observation of Neutron-Unstable Nuclei

Neutron-unstable nuclei cannot be detected as particles because they are unable to reach the usual detector owing to their short lifetime. Therefore, independent of the nature of the nuclear reaction in which their synthesis is realized, their observation is reduced to the study of multiple processes in which charged products and neutrons are formed. Such a many-particle reaction can be written in the general form

$$\mathrm{A} + \mathrm{B} \rightarrow \mathrm{C}_1 + \cdots + \mathrm{C}_K + \mathrm{n}_1 + \cdots + \mathrm{n}_i \qquad (2.4)$$

where $\mathrm{C}_K$ denotes the reaction charged products and $\mathrm{n}_i$ are neutrons. The problem to be solved is to identify a metastable state in the subsystem of one or several $C_K$ and $n_i$. The following reactions can serve as examples to be discussed further:

1. $^7Li + {}^7Li \to {}^8B + {}^3H + n + n + n$, a search for the state in the system $^3H + 3n(^6H)$,
2. $^3He + \pi^- \to \pi^+ + n + n + n$, a search for the state in the system $3n(^3n)$,
3. $^7Li + {}^{14}C \to {}^8B + {}^{12}Be + n$, a search for the state in the system $^{12}Be + n(^{13}Be)$.

The problem considered is, in essence, identical to that solved in elementary-particle physics in searching for resonances. The method of multiple-process analysis developed for that purpose is based on a statistical approach assuming that phase-space considerations are very important.

The probability of a transition from the initial state 0 to the final one F is

$$W_F = (2\pi/\hbar)|M|^2\rho_F \tag{2.5}$$

where $M$ is the transition matrix element and $\rho_F$ is the statistical weight (phase volume) of the final state.

The analysis of the phase volume available in the reaction is the first step in trying to explain the nature of many-particle reactions. It is assumed that the phase volume is a kinematical background. Deviations from this background suggest the existence of certain correlations between the particles formed due to their interaction.

The basic equations for the phase space in a many-particle reaction are well known. If a system disintegrates into $N$ particles with momenta lying between $\mathbf{p}_i$ and $(\mathbf{p}_i + d\mathbf{p}_i)$, according to the law of conservation of momentum

$$\sum_{i=1}^{N} \mathbf{p}_i = 0 \tag{2.6}$$

It follows from this conservation law that only $(N - 1)$ values of momentum are independent. In view of the fact that each particle can have only one final state in a phase-space element equal to $(2\pi\hbar)^3$ [at zero spin; if the spin $s$ is other than zero, the number of states is multiplied by $(2s + 1)$], the number of states of the system under consideration per unit energy interval equals

$$\rho_F = (1/dE_0) \prod_{i=1}^{N} d^3p_i/(2\pi\hbar)^3 \tag{2.7}$$

where $E_0 = \sum_{i=1}^{N} E_i$ is the total energy available in this process.

The spectrum of momenta of some particle, such as the first, can be obtained by integrating over the momenta of the other particles:

$$\rho_F(p_1)dp_1 = [4\pi p_1^2 dp_1/(2\pi\hbar)^3]\frac{d}{dE_0}\prod_{i=2}^{N}\int d^3p/(2\pi\hbar)^3 \tag{2.8}$$

If the particles are charged, the outgoing particles should overcome the Coulomb potential barrier. As a result, a strongly energy-dependent penetrability factor appears in the equations and conceals the role played by purely statistical factors.

In the simplest case of three particles in the final state there exists a highly effective method of analysis, allowing one to find out if the observable distributions of particles differ from the predicted ones in the phase volume only.

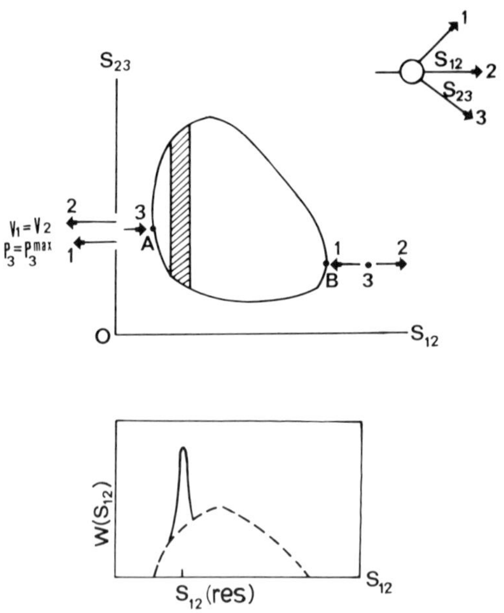


**Figure 11.** The Dalitz diagram for the case of two particles having a resonant interaction in the exit channel (see text).

This method employs Dalitz diagrams and was thoroughly examined (although in a relativistic formulation not very convenient for us) in a number of monographs on kinematics of nuclear processes [see, for example (Ba 68, By 73)]. The Dalitz diagram is defined as a physical region for the three-particle process limited by the conservation laws in the plane of kinetic energies of any pair of particles $E_i$, $E_j$ $(i, j = 1, 2, 3)$. Usually, instead of the energies the squares of the so-called effective masses of any two (of the three possible) pair combinations of secondary particles are measured off along the axes (Figure 11). The effective masses squared are connected linearly with the energies and they characterize the excitation energy of the given pair in the nonrelativistic approximation. Thus, if we consider the distribution in the effective mass $S_{12}$, the minimum value of $S_{12}$ corresponds to the "particle" formation threshold (1.2). In this case the momentum of the third particle reaches the maximum possible value and particles 1 and 2 move in the opposite direction with identical velocities as a unit. On the other hand, the maximum value $S_{12}$ corresponds to the third particle being stationary and particles 1 and 2 moving in opposite directions with equal momenta.

Experimental events are shown in a Dalitz diagram by individual points. Their density is proportional to the reaction matrix element squared:

$$d2N/dE_1\, dE_2 = \text{const}|M(E_1, E_2)|^2 \tag{2.9}$$

If the process probability is determined only by the respective phase volume, the matrix element is constant and

$$d^2N/dE_1\, dE_2 = \text{const} \tag{2.10}$$

Thus, the phase volume predicts a uniform distribution of events in the Dalitz diagram. This property of the diagrams allows us to connect any deviation of the experimental point density from a uniform one with the presence of a residual interaction of particles in the final state. Depending on the interacting pair of particles, there appears a condensation of points in the band parallel to one of the coordinate axes or making an angle of 45° with them. At the interaction of particles 1 and 2 the condensation of points takes place in the band parallel to the $S_{23}$ axis, and in the projection on the $S_{12}$ axis there appears a maximum lying above the statistical distribution (Figure 11).

The analysis of three-particle reactions with the help of Dalitz diagrams assumes the arrangement of an experiment with the detection of coincidences of two out of the three particles formed. The extremely low formation cross-sections of the neutron-decaying nuclei and the impossibility, in most cases, of using large solid angles explain why such measurements cannot be realized, as a rule, at the present time. In practice the energy spectrum of one of the particles formed in the reaction is measured; this corresponds to the projection of the two-dimensional distribution on one of the axes. As applied to the example shown in Figure 11, it means the measurement of the distribution in $E_3$ and, respectively, in the effective mass $S_{12}$, which allows the special features in the interaction of particles 1 and 2 to be detected.

Three situations are possible. First, particles 1 and 2 can form a stable nucleus. In this case the three-particle reaction goes over into a two-particle one and its kinematics becomes completely definite. Particles 3 are grouped into a line with an instrumental width, which is separated from the three-particle reaction threshold, i.e., the decay of $(1, 2)$ $1 + 2$, by the value of the stable nucleus binding energy $(1, 2)$. Thus, using the spectrum of particles 3 one succeeds in establishing the fact of the nucleus $(1, 2)$ existence and also in measuring its mass.

Second, particles 1 and 2 can form a metastable system. In this case the spectrum of particles 3 will have the form shown in Figure 11. A group with width $\Gamma$ corresponds to the excitation of the metastable state. The width, as usual, is related to the lifetime of the nucleus $\tau = \hbar/\Gamma$ and is separated from the breakup threshold by the value of the decay energy. This group is superimposed on a continuous distribution reflecting competing three-particle processes as well as other many-particle channels as they become open. Thus, in this case, two most important characteristics of the metastable state are determined: the decay energy (i.e., the unstable nucleus mass) and the lifetime.

Finally, the spectrum of particles 3 can have no special features with respect to the phase volume. In such cases it is usually concluded that the interaction of particles 1 and 2 is not manifested in the given reaction. Sometimes, even a more categorical statement of nonexistence of nucleus $(1, 2)$ can be made, since there exists a definite relationship between the intensities of the continuous spectrum and the quasi-discrete group.

Let us consider, following Baz *et al.* (Ba 66), the relationship between the yields of the two- and three-particle reactions. All the particles are treated as spinless and having no charge. For the three-particle reaction the following well-known Fermi spectrum (Figure 11) is obtained from equation (2.8) at low energies when the contribution of terms with $l > 0$ is negligible and under the condition that the formation of a metastable particle X is impossible:

$$j(E)\, dE = A_0\sqrt{E(E_0 - E)}(dE/\varepsilon), \qquad \varepsilon = \hbar^2/2\mu R^2 \tag{2.11}$$

where $R$ is the radiation source radius and $A_0$ is an energy-independent constant.

The total yield of the three-particle reaction equals

$$j = \int_0^E j(E)\, dE = A_0(\pi/16\varepsilon) \tag{2.12}$$

If the formation of the unstable particle $X$ with width $\Gamma$ is possible energetically, the yield (at $l = 0$) will be

$$j(E)\, dE = A_0(2/\pi)(\varepsilon/E_0)^{3/2}\frac{2E_0}{\Gamma}\frac{\Gamma^2/4}{(E-E_0)^2+(\Gamma^2/4)}\frac{\sqrt{F(E_0-\varepsilon)}}{2\varepsilon}\, dE \tag{2.13}$$

The appearance of the resonance factor results in a sharp increase of the yield at $E \simeq E_0$. The energy region $\Delta E$, where the reaction involving formation of the metastable nucleus dominates, is given by

$$\Delta E = \sqrt{E_0/2\Gamma}\,\Gamma \tag{2.14}$$

The total yield $j_X$ of particles X at energy $E$ close to $E_0$ will be

$$j_X = \int_{E_0-\Delta E}^{E} j(E)\, dE \simeq A_0\sqrt{\varepsilon/2}[\sqrt{(E-E_0)^2+(\Gamma^2/4)}+(E-E_0)]^{1/2} \tag{2.15}$$

It should be noted that the process taking place is considered only schematically. In practice, other channels often contribute to the continuous spectrum as well. Therefore, the relations between the quasi-discrete group and the continuous spectrum are distorted.

Thus, one can see that the measurement of energy distributions of particles accompanying the formation of neutron-unstable nuclei turns out to be much more informative than the direct detection of the neutron-rich nuclei themselves.

However, it should be mentioned that the replacement of the kinematically complete experiment, as a result of which the Dalitz diagram is obtained, by the measurement of the line spectrum of one of the three particles may lead to an ambiguous interpretation. If particles 2 and 3, and 1 and 3 interact in addition to 1 and 2, the three channels result in one and the same final state:

$$\mathrm{A}+\mathrm{B} \begin{array}{c} \nearrow 1+(2,3) \searrow \\ \rightarrow 2+(1,3) \rightarrow \\ \searrow 3+(1,2) \nearrow \end{array} 1+2+3 \tag{2.16}$$

Two more bands of the point concentration will then be observed in the Dalitz diagram. Being projected onto the axis $S_{12}$ they can yield additional maxima simulating resonances in the system (1, 2).

The situation is even less definite with reactions having more than three particles in the final state when the required nucleus decays with the emission of more than one neutron. At the current experimental state-of-the-art, measurements involved in the search for such nuclei are bound to be inclusive. The kinematical analysis may turn out to be rather complicated. These problems are concrete by their very nature and go beyond the scope of the present work. Some considerations will be given, however, in our description of the experiment on the detection of the $^6$H nucleus.

Thus, the inclusive process

$$\mathrm{A}+\mathrm{B} \rightarrow \mathrm{C}+\mathrm{X} \tag{2.17}$$

is practically the only method of search for, and investigation of, neutron-decaying nuclei at the present time. Here X is the required metastable nucleus.

Serious limitations on the use of this method stem from the requirements imposed upon the nuclei C.

It is clear that the greater the neutron excess of the required nuclei X, the more proton-rich should be the nuclei C. For the nuclei C to be reliably identified against a vast background of other nuclei, neighboring isotopes included, they should be comparatively light. Apparently, in using existing techniques, the limiting value of mass number $A$ is in the region of 20. It is also desirable that the nuclei C have no levels below the decay threshold for the emission of nucleons. Otherwise, these states will manifest themselves in the spectrum, complicating its decoding.

The $^{8}$B, $^{9}$C, $^{13}$O, and $^{14}$O nuclei, having no excited states below the nucleon-emission threshold and, to a lesser degree, the $^{10}$C nucleus, whose nearest level is positioned comparatively high (3.35 MeV), meet these requirements. No excited states have been observed for the $^{17}$Ne and $^{20}$Mg nuclei which may also turn out to be promising for this purpose.

In the next section we consider some of the reactions with emission of the above-mentioned nuclei.

## 2.7. Charge-Exchange Reactions

Charge-exchange reactions with $\pi$ mesons are widely used for the synthesis and measurement of the masses of weakly bound and unstable nuclei. Replacement of one or several neutrons of the nucleus by one or several protons takes place in the charge-exchange reaction. Therefore, in these reactions the nucleus charge changes by one or several units but its mass remains the same. The double charge-exchange reaction induced by $\pi$ mesons on nuclei occupies a particular place among the known nuclear reactions. Its unique property reflects the possibility of obtaining, as a result of the reaction, nuclei, the isospin $z$ component of which differs by two units from that of the initial nuclei. This is possible as a consequence of a double isospin-flip of the $\pi$ meson whose isospin equals unity and which has three charge states: +1, 0, and −1. Experimentally, this process was discovered in Dubna in 1964 (Ba 64). The mechanism of the $\pi$-meson double charge exchange on nuclei was treated in detail elsewhere (Ji 85a). The simplest diagram of the two-step process is given in Figure 12. For the

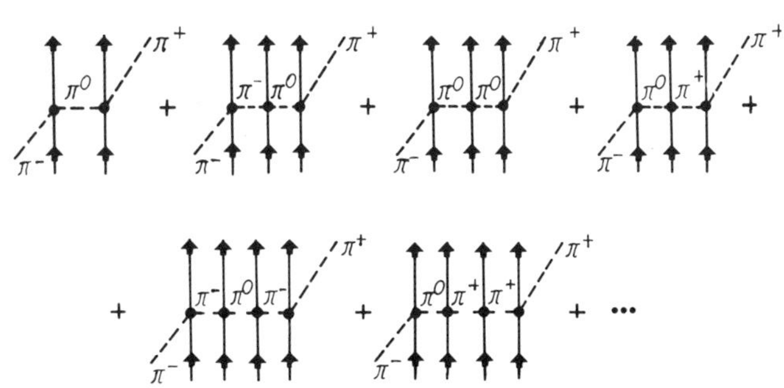

**Figure 12.** Diagrammatic decomposition of the double charge exchange ($\pi^-\pi^+$) reaction amplitude for a two-step process.

**Table 2. Cross-Sections for $(\pi^+\pi^-)$ Double Charge Exchange Reactions**

| Target | $\pi^+$ Kinetic energy (MeV) | Reaction cross-sections (mb) | Reference |
|---|---|---|---|
| $^4$He | 100 | $0.30 \pm 0.15$ | (Fa 74) |
| $^4$He | 145 | $0.34 \pm 0.17$ | (Fa 76) |
| $^4$He | 200 | $0.398 \pm 0.0626$ | (St 81) |
| $^4$He | 486 | $1.20 \pm 0.21$ | (Ca 68) |
| $^4$He | 1586 | $0.24 \pm 0.08$ | (Ga 77) |
| $^4$He | 1725 | $0.14 \pm 0.07$ | (Ga 77) |
| Emulsion | 80 | 0.500 | (Ba 64) |
| Li | 195 | 0.090 | (Gi 64) |
| Be | 195 | 0.010 | (Gi 64) |
| Na | 195 | 0.0014 | (Gi 64) |

synthesis of neutron-rich nuclei the reaction $(\pi^-\pi^+)$ is of particular interest. It was used to investigate systems of three and four neutrons (on $^3$He and $^4$He targets, respectively) of $^7$H, $^9$He, etc. Characteristic cross-sections of the double charge-exchange reactions of $\pi$ mesons usually occupy a micrȯbarn range. Some of them are presented in Table 2.

Recently, charge-exchange reactions with heavy ions have begun to be used, the cross-sections of which may reach values up to $10^{-2}$ mb/sr.

As far as the synthesis of the isotopes of the lightest element is concerned, multiple charge-exchange reactions are particularly interesting. These are charge-exchange processes whose mechanisms is analogous to that of the Gamow–Teller resonance excitation and which can proceed only in the interaction of complex nuclei. First experiments on single or double charge exchange in reactions with heavy ions were made in 1980–1981 by research groups from Darmstadt (Br 81) and Los Alamos (Dr 80). It was shown (Br 81) that at bombarding ion energies $E > 10$ MeV/nucleon the double charge-exchange reactions $^{42}$Ca($^{48}$Ti, $^{48}$Ca)$^{42}$Ti and $^{42}$Ca($^{48}$Ti, $^{48}$Cr)$^{42}$Ar can take place with a relatively high cross-section ($\sim$10 mb/sr) (see Table 3). Moreover, the charge-exchange reaction cross-section grows with increasing energy of the bombarding ions. The angular distribution of the products is directed sharply forward, a characteristic of all direct processes. As shown elsewhere (Na 82, Fi 82), charge-exchange reactions result in formation of final products in their ground states. Therefore, they can be used to obtain weakly bound nuclei in the vicinity of the nuclear-stability boundary. Table 3 represents the basic experimental data on the double charge-exchange reactions obtained thus far in various scientific centers. It follows from the totality of these experimental data that cross-sections of the double charge-exchange reactions may differ greatly for various target–projectile combinations. Figure 8 gives the cross-section of these reactions as a function of the $Q$-reaction value. It is clear that the charge-exchange reaction cross-sections correlate with the $Q$ value of the reaction. The last column of Table 3 presents the experimental data on the mass of the unstable nucleus $^9$He in charge-exchange reactions with accelerated ions of $^{14}$C, as obtained in Dubna.

**Table 3. Cross-Sections for Heavy-Ion-Induced Double Charge-Exchange Reactions**

| Reaction | Bombarding ion energy (MeV) | Registration angle (deg) | Cross-section for ground-state production ($\mu$b/sr) | Reference |
|---|---|---|---|---|
| $^{24}Mg(^{18}O, ^{18}Ne)^{24}Ne$ | 184 | 8 | 0.8 | (Na 82) |
| $^{26}Mg(^{18}O, ^{18}Ne)^{26}Ne$ | 124 | 8 | <0.1 | (Na 82) |
| $^{40}Ca(^{14}C, ^{14}O)^{40}Ar$ | 51 | 10–30 | 10 | (Dr 80) |
| $^{48}Ca(^{18}O, ^{18}C)^{48}Ti$ | 100 | 4–8 | 0.04 | (Na 82) |
| | 112 | 5 | 0.4 | (Fi 82) |
| $^{48}Ca(^{14}C, ^{14}Be)^{48}Ti$ | 87.4 | 4–8 | <0.04 | (Na 82) |
| $^{42}Ca(^{48}Ti, ^{48}Ca)^{42}Ti$ | 226 | 25 | 2 | (Br 81) |
| | 288 | 16 | 4.5 | |
| | 385 | 7 | 60 | |
| | | 8.5 | 25 | |
| | | 11 | 10 | |
| | | 13 | 5 | |
| $^{42}Ca(^{48}Ti, ^{48}Cr)^{42}Ar$ | 226 | 25 | 2 | (Br 81) |
| | 288 | 16 | 4.5 | |
| | 385 | 7 | 400 | |
| $^{9}Be(^{14}C, ^{14}O)^{9}He$ | 160 | 5 | 0.40 | (Be 84) |

These experiments have shown that the double charge-exchange reaction cross-sections with $^{14}C$ ions turn out to be an order of magnitude higher in cross-section than those induced by $\pi$ mesons, i.e., 0.4 $\mu$b/sr. Their further use is suggested for measuring the masses of unstable nuclei. Thus, it was shown (Be 84a) that the double charge-exchange reaction $^{10}Be(^{14}C, ^{14}O)^{10}He$ could be used to measure the $^{10}He$ mass. The cross-section of such a reaction may amount to ~0.1 $\mu$b/sr.

## 2.8. Light Nuclear Rearrangement Reactions

Such a general heading unites a number of reactions induced by intermediate-mass heavy ions (from $^{7}Li$ to $^{18}O$) whose mechanism can be treated, at least formally, as a multistage one. It includes pickup reactions involving three or more nucleons which form a quasi-stationary cluster, etc. The effectiveness of using the ($^{7}Li$, $^{8}B$) reaction has recently been demonstrated. Unstable nuclei such as $^{6}H$ and $^{13}Be$ were synthesized and excited states of $^{8}He$ observed with the help of this reaction (Al 82b, Al 83, Al 84).

The ($^{7}Li$, $^{8}B$) reaction itself corresponds to a rather complicated rearrangement of nucleons in the colliding nuclei, i.e., it can be treated as a pickup of two protons and transfer of one neutron to the target nucleus or as a proton pickup by a $^{7}Li$ nucleus with a simultaneous charge exchange of the neutron with a proton in it. Very low values of $Q$—reaching (35–40) MeV (Table 4)—are characteristic of the ($^{7}Li$, $^{8}B$) reaction on light nuclei.

Figure 13 shows the characteristic form of the spectra observed in these reactions. As expected from semiclassical theory (Br 72) predicting an exponential decrease in

**Table 4. Ground-State Production Cross-Sections for Different Nuclei Formed in the ($^7$Li, $^8$B) Reaction Measured at 10° (Al 85)**

| Target | Residual nucleus | $E_{lab}$ (MeV) | $Q$ (MeV) | Cross-section ($\mu$b/sr) |
|---|---|---|---|---|
| $^7$Li | $^6$H | 82 | −35 | 0.06 |
| $^9$Be | $^8$He | 82 | −28.3 | 0.1 |
| $^{12}$C | $^{11}$Be | 82 | −28.2 | 0.6 |
| $^{14}$C | $^{13}$Be | 82 | −39.8 | |
| $^{16}$O | $^{15}$C | 82 | −22.6 | 0.5 |
| $^{26}$Mg[a] | $^{25}$Ne | 79 | −22.1 | 0.3 |

[a] From (Wi 73).

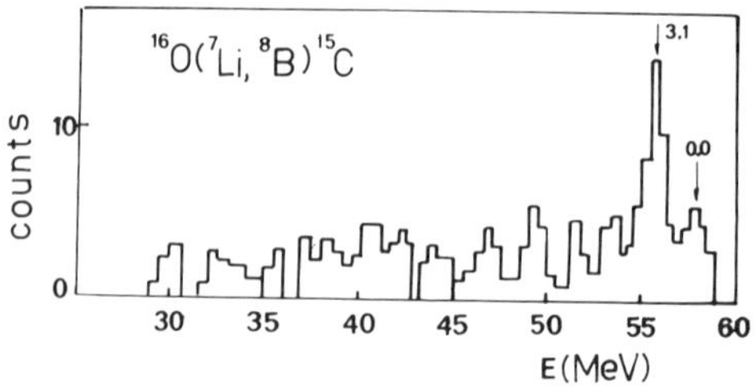


**Figure 13.** Energy spectra of $^8$B nuclei formed in the reaction $^{16}$O ($^7$Li, $^8$B)$^{15}$C at a $^7$Li energy of 82 MeV (registration angle $\theta_{lab}$ = 10°) (Al 85).

the cross-section, transitions are observed with a noticeable probability only to states near the ground states of the residual nuclei (in the given case, $^{15}$C).

Table 4 presents measured cross-sections of the ($^7$Li, $^8$B) reaction for transitions to the ground states of residual nuclei as measured at an angle of 10° in the laboratory frame. These cross-sections are very small as compared with cross-sections of conventional reactions in this region of collision masses and energies (Table 5). The cross-

**Table 5. Cross-Sections for Selected Reactions on Light Nuclei (An 74)**

| Reaction | | Cross-Section ($\mu$b/sr) |
|---|---|---|
| One-nucleon transfer | (n or p) | 1000 |
| Two-nucleon transfer | np | 1000 |
| | nn | 20 |
| | pp | 20 |
| Three-nucleon transfer | 2p1n | 500 |
| | 3n | 3 |
| Four-nucleon transfer | 2p2n | 1000 |
| Five-nucleon transfer | | 20 |
| Charge exchange: | ($^6$Li, $^6$He) | 100[a] |
| | ($^7$Li, $^7$Be) | 800[a] |

[a] From (Gl 84).

sections of other reconstruction reactions were found to be of the same order of magnitude, for example, $^{7}Li(^{7}Li, ^{10}C)4n$ (a pickup of three protons). Nevertheless, such reactions may happen to be practically the only means of obtaining some unstable neutron-rich nuclei. For example, presently there is only one available method of obtaining $^{10}He$ as a residual nucleus, i.e., the reaction

$$^{14}C(^{9}Be, ^{13}O)^{10}He, \qquad Q = -54.6 \text{ MeV}$$

which can be treated as a pickup of four protons.

It is very difficult to estimate the cross-sections of such exotic reactions. Some information on the orders of magnitude can be obtained using the model of Brink (Br 72) based on a quasi-classical approach. Within its framework the cross-section value is determined by the degree of linear and angular momenta nonconservation of the transferred nucleon in the initial and final nuclei. It is predicted that the cross-section should decrease exponentially with a hindrance factor given by

$$H = \Delta K/\Delta K_0 + \Delta L/\Delta L_0 \qquad (2.18)$$

Figure 14 presents a data compilation on the cross-sections of two-, three-, and four-nucleon transfer reactions taken from (De 83a). It is clear that there exists a general tendency but the spread reaches an order of magnitude.

## 2.9. Coulomb Shock Reactions

This review of nuclear reactions resulting in the formation of neutron-rich nuclei suggests that, despite the variety of possibilities, the cross-sections turn out to be rather small. It thus stimulates a search for new mechanisms. In the present section we consider a hypothetical process suggested by one of us (AAO) and called a Coulomb shock (Og 76, Og 79).

The idea is to arrange a peripheral collision of two heavy nuclei, in which a neutron-enriched region appears because the Coulomb repulsion of the protons present. The neutron enrichment will depend on the balance between the Coulomb repulsive

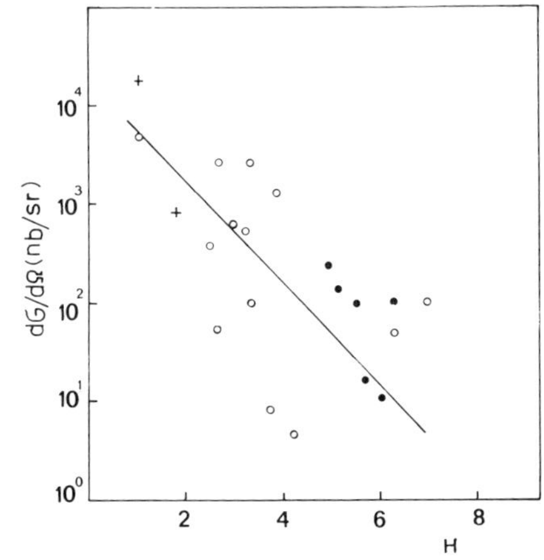


**Figure 14.** Comparison of calculated and experimental cross-sections for two- (+), three- (○), and four- (●) nucleon transfer vs. the hindrance coefficient $H$ (De 83a).

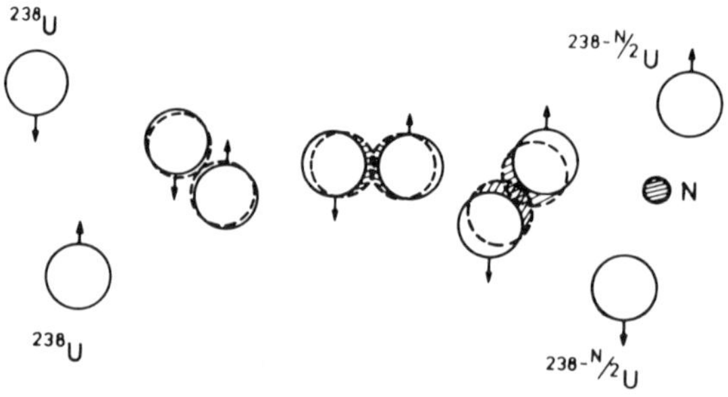


**Figure 15.** A schematic representation of the Coulomb shock in the case of two colliding uranium nuclei.

force and the restoring one determined by the symmetry energy

$$K\frac{(N-Z)^2}{A} \tag{2.19}$$

of the semiempirical mass formula. As a model of such a process it can be assumed that both nuclei consist of proton and neutron liquids which overlap in the ground state and can be separated partially by the mutual Coulomb field of colliding nuclei. Thus, this model is closely related to the hydrodynamic theory of the giant dipole resonance.

It is natural to assume that the proton–neutron separation maximum could be achieved if the relative velocity of the colliding nuclei matched the giant dipole resonance frequency. As only the first half-period of the oscillations has any practical meaning for the separation, this process is called the Coulomb shock. It is shown schematically in Figure 15.

For a rough estimate, let us consider a mechanical model of forced oscillations under the effect of an external force. The forced oscillation equation is

$$\ddot{X} + \omega_0^2 X = \frac{1}{M}F(t) \tag{2.20}$$

where $\omega_0$ is the internal oscillation frequency of the proton liquid with respect to the neutron one, i.e., the giant dipole resonance frequency is

$$\omega_0 \equiv \omega_{\mathrm{GD}} = \frac{2\pi}{72}\left(\frac{100}{A}\right)^{1/3} c/f_{\mathrm{m}} \tag{2.21}$$

$F(t) = Z_1 Z_2 e^2/D(t)$, $D(t)$ being the distance between the centers of the nuclei.

For simplicity, we neglect the oscillation damping, the final dimensions of the colliding nuclei, deviations of the mass coefficient from the reduced mass, and assume that the repulsive force has the form of a sinusoid. The solution of the equation

$$\ddot{X} + \omega_0 X = \frac{F_0}{M}\sin\omega_t \tag{2.22}$$

at $\omega = \omega_0$ is

$$X = \frac{F_0}{M\omega_0^2}(\sin\omega_0 t - \omega_0 t\cos\omega_0 t) \tag{2.23}$$

Quantity $F_0/M\omega_0^2$ is a static displacement caused by the constant force $F_0$.

The maximum proton-to-neutron displacement occurs at

$$\omega_0 t = \pi \tag{2.24}$$

and its value is

$$X_{\max} = \frac{F_0}{M\omega_0^2}\pi \tag{2.25}$$

For uranium nuclei $X_{\max} \simeq 0.8$ fm. Thus, the center of gravity of the protons would be displaced with respect to that of the neutrons by 0.8 fm in each uranium nucleus. Reflecting numerous approximations made in these calculations, this value can be treated only as a very rough estimate.

In any case, one can expect that a neutron layer of thickness about 2 fm is formed on the surface of the nuclei (at $2X_{\max} = 2$ fm, $N \sim 25$). In the separation of the nuclei it can either join one of the nuclei, so forming an isotope with a large neutron excess, or become separated from both the nuclei in a process resembling ternary fission and result in the formation of a light nucleus with an extremely large neutron excess.

The optimal energy for such a process can easily be determined from the condition

$$\tau_0 = \omega_0/\pi \tag{2.26}$$

where $\tau_0$ is the collision time.

For uranium nuclei we have $E/A \simeq 25$ MeV/nucleon. There are various ways in which it might be possible to identify the Coulomb shock mechanism, but we shall not dwell on them here.

## 3. Methods of Measurement and Identification of Neutron-Rich Light Nuclei

In the present section a far from complete review of the methods of neutron-rich nuclear identification is given. The first three parts consider the procedures connected with the detection of nuclei as charged particles: magnetic methods, systems of $\Delta E$-$E$ counters, and time-of-flight measurements. In the fourth part the problems concerning the detection of neutron nuclei are treated.

### 3.1. Mass Spectrometers and Magnetic Analyzers "On-Line" with Heavy-Ion Accelerators

Magnetic systems used to separate and identify isotopes of the lightest elements should be sufficiently rapid, have high effectiveness of nuclear separation, and high mass resolution. The creation of such a system with an ion source for the separation of the isotopes of the lightest elements is a rather complicated technical problem. Therefore, at present only isolated experiments, which use magnetic systems with ion sources, have been reported in which neutron-rich isotopes of the lightest nuclei are separated and their properties studied. The use of a mass spectrometer for alkali elements with an ion source developed at the R. Bernas Laboratory (Orsay, France) was most successful. This spectrometer was used effectively with high-energy proton beams to study reactions in which a number of new isotopes were obtained, including

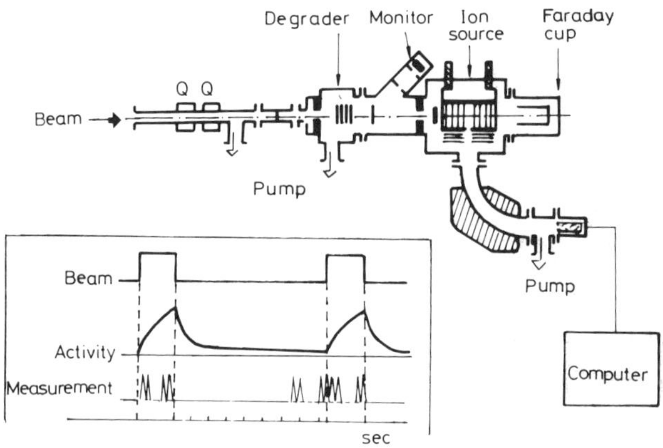


**Figure 16.** The experimental arrangement of an on-line mass separator (Kl 67) for use with heavy-ion beams. In the lower part the timing of the separator is shown. The upper curve shows the timing of the target irradiation, the next shows the variation in the current out of the ion source, and the lower one indicates the measurement timing.

$^{11}$Li (Kl 67). The spectrometer was also employed successfully in a joint Dubna–Orsay experiment using heavy-ion beams. The experimental layout of such a spectrometer is given in Figure 16. Its total efficiency for alkali elements (Li, Rb, Cs, Fr) amounted to 2%. The lifetime range, defined by the atomic diffusion from heated graphite, allowed the measurement of isotopic yields of these elements with lifetimes up to 0.1 s.

The ion source was a thin-walled graphite cylinder electrically heated to 1500 °C. Nuclear reaction products knocked out of the barget were decelerated by a wall of thin graphite foils positioned inside the ion source. The ions selected from the source and accelerated to an energy of 10 keV were mass separated by an electromagnet. The mass resolution $M/\Delta M \approx 500$.

The ions were detected in the magnet focal plane using a secondary electron multiplier. The measurements were carried out as follows. A certain regime of radiation and measurement was selected (Figure 16). By scanning the spectrometer accelerating voltage, two isotopic spectra were collected in a definite mass region. The first was collected during the irradiation and the second immediately upon its completion. The former spectrum reflected the independent yields of the isotopes under investigation. The latter spectrum was collected prior to the beginning of the next cycle of irradiation. It reflected radiation from a slowly diffusing part of the isotopes under investigation, as well as some established contribution of cumulative yields and natural impurities. The difference between these two spectra thus yielded relative independent yields of the alkali elements under investigation.

In spite of the fast response and high mass resolution, application of this spectrometer is restricted to studies of the alkali element isotopes.

One of the most effective methods of determining nuclear masses involves the use of two-particle reactions in which the mass of the investigated nucleus is determined from the reaction $Q$ value corresponding to the formation of two final nuclei in their ground states.

The $Q$ value is obtained by measuring the energy spectrum of the conjugated product of known mass. The presence of peaks in this spectrum suggests the formation of final nuclei in a definite energy state.

Such an approach is very important in the case of determining nucleon stability of unknown nuclei and studying their virtual and quasi-stationary states when the detection of the nuclei themselves is impossible.

Reactions resulting in the formation of exotic nuclei have the following features: a large negative $Q$ value, 30–60 MeV, which determines a corresponding small cross-section and a strong directionality of the reaction products to forward angles where the elastic scattering cross-section is large. These circumstances condition the selection of experimental procedures for studies in this area. On-line magnetic analyzers are typically employed making it possible to eliminate a considerable part of the background and to perform precision measurements of the energy spectra. Such analyzers are used successfully in many laboratories where the properties of the lightest elements are investigated, e.g., the spectrometer LISE at GANIL (Caen, France) (An 86), the double focusing spectrometer at the Institut de Physique Nucleaire (Orsay, France) (Ro 78), the spectrometer EPICS (Los Alamos, USA) (Se 78), the spectrometer QMG/2 (Daresbury, Gr. Britain) (Dr 74), the wide-range magnetic analyzer with stepped poles MSP-144 (Dubna, USSR) (Ba 75), etc.

Without considering the characteristics of each of these analyzers, which are treated in the literature, we shall briefly describe an experimental installation for measuring the energy spectra of nuclei formed in reactions with heavy ions. This installation is based on the use of the magnetic analyzer MSP-144 (Be 85). An outline of this installation is given in Figure 17. Besides the magnetic analyzer the installation comprises a scattering chamber, a collimation system, and a position-sensitive ionization chamber for reaction-product detection.

The parameters of the magnetic analyzer are given in Table 6. The following characteristics of the analyzer are important: focal-plane linearity throughout its whole length (1.5 m), as well as linear dependence of the dispersion and resolution on position along the focal plane. The energy is determined by the expression

$$E = k(Bx)^2 q^2 / A \tag{3.1}$$

where $E$ is the detected nucleus energy (MeV), $B$ the magnetic-field value across the large analyzer gap, $x$ the coordinate of the impact of the nucleus onto the focal plane

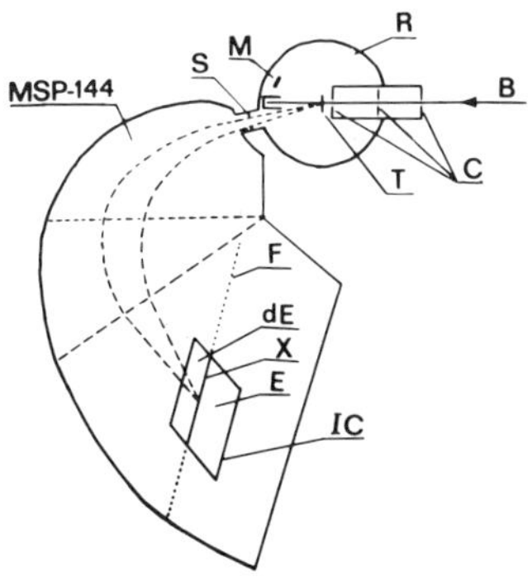


**Figure 17.** Schematic view of the MSP-144 stepped pole magnetic spectrograph (Be 85): C denotes collimators, T target, M detector for beam monitoring, IC position-sensitive (X) double ($\Delta E$–$E$) ionization chamber for detection of reaction products.

**Table 6. Main Parameters of the MSP-144 Magnetic Spectrograph**

| | | |
|---|---|---|
| Energy range | $E_{max}/E_{min}$ | 5.2 |
| Dispersion | $D(M)$ | 1.9 |
| Horizontal magnification | $M_{hor}$ | 0.3 |
| Vertical magnification | $M_{vert}$ | 2.7 |
| Deviation angle | deg | 110.7 |
| $A \cdot E/q^2$ | MeV | 80 |
| Maximal radius | m | 1.25 |
| Solid angle | msr | 5 |
| Energy resolution | $\Delta E/E$ | $5 \times 10^{-4}$ |

(m), $q$ the ionic charge, A the mass (amu), and $k$ a constant coefficient for a given analyzer.

To detect the products, use was made of a gridded ionization chamber. The chamber anode is divided into several sections which detect the energy losses ($\Delta E$) and residual energy ($E$). In the gap between the anode plates there is a proportional counter with a delay line to measure the coordinate ($x$) along the focal plane. The accuracy of the energy determination amounted to ±0.33% and, respectively, the accuracy of determining the $Q$ value and the mass of the light nuclei in the given experiment performed at the U-300 cyclotron of the JINR amounted to 280 keV. This precision was sufficient for a further investigation, using this spectrometer, of resonances in the region of the lightest neutron-rich nuclei and for the determination of the nucleon stability of these nuclei.

## 3.2. Systems of $\Delta E$–$E$ Detectors

The method of simultaneous measurement of total ionization ($E$) and specific ionization losses ($\Delta E$) by nuclei has become widely used to detect isotopes of the lightest elements.

### *3.2.1. Total Ionization Measurement*

The total ionization measurement is realized by using gas-filled ionization chambers, semiconductor detectors, and scintillators. Gas-filled ionization detectors are widely used for these purposes. This follows because in the gas detectors the number of freed electrons is, to a high degree of accuracy, a function of the energy losses in the gas; this holds true for a large range of particle charges and velocities. The linear characteristic of these detectors is determined by the fact that the average energy $W$, required to form an electron–ion pair, is constant. It usually equals 25 eV for gases (in silicon detectors this value is an order of magnitude lower). It has been shown experimentally that $W$ is practically independent of the type and energy of the radiation. Another advantage of the gas detectors is the absence of radiation damage. The effective thickness of the gas detector is easily varied by a replacement of gas or by gas pressure change. A basic virtue of gas detectors is the possibility of detecting reaction products over large solid angles. There are two types of ionization chamber: with the electric field perpendicular to the particle trajectory (Se 81b) and with it parallel to the trajectory (Ki 83). The ionization chambers of the second type ensure better energy resolution in systems having a large-area inlet opening, though they are not widely utilized. The use of an axial ionization chamber gives one more advantage. The anode current $I_A(t)$ is proportional to the total number of electrons distributed along the track for each time $t$. Measurement of the $I_A(t)$ value yields information on the distribution of specific ionization losses along the track. Consequently, they are called Bragg chambers (Ma 84). Measurement of the Bragg curves allows the $Z$ identification of nuclei and simultaneous measurement of their total energy.

Energy resolution for gas-filled detectors is typically $\Delta E/E \simeq 0.6\%$ (Qu 79).

To measure the total ionization produced by long-range particles and nuclei, much use has been made of semiconductor detectors (Go 75). However, their application is limited due to their small size, their amplitude defect (Fano factor), and the radiation damage to which they are susceptible.

The scintillation method is widely used for nuclear reaction spectrometry. This method is based on the detection of short light flashes, i.e., scintillations appearing in some substances after the passage of charged particles or photons through them. The widespread use of scintillators to measure total ionization of charged particles and light nuclei follows primarily from the possibility of making these scintillators of any size. Thus, for example, the dimensions of scintillators based on liquid and solid solutions, as well as on gases, are limited only by light collection requirements. The attainable energy resolution for charged particles and light nuclei in the 10 MeV energy range is 6–10% for inorganic scintillators and about 15% for organic ones. For detectors based on inert gases the energy resolution can amount to a few percent. The good timing characteristics of organic scintillators make them particularly useful in time-of-flight spectrometry.

### *3.2.2. Measurement of Light Nuclear Charges using the Specific Ionization Loss Method*

According to the Bethe–Bloch equation, the specific ionization losses of charged particles are directly proportional to the square of their ion charge and to the particle mass and inversely proportional to their kinetic energy, $dE/dx \sim q^2M/E$. Thus, the product of the energy and specific ionization losses depends on the charge and mass number of the fragment,

$$E\frac{\Delta E}{\Delta x} \sim Aq^2$$

A simultaneous measurement of the energy loss

$$\Delta E = \left(\frac{dE}{dx}\right)\Delta x$$

in the thin detector and the residual energy $E$ in a total ionization detector allows the nuclear charge $Z$ to be determined. To this end, the so-called identification parameter $\mathrm{PI} \sim (\Delta E + E_{\mathrm{res}}) \cdot \Delta E$ (Go 75) is introduced, on the basis of which each pair of pulses is associated with some charge value. For these measurements use is made of scintillators (which are optically connected to photoelectric multipliers), semiconductor $\Delta E \cdot E$

telescopes, as well as gas $\Delta E$ detectors together with various total ionization detectors. The accuracy of the charge determination usually corresponds to $Z/\Delta Z \sim 30$–$50$. There exist other methods of measuring the atomic numbers of light nuclei.

Thus, Clerc *et al.* (Cl 75) describe an experimental device in which the particle charge is determined by detecting secondary electrons emitted during its passage through a thin electron emitter. The number of secondary electrons emitted per passing particle is proportional to the specific ionization losses. The accuracy of the charge determination here is $Z/\Delta Z \simeq 35$. Moreover, this secondary electron emission effect is widely used to obtain time triggers in measuring the mass of light nuclei by time-of-flight methods (see below).

A typical system for light-nucleus detection and identification employing the $(dE/dx) \cdot E$ method is described elsewhere (Be 85), where an ionization chamber with an anode divided into two sections was used. The first section registered the energy losses ($\Delta E$) and the second, the residual energy ($E$). A proportional counter, with a delay line for measuring the spatial coordinate, was positioned in a gap between the two anode plates. The coordinate was then measured by the time difference between the arrival of signals at the delay line ends. The chamber was filled with the vapors of pentane ($C_5H_{12}$) or isobutane ($C_4H_{10}$). In the experiment described, the $\Delta E$ and $E$ resolutions were determined, mainly by the angular spread of ions at the entrance into the ionization chamber, and amounted to 3 and 2%, respectively. The coordinate resolution was 0.7 mm. For a more unequivocal identification of nuclei with respect to $Z$, ionization chambers with three or more $\Delta E$ sections (Sh 75) are typically used. High-energy resolution for both $\Delta E$ and $E$, for light nuclei, is obtained by using telescopes of semiconductor detectors. The number of detectors in a typical telescope may range up to ten. Thus, in work by Gornov *et al.* (Go 84) a telescope consisting of eight semiconductor Si(Au) and Si(Li) detectors was used to detect high-energy light particles. The energy of the particles detected in the case of total absorption is determined by summing the energy loss in all detectors (the "total absorption" case).

The nuclear identification is usually performed with the help of a $\chi^2$ criterion. It was shown (Go 81) that an acceptable result is obtained using a functional of the following type:

$$\chi_n^2 = \sum_{i=1}^{n} \left( \frac{R_i \exp - \Delta R_i}{\sigma(\Delta R_{i\,\exp})} \right)^2 \tag{3.2}$$

where $\Delta R_i$ is the thickness of the $i$th detector, $\Delta R_{i\,\exp}$ is the thickness of the $i$th detector determined by measurement of the specific energy loss for a concrete assumption as to the particle type, $\sigma(\Delta R_{i\,\exp})$ is the respective measurement error, and $n$ is the number of detectors.

Such a system of multilayer semiconductor detectors allows one to identify reliably the light nuclei up to isotopes of beryllium. However, this method of detection using a $\Delta E$–$E$ system does not allow high experimental sensitivity due to loading of the detectors by elastically scattered ions. To eliminate this effect, use is usually made of a magnetic analysis system in combination with $\Delta E$–$E$ detectors (Ar 70b). Such a system makes it possible to measure the nuclear reaction products being formed with the cross-sections down to $10^{-33}$ cm$^2$ (Bo 82) at forward-scattering angles. In this case the identification of the reaction products is made using the known ratios $\Delta E \cdot E \sim mZ^2$ and $m/q^2 \sim (Bx)^2/E_{\text{tot}}$, where $m$ is the mass number, $Z$ the nuclear charge, and $q$ the charge state of the ion after leaving the target. As a result of such a treatment of

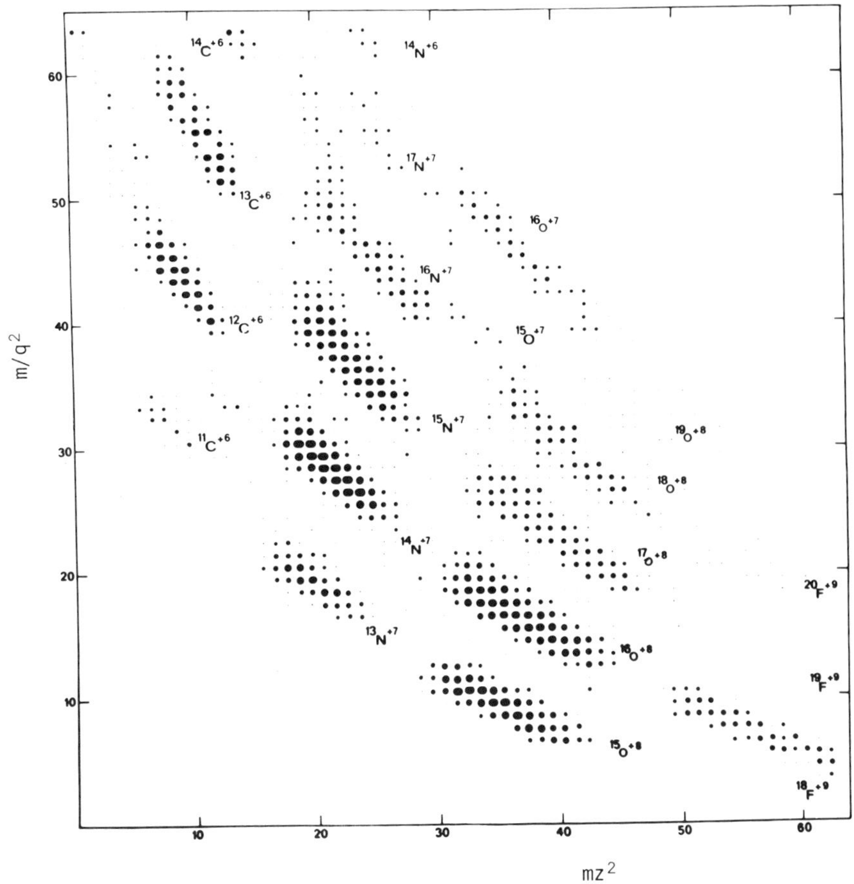


**Figure 18.** Identification matrix for the products of the reaction $^{14}C$ (158 MeV) + $^{9}Be$ recorded by the MSP-144 magnetic spectrograph in conjunction with the position-sensitive ionization chamber (Be 85).

experimental data, so-called identification matrices are obtained, which represent the product yield as functions of the parameters $mZ^2$ and $m/q^2$. Figure 18 shows such an identification matrix. One can see that the events group together to form peaks corresponding to different $m$, $Z$, and $q$ and are spaced sufficiently far from one another to provide a reliable identification of ions with mass $A \leqslant 20$.

## 3.3. Time-of-Flight Measurements

To obtain a more reliable identification of isotopes formed in nuclear reactions, as well as to measure their mass, a time-of-flight method is often used in combination with magnetic analysis and $\Delta E$–$E$ measurements.

In order to identify the products, the time triggers are usually taken from the acceleration RF system (the "start" signal) and from the particle detector (the "stop"

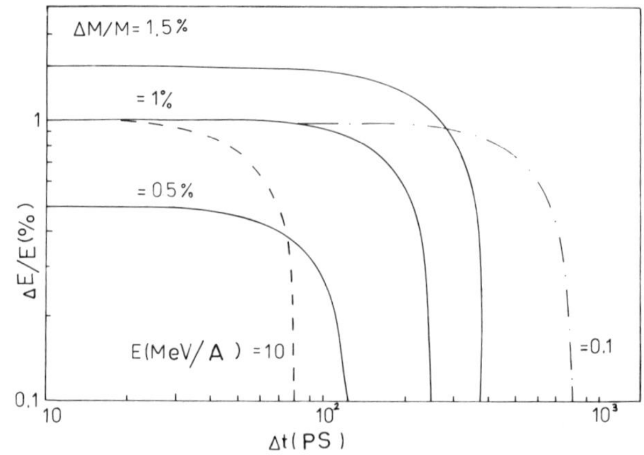


**Figure 19.** Mass resolution vs. time and energy resolution for detection of reaction products with $A = 100$ and 1 MeV/A energy after a time-of-flight path of 1 m. The dotted curve shows this dependence for products with 10 MeV/A energy and the dashed curve for products with 0.1 MeV/A energy.

signal). Such systems typically yield time resolutions of about 1 ns (An 86) which is quite sufficient for product identification. However, higher resolution is required to measure the nuclear mass. The mass resolution obtained in time-of-flight measurements is determined primarily by the energy and time resolution of the system if any error in the path-length measurement is neglected:

$$\Delta M/M = [(\Delta E/E)^2 + 4(\Delta T/T)^2]^{1/2} \tag{3.3}$$

It is clear from Figure 19, presenting mass resolution vs. time and energy, that for particles with energy of a few MeV/nucleon a time resolution of 250 ns is necessary (at a path length of 1 m) to obtain a mass resolution of $\Delta M/M = 1\%$ at $\Delta E/E = 0.5\%$. As an example of the use of a $\Delta E$-$E$-$T$ system, we consider the experimental setup employed by Alexandrov *et al.* (Al 82, Al 83, Al 84) to detect $^6$H and $^{13}$Be nuclei.

The general experimental system is shown in Figure 20. A 87-MeV $^7$Li ion beam, accelerated by the cyclotron, was injected into a scattering chamber, and struck a target

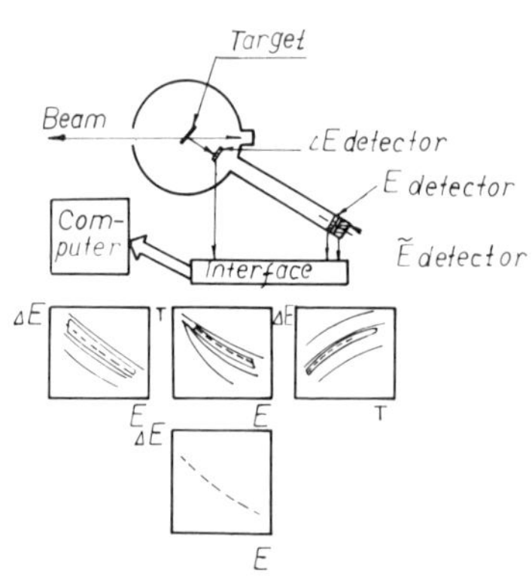


**Figure 20.** Experimental arrangement for recording reaction products by means of a ($\Delta E$-$E$) solid-state telescope and a time-of-flight measurement between the two telescope components. The schematic parameter dependence as obtained from experiment is shown in the squares (Al 82).

made of $^7$Li (in the search for $^6$H) or of $^{14}$C (in the search for $^{13}$Be). The problem was to observe and measure the spectra of $^8$B nuclei whose formation was extremely rare and took place against a vast background of other charged particles. Three parameters were measured simultaneously: the specific energy loss $\Delta E$, total energy $E$, and time of flight $T$ from the $\Delta E$ to the $E$ detector. Moreover, the $E$ counter, set in an anticoincidence mode, prohibited the detection of particles with a mean free path exceeding the maximum for the $^8$B nuclei. To eliminate side effects in the $\Delta E$ and $E$ counter channels, pile-up rejectors were also installed.

The mass separation in the two-dimensional spectra of $\Delta E - E$ was no worse than 8% and the time-of-flight resolution was 0.4–0.5 ns throughout the total energy range of the $^8$B nuclei detected. This allowed reliable separation of the $^8$B nuclei from all others, in particular, from $^{10}$B (Figure 21), the most probable contaminant. Three parameters ($E$, $\Delta E$, $T$) were entered into an "on-line" computer for each detected particle; three two-dimensional spectra $\Delta E - E$, $E - T$, $\Delta E - T$ were formed. Each spectrum covered a region in which the $^8$B nuclei were expected. Using all three programs in coincidence a final energy spectrum was formed. It was shown schematically in Figure 20.

One of the best methods of obtaining the time trigger in measuring velocities is the detection of secondary electrons from a thin foil. The yield of these secondary electrons takes place from a depth of several hundreds of Å, which makes it possible to use very thin films (a few tens of $\mu$g/cm$^2$) as an emitter. To detect the secondary electrons, use is made of microchannel plates (MCP) having a high efficiency for electron detection (~90%), a high amplification coefficient ($10^6$–$10^7$), and good time characteristics (~100–150 ns) (Dm 82a). However, in studying light nuclei formed in nuclear reactions with 5 MeV/nucleon energies, it is practically impossible to use conventional emitters (organic or carbon foils) due to their extremely low (less than 1) secondary emission coefficient. Thus, in detecting $^4$He ions with 7.5 MeV/nucleon

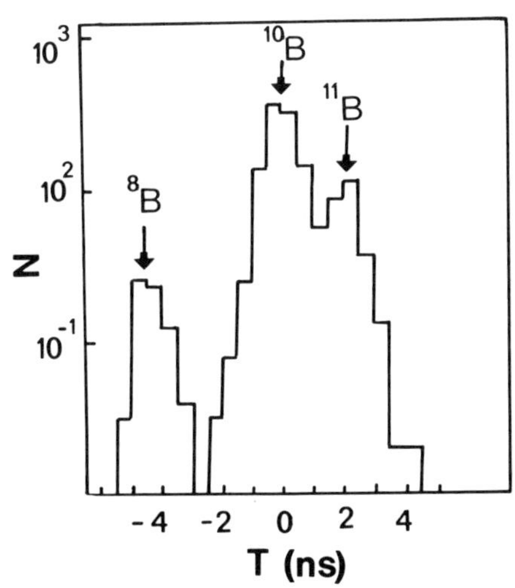


**Figure 21.** Time-of-flight spectrum for boron isotopes originating from the $^6$Li (68 MeV) + $^6$Li reaction (Al 85).

energy the detection efficiency of the emitter–MCP system amounted to only a few percent (Dm 82b).

The most effective and, at the same time, the simplest method of increasing the detection efficiency for light energetic nuclei is the use of emitters having controlled secondary electron emission (CSEE) based on use of porous dielectrics (Ka 84). The CSEE involves electron emission from porous dielectric layers when a strong electric field $E \geqslant 10^4$ V/cm is created in them with the help of a layer of electrons positioned on both sides of the foil. The porous dielectric layers are obtained by vacuum spraying and deposition on a substrate, which is simultaneously one of the electrodes, and are made from salts such as CsI, KCl, LiF, or magnesium oxide (MgO). The investigations of CSEE have shown that the coefficient of secondary electron emission from the porous layer exceeds that of a solid substance by more than $10^2$ for all investigated dielectrics at electric field $E \geqslant 3 \times 10^4$ V/cm. That follows because the electron-emission process takes place throughout the total volume of the porous dielectric with subsequent breeding in the dielectric pores under the effect of the strong electric field. The CSEE mechanism is thus analogous to electron multiplication in channel multipliers and MCP. Figure 22 presents a schematic picture of such a CSEE detector developed in Dubna and used at the GANIL accelerator to measure masses of exotic nuclei (Lu 86). The detector consists of a CSEE emitter with a porous layer of MgO, an acceleration grid, an electrostatic mirror, a chevron MCP, and a flat anode. The accelerated secondary electrons from the emitter are reflected at 90° by the electrostatic mirror and hit the MCP assembly where $10^6$–$10^7$-fold amplification takes place. An electric signal is taken from the flat anode positioned behind the MCP assembly. The first measurements performed with the LISE spectrometer at GANIL have shown that such a detector has unique characteristics (an efficiency of ~70%, a time resolution of ~200 ns) and is promising for use in time-of-flight spectrometers for detecting light nuclei with energies of a few tens of MeV/nucleon.

Meanwhile, in detecting isotopes of the lightest elements for the synthesis of which (as has been noted above) direct reactions with heavy ions at low energies

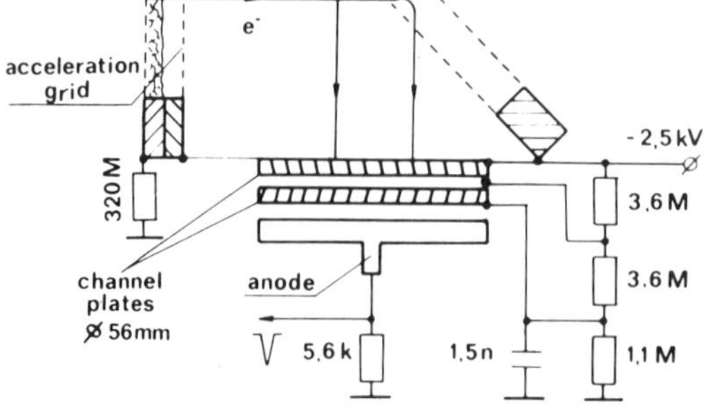


**Figure 22.** Schematic view of the time-zero detector based on microchannel plates and a porous dielectric emitter (Lu 86).

(~10 MeV/nucleon) and fragmentation at intermediate and high energies (~30 MeV/nucleon) are used, the detectors are positioned as close as possible to 0° with respect to the primary beam direction. Naturally, this results in heavy loading of the detecting systems by the scattered beam. In the most sensitive experiments on the synthesis of new isotopes of the lightest elements, combinations of all the above-mentioned methods are used together with magnetic analysis, making it possible to perform a space separation of the reaction products and elastically scattered ions.

### 3.4. Detection of Nuclei Consisting only of Neutrons

The above-described procedures are suitable for the detection and identification of neutron-rich nuclei having non-zero charge, but they are not applicable in the case of the detection of nuclei comprising only neutrons. So far two basic methods have been used in the latter case: the activation one and that based on time-of-flight detection.

The activation method was widely used in the past, when the possible existence of a bound dineutron was not excluded. We consider this method using a recent work (Ag 85) as an example. It reports that an effect was observed which, according to the authors, might be evidence for a nucleon-stable tetraneutron.

A uranium target was bombarded by a 100 MeV beam of $\alpha$ particles with 10 $\mu$A intensity. A copper shield was placed behind the target, with a thickness selected so that any charged particles formed in the target were stopped in the shield.

A sample made of $^{25}$Mg and $^{26}$Mg isotope oxides, with enrichments of 96.9 and 97.2%, respectively, was employed as a detector of neutron nuclei. If such nuclei were produced in the uranium target, they would be identified by the appearance of $^{28}$Mg activity as a result of the reactions $^{25}\mathrm{Mg}(^{x}\mathrm{n}, {}^{(x-3)}\mathrm{n})^{28}\mathrm{Mg}$ and $^{26}\mathrm{Mg}(^{x}\mathrm{n}, {}^{(x-2)}\mathrm{n})^{28}\mathrm{Mg}$. After bombardment of the samples they were subjected to radiochemical separation of $^{24}$Na, formed in (n, p$x$n) reactions on the magnesium isotopes, over a period of 10–15 h.

In the spectra of $\gamma$ rays (Figure 23) measured with a Ge(Li) detector having a 90 cm$^3$ volume, a characteristic line at 1342 keV, accompanying the decay of $^{28}$Mg, was observed as well as a line at 1779 keV from the decay of the daughter nucleus $^{28}$Al:

$$^{28}\mathrm{Mg} \xrightarrow[\beta^-,\,\gamma]{21\,h} {}^{28}\mathrm{Al} \xrightarrow[\beta^-,\,\gamma]{2.7\ \mathrm{min}} {}^{28}\mathrm{Si}$$

The neutron nucleus capture can be simulated by reactions on silicon impurities in the sample: $^{28}\mathrm{Si(n, p)}^{28}\mathrm{Al}$, $^{29}\mathrm{Si(n, 2p)}^{28}\mathrm{Mg}$, $^{30}\mathrm{Si(n, 2pn)}^{28}\mathrm{Mg}$, and reactions induced by secondary particles formed under exposure to fast neutrons:

$$(\mathrm{n}, \alpha) - {}^{26}\mathrm{Mg}(\alpha, 2\mathrm{p})^{28}\mathrm{Mg}$$

$$(\mathrm{n, t}) - {}^{26}\mathrm{Mg(t, p)}^{28}\mathrm{Mg}$$

$$(\mathrm{n}, {}^{6}\mathrm{He}) - {}^{25}\mathrm{Mg}({}^{6}\mathrm{He}, 2\mathrm{p})^{28}\mathrm{Mg}$$

The main problem in experiments of this type is the necessity of convincing test experiments which preclude contamination of the data from such secondary events. For this purpose measurements of the $^{28}$Mg yield were performed (Ag 85) using a quartz target instead of the magnesium one (to estimate the rate of impurities) and also with a magnesium target irradiation at an angle of 65° with respect to the $\alpha$-particle beam (to check for the existence of a correlation between the $^{28}$Mg activity yield and

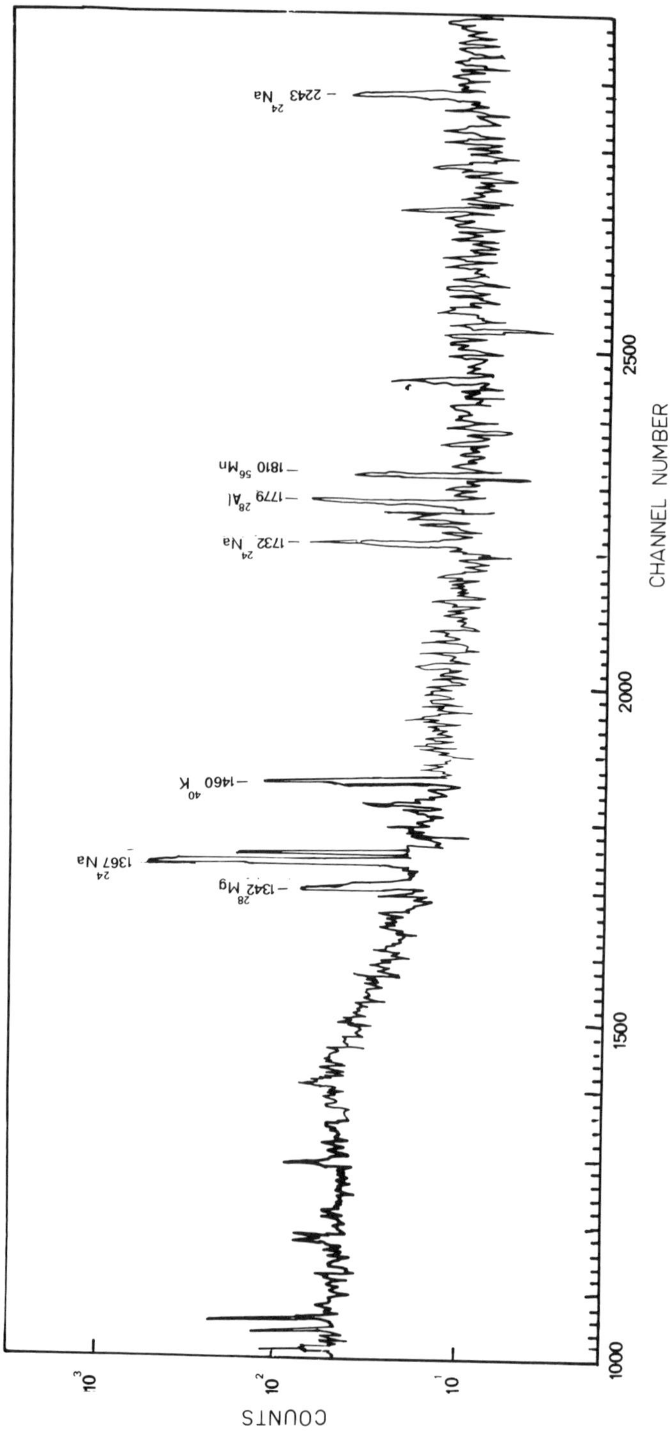


**Figure 23.** Part of the γ-ray energy spectrum from the reaction $^{26}Mg(^{x}n, ^{(x-2)}n)^{28}Mg$ resulting in Mg target activation with multineutrons (Ag 85).

the angular distribution of neutrons). Moreover, calculations of expected yields of secondary reactions were made based on available data on their cross-sections.

Without analyzing the results of these test experiments by Ageyer *et al.* (Ag 85), we would only note that it is very difficult to estimate their conclusiveness and reliability.

A method of neutron nucleus detection, different in principle, was suggested by Brill *et al.* (Br 64). Its essence is as follows. We assume that, as a result of one of the above-discussed nuclear reactions, a neutron nucleus is formed. It is reasonable to suppose that it contains a number of neutrons close to the minimum at which the neutron nucleus can exist, i.e., that the binding energy would be comparatively low. As a result, in interaction with other nuclei a disintegration into $x$ neutrons would be expected as one of the main mechanisms:

$$x_{\mathrm{n}} \rightarrow \mathrm{n}x$$

Thus, the problem of detecting the $x_{\mathrm{n}}$ nucleus is reduced to the detection of a number of neutrons simultaneously (Figure 23).

Brill *et al.* (Br 64) used a telescope comprising two scintillation counters, the first serving as a target in which neutron nuclei caught by it could dissociate. To obtain these neutron nuclei, the $^{48}$Ca + $^{12}$C reaction was used with 72 MeV carbon ions. The formation of $^{6}$n nuclei was energetically possible. The detecting system measured the energy and time-of-flight spectra, which made it possible to select events corresponding to simultaneous detection of several neutrons in the presence of a high background of solitary neutrons and $\gamma$ rays. An example of the resulting distribution of double coincidences is given in Figure 25. The arrows show the spectrum regions where the neutron nuclei ($^{2}$n and $^{4}$n) would be expected. All the observed events corresponded to random coincidences.

Such a measuring system could now be made considerably better than was possible in the quoted work (Br 64). It is essential to note that the number of chance coincidences could be decreased to any desired limit if a greater number of counters were used in the telescope, since this decreases the number of chance coincidences in any given experimental setup.

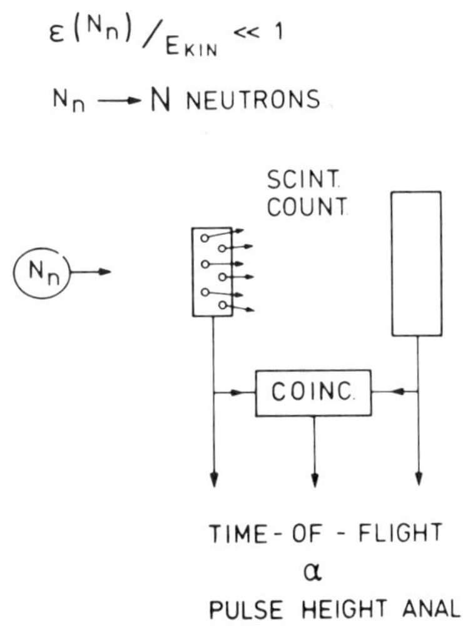


**Figure 24.** Schematic representation of an experimental arrangement for recording neutron nuclei by a time-of-flight technique (Og 79).

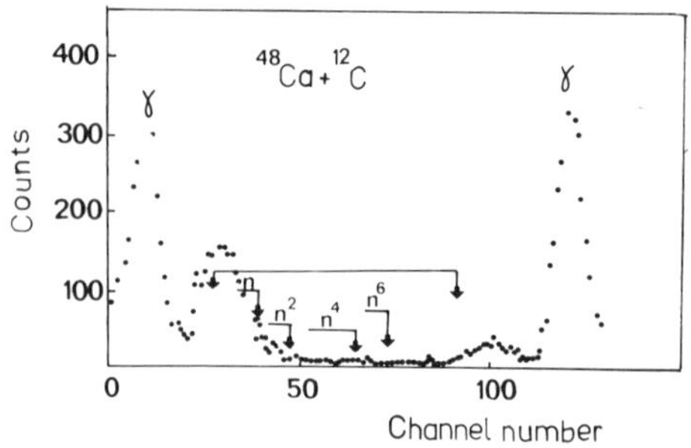


**Figure 25.** The time-of-flight coincidence spectrum from the $^{48}Ca + ^{12}C$ reaction measured with the system shown in Figure 24 (Br 64). The arrows indicate the $^2n$, $^4n$, and $^6n$ detection thresholds.

Naturally, the above method could be employed in combination with a charged particle detection system to observe nucleon-stable nuclei with a large excess of neutrons. For example, a hypothetical $^7H$ nucleus would be expected to dissociate into a triton and four neutrons.

# 4. Systematics of Neutron-Rich Isotopes of the Lightest Elements

## 4.1. The $N$–$Z$ Diagram in the Lightest Nuclear Region

We consider the region of the lightest nuclei. Figures 5 and 26 present relevant sections of the $N$-$Z$ diagram shown in Figure 1. Stable nuclei are found approximately along the $N = Z$ line as would be expected in the absence of Coulomb forces.

With increase in the proton number there appear radioactive nuclei with associated $\beta^+$ decay or K capture. The smallest neutron excess, equal to $\eta = (N - Z)/A = -0.33$, belongs to $^9C$; $^8B$ has $\eta = -0.25$ and for $^{13}O$, $\eta = -0.23$.

With further decrease in neutron excess, nuclei emitting protons from their ground states are observed. Such proton-decaying isotopes are observed for all elements in the above-mentioned section of the $N$-$Z$ diagram. These are $^2He$, $^{4,5}Li$, $^6Be$, $^{7,9}B$, $^8C$, $^{11}N$, $^{12}O$, $^{15,16}F$, and $^{16}Ne$. Analysis of their properties allows one to state that for the lightest nuclei the proton drip-line has been established.

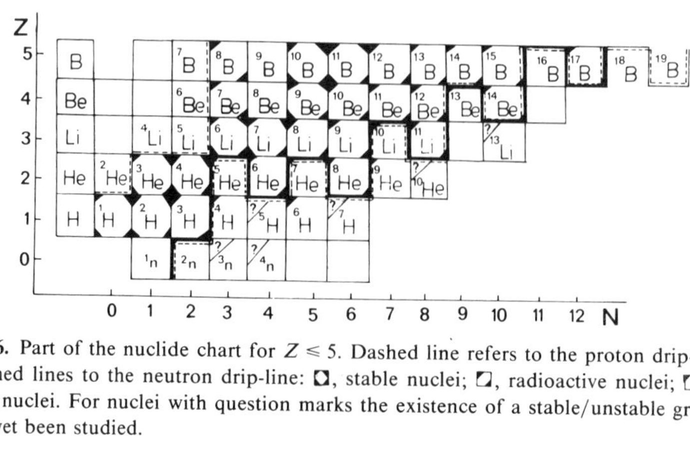

**Figure 26.** Part of the nuclide chart for $Z \leqslant 5$. Dashed line refers to the proton drip-line, thick and dashed lines to the neutron drip-line: ◘, stable nuclei; ◪, radioactive nuclei; ◰, neutron unstable nuclei. For nuclei with question marks the existence of a stable/unstable ground state has not yet been studied.

To the right of the $N = Z$ line, with increasing neutron excess a region of $\beta^-$-radioactive nuclei appears. Among them $^8$He is a record holder in terms of the $\eta$ value: $\eta = 0.50$. There are only four nuclei having $\eta > 0.4$: $^{11}$Li, $^{14}$Be, $^{17}$B, and $^{19}$B.

Finally, eight nuclei have been found up to now that decay by emitting neutrons: $^2$n, $^{4,6}$H, $^{5,7,9}$He, $^{10}$Li, and $^{13}$Be.

Several attempts have been made to synthesize such nuclei as $^5$H, $^7$H, and those consisting of a few neutrons, such as $^3$n and $^4$n. None have been reliably detected even in the form of very short-lived nuclear systems, i.e., resonances. The same is true for the very proton-rich nuclei, such as $^3$Li and $^5$Be. It is possible that the instability sea begins from these nuclear systems.

## 4.2. Types of Nuclear Instability

Since we deal with nucleon-decaying nuclei, we consider the main types of short-lived nuclear systems briefly. There is no unique definition of a metastable state but two are usually considered (Ph 67): (1) a definition on the basis of resonance scattering theory, which assumes that the scattering phase passage through $\pi/2$ signals the existence of a metastable state, and (2) a definition proceeding from the delay-time concepts according to which the metastable state appears in that energy region where the phase shift shows a maximum positive energy derivative (Sm 60). The latter definition is considerably broader, since it includes interactions of the final state. As the neutron-decaying lightest nuclei have widths reaching 1 MeV and more, corresponding to very short lifetimes even on the nuclear scale, such a general approach is possibly more justified. First, we shall consider the minimum permissible times for which it is possible to use the concept of a nucleus at all.

A general expression for the mean time $t$ spent by a particle in a region with radius $R$ during its scattering from a potential well was obtained and analyzed (Ba 66). This time is given by

$$t = (2/v)[d\delta/dK + R - (1/2K)\sin 2(KR + \delta)] \tag{4.1}$$

where $v$ and $K$ are the velocity and wave number of the relative motion of the particle in the potential region, and $\delta$ is the scattering phase.

For a particle with angular momentum $l = 0$

$$t = (2/v)[R - (1/2K)\sin 2KR] \tag{4.2}$$

If $KR \gg 1$, i.e., the wavelength $\lambda$ of the particle is much less than the dimension of the scattering region the second term can be neglected as compared with the first and the classical result is obtained, i.e.,

$$t = T_{\text{cl}} = 2R/v \tag{4.3}$$

This delay time $t$ simply equals the time of flight of the particle through the sphere with a trajectory going through its center. The $T_{\text{cl}}$ value is usually made equal to a period of the radial motion of the particle with orbital momentum $l = 0$ in a well of radius $R$ (Bo 69):

$$\tau_{e=0} = \frac{2\mu R}{\hbar K_0} = T_{\text{cl}} \tag{4.4}$$

The corresponding decay width is

$$\Gamma = \hbar P/\tau \tag{4.5}$$

where $\tau$ is a period of the particle radial motion in the potential well and $P$ is the barrier penetrability. For $l = 0$, in a square well model, we have

$$P_{l=0} = 4KK_0/(K + K_0)^2 \tag{4.6}$$

where $K_0$ is the wave number inside the well and $K$ outside it. Substituting into equation (4.5) the expressions for $\tau_{l=0}$ and $P_{l=0}$ at $K \ll K_0$ we obtain

$$\Gamma_s(l = 0) = 2\hbar^2\mu R/\mu R^2 \tag{4.7}$$

Since $\Gamma_s(l = 0)$ is the greatest possible model value of the decay width, it is natural to take this value as a conditional upper limit of the nuclear-state width. It is senseless to speak about the existence of a nuclear system when this limit is exceeded. Thus,

$$\Gamma_{\text{lim}} = \frac{83.6}{\mu R^2} KR \text{ (MeV)} \tag{4.8}$$

or, having expressed $K$ in terms of the decay energy $\varepsilon$;

$$\Gamma_{\text{lim}} = 18.31 \frac{\sqrt{\varepsilon}}{R\sqrt{\mu}} \text{ (MeV)} \tag{4.9}$$

The corresponding lower limit for the lifetime of the nuclear system is

$$T_{\text{lim}} = \frac{\hbar}{\Gamma_{\text{lim}}} = 0.36 \times 10^{-22} \frac{R\sqrt{\mu}}{\sqrt{\varepsilon}} \text{ s} \tag{4.10}$$

In equations (4.10), (4.11), and (4.12) the radius $R$ is taken in units of $10^{-13}$ cm, and the decay energy $\varepsilon$ in MeV.

Naturally, these limiting values of $\Gamma$ and $T$ are approximate, since they are obtained within the framework of a particular model. Moreover, one should bear in mind that at decay times comparable with the reaction time the characteristics of the state under consideration can be distorted by the presence of other particles. This problem is rather complicated and we shall touch upon it only in connection with the case of two-nucleon decay (Section 5.7). We note that presently there is a tendency to treat any enhancement over the phase volume as a manifestation of a metastable state—attributing the observed width to this state without analyzing its origin.

Nevertheless, one should realize that such an approach can result in mistakes. Therefore, it is useful to consider possible alternate sources of delay in particle scattering. For this classification it is convenient to make use of notions connected with the theory of final-state interactions (FSI) and the analytic properties of the scattering matrix.

The theory of FSI can be applied to reactions in which the potential acting between the particles can be divided into two parts: $V = V_R + V_I$. The reaction is caused by the potential $V_R$ (the primary potential) while the potential $V_I$ represents an interaction which somewhat changes the reaction caused by the potential $V_R$. In particular, in compliance with Section 2.6, in the case of a system which decays into several particles only two of which are interacting, $V_R$ corresponds to the phase volume and $V_I$ to the interaction between this interacting pair.

Using formal language [see numerous books on the theory of scattering, for example (Ta 72, Si 71)], the $T$-matrix element for the transition under consideration can be written as

$$t(p \leftarrow K) = t_0(p \leftarrow K)/f(p) \tag{4.11}$$

where $t_0(p \leftarrow k)$ is used to denote the matrix element for the same transition in the total absence of interaction in the final state and $f(p)$ is the so-called Jost function corresponding to a true interaction in the final state. An increase (or decrease) in the reaction amplitude results from the interaction in the final state, which corresponds to a change of probability of the particle presence in the region where the primary interaction is manifested. It is possible to show that for an attractive field $|f(p) \leqslant 1$, while in the case of a repulsion $|f(p)| \geqslant 1$.

Thus, the FSI of the attracting particles serves as an amplification factor for the process and increases the time the particles spend near one another.

The Jost function is expressed in terms of the scattering matrix and their analytical properties are connected with one another. In particular, the zeros of the Jost function correspond to the poles of the scattering matrix. It is obvious that, according to equation (4.11), one should expect, at these points, a strong increase in the cross-section.

On a plane of the complex linear momentum $p$ the zeros of the Jost function, lying on the imaginary axis in the upper half-plane (Figure 27), correspond to the bound states. The zeros in the lower half-plane (at the points designated by $p$) correspond to the resonances. At physical values of the variable $p$ the phase shift equals the phase of the $f(p)$ function with a negative sign. Its resonance part

$$\delta_{\text{res}}(p) = -\arg(p - \bar{p}) \tag{4.12}$$

equals the angle shown in Figure 27. One can see that when the variable $p$ passes (while increasing) a point on the real axis, which corresponds to the zero of the Jost function, the phase $\delta_{\text{res}}(p)$ passes through $\pi/2$ and the respective partial cross-section

$$\sigma_l(p) = \frac{4\pi(2l+1)}{p^2} \sin^2 \delta_l(p) \tag{4.13}$$

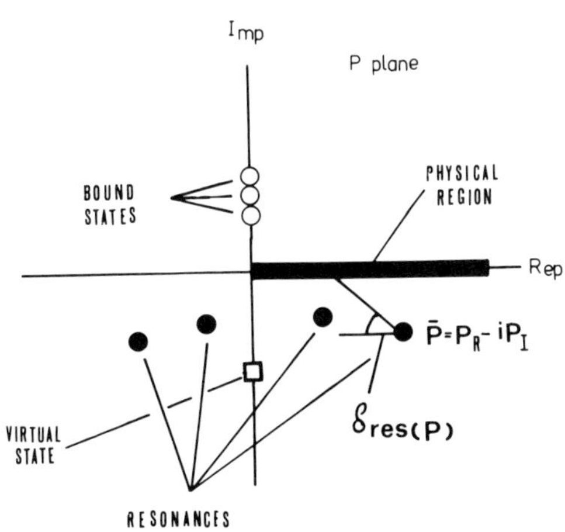


**Figure 27.** The complex momentum plane. Open circles represent zeros of the Jost function (poles of the $S$ matrix) that correspond to bound states; the square represents a virtual state, the full circles represent resonant states. The resonant phase shift $\delta_{\text{res}}(P)$ is shown for the point $P$ corresponding to the resonant state.

reaches a maximum. The passage of the phase of the $l$th partial wave through $\pi/2$ is the criterion for the occurrence of resonance.

If the Jost function is considered as a function of $E$, the zero at point $\bar{p}$, which corresponds to the resonance, becomes a zero in a nonphysical sheet at the point $E = E_R - i\Gamma/2$ (where $\Gamma/2$ is traditionally used to denote an imaginary part of the $\bar{E}$ value) and the resonance phase then has the form

$$\sin \delta_{\rm res}(E) = \frac{\Gamma/2}{[(E - E_R)^2 + (\Gamma/2)^2]^{1/2}} \tag{4.14}$$

The resonance cross-section is described by a Breit–Wigner formula and, as the energy function, has a maximum at the point $E_R$ with width $\Gamma$ which corresponds to a system lifetime $\tau = \hbar/\Gamma$.

It is clear from Figure 27 that the closer the Jost-function zero is to the real axis, the more distinct the resonance peak and the smaller its width. A considerable departure from the real axis will result in a gradual disappearance of any observable effects, which corresponds to the introduction of a limiting decay width.

The main physical sources of the width for a nuclear system are as follows:

1. The presence of a barrier, either Coulomb or centrifugal.
2. The action of various selection rules.
3. The structural hindrances connected with the necessity to reconstruct the nucleon configurations to realize the decay.

The Jost-function zeros positioned on the imaginary axis $p$ in the lower half-plane correspond to so-called virtual states. They are, in fact, resonances with $l = 0$. Sometimes they are called antibound states as they are analogs of the bound states with $l = 0$ in the attractive field which is somewhat weaker than would be necessary for the appearance of a discrete level. If the attractive field in which the discrete level exists is decreased, the Jost-function zero moves downward along the imaginary axis. Two situations are possible in crossing the real axis. If the discrete state has angular momenta $l > 0$, the Jost-function zero will split and leave the imaginary axis having taken a symmetrical position in the third and fourth quadrants. That will correspond to the true resonance. At $l = 0$, the zero of the Jost function is simple and it remains on the imaginary axis, corresponding to the virtual state.

In the presence of the virtual state the scattering phase $\delta$ is defined as

$$\operatorname{ctg} \delta = -1/aK \tag{4.15}$$

where $a$ is the scattering length ($a < 0$). At $kR \ll 1$ and $R \ll a$ it follows from equation (4.1) that the virtual state lifetime is

$$T_{\rm virt} = \frac{2R}{v} \frac{2a^2K^2}{1 + a^2K^2} \tag{4.16}$$

The time delay of the particles near one another can be rather long.

It should be noted that the virtual state is not the only type of $S$ resonance. In the case of a very deep well there are low-energy $S$-wave resonances whose width is determined by expression (4.7). The zeros in the fourth quadrant (Figure 27) correspond to such resonances.

Thus, one can see that final-state interactions in general and involving different types of short-lived states are characterized by different singularities and, in principle, it is possible to choose between them. However, for this purpose a comparison of the experimental data with the calculations using a specific type of interaction, which is, as a rule, unknown, is required. In some cases involving low-nucleon number systems one succeeds with a certain accuracy. Then a conclusion can be drawn that, for example, it is the FSI and not the resonance (Ji 85) that taks place in the system of three neutrons. However, in spite of such a possibility, the primary treatment of the experimental data is usually reduced to the extraction of effective resonance parameters for the maximum observed over the phase volume.

To determine the decay widths, use is usually made of the $R$-matrix theory whose detailed description can be found elsewhere.

We will include only a summary of the equations required to calculate the decay width with neutron emission.

The partial decay width for channel $\alpha$ with orbital momentum $l$ and spin $S$ has the form

$$\Gamma_{\alpha Sl} = \frac{2P_{\alpha l}(\rho)\gamma^2_{\alpha Sl}}{1+\sum_{Sl}\gamma^2_{\alpha Sl}\, dS_{dl}/d\rho}, \qquad \rho = KR \tag{4.17}$$

where $P_{\alpha l}$ is the barrier penetrability, $\gamma^2_{Sl}$ a reduced width, and $S_{\alpha l}$ the shift function. Expressions for $P_l(\rho)$ and $S_l(\rho)$ are given in Table 7 for some values of the momenta in the case of neutron emission.

## 4.3. Prediction of the Masses of the Lightest Nuclei

Prior to considering specific lightest neutron-rich nuclei, we shall very briefly and purely qualitatively discuss the way of predicting their stability or instability, i.e., the methods of calculating their masses.

We shall not discuss approaches based on macroscopic models of different kinds in view of their poor applicability to the lightest systems. To calculate the masses of light nuclei, shell calculations and considerations based to some extent on the isotopic invariance of nuclear forces turn out to be most suitable. This invariance allows the

**Table 7. Penetrability and Shift Functions for Neutrons[a]**

| $l$ | $P_1(\rho)$ | $S_l(\rho)$ |
|---|---|---|
| 0 | $\rho$ | 0 |
| 1 | $\dfrac{\rho^3}{1+\rho^2}$ | $\dfrac{-1}{1+\rho^2}$ |
| 2 | $\dfrac{\rho^5}{9+3\rho^2+\rho^4}$ | $\dfrac{-3(6+\rho^2)}{9+3\rho^2+\rho^4}$ |
| 3 | $\dfrac{\rho^7}{225+45\rho^2+6\rho^4+\rho^6}$ | $\dfrac{-3(225+30\rho^2+2\rho^4}{225+45\rho^2+6\rho^4+\rho^6}$ |

[a] From (Wi 63).

concept of isospin to be introduced and nuclei with identical $A$ to be treated as different states of a system of $A$ nucleons.

In compliance with the above, the total energy of the $A$ nucleus can be presented as a sum of three terms: a nuclear part $E^A_{\text{nucl}}(T)$ specified by interaction between the nucleons and depending only on the isotopic spin; the energies of the Coulomb interaction between protons $E^A_{\text{Coul}}(Z)$ and the masses of the neutrons and protons:

$$E^A_{\text{tot}} = E^A_{\text{nucl}}(T) + E^A_{\text{Coul}}(Z) + Nm_{\text{n}} + Zm_{\text{p}} \tag{4.18}$$

The expression for the energy of a uniformly charged sphere of radius $R = r_0 A^{1/3}$ is assumed to be the Coulomb energy of the nucleus:

$$E^A_{\text{Coul}} = \frac{3}{5}\frac{e^2 Z(Z-1)}{r_0 A^{1/3}} \tag{4.19}$$

It is impossible to calculate the Coulomb energy exactly since, as a rule, the wave functions of the nuclear state are not known. Therefore, practical approaches to the calculation of the nuclear masses are based, to a maximum degree, on empirical data on the neighboring nuclei. One of the simplest approaches is the three-term mass formula (Wi 64).

Since $T_Z$ is the isotopic spin projection and is given by $T_Z = A/2 - Z$, the Coulomb energy contains $T_Z$ raised to no higher than the second power and the total nuclear energy can be presented in the form of a second-degree polynomial in $T_Z$:

$$E^A_{\text{tot}} = a(T) + bT_Z + cT_Z^2 \tag{4.20}$$

This is a familiar mass formula applicable to identical states of an isotopic multiplet, allowing one to calculate the masses of nuclei if the coefficients $a$, $b$, and $c$ are known from independent data.

The most successful and widely used method of calculating the light nuclear masses uses the Garvey–Kelson relations (Ga 66, Ke 66) connecting the data on mass difference of a number of nuclei which differ by a unit in $T_Z$. These relations have the form

$$\sum_{i=1}^{\alpha} C_i M(N_i Z_i) = 0, \qquad |C_i| = 1 \tag{4.21}$$

The most generally used expressions for $\alpha = 6$ are as follows:

$$\begin{aligned} &M(N+2, Z-2) - M(N, Z) + M(N, Z-1) - M(N+1, Z-2) \\ &\quad + M(N+1, Z) - M(N+2, Z-1) = 0 \end{aligned} \tag{4.22}$$

$$\begin{aligned} &M(N+2, Z) - M(N, Z-2) + M(N, Z-1) - M(N+1, Z) \\ &\quad + M(N+1, Z-2) - M(N+2, Z-1) = 0 \end{aligned} \tag{4.23}$$

Predictions for some unknown nuclei will be given in the subsequent sections. Here we note that, as shown by experience, the Garvey–Kelson method not infrequently predicts nuclei to be less stable than they turn out to be in fact, particularly when the calculations were made for the lightest nuclei with large neutron excess. Thus $^8$He, $^{11}$Li, and $^{19}$C turn out to be bound while the Garvey–Kelson calculations predicted their instability.

A purely empirical extrapolation of the dependence of the binding energy of one or two neutrons on the number of protons at a constant number of neutrons has been

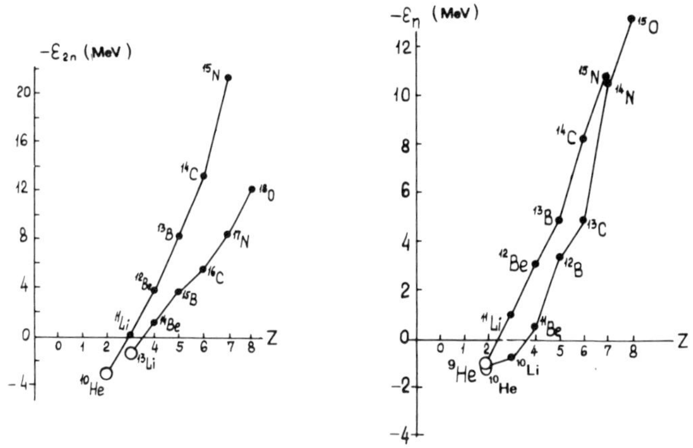


**Figure 28.** Dependence of binding energy for one and two neutrons in light nuclei on the number of protons for a constant number of neutrons. This is extrapolated into the region of unknown nuclei.

shown to be rather successful in predicting the stability of the neutron-rich lightest nuclei,

$$\varepsilon_n = f_1(Z)|_{N=\text{const}}, \qquad \varepsilon_{2n} = f_2(Z)|_{N=\text{const}} \tag{4.24}$$

Examples of such extrapolations are given in Figure 28. In particular, for $^{10}$He an instability of $Q_{2n} = 3$ MeV is predicted against the decay $^{10}\text{He} \rightarrow {}^{8}\text{He} + 2\text{n}$ and $Q_{1n} = 0.8$ MeV against the decay $^{10}\text{He} \rightarrow {}^{9}\text{He} + \text{n}$. Values were calculated using the experimental value of the $^9$He mass and amount to $Q_{2n} = 1.66$ MeV and $Q_{1n} = 0.52$ MeV, respectively. It should be noted that the observed decay energy of $^9$He turned out to be considerably lower than was predicted by any of the methods.

Though it has not been fundamentally established, the extrapolation method is apparently applicable to predicting the masses of unstable nuclei if $\varepsilon_n$ is assumed to be not the decay energy with the emission of not necessarily only one neutron but rather a total decay energy. Thus, the $^6$H mass (Og 79) predicted in this way was in good agreement with the experimental data (Al 84); $\varepsilon_n$ extrapolation in the $^{11}$B–$^{10}$B–$^9$Li–$^8$He–$^7$H chain predicts that $^7$H is also positioned in the neutron drip-line region.

## 4.4. Multineutrons

### *4.4.1. Dineutrons*

The stability of a system consisting of two neutrons was studied in a great number of experiments and as far back as the 1950s it became clear that there existed no bound dineutron. It was a natural conclusion from the charge invariance of the nuclear forces. For the first time, the unstable dineutron, i.e., a system of two neutrons in a singlet state, was observed by Voitovetsky with his collaborators (Vo 64) in the $^2$H(n, p)2n reaction at $E = 14$ MeV from observations on the spectrum of the protons (Figure 29). The characteristics of the two-neutron interaction were determined in that and in

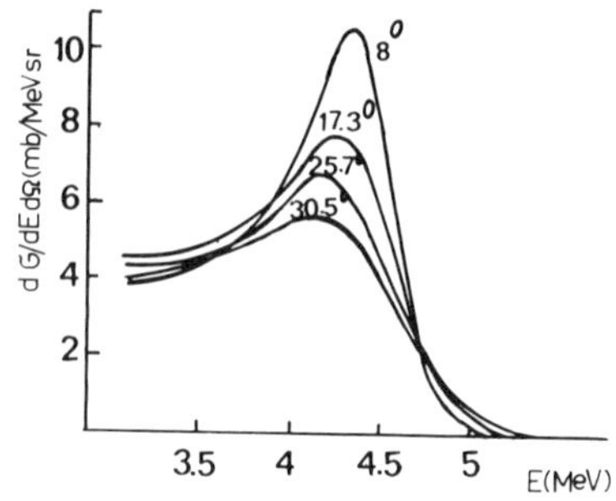


**Figure 29.** Energy spectra of protons from the $^2H(n, p)2n$ reaction at $E_n = 14$ MeV (Vo 64). In this work, for the first time, a final-state interaction in the 2n system was observed.

numerous subsequent studies. Without considering various methods of experimental-data analysis, which are sometimes very complicated, we note that they are all based on the theory of interaction of two nucleons in the final state (Mi 50, Wa 52). In the simplest case the distribution of neutrons as a function of the energy of their relative motion has the form

$$W(E) = \text{const}\,\frac{\sqrt{E}\,dE}{E + |\varepsilon|} \tag{4.25}$$

and the correlation function of the two neutrons

$$W(\theta) = \text{const}\,\frac{\theta d\theta}{\sqrt{\theta^2 + 4|\varepsilon|/E_0}} \tag{4.26}$$

Here $E_0$ is the center-of-mass energy if the two neutrons, $E$ is the energy of their relative motion, and $\varepsilon$ is the virtual level energy. The third particle spectrum (in the above-mentioned case it is that of the proton) can be obtained from equation (4.25):

$$W(E_p) = \text{const}\,\frac{\sqrt{E_p^m - E_p}\,dE_p}{\varepsilon + \frac{3}{2}(E_p^m - E_p)} \tag{4.27}$$

where $E_p^m$ is the minimum possible proton energy corresponding to $E = 0$.

As noted in Section 4.2, the FSI characteristics can be written in terms of resonance parameters. The resonance energies in the $A = 2$ system and their widths were estimated (Be 84) proceeding from a spectral distribution form

$$\frac{\sin^2 \delta(k)}{k} \tag{4.28}$$

where $\delta(k)$ is the phase shift and $k$ the relative linear momentum transfer. A modified form was used for $^2$He, i.e.,

$$\frac{\eta}{e^{2\pi\eta} - 1}\,\frac{\sin^2 \delta(k)}{k} \tag{4.29}$$

where $\eta$ is the Sommerfeld parameter.

The resonance parameters thus obtained are given in Table 8. The excitation level spectrum of the nuclei with $A = 2$ is presented in Figure 30.

**Table 8. Resonance Parameters for Two Nucleon Systems (Diproton, Deutron in the Singlet State, Dineutron) (Be 84b)**

| Resonance | Decay | Resonance energy $\varepsilon$ (MeV) | Resonance width $\Gamma$ (MeV) |
|---|---|---|---|
| $^2$He | p + p | 0.8 | 2.8 |
| $^2$H* | p + n | 0.07 | 0.5 |
| $^2$n | n + n | 0.12 | 1.0 |

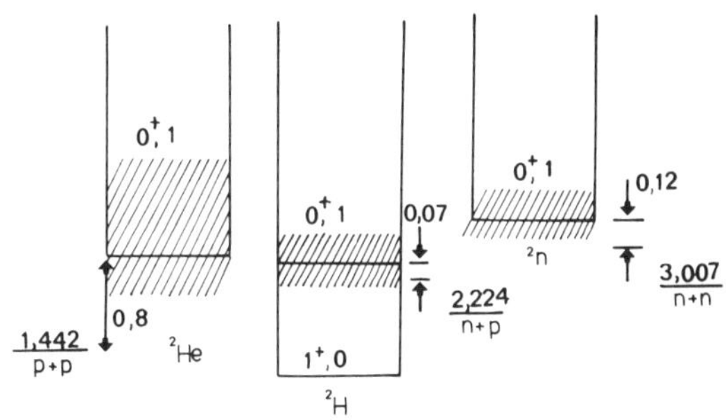


Figure 30. The level spectra for $A = 2$ nuclei.

## 4.4.2. Trineutrons

The reported observation of a bound trineutron (Aj 65) was not confirmed by more recent measurements (Th 66) and the generally accepted viewpoint is now that the trineutron is not bound. The upper limits for the bound trineutron formation cross-section in the $^3$He($\pi^-$, $\pi^+$)3n reaction amounts to 0.12 $\mu$barn/sr (for angles of 20–40°) (Sp 74), in the $^7$Li($^7$Li, $^{11}$C)3n reaction to 70 nbarn/sr (Ce 74), and in the $^7$Li($^{11}$B, $^{15}$O)3n reaction to 10 nbarn/sr (Be 86a).

The situation with respect to resonance states in the system of three neutrons is less unambiguous. A characteristic grouping of pions (Sp 70, Sp 74) was observed in the spectrum of the $^3$He($\pi^-$, $\pi^+$)3n reaction (Figure 31), which at first was explained as a result of the interaction of three neutrons in their final state (Sp 70). However, it followed from other calculations (Ph 70) that the observed form of the spectrum could be reproduced on the assumption of FSI involving only two neutrons (Figure 31a).

A more detailed analysis of the above-mentioned reaction was carried out by Jibuti and Kezerashvili (Ji 85a). Their main results are given in Figure 31. It is clear from this figure that a plane-wave approximation (i.e., a two-step mechanism without a FSI between three neutrons—the diagrams are presented in Figure 11) does not describe the experimental data even for a wide variety of nucleon-nucleon potentials. At the same time, comparison of these curves gives evidence of a strong sensitivity of the reaction cross-section to the initial-state wave function. In the microscopic approach developed by these authors (Ji 85b) the orthogonality of the wave functions of the

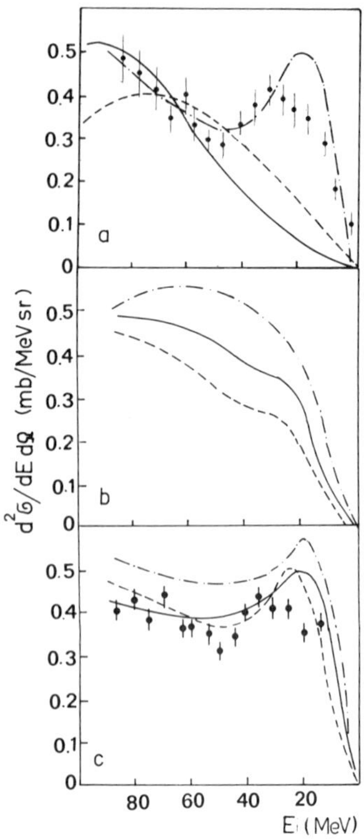


**Figure 31.** Energy spectrum of the DCX reaction $^3He(\pi^-\pi^+)3n$. (a) The solid curve is for four-body phase space and the dashed curve for four-body phase space distorted by an $^{15}O$ interaction between two of the neutrons. Both curves are normalized to the experimental data in the energy range 50–85 MeV. The dot-dash curve is the result of a calculation by Phillips. Experimental data from (Sp 70). (b) Calculations with the PWBA approach and a two-step mechanism. Different curves correspond to different potentials used (Ji 85b). (c) Calculations for a two-step mechanism in which FSI among the three neutrons is considered. Different curves correspond to different potentials used (Ji 85b). Experimental data from (Sp 74).

initial and final nuclear states is provided for, which was not the case, for example, in Phillips' work (Ph 70).

It follows from Figure 31c that the account taken of the FSI of the three neutrons changes the picture qualitatively, as compared with the plane-wave approximation, resulting in a strong increase of the predicted cross-section in that very region where it was observed in the experiment. The result depends weakly on the nucleon–nucleon potentials used.

Thus, we see that the explanation of the wide maximum observed in the $^3He(\pi^-, \pi^+)3n$ reaction is not unambiguous: it can be achieved with account taken of the FSI of two neutrons as well as three.

At the same time, in reactions with heavy ions no deviation from phase space, and consequently no manifestation of the FSI between three neutrons, has been observed (Figure 32). The trineutron system was investigated in the $^7Li(^7Li, ^{11}C)3n$ (Ce 77) and $^3H(^7Li, ^7Be)3n$ (Al 87) reactions.

### *4.4.3. Tetraneutrons*

Calculations of the binding energy of four neutrons by a variation method (Ta 65), by the methods of resonant groups (Th 70) and using hyperspherical functions (Ba 67), as well as by the Hilbert-Schmidt method (Ba 85) predict that there exists no bound

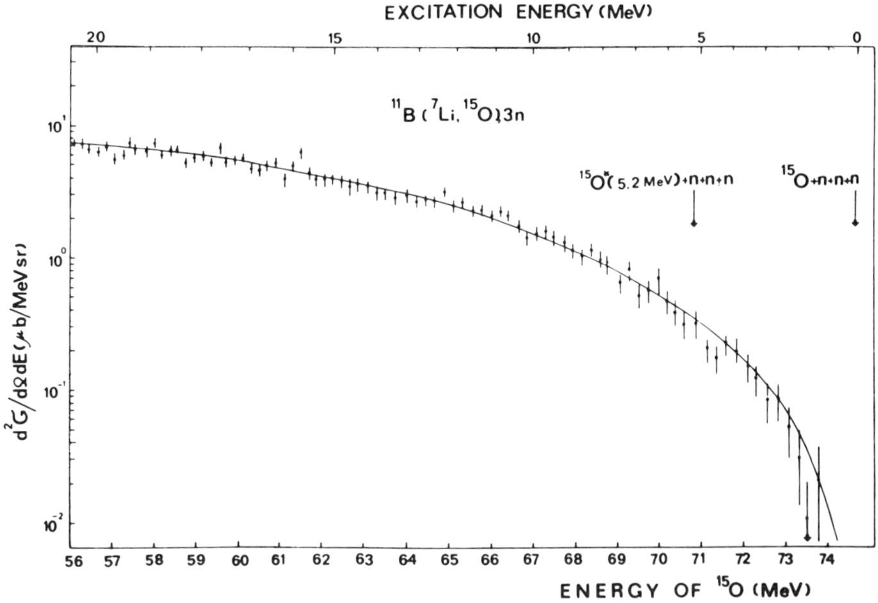


**Figure 32.** Energy spectrum of $^{15}O$ from the $^{7}Li(^{11}B, ^{15}O)3n$ reaction (Be 86). The full curve is a phase-space calculation including contributions from the two exit channels: $^{15}O + (n + n + n)$ and $^{15}O^*$ (5.2 MeV) + (n + n + n).

state of four neutrons. At the same time, these calculations do not prove the absence of such a state. One can reliably state that the upper limit of the tetraneutron drip-line equals 3.1 MeV; in the opposite case the $^{8}He$ nucleus would have undergone not $\beta$ decay but four-neutron decay $^{8}He \rightarrow {}^{4}He + 4n$.

Presently, there is only one experiment suggesting the existence of a stable tetraneutron (Ag 85). It was carried out using an activation method and requires independent verification.

Recently, the system of four neutrons in a final state has been studied in the following reactions: $^{4}He(\pi^-, \pi^+)4n$ (Un 84), $^{7}Li(^{7}Li, ^{10}C)4n$ (Ce 74), and $^{7}Li(^{11}B, ^{14}O)4n$ (Be 86a). Elsewhere (Un 84, Ce 77) a very small number of events was observed near the threshold but, according to the authors, they were of an instrumental origin. In other work (Be 86a) several events of unknown origin were observed in the threshold region but, as noted by the authors, they required verification. The upper limit on the formation cross-section of the bound tetraneutron in the $^{7}Li(^{7}Li, ^{10}C)$ reaction amounts to 30 nbarn/sr, and in the $(\pi^-, \pi^+)$ reaction to approximately 10 nbarn/sr ($E_\pi = 156$ MeV, the observation angle $\theta = 0°$). The highest sensitivity was achieved in the $^{7}Li(^{11}B, ^{14}O)4n$ reaction and amounted to 1 nbarn/sr (Figure 33).

In the above-mentioned reactions with heavy ions the spectrum form is satisfactorily described by a five-body phase space.

In the $(\pi^-, \pi^+)$ reaction a deviation from the phase-space prediction is observed. Figure 34a presents the spectrum of $\pi^+$ mesons measured by Ungar *et al.* (Un 84) and its comparison with calculations (Ke 86). These latter calculations were carried out

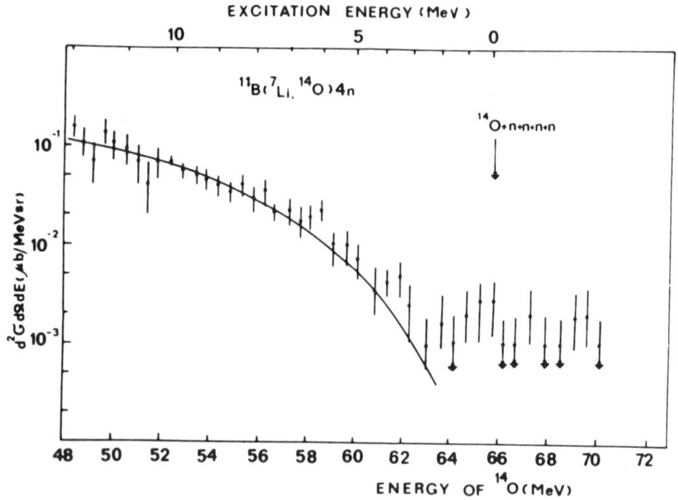


**Figure 33.** Energy spectrum of $^{14}$O from the $^{7}$Li ($^{11}$B, $^{14}$O)4n reaction (Be 86). The full curve is a phase-space calculation for the five-body decay in the exit channel ($^{14}$O + n + n + n + n).

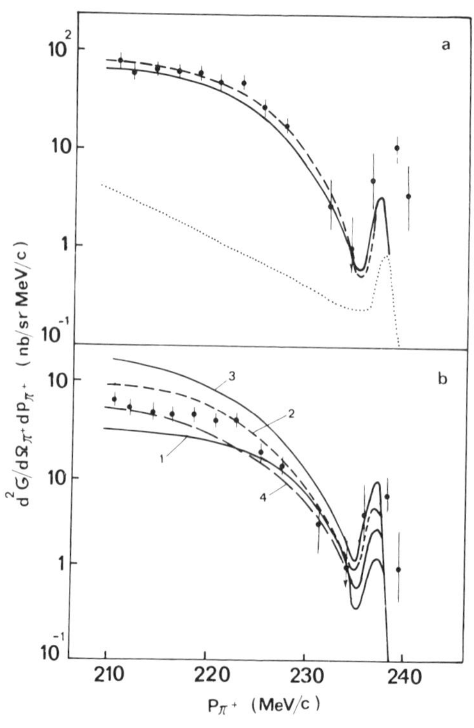


**Figure 34.** The $\pi^+$ spectrum from the $^{4}$He $(\pi^-\pi^+)$4n reaction at $E_\pi = 165$ MeV and $\theta = 0°$. (a) Experimental points from (Un 84); calculation (Ke 86) accounts for the final-state interaction (FSI) among the four neutrons in the exit channel. The three points forming a maximum near the threshold are uncertain. (b) Comparison with the calculation accounting for the FSI between two neutrons.

using the same technique as in Jibuti and Kezerashvili (Ji 85). It is clear that the FSI between all the neutrons results in a maximum of the predicted cross-section in the small momentum-transfer region regardless of the reaction mechanism (see Figure 12). It is interesting that this maximum is rather narrow and manifests itself near the threshold. However, no experimental confirmation has so far been obtained.

Simultaneously, as shown by Ungar *et al.* (Un 84), the same experimental data can be explained without using the FSI of the four neutrons but on the assumption that the FSI takes place involving each of the two pairs of neutrons separately (Figure 34b).

The necessity of taking into account the FSI between all the four neutrons is possibly suggested by the analysis performed (Ji 85a) of the data obtained by Gilly *et al.* (Gi 65) on the dependence of the differential cross-section of the $^4He(\pi^-, \pi^+)4n$ reaction on the $\pi^-$-meson energy. The account taken of the FSI considerably improves the agreement with experiment, as compared with that obtained using a plane-wave approximation. In so doing, the main contribution is made by the state with momentum $l = 0$ and the state with $l = 2$ is strongly suppressed.

The possible existence of a stable tetraneutron and its characteristics can be connected with the $T = 2$ resonance in the system of four nucleons. Therefore, a number of experiments was carried out in order to detect this $T = 2$ level in nuclei with $A = 4$, for example, in investigating the interaction of $P + {}^3He$ (Ba 72). In all these experiments no such resonance was found.

The data available at present on multineutrons show rather convincingly that there exist no stable neutron nuclei with $A$ in the range from 2 to 4 inclusive. A resonance (virtual) state is observed in the system of two neutrons. The situation of the metastable states of three and four neutrons is somewhat less clear. It is most probable that there exist no narrow decaying states. However, one cannot exclude the existence of a very short-lived system corresponding to the FSI of three and four neutrons, since there remain certain contradictions in the experimental data.

### 4.4.4. *Multineutrons with $n > 4$*

No experiments on the search for heavier neutron systems have been carried out as yet, though their importance is evident. A pentaneutron $^5n$ is expected to be unstable by 10 MeV (Be 81). The extrapolation method given in the previous section predicts that the $^5n$ decay energy can be approximately 5 MeV. In this case, however, the centrifugal barrier might result in some delay in the decay.

The multineutron systems with $N \geqslant 5$ in the final state can be obtained in nuclear reactions such as those presented in Table 9.

## 4.5. Superheavy Isotopes of Hydrogen

### 4.5.1. $^4H$

The great number of experimental investigations reported thus far have detected no nucleon-stable isotope of hydrogen with $A > 3$. Quasi-stationary states have been observed in the systems $^4H$ and $^6H$. The existence of a resonance in the $^4H$ system, which decays to product $n + T$, was first reported by Baz *et al.* (Ba 72) in which the $\gamma + {}^4He$ and $^6Li + \pi^-$ reactions were studied. A neutron-unstable state of $^4H$ with

**Table 9. Some Nuclear Reactions for the Production of Multineutron Systems with $n > 5$ in the Final State**

| System | Reaction | $Q$ (MeV) |
|---|---|---|
| 5n | $^{7}Li + {}^{7}Li \rightarrow {}^{9}C + 5n$ | −39.4 |
| | $^{7}Li + {}^{11}B \rightarrow {}^{13}O + 5n$ | −39.9 |
| | $^{9}Be + {}^{9}Be \rightarrow {}^{13}O + 5n$ | −40.8 |
| 6n | $^{9}Be + {}^{10}Be \rightarrow {}^{13}O + 6n^{a}$ | −39.5 |
| | $^{9}Be + {}^{14}C \rightarrow {}^{17}Ne + 6n^{a}$ | −50.5 |
| 7n | $^{9}Be + {}^{18}O \rightarrow {}^{20}Mg + 7n$ | −63.5 |
| 8n | $^{10}Be + {}^{18}O \rightarrow {}^{20}Mg + 8n^{a}$ | −70.3 |

[a] Reactions in which either the target or bombarding ion is radioactive.

$5.1 \pm 1.5$ MeV decay energy and a width $\Gamma \leqslant 3$ MeV was found in the reaction. Later on, interaction in the final state of n + T was observed in various simple stripping and pickup reactions at small angles. The $^{4}H$ nucleus characteristics were for the first time determined from the phase analysis of the n + $^{3}H$ scattering and interpreted as p-wave resonances corresponding to the $2^{-}$ and $1^{-}$ states at energies of 3.4 and 5.1 MeV, respectively, relative to the system n + $^{3}H$.

In the inclusive spectra obtained in reactions with heavy ions the $^{4}H$ nucleus usually appears as a very wide peak ($\Gamma \sim 3$ MeV). Figure 35 shows the spectrum of $^{8}B$ nuclei in the $^{6}Li(^{6}Li, {}^{8}B)^{4}H$ reaction (Ce 74). Analogous data were obtained in other reactions. One of the recent results is the spectrum of $^{16}O$ nuclei from the $^{9}Be(^{11}B, {}^{16}O)^{4}H$ reaction at a $^{11}B$ ion energy of 88 MeV (Be 86b), and is given in Figure 36. The wide range of excitation energies of the $^{4}H$ nucleus obtained is characteristic of these measurements. The solid curve is used to display the sum of the phase curves obtained from all possible decay channels ($^{3}H$ + n, $^{2}H$ + n + n, p + n + n + n). The observed excess over the phase space prediction is shown in the insert and is a peak with the width of ~4 MeV in the center-of-mass system. It is not yet excluded that this may be interpreted as compressing two peaks: one with energy of $3.5 \pm 0.5$ MeV above the threshold of n + $^{3}H$ and with width $\Gamma \sim 1.0$ MeV (the ground state of $^{4}H$) and the second with energy of 5 MeV (an excited state of $^{4}H$).

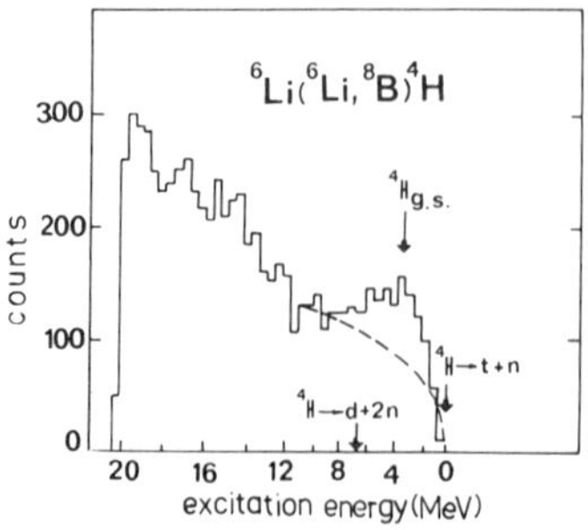


**Figure 35.** Energy spectrum of $^{8}B$ from the $^{6}Li(^{6}Li, {}^{8}B)^{4}H$ reaction. The histogram represents the experimental data (Ce 74). The broad group near threshold is interpreted as evidence for the appearance of metastable $^{4}H$ nuclei. The dotted line represents a phase-space calculation for the three-body breakup in the exit channel ($^{8}B$ + t + n).

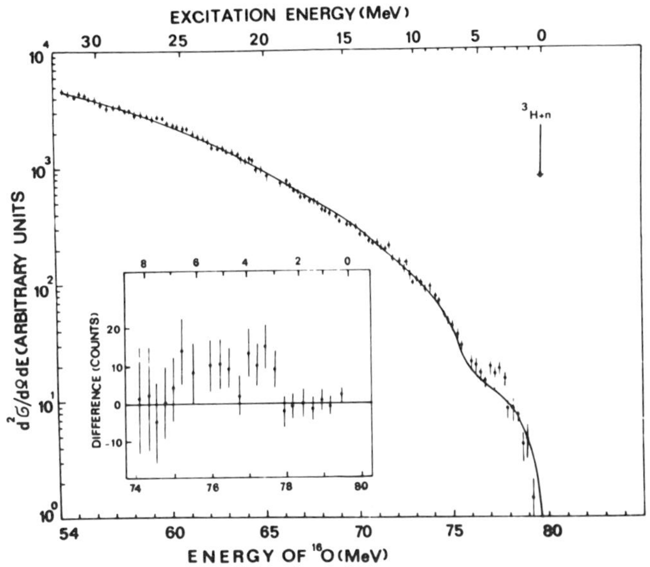


**Figure 36.** Energy spectrum of $^{16}$O from the $^9$Be($^{11}$B, $^{16}$O)$^4$H reaction (Be 85). The full line is a phase-space calculation that takes into account the following exit channels: $^{16}$O + $^3$H + n, $^{16}$O* + $^3$H + n, $^{16}$O + $^2$H + n + n, and $^{16}$O + H + n + n + n. In the inset, the difference between the phase-space curve and the experimental points is shown on a linear scale.

### *4.5.2.* $^5$H

Some time ago it was suggested that the $^5$H nucleus might lie near the breakup threshold and even be stable (Bl 64). The considerations were as follows. A level with excitation energy of 16.7 MeV and having quantum numbers $I = \frac{3}{2}^+$, $T = \frac{1}{2}$ and structure $(1s)^3(1p)^2$ is well known in the $^5$He nucleus. One can visualize it as a combined triton and deuteron. The singlet state of the deuteron with $T = 1$ lies, as known, higher than the triplet ground state by 2.3 MeV. Therefore, one may assume that $^5$He has a state with $T = \frac{3}{2}$ of the triton-plus-singlet deuteron structure and lying approximately 2.3 MeV higher than this 16.7 MeV state. Proceeding from the above, it was proposed (Bl 64) that the $^5$H nucleus could be stable with binding energy of approximately 0.4 MeV. In estimating the $^5$H energy one can, however, act otherwise (Ba 72) proceeding from analogy with a system of an $\alpha$-particle plus a deuteron, i.e., the $^6$Li nucleus. The state with $T = 1$, $I = 0^+$ lying at energy 3.56 MeV is known in the $^6$Li nucleus, and can be treated as an $\alpha$ particle plus a singlet deuteron. In this case the corresponding energy of the $T = \frac{3}{2}$ level in $^5$He would become approximately 20.3 MeV and $^5$H would already be unstable against decay into $^3$H + 2n by 0.8 MeV.

The search for $^5$H was undertaken in many laboratories and this work is reviewed elsewhere (Be 86b, Ba 72). The main conclusion is that there exists no stable $^5$H.

No levels corresponding to unbound states of $^5$H were detected in experiments which studied the $^3$H + 2n system in the final state.

The spectrum shown in Figure 37 is a typical result. It was measured (Be 86b) from the $^9$Be($^{11}$B, $^{15}$O)$^5$H reaction. This spectrum is reproduced satisfactorily by phase

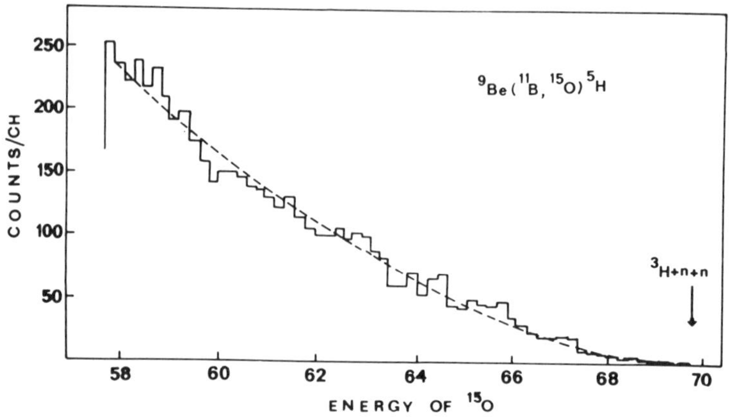


**Figure 37.** Energy spectrum of $^{15}O$ from the $^9Be(^{11}B, ^{15}O)^5H$ reaction (Be 85). The full line is a phase-space calculation for the four-body decay ($^{15}O + ^3H + n + n + n$) in the exit channel.

space appropriate to the four-particle reaction. No narrow peaks are observed in the spectra. Analogous results were obtained in studying the $^7Li(^6Li, B^8)$ (Al 85), $^3H(t, p)$, and $^9Be(\alpha, ^8B)$ reactions (Ba 72).

The results obtained with negatively charged projectiles differ from those with heavy ions. In the $^6Li(\pi^-, p)^3H$, 2n reaction at $E_\pi = 125$ MeV a deviation from the four-particle phase space (Figure 38) was observed (Se 81a) and interpreted by the authors as the $^5H$ ground state with decay energy $11 \pm 1.5$ MeV and width 14 MeV. The statistical accuracy was rather high in these measurements. The maximum excess over phase space amounts to a factor of 2, while in the energy range 25–55 MeV the experimental data coincide with the phase-space curve within $\pm 3\%$.

At the same time, the result obtained (Se 81a) does not completely agree with other data of the same group. In the search for $^7H$ in the $^7Li(\pi^-, \pi^+)^3H$, 4n reaction the continuous spectrum of pions was well described by the phase volume corresponding

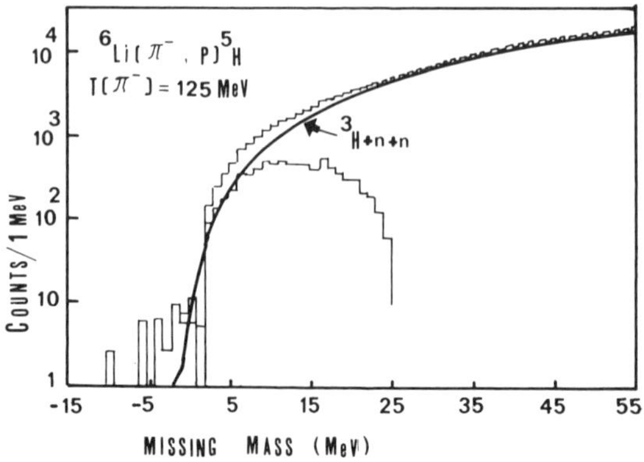


**Figure 38.** Proton spectrum from the $^6Li(\pi^-, p)^5H$ reaction (Se 81a).

to the $^7Li(\pi^-, \pi^+)^5H$, 2n process (Figure 42), which assumed the formation of an almost stable $^5H$. The curve corresponding to $^5H$ with energy 11 MeV higher than the threshold would have been considerably shifted to the right.

The identification of a nucleus from such phase-volume deviation—without observing an explicit maximum—is rather risky. Therefore, we hold that the data from both these studies require more detailed analyses that, in particular, take into account the n–n interaction in the final state.

The question of a $^5H$ resonant state remains open at the present time. However, as far as its decay width is concerned, all the data suggest that it is very large, and certainly not less than 5 MeV. In the next section some considerations will be formulated allowing so great a width to be understood qualitatively, even when the decay energy is comparatively low.

### *4.5.3.* $^6H$

In view of the absence of at least a quasi-stationary state of $^5H$, the search for a resonance in the $^3H + 3n$ system (the $^6H$ nucleus) may turn out to be pointless, since all experience in nuclear physics suggests that odd nuclei are always less stable than even ones. Nevertheless, the binding-energy extrapolation predicts a relatively low decay energy for $^6H$ (2–3 MeV) and in this case the centrifugal barrier could slow down the decay.

$^6H$ was synthesized in the $^7Li(^7Li, ^8B)^6H$ reaction (Al 84). The corresponding spectrum of $^8B$ is given in Figure 39. The test experiments and analyses have shown that the possible background sources (such as chemical impurities in the target, incomplete suppression of pulse pile-up, formation of $^8B$ in the five-particle reaction $^7Li + ^7Li \rightarrow ^8B + ^3H + 3n$) do not contribute appreciably in the region of the peak attributed to $^6H$ formation.

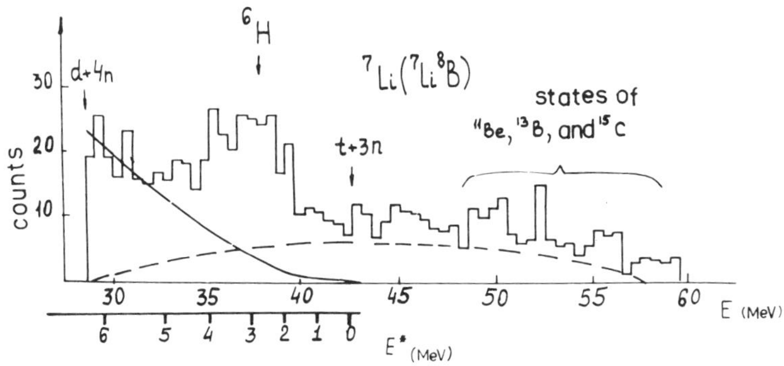


**Figure 39.** Energy spectrum of $^8B$ from the $^7Li(^7Li, ^8B)^6H$ reaction (Al 84). The histogram represents the experimental data. The full line is a phase-space calculation for the five-body decay in the exit channel. The dotted line shows the pile-up distribution. The peaks on the right side of the threshold may be identified as coming from reactions on $^{12}C$, $^{14}N$, and $^{16}O$ impurities. The broad maximum at the left of the threshold for t + 3n production is identified as a metastable state of $^6H$.

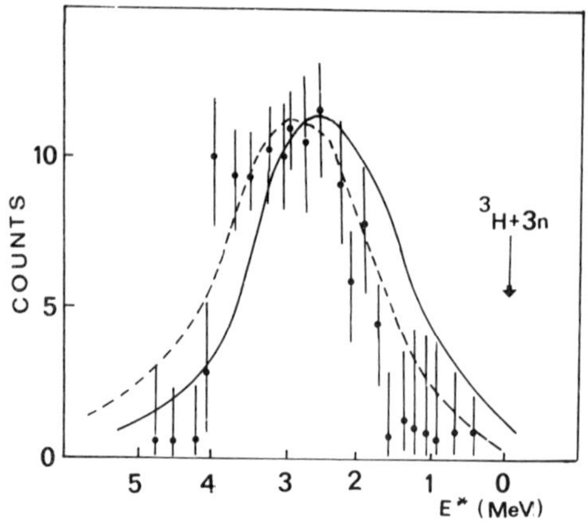


**Figure 40.** Energy spectrum of $^8$B from the $^7$Li($^7$Li, $^8$B)$^6$H reaction obtained after subtracting all backgrounds mentioned in the caption to Figure 39. The curves represent calculations with a Breit-Wigner formula and the following parameters: $E_c$ = 2.7 MeV, $\Gamma$ = 1.8 MeV—full curve; $E_0$ = 3.0 MeV, $\Gamma$ = 1.8 MeV—dotted curve.

Figure 40 shows the spectrum of the $^8$B nuclei after subtraction of background of all types. It is clear that the peak is satisfactorily approximated by a resonant curve of Breit-Wigner form. Alexandrov *et al.* (Al 84) gave the following resonance parameters: resonance energy $E = 2.7 \pm 0.4$ MeV and width $\Gamma = 1.8 \pm 0.5$ MeV. Possibly, the $E$ = 3.0 MeV value (Al 85) is in somewhat better agreement with experimental data. As for the resonance width, the $\Gamma$ = 1.8 MeV value is obtained in this treatment as well, and the quoted error value of $\pm 0.5$ MeV may even be overestimated.

As mentioned in the previous section, in the inclusive spectra it is possible, in principle, to observe groups with widths of several MeV reflecting the presence of consecutive two-particle processes. These groups can simulate the formation of a wide state of the residual nucleus [see, for example, Venikov *et al.* (Ve 81)]. In this case a careful kinematic analysis is required. However, in the experiment with $^6$H this mechanism for formation of the wide group is impossible, since the $^8$B nuclei are always formed in the final state with a number of particles greater than three in the $^7$Li + $^7$Li interaction. For example, the reaction $^7$Li($^7$Li, t)$^{11}$B* might take place, in principle, with $^{11}$B formation in a discrete state from which the $^8$B nucleus would be emitted but, in this case, three neutrons would be produced. Two-particle decay may take place in the case of $^9$B: $^9\text{B} \rightarrow {}^8\text{B} + \text{n}$. However, the $^9$B itself is formed only in the five-particle reaction $^7\text{Li} + {}^7\text{Li} \rightarrow {}^9\text{B} + {}^3\text{H} + 2\text{n}$. In all cases a continuous spectrum of $^8$B nuclei will be observed which cannot simulate the observed group.

The measurements made by Belozyorov *et al.* (Be 86b) using the $^9$Be($^{11}$B, $^{14}$O)$^6$H reaction have confirmed the presence of a $^6$H resonant state. The pertinent spectrum is given in Figure 41. The value of the $^6\text{H} \rightarrow {}^3\text{H} + 3\text{n}$ decay energy obtained, $E = 2.6 \pm 0.5$ MeV, practically coincides with the results of Alexandrov *et al.* (Al 84).

The $^6$H nucleus is the most neutron-rich nuclear system known at the present time: one proton binds 5 neutrons.

### 4.5.4. $^7$H *and Heavier Isotopes of Hydrogen*

It follows from extrapolations of various types, for example Ogloblin (Og 79), that the $^7$H nuclei should be not far from the neutron drip-line. Using the $^6$H mass, obtained elsewhere (Al 84), one can calculate the $^7$H mass using the Garvey-Kelson method. In

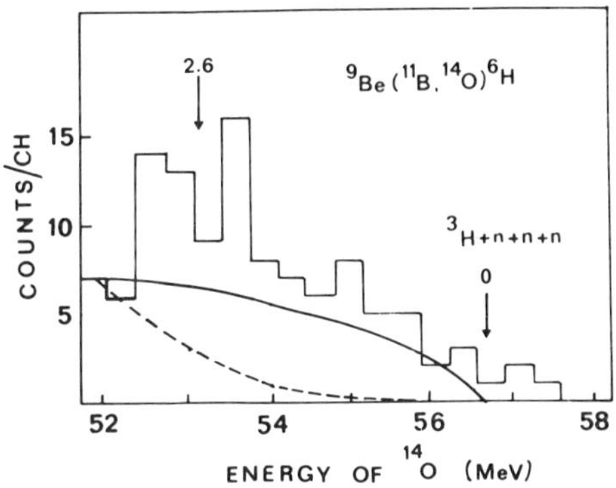


**Figure 41.** Energy spectrum of $^{14}O$ from the $^9Be(^{11}Be, ^{14}O)^6H$ reaction (Be 86). The dashed line is a phase-space calculation for the five-body breakup ($^{14}O + {}^3H + h + n + n$); the full line is a phase-space calculation that also takes into account the three-body breakup ($^{14}O + {}^5H + n$ on $^{14}O + {}^4H + 2n$ or $^{14}O + {}^3H - 3n$).

this case $^7H$ turns out to be unstable against decay into $^3H + 4n$ by 1.64 MeV. The search for nucleon-stable $^7H$ was mentioned in Section 2. Recent experiments on the search for $^7H$ in the spontaneous fission of $^{252}Cf$ and in the $^7Li\ (\pi^-, \pi^+)H$ reaction did not result in the observation of a stable state. The single experiment in which the search for $^7H$ in the final state was performed—in the $^7Li(\pi^-, \pi^+)^7H$ reaction (Se 81a)—yielded no indications of the existence of a metastable $^7H$ (Figure 42).

Thus, the experimental data on superheavy isotopes of hydrogen ($A > 3$) show that quasi-stable states exist for systems with an odd number of neutrons ($^4H$ and $^6H$), while systems with an even number of neutrons, if they exist at all, have a very large width.

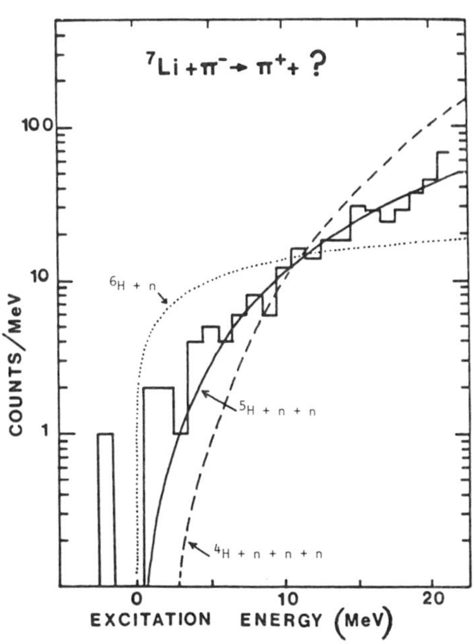


**Figure 42.** Excitation energy spectrum from the $^7Li(\pi^- \pi^+)$ reaction (Se 81).

**Table 10. Some Nuclear Reactions for the Production of Heavy Hydrogen Isotopes with $A > 6$**

| Nucleus | Reaction | $Q$ (MeV) |
|---|---|---|
| $^{7}$H | $^{7}Li + ^{9}Be \rightarrow ^{9}C + ^{7}H$ | −49.9 |
| | $^{9}Be + ^{11}B \rightarrow ^{13}O + ^{7}H$ | −50.3 |
| $^{8}$H | $^{7}Li + ^{14}C \rightarrow ^{13}O + ^{8}H^{a}$ | −52.4 |
| | $^{7}Li + ^{18}O \rightarrow ^{17}Ne + ^{8}H$ | −66.2 |
| | $^{11}B + ^{14}C \rightarrow ^{17}Ne + ^{8}H^{a}$ | −69.4 |
| | $^{11}B + ^{10}Be \rightarrow ^{13}O + ^{8}H^{a}$ | −66.4 |
| $^{9}$H | $^{14}C + ^{15}N \rightarrow ^{20}Mg + ^{9}H^{a}$ | −87.1 |

[a] Reactions in which either the target or bombarding ion is radioactive.

Table 10 presents some nuclear reactions which can be used to search for superheavy hydrogen isotopes with $A > 6$.

## 4.6. Neutron-Rich Isotopes of Helium

### 4.6.1. $^{5}$He, $^{6}$He

The $^{5}$He nucleus is a nucleon-unstable one with decay energy of 0.89 MeV, though it has a comparatively small neutron excess: $\eta = (N - Z)/A = 0.20$. It is evident that its instability is connected not with the neutron excess but with the properties of the interaction of $\alpha$ particles with the neutron. This nucleus is interesting for us because its resonance parameters are well known and it is convenient to compare those of the decay of other neutron-unstable nuclei with them.

The $^{6}$He nucleus is also well known. It is $\beta$ radioactive, with a binding energy of 0.98 MeV against decay into $^{4}He + 2n$. The stability of this nucleus is due to pairing, i.e., it is provided for by the adding of an extra neutron to the unstable odd neutron number $^{5}$He. $^{6}$He has its first excited state at 1.80 MeV, which is higher than the decay threshold for the emission of two neutrons. The investigation of the decay of this level is of great interest in connection with the question of the way the neutrons are emitted in such cases: in the form of a correlated pair ("dineutron") or sequentially via an intermediate state (in this case via $^{5}$He). This problem will be discussed below.

Both $^{5}$He and $^{6}$He have highly excited states connected with the promotion of one of the nucleons from the 1s to the 1p shell. As a result, clearly manifested cluster structures of the d + t and t + t types are formed, respectively, but this goes beyond the scope of the present review.

### 4.6.2. $^{7}$He

The $^{7}$He nucleus is unstable against the decay $^{7}He \rightarrow ^{6}He + n$. The decay energy amounts to $0.44 \pm 0.03$ MeV, and the width is $160 \pm 30$ keV. In view of our remarks about $^{5}$He, the $^{7}$He nucleus should apparently be considered as the first "real" neutron-

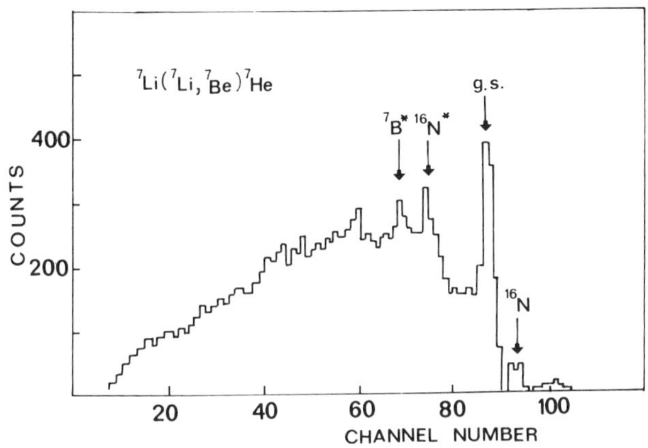


**Figure 43.** Energy spectrum of $^{7}$Be from the $^{7}$Li($^{7}$Li, $^{7}$Be)$^{7}$He reaction. A sharp peak is observed, corresponding to the ground state of $^{7}$He. No peaks corresponding to excited states of $^{7}$He were observed.

unstable isotope of helium. $^{7}$He manifests itself in various nuclear reactions as a clearly pronounced resonance. One of the recent observations (Al 85) is given in Figure 43. It shows the spectrum obtained in the $^{7}$Li($^{7}$Li, $^{7}$Be)$^{7}$He reaction with the formation of $^{7}$He in the final state. The $^{9}$Be($^{6}$Li, $^{8}$B)$^{7}$He reaction was also studied (Al 85) and a spectrum of the same type obtained. The main aim of these measurements was to study the excited states in this nucleus, since controversial data were available on their existence (Ce 74, We 77). No maxima with widths less than ~2 MeV were detected by Alexandrov *et al.* (Al 85) up to excitation energies of ~10 MeV. Since the ($^{6}$Li, $^{8}$B) and ($^{7}$Li, $^{7}$Be) reactions are absolutely different both in their mechanisms (a pickup of two protons and n → p charge exchange, respectively) and cross-section values, this result is a strong argument that $^{7}$He has no excited states—at least comparatively narrow ones.

### *4.6.3.* $^{8}$He

The $^{8}$He nucleus continues to attract our attention as the most neutron-rich nucleus which is stable against nucleon decay. Its stability was predicted in 1960 (Ze 60, Go 60) and its first observation dates back to 1961 (Lo 61). According to current data, $^{8}$He is stable against decay into $^{6}$He + 2n by 2.137 MeV. This exceeds the stability of $^{6}$He by 1.4 MeV. Thus, the addition of two neutrons (to $^{6}$He) does not decrease the binding energy, as is the case for almost all nuclei, but increases it considerably. The "helium anomaly" to be discussed below is based on this fact.

A search for excited states of $^{8}$He was performed (Al 82b, Al 85) using the $^{9}$Be($^{7}$Li, $^{8}$B)$^{8}$He reaction. Two series of measurements were made, the results of which are given in Figure 44. From an experimental point of view the series differ somewhat from one another (in one case a $\Delta E$-$E$ system was used; in the other, a $\Delta E$-$T$-$E$ system). To estimate the reliability of these data, it is important to note that the backgrounds due to pulse pile-up in the two series was different, despite the low statistics. One set of measurements (Al 82b), in Figure 44a, were interpreted as an observation of an excited state in $^{8}$He with energy 2.6 ± 0.2 MeV, while the other set

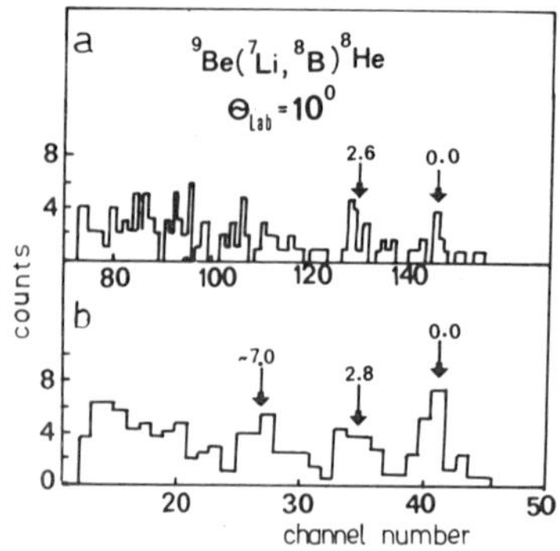


**Figure 44.** Energy spectra of $^8$Be from the reaction $^9$Be($^7$Li, $^8$B)$^8$He: (a) from (Al 82a); (b) from (Al 85).

(Al 85), in Figure 44b, show a corresponding energy of 2.8 ± 0.4 MeV, i.e., in good agreement. Some incompatibility between the widths of the maxima in the two spectra is due, first, to the low statistics and, second, to a large instrumental width in the case of Figure 44b.

This state at 2.7 MeV, in the same way as the excited state of $^6$He, may decay with the emission of two neutrons.

Apparently, in $^8$He, one more, rather wide, state was observed with energy of about 7 MeV, above the decay threshold for $^8\text{He} \rightarrow {}^4\text{He} + 4\text{n}$ and for which the channel with emission of a "tetraneutron" is, consequently, open.

### *4.6.4.* $^9$He

The $^9$He nucleus was found in the $^9$Be($\pi^-$, $\pi^+$)$^9$He reaction (Se 81a). The spectrum of $\pi^+$ mesons obtained in that work is given in Figure 45. $^9$He is unstable against decay into $^8$He + n by 1.14 MeV. It is interesting that all the reports predicting the $^9$He mass yielded a considerably higher instability (Table 11). Thus, similar to the case of $^8$He, the $^9$He stabilityhas turned out to be higher than expected. The results of an experiment on determining the $^9$He nucleus mass in the double charge exchange reaction $^9$Be($^{14}$C, $^{14}$O)$^9$He have been reported (Be 84a). The excess above the phase-space curve observed in the energy spectrum of $^{14}$O corresponding to $^9$He decay into $^8$He + n yielded a value $\Delta M(^9\text{He}) = 42.2 \pm 1.1$ MeV. For this value of the mass defect $B_n(^9\text{He}) = 2.5 \pm 1.1$ MeV. These values agree with the results of Seth (Se 81a).

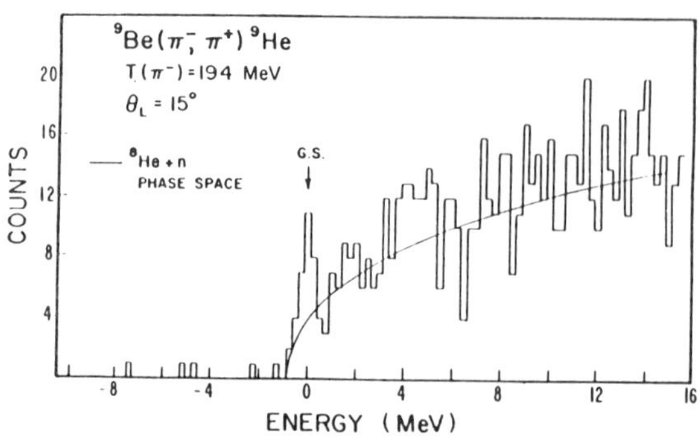


**Figure 45.** The $\pi^+$ energy spectrum from the $^9$Be($\pi^-\pi^+$)$^9$He reaction (Se 81). The full line indicates the phase-space calculation for three-body decay in the exit channel: ($\pi^+ + {}^8\text{He} + \text{n}$).

**Table 11. $^9$He Binding Energies Obtained in Different Studies**

| Author | Mass excess (MeV) | B(n) (MeV) | Reference |
|---|---|---|---|
| Seth | 40.98 (20) | −1.31 (20) | (Se 81a) |
| Janecke | 42.03 | −2.36 | (Wa 76) |
| Jelly *et al.* (GK) | 42.61 | −2.94 | (Je 75) |
| Jelly *et al.* (MSM) | 43.49 | −3.82 | (Je 75) |
| Thibault and Klapisch | 42.75 | −3.08 | (Th 74) |

### *4.6.5.* $^{10}$He

The anomalous stability of these nuclei, $^8$He in particular, results in continued interest in the search for $^{10}$He. It seemed quite natural to search for nucleon-stable $^{10}$He in view of the stability increase from $^6$He to $^8$He. Experiments on the ternary fission of $^{252}$Cf performed to this end were mentioned in Section 2. $^{10}$He was not detected either in bombarding $^{232}$Th with relativistic (4.8 GeV/c) deuterons (Be 79) (the cross-section upper limit was established at 2 $\mu$barn). The most sensitive experiment on the search for stable $^{10}$He was carried out by Oganessian *et al.* (Og 82): in the interaction of $^{10}$B, $^{11}$B, and $^{22}$Ne ions with energy from 8 to 10 MeV/nucleon with a thorium target, $^{10}$He nuclei were not observed at a cross-section level of $10^{-4}$ $\mu$barn/sr.

The estimate of the $^{10}$He mass using the Garvey–Kelson method and the measured $^9$He mass predicts that $^{10}$He is unstable against decay into $^8$He + 2n by 1.66 MeV. The extrapolation estimate yields a close value of 1.4 MeV (Og 79). All this makes the existence of $^{10}$He highly probable as a resonance. No experiments on the search for $^{10}$He with detection of its stable partner has been reported as yet.

### *4.6.6. Isotopes of Helium with $A > 10$*

Presently, there are no data on heavier isotopes of helium. A very rough (since there are no data on $^{13}$Li) extrapolation of function (4.24) along the chain $^{17}$N–$^{16}$C–$^{15}$Be–$^{13}$Li–$^{12}$He shows that the decay energy of $^{12}$He can be in the region of 2–3 MeV. Table 12 presents the nuclear reactions which may be used for the synthesis of helium isotopes with $A \geqslant 10$.

## 4.7. Neutron-Rich Isotopes of Lithium

### *4.7.1.* $^{10}$Li

A continuous chain of nucleon-stable isotopes of lithium ends with $^9$Li. The next isotope, $^{10}$Li, is known to decay into $^9$Li + n with a decay energy of 0.80 ± 0.25 MeV (Wi 75). $^{10}$Li was found in the $^9$Be($^9$Be, $^8$B)$^{10}$Li reaction with a cross-section of 30 kbarn/sr ($\theta_{lab} = 14°$) (Figure 46). The resonance width amounts to $\Gamma$ = 1.2 ± 0.3 MeV. An assumption was formulated (Ab 79, Ab 80) on the basis of the possible position of a $T = 2$ level in $^{10}$Be that it was an excited state of $^{10}$Li and not its ground state that had been observed in the above-mentioned work. Since (Wi 75)

**Table 12. Some Nuclear Reactions for Superheavy He Isotope ($A \geqslant 10$) Production**

| Nucleus | Reaction | $Q$ (MeV) |
|---|---|---|
| $^{10}$He | $^{10}Be + ^{14}C \rightarrow ^{14}O + ^{10}He^{a}$ | −40.1 |
| | $^{9}Be + ^{18}O \rightarrow ^{17}Ne + ^{10}He$ | −53.7 |
| | $^{9}Be + ^{14}C \rightarrow ^{13}O + ^{10}He^{a}$ | −56.5 |
| | $^{10}Be + ^{20}Ne \rightarrow ^{20}Mg + ^{10}He^{a}$ | −59.8 |
| $^{11}$He | $^{10}Be + ^{18}O \rightarrow ^{17}Ne + ^{11}He^{a}$ | −51.9 |
| | $^{10}Be + ^{14}C \rightarrow ^{13}O + ^{11}He^{a}$ | −63.3 |
| | $^{9}Be + ^{22}Ne \rightarrow ^{20}Mg + ^{11}He$ | −70.1 |
| $^{12}$He | $^{14}C + ^{18}O \rightarrow ^{20}Mg + ^{12}He^{a}$ | −79.2 |

[a] Reactions in which either the target or bombarding ion is radioactive.

is the only reference in which $^{10}$Li was observed and, as shown in Figure 46, the background was rather high, it is important to corroborate this result using some other reaction.

### *4.7.2.* $^{11}$Li

$^{11}$Li is stable with a binding energy of $156 \pm 80$ keV with respect to $^{9}$Li + 2n. It is one of the least bound nuclei known. Possibly, this explains the anomalously large radius of this nucleus, which was measured in experiments with radioactive beams (Ta 85b).

It should be noted that, prior to the discovery of $^{11}$Li, most of the calculations predicted this nucleus to be unstable against the nucleon decay.

The recently aroused interest in $^{11}$Li is because of its $\beta$ decay, which is accompanied by the emission of one, two, or even three neutrons, as well as tritons. These phenomena will be treated in the next section.

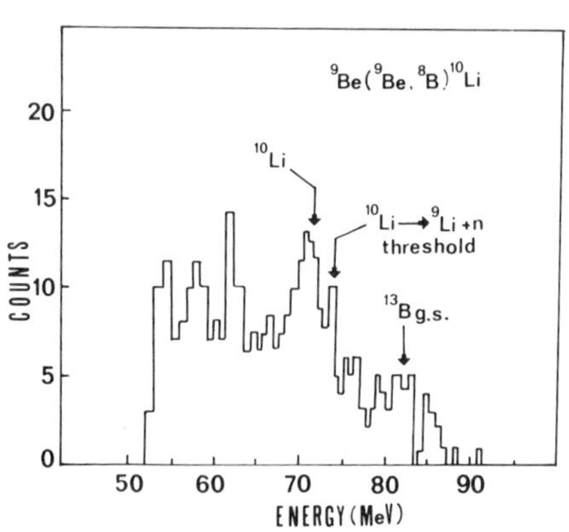


**Figure 46.** Energy spectrum of $^{8}$B from the $^{9}Be(^{9}Be, ^{8}B)^{10}Li$ reaction measured at an angle of 14°. The beam energy was 120 MeV (Wi 75).

**Table 13. Predicted Decay Energies for Very Heavy Li Isotopes**

| Nucleus | Mass excess (MeV) | Decay | Decay energy (MeV) |
|---|---|---|---|
| $^{12}Li$ | 52.93 | $^{11}Li + n$ | 3.92 |
| | | $^{9}Li + 3n$ | 3.76 |
| $^{13}Li$ | 60.34 | $^{11}Li + 2n$ | 3.26 |
| $^{14}Li$ | 72.29 | $^{13}Li + n$ | 3.88 |
| | | $^{11}Li + 3n$ | 7.14 |
| $^{15}Li$ | 81.60 | $^{14}Li + n$ | 1.24 |
| | | $^{11}Li + 4n$ | 8.38 |

[a] From (Aj 85, Aj 86).

### 4.7.3. *Isotopes of Lithium with A > 11*

There are no reliable experimental data on heavier isotopes of lithium. $^{12}Li$ was not observed in bombarding a uranium target by 4.8 GeV protons (Th 75). In view of the data on the masses of $^{10}Li$ and $^{11}Li$, one can say with sufficient confidence that $^{12}Li$ is not stable.

As for $^{13}Li$, the situation is less definite. This nucleus was, possibly, observed as a nucleon-stable one in bombarding $^{124}Sn$ by 6.7 GeV protons (Bo 80). However, the statistics were low in the region of interest and the authors (Bo 80) did not claim detection of $^{13}Li$. The predictions of the $^{13}Li$ mass are highly variable. It follows from other work (Aj 86) that $^{13}Li$ is unstable against decay into $^{11}Li + 2n$ by 3.26 MeV. Extrapolation (4.24) predicts instability of $^{13}Li$ with a decay energy of only about 1 MeV. Thus, the search for $^{13}Li$ as a residual nucleus in a binary reaction is undoubtedly justified. Table 13 presents expected decay energies for the $^{12-15}Li$ nuclei taken from the reviews (Aj 85, Aj 86).

Table 14 contains nuclear reactions for obtaining $^{10}Li$, $^{12}Li$, and $^{13}Li$ nuclei as a residual nucleus.

**Table 14. Some Nuclear Reactions for Heavy Li Isotope Production**

| Nucleus | Reaction | $Q$ (MeV) |
|---|---|---|
| $^{10}Li$ | $^{11}B + ^{13}C \rightarrow ^{14}O + ^{10}Li$ | −30.0 |
| | $^{7}Li + ^{13}C \rightarrow ^{10}C + ^{10}Li$ | −31.5 |
| $^{12}Li$ | $^{14}C + ^{15}N \rightarrow ^{17}Ne + ^{12}Li$[a] | −53.8 |
| | $^{7}Li + ^{18}O \rightarrow ^{13}O + ^{12}Li$ | −58.0 |
| | $^{7}Li + ^{14}C \rightarrow ^{9}C + ^{12}Li$[a] | −60.0 |
| | $^{11}B + ^{14}C \rightarrow ^{13}O + ^{12}Li$[a] | −60.4 |
| $^{13}Li$ | $^{15}N + ^{18}O \rightarrow ^{20}Mg + ^{13}Li$ | −75.3 |

[a] Reactions in which either the target or bombarding ion is radioactive.

## 4.8. Neutron-Rich Isotopes of Beryllium

For a long time $^{12}$Be has been the heaviest known isotope of beryllium. It is nucleon-stable with a rather high neutron binding energy $\varepsilon_n = 3.17$ MeV.

### *4.8.1.* $^{13}$Be

As for the $^{13}$Be nucleus, the calculations predicted (Th 74) that it should decay with the emission of a neutron and a decay energy of 1.5–2 MeV. An unstable $^{13}$Be nucleus was found (Al 83); repeated measurements are described elsewhere (Al 85). The $^{13}$Be nucleus was synthesized using the $^{14}$C($^7$Li, $^8$B)$^{13}$Be reaction with 82 MeV $^7$Li ions. The spectrum of the $^8$B nuclei obtained by summing three series of measurements (Al 85) is shown in Figure 47; the background due to the reaction on natural carbon has been subtracted. In the region 2 MeV above the threshold one can clearly observe a grouping of events (about 70 events in all). An insignificant excess over the background (within the limits of statistical accuracy) to the right of the reaction threshold is explained, apparently, by residual superimposed signals, the number of which should reach a maximum just in that region.

The observed maximum has a clearly manifested slope to the left, therefore it cannot result from the three-particle reaction $^{14}\text{C} + {}^7\text{Li} \rightarrow {}^{12}\text{Be} + {}^8\text{B} + \text{n}$. It is described by a resonance curve with parameters $E_0 = 2$ MeV and $\Gamma = 1.5$ MeV. The results of the first series of measurements (Al 83) yielded the values $E_0 = 1.8 \pm 0.5$ MeV and $\Gamma = 0.9 \pm 0.5$ MeV. Clearly, the summed result does not change the $E_0$ value. The procedure of spectrum summation coarsens the structure and does not allow widths to be reliably determined. Nevertheless, the value of $\Gamma = 0.9$ obtained from the first measurements is apparently somewhat underestimated.

The measured $^{13}$Be decay energy is in good agreement with the above-mentioned predictions. It should be noted that interest in the search for $^{13}$Be is mainly due to an urge to detect the heaviest unstable nucleus of those known up to now. One might expect that the heavier the nucleus, the less random the character of the deviations of its mass from the calculated values. From this point of view mass measurements assume particular importance. The obtained agreement of the calculations with the experimental

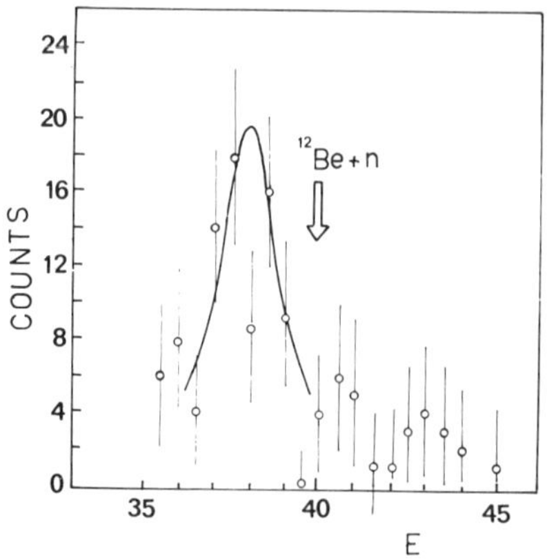


**Figure 47.** Energy spectrum of $^8$B from the $^{14}$C($^7$Li, $^8$B)$^{13}$Be reaction (Al 85). The full curve represents a calculation with a Breit–Wigner formula; the resonance parameters are $E_0 = 2$ MeV, $\Gamma = 1.5$ MeV.

data suggests that in the case of the heaviest known neutron-decaying nucleus no surprises have turned up as yet.

However, there are some considerations which make one think that, perhaps, it is the excited state of $^{13}Be$ and not its ground state that is in fact observed. This hypothesis is based on a spectrometric analysis to be considered in the next section. If it is confirmed, the question of the agreement between the calculation and the data will have to be reconsidered.

### *4.8.2.* $^{14}Be$

The first experiments on $^{14}Be$ synthesis used the heavy-ion reaction $^{15}N(145\ MeV) + {}^{232}Th(Ar\ 70)$ and yielded only an upper limit for the formation cross-section, on the basis of which it was concluded that $^{14}Be$ was not stable. However, subsequent experiments bombarding uranium targets with protons of 4.8 GeV have shown that $^{14}Be$ is stable against neutron decay. More recent experiments on $^{14}Be$ in the double charge exchange reaction $^{14}C(\pi^-, \pi^+)^{14}Be$ (Gi 84) allowed its mass to be determined. It corresponds to a binding energy of 1.12 MeV with respect to $^{12}Be + 2n$. Not long ago, at Michigan State University, a beam of $^{14}Be$ nuclei from the $^{18}O(30\ MeV/A) + {}^{181}Ta$ reaction was formed using a mass separator (RPMS) and the $^{14}Be$ half-life was measured to be $4.2 \pm 0.7$ ms (Cu 86).

### *4.8.3. Isotopes of Beryllium with $A > 14$*

No heavier isotopes of beryllium have been observed. It is predicted that they all are unstable. The expected decay energies taken from reviews (Aj 86, Aj 82, Aj 83) are given in Table 15.

Table 16 presents possible nuclear reactions for obtaining isotopes of beryllium with $A > 14$.

Our review of the systematics of the lightest neutron-rich nuclei is concluded here. No nucleon-unstable nucleus has so far been detected experimentally among the isotopes of boron, though there are grounds for assuming that $^{16}B$ is neutron-unstable while $^{15}B$, $^{17}B$, and $^{19}B$ are expected to be nucleon-stable.

In the following section we shall consider the conclusions that can be drawn from the available experimental data.

**Table 15. Predicted Decay Energies for Very Heavy Be Isotopes[a]**

| Nucleus | Mass excess (MeV) | Decay | Decay energy (MeV) |
|---|---|---|---|
| $^{15}Be$ | 51.18 | $^{14}Be + n$ | 2.42 |
| $^{15}Be$ | 59.22 | $^{14}Be + 2n$ | 2.4 |
| $^{17}Be$ | 70.67 | $^{16}Be + n$ | 3.37 |
| $^{18}Be$ | 78.43 | $^{16}Be + 2n$ | 3.1 |
| | | $^{14}Be + 4n$ | 5.45 |
| | | $^{12}Be + 6n$ | 4.92 |
| | | $^{10}Be + 8n$ | 1.25 |

[a] Data from (Aj 82, Aj 83, Aj 86).

**Table 16. Some Nuclear Reactions for Heavy Be Isotope Production**

| Nucleus | Reaction | $Q$ (MeV) |
|---|---|---|
| $^{15}$Be | $^{14}C + ^{18}O \rightarrow ^{17}Ne + ^{15}Be$ | −63.3 |
| $^{16}$Be | $^{14}C + ^{22}Ne \rightarrow ^{20}Mg + ^{16}Be$ | −79.7 |

## 5. Some Properties of Neutron-Rich Nuclei

In Section 1 we formulated some conclusions about the properties of neutron-rich nuclei, based upon our knowledge of nuclei near the stability valley. Let us now analyze these conclusions from the standpoint of the experimental data considered in the previous section.

### 5.1. Stability of the Lightest Nuclei as a Function of Neutron Excess

The first, and possibly most important, question is what one can say about the stability of nuclei as a function of the neutron excess $\eta = (N - Z)/A$. The averaged theoretical dependence determined (as noted in Section 1) by the symmetry energy is directly connected with the $\eta$ value. Even–odd effects and shell corrections manifested in specific experimental values of the binding energy modify this dependence. However, the primary dependence is expected to be that upon the neutron excess.

In the present section we wish to generalize the data considered in the previous section on the stability of the lightest nuclei and thus to formulate a prediction for heavier nuclei. It bears emphasis that such an extrapolation may be unreliable in view of the clearly defined individuality of the light nuclei.

All three types of nuclear systems mentioned in Section 1 are observed in the region of the lightest neutron-rich nuclei: radioactive systems stable against neutron decay, neutron-radioactive nuclei with finite observable decay widths, and neutron-decaying systems whose width is so large that it is effectively unmeasurable.

We consider these three types of nuclei in greater detail. Table 17 presents data on neutron binding energies in the most neutron-rich isotopes of each element from hydrogen to boron, separately for nuclei with even and odd neutron numbers. The neutron excess for the last stable nucleus with an odd $N$ in the case of lithium, beryllium, and boron turns out to be rather small ($\eta = 0.25$–$0.29$) and, as follows from Figure 5, does not apparently coincide with those values of $\eta$ which are observed for heavier nuclei (for a quantitative comparison Table 17 presents a value of $\eta = 0.37$ for the heaviest nucleon-stable isotope of carbon $^{19}$C). The steep decrease of data for the lightest nuclei is due to the features of their individual structure, i.e., the above-mentioned phenomenon which should be always taken into account in any extrapolations to heavier nuclei.

The above comments imply that the data on the odd nuclei agree, on the whole, with the prediction that neutron instability takes place at $\eta \gtrsim 0.36$.

The data on the lightest nuclei with even numbers of neutrons show convincingly that such nuclei can have a neutron excess of the order of $\eta = 0.45$–$0.50$. Here the key

**Table 17. Neutron Excess and Binding Energy for the Heaviest Nucleon Stable Isotopes of Very Light Elements**

| | Isotopes with odd neutron number | | | Isotopes with even neutron number | | |
|---|---|---|---|---|---|---|
| Element | Isotope | $\eta = (N - Z)/A$ | $\varepsilon_n$(MeV) | Isotope | $\eta = (N - Z)/A$ | $\varepsilon_n$(MeV) |
| H | $^2$H | 0 | 2.22 | $^3$H | 0.33 | 6.26 |
| He | | | | $^8$He | 0.50 | 2.14 |
| Li | $^8$Li | 0.25 | 2.03 | $^{11}$Li | 0.45 | 0.16[b] |
| Be | $^{11}$Be | 0.23 | 0.50 | $^{14}$Be | 0.43 | 1.12[b] |
| B | $^{14}$B | 0.29 | 0.97 | $^{19}$B | 0.47 | 0.4[a,b] |
| C | $^{19}$C | 0.37 | 0.53[a] | | | |

[a] Calculated value.
[b] Binding energy with respect to two-neutron decay.

problem remains the same: to what degree can these results be extrapolated to high $A$. The value of $\eta = 0.45$ corresponds to such nuclei as $^{72}$Ca, $^{182}$Sn, and $^{334}$U and in general complies with some predictions; see, for example, work by Bogatin *et al.* (Bo 80) where the neutron-drip-line going through $^{70}$Ca, $^{176}$Sn, and $^{302}$U is presented. To what extent the abnormal stability of $^8$He having $\eta = 0.5$ is an exception or a rule should be verified using other light isotopes. Such nuclei as $^{16}$Be, $^{24}$C, $^{32}$O, etc., correspond to a value of $\eta = 0.5$. However, nothing is known about their existence at present.

We now turn to metastable nuclei. Table 18 presents the main properties of all presently known nuclei that can be treated as neutron-radioactive ones. The minimum value of the neutron excess at which this instability is observed is $\eta = 0.40$ ($^{10}$Li), and if we do not take into consideration $^5$He, the maximum value is $\eta = 0.67$ ($^6$H); we also exclude the dineutron from these considerations.

**Table 18. Nuclei Metastable with Respect to Neutron Decay**

| Nucleus | $\eta = (N - Z)/A$ | Decay mode | Angular momentum of emitted neutron ($l$) | Decay energy (MeV) | Width $\Gamma$ (MeV) |
|---|---|---|---|---|---|
| $^2$n | 1.0 | n + n | 0 | 0.12 | 1.0 |
| $^4$H | 0.50 | $^3$H + n | 1 | 3.4 | ~3 |
| $^6$H | 0.67 | $^3$H + 3n | | 2.7 ± 0.4 | 1.8 ± 0.5 |
| $^5$He | 0.20 | $^4$He + n | 1 | 0.89 | 0.60 ± 0.02 |
| $^7$He | 0.43 | $^6$He + n | 1 | 0.44 ± 0.03 | 0.16 ± 0.03 |
| $^9$He | 0.55 | $^8$He + n | 1 | 1.14 ± 0.12 | ≤1.0[a] |
| $^{10}$Li | 0.40 | $^9$Li + n | 1 | 0.80 ± 0.25 | 1.2 ± 0.3 |
| $^{13}$Be | 0.38 | $^{12}$Be + n | (0.2) | 1.8 ± 0.5 | 0.9 ± 0.5 |

[a] Estimated.

Such a spread in the neutron-excess values is very large. Indeed, values of $\eta = 0.4$ or somewhat higher are in good agreement with our earlier conclusions regarding neutron-stable nuclei.

Quite a different situation occurs if we assume that there can exist metastable neutron-decaying nuclei with the same neutron excess as that of $^6H$. Such nuclear systems as $^{121}Ca$, $^{303}Sn$, and $^{558}U$ correspond to $\eta = 0.67$. It is difficult to imagine the possibility of their existence, however, even as a very short-lived resonance state.

In any case, if the values of the neutron excess, at which the neutron-decaying lightest nuclear systems are observed, are realized in the region of heavy nuclei, the neutron drip-line should instead be a very wide region in which the existence of hundreds or even thousands of neutron-radioactive nuclei is possible (Figure 2).

We now consider the third group of the lightest nuclear systems, namely $^4n$, $^5H$, and $^7H$, which have not been observed as residual nuclei. Naturally, the failure to observe them under the experimental conditions discussed in Section 4 does not mean for certain that these nuclei do not exist even as short-lived unstable systems. Nevertheless, one can apparently state that their widths are considerably larger than those given in Table 18 and amount to no less than 5–10 MeV. The estimates of the limiting width (4.9) will also fall within this energy range. Consequently, there are grounds for considering that the above nuclear systems already fall in the "neutron sea of instability." In Sections 5.5 and 5.7 we shall try to answer the question as to why $^6H$ is observed and $^5H$, having a smaller value of $\eta$, is not observed.

## 5.2. Helium Anomaly

We consider the problem of the stability of neutron-rich nuclei from another point of view. According to the semiempirical mass formula, the binding energy of the neutron in the nucleus should decrease smoothly with the number of neutrons. As a result of nucleon pairing this dependence should be treated separately for nuclei with even and odd numbers of neutrons, and shell effects cause breaks in the resulting monotonic behavior.

A universal dependence, however, appears; stability is always decreased with the addition of a pair of neutrons. An abnormal behavior of $\varepsilon_n$, such as an increase and not a decrease with increasing number of neutrons, is rarely observed. The abnormal dependence generally falls within the experimental error limits, but two or three cases are known (for example, the $^{15}N$–$^{17}N$ pair) where the stability increase is definite though very small (about 100 keV).

The strongest stability increase with number of neutrons is observed for the $^6He$–$^8He$ pair, where is exceeds 1 MeV. If we consider the binding energies of odd isotopes of helium, it turns out that the transition from $^5He$ to $^9He$, i.e., the addition of four neutrons, does not effectively change the binding energy. Thus, we deal with an effect which was called (Al 84) the "helium anomaly" (Figure 48). It is interesting primarily because it takes place for nuclei lying on the stability line. Since a microscopic explanation of this unusual stability of the superheavy isotopes of helium is not available, there arises the question as to whether it is a reflection of the fact that these nuclei have a neutron excess much larger than is the case in conventional nuclei. Or is it a qualitatively new phenomenon.

Presently, it is not possible to test any such hypothesis on other nuclei. The only pair of nuclei with known masses, which have a very large neutron excess, are the

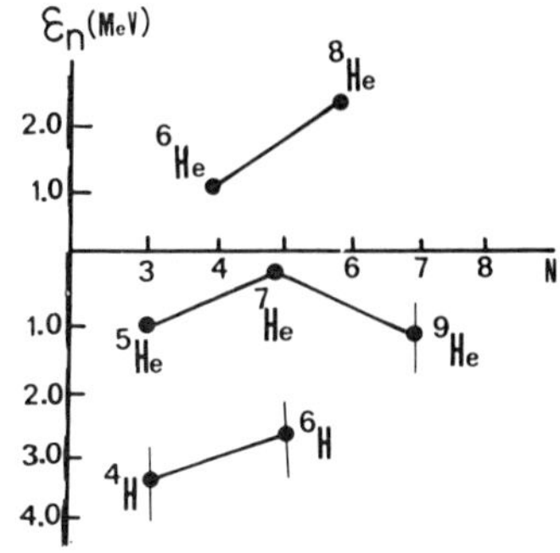


**Figure 48.** Neutron binding-energy dependence on neutron number for H and He isotopes.

superheavy unstable isotopes of hydrogen, $^4H$ and $^6H$. Their decay energies are also given in Figure 48. Though the errors are rather large, one can see that in the transition from $^4H$ to $^6H$ the instability, at least, does not increase, i.e., the same effect as for the isotopes of helium is observed.

Certainly, the helium anomaly can be the result of some fluctuations in the interaction. For example, it may turn out that it is not the abnormal stability of $^8He$ that takes place but an abnormal instability of $^6He$ due to the specific nature of the interaction of the $\alpha$ particle with two neutrons. A quantitative explanation of the nature of the helium anomaly is, we believe, one of the most important problems in the theory of the lightest neutron-rich nuclei. It should be noted that no matter whether a new phenomenon or some fluctuation is involved here, the very fact of the existence of an abnormal increase in stability with number of neutrons strongly enhances the possibility of observing separate stable nuclei with very large neutron excess or even islands containing a great number of neutron-radioactive nuclei.

## 5.3. Dimensions of Neutron-Rich Nuclei

The neutron-rich nuclei near the stability line have a comparatively low binding energy (we refer here to neutron stable nuclei). Their dimensions can be considerably larger than those of conventional nuclei. The small binding energy of the last neutron $E_n$ results in an elongated tail of the wave function of relative motion of this neutron. For example, if the neutron is in a p state, the wave function has the following form outside the interaction region (Ba 60):

$$R_n(r) = \left(1 + \frac{1}{n^r}\right) e^{-\chi_n r}, \qquad \chi_n = \left(\frac{2\mu\varepsilon_n}{\hbar^2}\right)^{1/2} \tag{5.1}$$

At $\varepsilon_n = 1$ MeV (assuming $\mu \simeq 1$) the tail of the neutron wave function is spread out over a distance of ~5 fm, while at $\varepsilon_n = 0.1$ MeV it reaches to 15 fm. Evidently, the radius of such a neutron-rich nucleus may differ strongly from the standard $R = r_0 A^{1/3}$ value.

Until recently, it was not possible to measure the radii of radioactive nuclei, such as $^8He$ or $^{11}Li$; recently, however, measurements were carried out with the availability of radioactive beams (Ta 85b). Beams of secondary nuclei, including those of various isotopes of helium and lithium up to $^8He$ and $^{11}Li$, were obtained in projectile fragmentation reactions, for example, using $^{11}B$ and $^{22}Ne$ ions with energy about

800 MeV/nucleon. By measuring the total cross-sections $\sigma_{tot} = \pi(R_1 + R_2)^2$ the values of the radii were obtained. It turned out that the radii of helium nuclei, $^8$He included, are in good agreement with the $A^{1/3}$ dependence. For $^{11}$Li it was observed that there was a marked increase in the radius as compared with that expected according to the $A^{1/3}$ law. It should be noted that the binding energy of the last neutron in $^{11}$Li amounts to only 0.16 MeV.

Though this observation is so far the only one, it is a strong argument in favor of the fact that weakly bound nuclei near the neutron drip-line have abnormally large dimensions. At the same time the data on $^6$He and $^8$He show that, at binding energies $\varepsilon_n \gtrsim 1$ MeV, the radius increase is negligible.

## 5.4. Decay Widths of the Lightest Neutron-Rich Nuclei

We return to Table 18 and ask which factors determine the widths observed and, consequently, the lifetimes of the neutron-decaying nuclei. As noted in Section 1, the main physical reason for the delay of the neutron emission with $l > 0$ is the centrifugal barrier. The angular momentum of the escaping neutron is known for five nuclei (presented in Table 18) that undergo single-nucleon decay and equals $l = 1$ in each case. The width can be estimated by using the formulae of $R$-matrix theory, (4.10). The parameters $\gamma^2$ and $r_0$ in the expression for the radius $R = r_0(A_1^{1/3} + A_2^{1/3})$ were determined on the basis of the data concerning $^5$He and $^5$Li which, in their turn, were obtained by approximating the scattering phase $\delta_{3/2}^-$ in the energy range 0–10 MeV for the systems $n + \alpha$ and $p + \alpha$. These and the calculations given below on the widths were made by one of us (AAO) in collaboration with L. V. Chulkov. The best agreement was obtained for values of $r_0 = 1.15$ fm and $\gamma^2 = 1.35\hbar^2/\mu R^2$, i.e., for $\gamma^2$ rather close to the Wigner limit.

The results of these calculations are given in Table 19. For comparison the same table presents data single-proton decay of some nuclei in this same region. It is clear from Table 19 that the experimental neutron widths are reproduced well with accuracy to a factor of 1.5–2.0. Comparison with proton-decay data makes this agreement still more convincing. The accuracy of calculating the proton widths turns out to be approximately the same and, moreover, it is clear that in some cases (such as $^{15}$F decay from the ground and excited states) it is the penetrability of the centrifugal barrier and not that of the Coulomb barrier that is crucial in determining the lifetime.

The difference between the characteristics of neutron and proton decaying nuclei is quantitative, reflecting the absence of the Coulomb barrier for the neutron-radioactive nuclei. In the lightest nuclei, where Coulomb forces are not large, the difference between the proton and neutron decays is not as significant as it is in heavy nuclei. If it ever becomes possible to reach the neutron drip-line in the region of heavy nuclei, where external neutrons can be in states with orbital momenta up to $l = 7$ (the $j$ shell), the penetrability of the centrifugal barrier might become of the order of $10^{-13}$ at decay energies of the order of a few hundred keV and, consequently, the lifetimes of the neutron-radioactive nuclei would reach values where direct measurements are feasible.

## 5.5. Nuclear Systems with Even Numbers of Neutrons

Another extremely interesting fact which follows from studying the systematics of unstable neutron-rich nuclei is the fact that all nuclei observed up to now, except the

**Table 19. One-Nucleon Decay Widths of Lightest Nuclei**

| Nucleus | Decay mode | Angular momentum of emitted neutron or proton ($l$) | Decay energy (MeV) | $\Gamma_{exp}$ (MeV) | $\Gamma_{calc}$ (MeV) |
|---|---|---|---|---|---|
| | | | Neutron decay | | |
| $^4H$ | $^3H+n$ | 1 | 3.4 | 3 | 4.0 |
| $^5He$ | $^4He+n$ | 1 | 0.89 | $0.60 \pm 0.02$ | 0.80 |
| $^7He$ | $^6He+n$ | 1 | 0.44 | $0.16 \pm 0.03$ | 0.30 |
| $^9He$ | $^8He+n$ | 1 | 1.14 | $\leqslant 1.0$ | 1.47 |
| $^{10}Li$ | $^9Li+n$ | 1 | 0.80 | $1.2 \pm 0.3$ | 0.87 |
| | | | Proton decay | | |
| $^5Li$ | $^4He+p$ | 1 | 1.97 | 1.50 | 1.76 |
| $^9Be$ | $^8Li+p$ | 1 | 0.09 | $0.47 \times 10^{-3}$ | $0.2 \times 10^{-3}$ |
| $^9B$ | $^8Be+p$ | 1 | 0.18 | $0.54 \times 10^{-3}$ | $0.9 \times 10^{-3}$ |
| $^{11}N$ | $^{10}C+p$ | 1 | 2.31 | $0.74 \pm 0.20$ | 1.21 |
| $^{15}F$ | $^{14}O+p$ | 0 | 1.47 | $1.0 \pm 0.2$ | 0.83 |
| $^{15}F$ | $^{14}O+p$ | 2 | 2.43 | $0.24 \pm 0.03$ | 0.18 |
| $^{16}F$ | $^{15}O+p$ | 0 | 0.55 | $0.04 \pm 0.02$ | 0.02 |

dineutron, have an odd number of neutrons. Systems with an even number of neutrons, such as $^4n$, $^5H$, and $^7H$, have not been detected. At first glance this seems strange, as experience gained in nuclear physics suggests that, because of nucleon pairing, the even nuclei are always more stable than the odd ones. The considerations and estimates presented in Section 4 indicate, in fact, that $^5H$ and $^7H$ are close to the decay threshold.

Certainly, it may turn out that the decay energy of the above-mentioned nuclear systems is still high due to some unknown reasons. However, it is natural to assume that a specific effect takes place which, in spite of the low decay energy, results in an unobservably large width. It was assumed (Al 84) that this effect is the pairing of two nucleons to form a dinucleon which then may have zero orbital momentum with respect to the residual nucleus and, consequently, it can be emitted without overcoming the centrifugal barrier. This assumption corresponds to the ideas (Go 60) about two-proton decay of nuclei.

Consider a nucleon being at a quasi-stationary level $E_1$ and blocked by the centrifugal barrier (Figure 49). The addition of a second nucleon brings about a change in the energy of this state. It will become approximately equal to

$$E_2 = 2E_1 - \varepsilon \tag{5.2}$$

where $\varepsilon$ is the pairing energy.

As a result, two situations are possible. If the energy gain due to pairing is high enough, the new two-nucleon state can go over into the discrete spectrum. Then the even (with respect to neutrons) nucleus will become stable. This effect is observed in transition from $^5He$ to $^6He$, from $^7He$ to $^8He$, and in a number of other cases. However, another situation is also possible, in which the energy gain due to pairing is not sufficient

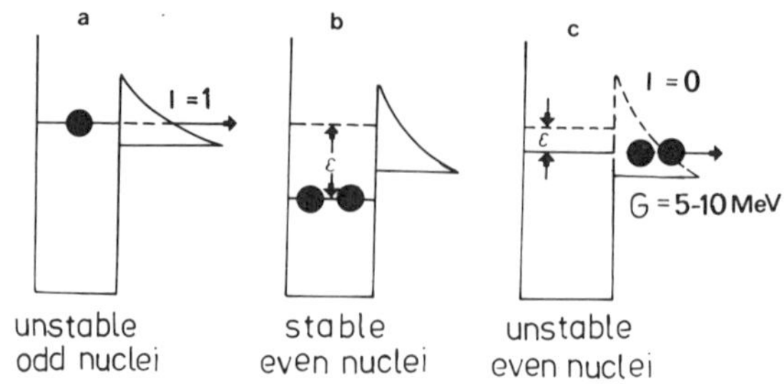


**Figure 49.** Schematic representations of the centrifugal-barrier influence on the decay of nuclei with odd and even numbers of neutrons in $l = 1$ states. (a) One neutron overrides the centrifugal barrier. (b) When a neutron is added to make an even number, the pairing energy is high enough and the system has a discrete spectrum. (c) When a neutron is added to make an even number, the system has a continuous spectrum and disintegrates by two-neutron emission. After pairing, a "dineutron" may appear with $l = 0$ relative to the residual nucleus and the decay takes place without penetration through the centrifugal barrier.

to bind the nucleus and the two-nucleon level remains in the continuous spectrum. The formed nucleus should be susceptible to many-particle (in the simplest case, a three-particle) decay. Such a situation takes place in transition from $^4$H to $^5$H, from $^6$H to $^7$H, and apparently from $^9$He to $^{10}$He. The realization of one of the two possibilities depends on the position of the single-particle level, the pairing energy, the decrease in well depth due to the increase in the neutron excess, i.e., it is a detailed quantitative problem.

In fact, in this case we deal with a three-particle problem in which we would like to know to what extent the interaction of two neutrons in the final state changes the probability of their escape. This is a very complicated problem in its general form. We shall consider decays with the emission of two nucleons in greater detail below. Presently, we only note that the existence of unstable nuclear systems with an even number of neutrons turns out to be immediately connected with the problem of the mechanism of two-nucleon decays.

## 5.6. Spectroscopy of Nuclei at the Stability Line

Spectroscopic studies of nuclei close to the neutron drip-line are particularly interesting. First, because the structure of these nuclei may differ strongly from that of other nuclei, they can have an unusual spectrum of levels. Second, the lower levels of nuclei with a large neutron excess are inevitably near the neutron-emission thresholds and the neutron decay channels are, effectively, the only open ones. In this respect, the excited states of neutron-rich, stable nuclei simulate the ground states of unstable ones and one can obtain information on the structure of the latter by studying the former.

At present, very little is known about the levels of even nuclei with large neutron excess. Spectroscopic information is available for some of the heaviest stable nuclei as presented in Table 17. Proceeding from it, we shall consider one example for each of the above-mentioned points of interest. We start with the decay modes. It was mentioned in Section 1 that if a competition among different decay modes was observed near the proton drip-line, no such diversity was expected close to the neutron drip-line and decays with the emission of neutrons would prevail there. But it becomes possible to study the many-neutron correlations, which in many respects determine the properties of these nuclei.

It has been mentioned in the previous section that the lifetime of even neutron-decaying nuclei is directly connected with the mechanism of two-neutron decay, which can be studied at the present time only via the excited states of stable nuclei.

If we return to the nuclei presented in Table 17, we see that excited states are observed for $^{8}$He, $^{11}$Be, and $^{14}$B, including those above the neutron-emission threshold. For $^{11}$Be this threshold is very low (0.50 MeV) and about 10 higher-lying levels are known, including rather narrow ones ($\Gamma \sim 10$ keV). The $^{8}$Li states lying above the threshold have, on the other hand, widths of the order of 1 MeV, with the exception of one state. For $^{11}$Be the 10.59 MeV state is known to have a width of about 100 keV, and lies higher than not only the $^{11}$Be $\rightarrow$ $^{10}$Be + n threshold but also higher than the $^{11}$Be $\rightarrow$ $^{9}$Be + 2n and $^{11}$Be $\rightarrow$ $^{8}$Be + 3n ones. Thus, the factors slowing down the decay are of great importance; but their nature is unknown. It should be mentioned that the neutron excess of these nuclei is not large and from this point of view they are of no particular interest. However, for study of the decay mechanisms the study of levels of such nuclei can be very informative, as we show in the next section.

Among the even nuclei listed in Table 17 excited states have so far been detected only for $^{8}$He. Their spectrum, according to the data of Alexandrov *et al.* (Al 82, Al 85), is presented in Figure 50. The relatively small width of the 2.6 MeV level is apparently connected with the low decay energy into available channels, which is the main difference between $^{8}$He and $^{7}$He, for which the predicted levels are located beginning, approximately, 2 MeV above the $^{6}$He + n threshold.

The question of whether there really exists a $^{8}$He level in the region of 7 MeV excitation energy is of great interest. Though the level has a comparatively low excitation energy, all four possible neutron channels are open for its decay and there are no other

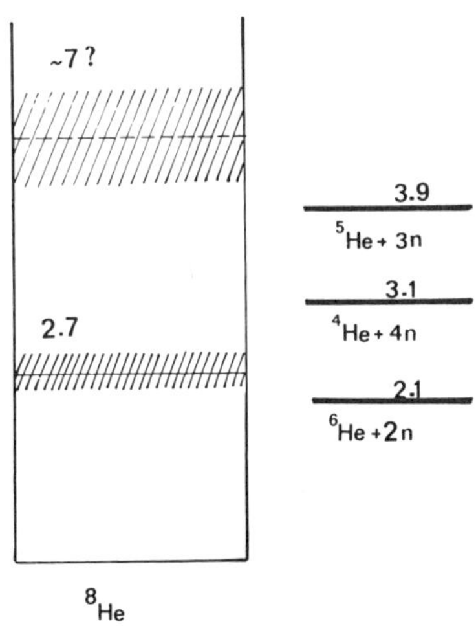


**Figure 50.** The level spectrum for $^{8}$He (Al 85).

open channels. Knowledge of the width of this level (quite apart from its decay modes) would allow one, in principle, to obtain information on multineutron correlations. The single- and two-neutron decays (in the absence of other open channels) with energy >1–2 MeV do not affect the finiteness of the width, which is demonstrated by the data on the spectrum of the $^{12}Be$ nucleus for which there are two wide states lying above the respective barriers.

Thus, the spectroscopy of neutron-rich nuclei can yield information on many-neutron correlations in nuclei, which is difficult to obtain otherwise. $^{14}Be$ is promising for the study of many-neutron decays, in addition to $^{8}He$ and, partly, $^{12}Be$. $^{14}Be$ has decay thresholds for 4, 5, and 6 neutrons at 4.8, 11.6, and 13.3 MeV, respectively, and the nearest nonneutron channel, $^{14}Be \to {}^{10}He + \alpha$, is no lower than 10.1 MeV. Certainly, it is possible to find other interesting cases as well.

We now consider the possible existence of nuclei "without a ground state" as an example of another phenomenon possible at the neutron drip-line. It is well known that the $^{5}He$ and $^{5}Li$ ground states have a considerable width (~1 MeV) determined by the probability for single-nucleon decay and, at the same time, in the region of excitation of ~16 MeV these nuclei have comparatively narrow levels corresponding, for example, to the d + t structure ($^{5}He$). For unstable neutron-rich nuclei it may turn out that the ground state will have $l = 0$, so that its decay width may be very large and thus be unobservable. At the same time, the excited states may have comparatively narrow widths.

It is possible that $^{13}Be$ is such a nucleus. It has 4 protons and 9 neutrons. Consequently, its configuration is $(1s)^2(1p)^2$ for protons and $(1s)^2(1p)^6(2sd)^1$ for neutrons. Then, it is natural to expect that the spin of $^{13}Be$ will be either $\frac{1}{2}^+$ or $\frac{5}{2}^+$. To determine which of the two values is most probable, consider the change of the spin

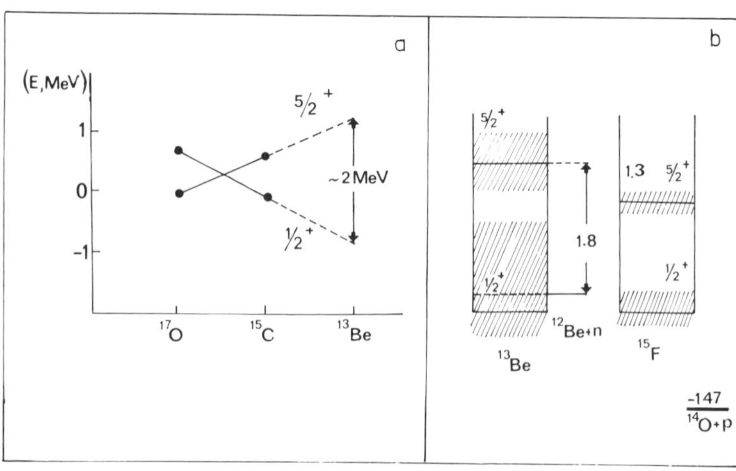


**Figure 51.** Supposed level scheme for $^{13}Be$. (a) Position of $\frac{1}{2}^+$ and $\frac{5}{2}^+$ levels in nuclei with 9 neutrons; extrapolation shows that $^{13}Be$ may have a $\frac{1}{2}^+$ ground state and a $\frac{5}{2}^+$ first excited state. (b) Comparison of the supposed level scheme of $^{13}B$ with the known levels of $^{15}F$ that has 9 protons. The $^{15}F$ g.s. has $J^{\pi} = \frac{1}{2}^+$ and, as a result, its width is greater than the width of the excited state.

for nuclei with the same neutron configuration but having one or two proton pairs more (Figure 51). We see that inversion takes place with decreasing number of protons: in the transition from $^{17}$O to $^{15}$C, the configurations of the ground and excited states are exchanged and the spin becomes $\frac{5}{2}^+$ instead of $\frac{1}{2}^+$. If this tendency is preserved for $^{13}$Be as well, its ground state should have spin $\frac{1}{2}^+$ and the $\frac{5}{2}^+$ state will have an excitation energy of about 2 MeV. The data on the spectrum of $^{11}$Be having, as does $^{13}$Be, 4 protons but 2 neutrons less, suggest that this is quite possible. The ground state of this nucleus is $\frac{1}{2}^+$ (i.e., the last neutron is the 2s shell, like that of $^{13}$Be), while the $\frac{5}{2}^+$ level lies at an energy of 1.78 MeV. Another comparison, namely, with $^{15}$F, also supports this hypothesis. $^{15}$F has the same number of protons (9) as that of neutrons for $^{13}$Be (and an even number of neutrons, i.e., 6). The ground state of $^{15}$F is $\frac{1}{2}^+$ and the first excited state is $\frac{5}{2}^+$ at an energy of 1.3 MeV.

Thus, there are grounds to expect that $^{13}$Be has spin $\frac{1}{2}^+$ in the ground state and, moreover, an excited state at an energy of 1.5–2 MeV with spin $\frac{5}{2}^+$. We immediately encounter difficulties in explaining the width of the observed state of $^{13}$Be, however. If it is a $\frac{1}{2}^+$ state, with the emission of an $l = 0$ neutron and a decay energy equal to 1.8 MeV (Al 84), the width will be considerably larger than the measured value of 0.9 MeV. This can be seen from comparison with $^{15}$F: at a decay energy of 1.5 MeV the width amounts to 1 MeV, in spite of the presence of the Coulomb barrier which is absent in the case of $^{13}$Be.

But there exists an alternate hypothesis. It is possible that in the experiments on $^{13}$Be detection we observe the $\frac{5}{2}^+$ excited state and not the ground state. The width of its decay is determined by the penetrability of the centrifugal barrier for $l = 2$ neutrons. The ground state possibly has an unobservably large width. Comparison with $^{15}$F once again suggests such a situation, i.e., the excited state of this nucleus has a width that amounts to one fourth that of the ground state. In this case $^{13}$Be may turn out to be unique: it has an excited state but does not have a ground state. More accurate experimental data are required.

## 5.7. Two-Nucleon Decays

We turn our attention to two-nucleon decays. Goldansky was the first to emphasize their importance in the understanding of phenomena near the stability line (Go 60). He predicted a new type of radioactive decay, i.e., double-proton radioactivity. The physical grounds for this process (which has not been discovered) are based on the assumption of the underbarrier correlation of the two emitted protons into a "diproton," which can considerably accelerate the decay as compared with a simultaneous emission of two single protons.

Since Goldansky's suggestion concerning double-proton radioactivity forms the basis of an assumption (Al 84) that two-neutron decaying nuclei might have very large widths, we consider briefly, following Baz *et al.* (Ba 72), the main model relations obtained for double-proton decay.

The probability of emission of a diproton as a unit with $l = 0$, from a nucleus with charge $Z$ and mass number $A$, equals

$$\lambda_{2p} \simeq K_{2p} \exp(-G_{2p}) \tag{5.3}$$

where the preexponential factor $K_{2p}$ characterizes the formation probability of the double-proton configuration while $G$ is the quasi-classical penetrability given by

$$G = \frac{(8\mu)^{1/2}l^2}{\hbar} \frac{2(Z-2)}{(Q-\varepsilon)^{1/2}}[\arccos X^{1/2} - X^{1/2}(1-X)^{1/2}] \tag{5.4}$$

with $\mu = 2(A-Z)/A$; $X = (Q-\varepsilon)/B_{2p}$ is the ratio of the diproton decay energy to the Coulomb barrier value for the diproton. Assuming the diproton to be a particle, we must decrease the total energy of the double-proton decay $Q$ by the value of the "diproton resonance" $\varepsilon$ (see Table 8).

The probability of independent simultaneous emission of two protons at the same total energy of the double-proton decay $Q$ will be

$$\lambda_{p_1p_2} \simeq K_p \exp(-G_{p_1}) \exp(-G_{p_2}) \tag{5.5}$$

where all the values refer to single protons. It has been shown (Ba 72) that $\lambda_{p_1p_2}$ is largest when the protons escape with the same energy equal to $Q/2$; then

$$G_{2p} = \text{const}/(Q-\varepsilon)^{1/2} \qquad \text{and} \qquad G_{pp} = \text{const}/Q^{1/2} \tag{5.6}$$

The emission of a diproton with energy $(Q-\varepsilon)$ has the same probability as that of the simultaneous independent emission of two protons with energy $Q$, with accuracy up to the preexponential factor. Since we look at the competition between the two processes at a preset value of $Q$, the escape of single protons with momentum $l = 0$ turns out to be more probable.

The situation changes considerably if the protons are in an $l > 0$ state. In this case a single proton must overcome a centrifugal barrier in escaping. This barrier corresponds to the orbital momentum $l$. The set of possible momenta for the diproton is determined by the quantum numbers and the structure of the state under consideration. If a combination of two protons possessing momentum $l = 0$ with respect to the daughter nucleus is permitted, one can speak about the emission of the diproton involving only the Coulomb barrier. The pairing energy for the diproton is

$$R_d = \frac{\hbar}{(m\varepsilon)^{1/2}}[l(l+1)]^{1/2} \tag{5.7}$$

If $R_d < b = 2(Z-2)e^2/Q$, the diproton pair dissociates inside the Coulomb barrier and the protons go the rest of the way as single ones; if $R_d > b$, the diproton goes all the way as a single particle.

The estimates show that for $l > 0$ the formation of a diproton can make the decay faster by many orders of magnitude as compared with the emission of single protons.

Certainly, all the above considerations are based on the assumption that the decay of a specific nuclear state may take place by emitting a diproton or, in a general case, a dinucleon treated as a unified particle.

In the case of actual nuclear decays the situation becomes more complicated, since we have to consider competition between the emission of the dinucleon and two single nucleons, as well as that between the simultaneous decay to which both the above-mentioned cases belong and the successive decay process. Figure 52 shows different versions of the decay. The nucleus may emit two nucleons (now we speak about neutrons for the sake of definiteness) simultaneously making a transition from the $N,Z$ state directly into the $(N-2), Z$ state, or, if there exists an energy-available intermediate state $N-1, Z$, successively, via this state. So far, we have been considering only the

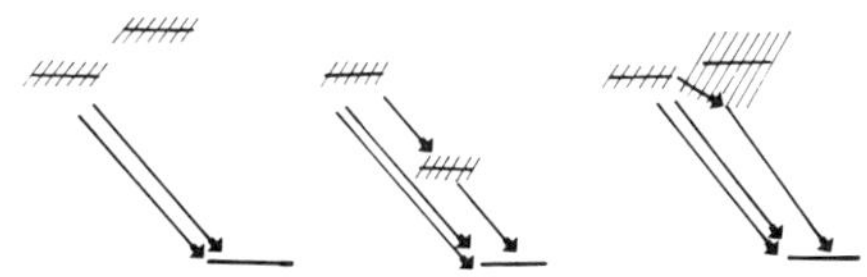


**Figure 52.** Various possibilities for two-nucleon decay: (a) Only the synchronous emission of the two nucleons is possible. (b) Synchronous as well as sequential emission of the two nucleons is possible. (c) Decay through an intermediate state is possible. The separation into synchronous and sequential emission is conditional.

situation shown in Figure 52a, in which the $(N-1)$, Z nucleus level lies higher than the decaying nucleus level $N$ and cannot be formed in the successive decay process except in a virtual process. Actually, however, the situation shown in Figures 52b and 52c is realized when the intermediate state can be populated as a result of the emission of a single nucleon: this state lies either lower than the decaying two-nucleon state or higher, but it has such a large width that it can still participate in the decay. In this case, if the widths of the two-nucleon and intermediate decays are small compared with the decay energy, as shown in Figure 52b, one can differentiate between the simultaneous and successive emission of nucleons; in the case of large widths or the situation shown in Figure 52c, such a separation becomes rather artitrary. It will be shown that precisely this case occurs in the light nuclei and hence considerably complicates the analysis.

It has already been mentioned in Section 5.5 that to clarify the nature of two-nucleon decays, one should solve the problem of three bodies from both the experimental and the theoretical points of view. The first steps in this direction, which propose the performance of a kinematically complete experiment for the decay of nuclei with $A = 6$ and analysis of the results using the method of $K$-harmonics, are being taken at present but it is too early to present them here. Rather we simply formulate the problem as follows: can the available data on nuclei capable of decay via two-nucleon emission yield information as to how this decay happens and, in particular, it is possible to consider any features of these decays as a qualitative indication of the emission of the dinucleon.

First, consider the data on the widths of states in which the two-nucleon decay is possible. Table 20 presents the characteristics of such states taken from reviews (Aj 84, Aj 85, Aj 82). They belong to the type shown either in Figure 52b or in Figure 52c. "Pure" types of two-nucleon decay (Figure 52a) are not observed in practice. Two such cases are given in Table 20, but one should bear in mind that competing single-nucleon decays are not absolutely prohibited.

The model analysis of the widths of these nuclei was performed by L. V. Chulkov and one of the authors (AAO). Without going into details, they failed to arrive at a definite conclusion regarding dinucleon emission. In some cases the total widths are well reproduced on the assumption of predominance of one type of decay: for example, in the case of $^{9}C^{*}$ and $^{7}B$ it is the successive emission of protons while for $^{6}He^{*}$ it is dineutron emission that appears to dominate. The diproton escape and the emission of single protons play comparable roles in the $^{6}Be^{*}$ decay.

Table 20. Nuclei that Undergo Two-Nucleon Decay

| Decay mode[a] | Nucleus | $J^\pi, T$ | $E^*$ (MeV) | Width (keV) | Decay channel | $Q$ (MeV) |
|---|---|---|---|---|---|---|
| a | $^8$B | $0^+, 2$ | 10.62 | 60 | p + $^7$Be ($T = \frac{3}{2}$) | −0.39 |
| | | | | | $^2$He + $^6$Li ($T = 1$) | 1.31 |
| | $^8$Be | $0^+, 2$ | 27.49 | 5.5 ± 2.0 | p + $^7$Li ($T = \frac{3}{2}$) | −1.00 |
| b | $^6$Li | $2^+, 1$ | 5.37 | 540 ± 20 | p + $^5$He | 0.78 |
| | | | | | $^2$H + $^4$He | 1.67 |
| | $^6$Be | $2^+, 1$ | 1.6 | 1160 ± 60 | p + $^5$Li | 1.08 |
| | | | | | $^2$He + $^4$He | 3.04 |
| | $^7$B | $\frac{3}{2}^-, \frac{3}{2}$ | 0 | 1300 ± 200 | p + $^6$Be | 2.21 |
| | | | | | $^2$He + $^5$Li | 1.61 |
| | $^9$C | $\frac{1}{2}^-, \frac{3}{2}$ | 2.22 | 100 ± 20 | p + $^8$B | 0.92 |
| | | | | | $^2$He + $^7$Be | 0.78 |
| c | $^6$Be | $0^+, 1$ | 0 | 92 ± 6 | p + $^5$Li | −0.59 |
| | | | | | $^2$H + $^4$He | 1.37 |
| | $^6$He | $2^+, 1$ | 1.8 | 113 ± 20 | n + $^5$He | −0.07 |
| | | | | | $^2$n + $^4$He | 0.825 |
| | $^8$C | $0^+, 1$ | 0 | 230 ± 50 | p + $^7$B | − 0.13 |
| | | | | | $^2$He + $^6$Be | 2.14 |
| | $^{12}$O | $0^+, 2$ | 0 | 400 ± 250 | p + $^{11}$N | −0.41 |
| | | | | | $^2$He + $^{10}$C | 1.79 |
| | $^{16}$Ne | $O^+, 2$ | 0 | 110 ± 40 | p + $^{15}$F | 0.78 |
| | | | | | $^2$He + $^{14}$O | 1.43 |

[a] Decay mode chosen as shown in Figure 52.

The problem is, however, that for some decays one fails to obtain any self-consistent picture at all. One can see this particularly in comparing $^6$Be (ground) and $^{16}$Ne. These nuclei have practically the same energies for double-proton decay (1.37 and 1.43 MeV, respectively) and (within error) identical width. In both cases the orbital angular momentum of the diproton equals zero. At the same time, the charges of the nuclei resulting from diproton escape differ by a factor of four so that the Coulomb penetrabilities differ greatly. Indeed, the calculation predicts that the diproton widths of $^6$Be and $^{16}$Ne should differ by three orders of magnitude. For the emission of one proton the widths are small in both cases and amount to approximately one percent of the total width. If we assume, though it is not obvious for three-particle decay, that the calculation of the single-nucleon widths does not contain a large error (it was clear from Section 5.4 that agreement with experiment is within a factor of 2) this result means that there are some difficulties in principle in the description of the observed widths using a dinucleon escape model.

In this connection, it does not seem possible at present to answer the question as to whether decays with dinucleon escape are observed or not.

J. Cerny with his collaborators (Ca 84) carried out direct measurements from which such information could be obtained. They searched for double-proton decay of $^{22}$Mg and $^{26}$Si. After the first report that simultaneous emission of two protons had been observed, more accurate measurements and analyses showed that successive emission of protons actually took place.

Since all questions concerning two-nucleon decays are, as one can see, very important for solving the question of even nuclei at the neutron drip-line, of the double-proton radioactivity, and, in general, of the nuclear three-body problem, a cycle of investigations of such decays in $A = 6$ nuclei was initiated at the I. V. Kurchatov IAE (Al 87).

Isobaric multiplets of states of nuclei unstable against two-nucleon emission are extremely interesting subjects for study and for the possible provision of information concerning two-nucleon decays. In this case, when the n–n, p–p, and p–n pairs are emitted, the quantum characteristics and nuclear interaction of the decay particles are identical while differences are present only in the Coulomb interaction which results, in particular, in different values for the decay barriers. The isospin triplet with $A = 6$ ($I^\pi = 2^+$) is particularly convenient, since in this case the third product of the decay is the $\alpha$ particle whose interaction with nucleons is well known, and since excitation of its intrinsic degrees of freedom can be neglected. Though the decay of states of this triplet was studied previously (Ab 70, Ko 78, Ar 73, Wi 75), the available data were not complete and contradicted one another.

In the above-mentioned references nuclei with $A = 6$ were obtained in the $^6Li(^3He, t)^6Be$, $^7Li(d^3, He)^6He$, and $^7Li(d, t)^6Li$ reactions (Figure 53). The tritons or $^3He$ nuclei formed in the reactions were detected in coincidence with $\alpha$ particles appearing as a result of the $^6Be \to \alpha + 2p$, $^6He \to \alpha + 2n$, and $^6Li \to \alpha + p + n$ decays. The detection of $^3H(^3He)$ in the required energy range indicated the formation of the nucleus in the $2^+$ state (in the case of $^6Be$, in the $0^+$ state as well), while the spectrum of $\alpha$ particles yielded information on the character of the processes taking place. A special analysis has shown that, despite the comparable decay times of the nuclei being formed and of their fly-off times, the recoil nucleus accompanying the formation of the nucleus with $A = 6$ does not affect its decay.

In accord with the above discussion of various possible decays and the level scheme of the nuclei under investigation (Figure 53), it should be expected that the $2^+$ states can decay both successively via intermediate nuclei with $A = 5$ and directly via two-nucleon emission (the three-particle decay).

The spectrum of $\alpha$ particles in the case of $^6He \to \alpha + 2n$ decay is shown in Figure 54 as an example. It is clear that the phase volume (the dotted line) cannot account for the grouping of the $\alpha$ particles in the high-energy part of the spectrum. The same result was obtained for $^6Be$ and $^6Li$.

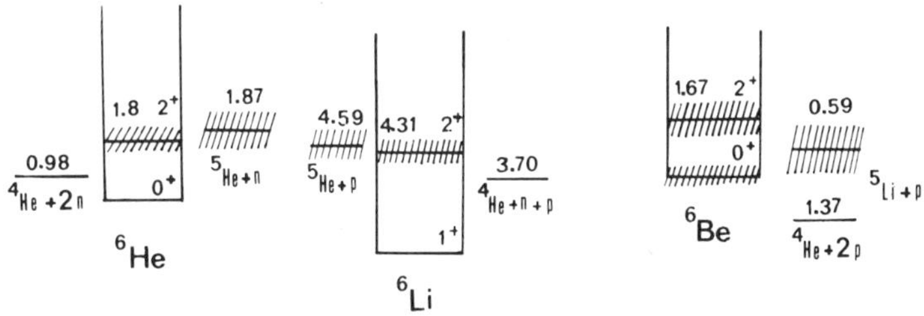


**Figure 53.** Level spectra for $A = 6$ nuclei.

In the case of $^6$He successive decay is possible through the tail of the wide state of $^5$He: $^6He(2^+) \rightarrow n + {}^5He \rightarrow n + n + \alpha$. The dynamic part of the amplitude $f_s^d$ of this process, which has two stages, is represented in a factorized form

$$f_s^d = f_I \cdot f_{II} \tag{5.8}$$

The amplitudes $f_I$ and $f_{II}$ of the two-particle decays at the first ($^6A \rightarrow N + {}^5A$) and second ($^5A \rightarrow N + \alpha$) stages were represented by Breit–Wigner expressions. Taking into account the spins of the particles involved in the process, the successive decay amplitude $f_s$ has the form

$$f_s = \sum f_s^d G \tag{5.9}$$

where $G$ is the angular part of the amplitude, which has a kinematic origin and is a combination of spherical harmonics and vector-addition coefficients providing for conservation of total angular momentum and taking into account that the intermediate system ($^5$He, $^5$Li) has $I^\pi = \frac{3}{2}^-$. The results of calculations on successive decay, normalized with respect to the experimental data, are shown in Figure 54. It is clear (and it holds true for the other two nuclei) that the successive decay model satisfactorily reproduces the observed spectra in the range of small $\alpha$-particle energies. However, in the region of the high-energy peak the theoretical prediction decreases monotonically with increasing energy.

Thus, neither the three-particle nor the successive decay model accounts for the appearance of marked peaks in the high-energy part of the $\alpha$-particle spectrum. Effects connected with the existence of a singlet virtual state in the $N$–$N$ system may be responsible for them.

The final-state interaction (FSI) was studied using a Migdal–Watson approach (Mi 55, Wa 52). The total amplitude of the process was presented in the form

$$|f_i|^2 \sim |f_r|^2 \cdot |\phi|^2 \tag{5.10}$$

where $f_r$ is the amplitude for the reaction-product formation. Among these products are two nucleons with relative motion wave function $\phi$. It is assumed, as often happens in studying the FSI in such reactions, that $|f_r|^2 = \text{const}$. In such cases, equations like (4.1)–(4.3) are obtained. For the case of p–p interaction the wave function $\phi$ includes account of Coulomb interaction. It is clear from Figure 54 that allowance for the FSI, being in fact the phase volume of the three-particle decay modulated by the function $\phi$, results in a wide distribution and does not explain the narrow peaks observed in the experiment. From a formal point of view, the calculated distributions in the high-energy part of the spectrum can be made narrow enough, if one takes very long (with respect to the modulus) scattering lengths ($a_{nn} = -50$ fm, $a_{pp} = -25$ fm instead of the known values of $-16$ and $-7.7$ fm, respectively). This would mean (naturally, in the framework of the $|f_r|^2 = \text{const}$ hypothesis) a decrease in the virtual energy of the pair of nucleons, i.e., their stabilization. However, the escape of a stable dinucleon naturally requires the selection of such $f_r$ which describe the two-particle decay.

These considerations suggest that it would be useful to perform calculations in which $f_r$ is preset as an amplitude for two-particle decay, i.e., by a Breit–Wigner expression. In this expression the penetrability for the system of $\alpha$ particles plus a "dinucleon" is the most significant factor. The function for the relative motion of the two nucleons was calculated for conventional scattering lengths, while the channel radius was preset as the sum of the $\alpha$-particle radius and an effective radius for the

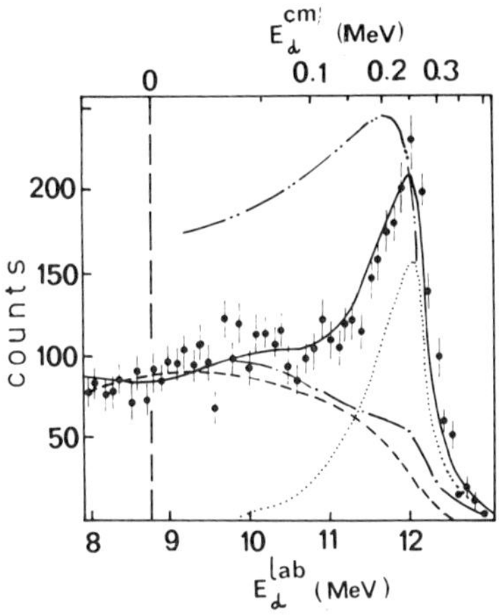


**Figure 54.** The $\alpha$-particle spectra from the decay of $^6He^*(2^+)$ nuclei formed in the $^7Li(d, ^4He)^6He^*$ reaction at $E_d = 30.7$ MeV ($\theta_\alpha = 39.9°$, $\theta_{^3He} = -80.0°$); $E_\alpha^{lab}$ is the $\alpha$-particle energy in the laboratory system and $E_\alpha^{cm}$ the $\alpha$-particle energy in the rest frame of $^6He^*(2^+)$. Different calculations are represented as follows: · —— · sequential decay of $^6He^*$ through $^5He$ (3/2), · · · "dineutron" emission, —— their sum, - - - phase-space calculation, · · —— · · final-state interaction of the two neutrons.

pair of nucleons. The results of such a calculation are given in Figure 54 by the dotted line. In all three cases the theoretical dependences satisfactorily reproduce the characteristic peaks in the high-energy part of the spectrum. It is interesting that this result turned out to be stable against severalfold variation in the channel radius, from $R = R_\alpha = 2$ to $R = R_\alpha + |a|/2$, despite the strong dependence of the magnitudes of the penetrability on $R$. This means that it is the two-particle character of the process itself and not its details that turn out to be significant for the reproduction of the spectrum form. The escaping dinucleon is not equivalent to interacting free nucleons but is a product of the effect of the third particle on them (in the given case this is the $\alpha$ particle).

Hence, these studies have shown that in the two-nucleon decay process in the lightest nuclei there is a noticeable component which can be treated as a two-body one with the escape of a resonance dinucleon. Within the framework of the considered model approach one can attribute approximately 50, 65, and 20% of the $^6He(2^+)$, $^6Li(2^+)$, and $^6Be(2^+)$ decay, respectively, to such emission.

Presently, it is not clear to what extent the character of the result obtained on the emission of the dinucleon is general. For now it should apparently be assumed that such a mechanism can play a significant part in the decay properties of the lightest nuclei and in many respects can determine their lifetimes. The detection of $^{10}He$ in its ground state would be of great importance for checking the hypothesis that dineutron emission sharply decreases the lifetimes of the neutron-radioactive nuclei with even numbers of neutrons ($^4n$, $^5H$, $^7H$). The calculations predict (see Section 4.3) that $^{10}He$ is not stable against single-nucleon decay $^{10}He \rightarrow n + {}^9He$ by 0.5 MeV and, according to the data of Table 19, this decay channel would have resulted in a width $\Gamma(1n) \sim$ 0.5 MeV. The emission of a dineutron with $l = 0$ and energy 1.66 MeV (according to the predictions) would have increased the width severalfold.

## 5.8. Delayed Decays

The study of neutron-rich isotopes of the lightest elements at the stability line makes it possible to observe different types of delayed processes. For these nuclei the

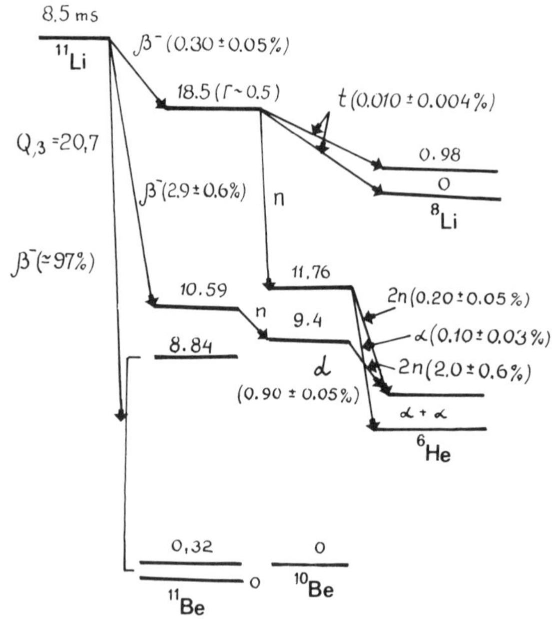


**Figure 55.** The $^{11}$Li decay scheme (La 84).

difference between the masses of the neighboring isobars (the ($\beta$-decay energy) may reach 2–3 tens of MeV and, as a result of $\beta$-decay levels with a high excitation energy, can be populated, which leads to a wide range of possibilities for different types of the decay: $\beta$-delayed two-neutron and three-neutron decay (Az 79, Az 80), emission of tritons, $\alpha$ particles, and heavier particles (La 84, De 83). From this point of view, the $^{11}$Li and $^{8}$He isotopes with a high $\beta$-decay energy are particularly interesting. For the $^{11}$Li isotope the value of $Q_\beta = 20.7$ MeV, so that all the above-mentioned types of decays are energetically possible. The emission of the $\beta$-delayed triton is energetically favorable for this nucleus and has a threshold at 15.72 MeV for the decay of $^{11}$Be into $^{8}\text{Li} + \text{t}$. This new type of decay was predicted for light nuclei far from the $\beta$-stability line (Be 70) and was detected experimentally (La 84). The $^{11}$Li decay scheme is shown in Figure 55.

Recently, a delayed escape of a triton after the $\beta$ decay of $^{8}$He was observed (Bo 86). The $\beta$-decay energy for this nucleus is $Q_\beta = 10.653$ MeV and the breakup thresholds for $^{8}$Li into $\alpha + \text{t} + \text{n}$ and $^{5}\text{He} + \text{t}$ equal 4.50 and 5.39 MeV, respectively. Both the breakup branches following the $\beta$ decay of $^{8}$He into a lithium level with energy 8.8 MeV and width $\Gamma \simeq 1$ MeV have been identified.

The processes accompanying the $\beta^-$ decay of neutron-rich nuclei were treated in detail by Lutostansky (Lu 86b). Their theoretical description is based on calculations of the $\beta$-decay strength function $S_\beta(E)$ which determines the population of the daughter nucleus levels in terms of an integral with the Fermi function. The function $S_\beta(E)$ has a complex structure of resonance type, which is based on the isobaric states (Figure 56). The most important part is played by the Gamow–Teller resonance tails and states of the core polarization type which are determined by the "proton–neutron hole"

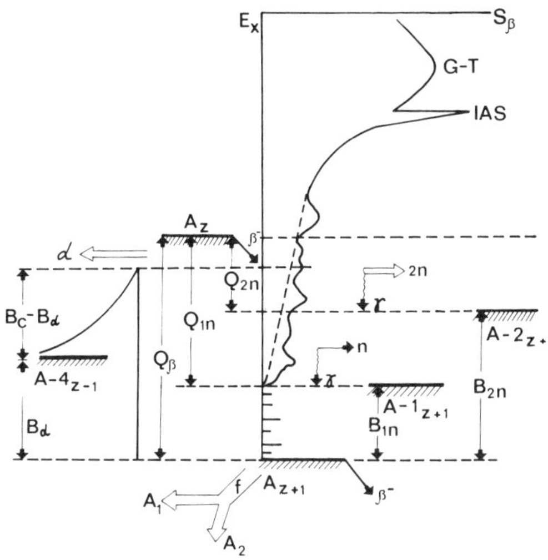


**Figure 56.** A schematic showing processes accompanying $\beta^-$ decay (Lu 86b).

configurations and connected with the total momentum flip. The latter are usually positioned in the $\beta$-decay range.

The possibility of the $(\beta^-, kn)$ process $(k = 1, 2, \ldots)$ is determined by the condition $Q_\beta > B_{kn}$, where $B_{kn}$ is the escape energy of $k$ neutrons in the daughter nucleus (see Figure 56). The analysis performed by Lutostansky (Lu 86b) shows that the range of nuclei undergoing delayed decay with emission of two neutrons is rather close to the boundary of already identified nuclei. For some of the detected nuclei a prediction of the three-neutron decay probability has been made.

The probability of $k$-neutron emission is defined by the relation

$$P_{kn} = \int_{B_{kn}}^{Q_\beta} \int_{0}^{Q_{kn}} I_\beta(U)\, W_n(U, E)\, dU\, dE \tag{5.11}$$

where $I_\beta(U)$ is the probability of isobaric state population in the daughter nucleus, and $Q_{kn} = Q_\beta - B_{kn}$. The internal integral over $U$ yields the spectrum of emitted neutrons determined by the probability of their escape $W_n(U, E)$, which is calculated using a statistical theory.

For rough estimates the following expression is applicable:

$$P_{kn} \approx 125(Q_{kn}/Q_\beta)^{4.35}(\%), \qquad k \geqslant 2 \tag{5.12}$$

The even–odd effects are disregarded in this expression, since at $k \geqslant 2$ they are already negligible.

The most comprehensive study of multineutron $\beta$-delayed emission has been performed for the Na isotopes. The experimental data and comparison with calculations were reviewed elsewhere (Lu 86b).

One of the most important questions is related to the possibility of emitting a correlated pair of neutrons (a dineutron). In $\beta$-delayed decays such a process has not yet been observed. The results of Lutostansky's calculations (Lu 86b) predict the probability of dineutron emission, $P_{2n}$, to be lower by several orders than that of the cascade mechanism $P_{n+n}$ (for nuclei in the region of $^{30}Na$–$^{55}K$). These calculations are purely statistical and do not take into account the possibility of neutron correlations in the surface, which may result in amplification of the process. At the present time, it does not seem possible to correlate these calculations even qualitatively with experimental data on two-nucleon decays examined in the previous section.

## 6. Conclusions

There have been several stages in the study of light neutron-rich nuclei during which interest in this problem passed through peaks and valleys.

The first stage dates back to the end of the fifties and beginning of the sixties. Already at that time, the basic theoretical approaches to describing the structure of the known light nuclei were developed and it was natural to use them to predict the stability of new ones. The first methods for calculating masses of unknown light nuclei appeared at that time: the extrapolation method, the approaches based on the use of isotopic invariance, and, finally, the Garvey–Kelson method. Predictions of the stability of $^{5}$H (Bl 64), of the tetraneutron, and, in particular, of $^{8}$He (Ze 60) played an important part in attracting attention to the lightest neutron-rich nuclei. Works devoted to the neutron-deficient nuclei, particularly the prediction of double-proton radioactivity (Go 60), also suggested the possibility of new phenomena appearing in the region of the stability line.

These and other theoretical studies stimulated numerous experiments on the lightest nuclear systems. The above stage ended in the detection of the bound state of $^{8}$He [the first observation, final confirmation, and measurement of the characteristics (Lo 61)] and the observation of two neutrons in the final state with the determination of the constant of their interaction (Vo 64). The synthesis of $^{6}$Be (Bo 57), which happened to be not only the first unstable nucleus outside the stability line but also the first nuclear system undergoing two-nucleon decay, should have been mentioned here. To our regret, this work was not properly recognized at that time.

The second stage of investigation lasted until the end of the seventies. Among the experimental work, one should mention that carried out in Berkeley and in Dubna, during which a number of proton- and neutron-decaying nuclei were synthesized and thus the proton and neutron drip-lines were determined in the region of the lightest nuclei. Such nuclei as $^{7}$He, $^{10}$Li, $^{11}$Li, and $^{14}$Be were found. However, at the same time there also appeared a certain pessimistic tendency as far as the prospects of further studies were concerned, since attempts to synthesize the tetraneutron, $^{5}$H, and $^{10}$He, failed. It became clear that experiments on the search for new nuclei were extremely difficult and the fact that after many long-run measurements a negative result was obtained did not help to make these investigations popular. The calculations, whose methods were constantly developed, did not yield encouraging predictions either and, in any case, they did not stimulate the experimentalist into performing measurements which could even qualitatively change the gradually formed idea that all important and available studies in the region of light neutron-rich nuclei had already been done.

By the beginning of the eighties, however, three independent programs of investigations had been started. Their realization has resulted in advances. First is a series of studies on reactions with $\pi^-$ mesons at the Los Alamos Meson Physics Facility which demonstrated the success of this new approach, in particular, the value of the charge-exchange reactions $(\pi^-, \pi^+)$. As a result, unstable $^{9}$He was discovered, much closer to the decay threshold than had been assumed. Second was the development and use of a new method of obtaining nucleon-stable nuclei in fragmentation reactions at

Berkeley and at GANIL. The appearance of beams of radioactive nuclei should be treated as a natural development of this method and is one of the most promising means of studying neutron-rich nuclei in the future.

Finally, a program of investigations of neutron-decaying nuclei has been undertaken at the cyclotron of the I. V. Kurchatov IAE in Moscow and at JINR, Dubna. This program's main results have been:

1. The synthesis of two new unstable nuclei, $^6H$ and $^{13}Be$, the first of which is the most neutron-rich while the second is the heaviest of the presently known neutron-radioactive nuclei.
2. The observation of decays with the emission of the unstable dinucleon, in particular, the dineutron as a unit entity (Bo 85).

It is important that these data not only considerably expanded our knowledge of neutron-decaying nuclear systems, but they allowed the formulation of a number of general questions pertaining to the existence of nuclei with large neutron excess.

All the main phenomena expected for heavier nuclei have also been observed in the region of the lightest ones. The proton drip-line has been reliably established. The neutron drip-line has been reached, in the sense that the existence of neutron-radioactive nuclei has been found. The lifetimes of unstable nuclei undergoing single-nucleon decay are satisfactorily explained by penetrability of the Coulomb and (or) centrifugal barriers. Neutron instability takes place approximately at the expected neutron-excess values. The neutron drip-line itself is a zone where decaying nuclei with odd numbers of neutrons alternate with stable even ones.

At the same time, unexpected phenomena have been observed for which nothing can yet be said about their analogs in heavier nuclei.

First, at large neutron excess $\eta = (N - Z)/A$ which is presently not available anywhere except in the lightest nuclei, a discontinuity was observed in the dependence of the binding energy on $\eta$ (Figure 57). The existence of this discontinuity does not stem from any theoretical concepts and it is not clear whether it is a characteristic of the lightest nuclei or whether its character is more general.

Second, the detection of $^6H$ shows that one proton can hold five neutrons during a period which exceeds severalfold the characteristic free fly-off time of the system. If this holds true for heavier nuclear systems as well, the existence of a great number of unstable neutron-decaying nuclei can be expected. As a result, the neutron drip-line turns into a zone comparable in its dimensions with the region of stable and $\beta$-radioactive nuclei.

Third, unstable nuclei with even numbers of neutrons can undergo two-neutron decay, as a result of which both neutrons escape in the form of a unified particle. If dineutron escapes with $l = 0$ are possible, this can sharply decrease the lifetime of such systems. The possibility of such an effect for heavier nuclei would mean that the zone of the neutron-radioactive nuclei would consist mainly of nuclei with odd numbers of neutrons. In the region of the proton drip-line, the possibility of dinucleon escape is the physical analog of the phenomenon of double-proton radioactivity.

At present, the challenge is to choose the most promising directions of research and the best approaches.

We shall now briefly discuss the most important unsolved questions. The first is that one should understand the physical nature of the helium anomaly. Microscopic

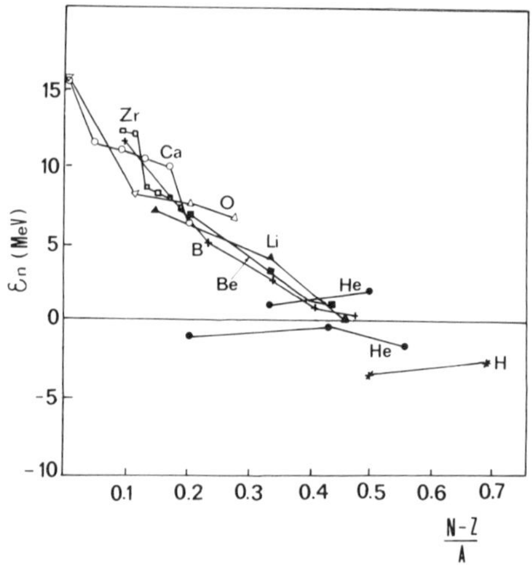


**Figure 57.** Mediated dependence of the neutron binding energy on the parameter $\eta = (N - Z)/A$, where $N$ is the neutron number, $Z$ the proton number, and $A$ the atomic mass of the isotope.

calculations are required so that one can decide whether it is a consequence of some random coincidences or whether the increased stability of the superheavy isotopes of helium and hydrogen is due to the fact that the neutron excess in them has reached some critical value.

The second problem is that one should determine experimentally whether helium and hydrogen nuclei with large values of neutron excess $\eta$ are an exception or a rule. The answer to this question implies the synthesis of nuclei with $\eta \geqslant 0.5$ for greater mass numbers. Presently, it is easiest to advance in this direction by using fragmentation reactions of moderately light nuclei.

The next question, which is also experimental, involves an advance to still larger neutron excesses than those reached at present. It means the search for unstable lightest isotopes which are still more neutron-rich than $^6$H. Proceeding from the analysis performed in the present review, special attention should be devoted to the synthesis of such nuclei from which the escape of neutrons can be slowed down by centrifugal barriers, i.e., those having an odd number of neutrons, such as $^8$H. It is evident that the fact that some nuclei with an even number of neutrons are not observed does not mean that the odd nucleus which follows does not exist.

It is natural that in searching for unstable nuclei the problem of multineutrons as purely neutron systems should occupy a special position. An experiment on the search for the pentaneutron $^5$n by using one of the nuclear reactions mentioned in Table 9 seems very important and quite realistic. It should be noted that the experiments performed using the procedures described in Section 3.4, which are aimed at searching for simultaneous escape of a great number of neutrons, are also feasible.

The question of how the two-neutron decay takes place is the next open problem. We have seen that the escape of two nucleons from the nuclei with $A = 6$ may be considered, in part, as the emission of a dinucleon. How does it take place for heavier nuclei? So far the widths of double-proton decay of a number of nuclei, such as $^6$Be and $^{16}$Ne, have failed to be explained. There are several ways of answering this question. One is to construct a theory of three-particle decay based on a kinematically complete experiment. Such work, as applied to nuclei with $A = 6$, is presently being carried out at the IAE. Another way is a direct study of decays of heavier nuclei. Probably, the search for $^{10}$He in the final state is the most important experiment to be conducted at present.

The question of explaining the decay of nuclei with the escape of three or more neutrons is absolutely open.

We have stressed the importance of spectroscopic investigations for understanding the structure of neutron-rich nuclei and have given several examples. We very briefly mention the prospects for new experimental investigations.

Two main trends in the search for, and study of, new nuclei have emerged: study of those which are stable against the emission of neutrons and those which are unstable.

Presently, the most promising area in the search for new stable nuclei comprises projectile fragmentation reactions at bombarding energies of tens and hundreds of MeV/nucleon. The use of beams of secondary radioactive nuclei is a logical development of this trend.

Various reactions with heavy ions and $\pi$ mesons are now used in the search for unstable nuclei. Their ability to compete depends to a considerable extent on the intensity of the respective beams. Without insisting on the cross-section values as well, which are higher in many cases in reactions with heavy ions, it is possible to show that in binary reactions with heavy ions there are many more combinations of projectiles and targets available for obtaining unstable nuclei and the respective conjugate products in the final state than in the case of $\pi$ mesons.

Using heavy-ion beams in the region of intermediate energies, one can also determine the properties of unstable nuclei by employing the missing mass method, performing kinematically complete experiments with the aid of highly effective detectors of particles in a $4\pi$ geometry.

Finally, secondary beams of radioactive nuclei open up great possibilities for these investigations. Their use will make possible a considerable advance into the region of higher neutron excess for the lightest elements (see Table 21). Here the question of the intensity of the primary beam is of primary importance. From our viewpoint, this

**Table 21. Some Nuclear Reactions with Secondary Beams of Radioactive Nuclei That May be Used to Obtain Neutron-Rich Isotopes of Lightest Elements**

| Nucleus | Reaction | $Q$ (MeV) |
|---|---|---|
| $^{6}$n | $^{6}\mathrm{He} + {}^{14}\mathrm{C} \rightarrow {}^{14}\mathrm{O} + {}^{6}\mathrm{n}$ | −35.8 |
| | $^{8}\mathrm{He} + {}^{14}\mathrm{C} \rightarrow {}^{16}\mathrm{O} + {}^{6}\mathrm{n}$ | −9.1 |
| $^{7}$n | $^{13}\mathrm{B} + {}^{7}\mathrm{Li} \rightarrow {}^{13}\mathrm{O} + {}^{7}\mathrm{n}$ | −48.1 |
| $^{8}$n | $^{8}\mathrm{He} + {}^{14}\mathrm{C} \rightarrow {}^{14}\mathrm{O} + {}^{8}\mathrm{n}$ | −37.9 |
| $^{8}$H | $^{13}\mathrm{B} + {}^{9}\mathrm{Be} \rightarrow {}^{14}\mathrm{O} + {}^{8}\mathrm{H}$ | −35.4 |
| $^{9}$H | $^{13}\mathrm{B} + {}^{9}\mathrm{Be} \rightarrow {}^{13}\mathrm{O} + {}^{9}\mathrm{H}$ | −58.6 |
| $^{12}$He | $^{12}\mathrm{Be} + {}^{14}\mathrm{C} \rightarrow {}^{14}\mathrm{O} + {}^{12}\mathrm{He}$ | −43.9 |
| | $^{16}\mathrm{C} + {}^{9}\mathrm{Be} \rightarrow {}^{13}\mathrm{O} + {}^{12}\mathrm{He}$ | −62.0 |
| $^{14}$He | $^{12}\mathrm{Be} + {}^{22}\mathrm{Ne} \rightarrow {}^{20}\mathrm{Mg} + {}^{14}\mathrm{He}$ | −80.6 |
| $^{14}$Li | $^{13}\mathrm{B} + {}^{14}\mathrm{C} \rightarrow {}^{13}\mathrm{O} + {}^{14}\mathrm{Li}$ | −68.7 |
| $^{18}$Be | $^{24}\mathrm{Ne} + {}^{14}\mathrm{C} \rightarrow {}^{20}\mathrm{Mg} + {}^{18}\mathrm{Be}$ | −93.7 |

problem will be solved using a new generation of accelerators, such as UNILAC in Darmstadt or the cyclotron complex for heavy ions (U-400–U-400M) in Dubna for which a program of obtaining and using relatively intense secondary beams of radioactive nuclei is envisaged.

## Acknowledgments

The authors express their gratitude to Academician G. N. Flerov for his support of the present work and for useful remarks. We also wish to thank Drs. C. Borcea, L. M. Chulkov, A. A. Korsheninnikov, R. Kalpakchieva, B. G. Novatsky, and Nguen Hoai Chau for useful discussions.

## References

(Ab 70) S. M. Abramovich, L. A. Markin, V. A. Pereshivkin, and V. I. Serov, *Izv. Akad. Nauk SSSR, Fiz.* **34**:1727 (1970).

(Ab 80) S. N. Abramovich, A. I. Baz, and V. J. Guzhovski, *Yad. Fiz.* **32**:402 (1980).

(Ag 85) V. A. Ageyev, I. N. Vichnevsky, V. I. Gavrileyuk, V. A. Zheltonozhsky, T. N. Lashko, and N. V. Strilchuk, Report 84-4, KINR, Kiev (1985).

(Aj 65) A. Ajdačik, M. Cerineo, B. Laboreic, G. Paic, I. Slans, and P. Tomas, *Phys. Rev. Lett.* **14**:444 (1965).

(Aj 82) F. Ajzenberg-Selove, *Nucl. Phys. A* **375**:1 (1982).

(Aj 84) F. Ajzenberg-Selove, *Nucl. Phys. A* **413**:1 (1984).

(Aj 85) F. Ajzenberg-Selove, *Nucl. Phys. A* **433**:1 (1985).

(Aj 86) F. Ajzenberg-Selove, *Nucl. Phys. A* **449**:1 (1986).

(Al 82) D. V. Alexandrov, Yu. A. Glukhov, A. S. Demyanova, V. I. Duchanov, I. B. Mazurov, B. G. Novatsky, A. A. Ogloblin, and D. E. Stepanov, *Yad. Fiz.* **35**:277 (1982).

(Al 83) D. V. Alexandrov, Yu. A. Glukhov, A. S. Demyanova, V. I. Mazurov, B. G. Novatsky, A. A. Ogloblin, S. B. Sakuta, and D. N. Stepanov, *Yad. Fiz.* **35**:277 (1983).

(Al 84a) D. V. Alexandrov, E. A. Ganza, Yu. A. Glukhov, B. G. Novatsky, A. A. Ogloblin, and D. N. Stepanov, *Yad. Fiz.* **39**:513 (1984).

(Al 84b) D. V. Alexandrov, E. A. Ganza, Yu. A. Glukhov, B. G. Novatsky, A. A. Ogloblin, and D. N. Stepanov, *Zh. Eksp. Teor. Fiz. Pis'ma* **40**:152 (1984).

(Al 85) D. V. Alexandrov, E. A. Ganza, Yu. A. Glukhov, E. Yu. Nikolsky, B. G. Novatsky, A. A. Ogloblin, and D. N. Stepanov, *Vopr. At. Nauki Tekh.* **2**(31):3 (1985).

(Al 87) D. V. Alexandrov, E. A. Ganza, Yu. A. Glukhov, B. G. Novatsky, A. A. Ogloblin, and D. N. Stepanov, *Yad. Fiz.* **45**:132 (1987).

(An 74) N. Anyas-Weiss, J. C. Cornell, P. S. Fisher, P. N. Hudson, A. Menchaca-Rocha, D. J. Millener, A. D. Panagiotou, D. K. Scott, D. Strottman, D. N. Brink, B. Buck, P. J. Ellis, and T. Engeland, *Phys. Rev.* **12**:201 (1974).

(An 86) R. Anne, D. Bazin, and A. C. Mueller, GANIL report P86-23, Caen (1986).

(Ar 70a) A. G. Artukh, V. V. Avdeichikov, J. Erö, G. F. Gridnev, V. L. Mikheev, V. V. Volkov, and J. Wilczynsky, *Phys. Lett.* **33B**:407 (1970).

(Ar 70b) A. G. Artukh, V. V. Avdeichikov, J. Erö, G. F. Gridnev, V. L. Mikheev, and V. V. Volkov, *Nucl. Instrum. Methods* **83**:72 (1970).

(Ar 71) A. G. Artukh, V. V. Avdeichikov, G. F. Gridnev, V. L. Mikheev, V. V. Volkov, and J. Wilczynsky, *Nucl. Phys. A* **168**:321 (1971).

(Ar 73) K. P. Artemov, V. Z. Goldberg, V. P. Petrov, V. P. Rudakov, I. N. Serikov, and V. A. Timofeev, *Yad. Fiz.* **17**:225 (1973).

(Au 79) P. Auger, T. H. Chiang, J. Gazin, B. Gatty, D. Guerreau, E. Nolte, J. Pouthas, X. Tarrago, and J. Girard, *Z. Phys. A* **289**:255 (1979).

(Az 79) R. E. Azuma, L. C. Carraz, P. G. Hansen, B. Jonson, K. L. Kratz, S. Mattsson, G. Nyman, H. Ohm, H. L. Ravn, and A. Schröder, *Phys. Rev. Lett.* **43**:1652 (1979).

(Az 80) R. E. Azuma, T. Björnstod, H. A. Gustafsson, P. G. Hansen, B. Jonson, S. Mattsson, G. Nyman, A. M. Poskanzer, and H. L. Ravn, *Phys. Lett.* **96B**:31 (1980).

(Ba 60) A. I. Baz, V. N. Goldansky, and Ya. B. Zeldovich, *Usp. Fiz. Nauk* **72**:211 (1960).

(Ba 64) Yu. A. Batusov, S. A. Bunyatov, V. M. Sidorov, and V. A. Yarba, *Zh. Eksp. Teor. Fiz.* **46:**817 (1964).

(Ba 66) A. I. Baz, Ya. B. Zeldovich, and A. M. Perelomov, in: *Scattering, Reactions, and Decay in Nonrelativistic Quantum Mechanics* (in Russian), Nauka, Moscow (1966).

(Ba 67) A. M. Badalyan, E. S. Galperin, V. N. Lyakhovicki, V. V. Pustovalov, Yu. A. Simonov, and E. L. Surkov, *Yad. Fiz.* **6**:473 (1967).

(Ba 68) A. M. Baldin, V. N. Goldansky, V. M. Maksimenko, and I. L. Rosental, in: *Kinematics of Nuclear Reactions* (in Russian), Atomisdat, Moscow (1968).

(Ba 70) A. I. Baz, A. M. Gorbatov, V. F. Demin, and I. M. Pasinkov, *Zh. Eksp. Teor. Fiz. Pis'ma* **12**:151 (1970).

(Ba 71) A. I. Baz and M. V. Jukov, *Proceedings of the Problem Symposium on Nuclear Physics (Novosibirsk, USSR, 1970)*, Nauka, Moscow (1971).

(Ba 72) A. I. Baz, V. I. Goldansky, V. Z. Goldberg, and Ya. B. Zeldovich, in *Light and Medium Heavy Nuclei Close to the Border of Nuclear Stability* (in Russian), Nauka, Moscow (1972).

(Ba 75) Yu. G. Bassargin, N. I. Boldin, L. E. Korolev, V. G. Levchenko, and Yu. P. Servergin, *Nucl. Instrum. Methods* **126**:413 (1975).

(Ba 81) R. Bayer, Z. Dlouhy, J. Svanda, and I. Wilhelm, *Czech. Phys.* (*Engl. Transl.*) *B* **31**:1273 (1981).

(Ba 85) A. M. Badalyan, T. I. Belova, N. B. Kopnekhova, and V. D. Efros, *Yad. Fiz.* **41**:1460 (1985).

(Be 70) E. Y. Berlovich, Proc. Int. Conf. on the Properties of Nuclei Far From the Region of $\beta$-Stability (Leysin, Sw. 1970), CERN report 70-30:497, Geneva (1970).

(Be 71) H. A. Bethe, *Annu. Rev. Nucl. Sci.* **21**:93 (1971).

(Be 81) B. Bevelacqua, *Phys. Lett.* **102B**:79 (1981).

(Be 84a) A. V. Belozyorov, C. Borcea, E. Gierlik, Nguyen, Choai Chau, V. B. Kutner, R. Kalpakchieva, A. M. Kalinin, Yu. Ts. Oganessian, Yu. E. Penionzkevich, N. V. Pronin, and G. N. Flerov, Proc. 34 Conf. on Nucl. Spectroscopic (Alma-At, USSR, 1984), Nauka, Leningrad (1984), p. 379.

(Be 84b) M. A. Bernstein and W. A. Friedman, *Phys. Rev. C* **29**:132 (1984).

(Be 85) A. V. Belozyorov, C. Borcea, Z. Dlouhy, A. M. Kalinin, R. G. Kalpakchieva, Ngueyn Hoai Chau, Yu. E. Penionzhkevich, and N. K. Skobelev, Report 13-85-535 JINR, Dubna (1985).

(Be 86a) A. V. Belozyorov, C. Borcea, Z. Dlouhy, A. M. Kalinin, Nguyen, Hoai Chau, and Yu. E. Penionzhkevich, Report P7-86-309 JINR, Dubna (1986).

(Be 86b) A. V. Belozyorov, C. Borcea, Z. Dlouhy, A. M. Kalinin, R. Kalpakchieva, Nguyen Hoai Chau, Yu. Ts. Oganessian, and Yu. E. Penionzhkevich, *Nucl. Phys. A* **460**:352 (1986).

(Bi 85) B. Bimbot, S. Della-Negra, P. Aguer, G. Bastin, R. Anne, H. Delagrange, and F. Hubert, *Z. Phys. A* **322**:443 (1985).

(Bi 86) R. Bimbot, S. Della-Negra, M. Manasijevic, P. Aguer, G. Bastin, R. Anne, H. Delagrange, Y. Schutz, F. Hubert, Y. Gono, K. Hatanaka, Proc. Int. Conf. on Heavy Ion Nuclear Collisions in the Fermi Energy Domain (Caen, France, 1986), *J. Phys.* (*Paris*) **C4–8**:241 (1986).

(Bl 64) C. H. Blanchard and R. G. Winter, *Phys. Rev.* **107**:774 (1964).

(Bo 57) G. F. Bogdanov, N. A. Vlasov, S. P. Kalinin, B. V. Rybakov, and V. A. Sidozov, *At. Ener.* **3**:204 (1957).

(Bo 69) A. Bohr and B. Mottelson, in *Nuclear Structure*, Benjamin, New York, (1969) Vol. 1, Ch. 2.

(Bo 80) V. I. Bogatin, E. A. Ganza, O. V. Lozhkin, Yu. A. Murin, V. S. Oplavin, H. A. Perfilov, and Yu. P. Yakovlev, *Yad. Fiz.* **32**:27 (1980).

(Bo 82) C. Borcea, E. Gierlik, A. M. Kalinin, R. Kalpakchieva, Yu. Ts. Oganessian, T. Pavlat, Yu. E. Penionzhkevich, and A. V. Rykhlyuk, *Nucl. Phys. A* **391**:520 (1982).

(Bo 83) V. Borrel, D. Guerrlau, J. Galin, B. Gatty, D. Jacquet, and Y. Tarrayo, *Z. Phys. A* **314**:191 (1983).

(Bo 84a) C. Borcea, E. Gierlik, R. Kalpakchieva, Nguyen Hoai Chau, Yu. Ts. Oganessian, T. Pavlat, and Yu. E. Penionzhkevich, *Nucl. Phys. A* **415**:169 (1984).

(Bo 84b) O. V. Bochkarev, A. A. Korcheninikov, E. A. Kuzmin, I. G. Mukha, A. A. Ogloblin, and L. V. Chulkov, *Zh. Eksp. Teor. Fiz. Pis'ma* **40**:204 (1984).

(Bo 85a) O. V. Bochkarev, A. A. Korcheninikov, E. A. Kuzmin, I. G. Mukha, A. A. Ogloblin, and L. V. Chulkov, *Zh. Eksp. Teor. Fiz. Pis'ma* **42**:303 (1985).

(Bo 85b) O. V. Bochkarev, A. A. Korcheninikov, E. A. Kuzmin, I. G. Mukha, A. A. Ogloblin, and L. V. Chulkov, *Zh. Eksp. Teor. Fiz. Pis'ma* **42**:305 (1985).

(Bo 86) M. J. G. Burge, M. Epherre-Rey-Campagnolle, D. Guillemaud-Mueller, B. Jonson, M. Langevin, G. Nyman, and C. Thibault, CERN report EP/86-67, Geneva (1986).

(Br 61) H. C. Britt and A. R. Quinton, *Phys. Rev.* **124**:877 (1961).

(Br 64) O. D. Brill, N. I. Venikov, A. A. Kurashov, A. A. Ogloblin, V. M. Pankratov, and V. P. Rudakov, *Phys. Lett.* **12**:51 (1964).

(Br 72) D. M. Brink, *Phys. Lett.* **40B**:37 (1972).

(Br 80) D. A. Bromley, *Science* **209**:110 (1980).

(Br 81) C. Brendel, V. Beck, J. Canter, G. Delic, M. Marinescu, A. Richter, G. Schreider, F. Busch, D. Schüll, and M. Shen, Proc. 4th Conf. on Nuclei Far From Stability (Helsingör, Denmark, 1981), CERN report 81-09:664, Geneva (1981).

(By 73) E. Byckling and K. Kajantie, *Particle Kinematics*, John Wiley and Sons, London, New York, Sydney, Toronto (1973).

(Ca 68) N. Carayannopoulos, J. Head, N. Kwak, J. Manweiler, and R. Stump, *Phys. Rev. Lett.* **20**:1215 (1968).

(Ca 84) M. D. Cable, J. Honkanen, E. C. Schloemer, M. Ahmed, J. E. Reiff, Z. Y. Zhou, and J. Cerny, *Phys. Rev. C* **30**:1276 (1984).

(Ce 74) J. Cerny, *Proceedings of the International Conference on Reactions Between Complex Nuclei (Nashville, USA, 1974)*, Vol. 2, North-Holland Publ. Co., Amsterdam (1974), p. 483.

(Ce 77) J. Cerny, R. B. Weisenmiller, N. A. Jelley, K. H. Wilcox, and G. J. Wozniak, *Phys. Lett.* **53B**:247 (1977).

(Cl 75) H.-G. Clerc, K.-H. Schmidt, H. Wohlfarth, W. Lang, H. Schrader, K. E. Pferdekamper, R. Jungmann, M. Asghar, J. P. Bocquet, and G. Siegert, *Nucl. Phys. A* **247**:74 (1975).

(Co 67) S. W. Cosper, J. Cerny, and R. C. Gatly, *Phys. Rev.* **154**:1193 (1967).

(Cu 86) M. S. Curtin, L. H. Harwood, J. A. Nolen, B. Sherrill, Z. Q. Yie, and B. A. Brown, *Phys. Rev. Lett.* **56**:34 (1986).

(De 77) C. Detraz, *Phys. Lett.* **66B**:333 (1977).

(De 83a) C. Detraz, *Nucl. Phys. A* **409**:353c (1983).

(De 83b) C. Detraz, M. Langevin, D. Guillemaud-Mueller, A. Mueller, C. Thibault, F. Touchard, G. Klotz, C. Miehe, and G. Walter, *Nucl. Phys. A* **402**:301 (1983).

(De 86) C. Detraz, GANIL report P86-20, Caen (1986).

(Dm 82a) V. D. Dmitriev, S. M. Lukyanov, Yu. E. Penionzhkevich, and D. K. Sattarov, *Prib. Tekh. Eksp.* **2**:7 (1982).

(Dm 82b) V. D. Dmitriev, H. Sodan, A. M. Kalinin, S. M. Lukyanov, Yu. Ts. Oganessian, Yu. E. Penionzhkevich, and T. S. Salamatina, *Yad. Fiz.* **35**:1112 (1982).

(Dr 74) A. G. Drentje, H. A. Enge, and S. B. Kowalski, *Nucl. Instrum. Methods* **122**:485 (1974).

(Dr 80) D. M. Drake, J. D. Moses, J. C. Peng, S. Nelson, and J. W. Sunier, *Phys. Rev. Lett.* **45**:1765 (1980).

(Du 84) E. Duek, N. N. Ajitanano, J. M. Alexander, P. Logan, M. Kildir, L. Kowalski, L. C. Vaz, D. Guerreau, M. S. Zisman, M. Kaplan, and D. J. Moses, *Z. Phys. A* **317**:83 (1984).

(Fa 74) I. V. Falomkin, M. M. Kulyukin, V. I. Lyashenko, G. B. Pontekorvo, and Yu. A. Shcherbakov, *Nuovo Cimento* **22A**:333 (1974).

(Fa 76) I. V. Falomkin, V. I. Lyashenko, G. B. Pontecorvo, Yu. A. Shcherbakov, M. Alev, T. Angelescu, O. Balea, B. Mihul, F. Nichitin, A. Seraru, F. Balestra, R. Garfagini, G. Piragino, C. Guaraldo, and R. Schrimaglio, *Lett. Nuovo Cimento* **16**:525 (1976).

(Fi 82) L. K. Fifield, J. L. Durell, M. A. C. Hotchiks, J. R. Leigh, T. R. Ophel, and D. C. Weisser, *Nucl. Phys. A* **385**:505 (1982).

(Fl 68) G. N. Flerov, *At. Energy* **24**:5 (1968).

(Ga 66) G. T. Garvey and I. Kelson. *Phys. Rev. Lett.* **16**:197 (1966).

(Ga 77) F. Gaille, S. Mureramanzi, M. Bogdanski, and J. B. Fecenneret, *Nuovo Cimento* **A40**:31 (1977).

(Gi 64) L. Gilly, M. Jean, R. Meunier, M. Spigheb, J. P. Stroot, P. Duteil, and A. Rode, *Phys. Lett.* **11**:244 (1964).

(Gi 65) L. Gilly, M. Jean, R. Meunier, N. Spignel, Y. P. Stroot, and P. Duteil, *Phys. Lett.* **19**:335 (1965).

(Gk 84) H. Gilman, H. T. Fortune, L. C. Bland, Rex R. Kuziah, C. Fredmoore, P. A. Seidl, C. L. Morris, and W. B. Cottingame, *Phys. Rev. C* **30**:958 (1984).

(Gl 84) Yu. A. Gluchov, A. S. Demyanova, I. Lazor, A. A. Ogloblin, S. B. Sakuta, and V. B. Sukharevsky, *Yad. Fiz.* **40**:62 (1984).

(Go 60a) V. I. Goldansky, *Zh. Eksp. Teor. Fiz.* **38**:1637 (1960).

(Go 60b) V. I. Goldansky, *Zh. Eksp. Teor. Fiz.* **39**:497 (1960).

(Go 75) F. S. Goulding and B. G. Harvey, *Annu. Rev. Nucl. Sci.* **25**:167 (1965).

(Go 81) M. G. Gornov, *Prib. Tekh. Eksp.* **2**:42 (1981).

(Go 84) M. G. Gornov, Yu. B. Gurov, A. S. Lukin, P. V. Morochov, K. O. Oganessian, B. P. Osipenko, V. A. Pechkurov, A. P. Pichugin, V. I. Saveliev, F. M. Sergeev, V. G. Sandukovsky, A. A. Chamutov, and P. P. Chafigullin, *Nucl. Instrum. Methods* **225**:42 (1984).

(Gr 86) G. Gregoire and B. Tamain, GANIL report P86-10 (1986).

(Gu 80) D. Guerreau, J. Galin, B. Gatty, Y. Tarrago, J. Girard, R. Lucas, and C. Ngô, *Z. Phys. A* **295**:105 (1980).

(Gu 86) D. Guerreau, Proc. Int. Conf. on Heavy Ion Nuclear Collisions in the Fermi Energy Domain (Caen, France, 1986), *J. Phys.* (*Paris*) **C4–8**:205 (1986).

(Ir 75) J. M. Irvine, in *Heavy Nuclei, Superheavy Nuclei and Neutron Stars*, Clarendon Press, Oxford (1975).

(Je 75) N. A. Jelly, J. Cerny, D. P. Stahel, and K. H. Wilcox, *Phys. Rev. C* **11**:2049 (1975).

(Ji 85a) R. I. Jibuti and R. Ya. Kezeraçhvili, *Part. Nucl.* **16**:1173 (1985).

(Ji 85b) R. I. Jibuti and R. Ya. Kezeraçhvili, *Nucl. Phys. A* **437**:687 (1985).

(Jo 81) B. Johson, Proc. Int. Conf. on Nuclear Physics (Berkeley, USA, 1980), *Nucl. Phys. A* **354**:77 (1981).

(Ka 69) S. A. Karamyan, Yu. Ts. Oganessian, Yu. E. Penionzhkevich, and B. I. Pustylnik, *Yad. Fiz.* **9**:715 (1969).

(Ka 84) R. L. Kavalov, S. M. Lukyanov, Yu. L. Markarian, S. G. Gasparian, Yu. E. Penionzhkevich, and G. G. Chubarian, *Prib. Tekh. Eksp.* **3**:46 (1984).

(Ke 86) R. Ya. Kezerachvili, *Yad. Fiz.* **44**:842 (1986).

(Ki 83) K. Kimora, *Nucl. Instrum. Methods* **212**:227 (1983).

(Kl 67) R. Klapisch, J. Chaumont, C. Philippe, I. Amarel, R. Fergeau, M. Salome, and R. Bernas, *Nucl. Instrum. Methods* **5**:216 (1967).

(Ko 78) R. Kolalis, Proc. Int. Conf. Probl. petite nombre corp. Phys. noyaux et part. elem. (Quebec City, Canada, 1977) (Les Presses de L'Université Labal, Quebec, 1978), p. 406.

(La 84) M. Langevin, C. Detraz, M. Epherre, D. Guillemaud-Mueller, B. Jonson, C. Thibault, and the ISOLDE Collaboration, *Phys. Lett.* **146B**:176 (1984).

(Lo 61) O. V. Lozhkin and A. S. Rimski-Korsakov, *Zh. Eksp. Teor. Fiz.* **40**:1519 (1961).

(Lo 83) O. V. Lozhkin, V. S. Oplarin, and Yu. P. Yakovlev, preprint RI-168, Rad. Inst., Leningrad (1983).

(Lu 86a) S. M. Likyanov, M. Levitowicz, Yu. E. Penionzhkevich, G. G. Chubarian, D. Bazin, D. Guillemaud-Mueller, A. C. Mueller, and M. G. Saint-Laurent, GANIL report P86–21, Caen (1986).

(Lu 86b) Yu. S. Lutostansky, *Izv. Akad. Nauk SSSR, Fiz.* **50**:834 (1986).

(Ma 84) R. J. Macdonald, L. G. Sobotka, Z. Q. Yao, G. J. Wozniak, and G. Guarino, *Nucl. Instrum. Methods* **219**:508 (1984).

(Mc 68) R. L. McGrath, J. Cerny, and S. W. Cosper, *Phys. Rev.* **165**:1126 (1968).

(Mi 55) A. B. Migdal, *Zh. Eksp. Teor. Tiz.* **28**:3 (1955).

(Mi 78) A. B. Migdal, in *Fermions and Bosons in Strong Fields* (in Russian), Nauka, Moscow (1978), p. 246.

(Mi 84) V. L. Mikcheev, A. G. Artukh, V. V. Volkov, G. F. Gridnev, M. G. Gruchecky, V. Karch, A. M. Mezentzev, V. M. Morozov, and L. Pomorski, report P7-84-614, JINR, Dubna (1984).

(Na 82) F. Naulin, C. Detraz, M. Roy-Stephan, M. Bernas, J. Boer, D. Guillemaud, M. Langevin, F. Pougheon, and P. Roussel, *Phys. Rev. C.* **25**:1074 (1982).

(Ni 84) J. M. Nitschke, Proc. Workshop on Prospects for Research with Radioactive Beams from Heavy Ion Accelerators (Washington, USA, 1984), LBL report 18187:48, Berkeley (1984).

(Og 70a) Yu. Ts. Oganessian, Yu. E. Penionzhkevich, and A. O. Chamsutdinov, *Yad. Phys.* **14**:54 (1970).

(Og 70b) Yu. Ts. Oganessian and Yu. E. Penionzhkevich. *J. Phys.* **31**:259 (1970).

(Og 73) Yu. Ts. Oganessian, Yu. E. Penionzhkevich, Nguyen Tac Ank, D. M. Nadkarni, K. A. Gavrilov, Kim De En, and M. Hussonois, *Yad. Fiz.* **18**:734 (1973).

(Og 76) A. A. Ogloblin, Proc. Int. Conf. on Nuclear Physics with Heavy Ions (Caen, France, 1976) (1976), p. 10.

(Og 78) Yu. Ts. Oganessian, V. I. Manko, Yu. E. Penionzhkevich, and V. N. Poljansky, *Nucl. Phys. A* **303**:259 (1978).

(Og 79) A. A. Ogloblin, GSI report 79–12, Darmstadt (1979).

(Og 82) Yu. Ts. Oganessian, Yu. E. Penionzhkevich, E. Gierlik, R. Kalpakchieva, T. Pavlat, C. Borcea, A. V. Belozyorov, Yu. P. Kharitonov, S. P. Tretyakova, V. G. Subbotin, S. M. Lukyanov, N. V. Pronin, and A. A. Bykov, *Zh. Eksp. Teor. Fiz. Pis'ma* **36**: 104 (1982).

(Og 85) A. A. Ogloblin, Proc. Int. School on Nuclear Structure (Alushta, USSR, 1985), report P-352, JINR, Dubna (1985).

(Pe 86) Yu. E. Penionzhkevich, E. Gierlik, V. V. Kamanin, and C. Borcea, *Part. Nucl.* **17**:165 (1986).

(Ph 67) G. C. Phillips, Proc. Int. Conf. Nuclear Physics (Gatlinburg, Tennessee, 1966) (Academic, New York, 1967), p. 949.

(Ph 70) G. C. Phillips, *Phys. Lett.* **33B**:260 (1970).

(Po 71) A. M. Poskanzer, G. W. Butler, and E. K. Hyde, *Phys. Rev. C* **3**:881 (1971).

(Qu 79) U. Quade, K. Rudolph, and G. Siegert, *Nucl. Instrum. Methods* **164**:435 (1979).

(Ra 72) M. Rajagopalam and T. D. Thomas, *Phys. Rev. C* **5**:1402 (1972).

(Ro 78) R. Roussel, M. Bernas, F. Diaf, F. Naulin, F. Pougheon, G. Rutbard, and M. Roy-Stephan, *Nucl. Instrum. Methods* **153**:111 (1978).

(Ro 84) H. J. Rose and G. A. Jones, *Nature* **307**:245 (1984).

(Se 78) K. K. Seth, H. Nann, S. Inversen, M. Kaletka, J. Hird, and H. A. Thiessen, *Phys. Rev. Lett.* **41**:1589 (1978).

(Se 81a) K. K. Seth, Proc. 4th Int. Conf. on Nuclei Far From Stability (Helsingør, Denmark, 1981), CERN report 81-09:655, Geneva (1981).

(Se 81b) V. Seidel, H. Sodan, S. M. Lukyanov, P. Manfras, Yu. E. Penionzhkevich, F. Stary, and K. D. Schilling, report P7-81-807, JINR, Dubna (1981).

(Sh 75) D. Shapiro, R. M. Devries, H. W. Fulbright, J. Toke, and N. R. Clover, *Nucl. Instrum. Methods* **129**:123 (1975).

(Si 69) Yu. A. Somonov, *Yad. Fiz.* **23**:1210 (1969).

(Si 71) A. G. Sitenko, in *Lectures on Diffusion Theory* (in Russian), Vishchaya Shkola, Kiev (1971).

(Sm 60) F. T. Smith, *Phys. Rev.* **118**:349 (1960).

(So 85) M. Sowinsky, M. Levitovich, R. Kupchak, A. Jankovski, N. K. Skobelev, and S. Choinacki, report P7-85-377, JINR, Dubna (1985).

(Sp 70) J. Sperinde, D. Fredrickson, R. Hinkins, V. Perez-Mendez, and B. Smith, *Phys. Lett.* **32B**:185 (1970).

(Sp 74) J. Sperinde, D. Fredrickson, and V. Perez-Mendez, *Nucl. Phys. B* **78**:345 (1974).

(St 81) A. Stetz, L. W. Swenson, J. Davis, J. Kallne, R. C. Minehart, R. R. Whitney, V. Perez-Mendez, A. Sagle, J. Carroll, J. B. McClelland, and J. Faucett, *Phys. Rev. Lett.* **47**:782 (1981).

(Sy 79) T. J. M. Symons, *Proc. Int. Conf. on Atomic Masses (East Lansing, USA, 1979)*, Plenum Press, New York and London (1979), p. 61.

(Ta 65) V. C. Tang and B. F. Bayman, *Phys. Rev. Lett.* **15**:165 (1965).

(Ta 72) J. R. Taylor, in *Scattering Theory*, John Wiley and Sons, New York (1972).

(Ta 85a) I. Tanihata, *Hyperfine Interact.* **21**:251 (1985).

(Ta 85b) I. Tanihata, H. Hamagaki, O. Hashimoto, Y. Shido, N. Yoshikawa, K. Sugimoto, O. Yamakawa, T. Robayashi, and N. Takahashi, LBL report 19904, Berkeley (1985).

(Th 66) G. S. T. Thoronton, J. K. Bair, C. M. Jones, and H. B. Willard, *Phys. Rev. Lett.* **17**:701 (1966).

(Th 70) D. R. Thompson, *Nucl. Phys. A* **143**:304 (1970).

(Th 74) C. Thibault and R. Klapisch, *Phys. Rev.* **9**:793 (1974).

(Th 75) C. Thibault, R. Klapisch, C. Rigaud, A. M. Poskanzer, P. Prieels, L. Lessard, and W. Reisdorf, *Phys. Rev. C* **12**:644 (1975).

(Un 84) J. E. Ungar, R. D. McKeown, D. F. Geesaman, R. T. Holt, T. R. Specht, K. E. Stephenson, B. Zeidman, and C. L. Morris, *Phys. Lett.* **144B**:333 (1984).

(Ve 81) N. I. Venikov, Yu. A. Glukhov, A. S. Demyanova, S. I. Drozdov, B. G. Novatsky, A. A. Ogloblin, S. B. Sakuta, D. N. Stepanov, B. N. Unezhov, and Yu. L. Yuzhinov, *Yad. Fiz.* **33**:875 (1981).

(Vo 64) V. K. Voitovetsky, I. L. Korsunsky, and Yu. F. Pazhin, *Zh. Eksp. Teor. Fiz.* **47**:1628 (1964).

(Vo 69) A. A. Vorobiev, D. M. Seleverstov, V. T. Grachov, I. A. Kondurov, A. M. Nikitin, A. I. Yegorov, and Yu. K. Zalite, *Phys. Lett.* **30B**:332 (1969).

(Vo 71) V. V. Volkov, *Part. Nucl.* **2**:287 (1971).

(Vo 75) V. V. Volkov, *Part. Nucl.* **6**:1040 (1975).

(Vo 82) V. V. Volkov, in *Nuclear Reactions of Deep Inelastic Transfer* (in Russian), Energoizdat, Moscow (1982).

(Wa 52) K. M. Watson, *Phys. Rev.* **88**:1153 (1952).

(Wa 76) A. H. Wapstra and K. Bos, in *At. Data Nucl. Data Tables* **17**:474 (1976).

(We 77) R. B. Weisenmiller, N. A. Jelley, D. Ashery, K. H. Wilcox, G. J. Wozniak, M. S. Zisman, and J. Cerny, *Nucl. Phys. A* **280**:217 (1977).

(We 79) G. D. Westfall, T. J. M. Symons, D. E. Greiner, H. H. Heckman, P. J. Lindstrom, J. Mahoney, A. C. Shotter, D. K. Scott, H. J. Crawford, C. McParland, T. C. Awes, C. K. Gelbke, and T. M. Kidd, *Phys. Rev. Lett.* **43**:1859 (1979).

(Wh 67) S. L. Whetstone, Jr. and T. D. Thomas, *Phys. Rev.* **54**:1174 (1967).

(Wi 64) D. H. Wilkinson, *Phys. Lett.* **11**:243 (1964).

(Wi 67) J. Wilczynski, V. V. Volkov, and P. Dezovski, *Yad. Fiz.* **5**:942 (1967).

(Wi 73) K. H. Wilcox, N. A. Jelley, G. J. Wozniak, R. B. Weisenmiller, H. L. Harney, and J. Cerny, *Phys. Rev. Lett.* **30**:866 (1973).

(Wi 75) K. H. Wilcox, R. B. Weisenmiller, G. J. Wozniak, N. A. Jelley, D. A. Shery, and J. Cerny, *Phys. Lett.* **59B**:142 (1975).

(Wi 80) J. Wilczynski, K. Siwek-Wilczynska, and J. Van Driel, *Phys. Lett.* **45**:606 (1980).

(Yo 68) P. G. Young and R. H. Stokes, *Phys. Rev. C* **4**:1597 (1971).

(Ze 60) Ya. B. Zeldovich, *Zh. Eksp. Teor. Fiz.* **38**:1123 (1960).

# 5

HELGE L. RAVN (left) is senior applied physicist in the Experimental Physics division of the European Center for Nuclear Research, CERN, in Geneva, Switzerland. He has worked extensively on targets and ion sources for the ISOLDE on-line mass separator facility. He received his M.Sc. in chemical engineering at the Technical University of Copenhagen in 1965. From 1965 through 1967 he held a CERN fellowship and studied high-energy nuclear reactions. From 1967 through 1971 he took up a scientific post at the Institute of Physics, University of Aarhus and participated in experiments in nuclear spectroscopy and solid-state hyperfine interactions of ion-implanted nuclei. In 1980 he received his Dr. Sci. Nat. from the University of Aarhus. He serves on editorial boards, conference committees, and advisory and visiting committees at universities and national laboratories. He is a member of the Danish Academy of Technical Sciences.

BRIAN W. ALLARDYCE (right) was trained in nuclear physics at the University of Oxford where he gained a D.Phil. for his work on statistical fluctuations in compound nuclear reactions in 1965. He then participated in a series of experiments on the reaction of pions with nuclei, working both on the synchrocyclotron at CERN, Geneva and on the proton synchrotron at the Rutherford Laboratory in Great Britain. In 1973 he moved to CERN as a Fellow and worked on technical aspects of the improvement program of the CERN synchrocyclotron (SC). This led to an interest in accelerator developments, and he was prominent in the work of converting the SC to accelerate light ions to energies up to 86 MeV/U and in promoting the use of these beams before other dedicated accelerators came on-line in this energy range. He became involved with ISOLDE in 1982 when he became the project leader for the new separator while continuing to be in charge of the synchrocyclotron accelerator.

# On-Line Mass Separators

HELGE L. RAVN AND BRIAN W. ALLARDYCE

## 1. Introduction

### 1.1. Definition

Ever since the discovery of artificially produced radioactivity in nuclear reactions, a major experimental challenge has been to separate the many different reaction products from the bulk target material in order to be able to study the properties of particular product nuclei under sufficiently clean conditions. One of the most efficient methods, which has now evolved to a high degree of sophistication, is the on-line mass separator, also called the on-line isotope separator or ISOL. Such an instrument is essentially a target, an ion source, and an electromagnetic mass analyzer coupled in series. The apparatus is said to be on-line when the material analyzed is directly the target of a nuclear bombardment, where the reaction products of interest formed during the irradiation are slowed down and stopped in the system. Chemical and physical methods are then used continuously to separate the various elements and transfer them into the ion source.

On-line mass separation was originally seen as a way to study nuclei far from $\beta$ stability, because these nuclei have short half-lives and an on-line analysis can be made much more speedily than an off-line one. Results have shown that one can indeed study nuclei far from stability with these instruments, in many cases out to the physical limits of nuclear existence, the so-called neutron or proton drip-lines. Detailed overviews of such studies can be found in the proceedings of the series of conferences on nuclei far from stability started in Lysekil (Ly 66) and followed by Leysin (Le 70), Cargèse (Ca 76) and Helsingør (He 81). There have also been excellent reviews by Klapisch (Kl 69), Hansen (Ha 79), and Epherre (Ep 84).

However, modern instruments have reached such a high level of performance, in intensity, beam purity, etc., that they are employed not only in the difficult region away from $\beta$ stability, but also for studies close to stability, where other techniques are in competition. Progress in the field of isotope separators is recorded in the series of

HELGE L. RAVN AND BRIAN W. ALLARDYCE • European Center for Nuclear Research (CERN), Geneva, Switzerland.

International Conferences on Electromagnetic Isotope Separators or EMIS, the most recent of which are (Em 65, Em 70, Em 73, Em 76, Em 80, Em 87).

Alternative techniques which have produced important results in the study of radioactive nuclei consist of in-beam $\gamma$-ray spectroscopy (Di 70), fast radiochemical separation (He 69, He 86), and helium jet transport (Ma 63, Wo 76, Mo 81b). Of these only the helium jet technique coupled to a mass analyzer will be mentioned in this paper. These methods are not highly selective when compared to the on-line mass separator and are therefore most powerful in cases where the nuclear reaction has a large cross-section and few reaction channels, and the nuclei of interest have characteristic decay modes, such as delayed particle emission. Furthermore, large amounts of stable material have to be tolerated in the samples.

Another alternative technique to the on-line mass separator is the recoil analyzer treated by Enge in Vol. 7 of this Treatise (En 85). Here the reaction products of the irradiation recoil from the relatively thin target material and are mass analyzed directly. Such separators clearly are very fast and can be used successfully for the shortest nuclear lifetimes. However, they suffer from low intensities and have rather poor mass resolution.

## 1.2. The On-Line Isotope Separator Concept

The decisive point in the production of nuclei far from stability is that the product of interest is usually formed with a very low cross-section while at the same time other nuclear reactions (and often very many and very complex nuclear reactions) are producing a variety of other product nuclei with much higher yields. Thus, the nucleus of interest is hidden in a huge sea of other products. This is true for high-energy proton reactions, for fission, and in high-energy heavy-ion reactions. Figure 1 shows a typical example of the yields for high-energy proton spallation reactions on a La target. The figure shows the cross-section calculated from the semiempirical formula of Rudstam (Ru 66), which agrees well with experiment. For each element produced, the curves show the evolution of the cross-section as a function of mass, and it can be seen that many orders of magnitude are involved as one goes from the peaks to regions far from stability.

In the study of nuclei far from stability it is clear therefore that it is essential to achieve sufficiently high selectivity, and that a separation not only in $A$ but also in $Z$ is necessary. Consider, for example, the case of $^{114}$Cs, produced according to Figure 1 with a cross-section of $\sim$1 nb. After a mass analysis, the $A = 114$ products are dominated by the much more intense lighter elements from Xe down to Cd or below, and $^{114}$Cs would be completely undetectable. However, in combination with a chemical separation, it is indeed possible to obtain the clean conditions to observe $^{114}$Cs.

Today, new rapid separation methods based on thermochemical and physical principles allow radioactive nuclei of practically all elements to be transferred from a bombarded target and converted into an ion beam in a continuous way. The devices using these techniques, often referred to as target and ion-source systems, serve as an injector to the electromagnetic mass separator, with its associated beam optical system. By their performance the target and ion-source systems entirely determine the quality of the secondary ion beams that are available at a given on-line mass separator, and together with the bombarding particles, the intensity of these beams. In this context the word "performance" refers to the number of elements available, the purity of the

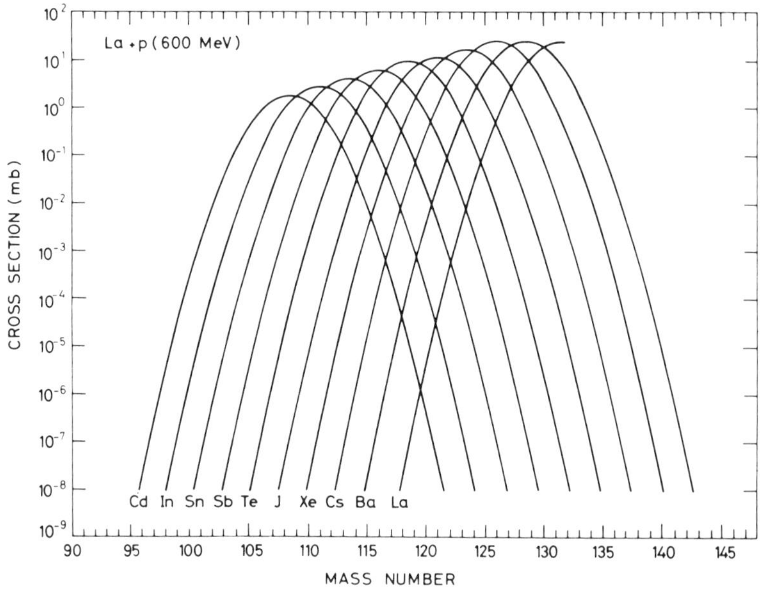


**Figure 1.** Calculated cross-sections for high-energy spallation reactions on a La target bombarded by 600 MeV protons.

beams, the speed, the yield, and the emittance. The mass transfer and ionization processes are discussed in detail in Section 2.

Although the various target and ion-source systems have been developed with particular bombarding particles in mind, they present a number of common features (Sections 3 and 4). The reaction products formed in a target are transferred to an ion source from which they are injected into the acceleration stage of the mass separator. This process requires that the nuclei be thermalized in the system. However, the decay losses for short-lived species can be kept reasonably small as seen in Figure 2 (Ra 76a), which displays the decay loss as a function of the lifetime of the nucleus. Despite the losses, radioactive ion-beam intensities of up to 3% of the primary bombarding beam intensity have been achieved (Ra 85).

The many target and ion-source systems developed over the years have been combined with different mass analyzing systems to create essentially three different types of instrument: the mass spectrometer, the low current mass separator, and the medium current mass separator. These are discussed in the following sections, but full details of individual machines are recorded in the EMIS series of conferences.

## 1.3. From the First Mass Separation to Today's Instruments

Soon after the application of accelerators to produce new radioactive isotopes, the mass spectrometer principle was used for identification of their mass numbers (Ya 41). But it was only in 1944 when reactor-produced fission-product mixtures of

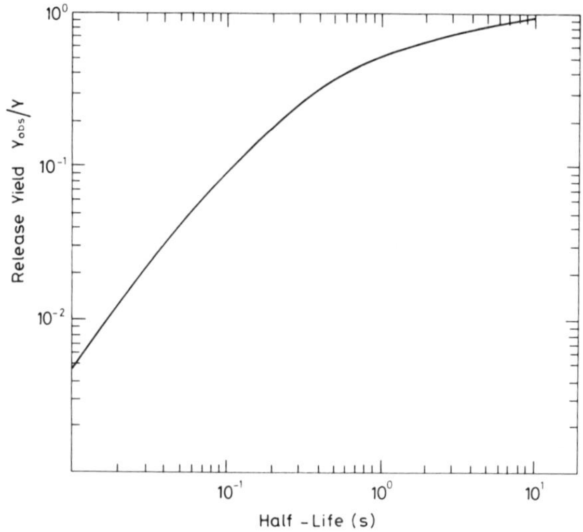


**Figure 2.** Release yield or decay loss of Rb from a Nb powder target as a function of half-life.

high specific activity became available that this principle turned into a general method of analysis and study of radioactive nuclei.

The early experiments of Dempster (De 44), in which fission products were separated in a mass spectrograph, used a hot spark for ion production with an efficiency of $10^{-8}$. These experiments encouraged the development of more efficient methods for the separation of nuclear reaction products. The fundamental data accumulated in the $^{235}U$ separations by the wartime Calutrons served as the starting point for developments which led to the construction of the different types of laboratory machine in use today. These instruments not only became important for separation of stable and radioactive nuclei, but were used as general-purpose low-energy heavy-ion accelerators with applications in many fields of physics and technology. These applications were reviewed by Koch *et al.* (Ko 58) and Karmohapatro (Ka 76).

The birth of on-line isotope separation was reported 36 years ago by Kofoed-Hansen and Nielsen (Ko 51). They irradiated a target consisting of an uranium oxide/baking powder $\{(NH_4)_2CO_3\}$ mixture with thermal neutrons from the Copenhagen cyclotron. The noble gases which were selectively emanated at room temperature were carried through a 15-m-long tube to the ion source of the electromagnetic isotope separator as shown in Figure 3 by means of the baking powder's radiation decomposition products, $NH_3$, $H_2O$, and $CO_2$. The most neutron-rich product which this technique allowed one to study was the 10-second $^{91}Kr$ isotope. The contrast between the technique of the pioneering experiment in 1950 and that of nuclear physics 25 years later has been pointed out in a very personal as well as informative account by Kofoed-Hansen (Ko 76). The experiments were interrupted in 1952 when the cyclotron was moved to a new building and were never resumed.

In the mid-sixties, experiments needing on-line mass separation regained interest, led by new developments in accelerators and in high-resolution Ge (Li) $\gamma$-spectrometers, and by more powerful techniques in nuclear instrumentation like fast electronics and

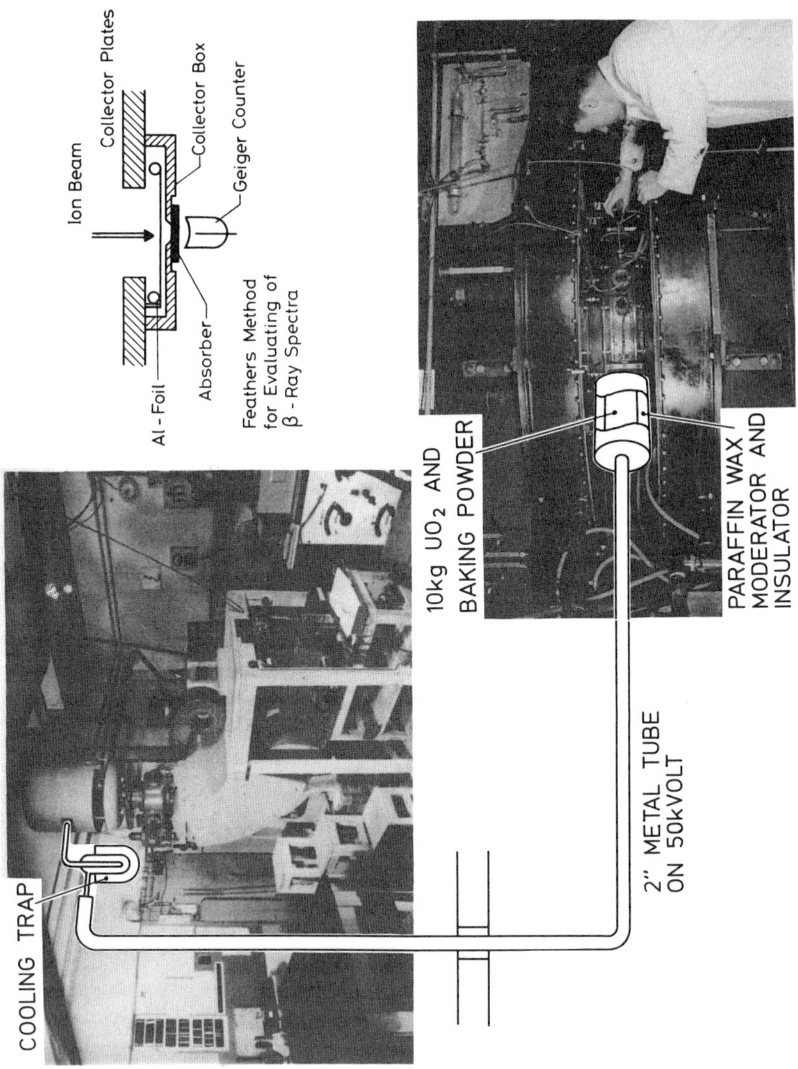


**Figure 3.** The Copenhagen isotope separator.

computers. In addition, there was now an understanding of nuclear theory that allowed the prediction of many effects specific to nuclei far from stability. At the Lysekil conference (Ly 66) a number of very good and precise physics questions were formulated. The search for new radioactive-decay modes, the refinement of the mass formulas, and the prediction of new regions of stable deformation and magic numbers formed the motivation for reviving the technique of Kofoed-Hansen and Nielsen. The first group to resume on-line mass separation and extend it to other elements was that of Klapisch and Bernas (Kl 65). They used a mass spectrometer equipped with an integrated target and surface-ionization source, mainly for nuclear-reaction studies. This application was reviewed by Klapisch (Kl 69). A large number of new on-line mass separators based at different accelerators then began to take shape, but the merits of the various bombarding particles were not clearly understood and the target and ion-source technique for most elements was still somewhat primitive. Since then, a large variety of on-line separators has been constructed, many of which are listed in Table 1.

In the following it is not intended to describe all these separators in detail, but rather to extract the essential features and illustrate them by means of selected examples.

**Table 1. On-Line Mass Separators Classified According to Their Type and the Bombarding Particle**

| At heavy ion accelerators | At reactors or neutron generators | At high-energy proton accelerators |
|---|---|---|
| *Mass Spectrometers* | | |
| *Montreal (Le 75) | GODIVA—Los Alamos (Ba 73) | *ORSAY—CERN (Sa 81) |
| *Orsay (Kl 65) | SOLAR—Richland (St 74) | |
| Tokyo (Yo 76) | OSTIS—Grenoble (Wü 78) | |
| *Low Current Separators* | | |
| DOLIS—Daresbury (Ge 83) | *Copenhagen (Ko 51) | IRIS—Leningrad (Be 73) |
| GSI—Darmstadt (Br 81) | *ARIEL—Grenoble (Bo 70) | ISOLDE 2—CERN (Ra 79) |
| Jyväskulä (Ar 86b) | HELIOS—Mainz (Br 85) | |
| LISOL—Leuven (Ve 81) | Kyoto (Ok 81) | |
| OASIS—Berkeley (Ni 83) | OSIRIS—Studsvik (Ru 76) | |
| *PINGIS—Stockholm (Fr 73b) | SOLIS—Soreq (Or 70) | |
| *Stockholm (Bo 65) | TRISTAN I—Ames (Ta 76) | |
| *POLARIS—Princeton (An 75) | TRISTAN II—Brookhaven (Ta 79) | |
| RAMA—Berkeley (Mo 81a) | Mainz (Re 80) | |
| SARA—Grenoble (Pl 87) | INEL—Idaho Falls (An 87) | |
| Tohoku (Fu 81) | | |
| Tokai (Ta 83) | | |
| UNISOR—Oak Ridge (Sp 81) | | |
| *Medium Current Separators* | | |
| Chalk River (Sc 81b) | IAELE—Buenos Aires (Ac 70) | ISOCELE—Orsay (Pa 81) |
| BEMS 2 Dubna (Ka 74) | SIRIUS—Strasbourg (Zi 73) | ISOLDE 3—CERN (Al 86) |
| EMSONHIB Dubna (Ta 70) | | TISOL—TRIUMF (Ox 87) |

* Separators no longer in operation.

## 2. Production of Nuclei Far From Stability

### 2.1. Nuclear Reactions

The particular nucleus which it is desired to study may be produced by a range of nuclear reactions, depending on how far it is from stability and in which region of the periodic chart it is situated. One can calculate the potential intensity of the product beam in the on-line separator knowing the flux of the bombarding particles, the target thickness, and the cross-section for the reaction involved. These considerations quickly establish whether or not a particular experiment is feasible at a given on-line separator installation. The most important point turns out to be the choice of the bombarding particle and its energy. For the different particle/energy combinations, various reaction mechanisms are used to produce the desired product nucleus, and several examples are discussed below. A fuller discussion will be found in Chapter 2 of this volume by Volkov (Vo 87a). Table 1 lists many of the installations in use today, grouped by the type of bombarding particle.

### 2.2. Thermal Neutrons

The thermal neutron-induced fission of $^{235}U$ was the first reaction to be used in an on-line mass separator (Ko 51). Since this reaction may reach cross-sections as high as 20 b for the formation of individual fission products, it is very attractive for the production of the heavy neutron-rich nuclei within the fission mass distribution shown in Figure 4 (Wo 77). The reaction is quite complex and requires a very high selectivity of the separator system in order to obtain clean conditions. Although only a limited

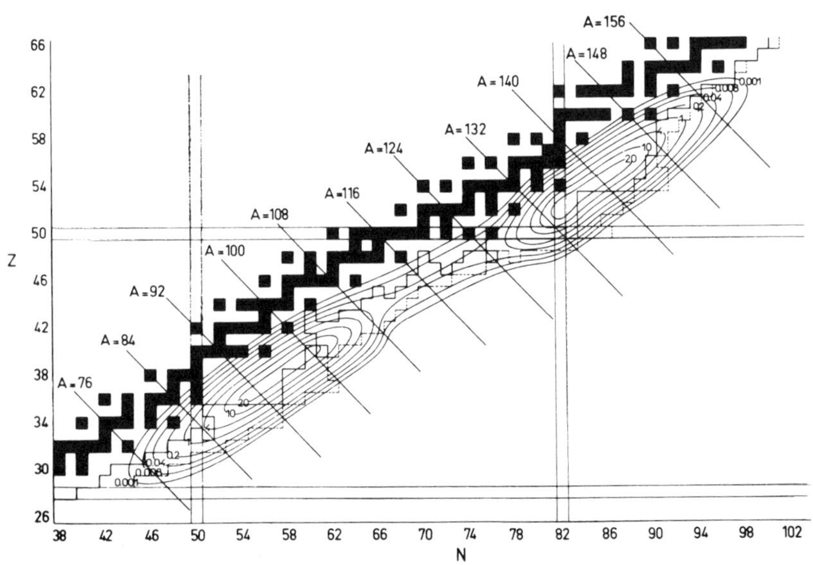


**Figure 4.** A section of the chart of the nuclides showing independent cross-section contours for thermal fission of $^{235}U$ in barns.

region of elements is possible, long series of isotopes of a given $Z$ may be studied. The high momentum imparted to the products due to the kinematics of the fission process may conveniently serve to let the nuclei recoil out of the bulk target material into a catcher with more favorable diffusion or transport properties.

Most of the 13 thermal-fission-based experimental programs listed in Table 1 are on-line to reactors. In the most favorable cases the target is situated in the high-flux region close to the reactor core. Only a few groups (Ru 76) have solved this technologically difficult problem and have used a neutron flux of $10^{11}$ neutrons/cm$^2 \cdot$ s. The remaining setups work with the much more convenient external neutron beams, where the flux is reduced by several orders of magnitude. The typical target mass that can be used with thermal neutrons is three to ten grams of $^{235}$U, before the neutrons are absorbed: this results in maximum intensities of $\sim 10^8$ atoms/s for the products.

Radioactive nuclei have also been produced by fission induced by neutrons from neutron generators (Ac 70), or by the spontaneous fission of $^{252}$Cf (An 87, Si 65) but with much lower yields. However, these two methods can be very useful to study target and ion-source techniques.

## 2.3. Light Particles with Relativistic Energies

Due to the limited number of accelerators at which light, relativistic projectiles are available in high intensities, only four experimental programs using these particles are found in Table 1: at CERN—ISOLDE with 600 MeV protons and 910 MeV $^3$He, at ORSAY—ISOCELE with 156 MeV protons and 210 MeV $^3$He, at Leningrad—IRIS with 1 GeV protons, and at TRIUMF with 450 MeV protons. The three principal reactions used here, spallation, fission, and fragmentation, which cover almost the entire chart of the nuclides, are discussed below.

### *2.3.1. Spallation Reactions*

This two-step reaction which becomes dominant at proton, neutron, or light-ion ($^3$He, $^4$He) energies from 100 MeV/U and upward is highly complex. Here the emission of charged particles followed by evaporation of neutrons limits the production of the lightest isotopes of a given element. The reaction leads to a mixture of a large variety of different product nuclei with formation cross-sections up to 20 mb as illustrated in Figure 1. The lower cross-sections and the complexity of this reaction are strongly offset by the long range of the bombarding particles, which allows the use of very thick targets. At 600 MeV, a target thickness of 150 g/cm$^2$ corresponds to one interaction length at mass $\sim$140, and the incoming beam is used up principally through nuclear reactions. With a proton-beam intensity of 1 $\mu$A the production rate in such a target becomes $10^{11}$ atoms/s for a nuclide with a formation cross-section of 20 mb. Present technology would allow the use of proton-beam intensities of 10–100 $\mu$A (Ea 86).

Theoretical calculations do not permit the estimation of the cross-sections with any precision. However, semiempirical formulas have good predictive power (Ru 66, Si 73) and describe well the yields of the neutron-deficient products all the way out to products like $^{114}$Cs produced in the reaction $^{139}$La(p, 3p23n)$^{114}$Cs as shown in Figure 1. The optimum bombarding energy for this type of reaction is $\sim$1 GeV as seen from the typical excitation curve in Figure 5, which shows the wide maximum around this proton energy (Re 77a).

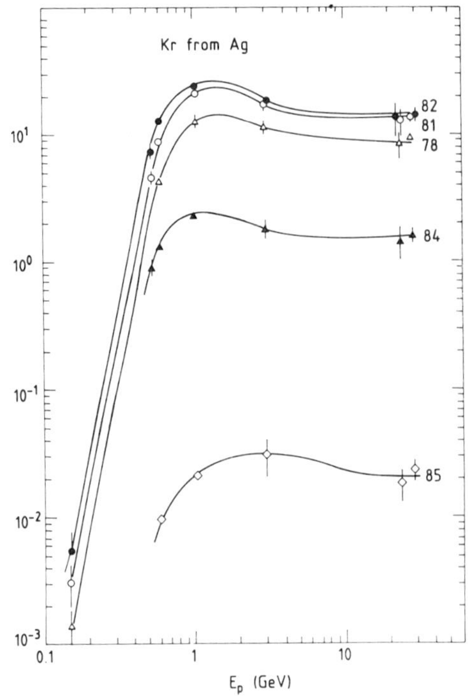


**Figure 5.** Excitation functions for the production of Kr isotopes by proton spallation of Ag.

An interesting possibility is to use spallation induced by a 910 MeV $^3$He beam instead of 600 MeV protons as done by Bjørnstad *et al.* (Bj 81a, Bj 81b). Compared to the proton beam, which favors close spallation, the $^3$He beam strongly enhances the deep spallation yields and gives interesting gains far from stability, as shown in Figure 6. The higher total energy transferred to the system favors the evaporation of many particles while the range of $^3$He still allows the use of thick targets. This is no longer the case when a 1 GeV $^{12}$C beam is used, however. The low yield ratios (Bj 81b) shown in Figure 7 are not only a consequence of the reaction mechanism, but also reflect the small effective target thickness for the $^{12}$C which has a range of only a few mm in the target.

### *2.3.2. Target Fragmentation and Fission*

As the particle energy rises toward the GeV region, the cross-section curve as a function of product mass formed in a heavy target changes as shown in Figure 8. In addition to the spallation products one sees the onset of symmetric fission represented by the product peak in the medium mass region, and of target fragmentation which is responsible for the nuclei occurring in the $Z < 20$ region. These two processes, whose mechanism is still under study, very efficiently produce the neutron-rich nuclei in their respective mass regions as shown in the figure. It should be noted that the high-energy fission of $^{232}$Th or $^{238}$U leads to a much broader mass yield curve than that of thermal neutron fission of $^{235}$U (compare with Figure 4).

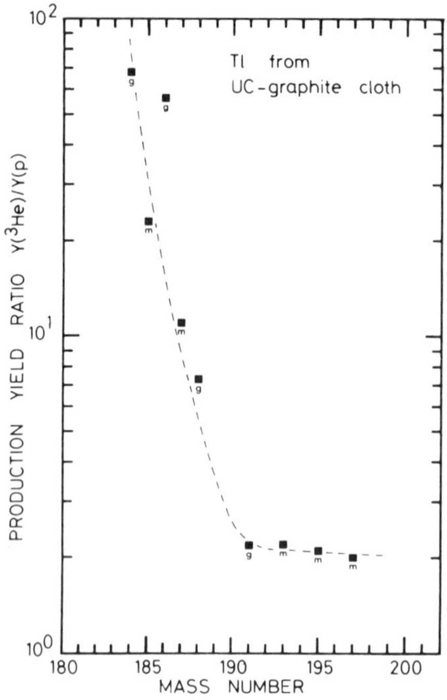


**Figure 6.** Ratios of the $^3$He- to the proton-induced saturation yields of Tl isotopes normalized to the same beam intensity.

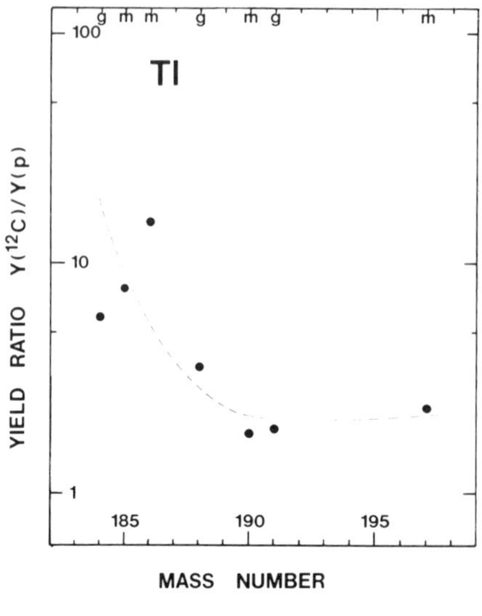


**Figure 7.** Ratios of the $^{12}$C- to the proton-induced saturation yields of Tl isotopes produced in a UC target and normalized to the same beam intensity. Letters m and g indicate the metastable and ground state, respectively.

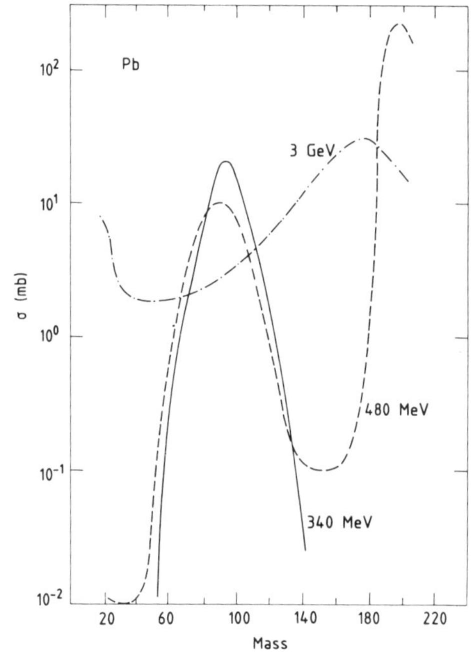


**Figure 8.** Cross-sections for the formation of products by proton bombardment of Pb as a function of product mass and proton energy.

As with the 910 MeV $^3$He-induced spallation reaction, the $^3$He *fragmentation* reaction is also enhanced relative to 600 MeV protons. However, in the fission region the cross-sections are hardly affected by the change to $^3$He (Bj 81b).

## 2.4. Heavy Ions

As seen from Table 1 many experimental programs are based on heavy-ion accelerators where often slow, light ions are also available. The large variety of reactions possible with heavy ions gives access to all regions of unstable nuclei as discussed below.

### *2.4.1. Fusion-Evaporation Reactions*

The heavy-ion reaction that has been the mainstay in on-line mass-separator work is the fusion-evaporation mechanism which involves slow, light and heavy ions up to ~10 MeV/U.

This reaction leads to extremely neutron-deficient nuclei and is strongly peaked in a few reaction channels at a given bombarding energy, so that considerable intrinsic selectivity is assured, as illustrated in Figure 9 (Ha 82). The variety of ions available as projectiles can be combined with all possible target elements, including isotopically enriched ones, in order to find ideal conditions to produce a given nuclide. A special advantage is the forward momentum imparted to the products in this reaction. It may

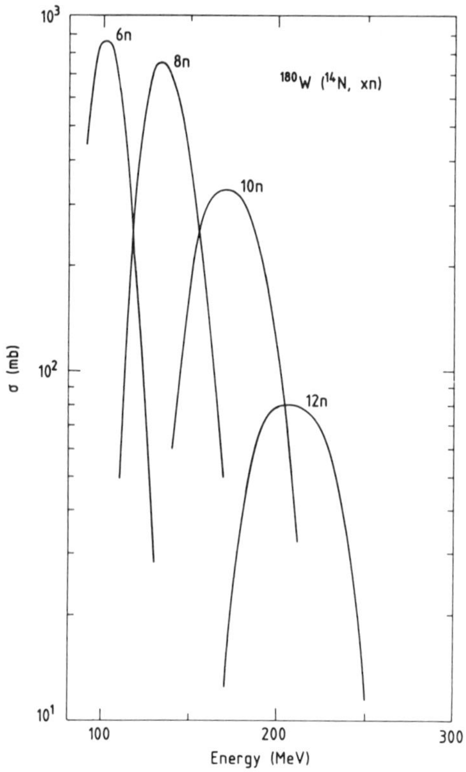


**Figure 9.** Excitation functions for various fusion-evaporation reactions on a W target.

be used to separate them from the bulk target and let them recoil into a catcher at high temperature, where the diffusion and release properties are favorable. Furthermore, the large angular momentum brought in by the heavy projectile very efficiently populates high-spin states.

Only thin targets, ~0.1 g/cm$^2$, can be used in these experiments, resulting in low maximum production rates ~$10^6$ atoms/s which, for very unstable products, may be compensated by the high cross-sections seen in Figure 9. The production cross-sections can be predicted by the Alice code (Bl 78) which is a cascade calculation. The predicted ion intensities in the focal plane of the Chalk River isotope separator (Sh 87), shown in Figure 10, are based on the calculated cross-sections and the typical performance of existing target and ion sources.

A unique application of the neutron-rich projectiles such as $^{54}$Cr or $^{58}$Fe with energies ~5 MeV/U and a $^{209}$Bi target has been in the study of transuranium elements like element 107, reached in the reaction $^{54}$Cr($^{209}$Bi, n)$^{262}$107. This type of reaction has only rarely been used with on-line mass separators (Ni 83) but has been studied by means of the recoil velocity filter separator SHIP (Mü 82). However, the fusion-reaction mechanism has been particularly well studied and used by the GSI on-line mass-separator group which has found the optimum conditions for production of a number of new nuclei situated at or near the proton drip-line in the $Z$ = 50–70 region (Sc 81a).

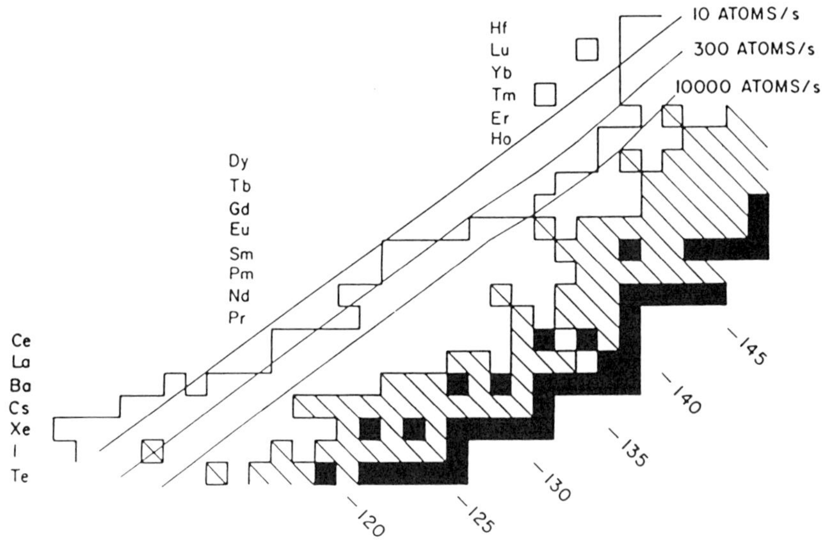


**Figure 10.** Section of the chart of nuclides indicating the predicted production rates obtained in the focal plane of the Chalk River heavy-ion on-line mass separator.

Stable target nuclei in the narrow part of the stability valley below $Z = 20$ follow the $Z = N$ line, and light-ion fusion reactions in this region already lead to very neutron-deficient nuclei. The Berkeley group (Ay 80b) have used their high-intensity (up to 10 $\mu$A for 110 MeV $^3$He) light-ion beams to study a number of new nuclei which belong to the isospin projection families of $T_z = (N - Z)/2 \geqslant -\frac{3}{2}$.

### 2.4.2. *Deep-Inelastic Transfer Reactions*

Volkov's group (Vo 87a) has pioneered the use of deep-inelastic transfer reactions to produce nuclei far from stability. By bombarding a target of high $N/Z$ ratio with a projectile a few protons heavier and an incident energy only slightly above the interaction threshold, it is possible to produce fragments with $N/Z$ higher than that of the projectile. A variety of neutron-rich nuclei outside the region of thermal-neutron fission has been produced in such multinucleon transfer reactions. Recently, the GSI on-line mass-separator group has studied this process and used it to produce a number of new neutron-rich nuclei. A typical example of these thin target production yields is shown in Figure 11 (Ru 85a).

### 2.4.3. *Projectile Fragmentation*

When the ion energy rises above ~40 MeV/U, the fusion-evaporation reaction turns into the spallation-like process discussed previously. At the same time, the projectile fragmentation channel opens up and becomes interesting for production of new nuclei on both sides of stability. This process has not been used in on-line mass

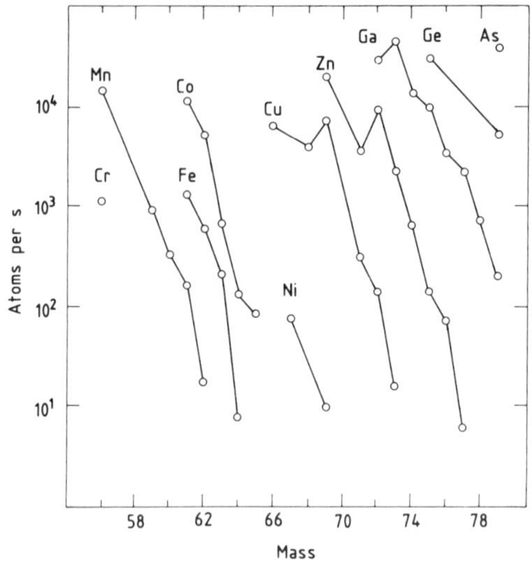


**Figure 11.** Production yields of a variety of elements by multinucleon transfer from a GSI heavy-ion beam, using thin targets.

separators but often in recoil mass spectrometers. The typical cross-sections shown in Figure 12 (Ep 84) are comparable to those of high-energy proton fragmentation of $^{238}U$ which are also plotted. The compulsory thin-target technique limits the maximum ion-beam intensities to about $10^{-6}$ of that of the projectile. This production method has had great success in identifying a number of new nuclei (De 86), while determination

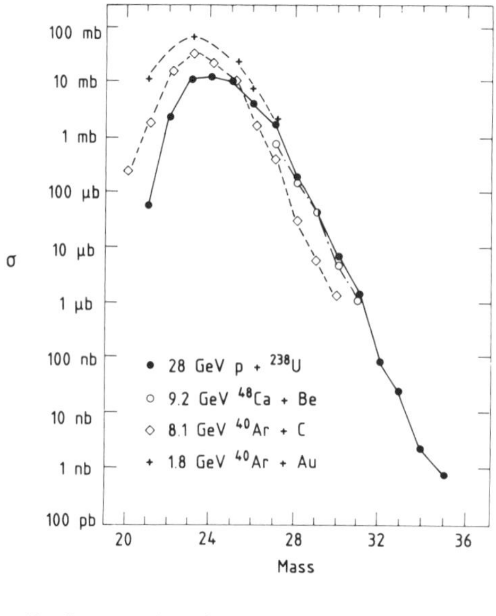


**Figure 12.** Cross-sections for the production of Na isotopes.

of decay properties appears to be more difficult, as demonstrated in the $^{31}$Ar work at the Lise recoil spectrometer at GANIL (La 86).

## 2.5. Status of Beams Available

As is seen from the discussion above, each of the different bombarding particles used in connection with on-line mass separators has its own merits and, to a large extent, the instruments at the different laboratories are complementary.

However, the thick-target technique used in conjunction with high-energy light bombarding particles offers the highest possible yields for almost all products. But the radioactivity induced in the target and the surrounding equipment requires elaborate safety and handling precautions. The separation of the products from the bulk target material raises a number of technological problems, and systems dedicated to each single element have to be developed in order to efficiently exploit the high production rates and to obtain sufficient selectivity. These problems have been solved for many elements. The beam intensities obtained and the broad isotopic distributions are unmatched by any other technique and allow very detailed studies of nuclear and atomic properties to be accomplished.

The heavy-ion fusion reaction which employs the thin-target technique has the advantage of only a few open reaction channels providing considerable selectivity and sensitivity in the very neutron-deficient region of the nuclear chart. The large recoil energy of the products offers a universal means to separate them from the target material independent of their physical and chemical properties. One is therefore less restricted in the choice of target element or compound compared to the thick-target technique. The heavy-ion-based systems seem to show an advantage in speed, although much of the advantage is lost in the systems to be discussed here, where the recoils are stopped in a catcher in the ion source from where they have to diffuse and evaporate off. Together with the heavy-ion fragmentation and deep-inelastic reactions these systems have proven to be very efficient for identification of new nuclei far from stability.

It is interesting to note that for neutron-rich products, higher intensities can be obtained from high-energy fission of natural Th and U than with reactor neutron fission of $^{235}$U. The main advantage of reactor-based on-line mass separators seems to be in the long running periods available as compared to accelerators, where beam time is more precious.

As a result of the systematic target and ion-source developments undertaken by the many on-line mass-separator groups, discussed in detail in Section 3, 82 of the chemical elements shown in Figure 13 are available as radioactive ion beams.

They can be divided into two groups:

1. About 23 elements that are successfully separated with good chemical selectivity, and with overall yields of about 10%.
2. 59 elements that are observed as on-line beams but still have need of improvements in yields, selectivity, delay time or range of isotopes than can be produced.

It is possible to classify the elements broadly according to their behavior in the target and ion source even though each one presents specific problems. Of major practical importance is the temperature required in the target and ion source in order to keep the element or a suitable compound in the gas phase so that it can be transferred

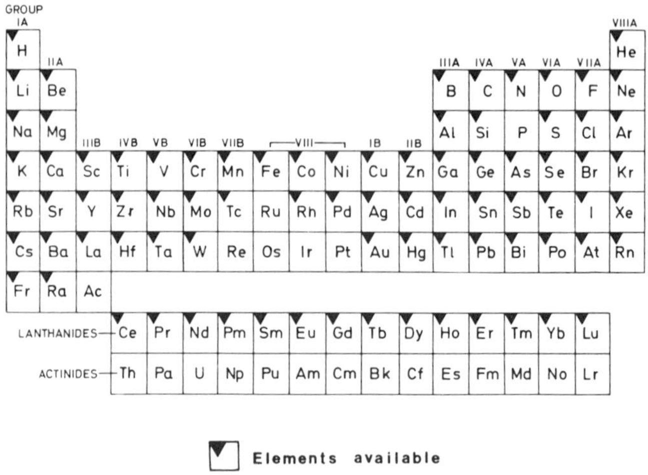


**Figure 13.** Periodic table of the elements showing those available as beams at on-line mass separators.

and ionized efficiently. The general physical and chemical properties must also be taken into account, so in the following the elements are grouped roughly as in the periodic table.

### 2.5.1. *The Alkali Elements Li, Na, K, Rb, Cs, and Fr*

Due to their high vapor pressure and simple chemical behavior, these elements are the easiest to handle in an on-line mass separator. Their low ionization potential has permitted many laboratories to produce element-selective beams very efficiently. Figure 14 shows the Na yields obtained by the ORSAY group (To 81), which has pioneered the production of the light alkali beams. They have reached the 1.5 ms isotope $^{35}$Na (La 83a), which is considered to be the heaviest bound isotope of Na and is the fastest $\beta$ decay observed in a nuclear system. The mass yield curve of cesium produced at ISOLDE (Ra 79) is shown in Figure 15. Note that all 38 known Cs isotopes are available. On the proton-rich side, $^{114}$Cs was shown by D'Auria *et al.* (Da 78) to be situated at the proton drip-line. This was confirmed by Faesterman *et al.* (Fa 84) who observed the proton-unstable 1 $\mu$s $^{113}$Cs.

### 2.5.2. *The Alkaline Earth Metals Be, Mg, Ca, Sr, Ba, and Ra*

Only beams of the three heaviest and most volatile elements of this group are produced efficiently on-line, the lighter ones still needing improvement. Despite low production yields the RAMA Group (Mo 81a) could identify the heaviest and lightest nuclei with isospin projection $T_Z = -2$, $^{36}$Ca and $^{20}$Mg respectively. The Ra-beam intensities of $10^9$ atoms/s at ISOLDE (Ra 87) have allowed the identification of two more cases of $^{14}$C emission (Pr 85).

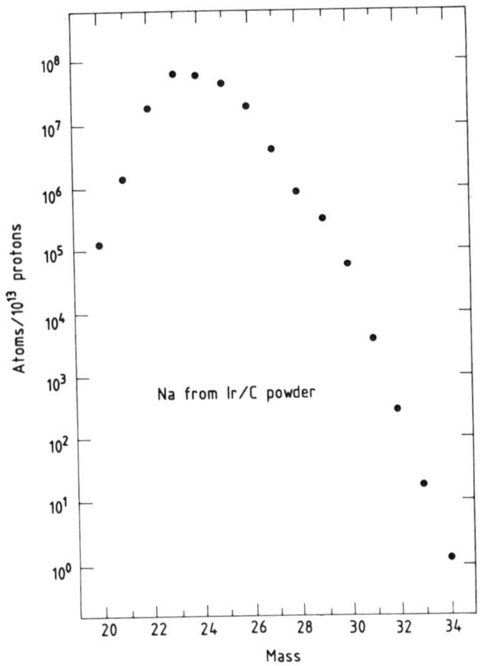


**Figure 14.** Production yields of Na isotopes in a Ir target bombarded by 20 GeV protons, obtained by the Orsay group.

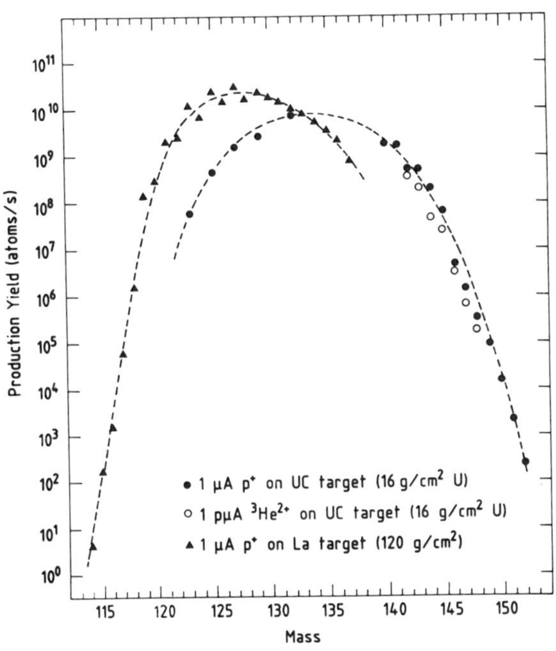


**Figure 15.** Production yields of Cs isotopes at ISOLDE.

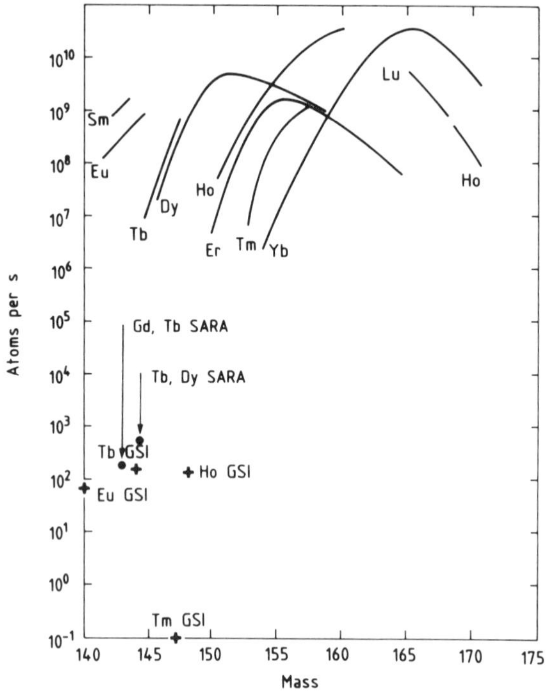


**Figure 16.** Production yields of the lanthanides obtained at ISOLDE by bombardment of a 122 g/cm$^2$ Ta target with a beam of 1 $\mu$A of 600 MeV protons. As a comparison, some yields obtained at the GSI and SARA on-line mass separators are also plotted for nuclei far from stability.

### *2.5.3. The Trivalent Metals Al, Sc, Y, and the Rare Earths*

The two lightest members of this group are only available in low yields but many groups successfully produce the rare earths. The yields of lanthanides obtained at ISOLDE (Bj 86b) and shown in Figure 16 have mainly been used in an extensive laser-spectroscopy program by the Mainz Group (Ot 87): the displayed yields correspond to an overall efficiency of 10–50%. The reason for not using these beams for nuclear spectroscopy purposes is the lack of element selectivity. The limits of stability in this region have been probed by the GSI Group (La 83b) whose beam intensities are also plotted in Figure 16. In the reaction $^{58}$Ni + $^{92}$Mo at a bombarding energy of 5 MeV/U, the ground-state proton emitter $^{147}$Tm with a half-life of 0.56 s was produced at a rate of 0.1 atom/s. The similar production rates of other rare earths far from stability which have been obtained by the SARA group (Pl 87) are also shown in Figure 16. They have elegantly used the high gas throughput of a medium-intensity mass separator to obtain 2% overall efficiency of an He-jet coupled ion source.

### *2.5.4. The Actinides Ac, Th, Pa, U, Np, Pu, Am, Cm, Bk, Cf, Es, Fm, Md, No, and Lw*

The heavy members of this group can only be produced in heavy-ion fusion reactions and the cross-sections for their formation are quite low, since the preferred de-excitation mode of the compound nucleus is fission. At the OASIS separator at Berkeley they have succeeded in making a $^{250}$Fm beam of 150 atoms/h (Ni 83). The overall efficiency of the separator in the case of Fm formed in the reaction $^{238}$U($^{16}$O, 4n)Fm was 10%.

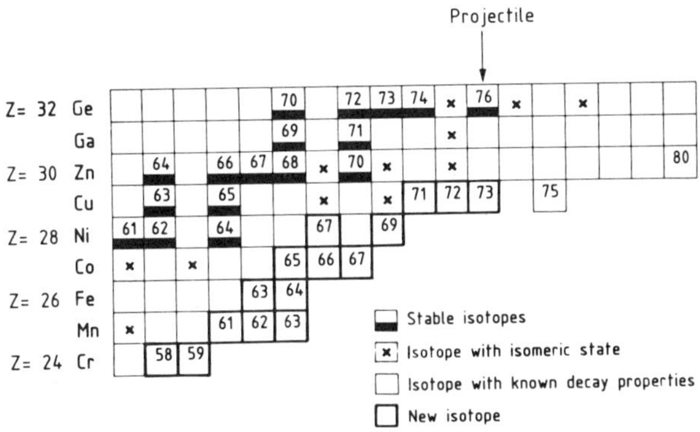


**Figure 17.** A section of the chart of the nuclides showing the neutron-rich nuclei produced at GSI by bombarding a U target with a 9 MeV/U beam of $^{76}$Ge.

## *2.5.5. The Iron Group Cr, Mn, Fe, Co, and Cu*

The proton-rich isotopes of this series of elements are difficult to produce by means of spallation reactions on thick targets, since no refractory target material with high reaction cross-section has been found. The neutron-rich nuclei fall in the minimum region of the cross-section curve for production by means of high-energy fission and fragmentation.

These elements have mainly been studied in heavy-ion reactions. The part of the chart of nuclei shown in Figure 17 illustrates the region of the neutron-rich nuclei produced at GSI (Ry 85) by means of multinucleon transfer reactions induced by irradiating a $^{nat}$W target with 9 MeV/U $^{76}$Ge ions. The beam intensities obtained were sufficient to establish the decay schemes of the new nuclei indicated in Figure 17.

## *2.5.6. Refractory Metals Ti, V, Zr, Nb, Mo, Hf, Ta, and W*

The very low vapor pressure of these transition metals and many of their compounds has for a long time prevented their on-line study. In order to solve this problem several groups have taken up the chemical evaporation technique introduced by Sidenius and Skilbreid (Si 61) for stable beam production. In this method reactive gas is introduced into the target and ion source so that volatile compounds of the elements in question are synthesized. These molecules serve to transport the nuclear reaction products to the ion source and keep them in the gas phase. The ISOCELE group has made radioactive ion beams of the elements Nb, Zr, Hf, Ta, and W (Sa 84) by introducing $CF_4$ in the ion source. Since this method gives rise to high gas pressures in the ion source and consequently high beam intensities, their use of a medium-current mass separator was a necessary condition for success. An example of the beam intensities obtained is shown in Figure 18. Note that the Hf is separated in the form of $HfF_3^+$ ions which, in addition, assured the isobaric separation so that the group could extend our knowledge to 7 new Hf and Ta nuclei.

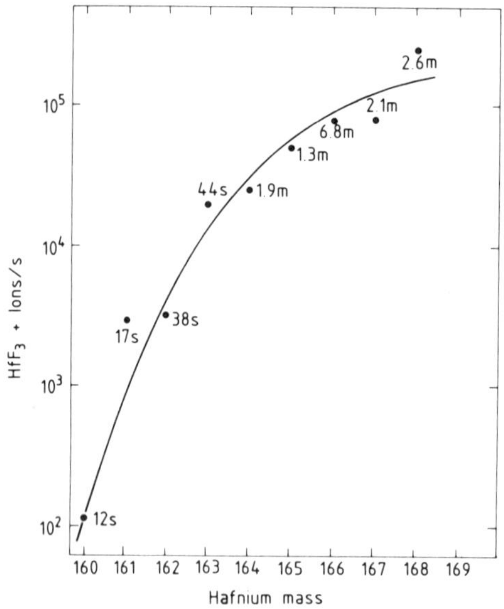


**Figure 18.** Production yields of Hf isotopes by the ISOCELE group using the technique of $CF_4$ gas addition to synthesize volatile Hf compounds.

### *2.5.7. Platinum-like Metals Tc, Ru, Rh, Pd, Re, Os, Ir, and Pt*

Like the refractory elements discussed in Section 2.5.6 it is difficult to make beams of these noble metals, since very high temperatures are required to keep them in the gas phase. It has often been claimed that He-jet systems may do the job in such cases. This may be true up to the ion source, but there they are unfortunately ionized with very low yields due to condensation. An elegant way to avoid this problem has been devised by the Jyväskylä group (Ar 86b). In their He-jet ion guide system shown in Figure 19 they make use of the primary recoil ions produced in a heavy-ion reaction. After charge equilibration in the He gas the $1^+$ ions are injected via a He skimmer into the separator without the use of a conventional ion source. The element-dependent overall efficiency of the process is 0.03–10% and the delay half-time is below 1 ms. With this technique they have made beams of a number of refractory elements including Tc. In a recent experiment with proton-induced fission the 1 s isotope $^{116}Rh$ was identified (Ay 80a).

### *2.5.8. Group 2B Elements Zn, Cd, and Hg*

The volatility of these elements allowed intense beams to be made at an early stage of the on-line mass-separation era. The high yields of Hg opened up the possibility to study nuclear ground states by techniques of atomic physics. The pioneering experiment by the Mainz group (Bo 72) on the nuclear radii of the Hg isotopes revealed an unexpected sharp change between $^{187}Hg$ and $^{185}Hg$. This triggered many groups to study the level structure of the light Hg nuclei and led to the discovery of shape coexistence in this region.

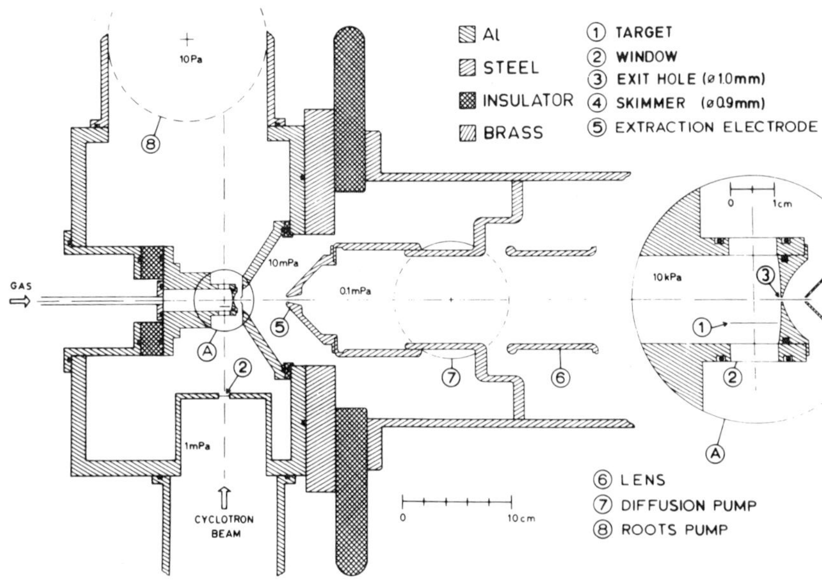


**Figure 19.** The He-jet ion guide system of the Jyväskylä group.

### 2.5.9. *Group 3A Elements Ga, In, and Tl*

The low ionization potential and the relatively high vapor pressure has allowed a number of groups to make selective and intense beams of these elements. Recently, the element In has been extensively studied since the undiscovered but presumably doubly magic $^{100}Sn$ decays to it, and the doubly magic $^{132}Sn$ can be studied via the decay of $^{132}In$ (Bj 86a). In this context helium refrigerators for nuclear orientation work have been placed on-line with isotope separators at Leuven and Daresbury. This technique has been applied to spin and nuclear moment measurements in the region from $^{105}In$ to $^{108}In$ (Va 82).

The studies of the decay of the Tl isotopes of mass 184, 186, 188 at UNISOR (Wo 84) revealed two coexisting bands in the even Hg isotopes, the strongly deformed one being built on a low-lying $0^+$ state.

### 2.5.10. *Elements which form Refractory Compounds B, C, N, O, Si, S, and P*

Although Figure 13 shows that beams of several of these elements have been observed, the overall efficiencies are typically 0.01% and they clearly still need much more development work. Recent interest in post-acceleration of these elements for nuclear astrophysics purposes has renewed the interest in high-efficiency methods as discussed elsewhere (Ra 85).

### 2.5.11. *Group 4A, 5A and 6A Elements Ge, Se, Te, Sn, As, Sb, Pb, Bi, and Po*

The neutron-deficient nuclei of the six lightest members of this group are at present best produced in heavy-ion fusion reactions. The GSI mass-separated beams of Te, Sn, and other elements (Ki 77, Ti 81) permitted the discovery of a new island of $\alpha$ emission. The lightest $\alpha$ emitter known so far, $^{106}$Te, gives insight into the $Z = 50$ shell strength through $Q$-value and mass measurements (Sc 81a).

The heavy members of this group and the neutron-rich isotopes of the lightest ones are best produced in spallation and fission of U or Th respectively. The Studsvik (Ru 85b) group have used an integrated target and ion source containing 3 g of $^{235}$U, situated near the core of their reactor in a flux of thermal neutrons of $4.10^{11}$ n/cm$^3 \cdot$ s in a very careful study of production cross-sections. They not only give the production yields of about 170 nuclides but also the delay-time distribution and ion-source efficiency of each element separated. They have recently introduced a new system (Ja 87) in which they have increased the target temperature from 1500 °C to 2500 °C. The resulting shorter delays give the improved overall efficiencies shown in Figure 20.

### 2.5.12. *The Halogens F, Cl, Br, I, and At*

Characterized by their volatility and their high electron affinity the halogens heavier than F may be produced almost as intensely and selectively as the alkalies. The chemical selectivity has been obtained at ISOLDE by means of negative surface ionization; typical examples (Kl 86, Vo 81) are the At yields shown in Figure 21. The new isotopes $^{226-228}$At have at present only been identified as radioactive ion beams, and their decay

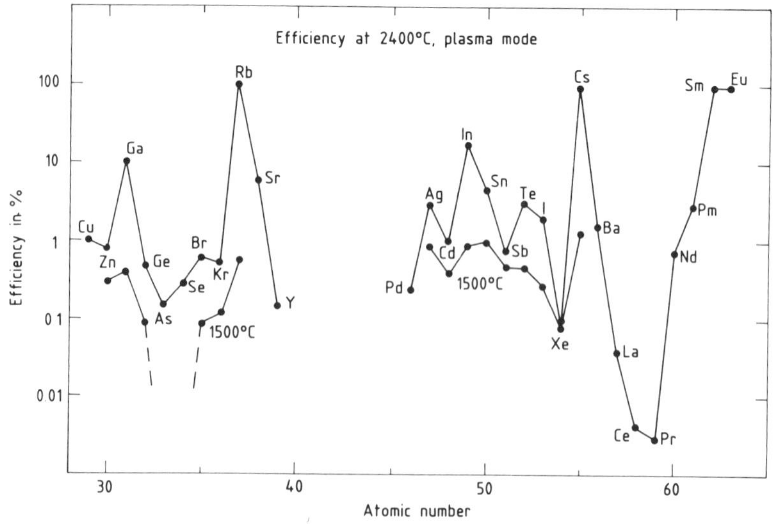


**Figure 20.** Ion-source efficiencies of products from thermal neutron fission of $^{235}$U obtained at Studsvik with ion sources working at two different temperatures.

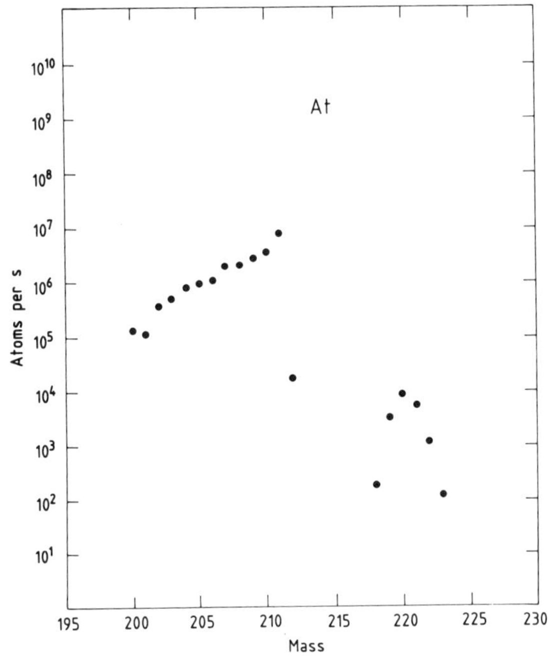


**Figure 21.** Production yields of At isotopes obtained at ISOLDE by bombardment of a 47 g/cm$^2$ $ThO_2$ target with a beam of 1 $\mu$Å of 600 MeV protons.

properties will be studied later. The $^{72}$Br beam at the Chalk River high-resolution mass separator was used to demonstrate the feasibility of the on-line separator itself for direct mass measurements (Sh 84).

### *2.5.13. The Rare Gases He, Ne, Ar, Kr, Xe, and Rn*

These noble gases are the easiest elements to bring on-line since they form no chemical bonds and diffuse out of a number of target materials, even at room temperature in some cases. One of the first targets (Pa 70) used at ISOLDE in 1967 was $ThO_2 \cdot XH_2O$, which emanated Rn at room temperature. Since then, all known isotopes of these elements are available as beams at ISOLDE in high yields and with high chemical selectivity, and they are continuously being improved. Recently, new targets consisting of alkaline-earth oxides have been tried for the production of the neutron-deficient isotopes of Ne, Ar, and Kr (Ra 87). They have been found to be the fastest ISOLDE targets with delay half-times of ~0.5 s. As an example the Ar yields are shown in Figure 22. The three orders of magnitude increase in $^{32}$Ar yield by using a CaO target, and its dominant $\beta$-delayed proton-decay mode allowed (Bj 85) a precise evaluation of the axial-vector (Gamow–Teller) strength relative to the free-nucleon value. The isotope $^{31}$Ar, which according to most updated mass formulas is stable or nearly stable against direct two-proton emission, was not observed in this experiment but has recently been observed (Po 86) at the LISE recoil spectrometer at GANIL.

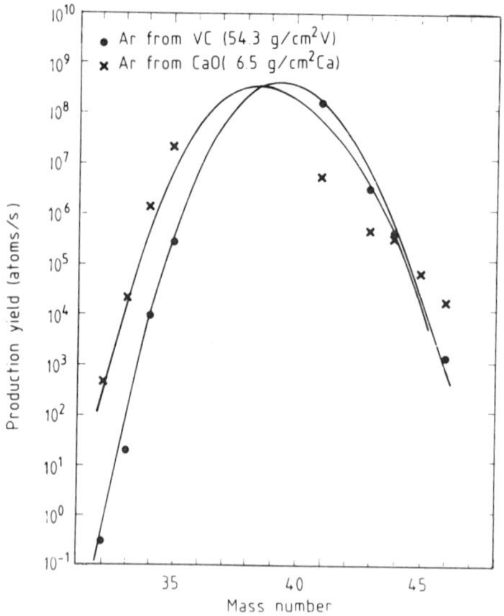


**Figure 22.** Yields of Ar isotopes produced at ISOLDE by means of VC and CaO targets irradiated with a beam of 1 $\mu$Å of 600 MeV protons.

## 3. Targets for On-Line Mass Separators

### 3.1. Release Kinetics of Nuclear Reaction Products from the Bulk Target Material

In the on-line mass separators discussed in this paper the reaction products are always brought to rest before being ionized. The general target problem becomes one of mass transfer, i.e., how to separate continuously the product nuclei from up to $10^{20}$ times more abundant target or catcher nuclei and transfer them into the gas phase of the ion source. This can only be done by heating a low vapor pressure target or catcher material to a sufficiently high temperature at which the nuclei of interest are released by diffusion and desorption processes.

In Figure 23, the different steps of this mass transfer are illustrated schematically together with an indication of the possible losses from various effects. These determine the overall yield (i.e., the ratio of the number of atoms of a given mass obtained in the focal plane of the separator, to the amount produced in the target) which is the most important single parameter of the system.

In the following it is shown that this transport process is governed entirely by element-specific diffusion, desorption and chemical processes and that the large amount of data which exists allows predictions to be made for many systems. Each of the steps is discussed separately and typical examples of on-line targets are given. Methods for the determination of the resulting overall decay losses of actual target and ion-source systems are discussed in Section 5.1.6.

#### *3.1.1. Reaction Recoil Transport*

It has already been noted that a major advantage of using heavy ions as projectiles is the large recoil energy imparted to the target nucleus, which makes it possible to get

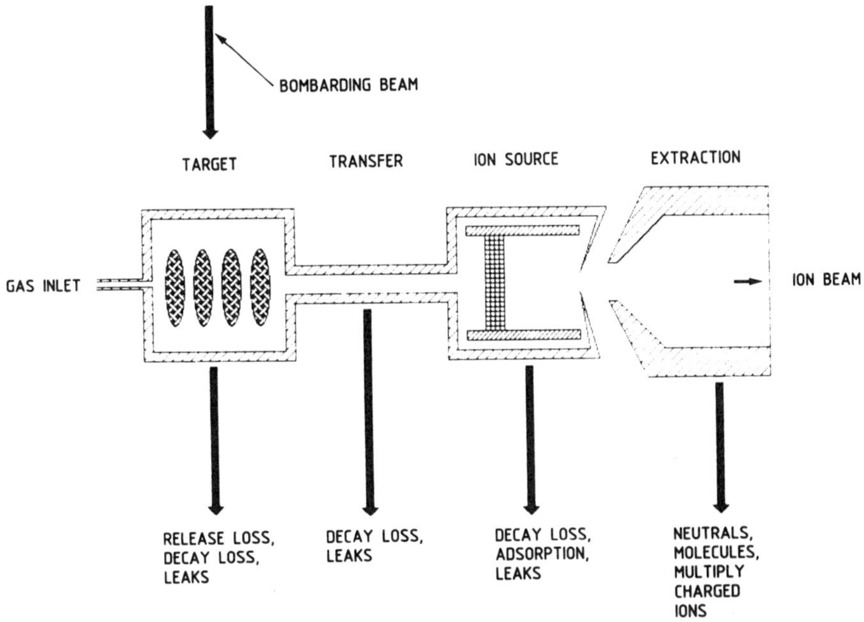


**Figure 23.** The steps in the production of an ion beam, with the losses which may occur at each stage.

the reaction product out of the thin target in times of the order of $10^{-6}$ s. Compared to the recoil mass separators this extraction time advantage is almost lost, however, in on-line mass separators. Here the recoils are brought to rest by implantation into a catcher, from where they have to be released via element-dependent diffusion and desorption processes. The real usefulness of this technique when used with on-line mass separators is that there is an almost free choice of target material, unrestricted by melting-point and vapor-pressure considerations, since the target can be situated at a well-cooled location outside the ion source; then the products recoil through a thin window into the catcher situated inside the ion source. This was demonstrated by Karnaukhov *et al.* (Ka 74) and is now adopted as a standard technique by other heavy-ion-based on-line separators. The catcher can be chosen to have favorable diffusion and desorption properties for the element in question and the source stays undisturbed by vapors released from the target.

The effective thickness of these heavy-ion targets is governed by the range of the projectile and of the recoiling product atoms, both in the target and in the entrance window, as discussed elsewhere (Ar 86b and Ad 85). A particular problem of these thin targets is the damaging effect of high power densities (lattice damage and sputtering), which limits the lifetime of targets and windows and therefore imposes an upper limit on the bombarding intensity (Ad 85).

The recoiling products may also be stopped in a gaseous catcher. This is done as shown in Figure 24 in the He-jet (Sh 87) technique, where the products are thermalized in helium doped with clusters of heavy molecules, onto which they attach themselves.

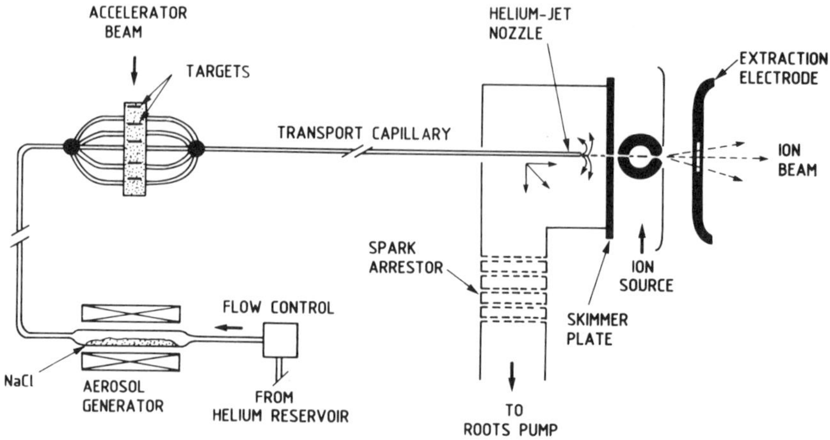


**Figure 24.** The He-jet technique.

The He flow now serves as a vehicle to assure the transport away from the target region. Downstream, the He is pumped away in a skimmer arrangement which allows the clusters to continue their flight into a suitable collection device. This technique has been used extensively in the studies of nuclei far from stability as reviewed by Herrmann and Denschlag (He 69).

The coupling of this transport technique to an on-line mass separator was originally proposed by Nitchke (Ni 70) and later successfully achieved by a number of groups, as seen in the review of Moltz (Mo 81b). The major problem here is to efficiently transfer the recoils into the sensitive volume of the ion source and avoid it being disturbed by the He flow. Once inside the ion source it is the surface adsorption-desorption process that for most elements controls the overall speed.

The atoms thermalized in He remain in the singly charged state until there is a collision with a wall or an impurity atom. This effect was used by Ghiorso (Gh 59) for the detection of $\alpha$-particle emitters produced in heavy-ion reactions. The Jyväskylä group (Ar 86a), followed by several other groups (Em 87), has successfully used this effect to couple the He jet directly to an on-line mass separator without using an ion source. The transfer efficiency of the recoils to the ion beam is determined by their range in He and subsequent diffusion and flow toward the skimmer region (Ar 86b).

### 3.1.2. *Thermal Diffusion*

A rapid escape of nuclear-reaction products embedded in a solid catcher or target material implies that a diffusion process brings the element of interest out to the surface. The rate of diffusion in solids can be found by solving Fick's second diffusion law (Cr 57) for the particular geometrical shape of the target material. The time dependence is then described in terms of the fractional activity $F(t)$, the fraction of a nuclide remaining in the target material as a function of heating time, as shown in Figure 25. For a spherical particle with radius $r$ and an initial homogeneous activity distribution,

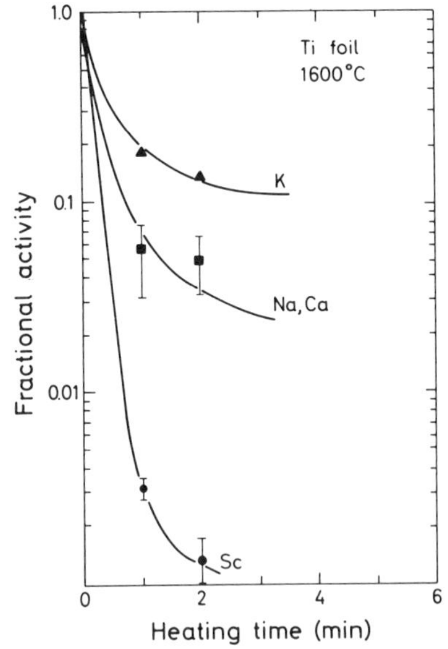


**Figure 25.** Diffusion of nuclear-reaction products to the surface of a Ti-foil target plotted as the fractional activity, as a function of heating time.

the following typical expression can be derived:

$$F(t) = \frac{6}{\pi^2} \sum_{n=1}^{\infty} \frac{1}{n^2} \exp(-n^2\pi^2 Dt/r^2) \tag{3.1}$$

where $D$ is the diffusion coefficient. The temperature dependence of $D$ is described by the Arrhenius relation

$$D = D_0 \, e^{-Q/RT} \tag{3.2}$$

where $Q$ is the activation energy for diffusion, $R$ the gas constant, and $T$ the absolute temperature.

It is seen from equations (3.1) and (3.2) that the idea behind using refractory target materials is to combine a short diffusion path and a high temperature in order to bring the products out to the surface rapidly. It should be noted that the large spread in fractional activity seen in Figure 25 in many cases contributes strongly to the element selectivity of the combined target and ion-source system. In general, the temperature should be as high as possible in order to enhance the diffusion without exceeding the melting point or producing intolerable evaporation of target material. These conditions are fulfilled by a number of metals, borides, carbides, nitrides, oxides, and sulfides. A rough estimate of the lower limit of $D$ is often obtained by looking at the self-diffusivities (No 75). A useful empirical relationship has been pointed out by LeClaire (Le 49):

$$Q = 38\, T_{\rm m} \tag{3.3}$$

where $Q$ is the activation energy (cal/mol) and $T_{\rm m}$ the melting point of the material in K. Today, a large amount of diffusion data for systems of interest to on-line mass separators has been determined experimentally and consequently the performance of a given system may be predicted. Reference to a number of such cases is given in Table 2, which lists target materials for particular product nuclei.

**Table 2. Target Materials Which Have Been Used for Particular Products**

| Product nucleus | Reference | Target material |
|---|---|---|
| Li | (Th 71) | BC |
| Be | (Ho 84) | SiC, TiC, VC, ZrC, Ru + C |
| C | (Ra 87) | MgO, CaO |
| | (Ho 84) | $ZrO_2$, TaO |
| N | (Ra 87) | MgO, CaO |
| Ne | (Ra 87) | MgO |
| Na | (To 81) | SiC, Ta, Ir, UC |
| | (Ho 84) | SiC, $Si_3Hf_5$, $Si_3Ta_5$ |
| | (Bj 86b) | Ti |
| | (Ca 79) | UC |
| S | (Ca 78) | VC |
| Cl | (Ca 78) | VC |
| Ar | (Wö 82) | C |
| | (Bj 86b) | CaO |
| | (Ca 78) | VC |
| K | (Be 77a) | Ti |
| | (Bj 86b) | Ti |
| | (Ca 78) | TiC, VC |
| | (Ho 84) | TiC |
| Ca | (Be 77a) | Ti |
| | (Bj 86b) | Ti |
| | (Ca 78) | VC |
| | (Ho 84) | TiC |
| Sc | (Be 77a) | Ti |
| | (Bj 86b) | Ti |
| | (Ho 84) | TiC, VC |
| Ti | (We 74) | C |
| V | (We 74) | C |
| Cr | (We 74) | C |
| Mn | (Ho 84) | $Cr_2O_3$, $Hf_5Ge_3$, $Zr_5Ge_3$ |
| Fe | (We 74) | C |
| Co | (Ho 84) | $Zr_5Ge_3$ |
| Ni | (We 74) | C |
| Cu | (We 74) | C |
| Zn | (Re 77b) | Ni |
| | (Ha 70) | Ge |
| | (Ho 84) | $Hf_5Ge_3$, $Zr_5Ge_3$ |
| | (Be 77a) | Zr |
| | (Be 76f) | Nb |
| | (Be 76d) | Mo |
| | (Ru 85b) | $UO_2$ |
| Ga | (Re 77b) | Ni |
| | (Ho 84) | $Hf_5Ge_3$, $ZrO_2$, $Zr_5Ge_3$, Ru, Ru + C |
| | (Ru 85b) | $UO_2$ |
| Ge | (Ho 84) | Ru |
| As | (Co 57) | C, MgO |
| | (Ho 84) | ZrC, $ZrO_2$, $Zr_5Ge_3$, Ru, Ru + C |
| | (Be 76c) | Ta |

**Table 2.** *(Continued)*

| Product nucleus | Reference | Target material |
|---|---|---|
| Se | (Re 77b) | Ni |
| | (Ho 84) | ZrC, $ZrO_2$, Ru, Ru + C, UC |
| | (Be 76f) | Nb |
| | (Be 76d) | Mo |
| | (Ca 78) | Nb, Mo, UC |
| Br | (Co 57) | C |
| | (Ho 84) | ZrC, $ZrO_2$, $Zr_5Ge_3$, Ru, Ru + C |
| | (Ru 85b) | $UO_2$ |
| Kr | (Ho 84) | Ru |
| | (Tu 77) | $UO_2$ |
| | (Ru 85b) | $UO_2$ |
| Rb | (Ra 75) | YLa |
| | (Be 77a) | Zr |
| | (Ho 84) | ZrC, $ZrO_2$, $Zr_5Ge_3$, Ru + C |
| | (Be 76f) | Nb |
| | (Ca 78) | Nb, Mo |
| | (Be 76d) | Mo |
| | (Be 76e) | Hf |
| | (Be 76c) | Ta |
| | (Ca 79) | UC |
| | (Ru 85b) | $UO_2$ |
| Sr | (Co 57) | C |
| | (Be 77a) | Zr |
| | (Ho 84) | ZrC, Ru, Ru + C |
| | (Be 76f) | Nb |
| | (Ca 78) | Nb, Mo |
| | (Be 76d) | Mo |
| | (Be 76e) | Hf |
| | (Be 76c) | Ta |
| Y | (Be 77a) | Zr |
| | (Be 76f) | Nb |
| | (Be 76d) | Mo |
| | (Be 76c) | Ta |
| | (Be 76g) | W |
| Zr | (Co 57) | C, MgO |
| | (We 74) | C |
| Nb | (We 74) | C |
| Mo | (We 74) | C |
| Ru | (Co 57) | C |
| Rh | (Co 57) | C |
| Pd | (Co 57) | MgO |
| Ag | (Co 57) | C, MgO |
| | (Ho 84) | $Hf_5Sn_3$, $ThO_2$ |
| | (Ca 78) | UC |
| | (Ru 85b) | $UO_2$ |
| Cd | (Co 57) | C, MgO |
| | (Ha 70) | Sn |
| | (Ca 78) | UC |
| | (Ru 85b) | $UO_2$ |

*(continued)*

**Table 2.** *(Continued)*

| Product nucleus | Reference | Target material |
|---|---|---|
| In | (Ho 84) | $Hf_5Sn_3$ |
| | (Ca 78) | UC |
| | (Ru 85b) | $UO_2$ |
| Sn | (Co 57) | MgO |
| | (Ca 78) | UC |
| | (Ru 85b) | $UO_2$ |
| | (Ho 84) | $ThO_2$ |
| Sb | (Co 57) | C, MgO |
| | (We 74) | C |
| | (Ca 78) | BaZrO |
| | (Ru 85b) | $UO_2$ |
| | (Ho 84) | ThO |
| Te | (Co 57) | C, MgO |
| | (Ca 78) | CeS, $BaZrO_3$, UC |
| | (Ru 85b) | $UO_2$ |
| | (Ho 84) | $ThO_2$ |
| I | (Co 57) | C, MgO |
| | (Ca 78) | CeS, $BaZrO_3$, UC |
| | (Ru 85b) | $UO_2$ |
| | (Ho 84) | $ThO_2$ |
| Xe | (Ca 78) | CeS |
| | (Ra 87) | BaO, ThC |
| | (Ru 75) | LaC |
| | (Ca 79) | LaC |
| | (Ha 70) | Pb |
| | (Tu 77) | $UO_2$ |
| | (Ru 85b) | $UO_2$ |
| | (Ho 84) | $ThO_2$ |
| Cs | (Ca 78) | CeS, $BaZrO_3$, UC |
| | (Ra 75) | LaC |
| | (Be 76e) | Hf |
| | (Be 76c) | Ta |
| | (Ho 84) | UC |
| | (Ca 79) | UC |
| | (Ru 85b) | $UO_2$ |
| Ba | (Ca 57) | MgO |
| | (Ca 78) | CeS, UC |
| | (Ra 75) | LaC |
| | (Be 76e) | Hf |
| | (Be 76c) | Ta |
| | (Be 76g) | W |
| | (Ho 84) | UC |
| RaE | (Or 61) | C |
| | (Co 57) | C |
| | (Ho 84) | $Hf_5Ge_3$, $Hf_5Sn_3$, TaC, Re, Re + C, Os, Ir, Ir + C, UC |
| | (Ra 75) | GdLa |
| | (Be 76e) | Hf |

Table 2. *(Continued)*

| Product nucleus | Reference | Target material |
|---|---|---|
| | (Ca 78) | Hf, Ta, W |
| | (An 67) | Ta |
| | (An 65) | Ta |
| | (Be 76c) | Ta |
| | (Be 76g) | W |
| | (La 74) | Re |
| Hf | (Be 76c) | Ta |
| | (Be 76g) | W |
| Hg | (Re 77b) | C, Cu, Au |
| | (Ha 70) | Pb, Bi |
| Tl | (Re 77b) | C, Ni, Cu, Au |
| | (Ra 75) | ThLa |
| Pb | (Re 77b) | C, NiCu, Au |
| | (Ru 75) | ThLa |
| | (Ho 84) | ThC |
| Bi | (We 74) | C |
| | (Re 77b) | C, Cu, Au |
| | (Ra 75) | ThLa |
| | (Ho 84) | ThC |
| Po | (Re 77b) | C |
| | (Ha 67) | BiO |
| | (Ru 75) | ThLa |
| | (Ho 84) | ThC |
| Rn | (Ra 75) | ThLa |
| Fr | (Ca 79) | UC |
| Ra | (Ra 75) | ThLa |
| Pa | (Mc 56) | U |
| actinides | (Or 61) | C |

### *3.1.3. Surface Desorption*

Once the product has reached the surface of the target or catcher material it has to obtain sufficient energy to overcome the surface desorption enthalpy of the material or of any other surface encountered on its way to the ion-source exit. In a few cases, like the rare gases or the alkali metals, where $\Delta H_d \sim 1$ eV, the delay caused by the multiple absorption/desorption cycles on the walls of the enclosure may be neglected. For these "ideal" gases, there remains, however, a small intrinsic delay determined by the vacuum flow conditions of the particular system. This has been shown by Rudstam (Ru 65) and Kirchner (Ki 81a) to have half-times of the order of 1–300 ms and to agree well with standard vacuum flow calculations for the particular geometry involved (Ho 74). In all other cases the atom lingers a time $\tau$ on the surface after each collision, as given by the Frankel equation (De 68):

$$\tau = \tau_0 \exp(-\Delta H_d / RT) \tag{3.4}$$

where $\tau_0$ is a constant of the order of $10^{-13}$ s. A measurement of $\tau$ allows the effective diffusion constant for a given system to be calculated. The mathematical treatment of the flow to and through the ion source becomes very similar to the case of solid-state diffusion as shown by Carraz *et al.* (Ca 78).

In spite of a large amount of surface desorption data available in the literature, precise predictions are often difficult due to the often undefined surface conditions, influenced by vacuum rest-gas pressures. However, in recent years a number of mass-separator groups have measured adsorption enthalpies of surface-desorption-controlled diffusion parameters under conditions relevant to on-line target and ion-source systems.

It was realized by the Osiris group that the large variations in $\Delta H_d$ could be used to obtain chemical selectivity in on-line mass separators. They applied a thermo-chromatograph at the collector end of the mass separator to separate a number of isobaric nuclei (Ru 81). Similarly, the experiments of Beyer *et al.*, based on the experimental determination of $\tau$ values for a number of elements on refractory metal surfaces, demonstrated off-line separation in the focal plane (Be 81) and in the ion source (Be 76a, Be 76b, Be 77b).

With special reference to negative surface ion sources Mårtenson and Wilhelmsson studied the desorption energies of bromine and iodine on the ionizer material $LaB_6$ (Må 85), while Kirchner *et al.* (Ki 86, Ki 87) made a systematic study of the desorption enthalpies of a number of elements in their FEBIAD ion source and compared them to the semiempirical $\Delta H_d$ values of Eichler *et al.* (Ro 84, Ei 85, Ei 87). They used these data not only to predict the delay-time distributions, but to make an on-line element-selective bunched-beam release as shown in Figure 26.

A study of both the diffusion and desorption parameters of a number of fission products was made by Rudstam *et al.* (Ru 85b) in their target and ion source typical of nuclear-reactor-based on-line mass separators. Their data allowed them to correct the measured fission yields for the decay of the short-lived species which occurred between production and measurement in order to derive the production cross-sections.

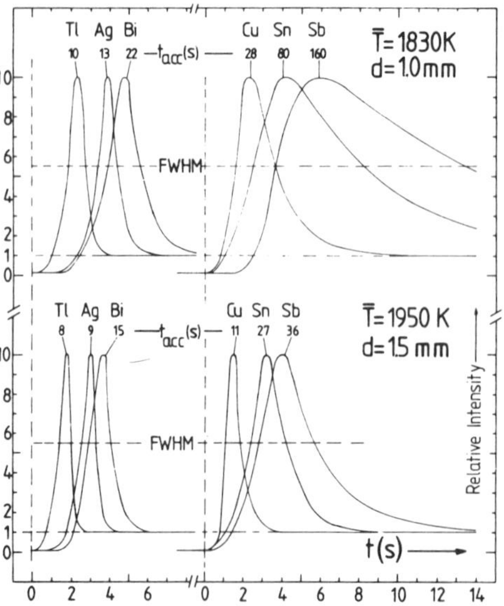


**Figure 26.** Bunched release of beams of various elements obtained from the GSI FEBIAD by cyclic heating of a cooled adsorption trap mounted in the wall of the ion source as shown in Figure 38.

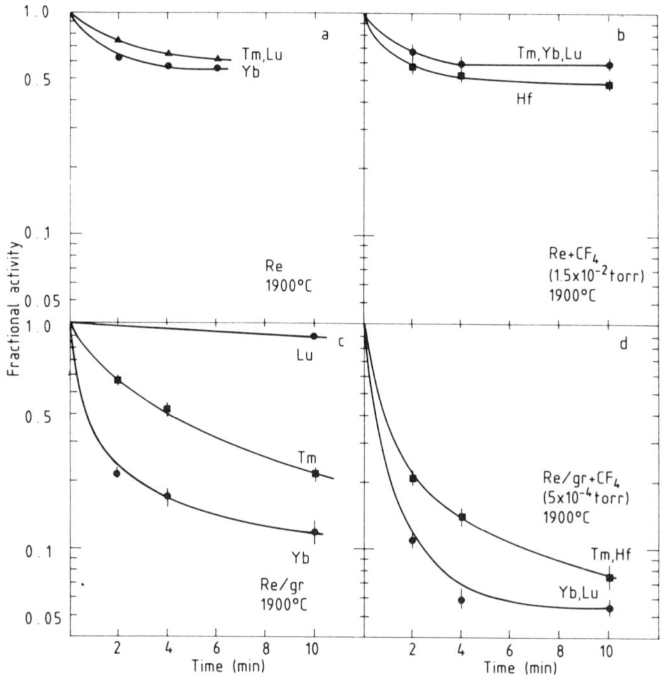


**Figure 27.** Effect of $CF_4$ addition to a Re and a Re/graphite target. Fractional activity measured as a function of heating time without $CF_4$ addition shows no release of Hf in (a) and (c). Fast release of Hf under influence of $CF_4$ is seen in (d) where the graphite admixture prevents the Re from sintering.

Many refractory elements are not released from the target material even at the highest, experimentally possible temperatures. An example of this is seen in Figure 27 by the high fractional activity of Lu remaining in the Re target after 10 min of heating, and Hf which is not released at all. Most often this is due to too slow a surface desorption, as shown by Carraz *et al.* (Ca 79) and Hoff *et al.* (Ho 84). This group of elements can be volatilized by introduction of a reactive gas into the target region, which forms a volatile compound of the element in question in a reversible reaction. The high volatility and thermodynamic stability of gaseous metal chlorides was used by Sidenius and Skilbreid (Si 61) for off-line mass separation of a number of elements evaporated by introducing $CCl_4$ vapor into the ion source. A number of on-line mass-separator groups are now using $CF_4$ addition for the case of the refractory elements. Figure 28 shows the approximately 20 orders of magnitude rise in partial pressure of Zr atoms obtained by fluorine addition. The thermochemical data which allows the calculation of this partial-pressure rise exists for a number of refractory elements and has been reviewed by Hildenbrand (Hi 81).

The effect of $BF_3$, $CF_4$, and $SF_6$ on the fractional release has been studied by Hoff *et al.* (Ho 84) for conditions which can be tolerated by ion sources. A typical example of the effect of $CF_4$ addition is seen in Figure 27. Here the fractional activity of Hf left in the Re/graphite target becomes very small when $CF_4$ is added.

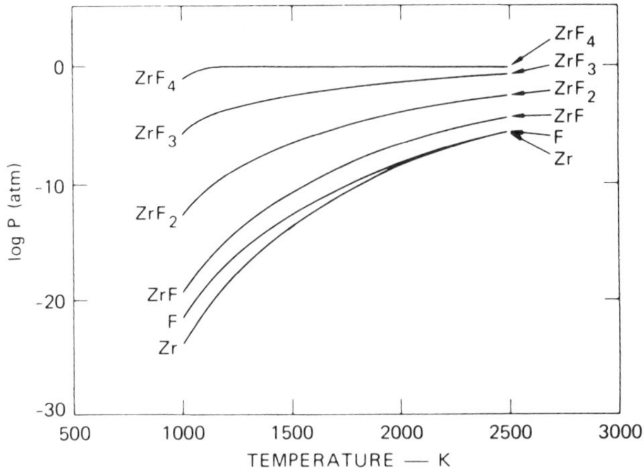


**Figure 28.** Partial pressure of Zr–X species in equilibrium with metallic Zr showing the effect of F addition.

### 3.2. Containment of High-Temperature Target Materials

Graphite, molybdenum, tantalum, tungsten, and rhenium are practically the only materials suitable for the construction of containers and ion sources which can withstand the deleterious effects of prolonged contact with the target materials listed in Table 2. Ceramic materials like BeO, BN, $Al_2O_3$, and $SiO_2$ are usually less favorable due to their much higher impurity outgassing rate and chemical reactivity. General information on refractory materials, their use in constructing containers, and their compatibility with different elements can be found elsewhere (Ca 67).

## 4. Ion Sources for On-Line Mass Separators

### 4.1. Performance Criteria

Once the products have been released from the target or catcher surface they must be transferred as fast as possible to an ion source in order to convert them into singly or multiply charged ions. The successful operation of an on-line mass separator depends to a large extent on the performance of this combination of target and ion source. The widespread use of ion sources for the production of heavy-ion beams both around accelerators and in industry has led to the development of a large variety of ion sources. Unfortunately, most often only very little emphasis has been put on the performance criteria dictated by the complex mixture of rare and short-lived nuclei which an on-line target provides as feed material. Because of this it is not possible to define a universal ion source, so that for on-line mass separation the source has to be developed specifically. Most of the inspiration for this development comes from the research fields mentioned above which are reviewed by Freeman and Sidenius (Fr 73a) and are the subject of a number of conference proceedings (Em 65, Em 70, Em 73, Em 80, Em 87, Io 72, Io 83). High-efficiency sources dedicated to on-line mass separation are much rarer and these have been reviewed by Ravn (Ra 76b) and Kirchner (Ki 81b).

Certain of the problems of radioactive ion production are technological, while others are associated with an incomplete understanding of the complex ionization phenomena in the electrical discharges. A discussion of these phenomena can be found in the literature (De 73, Wi 73, Cl 87). Most of the progress in ion-source development will continue to be empirical, but novel techniques such as ECR and laser ionization will play an increasingly important rôle. In general, each element and each experiment requires a somewhat different form of source or at least places a different emphasis on the performance characteristics.

In the following we will discuss the performance criteria that are of greatest importance to on-line separators and then describe some sources which best illustrate the particular problems of radioactive ion production and show the state of the art in this field. The various losses encountered in producing a radioactive ion beam have already been illustrated schematically in Figure 23. The following design criteria have also to be taken into account.

### 4.1.1. *Ionization Efficiency*

The useful reaction rate in the target varies from ~0.01 atom/s for the most exotic species to $\sim 10^{12}$ atoms/s for those close to stability. Since the amount of feed material is so strongly limited, most emphasis has been put on achieving a high ionization efficiency. This is defined here as the ratio of singly charged ions extracted from the source to the number of atoms brought into the source, assuming no decay losses. The values obtained vary strongly from element to element and are often difficult to measure with any precision, since they depend critically on the mechanical tolerances and the type of ion source, the amount and chemical composition of the gases which accompany the radioactivity, and the chemical affinity of the atoms in question to the source construction materials. A number of measured values are given in the examples discussed below. Comparisons of measured and theoretical efficiencies have been published (Ra 79, Ki 81b, Ki 76, Ni 85).

For plasma discharge sources, the efficiency is often given only for the noble support gas that is admitted to the source through a calibrated leak. Apart from sustaining the plasma the intensity of the noble gas beam allows the source performance to be monitored, and a high gas efficiency usually indicates a good efficiency also for other elements present in the plasma. The more refractory an element, the higher the temperature required in order to keep it in the gas phase; but only very few sources work reliably above 2000 °C, since the coldest spot in the source enclosure determines the condensation or absorption losses. Contrary to ion sources used in other fields the gas tightness of the on-line target and ion source plays an important rôle. This is because the losses caused by escape through leaks are at best proportional to the ratio of the leak area to the source outlet area. If there is any vacuum flow impedance between the target and the ion-source outlet, the loss is even larger.

### 4.1.2. *Selectivity*

As pointed out in Section 1.2, a combination of mass separation with chemical separation is essential for the study of nuclei far from stability. However, only a limited number of elements are selectively produced by present target and ion-source systems. In many cases the desired selectivity has been obtained in the ionization stage either by surface ionization (Ra 76a), ionization of molecular sidebands (Ho 80, Ho 81, Sa 84), or thermochromatographic separation in the ion source (Ki 86, Ki 87). A very promising technique is the two-step laser photoionization source which may provide elementally pure beams with both high efficiency and selectivity, as discussed in Section 4.6.

### 4.1.3. Speed

The losses caused by decay during the time the products spend in the ion source play a more important role for the short-lived nuclei than the trivial losses through leaks in the system. This delay loss varies from the minimal intrinsic delay (1–300 ms) discussed in Section 3.1.3 to quite large values for elements with high chemical affinity for the construction materials of the ion source. For the short-lived elements it is most important to minimize the intrinsic delay by a reduction of the source volume, an increase in the temperature, an increase in the outlet orifice area, and, if possible, a judicious choice of the construction materials (minimum desorption enthalpy for the product elements in question). In addition, it is desirable to have a high pumping speed through the source outlet in those cases where high partial pressures of target material or their compounds occur as a result of the chemical evaporation technique or the use of volatile target materials.

### 4.1.4. Emittance and Beam Shape

The angular divergence and physical size of a beam are conveniently discussed in terms of the emittance. A small emittance value has become increasingly necessary as the experiments have moved away from the focal plane of the analyzing magnet, which is where the first experiments with on-line separators took place. The low-energy (40–60 keV) ion beam is now often transported over long distances to the experiments via a complex beam-transport system, and stringent requirements are often made on the precision with which the ion beam must be aligned and focused inside the experimental equipment. Emittance and yield do not seem to have an optimum at the same operating parameters of plasma discharge sources, although no measurements exist for on-line conditions. Kirchner (Ki 78) has shown that the mass resolving power and ion-source efficiency are inversely correlated for the FEBIAD type source; a 10-fold loss of yield was found when the source gave the best resolving power. Off-line measurements by Dangaard and Heinemeyer (Da 81) show that the emittance of a similar ion source (Ca 79) may vary from 7 to 100 mm · mrad depending on the operating conditions.

The requirement for a large source outlet area discussed in Section 4.1.3 is, from a beam optical point of view, best fulfilled by means of ion sources with a slit-shaped outlet opening. Puteaux *et al.* (Pu 87) have compared the emittances of several plasma discharge sources commonly used in on-line mass separators. They found typical emittance values around 20 mm · mrad.

### 4.1.5. Energy Spread

The distribution of initial energies of the ions extracted from the source should, of course, be a minimum in order to obtain a high resolution in the mass separator.

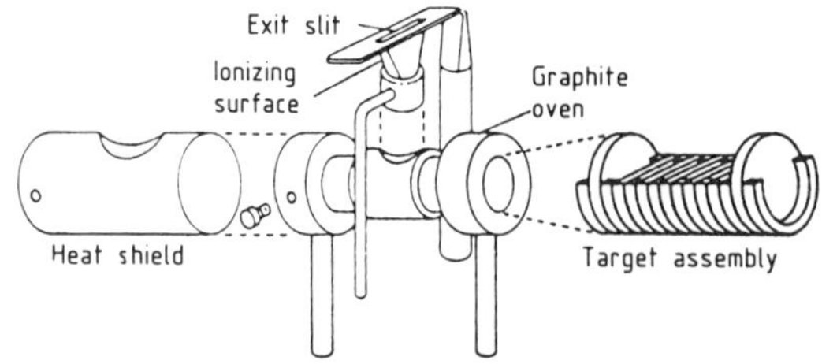


**Figure 29.** Target and surface ionization source assembly with slit geometry as used by Chan *et al.* (Ch 77).

This is also a requirement from collinear laser spectroscopy experiments, where the resolution is limited only by the Doppler broadening of the optical resonance lines caused by any velocity spread of the ions. In fact, the collinear laser setup is ideally suited to a study of energy spreads of ion sources as discussed by Neugart (Ne 86). Investigations by this technique (Ne 82) of the variations in potential of the ions in the plasma of a FEBIAD-type ion source showed that 10 V was not an unusual value. Depending on the operating mode of the source, variations down to 3 V were observed. For surface ionization sources, energy distributions of 3 eV (Bu 82) and 0.5 eV (Kl 79) are reported.

## 4.2. Surface Ionization Sources

Surface ionization is a process in which an atom hits a heated surface, where it loses or gains an electron before re-evaporating as an ion. Most elements and many molecules can be ionized in this way. Efficiencies of interest in on-line mass separation (from 0.1 to 100%) can only be obtained for the roughly 50 elements with electron affinity $>1.5$ eV or ionization potential $<7$ eV. The ionization efficiency is given by the Saha-Langmuir equation as discussed in detail elsewhere (De 73, Wi 73, Ka 83). Sources based on this principle often allow a high degree of selective element ionization. This was first exploited on-line, for the alkali metals, by Amarel *et al.* (Am 67), and has since become a standard technique in the field. The simplicity of such a source is seen in Figure 29, which shows the target and ion-source assembly used by Chan *et al.* (Ch 77) for ionization of In and Ga. The ionization takes place on the inside surface of a "chimney" made of rhenium from which the ions are extracted through a slit. At ISOLDE (Bj 86b) this type of source has been developed in conjunction with thick targets for ionization of many more elements. By means of a tubular ionizer, which forms a constriction at the end of the transfer line as shown in Figure 30, the temperature can be raised toward 3000 K and materials with work functions of up to 6.5 eV may be used for the constriction. It should be noted that this kind of ion source has short delays due to its very high temperature and is insensitive to the impurity vapors which may accompany the products. This makes it quite attractive in combination with He-jet transport, as done by Brügger *et al.* (Br 85) with the system shown in Figure 31.

More recently, the surface ionization principle was extended to the on-line negative-ion formation of the high electron affinity halogens (Vo 81, Bj 86b). The principle and construction of such a source is shown in Figure 32. The negative-ion emitter onto which the product flow impinges consists of a small $LaB_6$ pellet with a work function of 2.7 eV, situated in the constriction of the transfer line. In order to assure adequate pumping speed of the target material, an annular opening is left around the pellet

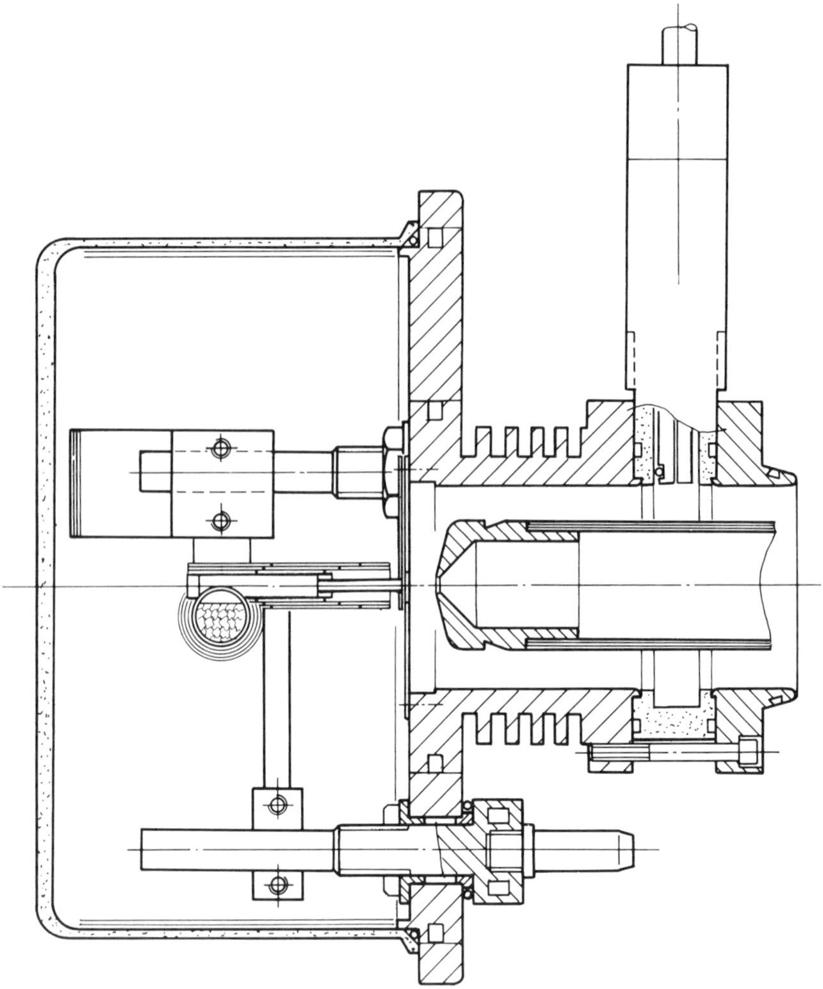

**Figure 30.** ISOLDE-type target with a tubular surface ionizer mounted in its remotely handled vacuum container with quick couplings.

through which about 50% of the product atoms escape without hitting the ionizer. The abundantly emitted electrons are deflected onto a separately powered electrode by means of a permanent magnetic field. Selective ionization with efficiencies of up to 40% for Cl, Br, I, and At have been obtained. An as-yet unused potential of this source is the ionization of volatile chemical compounds with high electron affinity.

### 4.3. High-Temperature Cavity Ion Sources

Beyer *et al.* (Be 71) and Johnson *et al.* (Jo 73) observed that a tungsten cavity, similar to the surface ionization source but with a small orifice, was a very efficient

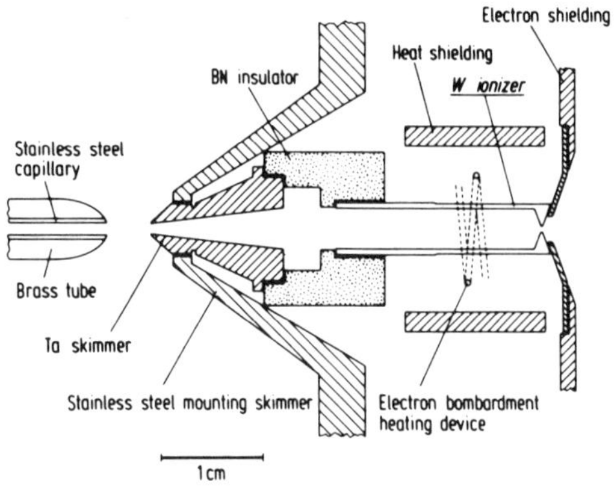


**Figure 31.** The tubular surface ionizer coupled via a skimmer to a He-jet transport system as used by Brügger *et al.* (Br 85).

thermal ionizer that could give higher efficiencies than expected from surface ionization alone. Figure 33 shows such a source. As discussed by a number of authors and reviewed by Kirchner (Ki 81b) and Huyse (Hu 83), a quasi-neutral thermal plasma may be created under certain conditions of surface ionization and electron emission from the walls. Depending on the total pressure, the partial pressure of the impurities, and the total extracted current (or outlet area), the efficiency rises from that of surface ionization described by the Langmuir equation to that of thermal-equilibrium plasma ionization described by the Saha equation. This is shown in Figure 34, plotted as a function of ionization potential. Due to the sensitivity of the source to impurities and gas pressure, a high efficiency is best obtained in conjunction with thin targets which allow only the recoiling products to enter the cavity. Such an ion source was first developed by Karnaukhov *et al.* (Ka 74) and is now used frequently at heavy-ion-based on-line mass separators. A typical example taken from the work of Kirchner *et al.* (Ki 81a) is shown in Figure 35.

Recently Piotrowski *et al.* (Pi 84) have developed this source concept for use in conjunction with thermal neutron fission of $^{235}$U. By accommodating 5 g of $^{235}$U in the form of carbide inside the cavity as shown in Figure 36, they demonstrated the ionization of a number of elements with high ionization potential.

## 4.4. Hot Cathode Arc-Discharge Ion Sources

The elements with ionization potential $>7$ eV are most efficiently ionized in a low-pressure gas-discharge ion source. The common feature of such ion sources is that electrons are extracted from a heated cathode, guided by a suitably oriented magnetic field, and accelerated into a gas-filled chamber where a plasma is created. The products to be ionized are fed into the plasma by diffusion, recoil, or He-jet technique as discussed in Section 3. A large variety of ionization-chamber geometries and cathode configurations have been used in on-line mass separators, but the sources broadly fall into two categories. Extraction of the ions may be effected from a circular orifice in the direction of the plasma column and electron flow, or from a slit perpendicular to the discharge. There is no fundamental difference between the two types of source, but for practical purposes the distinction is real since a number of performance parameters are different in the two geometries, as will be seen from the following discussion.

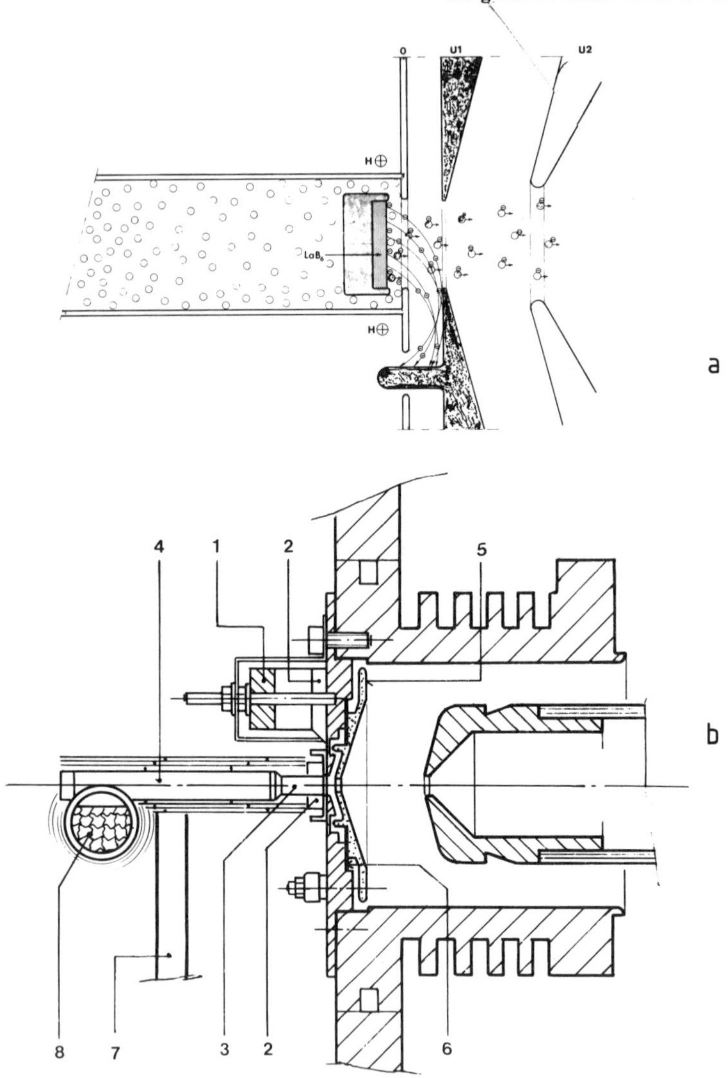


**Figure 32.** The principle (a) and technical details (b) of the ISOLDE negative surface-ionization source: (1) permanent magnet, (2) pole piece, (3) ionizer constriction, (4) product transfer line, (5) electron catcher electrode, (6) insulator, (7) terminal for line and ionizer heating current, (8) target.

### *4.4.1. Axial Extraction Sources*

The first arc-discharge source to be used on-line was the axial extraction ion source which was developed by Nielsen (Ni 57) originally to meet the requirements of low-current laboratory isotope separators of the Scandinavian type. A typical example of the Nielsen source is seen in Figure 37 taken from the on-line studies of thermal neutron

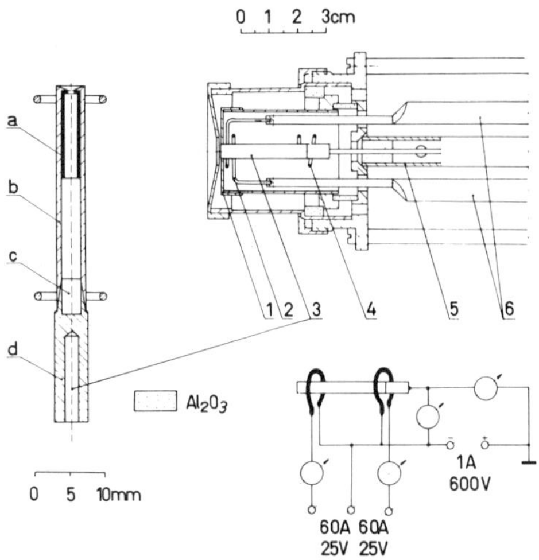


**Figure 33.** Hot cavity ion source and its power supply scheme: (1) heat shield and electron reflector, (2) cathode of ionizer, (3) ionizer and vaporizer (a, Re foil; b, ionizer; c, vaporizer; d, support rod), (4) vaporizer cathode, (5) adjustable support, (6) supports and cathode terminals.

fission products of Talbert *et al.* (Ta 79). The $^{235}U$-loaded graphite-cloth target was situated inside the anode cylinder as developed by Rudstam *et al.* (Ru 85b), who also mapped the on-line ionization efficiencies of the elements shown in Figure 20. The operating parameters, in particular the gas pressure, were very critical when keeping this Nielsen source running at a high efficiency, and its lifetime was limited by reactions between the filament and the impurities released by the target material.

These drawbacks inspired Kirchner and Roeckl (Ki 76) to develop a new source specifically for on-line mass separators. Their Forced Electron Beam Induced Arc Discharge (FEBIAD) source is shown in Figure 38. The essential changes are that it is equipped with two anodes and the nude filament is here replaced by an indirectly

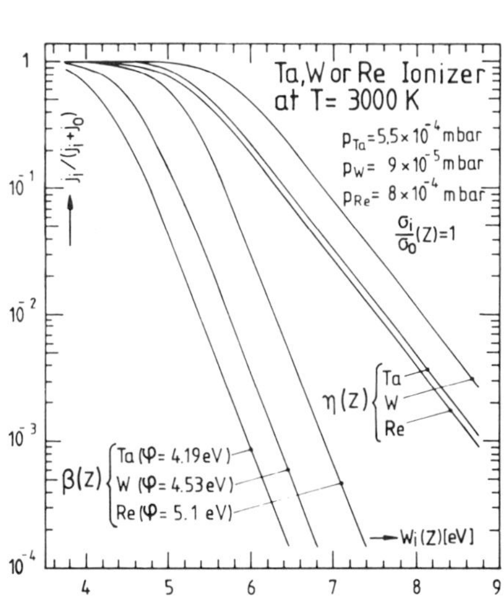


**Figure 34.** Theoretical ionization efficiencies as a function of ionization potential $W_i$ for surface ionization $\beta(z)$ and thermal-equilibrium plasma ionization $\eta(z)$.

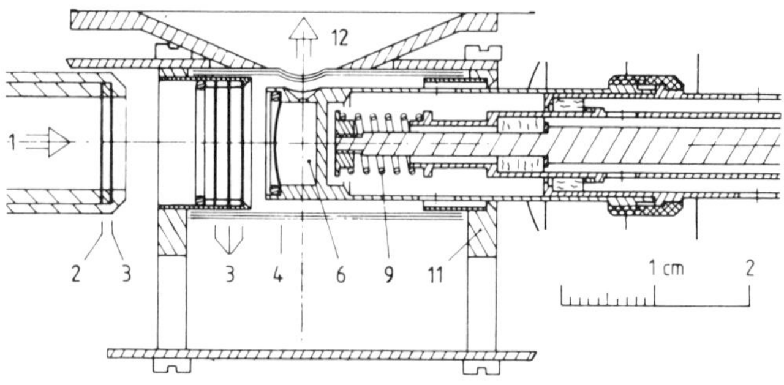


**Figure 35.** GSI hot cavity ion source: (1) heavy-ion beam, (2) target foil, (3) heat shields, (4) window, (6) ionizer volume, (9) filament for electron bombardment, (11) support, (12) ion beam.

heated disk-shaped cathode from where the electrons are extracted into the plasma by means of a grid. These improvements resulted in a longer lifetime, a decoupling of the ion-source parameters for ease of operation, and an expansion of the regime of efficient operation into the $10^{-5}$ torr region of support-gas pressure. The version of the source shown in Figure 38 is for heavy-ion recoil transport and incorporates the cooling device which allows bunching of the ion beam and thermochromatographic separation as discussed in Section 3.1.3. FEBIAD sources have now been widely adopted and Figure 39 shows a version for use with a high-temperature, thick ISOLDE target. In order to avoid adsorption on cold spots, the products are fed into the source via the cathode which is directly heated by means of the current, which also heats the transfer line.

Such ion sources work best with gas pressures $\sim 10^{-3}$ torr or below. In cases where the source is exposed to a higher gas load but, for beam optics reasons, a circular aperture is required, the hollow cathode source is a solution. This type of source was developed by Sidenius (Si 70) and works with high efficiency at a very high pressure, but the resulting beam intensity can still be handled by low-current separators. Furthermore, the working temperature of close to 2000 °C allows the efficient ionization of low vapor pressure elements. These properties make it the preferred choice of many groups for coupling a He-jet transport system to a mass separator. The source developed by Anderl *et al.* (An 87) shown in Figure 40 incorporates the interesting feature of an indirectly heated cathode which most certainly increases the lifetime of the source.

Sources with a similar geometry of an indirectly heated cylindrical cathode, but with cylindrical anode grids for electron injection into the plasma, have been reported by Nitchke (Ni 83) for heavy-ion recoil products (Figure 41) and by Gill (Gi 84) for fission products (Figure 42). These new concepts deserve attention, since they are claimed to work at temperatures above 2500 °C without problems with their insulators, usually a critical feature of the design.

### *4.4.2. Lateral Extraction Sources*

Medium-current mass separators usually work with a vertical arc perpendicular to the extracted ion-beam direction. Apart from the high resolution obtained from the narrow-slit extraction geometry, their increased throughput due to the large slit area is a major advantage as discussed in Section 3.1.3. The ISOCELE group (Pu 81) has

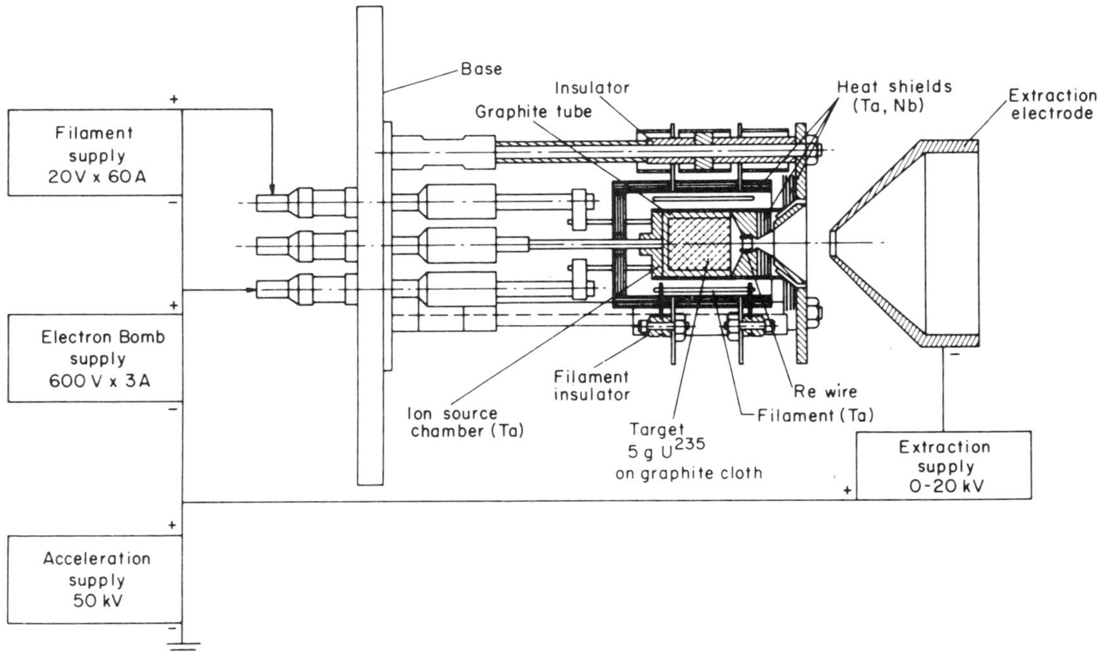


**Figure 36.** The Brookhaven hot cavity target and ion source and associated power supplies.

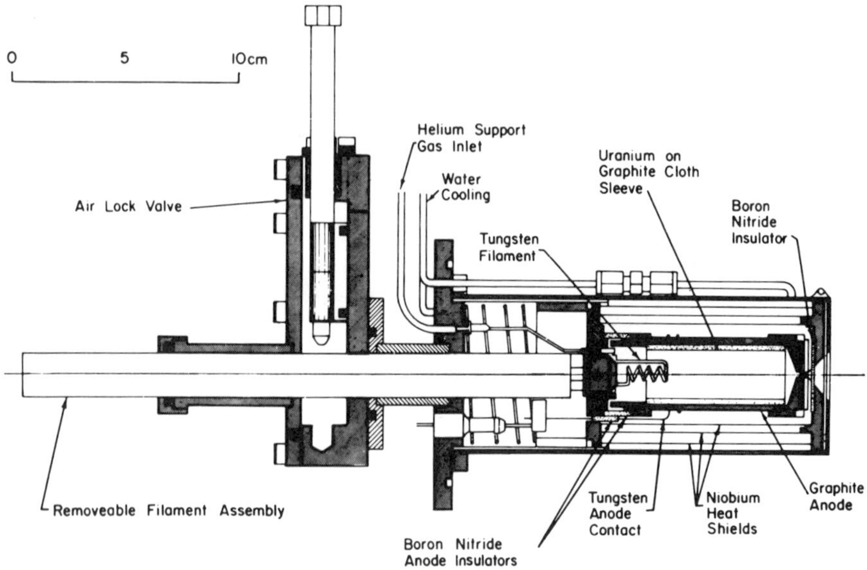


**Figure 37.** Nielsen-type plasma-discharge ion source with integrated $^{235}$UC target.

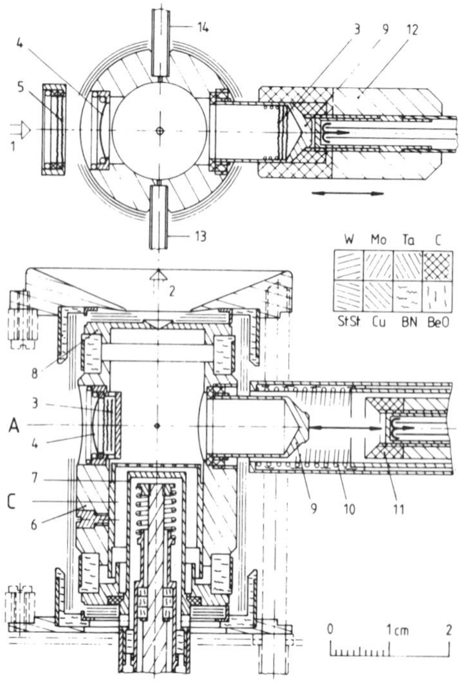


**Figure 38.** GSI-type FEBIAD ion source with cooling pocket for beam bunching: (1) target in heavy-ion beam, (2) mass-separator beam, (3) catcher stack of thin tantalum foils, (4) entrance window, (5) window heat-shields, (6) anode, (7) cathode, (8) outlet plate, (9) accumulation pocket, (10) extra heater, (11) weak cooler, (12) 15 $\mu$m tantalum foil (cold trap), (13) associated cooler, (14) gas line, (15) oven line.

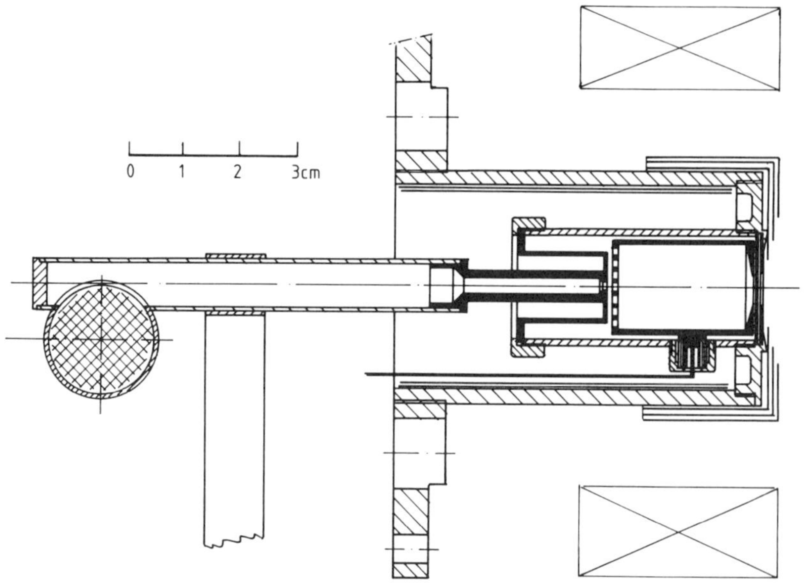


**Figure 39.** ISOLDE-type FEBIAD ion source with combined DC heating of the product transfer line and the cathode.

taken advantage of this by introducing the consumable target technique and by developing the chemical evaporation technique. A schematic view of their Bernas–Nier-type ion source in Figure 43 shows that their few $g/cm^2$ thick target is located inside the ion source and is heated by the arc itself. In fact they may choose volatile target materials or such a high temperature that a major part is lost through the extraction slit during operation.

The large throughput of this source geometry has also attracted the users of He-jet transport. Schmeing *et al.* (Sc 87) developed a high-temperature version of the slit source with a long-lived indirectly heated cathode and He-jet catcher region as seen in Figure 44. They report efficiencies of up to 5% for a number of less volatile elements. The He-jet coupled Bernes–Nier source of Planetier *et al.* (Pl 87) was successfully used for production of the rare-earth beams shown in Figure 16.

If the more compact axial extraction source is equipped with a slit, Puteaux *et al.* (Pu 87) have shown that it is no different to the lateral extraction source with respect to beam optical properties, as well as current handling and throughput capabilities, a feature which will be exploited in the new separator at ISOLDE.

## 4.5. Electron Cyclotron Resonance Sources

Due to their fundamental simplicity, ion sources where the plasma is excited by means of microwaves now often replace hot-cathode sources at accelerators. In particular, the single-stage ECR source where the cyclotron resonance of electrons in a magnetic field is used for plasma generation has attracted interest as a possible ion

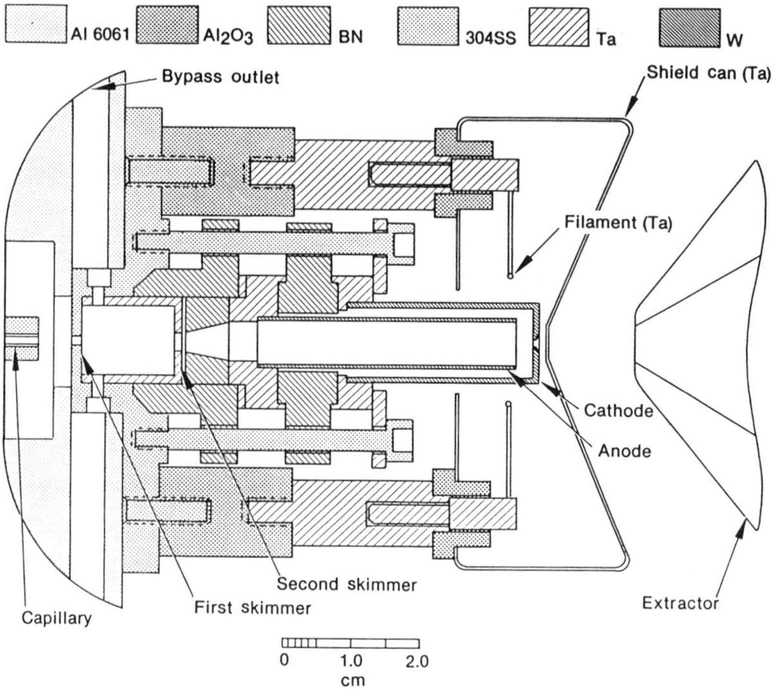


**Figure 40.** Hollow cathode ion source of Anderl *et al.* (An 87) connected to a He-jet transport system via two skimmers for He removal.

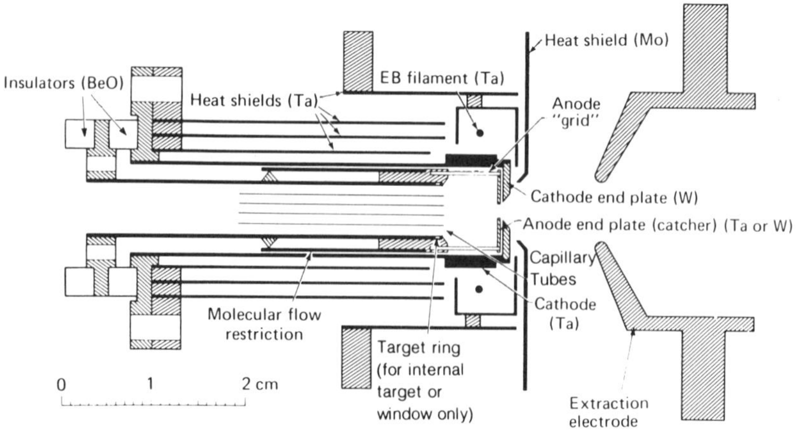


**Figure 41.** OASIS ion source with indirectly heated cathode and molecular back-flow restriction for the ionization of heavy-ion recoil products.

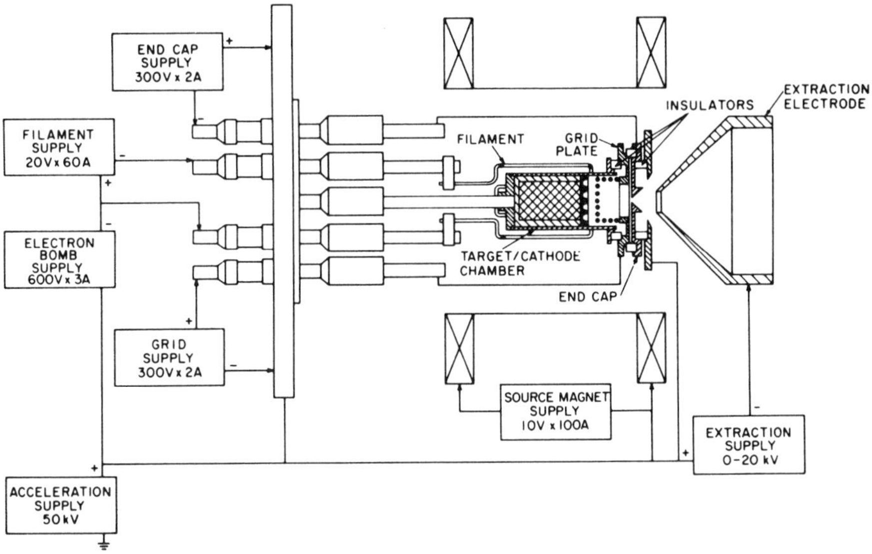


**Figure 42.** TRISTAN ion source with integrated $^{235}U$ target and indirectly heated cathode for the ionization of fission products.

source for on-line mass separators. The advantages of this application are that it has a high gas throughput since it can work at high pressure, has no hot parts that can wear out, and gives higher yields of singly charged ions because the electrons and ions are much better confined in its magnetic well than they are in the conventional plasma-discharge ion sources. Especially for light gaseous elements or compounds, the

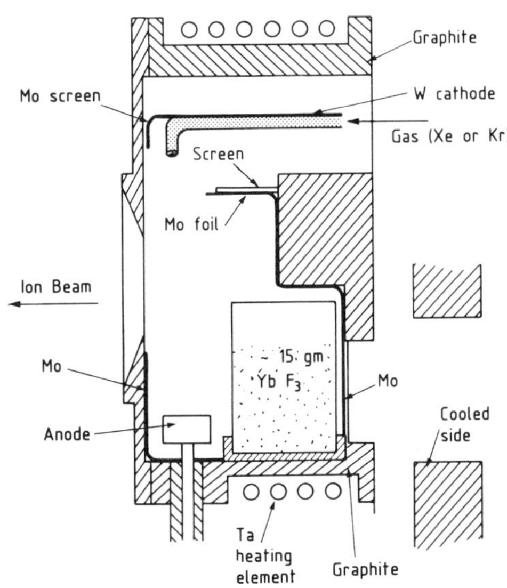


**Figure 43.** ISOCELE ion source wth integrated target and slit extraction geometry.

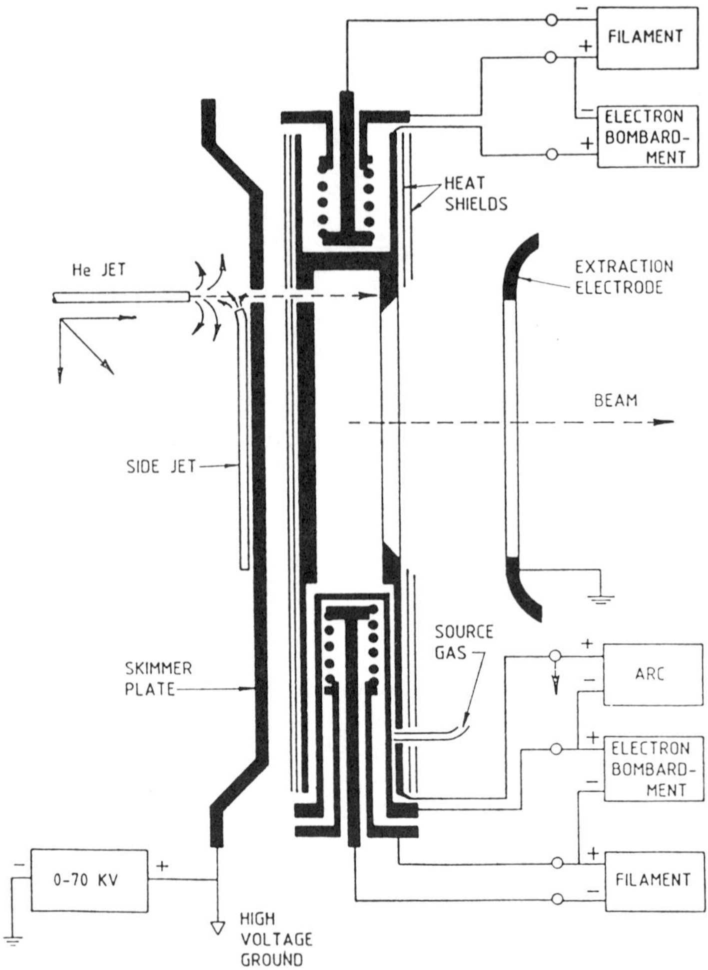


**Figure 44.** Chalk River ion source with indirectly heated cathode and slit extraction geometry coupled to a He-jet transport system.

measured yields are very high as shown by Bechtold and Sheikh (Be 85) by means of the source configuration shown in Figure 45. The radiofrequency sources seem to have some unused potential for ionization also for less volatile elements, because a high-temperature version could be developed. A number of groups (Wa 87, Ho 86, Bu 87) are at present pursuing developments along these lines.

## 4.6. Laser Resonance Photoionization Sources

A promising new scheme of ionization is the stepwise excitation by laser radiation leading to photoionization in the last transition. The high selectivity of this process

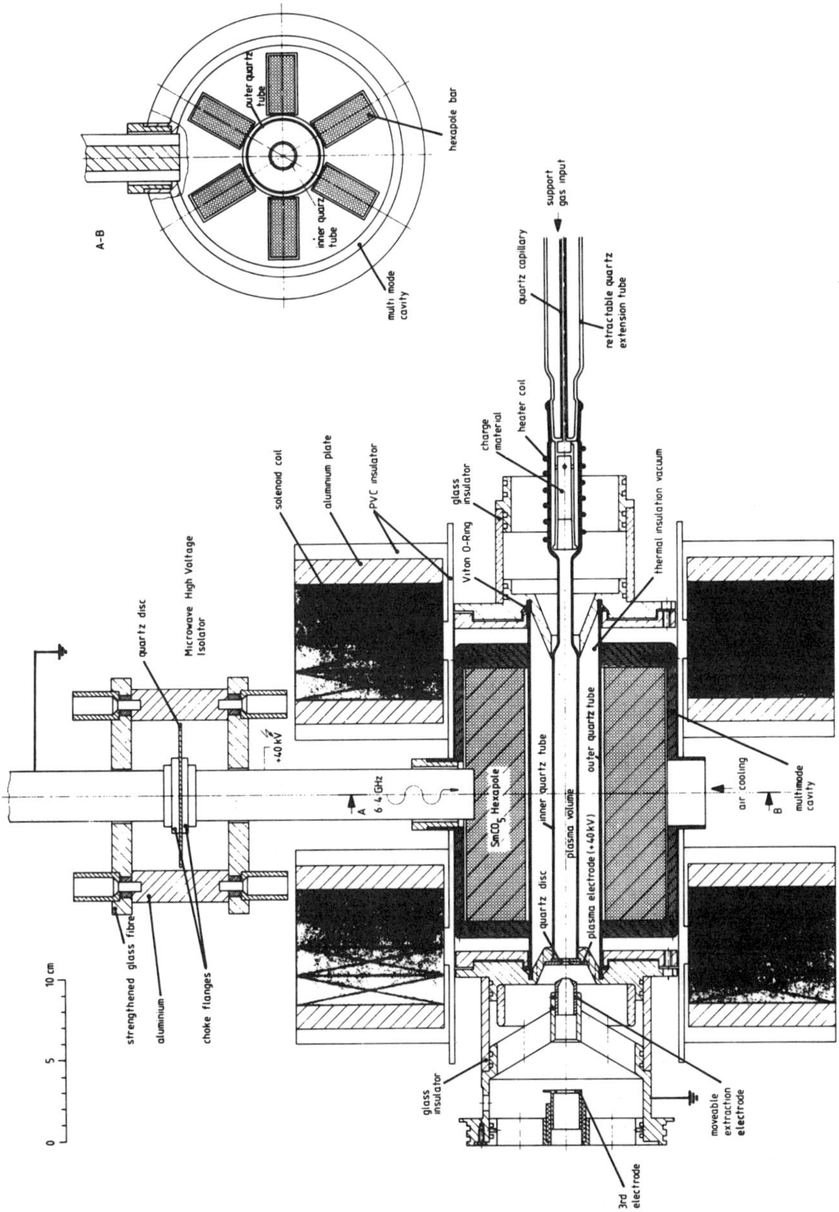


**Figure 45.** Electron cyclotron resonance ion source optimized for high yields of singly charged ions.

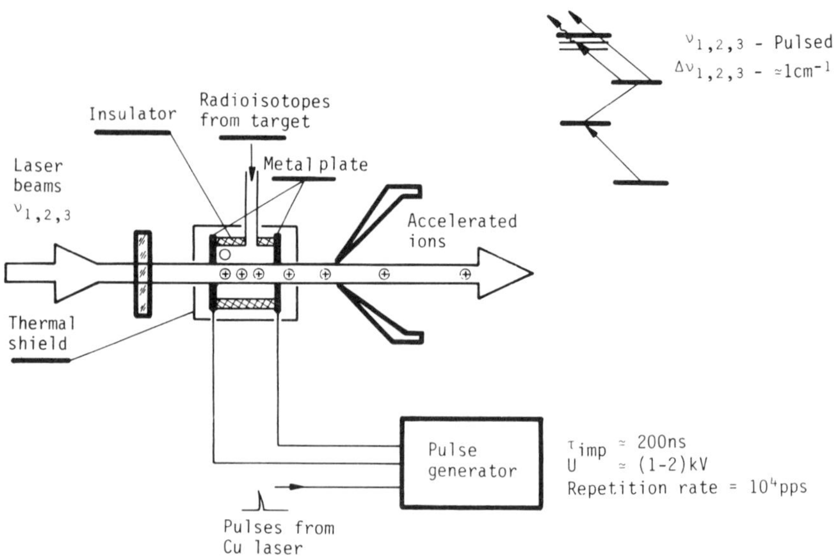


**Figure 46.** Layout of the proposed laser ion source.

enables an efficient discrimination against isobars and, if desired, against isotopes. A typical example of such an application is in the rare-earth region where the production rates at ISOLDE are very high as shown in Figure 16, but due to the chemical similarities no isobaric separation has been achieved. The atoms in the molecular flow of nuclear reaction products which leaves the high-temperature target seem to be ideally suited to stepwise ionization by means of radiation from several lasers as proposed by Letokhov (Le 85) and Kluge *et al.* (Kl 85). In fact the scheme shown in Figure 46 has been shown by Andreev *et al.* (An 86) to be able to ionize Sr with 17% efficiency. There will certainly be future developments in this domain (Cr 87).

## 5. On-Line Mass-Separator Systems

### 5.1. General Properties of Separators

The ion beams of radioactive nuclei emitted by the target and ion-source systems described above can be mass separated by means of one of the electromagnetic mass separators listed in Table 1, where the different machines are classified according to the beam current which may be transported. From the standpoint of ion optics, these machines are identical to, or further developments of, the widely used off-line mass spectrometers or isotope separators. However, for on-line machines there are additional considerations, and it is useful to note some of the common terms employed in this context.

### 5.1.1. *Maximum Current which can be Transported*

The classification has been made here and in Table 1 into three types of isotope separator. *Mass spectrometers*, where $I < 1\ \mu$A, have mostly been used on-line for nuclear-reaction studies and more rarely for radioactive sample production and studies. *Low-current separators*, with $I < 100\ \mu$A, were originally developed in the Scandinavian laboratories (An 64, Ko 58) with off-line radioactive sample preparation in mind. They are now widely used as general-purpose low-energy heavy-ion accelerators. *Medium-current separators*, with $I < 10$ mA, were developed mainly in France and Britain (Be 61, Fr 63) with the aim of producing stable isotopes in weighable amounts, and have since formed the basis for the implantation machines used in the semiconductor industry. They should be distinguished from the very high current Calutron-type machines which are not discussed here.

The need for higher intensity is not so obvious when one wants to separate carrier-free radioactive isotopes and, as seen from Table 1, the low-current mass separator has been the most frequently used type for on-line use. However, the medium current separator has later shown its advantage in on-line use, especially when it is necessary to handle high vapor pressures in the ion source with the resulting high stable beam current accompanying the desired radioactive beam. Classified under this heading is the last generation of on-line mass separators which have more complex beam optical systems and can handle higher beam currents.

### 5.1.2. *Mass Range Simultaneously Available*

Most on-line mass separators have been built for a large range of masses to be simultaneously accessible in the focal plane in order to be able to direct several beams to different experiments at the same time, thus making for an extremely efficient use of the installation. Defined as $(M_{\max} - M_{\min})/M_c$, where $M_{\min}$ is the lowest mass focused in the focal plane, $M_{\max}$ is the largest one, and $M_c$ is the central mass, this mass range has reached up to ±20% in some separators. Beams may be selected by electrostatic deflectors situated in or near the focal plane of the analyzing magnet. The switchyard at ISOLDE described by Andersen (An 73) allowed up to four beams to be selected simultaneously from a ±15% mass range, each beam being sent out into the experimental hall. The simultaneous use of beams at ISOLDE has become less frequent, however, as the complexity of experiments has increased: the average number of simultaneous experiments at ISOLDE was 2.2 in 1980, and dropped to 1.1 in 1985.

### 5.1.3. *Resolving Power of the Instrument*

This quantity is defined as $R = \Delta M/M$, where $\Delta M$ is the full width at half maximum of a beam of ions with mass $M$ in the focal plane. The purpose of the on-line setup is to enhance the abundance of an isotope of one element from a mixture of many elements. In principle, this selectivity can be performed by means of the beam optical system if it is given sufficient resolution to separate isobaric masses. The resolving power $R$ needed for this is seen in Figure 47 taken from work of Sharma *et al.* (Sh 87). Obviously, isobaric mass separation near $\beta$ stability is difficult since the $Q_\beta$ values are below 1 MeV which, for masses around 100, would require a resolving power of the order of $10^5$. But for nuclei far from stability with $Q_\beta$ values of 3–10 MeV, resolution

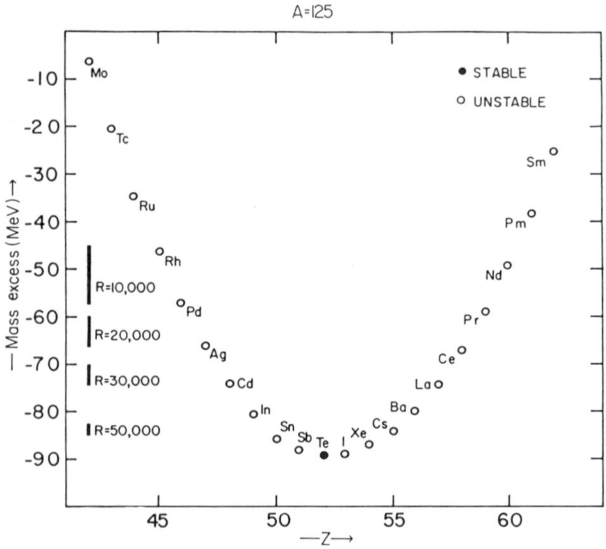


**Figure 47.** Mass excess of isobars with $A = 125$. The bars to the left indicate by what mass excess two neighboring isobars must differ for them to be resolved at the labeled resolution.

of 1000 to 30,000 is sufficient and is technically within reach. A high resolution may furthermore reduce the contamination from molecular ions and multiply charged ions which can be particularly disturbing in on-line mass separators. The resolving power of a mass separator can be calculated and reflects the beam optical properties like source emittance, image aberrations, and the stability of the power supplies. The typical value for many on-line separators is around 1000.

### *5.1.4. Dispersion in the Focal Plane*

The dispersion is given by $D = Rd$, where $d$ is the separation in mm between focused ion beams of mass $M$ and $M + \Delta M$ in the focal plane of the separator, measured in the direction perpendicular to the beam direction.

### *5.1.5. Enhancement Factor*

The enhancement factor is a measure of how well the separator succeeds in delivering the desired nuclide to the focal plane, uncontaminated by unwanted nuclides. It has two elements; the efficiency of delivering the required nuclide, and the efficiency of rejecting the unwanted species. If $C_{\text{iw}}$ and $C_{\text{fw}}$ are the initial and final concentrations of the wanted nuclide, and $C_{\text{iu}}$ and $C_{\text{fu}}$ are the initial and final concentrations of the unwanted contaminant, then

$$E = \frac{C_{\text{fw}}}{C_{iw}} \cdot \frac{C_{\text{iu}}}{C_{\text{fu}}}$$

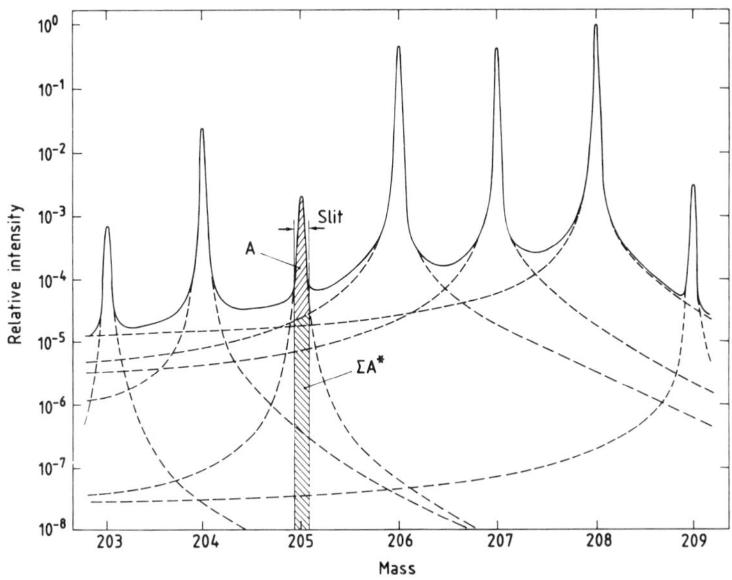


**Figure 48.** Mass scan of Pb beams showing the contamination of mass 205 by the tails of the other beams.

The real figure of merit of a separator is its enhancement factor $E$, where in particular all nuclides are considered, not just the isotopes of one element (which is often the case in off-line mass separation where chemical selectivity is assured by the purity of the feed material). The dependence of $E$ on the separator properties has been extensively studied by Camplan and co-workers (Ca 81b). Some typical beam profiles of a mass-separator beam are illustrated in Figure 48 (Fa 65), which shows that $E$ is mainly determined by the tails of the beams. With reference to Figure 48 the enhancement factor can be expressed as

$$E = \frac{A}{\sum A^*} \cdot \frac{\sum C_{\mathrm{iu}}}{C_{\mathrm{iw}}}$$

where $\sum A^*$ is the sum of the unwanted nuclides in the focal plane and $A$ is $C_{\mathrm{fw}}$. It is clear that much effort should be spent on reducing the image width, and hence the slit width, in order to reduce the intensity of the tails. Typically $E$ varies from $10^2$–$10^5$ depending on mass region and mass difference. The origin of the tails, as discussed by Camplan (Ca 81b), is collisions with rest gas at various places in the separator. These give rise to ions or neutrals with the wrong mass and wrong momentum that are scattered through the slit, or ions of wrong mass and correct momentum which pass the slit because they have not been fully accelerated. Bernas *et al.* (Be 60) showed that, by means of a second magnetic analyzing field, the contribution from the scattered particles could be reduced by a factor of 10 to 100 but since the not-fully-accelerated ones dominated, the expected enhancement factor equal to the product of the two stages was not obtained. In a recent paper Wollnik and Becker (Wo 85) rediscussed the origin of the contaminations and suggested suppressing the contribution of the

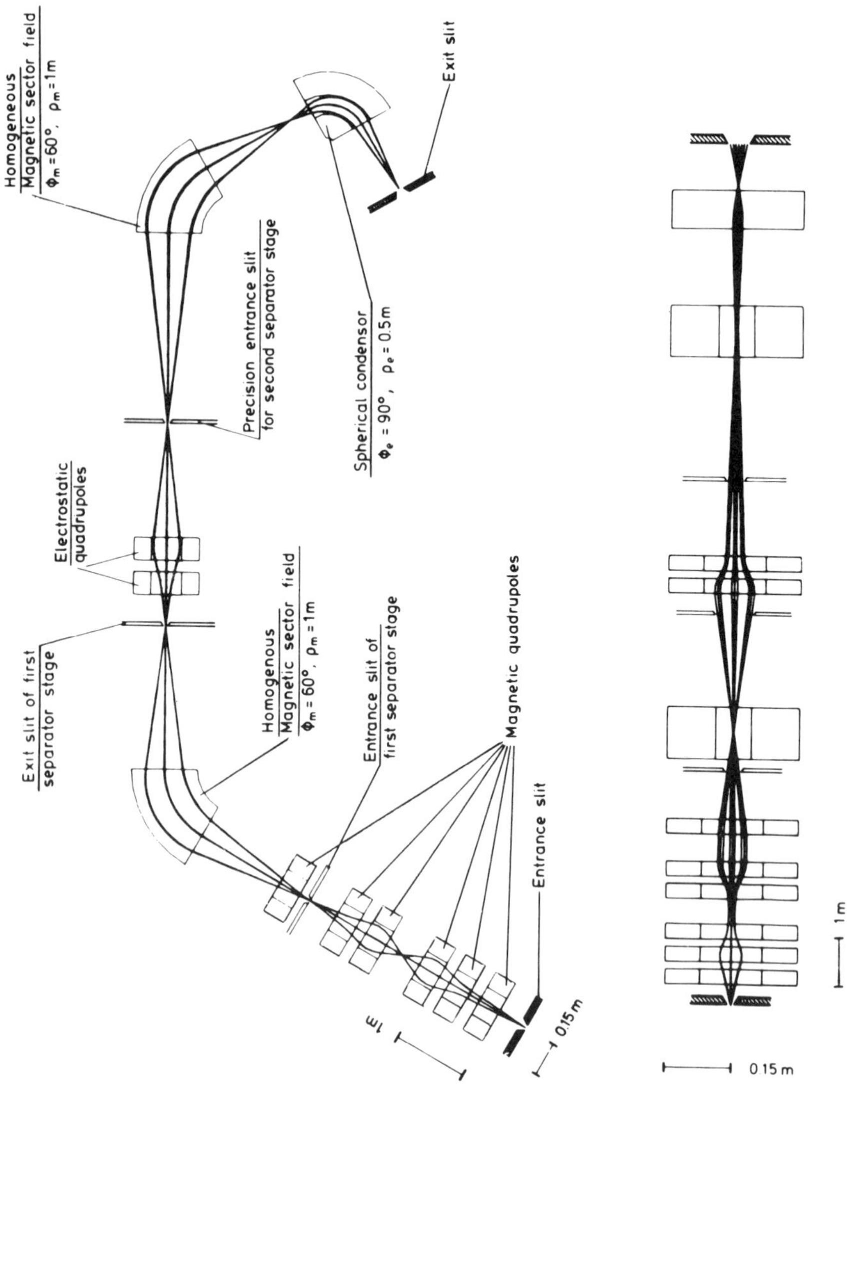


**Figure 49.** Beam optical layout of a two-stage high-resolution on-line mass separator.

not-fully-accelerated ions by placing the vacuum chamber of the analyzing magnet at a potential. In the same paper they present an on-line two-stage mass-separator concept optimized for high resolution and enhancement by means of magnetic quadrupole lenses and an electrostatic sector field as shown in Figure 49. A more general discussion of the optimal beam-optical layout of a high-resolution on-line mass separator with high transmission was made by Kornahl *et al.* (Ko 75). Pole-face correction coils as shown in Figure 50 have been used by Camplan and Meunier (Ca 81a) for adjustment of the beam-waist position after the analyzing magnet and to improve the resolution by reducing the aberrations. Geisse *et al.* (Ge 87) use systematically pole-face coils and electrostatic multipole elements in their design of the new high-resolution ISOLDE separator.

### 5.1.6. Delay Time

As already discussed in Section 3.1, there is a delay between the production of a nuclide and its arrival in the focal plane which is controlled by the diffusion through the target or catcher material, the release from the surfaces, and the intrinsic delay in the gaseous phase of the target and ion-source system (cf. Figure 23). The ionic-transport time from the source to the measuring equipment is short and can be neglected here. The delay losses of short-lived nuclei, expressed as the fraction $(Y_{obs}/Y)$ of nuclei which survived the time lag between production and measurement, is a particularly important parameter when the separator is used for reaction cross-section measurements. In addition, it is a most useful figure of merit for a given on-line mass-separated beam.

The general way to treat this delay is to define a delay-time distribution $P(t)$, which is the probability of a delay between 0 and $t$ after the production of a given nuclide. It is shown (Ca 78) that this delay-time function is most often determined by one of the delay-causing mechanisms mentioned above. It can therefore be expressed as a sum of exponential terms similar to the $F(t)$ functions defined in Section 3.1. The

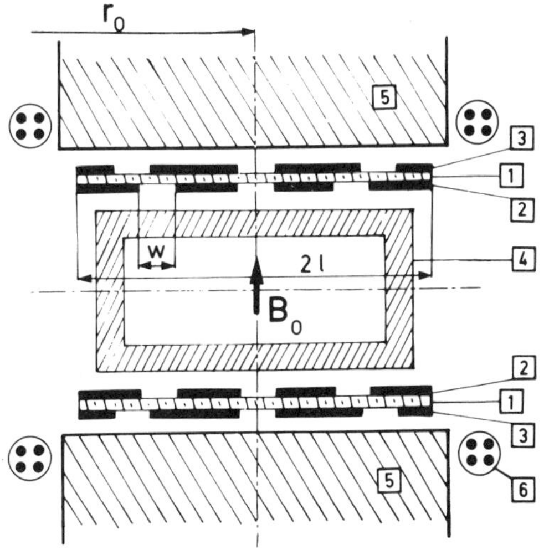


**Figure 50.** Cross-section of a set of $\alpha$ and $\beta$ coils machined into circuit boards. For clarity, conductors are shown almost as thick as the circuit board and each coil is assumed to have only two pairs of conductors: (1) insulating part of the circuit board, (2) $\alpha$ coil, (3) $\beta$ coil, (4) vacuum chamber, (5) pole piece, (6) returning conductors.

relation between the observed and decay-loss-corrected production yields may now be written as

$$Y_{obs}/Y = \int_0^\infty \lambda\, e^{-\lambda t} P(t)\, dt$$

where $\lambda$ is the decay constant of the nuclide in question. So far the fastest systems developed are based on high temperature, solid targets or catchers. For the thick powder target discussed in Section 3.1.2 the yield becomes

$$Y_{obs}/Y = (3/r)(D/\lambda)^{1/2}$$

These values are shown in Figure 2 as a function of nuclide half-life. The fact that the yield decreases only as the square root of $1/\lambda$ implies that such target systems are extremely favorable for production of short-lived nuclei. Similar high release rates are obtained in mass-separator systems on-line to reactor neutrons or heavy ions as discussed by Rudstam (Ru 85b) and Klepper (Kl 81), respectively.

Instead of referring to the delay-time distribution $P(t)$ or the release yield $Y_{obs}/Y$ as the figure of merit, many groups often only give a delay half-time, i.e., the time required for the beam intensity to drop by a factor of two when the bombarding beam is switched off.

### *5.1.7. Overall Efficiency*

This is, of course, the product of all the partial efficiencies involved in transporting the required nuclide through the separator, such as target release, transfer to the source, ionization in the source, and transport in the beam optics to the focal plane.

## 5.2. Mass Spectrometers

The mass spectrometer is an analytical instrument where the individual mass-separated ions are counted as the magnetic or electric field is varied in order to determine a mass spectrum. The ion sources for on-line use have been limited to surface ionization types, since the maximum current the beam optics can handle is of the order of 1 $\mu$A. The accelerating voltage is 5–10 kV and the ions are extracted from a slit opening. With the following entrance slit wide open the resolving power is ~400 and the transmission ~20%. By closing down the slit the apparatus can reach resolving powers of $\sim 10^5$, enough for on-line mass measurements.

The ORSAY group (Sa 81) has used this type of instrument at a variety of accelerators for nuclear-reaction studies (Kl 69) and later for on-line mass measurements and nuclear-structure studies of the light alkali metals (cf. Section 2.5.1). The layout of the setup at the 20 GeV CERN-PS proton beam is seen in Figure 51. It should be noted that this group exploited a 100 ms delay-time component in the release from their target to use the pulse structure of the PS beam ($10^{13}$ protons in 2 $\mu$s every few seconds) to obtain a bunched release of the nuclei of interest.

Despite the element limitations these compact and mobile instruments have at an early stage made decisive contributions to on-line mass-separator physics. Their rôle

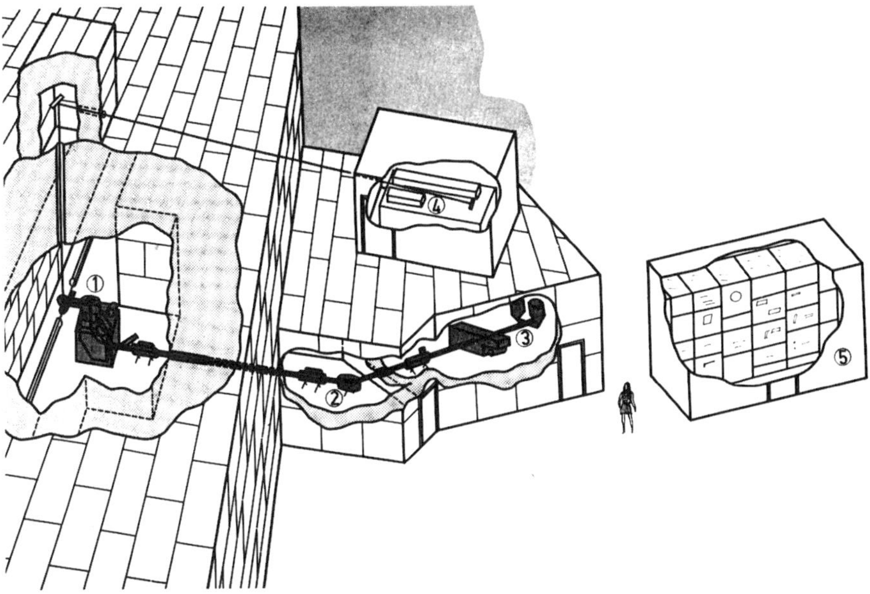


**Figure 51.** General layout of the Orsay on-line mass spectrometer at the CERN 20 GeV PS proton beam: (1) mass spectrometer, (2) ion-beam line, (3) cadmium-cladded counting room, (4) laser, (5) experimental control room.

has today been taken over by the low-current mass separator. But the mass spectrometer continues to be used now on the experimental side of on-line mass separators (To 81).

## 5.3. Low-Current Isotope Separators

This type of separator is the most frequently used on-line, as seen from Table 1. It allows the use of ion sources which deliver ion-beam currents even above 100 $\mu$A. The ions are accelerated to ~60 kV in two steps and sent into the typical 55°, 1.5 m bending radius sector-field magnet as a parallel beam. The fringe field of the magnet is used for vertical focusing and gives spot images of diameter 2 mm in the focal plane, where the mass range is ±20% of the central mass and the dispersion is (1500/M) mm. The resolving power is 500–1500, depending on the diameter of the source outlet which may vary from 3 mm downward. The enhancement factor obtained is at best $10^4$ as measured in an external beam line after a 3 mm slit and electrostatic deflection. Many configurations are now in use at accelerators and reactors.

The ISOLDE facility on-line to the CERN 600 MeV proton synchrocyclotron (SC) has in many respects been a model for the others. It enjoys the collaboration of more than 200 physicists from 50 laboratories. With a physics program (Kl 86) which ranges from the study of nuclei far from stability over atomic, solid-state, surface, and elementary particle physics, to medical applications, it is by far the most ambitious project in the field. Figure 52 gives the general layout of the ISOLDE low-current separator and its external beam line system, which feeds roughly 10 experimental

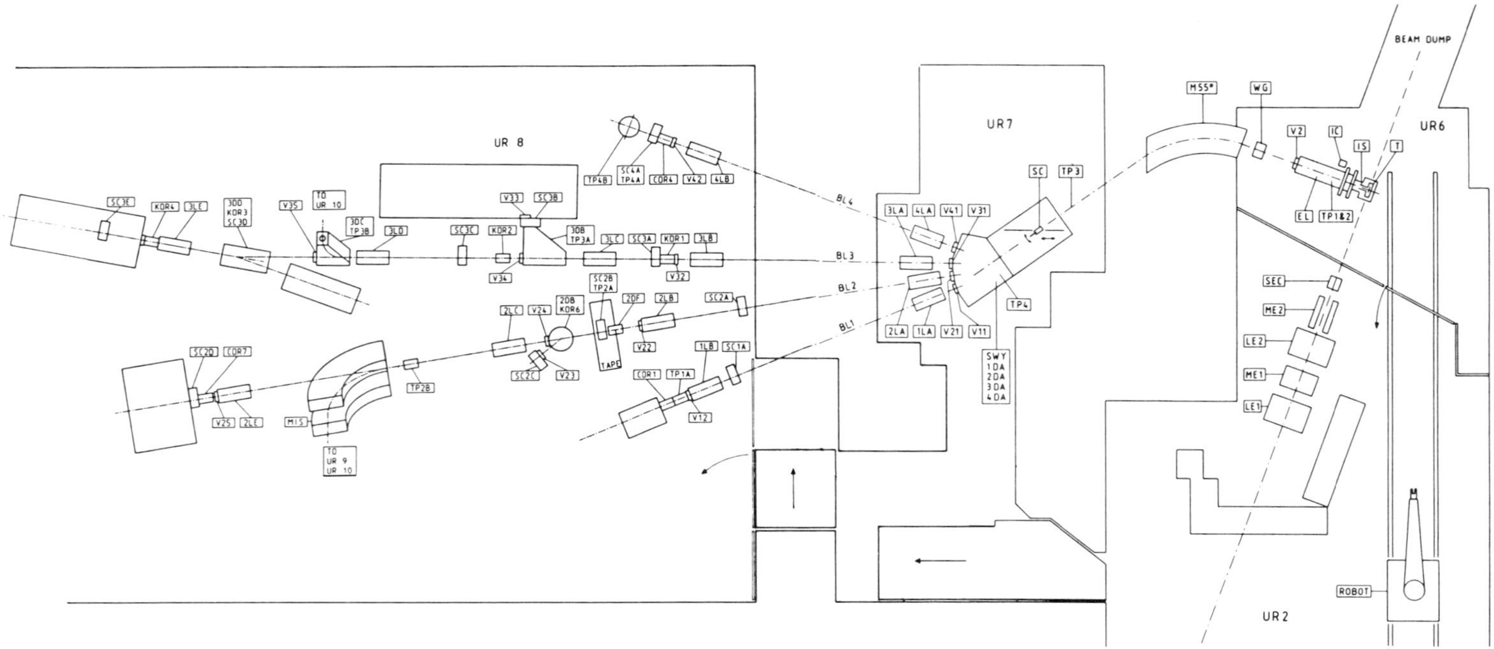
BEAM DUMP
UR6
UR7
UR 8
UR2
M55°
WG
V2
IC
IS
T
EL
TP1&2
SC
TP3
TP4
SEC
ME2
LE2
ME1
LE1
ROBOT
SWY
1DA
2DA
3DA
4DA
3LA
4LA
V41
V31
2LA
1LA
V21
V11
BL4
BL3
BL2
BL1
TP4B
SC4A
TP4A
COR4
V42
4LB
SC3E
KOR4
3LE
3DD
KOR3
SC3D
V35
TO
UR 10
3DC
TP3B
3LD
SC3C
KOR2
V33
SC3B
3DB
TP3A
3LC
SC3A
KOR1
3LB
V34
V32
SC2B
TP2A
2DF
2LB
SC2A
2LC
V24
2DB
KOR6
V22
TAPE
SC2C
V23
SC2D
COR7
V25
2LE
MIS
TO
UR 9
UR 10
TP2B
COR1
TP1A
1LB
SC1A
V12

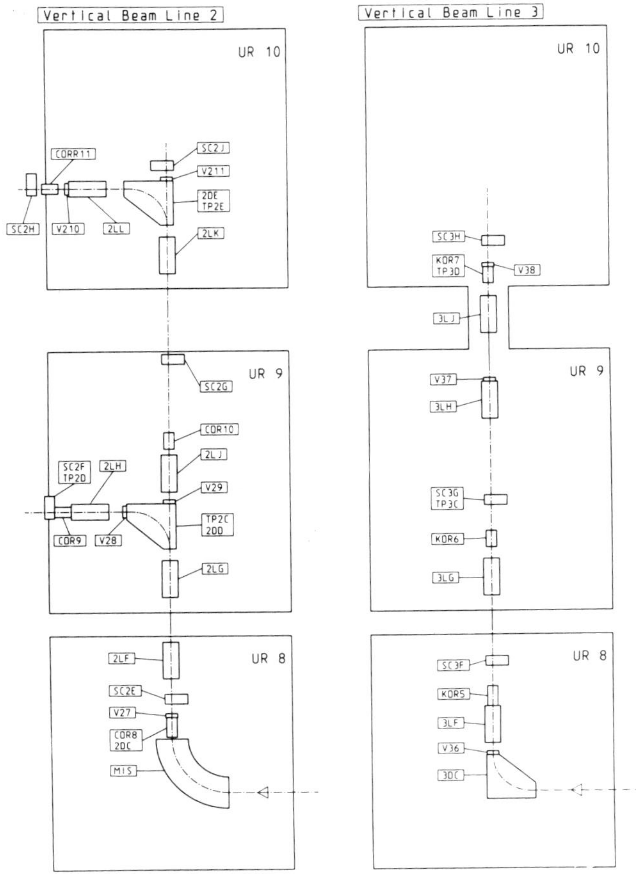


**Figure 52.** Layout of the ISOLDE low-current mass separator and its beam-distribution system: COR, electrostatic dipole correction element; D, electrostatic dipole deflector; EL, electrostatic einzel-lens triplet; IS, ion source; IC, ionization chamber for proton-beam monitoring; L, electrostatic quadrupole triplet lens; M, analyzing magnet; ME, magnetic-dipole deflector; LE, magnetic quadrupole lens; SC, beam observation box with $X$-$Y$ scanner and Faraday cup; SEC, proton beam observation box; T, target; TP, turbomolecular pump; V, valve.

stations up to 25 m away from the target. One of the most remarkable features of ISOLDE is the broad range of elements and nuclei available for experiments. The ISOLDE beam catalogue (Kl 86) lists the intensities of ~600 isotopic beams of ~66 elements. The ion-source efficiencies range from 1 to 95%, the delay half-times from 0.5 to 30 s and the radioactive beam intensities reach 60 nA (i.e., $4 \times 10^{11}$ atoms/s). The resolving power is ~1500 when a plasma ion source of 0.5 mm opening is used. As seen from Figure 15, one can study a given nuclear property over a very large mass range. The enhancement factor is further improved in one beam line where a second analyzing magnet is used to deflect the beam vertically to the low-background nuclear-spectroscopy experimental area two floors above.

The price paid for using a high-energy proton beam at ISOLDE is counted in radiation-dose effects. The ionization of the air in the target region gives rise to a pulsed current load of 6 mA per pulse on the accelerating high-voltage power supply. A series-tube regulator developed by Vosicki (Vo 87b) allows the voltage to be stabilized to ±1 V. The laser-spectroscopy groups further reduce the ripple on the beam to ±0.1 V by applying a correction voltage derived from the 60 kV supply to their charge exchange cell (Ge 81).

The amount of beam time given to the users by this facility is about 2000 h/yr, which is close to saturation. With the present manpower available and taking into account the necessary maintenance work, radioactive cooling prior to any interventions, target and ion-source exchanges, etc., it would hardly be possible to give more time to physics.

A typical example of an ISOLDE-type low-current separator on-line to a nuclear reactor is the OSIRIS facility in Studsvik (Ru 76) of which the layout is shown in Figure 53. This machine features an ion source located in the high-flux ($4 \times 10^{11}$ n/cm$^2 \cdot$ s) region at the inside wall of the reactor pool. The target material is 3 g of $^{235}$U as a solid solution in graphite cloth, and it is situated on the inner surface of the anode cylinder of a plasma-discharge ion source, or more recently in a thermal ion source (Ja 87), similar to the one developed at TRISTAN, shown in Figure 36. The element range at OSIRIS covers practically all the fission-product distribution shown in Figure 4. Since the ion source has no chemical selectivity, this group has developed a thermochromatographic element separation (Ru 81) after the focal plane where individual ion-beam intensities reach $10^8$ atoms/s. A method for element separation by means of molecular ions (Ho 81) has also been developed. Recently, this separator was used for the nuclear-reaction studies (Ru 85b) discussed in Section 2.5.11 during which delay half-times of 0.4 to 4 s and ion-source efficiencies of 0.1 to 1.0% were measured. The relatively low yields reflect the conservative running mode of the ion source which achieved a lifetime equal to the typical experimental period of five months.

The development of experimental techniques for heavy-ion-based mass separators has been particularly active in the groups at GSI (Br 81), Louvain-La-Neuve (Ve 81), and Jyväskylä (Ar 86b) all of which use ISOLDE-type mass separators. As an example of such a separator, the layout of LISOL (Hu 87) on-line to the Louvain-La-Neuve isochronous cyclotron is shown in Figure 54. Here the bombarding beams of light ions vary from 90 MeV $^3$He to 240 MeV $^{20}$Ne. The beam optical performance of the separator has been improved by including a quadrupole doublet lens in front of the magnet. It was shown by Nielsen (Ni 81) that the better matching between the source and the magnet obtained in this way improved the resolving power by a factor of four. A large

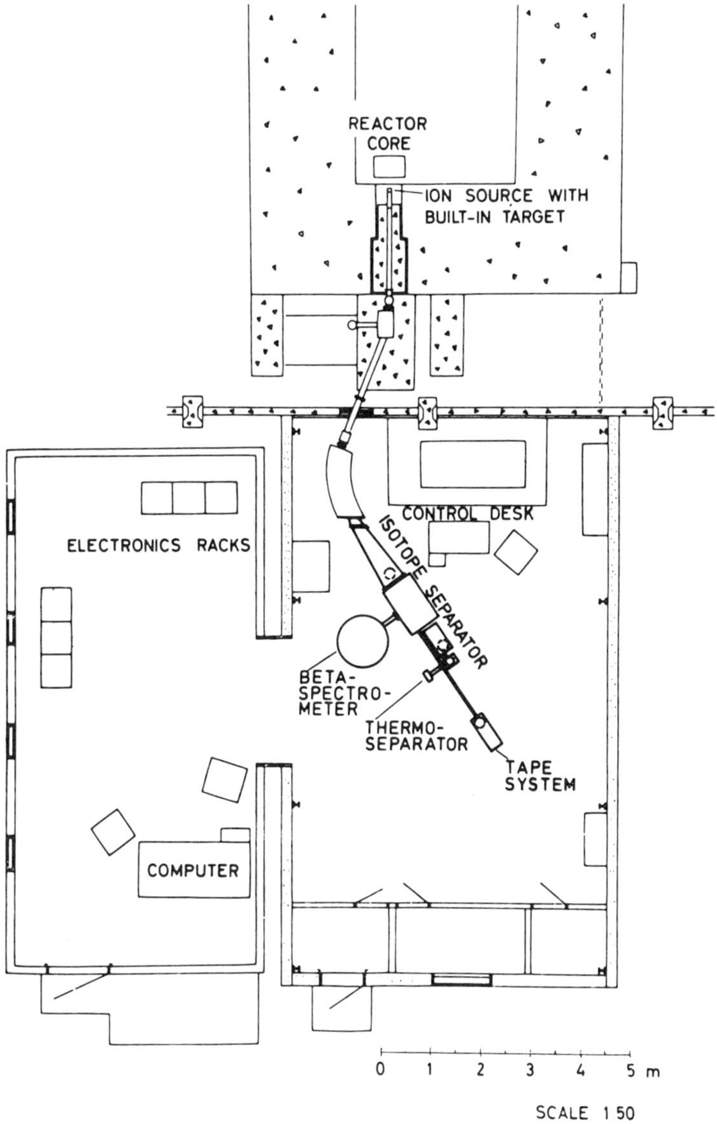


**Figure 53.** Layout of the OSIRIS low-current mass separator on-line at the Studsvik reactor.

number of target and ion-source techniques has been adapted to this machine, ranging from FEBIAD ion sources to the He-jet ion guide (cf. Figure 19), and there are plans for an ECR ion source. Beam intensities reach $\sim 10^6$ atoms/s and beams of rather inaccessible elements like fluorine have been developed. In the physics program of this facility one notes a pioneering effort in on-line measurement of nuclear moments of nuclei oriented by implantation into a cold target.

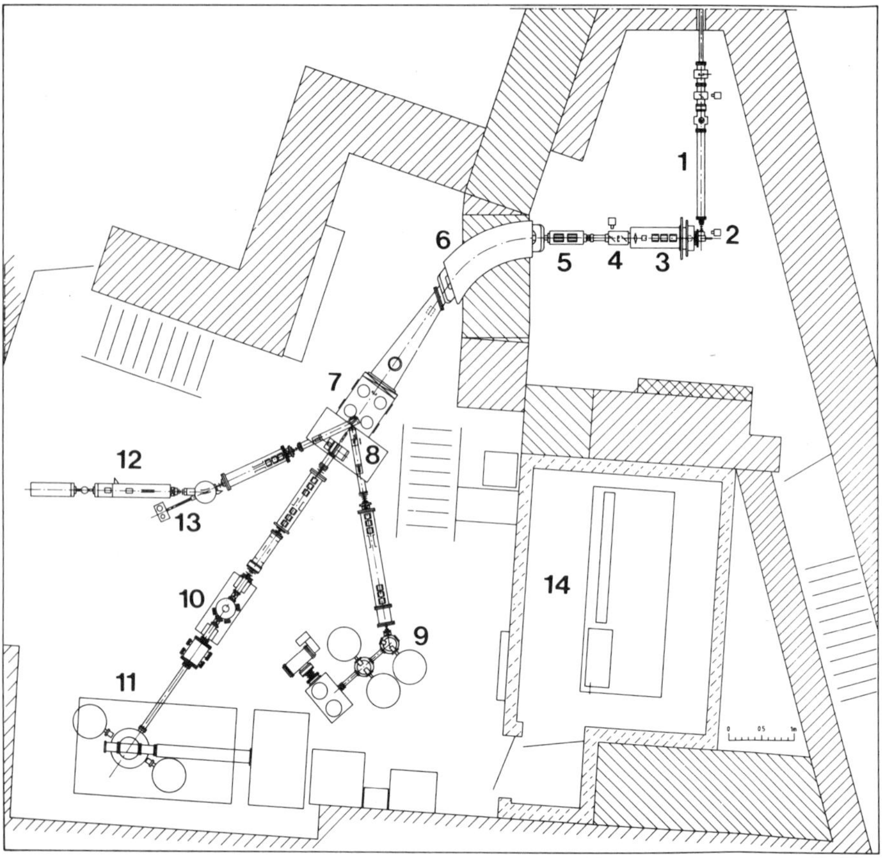


**Figure 54.** Layout of the LISOL low-current mass separator on-line at the Leuven cyclotron: (1) cyclotron beam, (2) ion source, (3) einzel lens, (4) beam diagnostics, (5) quadropole doublet, (6) magnet, (7) collector chamber, (8) beam switch yard, (9) tape system, (10) scattering chamber, (11) $^3$He–$^4$He dilution refrigerator, (12) laser-interaction chamber, (13) minitape system, (14) laser apparatus.

## 5.4. Medium-Current Isotope Separators

The wedge-shaped beam geometry obtained from the lateral beam extraction sources used in conjunction with medium-current separators offers a number of advantages apart from reduction of the transport times discussed in Section 4.4.2. The line beam shape in one plane allows the electrostatic beam elements in front of the magnet to be omitted, so that space-charge compensation is preserved and aberrations are minimized. The strongly divergent beam obtained after the extraction electrode, kept only 200 V below ground potential, maximizes the resolving power by allowing the beam to occupy the largest possible pole area in the magnet. Furthermore, the slit geometry gives a lower density of neutrals in the extraction gap, for a given beam

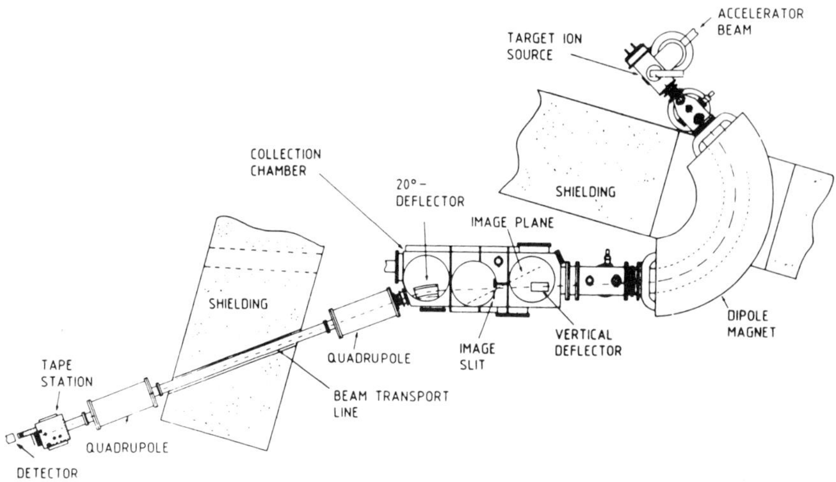


**Figure 55.** Layout of the Chalk River medium-current mass separator on-line at the tandem/superconducting cyclotron.

intensity, thus reducing the tails of the mass-separated beams. The resolving power reaches typically 6000–8000.

Designed strictly according to the above-mentioned principles, the Chalk River separator (Sc 81b) has reached an excellent performance. The layout of this machine, on-line at the tandem/superconducting cyclotron heavy-ion facility, is seen in Figure 55. Note the short distance between the source and the magnet where there are no lenses. By means of pole face windings the resolving power can be pushed to more than 8000. From the resulting beam profile shown in Figure 56 an enhancement factor of $>10^6$ can be deduced. The separator can use axial-extraction FEBIAD-type sources, and Bernas–Nier-type sources the efficiencies of which vary from 0.01 to 50%, depending on the element. When they (Sc 87) use the He-jet coupled ion source shown in Figures 24 and 45, overall efficiencies of the order of 5% have been obtained for the elements Si, As, Sc, Ag, and Sn which are remarkably good.

When the radioactive-beam intensity and radiation level are increased by bombardment with light particles in the 100 MeV and $\mu$A range, more elaborate shielding is needed between the target in the primary beam area, and the experimental hall. In order to obtain the necessary space, an image of the ion-source outlet is produced by means of lenses, at some distance from the target in front of the analyzing magnet. The ISOCELE-2 separator shown in Figure 57 (Pa 81) on-line to the 200 MeV, 2.5 $\mu$A proton beam and 283 MeV $^3$He beam of the ORSAY synchrocyclotron is designed accordingly. An electrostatic quadrupole triplet and a singlet are interposed between the rectangular slit ion source and the magnet. This configuration allows beam currents of up to 2 mA to be handled when proper electrostatic guard rings are installed at the boundaries of the quadrupoles in order to preserve space change compensation. The ISOCELE group has exploited the high throughput of their ion source and separator to develop beams of a number of refractory elements (Sa 84) as discussed in Section 3.1.3 and illustrated in Figure 18.

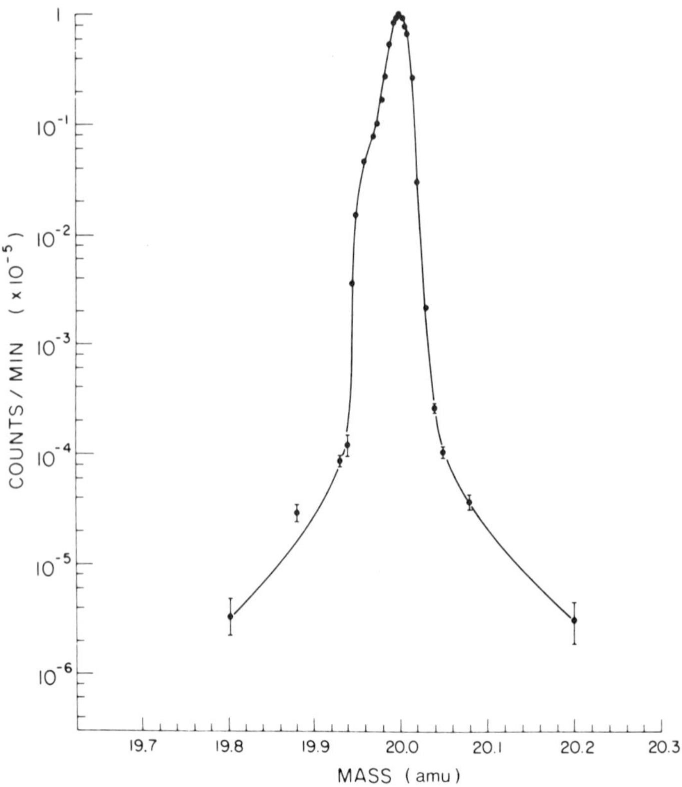


**Figure 56.** Beam profile for $^{20}Na$ measured at the Chalk River mass separator. The shoulder at the 5% level on the low-mass side indicates that the $\beta$-pole-face coil is not perfectly adjusted. The enhancement factor is larger than $10^6$.

The increased demand for beam time and space at the low-current ISOLDE separator led to the decision in 1983 to expend the CERN ISOLDE facility by the addition of a second primary proton target station, together with a second on-line isotope separator and experimental zone, thereby reserving the CERN synchrocyclotron almost exclusively for ISOLDE use in the future. By alternating the operation of the two separators, it is planned to give 5000 h/yr of radioactive beam time to the users. As seen from the perspective view of the SC-ISOLDE on-line mass-separator facility shown in Figure 58, an economical solution was chosen by incorporating the new separator in the existing buildings. The target station was put inside the cyclotron vault close to the location of the former pion production target, and the experiments will be housed in an existing experimental hall. The design parameters of this new machine are listed in Table 3 and are based on experience from both the ISOLDE low-current separator and the ISOCELE medium-current separator. The beam optical layout, with the exception of the last analyzing stage, is as shown in Figure 49, which has been calculated by the Giessen group (Ge 87). The layout of the new ISOLDE separator is shown in Figure 59. The ion beam which is accelerated in one step to 60 keV enters a

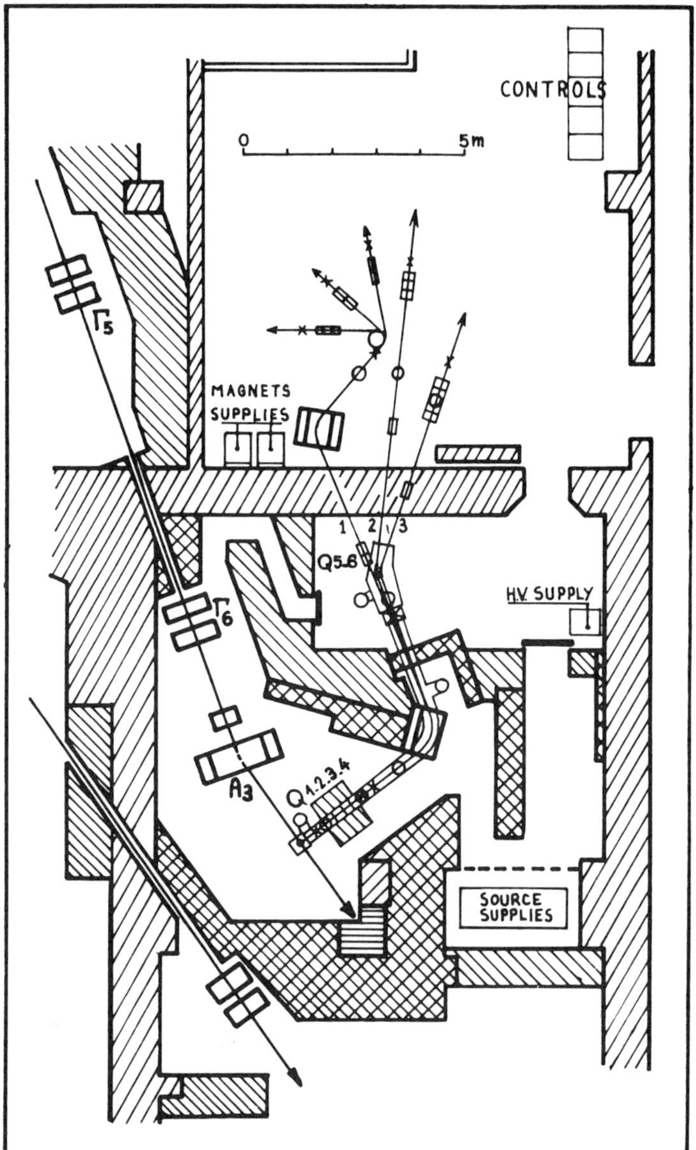


**Figure 57.** Layout of the ISOCELE medium-current mass separator on-line at the Orsay synchrocyclotron. Q1, 2, 3, 4 denote electrostatic quadrupole elements situated between the source and the magnet.

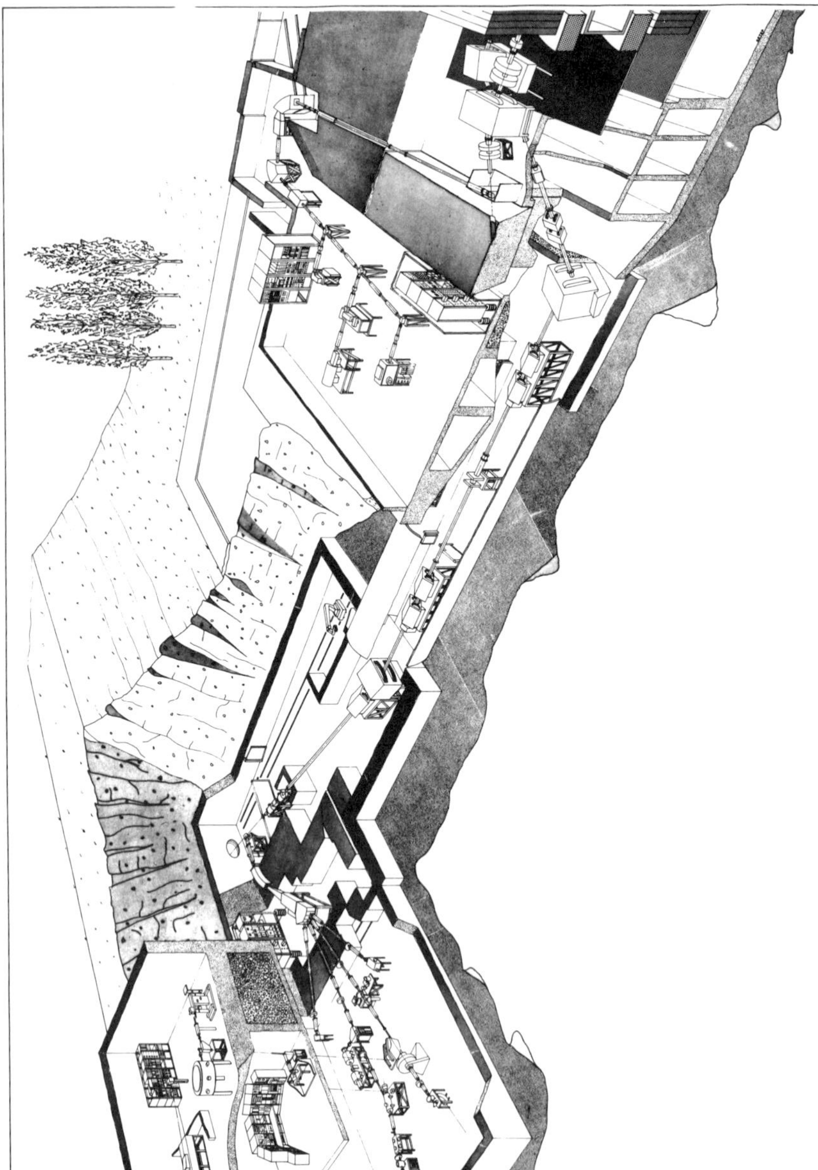

**Figure 58.** Perspective view of the SC-ISOLDE facility showing both the underground low-current separator and the new medium-current separator with their corresponding experimental areas.

**Table 3. Parameters of the ISOLDE Medium-Current Mass Separator**

| | |
|---|---|
| Acceleration voltage | 60 kV |
| Stability of high-voltage supplies | 1 in $10^5$ |
| Maximum stable ion current | 3 mA |
| Ion-source outlet geometry | circular and slit |
| Mass dispersion in 1st focal plane | 18 mm at mass 100 |
| Mass dispersion in 2nd focal plane | 35 mm at mass 100 |
| Resolving power in 1st focal plane | 5000 |
| Resolving power in 2nd focal plane | 7000 at 98% transmission; up to 30,000 with slits |
| Enhancement factor | $10^5$–$10^8$ |
| 600 MeV proton beam intensity | 4 $\mu$A |
| 910 MeV $^3$He beam intensity | 0.5 $\mu$A |
| Number of analyzing magnets | 2 |
| Magnet angle | 90° and 60° |
| Magnet radius | 1.0 m |
| Magnet polegap | 110 mm |
| Magnet geometry | C type |
| Maximum magnetic field | 0.7 T |
| Number of magnet pole-face coils | 2 on each |
| 90° magnet exit poletip | inclined 20° |
| 60° magnet exit poletip | curved, 4 m radius |
| Magnification in 1st focal plane | 1–1.7 |
| Magnification in 2nd focal plane | 1.3–2.1 |
| Vertical beam size in 1st focal plane | <4 mm |
| Mass range in 1st focal plane | ±12% of central mass |
| Beam transport | Electrostatic quadrupoles |
| Aberration correction elements | Electrostatic multipoles |
| Vacuum system | Turbomolecular pumps |
| Area for experiments | ~200 m$^2$ |

transfer section which consists of five electrostatic quadrupoles. They produce an almost source-independent image in front of the first magnet situated outside the highly radioactive target zone. The first separator stage uses a quadrupole to fill the 90° magnet so as to obtain a high resolving power. In order to add the dispersion of the two magnets, the second stage includes a doublet that creates a second focus between the two magnets. A last singlet makes a final focus from where the beam is taken over by the external beam lines that distribute it to about ten experimental stations. In addition to the pole-face correction coils to be fitted on both magnets, aberrations are further corrected by means of three electrostatic multipole elements of novel design (Ge 87). They consist of 32 rods aligned parallel to the optic axis as shown in Figure 60. Each rod has its own voltage supply cntrolled by a microprocessor, which allows the element to become anything from a dipole to a high-order multipole. Finally, the magnet vacuum chambers are electrically insulated so that they can be kept at a high voltage in order to achieve the increase in enhancement factor discussed in Section 5.1.5.

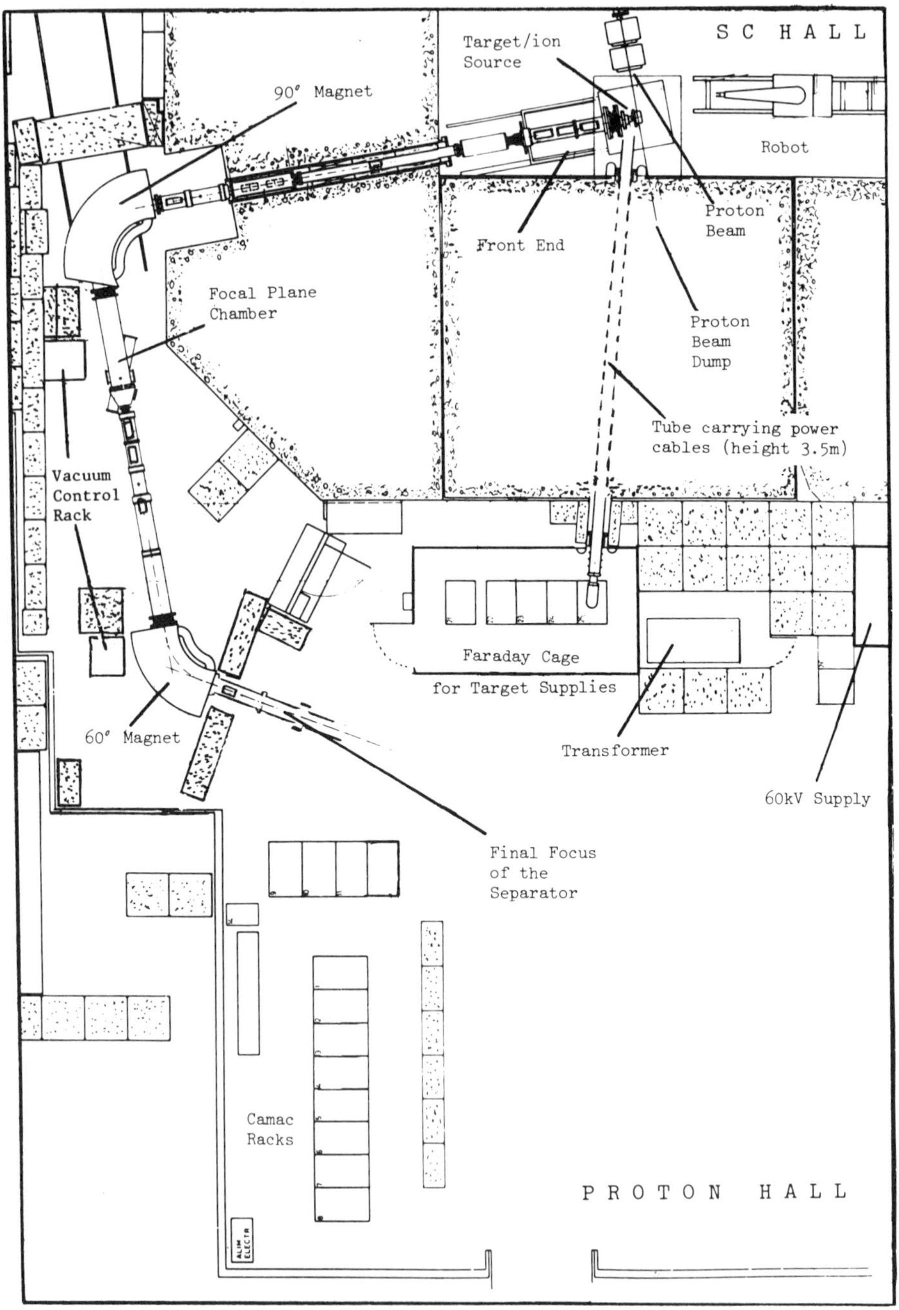


**Figure 59.** Layout of the new ISOLDE medium-current mass separator.

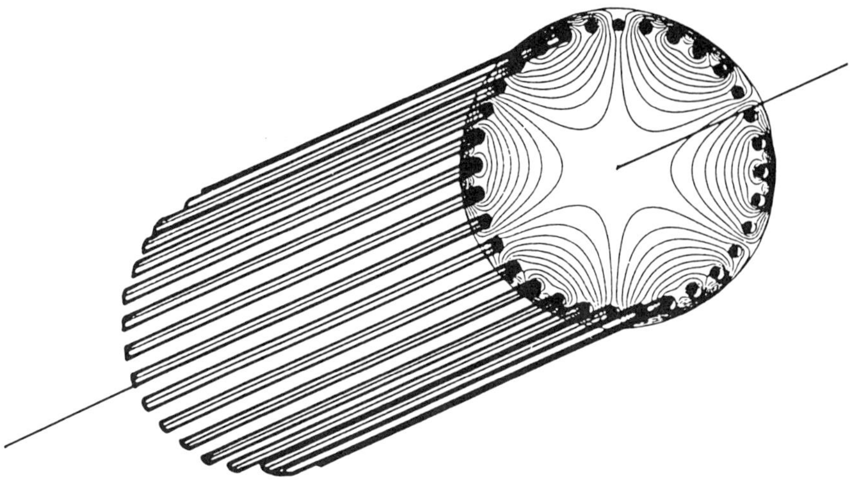

**Figure 60.** Squirrel-cage-type electrostatic lens which allows one to generate multipoles from a dipole and upward by separate voltage control of each of the 32 rods, here shown in the hexapole mode.

## 6. Concluding Remarks

The intention of this chapter has been to summarize the state of the art in the field of on-line isotope separators. Some pointers to the future have also been given. The limited development effort invested by a relatively small number of groups has led to the present variety of intense radioactive beams which make many new experiments possible.

It should be realized that a good understanding of the behavior of the different chemical elements under on-line conditions has accumulated over the years, so that it is now a relatively simple but interesting task to develop further beams. The authors hope that this overview together with the references therein has convinced the reader of the importance of new developments and has whetted his appetite to join forces with groups making developments in this field, or to become a user of the end product—the radioactive ion beams.

## References


(Ac 70) E. O. Achterberg, E. Y. de Aisenberg, F. D. Iglesias, A. E. Jech, J. A. Moragues, D. Otero, M. L. Perez, A. N. Proto-Otero, R. Requejo, J. J. Rossi, W. Scheuer, and J. F. Suarez, Proceedings of the International Conference on Electromagnetic Isotope Separators and the Techniques of their Application, Marburg, 1970 (Bundesministerium fur Technologie und Wissenschaft, rep. K 70-28) (1970), p. 108.

(Ad 85) H. L. Adair and E. H. Kobisk, *Treatise on Heavy-Ion Science* (D. Allan Bromley, ed.), Vol. 7, Plenum Press, New York (1985), p. 119.

(Al 86) B. W. Allardyce and H. L. Ravn, *Nucl. Instrum. Methods* **B26**:112 (1987).

(Am 67) I. Amarel, R. Bernas, J. Chaumont, R. Foucher, J. Jastrezbski, A. Johnson, R. Klapisch, and J. Teillac, *Ark. Fys.* **36**:77 (1967).

(An 64) G. Anderson, *Nucl. Instrum. Methods* **28**:245 (1964).

(An 65) M. L. Andersen, O. B. Nielsen, and B. Scharff, *Nucl. Instrum. Methods* **38**:303 (1965).

(An 67) M. L. Andersen and B. Scharff, *Ark. Fys.* **36**:151 (1967).

(An 73) G. Anderson, Proceedings of the 8th International EMIS Conference on Low Energy Ion Accelerators and Mass Separators, Skövde (1973), Chalmers Univ. of Technology and Univ. of Gothenburg (G. Andersson and G. Holmén, eds.), p. 356.

(An 75) E. J. Anasaldo, G. T. Garvey, J. M. Lind, F. W. Loeser, and F. A. Naumann, *Nucl. Instrum. Methods* **125**:429 (1975).

(An 86) S. V. Andreev, V. I. Mishin, and V. S. Letokhov, *Opt. Commun.* **57**:317 (1986).

(An 87) R. A. Anderl, J. D. Cole, and R. C. Greenwood, *Nucl. Instrum. Methods* **B26**:333 (1987).

(Ar 86a) J. Arje, J. Aystö, H. Hyvönen, P. Taskinen, V. Koponen, J. Honkanen, K. Valli, A. Hautojärvi, and K. Vierinen, *Nucl. Instrum. Methods A* **247**:431 (1986).

(Ar 86b) J. Årje, University of Jyväskylä, Research Report No. 3/86 (1986), p. 24.

(Ay 80a) J. Aystö, private communication (1980).

(Ay 80b) J. Aystö and J. Cerny, in: *Future Directions in Studies of Nuclei Far From Stability* (J. H. Hamilton, E. H. Spejewski, C. R. Bingham, and E. F. Zganjar, eds.), North-Holland, Amsterdam (1980), p. 257.

(Ba 73) S. J. Balestrini and L. Forman, Proceedings of the International EMIS Conference on Low Energy Accelerators and Mass Separators, Skövde, 1973, Chalmers University of Technology and University of Gothenburg (G. Anderson and G. Holmén, eds.) (1973), p. 359.

(Be 60) R. Bernas, J. L. Sarrouy, and J. Camplan, *Le Journal de Physique et le Radium Physique Appliquée* **21**:191, Suppl. 11 (1960).

(Be 61) R. Bernas, *Rev. Sci. Instrum.* **32**:96 (1961).

(Be 71) G. J. Beyer, E. Herrmann, A. Piotrowski, V. J. Raiko, and H. Tyrroff, *Nucl. Instrum. Methods* **96**:437 (1971).

(Be 73) E. Ye. Berlovich, E. I. Ignatenko, and Yu. N. Novikov, Proceedings of the International EMIS Conference on Low Energy Accelerators and Mass Separators, Skövde, 1973, Chalmers University of Technology and University of Gothenburg (G. Anderson and G. Hölmén, eds.) (1973), p. 349.

(Be 76a) G. J. Beyer and A. F. Novgorodov, Akademie der Wissenschaften der DDR, Zentralinstitut für Kernforschung, Rossendorf bei Dresden, Report ZKF 305 (1976).

(Be 76b) G. J. Beyer, A. F. Novgorodov, O. Knotek, M. Jachim, and J. Juschkewitsch, ZKF 307 (1976).

(Be 76c) G. J. Beyer, W. D. Fromm, and A. F. Novgorodov, Zentralinstitut für Kernforschung, Rossendorf bei Dresden, Report ZkF 309 (1976).

(Be 76d) G. J. Beyer and A. F. Novgorodov, ZfK 317 (1976).

(Be 76e) G. J. Beyer and A. G. Novgorodov, ZfK 303 (1976).

(Be 76f) G. J. Beyer and A. F. Novgorodov, Radiochem. Radioanal. Lett. **27**(5–6):341 (1976).

(Be 76g) G. J. Beyer, A. F. Novgorodov, and W. D. Fromm, ZfK 310 (1976).

(Be 77a) G. J. Beyer and A. F. Novgorodov, Akademie der Wissenschafter der DDR, Zentralinstitut für Kernforschung, Rossendorf, Report ZfK 328 (1977).

(Be 77b) G. J. Beyer, O. Knotek, M. Jachin, J. Jushkevic, and A. F. Novgorodov, ZfK 332 (1977).

(Be 81) G. J. Beyer, A. F. Novgorodov, A. Kovalev, F. Prazak, V. A. Khalkin, and Yu. Yushkevich, *Nucl. Instrum. Methods* **186**:401 (1981).

(Be 85) V. Bechtold and S. Shiekh, Proc. Accelerated Beams Workshop, Parksville B.C. 1985 (L. Buchmann and J. M. D'Auria eds.), Triumf Report TRI-85-1, p. 110.

(Bj 81a) J. Bjørnstad, L. C. Carraz, H. A. Gustafsson, J. Heinemeier, B. Jonson, O. C. Jonsson, V. Lindfors, S. Mattsson, and H. L. Ravn, *Nucl. Instrum. Methods* **186**:391 (1981).

(Bj 81b) T. Bjørnstad, H. -A. Gustafsson, B. Jonson, O. C. Jonsson, V. Lindfors, S. Mattsson, A. M. Poskanzer, H. L. Ravn, and D. Schardt, *Z. Phys.* **303A**:227 (1981).

(Bj 85) T. Bjørnstad, M. G. J. Borge, P. Dessagne, R. -D. von Dincklage, G. T. Ewan, P. G. Hansen, A. Huck, B. Jonson, G. Klotz, A. Knipper, P. O. Larsson, G. Nyman, H. L. Ravn, C. Richard-Serre, K. Riisager, D. Schardt, and G. Walther, *Nucl. Phys. A* **443**:283 (1985).

(Bj 86a) T. Bjørnstad, M. J. G. Borge, J. Blomquist, R. D. von Dincklage, G. T. Ewan, P. Hoff, B. Jonson, K. Kawade, A. Kerek, O. Klepper, G. Løvhøiden, S. Mattsson, G. Nyman, H. L. Ravn, G. Rudstam, K. Sistemich, and O. Tengblad, *Nucl. Phys. A* **453**:463 (1986).

(Bj 86b) T. Bjørnstad, E. Hagebø, P. Hoff, E. Kugler, H. L. Ravn, S. Sundell, and B. Vosicki, *Phys. Scr.* **34**:578 (1986).

(Bl 78) M. Blann, University of Rochester Report URNSRL-181 (1978).

(Bo 65) S. Borg, U. Fägerquist, G. Holm, and F. Kropff, *Nucl. Instrum. Methods* **38**:296 (1965).

(Bo 70) J. P. Bocquet, R. Brissot, J. Crancon, C. Ristori, and A. Moussa, *Nucl. Instrum. Methods* **88**:51 (1970).

(Bo 72) J. Bonn, G. Huber, H. J. Kluge, L. Kugler, and E. W. Otten, *Phys. Lett.* **38B**:308 (1972).

(Br 81) C. Bruske, K. H. Burkhard, W. Hüller, R. Kirchner, O. Klepper, and E. Roeckl, *Nucl. Instrum. Methods* **186**:61 (1981).

(Br 85) M. Brügger, N. Hildebrand, T. Karlewski, N. Trautmann, A. K. Mazumdar, and G. Herrmann, *Nucl. Instrum. Methods A* **234**:218 (1985).

(Bu 82) F. Buchinger, A. C. Mueller, B. Schinzler, K. Wendt, C. Ekström, W. Klempt, and R. Neugart, *Nucl. Instrum. Methods* **202**:159 (1982).

(Bu 87) L. Buchmann, J. L. D'Auria, M. Dombsky, P. Schmor, and J. Vincent, Proceedings of the Eleventh International Conference on Electromagnetic Isotope Separators and Techniques Related to their Applications, Los Alamos, August 1986 (W. Talbert, ed.), *Nucl. Instrum. Methods* **B26**:253 (1987).

(Ca 67) I. E. Campbell and E. M. Sherwood, *High Temperature Materials and Technology*, Wiley, New York (1967).

(Ca 76) Proceedings of the 3rd International Conference on Nuclei Far From Stability, Cargèse, CERN 76-13 (1976).

(Ca 78) L. C. Carraz, I. R. Haldorsen, H. L. Ravn, M. Skarestad, and L. Westgaard, *Nucl. Instrum. Methods* **148**:217 (1978).

(Ca 79) L. C. Carraz, H. L. Ravn, M. Skarestad, S. Sundell, and L. Westgaard, *Nucl. Instrum. Methods* **158**:69 (1979).

(Ca 81a) J. Camplan and R. Meunier, *Nucl. Instrum. Methods* **186**:445 (1981).

(Ca 81b) J. Camplan, *Nucl. Instrum. Methods A* **187**:157 (1981).

(Ch 77) K. C. Chan, B. P. Pathak, L. Nikkinen, L. Lessard, and I. S. Grant, *J. Inorg. Nucl. Chem.* **39**:1915 (1977).

(Cl 87) D. J. Clark, *Treatise on Heavy-Ion Science* (D. Allan Bromley, ed.), Vol. 7, Plenum Press, New York (1987), p. 3.

(Co 57) G. A. Cowan and C. J. Orth, Proceedings of the 2nd Conference on Peaceful Uses Atomic Energy, Geneva, Vol. 7, United Nations, Geneva (1958), p. 328.

(Cr 57) J. Crank, *The Mathematics of Diffusion*, Clarendon Press, Oxford (1957).

(Cr 87) J. E. Crawford, J. K. P. Lee, V. Raut, S. Savard, and G. Thekkadath, *Nucl. Instrum. Methods* **B26**: 128 (1987).

(Da 78) J. M. D'Auria, J. W. Grüter, E. Hagberg, P. G. Hansen, J. C. Hardy, P. Hornshøj, B. Jonson, S. Mattsson, H. L. Ravn, and P. Tidemand-Petersson, *Nucl. Phys. A* **301**:397 (1978).

(Da 81) S. Dangaard and J. Heinemeyer, Annual Report 1981–1982, Institute of Physics, University of Aarhus, Denmark.

(De 44) A. J. Dempster, Metall. Rep. CP 1507:25 (1944).

(De 68) J. H. DeBoer, The Dynamical Character of Adsorption, Clarendon Press, Oxford (1968), p. 232.

(De 73) G. Dearnaly, J. H. Freeman, R. S. Nelson, and J. Stephen, *Ion Implantation*, North-Holland, Amsterdam (1973).

(De 86) C. Detraz, GANIL report 86-20, Proceedings of the International Nuclear Physics Conference, Harrogate 1986, to appear.

(Di 70) R. M. Diamond, Proceedings of the International Conference on the Properties of Nuclei Far From the Region of Beta Stability, Leysin (1970), CERN 70-30 (1970).

(Ea 86) T. Eaton and H. L. Ravn, *Nucl. Instrum. Methods* **B26**:190 (1987).

(Ei 85) B. Eichler, S. Hübner, and H. Rossbach, ZfK Rossendorf Reports 560 (1965) and 561 (1985).

(Ei 87) B. Eichler and H. Rossbach, *Kernenergie*, submitted.

(Em 65) Proceedings of the International Conference on Electromagnetic Isotope Separators and Related Ion Accelerators and their Application to Physics (J. Koch and L. Sarholt-Kristensen, eds.), *Nucl. Instrum. Methods* **38**:1 (1965).

(Em 70) Proceedings of the International Conference on Electromagnetic Isotope Separators and the Techniques of their Applications, Marburg (1970) (H. Wagner and W. Walcher, eds.), Bundesministerium für Bildung and Wissenschaft Forschungsbericht BMBW-FB K 70-28 (1970).

(Em 73) Proceedings of the 8th International EMIS Conference on Low-Energy Ion Accelerators and Mass Separators, Skövde (1973), Chalmers University of Technology and University of Gothenburg (G. Anderson and G. Hölmén, eds.) (1973).

(Em 76) Proceedings of the 9th International Conference on Electromagnetic Isotope Separators and Related Ion Accelerators, Kiryat Anavim (1976) (S. Amiel and G. Engler, eds.), *Nucl. Instrum. Methods* **139**:1 (1976).

(Em 80) Proceedings of the 10th International EMIS Conference, Zinal, Switzerland (1980) (E. Kugler, H. L. Ravn, and S. Sundell, eds.), *Nucl. Instrum. Methods* **186**:1 (1981).

(Em 87) Proceedings of the Eleventh International Conference on Electromagnetic Isotope Separators and Techniques Related to their Applications, Los Alamos, August 1986 (W. Talbert, ed.), *Nucl. Instrum. Methods* **B26**:1 (1987).

(En 85) H. A. Enge, *Treatise on Heavy-Ion Science* (D. Allan Bromley, ed.), Vol. 7, Plenum Press, New York (1985), p. 403.

(Ep 84) M. Epherre, *Ann. Phys. Fr.* **9**:717 (1984).

(Fa 65) H. Fabricius, K. Freitag, and S. Göring, *Nucl. Instrum. Methods* **38**:64 (1965).

(Fa 84) T. Faesterman, A. Gillitzer, K. Hartel, P. Kienle, and E. Nolte, *Phys. Lett.* **137B**:23 (1984).

(Fr 63) J. H. Freeman, *Nucl. Instrum. Methods* **22**:306 (1963).

(Fr 73a) J. H. Freeman and G. Sidenius, *Nucl. Instrum. Methods* **107**:477 (1973).

(Fr 73b) K. Fransson, M. af Ugglas, and A. Engström, Proceedings of the International EMIS Conference on Low Energy Accelerators and Mass Separators, Skövde, 1973, Chalmers University of Technology and University of Gothenburg (G. Anderson and G. Holmén, eds.) (1973), p. 330.

(Fu 81) M. Fujioka, T. Shinozuka, E. Tanaka, Y. Arai, S. Hayashibe, and T. Ishimatsu, *Nucl. Instrum. Methods* **186**:121 (1981).

(Ge 81) K.-H. Georgi, *Nucl. Instrum. Methods* **186**:271 (1981).

(Ge 83) W. Gelletly, *Nucl. Instrum. Methods* **211**:89 (1983).

(Ge 87) C. Geissen, H. Wollnik, B. Allardyce, and E. Kugler, *Nucl. Instrum. Methods* **B26**:120 (1987).

(Gh 59) A. Ghiorso, *At. Energ. (USSR)* **7**:338 (1959); *Soviet J. At. Energy* **7**:819 (1959).

(Gi 84) R. L. Gill, Proceedings of the Triumf-ISOL Workshop, Mount Gabriel (J. Crawford and J. M. D'Auria, eds.), TRIUMF report TRI-84-1, p. 99.

(Ha 67) R. A. Hasty, *J. Inorg. Nucl. Chem.* **29**:2679 (1967).
(Ha 70) E. Hagebø, Proceedings of the International Conference on Electromagnetic Isotope Separators and the Techniques of their Applications, Marburg, 1970, Report BMBW-FB K 70-28 (1970), p. 146.
(Ha 79) P. G. Hansen, *Annu. Rev. Nucl. Part. Sci.* **29**:69 (1979).
(Ha 82) J. H. Hamilton, *Heavy Ion Collisions*, Vol. 3 (R. Bock, ed.), North-Holland, Amsterdam (1982), p. 574.
(He 69) G. Herrmann and H. O. Denschlag, *Annu. Rev. Nucl. Sci.* **19**:1 (1969).
(He 81) Proceedings of the 4th International Conference on Nuclei Far From Stability, Helsingør, Denmark (1981), CERN 81-09 (1981).
(He 86) G. Herrmann, GSI 86-53 (1986).
(Hi 81) D. L. Hildenbrand, *Nucl. Instrum. Methods* **186**:357 (1981).
(Ho 74) L. Holland, W. Steckelmacher, and J. Yarwood, *Vacuum Manual*, E. and F. N. Spon, London (1974).
(Ho 80) P. Hoff, L. Jacobson, B. Johanson, P. Aagaard, G. Rudstam, and H.-U. Zwicky, *Nucl. Instrum. Methods* **172**:413 (1980).
(Ho 81) P. Hoff, K. Aleklett, E. Lund, and G. Rudstam, *Z. Phys. A* **300**:289 (1981).
(Ho 84) P. Hoff, O. C. Jonsson, E. Kugler, and H. L. Ravn, *Nucl. Instrum. Methods* **221**:313 (1984).
(Ho 86) M. Hofer, S. Sundell, H. L. Ravn, A. Ruf, and B. Vosicki, Letter of Intent to the ISOLDE Committee, CERN, May 1986, unpublished.
(Hu 83) M. Huyse, *Nucl. Instrum. Methods* **215**:1 (1983).
(Hu 87) M. Huyse, K. Deneffe, J. Gentens, P. Van Duppen, and D. Wouters, *Nucl. Instrum. Methods* **B26**:105 (1987).
(Io 72) Proceedings of the International Conference on Ion Sources, Vienna 1972 (F. Viehbock, H. Winter, and M. Brack, eds.), SGAB, Vienna (1972).
(Io 83) Proceedings of the International Ion Engineering Congress, Tokyo 1983 (K. Takagi, ed.), Institute of Electrical Engineers, Japan (1983).
(Ja 87) L. Jacobson, B. Fogelberg, B. Erkström, and G. Rudstam, *Nucl. Instrum. Methods* **B26**:223 (1987).
(Jo 73) P. G. Johnson, A. Bolson, and C. M. Henderson, *Nucl. Instrum. Methods* **106**:83 (1973).
(Ka 74) V. A. Karnaukhov, D. D. Bogdanov, A. V. Demyanov, G. I. Koval, and L. A. Petrov, *Nucl. Instrum. Methods* **120**:69 (1974).
(Ka 76) S. B. Karmohapatro, *Adv. Electron. Electron Phys.* **42**:113 (1976).
(Ka 83) H. Kawano, Y. Hidako, and F. M. Page, *Int. J. Mass Spectrom. Ion Phys.* **50**:1 (1983).
(Ki 76) R. Kirchner and E. Roeckl, *Nucl. Instrum. Methods* **133**:187 (1976).
(Ki 77) R. Kirchner, O. Klepper, G. Nyman, W. Reisdorf, E. Roeckl, D. Schardt, N. Kafrell, P. Peuser, and K. Schneeweiss, *Phys. Lett.* **70B**:150 (1977).
(Ki 78) R. Kirchner, Proceedings of the International Conference of Low Energy Ion Beams (K. G. Stephens *et al.*, eds.), Inst. Phys. Conf. Ser. No. 38 (Institute of Physics, Bristol) (1978), p. 29.
(Ki 81a) R. Kirchner, K. H. Burkard, W. Hülle, and O. Klepper, *Nucl. Instrum. Methods* **186**:295 (1981).
(Ki 81b) R. Kirchner, *Nucl. Instrum. Methods* **186**:275 (1981).
(Ki 86) R. Kirchner, O. Klepper, D. Marx, G. E. Rathke, and B. Sherril, *Nucl. Instrum. Methods A* **247**:265 (1986).
(Ki 87) R. Kirchner, Proceedings of the Eleventh International Conference on Electromagnetic Isotope Separators and Techniques Related to their Applications, Los Alamos, August 1986 (W. Talbert, ed.), *Nucl. Instrum. Methods* **B26**:204 (1987).
(Kl 65) R. Klapisch and R. Bernas, *Nucl. Instrum. Methods* **38**:291 (1965).
(Kl 69) R. Klapisch, *Annu. Rev. Nucl. Sci.* **19**:33 (1969).
(Kl 79) W. Klempt, J. Bonn, and R. Neugart, *Phys. Lett.* **82B**:47 (1979).

(Kl 81) O. Klepper, *Nucl. Instrum. Methods* **186**:385 (1981).

(Kl 85) H.-J. Kluge, F. Ames, W. Ruster, and K. Wallmeroth, Proceedings of the Accelerated Beams Workshop, Parksville, B.C. (L. Buchmann and J. M. D'Auria, eds.), TRIUMF Report TRI-85-1 (1985), p. 119.

(Kl 86) H.-J. Kluge, ISOLDE User's Guide, CERN 86-05 (1986).

(Ko 51) O. Kofoed-Hansen and K. O. Nielsen, *K. Danskė Vidensk. Selsk. Mat.-Fys. Medd.* **26**, No. 7 (1951).

(Ko 58) J. Koch, R. H. V. M. Dawton, M. L. Smith, and W. Walcher, *Electromagnetic Isotope Separators and Applications of Electromagnetically Enriched Isotopes*, North-Holland, Amsterdam (1958).

(Ko 75) W. Kornahl, H. Wagner, and W. Walcher, GSI-Report, Pk-1-75.

(Ko 76) O. Kofoed-Hansen, Proceedings of the 3rd International Conference on Nuclei Far From Stability, Cargèse, CERN 76-13 (1976), p. 65.

(La 74) A. Latuszynski, JINR E6-8057 (1974).

(La 83a) M. Langevin, C. Detraz, D. Guillemaud-Mueller, A. C. Mueller, C. Thibault, F. Touchard, F. Klotz, G. Miche, G. Walther, M. Epherre, and C. Richard-Serre, *Phys. Lett.* **130B**:251 (1983).

(La 83b) P. O. Larsson, T. Batsch, R. Kirchner, O. Klepper, W. Kurcewicz, E. Roeckl, D. Schardt, W. F. Feix, G. Nyman, and P. Tidemand-Petersson, *Z. Phys. A* **314**:9 (1983).

(La 86) M. Langevin, A. C. Mueller, D. Guillemaud-Mueller, M. G. Saint-Laurent, R. Anne, M. Bernas, J. Galin, D. Guerreau, J. C. Jacmart, S. D. Hoath, F. Naulin, F. Pougheon, E. Quiniou, and C. Détraz, *Nucl. Phys. A* **455**:149 (1986).

(Le 49) A. D. LeClaire, *Prog. Metal. Phys.* **306**:1 (1949).

(Le 70) Proceedings of the International Conference on the Properties of Nuclei far from the Region of Beta Stability, Leysin (1970), CERN 70-30 (1970), 2 vols.

(Le 75) J. K. P. Lee, G. Pilar, B. L. Tracy, and L. Yaffe, *J. Inorg. Nucl. Chem.* **37**:2035 (1975).

(Le 85) V. S. Letokhov, On-line in 1985 and beyond, A workshop on the ISOLDE programme, Zinal, Switzerland (1984), Abstracts, unpublished.

(Ly 66) Proceedings of the International Symposium on Why and How should we Investigate Nuclides Far Off the Stability Line, Lysekil, Sweden (1966) (W. Forsling, C. J. Herrlander, and H. Ryde, eds.), *Art. Fys.* **36**:1 (1967).

(Ma 63) R. D. Macfarlane and R. D. Griffioen, *Nucl. Instrum. Methods* **24**:461 (1963).

(Må 85) B. M. Mårtenson and S. O. Wilhelmsson, *Surf. Sci.* **161**:181 (1985).

(Mc 56) D. E. McKenzie, *Can. J. Chem.* **34**:515 (1956).

(Mo 81a) D. M. Moltz, J. Aystö, M. D. Cable, R. F. Parry, P. E. Haustein, J. M. Wouters, and J. Cerny, *Nucl. Instrum. Methods* **186**:141 (1981).

(Mo 81b) D. M. Moltz, *Nucl. Instrum. Methods* **186**:135 (1981).

(Mü 82) G. Münzenberg, P. Armbruster, F. P. Hebberger, S. Hofmann, K. Poppensieker, W. Reisdorf, H. S. Schneider, K. H. Schmidt, C. C. Sahm, and D. Vermeulen, *Z. Phys. A* **309**:89 (1982).

(Ne 82) R. Neugart, private communication.

(Ne 86) R. Neugart, *Progress in Atomic Spectroscopy* (H. Kleinpopper and H. J. Beyer, eds.), Plenum Press, New York and London (1987).

(Ni 57) K. O. Nielsen, *Nucl. Instrum. Methods* **1**:289 (1957).

(Ni 70) J. M. Nitchke, Proceedings of the International Conference on the Properties of Nuclei Far From Stability, Leysin 1970, CERN Report 70-30 (1970), p. 153.

(Ni 81) B. R. Nielsen, *Nucl. Instrum. Methods* **186**:457 (1981).

(Ni 83) J. M. Nitschke, *Nucl. Instrum. Methods* **206**:341 (1983).

(Ni 85) J. M. Nitschke, *Nucl. Instrum. Methods A* **236**:1 (1985).

(No 75) A. S. Norwick and I. J. Burton, *Diffusion in Solids: Recent Developments*, Academic Press, London (1975).

(Ok 81) K. Okano, Y. Kawase, K. Kawade, Y. Yamamoto, M. Hanada, T. Katoh, and I. Fujiwara, *Nucl. Instrum. Methods* **186**:115 (1981).

(Or 61) C. J. Orth, *Nucl. Sci. Eng.* **9**:417 (1961).

(Or 70) M. Oron and S. Amiel, Proceedings of the International Conference on Electromagnetic Isotope Separators and the Techniques of their Application, Marburg (Bundesministerium fur Technologie und Wissenschaft), rep. K70-28 (1970), p. 87.

(Ot 87) E. W. Otten, *Treatise on Heavy-Ion Physics* (D. Allan Bromley, ed.), Vol. 8, Plenum Press, New York (1988), this volume.

(Ox 87) K. Oxhorn, J. E. Crawford, H. Dautet, J. K. P. Lee, R. B. Moore, L. Nikkinen, L. Buchmann, J. M. D'Auria, R. Kokke, A. J. Otter, and J. Vincent, Proceedings of the Eleventh International Conference on Electromagnetic Isotope Separators and Techniques Related to their Applications, Los Alamos, August 1986 (W. Talbert, ed.), *Nucl. Instrum. Methods* **B26**:143 (1987).

(Pa 70) P. Patzelt, Proceedings of the International Conference on Electromagnetic Isotope Separators and the Techniques of their Application, Marburg (Bundesministerium fur Technologie und Wissenschaft), rep. K70-28 (1970), p. 158.

(Pa 81) P. Paris, C. F. Liang, D. Lecouturier, M. Arianer, J. Obert, A. Caruette, A. Ferro, J. Fournet, J. C. Putaux, J. L. Sarrouy, and the ISOCELE Collaboration, *Nucl. Instrum. Methods* **186**:91 (1981).

(Pi 84) A. Piotrowski, R. L. Gill, and D. C. McDonald, *Nucl. Instrum. Methods A* **224**:1 (1984).

(Pl 87) A. Plantier, R. Beraud, T. Olliver, A. Charvet, R. Duffait, A. Emsallem, N. Redon, V. Boninchi, S. Vanzetto, A. Gizon, J. Treherne, N. Idrissi, J. Genevey, and J. C. Vieux-Rochaz, Proceedings of the Eleventh International Conference on Electromagnetic Isotope Separators and Techniques Related to their Applications, Los Alamos, August 1986 (W. Talbert, ed.), *Nucl. Instrum. Methods* **B26**:314 (1987).

(Po 86) F. Pougheon, R. Anne, M. Bernas, C. Detraz, J. Galin, D. Guerreau, S. Hoath, D. Guillemaud-Mueller, J. C. Jacmart, M. Langevin, A. C. Mueller, F. Naulin, M. G. Saint-Laurent, and E. Quiniou, IPNO DRE 86-01 (1986).

(Pr 85) P. B. Price, J. D. Stevenson, and S. W. Barwik, *Phys. Rev. Lett.* **54**:297 (1985).

(Pu 81) J. C. Putaux, J. Obert, L. Kotfila, B. Roussiere, J. Sauvage-Letessier, C. F. Liang, A. Peghaire, P. Paris, and J. Giroux, *Nucl. Instrum. Methods* **186**:321 (1981).

(Pu 87) J. C. Putaux, J. Obert, G. Boisser, and P. Paris, Proceedings of the Eleventh International Conference on Electromagnetic Isotope Separators and Techniques Related to their Applications, Los Alamos, August 1986 (W. Talbert, ed.), *Nucl. Instrum. Methods* **B26**:213 (1987).

(Ra 75) H. L. Ravn, S. Sundell, and L. Westgaard, *Nucl. Instrum. Methods* **123**:131 (1975).

(Ra 76a) H. L. Ravn, L. C. Carraz, J. Denimal, E. Kugler, M. Skarestad, S. Sundell, and L. Westgaard, *Nucl. Instrum. Methods* **139**:267 (1976).

(Ra 76b) H. L. Ravn, *Nucl. Instrum. Methods* **139**:281 (1976).

(Ra 79) H. L. Ravn, *Phys. Rep.* **54**:203 (1979).

(Ra 85) H. L. Ravn, J. M. D'Auria, M. J. G. Borge, Y. Huang, O. C. Jonsson, L. Nikkinen, and S. Sundell, Proceedings of the Accelerated Beams Workshop, Parksville B. C. (L. Buchmann and J. M. D'Auria, eds.), Triumf Report TRI-85-1 (1985), p. 94.

(Ra 87) H. L. Ravn, T. Bjørnstad, P. Hoff, O. C. Jonsson, E. Kugler, S. Sundell, and B. Vosicki, Proceedings of the Eleventh International Conference on Electromagnetic Isotope Separators and Techniques Related to their Applications, Los Alamos, August 1986 (W. Talbert, ed.), *Nucl. Instrum. Methods* **B26**:183 (1967).

(Re 77a) S. Regnier, Thesis, IN2P3, University of Bordeau-Gradignan, CENBG 7726 No. 557 (1977).

(Re 77b) T. Reetz, B. Eichler, H. Gäggeler, and I. Zvara, *Radiochim. Acta* **24**:69 (1977).

(Re 80) L. von Reisky, J. Bonn, S. L. Kaufman, L. Kugler, E. W. Otten, J.-M. Rodriguez-Giles, K. P. C. Spath, and D. Weskott, *Nucl. Instrum. Methods* **172**:423 (1980).

(Ro 84) H. Rossbach and B. Eichler, ZfK Rossendorf Report 527 (1984).

(Ru 65) G. Rudstam, *Nucl. Instrum. Methods* **38**:282 (1965).

(Ru 66) G. Rudstam, *Z. Naturforsch.* **21A**:1027 (1966).

(Ru 76) G. Rudstam, *Nucl. Instrum. Methods* **139**:239 (1976).

(Ru 81) G. Rudstam, P. Aagaard, P. Hoff, B. Johansson, and H.-U. Zwicky, *Nucl. Instrum. Methods* **186**:365 (1981).

(Ru 85a) E. Runte, K.-L. Gippert, W.-D. Schmidt-Ott, P. Tidemand-Petersson, L. Ziegeler, R. Kirchner, O. Klepper, P. O. Larsson, E. Roeckl, D. Schardt, N. Kaffrell, P. Peuser, M. Bernas, P. Dessagne, M. Langevin, and K. Rykaczewski, *Nucl. Phys. A* **441**:237 (1985).

(Ru 85b) G. Rudstam, P. Aagaard, and H.-U. Zwicky, Studsvik Science Research Laboratory, Sweden, Rep. NFL-42 (1985).

(Ry 85) K. Rykaczewski, R. Kirchner, O. Klepper, V. T. Koslowsky, P. O. Larsson, G.-E. Rathke, E. Roeckl, D. Schardt, E. F. Zganjar, U. Bosch, W. Eschner, K.-L. Gippert, E. Runte, W.-D. Schmidt-Ott, P. Tidemand-Petersson, N. Kaffrell, P. Peuser, W. Kurcewicz, M. Bernas, P. Dessagne, M. Langevin, H. Beer, and G. Walter, Contribution to the XXIII International Winter Meeting on Nuclear Physics, Bormio, Italy (January 1985).

(Sa 81) M. de Saint-Simon, R. Fergeau, M. Jacotin, J. F. Kepinski, R. Klapisch, C. Thibault, and F. Touchard, *Nucl. Instrum. Methods* **186**:87 (1981).

(Sa 84) J. Sauvage, C. Bourgeois, A. Caruette, P. Kilcher, J. Obert, J. Oms, J. C. Puteaux, B. Roussière, A. Ferro, J. Fournet, L. Kotfila, C. F. Liang, P. Paris, M. G. Porquet, and G. Cornudet, Proc. TRIUMF-ISOL Workshop, Mount Gabriel, Quebec (J. Crawford and J. M. D'Auria, eds.), Triumf Report TRI-84-1 (1984), p. 161.

(Sc 81a) D. Schardt, T. Batsch, R. Kirchner, O. Klepper, W. Kurcewicz, E. Roeckl, and P. Tidemand-Petersson, *Nucl. Phys. A* **368**:153 (1981).

(Sc 81b) H. Schmeing, J. C. Hardy, E. Hagberg, W. L. Perry, and J. S. Wills, *Nucl. Instrum. Methods* **186**:47 (1981).

(Sc 87) H. Schmeing, J. S. Wills, E. Hagberg, J. C. Hardy, V. T. Koslowsky, and W. L. Perry, Proceedings of the Eleventh International Conference on Electromagnetic Isotope Separators and Techniques Related to their Applications, Los Alamos, August 1986 (W. Talbert, ed.), *Nucl. Instrum. Methods* **B26**:321 (1987).

(Sh 84) K. S. Sharma, H. Schmeing, H. C. Evans, E. Hagberg, J. C. Hardy, and V. T. Koslowsky, Proc. TRIUMF-ISOL Workshop, Mount Gabriel, Quebec (J. Crawford and J. M. D'Auria, eds.), Triumf-ISOL Report TRI-84-1 (1984), p. 70.

(Sh 87) K. S. Sharma, R. C. Barber, H. E. Duckworth, G. R. Dyck, C. A. Lander, H. Schmeing, G. Audi, J. S. Guo, E. Hagberg, J. C. Hardy, and V. T. Koslowsky, Proceedings of the Eleventh International Conference on Electromagnetic Isotope Separators and Techniques Related to their Applications, Los Alamos, August 1986 (W. Talbert, ed.), *Nucl. Instrum. Methods B*, to appear.

(Si 61) G. Sidenius and O. Skilbreid, *Proceedings of the International Symposium on Electromagnetic Separation of Radioactive Isotopes, Vienna (1960)*, Springer-Verlag. Vienna (1961), p. 234.

(Si 65) G. Sidenius, R. M. Gammon, R. A. Naumann, and T. D. Thomas, *Nucl. Instrum. Methods* **38**:299 (1965).

(Si 70) G. Sidenius, Proceedings of the International Conference on Electromagnetic Isotope Separators and the Techniques of their Applications, Marburg (1970) (H. Wagner and W. Walcher, eds.), Bundesministerium für Bildung und Wissenschaft Forschungsbericht, BMBW K, 70-28 (1970), p. 423.

(Si 73) R. Silberberg and C. H. Tsao, Naval Research Laboratory, Washington, D.C., NRL Report 7593 (1973).

(Sp 81) E. H. Spejewski, R. L. Mlekodaj, and H. K. Carter, *Nucl. Instrum. Methods* **186**:71 (1981).

(St 74) J. J. Stoffels, *Nucl. Instrum. Methods* **119**:251 (1974).

(Ta 70) N. I. Tarantin, A. P. Kabachenko, I. V. Kuznetsov, Yu. A. Dyachihin, N. S. Ivanov, and Om Zai Khun, Proceedings of the International Conference on Electromagnetic Isotope Separators and the Techniques of their Application, Marburg 1970 (Bundesministerium fur Technologie und Wissenschaft), rep. K 70-28 (1970), p. 59.

(Ta 76) W. L. Talbert, Jr., F. K. Wohn, A. R. Landin, J. C. Pacer, R. L. Gill, M. A. Cullison, G. A. Sheppard, K. A. Burke, K. L. Malaby, and A. F. Voigt, *Nucl. Instrum. Methods* **139**:257 (1976).

(Ta 79) W. L. Talbert, Jr., F. K. Wohn, J. C. Hill, A. R. Landin, M. A. Cullison, and R. L. Gill, *Nucl. Instrum. Methods* **161**:431 (1979).

(Ta 83) T. Tamura, private communication (1983).

(Th 71) C. Thibault, Thèse à la Faculté des Sciences d'Orsay, Université de Paris, CNRS AO 5862/No 783 (1971).

(Ti 81) P. Tidemand-Petersson, R. Kirchner, O. Klepper, W. Kircewicz, E. Roeckl, and E. F. Zganjar, *Z. Phys. A* **302**:343 (1981).

(To 81) F. Touchard, J. Biderman, M. de Saint-Simon, C. Thibault, G. Huber, M. Epherre, and R. Klapisch, *Nucl. Instrum. Methods* **186**:329 (1981).

(Tu 77) J. A. Turnball, C. A. Friskney, F. A. Johnson, A. J. Walter, and J. R. Findlay, *J. Nucl. Mater.* **67**:301 (1977).

(Va 82) D. Vandenplassche, E. van Walle, C. Nuytten, and L. Vanneste, *Phys. Rev. Lett.* **49**:1390 (1982).

(Ve 81) J. Verplancke, B. Brijs, K. Cornelis, J. Gentens, M. Huyse, and G. Lhersonneau, *Nucl. Instrum. Methods* **186**:99 (1981).

(Vo 81) B. Vosicki, T. Bjørmstad, L. C. Carraz, J. Heinemeyer, and H. L. Ravn, *Nucl. Instrum. Methods* **186**:307 (1981).

(Vo 87a) V. V. Volkov, *Treatise on Heavy-Ion Science* (D. Allan Bromley, ed.), Vol. 8, Plenum Press, New York (1988), this volume.

(Vo 87b) B. Vosicki, private communication.

(Wa 87) E. Van Walle, B. Brijs, K. Deneffe, M. Huyse, N. Severijns, L. Vanneste, D. Vandeplassche, and J. Wouters, *Nucl. Instrum. Methods* **B26**:345 (1987).

(We 74) W. Weisweiler and G. D. Nayeshwar, *Carbon* **13**:175 (1975).

(Wi 73) R. G. Wilson, *Ion Beams with Applications to Ion Implantation*, Wiley, New York (1973).

(Wo 76) H. Wollnik, *Nucl. Instrum. Methods* **139**:311 (1976).

(Wo 77) F. K. Wohn, Iowa State University, Report IS-4270 (1977), p. 32.

(Wö 82) R. Wölfle, S. M. Qaim, and R. Nider, *Radiochim. Acta* **31C**:165 (1982).

(Wo 84) J. L. Wood, Proc. TRIUMF-ISOL Workshop, Mount Gabriel, Quebec (J. Crawford and J. M. D'Auria, eds.), Triumf Report TRI 84-1 (1984), p. 87.

(Wo 85) H. Wollnik and K. Becker, *Nucl. Instrum. Methods A* **238**:206 (1985).

(Wü 78) K.-D. Wünsch, *Nucl. Instrum. Methods* **155**:347 (1978).

(Ya 41) S. Yamaguchi, *Proc. Phys. Math. Soc. Jpn.* **23**:264 (1941).

(Yo 76) Y. Yoshizawa, H. Noma, T. Horiguchi, T. Katoh, S. Amemiya, M. Itoh, K. Hisatake, M. Sekikawa, and K. Chida, *Nucl. Instrum. Methods* **134**:93 (1976).

(Zi 73) J. P. Zirnheld and L. Schutz, Proceedings of the International EMIS Conference on Low Energy Accelerators and Mass Separators, Skövde, 1973, Chalmers University of Technology and University of Gothenburg (G. Anderson and G. Holmén, eds.) (1973), p. 469.

# 6

ISAO TANIHATA is Principal Scientist and Head of the Linear Accelerator Laboratory at The Institute of Physical and Chemical Research (RIKEN), Wako, Saitama, Japan and has held this post since 1986. He received his undergraduate and graduate education at Osaka University, receiving the Ph.D. degree in 1975. During the periods 1978–79 and 1983–85, he was Postdoctoral Fellow and Research Scientist, respectively, at the Lawrence Berkeley Laboratory and during 1979–85 and 1985–86 he was Assistant and Associate Professor, respectively, at the Institute for Nuclear Study at the University of Tokyo. During 1976–78, he held the Miller Fellowship awarded by the Lawrence Berkeley Laboratory.

# On the Possible Use of Secondary Radioactive Beams

ISAO TANIHATA

## 1. Introduction

Since the first experiment on $\alpha$ scattering by Rutherford, nuclear reactions using accelerated beams of electrons, protons, deuterons, alphas, and so on, have been successfully used in the study of nuclear structure and reaction mechanisms. The use of accelerated heavy ions during the past ten years has opened a wide range of possibilities for studies in nuclear physics and other fields of research. However, until a few years ago experiments had been performed using only ions of stable isotopes.

Recently, beams of short-lived radioactive nuclei have been produced as secondary beams at the Berkeley BEVALAC (Ta 85). In the first experiment, interaction cross-sections of high-energy reactions using beams of radioactive nuclei were measured, and matter radii of those radioactive nuclei were determined (Ta 85a, Ta 85b). Although the radii of stable nuclei have been studied extensively by electron and proton scattering and by muonic X-rays, radii of unstable nuclei have not been studied except for a small number of cases that were determined in isotope-shift measurements. The use of beams of radioactive nuclei opened new possibilities for studying nuclear structures which were difficult to reach by conventional methods. Several experiments other than the study of nuclear radii are also in progress.

In this chapter, we review methods for producing beams of radioactive nuclei and then present examples of studies which can be usefully developed through the use of these radioactive beams.* Since the use of radioactive nuclei is still very new, only a few experiments have been completed. Many more are in progress or being developed

* The isotopes produced by ISOL (Isotope Separator On Line) are also often termed "beams of radioactive nuclei." Because the studies developed using ISOL facilities are discussed in another chapter, it is not the author's intention to cover them here. The discussion will therefore be restricted to radioactive beams of energy higher than 1 MeV/nucleon.

ISAO TANIHATA • Institute of Physical and Chemical Research (RIKEN), Wako, Japan.

and many others are currently under discussion. Therefore a considerable part of the discussion presented in this chapter has no practical realization at this present moment. However, it would be appropriate to review current production methods and uses of radioactive beams because the author believes that the use of radioactive beams will open a new era in the history of nuclear physics.

To the best of the author's knowledge, only two workshops dedicated to the possibility of producing and using radioactive beams have been held in recent years. Because the author simply cannot cover all the currently available ideas in detail, interested readers are referred to the proceedings of these workshops (Pr 84, Pr 85). Formal proposals for new facilities at TRIUMF (Pr 85a), GSI (Pr 82), and LBL (Pr 86) also include relevant information.

## 1.1. Why and What

Some 8000 species of nuclides are predicted by current nuclear theory, to be bound or to have lifetimes longer than microseconds. Among these bound nuclei only about 2000 nuclides have as yet been identified. About 270 of these are stable and lie along the stability line in an $N$-$Z$ nuclear chart as shown in Figure 1. The remaining species are unstable and decay spontaneously by $\alpha$ emission, $\beta$ emission, nuclear fission, proton emission, or other decay modes. The nuclei that have so far been studied almost all lie near the stability line; nuclei far from that line have not as yet been synthesized over a wide region. Moreover, properties of nuclei such as spins, radii, electromagnetic moments, and decay modes have only been investigated for a small fraction of the known nuclei and these, too, are predominantly near the stability line.

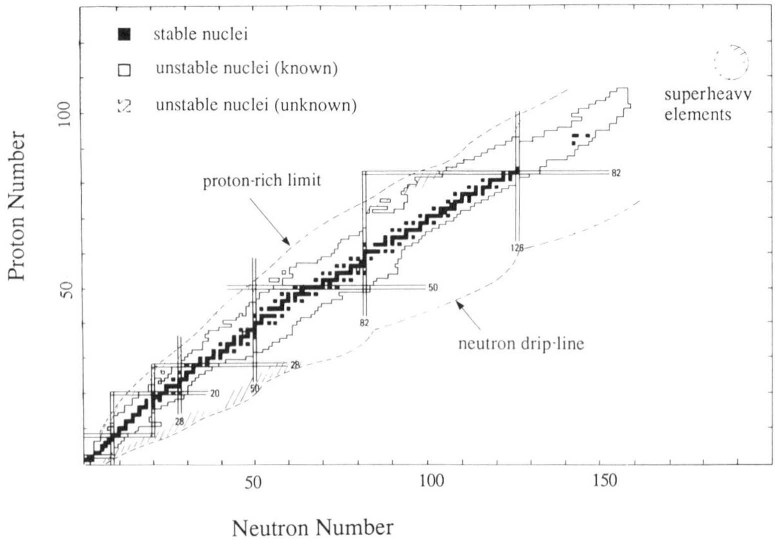


**Figure 1.** The nuclear chart. Solid squares in the figure indicate stable nuclei. Nuclei within the jagged solid line are unstable nuclei which have been observed. Theoretically predicted limits of the existence of nuclei are shown by dashed lines with shadows. Only about one quarter of predicted nuclei have been observed so far.

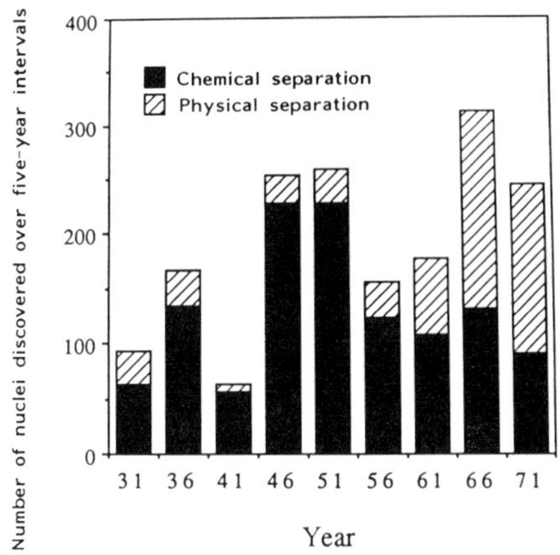


**Figure 2.** Numbers of nuclides discovered during five-year periods. The recent increase shows that physical separation has become a major tool for such discoveries.

It is then no exaggeration to say that our present knowledge of nuclear structure is based essentially on experiments on nuclei near the stability line. It is, therefore, particularly desirable to extend these studies to nuclei far from stability in order to obtain a deeper understanding of the rich spectrum of properties of nuclei and of nuclear structure. Moreover, it is quite possible that we may encounter entirely new phenomena in such an unstudied region of nuclei.

The main reason why studies so far have been restricted to the domain near stability reflects the difficulty of producing and preparing the radioactive species. In this respect two important techniques have been developed in the past twenty years: one uses the ISOL (Isotope Separator On-Line) facilities and the other is based upon use of heavy-ion accelerators. Figure 2 shows the number of nuclides that were discovered, over five-year intervals, in the past. Three distinct peaks reflect important developments of technique in nuclear physics. The first peak during 1936–1940 reflects work using neutron beams and also the invention of accelerators such as the cyclotron. The evolution in accelerator techniques together with the development of new radiation detectors—especially the scintillation counter [NaI(Tl)]—produced a second peak in the 1950s. Nuclides were separated mainly by chemical methods during these two peak intervals. On the other hand, physical separation methods, which use electromagnetic separators, became more important during the third peak after the 1960s. This last peak, in fact, reflects the development of the ISOL and of heavy-ion accelerators. The physical-separation method, incorporating sophisticated beam optics and supported by advanced radiation-detector systems, provided the extremely sensitive tools required for identifying rare nuclei. Recently, a number of new isotopes have been discovered through projectile fragmentation studies with intermediate-energy and high-energy heavy-ion beams at Berkeley and GANIL (La 85, Gu 86, Mu 84, and elsewhere). The use of high-intensity heavy-ion accelerators is expected to lead to the widespread production of nuclei situated further from stability.

If one is able to use radioactive nuclei in incident beams and induce nuclear reactions with them, many important measurements that would otherwise be impossible can be executed. Experiments at the Berkeley BEVALAC have demonstrated the feasibility of this, and have suggested wide applications for radioactive beams. The use of radioactive beams can provide powerful methods for exploring nuclei under extreme conditions as will be summarized below.

### *1.1.1. How far from the Stability Line are Nuclei Bound?*

Studying the limits of existence of nuclei is important not only for its own sake but also for a precise understanding of the effective interactions in nuclei. For example, the nuclear mass formulas are developed semiempirically using masses of known nuclei. However, they give considerably different predictions when extrapolated to nuclei far from stability. In the region near the particle drip-lines, the application of such formulas is still of doubtful utility. This indicates the importance of inadequately known effective interactions such as the isospin-dependent interactions and many-body effects. The production of extremely exotic nuclei can be expected if one has available high-intensity radioactive beams. Mass measurements on these exotic nuclei using radioactive-beam techniques are expected to lead to important new input in the field of effective nuclear interactions.

### *1.1.2. Phenomena in High-Isospin Nuclei*

In a stable nucleus the numbers of protons and neutrons are balanced but nuclei far from stability are always either neutron or proton rich and thus provide unbalanced nucleon environments. It is important to know the properties of such nuclei. As an example, a possible change of the shell-model magic numbers in such unbalanced nuclei is an interesting open question. Studies of the nucleus size, its electromagnetic moments, its decay characteristics, and its Coulomb excitation are all important when these nuclei are radioactive.

### *1.1.3. High-Spin and Highly Exited States*

One of the most important contributions of heavy-ion beams, thus far, has been in providing great scope for the study of high-spin and highly excited states. It is known that the production rate of high-spin states is strongly dependent on the proton and neutron numbers of both projectile and target nuclei. The use of radioactive beams provides greater selection in both numbers.

### *1.1.4. Nuclear Astrophysics*

It is belived that all elements existing in our universe were produced by nucleosynthesis in stars. In hot stars and in novas, low-energy nuclear reactions between unstable nuclei are known to be very important. However the reaction cross-sections on such unstable nuclei have not been determined because of obvious experimental difficulties. The use of low-energy radioactive beams permits the measurement of cross-sections in the laboratory. In his Nobel lecture W. A. Fowler emphasized the importance of radioactive beams:

> The production of radioactive ion beams hold great promise for the future . . . Continued development and application of radioactive ion-beam techniques could bring the most exciting results in laboratory nuclear astrophysics in the next decades.

He also pointed out that the possibility of using beams of isomers is also important (Fo 84).

## 1.2. Radioactive Beams or Radioactive Targets?

The second possible method of studying reactions by means of radioactive nuclei is to use the radioactive nuclei as targets. In this application, one can then use a high-intensity beam of stable nuclei as projectiles. One may obtain high reaction rates if sufficiently thick targets are available. In fact, this technique was developed twenty years ago for the production of nuclides in the transuranium region (Se 68).

But let us now compare the reaction rates obtainable in radioactive-beam applications and in radioactive nuclear-target applications. A comparison will be made only for nuclei which are not stable or which do not exist in nature because of their short lifetimes. A naturally existing radioactive nucleus such as $^{14}C$ can be made into a thick target just as can stable nuclei. Therefore our discussion will be different for much shorter-lived radioactive nuclei which have to be produced by an accelerated beam. For radioactive-beam applications, the number of nuclear reactions $I_p(T)$ that occur between a radioactive projectile and a target nucleus within a time interval $T$ can be written as

$$I_p(T) = n_r n_t \sigma T \tag{1}$$

where $n_r$ is the number of radioactive nuclei per unit time in the beam, and $n_t$ is the number of target nuclei in unit area. The reaction cross-section of interest is denoted as $\sigma$.

If one uses the radioactive nuclei as target, the maximum reaction rate is obtained when a beam of stable nuclei is simultaneously injected into the target area where the radioactive nuclei are being accumulated. The number of collected radioactive nuclei $N_r(t)$, which can be used as target at time $t$, is given by

$$N_r(t) = \tau n_r (1 - e^{-t/\tau}) \tag{2}$$

The thickness of the radioactive target is then $N_r(t)/a$, where $a$ is the area over which these radioactive nuclei are collected. The number of reactions $I_t(T)$ in a time interval $T$, when a stable nuclear beam of intensity $n_s$ per unit area is simultaneously injected, is then given by

$$I_t(T) = \int_0^T \sigma n_s \frac{N_r(t)}{a}\, dt \tag{3}$$

$$= \frac{n_s n_r}{a} \sigma [T + \tau(e^{-T/\tau} - 1)] \tag{4}$$

If the experiment continues long enough to satisfy $T/\tau \gg 1$, then

$$I_t(T) \approx \frac{n_s n_r}{a} \sigma T \tag{5}$$

The ratio of the numbers of reactions observable in the two methods can then be written

$$\frac{I_t(T)}{I_p(T)} = \frac{1}{a} \frac{n_s}{n_t} \tau \tag{6}$$

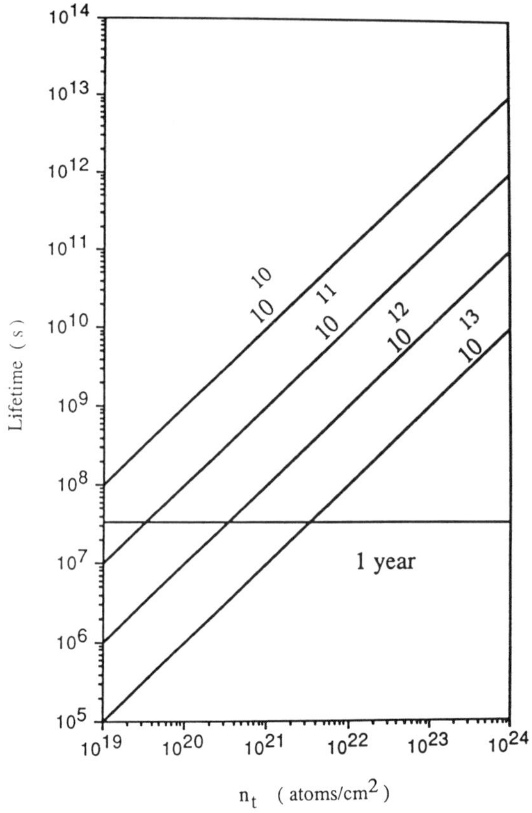


**Figure 3.** Lines show relations between lifetime and target thickness $n_t$ under the condition that the radioactive-target and radioactive-beam methods give the same reaction rates. Numbers along the lines indicate the intensities of stable nuclear beams in particles per second ($n_s$).

The ratio is independent of $T$, the reaction cross-section $\sigma$, and the intensity of the radioactive nuclei $n_r$. Lines in Figure 3 show the relation between $n_t$ (the thickness of the target when the radioactive nucleus is used as projectile) and $\tau$ (the lifetime of the nucleus) under the condition that the radioactive-beam application and the radioactive-target application give equal reaction rates. When the lifetime of the nucleus is longer than the value indicated by the line, the radioactive-target method provides a higher reaction rate than does the radioactive-beam method. From the figure, one can see that the radioactive-beam technique is preferable for studying nuclear reactions with radioactive nuclei of lifetime shorter than about one year, in most conditions. Also, the radioactive nuclear beam gives a lower background environment than does the radioactive nuclear target for two reasons. One is that the radioactivity of the target nucleus itself (which is stronger when one uses a thicker target) is usually a problem in terms of detector overload or damage. The other is that contaminant reactions necessarily occur in the stopping material that is used to accumulate the radioactive nuclei.

The above discussion is valid only when the radioactive nuclei are produced by the same nuclear reaction in both methods. Nuclei which are produced by other reactions, such as slow neutron capture, may also be used in large amounts as a target. The comparison in that case is different and cannot be generalized.

From our discussion thus far, it is clear that use of radioactive beams provides a higher reaction rate for short-lived nuclei. The effectiveness of the method, however, depends strongly on the characteristics of the detection system. Thus its selection has to be made carefully for each individual reaction to be studied. A more-detailed discussion can be found elsewhere (Ha 85). Many kinds of radioactive targets have been used for the study of heavy nuclei as well as the synthesis of new elements and isotopes. In this review, however, no further discussion of the use of radioactive targets will be given.

## 2. Production of Beams of Radioactive Nuclei

Two, basically different, methods can be used for producing the radioactive beams. In either method, the radioactive nuclei have to be produced through nuclear reactions induced by an accelerated beam. The two methods differ in terms of the extraction of the ions of the radioactive nuclei. One method uses a so-called isotope separator on line (ISOL) for selecting the product nuclei. In this method, nuclei are extracted from a production target by thermal diffusion and then ionized and accelerated into a beam at a low energy ($\approx$50 keV). This selected beam is then accelerated to the energy necessary for the experiment. The second method does not employ an ion source but uses, instead, recoil radioactive nuclei directly from the production target, and a suitable recoil separator is used to select the desired nuclide. The recoil energy can be used directly in many cases although both acceleration or deceleration can readily be applied.

Elements so far available as ion beams from ISOLDE at CERN are shown in Figure 4. It is impressive to see that many elements have been extracted as ions by this method. In the ISOL method, an ion source located near the production target supplies ions of radioactive nuclei accelerated from thermal energy. Further acceleration of these nuclei is essentially identical to that of heavy ions of stable nuclei. Important requirements specific to the acceleration of the radioactive nuclei arise because the radioactive nuclei have a finite lifetime and because their intensity is generally many orders of magnitude weaker than typical intensities of ions of stable nuclei. Their finite lifetime requires the ion extraction and acceleration to be more rapid than the decay of the nucleus. In general, the necessary time for acceleration does not depend on the nuclear species. However, the extraction time for ions from the target material does depend strongly on the chemical character of the element. For example, noble gas and alkali-metal nuclei can be extracted in as short a time as a few milliseconds. On the other hand, elements such as oxygen and nitrogen are known to be very difficult to extract. In addition, a high capturing efficiency of the ions into the acceleration phase is very important for efficient use of the radioactive ions produced.

Various types and energies of primary beam have been used for ISOL studies. For production of exotic nuclei, a high-energy proton beam (500–1000 MeV) accelerated by a cyclotron or linac is advantageous for several reasons. First, high beam intensities are easily available. Second, a very thick target can be used because of the long stopping

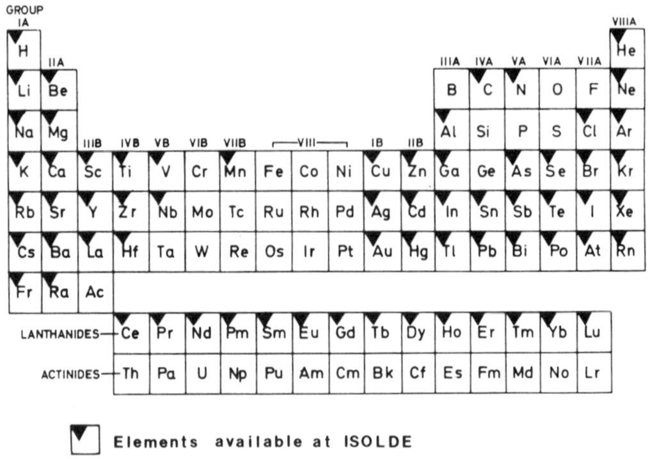


**Figure 4.** Periodic table of the elements, indicating the elements available at ISOLDE (provided by H. L. Ravn).

range of high-energy protons. Also, production cross-sections for exotic nuclei are large if one uses spallation reactions for proton-rich nuclei or induced-fission reactions for neutron-rich isotopes. The production cross-sections for radioactive nuclei will be discussed in a later part of this chapter. Further discussion of the acceleration of an ISOL beam is given in Section 2.2.

The recoil-separator method has several advantages when applied to the separation of nuclei produced in heavy-ion reactions. In such a reaction, the velocity of the center-of-mass system is, in general, large. Therefore product nuclei are emitted primarily into a small forward cone. In particular, nuclei which are produced by projectile fragmentation are emitted in the direction of the primary beam with almost constant velocity.

For production and separation of radioactive nuclei, heavy ions of energy higher than a few hundred MeV/nucleon give the best results because the product separation can be made with high collection efficiency. The separated secondary beams, at that energy, have a typical momentum broadening of a few percent, depending on the separator used. For an experiment which does not require good energy resolution, such separated nuclei can be used directly, to induce further reactions. These methods will be discussed in Sections 2.3 and 2.4.

In an experiment which requires better momentum resolution, the separated beam may be injected into a storage-cooler ring in which the radioactive beam is cooled down to a relative momentum broadening ($\Delta P/P$) of less than $10^{-3}$. The cooled beam may be extracted for use in an experiment or may be used with an internal target. The cooler ring can also be used to decelerate or to accelerate the radioactive beam for experiments of various energies. An example of such a storage-cooler ring will be discussed in Section 2.5.

A storage ring can also be used for production of secondary nuclei using a thin internal target. In an external target system, the most efficient target thickness for production of a secondary radioactive beam is not determined by the thickness corresponding to the maximum production yield (see Section 2.4). The most efficient thickness of target is usually much thinner than the reaction mean-free-path of the primary beam. Therefore, when the external beam is injected onto such a target, a considerable fraction of the primary nuclei are transmitted without inducing any nuclear reaction. On the other hand, in an internal-target application, noninteracting nuclei are recirculated and used repeatedly until they fragment, or until they produce the desired product. The number of product nuclei ($I_p$) is then given in terms of the number of initial circulating nuclei ($I_i$) by

$$I_p = I_i \sigma_p / \sigma_r \tag{7}$$

where $\sigma_p$ is the production cross-section for the desired nuclide and $\sigma_r$ is the total reaction cross-section. The nuclei produced are not attenuated by a thin target and, therefore, use of an internal target in a recirculation mode provides the most efficient use of the primary beam for production of radioactive nuclei.

In the following sections methods of producing radioactive beams will be discussed in greater detail with reference to specific examples.

## 2.1. General Considerations

In all methods of producing radioactive beams, the steps listed below are necessary before a radioactive beam can be delivered:

1. Production of the radioactive nuclei.
2. Extraction of the ions of the desired nucleus.
3. Selection of the nuclide for experimental use.
4. Acceleration or deceleration of the beam of these nuclides to a suitable energy for the experiment.

Of course, the required intensity and quality of the beam depend on the experiment. For a measurement of the interaction cross-sections at 1 GeV/nucleon (see Section 3.2.2), for example, a beam intensity of about $10^3$/s was sufficient. There, an admixture of other isotopes of similar intensity in the beam was tolerable because each incident nucleus was identified and reactions were observed for each incident nucleus. On the other hand, measurement of a (p, $\gamma$) reaction cross-section (see Section 3.2.6) requires a low-energy beam (less than a few MeV/nucleon) with an intensity as high as $10^8$/s and high purity.

The intensity of the radioactive beam ($I_r$) depends on many factors and can be written

$$I_r = \varepsilon \cdot N_t \cdot \sigma_p \cdot I_1 \tag{8}$$

where $I_1$ is the intensity of the primary beam, $\sigma_p$ is the production cross-section for the desired nuclide, $N_t$ is the number of target nuclei per unit area, and $\varepsilon$ is the extraction efficiency for the desired radioactive nuclei. Here $\varepsilon$ includes the extraction efficiency of the ions from a target, the transmission efficiency of the nuclear separation, the capture and transmission efficiency for acceleration, and the survival ratio of the nucleus from its decay before it reaches its ultimate target. These factors depend strongly on details of the method used and will therefore be discussed in the following subsections, each devoted to one of the methods.

Next we discuss the available intensities of primary beams and the production cross-sections for radioactive nuclei as the basis for an overall comparison of the methods discussed in the following sections. For comparison, the following classification of the methods is applied:

1. The use of high-energy proton–nucleus reactions followed by an ISOL and a heavy-ion accelerator.
2. The use of low-energy ($\leqslant$20 MeV/nucleon) heavy-ion reactions followed by a recoil mass separator.
3. The use of medium-energy (30–150 MeV/nucleon) heavy-ion reactions followed by a separator.
4. The use of high-energy ($\geqslant$200 MeV/nucleon) heavy-ion reactions followed by a separator. (Use of a storage ring for production of radioactive isotopes followed by a separator is included here as well.)

Final beam quality depends strongly on the separation methods and details of these methods are discussed in each of the following sections.

### *2.1.1. Primary-Beam Intensities*

Intensities of primary beams depend on the type of accelerator in use. In general, a proton accelerator provides the highest-intensity beam ($10^{13}$/s) up to high energies ($\geqslant$500 MeV). Even higher intensity (up to $10^{15}$/s) is expected in the TRIUMF proposal (Le 85). As an example of current capability, ISOLDE at CERN use the 600 MeV proton beam accelerated by a synchro-cyclotron (SC) with $10^{13}$/s intensity. It would be possible to obtain an even higher intensity beam from the SC, however the usable intensity at ISOLDE is limited by current radiation-control regulations.

Low-energy heavy-ion beams are provided by tandem Van de Graaffs or heavy-ion LINACS and cyclotrons, and their intensity can be as high as $10^{15}$/s for elements such as Ne or Ar. The intensity for other elements is generally lower but $10^{13}$/s is attainable for ions of many kinds of elements.

At intermediate energy ($\leqslant$150 MeV/nucleon), cyclotrons can also be used for the acceleration of heavy ions. At present in cyclotrons, the intensities of heavy-ion beams are about $10^{13}$/s. Equally high intensities can also be obtained with high duty factor (100%), while this duty factor is usually much lower for higher-energy accelerators such as synchrotrons.

High-energy heavy ions (above a few hundred MeV/nucleon) are currently accelerated by synchrotrons such as the Berkeley BEVALAC and the Synchrophasatron at Dubna. With those older generation machines, a heavy-ion intensity of about $10^9$/s for light ions such as Ne and $10^7$/s, and less for heavier ions, is standard. However, new-generation synchrotrons which are under construction (Ki 85a, Ki 85b), or at the proposal stage (Pr 86), are expected to give a $4 \times 10^{12}$/s intensity for light elements and $3 \times 10^{11}$/s for uranium.

When applying a storage ring for secondary nuclei, the number of accumulated nuclei is essentially the same as the number of nuclei which can be accelerated by a synchrotron ($\approx 10^{11}$–$10^{12}$/s). However, the effective beam intensity on a target has to be multiplied by the revolution frequency of the beam (usually about 1 MHz). The *effective* beam intensity of the accumulated ions is then $10^{17}$–$10^{18}$/s. This, in turn, leads to a very high production rate even when using an extremely thin target as required for circulating-beam operation.

### 2.1.2. Production Cross-Sections for Radioactive Nuclei

Production cross-sections for radioactive nuclei at low energy depend strongly on the reaction used; at higher energy the production cross-sections show a systematic trend. Production cross-sections for radioactive nuclei, using a high-energy proton beam, are discussed in detail by Ravn (Ra 79) and by Talbert *et al* (Ta 84, Ta 84a). The primary reactions used are spallation and induced fission. Figure 5a shows the

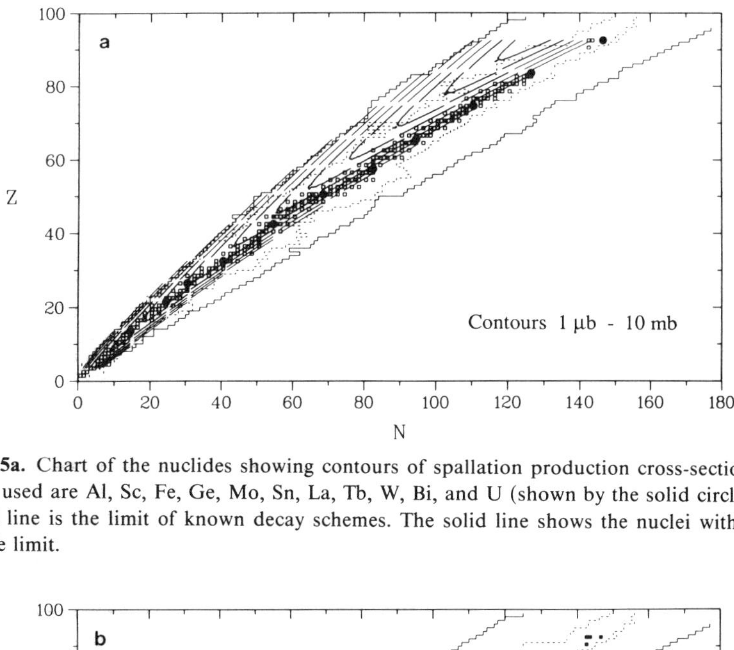


**Figure 5a.** Chart of the nuclides showing contours of spallation production cross-sections. The targets used are Al, Sc, Fe, Ge, Mo, Sn, La, Tb, W, Bi, and U (shown by the solid circles). The dashed line is the limit of known decay schemes. The solid line shows the nuclei with 300 ms half-life limit.

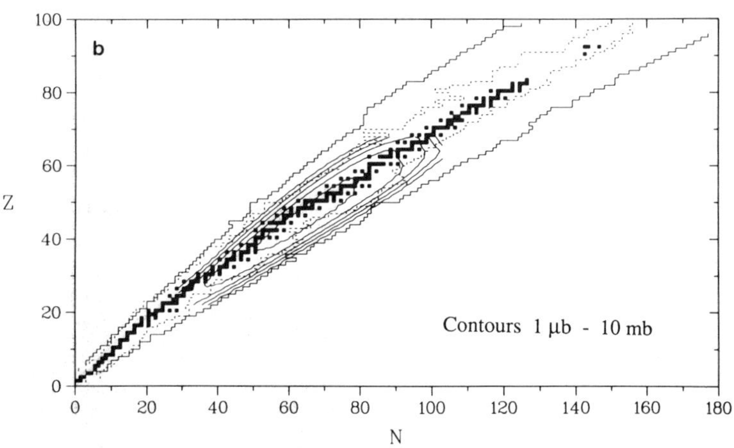


**Figure 5b.** Chart of the nuclides showing contours of $^{238}U(p, f)$ production cross-sections (Ta 84).

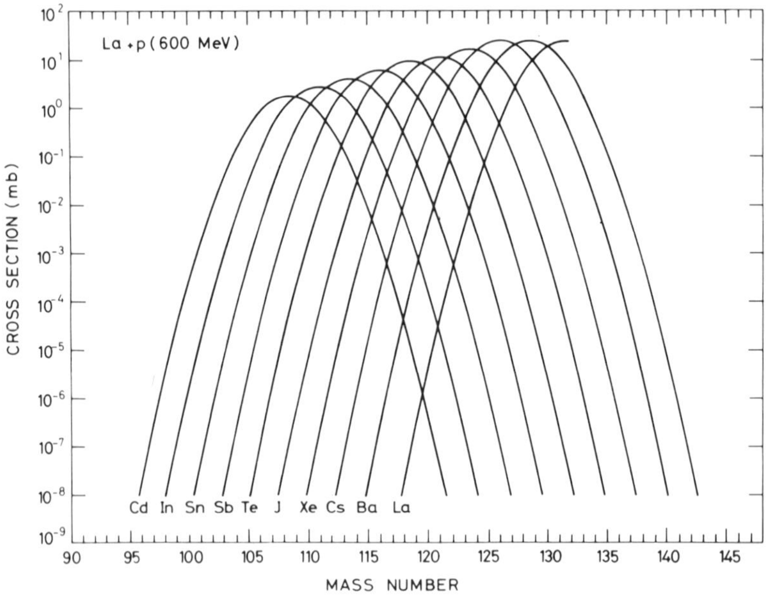


**Figure 6.** Isotopic yield distribution for the spallation of La by 600 MeV protons, calculated (Ra 79) according to the semiempirical formula of Rudstamm (Ru 66).

estimated production cross-sections for nuclei using spallation reactions (Ta 84). As is well known, primarily proton-rich nuclei are produced in these reactions. As shown in this figure, nuclei with half-life longer than 300 ms are produced with cross-sections larger than 1 $\mu$b. The production cross-sections using induced fission, e.g., $^{238}$U(p, f), are shown in Figure 5b—neutron-rich nuclei are produced in this reaction. Again, 1 $\mu$b cross-sections reach the 300 ms half-life limit. With both reactions, the production cross-sections are about 10 mb in the most favorable cases. Figure 6 shows the isotopic dependence of the production cross-sections in the 600 MeV La + p reaction (Ra 79) calculated using a semiempirical formula of Rudstam (Ru 66). It should be noted, for example, that Sn isotopes having about ten neutrons less than the stable isotopes can be produced at about 10 pb, which still provides a yield of these nuclei of about 10/s with a $10^{13}$/s proton beam on a $6 \times 10^{23}$ atoms/cm$^2$ thick target.

Production cross-sections for neutron-rich nuclei in reactions induced by low-energy heavy-ion beams were studied by Breuer *et al.* (Br 80, Br 83, Br 83a). Their data was taken on the production cross-sections for the neutron-rich isotopes using a 8.3 MeV/nucleon $^{56}$Fe + $^{238}$U reaction. Nuclei with $A/Z$ ratios similar to that of the nucleus formed by a compound of the projectile and the target are produced with a few hundred $\mu$b. Figure 7 shows the energy-loss ($E_L$) dependence of the cross-section. Although the cross-sections for production of nuclei far from stability increase with $E_L$, they are still small compared to those obtained in higher-energy reactions. Therefore, application of low-energy heavy-ion beams for production of secondary radioactive beams will be restricted to production of low-energy beams of nuclei relatively near to stability.

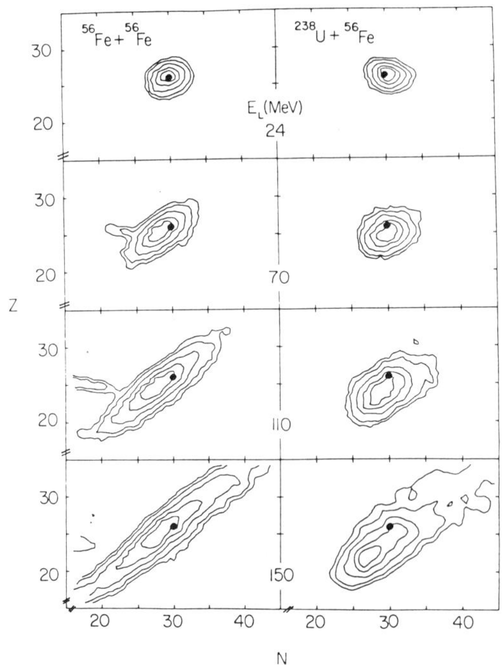


**Figure 7.** Contour lines of measured cross-sections in the $N$–$Z$ plane obtained in the $^{56}Fe + ^{56}Fe$ reaction at 8° and the $^{238}U + ^{56}Fe$ reaction at 45° for energy losses $E_L$ of $24 \pm 2$, $70 \pm 5$, $110 \pm 5$, and 150 MeV. The contour lines represent 75%, 50%, 25%, 10%, and 5% of the maximum cross-section of each distribution. Filled circle indicates the location of the projectile $^{56}Fe$. The low-$N$ tail near $Z = 26$ in the Fe + Fe reaction is due to elastically scattered projectiles which have been subsequently rescattered from the experimental structure (Br 83).

In intermediate-energy or high-energy heavy-ion collisions, target fragmentation leads to products similar to spallation and it can, in principle, be used as a source of radioactive nuclei. Because the primary-beam intensity is comparable to (in the most favorable cases), or weaker than, that of protons and because the usable targets are thinner, the use of target fragmentation for production of radioactive beams is of no advantage.

On the other hand, projectile fragmentation in heavy-ion collisions provides an excellent opportunity to produce secondary radioactive nuclear beams because the extraction of the product nuclei is extremely easy. The production cross-sections for radioactive nuclei by projectile fragmentation have been studied by many groups (Go 78, Vi 79, We 79, St 81, Mu 84, La 85). It has been found that exotic isotopes, neutron rich as well as proton rich, are produced with large cross-sections. Figure 8 shows the

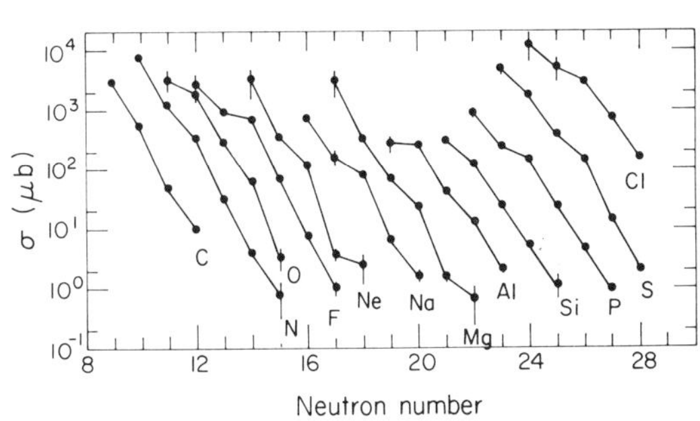


**Figure 8.** Production cross-sections for elements observed in the fragmentation of 212 MeV/nucleon $^{48}Ca$ by a beryllium target. The lines are to guide the eye (We 79).

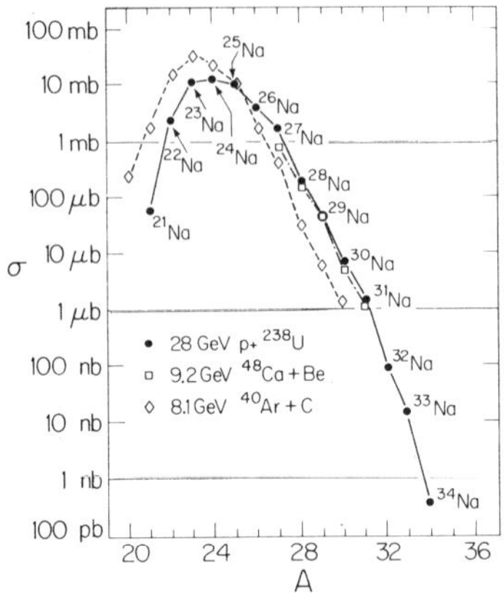


**Figure 9.** Measured production cross-sections (We 79) for Na isotopes from the reactions 9.2 GeV $^{48}Ca + {}^{9}Be$ (We 79); 8.1 GeV $^{40}Ar + C$ (Sy 79); and 28 GeV p + U(De 79).

cross-sections for projectile fragmentation of 212 MeV/nucleon $^{48}Ca$ (We 79). Many neutron-rich nuclides are produced with cross-sections larger than 1 $\mu$b. Comparison of the production cross-sections for sodium isotopes with different reactions is presented in Figure 9. From this comparison of the cross-sections with $^{40}Ar$ and $^{48}Ca$ beams, it follows that more neutron-rich isotopes are produced from a neutron-rich primary. It also follows that the production cross-sections of neutron-rich isotopes from a $^{48}Ca$ beam are almost identical to those obtained in reactions induced by high-energy protons. Proton-rich isotopes can also be produced by projectile fragmentation of proton-rich primaries. The production cross-sections for heavy nuclei ($A \geq 100$) have not yet been studied in projectile fragmentation; it is therefore not known whether proton-rich or neutron-rich nuclei are produced with larger cross-sections.

However, it is known that projectile fragmentation cross-sections are almost independent of the incident-beam energy. Figure 10 shows production cross-sections for light nuclei by a 44 MeV/nucleon $^{40}Ar$ beam; the same cross-section behavior as at high energy is seen. Figure 11 shows the results of a semiempirical calculation of the fragmentation cross-sections. It is evident that the isotope distribution is strongly correlated to the $N/Z$ ratio of the incident projectile, at this energy as well.

Table 1 is a summary of known production rates for radioactive nuclei. The thickness of the target depends on details of the separation methods and therefore only typical target thicknesses are listed. Further, only typical values of the extraction efficiencies $\varepsilon$ are shown because they also depend on details of the method in use. These details will be discussed in the following subsections.

## 2.2. Use of High-Energy Proton Reactions

It is well known that many unstable isotopes are produced through spallation or induced fission in high-energy proton–nucleus collisions. Many ISOL facilities use these reactions with proton energies of several hundred MeV. The advantages of using a proton beam for the production of unstable nuclei are as follows.

1. High-intensity beams of up to 100 $\mu$A are available. These are several orders of magnitude greater than the beam intensities available for heavy ions.

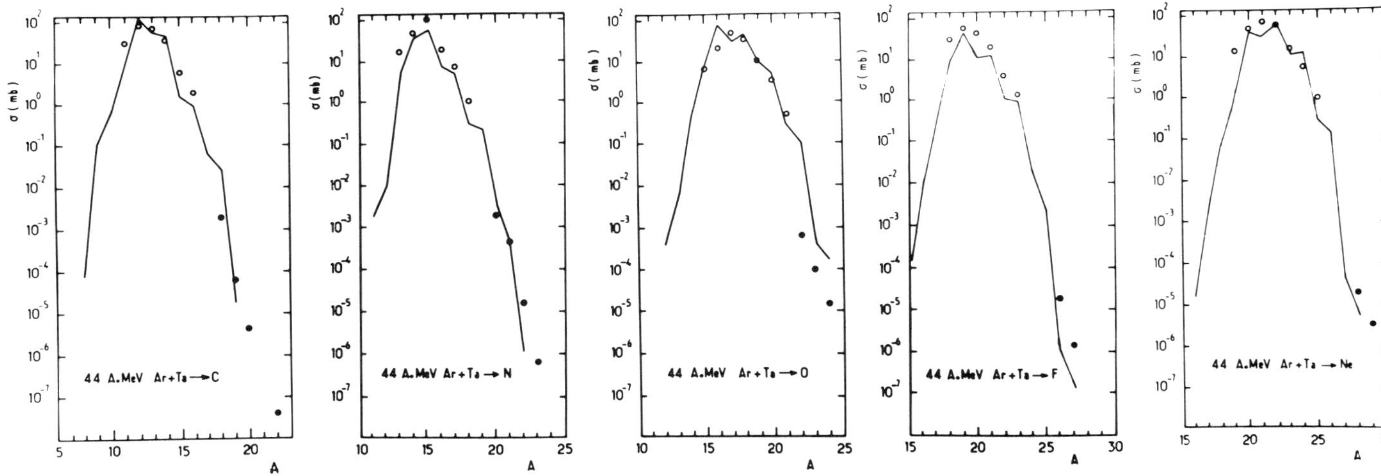


**Figure 10.** Calculated final isotopic distributions for the reaction 44 MeV/nucleon $^{40}$Ar + Ta (solid lines). Comparison is made with results of Guerreau *et al.* (Gu 83) (open circles) and of Langevin *et al.* (La 85) (solid circles) (Gu 86).

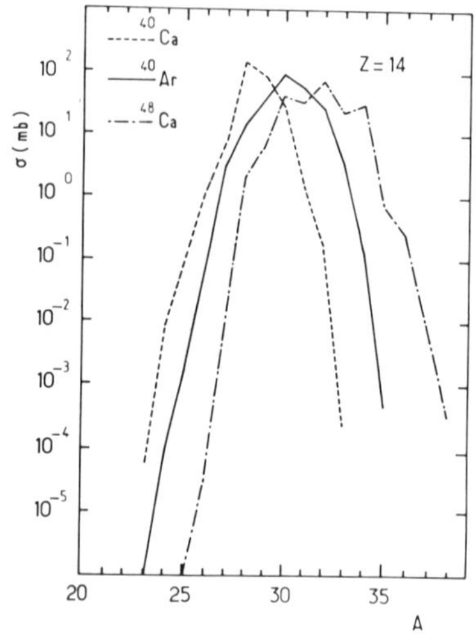


**Figure 11.** Calculated final distributions for $Z = 14$ isotopes produced with different projectiles. The influence of the $N/Z$ ratio of the projectile appears very clearly ($E_{inc}/A = 44$ MeV/nucleon, Ta target) (Gu 86).

2. Thick targets can be used because the range of protons is much longer than that of heavy ions of the same energy.

Because of these characteristics, one usually obtains the largest production yield of exotic nuclei by using high-energy proton reactions. If one wishes to use these product nuclei as secondary beams, however, several difficulties arise. First, because the recoil energies of the products are very low, extraction of exotic isotopes is not a trivial process but requires very advanced technology. In fact, this is the main area of development at the ISOL facilities. Attempted improvements at ISOL have been reported in many places (Ra 79), but because it is not the purpose of this chapter to

**Table 1. Comparison of Radioactive-Beam Production Methods**

| Method | 1 | 2 | 3 | 4 | 5[a] |
|---|---|---|---|---|---|
| Primary beam | proton | heavy ion | heavy ion | heavy ion | heavy ion |
| Energy (MeV/nucleon) | ⩾500 | ⩽20 | 30–150 | ⩾200 | >200 |
| Intensity (present) (particles/s) | $10^{13}$ | $10^{13}$ | $10^{13}$–$10^{15}$ | $10^{9}$ | $10^{7}$–$10^{9}$ |
| Intensity (future) (particles/s) | $10^{15}$ | $10^{15}$ | $10^{14}$ | $10^{11}$–$10^{12}$ | $10^{11}$–$10^{12}$ |
| Target thickness | $6 \times 10^{23}$ | $5 \times 10^{21}$ | $2 \times 10^{22}$ | $3 \times 10^{23}$ | $10^{16}$ |
| Production rate for 1 mb cross-section | $6 \times 10^{11}$ | $5 \times 10^{9}$ | $2 \times 10^{9}$ | $3 \times 10^{7}$–$10^{9}$ | $10^{8}$–$4 \times 10^{9}$ |
| Extraction efficiency (typical) | $10^{-4}$ | 0.1 | 0.1 | 0.5 | 0.5 |

[a] Use of a storage ring for production of radioactive nuclei followed by an isotope separator.

review ISOL facilities, as such, no detailed discussion will be given here. This method is of limited applicability because the extraction efficiency depends strongly on the element and it is often very difficult to extract the nuclei of interest. It is usually easy to extract alkaline elements with high efficiency, but in the case of a gas such as nitrogen the efficiency is typically much lower. As an example, the light-ion beams available at ISOLDE (at CERN) are listed in Table 2 (Ra 85). As can be seen, the source yield is strongly dependent on the element. Isotopes of K, Na, and Li are extracted very efficiently; in contrast, isotopes of C, N, and Mg give extremely small source yield.

The second difficulty in the extraction of a secondary beam arises in the case of short-lived nuclei because the time for extraction of the isotopes is generally longer than 10 ms. The third difficulty is that after extraction of a nucleus at an ISOL, the energy of the extracted beam is still too low (<100 keV) for direct use in the study of nuclear reactions. Therefore reacceleration of these secondary nuclei is required.

Even given the restrictions noted above, reaccelerated secondary beams of radioactive isotopes are still, in many cases, useful for research. Nuclear astrophysics has been one of the most exciting applications of this method. In fact, a facility for accelerated radioactive beams has been proposed at TRIUMF, the meson facility in Vancouver, Canada. The TRIUMF group propose to construct an ISOL at its cyclotron facility and to reaccelerate the secondary radioactive nuclei. Considerations in building such a facility were discussed at an accelerated radioactive-beam workshop in 1985 and summarized in its proceedings (Pr 85). The following are a few brief remarks on the project.

**Table 2. Light-Ion Beams at ISOLDE (Ra 85)**

| Ion beam | Intensity (atoms/s · $\mu A$) | Target thickness ($g/cm^2$) | Source yield (%) |
|---|---|---|---|
| $^{6}He$ | $7.3 \times 10^5$ | 122 | 0.5 |
| $^{8}He$ | $8.4 \times 10^6$ | 13 | 0.5 |
| $^{8}Li$ | $3.9 \times 10^8$ | 122 | 50 |
| $^{9}Li$ | $3.6 \times 10^7$ | 122 | 50 |
| $^{11}Be$ | $1.8 \times 10^4$ | 122 | 0.1 |
| $^{10}C$ | $5.4 \times 10^5$ | 2.5 | 0.23 |
| $^{11}C$ | $1.3 \times 10^7$ | 2.5 | 0.23 |
| $^{13}N$ | $4.1 \times 10^5$ | 2.5 | 0.01 |
| $^{16}N$ | $1.4 \times 10^4$ | 2.5 | 0.01 |
| $^{19}Ne$ | $3.0 \times 10^7$ | 3 | 0.7 |
| $^{24}Ne$ | $2.0 \times 10^5$ | 122 | 1 |
| $^{26}Na$ | $4.0 \times 10^6$ | 122 | 50 |
| $^{27}Mg$ | $8.5 \times 10^5$ | 122 | 0.1 |
| $^{23}Mg$ | $7.0 \times 10^4$ | 3.2 | 0.1 |
| $^{29}Al$ | $2.1 \times 10^6$ | 13 | 0.1 |
| $^{38}Cl$ | $6.0 \times 10^6$ | 41 | 1 |
| $^{35}Ar$ | $8.8 \times 10^6$ | 6.5 | 5 |
| $^{37}Ar$ | $4.5 \times 10^8$ | 38 | 5 |
| $^{38}K$ | $7.5 \times 10^7$ | 3.2 | 100 |
| $^{42}K$ | $1.4 \times 10^9$ | 3.2 | 100 |
| $^{43}K$ | $5.0 \times 10^7$ | 122 | 100 |

Special requirements exist for acceleration of ions delivered from an ISOL. They are the following:

1. Generally, singly charged ions are most efficiently delivered from an ISOL. Therefore the reaccelerator should be designed to accept singly charged ions at injection.
2. To keep the ISOL ion source reasonably stable and reliable, the energy of the ions exiting from it is kept to less than 100 keV.
3. Secondary radioactive beams are not expected to be of high intensity. Therefore it is essential for the reaccelerator to have a high duty factor as well as high transmission.

In the light of these requirements several types of accelerators were examined. Here we summarize some of the most useful considerations in selecting the reaccelerator.

### 2.2.1. *Tandem Van de Graaffs*

The principal advantage of DC accelerators such as Van de Graaff and Tandem Van de Graaff units is that they have a 100% duty cycle and accept ions of any charge-to-mass ratio. However an ordinary Van de Graaff requires that an ISOL or secondary target be at the high-voltage terminal at voltages of up to 20 MV. This kind of configuration would not be impossible to construct but the size of the facility and the difficulty of maintenance would not be desirable features. A Tandem Van de Graaff does not have this problem, because it has both its injection and extraction sides at ground potential. A terminal voltage of about 8 MV would be enough to obtain the required secondary beam energy: $A = 60$ with 1.5 MeV/nucleon would satisfy the conditions required for the study of nuclear reactions of astrophysical interest. An unfortunate aspect of the Tandem Van de Graaff is that it requires negative ions for injection. A charge-exchange system is therefore needed for ions of most elements. A typical charge-exchange efficiency would be expected to be around $10^{-2}$ for most heavier ions and, in some cases, even less than that. Therefore the intensity of secondary beams accelerated in a tandem machine would be strongly limited. The major advantage of the DC machines and one not immediately applicable in our field is their unmatched precision in energy and in beam quality.

### 2.2.2. *Cyclotrons*

An advantage of the cyclotron is that injection can be done continuously and a 100% duty factor can be obtained. However, the main disadvantage of a cyclotron for use with heavy ions arises because of the fact that its diameter scales with $A/q$ (mass to ion-charge ratio) for a given energy per nucleon. Since we would expect to have $q = 1^+$ from the source, it therefore scales with $A$. A cyclotron which accelerates ions of $A = 60$ to 1.5 MeV/nucleon would have a 20 m diameter, which is larger than the meson factory cyclotron at TRIUMF. Even if a superconducting magnet were used, we would still require a diameter larger that that of the existing modern cyclotron at the National Cyclotron Laboratory at Michigan State University! A more reasonable solution would involve a small preaccelerator to get the radioactive ions stripped to higher charge states, followed by a considerably smaller cyclotron. The circular-machine options are reviewed, in detail, by Mackenzie (Ma 85a) to whom interested readers are referred.

### *2.2.3. LINACs*

LINACs are often considered to be the best accelerators for low-energy heavy ions and have been used in many laboratories, e.g., the HILAC used at the Lawrence Berkeley Laboratory and the UNILAC used at GSI. It allows ground potential for both injection and extraction, and the extraction of beams from such accelerators is very easy. Because a LINAC is essentially a constant-velocity machine, it can deliver a certain energy per nucleon up to the heaviest nuclei that can be accepted for acceleration. Charge strippers can easily be accommodated between sections of the LINAC so that the charge state of the accelerated beam can be raised to enhance the acceleration efficiency. However, a LINAC is generally very power consuming and thus requires a pulsed mode of operation with a duty factor of not more than 10–25%. For the present application, it would be desirable to have a LINAC duty cycle of 100%, which is certainly possible. In fact, as an example, the heavy-ion LINAC (RILAC) at RIKEN in Japan is now operating at 100% duty factor (Od 84, Mi 86a) and with variable frequency and is thus providing variable-energy (2–4 MeV/nucleon) heavy ions up to Au.

An RFQ (radio-frequency quadrupole) type of LINAC appears to have desired injection properties (Mu 85). It uses an RF electric field for simultaneous acceleration and focusing of the beam; it also provides self-bunching of the injected DC beam so that the acceptance for the secondary beam would be as large as that of a DC accelerator. An RFQ LINAC has been used as an injector for the BEVALAC at Berkeley and it is also planned for it to be used as an injector of heavy ions at both CERN and GSL (Mu 85).

An RFQ LINAC has also been proposed for the TRIUMF low-energy radioactive beam facility. In the proposal beams of radioactive isotopic ions of up to $10^{12}$ atoms/s are anticipated. Specifications for the post-accelerator are listed in Table 3 (Bu 85a). The estimated beam intensities for selected nuclei are listed in Table 4.

**Table 3. Specifications of the TRIUMF–ISOL Post-Accelerator (Bu 85a)**

| Parameters | Range | Comment |
|---|---|---|
| Ion | $A \leqslant 60$ | singly charged ions (+or − ) delivered by ISOL |
| Input energy | $E_{\text{total}} = 60$ keV | maximum ISOL voltage, may be raised by biasing |
| | $(E/A \leqslant 1.5$ keV/nucleon) | the first section of the accelerator |
| Output energy | $E_{\text{total}} = 90$ MeV | continuously variable range of |
| | $(E/A \leqslant 1.5$ MeV/nucleon) | $0.2 \leqslant E/A \leqslant 1.5$ MeV/nucleon |
| Input energy spread | $\Delta E/E \leqslant 10^{-4}$ | delivered by ISOL |
| Output energy spread | $\Delta E/E \leqslant 10^{-3}$ | debunching after the accelerator |
| Duty factor | 100% | due to low currents delivered by ISOL |
| Transmission | ≈50% | |
| Beam current | $\leqslant 10^{12}$/s | low ISOL currents |

**Table 4. Estimated Radioactive Nuclear-Beam Intensities from the TRIUMF Proposal (Bu 85a)[a]**

| Beam | Estimated intensity (atoms/s) |
|---|---|
| $^{13}$N | $2 \times 10^{11}$ |
| $^{15}$O | $1 \times 10^{12}$ |
| $^{18}$F | $5 \times 10^{11}$ |
| $^{19}$Ne | $2 \times 10^{11}$ |
| $^{21}$Na | $9 \times 10^{11}$ |

[a] Estimated losses of the beam are 30% in the target, 50% in the ion source, and 60% in the post-accelerator. A primary proton intensity of 50 $\mu$A was assumed.

## 2.3. Use of Low-Energy Heavy-Ion Reactions

For studying low-energy reactions such as (p, $\gamma$), (p, $\alpha$), and Coulomb excitations (which are of great astrophysical interest; see Section 3.2.6), low-energy heavy-ion beams can be used in reversed geometries. Such developments are under way at the Lawrence Livermore National Laboratory (LLNL), at the Ohio State University, at the Rochester Nuclear Structure Research Laboratory (NSRL), and at Yale's A. W. Wright Nuclear Structure Laboratory. All use Tandem Van de Graaff accelerators for supplying heavy-ion beams. A heavy-ion primary has several advantages in these applications. First, it permits the study of reactions on short-lived radioactive nuclei. Second, it results in a kinematic compression of the reaction products into a forward cone, resulting in a large secondary beam-intensity enhancement. Third, the center-of-mass energies for the reactions of greatest interest lie in the few MeV range.

### *2.3.1. Quadrupole Sextuplet at the LLNL*

At the LLNL a beam of $^7$Be has been produced using the Quadrupole Sextuplet Beam Transport System (QSBTS) (Ha 83, Ha 83a, Ma 84) shown in Figure 12. Primary beams of $^7$Li, accelerated by the LLNL EN Tandem Van de Graaff, were used to

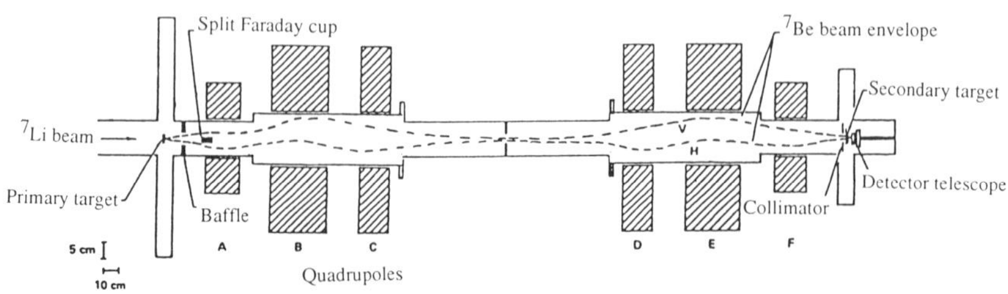


**Figure 12.** Configuration of the Quadrupole Sextuplet Beam Transport System (QSBTS) at the Lawrence Livermore National Laboratory. By providing a focus at the center, the background is significantly reduced [modified from (Ma 84)].

produce $^7$Be through the $^1$H($^7$Li, $^7$Be)n reaction. Because of the motion of the center of mass, the $^7$Be nuclei are all produced in the forward direction. The $^7$Li that does not interact with the primary target stops in a shadow bar, which serves as a split Faraday cup in order to monitor the primary-beam intensity and location.

The secondary particles that are emitted close to an angle of zero degrees but that do not strike the shadow bar are then focused onto a secondary target by the QSBTS which has a large solid angle acceptance. Because this system is chromatic, most of the scattered primary beam that misses the shadow bar is also eliminated. Furthermore, the chromaticity can be useful in defining the energy of the desired secondary-beam particles, and the time-of-flight of the particles from the primary target to the secondary one can be used to improve the purity of the secondary beam. To achieve this, the primary beam was pulsed with good time resolution (a few ns). The flight path was 2.0 m and the trajectories for the desired particles were nearly isochronus. It was therefore easy to discriminate between the foreground secondary-beam particles and other particles that had the same trajectories but different velocities.

The beam of 18.6 MeV $^7$Be was obtained from a 23 MeV $^7$Li primary beam. A particle-identification spectrum of the secondary beam is shown in Figure 13. The

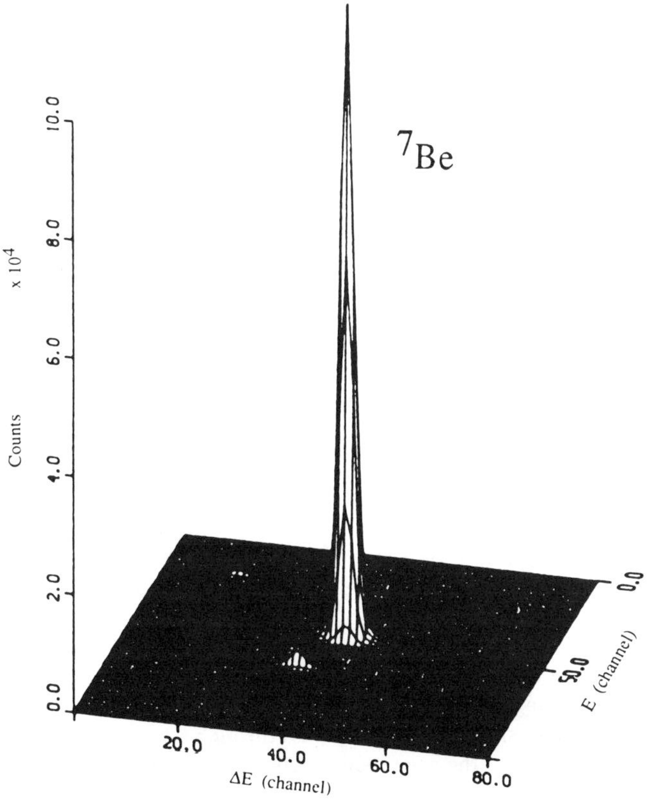


**Figure 13.** A $\Delta E$ *vs.* $E$ spectrum showing the profile of a 14 MeV $^7$Be beam produced by the $^1$H($^7$Li, $^7$Be)n reaction in the QSBTS [modified from (Ma 84)].

**Table 5. Radioactive Beams Studied at LLNL and Their Characteristics**

| Beam | Production reaction | Primary beam energy (MeV) | $E$ (FWHM) (MeV) | Conversion efficiency[a] | (d, n) rate (cts/hr) | (p, $\gamma$) rate (cts/day) |
|---|---|---|---|---|---|---|
| $^{7}$Be | $^{1}$H($^{7}$Li, $^{7}$Be)n | 24 | 20 (0.5) | $3 \times 10^{-6}$ | ≈100 | 0.1 |
| $^{13}$N | $^{2}$H($^{12}$C, $^{13}$N)n | 36 | 34 (1.6) | $1 \times 10^{-7}$ | ≈100 | 6 |
| $^{15}$O | $^{2}$H($^{14}$N, $^{15}$O)n | 36 | 16 (2.1) | $1 \times 10^{-7}$ | — | — |

[a] The ratio between the number of secondary-beam particles to the number of primary-beam particles (Ma 84).

largest background came from the scattered $^{7}$Li and the ratio of $^{7}$Be to $^{7}$Li was 20:1. The energy resolution of the secondary $^{7}$Be beam was about 500 keV or 2.5%.

Beams of $^{13}$N were also obtained using the $^{2}$H($^{12}$C, $^{13}$N)n reaction (Ma 85). These beams were contaminated with scattered $^{12}$C of intensity about equal to that of the $^{13}$N. However, by demanding a time-of-flight in the proper range, a reasonably well isolated beam was obtained. A list of the radioactive beams produced at LLNL is shown in Table 5 with their conversion efficiencies.

Using $^{7}$Be and $^{13}$N beams, the LLNL group is currently measuring (p, $\gamma$) and (d, n) reaction cross-sections, which are important for an astrophysical understanding of nuclear synthesis (Ha 83). Their experiments will be discussed in Section 3.2.6. The important branching ratio in the decay of $^{7}$Be has also been measured using this system (Ma 83).

### 2.3.2. *RISOAR at the Ohio State University*

In contrast to the quadrupole system used by the LLNL group, a dipole magnet is used in the Ohio State University system (Bo 83, Bo 84, Bo 85a). This system, called RISOAR (Radioactive Ion Source for Astrophysical Research), is designed to study both the radiative capture of protons and $\alpha$ particles, and particle-transfer reactions involving low-energy radioactive ions such as $^{7}$Be, $^{11}$C, $^{13}$N, $^{15}$O, and $^{19}$Ne with intensities of $10^{6}$ ions/s and energies of 8 to 16 MeV. The broad features of the facility are shown in Figure 14 in the specific case (for purposes of illustration) of the reaction $^{4}$He($^{15}$O, $\gamma$)$^{19}$Ne. Initially, a beam of $^{16}$O ions from the OSU Van de Graaff accelerator is incident on a $^{3}$He cell, where a part of the incident beam is converted to the desired $^{15}$O ions via the $^{3}$He($^{16}$O, $^{15}$O)$^{4}$He reaction. Because the incident projectile is much heavier than the target, all of the $^{15}$O ions are emitted within a laboratory scattering angle of about 25°. Thus a large number of $^{15}$O ions emerging from the $^{3}$He gas cell (Target I) can be collected and focused, by a large dipole magnet, onto Target II which, in this example, is a $^{4}$He target. Just prior to arriving at Target II, each ion passes through a pair of thin foils. Electrons knocked out of the foils by the $^{15}$O ions are detected, thus providing a start and stop signal for a time-of-flight determination on each ion. Since the ions slow down as they pass through the gas in Target II, they map out a significant fraction of the excitation function for the reaction being studied. The time between the arrival of the $^{15}$O ions at the entrance to Target II and the detection of the reaction $\gamma$ ray thus determines the energy at which the reaction takes place. The intensity of the radioactive beam at Target II is generally around $10^{6}$ ion/s, resulting

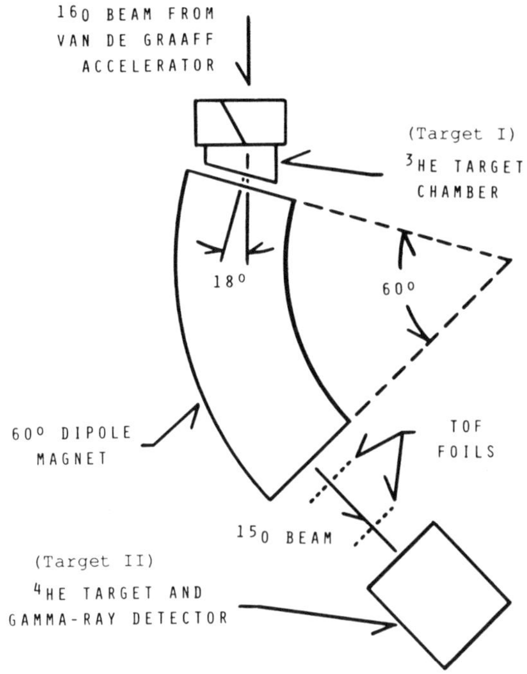


**Figure 14.** Basic components of RISOAR at Ohio State University—the two-stage facility for the study of nuclear reactions on short-lived radioactive nuclei (Bo 85).

in a count rate for a typical radiative-capture reaction study of several hundred counts per day. Optimization of the system has been discussed elsewhere (Bo 85a).

### *2.3.3. University of Rochester Recoil Mass Spectrometer*

At the University of Rochester Nuclear Structure Research Laboratory (NSRL), the production of secondary beams is being planned using the Rochester recoil mass spectrometer (RMS) (Co 81, Co 83). Figure 15 shows the RMS. The magnetic-quadrupole triplet $Q_1$, $Q_2$, and $Q_3$ produces point-to-parallel transport for reaction products. The reaction products and the diverging primary beam then enter a 15° electrostatic deflector, $E_1$. The primary beam strikes the positive plate of $E_1$ while reaction products with energies $E_0 \pm 15\%$ are deflected through a 15° angle. The next active element is a 36.2° dipole magnet which, together with the $E_1$ unit, constitutes an Aston-type mass spectrometer (As 19). The second electrostatic deflector $E_2$ works essentially as a background rejector. Scattered beam particles, originating at the beam stop position on the $E_1$ plate, have trajectories which intersect the positive plate of $E_2$. Therefore, primary-beam particles can reach the focal plane only after scattering at least twice as well as satisfying momentum selection in the dipole. The last quadrupole triplet focuses the selected secondary beam onto a secondary target.

Figure 16 shows the mass spectrum observed for a typical evaporation residue from the $^{35}Cl + ^{58}Ni$ reaction. The mass resolution is about $\Delta m/m = 600$ and was shown to be insensitive to the geometric solid angle. At a larger solid angle (4 msr) the absolute detection efficiency was about 20%. The background, which results from

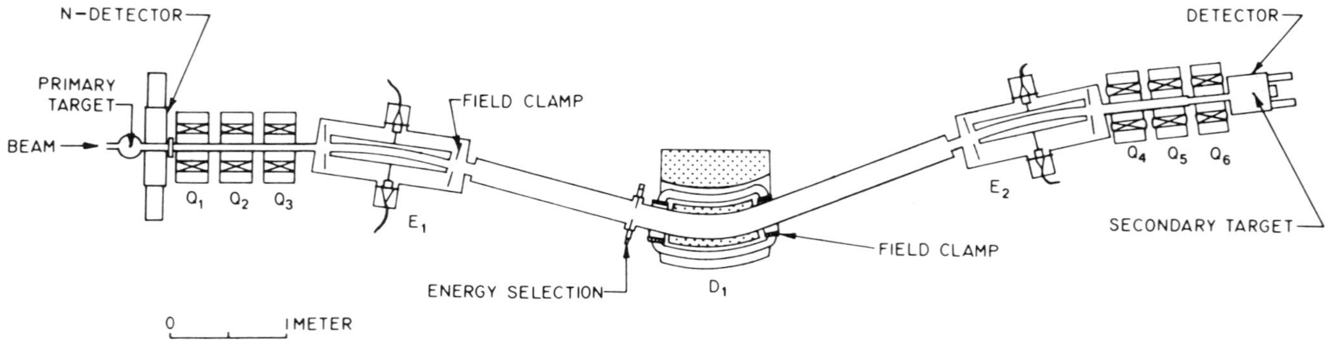


**Figure 15.** Schematic diagram of the NSRL RMS facility showing the neutron-multiplicity detector, the recoil-mass separator, and the focal-plane detector (Pi 85).

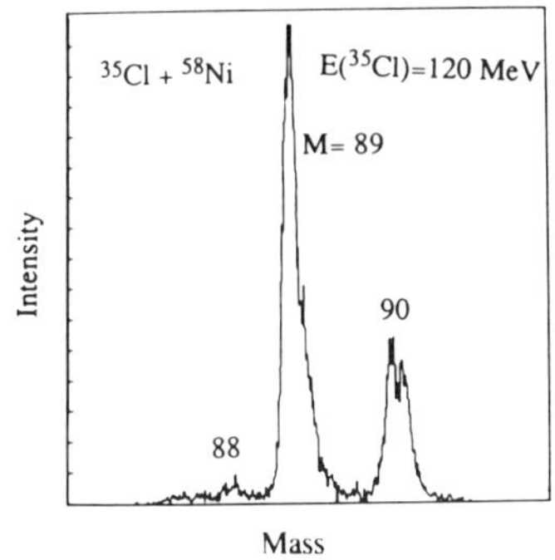


**Figure 16.** A typical evaporation-residue mass spectrum observed in the $^{35}Cl + ^{58}Ni$ reaction. The mass resolution was insensitive to the geometric solid angle. At the larger solid angle (4 msr) the absolute detection efficiency was 20% (Co 83).

internal double scattering and atomic-charge exchange, is irrelevant above a cross-section level of ~0.1 nb for virtually any mass. The flight path in the spectrometer is such that nuclei with lifetimes of the order of microseconds can be easily transported to the focal plane.

Recently, the addition of a PANDA neutron-multiplicity detector (Co 83, Pi 84), located at the primary target, was used to further enhance the selection of neutron-emitting reaction channels. The PANDA is a five-segment liquid-scintillation detector which can detect single neutrons with efficiencies in excess of 50% and can measure the neutron multiplicity while discriminating against gamma-ray events. Preliminary indications are that, for many nuclei, beams can be produced with high purity at intensities of over $10^6$ ions per second.

The studies to which this system has been directed will be described in Section 3.

## 2.4. Use of Intermediate-Energy and High-Energy Heavy-Ion Reactions

As has already been described in this chapter, heavy-ion reactions have several advantages for the production of beams of nuclei far from stability. The most important is that recoil-mass separators can be used to produce secondary beams of radioactive nuclei without need for an ion source. The mass separator is especially effective when it is used for the separation of projectile fragments from intermediate-energy or high-energy heavy-ion reactions. The main reason for this is the fact that the projectile fragments are kinematically emitted into a very narrow forward cone with little momentum spread, thus leading to high selectivity and transmission efficiencies.

In the next section, properties of projectile fragmentation are discussed from the viewpoint of secondary-beam production and then the principles of a nuclear-separation technique are described. Examples of such systems are then presented.

### *2.4.1. Characteristics of Projectile Fragmentation*

From measurements on nucleus–nucleus collisions, it is known that the fragmentation cross-section is almost independent of the incident energy if it is above several hundred MeV/nucleon (Go 78). It has also been found that nuclei far from stability are produced with large cross-sections. Since these production cross-sections were discussed in Section 2.1.2, we shall limit ourselves here to discussion of the momentum distribution of the fragments.

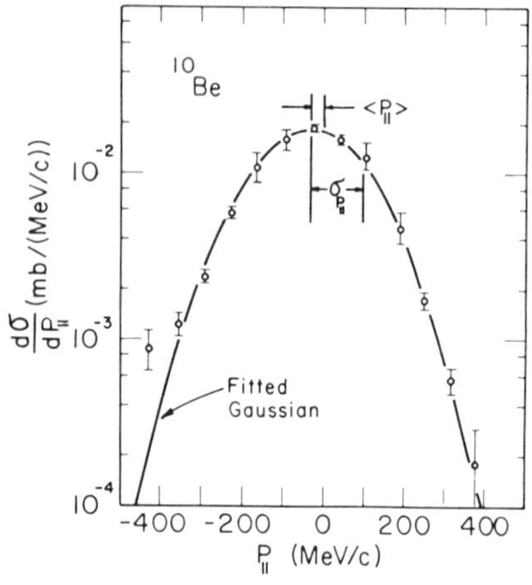


**Figure 17.** The projectile-frame parallel-momentum distribution for $^{10}$Be fragments from $^{12}$C at 2.1 GeV/nucleon on a Be target. The mean momentum $\langle P_{\parallel} \rangle = -30$ MeV/c and standard deviation $\sigma(P_{\parallel}) = 129$ MeV/c are indicated (Gr 75).

Figure 17 shows the momentum distribution, in the projectile rest frame, for $^{10}$Be fragments from $^{12}$C fragmentation at 2.1 GeV/nucleon (He 72, Gr 75). As can be seen, a Gaussian shape provides a good fit to the spectra observed for all isotopes regardless of beam, energy, or target. A spectrum is characterized by its central momentum $\langle P_{\parallel} \rangle$ and standard deviation $\sigma(P_{\parallel})$. The value of $\langle P_{\parallel} \rangle$ was found to be in the range of $-10$ to $-130$ MeV/c for various fragments. (A negative sign for $\langle P_{\parallel} \rangle$ indicates that the fragment speed is less than that of the projectile.) In effect, the fragments have almost the same velocity as the incident projectile nucleus.

The width of the momentum spread $\sigma(P_{\parallel})$, as well as $\langle P_{\parallel} \rangle$, are found to be essentially independent of target mass and beam energy but do depend on the mass number of the projectile ($A_P$) and of the fragment ($A_F$). The dependence of $\sigma(P_{\parallel})$ on $A_P$ and $A_F$ can be expressed as

$$\sigma(P_{\parallel}) = \sigma_0 \sqrt{A_F(A_P - A_F)/(A_P - 1)} \tag{9}$$

where $\sigma_0 \approx 90$ MeV/c (Go 78); $\sigma(P_{\parallel})$ takes its maximum value when $A_F = A_P/2$. The width $\sigma(P_t)$ of the transverse momentum distribution of the fragment is found to be equal to $\sigma(P_{\parallel})$, consistent with isotropic production of fragments in a frame moving at $\beta_{\parallel} \approx -\langle P_{\parallel} \rangle / E$ in the projectile frame. The value of $\sigma(P_{\parallel})$ is known to remain constant down to energies approaching as low as 20 MeV/nucleon (Ge 76). It is found that $\sigma(P_t)$ is almost constant with a value of $\approx$200 MeV/c. Empirically, it was shown that, at projectile energies below 200 MeV/nucleon, the $\sigma(P_t)$ is fitted by (Bi 79)

$$\sigma^2(P_t) = \sigma_0^2 A_F(A_P - A_F)/(A_P - 1) + \sigma_1^2 A_F(A_F - 1)/A_P(A_P - 1) \tag{10}$$

where $\sigma_1 \approx 200$ MeV/c.

In order to minimize broadening of the product momentum distribution it is important to use a primary beam of mass number close to that of the desired nucleus but, especially at high energy, to have both the proton and the neutron numbers of the projectile nucleus larger than those of the desired product nucleus.

Figure 18 shows energy dependences of the relative-momentum width of the parallel momentum $[\sigma(P_{\parallel})/P]$ and the transverse momentum $[\sigma(P_t)/P]$ for several mass regions. The numbers indicated by 20 → 16, for example, identify the fragmentation

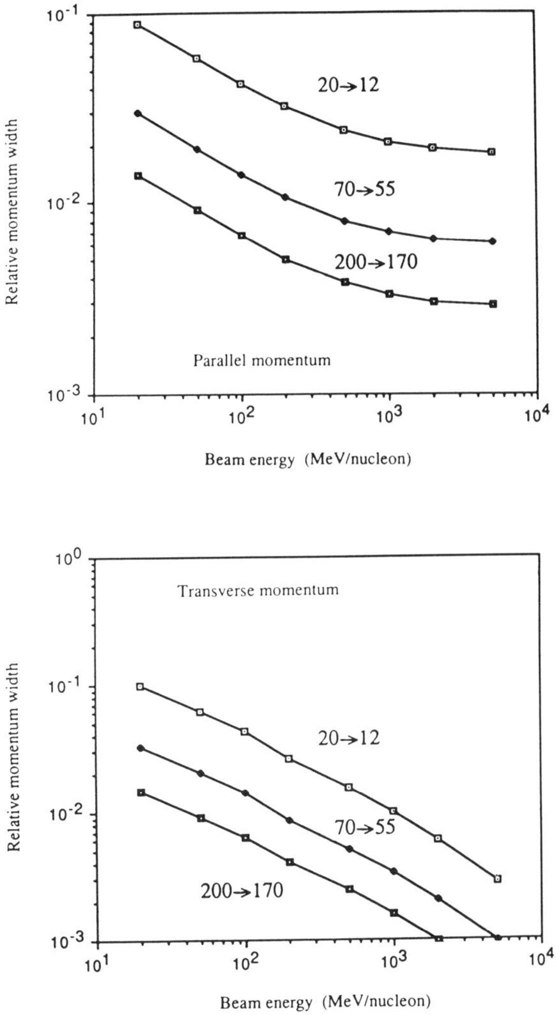


**Figure 18.** Energy dependences of the relative-momentum width for projectile fragments from several different mass combinations of the primary and the fragment.

of a projectile nucleus with mass number 20 into a secondary nucleus with mass number 16. One can see that the momentum spread ranges from 10% to 0.3% depending on beam energy and mass combination. For a projectile energy higher than 200 MeV/c, the momentum spread is less than 3% in all mass ranges.

The angular spread is 6° for the lightest projectile mass at an incident energy of 20 MeV/nucleon, but it converges to less than a few degrees when the energy—or the mass number—becomes large. From these values one can see that high transmission efficiency can be obtained using a recoil-mass separator which has a momentum acceptance larger than a few percent and an angular acceptance of a few degrees.

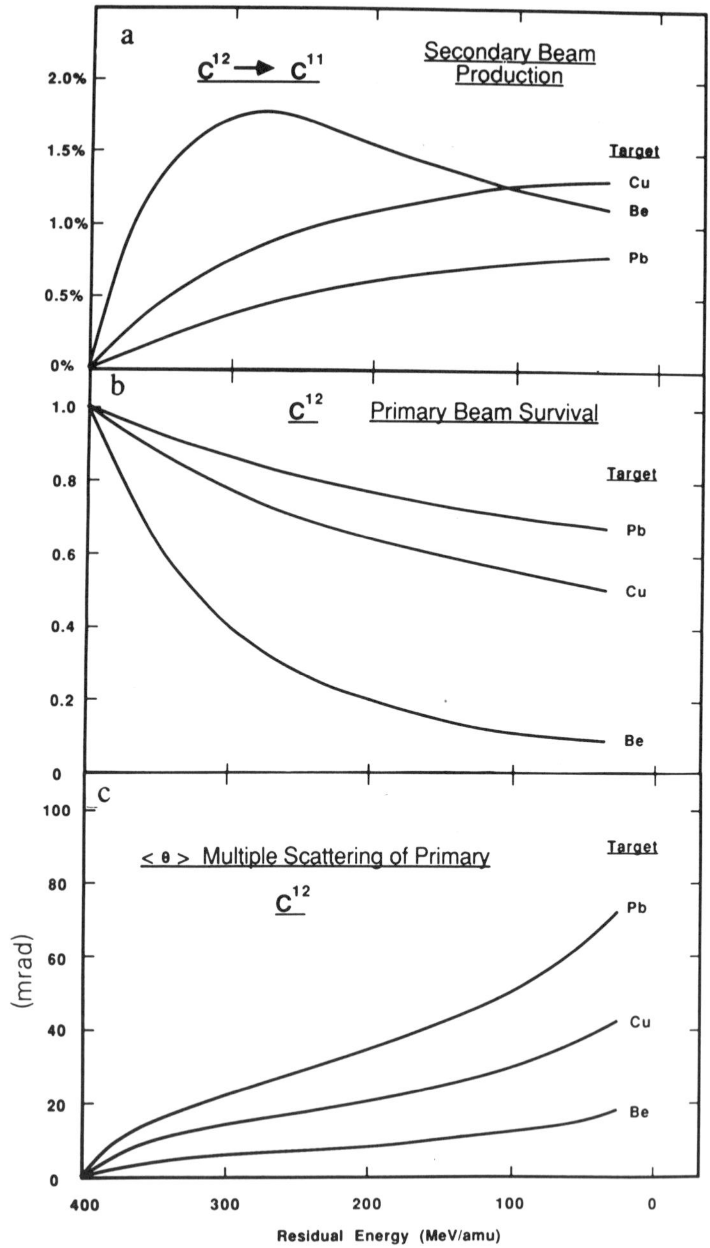


**Figure 19.** (a) Production efficiency (as a percent of the total primary beam) for the single-neutron-loss projectile fragment $^{11}$C for different target materials, plotted against primary-beam residual energy. (b) Survival of the primary $^{12}$C beam vs. energy lost in different targets. Attenuation is calulated (Al 79) from the total nuclear-reaction cross-sections measured by Lindstrom *et al.* (Li 75). (c) Multiple Coulomb scattering of the primary $^{12}$C beam vs. energy lost for different targets. This angle must be taken together with the spreading due to the nuclear reaction in order to calculate the emittance of the secondary $^{11}$C beam (Al 79).

### 2.4.2. Selection of the Target

Usable secondary-beam intensity and momentum spread of the fragments depend, in large measure, on the choice of target material as well as on the properties of the production reaction as discussed in the previous section. Figures 19a, b, and c show results of a study of $^{11}C$ production efficiency, primary-beam survival, and multiple scattering for a primary 400 MeV/nucleon $^{12}C$ beam passing through beryllium, copper, and lead targets (Al 79). The horizontal axis in the figure is the remnant primary-beam energy. Since the $dE/dx$ value is lower for a low-$Z$ target, a given beam-energy loss can also result from passage through a much thicker low-$Z$ target.

The other factor in yield considerations reflect the target mass-number dependence of the fragmentation cross-sections, which are generally proportional to $A^{2/3}$. Because the number of nuclei in a target of given thickness (in g/cm$^2$) is proportional to $1/A$, the production yield is proportional to $A^{-1/3}$ and therefore is larger for targets of the lighter nuclei. The fall-off in the $^{11}C$ yield from the beryllium target below 300 MeV/nucleon, in Figure 19, results from the loss of $^{11}C$ due to nuclear reactions and lower production rates because of a dwindling supply of primary nuclei throughout the target.

From Figure 19 one sees that low-$Z$ materials make the best targets for secondary-beam production via projectile fragmentation reactions. Production is higher, primary beams attenuate more rapidly, and there is less multiple scattering (hence less beam-emittance growth).

Selection of the optimum target thickness is also important for production of secondary beams. The number of nuclei produced, $N_f(x)$, can be calculated using

$$dN_f(x) = (N(x)\sigma_f - N_f(x)\sigma_{0,f})\, dx \tag{11}$$

where $\sigma_f$ is the production cross-section for the fragment f, $\sigma_{0,f}$ is the total fragmentation cross-section for the fragment nucleus itself, and $N(x)$ is the number of primary nuclei at a target depth $x$,

$$N(x) = N(0)\exp(-\sigma_{0,p}x) \tag{12}$$

Here the number of incident nuclei is shown as $N(0)$ and the total fragmentation cross-section of the projectile nucleus is $\sigma_{0,p}$. Figure 20 shows the target-thickness dependence of the production yield. In this calculation the ratio of the production cross-section to the total fragmentation cross-section ($\sigma_f/\sigma_{0,f}$) is assumed to be 0.2. Two curves are shown for the cases $\sigma_{0,f}/\sigma_{0,p} = 0.8$ and 1, respectively, indicating that the maximum yield is obtained when the target thickness is about equal to the reaction mean-free-path ($\lambda = 1/\sigma_{0,p}$) of the incident nuclei.

Because the total fragmentation cross-section is about a barn, the mean-free-path of the incident nuclei is a few tens of g/cm$^2$; the range of these ions, however, is shorter than the mean-free-path for energies below several hundred MeV/nucleon, and therefore one must use a thinner target at those energies.

If all the product nuclei can be used as secondary beams, the thickness of the target is determined by maximization of the product yield. In practice, however, a beam separator has limited acceptance in both momentum and angle. In addition to momentum and angular broadening arising from the production mechanism, energy loss and multiple scattering make the broadening even greater. As illustrated in Figure 21, a difference in energy loss between the projectile and the product nucleus increases

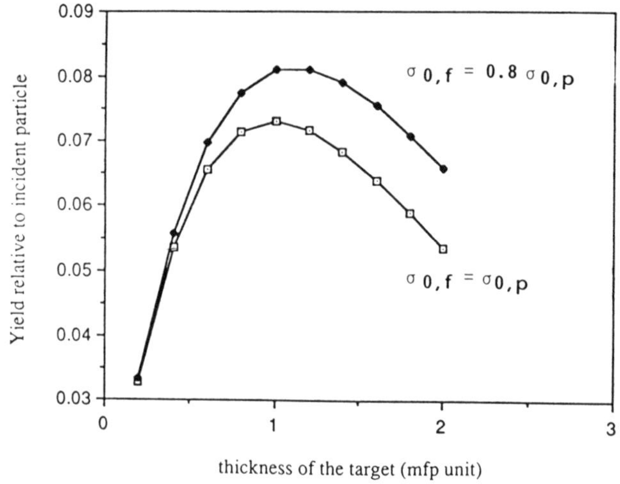


**Figure 20.** Target-thickness dependence of the fragment yield. The maximum yield is obtained for a thickness of about a mean-free-path of the projectile.

the momentum spread of the product nuclei coming out of the target. As shown in the figure, if the reaction occurs at the front surface of the target, the energy loss is determined by $dE/dx$ of the fragment; on the other hand, if the reaction occurs at the back surface, the energy loss is determined by $dE/dx$ of the primary beam. Figure 22 shows a comparison of causes of momentum broadening for the specific case of $^8$He production from a 800 MeV/nucleon $^{12}$C primary. The effect of the energy-loss difference ($\Delta dE/dx$ in the figure) contributes to the momentum spread. On the other hand, multiple Coulomb scattering affects mainly the transverse momentum. Multiple Coulomb scattering sets no limit on the target thickness at this energy.

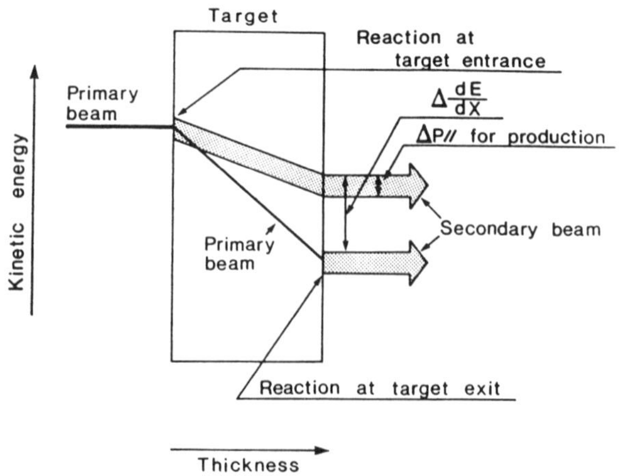


**Figure 21.** A schematic diagram for energy loss in the target.

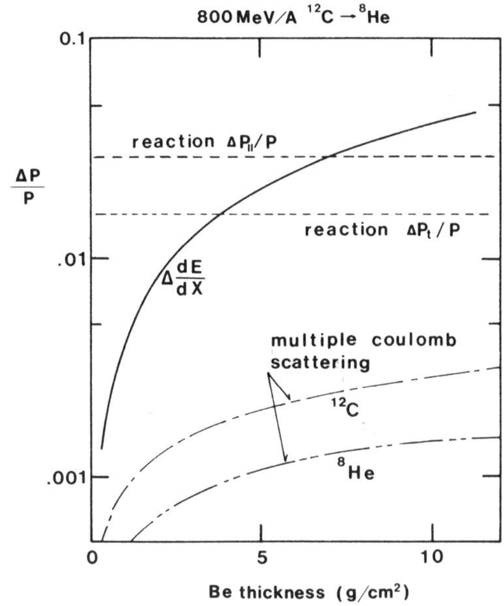


**Figure 22.** The momentum spread of $^{8}$He secondaries. The difference in $dE/dx$ results in additional broadening of $P_{\parallel}$. The multiple Coulomb scattering affects $P_t$ (Ta 85).

The momentum broadening owing to the energy-loss difference becomes comparable to that intrinsic to the reaction when the target is thicker than 5 g/cm$^2$. Unless the momentum acceptance of the separator is much larger than a few percent, a thicker target will not increase the intensity of the secondary beam. In practice, it is very difficult to design a separator with an acceptance larger than a few percent; therefore, the target thickness is ultimately determined by the relation between the momentum broadening intrinsic to the reaction and the energy-loss difference in the target. In general, the thickness of target which equates these effects is the optimum one. This special thickness of target is shown in Figure 23 as a function of the incident-beam energy. The three curves in this figure show secondary beams of $^{16}$C, $^{68}$Ni, and $^{190}$W from the reactions (1) $^{20}$Ne + Be → $^{16}$C, (2) $^{76}$Ge + Be → $^{68}$Ni, and (3) $^{208}$Pb + Be → $^{190}$W, respectively, as arbitrarily chosen examples. The usable target thickness varies with the energy as $E^a$, where $a$ ranges from 1 to 1/2 depending on the specific combination of projectile and product nucleus involved.

The nuclear separation necessary depends on the specific experiment. For example, if one wishes to measure the reaction cross-section at high energy, the admixture of other nuclei in the secondary beam can be effectively rejected by identifying the secondary reaction products. On the other hand, if one wishes to study low-energy reactions having small cross-sections, a single-nuclide beam is essential.

Several recoil-mass separators have already been built and many are currently under construction. We now consider several of these recoil-mass separators that are used with radioactive beams.

### *2.4.3. Secondary-Beam Lines at the Berkeley BEVALAC*

As shown in previous sections, per-nucleon momentum $p$ of fragments is almost constant. In addition, light ions are fully stripped at high energy. Therefore, separation

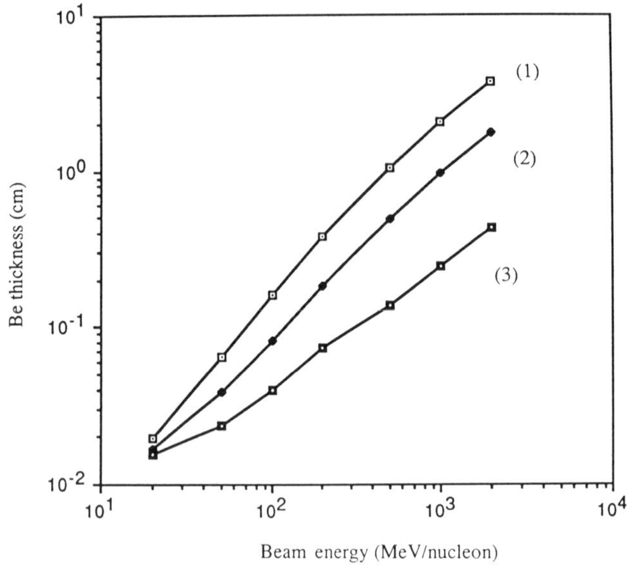


**Figure 23.** The target thickness which gives a momentum broadening as a result of energy loss equal to that intrinsic to the production reaction: (1) $^{20}Ne + Be \rightarrow {}^{16}C$, (2) $^{76}Ge + Be \rightarrow {}^{68}Ni$, and (3) $^{208}Pb + Be \rightarrow {}^{190}W$.

of the fragments in terms of their rigidity $R$ provides a separation in $A/Z$ because

$$R = pA/Z \tag{13}$$

where $p$ is constant.

Several nuclear-physics experiments have already been carried out at Berkeley using radioactive nuclear beams produced through the projectile fragmentation of high-energy heavy ions accelerated by the BEVALAC. Two beam lines on the BEVALAC experimental floor can be used for delivering radioactive secondary beams. One of the beam lines provides the simplest separation—only in terms of $A/Z$. Figure 24 shows such a beam line, which has been used for measurement of interaction cross-sections using radioactive beams of all the p-shell nuclei (Ta 85, Ta 85a, Ta 85b, Ta 85c, Ta 85d). In these experiments each beam nucleus is identified by a set of

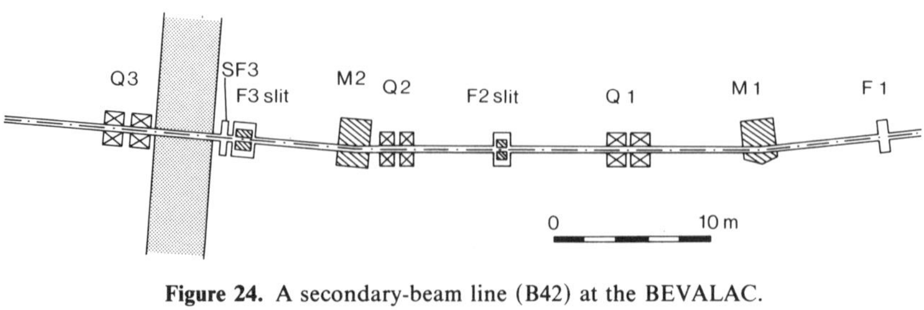


**Figure 24.** A secondary-beam line (B42) at the BEVALAC.

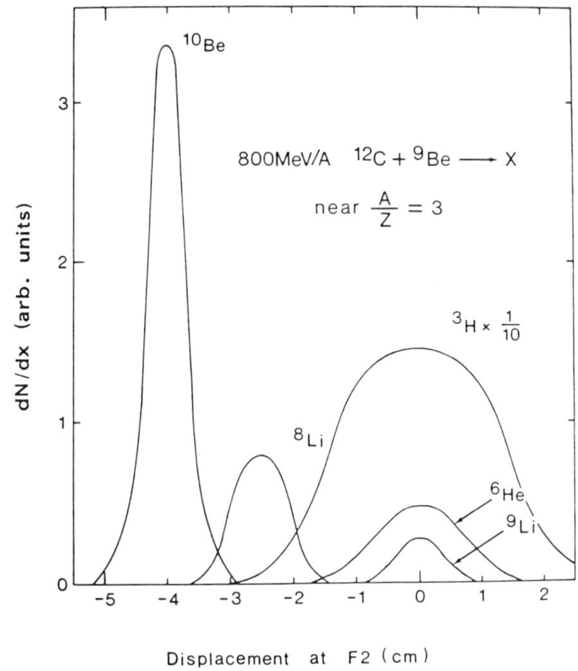


**Figure 25.** Example of the fragment distributions at F2. The beam line is set for nuclei with $A/Z = 3$.

counters before it is incident on a secondary reaction target, so that an admixture of other nuclei in the beam causes no essential problems.

The beam line operates as follows. Primary heavy ions at 800 MeV/nucleon, accelerated by the BEVALAC, are first focused at F1. A production target made of 5 g/cm$^2$ Be is used. Fragments produced at the target are analyzed at F2—a focus with large momentum dispersion. Figure 25 shows an example of the distribution of fragments from $^{11}$B projectiles for a setting of $A/Z = 3$; nuclei of different $A/Z$ are well separated. A separation slit is located at F2 and a pair of copper absorber reduces the energy of nuclei which have different rigidity from those of interest. One difference between the separation achieved in this beam line and that of the other separator, described previously, is that the unwanted nuclei cannot be easily stopped by absorbers or collimators because it would require many hundreds of g/cm$^2$ of material to stop beams at 1 GeV/nucleon and their fragments. On the other hand, the width of the slit is 2–5 mm for the presently available dispersion of the beam line. Consequently, use of thick material for collimation produces large amounts of background as a result of edge scattering. To prevent this, a relatively thin collimator is used. The opening of the copper collimator jaws was determined such that nuclei passing through the collimator can be well separated, positionally, at the next focus F3. F3 provides an achromatic focus for those nuclei which passed F2 without energy loss but gives a momentum dispersion for those which pass through the collimator. The characteristics of the F2 collimator was determined using an 800 MeV/nucleon $^{12}$C primary beam and

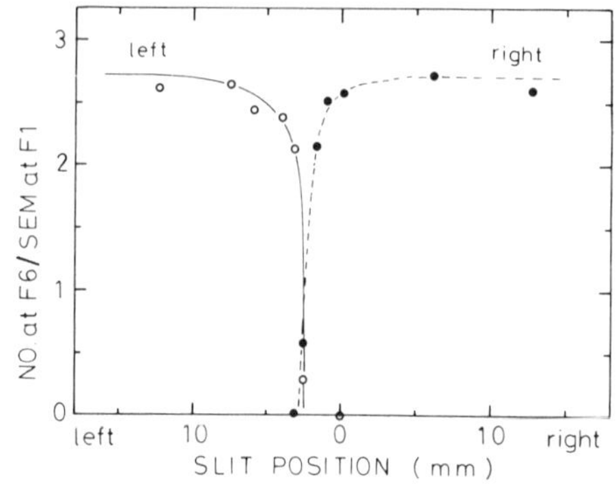


**Figure 26.** The characteristics of the slit at F2 measured using a $^{12}C$ primary beam.

is shown in Figure 26. The horizontal size of the primary beam at F2 is about 1 mm. Sharp clipping characteristics are evident in this figure. Separated secondary nuclei are finally transmitted to the HISS (heavy-ion superconducting spectrometer) experimental area using an ordinary beam line system.

Figure 27 shows a particle-identification spectrum of the secondary beams after the $A/Z = 3$ selection in this beam line. The figure shows a scatter plot between the pulse height of a scintillation counter and the time-of-flight. A primary beam of $^{11}B$ at 800 MeV/nucleon was used in this case. No admixture of nuclei other than $^{3}He$, $^{6}He$, and $^{9}Li$, which all have $A/Z = 3$, was found. For measurements of interaction cross-sections, each incident nucleus was identified in this way before it was incident on a reaction target. Using this beam line, all the p-shell nuclei and some sd-shell nuclei were transmitted for reaction measurements (see Section 3.2.2).

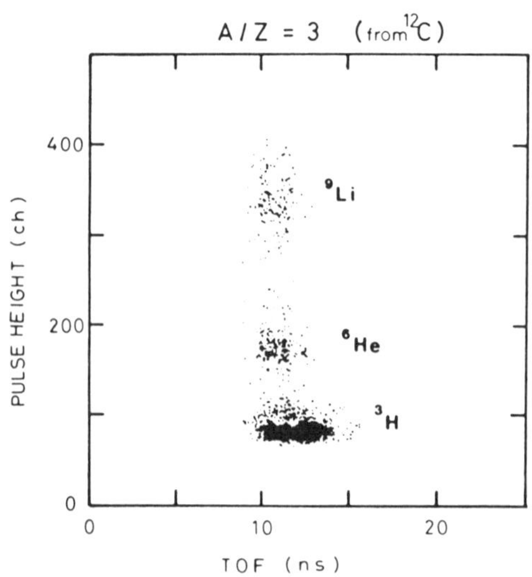


**Figure 27.** Identification spectra of the secondary beam.

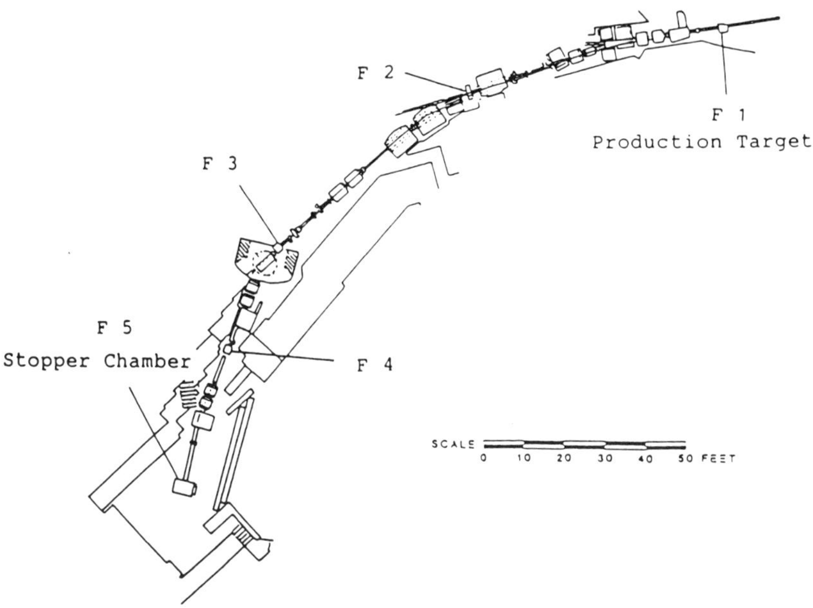


**Figure 28.** The beam transport system B44 at the BEVALAC.

Another beam line recently constructed at the Berkeley BEVALAC was developed to deliver lower-energy radioactive beams. Using this beam line, measurements of nuclear moments and precise *ft* values for $\beta$-emitting mirror nuclei in the f-shell are currently in progress (Mi 84a). Because the latter measurements are performed by detecting $\beta$ particles emitted after stopping the nuclei in some suitable material, it is necessary to avoid stopping unwanted nuclei, as well, in this stopping material. It is therefore necessary to separate nuclei not only in terms of their $A/Z$ value but also in terms of their mass or charge. Because the per-nucleon energy loss of a heavy ion with constant velocity is approximately proportional to $Z^2/A$ at high energy, a different sort of selection from that provided by magnetic separation ($A/Z$) can be obtained using an absorber. In order to effect this second separation, this beam line uses two complete magnetic analyzing systems as shown in Figure 28. The first selection in terms of $A/Z$ is made at F2 and then selected nuclei are achromatically focused on F3.

The absorber is positioned at F3. Because of the difference in energy loss, nuclei with different charges are spatially separated at F4. Using a collimator at F4, a high-purity separated beam is focused at F5 where a stopping medium is placed for measurements such as those on $\beta$ decays and on magnetic moments. If thin stopping material is used, the difference in the range of the different nuclei transmitted can also be used for even further purification of the effective secondary beam.

Recently beams of $^{20}$F, $^{21}$F, and $^{39}$Ca were produced from 200 MeV/nucleon $^{22}$Ne and $^{40}$Ca primary, respectively, and stopped in an Al foil of 0.1 mm thickness. Extremely clean decay-time sepctra were obtained as will be discussed in Section 3.1.2.

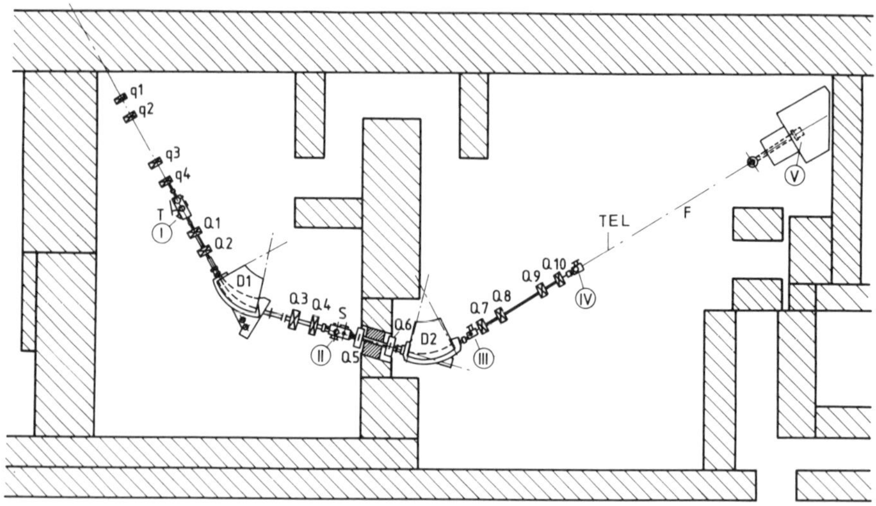


**Figure 29.** The LISE projectile-fragment separator at GANIL. The symbols q1 to q4, and Q1 to Q10 refer to quadrupoles; and D1, D2 to dipoles. The target is denoted by T, the slits by S, the achromatic focal point by F, and the telescope by TEL. Faraday cups are placed at points I, II, IV, and V, and beam diagnostics (secondary emission and ionization chambers) at points I, II, III, IV, and V (Bi 85).

Still another beam line which can deliver radioactive beams has been used at the BEVALAC for biomedical use. It shall not be described further because the separation technique used is essentially the same as that discussed above. Interested readers are referred elsewhere (Ch 82), (Al 79), (Al 84).

### *2.4.4. LISE at GANIL*

At intermediate energy the use of an absorber is efficient in terms of separation of the secondary nuclei as discussed in the last section. The LISE system at GANIL uses an absorber in a much more refined way at 30 to 100 MeV/nucleon (Bi 85, Du 86). The LISE system is shown in Figure 29. The primary beam is focused on a production target (T). Projectile fragments produced in the target are predominantly emitted into a scattering angle close to zero degrees. These secondary beams are analyzed by the momentum-dispersive part of the line, composed of a magnetic dipole D1 and quadrupoles Q1 to Q4. At the first focal plane where the beam focuses on different positions, depending on its rigidity, a slit S of variable aperture is located and is used to limit the rigidity acceptance of the line, hence achieving a selection on $A/Z$. A wedge-shaped absorber is positioned here for separation in terms of charge. The second dipole D2 associated with quadrupoles Q5 to Q6 compensates for the dispersion in the first section and ensures double achromatism (in angle and position) of the line. Finally, quadrupoles Q7 to Q10 are used to focus the secondary beam to the proper position in the experimental area.

We now discuss the principles of this separation in greater detail. It is known that at intermediate energy the range $\mathfrak{R}$ of the $^{A}Z$ nucleus having an energy per nucleon

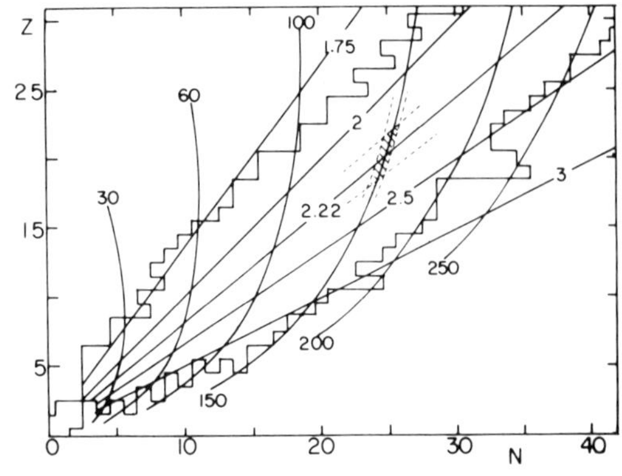


**Figure 30.** In this part of the chart of the nuclides, the ions selected by the first dipole are located on the straight lines corresponding to particular $A/Z$ ratios (here shown in the range from 1.75 to 3) for several $B_1\rho$ (rigidity) settings. The ions selected by the second dipole are located on the curves corresponding to particular $A^{2.5}/Z^{1.5}$ ratios (here shown in the range 30 to 250) for several $B_2\rho$ settings. Each selection has a given width in the $(N, Z)$ plane (locally represented by the dotted lines) and the nuclei transmitted by the spectrometer are located in a $(N, Z)$ domain such as the one represented by the lightly hatched area. This domain usually includes only one isotope, but in less-favorable cases two or three may include (Du 86).

$E$ can be empirically expressed as

$$\Re = kE^{\gamma}A/Z^2 + CA \tag{14}$$

where $k$, $\gamma$, and $C$ are constants depending only on the stopping material. After some manipulation of this equation, one can calculate the rigidity $R_2$ of the nucleus after passage through an absorber as

$$R_2(A, Z, d, R_1) = R_1\left(1 - \frac{dA^{2\gamma-1}}{kZ^{2\gamma-2}}R_1^{-2\gamma}\right)^{1/2\gamma} \tag{15}$$

The ions not separated in terms of $R_2$ are those having the same ratio $A^{2\gamma-1}/Z^{2\gamma-2}$. Using $\gamma = 1.75$ as an example (the value for an aluminum absorber) this ratio may be written as $A^{2.5}/Z^{1.5}$. This system thus provides a selection on $A$ and $Z$ that is quite different from the first selection which is sensitive only to $A/Z$. This difference in selectivity is shown in Figure 30. One can see that any single nucleus can, in principle, be separated using this approach.

If an absorber were placed at the first dispersive focus of the separator, it would be interesting to follow how the second part of the separator functions. If the first part of the separator has dispersion $D_1$, the relation between the rigidity of a nucleus and its position $x$ can be written as

$$R_1 = R_0(1 + x/D_1) \tag{16}$$

As an example, if one use an absorber of uniform thickness, the rigidity loss of the nucleus depends on its position because of the term which includes $R_1^{-2\gamma}$ in equation (15). The change in the dispersion introduced by the absorber is characterized by $[\Delta R_2(x)/R_2]/[\Delta R_1(x)/R_1]$, and is calculated to be

$$1/[1 - d/(\Re - CA)] \tag{17}$$

This means that the separator, which was achromatic before insertion of the absorber, cannot focus the selected beam; it also means that the separation power is no longer optimized. If the second part of the spectrometer is designed to have variable dispersion, optimization can be retrieved, however this adjustment has to be made directly, using precious radioactive beam. Therefore the selection of an absorber, which retains the achromatism of the separator, is preferable.

Such an absorber is, in fact, realized if $R_2$ is made proportional to $R_1$ for all $x$ values. This condition is fulfilled if $[d(x) \times R_1^{-2\gamma}]$ is independent of $x$. The shape of absorber which preserves achromatism is thus given by

$$d(x) = d_0(1 + x/D_1)^{2\gamma} \tag{18}$$

Because $\gamma$ is almost independent of $A$, $Z$, and $E$, in the intermediate energy range, it is a remarkable fact that the one type of absorber can conserve the achromatism of the separator, independently of $A$, $Z$, $E$, $R_1$, and $R_2$.

One other useful absorber is the monoenergetic wedge. This is the type of wedge which produces monoenergy (or uniform rigidity) after the absorber. Although the achromatism of the separator is destroyed, a uniformity in the range of the secondary nuclei provides further selectivity, useful in experiments which use stopped nuclei in material. This type of wedge is being used at Berkeley in beam line B44 (see Figure 28), at F4, for magnetic-moment measurements where range uniformity is essential in producing polarized nuclei by the tilted-foil technique (see Section 3.1.3).

Production of beams of $^{39}$Cl was demonstrated using an $^{40}$Ar projectile at 44 MeV/nucleon without applying energy absorption (Bi 85). At this energy it was found that Ar or other product nuclei are essentially fully-stripped after passing through the Be production target. In contrast, admixtures of significant amounts of lower-charge states were observed when a Au target was used. Figure 31 shows a spectrum from a $\Delta E$ counter at position F in Figure 29; with this optimized setting for a $^{39}$Cl beam, the admixture of other nuclei was less than 16%. The size of the beam spot was also reasonably small (10 mm horizontal and 4 mm vertical) at position F. The conversion ratio of the primary to the secondary beam was about $2 \times 10^{-4}$.

During production of a $^{38}$S beam, a greater admixture of other isotopes was observed, as shown in Figure 32. A monoenergy-type absorber was used in order to obtain this better separation. Figure 32c shows the result of setting the D2 field such as to favor the most neutron-rich beams like $^{36}$P, $^{39}$S, and $^{40}$Cl. A significant increase in the intensity of these nuclides was observed.

The best purified $^{38}$S beam obtained is shown in Figure 32d. The admixture of other nuclei was about 28%; the conversion ratio was about $2 \times 10^{-6}$.

Further details regarding the principles of separation and examples of the nuclear separation achieved can be found elsewhere (Du 86).

A similar type of radioactive-beam separator is currently being designed at RIKEN in Japan. It will use an intermediate absorber without destroying the achromatism of the beam line. This beam line will be used for producing secondary beams from primary heavy ions at up to 135 MeV/nucleon which will be accelerated by the ring cyclotron at RIKEN.

### 2.4.5. CERN Synchrocyclotron

Production of $^{17}$N and $^{16}$C secondary beams, using an 84 MeV/nucleon $^{18}$O beam accelerated by the synchrocyclotron at CERN, has been reported by Fleury *et al.* (Fl 85). The separation procedure was a simple magnetic analysis with one dipole magnet. Conversion ratios were found to be $5 \times 10^{-5}$ for $^{17}$N and $7 \times 10^{-6}$ for $^{16}$C.

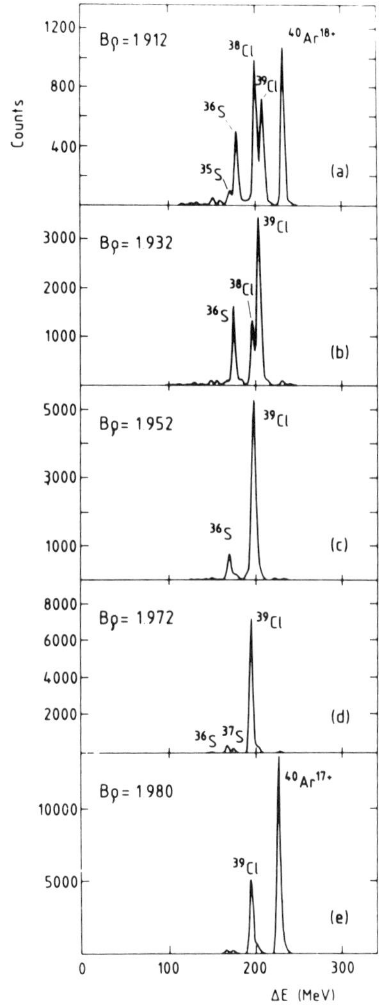


**Figure 31.** Effect of a Bρ variation on the $^{39}$Cl secondary-beam purity. The beam compositions (by $\Delta E$ spectra) are shown for five different Bρ values (in Tm). The ordinate scale is in arbitrary units. The most abundant isotopes in the secondary beams are indicated close to their corresponding peaks (Bi 85).

### *2.4.6. Reaction-Product Mass Separator at MSU*

The separation of exotic secondary beams has been reported by a group at the Michigan State University (MSU) superconducting cyclotron (Ha 84, Cu 86). They used a reaction-product mass separator (RPMS) for separating $^{9}$Li, $^{11}$Li, $^{12}$Be, $^{14}$Be, $^{12}$B, $^{14}$B, $^{15}$B, and $^{17}$C secondary beams, and measured the half-lives of these isotopes—including the previously unknown half-lives of $^{14}$Be and $^{17}$C. Although they have so far used the RPMS only for half-life measurements, it is clearly capable of providing high-purity radioactive beams for reaction studies as well (Ha 84).

Figure 33 shows a side view of the RPMS. The RPMS combines a Wien filter (velocity selector) and a magnetic dipole with quadrupole magnets so as to achieve a system which has a focal plane in which the position of the particle depends solely on

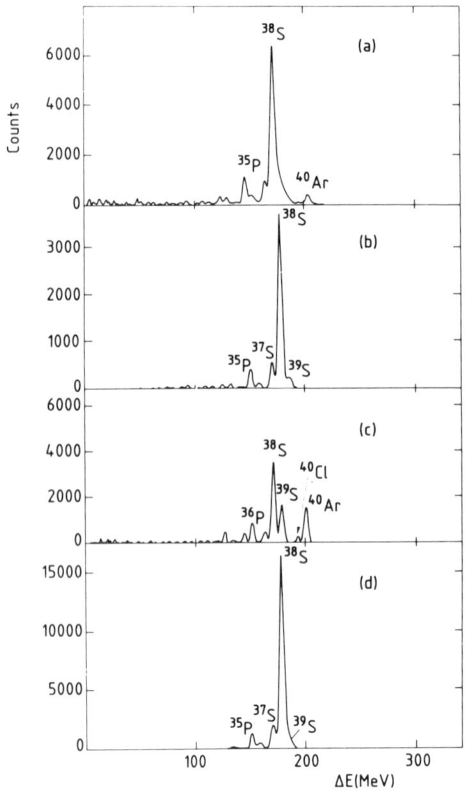


**Figure 32.** Influence of an intermediate absorber on the purity of a $^{38}$S secondary beam. The $\Delta E$ spectrum is plotted for $B_1\rho$ = 2.055 Tm and for the following experimental conditions: (a) No absorber, $\Delta B_1\rho/B_1\rho = \pm 0.56\%$, $B_2\rho = B_1\rho$; (b) Al absorber, $\Delta B_1\rho/B_1\rho = \pm 2.8\%$, $B_2\rho$ = 2.005 Tm; (c) Al absorber, $\Delta B_1\rho/B_1\rho = \pm 2.8\%$, $B_2\rho$ = 2.051 Tm; (d) Al absorber, $\Delta B_1\rho/B_1\rho = \pm 0.4\%$, $B_2\rho$ = 2.005 Tm (Bi 85).

its mass-to-charge ratio $A/q$. It has unit-mass resolution and therefore would appear to deliver a clean secondary beam. However this separator has only $A/q$ separation, and thus the separation of a nuclide with the same, or nearly the same, $A/q$ cannot be achieved. Introduction of an absorber after the RPMS, followed by another separator, is being considered in order to achieve $A/q$ separation.

## 2.5. Storage Rings

Because the intensity of radioactive beams is generally weak, efficient use of these beams is essential. As described in Section 2.1, secondary beams produced by projectile

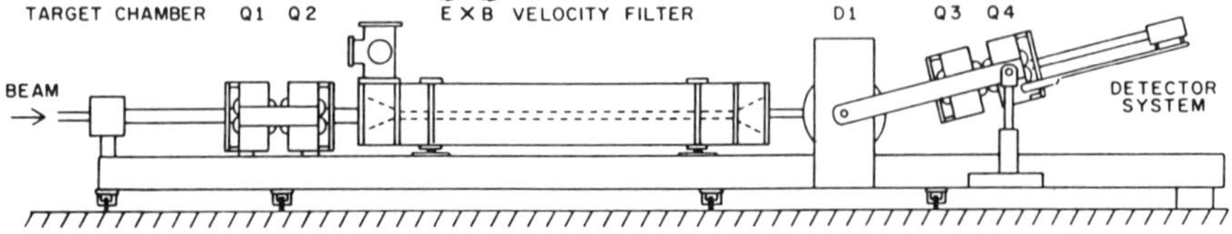


**Figure 33.** The reaction-product mass separator at the Michigan State University Cyclotron.

fragmentation in high-energy heavy-ion collisions have a momentum broadening of only a few percent. Therefore, it is possible to store these secondary beams in a storage ring. In fact, such a facility has been proposed in several places including INS at the University of Tokyo, GSI in Darmstadt, and LBL in Berkeley. If a storage ring is used with a rapid cycling heavy-ion synchrotron, it provides high-intensity, high-quality cooled beams of radioactive nuclei, for a broad spectrum of experiments, that would otherwise be very difficult, if not impossible. A storage ring also provides superior usage of the primary accelerated beams such as the stretching of the heavy-ion beam for a high-duty-factor experiment and internal-target experiments for small cross-section or high-resolution studies.

Several modes of operation can be considered in using a storage ring for radioactive-beam experiments.

1. *Measurements of nuclear mass.* Initially a primary beam, accelerated by a synchrotron, is incident on a production target; after a rough separation, secondary nuclei of interest are injected into the storage ring. If the storage ring is designed so that the circulation time of the nucleus depends only on its mass (isochronus), the mass of the stored nuclei can be obtained with high accuracy from a time measurement taken over several thousand revolutions. If cooling of the beam is applied, the accuracy of this measurement is further increased. However the necessary cooling time restricts the range of nuclear lifetimes that can be studied.

2. *Reaction studies with cooled beams.* After accumulation of radioactive nuclei, nuclear reactions induced by them can be studied using an internal target. For such a study, beam cooling is also essential to prevent phase space broadening and loss of the beam by multiple Coulomb scatterings. In the usual fixed-target experiments with an extracted beam, most of the incident nuclei leave the target without inducing any nuclear reactions. On the other hand, in internal-target experiments, all the stored nuclei can be used until they disappear as a result of nuclear interactions. This mode of operation, therefore, permits the study of reactions having relatively small cross-sections.

3. *Reaction studies with low-energy radioactive beams.* After injection and cooling of the radioactive nuclei, the storage ring can be used to decelerate the beam down to around a few 10 MeV/nucleon. Then this low-energy beam can be extracted from the storage ring and used for low-energy nuclear-reaction studies or even for structure studies with a stopped beam.

4. *Production of high-intensity radioactive beams.* When an external target is used for production of secondary beams, the target thickness is limited because of momentum broadening considerations, as described in Section 2.4.2. With a thickness of target which gives a usable momentum spread, only 10–20% of the primary nuclei fragment in the target. If the primary beam is stored in a storage ring, and a thin internal target is used for production of the radioactive isotope, the production efficiency is maximized. The yield $Y$ of a certain nucleus can, in this case, be written as $Y = N_0\sigma/\sigma_0$, where $\sigma_0$ is the total nuclear cross-section, $\sigma$ is the production cross-section for the desired nucleus, and $N_0$ is the number of nuclei circulating. Moreover, the momentum spread of the secondary beam is minimized (i.e., the spread is due to the intrinsic production mechanism only). Because the nucleus that is of interest has, generally, a different rigidity from that of the circulating beam, its extraction is possible. Fortunately, beam extraction is easier for exotic nuclei, for which the production cross-sections are generally small, and therefore efficient production is most important. It is even possible,

in principle, to inject this extracted beam into yet another storage ring, for accumulation.

At INS, a storage ring called TARN II is being constructed following a successful stochastic beam-cooling test in the test ring TARN (Ka 84, Ka 86). This ring design includes plans to accelerate protons up to 1.3 GeV/nucleon and up to 0.45 GeV/nucleon for ions of charge-to-mass ratio $\frac{1}{2}$. In this ring, both stochastic and electron beam-cooling methods are planned in order to obtain both high resolution and small emittance. At present, the injector used is the sector focusing cyclotron, with $K = 67$, which accelerates ion beams ranging from protons to neon. The layout of the TARN II system is shown in Figure 34.

A storage ring, called ESR (Experimental Storage Ring), has been proposed at GSI as an important part of the high-energy heavy-ion accelerator project, SIS 18. It features a synchrotron for acceleration of heavy ions up to 2 GeV/nucleon. Figure 35 shows the layout plan of SIS 18 and the ESR. The latter will be used for accumulation, storage, and phase-space-density increase of heavy ions with energies ranging between 834 MeV/nucleon ($Ne^{10+}$) and 556 MeV/nucleon ($U^{92+}$). A heavy ion accelerated in the UNILAC up to 20 MeV/nucleon, and stripped to an adequately high charge state, will be injected into SIS 18 over an interval of 10 to 30 turns and accelerated to the desired energy. The synchrotron is designed to operate with a cycling rate of 3 Hz at up to 1.2 T, and of 1 Hz at its highest energy. Fast extraction is used over a single turn for transfer of the bunch structure into the ESR. Because the ESR can be operated at $\Delta P/P = \pm 2\%$ and is designed for a radial acceptance of about $140\pi$ mm mrad, radioactive beams can be accumulated efficiently.

Figure 36 shows the layout plan of the ESR. It has two 9.5 m long straight experimental sections, in one of which an electron cooling device will be installed. The other four straight sections will be used for installation of rf cavities and slow and fast extraction elements. The rf cavities are used for acceleration, deceleration and, especially, together with the longitudinal phase-space electron cooling system, for bunching of the beam. With a fast extraction system one could hope to transfer a highly ionized and cooled beam back to SIS 18 for further acceleration or deceleration. The most important facilities of the ring are various complementary cooling devices which can be operated simultaneously. For low phase-space-density secondary beams, stochastic precooling may be used. For cooling to a very high phase-space density, electron cooling of completely stripped heavy ions is planned using an interaction zone of 2 m length.

While the cooled beam circulates in the ring, it may be used in the second straight section, for the study of collision processes, using internal targets which may be atomic beams, electron beams, gas jets, or fibers. For all experiments which need thin targets, a high gain in luminosity may be achieved, as compared with a single-pass experiment, because of the increase in circulating beam current (an increase of about $2 \times 10^6$). Also, the interaction of collinear laser beams with circulating ions of high intensity and small momentum spread can be studied effectively in this geometry.

The proposal for the BEVALAC upgrade at LBL also includes a storage ring for secondary radioactive-beam experiments and other high-resolution or high-luminosity experiments. This storage ring will include stochastic and electron-beam cooling, which will permit the use of internal targets with the circulating beams and the storage of high-intensity beams of radioactive nuclei (Pr 86).

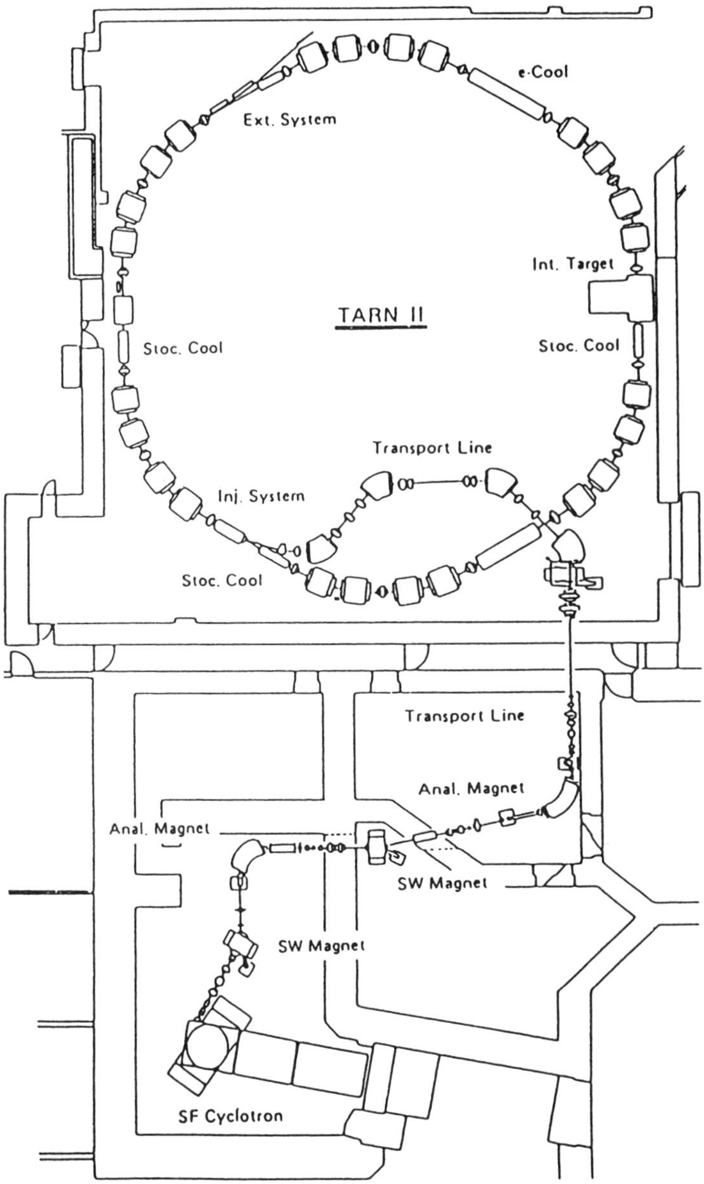


**Figure 34.** A proposed layout of TARN II at the Institute for Nuclear Study, University of Tokyo (Ka 86).

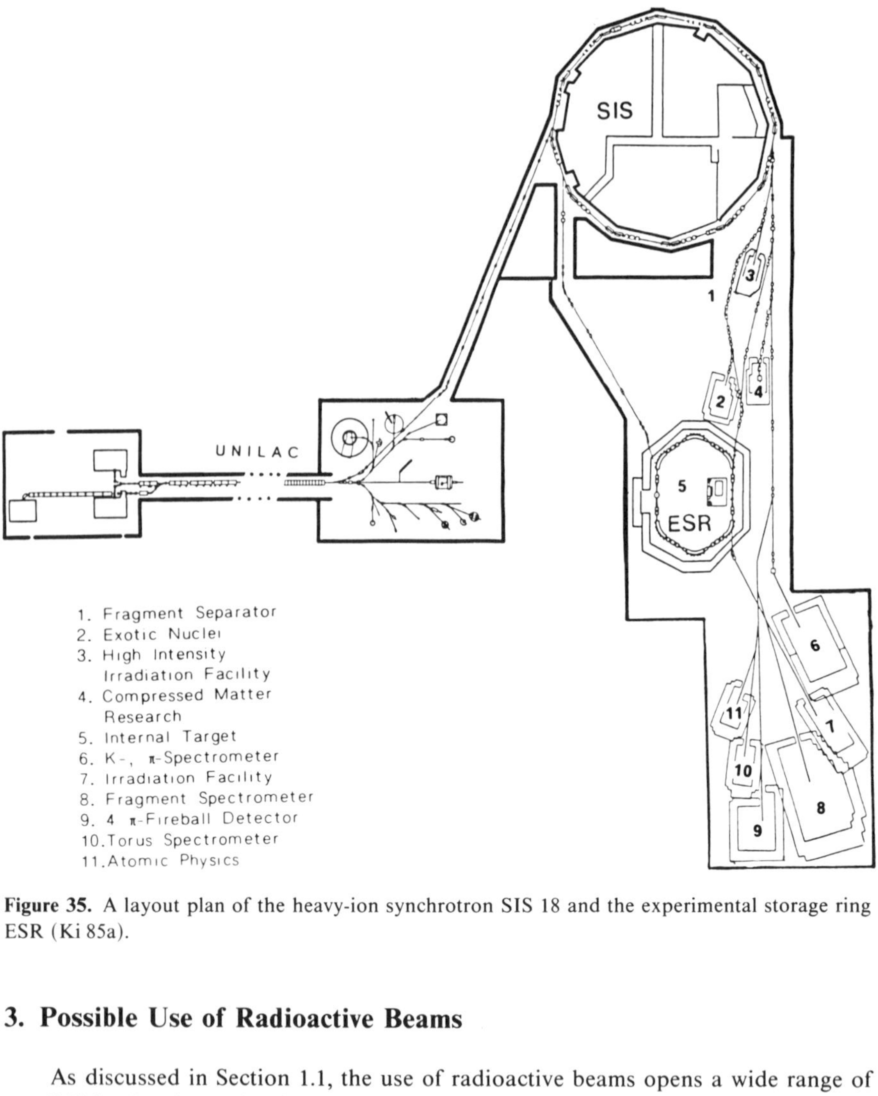


**Figure 35.** A layout plan of the heavy-ion synchrotron SIS 18 and the experimental storage ring ESR (Ki 85a).

# 3. Possible Use of Radioactive Beams

As discussed in Section 1.1, the use of radioactive beams opens a wide range of possibilities for the study of nuclei far from stability. One of the most important goals of such work is that of determining the systematics of nuclear behavior over the widest possible range of nuclei. In this section, several examples of studies which benefit from the use of radioactive beams are presented.

## 3.1. Properties of Exotic Nuclei

### *3.1.1. Masses of Exotic Nuclei*

The study of nuclear masses has long been an important field of research in nuclear physics. The mass and binding energy are global properties of a nucleus that depends

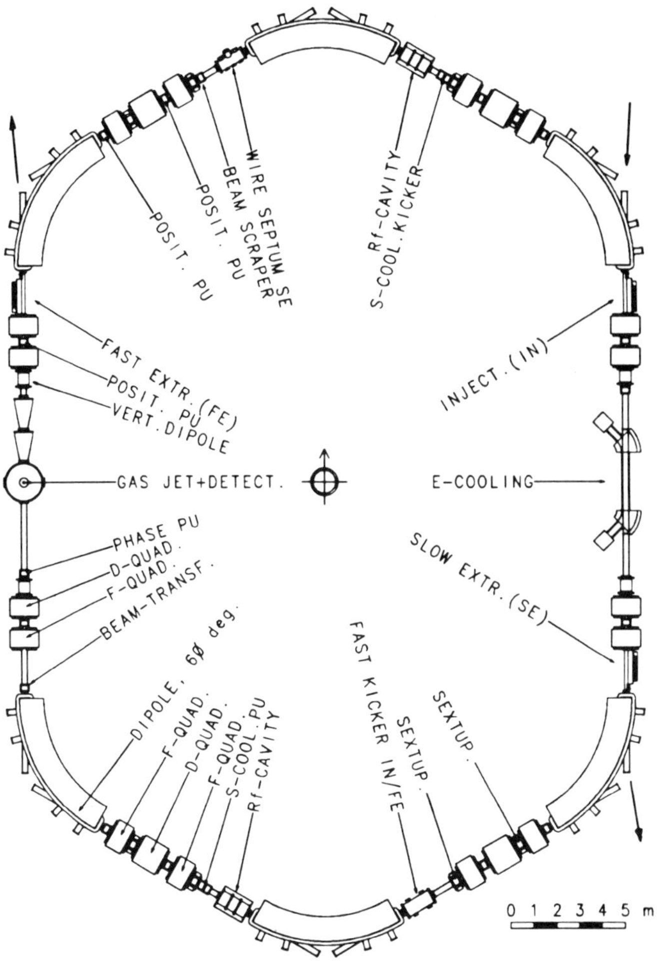


**Figure 36.** A layout plan of the experimental storage ring ESR with a circumference of 103.2 m. In two straight sections, each of length 9.5 m, the experimental setup and the facility for cooling can be mounted (Ki 85b).

on its structure, i.e., on such properties as single-particle levels and collective effects. The determination of nuclear masses over a wide range of proton–neutron ratios would provide a stringent test for nuclear models. Most of the masses of the nuclei far from stability have been determined by measurements of $\beta$-ray end-point energies and $\gamma$-transition energies; many masses have also been determined using mass spectrometers combined with ISOL facilities. Methods of, and the importance of, the mass measurements have been discussed in many places, and recent reviews can be found elsewhere (Wa 80, Ha 82a). Here we focus on several methods for determining nuclear masses of exotic nuclei, using beams of radioactive nuclei.

Direct measurement of a nuclear mass, using radioactive nuclear beams, is possible. At GANIL, using the spectrometer SPEG, masses of $^{20}N$, $^{21}N$, $^{23}O$, $^{24}F$, $^{25}F$, and $^{26}F$

have recently been measured (Mi 86). Projectile fragmentation was used for producing these nuclei and the beam character was used directly. The masses were determined by measurements of the magnetic rigidity $R$ and the time of flight $\tau$, where

$$R = \beta\gamma m/Z \qquad \text{and} \qquad \tau = l/\beta c$$

with $\gamma = 1/(1-\beta^2)^{1/2}$. The mass resolution was determined by the resolutions of the magnetic rigidity and of the time-of-flight measurements. The SPEG provides a rigidity resolution of $10^{-4}$ and, therefore, the resolution of the time of flight determined the mass resolution. Because of kinematic focusing of the projectile fragments, it was possible to use an entire beam line, from the cyclotron to the spectrometer, as the flight path—116 m long in all. With a time resolution of about 500 ps it was possible to obtain a mass resolution of about 500 keV. This example shows the usefulness of a high-energy secondary beam line in a mass measurement in which the accuracy of the measurement is limited only by the resolution of the time of flight. Thus the development of detectors with better time resolution is very important.

A more elaborate method for measuring nuclear masses is currently being considered as an application of a storage ring. As described in Section 2.5, radioactive isotopes produced through projectile fragmentation can be accumulated in a storage ring and, if the storage ring is designed so that ions with equal masses have equal orbital frequencies (isochronus), time-of-flight measurements taken over many turns in the storage ring provide a high-resolution measurement of the mass. Beam cooling may be applied to obtain reduced momentum spread and thus even better mass resolution. Another possibility is deceleration of the nuclei followed by a combined measurement of the energy, the momentum, and the time of flight through a spectrometer.

### *3.1.2. Decay Characteristics of Exotic Isotopes*

Exploration of new regions of the nuclear chart is expected to reveal nuclei with unusual sizes, shapes, and level structures. Many of these nuclei, including many near the proton and neutron drip-lines, also exhibit unusual modes of radioactive decay. There is widespread interest in p, 2p, and $\alpha$ emission near the proton drip-line, both prompt and $\beta$-delayed, and, from studies of the latter, in determination of $\beta$-strength functions. By studying long sequences of isotopes, the underlying symmetries of the nuclear structure and their dependence on the neutron and proton numbers can be determined. The determination of gross $\beta$-decay and mass-surface properties for very neutron-rich nuclei of above $N = 50$, is of special importance for astrophysics since the r-process path may be reachable. Studies of those decay characteristics benefit from use of the secondary-beam technique because of the background-free conditions which can be obtained for measurements on rarely produced nuclei.

Half-lives of the heaviest Be isotopes, $^{14}$Be, and of $^{17}$C were recently determined for the first time at the MSU reaction-product mass separator (see Section 2.4) (Cu 86). The decay curves for these and other nuclei are shown in Figure 37, and comparison with the theoretical predictions is shown in Table 6. Half-lives of neutron-rich nuclei in the sd-shell, such as $^{40}$S, $^{38}$P, $^{37}$P, $^{36}$Si, $^{35}$Si, $^{34}$Al, $^{26}$Ne, $^{24}$F, and $^{19}$N, have been determined for the first time using the LISE spectrometer (see Section 2.4) (Hu 86). The decay curves are shown in Figure 38. The predictive powers of several theoretical models was examined against these new data, as shown in Table 7. It was found that,

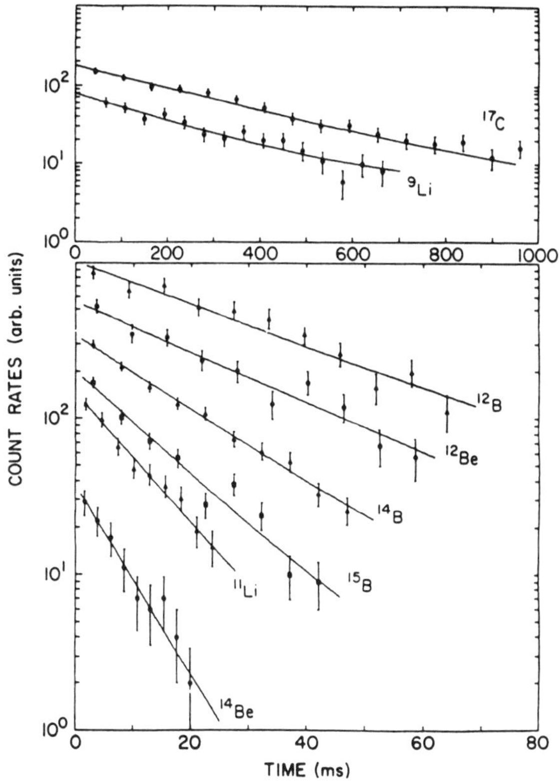


**Figure 37.** $\beta$-decay half-life data from an experiment at Michigan State University shown with the best-fit curves (Cu 86).

**Table 6. $\beta$-Decay Half-Lives in Milliseconds**[a]

| Isotope | This measurement | Previous measurement | Theoretical prediction |
|---|---|---|---|
| $^{9}$Li | 173 ± 14 | 175 ± 1 | 76 |
| $^{11}$Li | 7.7 ± 0.6 | 8.5 ± 0.2 | 2.3 |
| $^{12}$Be | 21.3 ± 2.2 | 24.4 ± 3.0 | 8.6 |
| $^{14}$Be | 4.2 ± 0.7 | — | 2.8 |
| $^{12}$B | 20.0 ± 1.5 | 20.41 ± 0.06 | 13.6 |
| $^{14}$B | 12.8 ± 0.8 | 16.1 ± 1.2 | 11.1 |
| $^{15}$B | 8.8 ± 0.6 | 11.0 ± 1.0 | 7.4 |
| $^{17}$C | 202 ± 17 | — | (422, 297, 242) |

[a] $\beta$-decay half-lives for the neutron-rich isotopes measured in this experiment compared with previous measurements (when they exist). Theoretical predictions for Gamow–Teller $\beta$ decay are also provided for comparison. The three theoretical half-lives for $^{17}$C correspond to the assumptions in the calculation of $\frac{1}{2}^+$, $\frac{3}{2}^+$, and $\frac{5}{2}^+$ ground-state $J^\pi$ values, respectively, for $^{17}$C (Cu 86).

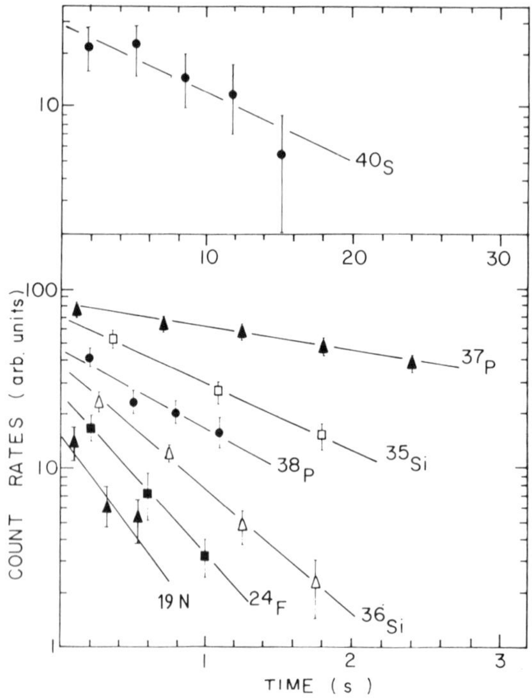


**Figure 38.** $\beta$-decay half-life data from an experiment at GANIL shown with the best-fit curves (Hu 86).

as far as the half-lives are concerned, the predictions of Wildenthal gave the best reproduction of the data. Measured $\beta$–$\gamma$ coincidences in some of these nuclei, such as $^{22}O$ and $^{17}C$ and other sd-shell nuclei, have also been reported.

The background-free conditions also provide an opportunity for precise measurement of the decay characteristics of nuclei close to the stability line. Measurements of $\beta$-decay lifetimes and $\beta$–$\gamma$ studies have recently been reported using particle separation following heavy-ion reactions. Using a quadrupole-sextuplet beam-transport system, as shown in Section 2.2, Mathews *et al.* (Ma 83) measured the branching ratio for $^7Be$

**Table 7. Lifetimes of Neutron-Rich Nuclei (Hu 86)**

| Isotope | Measured half-life (s) | Predicted half-life (s) (Ta 73) | (Ko 85) | (Kl 84) | (Wi 83) |
|---|---|---|---|---|---|
| $^{40}S$ | 7.6 (25) | 30.0 | 18.0 | 13.6 | |
| $^{38}P$ | 0.70 (15) | 0.7 | 1.0 | 0.146 | |
| $^{37}P$ | 2.3 (2) | 4.0 | 4.4 | 0.744 | |
| $^{36}Si$ | 0.40 (6) | 5.0 | 3.0 | 0.14 | |
| $^{35}Si$ | 1.0 (3) | 2.0 | 2.0 | 0.305 | |
| $^{34}Al$ | 0.05 (2) | 0.2 | 0.26 | 0.076 | |
| $^{26}Ne$ | $<0.3$ | 1.0 | 0.6 | 0.043 | 0.170 |
| $^{24}F$ | 0.34 (8) | 0.1 | 0.14 | 0.577 | 0.171 |
| $^{19}N$ | 0.27 (7) | 0.4 | 0.46 | 0.591 | 0.321 |

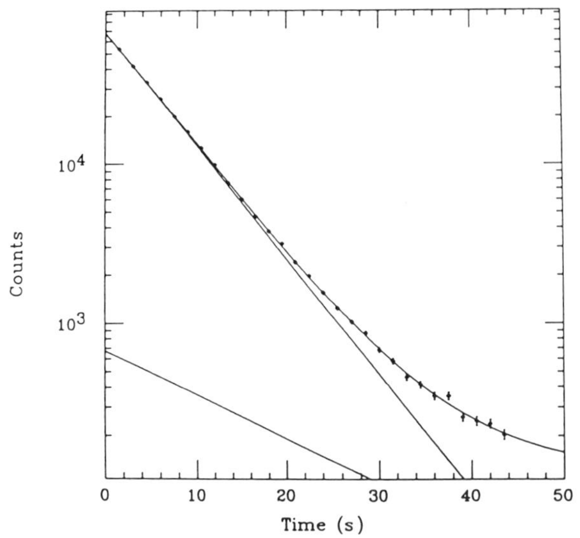


**Figure 39.** A $\beta$-decay time spectrum of $^{21}F$ measured at the BEVALAC. The two straight lines in the figure show the decomposition of the spectrum by two components.

electron-capture decay to the first excited state in $^{7}Li$. Very clean $\gamma$-ray spectra were obtained in the branching-ratio measurement.

The half-life of $^{21}F$ was measured using the secondary-beam line at the BEVALAC (No 86). The intensity of the stopped $^{21}F$ at the end of the beam line was greater than $10^5$ /s when a $10^9$ /s primary beam of $^{22}Ne$ at 200 MeV/nucleon was used. A measurement over about 20 min gave the clean spectrum shown in Figure 39. The background from longer lifetimes shown in this spectrum is considered to come from the decay of $^{20}F$ which is produced in the stopping material through nuclear reactions induced by $^{21}F$. The half-life of $^{21}F$ was determined to be $4.21 \pm 0.03$ s.

The secondary-beam technique thus shows marked advantages for the measurement of nuclear decay characteristics, not only for nuclei far from stability but for nuclei in general.

### *3.1.3. Electromagnetic Moments of $\beta$-Unstable Nuclei*

Electromagnetic moments of $\beta$-unstable nuclei have been measured using several different methods. Among them are the atomic-beam resonance method, using an ISOL facility; another important method, which can be applied to radioactive nuclei, is NMR (Nuclear Magnetic Resonance) detected by nuclear radiation from $\beta$-unstable nuclei. Such an experiment is currently in progress at the BEVALAC (Mi 84a, No 86). Measurements on magnetic moments of short-lived partners of mirror nuclei in the f-shell, such as $^{43}Ti$, $^{45}V$, . . . , $^{55}Ni$, and $^{57}Cu$, are planned.

Magnetic moments of mirror nuclei have been intensively studied in p-shell and sd-shell nuclei (Ho 73, Ha 85a). The symmetry of the mirror nuclei provides a unique opportunity for studying details of the nuclear wave functions as well as the renormalizations of the nucleon magnetic moments and the Gamow–Teller coupling constant in

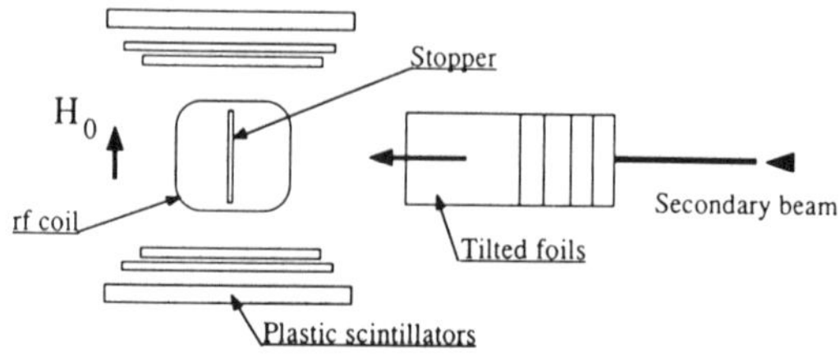


**Figure 40.** An experimental system for magnetic moment measurements.

nuclei. The determination of magnetic moments and the precise measurements of *ft*-values for f-shell mirror nuclei are expected to give important systematic information (Su 73).

The experimental method used in these measurements is shown in Figure 40 (Mi 84a). The experiment proceeds in several steps:

1. The desired nucleus is produced by projectile fragmentation of heavy ions at around 200 MeV/nucleon. These nuclei are then separated in a secondary beam line which includes an absorber and a magnetic analyzer. (The beam line was shown in Figure 28.)
2. The separated radioactive nuclei are decelerated by the absorber before being incident on the detection apparatus.
3. The nuclei—slowed down to a few MeV/nucleon—then pass through thin tilted foils, thus becoming polarized.
4. The polarized nuclei are implanted into some stopping material which is located in a strong magnetic field, $H_0$, parallel to the polarization axis. The stopping material is selected such that this nuclear polarization can be retained during the entire nuclear lifetime.
5. Beta rays emitted from the polarized nuclei are detected by a pair of scintillation-counter telescopes positioned above and below the stopper, respectively. The nuclear polarization is measured by the asymmetric emission of $\beta$-rays.
6. An rf-magnetic field is applied to the stopper and the resulting NMR is detected by a change in the asymmetry in the $\beta$-ray emission. The magnetic moment of the nucleus is determined from the resonance frequency of the NMR.

This method has several advantages. First, one can study a very short-lived nucleus (of lifetime as short as a millisecond) because no ion source is involved in the separation procedure. For the same reason, this method can be applied to elements of any chemical character. (One possible difficulty of the experiment is polarization by the tilted foils.) This technique has been demonstrated to be powerful for low-energy ions where the charge state of the ions is not high (No 86a). However, in the present application, the momentum spread of the secondary beam sets a lower limit on the energy of the nuclei passing through the tilted foils even when the momentum spread is minimized by use of a wedge-shaped absorber (see Section 2.4.4). The degree of polarization one can get when highly ionized nuclei are used has not yet been reliably determined. If the secondary beam is cooled in a storage ring, the momentum spread becomes negligibly small and thus such experiments will benefit highly from use of a storage ring.

It is known that the momentum transfer in projectile fragmentation at high energy is relatively small. However, at energies below 200 MeV/nucleon, the effect of nuclear friction has been observed to result in broadening of the transverse-momentum distribution of a fragment (Bi 79). Therefore one would expect there to be some degree of nuclear polarization produced directly in the fragmentation reaction. If this is so, then it might be possible to produce a polarized nuclear beam by selecting fragments at a certain emission angle, with the obvious disadvantage of loss of intensity of the secondary nuclei. But if reasonably large polarization should be possible, this would be an attractive technique.

Present techniques, of course, can be applied to other $\beta$-decaying nuclei of interest. A systematic study of the magnetic moments of long-isotope chains would give important new information on nuclear structure. The method can also be extended to measurement of electric-quadrupole moments of radioactive nuclei by using a stopping material which provides a strong electric-field gradient.

## 3.2. Nuclear Reactions with Radioactive Beams

### *3.2.1. Production of Exotic Isotopes and New Elements*

Most unstable nuclei have been synthesized in nuclear reactions using beams of stable particles, or by fission of heavy elements such as uranium. Recent developments show that intermediate-energy and high-energy heavy-ion beams are very effective for the production of exotic nuclei. Such use of radioactive beams is expected to provide further versatility in the production of exotic nuclei and exotic states such as superdeformed and very high spin states. Here we review several advantages of the use of radioactive beams in the production of exotic nuclei.

*3.2.1a. Superheavy Elements.* The tentative prediction (My 66) of an island of stability around $N = 184$ and $Z = 114$–$126$ initiated intensive experimental effort to discover or synthesize such superheavy elements. Extensive theoretical study continues to predict the existence of such an island, but all experimental attempts to find it have failed. These studies thus cast doubt on the reliability of the theoretical estimates, which in other regions of the periodic table, however, have been eminently successful. This confusing situation has lasted for a number of years, but it appears to be changing dramatically as a result of experimental and theoretical developments over recent years.

Detailed study of the production cross-sections for isotopes of the heaviest elements (leading to the synthesis at GSI of elements 107, 108, and 109 with $N$ of up to 157), together with theoretical predictions of an entrance-channel dynamic hindrance to fusion, have provided an explanation for the smallness of the cross-sections characterizing the production of the postulated new elements. The picture has been further clarified by recent refinements in the theoretical treatment of nuclear masses and deformations (Le 84, Mo 84). In particular, the theoretically predicted masses track—with remarkable accuracy—the experimental ground-state masses of the newly discovered isotopes in the region $Z = 106$–$109$.

As a result of all these developments, there is now no reason to dismiss the theoretical predictions as a sensible guide to what may lie beyond the upper end of the periodic table. The latest theoretical predictions are qualitatively similar to the earlier ones, but they include some new features. An island of spherical superheavy nuclei, with $Z \cong 114$, is once again predicted. The island spreads out below the magic number $N = 184$. In addition, there appears a small secondary islet of deformed nuclei centered around $Z = 110$ and $N = 162$. According to these predictions two research approaches are important in searching for new heavy and superheavy nuclei. One is (the already intense effort) to characterize more quantitatively the magnitude of fusion cross-sections at the upper end of the periodic table, paying particular attention to the role of the dynamic entrance-channel hindrance. The other is a direct attack, with improved targets, projectiles, and detection techniques, on the heavy islet around $Z = 110$, $N = 162$ and on the superheavy island around $Z = 114$, $N \approx 178$.

In both of these approaches, the availability of a wide range of projectiles—including low-energy extremely neutron-rich (radioactive) nuclei, such as $^{50,52}$Ca—will be of great value and could be decisive. An additional, very attractive, possibility is to use such beams in conjunction with a target of highly radioactive $^{254}$Es, or even $^{255}$Es, production of which is planned as part of the LEAP (Large Einsteinium Activation Project) at LBL (Pr 86).

*3.2.1b. Production of New Isotopes.* To synthesize new neutron-rich isotopes of the heaviest elements there is a need for high-intensity beams of neutron-rich projectiles, with energies from near the Coulomb barrier to about twice that value, for bombarding the heaviest possible targets. Recent studies suggest that there is a host of new neutron-rich heavy-element isotopes that can be produced in quantities large enough for their identification and study. Many of these should have relatively long spontaneous-fission half-lives, reflecting hindrance factors associated with odd protons or neutrons. For production of neutron-rich nuclei the most neutron-rich stable beams such as $^{18}$O and $^{22}$Ne have been used, but the cross-sections are still small. In order to produce very neutron-rich heavy-element isotopes, it is therefore necessary to use projectiles that are even more neutron-rich and hence radioactive.

In the region between Ra ($Z = 88$) and the heaviest elements, many new neutron-rich isotopes can be produced in compound-nucleus or transfer reactions involving neutron-rich, light-ion radioactive beams such as $^6$He and $^8$He. The study of these isotopes should reveal a number of interesting properties, such as the $\beta$-delayed fission decay mode. These studies will contribute to the development of dynamic models of nuclear fission and to the proper treatment of $\beta$ decay and $\beta$-delayed fission in astrophysical models of nucleosynthesis.

Radioactive nuclear beams of proton-rich nuclei, such as $^8$B, $^9$C, and $^{58}$Ni, can be effectively used for synthesizing proton-rich isotopes of light and medium-heavy nuclei. Near the proton drip-line new decay modes such as direct proton decay, $\beta$-delayed proton decay, and $\alpha$ emission can be studied.

Nuclei having double closed shells provide important information about nuclear structure because their theoretical treatment is relatively simple and details of their structure can be calculated. For synthesizing doubly closed nuclei far from stability, such as $^{78}$Ni, $^{100}$Sn, $^{132}$Sn, $^{164}$Pb, and subshell-closure nuclei, such as $^{76}$Sr and $^{80}$Zr, use of compound nuclear reactions with radioactive beams, such as those listed below, are expected to be useful:

$$^{70}\mathrm{Se} + {}^{9}\mathrm{Be} \rightarrow 3\mathrm{n} + {}^{76}\mathrm{Sr}$$

$$^{75}\mathrm{Rb} + {}^{7}\mathrm{Li} \rightarrow 2\mathrm{n} + {}^{80}\mathrm{Zr}$$

$$^{95}\mathrm{Ag} + {}^{7}\mathrm{Li} \rightarrow 2\mathrm{n} + {}^{100}\mathrm{Sn}$$

*3.2.1c. High-Spin and Highly Deformed States.* In a recent study at Daresbury a superdeformed band in $^{152}$Dy was discovered (Tw 86). The experiment showed that this superdeformed band extends from a spin of 22 $\hbar$ up to 60 $\hbar$ and has a deformation $\delta_2 \approx 0.6$. High-spin states of $^{152}$Dy have been studied extensively using the $^{124}$Sn + $^{32}$S reaction. However, in the Daresbury study the $^{106}$Pd + $^{48}$Ca $\rightarrow$ $^{152}$Dy + 4n reaction was used. The reason why high-spin states are observed in this reaction can be understood in terms of the excitation energy $E^*$ and angular momentum $l$ involved in the reactions. For $^{124}$Sn + $^{32}$S, $E^* = 83$ MeV and $l = 60$ $\hbar$; and for $^{106}$Pd + $^{48}$Ca, $E^* = 79$ MeV and $l = 71$ $\hbar$. The angular momenta were calculated using the classical methods of Bass (Ba 73). It follows that the use of the neutron-rich $^{48}$Ca beam brings a higher angular momentum into the compound nucleus together with a smaller excitation energy, demonstrating that the selection of the incident beams is extremely critical in the production of high-spin and highly deformed nuclei.

By extending the method to radioactive and isomeric-state nuclear beams, it is expected that one will find the optimum combinations for producing high-spin states near the nuclear yrast line. For example, if $^{107m}$Pd($11/2^-$) is used as an incident beam on a $^{48}$Ca target, angular momenta as high as 76 $\hbar$ are involved.

The heaviest Yb nucleus that can currently be made at the limiting spin ($\sim$60 $\hbar$) is $^{168}$Yb, produced in the fusion reaction $^{48}$Ca($^{124}$Sn, 4n)$^{168}$Yb. Although it is not possible to increase the fission barrier of a given nucleus, one can produce different nuclei having higher fission barriers. The fissility of a nucleus depends largely on $Z^2/A$, and in the ytterbium ($Z = 70$) region, for example, an increase of mass number by one unit will produce a nucleus holding about one additional unit of angular momentum before fissioning. Thus to make Yb nuclei that sustain more angular momentum, it is necessary to make heavier isotopes. For example, the reaction with radioactive Sn nuclei, $^{132}$Sn($^{48}$Ca, 4n)$^{176}$Yb, is considered to produce a reasonable candidate. The $^{176}$Yb nucleus should sustain about 8 $\hbar$ more angular momentum than $^{168}$Yb, giving about 10% more angular momentum, and correspondingly larger Coriolis and centrifugal forces. This will greatly help in the understanding of the entire virgin region above 49 $\hbar$, and it may well provide access to some entirely new phenomena.

*3.2.1d. Highly Excited States of Neutron-Rich Nuclei.* In compound nuclear reactions using neutron-rich nuclei, such as $^{6,8}$He and $^{9}$Li, on stable targets, highly excited states of neutron-rich nuclei are populated (Bu 85). Nearly all known, and even some unknown, neutron-rich nuclei can be reached in this way. Some of these reactions are listed in Table 8. The excitation energies of these compound nuclei can be varied over a wide range by using different target–projectile combinations; in particular by using higher $Z$ neutron-rich projectiles, the reaction Q-value can be lowered. The Q-values attainable in some of these compound nuclear reactions is so high that even the excitation of giant resonances or of stretched states seems possible. As yet, nothing is known about giant resonances or highly excited states in extremely neutron-rich light nuclei.

### *3.2.2. Interaction Cross-Sections and Sizes of Exotic Nuclei*

Many detailed studies of the size and shape of stable nuclei have been made using electromagnetic and nuclear probes. Among them, electron scattering gives the most accurate information about the charge distribution over a wide range of stable nuclei.

**Table 8. Fusion Reactions and their Q-Values (Bu 85)**

| Compound Nucleus | Reaction | $Q$-value (MeV) |
|---|---|---|
| $^{9}$Li | $^{6}$He + t | 6.4 |
| $^{11}$Li | $^{8}$He + t | 3.3 |
| $^{10}$Be | $^{9}$Li + p | 19.6 |
| | $^{6}$He + $^{4}$He | 7.4 |
| $^{11}$Be | $^{9}$Li + d | 17.9 |
| | $^{8}$He + $^{3}$He | 26.4 |
| $^{12}$Be | $^{9}$Li + t | 15.0 |
| | $^{8}$He + $^{4}$He | 9.1 |
| | $^{11}$Li + p | 25.6 |
| $^{13}$Be | $^{11}$Li + d | 20.7 |
| $^{14}$Be[a] | $^{11}$Li + t | 14.5 |
| $^{12}$B | $^{6}$He + $^{6}$Li | 18.3 |
| | $^{8}$Li + $^{4}$He | 10.0 |
| | $^{9}$Li + $^{3}$He | 26.5 |
| $^{13}$B | $^{6}$He + $^{7}$Li | 27.1 |
| | $^{9}$Li + $^{4}$He | 10.8 |
| $^{14}$B | $^{8}$He + $^{6}$Li | 21.5 |
| | $^{9}$Li + $^{4}$He | 34.0 |
| $^{15}$B | $^{8}$He + $^{7}$Li | 17.1 |
| | $^{11}$Li + $^{4}$He | 16.3 |

[a] Mass assumed: $^{13}$Be + n.

Figure 41 shows the half-density radius $r_{0.5}$ of the nuclear charge density (Co 67). The mass-number dependences of the half-density radii are, on the whole, well-fitted by the formula

$$r_{0.5} = 1.18A^{1/3} - 0.48 \quad \text{(fm)} \tag{19}$$

as shown by the solid line in the figure.

Although they give precise information, electron scattering and other electromagnetic probes, such as muonic X-rays and isotope-shift determinations, are primarily sensitive to the charge distribution and thus to the proton distribution. The matter distribution and the neutron radius have, therefore, to be studied through strongly interacting probes. A determination of the radius from an experiment with strongly interacting probes is generally less accurate, due to difficulties in the analyses. So far, the most reliable results on nuclear matter distributions have been obtained from the elastic scattering of high-energy ($\sim$1 GeV) protons. An advantage of this method is that at high energy the mechanism of scattering is simple and there exists a reliable scattering theory which allows one to relate the measured cross-sections to the free proton–nucleon amplitudes and nuclear densities under investigation. It has been shown that the differences between the neutron and proton distributions are small for stable

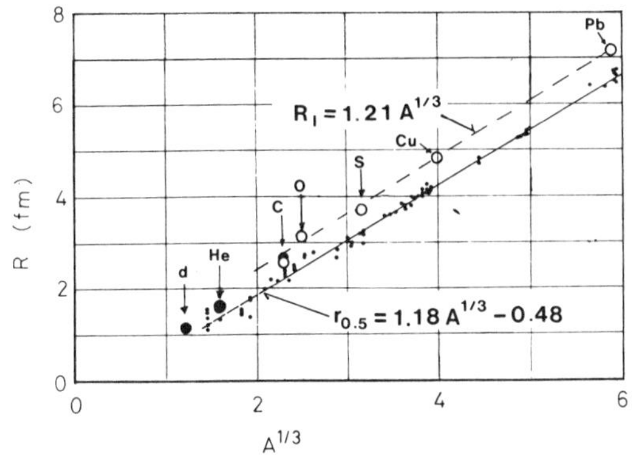


**Figure 41.** Half-density radii measured by electron scattering (small dots). The interaction nuclear radii $R_I$ defined by equation (20) are indicated by larger circles.

nuclei. Small but definite differences between the neutron and proton root-mean-square (rms) radii have been determined for several nuclei, e.g., that for $^{48}$Ca is about 0.1 fm (Al 76, Ch 78).

Nuclear radii have, however, so far been studied mainly for stable nuclei. To date, the radii of unstable nuclei have been measured only by isotope-shift measurements using on-line mass separators. Isotope-shift measurements have been extended to unstable isotopes, especially at the ISOLDE facility at CERN (Ot 81). (See also Chapter 7 of this Volume.) Interesting data have been obtained on the Hg-isotopic chain, and on the alkali isotopes (Bo 73). The Hg data have revealed a shape transition to strong deformation at $A = 185$, 183, and 181. Also, for the isotopes in the vicinity of the magic neutron number $N = 82$, the data reveal a clear indication of change of shape. Although it provides a challenging test to our understanding of nuclear structure in the region far from stability, the applicability of the isotope-shift method is restricted to the nuclei of certain elements only, because of the necessary use of an ion source.

Recently, new experiments which determine nuclear-matter size using a high-energy radioactive beam have been reported. Here we present the principle of this experiment and some recent results from it.

Interaction cross-sections $\sigma_I$ for heavy-ion collisions have been extensively measured at the BEVALAC in Berkeley for incident energies from several hundred MeV per nucleon up to 2 GeV/nucleon. Here, $\sigma_I$ is defined as the total cross-section for the change of proton and/or neutron number in the incident nucleus. It has for some years been known that $\sigma_I$ between stable nuclei is essentially independent of the beam energy when above a few hundred MeV/nucleon (Go 78, Ja 78). Also, the nucleon–nucleon cross-section reaches a saturation value at an energy of around 800 MeV. It is, therefore, not surprising that $\sigma_I$ reflects a well-defined nuclear size. One can define an interaction nuclear radius, assuming the separability of the projectile and the target radius, as follows:

$$\sigma_I(\mathrm{p}, \mathrm{t}) = \pi[R_I(\mathrm{p}) + R_I(\mathrm{t})]^2 \qquad (20)$$

where $R_I(\mathrm{p})$ and $R_I(\mathrm{t})$ are the interaction radii of the projectile and the target nuclei, respectively. The systematic change of $R_I(\mathrm{p})$, for example, can be obtained from a series of $\sigma_I$ measurements with the same target. The absolute value can be obtained from the $\sigma_I$ for collisions between identical nuclei.

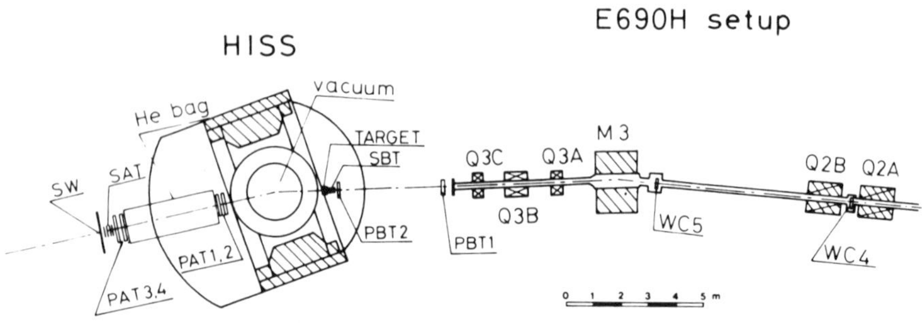


**Figure 42.** Experimental arrangement for the interaction cross-section measurements. All the scintillation counters (SBTs, and SATs) were 3 mm in thickness. Wire chambers PAT1–PAT4, PBT1, and PBT2 were used for tracing nuclei for purposes of their identification.

In the definition of the interaction radius $R_I$, a separability of the projectile and the target radius was the only assumption introduced. It should be noted that a black-sphere model satisfies this assumption although it is not *necessary* to have a black sphere nucleus in order to satisfy it. Open circles in Figure 41 show the interaction radii obtained from $\sigma_I$ between stable isotopes. As shown in this figure, the $R_I$ lie about 0.6 fm above the half-density radii obtained from the electron-scattering measurement. This shows that the $R_I$ values are sensitive to the peripheral density distribution of nuclei. Before discussing the meaning of the $R_I$ in greater detail, let us examine the data obtained with radioactive nuclei.

At the BEVALAC, interaction cross-sections were determined at 790 MeV/nucleon by a transmission measurement using a large-acceptance spectrometer. The experimental arrangement is shown in Figure 42. The individual nuclear species were identified before incidence on a secondary reaction target. Their identification and selection were achieved using scintillation counters (SBTs) and multiwire proportional chambers as well as time-of-flight measurements between scintillation counters located 50 m upstream of the target and the SBTs. The total fraction of all other isotopes in such a selected-isotope beam was considerably less than $10^{-3}$. The intensity of the beam was about 1000 $s^{-1}$ and was limited by the speed of the data-taking system. For nuclei near the particle drip-line, such as $^{8}$He ($\approx$500 $s^{-1}$), $^{11}$Li ($\approx$200 $s^{-1}$), beam intensities were smaller but adequate for measurements.

When the same nuclei were detected both before and after the target, the event was counted as noninteracting. The cross-section was then calculated from the equation

$$\sigma_I = (1/N_t)\log(\gamma_0/\gamma) \tag{21}$$

where $\gamma$ is the ratio of the number of noninteracting nuclei to the number of incoming nuclei for a target-in run, and $\gamma_0$ is the same ratio for a target-out run; and $N_t$ is the number of target nuclei per unit area. By taking the ratio $\gamma_0/\gamma$, uncertainties due to counter inefficiencies and reactions occurring anywhere other than in the target were eliminated. Interaction cross-sections were determined for all He, Li, and Be isotopes using three target nuclei: Be, C, and Al.

The assumption of separability in equation (20) was examined using $\sigma_I$ for the various projectile–target combinations. Figure 43 shows the $R_I$ obtained from different

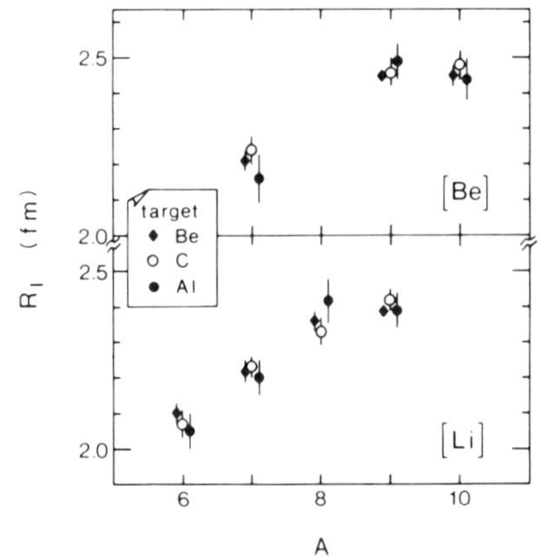


**Figure 43.** Interaction radii $R_I$ for Li and Be isotopes. The values obtained using three different targets agree, showing the separability of projectile and target contributions to $R_I$ (Ta 85b).

targets where the absolute scale for the radius is determined from a least-squares fitting of $^4He + ^4He$, $^9Be + ^9Be$, $^{12}C + ^{12}C$, $^4He + ^{12}C$, and $^9Be + ^{12}C$ interaction cross-sections (Ta 85b). It is evident that the projectile radius is independent of the target material. The validity of the separability assumption for He isotopes was also demonstrated (Ta 85a), confirming that it holds well for the light nuclei studied.

As shown above, $R_I$ is a well-defined experimental size parameter. A comparison of $R_I$ with the calculated matter distributions is, however, remote at this point in time. Therefore, Glauber-type calculations were made to determine the rms radius from the measured $\sigma_I$. It was found that the rms radius for the nuclear-matter distributions could be determined essentially independently of the selection of model density distribution used in the Glauber calculation. This was demonstrated using three different distributions: Gaussian (Ka 75), harmonic-oscillator shell model (El 61), and droplet model with a Yukawa folding function (My 83).

Table 9 shows the values of the rms radii deduced from the measured $\sigma_I$ using these Glauber model calculations. The rms radii of the charge distributions were calculated by folding the charge distribution of the proton into a point-proton distribution, i.e.,

$$R^c_{rms} = [(R^p_{rms})^2 + (0.8)^2]^{1/2} \quad (\text{fm}) \tag{22}$$

where 0.8 fm is the rms charge radius of a proton. The proton rms radius was assumed to be equal to that of the neutron in the case of the Gaussian distribution. A comparison of the rms charge radii with those determined by electron-scattering experiments is shown in Figure 44 for stable nuclei. It can be seen that they agree very well with one another. Even the difference between the radii of $^6Li$ and $^7Li$ is reproduced by the harmonic-oscillator distribution (because of the occupation-number difference between protons and neutrons).*

* It has been reported (Ta 85b) that the reduction factor (80%) for the nucleon–nucleon cross-section used in the Glauber calculation was chosen to give the best fit to the electron-scattering data. However, it was then found that there was some confusion between the point-nucleon distribution and the folded (by the nucleon form-factor) distribution in the calculation. After correction of this mistake in the calculation, it was found that the *free nucleon–nucleon cross-section* gives a good fit, as shown in Figure 44.

**Table 9. Interaction Nuclear Radii and rms Radii, in Fermis**

| | | $R_{rms}$ | | | |
|---|---|---|---|---|---|
| | | Gaussian | Harmonic osc. | | Electron scatt. |
| | $R_I$ | (matter) | (matter) | (charge)[a] | (charge) |
| $^4$He | 1.41 ± 0.03 | 1.59 ± 0.05 | 1.57 ± 0.05 | 1.76 ± 0.04 | 1.67 ± 0.01 |
| $^6$He | 2.18 ± 0.02 | 2.52 ± 0.03 | 2.45 ± 0.03 | 2.35 ± 0.03 | |
| $^8$He | 2.48 ± 0.03 | 2.55 ± 0.03 | 2.52 ± 0.03 | 2.30 ± 0.03 | |
| $^6$Li | 2.09 ± 0.02 | 2.35 ± 0.03 | 2.32 ± 0.03 | 2.46 ± 0.03 | 2.56 ± 0.10 |
| $^7$Li | 2.23 ± 0.02 | 2.35 ± 0.03 | 2.33 ± 0.03 | 2.40 ± 0.03 | 2.41 ± 0.10 |
| $^8$Li | 2.35 ± 0.02 | 2.38 ± 0.02 | 2.37 ± 0.03 | 2.40 ± 0.02 | |
| $^9$Li | 2.41 ± 0.02 | 2.32 ± 0.02 | 2.32 ± 0.02 | 2.32 ± 0.02 | |
| $^{11}$Li | 3.14 ± 0.16 | 3.11 ± 0.22 | 3.10 ± 0.23 | 2.97 ± 0.21 | |
| $^7$Be | 2.22 ± 0.02 | 2.33 ± 0.02 | 2.31 ± 0.03 | 2.49 ± 0.03 | |
| $^9$Be | 2.45 ± 0.01 | 2.38 ± 0.01 | 2.38 ± 0.01 | 2.47 ± 0.01 | 2.52 ± 0.01 |
| $^{10}$Be | 2.46 ± 0.02 | 2.28 ± 0.02 | 2.30 ± 0.02 | 2.38 ± 0.02 | |
| $^{12}$C | 2.61 ± 0.02 | 2.32 ± 0.02 | 2.35 ± 0.02 | 2.48 ± 0.02 | 2.45 ± 0.01 |

[a] Charge radii are obtained by folding the rms charge radius of the proton (0.8 fm) with its matter distribution.

Comparison of $R_I$ and the density distributions which reproduce the measured $\sigma_I$, show that $R_I$ represents the radius for which the matter density is about 0.05 fm$^{-3}$ for $A \geqslant 6$ nuclei. Because $R_I$ is thus an almost constant-density radius, it is reasonable that it shows a similar mass dependence to the half-density radius as shown in Figure 41.

Figure 45 shows the rms radii of nucleon distributions determined by use of the harmonic-oscillator distribution. Appreciable differences of radii are observed, for the first time, between pairs of isobars with different isospin, such as $^6$He–$^6$Li, $^8$He–$^8$Li, and $^9$Li–$^9$Be. The larger radii of the neutron-rich isotopes $^6$He and $^8$He, which have only two protons, suggest the existence of a thick neutron skin. However, this statement depends on the assumed density distribution and is thus model-dependent. Other measurements, such as those on charge distributions, are therefore necessary to confirm

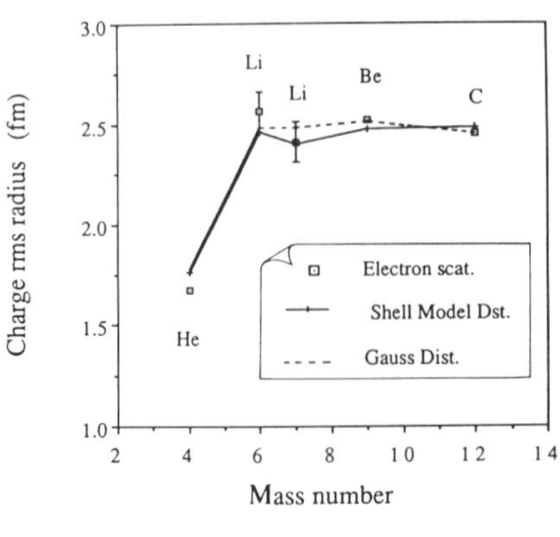


**Figure 44.** Charge rms radii. Squares indicate the charge rms radii $R_{rms}$ determined by electron-scattering experiments. The dashed line shows the $R_{rms}$ obtained by fitting $\sigma_I$ with a Glauber-type calculation employing a Gaussian density distribution. The solid line shows the result obtained using a shell-model harmonic-oscillator distribution.

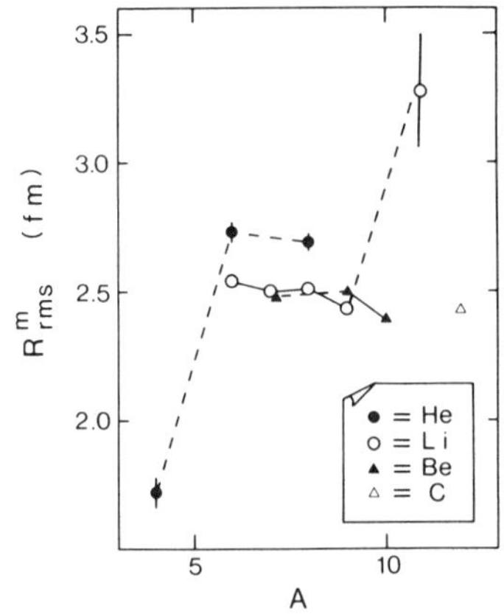


**Figure 45.** Matter rms radius $R^{m}_{rms}$. Lines connecting isotopes are only guides for the eye. Differences in radii are seen for isobars with $A = 6, 8$, and 9. The $^{11}$Li isotope has a much larger radius than other nuclei (Ta 85b).

this apparent difference of radii between proton and neutron distributions. The radii of a pair of mirror nuclei $^7$Li–$^7$Be, on the other hand, are identical.

It is interesting to note that $^{11}$Li, with neutron number 8—which is the p-shell closure in the naive shell model—shows a considerably larger radius than do neighboring nuclei. This may suggest the existence of a large deformation and/or of a long tail on the nucleon distribution reflecting the very weak binding of the outer nucleons. A weakly bound nucleon may increase the $\sigma_I$ because it could be knocked out of the nucleus with only a small momentum transfer. A rough estimate, based on the energy distribution of nucleons after a nucleon–nucleon scattering, shows that a change in separation energy from 10 MeV to zero affects the $\sigma_I$ by only about 3%. This binding-energy effect, therefore, seems inadequate to explain the bulk of the observed deviation. Further studies are clearly necessary to understand this situation.

As shown above, it has been demonstrated that measurements of the interaction cross-sections in high-energy nuclear collisions using radioactive nuclei provide a method for determining the nuclear radii of $\beta$-unstable nuclei as well as of stable nuclei. Extension of these measurements to a wider area of the nuclear chart will definitely provide basic and important information on the structure of nuclei far from stability.

### *3.2.3. Fragmentation of Exotic Nuclei*

It has been shown that the momentum distribution of projectile fragments reflects the motion of the nucleons in the nucleus (Go 74). In particular, the momentum distribution of the one-nucleon removal channel in projectile fragmentation was studied in detail by Fujita and Hüfner (Fu 79, Hu 80). They showed that the momentum distribution of a nucleon could be determined up to around 500 MeV/c. From an analysis of $^{16}\text{O} \rightarrow {}^{15}\text{O}$ fragmentation at 2.1 GeV/nucleon, it was also shown that the correlation of nucleons in the nucleus affects the momentum distribution.

When the momentum distribution of a nucleus in a two-nucleon removal channel is measured, the width (standard deviation) of the $P_\parallel$ distribution (see Figure 17 for definition of $P_\parallel$ and $\sigma$) in the fragment coordinate system $\sigma_{12}$ can be expressed as

$$\sigma_{12}^2 = \sigma_1^2 + \sigma_2^2 + 2/3\langle P_1 P_2 \rangle \tag{23}$$

where $\sigma_1$ and $\sigma_2$ indicate the widths (standard deviation) of the single-nucleon distribution of the two removed nucleons (proton or neutron), respectively, and $\langle P_1 P_2 \rangle$ is the correlation between the two nucleons. As an example, the momentum distributions of $^{14}$C, $^{14}$N, and $^{14}$O from the fragmentation of 2.1 GeV/nucleon $^{16}$O have been reported by Heckman *et al.* (Li 75, He 78). Their data:

$$\sigma_{pp} = 125 \pm 3 \text{ MeV/c} \quad \text{for} \quad ^{14}\text{C} \tag{24}$$

$$\sigma_{pn} = 112 \pm 3 \text{ MeV/c} \quad \text{for} \quad ^{14}\text{N} \tag{25}$$

$$\sigma_{nn} = 99 \pm 6 \text{ MeV/c} \quad \text{for} \quad ^{14}\text{O} \tag{26}$$

shows differences between the widths, depending on the combination of nucleons present. It has not yet been investigated in detail, but the difference of the pp, nn, and pn correlations would be expected to be important in understanding these differences.

The momentum distributions have been studied for stable nuclei using electron and proton scattering. Nucleon correlations were also studied, but only for stable nuclei, using proton scattering at high energy. In exotic nuclei, where the proton and neutron numbers are considerably different—and thus the binding energies of protons and neutrons are different—it is very important to know the momentum distributions of the nucleons. In particular, a comparison of the momentum distributions for isobars is very important. Projectile fragmentation of radioactive nuclei is certainly a promising tool for such studies.

### 3.2.4. Coulomb Excitation and the Giant Resonance

As described by de Boer in Volume 1 of this treatise (Bo 84b), Coulomb excitation has provided much spectroscopic information on the low-lying states of nuclei. Coulomb excitation has the advantage that its reaction mechanism is completely understood and thus model-independent parameters for the states involved can be found. Indeed, states with large electrical-quadrupole as well as electric-magnetic dipole moments are strongly excited in Coulomb excitation, thus probing mainly the collective properties of nuclei. Coulomb excitation using radioactive beams provide unique conditions for studies of nuclear structure, level densities, and deformation of nuclei far from stability.

Such an experiment has been proposed using the Rochester recoil separator (see Section 2.2.3). The plan is to detect de-excitation $\gamma$ rays emitted from Coulomb-excited levels (Pi 85). Since the probability of Coulomb excitation increases with nuclear deformation, an experiment to study neutron-deficient nuclei in the Kr–Sr–Zr region is proposed. The recent discovery of very large deformations (Pi 81) in this region, near $N = Z = 40$, has generated much new theoretical activity (Hi 83, Pe 83). Figure 46 shows the systematics of the $2^+ \rightarrow 0^+$ energies and B(E2) values in the light even–even Sr nuclei. It is evident that the lighter-mass isotopes are greatly favored in Coulomb excitation.

Thick-target Coulomb excitation calculations were made for $^{78}$Sr. It was found that the Rochester recoil mass spectrometer facility, a counting rate of 2000 counts/s of $\gamma$ rays can be detected from the $2^+ \rightarrow 0^+$ transition using $^{58}$Ni ions on a $^{24}$Mg target. This method has several advantages over in-beam spectroscopy measurements. Most importantly, the high and complex background produced by large gamma multiplicities, which characteristically accompany in-beam measurements, will be removed. Also, the

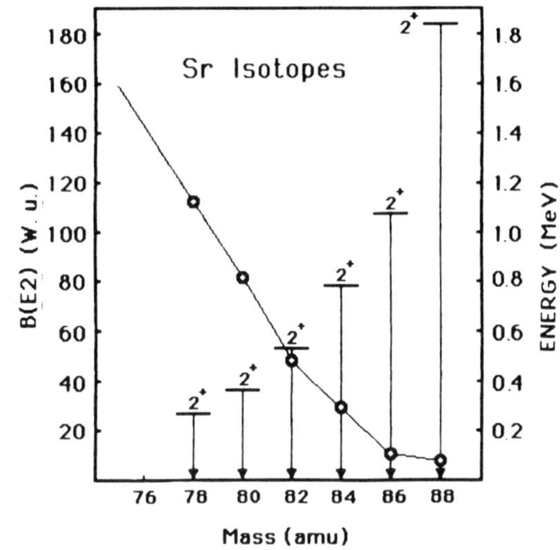


**Figure 46.** Systematics of the energies and transition probabilities for the first excited states in the light Sr nuclei (Pi 85).

technique strongly favors the population of states in deformed nuclei, which in this region are the ones farthest from stability. And finally, since compound nuclear reactions tend to populate states near or on the yrast line, this method also provides, for the first time, the possibility of studying low-spin collective states which are well above the yrast line in a nucleus with a lifetime shorter than a few hours.

At high energy, the electromagnetic dissociation of projectile nuclei has been studied (Ol 81, Ol 83) using 1.7 GeV/nucleon $^{18}O$ ions incident on a number of target materials ranging from Be to U. It was observed that the bombardment of a high-$Z$ target by beam nuclei produces a situation in which, for impact parameters greater than the range of nuclear forces, the projectile nucleus can pass through the electromagnetic field of the target nucleus and thus, in effect, see a virtual-photon target (almost entirely transverse for incident energies much larger than the Coulomb barrier). This phenomenon was observed as an enhancement of the fragmentation cross-sections in one-nucleon and two-nucleon removal channels. Figure 47 shows the target-mass dependence of the fragmentation cross-sections for the $^{18}O$ projectiles. The solid line indicates the cross-section expected from the reaction due to nuclear effects only. Large enhancements of the cross-sections due to electromagnetic dissociation are clearly present. Figure 48 shows the cross-sections for electromagnetic dissociation only (i.e., after subtraction of the nuclear contributions). The curves in this figure indicate the theoretical predictions based on an assumed virtual-photon spectrum and the measured photonuclear cross-sections ($\gamma$, n), ($\gamma$, 2n), and ($\gamma$, p). They show striking agreement. The important conclusion to be drawn is that we can deduce the photonuclear cross-sections for unstable nuclei from measurements of the electromagnetic dissociation using virtual-photon theory.

Figure 49 shows the theoretical photon spectrum for various incident energies. As shown in the figure, high-energy photons above 30 MeV may be present when a heavy ion of 1 GeV/nucleon or higher is involved. This therefore gives rise to the possibility of exciting giant resonances in such studies.

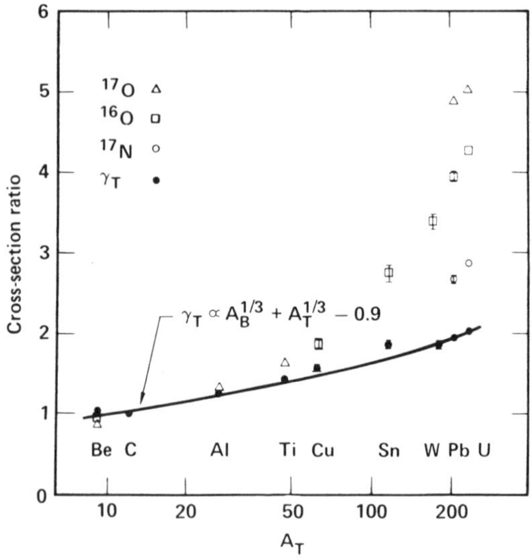


**Figure 47.** Target dependence of the fragment-production cross-section ratios $[\sigma(\mathrm{T, F})]/[(\sigma(\mathrm{Be, F})]$. The target factor $\gamma_T$ is obtained by averaging these ratios over eight different fragments. The curve is obtained by fitting the $\gamma_T$ points into the form $\gamma_T \propto A_T^{1/3} + A_B^{1/3} + X$, where $X$ is the fitting parameter. Note the enhancement for $^{17}$O, $^{16}$O, and $^{17}$N fragments resulting from the electromagnetic-dissociation process (Ol 81).

The isospin distribution of the giant dipole resonance requires the study of isobaric series of non-self-conjugate nuclei, only one of which is normally stable. To date, $^{18}$O is the only nucleus for which such an analysis of the isospin decomposition has been made (Be 76, Al 79a). In order to see whether the isospin composition of the giant dipole resonance varies with $N - Z$, it is necessary to study $^{18}$Ne, which is unstable. The other non-self-conjugate nuclei are $^{13}$C and $^{13}$N, and in this case we need to study the unstable nucleus $^{13}$N (Be 84).

Berman has also pointed out the possibility of using this mechanism for studying photonuclear properties of radioactive nuclei (Be 84). One question here is that of the

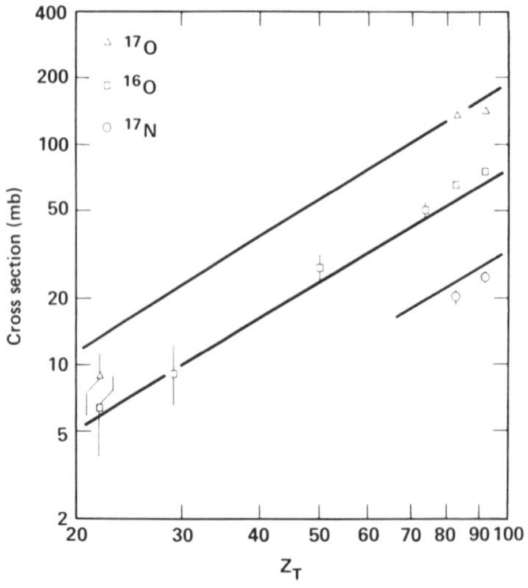


**Figure 48.** Electromagnetic-dissociation cross-section $\sigma_{em}$ vs. $\log(Z_T)$. The curves represent the theoretical predictions (Ol 81).

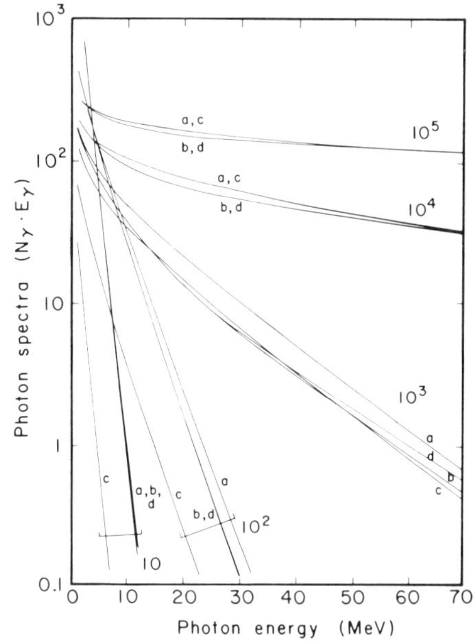


**Figure 49.** Virtual-photon spectra. The sets of curves are labeled according to beam energy, from 10 to $10^5$ MeV/nucleon. These are calculated for a U target with $b_{min} = 11.6$ fm. The labels a, b, c, and d indicate different parameter sets (Ol 81).

existence (and strength) of the short-range correlations in nuclear ground states which can be determined from the enhancement of the electric-dipole absorption strength over the sum-rule value. There is, currently, strong disagreement—particularly for $^{16}O$ and $^{40}Ca$—between the large values reported from total photon-absorption measurements and the small ones deduced from summing the measurements of all the partial photonuclear-reaction channels (Ah 75, Be 82). Since the giant-dipole-resonance strength for self-conjugate nuclei is entirely $T_>$ (=1), one would expect that the energy distribution of the dipole strength to be at least roughly the same, for all light self-conjugate nuclei. Hence systematic measurements on these nuclei would be of great value especially for the $\beta$-unstable species $^{18}F$, $^{22}Na$, $^{26}Al$, $^{30}P$, $^{34}Cl$, and $^{38}K$, in order to bridge the gap between $^{16}O$ and $^{40}Ca$.

### *3.2.5. Other Nuclear Reactions*

Besides the nuclear reactions described above, many other reactions can be studied using radioactive beams. Here we list such reactions together with some brief discussion.

*3.2.5a. Transfer Reactions.* Measurements on transfer reactions have been proposed for the TRIUMF facility (Bu 85). In the proposal these authors have stated that more states, as well as more nuclei, will be accessible in transfer reactions involving the direct use of a radioactive beam. Besides discovering the states (thus giving excitation energies and widths), other properties of these states can be derived. Table 10 shows $Q$-values of selected reactions together with $Q$-values for compound nuclei formed with helium and lithium isotopes as projectiles (Wa 71). All of the $Q$-values in the table are positive—a common property of reactions using nuclei far from stability as projectiles—thus allowing measurements made at energies close to the Coulomb barrier

**Table 10. *Q*-Values of Transfer/Fusion Reactions on Neutron-Rich Light Nuclei (Bu 85)**

| Final nucleus | Reaction | Compound nucleus | *Q*-Compound (MeV) | *Q*-Transfer (MeV) |
|---|---|---|---|---|
| $^4H$ | $^6He(d, \alpha)^4H$ | $^8Li$ | 9.8 | 2.4 |
| $^5H$ | $^8He(p, \alpha)^5H$ | $^9Li$ | 14.0 | 2.7 |
| $^6H$ | $^8He(d, \alpha)^6H$ | $^{10}Li$ | 9.4 | >3.2 |
| $^5He$ | $^6He(^3He, \alpha)^5He$ | $^9Be$ | 21.2 | 18.7 |
| $^6He$ | $^8He(p, t)^6He$ | $^9Li$ | 14.0 | 6.4 |
| | $^8He(^{16}O, ^{18}O)^6He$ | $^{24}Ne$ | 32.9 | 10.0 |
| $^8He$ | $^9Li(t, \alpha)^8He$ | $^{12}Be$ | 15.0 | 5.8 |
| | $^{11}Li(p, \alpha)^8He$ | $^{12}Be$ | 25.6 | 16.5 |
| $^8Li$ | $^9Li(^3He, \alpha)^8Li$ | $^{12}B$ | 26.5 | 16.5 |
| $^9Li$ | $^{11}Li(p, t)^9Li$ | $^{13}Be$ | 25.6 | 10.7 |
| $^{10}Li$ | $^{11}Li(^3He, \alpha)^{10}Li$ | $^{14}B$ | 34.0 | 20.5 |

still to reach highly excited states of the final residual nucleus. For example, a proton target bombarded with $^8$He at a few MeV incident energy can be used for the simultaneous study of $^6$He, $^7$He, and $^5$H, as well as the observation of elastically scattered particles, using a single detection system. In some cases, neutron or neutron-rich transfer reactions are possible [e.g., $^{11}$Li (t, p)$^{13}$Li] leading to extremely neutron-rich nuclei.

In these experiments, the most important factor is the beam intensity. Transfer reactions above the Coulomb barrier show cross-sections of about 1–100 mb for the population of one state, with cross-sections normally peaking in the forward direction. Thus, with a target thickness of $10^{19}$ atoms/cm$^2$, a beam intensity of $10^{10}$ ions/s, and a backward angle cross-section of 10 $\mu$b/sr, one would still expect 0.1 event/s in a detector with solid angle $(\Omega/4\pi)$ of $10^{-2}$. Intensities up to $10^{10}$ s$^{-1}$ can be delivered from ISOL + LINAC combinations for $^6$He and $^8$He as well as for $^8$Li and $^9$Li. A typical intensity for $^{11}$Li is expected to be $10^6$ ions/s (Ra 85). But nevertheless, the forward peaked cross-sections of about 1 mb/sr should result in 100 events per day, thus permitting the establishing of the structures of such nuclei as the nearly or totally unknown $^9$He and $^{10}$He. In addition, if one can design a counter with something more like $4\pi$ geometry, the yield would increase by at least a factor of ten.

*3.2.5b. Charge-Exchange Reactions Between Mirror Nuclei.* Charge-exchange reactions between a stable projectile and target, and single and double charge-exchange reactions with pions incident on stable targets, are well-established spectroscopic tools. Vary and Nagarajan (Va 85) suggested that charge-exchange reactions between mirror nuclei, such as $^{17}F(^{17}O, ^{17}F)^{17}O$, provide a unique opportunity for studying interesting and nontrivial wave-function components. These reactions between mirror nuclei are relatively simple to analyze in a coupled-channel formalism and hence may provide a powerful new spectroscopic tool. These authors indeed suggest that a resonance charge-exchange may occur.

The production of the $\beta$-unstable partner in mirror nuclei in the sd-shell is relatively easy and one could thus expect a high-intensity beam, making such studies quite feasible.

### 3.2.6. *Nuclear Astrophysics*

One of the primary goals of nuclear astrophysics is the determination of the location, time, and physical environment for the nucleosynthesis of the chemical elements and their isotopes. Experimental and theoretical nuclear physics supply the rates, as functions of temperature and density, for the strong and weak interactions which generate stellar energy through nuclear transmutation. The results of integrating nuclear physics within the framwork of theoretical astrophysics will provide an understanding of the origin of the chemical elements of which our planetary system is composed and on which life itself is based.

In ordinary stellar environments, such as the interior of our sun, nuclear burning occurs at such low temperatures that the nuclear reaction rates are slower than those associated with the radioactive decay. The species involved are then close to the line of $\beta$ stability. This type of cold nucleosynthesis (typically via a CNO cycle) has received a great deal of attention in the past. One of the important properties of the CNO cycle, however, is that it is a closed one and therefore no element heavier than fluorine is synthesized.

It is now widely recognized, however, that a variety of stellar environments are so hot and dense that the nuclear reaction rates are comparable with those of natural decay processes (Fo 84). Such conditions are believed to prevail on, for example, the surfaces of accreting neutron stars and in novae, supernovae, and massive stars. The high rates of nuclear reaction in such conditions break the closed CNO cycle and produce leak-out of such nuclei as $^{19}$Ne and subsequently $^{20}$Na through $^{15}$O($\alpha$, $\gamma$)$^{19}$Ne(p, $\gamma$)$^{20}$Na reactions. A complicated chain of (p, $\gamma$) and weak decays then leads to the production of iron-group nuclei. (This process is illustrated in Figure 50.) The flow path near the proton drip-line includes mainly nuclei with lifetimes less

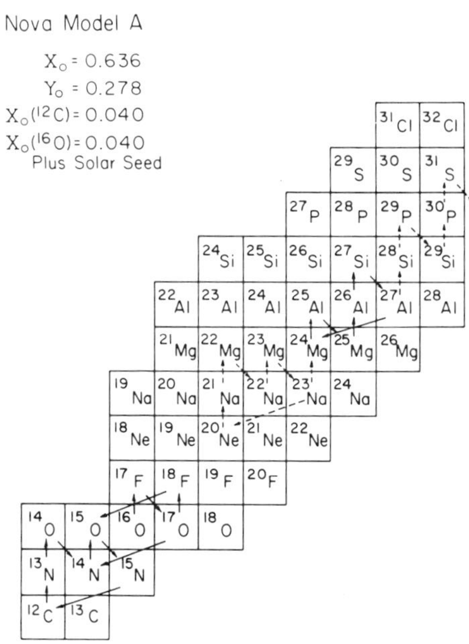


**Figure 50.** Flow path of the hot CNO cycle (Wa 84).

than 300 ms. Such an rp-process (Wa 81) can produce up to 100 times the energy generation rate of the $\beta$-decay limited CNO cycle at early stages of stellar evolution. However, it destroys seed nuclei, thus decreasing the energy generation at later stages. However, very few of the thermonuclear reaction rates required for this hot CNO cycle are known, and most are simply assumed to be averaged values. The actual rates could vary considerably from the estimated ones, and the results of the present hot CNO network calculations could differ appreciably from those which actually occur.

Another example involves the nuclear flows occurring during a nova outburst. The final abundance of $^{22}$Na (a $\gamma$-line astronomy candidate and source for the neon-E meteoritic anomaly) and $^{26}$Al (another important meteoritic anomaly) rely strongly upon the poorly known $^{21,22}$Na(p, $\gamma$)$^{22,23}$Mg and $^{26}$Al(p, $\gamma$)$^{27}$Si rates. Some of the published calculations have differed by a factor of $10^{10}$ in their estimate of the Na rates. There are further complications in $^{26}$Al reactions because of the existence of the isomer; the ground state of $^{26}$Al ($5^+$) has a half-life of $7.2 \times 10^5$ years and the $0^+$ isomeric state ($E_x = 0.228$ MeV) has a half-life of 6.34 s. Both levels contribute to the chain of nuclear reactions. A similar complication is found in $^{34}$Cl. A detailed discussion of these isomeric states is presented elsewhere (Wa 80a, Bo 85).

Many nuclear reaction cross-sections which involved $\beta$-unstable nuclei have to be measured if we are to fully understand the distribution of elements in the universe. The importance of the use of radioactive as well as isomer beams for nuclear astrophysics has been emphasized by Fowler (Fo 84).

At the Lawrence Livermore National Laboratory, the reaction $^7$Be(p, $\gamma$)$^8$B has been studied using a quadrupole sextuplet secondary transport system (see Section 2.3.1). This reaction produces $^8$B, the decay of which is believed to yield about 90% of the neutrinos from our Sun that are detectable using the $^{37}$Cl($\nu$, e$^-$)$^{37}$Ar reaction. Several groups have measured the proton capture cross-section by constructing targets of $^7$Be which has a 54-day half-life. There is a suggestion of disagreement among the results (Ba 80), perhaps because of the difficulty of working with such a highly radioactive target. An independent measurement would therefore be useful.

At LLNL beams of $^7$Be were produced using the $^1$H($^7$Li, $^7$Be)n reaction and a measurement of the $^7$Be(p, $\gamma$)$^8$B reaction cross-section was attempted. The attempt failed for two reasons. One difficulty arose from the fact that the unreacted $^7$Be struck the detectors. It is extremely difficult to detect the very weak yield of $^8$B produced by the (p, $\gamma$) reaction in the presence of the primary beam. Another difficulty was due to the deuterons contained in any natural hydrogen target. Although the deuteron content in hydrogen is about $1.5 \times 10^{-4}$, the cross-section of the (d, n) reaction, which produces exactly the same residual nucleus $^8$B, is about $10^4$ larger than that of the (p, $\gamma$) reaction. Therefore, this contamination can be of the same order of magnitude as the signal to be detected. The LLNL group is therefore trying to use a target made of *protium*, that is, of hydrogen depleted of deuterium, and they expect to obtain a deuteron-to-proton ratio of $10^{-6}$ in such a target.

A measurement of the (d, n) reaction is easier if one uses a deuteron target. In fact, the LLNL group measured the cross-section for $^7$Be(d, n)$^8$B using a $CD_2$ target. A spectrum measured in this experiment is shown in Figure 51 together with spectra from other targets. The signal is the $^8$B peak in Figure 51b. The secondary-beam intensity was too low (because it was limited to the amount of beam that the primary target could take) in this case to observe any $^8$B formed by proton capture from the $(CH_2)_n$ target. Therefore the spectrum of Figure 51a served as a background. The $^2$H($^7$B, $^8$B)n

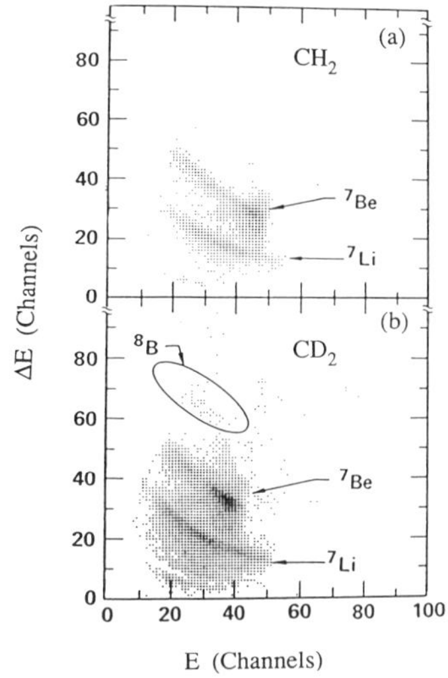


**Figure 51.** Scatter plots of $\Delta E$ *vs.* $E$ spectra for $^7$Be beams on various targets. The size of the points indicates the order of magnitude of the counts per channel: the smallest is for between 1 and 10 counts per channel, the largest for between $10^5$ and $10^6$. (a) Spectrum from a 17.6 MeV $^7$Be beam onto a ($CH_2$) target. (b) Spectrum from a 18.6 MeV $^7$Be beam onto a $(CD_2)_n$ target showing $^8$B particles produced by the $^2$H($^7$Be, $^8$B)n reaction (Ha 83a).

reaction cross-section was thus determined to be 58 ± 11 mb at a $^7$Be beam energy of 16.9 MeV.

Coulomb dissociation of high-energy ions, which was discussed in Section 3.2.4, may provide another method of determining the cross-sections for the (p, $\gamma$) reaction. As was shown in Section 3.2.4 it is possible to deduce the ($\gamma$, p) reaction cross-section from the Coulomb dissociation cross-section. The cross-section for the (p, $\gamma$) reaction is then calculated from the cross-section of its inverse ($\gamma$, p) reaction. A difficulty here is that this reaction cross-section is always given as a cross-section integrated over photon energy. However the cross-section needed for astrophysical purposes is, more or less, cross-section integrated over a certain energy range so that there is a reasonable possibility of obtaining a useful result.

## 4. Conclusion

In this chapter we have reviewed the production and the use of radioactive beams for nuclear studies. Although only a few experiments using radioactive beams have been realized as yet, they already show the necessity as well as the power of radioactive beams in the study of exotic nuclei. The use of radioactive beams provides an opportunity for studying, systematically, nuclear structure over the widest range of nuclei. The author believes that radioactive beam applications will provide a large amount of new nuclear information during the coming ten or more years. He hopes that radioactive beam facilities, which are under construction or in preparation, will become operational as soon as possible.

In this chapter, discussions were mostly devoted to the study of nuclear physics. However, radioactive beams are expected to provide a new tool of study in other fields of research, such as atomic physics, solid-state physics, and biomedicine. Indeed, biomedical studies have already been performed at the BEVALAC in Berkeley using beams of $^{19}$Ne (Ch 82).

The author would much appreciate receiving all information, in whatever field, that relates to the use of radioactive beams.

## Acknowledgments

The author wishes to thank Professor Shoji Nagamiya of Columbia University for his suggestion that he write this review, and for helpful discussions. He also thanks Professor Kenzo Sugimoto for many suggestions. Some of the ideas in the article were provided by the working group on the Japanese Hadron Facility (JHF); the author is grateful to the members of this working group and in particular to the members of the EXOTIC NUCLEI ARENA. He also wishes to express his great appreciation to his many colleagues who have provided important information and figures for use in this chapter.

## References

(Ah 75) J. Ahrens, H. Borchert, K. H. Czock, H. B. Eppler, H. Gimm, H. Gundrum, M. Köning, P. Riehn, G. Sita Ram, A. Zieger, and B. Ziegler, *Nucl. Phys. A* **251**:479 (1975).

(Al 76) G. D. Alkhazov, T. Bauer, R. Beurtey, A. Boudard, G. Bruge, A. Chaumeaux, P. Couvert, G. Cvijanovich, H. H. Duhm, J. M. Fontaine, D. Garreta, A. V. Kulikov, D. Legrand, J. C. Lugol, J. Saudinos, J. Thirion, and A. A. Vorbyov, *Nucl. Phys. A* **274**:443 (1976).

(Al 79) J. R. Alonso, A. Chatterjee, and C. A. Tobias, *IEEE Trans. Nucl. Sci.* **26**:3003 (1979).

(Al 79a) R. A. Alvarez, B. L. Berman, D. D. Faul, and D. Meyer, *Phys. Rev. C* **19**:1667 (1979).

(Al 84) J. R. Alonso and G. Krebs, *Proc. of the Workshop on Prospects for Research with Radioactive Beams from Heavy Ion Accelerators*, Washington, D.C., April 1984, LBL-18187 (1984), p. 153.

(As 19) F. W. Aston, *Philos. Mag.* **38**:707 (1919).

(Ba 73) R. Bass, *Phys. Lett.* **47B**:139 (1973).

(Ba 80) F. C. Barker, *Aust. J. Phys.* **33**:177 (1980).

(Be 76) B. L. Berman, D. D. Faul, R. A. Alvarez, and P. Meyer, *Phys. Rev. Lett.* **36**:1441 (1976).

(Be 82) B. L. Berman, R. Bergire, and P. Carlos, *Phys. Rev. C* **26**:304 (1982).

(Be 84) B. L. Berman, *Proc. of the Workshop on Prospects for Research with Radioactive Beams from Heavy Ion Accelerators*, Washington, D.C., April 1984, LBL-18187 (1984), p. 92.

(Bi 79) K. Van Bibber, D. L. Hendrie, D. K. Scott, H. H. Wieman, L. S. Schroeder, J. V. Geaga, S. A. Cessin, R. Treunhaft, J. Y. Grossiord, J. O. Rasmussen, and C. Y. Wong, *Phys. Rev. Lett.* **43**:840 (1979).

(Bi 85) R. Bimbot, S. Della Negra, P. Aguer, G. Bastin, R. Anne, H. Delagrange, and F. Hubert, *Z. Phys. A* **322**:443 (1985).

(Bo 73) J. Bonn, G. Huber, H.-J. Kluge, U. Köpf, and L. Kugler, *J. Phys. Soc. Jpn.*, Suppl. **34**:317 (1973).

(Bo 83) R. N. Boyd, *OSU Van de Graaff Lab Rep.* **2**:1 (1983).

(Bo 83a) R. N. Boyd, L. Rybarcyk, M. Wiescher, and H. J. Hausman, *IEEE Trans. Nucl. Sci.* **30**:1 (1983).

(Bo 84) R. N. Boyd, L. Rybarcyk, M. Wiescher, H. J. Hausman, and W. Kim, *Proc. of the Workshop on Prospects for Research with Radioactive Beams from Heavy Ion Accelerators*, Washington, D.C., April 1984, LBL-18187 (1984), p. 6.

(BO 84b) J. Boer, In *Treatise on Heavy-Ion Science*, Vol. 1 (D. Allan Bromley, ed.), Plenum Press, New York (1984), p. 293.

(Bo 85) R. N. Boyd, L. Rybarcyk, H. J. Hausman, W. Kim, and P. Schmalbrock, *Nucl. Instrum. Methods Phys. Res. B* **10/11**:378 (1985).

(Bo 85a) R. N. Boyd, *Proc. of the Accelerated Radioactive Beams Workshop*, Parksville, Canada, Sept. 1985, TRIUMF report TRI-85-1 (1985), p. 233.

(Br 80) H. Breuer, K. L. Wolf, B. G. Glagola, K. K. Kwiatkowski, A. C. Mignerey, V. E. Viola, W. W. Wilcke, W. U. Schröder, J. R. Huizenga, D. Hilscher, and J. Birkelund, *Phys. Rev. C* **22**:2454 (1980).

(Br 83) H. Breuer, A. C. Mignerey, V. E. Viola, K. L. Wolf, J. R. Birkelund, D. Hilscher, J. R. Huizenga, W. U. Schroder, and W. W. Wilcke, *Phys. Rev. C* **28**:1080 (1983).

(Br 83a) H. Breuer, N. R. Yoder, A. C. Mignerey, V. E. Viola, K. Kwiatkoski, and K. L. Wolf, *Nucl. Instrum. Methods* **204**:419 (1983).

(Bu 85) L. Buchmann, *Proc. of the Accelerated Radioactive Beams Workshop*, Parksville, Canada, Sept. 1985, TRIUMF report TRI-85-1 (1985), p. 149.

(Bu 85a) L. Buchmann, J. M. D'Auria, G. Mackenzie, H. Schneider, and R. B. Moore, *IEEE Trans. Nucl. Sci.* **32**:3330 (1985).

(Ch 78) A. Chanmeaux, V. Layly, and R. Schaffer, *Ann. Phys (N.Y.)* **116**:247 (1978).

(Ch 82) A. Chatterjee, W. Saunders, E. L. Alpen, J. Alonso, J. J. Scherer, and J. Llacer, *Radiat. Res.* **92**:230 (1982).

(Co 67) H. R. Collard, L. R. B. Elton, and R. Hofstafter, in *Nuclear Radii*, Vol. 2 (H. Schopper, Landolt Bornstein, and NS-Gi', eds.), Springer-Verlag, Berlin (1967).

(Co 81) T. M. Cormier and P. M. Stwertka, *Nucl. Instrum. Methods* **184**:423 (1981).

(Co 83) T. M. Cormier, M. G. Herman, B. S. Lin, and P. M. Stwertka, *Nucl. Instrum. Methods* **212**:185 (1983).

(Cu 86) M. S. Curtin, L. H. Harwood, J. A. Nolen, B. Sherrill, Z. Z. Q. Xie, and B. A. Brown, *Phys. Rev. Lett.* **56**:34 (1986).

(Du 86) J. P. Dufour, R. Del Moral, H. Emmermann, F. Hubert, D. Jean, C. Poinot, M. S. Pravikoff, A. Fleury, H. Delagrange, and K.-H. Schmidt, *Nucl. Instrum. Methods Phys. Res. A* **248**:267 (1986).

(El 61) L. R. B. Elton, *Nuclear Sizes*, Oxford University Press (1961), p. 21.

(Fl 85) A. Fleury, R. D. Moral, J. P. Dufour, H. Emmermann, F. Hubert, C. Poinot, M. S. Pravikoff, H. Delagrange, and R. Bimbot, *Nucl Instrum. Methods Phys. Res. B* **10/11**:369 (1985).

(Fo 84) W. A. Fowler, *Rev. Mod. Phys.* **56**:149 (1984).

(Fu 79) T. Fujita and J. Hüfner, *Nucl. Phys. A* **343**:493 (1980).

(Go 74) A. S. Goldhaber, *Phys. Lett.* **53B**:306 (1974).

(Go 78) A. S. Goldhaber and H. H. Heckmann, *Annu. Rev. Nucl. Sci.* **28**:161 (1978).

(Gr 75) D. E. Greiner, P. J. Lindstrom, H. H. Heckman, Bruce Cork, and F. S. Bieser, *Phys. Rev. Lett.* **35**:152 (1975).

(Gu 86) D. Guerreau, *J. Phys. (Paris)*, Suppl. #8, **C4**:207 (1986).

(Ha 82a) J. H. Hamilton, *Heavy Ion Collisions*, Vol. 3 (R. Bock, ed.), North-Holland, Amsterdam (1982), p. 574.

(Ha 83) R. C. Haight, *Nucl. Instrum. Methods* **212**:245 (1983).

(Ha 83a) R. C. Haight, G. J. Mathews, R. M. White, L. A. Aviles, and S. E. Woodard, *IEEE Trans. Nucl. Sci.* **30**:1160 (1983).

(Ha 84) E. Hagberg, J. C. Hardy, H. Schmeing, and G. Audi, *Proc. of the Workshop on Prospects for Research with Radioactive Beams from Heavy Ion Accelerators*, Washington, D.C., April 1984, LBL-18187 (1984), p. 71.

(Ha 85) E. Hagberg, J. C. Hardy, G. Audi, and H. Schmeing, *Proc. of the Accelerated Radioactive Beams Workshop*, Parksville, Canada, Sept. 1985, TRIUMF report TRI-85-1 (1985), p. 323.

(Ha 85a) S. S. Hanna and J. W. Hugg, *Hyperfine Interact.* **21**:59 (1985).

(He 72) H. H. Heckmann, D. E. Greiner, P. J. Lindstrom, and F. Bieser, *Phys. Rev. Lett.* **28**:926 (1972).

(He 78) H. H. Heckmann, D. E. Greiner, P. J. Lindstrom, and H. Shwe, *Phys. Rev. C* **17**:1735 (1978).

(Hi 83) T. Higo, S. Matsuki, and T. Yanabu, *Nucl. Phys. A* **393**:224 (1983).

(Ho 73) H. Horie and K. Sugimoto, *J. Phys. Soc. Jpn.*, Suppl. **34** (1973).

(Hu 80) J. Hüfner and M. C. Nemes, *Phys. Rev. C* **23**:2538 (1981).

(Hu 86) F. Hubert, J. P. Dufour, R. Del Moral, H. Emmermann, D. Jean, C. Poinot, M. S. Pravikoff, A. Fleury, H. Delagrange, H. Geissel, and K.-H. Schmidt, *J. Phys. (Paris)* Suppl. 47, **C4**:229 (1986).

(Ja 78) J. Jaros, A. Wagner, L. Anderson, O. Chamberlain, R. Z. Fuzesy, J. Gallup, W. Gorn, L. Schroeder, S. Shannon, G. Shapiro, and H. Steiner, *Phys. Rev. C* **18**:2273 (1978).

(Ka 75) P. J. Karol, *Phys. Rev. C* **11**:1203 (1975).

(Ka 84) T. Katayama, *Proc. of the 10th Int. Conf. on Cyclotron and their Appl.*, East Lansing, MI, May 1984, p. 595.

(Ka 86) T. Katayama, *Proc. of the 11th Int. Conf. on Cyclotrons and their Appl.*, Tokyo, Japan, Oct. 1986, to appear.

(Ki 85a) P. Kienle, GSI report GSI-85-16 (1985), p. 1.

(Ki 85b) P. Kienle, GSI report GSI-85-33 (1985), p. 1.

(La 85) M. Langevin, E. Quiniou, M. Bernas, J. Galin, J. C. Jacmart, F. Naulin, F. Pougheon, R. Anne, C. Détraz, D. Guerreau, D. Guillemaud-Mueller, and A. C. Mueller, *Phys. Lett.* **150B**:71 (1985).

(La 85a) F. Pougheon, R. Anne, C. Detraz, D. Guerreau, D. Guillemaud-Mueller, and A. C. Mueller, *Phys. Lett.* **150B**:71 (1985).

(Le 84) G. A. Leander, *Proc. of 7th International Conf. on Nucl. Mass* (1984), p. 466.

(Le 85) J. K. P. Lee, *Proc. of the Accelerated Radioactive Beams Workshop*, Parksville, Canada, Sept. 1985, TRIUMF report TRI-85-1 (1985), p. 77.

(Li 75) P. J. Lindstrom, D. E. Greiner, H. H. Heckmann, B. Cork, and F. S. Bieser, LBL report 3650 (1975), p. 1.

(Ma 83) G. J. Mathews, R. C. Haight, R. G. Lanier, and R. M. White, *Phys. Rev. C* **28**:879 (1983).

(Ma 84) G. J. Mathews, R. C. Haight, and R. W. Bauer, *Proc. of the Workshop on Prospects for Research with Radioactive Beams from Heavy Ion Accelerators*, Washington, D.C., April 1984, LBL-18187 (1984), p. 21.

(Ma 85) G. H. Mackenzie, *Proc. of the Accelerated Radioactive Beams Workshop*, Parksville, Canada, Sept. 1985, TRIUMF report TRI-85-1 (1985), p. 210.

(Ma 85a) G. J. Mathews, R. W. Bauer, R. C. Haight, and K. E. Sale, *Proc. of the Accelerated Radiactive Beams Workshop*, Parksville, Canada, Sept. 1985, TRIUMF report TRI-85-1 (1985), p. 241.

(Mi 84a) T. Minamisono, BEVALAC Proposal E732H (1984).

(Mi 86) W. Mittig, L. Bianchi, A. Cunsolo, B. Fernandez, A. Foti, J. Gastebois, A. Gillibert, C. Grégoire, Y. Schutz, C. Stephan, and A. Peghaire, GANIL report P.86-03 (1986), p. 1.

(Mi 86a) Y. Miyazawa, M. Hemmi, T. Kambara, M. Yanokura, M. Kase, E. Ikesawa, Y. Chiba, and I. Tanihata, to be published in the *Proc. of the 11th Int. Conf. on Cyclotrons and their Appl.*, Tokyo, Japan, Oct. 1986, to appear.

(Mo 84) P. Moller, *Proc. of 7th International Conf. on Nucl. Mass* (1984), p. 457.

(Mu 84) J. A. Musser and J. D. Stevenson, *Phys. Rev. Lett.* **53**:2544 (1984).

(Mu 85) R. W. Muller, *Proc. of the Accelerated Radioactive Beams Workshop*, Parksville, Canada, Sept. 1985, TRIUMF report TRI-85-1 (1985), p. 161.

(My 66) W. D. Myers and W. J. Swiatecki, *Nucl. Phys.* **81**:1 (1966).

(My 83) W. D. Myers and K.-H. Schmidt, *Nucl. Phys. A* **410**:61 (1983).

(No 86) Y. Nojiri, K. Matsuta, T. Minamisono, K. Sugimoto, K. Takeyama, H. Hamagaki, S. Nagamiya, K. Omata, Y. Shida, I. Tanihata, T. Kobayashi, S. Matsuki, S. Shimoura, J. R. Alonso, G. Krebs, and T. J. M. Symons, *Hyperfine Interact.*, **35**:1019 (1987).

(No 86a) Y. Nojiri, K. Takeyama, K. Matsuta, K. Asahi, and T. Minamisono, *J. Phys. Soc. Jpn., Suppl.* **55**:391 (1986).

(Od 84) M. Odera, Y. Chiba, T. Tonuma, M. Hemmi, Y. Miyazawa, T. Inoue, T. Kambara, M. Kase, T. Kubo, and F. Yoshida, *Nucl. Instrum. Methods* **227**:187 (1984).

(Ol 81) O. L. Olson, B. L. Berman, D. E. Greiner, H. H. Heckman, P. J. Lindstrom, G. D. Westfall, and H. J. Crawford, *Phys. Rev. C* **24**:1529 (1981).

(Ol 83) D. L. Olson, B. L. Berman, D. E. Greiner, H. H. Heckman, P. J. Lindstrom, and H. J. Crawford, *Phys. Rev. C* **28**:1602 (1983).

(Ot 81) E. W. Otten, *Nucl. Phys. A* **354**:471c (1981).

(Pe 83) A. Petrovici, A. Faessler, and Th. Koppel, *Z. Phys. A* **314**:227 (1983).

(Pi 81) R. B. Piercey, J. H. Hamilton, R. Soundranayagam, A. V. Ramayya, C. F. Maguire, X.-J, Sun, Z. Z. Zhao, R. L. Robinson, H. J. Kim, S. Frauendorf, J. Döring, L. Funke, G. Winter, J. Roth, L. Cleemann, J. Eberth, W. Neumann, J. C. Wells, J. Lin, A. C. Rester, and H. K. Carter, *Phys. Rev. Lett.* **47**:1514 (1981).

(Pi 84) R. B. Piercey, *Proc. Conf. on Instrumentation for Heavy-Ion Nuclear Research*, Oct. 1984, Oak Ridge, TN (1984).

(Pi 85) R. B. Piercey, T. M. Cornier, and M. Satteson, *Proc. of the Accelerated Radioactive Beams Workshop*, Parksville, Canada, Sept. 1985, TRIUMF report TRI-85-1 (1985), p. 250.

(Pr 82) SIS-Ein Beschleuniger fuer schwere lonen hoher Energie, GSI-Bericht 82-2; B. Franzke *et al.*, Zwischenbericht zur Planung des Experimentier-Speicherrings (ESR) der GSI, GSI-SIS-INT/84-5, Aug. 1984.

(Pr 84) *Proc. of the Workshop on Prospects for Research with Radioactive Beams from Heavy Ion Accelerators*, Washington, D.C., April 1984, LBL-18187 (1984).

(Pr 85) *Proc. of the Accelerated Radioactive Beams Workshop*, Parksville, Canada, Sept. 1985, TRIUMF report TRI-85-1 (1985).

(Pr 85a) The TRIUMF-ISOL Facility, A Proposal for an Intense Radioactive Beams Facility, June 1985.

(Pr 86) The BEVALAC Upgrade-A preliminary proposal, LBL PUB-5166, March 1986.

(Ra 79) H. L. Ravn, *Phys. Rep.* **54**:201 (1979).

(Ra 85) H. L. Ravn, J. M. D'Auria, M. J. G. Borge, Y. Huang, O. C. Jonsson, L. Nikkinen, and S. Sundel, '*Proc. of the Accelerated Radioactive Beams Workshop*, Parksville, Canada, Sept. 1985, **1** (1985), p. 94.

(Ru 66) G. Rudstam, *Z. Naturforsch.* **21a**:1027 (1966).

(Se 68) G. T. Seaborg, *Annu. Rev. Nucl. Sci.* **18**:53 (1968).

(St 81) J. D. Stevenson and P. B. Price, *Phys. Rev. C* **24**:2102 (1981).

(Su 73) K. Sugimoto and I. Tanihata, *J. Phys. Soc. Jpn., Suppl.* **34**:245 (1973).

(Ta 84) W. L. Talbert, Jr., M. E. Bunker, and J. W. Starner, Proc. of the TRIUMF-ISOL Workshop, Mont Rolland, Quebec (J. E. Crawford and J. M. D'Auria, eds.), TRIUMF Report TRI-84-1 (1984), p. 179.

(Ta 84a) W. L. Talbert, Jr. and M. E. Bunker, *Proc. of the Workshop on Prospects for Research with Radioactive Beams from Heavy Ion Accelerators*, Washington, D.C., April 1984, LBL-18187 (1984), p. 59.

(Ta 85) I. Tanihata, *Hyperfine Interact.* **21**:251 (1985).

(Ta 85a) I. Tanihata, H. Hamagaki, O. Hashimoto, S. Nagamiya, Y. Shida, N. Yoshikawa, O. Yamakawa, K. Sugimoto, T. Kobayashi, D. E. Greiner, N. Takahashi, and Y. Nojiri, *Phys. Lett.* **160B**:380 (1985).

(Ta 85b) I. Tanihata, H. Hamagaki, O. Hashimoto, Y. Shida, N. Yoshikawa, K. Sugimoto, O. Yamakawa, T. Kobayashi, and N. Takahashi, *Phys. Rev. Lett.* **55**:2676 (1985).

(Ta 85c) I. Tanihata, *Proc. of the 7th High-Energy Heavy Ion Study*, GSI Darmstadt, Oct. 1984, GSI-85, **10** (1985), p. 413.

(Ta 85d) I. Tanihata, *Proc. of the Accelerated Radioactive Beams Workshop*, Parksville, Canada, Sept. 1985, TRIUMF report TRI-85-1 (1985), p. 354.

(Tw 86) P. J. Twin, B. M. Nyako, A. H. Nelson, J. Simpson, J. F. Sharpey-Schafer, and G. Sletten, *Phys. Rev. Lett.* **57**:811 (1986).

(Va 85) J. P. Vary and M. A. Nagarajan, *Proc. of the Workshop on Prospects for Research with Radioactive Beams from Heavy Ion Accelerators*, Washington, D.C., April 1984, LBL-18187 (1984), p. 113.

(Vi 79) Y. P. Viyogi, T. J. M. Symons, P. Doll, D. E. Greiner, H. H. Heckman, D. L. Hendrie, P. J. Lindstrom, J. Mahoney, D.K. Scott, K. Van Bibber, G. D. Westfall, H. Wieman, H. J. Crawford, C. McParland, and C. K. Gelbke, *Phys. Rev. Lett.* **42**:33 (1979).

(Wa 71) A. H. Wapstra and N. B. Gpve, *Nucl. Data, Sect. A* **9**:267 (1971).

(Wa 80) A. H. Wapstra and K. Bos, in *Atomic Masses and Fundamental Constants 6* (J. A. Nolen and W. Benenson, eds.), Plenum Press, New York (1980), p. 547.

(Wa 80a) R. A. Ward and W. A. Fowler, *Astrophys. J.* **238**:266 (1980).

(Wa 81) R. K. Wallace and S. E. Woosley, *Astrophys. J., Suppl. Ser.* **45**:389 (1981).

(We 79) G. D. Westfall, T. J. M. Symons, D. E. Greiner, H. H. Heckman, P. J. Lindstrom, J. Mahoney, A. C. Shotter, D. K. Scott, H. J. Crawford, C. McParland, T. C. Awes, C. K. Gelbke, and J. M. Kidd, *Phys. Rev. Lett.* **43**:1859 (1979).

# 7

ERNST W. OTTEN has been Professor of Physics at the University of Mainz, Federal Republic of Germany, since 1972. He was educated at the Universities of Bonn and of Heidelberg. In 1962 he received his Ph.D. at Heidelberg and continued there as scientific assistant and lecturer until 1972. In 1969 and 1979 he spent sabbatical years at the European Center for Nuclear Research (CERN) in Geneva and participated in research of the ISOLDE Collaboration studying nuclei far from stability. He is a member of the Academy of Sciences and Literature at Mainz.

# 7

# Nuclear Radii and Moments of Unstable Isotopes

ERNST W. OTTEN

## 1. Introduction

The development of on-line mass separators, which provide long chains of isotopes extending far off stability, immediately raised the question of how to gain access to the basic nuclear ground-state properties of these exotic nuclei—their spins, moments, radii, and masses. In general, the amount of radioactive material produced at these facilities is too small to form beams or targets for any kind of scattering or nuclear-reaction experiment. On the other hand, the traditional spectroscopic methods, namely, atomic, nuclear, and mass spectroscopy, have turned out to be very suitable for this application since they are easily adapted to the special on-line conditions.

This chapter deals with radii, spins, and moments, which are all the domain of atomic spectroscopy, at least as far as nuclear ground states and long-lived isomers are concerned. The sources of information are the isotope shift (IS) and hyperfine structure (hfs) of spectral lines. While the evaluation of spins (I) and magnetic moments ($\mu_I$) from hfs poses no special problems, the precision obtainable in evaluating spectroscopic quadrupole moments ($Q_s$) and isotopic variations of mean-square (ms) charge radii from hfs and IS, respectively, gives cause for concern. It is therefore appropriate to first briefly summarize the basic concepts involved in analyzing IS and hfs data (Section 2). A major part of this review will be devoted to the description of the various spectroscopic methods which have been developed in this field (Section 3). Most of them are still highly competitive areas of research, and only a few have already exhausted their potential. During the past decade, development has focused on laser spectroscopic methods. These give optimum spectral resolution, i.e., free of Doppler broadening, together with very high sensitivity. The principal goal of present methodological development is to further increase the sensitivity obtainable. To achieve this, the basic laser spectroscopic devices are imbedded in complex, hybrid systems.

The discussion and interpretation of experimental results forms another major topic of this review (Section 4). For quite a number of elements, typically 20 or more

ERNST W. OTTEN • Institute of Physics, University of Mainz, 6500 Mainz, FRG.

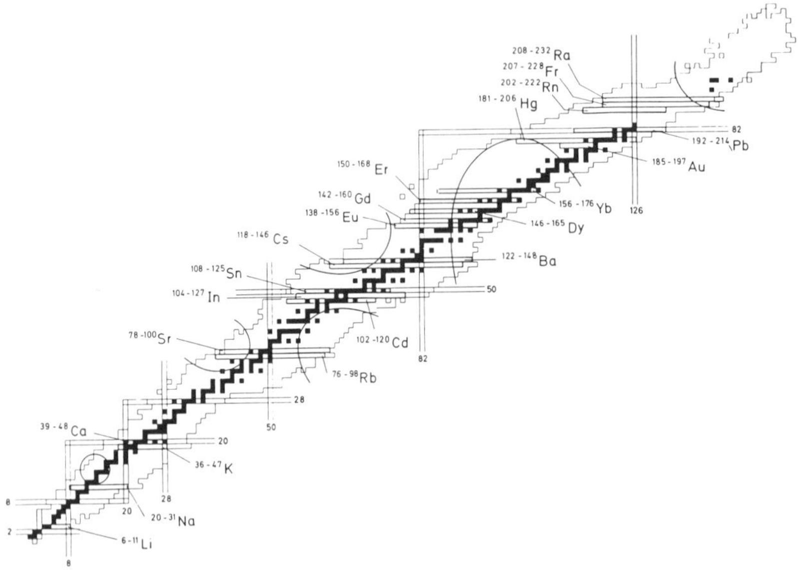


**Figure 1.** Chart of nuclides showing extended chains of isotopes which have been investigated by optical spectroscopy off stability.

isotopes have been investigated by means of atomic spectroscopy. In some regions of the nuclear chart the results already cover a dense network, for instance, from Cd to Sn, in the rare-earth region, from Au to Pb, and along the Rn–Fr–Ra sequence. Figure 1 provides an overview of the systematic work on long chains of isotopes that has so far been performed. The plot also includes some off-line work on longer chains of stable or long-lived isotopes such as, for example, the important series of Ca isotopes. These complete sets of nuclear ground-state data form a unique basis for a systematic overview and interpretation of the development of nuclear structure throughout the nuclear chart, linking distant regions of very different structure, e.g., spherical and deformed equilibrium shapes.

In this context the change of ms nuclear charge radii ($\delta\langle r^2\rangle$) plays a guiding role, very similar to that of mass. Both quantities are governed globally by the liquid-drop properties of the nucleus which leads to the expression in which the nuclear radius is proportional to $A^{1/3}$ and to the Bethe–Weizsäcker formula. If we define the deviations from the liquid-drop model as being the more interesting features of nuclear structure to be searched for, then we will find them in the deviations of the measured radii and masses from these two global laws. In isotope-shift studies such deviations are predominantly caused by nuclear deformation which obviously increases the observable $\langle r^2\rangle$. At shell closures we find characteristic changes in the average slope of the expansion of $\langle r^2\rangle$, and at shape transitions we meet abrupt discontinuities.

Since isotope-shift data are beginning to fill complete blocks in the $N$–$Z$ plane, one can combine results from different elements in order to extract isotonic and isobaric shifts of the ms charge radius.

While isotope shifts are sensitive to the mean-square deformation ($\langle\beta_2^2\rangle$) of a nucleus, spectroscopic quadrupole moments are proportional to the first-order term

($\langle\beta_2\rangle$) itself. Hence they show not only the size but also the sign of the deformation, provided that the coupling of the spin of the unpaired nucleons to the core is known. This interplay between collective and single-particle structure is another interesting and important aspect of these systematic investigations. In regions of transitional nuclei we also find strong differences between the two observables $\langle\beta_2^2\rangle_{IS}$ (from $\delta\langle r^2\rangle$) and $\langle\beta_2\rangle_{Q_s}^2$ (from $Q_s$), since the former includes zero-point vibrations and the latter does not. In the same sense, spins and magnetic moments may be used as fingerprints of collective nuclear structure by, for instance, characterizing a particular Nilsson state.

The relations of spins, moments, and ms radii to collective properties of nuclei justify their joint discussion in this chapter. Still, we do not aim at a complete discussion of all the spins and moments that have been measured off stability in recent years, but wish to focus on the systematics of ms radii. Their interpretation will be structured primarily according to different regions of nuclei—accompanied by introductions to the different phenomena observed, such as dynamic deformation near shell closures, shape transitions to static deformation, shape coexistence, isotone and isobar shift, odd-even staggering, and octupole effects.

The different sections of this chapter may be read almost independently of one another. The appendix consists of tables of spins, moments, and $\delta\langle r^2\rangle$ values measured in extended chains of isotopes by the techniques described in this chapter.

The field of study to be treated here has already been reviewed—at an early stage—by Jaquinot and Klapisch (Ja 79). Some of the techniques and results have also been summarized by Kluge (Kl 79, Kl 85), Neugart (Ne 85, Ne 87), and others (Ot 81). A collection of papers in the field of laser spectroscopy of unstable isotopes is also available (Be 82). Nuclear moments have been analyzed systematically by Ekström in the framework of the particle-plus-rotor model (Ek 81).

## 2. Introduction to Isotope Shift and Hyperfine Structure in Atomic Spectra

### 2.1. The Isotope Shift

The IS of a spectral line between two isotopes with mass numbers $A$ and $A'$ is defined to be

$$\mathrm{IS}^{AA'} = \nu^A - \nu^{A'} \tag{1}$$

In the case of hfs splitting, the transition frequency $\nu$ is taken to be the center of gravity (cg) of all hfs components of the line. It should be noted that the hfs interaction does not shift the cg in first-order perturbation theory (which is usually an excellent approximation). The origin of the IS is twofold, namely, the change of the nuclear mass $M^A$ and of the nuclear charge distribution.

#### *2.1.1. The Mass Shift*

In hydrogen-like spectra the mass shift ($\delta\nu_\mathrm{M}$) is trivially accounted for by introducing the reduced electron mass. To good approximation it is given by

$$\delta\nu_{\mathrm{MN}}^{AA'} = \frac{(M^A - M^{A'})m_e}{M^A M^{A'}}\,\nu^A \approx \frac{A - A'}{AA'}\,\frac{m_e}{\mathrm{amu}}\,\nu^A = N\frac{A - A'}{AA'} \tag{2}$$

where $m_e$ is the electron mass and amu denotes atomic mass unit; $N$ indicates the so-called normal mass shift. In complex spectra different electron momenta may be correlated, leading to a collective retroaction onto the nuclear motion. It causes the so-called specific mass shift:

$$\delta\nu_{\mathrm{MS}}^{AA'} \approx S\frac{A - A'}{AA'} \tag{3}$$

which still has the same $A$-dependence as the normal mass effect. When shielded d electrons and f electrons are involved in the transition, as for the transition elements and rare earths, $S$ can be of the order of $10N$ or more. But for simple s → p transitions, which are the most important ones in terms of probing the nuclear charge distribution, experimental evidence has demonstrated that $|S| \leqslant N$ (He 74). Hartree–Fock calculations of $S$ are very difficult; for some cases they are available and partially reproduce the experimental values. Since the mass shift is entirely a function of atomic and not of nuclear structure, we will not pursue it further here and instead refer the reader to the books of Kopfermann (Ko 58), Kuhn (Ku 69), and King (Ki 84) for more details. Methods of separating the mass effect from the field effect will be discussed below.

### *2.1.2. The Field Shift*

The influence of the finite extension of the nuclear charge distribution on the electronic binding energy, usually called the field shift ($\delta\nu_{\mathrm{F}}$), is easily derived in nonrelativistic, first-order perturbation theory. In order to derive it here, we write the electrostatic energy as an integral over the nuclear charge distribution $[\rho(\mathbf{r})]$ times the potential of the electron $[V(\mathbf{r})]$:

$$E_c = \int_0^{R_1} \rho(\mathbf{r})\, V(\mathbf{r})\, d\tau \tag{4}$$

where $R_1$ is the radius at which $\rho(\mathbf{r})$ drops to zero (see Figure 2). We will restrict ourselves to s electrons, which produce a spherically symmetric $V(r)$. We can separate $V(r)$ for $r \leqslant R_1$ into a constant part $V(R_1)$ that stems from the electronic charge outside $R_1$ and a variable part $v(r)$ arising from the inner charge. To good approximation we may take the nonrelativistic s wave function $\psi_s(r)$ of an electron as being constant over the nuclear volume and hence get the parabolic form for $v(r)$:

$$v(r) = -2\pi e|\psi_s(0)|^2(R_1^2 - r^2/3) \tag{5}$$

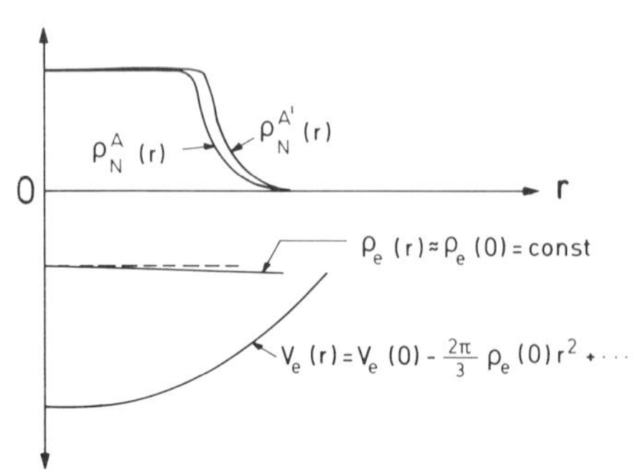


**Figure 2.** Schematic drawing of a nuclear charge distribution $\rho_N$ and an electronic charge distribution $\rho_e$, the latter for an s wave function. In addition, the potential $V_e$ of the electron is shown.

Further, recalling the relation

$$V(R_1) + v(0) = V(0) = -e\left\langle \frac{1}{r} \right\rangle_{\text{electron}} \tag{6}$$

we get the total potential inside $R_1$:

$$V(r) = -e\left\langle \frac{1}{r} \right\rangle_{\text{electron}} + \frac{2\pi}{3} e|\psi_s(0)|^2 r^2 \tag{7}$$

After inserting equation (7) into equation (4) it follows that

$$E_c = -Ze^2\left\langle \frac{1}{r} \right\rangle_{\text{electron}} + \frac{2\pi}{3} Ze^2|\psi_s(0)|^2\langle r^2\rangle_{\text{nucleus}} \tag{8}$$

The first term in equation (8) represents the monopole interaction with a point-like nucleus and the second the finite-size correction. The latter is the product of the electronic charge density at the origin and the second radial moment of the nuclear charge distribution.

We note that: (1) equation (8) was derived model independently with respect to $\rho(\mathbf{r})$ and thus holds for any nuclear shape; (2) the restriction to s electrons is justified since the others do not overlap the nucleus considerably; and (3) the theory also holds for the lightest muonic atoms; for heavier ones the muon wave function varies strongly over the range of $\rho(\mathbf{r})$. Therefore in the case of muons, the finite-size correction is characterized by more complicated radial moments. These are the so-called Barett moments, which depend on the particular muonic transition in question. Muonic results have been tabulated by Engfer *et al.* (En 74).

The finite-size effect in electronic spectra is too small to allow a direct determination of the total $\langle r^2\rangle$ of a nucleus from a given electronic-term value, even if this has been measured to the highest accuracy. This reflects the fact that the leading terms in the binding of many-electron atoms cannot be calculated to the precision (i.e., to order $10^{-6}$) at which the finite-size effect appears. Thus only a change of $\langle r^2\rangle$ within an isotopic series or between isomers can be measured, since the first term in equation (8) as well as other contributions to the binding cancel by taking the isotopic difference (except for the mass shift). The nonrelativistic field shift between isotopes or isomers in an optical transition is thus given by

$$\delta\nu_{F_{nr}}^{AA'} = -\frac{2\pi}{3} Ze^2\Delta|\psi(0)|^2\delta\langle r^2\rangle^{AA'} \tag{9}$$

where $\Delta|\psi(0)|^2$ is the change of the electron charge density at the nucleus between lower and upper states of the optical transition.

In the case of heavier elements it is necessary to proceed to a relativistic theory. In this case it is no longer adequate to use the Dirac s wave functions for a point-like nucleus as the starting point of a perturbation theory, as in equation (4), because they show a logarithmic divergence at the origin which leads to an unrealistic exponent of the radial moment of the charge distribution measured by the field shift. Instead, the Dirac equation is solved first near the origin in an extended, approximate nuclear charge distribution of liquid-drop dimension:

$$R_0 = r_0 A^{1/3} \tag{10}$$

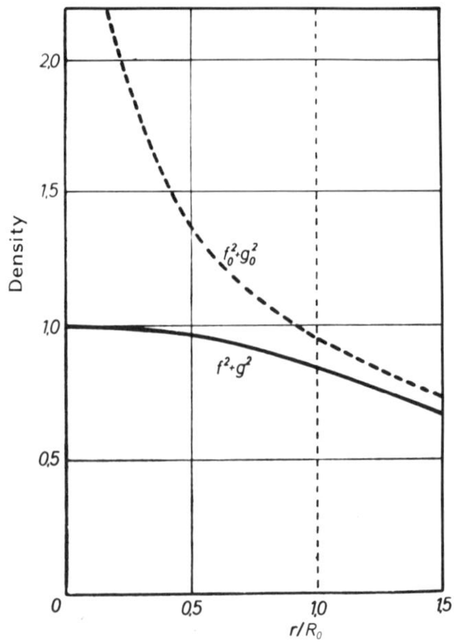


**Figure 3.** Density of an electronic Dirac s wave function in the potential of a point-like nucleus (broken line) and an extended nucleus of radius $R_0$ (solid line) for $Z = 90$ (Ko 58).

Figure 3, taken from Kopfermann's book (Ko 58), shows the difference between the two solutions for $Z = 90$. The solution for an extended charge is still fairly constant within the nuclear radius $R_0$. If we calculate the field shift with this wave function by perturbation theory, we can therefore expect that $\delta\nu_F$ will still be proportional, within reasonable accuracy, to the isotopic change of the second radial moment. Bodmer (Bo 59) has shown that the Dirac solution in an extended nuclear charge, and hence the field shift, are quite insensitive to details of the charge distribution. In fact, he has derived the relativistic formula

$$\delta\nu_F = -\frac{\pi a_0^3}{Z}\,\Delta|\psi(0)|^2\zeta(Z)N(Z)\langle r^2\rangle^{\sigma-1}\delta\langle r^2\rangle \tag{11}$$

where $\zeta$ and $N$ are known functions of $Z$ [see (Ki 84)], $\sigma = \sqrt{1-\alpha^2Z^2}$, $\alpha$ is the fine-structure constant, and $a_0$ is the Bohr radius. Since $\sigma$ is close to unity, $\delta\nu_F$ is still dominated by $\delta\langle r^2\rangle$ and in addition depends weakly on the total $\langle r^2\rangle$ of the nucleus. One can account for the latter effect with sufficient accuracy by inserting the liquid-drop estimate into equation (11):

$$\langle r^2\rangle = \tfrac{3}{5}r_0^2A^{2/3} \tag{12}$$

This allows further contraction of equation (11) to give

$$\delta\nu_F^{AA'} = -\frac{\pi a_0^3}{Z}\,\Delta|\psi(0)|^2f(Z)\delta\langle r^2\rangle^{AA'} = F(Z)\delta\langle r^2\rangle^{AA'} \tag{13}$$

where the field shift is now factorized into an electronic factor $F(Z)$ and a nuclear one $\delta\langle r^2\rangle$. A revised table of $F(Z)$ has recently been published by Blundell *et al.* (Bl 85).

In equation (13), $\Delta|\psi(0)|^2$ still describes the nonrelativistic change of the electronic charge density, which may be calculated from nonrelativistic, semiempirical formulas (see below). It serves to calibrate the amplitude of the Dirac solution of the electronic states in question by fitting the relativistic to the nonrelativistic wave functions at $r > R_1$, outside the nucleus.

For the very heaviest elements we can no longer claim the electronic wave function to be constant over the nuclear volume, and we thus require an expansion of the nuclear factor into a power series of radial moments (Se 69) which replaces $\delta\langle r^2\rangle$ by

$$\lambda^{AA'} = \delta\langle r^2\rangle^{AA'} + \frac{G(Z)}{F(Z)}\,\delta\langle r^4\rangle^{AA'} + \frac{H(Z)}{F(Z)}\,\delta\langle r^6\rangle^{AA'} + \cdots \tag{14}$$

For a spherical nucleus, $\lambda$ is about 6% smaller than $\delta\langle r^2\rangle$ at $Z = 80$. The shape dependence of the Seltzer correction [equation (14)] has been discussed by, e.g., Ahmad *et al.* (Ah 85, Ah 88).

According to equations (9), (12), and (16) (see below) the field effect varies approximately as $Z^2A^{-1/3}$. For $\delta A = 2$, it is of order 10 MHz in Li and 10 GHz in Fr.

### 2.1.3. Separation of Mass and Field Shift

In the past the IS has usually been measured in several electronic transitions, but data for the full isotopic series have rarely been taken for the particular transition in which the electronic mass-shift and field-shift factors can be calculated most reliably. However, different data sets can be related to each other by the so-called King plot [see King (Ki 84)], which fortunately separates field and mass shift and also controls the consistency of the data sets. For this purpose the measured IS in two different transitions, 1 and 2, given by

$$\mathrm{IS}_1^{AA'} = \frac{A - A'}{AA'}(N_1 + S_1) + F_1\lambda^{AA'} \quad \text{and} \quad \mathrm{IS}_2^{AA'} = \frac{A - A'}{AA'}(N_2 + S_2) + F_2\lambda^{AA'} \tag{15}$$

respectively, are multiplied by $AA'/(A - A')$ thereby transforming them into linear equations of the same argument $\lambda^{AA'}(A - A')/AA'$. These renormalized isotope shifts are plotted for all pairs $AA'$ in a $(x_1, x_2)$-plane where the data points form a straight line with slope $F_2/F_1$ and intersection $(N_2 + S_2) - (N_1 + S_1)F_2/F_1$ on the $x_2$-axis. Thus the ratio of field shifts and the difference of mass-shift factors (modified by $F_2/F_1$ in one term) can be extracted separately from the King plot. Figure 4 shows some King lines for the example of Ba. The alkali-like, ionic resonance transition ($6s\,^2S_{1/2} \rightarrow 6p\,^2P_{3/2}$, $\lambda = 455$ nm) in BaII (We 84), whose $F$-factor is favorably calculated, serves as reference on the horizontal axis. Within the limits of error the data points lie on straight lines as would be expected. The King line vs. the atomic-resonance transition ($6s^2\,^1S_0 \rightarrow 6s6p\,^1P_1$, $\lambda = 554$ nm) (Be 79, Mu 83) almost crosses the origin and hence shows no evidence for any specific mass shift in either of the two transitions. On the other hand, the transition ($5d\,^2D_{3/2}$—$6p\,^2P_{3/2}$, $\lambda = 585$ nm) (Si 80) has a pronounced specific mass shift $S \approx 4N$ and a negative $F$-factor with respect to the reference transition. Both effects are consequences of the inner 5d electron which orbits inside the core and thus gives rise to correlations with, and screening of, core electrons.

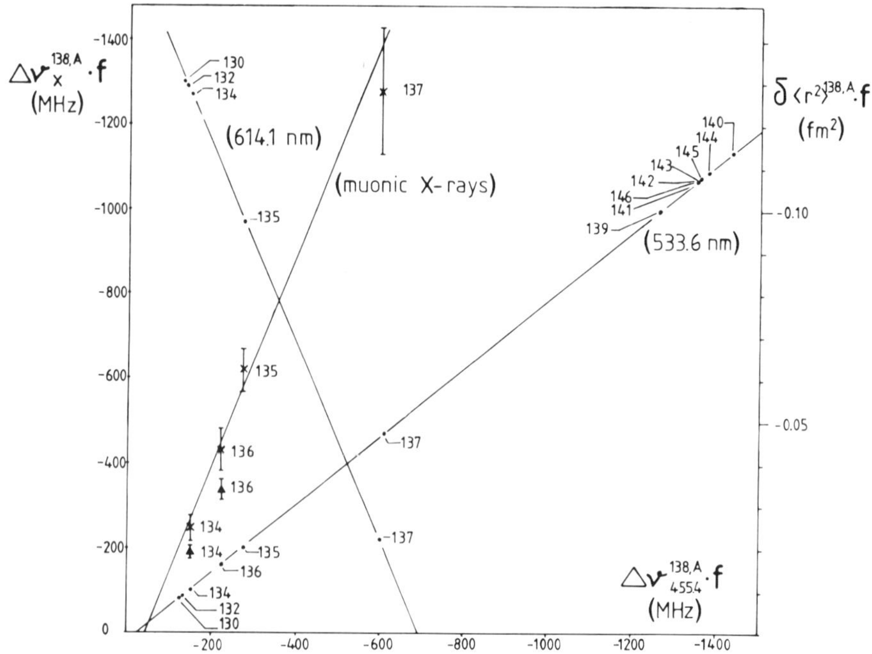


**Figure 4.** King-plot $IS^{A,138}$ for transitions in BaI and BaII with respect to the reference transition $\lambda = 455$ nm. The IS is renormalized by a factor $f = 2A/136(A\text{-}138)$. Also shown is the King line vs. the muonic isotope shift fitted to the results of Shera *et al.* (Sh 82) (crosses). Muonic results of Kunold *et al.* (Ku 83) are given for comparison (triangles). The vertical scale on the right applies to the muonic IS. For all transitions the normal mass shift has already been subtracted from the IS.

In preparing a King plot one needs at least three measured isotopes. But if the field shift in an isotopic series happens to be as regular as the mass shift, i.e., almost proportional to $A - A'$, then the data points converge toward a single point on the King line and the fit of its two parameters—slope and intersection—becomes rather poor. Unfortunately, a regular expansion of $\langle r^2 \rangle$ is the rule for sequences of stable, even isotopes; and these are more or less the only ones for which precise, absolute $\delta\langle r^2 \rangle$ determinations are available from either X-ray or muonic IS or from electron-scattering experiments. Therefore until now, these nonoptical measurements have enabled us in only very few cases to calibrate the optical field and mass shift on an absolute scale, via King plots (see Figure 4 and the following section). This unfortunate situation is to some extent saved by the fact that poorly fitted King parameters are still well correlated, and this correlation partly compensates their individual errors in the final evaluation of $\delta\langle r^2 \rangle$.

### 2.1.4. *Evaluation and Calibration of Field-Shift Constants*

An important factor in the evaluation of $\delta\langle r^2 \rangle$ via equation (13) is the determination of $\Delta|\psi(0)|^2$. Traditionally, it is calculated by the semiempirical Goudsmit–Fermi–Segré

(GFS) formula (Ko 58, Ar 71) which relates $|\psi(0)|^2$ of s electrons to the effective quantum number $n^* = n - \sigma = (Z_a^2 R_\infty / T_n)^{1/2}$ and its derivative $dn^*/dn = 1 - d\sigma/dn$, where $\sigma$ denotes the quantum defect, $Z_a$ outer charge (equal to one for neutral atoms), $R_\infty$ Rydberg constant, and $T_n$ is the term value. The GFS formula reads

$$|\psi(0)|^2 = ZZ_a^2\left(1 - \frac{d\sigma}{dn}\right) \Big/ n^{*3}\pi a_0^3 \tag{16}$$

In alkali-like spectra, for example, the favorable transitions, being sensitive to $\delta\langle r^2\rangle$, are those of the principal series $n$s to $n'$p. The p doublet is of zero density at the nucleus in the nonrelativistic limit, but the $p_{1/2}$ state becomes of finite density from relativistic effects [see Kopfermann (Ko 58)]. A semiempirical formula for the relativistic state has recently been proposed and proven experimentally by Wendt *et al.* (We 84). Besides its direct contribution to $|\psi(0)|^2$, a valence electron may also give rise to an indirect contribution by screening inner s-shells. Both effects are taken into account by a correction $\beta$ of the form

$$\Delta|\psi(0)|^2_{ns\rightarrow n'\mathrm{p}} = \beta|\psi(0)|^2_{ns} \tag{17}$$

where $\beta$ is of order 1.1 for an alkali-like $n$s $\rightarrow$ $n$p transition (He 74).

The quantity $|\psi(0)|^2$ also determines the magnetic hfs splitting constant $a_s$ (see below) of an s electron, through the so-called Fermi contact interaction:

$$a_s = \frac{8\pi}{3}\frac{\mu_0}{4\pi}2\mu_B\frac{\mu_I}{I}|\psi(0)|^2_{ns}F_{1/2}(Z)(1-\delta)(1-\varepsilon) \tag{18}$$

where $\mu_0$ is the permeability of vaccum, $\mu_B$ the Bohr magneton, $F_{1/2}$ the relativistic correction, $(1 - \delta)$ the Breit-Crawford-Schawlow correction taking into account the influence of the extended nuclear charge on the Dirac wave function as discussed above, and $(1 - \varepsilon)$ is the Bohr-Weisskopf effect stemming from the finite extension of nuclear magnetization. The corrections are discussed and tabulated elsewhere (Ko 58, Sc 55, Sc 57, Ar 71, Li 74).

Since the nuclear $g_I$-factor is usually known from an independent source (such as NMR) for at least one isotope of a given element, $|\psi(0)|^2$ may therefore also be determined from $a_s$ and $g_I$ through equation (18). This procedure is certainly correct for the direct contribution of the valence s electron to the IS and hfs. But for the indirect contributions induced by screening, core polarization, and so on, this is not necessarily true since the IS is sensitive to the charge density while $a_s$ is sensitive to the spin density. With respect to many-body effects these two may differ from one another because of the exchange interaction. It is astonishing that in spite of these complications by many-body effects equation (18) correctly reproduces $a_s$ to within about 5% in almost any known case (Ko 58). This success of the GFS formula is the main reason for its continued application in hfs as well as IS analysis. Its extension to configurations with more than one valence electron is in many cases possible and successful (Ko 58). Even in complex spectra, like those of rare earths and transition elements, the IS can be analyzed systematically in this way [see, e.g., Heilig *et al.* (He 74) and King (Ki 84)]. These analyses are subtle, cover many different lines in each spectrum, and in addition make use of Hartree-Fock calculations (see below) in order to fix screening corrections and so on.

An alternative way to analyze field shifts is by using straightforward relativistic Hartree-Fock calculations of $F(Z)$, which also include most of the many-body effects such as screening. It has been shown from many examples that such calculations reproduce ratios of $F$-factors quite well, even in rather complex spectra (Au 85, Au 86, He 74, Ki 84). For example, the ratio $F_{554}/F_{455}$ for Ba is determined from the King plot in Figure 4 to be 0.808 (2), while Torbohm *et al.* (To 85) have calculated a value of 0.783. However, the absolute precision of $F(Z)$ values obtainable from Hartree–Fock calculations is still debatable. They systematically lie 15–25% below those calculated semiempirically. King and Palmer have recently reviewed the situation and conjecture that their actual values lie somewhere in between (Ki 86). On the other hand, the systematic character of the discrepancy makes it clear that it is not a reflection of deficiencies in the calculation of individual valence states. It may, instead, be caused by perturbations located deep inside the core, thus reaching states of high momenta far out in the continuum, which would lead to a kind of renormalization of the field shift. The dilemma is that the experimental calibration of $F(Z)$ discussed below is difficult and cumbersome and only a few useful data sets have so far been published.

For calibrating $F(Z)$ one needs independent experimental information on $\delta\langle r^2\rangle$ or, more correctly, on $\lambda$. From the standpoint of pure physics, K X-ray isotope shifts are the more suitable experiments to perform since they also measure $\lambda$ directly. Moreover, the calculation of mass-shift and field-shift factors poses no serious problem in such experiments, because the inner-shell electrons are essentially unscreened. However, the experimental precision obtainable is low since $\delta\nu_{\mathrm{F}}$ is about a hundred times smaller than the linewidth. Only for the heaviest elements is satisfactory accuracy achieved. In the case of lead, for example, an accuracy of 5% was obtained, from very careful measurements, by Borchert *et al.* (Bo 83) who also compared their results systematically to those from optical and muonic IS and from electron scattering. X-ray IS have been compiled by Boehm and Lee (Bo 74).

Electron-scattering experiments offer an alternative means of access to nuclear charge radii through the form factor $F(q)$. For small momentum transfer $q$ it may be expanded as

$$F(q) = 1 - q^2\langle r^2\rangle/6 + \mathrm{O}(q^4)\langle r^4\rangle) \tag{19}$$

This method provides an excellent model-independent determination of the total $\langle r^2\rangle$ by taking the slope $dF/d(q^2)$ at $q^2 = 0$. However, the precise measurement of the small $\delta\langle r^2\rangle$ effect in an isotopic series is difficult and has therefore only been performed in a limited number of cases [see the compilations of de Jager and de Vries (Ja 74)]. The more recent experiments are designed to obtain not just the $\langle r^2\rangle$ term in equation (19) but also a complete measurement of $F(q)$ over as wide a $q$-range as possible. A Fourier–Bessel transformation of $F(q)$ leads to a determination of the monopole charge distribution $\rho(r)$. By extending these measurements over series of isotopes and isotones one obtains complete information about the local change $\delta\rho(r)$ at any $r$, and not merely $\delta\langle r^2\rangle$. Needless to say such measurements provide much more detailed information on nuclear structure than does the optical IS. Figure 5 gives an example for the isotope pair $^{48,40}$Ca (Em 83) which is discussed in Section 4.8 in more detail. Obviously $\delta\langle r^2\rangle$ can be determined directly from this measurement by taking the integral

$$\delta\langle r^2\rangle = (1/Z)\int_0^\infty r^2\delta\rho(r)\,d\tau \tag{20}$$

Again, examples of such measurements are rare since it would require extraordinary effort to perform them systematically throughout the nuclear chart.

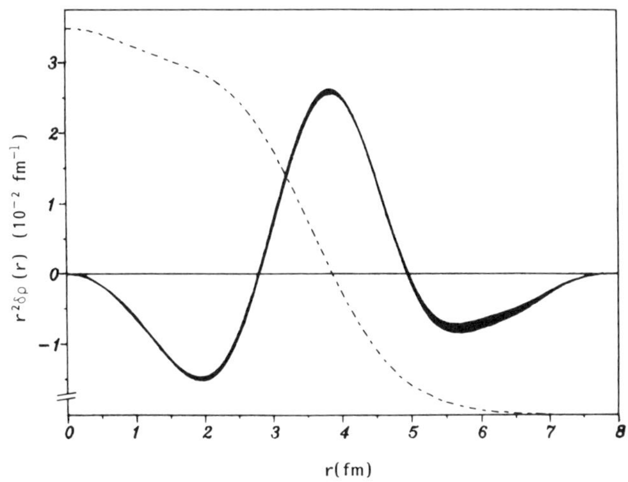


**Figure 5.** Plot of $r^2\delta\rho(r)$ for the isotope pair $^{48,40}Ca$. Charge is shifted from the center and the outer skin toward the surface region around the half-density radius $c$. The width of the band corresponds to the uncertainty. For comparison $\rho(r)^{40}$ is also shown on an arbitrary scale (dotted line) (Em 83).

An easier method of access to nuclear charge distributions and their isotopic dependence is provided by muonic spectra. As already mentioned, muonic spectra measure a radial moment which depends explicitly on the particular muonic transition being measured and which differs strongly from $\langle r^2 \rangle$ except for the lightest elements. In order to convert both observables into one another, one again needs information on $\delta\rho(r)$. Fricke *et al.* have therefore concluded that a combined analysis of electron scattering and muonic data probably offers the most effective method of calibrating the $\delta\langle r^2 \rangle$ scale (Fr 81, Fr 83, Bo 83). However, this conclusion presupposes not only precise measurements but also sufficiently correct calculations of nuclear-core polarization by the muon—which constitutes a large fraction of the muonic IS [reviewed, e.g., by Scheck (Sc 78)]. At present, it appears that the accuracy of these calculations is still a problem. For instance, the results of calculations of $\delta\langle r^2 \rangle$ from muonic Ba spectra by Kunold *et al.* (Ku 83) differ from those of Shera *et al.* (Sh 82) by more than 20% [compare Figure 4; see also Borchert *et al.* (Bo 83) for the case of lead].

Let us then summarize the situation as it appears at present. Since the mass shift decreases about as fast with mass number as the field shift increases, it can cause serious difficulties in the analysis of lighter isotopes up to $A \lesssim 150$ and it is of little importance for $A \gtrsim 150$.

The results of calculations of absolute field-shift factors by Hartree–Fock methods and semiempirical formulas, respectively, differ from one another and may be in error by 10–30%. Similar accuracy has been obtained by the various calibration methods. The error in $F$ is purely a scaling error and does not affect ratios of $\delta\langle r^2 \rangle$ in an isotopic series. If there remains a systematic error in $F$, it would appear not to scatter significantly but to be about the same for all the elements measured; otherwise, the slope $\delta\langle r^2 \rangle^{NN'}/(N - N')$ would scatter substantially for different isotones. This would necessarily lead to a crossing of the ms charge radii of different isotones somewhere off stability such that $\langle r^2 \rangle$ of a nucleus $(Z, N)$ would become smaller (at constant shape!) than that of a lighter isotone $(Z - x, N)$. Such an unphysical behavior has never been observed.

## 2.2. Atomic Hyperfine Structure

### *2.2.1. Introduction to Hyperfine Structure*

The hfs in atomic spectra is the classical example of static multipole interaction. Because of the small ratio of nuclear to atomic radius, it converges very rapidly with powers of this ratio. Thus, besides the leading electric-monopole interaction, only the magnetic-dipole and electric-quadrupole interactions are of practical importance. For the same reason the hfs interaction is separable into a product of a nuclear matrix element (the moment) and an electronic one (the hfs field). Further, we note that the nuclear matrix element is always diagonal for a state of definite spin and parity assignment $I^P$. Therefore, only moments of even parity contribute to hfs. The odd ones vanish identically. This fact answers, for instance, the question of why octupole deformation of the nuclear charge distribution cannot be observed through hfs, although it may be of finite size if measured in the intrinsic frame of the nucleus: It is because a measurement in the intrinsic frame never corresponds to a pure quantum state $I^P$.

The hfs coupling of nuclear and electronic angular momenta $I$ and $J$ to the total angular momentum of the atom $F$ is given by the well-known formula (Ko 58)

$$W(F) = AK/2 + B\frac{\frac{3}{4}K(K+1) - I(I+1)J(J+1)}{2(2I-1)(2J-1)I \cdot J} \tag{21}$$

where $K = F(F+1) - I(I+1) - J(J+1)$, $A = \mu_I H_e(0)/I \cdot J$ is the magnetic dipole, and $B = eQ_s\varphi_{jj}(0)$ is the electric-quadrupole coupling constant; $H_e(0)$ is the magnetic field and $\varphi_{jj}(0)$ the electric-field gradient produced by the electrons at the site of the nucleus. In the traditional notation of equation (21) the moments and fields are defined as the expectation values of the respective operators in the states of maximum projection $m_I = I$, $m_J = J$.

Figure 6 shows a schematic plot of a hfs multiplet. The magnetic hfs follows Lande's interval rule which states that two neighboring levels are split by $F_> \cdot A$, where $F_>$ is the larger of the two $F$-values. For $Q_s > 0$ the quadrupole interaction raises those levels with $I$ and $J$ parallel or antiparallel, and lowers those with perpendicular spins.

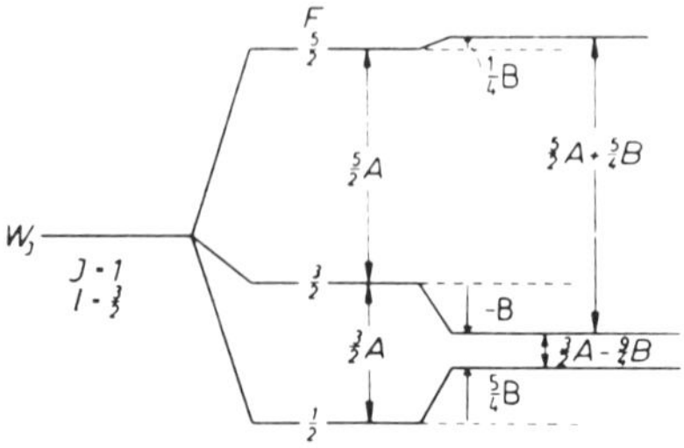


**Figure 6.** Hyperfine multiplet for $J=1$, $I=\frac{3}{2}$ with magnetic-dipole ($A$) and electric-quadrupole (B) splitting.

For unpaired s electrons the magnetic hfs usually exceeds the Doppler width of a spectral line of roughly 1 GHz; for p electrons it is at least larger than the natural linewidth which is typically 1-20 MHz. For a deformed nucleus of medium mass ($A \approx 100$) a p electron produces an electric-quadrupole splitting about equal to the magnetic-dipole splitting.

### 2.2.2. On the Evaluation and Calibration of Hyperfine Structure Fields

For a single valence electron in a state $|nlj\rangle$ the hfs coupling constants, which we denote now by lower-case letters $a_{nlj}$ and $b_{nlj}$, are given for $l > 0$, to a first approximation, by (Ko 58, Ar 71)

$$a_{nlj} = \frac{\mu_0}{4\pi} \frac{2\mu_B \mu_I}{I} \frac{l(l+1)}{j(j+1)} \langle r^{-3} \rangle_{nl} F_j(Z_i)(1-\delta)(1-\varepsilon) \tag{22}$$

and

$$b_{nlj} = \frac{e^2}{4\pi\varepsilon_0} Q_s \frac{2j-1}{2j+2} \langle r^{-3} \rangle_{nl} R_{lj}(Z_i) \tag{23}$$

where $F_j(Z_i)$ and $R_j(Z_i)$ are relativistic corrections factors (Ko 58, Sc 55, 57, Li 74), $Z_i$ is the nuclear charge seen by the inner part of the electronic wave function and is close to $Z$, e.g., $Z_i = Z - 4$ for p electrons. The $a$-factor for an s electron has already been given in Section 2.1.4 by equation (18).

The main problem in calculating the hfs coupling constants using equations (22) and (23) is the evaluation of the radial matrix element $\langle r^{-3} \rangle$. Semiempirically, it can be determined either from the effective quantum nuclear $n^*$ through Lande's formula:

$$\langle r^{-3} \rangle = \frac{Z_i Z_a}{\pi a_0^3 n^{*3} l(l+\frac{1}{2})(l+1)} \tag{24}$$

or from the fine-structure splitting $\Delta W_{FS}$ which is also proportional to $\langle r^{-3} \rangle$:

$$\Delta W_{FS} = \frac{\mu_0}{4\pi} 2\mu_B^2 (l+\tfrac{1}{2}) \langle r^{-3} \rangle H(Z_i) \tag{25}$$

where $H(Z_i)$ is another relativistic correction. In the p-series of alkalis, for example, both methods agree well. However, their success in this respect does not completely solve the problem since core-polarization effects are not included in equations (22) and (23). In the magnetic case these effects can exceed even the direct contribution; in the electric case they are of the order of 20% of the direct contribution and are known as Sternheimer corrections (St 50). In the case of simple spectra these two corrections are reasonably well understood. The hfs of alkalis is also reproduced quite well by *ab initio* calculations.

The analysis of the hfs of complex spectra is usually performed with the help of a relativistically correct parametrization of the hfs interaction, as given by Sandars and Beck (Sa 65). A certain number of radial matrix elements attached to angular-momentum operators are treated as free parameters and fitted to experimental results. They can also be computed *ab initio*, with limited success (Li 74, Bü 82).

Since $n^*$ is roughly constant ($\approx$1.5) for atomic ground states throughout the periodic system, the hyperfine fields increase approximately with $Z$, according to equation (24). The experimental calibration of magnetic hfs fields is straightforward since, for virtually any element, $\mu_I$ of at least one isotope has been measured directly in an external field by NMR or an equivalent method. The last major gap in our knowledge of $\mu_I$ was for the radioactive elements Rn, Fr, and Ra; it has recently been filled by experiments at ISOLDE (Ki 88, Ek 86, Ar 87). Electric hfs fields have only been calibrated for a limited number of elements, through measurement of the muonic quadrupole splitting. For higher excited states $n \geqslant 3$ this can safely be analyzed with respect to $Q_s$ (Sc 78, St 85). Muonic measurements have also shown that for well-deformed nuclei, the formula for the projection of the intrinsic quadrupole moment $Q_0$ onto the spectroscopic one $Q_s$ holds with remarkable precision (i.e., within a few percent) in its strong-coupling limit:

$$Q_s = \frac{I(2I-1)}{(I+1)(2I+3)} Q_0 \tag{26}$$

Hence this formula offers an alternative method of calibrating the quadrupole splitting for spectra of elements with well-deformed isotopes and well-measured $Q_0$.

## 3. Methods of Atomic Spectroscopy Far Off Stability

### 3.1. Early and Nonoptical Experiments Off Stability

Optical spectroscopy of unstable isotopes poses two basic problems: (1) how to obtain sufficient sensitivity in the tracing of optical signals from the minute samples of material available; and (2) how to achieve, or at least preserve, under the constraint of highest sensitivity, sufficiently high spectral resolution in order to resolve IS and hfs. In the course of this section we will show how laser spectroscopic methods, on-line with mass separators for radioactive beams (ISOL systems), solved these problems in several elegant and advantageous ways. But before discussing these more recent developments we will touch upon a few early experiments in the field of unstable isotopes, including nonoptical methods which provided knowledge of spins and moments but not of the isotope shift.

In traditional optical spectroscopy, emission spectra from hollow cathodes or other kinds of spectral lamps are analyzed with high-resolution gratings or Fabry–Perot interferometers. This method is, of course, Doppler limited but resolves hfs and IS of heavier nuclei ($A \geqslant 100$)—particularly if cooled sources are used. Applications to unstable isotopes necessitated off-line experiments, and were thus limited to half-lives $T_{1/2} \geqslant 1$ day. The classical example of this type of work is the measurement of a series of Hg isotopes $192 \leqslant A \leqslant 204$ by Tomlinson and Stroke (To 64). The lighter members of this chain were produced by $(p, x_n)$ reactions in a Au foil. Thereafter, Hg was diffused out of the foil *in vacuo* and distilled into a small quartz bulb which finally was sealed off. The spectrum was excited by a rf discharge and photographed with a diffraction-grating spectrograph. Figure 7 shows the IS in the Hg series together with results from neighboring Tl.

Since Hg is a rather volatile and inert element, this experiment worked for quantities as small as of the order of 10 ng. The demand for volatility and/or chemical stability seriously restricted the application of the traditional optical methods to unstable isotopes

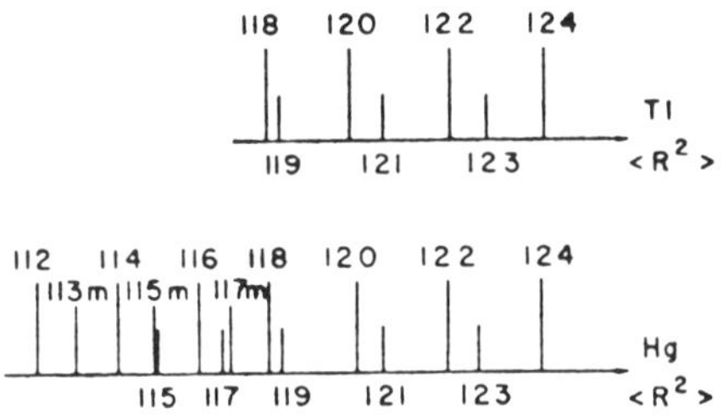


**Figure 7.** Mean-square (ms) charge radii of stable and long-lived Hg and Tl isotopes as determined by classical spectroscopy off-line. The scale is arbitrary [(To 64) and references cited therein].

and is still a crucial handicap in many of the more recent experimental methods that have been applied off stability.

Among the nonoptical methods we mention Rabi's famous atomic-beam magnetic-resonance method (ABMR). It could be adapted to the investigation of unstable isotopes in general, by tracing the radioactivity of the collected beam thereby taking advantage of the high sensitivity and selectivity of this detection technique. In this way a huge number of long-lived unstable nuclei have, in the past, been investigated off-line (Ni 57). The method is also a competitive one far off stability when implemented at a powerful isotope separator on-line (ISOL system), such as the ISOLDE at CERN (Ra 79). The first systematic measurements of spins and magnetic moments of a long series of alkalis, Rb (Ek 78), Cs (Ek 77), and Fr (Ek 86), were performed using ABMR at ISOLDE; measurements on Au (Ek 80), In, and Tl (Ek 84) followed. Later, laser spectroscopic experiments enabled the determination of $Q_s$ and $\delta\langle r^2\rangle$ values as well (see below). Under on-line conditions ABMR is sensitive down to beam intensities of about $10^5$ atoms/s.

Other rf methods, such as NMR and optical pumping, involve a sophisticated trick to monitor the rf resonance by radioactivity: since the nuclear rf resonance effectively results in a destruction of nuclear polarization, it can be monitored by a change in the parity-violating asymmetry of the angular distribution of $\beta$ decay. If excited nuclear levels are involved, the anisotropy of $\gamma$ radiation may be used instead. These methods belong to the vast field of angular-correlation techniques from which most of our knowledge of moments of excited states stems. The nuclear-orientation technique at low temperature (NOR) permitted the extension of such measurements to ground states of nuclei far from stability. On-line applications of this method have recently been successful, for example, in neutron-deficient Au and In isotopes (Wa 85, Va 85), and are developing rapidly [reviewed, e.g., by Stone (St 85b)]. Since these methods cannot be covered by this chapter we refer the reader to the topical literature (Ka 81, Ni 83, Ra 85). However, there was also an optical method, linked to this field, which played an important role in the early years of research far off stability; it is the RADOP method and it will be discussed in the following section.

## 3.2. Radioactive Detection of Optical Pumping (RADOP)

Optical Pumping (OP) is an efficient method for production of polarized atomic vapor by absorption of polarized resonance light. The angular momentum of the photons is first transferred to the valence electron and then, by hfs coupling, to the nuclear spin as well.

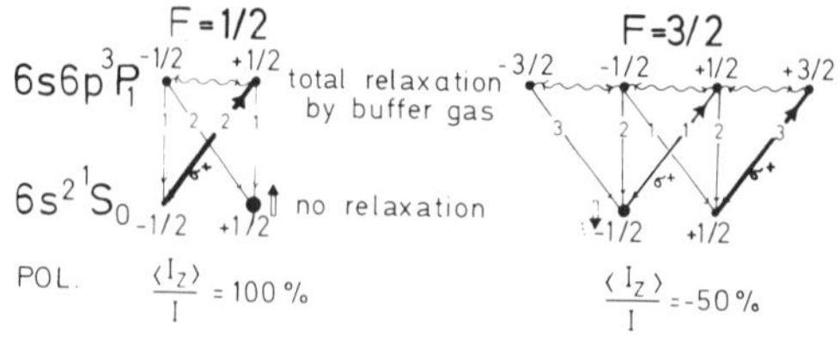


**Figure 8.** Optical pumping (OP) cycle for a mercury isotope with $I = \frac{1}{2}$ between states $J = 0$ and $J = 1$.

The detailed mechanism of OP is shown in Figure 8 for the example of a mercury isotope with $I = \frac{1}{2}$. The diamagnetic ground state $^1S_0$ has total angular momentum $F = I = \frac{1}{2}$ because $J = 0$. It splits into two nuclear Zeeman levels $m_I = +\frac{1}{2}$ and $-\frac{1}{2}$. The excited $^3P_1$ state has two hyperfine components with $F = \frac{1}{2}$ and $\frac{3}{2}$ which split into two and four Zeeman levels, respectively, as shown in Figure 8. Let us assume that the optical excitation via the resonance line excites only one of the upper $F$ levels, either the $F = \frac{1}{2}$ or the $F = \frac{3}{2}$ level. In Hg this is in general possible, since the hfs splitting is large as compared to the Doppler width of the spectral line. Right-handed $\sigma^+$-light leads to $\Delta m_F = +1$ absorption. The relative transition probabilities given in Figure 8, the pumping rate, and the relaxation rates in the excited and the ground state, govern the nuclear polarization $P_I = \langle I_z \rangle / I$ finally achieved. Collisions with buffer-gas atoms (usually a noble gas) depolarize the excited p state during its lifetime of about 100 ns because of the atomic quadrupole moment of the state with $l > 0$. Thus reemission is isotropic. In the spherically symmetric atomic ground state, however, relaxation rates are negligible as compared to the pumping rate. Under these conditions the steady state is reached when the product of population numbers and absorption probability is the same for all of the ground-state sublevels. This leads to $P_I = +100\%$ for pumping through the $F = \frac{1}{2}$ channel and to $-50\%$ through the other one.

The angular distribution of radiation emitted by an oriented ensemble of nuclei is given by

$$W(\vartheta) = \sum_k A_k f_k P_k(\cos \vartheta) \tag{27}$$

where $A_k$ is a correlation coefficient in the decay, $P_k$ is a Legendre polynomial, and $f_k$ is the $k$th moment of orientation defined by

$$f_k(I) = \binom{2k}{k}^{-1} I^{-k} \sum_m \sum_{\nu=0}^{k} (-1)^\nu \frac{(I-m)!(I+m)!}{(I-m-\nu)!(I+m-k+\nu)!} \binom{k}{\nu}^2 n_m \tag{28}$$

where $n_m$ is the normalized population number of the substrate $m_I = m$. While $f_0$ measures the degree of isotropic population, $f_1$ is the ordinary polarization $P_I$, relevant to $\beta$-decay asymmetry through the asymmetry coefficient

$$A_1 = \left\langle \frac{v}{c} A \right\rangle_{\text{spectrum}} \tag{29}$$

where $v$ is the velocity of the $\beta$ particle and $A$ is the asymmetry parameter, both averaged over the $\beta$-decay spectrum in question. For exotic nuclei this spectrum is usually very complex and not known in sufficient detail to estimate $A_1$. In $\beta$-RADOP applications competing channels with opposite signs of $A$ tend to diminish $A_1$ to unacceptably small values.

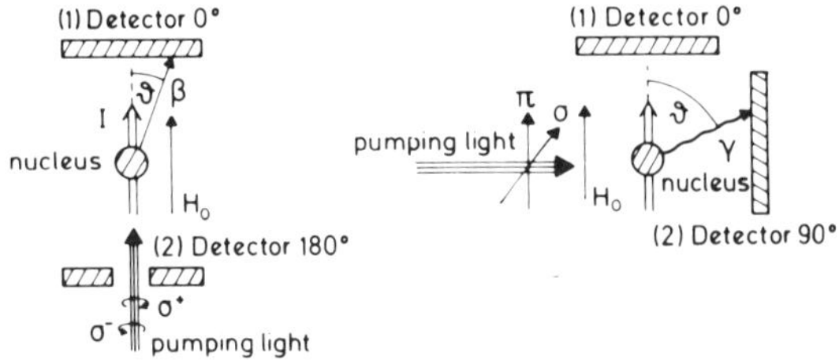


**Figure 9.** Geometries in $\beta$ RADOP and $\gamma$ RADOP with respect to an external field $H_0$.

The second moment

$$f_2 = (1/I^2)\sum_m m^2 \cdot n_m - \tfrac{1}{3}I(I+1) \tag{30}$$

is the so-called alignment and the most important moment in $\gamma$ RADOP, since $A_2$ dominates the anisotropy of $\gamma$ decay of oriented nuclei for all multipolarities.

Since $f_K \equiv 0$ for $I \leqslant k/2$, $\beta$ RADOP requires $I \geqslant \frac{1}{2}$ and $\gamma$ RADOP requires $I \geqslant 1$. These rules exclude, for instance, IS measurements on even nuclei, which obviously restricts the systematic application of RADOP. $\gamma$ RADOP is not limited to excited nuclear levels, since $\beta$ decay from the ground state transfers the alignment to excited levels of the daughter whose $\gamma$ decay may then be measured. Since $\beta$ decay is sensitive to odd and $\gamma$ decay to even moments of orientation, the pair of detectors observing the angular correlation are mounted at 0° and 180° in the former case and at 0° and 90° in the latter (see Figure 9).

A special advantage of OP is its capability of inverting the polarization, in a fast and very gentle manner, for example by switching the polarization of the incident light from $\sigma^+$ to $\sigma^-$ in $\beta$ RADOP. This enables the suppression of any kind of instrumental asymmetry by taking the relative difference signal of $\beta$-counting rates

$$S_\beta = (N_1^+N_2^- - N_1^-N_2^+)/(N_1^+N_2^- + N_1^-N_2^+) = 2A_1/(1 + A_1^2) \tag{31}$$

where $N_1^+$ is the counting rate in detector 1 at positive polarization, and so on. The detection limit of $S_\beta$ is determined by statistical accuracy only. Signals $\approx$0.1% have been observed unambiguously. In the case of $\gamma$ RADOP it is preferable to irradiate with the pumping light perpendicular to the external field $H_0$ and with linear polarization. Turning this polarization axis by 90° changes the sign of $f_2$, i.e., from polar to equatorial alignment. Hence we can define a signal $S_\gamma$ in analogy to $S_\beta$ [equation (31)].

The most serious problem in RADOP experiments arises from unpolarized radioactive material which may stick to the walls of the resonance vessel and hence does not participate in the OP cycle. This problem is avoided, or at least minimized, in on-line experiments on short-lived isotopes because the radioactive species is produced or deposited in the center of the vessel and already decays on its diffusion path through the buffer gas to the walls. The first $\beta$-RADOP experiments on short-lived alkalis strictly followed this scheme (Kö 69). A deuteron beam passed through a resonance vessel filled with, e.g., $^{36}$Ar giving rise to the reaction $^{36}\text{Ar(d, n)}^{37}\text{K} \xrightarrow[\beta^+,1\text{s}]{} {}^{37}\text{Ar}$. The magnetic moment of $^{37}$K and some other mirror nuclei were measured this way [reviewed, e.g., by Otten (Ot 85, 85b)].

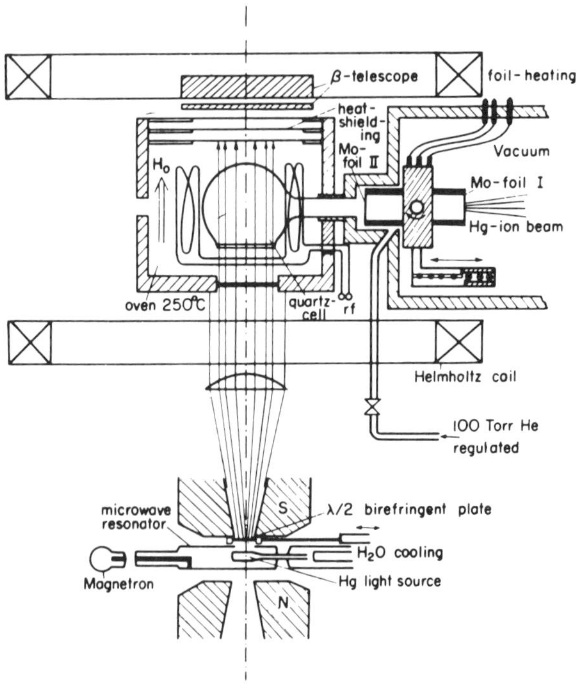


**Figure 10.** Apparatus used for RADOP experiments with on-line mass-separated, short-lived Hg isotopes (Hu 76).

Off-line experiments are only possible for a few noble and volatile elements which do not stick to the walls. The first $\gamma$-RADOP experiment which was conducted on $^{203}$Hg ($T_{1/2}$ = 47 days) is an example of this type (Ca 73).

Figure 10 shows the arrangement of the apparatus used for RADOP experiments on short-lived Hg isotopes at ISOLDE. These experiments opened the era of optical investigations at on-line mass separators (Bo 71, Hu 76). Collector foils are mounted two on each side of a flange and receive the mass-separated beam on one side and release it on the other after interchange of position and heating. OP is performed in the presence of He buffer gas by circularly polarized resonance light ($6s^2\,{}^1S_0 \to 6s6p\,{}^3P_1$, $\lambda = 254$ nm) from a spectral lamp containing an even stable isotope, e.g., $^{198}$Hg. Since hfs and IS of this spectral line are large compared to the Doppler width, a Zeeman component of the lamp isotope is tuned to the position of the hfs components of the unstable isotopes. This gives rise to scanning signals of the asymmetry $S_\beta$ as shown in Figure 11 for the case of $^{189}$Hg. The positions of the signals directly display hfs splitting and IS.

In addition to this information $g_I$ can also be directly determined by applying NMR to the oriented ensemble of nuclei in the resonance vessel. At the resonance point

$$\nu_0 = \gamma H_0 = g_I \mu_n H_0 / h \tag{32}$$

(where $\mu_n$ is the nuclear magneton and $H_0$ the external field) the orientation and hence the asymmetry signal $S_\beta$ is destroyed as shown in the insert of Figure 11. In the original

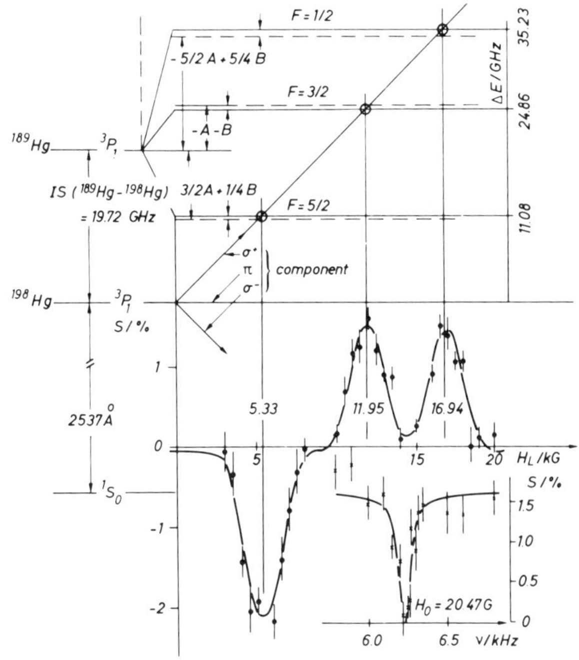


**Figure 11.** $\beta$ RADOP on $^{189}$Hg. The upper part of the diagram shows the various splittings of the resonance line for the lamp isotope $^{198}$Hg and for the unstable $^{189}$Hg, in their relative positions. Below, the scanning signal of $S_\beta$ is given. The insert shows the NMR signal of $^{189}$Hg (Hu 76).

experiment $\beta$-RADOP signals were obtained in the Hg series $181 \leqslant A \leqslant 205$. One particular surprise was the discovery of a shape transition from small-oblate to strong-prolate deformation in $^{185}$Hg, observed as a srong discontinuity of the IS (Bo 72, Bo 76) (see Section 4.7).

At the very ends of this chain the yield curve drops to about $10^3$ atoms/s. Since the lifetime is only a few seconds, the number of atoms in the vessel is of the same order. Still, a satisfactory signal-to-noise ratio is obtained at an asymmetry level of $S_\beta \approx 4\%$. For comparison, a usual fluorescence signal would optimize at a density of $10^{10}$–$10^{12}$ atoms/cm$^3$. As seen from the point of view of our current understanding these early $\beta$-RADOP experiments surpassed the sensitivity of many of the later laser spectroscopic experiments by order of magnitude; but their scope was fairly restricted. In the course of our discussion of the various methods it will become clear that the question of sensitivity and applicability of a particular method is always complex and subject to many experimental circumstances. General statements are therefore of little use and have poor predictive power.

An example of $\gamma$ RADOP is given in Figure 12. The vessel used contained $^{193}$Hg together with its isomer with $I = \frac{13}{2}$. Both the decay to $^{193}$Au and the $\gamma$ spectra were measured with a GeLi detector. The upper part of the figure shows the spectrum of the anisotropy signal (from one detector only), defined here as

$$S_\gamma = (N_1^+ - N_1^-)/(N_1^+ + N_1^-)^{1/2} \tag{33}$$

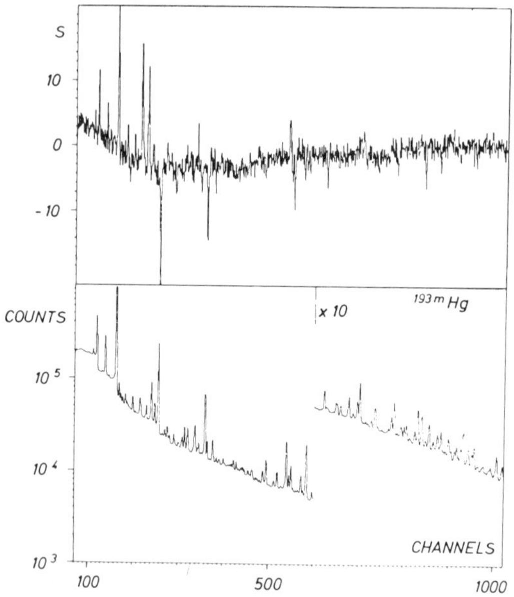


**Figure 12.** Upper part: normalized $\gamma$-RADOP signal in the decay spectrum of $^{193}Hg^m$. The anisotropy is calculated in each channel. Lower part: the decay spectrum itself (Hu 76).

The square root in the denominator of equation (33) normalizes $S_\gamma$ such that it is measured in units of one statistical standard deviation. Strong signals of opposite sign are seen in many transitions. These data contain a great deal of as yet unanalyzed information on the multipolarities of the decay scheme. The recently revived interest in detailed nuclear spectroscopy may well stimulate more work in this field, especially in the area of nuclear orientation (St 85b).

Figure 13 shows NMR curves from $\gamma$ RADOP on the (M1/E2) transition in $^{203}Hg$ (Ca 73). The width of the curve increases with the rf power, while its shape is characteristic of the orientation parameter $f_k$ and multipolarity involved (Ma 71).

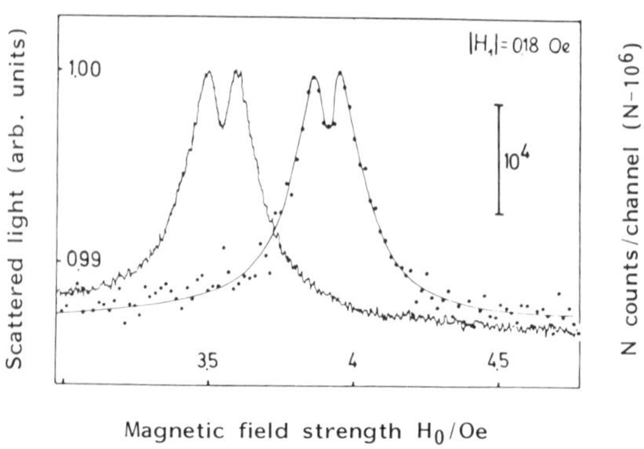


**Figure 13.** NMR resonance in the $\gamma$-RADOP signal of $^{203}Hg$ (smooth curve) in comparison with the ordinary optical pumping NMR signal of $^{201}Hg$ (jagged curve) (Ca 73).

A very interesting, but also difficult, problem was tackled in a RADOP experiment on the fission isomer $^{240m}$Am ($T_{1/2} = 1$ ms). It concerned the determination of the isomer shift (Be 79b). A huge shift of the order of 2 Å would be expected, reflecting the extraordinarily large deformation of the isomeric state in the second minimum of the potential-energy surface. Level structure and BE2 values for fission isomers indicate an ellipsoidal shape with a ratio of axis of 2:1. Figure 14a shows the experimental

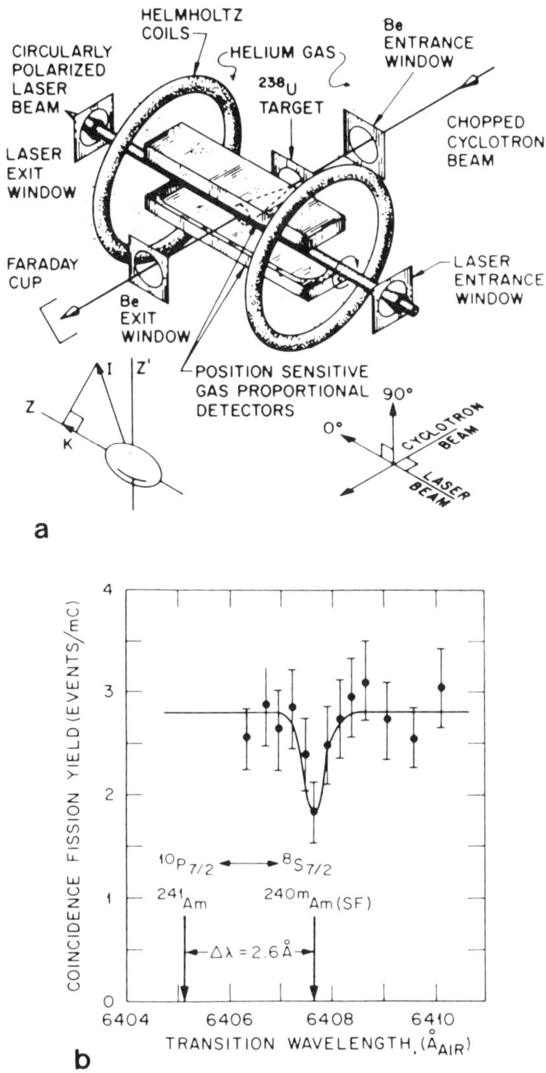


**Figure 14.** Fission-RADOP experiment on $^{240m}$Am (Be 79b): (a) experimental arrangement; (b) scan of the coincidence rate of the fission fragments as a function of laser frequency. Arrows indicate the IS with respect to $^{241}$Am.

system. $^{240m}$Am is produced by a Li beam incident on a thin $^{238}$U target from which the reaction product recoils into the He buffer gas in the resonance vessel, where it comes to rest. In contrast to earlier RADOP experiments, advantage was taken of a broad-band tunable dye laser (instead of a spectral lamp) for polarizing the Am atoms. Am is the only actinide with a spherically symmetric atomic ground state (the configuration is $5f^7 7s^2\ ^8S_{7/2}$) and hence is suited to RADOP since the relaxation rate of the polarization in the buffer-gas atmosphere is moderate. Broad-band running ($\Delta\nu$ = 0.5 Å) facilitated the search for the resonance line by taking large tuning steps. This was essential in view of the low event rate (about 200/h). On the other hand, the light intensity necessary to achieve a certain pump rate rises in proportion to the bandwidth. Only powerful lasers can meet these requirements. The RADOP signal was obtained in this particular case from the anisotropy of spontaneous fission with respect to the axis of orientation. This signal is absolutely free from background. Figure 14b shows a scan of the coincidence rate between the two fission fragments detected by two oppositely mounted proportional counters. A 30% dip is observed at the expected resonance position at which the nuclear spins should have lined up along the laser-beam axis. Using a somewhat different arrangement, whereby production and pumping sections are separated, this experiment is planned to be repeated by another group in order to strengthen the statistical confidence level of the resonance and eventually to resolve the hfs (Ba 86).

Another variant of RADOP was recently realized off-line by Calaprice *et al.* (Ca 85, 85b) in the case of the radioactive Xe isotopes 131$^{m}$, 133, and 133$^{m}$. The noble gas was spin-polarized by gas kinetic collisions with an optically pumped Rb vapor in a resonance vessel. During a collision the two atoms form a transient molecular state in which they share the spin polarization of the Rb valence electron that couples to the Xe spins and polarizes them. This polarization is monitored by $\gamma$-decay anisotropy. The extremely slow relaxation rate allows perfect NMR resolution. A linewidth of 0.5 Hz was reached, still dominated by power broadening. This experiment should eventually serve to set a better lower limit for the electric-dipole moment of a nucleus by measuring the resonance shift in an external electric field. Quite recently the group applied this method to $^{209}$Rn and performed the first direct determination of a $g_I$ factor of this element (Ki 88). This measurement enabled the calibration of hfs splittings in spectral lines of Rn which has recently been investigated for a long series of isotopes (Bo 87, 88).

In Section 3.7.2c we will come back to RADOP in connection with collinear laser spectroscopy. In summary, the RADOP method continues to be fruitful in certain, very specific experiments; but for systematic-survey experiments off stability it has been displaced by the more general and less cumbersome laser-spectroscopic methods discussed below.

## 3.3. Principles of Laser Spectroscopy

When tunable lasers were beginning to revolutionize optical spectroscopy in the early seventies, several laboratories were engaged in developing and testing methods with particular emphasis on the simultaneous achievement of high sensitivity and high resolution. Quite a number of those attempts turned out to be useful for investigating unstable isotopes and were able to be adapted to on-line conditions. Prior to discussing

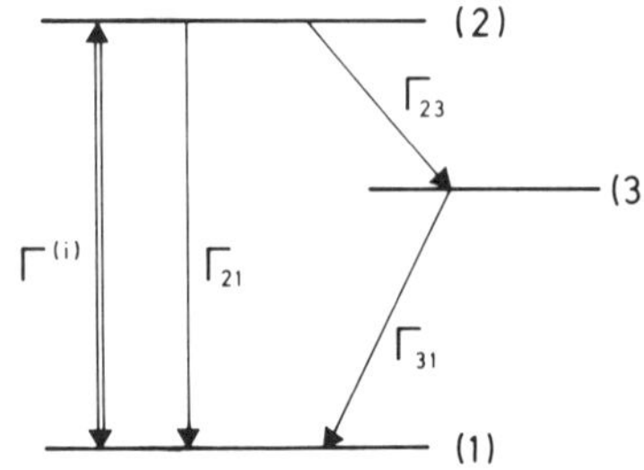


**Figure 15.** Schematic level diagram for laser spectroscopy. The laser drives transitions between levels 1 and 2 with the induced rate $\Gamma^{(i)}$. $\Gamma_{kl}$ are decay rates. Level 1 is assumed to be the atomic ground state.

experimental details the principal physical situation in laser spectroscopy will be briefly described.

Let us suppose that the laser radiation induces transitions between atomic levels 1 and 2 which may (or may not) couple to a third level (or a group of levels) by spontaneous emission or relaxation rates $\Gamma_{23}$, $\Gamma_{31}$ (refer to Figure 15). An atom with resonance frequency $\nu'$ in the laboratory frame will present to a photon of frequency $\nu$ an absorption (or induced emission) cross-section

$$\sigma(\nu - \nu') = \frac{\lambda^2}{2\pi} \frac{\Gamma_{21}/\Gamma_2}{1 + [4\pi(\nu - \nu')/\Gamma_2]^2} \tag{34}$$

where $\lambda$ is the wavelength of the photon and $\Gamma_2 = 2\pi\Delta\nu_h$ is the total decay rate of level 2 which determines the so-called homogenous (or natural) linewidth $\Delta\nu_h$ of the Lorentz profile equation (34). At the resonance point $\nu = \nu'$ and under the assumption that $\Gamma_2$ is comparable to $\Gamma_{21}$, the cross-section for an optical photon is of the order of $10^{-9}$ cm$^2$, i.e.,

$$\sigma(\nu = \nu') = \mathcal{O}(10^{-9}\ \text{cm}^2) \tag{35}$$

This is an enormous dimension for an elementary system and explains the effectiveness of laser spectroscopy. Let us assume, for example, a laser power of 10 mW/cm$^2$ which can be achieved by any continuous wave (cw) laser; this power corresponds to a photon beam intensity $j_{\text{Ph}}(\nu)$ of the order of $10^{17}$/cm$^2$ s; thus the induced transition rate $\Gamma^{(i)}$ reaches a level of $10^8$/s which is comparable to the spontaneous decay rate $\Gamma_{21}$ of allowed optical transitions. $\Gamma^{(i)} > \Gamma_2$ marks the regime of saturation where the populations in the excited level (level 2) and the ground level (level 1) are equalized in the absence of branching ($\Gamma_{23} = 0$).* A single atom excited to saturation will hence already

* Whereas equation (34) is a result of perturbation theory for a white photon spectrum, the interaction of a strong (i.e., classical) monochromatic field with a two-level system is described exactly by an analog to Bloch's equations for the motion of a spin-$\frac{1}{2}$ particle in an external magnetic field. At resonance the transient solutions show oscillations of the population numbers with the so-called Rabi frequency $\omega_1 = E\mu_e/\hbar$ ($E$ is the electric field strength of the laser and $\mu_e$ the transition electric dipole moment). The oscillation decays with $\Gamma_2$ toward the steady-state population. The linewidth of the resonance curve of the steady-state solution increases with $\omega_1$ according to $\Delta\omega_h \approx \sqrt{\Gamma_2^2 + \omega_1^2}$ (power broadening). Rabi frequency and power broadening are hence proportional to the square root of the laser power. The physical reality in laser spectroscopy does not usually correspond to either the white or the monochromatic solution but is complicated by the "inhomogeneous" Doppler width $\Delta\nu_D$ (Ab 61, Sh 76).

emit about $10^8$ fluorescence quanta per second, forming a strong signal. In an experiment on a single trapped Ba ion this fluorescence signal has actually been observed by the naked eye (Ne 78).

In most cases, however, branching leads to optical pumping (OP) into a third state (level 3) which is metastable for the time of observation $T$, that is, $T \cdot \Gamma_{31} \ll 1$. (This "mute" level can be, for example, one of the hfs levels of the ground state.) In this case the excitation emission cycle stops after an average of $1 + \Gamma_{21}/\Gamma_{23}$ fluorescence quanta per atom have been emitted. Usually this number is not much above unity. Therefore, many attempts have been made to find another signature of optical-resonance excitation which is more efficient and less sensitive to background than the detection of fluorescence quanta. RADOP was one such example and a number of others will be described below. Experience has shown, however, that in the absence of OP and for long observation times, the sensitivity of fluorescence detection is not easily improved on. Moreover, it is the easiest and, therefore, most generally used detection method.

In the case of Doppler broadening the cross-section is a convolution of a Lorentzian and Gaussian (so-called Voigt) profile:

$$\sigma_D(\nu) = \frac{\frac{\lambda^2}{2\pi}\Gamma_{21}/\Gamma_2}{\sqrt{\pi}\Delta\nu_D} \int_{-\infty}^{+\infty} \frac{\exp[-(\nu' - \nu_0)^2/(\Delta\nu_D)^2]}{1 + [4\pi(\nu - \nu')/\Gamma_2]^2} d\nu' \tag{36}$$

where $\nu_0$ is the resonance frequency for the atom at rest. For a thermal gas the (single) $1/e$ width

$$\Delta\nu_D = (\nu_0/c)(2kT/m)^{1/2} \tag{37}$$

is of the order of 1 GHz and exceeds the natural one by a factor of between 100 and 1000. The cross-section (equation 36) is flattened by the same factor, as compared to the opposite Doppler-free case ($\Delta\nu_D \ll \Delta\nu_h$). This implies that only a small fraction—of the order $\Delta\nu_h/\Delta\nu_D$—of the atoms can participate in the transition at a given excitation frequency $\nu$.

### 3.4. Laser-Excited Fluorescence in Resonance Cells

Our discussion of the individual experiments in this field will not follow their chronological order but will start instead with the conceptually simplest one.

A straightforward approach to laser-excited fluorescence spectroscopy on-line is to start from the RADOP arrangement of Figure 10. One replaces the spectral lamp by a tunable dye laser and the $\beta$ counters by a photomultiplier (PM). Figure 16 shows the modified apparatus used for measurements on a number of even Hg isotopes (Kü 77) as well as $i_{13/2}$ isomeric states (Da 78, Da 79) down to $^{184}$Hg and $^{185}$Hg$^m$, respectively. Both have evaded measurement by the earlier RADOP experiments, the even isotopes for reasons of principle ($I = 0$), and the isomers because of $\beta$-asymmetry parameters which were apparently too small.

The resonance wavelength of 254 nm for Hg cannot be generated directly in a dye laser but can be obtained by frequency doubling a laser beam of wavelength 508 nm in a birefringent crystal. Since this nonlinear effect is proportional to the square of the incoming intensity, it is preferable to choose a pulsed laser with high peak power instead of a cw laser. Here, a primary nitrogen pump laser yields ≈3 ns wide pulses

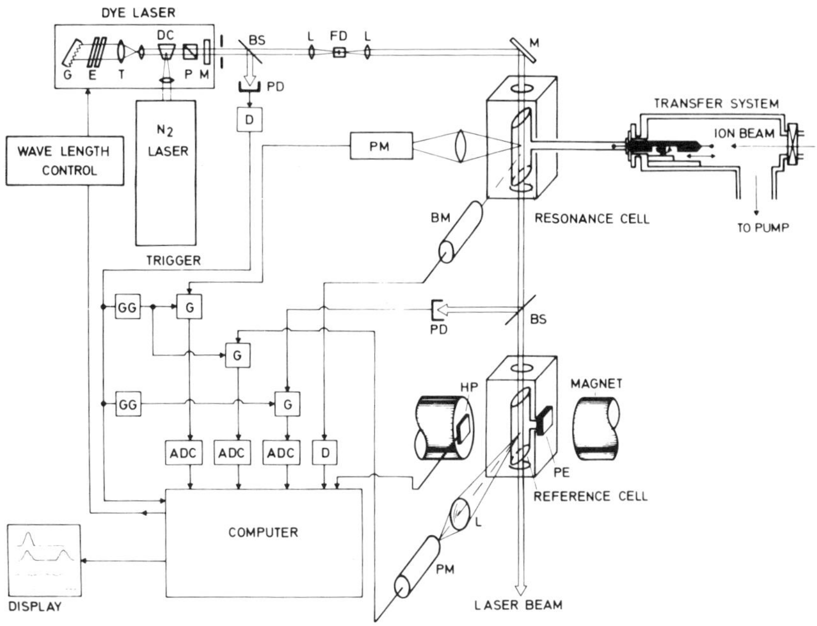


**Figure 16.** Experimental arrangement for fluorescence spectroscopy in a resonance cell on short-lived Hg isotopes. G: grating; E: etalon; T: beam-expanding telescope; DC: dye cell; P: polarizer; M: mirror; FD frequency-doubling crystal; L: lens; PM: photomultiplier; PD: photodiodes for monitoring laser intensity and triggering gates (G); ADC: analog-to-digital converters; BM: detector monitoring $\beta$-activity in the cell (Kü 77, Da 78).

of $\approx 100$ kW power at a repetition rate $\approx 100$ Hz; the frequency-doubled pulse has a power of the order of 1 kW and a spectral width of about 2 GHz which covers the Doppler width. This latter circumstance, together with the high power, results in effective saturation of the transition at resonance for the whole ensemble and maximizes the fluorescence peak.

Figure 17, taken from Kremmling *et al.* (Kr 79), shows a scan of the hfs spectrum of $^{187}$Hg containing components from the ground state as well as from the isomeric one. The laser frequency is calibrated by simultaneously scanning Zeeman components from stable isotopes at a fixed field (performed in the lower resonance cell of Figure 16.)

A modification of this method led to a direct determination of the nuclear spins of Hg isomers (Kr 79) by utilizing the fact that the laser excitation is performed using very short pulses and is followed by the slow decay of the atomic state with a lifetime of $10^{-7}$ s. If a small magnetic field is applied to the resonance vessel, the total spin $F = I + J$ performs a Larmor precession in the excited state, resulting in a rotation of the angular distribution of the emitted photons. Figure 18 shows the number of photons observed, as a function of time after the laser shot. For electric-dipole radiation, the modulation of the decay curve occurs at twice the Larmor frequency, which is connected to the spin via the $g_F$ factor:

$$g_F = g_J[F(F+1) + J(J+1) - I(I+1)]/2F(F+1) \tag{38}$$

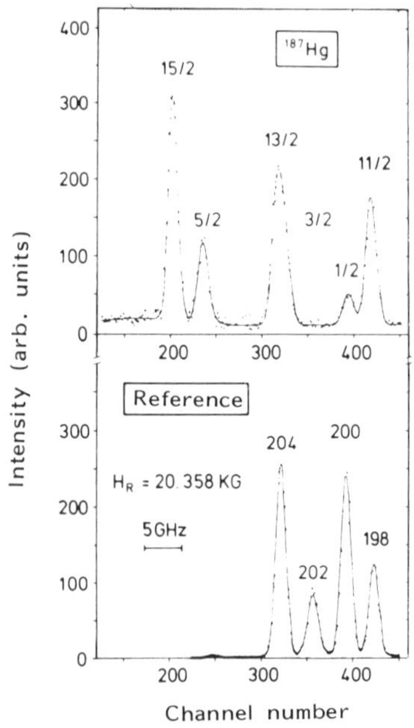


**Figure 17.** Upper part: hfs spectrum of the ground state ($I=\frac{3}{2}$) and the isomer ($I=\frac{13}{2}$) of the $6s^2\,{}^1S_0 \rightarrow 6s6p\,{}^3P_1$ ($\lambda = 2537$ Å) lines of $^{187}$Hg. Lower part: reference spectrum of the $\sigma^+$-Zeeman components of the even, stable Hg isotopes (Kr 79).

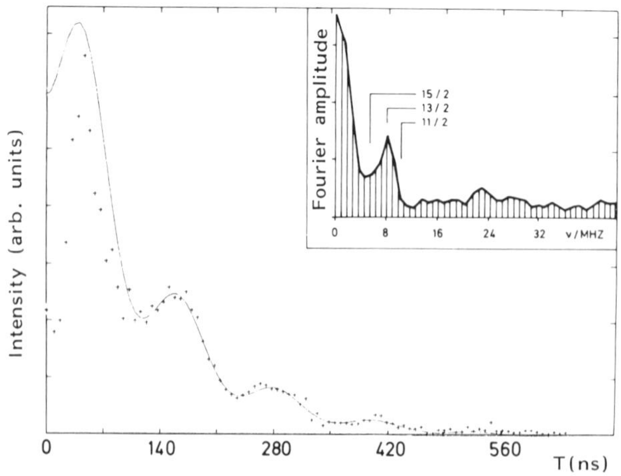


**Figure 18.** Time-dependent perturbed angular correlation in the fluorescence from the $F=I=\frac{13}{2}$ hfs level of $^{187}$Hg$^m$ in a magnetic field of 91.1 G. A damped oscillation is fitted to the data points. Owing to prompt stray light, the photomultiplier is overloaded in the first 80 ns. The inset shows the Fourier transform of the signal, peaking at the Larmor frequency expected for $I=\frac{13}{2}$ (Kr 79).

$I = \frac{13}{2}$ has been confirmed for all the isomers, including the previously questionable case of $^{185}Hg^m$. The method of laser-excited fluorescence in cells has also been applied to a series of Cd isotopes $102 \leqslant A \leqslant 120$ (Bu 81, Bu 87) and to some neutron-deficient Au isotopes (St 85c). The latter experiment was the first to be conducted on radioactive isotopes of a refractory element. It required special high-temperature technology, including sapphire cells heated to 1000 °C.

The original sensitivity of fluorescence spectroscopy with pulsed lasers was seriously limited by their low duty cycle ($\approx 10^{-6}$). About $10^9$ atoms were required in the cell in order to achieve a reasonable signal. Only a few percent of the atoms were in the region of observation; the others filled the dead volume. With modern copper-vapor or excimer lasers it is possible to gain a factor of 100 in duty cycle but even this is still far below the optimum obtained using a cw laser. From this point of view it would be preferable to choose for study elements with low-lying resonance lines, accessible to cw lasers. However, low excitation energies are accompanied by high chemical affinity (as in the cases of the alkalis and alkaline earths), which causes the atoms to stick to the wall on their first encounter with it.

## 3.5. Laser Spectroscopy on Collimated Atomic Beams

The experiments to be described in this section avoid the difficulties specific to resonance vessels by employing atomic-beam techniques instead. These also offer the chance to greatly reduce the Doppler width, as is necessary for application to lighter elements in order to resolve hfs and IS. The possibility of achieving high sensitivity together with high resolution in laser spectroscopy on atomic beams has been explored by Jaquinot and collaborators for various detection schemes (Ja 76). The experiments described in Sections 3.5.1. and 3.5.2 arose from considerations along these lines.

### *3.5.1. Laser-Excited Fluorescence from an Atomic Beam*

Excitation of a collimated, thermal atomic beam obviously leads to reduction of the Doppler width by a factor corresponding to the truncation of the atomic velocity in the direction of the light. It is also clear that resolution can only be gained at the cost of signal strength, since the collimation reduces the solid angle of the atomic beam observed. For a simple geometry, as sketched in Figure 19, the necessary integrations can be carried out analytically. In the absence of OP—and for moderate light intensities below the saturation point—the rate of fluorescence photons emerging from the illuminated volume $V$ is given by

$$R(\nu) = j_{\rm Ph}(\nu) n_{\rm V} V \sigma_{\rm Dc}(\nu) \tag{39}$$

where

$$n_{\rm V} \approx n_0 \frac{A}{4\pi r^2} \tag{40}$$

is the total density of atoms in volume $V$ and $n_0$ is that in the oven; $\sigma_{\rm Dc}(\nu)$ is the frequency-dependent cross-section (equation 36) but now taken for the "collimated" Doppler width,

$$\Delta\nu_{\rm Dc} = \Delta\nu_{\rm D} \sin\delta_0 = \frac{\nu_0}{c}(2kT/m)^{1/2} \sin\delta_0 \tag{41}$$

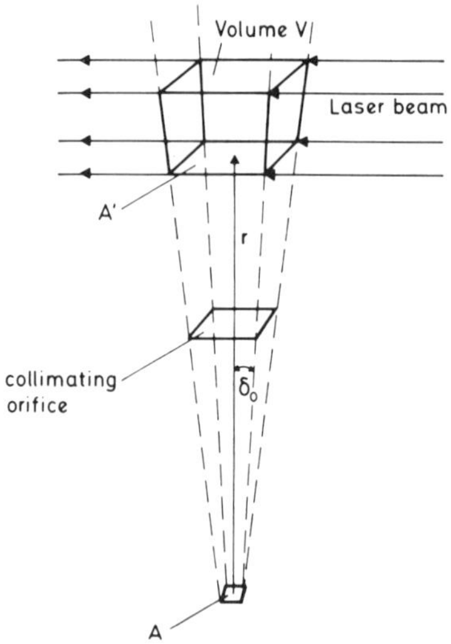


**Figure 19.** Geometry for exciting a collimated atomic beam in the volume $V$. The beam emerges from an oven with orifice A.

The remaining symbols are defined in Figure 19. The gain in resolution is given by the sine of the collimation angle $\delta_0$ and the loss in density by the solid angle through which the orifice of the oven is seen in the observation volume.

Let us now assume, more realistically, that in practice OP limits the number of fluorescence photons to about one per atom passing through $V$; furthermore, the laser intensity may be adjusted such that this limit is just reached at the resonance peak $\nu_0$. Then the peak count rate is given by

$$R(\nu_0) \approx I\frac{A'}{\pi r^2} \tag{42}$$

where $I$ is now the total current of atoms leaving the oven and $A'$ is the cross-section of the observation volume. Assuming a typical collimation of the beam to 25 mrad in both directions, equation (42) implies a geometric reduction factor of about 5000. Fluorescence photons may be detected within a solid angle of $\Delta\Omega/4\pi \approx 5\%$ and a quantum efficiency of the PM of 10%. From this one calculates a total detection efficiency per atom of only $10^{-6}$.

Nevertheless, this technique has proved to be very fruitful in off-line applications to the following isotopic chains: $^{40}$Ca to $^{48}$Ca (Be 80, An 82), $^{80-90}$Sr (Bu 85, An 86, 87), $^{110}$Sn to $^{125}$Sn (An 86b), $^{124}$Ba–$^{138}$Ba (No 78, Be 79), and $^{196}$Pb to $^{214}$Pb (Th 83, An 86c). These experiments have concentrated on the particularly interesting cases of magic proton elements $Z = 20$, 50, and 82. Figure 20 shows the system used by the Karlsruhe group for Ba and other elements (No 78). The atomic beam, evaporating from an off-line mass-separated sample of an unstable Ba isotope, is excited by a cw dye laser tuned periodically through the resonance. The laser frequency is controlled by heterodyning it with a second laser which is stabilized to the atomic resonance line of a stable Ba isotope.

Figure 21 shows a recording for the case of $^{126}$Ba ($T_{1/2} = 97$ min) (Be 79). The linewidth has shrunk to about 50 MHz which is an enormous gain over that of the cell experiments. About $3 \times 10^{11}$ atoms of $^{126}$Ba were produced; $6 \times 10^{10}$ of these were used

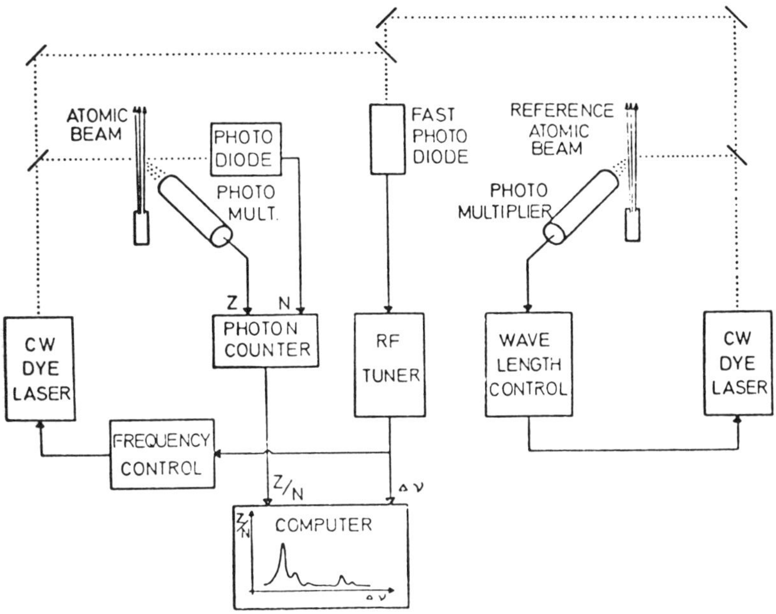


**Figure 20.** Experimental arrangement for fluorescence spectroscopy from a collimated atomic beam of off-line mass-separated unstable Ba isotopes excited by a cw dye laser. On the left is the unstable beam and on the right the stable reference beam (No 78).

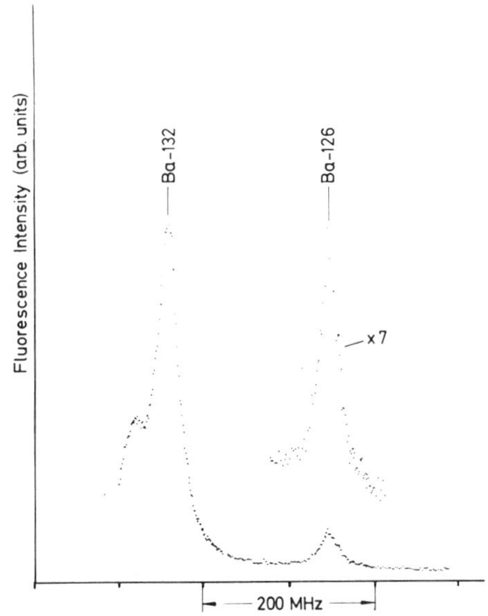


**Figure 21.** Fluorescence spectrum of the $6s^2\,{}^1S_0 \rightarrow 6s6p\,{}^3P_1$ ($\lambda = 5535$ Å) line taken from a sample of $^{126}$Ba. The resonance line is slightly power-broadened by high laser intensity in order to profit from multiple excitation of each atom (Be 79).

in a single recording over 6 min. From this one calculates that the average number of atoms present in the observation region was only about one. From this point of view, it is astonishing that the method can compete with the cell technique. But the 100% duty cycle of the cw laser compensated for most of the handicaps; some other factors, such as higher oscillator strength and better stray-light suppression, also helped.

### *3.5.2. Laser-Induced Optical Pumping (LIOP) of an Atomic Beam with a Stern–Gerlach Analyzer*

This experimental method was the first in the field of on-line laser spectroscopy. It was developed by a collaboration between the René Bernas and Aimé Cotton institutes, both at Orsay, and is particularly well suited to investigating alkalis. The pilot experiment was run on-line with the 150 MeV proton beam at Orsay and produced Na isotopes in the range $^{21}$Na to $^{25}$Na (Hu 75). Later, an improved apparatus was used on-line with the 24 GeV proton beam at CERN, which yielded extremely neutron-rich Na isotopes by fragmentation of U. Laser spectroscopy could be performed up to $^{31}$Na (Hu 78). Finally, the experiment was performed at the ISOLDE facility and all the other series of alkalis were measured, namely, $^{38}$K to $^{47}$K (To 82b), $^{76}$Rb to $^{98}$Rb (Th 81), $^{118}$Cs to $^{146}$Cs (Th 81b, Co 87), and $^{207}$Fr to $^{223}$Fr (Li 80, Co 85, Co 87).

The principle of the method can be explained using Figure 22, which shows the main elements of the apparatus used in the experiments at ISOLDE. The mass-separated ion beam is stopped in a hot Ta tube covered with Y in order to reduce the work function of the surface to below the ionization potential of alkalis. Thus the stopped particles reevaporate predominantly as neutral atoms and form an atomic beam which undergoes OP by the transverse laser beam. For the example shown in the level diagram in the lower part of Figure 22, pumping via transitions a and b depopulates the upper hfs level ($F = 2$) of the atomic ground state. For the spontaneous reemission from the excited state, channels c and d are also open and connect to the lower hfs level ($F = 1$). Thus the lower level is enriched at the cost of the upper. Pumping via channels c and d has the opposite effect.

After passing through the pumping section the beam enters a strong magnetic sextupole field where the Paschen–Back effect forms two groups of Zeeman levels with $m_j = \frac{1}{2}$ and $-\frac{1}{2}$, respectively. Only the rising group is focused by the Stern–Gerlach force of the field gradient; the other is defocused. Thus pumping via a and b decreases, and via c and d increases, the current to the detector thus giving rise to the pattern sketched in the figure. Selective and efficient detection of the atomic beam is achieved by an advanced version of the well-known Langmuir–Taylor detector. A hot tungsten tube is placed at the focus of the sextupole and converts, by surface ionization, alkali atoms into ions which are then mass-separated and counted by a multiplier.

Figure 23 displays the whole series of signals obtained from the chain of Na isotopes (Hu 78). The large splitting corresponds to the hfs of the 3s ground state and the small one to that of the excited 3p $^2P_{1/2}$ state. The center of gravity is displaced by the isotope shift which is dominated by the mass shift. If the small hfs splitting of the excited state is compared with the huge one in Hg (see Figure 11) one immediately recognizes the importance of reducing the Doppler width in the case of light elements.

The accuracy of determination of the ground-state splitting was recently improved by a factor of 1000 over that of these optical measurements by inducing rf transitions between the hfs levels in the region between the pumping zone and the Stern–Gerlach

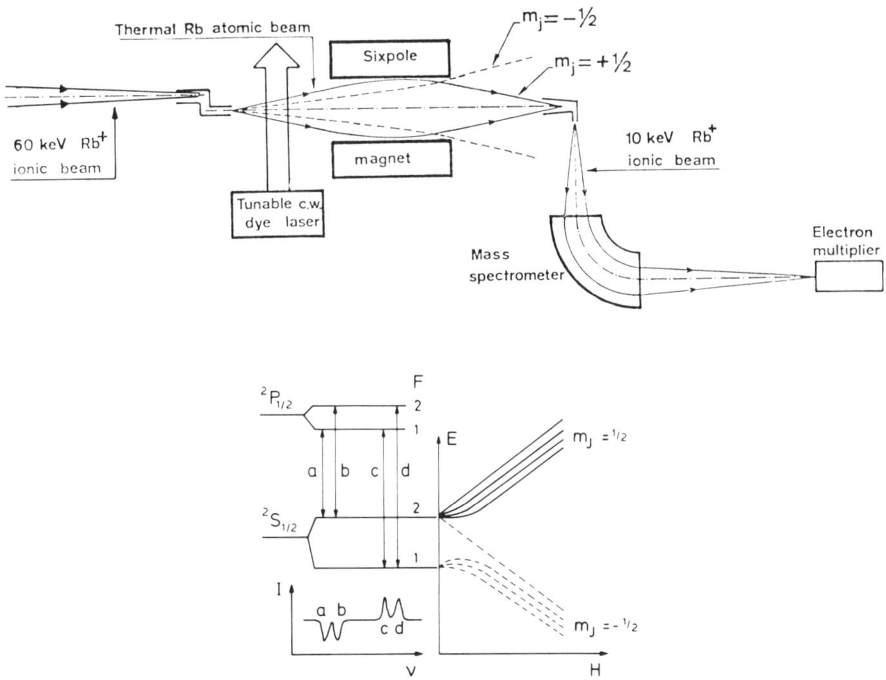


**Figure 22.** Upper part: experimental arrangement for laser-induced optical pumping (LIOP) of an atomic beam of alkalis detected by magnetic-state selection (Th 81). Lower part: hfs level diagram of the $D_1$ resonance line of an alkaline isotope with $I = \frac{3}{2}$. The inset shows schematically the spectrum of the four pumping signals.

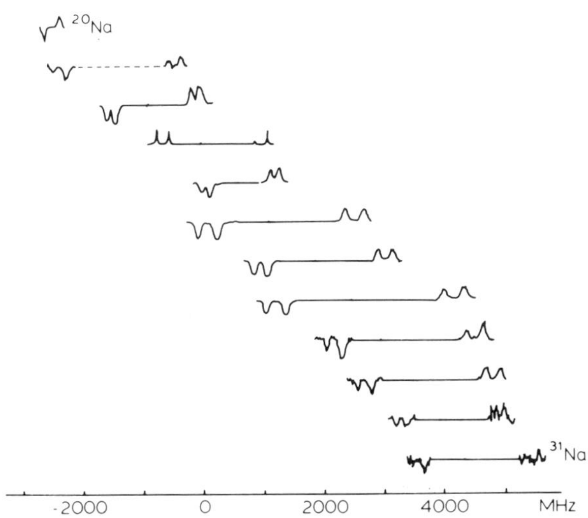


**Figure 23.** An entire series of Na signals in the $D_1$ resonance line obtained by LIOP (Hu 78).

analyzer; thus a type of Rabi apparatus was formed (Du 86). This high precision will enable the determination of the hfs anomaly once the $g_I$ factors have been measured directly.

Rf transitions were also applied to the excited $^2P_{3/2}$ state because its quadrupole splitting could frequently not be analyzed from the barely resolved hfs splitting in the optical D2-line. This double-resonance technique was perfectly suited to LIOP in the case of exciting the optical $\Delta F = 1$ hfs component $(^2S_{1/2}, F = I + \frac{1}{2}) \rightarrow (^2P_{3/2}, F = I + \frac{3}{2})$ which connects the states of highest angular momenta. Since the spontaneous decay obeys the selection rule $\Delta F \leqslant 1$, it cannot pump the atoms into the lower $F = I - \frac{1}{2}$ hfs level of the ground state. Hence, it cannot produce any LIOP effect unless the angular momentum of the excited state is lowered by one unit before it decays. This is done by the rf transition $F' = I + \frac{3}{2} \rightarrow F' = I + \frac{1}{2}$. Its linewidth $\Delta\nu = 1/\pi\tau$ is twice the natural one and is still narrower than the rest Doppler width of the optical transition (Du 82).

Of the many important results obtained by the LIOP method (see Section 4) we quote here a particular one which is of relevance to atomic physics. It concerns the first discovery of spectral lines in Fr. The first two doublets in the principal series have now been completely measured including their fine structure, hfs, and IS (Li 80, Co 87, Du 87). Fr attracted the interest of theoreticians because of the simplicity of the spectrum on the one hand (there is only one electron outside closed shells) and because of its highly relativistic character on the other. For example, the ground-state hfs field was predicted with an accuracy of about 5% (Dz 83, Ek 86). The experimental check of the prediction required the measurement of the hfs splitting by LIOP (Li 80) as well as of the $g_I$ factor by ABMR (Ek 86).

Under typical running conditions at ISOLDE the LIOP experiments produced results for beams down to the order of $10^5$ particles/s. Another characteristic number is the total transmission of the apparatus from the atomic-beam oven to the final ion multiplier which detects the signal; it was determined to be $10^{-5}$. The critical signal rate is thus of the order of one per second. For comparison we remark here that on-line fluorescence detection, as performed in collinear laser spectroscopy experiments for example, requires a minimum signal rate of 100/s to overcome background (see Section 3.7). But a better efficiency, i.e., a higher signal rate per incoming atom, compensates for the higher background. In any case, the success of LIOP makes it clear that particle detection in optical experiments is a powerful alternative to photon detection since it solves the problem of background.

### 3.6. Resonance Ionization Spectroscopy (RIS)

Ionization of atoms and molecules by stepwise resonant excitation with two or more laser quanta is a well-established and quite generally applied method in trace analysis (Le 84, Hu 79). The method first demonstrated its merits in the field of unstable nuclei, in an experiment at the Leningrad on-line mass separator on a series of Eu isotopes $141 \leqslant A \leqslant 150$ (Al 83, Fe 84). Figure 24 shows the relevant part of the level spectrum. Three photons from three different dye lasers lead from the atomic ground state via two relay states into the continuum. Ionization efficiency is optimized when all three transition are saturated. This requirement is easily met for the first two steps but poses a problem for the third one, since the cross-section for photoionization into the plane continuum is of the order of only $10^{-17}$ cm$^2$. Therefore, it would be expedient

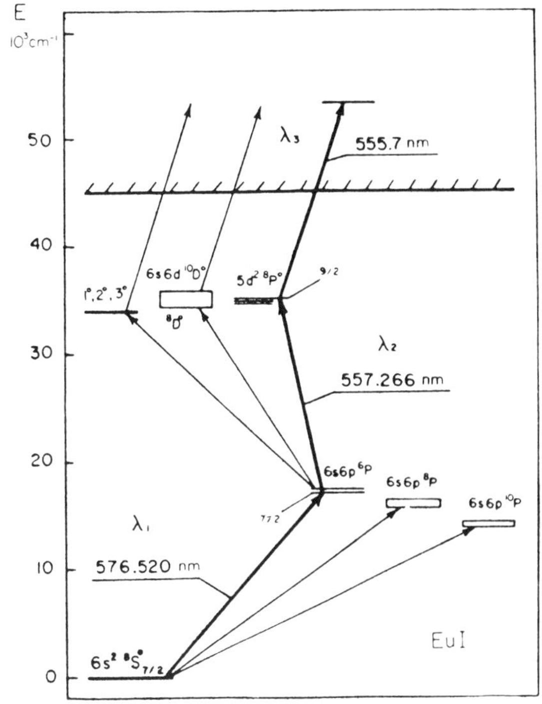


**Figure 24.** Partial scheme of Eu levels showing different ladders used for three-step laser photoionization. The right-hand ladder is the most efficient one and leads to an autoionizing state (Fe 84).

to tune this last step to an autoionizing state; in complex spectra many of these are present just above the ionization limit. From equation (36) we see that the excitation cross-section for such a state is higher, the smaller its autoionizing width ($\approx \Gamma_2$) is in comparison with its radiation width. One could, alternatively, excite a Rydberg state, just below the ionization limit, which can thereafter be easily field-ionized.

Hfs and IS are measured in the first step of RIS. The spectral width of the laser which serves this transition is therefore adjusted to its Doppler width. Since the latter depends on the collimation of the atomic beam, one again compromises between resolution and efficiency. The two other lasers are run broadband (a few GHz wide) so as to match the second and third steps for any of the isotopes, without requiring tuning.

Figure 25 shows some details of the apparatus used. The ion beam from the mass separator is stopped in an oven from which it evaporates as an atomic beam into the ionization region. The light intensity is amplified by a pair of mirrors that provide for multipassing of the laser beam through the atomic beam. The ions are then extracted, focused onto a channeltron detector, and counted as a function of laser frequency. Figure 26 shows a scan of $^{144}$Eu together with reference lines from stable isotopes and fringes from a calibrating interferometer. Interest in this experiment, from the point of view of nuclear physics, focused on the very neutron-deficient isotopes around and below the shell closure $N = 82$ (see Section 4). The group has recently extended these measurements to still-lighter Eu isotopes, down to $^{138}$Eu. Moreover, long isotopic chains of other rare-earth elements have been investigated this way, namely, Nd, Sm, Hm,

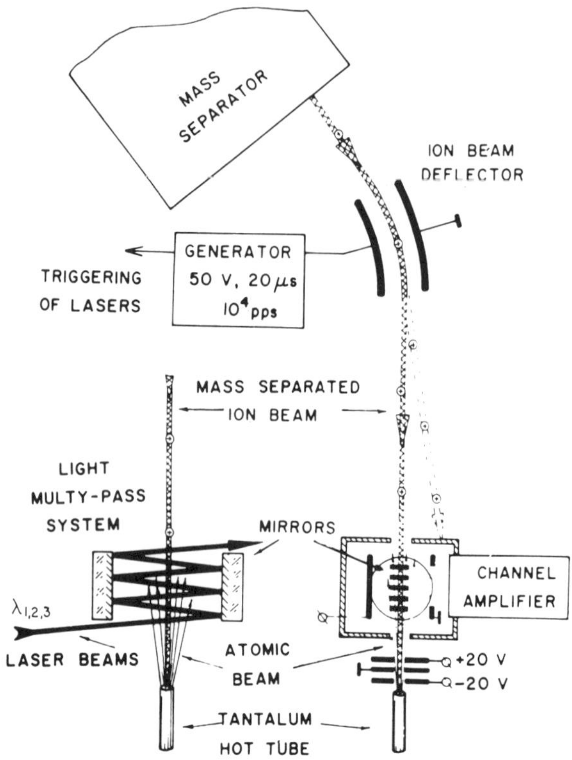


**Figure 25.** Experimental arrangement for on-line resonance ionization spectroscopy used at the Leningrad mass separator. The inset on the left shows an expanded view of the ionization region (Fe 84).

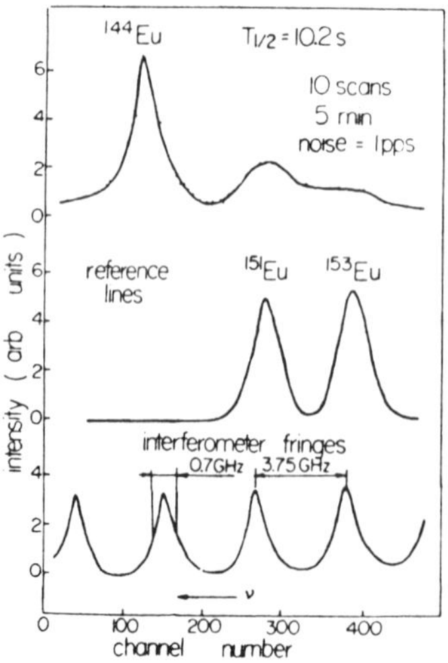


**Figure 26.** Upper curves: scan of the $^{144}$Eu resonance by the RIS technique, together with reference signals from stable isotopes. Bottom curve: interferometer fringes for calibrating the laser frequency (Fe 84).

and Tm (Al 87, Le 87, see the Appendix) proving the generality and effectiveness of this technique.

The RIS technique is usually based on pulsed, synchronized lasers, preferably with high repetition rate. There are five principal reasons for this choice:

1. The high power necessary to saturate the intermediate and ionizing transition with broadband irradiation is only available from pulsed lasers.
2. The pulse width (of the order of 10 ns) is smaller than, or at least comparable to, spontaneous decay times into any other atomic state. By running up the ladder very rapidly, one therefore prevents atoms from getting lost on the way, through OP.
3. The demand for a high pulse rate is self-explanatory in view of the high efficiency desired. In the experiment discussed above a copper vapor laser was used to pump the dye lasers. Its repetition rate, of the order of 10 kHz, already comes close to the optimum at which each atom experiences at least one laser pulse during its transit time through the interaction region. An overall efficiency of $3 \times 10^{-4}$ was reached in the Eu experiment (Fe 84), which is a very good number.
4. A sharp (delayed) coincidence between the laser pulse and the ion signal allows reduction of the background by a factor equal to the duty cycle of the laser; this was of the order of $10^{-4}$ in the Eu experiment and thus resulted in a noise level of 0.1 counts/s (Fe 84).
5. Pulsed operation provides the very important advantage of combining mass separation with RIS by a time-of-flight measurement. For this purpose the acceleration voltages and the geometry of the acceleration and drift regions of the ions are adjusted to one another such that a time focus is created at the site of the ion detector. At that focus the total time-of-flight is independent (to first or even higher order) of the spatial extension of the ionization region. Figure 27 shows the time-of-flight spectrum obtained in an experiment on neutron-deficient Au isotopes at ISOLDE (Wa 87). The peak for $^{186}$Au is clearly

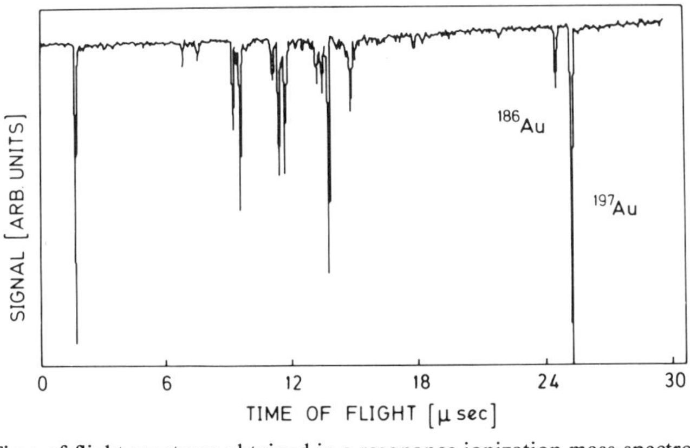


**Figure 27.** Time-of-flight spectrum obtained in a resonance ionization mass-spectrometry experiment (RIMS). $^{186}$Au was collected in the atomic-beam oven. Stable $^{197}$Au was added to serve as a mass marker. The medium-mass ions arise from photoionization of rest gas and can be suppressed by setting an appropriate time window (Wa 87).

separated from that corresponding to the stable $^{197}$Au and also from the background corresponding to lighter molecular masses. The latter separation was very important since the high laser power, necessary for the ionization step in Au, also led to nonresonant multiphoton ionization of rest-gas molecules. The Au series is very interesting from a nuclear physics standpoint since it repeats the shape transition that was observed earlier in light Hg isotopes (see Section 4).

## 3.7. Collinear Laser Spectroscopy

### *3.7.1. The Standard Method with Fluorescence Detection*

In the optical on-line experiments discussed thus far, an ion beam was converted into thermal atoms by stopping and reevaporation, thus forming either a vapor in a cell or a collimated beam. In the former case chemical stability is a restrictive condition and, in the latter, a large intensity factor is lost through collimation if high resolution is required. The original idea behind collinear laser spectroscopy was simply to avoid these problems by using the mass-separated beam directly as the spectroscopic sample. Since at typical ion velocities of $10^7$ cm/s the interaction time with the laser in a crossed geometry would shrink to about $10^{-8}$ s, a collinear superposition of laser and ion beams seemed favorable. In such a geometry, interaction time and length could easily be increased by a factor of 100 when observing fluorescence light from a pathlength of 10–20 cm. This can be achieved, e.g., by means of a cylindrical lens which forms an image of the beam on the entrance slit of a light pipe that in turn adapts the image to the geometry of the PM photocathode (see Figure 28).

A surprising and very important advantage of collinear geometry was discovered by Kaufman (Ka 76) and, independently, by Wing *et al.* (Wi 76). It is that the spread in kinetic energy ($\delta E$) of the beam remains unchanged under electrostatic acceleration,

$$\delta E = \delta(mv^2/2) = mv\delta v = (mc^2/\nu^2)\Delta\nu_{\rm D}\delta\nu_{\rm D} = \text{const} \tag{43}$$

Therefore, the product of the average velocity $v$ and the velocity spread $\delta v$, or equivalently, the product of the Doppler shift $\Delta\nu_{\rm D}$ and the Doppler width $\delta\nu_{\rm D}$ are constants of the motion. In other words, acceleration reduces the Doppler width along the beam direction from its original value in the ion source, by a large factor. Assuming ideal starting conditions, i.e., a thermal velocity distribution at the ion source, one observes—after acceleration—a reduced Doppler width of

$$\delta\nu_{\rm D} = \tfrac{1}{2}(kT/eU)^{1/2}\delta\nu_{\rm D}(0) \tag{44}$$

The reduction factor is about $10^3$ for a source temperature of 2000 K and an acceleration voltage of 60 kV. Under these conditions the residual Doppler width of, for example, the green barium resonance line at 535 nm would be about 1 MHz, far below the natural linewidth of 19 MHz. In practice the linewidth ranges from 10 to 50 MHz depending on the type and performance of the ion source, the stability of the acceleration voltage, and the quality of the beam optics. The angular emittances of the laser and ion beams, which reduce collinearity, do not contribute significantly to the linewidth as long as they are not greater than of the order of 1 mrad.

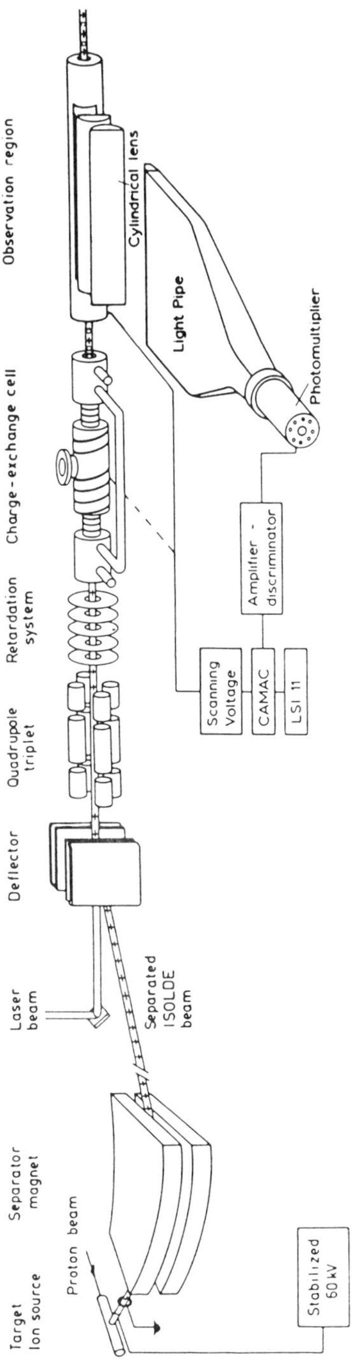


**Figure 28.** Schematic diagram of a standard collinear-laser-spectroscopy experiment using fluorescence detection (Mu 83).

The effect of velocity bunching increases not only the resolution but also the sensitivity: since the total area under the resonance curve remains unchanged, the peak intensity rises by the factor with which the linewidth narrows.

Another very useful feature of collinear geometry is the charge-exchange cell containing an alkali vapor which neutralizes the beam, in flight. By this means, laser spectroscopy is considerably facilitated since most ions do not have resonance lines in the visible region. It is assumed, of course, that the charge-exchange process does not disturb the velocity distribution of the beam. Actually, this is guaranteed by the large cross-section of about $10^{-15}$–$10^{-14}$ $cm^2$. Since it exceeds the kinetic cross-section by two orders of magnitude, most exchange collisions are peripheral. If the charge exchange is nonresonant then the beam energy is changed by (almost) exactly the energy defect $\Delta E$. [In forward scattering, the energy transferred to the target atom is of the order of only $(\Delta E)^2/eU$, which is negligible.] As shown below, charge exchange may also populate a metastable state of the atom which can then serve as the lower spectroscopic state from which laser light is absorbed. Collinear laser spectroscopy owes its currently wide application to these four aforementioned qualities, namely, (1) good adaptation to mass separators, (2) high resolution, (3) high sensitivity because of velocity bunching, and (4) great versatility because of the preparation of suitable spectroscopic states by charge exchange.

Figure 28 shows, in detail, the measurement procedure for the example of the collinear system at ISOLDE (Mu 83) which was built with the benefit of the experience gained from pilot experiments on stable Na (An 78) and on unstable fission isotopes of Rb and Cs (Sc 78b, Kl 79b). The determination of the atomic transition frequency requires a precise knowledge of the laser frequency, of the acceleration voltage, and of the atomic mass. Instead of measuring the two former quantities independently, it is safer and easier to run stable isotopes alternatively through the apparatus, for calibration. For this purpose the separator magnet is periodically switched from one mass to the other and an appropriate correction voltage is applied to the post-acceleration stage situated just in front of the charge-exchange cell. This generates a Doppler shift such that the resonances of the different isotopes coincide in the laboratory system, that is, for one and the same laser frequency. Finally, scanning of the resonances is achieved by adding another small voltage, stepwise, to the charge-exchange cell. Thus it is sufficient to run the laser at constant frequency in a stabilized but uncalibrated mode.

Figure 29 shows a set of resonances for three even Ra isotopes that was recorded in the manner described above (We 85). In Ra the atomic resonance leads from a single, diamagnetic ground state ($7s^2\ {}^1S_0$) to an excited ${}^1P_1$ state from where branching into a metastable ${}^1D_2$ state is only a few percent. Therefore, the atom emits many fluorescence quanta in the observation region before decaying into the "mute" ${}^1D_2$ state (see Section 3.3 and Figure 15 where the principle is explained). The total detection efficiency for an atom, under these conditions, has been found to be about 1%. At typical background counting rates of 10 kHz in the PM (from stray light, surrounding radioactivity, and so on), measurements were meaningful down to currents of about $10^4$ atoms/s. The resonances of ${}^{208,232}$Ra shown in Figure 29 were taken under these limiting conditions.

In complex spectra the sensitivity is reduced by the multiplicity of (metastable) fine structure and hfs levels. The example of the hfs spectrum of ${}^{151}$Dy with $J_1 = 8$, $J_2 = 9$, and $I = \frac{7}{2}$ shows, nevertheless, that even such cases can be mastered at an

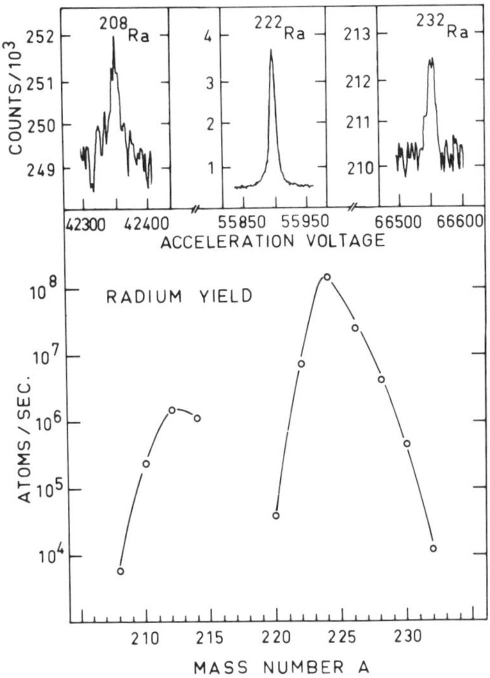


**Figure 29.** Lower part: yield curve of Ra isotopes separated at the ISOLDE facility at CERN. Upper part: scanning signals from collinear laser spectroscopy for some even Ra isotopes. The integration time is about 30 s/channel for $^{208,232}$Ra and a fraction of a second per channel for the abundant $^{222}$Ra (We 85).

on-line mass separator of sufficient intensity, although the detection efficiency per separated atom will have dropped to a range of $10^{-4}$–$10^{-5}$ (see Figure 30) (Ne 82).

The high resolution of the collinear method is also of benefit in cases where several isomers are produced and separated, simultaneously with the nuclear ground state. Figure 31 shows the example of $^{122}$In; 13 of the 15 hfs components are resolved, stemming from spins $I = 8$, 5, and 1 (Eb 87b). In this way it was possible to investigate a total of 37 nuclear states in the In chain $104 \leq A \leq 127$.

In number of cases one can omit neutralization by charge exchange because ion sources (especially plasma sources) also produce ions in metastable states which display resonance lines in a convenient spectral range. One of the first collinear experiments was conducted on a metastable $Xe^+$ beam (Me 77) in this way. The same technique was applied to a series of Eu isotopes at the Leningrad on-line mass separator (Dö 83, Dö 84).

Charge exchange with an alkali vapor also provides access to metastable states of noble elements. Let us consider, for instance, the case of the noble gas Rn which has a metastable state ($7s\,[\frac{3}{2}]_2$) bound by $-3.9$ eV. This is very close to the ionization potential of a Cs atom in its ground state (see Figure 32). Since charge exchange favors resonant conditions, the transfer rate into the metastable state is much greater than into the ground state which is bound by 10.7 eV. This metastable state is connected by

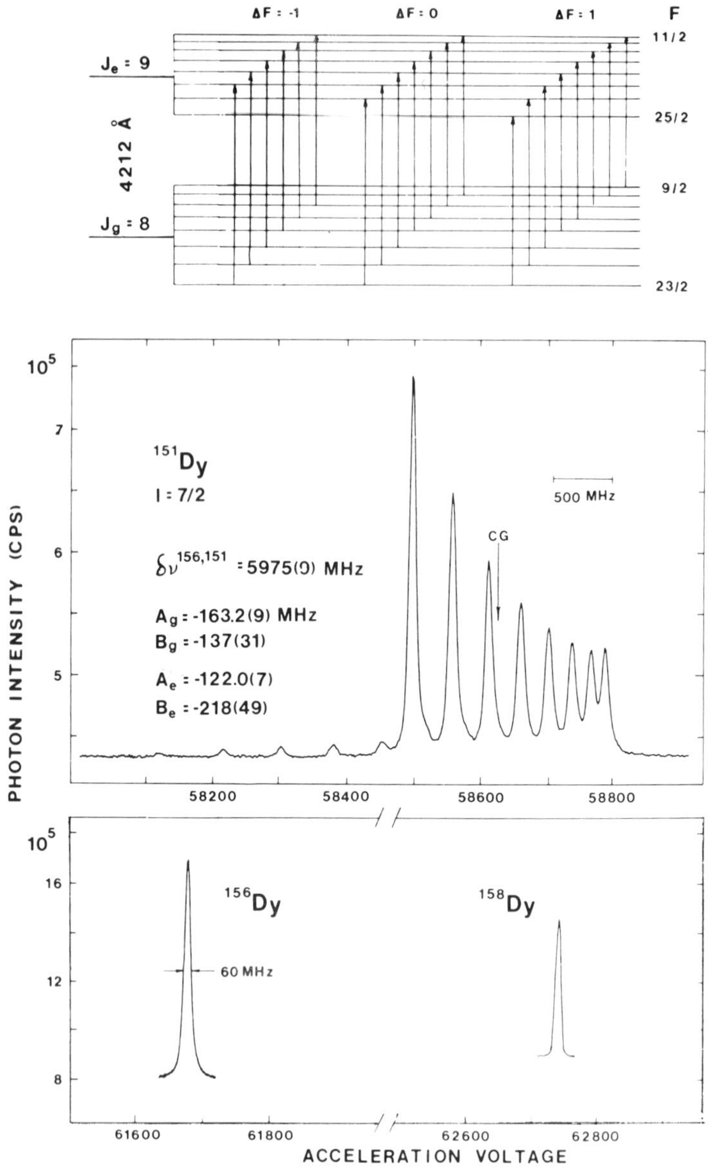


**Figure 30.** Center part: scan of the hfs of $^{151}$Dy by collinear laser spectroscopy. The strong group of resonances belongs to $\Delta F = +1$ transitions, the weak group to $\Delta F = 0$ (note the level scheme at the top of the figure). $\Delta F = -1$ transitions are two weak to be seen. Bottom part: reference signals from stable Dy isotopes (Ne 82).

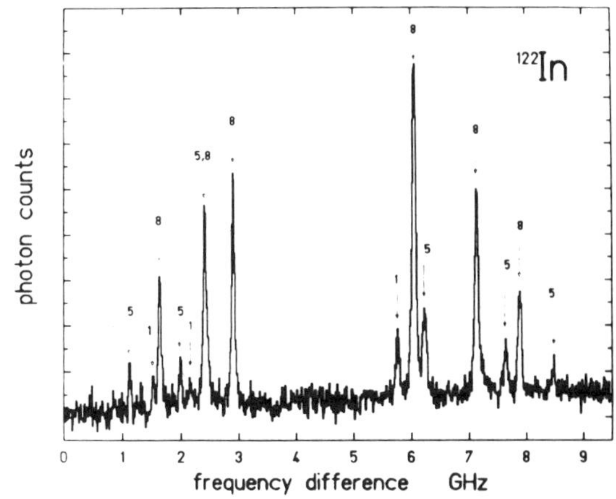


**Figure 31.** Scan of hfs of $^{122}$In in the transition $5p\,^2P_{3/2} \rightarrow 6s\,^2S_{1/2}$ ($\lambda = 451$ nm), obtained by collinear laser spectroscopy (Eb 87b). The hfs components of the nuclear ground state ($I = 1$) and the two isomers are marked by their spin values. Fluorescence is detected on the other doublet components down to the $5p\,^2P_{1/2}$ state ($\lambda = 410$ nm). Stray light from the laser can thus be cut off by a color filter.

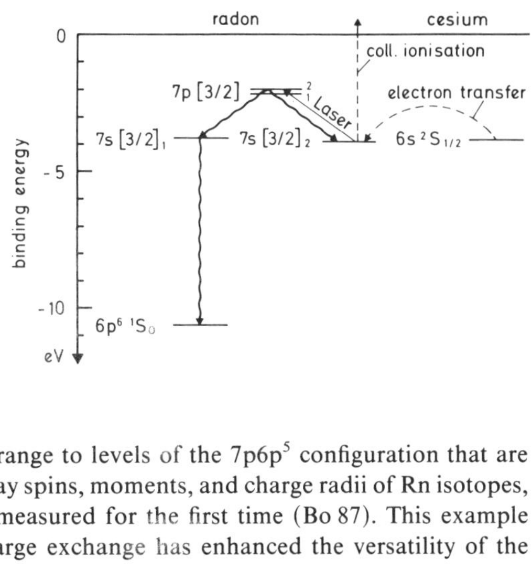


**Figure 32.** Level diagrams of Rn and Cs. The metastable Rn state at $-3.9$ eV is populated by resonant charge exchange with Cs. Excitation by a laser is then followed by branched fluorescence decay. Also shown is collisional ionization of the metastable state (see Section 3.7.2).

a group of lines in the red spectral range to levels of the $7p6p^5$ configuration that are easily excited by a cw laser. In this way spins, moments, and charge radii of Rn isotopes, ranging from $^{202}$Rn to $^{222}$Rn, were measured for the first time (Bo 87). This example is a further proof of how much charge exchange has enhanced the versatility of the method.

Thus far, collinear laser spectroscopy has been applied to unstable isotopes of the following elements: Li, Rb, Sr, In, Sn, Xe, Cs, Ba, Sm, Eu, Gd, Dy, Ho, Er, Yb, Hg, Tl, Pb, Rn, Fr, and Ra (see Appendix for references). Experience has shown that it may be applied to any element for which beams of good ion optical quality and an intensity exceeding $10^4$ to $10^7$ particles/s are available. (The lower intensity applies to the simplest, and the higher to very complex, atomic spectra.)

### *3.7.2. Variants of Collinear Laser Spectroscopy*

Very far off stability, as well as for refractory elements, present on-line mass separators cannot provide the intensity required for the standard collinear technique.

The separation yields from, for example, heavy-ion reactions—which are comparable to those from spallation very far from stability—lie in the critical region of $10^4$ to $10^7$/s mentioned above. Most of the methodological developments in recent years have therefore concentrated on increasing the sensitivity of on-line optical methods. This is true of three of the four variants of collinear laser spectroscopy that we shall discuss.

*3.7.2a. Fluorescent Atom Coincidence Spectroscopy* (*FACS*). In the standard approach the sensitivity is limited primarily by background from stray laser light, radioactivity, PM noise, and so on, all of which add up to typically 10,000/s. A group at the Daresbury on-line separator recently succeeded in suppressing this background by tagging true fluorescence photons with a coincidence signal from the incoming beam particle which was then detected downstream by a channeltron (see Figure 33a). In this way they were able to measure the resonance transitions for light Sr isotopes at a beam level as low as 60 particles/s (Ea 86, Ea 87). Figure 33b shows the coincidence count rates for $^{80}$Sr and $^{78}$Sr as functions of the scanning voltage.

This elegant method clearly requires a very clean beam. Contamination by other isotopes or isobars would bury the signal in the background owing to accidental coincidences, especially as the coincidence window has to be as wide as about 100 ns—

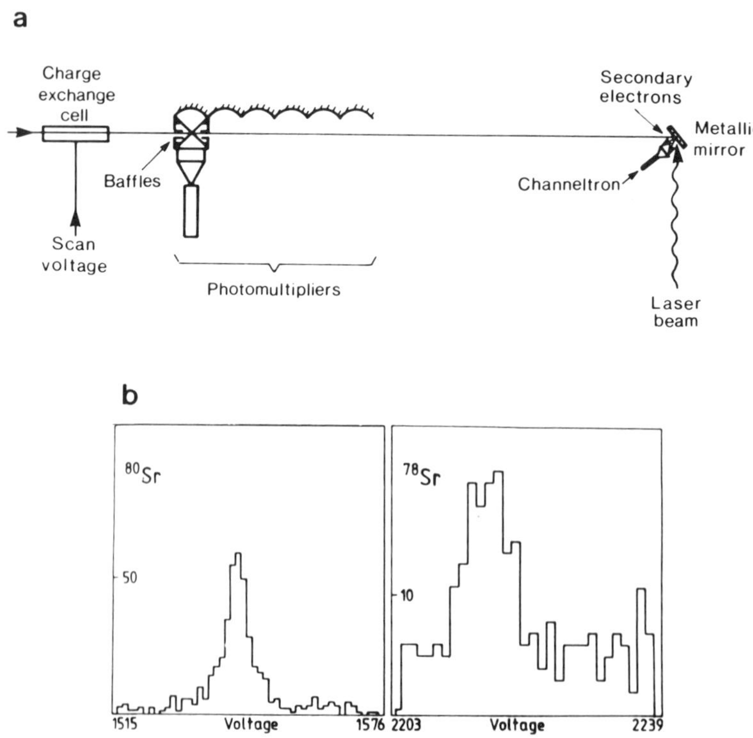


**Figure 33.** (a) Experimental arrangement for photon–atom coincidence detection in collinear laser spectroscopy (Ea 86). (b) Number of photon–atom coincidences from a fast beam of Sr atoms as a function of the voltage scanning through the resonance (Ea 87).

the time of flight of the particle through the observation region of the PM. In the Daresbury experiment this difficulty was surmounted by carefully conditioning the surface ionization source. The requirement may also be met by a mass separator of high resolution, capable of separating isobars. Another way of getting clean beams will, in the near future, be offered by laser ion sources, using the RIS technique. In a recent pilot experiment on stable Sr a total ionization efficiency of 10% was reached (An 86d). Since these laser ion sources are pulsed they allow triple coincidences between the times of ion birth $t_0$, of fluorescence detection $t_1$, and of ion stopping $t_2$. The foregoing discussion indicates the great impact which the development of new, sophisticated ion source-separator combinations will have on the progress of on-line optical spectroscopy in the future.

*3.7.2b. Detection by Collisional Ionization and Charge Exchange.* As pointed out earlier, detection efficiency as well as background rejection may be improved by changing from photon to particle counting. In the scheme under discussion here this is achieved by optical pumping into another electronic state of different binding energy. The OP effect is then sensed by the energy dependence of the collisional-ionization or charge-exchange cross-section. Such an experimental scheme has been developed by Neugart *et al.* (Ne 86) and is explained in Figure 32 for the case of Rn. The metastable $7s[\frac{3}{2}]_2$ state of Rn is populated by charge exchange with Cs. This state is then excited by a laser to a level of the 7p configuration whose decay branches into the tightly bound ground state by a cascade. Downstream of the OP region, the beam passes another gas-filled cell (containing, for preference, an electronegative gas) in which the remaining metastable fraction of the beam is ionized by collisions (see Figure 34a). The reionized fraction is finally deflected onto a detector, which gives the output signal. The pilot experiment was actually performed on the analogous case of a stable Kr beam for which the reionization signal is shown in Figure 34b together with the standard fluorescence signal. The former, which has the character of a flop-out signal, is a factor of 1000 stronger (Ne 86).

Quite recently the method was used on-line at ISOLDE and produced results for a series of Xe isotopes and some other Rn isotopes (see the Appendix). The least abundant isotope was $^{146}$Xe, which had a yield of about 400/s at the exit of the on-line separator (Bo 88).

A similar idea was realized in a reversed sense, independently, by Silverans *et al.* (Si 85). In their scheme $Sr^+$ ions were pumped from the 5s ground state via excitation to 6p and subsequent branching into a metastable 4d state (compare the basic level diagram in Figure 15). The pumping effect was monitored by the difference in neutralization probability of the two ionic states in a subsequent charge-exchange cell. This technique was recently applied at the collinear setup at ISOLDE and allowed the measurement of the Sr series up to $^{100}$Sr, far out on the wing of the yield curve (Si 88).

*3.7.2c. Collinear Laser RADOP.* This experiment carries the ideas of RADOP (Section 3.2) over to collinear fast-beam laser spectroscopy (see Figure 35a). A circularly polarized laser polarizes a fast beam of alkali atoms, in flight, by optical pumping. The beam is stopped in a suitable matrix exposed to an external magnetic field $B$. If relaxation times in this matrix are longer than, or at least comparable to, the nuclear lifetime then the nuclear polarization $P_I$ can be detected by the decay asymmetry measured by a pair of $\beta$ telescopes arranged at 0° and 180° to the field axis. A pilot

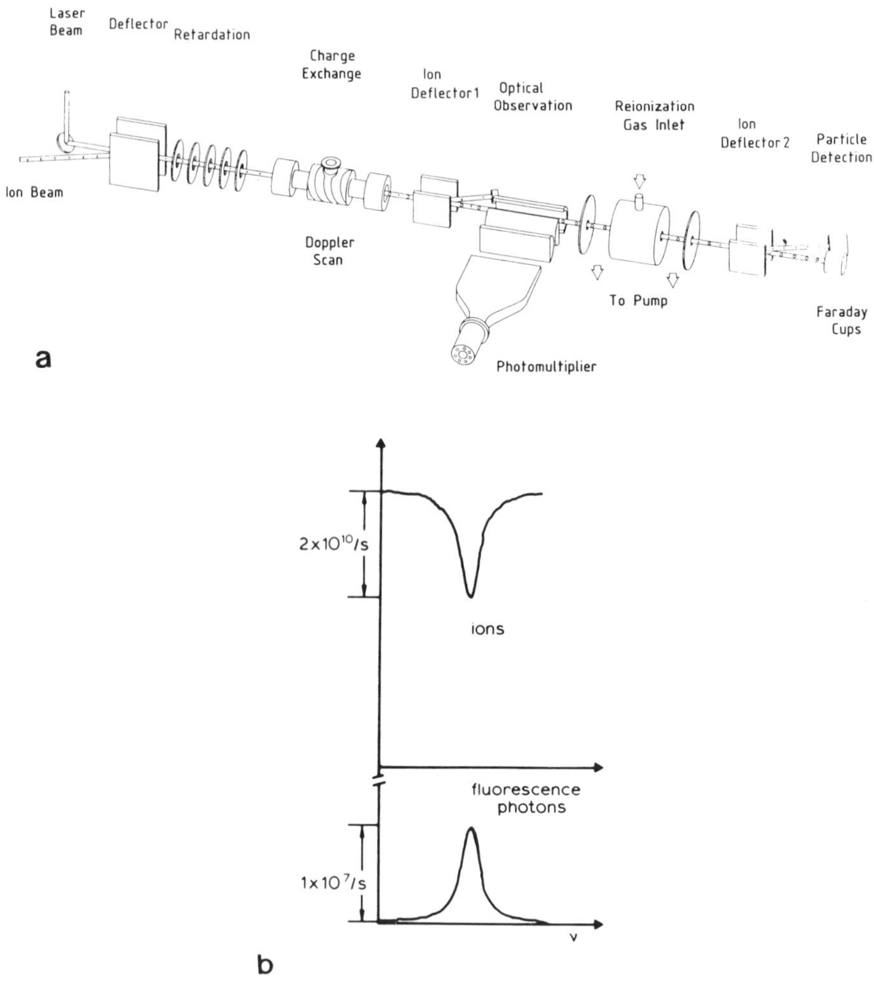


**Figure 34.** (a) Experimental arrangement for collisional-ionization detection in collinear laser spectroscopy. (b) Resonance signals from a beam of $5 \times 10^{11}$ krypton atoms/s; upper signal: reionized fraction of beam; lower signal: standard fluorescence signal (Ne 86).

experiment was conducted on some neutron-rich Rb isotopes at an on-line facility at the Mainz Triga reactor (Bo 85, Sp 85, Ge 86) and was followed by measurements on short-lived Li isotopes at ISOLDE. Figure 35b shows asymmetry signals from $^{9}$Li and $^{11}$Li. The signal is small, of the order of only 1%, although a beam polarization $P_I = 30\%$ was attained, as could be checked by fluorescence signals from stable $^{7}$Li in the same apparatus. The $^{11}$Li current was 600 atoms/s. Spin and magnetic moment were determined. The IS, however, is so strongly dominated by the mass shift that the field shift—which would be of great interest for such a light nucleus—could not be disentangled at the current level of sensitivity (Ar 87b).

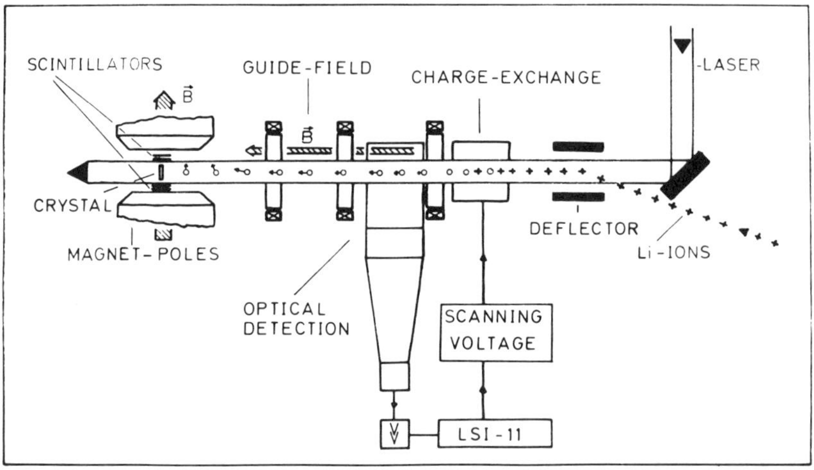


a

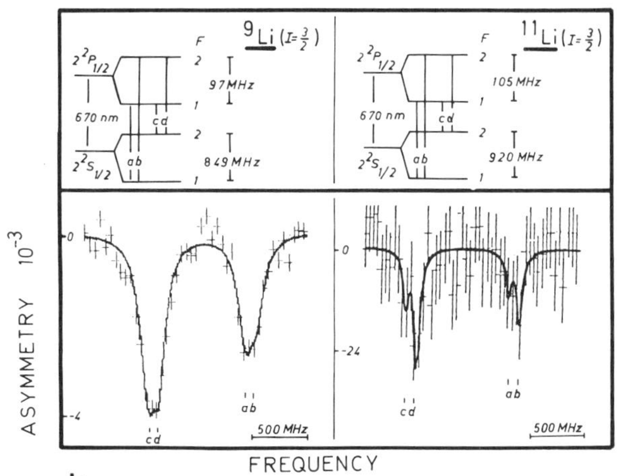


b

**Figure 35.** (a) Experimental arrangement for collinear laser RADOP experiment. (b) $\beta$-asymmetry signals from $^9$Li and $^{11}$Li obtained when scanning the hfs of the $D_1$ resonance line (lower part); and the corresponding level schemes (upper part) (Ar 87b).

*3.7.2d. $g_I$ Measurement by Spin Rotation.* This variant of collinear laser spectroscopy also makes use of spin polarization of a fast beam induced by optical pumping (see Figure 36a). After neutralization in a charge-exchange cell the beam passes a pumping zone I which is followed by a region of length $L$ of a perpendicular magnetic field $H_0$ in which the oriented spins precess by an angle $\vartheta = \gamma H_0 L/v$, where $v$ is the velocity of the beam. Note that $H_0$ is strong enough to decouple the atoms by Zeeman splitting from interaction with the laser during the precession phase. The precession angle can be measured downstream, in a field-free zone III, by the modulation of the

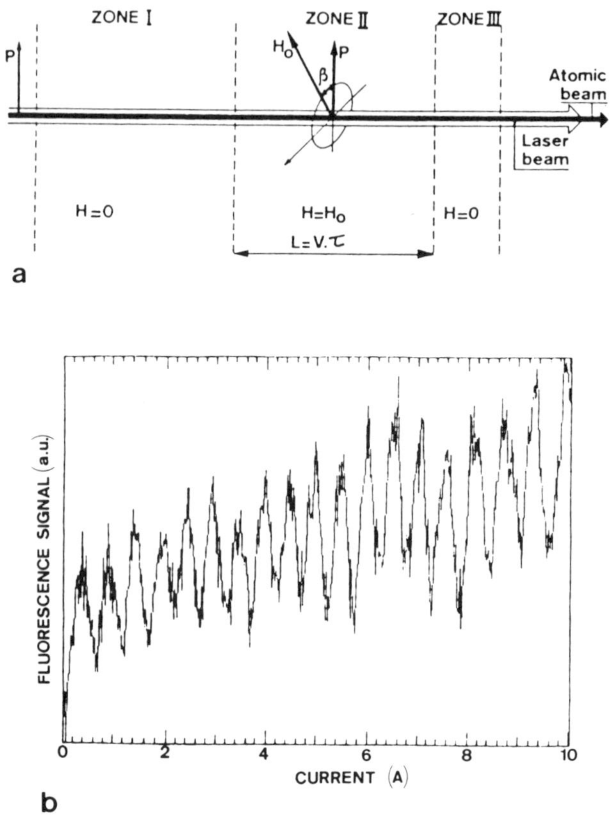


**Figure 36.** (a) Experimental arrangement for observation of spin precession in collinear laser spectroscopy (Be 84c); I = pumping zone, II = precession zone, III = detection zone. (b) Fluorescence intensity of $^{213}$Ra observed in zone III as a function of field $H_0$ in zone II (Ar 87).

fluorescence intensity as a function of $H_0$, because excitation by a polarized laser beam depends strongly on the angle of orientation of the ground-state spin. If the atom in question has a diamagnetic ground state, as in the case of alkaline earths, it is the nuclear spin which precesses and the measurement of $\vartheta$ directly determines $g_I$. The method described was developed by Vialle and collaborators for the examples of Na (Be 84c) and Ba (Ca 86b). The method has since been applied to Ra at the collinear setup at ISOLDE. It yielded the first direct $g_I$ measurements for certain isotopes of this element, namely, $^{213}$Ra and $^{225}$Ra (Ar 87). Figure 36b shows the precession signal of $^{213}$Ra. This measurement also served to calibrate hyperfine fields in Ra spectra from measured $A$-factors (Ah 83, We 87) and to allow Hartree–Fock calculations for this highly relativistic case to be tested against experiment. The hyperfine field of the 7s ionic ground state was determined in this way to be 1226 T as compared to the following theoretical values: 1293 T (Da 86), 1304 T (He 85), and 1266 T (Dz 85). The simple semiempirical Goudsmit–Fermi–Segrè formula yields 1285 T (We 87).

### 3.8. Summary of Experiments

In the 20 years of on-line optical spectroscopy, experimental methods have shifted radically from the use of traditional light sources to the use of tunable lasers. In fact, some of the new techniques, such as resonance ionization and collinear laser spectroscopy, would be simply unthinkable without lasers. Most of the methods discussed have yielded large numbers of results; in the Appendix about 500 nuclei are listed whose spins, moments, or charge radii have been determined in such experiments. But we should also emphasize the importance and the elegance of some peripheral developments—variants of the main methods—dedicated to the solution of specific problems.

At present methodological work is progressing rapidly toward higher sensitivity, thus pushing the exploration of nuclear structure by optical means further off stability and toward refractory elements. At the same time, new ion-source and separation techniques are being installed, which enrich the scope of separable elements and better fulfill the demands of on-line optical methods with regard to isobaric purity, beam quality, beam pulsing, and so on. It is also foreseeable that laser spectroscopy of ions, stored in traps, will be used in on-line experiments. In view of all this vigorous activity we expect that our knowledge of spins, moments, and charge radii of nuclear ground states will be essentially complete by the end of the next decade.

## 4. Discussion of Nuclear Moments and Radii in Extended Chains of Isotopes

As has already been mentioned, the results of optical investigations off stability, described in the preceding section, cover around 500 nuclear states. These results are collected in the Appendix where spins, moments, and $\delta\langle r^2\rangle$ values are listed. It is not the object of this chapter to discuss this vast amount of new material, nucleus by nucleus. Instead, we will try to gain some insight into this field (1) by focusing on exemplary and particularly interesting cases and (2) by analyzing systematic features of nuclear structure which are found in these data. The latter aim refers particularly to nuclear radii and shapes; the single-particle structure, displayed by spins and magnetic moments, is treated only briefly and occasionally.

### 4.1. Spins and Moments

#### *4.1.1. Interpretation in Terms of a Particle-Plus-Rotor Model*

The quantitative interpretation of spins and moments requires, in each case, not only knowledge of the gross and collective structure but also a detailed analysis of the single-particle character of the particular nuclear state in question. This involves a large amount of work. In fact, the published literature on systematic calculations of moments far off stability is sparse. Ekström's review article covers the material available up to 1980 (Ek 81). As mentioned above, he calculates magnetic and quadrupole moments using a particle-plus-rotor model (La 78) based on the Nilsson model. With respect to magnetic moments, the model naturally works better, the clearer the state in question can be identified with an asymptotic Nilsson state in a well-deformed nucleus. In the region of intermediate coupling, which applies to moderately deformed nuclei, the model permits calculation of the wave function of an odd nucleus as a superposition of individual single-particle Nilsson states. In this region the model still produces qualitatively correct results with respect to magnitudes and trends of magnetic moments, although its predictions of level ordering and its identification of the ground state tend to be less reliable. The model may be extrapolated to spherical nuclei as long as their states correspond to single-particle shell-model configurations and not to complex ones. Core polarization and mesonic effects on $\mu_I$ are taken into account schematically by introducing fixed, effective nucleon $g$-factors.

### 4.1.2. Trends of Magnetic Moments near Shell Closures

The kind of systematic data currently obtained is demonstrated in Figure 37 for a few selected chains of magnetic moments; they were not yet, or only partly, known at the time of Ekström's review (Ek 81). The Li, Eu, and Tl moments in Figure 37a

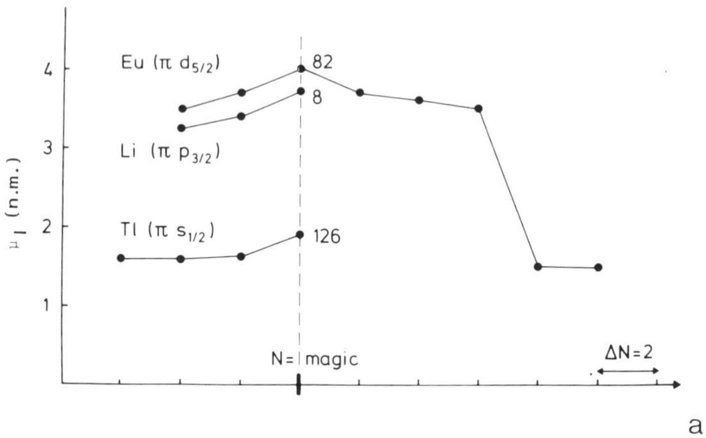


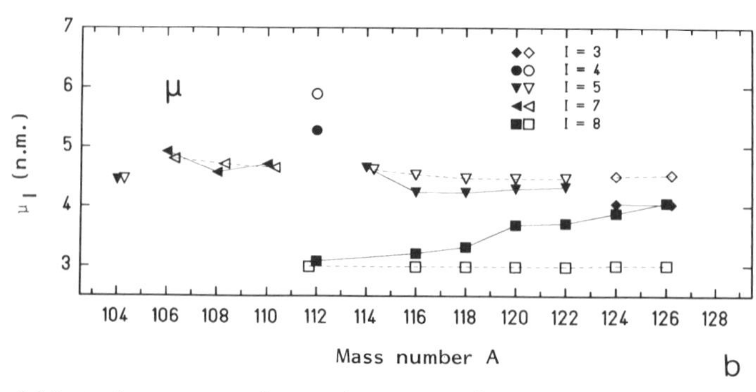


**Figure 37.** (a) Isotopic sequences of magnetic moments of some single-particle proton states near shell closures. Numbers indicate neutron numbers. The $\pi 3s_{1/2}$ moments in Tl are known down to $N = 114$ and remain constant at a value of 1.6 nm (data and references are given in the Appendix). (b) $\mu_I$ values of odd-odd In isotopes (Eb 87b). Equal spins are connected by lines. Solid symbols represent experimental values and open symbols semiempirical values.

are sequences of constant spin which should always represent the same dominant single-particle proton state (or hole) just above (or below) a closed proton shell. The moments always peak at the neutron shell closure at which 100% of the Schmidt value is reached in the case of $^{11}$Li, 80% for $^{145}$Eu, and 70% for $^{207}$Tl. The corresponding states are $1p_{3/2}$, $2d_{5/2}$, and $3s_{1/2}$, respectively.

The moments of direct neighbors of double magic nuclei have always attracted a great deal of theoretical interest. For the case of $^{207}$Tl, Neugart *et al.* (Ne 85b) have discussed the various theoretical results in comparison to their experimental values. The increase of the moment at the magic number is explained by two effects: (1) First-order core polarization from the $\nu(p_{3/2}^{-1} \otimes p_{1/2})\, 1^+$ particle-hole excitation vanishes because of the closed neutron shell. (2) Admixtures of collective excitations like $(2^+ \otimes d_{3/2}^{-1})\, 1/2^+$ are suppressed by the sharp rise of the excitation energy at the magic number. The closest theoretical results lie about 3% away from the experimental values (Hy 80, Ba 73).

### 4.1.3. *Checking the Odd-Group Coupling Rule*

Figure 37b shows the series of odd–odd magnetic moments in In (Eb 87b). Since the moments of most of the relevant states in neighboring isotopes and isotones (in Cd and Sn) are known as well, this complete set of data offers an excellent opportunity of systematically checking the odd-group coupling model. It predicts $\mu_{I,\text{odd,odd}}$ to be the vector sum of the moments of the odd-proton group (p) in the adjacent odd–even nucleus and of the odd–neutron group (n) in the adjacent even–odd nucleus, provided that the configurations are the same:

$$\mu_{I,\text{odd,odd}} = \frac{I}{2}\left[ g_p + g_n + (g_p - g_n)\frac{I_p(I_p+1) - I_n(I_n+1)}{I(I+1)} \right] \tag{45}$$

These semiempirical moments are plotted together with the experimental ones in Figure 37b (Eb 87b). The configurations involved are $(\pi g_{9/2})^{-1} \otimes (\nu d_{5/2})$, $(\nu g_{7/2})$, $(\nu s_{1/2})$, $(\nu d_{3/2})$, and $(\nu h_{11/2})$. For $I < 8$ the agreement varies from fair ($I = 3$) to very good ($I = 5$). The discrepancy at $I = 8$ is explained below.

The odd-group coupling model may be applied to spectroscopic quadrupole moments as well as to magnetic ones, although the quadrupole case is hardly discussed in the literature. The odd-group moments $Q_{s,p}$ and $Q_{s,n}$, with their tensor coupling coefficients $q_1$ and $q_2$, respectively [defined elsewhere (Bo 69)], are added to give a resulting moment

$$Q_{s,\text{odd,odd}} = q_1 Q_{s,p} + q_2 Q_{s,n} \tag{46}$$

The near-perfect agreement of these semiempirical moments with the measured ones, as shown in Figure 38, came as a great surprise (Eb 87b).

### 4.1.4. *Quadrupole Moments of Decoupled States*

The $Q_s$-values of the $I = 8$ states again form an exception to the odd-group coupling rule. The experimental moments rise much faster than the semiempirical ones. In both cases the experimental odd-neutron-group moments were taken from Sn isotones since they are not available for Cd (except for $N = 67$). The discrepancy might therefore be

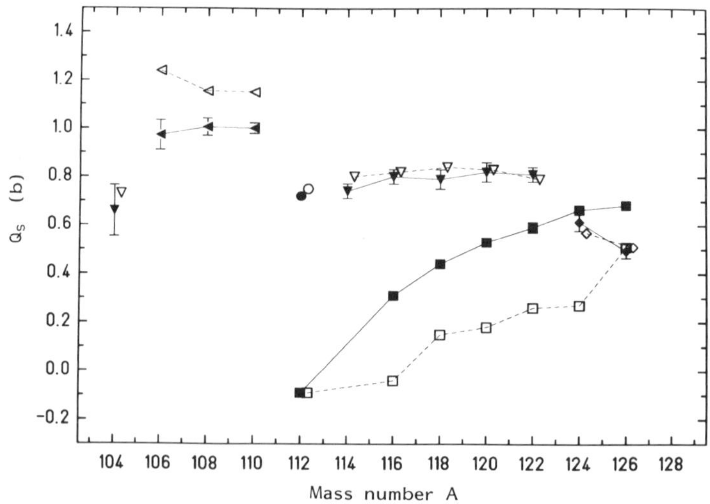


**Figure 38.** Experimental and semiempirical $Q_s$ values of odd-odd In isotopes (Eb 87b). Solid symbols: experiment; open symbols: semiempirical calculation.

connected with the strong polarizing power of the $\nu(h^n_{11/2})$ group which affects the In core more than the proton-magic Sn core. A change in nuclear structure would, of course, invalidate the odd-group coupling model.

Another very interesting phenomenon shown in Figure 38 is the steady increase of $Q_s$ from 0 to 0.6 barn in the $I = 8$ series. It can hardly be ascribed to a corresponding change in deformation since $Q_s$ remains reasonably constant for other spin values. It is explainable if one assumes that the high-spin states decouple from the core (Eb 87b).

The interpretation of spectroscopic quadrupole moments of intermediate nuclei in terms of decoupled states, and its quantitative proof by calculations with a particle-plus-rotor model, has thrown light onto this hitherto very obscure problem (Vo 77, Da 79, Ra 80, Fa 81, Ek 81). If we let $\Omega$ be the projection of the spin onto the intrinsic nuclear symmetry axis, then $Q_s$ varies with $\Omega$ as

$$Q_s = \frac{3\Omega^2 - I(I+1)}{(I+1)(2I+3)} Q_0 \tag{47}$$

where $Q_0$ is the intrinsic quadrupole moment. At moderate deformation a single nucleon in a high-$j$ shell decouples from the intrinsic axis and orients itself such that the wave function has maximum overlap with the core. For an oblate shape ($Q_0 < 0$) this is achieved at $\Omega = j = I$, for a prolate one ($Q_0 > 0$) at $\Omega = \frac{1}{2}$. In both cases $Q_s$ will be negative, according to equation (47). As $\Omega$-levels get filled, $Q_s$ will increase, pass zero in the middle of the shell, and finally reach its maximum positive value for the single-hole state.* $Q_s$-values of $i_{13/2}$ isomers of Hg were first interpreted this way; they represent

* It should be remembered that in the case of a $(j)^n$ shell-model configuration, $Q_s$ shows the same qualitative behavior as a function of $n$ as is ascribed to the decoupled states above. The choice between these two interpretations has to be made on quantitative grounds and with the help of additional spectroscopic information.

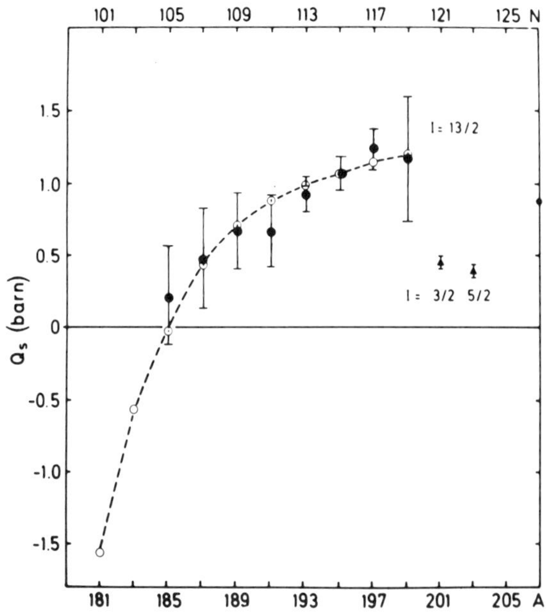


**Figure 39.** $Q_s$-values of $i_{13/2}$ isomers of Hg isotopes; solid symbols: experiment (Ul 86); open symbols: particle-plus-rotor calculation (Ra 80).

a favorable example since the intrinsic deformation of this series of nuclei is known to be fairly constant. Figure 39 shows the results of the first quantitative theoretical treatment of this problem within the framework of a particle-plus-rotor model by Ragnarsson. An equilibrium deformation of $\varepsilon = -0.125$ was chosen throughout the series (Ra 80). The agreement with experiment is quite satisfactory.

Ekström extended these calculations, systematically, to other regions of nuclei and was thus able to interpret many other $Q_s$ values of intermediate nuclei in this manner (Ek 81). The course of the $8^-$ In moments in Figure 38 strongly resembles that of the Hg moments; therefore the same cause is advocated, namely, the filling up a shell of decoupled high-spin states.

Although our general discussion of nuclear moments has been completed, we will again refer to them at particular points in the following discussion of nuclear radii and shapes where spins and moments help to identify collective structure.

## 4.2. Decomposition of ms Charge Radii into Volume and Shape Effect

Since the field effect of the isotope shift measures the change of the *mean*-square charge distribution, one would expect it to be governed by gross nuclear properties rather than by fine details of the single-particle structure. According to any nuclear model with saturated short-range binding, the nuclear volume should increase in proportion to the mass number and hence $\langle r^2 \rangle$ in proportion to $A^{2/3}$ [equation (12)]. A brief look at plots of isotope shifts from Section 4.5 onward shows, on the other hand, that $\delta\langle r^2 \rangle$ is not a smooth function of $A$. Obviously this cannot be explained by a liquid-drop estimate of the change of nuclear volume alone, and, one has to consider the influence of nuclear shape in addition; the latter has a strong and sometimes even abrupt $A$-dependence as a consequence of shell structure. An interpretation of isotope shifts must therefore start with a discussion of some general relationships between nuclear volume and shape, and $\delta\langle r^2\rangle$. This field has been reviewed by Barret (Ba 74) and by Brix (Br 77).

### 4.2.1. *Models of Charge Distributions for Spherical Nuclei*

The assumption of a uniformly charged, spherical liquid drop [equations (10) and (12)] leads to the so-called standard or uniform shift

$$\delta\langle r^2\rangle_{\text{unif}}^{A,A'} = \tfrac{2}{5}r_0^2 A^{-1/3}(A - A') \tag{48}$$

A choice of $r_0 = 1.2$ fm was supported by early electron-scattering experiments which analyzed the form factor in terms of the two-parameter Fermi distribution

$$\rho(r) = \rho(0) \Big/ \left[1 + \exp\left(4 \ln 3 \frac{r - c}{t}\right)\right] \tag{49}$$

where the two parameters are the skin thickness $t$ and the half-density radius $c$. In this model $\langle r^2\rangle$ is given, to very good approximation, by (En 74)

$$\langle r^2\rangle = \tfrac{3}{5}c^2\left[1 + \frac{7}{3}\frac{\pi}{4\ln(3c/t)}\right] \tag{50}$$

More recent results, including those from muonic spectra, have shown that nuclear expansion along the stable valley is somewhat weaker than $A^{1/3}$; instead the parametrization

$$c = (1.146\, A^{1/3} - 0.784\, A^{-1/3} + 0.122\, A^{-5/3})\ \text{fm} \qquad \text{and} \qquad t = 2.3\ \text{fm} \tag{51}$$

gives a better fit to the data (En 74). Equation (51) implies a reduction in the standard shift (equation 48) of about 10% for the heaviest nuclei.

However, neither of these formulas are valid for the isotope shift: it is on average a factor of two smaller than the standard shift. This deficit in the isotope shift must be compensated by a correspondingly larger isotone shift in order to recover the validity of the $A^{1/3}$ law or, more precisely, of equation (51) along the stable valley.

These facts immediately tell us that protons and neutrons cannot be equally distributed over the nucleus such that the ratio of their densities is everywhere $Z/N$. Thus a more realistic substitute for the uniform model was the spherical-droplet model of Myers *et al.* (My 69, My 77, My 83); it turned out to be a very satisfactory gross model, not only for the interpretation of nuclear masses but also of the average isotope shift, i.e., for its volume-dependent part, measured at constant nuclear shape. This success is mainly due to an additional degree of freedom present in this model: proton and neutron distributions are allowed to vary with respect to one another in order to balance isospin-dependent forces (Coulomb and symmetry forces) against isospin-independent ones. Therefore, the model discriminates inherently between isotope and isotone shifts. Its predictions for isotone and isobar shifts will be discussed in Section 4.10.

In the following, $\langle r^2\rangle_s$ and $\delta\langle r^2\rangle_s$ define the ms radius and its change for any spherical charge distribution; $\delta\langle r^2\rangle_s$ will be called the volume contribution to $\delta\langle r^2\rangle$.

Quantities $\langle r^2 \rangle_{sD}$ and $\delta\langle r^2 \rangle_{sD}$ are the corresponding values calculated from the spherical-droplet model with parameters as given elsewhere (My 83).

### 4.2.2. *The Shape Effect on the ms Radius*

The surface of an axially symmetric, deformed, and uniformly charged nucleus may be obtained from a spherical one by the transformation (My 83)

$$R_d = R_0 n \left[ 1 + \sum_i \beta_i Y_{i0}(\vartheta) \right] \tag{52}$$

where the parameters $\beta_i$ are a measure of the multipole deformations of order *i*. The normalization factor *n* is a function of the $\beta_i$ and ensures volume conservation. The lowest-order contribution of the $\beta_i$ to $\langle r^2 \rangle$ is given by

$$\langle r^2 \rangle_d = \langle r^2 \rangle_s \left( 1 + \frac{5}{4\pi} \sum_i \langle \beta_i^2 \rangle \right) \tag{53}$$

and hence, to first order,

$$\delta\langle r^2 \rangle_d = \delta\langle r^2 \rangle_s + \langle r^2 \rangle_s \frac{5}{4\pi} \sum_i \delta\langle \beta_i^2 \rangle \tag{54}$$

with $\langle r^2 \rangle_s = \frac{3}{5} R_0^2$. In the literature it has become standard to use for $R_0$ either the droplet value or the simple liquid-drop value $1.2\,A^{1/3}$ fm. The difference is tolerable in evaluating $\delta\langle \beta^2 \rangle$ from equation (54.). With equations (53) and (54) we have now found the sought-for separation of $\delta\langle r^2 \rangle$ into volume and shape effects. Myers and Schmidt (My 83) have also given higher-order corrections to equation (53) and here we quote the third-order contribution to the quadrupole deformation. It is equal to

$$\langle r^2 \rangle_s \frac{10}{21} \left( \frac{5}{4\pi} \right)^{3/2} \langle \beta_2^3 \rangle \tag{55}$$

and distinguishes between prolate and oblate shapes. The term (55) may be generalized to include a triaxial shape, characterized by the parameters $\beta$ and $\gamma$ [defined, e.g., by Eisenberg *et al.* (Ei 70)]; it is then (Gr 60) (corrected for a numerical factor of five) equal to

$$\langle r^2 \rangle_s \frac{3}{8\pi} \beta^3 \cos \gamma (1 - 4 \sin^2 \gamma) \tag{56}$$

For a nucleus with $\gamma = 30°$—which is neither prolate nor oblate—the third-order term vanishes. At $\beta = 0.3$ and $\gamma = 0°$ or $60°$, it contributes $\pm 9\%$ to the total shape effect but is nevertheless neglected in most analyses. This procedure is consistent inasmuch as in the evaluation of deformation parameters from other collective observables, such as BE2 or $Q_s$ values, higher-order corrections are also neglected in the standard literature. One should keep in mind, however, that this procedure is not satisfactory at large deformation, although the validity of the model, by itself, is subjected to certain accuracy limits.

In droplet-model calculations one considers a redistribution of the nucleon density from its primary droplet ansatz which allows residual forces to relax toward maximum binding energy. The effect of redistribution on $\langle r^2 \rangle$ is calculated for spherical as well as for deformed nuclei. In analyses of isotope shifts using equation (54) one only takes the spherical redistribution into account since the deformation is not known *a priori* but is extracted from the IS. This deficiency could, in principle, be avoided by iteration, but it is not significant in comparison to other shortcomings of the analyses.

All deformation parameters contribute with equal weight to equations (53) and (54), and always according to their mean-square values. Therefore, the IS measures not only the static deformation $\langle \beta \rangle$, as $Q_s$ does, but its zero-point fluctuation $\langle (\langle \beta \rangle - \beta)^2 \rangle$ as well. This will play a fundamental role in our discussion of the results.

A typical quadrupole deformation, $\beta_2 = 0.3$, causes an increase in $\langle r^2 \rangle$ of about 4% [according to equation (53)]. At first sight this seems to be a small effect. But it is, in fact very large compared to the volume effect caused by the addition of one neutron. The latter is of order

$$\delta \langle r^2 \rangle_s^{N,N-1} / \langle r^2 \rangle_s \approx 1/3A \tag{57}$$

which is 0.3% at mass $A = 100$. This example illustrates how sensitive the isotope shifts are to a change in nuclear shape.

### 4.2.3. *Influence of Surface Diffuseness*

Equations (53) and (54) were derived under the assumption of a sharp nuclear surface. In fact, they remain valid if instead of the awkward Fermi function, the Helm model (He 56) is used to represent the diffuseness of the nuclear surface. It consists of a folding of a sharp-edged distribution with a Gaussian one

$$G(r) = (2\pi\sigma)^{-3/2} \exp(-r^2/2\sigma^2) \tag{58}$$

of width $\sigma$. It has the following mathematical advantages (My 83): (1) All multipole moments of $\rho(r)$, i.e., the total charge, the quadrupole moment, and so on, are independent of $\sigma$ and hence the same as for the original, sharp-edged distribution. (2) The deformation part of $\delta \langle r^2 \rangle$ remains unchanged, as noted above. (3) The increase of the total $\langle r^2 \rangle$ due to diffuseness is given by the very simple formula

$$\Delta \langle r^2 \rangle_\sigma = 3\sigma^2 \tag{59}$$

The Helm model is certainly as good an approximation to the surface diffuseness as is the Fermi function. Friedrich and Voegler produced quite a good fit to form factors from electron scattering data, from droplet charge distributions folded with a Gaussian of width $\sigma = 0.99$ fm (Fr 81b, Fr 82).

## 4.3. The Two-Parameter Model for $\delta \langle r^2 \rangle$ and its Relation to other Collective Observables and Gross Properties

In many cases the shape effects on $\langle r^2 \rangle$ will be dominated by the quadrupole mode $\beta_2$ compared to which all others are negligible. With this simplification, equations (53) and (54) are known as the two-parameter model. The change in deformation $\delta \langle \beta_2^2 \rangle$ may then be extracted from experimental $\delta \langle r^2 \rangle$ values with the help of a model assumption for $\delta \langle r^2 \rangle_s$ (e.g., the droplet model) or *vice versa*.

For very heavy elements the simple decomposition into volume and shape effects is complicated by the Seltzer expansion [equation (14)]. The higher radial moments react more strongly to a change of shape than to one of volume. But since their contributions are small, one may expand them in terms of $\delta\langle r^2\rangle_s$ and $\delta\langle\beta_2^2\rangle$ to obtain

$$\lambda = (1+x)\delta\langle r^2\rangle_s + (1+y)\frac{5}{4\pi}\langle r^2\rangle_s\delta\langle\beta_2^2\rangle \tag{60}$$

which is now the two-parameter model for $\lambda$. The correction factors

$$x = \frac{10}{7}\frac{G}{F}R_0^2 + \frac{5}{3}\frac{H}{F}R_0^4 \tag{61}$$

and

$$y = 2\frac{G}{F}R_0^2 + 3\frac{H}{F}R_0^4 \tag{62}$$

are of the order of 10% for the heaviest nuclei (Ah 85, 88). Equation (60) can also be used for a conversion of the experimental $\lambda$ into $\delta\langle r^2\rangle$ values, in which model dependence enters only with the small weights of $x$ and $y$. In the Appendix $\delta\langle r^2\rangle$ is listed instead of $\lambda$, since the former is more convenient for comparison with theory.

### *4.3.1. Relation to the Geometrical Model*

It is clearly of interest to relate $\langle\beta_2^2\rangle_{IS}$ or $\delta\langle\beta_2^2\rangle_{IS}$ values, extracted from the isotope shift, to other collective observables. A rigorous relation to quadrupole transition strengths can be derived from the closure rule of quantum mechanics yielding

$$\langle\beta_2^2\rangle = \left(\frac{4}{3ZR_0^2}\right)^2 \sum_i \mathrm{B}(\mathrm{E2}, 0^+ \to 2_i^+) \tag{63}$$

Equation (63) should be model independent except for basic assumptions about size and shape of the geometrical model of the nucleus. The sum is already exhausted by its first term in two limits of collective motion: (1) the harmonic vibrator, and (2) the rigid rotor (Ku 72). Since in the latter case $\langle\beta_2^2\rangle = \langle\beta_2\rangle^2$, one may introduce the concept of the intrinsic quadrupole moment, related to $\langle\beta_2\rangle$ by

$$Q_0 = 3ZR_0^2\langle\beta_2\rangle/\sqrt{5\pi} \tag{64}$$

In the limit of a well-deformed rigid rotor, the strong coupling rule $\Omega = I$ is also expected to be valid. Hence one can project $Q_0$ onto $Q_s$ via equation (26) and finally relate the shape effect of the IS to the spectroscopic quadrupole moment $Q_s$.

### *4.3.2. Relation to the Interacting Boson Model*

An expression similar to the two-parameter model [equations (53) and (54)] has also been derived from the interacting boson model (IBA). The ms radius is given by it as that of the closed-shell nucleus $\langle r^2\rangle^{cs}$ plus four more terms which are proportional to the numbers of pairs of valence protons $N_\pi$, of valence neutrons $N_\nu$, of proton bosons $\langle n_{d\pi}\rangle$, and of neutron bosons $\langle n_{d\nu}\rangle$ (Ia 81):

$$\langle r^2\rangle = \langle r^2\rangle^{cs} + \gamma_{\pi\pi}N_\pi + \gamma_{\pi\nu}N_\nu + \beta_{\pi\pi}\langle n_{d\pi}\rangle + \beta_{\pi\nu}\langle n_{d\nu}\rangle \tag{65}$$

The prefactors $\gamma$ and $\beta$ are free parameters of the model. The first and second terms describe the isotone and isotope shift, respectively, for a spherical nucleus. Since the expectation value of the number of d bosons $\langle n_d\rangle$ is proportional to $\langle\beta_2^2\rangle$, equation (65) is equivalent to the two-parameter model [equation (53)] except that the deformation part is split into separate contributions from the proton and neutron d bosons (the latter is usually neglected). In cases where $\delta\langle r^2\rangle$ values are to be compared to BE2 values only, the IBA equation is short-circuited by equation (53) and yields no additional information. It only becomes useful when $\delta\langle r^2\rangle$ values are included and are used within the framework of a general fit of IBA parameters from spectroscopic data. Some analyses in regions around $N = 82$ have been published (Ia 83).

Alonso *et al.* (Al 85) extended the model to odd nuclei and applied it to neutron-deficient Cs and Xe isotopes. Focusing on differential isotope shifts, $\delta\langle r^2\rangle^{N,N-2}$ and $\delta\langle r^2\rangle^{N,N-1}$, they were able to reproduce the odd–even staggering in Xe and, in particular, the isotope shift of intruder states in Cs (see also Sections 4.5.2 and 4.11).

## 4.4. Effects of Nuclear Core Polarization on the Optical Isotope Shift

It is well known that muonic spectra are seriously affected by the so-called dynamical muon-nucleus interaction which describes nuclear excitation by the muon. At first sight it would seem very unlikely that the much weaker matrix elements of the electronic interaction could cause any measurable second-order effect of that kind. However, in comparison to the atomic IS, which is itself a tiny effect, the effect of nuclear polarization is not completely negligible, as has been pointed out by Reiner and Wilets (Re 62). Their results have recently been confirmed and extended by Hoffmann *et al.* (Ho 84) in a more general second-order perturbation treatment of the problem using quantitative models for the nuclear polarizability. They show explicitly that the interaction strength is centered far out in the electron continuum, at energies of 20–100 MeV, in contrast to the dynamical muon-nucleus interaction which is strongest among neighboring fine-structure levels.

The high electron excitation energy has three consequences: (1) The nuclear excitation energy may be neglected, as compared to the electronic one, in the denominator of the perturbation energy which may formally be written as a multipole expansion

$$\Delta E = -\sum_{i,j\neq 0,k=0}^{\infty} \frac{|\langle\psi_i^{(e)}\psi_j^{(n)}|\mathbf{F}_k^{(e)} \otimes \mathbf{M}_k^{(n)}|\psi_0^{(e)}\psi_0^{(n)}\rangle|^2}{E_i^{(e)} + E_j^{(n)} - E_0^{(e)} - E_0^{(n)}} \tag{66}$$

(The $\mathbf{F}_k^{(e)}$ and $\mathbf{M}_k^{(n)}$ terms are operators of the electronic multipole fields and nuclear multipole moments, respectively.) With the neglect of $E_j^{(n)} - E_0^{(n)}$, one can approximate the sum over nuclear matrix elements by $\langle(\mathbf{M}_k^{(n)})^2\rangle$ via the closure relation, that is, by the ms zero-point amplitude of the mode $k$, as in equation (63). (2) The de Broglie wavelengths of the admixed electronic states are of the order of nuclear dimensions. Therefore, the electronic wave function is reshaped at the site of the nucleus and remains unchanged further out. (3) $\Delta E$ is proportional to the electron density at the nuclear site as in the case of the ordinary IS.

From these observations we see that nuclear excitation effects are very similar to those which led to the corrected field-shift formulas [equations (11) and (13)] discussed in Section 2.1: There, the Dirac wave function was readjusted from the point-charge solution to the one obtained for an extended, static monopole charge distribution. The correction to the wave function considered here also accounts for the "dynamic" charge distribution of any multipole shape. Both effects are of second order and hence *a priori* of comparable size. Hoffmann *et al.* have calculated the effect for some characteristic cases (Ho 84): (1) For $^{208}$Pb the *total* nuclear polarization effect $\Delta E$ amounts to about 5% of the ordinary $\mathrm{IS}^{206/208}$. For such a stiff nucleus all of the excitation strength lies in the giant resonances which have a weak and smooth $A$-dependence. Hence the differential contribution $\delta(\Delta E)^{AA'}$ to the $\mathrm{IS}^{AA'}$ is negligible. (2) The sudden onset of deformation in the Sm chain also leads to a sharp increase in the nuclear quadrupole polarizability. While the expansion of the nuclear charge distribution lowers the electronic binding energy in first-order perturbation theory, the enhanced polarizability strengthens it in second-order theory. In that way the IS in the pair $^{150,152}$Sm is reduced by 5%.

In summary, contributions from nuclear polarization to the IS are in most cases negligible but in the region of shape transitions are worth considering. Although this effect was eventually proven to be small, we wished to emphasize it here since the question remained open or perhaps was simply ignored for many years.

## 4.5. The Exemplary Region of Rare Earths and its Neighborhood

### 4.5.1. *Europium Isotope Shift*

Many characteristic features of isotope shifts and their relation to nuclear structure can be demonstrated from the well-investigated region of rare earths. Figure 40 shows a plot of $\delta\langle r^2\rangle^{A,145}$ for Eu isotopes in the range $138 \leqslant A \leqslant 156$ (He 71, Dö 83, Dö 84, Al 83, Fe 84, Ah 85, Al 87). The reference isotope $^{145}_{63}\mathrm{Eu}^{82}$ has a closed neutron shell and the proton number is one below the semimagic $Z = 64$ of Gd. Although the proton shell gap at $Z = 64$ is not as wide as at the major shell closures, $Z = 50$ and 82, together with the closed neutron shell it should establish a spherical ground state for $^{145}$Eu. A straight line $\langle\beta_2^2\rangle^{1/2} = 0$ with the slope of the spherical-droplet model $\delta\langle r^2\rangle_{\mathrm{sD}}/\delta A$ has therefore been plotted through that reference point. The assumption of zero deformation for closed-shell nuclei cannot be strictly valid since at least the zero-point motion of giant resonances is still present at the shell closures. Its smooth $A$-dependence cannot be verified in an analysis of $\delta\langle r^2\rangle$. Moreover, the strength of the shell-dependent low-energy collective motion does not completely die out at shell closures, especially not for the octupole mode (see Section 4.8.1) which we do not consider for the moment. On the other hand, a small residual deformation at the magic reference point will hardly influence the determination of $\langle\beta_2^2\rangle^{1/2}$ through equation (54) for another isotope, if this is well deformed. This is easily seen from the progressive distance of equideformation lines in Figure 40. Only close to $\langle\beta_2^2\rangle^{1/2} = 0$ is the scale critical.

*4.5.1a. The Kink in $\delta\langle r^2\rangle$ at the Shell Closure.* The point of shell closure is marked by a characteristic kink in the $\delta\langle r^2\rangle$ plot. [Actually the kink is somewhat rounded by the strong odd–even staggering at $N = 81$, which may be ignored for the moment (see Section 4.11).] Below $N = 82$ the ms radius remains almost constant; above $N = 82$

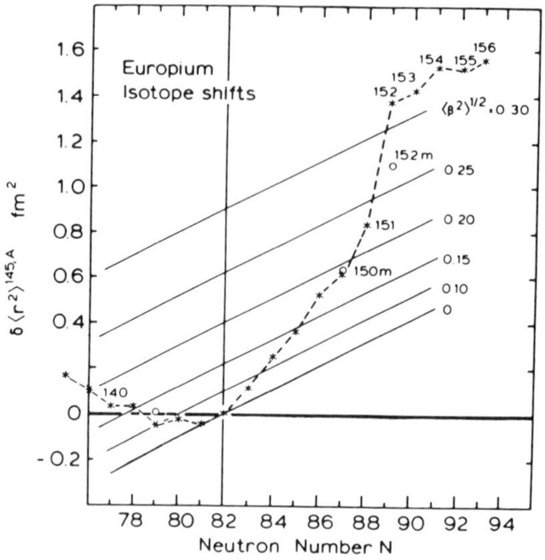


**Figure 40.** $\delta\langle r^2\rangle^{A,145}$ values in the series $^{138-156}$Eu with respect to the semimagic $^{145}$Eu. Isomers are indicated by open circles. The straight lines are equideformation lines with the slope of the droplet model at $N = 82$. The weak curvature of the correct droplet line is neglected in the plot (data and references are given in the Appendix).

it rapidly expands at twice the rate of the spherical-droplet model. Although this kink signals a strong and very pertinent shell effect (see also the figures below), it can hardly be explained within the limits of a strictly spherical shell model. One could argue that the neutron distribution expands faster above the shell closure as a consequence of the new particles in the outer shell. The proton distribution would have to follow suit because of the neutron–proton force. From the viewpoint of gross nuclear structure, however, it would take a substantial amount of energy to lower the central density in favor of the surface because of the high compression modulus of nuclear matter. Seen this way, it would seem more energy-effective to increase $\langle r^2\rangle$ by deformation rather than by volume expansion. Adopting the latter interpretation of the kink, the ms deformation should increase steadily on both sides of the magic number at about equal rates. If one chooses $\delta\langle r^2\rangle_{\text{unif}}$ from equation (48) as basis for the volume effect of $\delta\langle r^2\rangle$, then the equideformation lines in Figure 40 would be a factor of two steeper. That would imply the unphysical consequences of observing no deformation on the neutron-rich side but an extremely large one on the neutron-deficient side.

*4.5.1b. The Shape Transition at $N = 88$.* Another prominent feature in Figure 40 is the sharp increase in $\delta\langle r^2\rangle$ at $A = 151$. The two stable isotopes $^{151}$Eu and $^{153}$Eu have for a long time represented the classical examples of a nuclear shape transition, which was first discovered in the spectroscopic quadrupole moments (Sc 35, Ca 35) and later also found to manifest itself in an exceptionally large isotope shift (Br 52) [reviewed by Brix (Br 85)]. A later off-line measurement on $^{152}$Eu (He 71) made it clear that the shape transition actually occurs between the 88th and 89th neutron, in a single step. The intermediate position of the $0^-$-isomer $^{152m}$Eu is very interesting. The authors conclude that it is a permanently deformed, prolate nucleus but with a single-particle configuration, completely different to that of the ground state (Dö 84, Eg 78). However, that interpretation does not provide a clear argument in favor of the intermediate position which is never observed at other shape transitions unless they are in any case washed out (see below).

Beyond the point of transition, nuclear shape seems to be stabilized; the rms deformation remains constant (apart from odd–even staggering) at a level $\langle\beta_2^2\rangle_{\rm IS}^{1/2} = 0.32$. This value is to be compared with $\langle\beta_2^2\rangle_{\rm BE2}^{1/2} = 0.31$ obtained from $\mathrm{BE2}(0^+ \to 2_1^+)$ values (Lö 70) and with $\langle\beta_2\rangle_{Q_s} = 0.32$, derived from the spectroscopic quadrupole moment of $^{153}$Eu $Q_s = 2.414(21)$ (Ta 83) via equations (26) and (64). The excellent consistency seems to confirm, quantitatively, the assumptions and hypotheses involved in the different analyses, namely: (1) the evaluation of the field-shift constant $F$ [equation (13)]; (2) the two-parameter model and the droplet prediction for $\delta\langle r^2\rangle_s$; (3) the closure relation (63) and in particular its truncation to the first BE2 value; (4) the static character of the deformation; (5) the evaluation of $Q_s$ from muonic spectra; and (6) the projection formulas [equations (26) and (47)] with the strong coupling rule $\Omega = I$ for well-deformed nuclei.

It must be admitted, however, that in other regions of nuclei this level of consistency is not usually reached. Moreover, a consistency on the level of accuracy of a few percent appears somewhat fortuitous in view of the neglect of higher-order corrections (see Section 4.2.2). But it is still nevertheless true that the two-parameter model in general works best when it bridges large distances between rigid spherical and rigid deformed nuclei.

### 4.5.2. *Isotope Shifts in Xe, Cs, and Ba*

As a result of systematic measurements during recent years, the neighborhood of Eu has now been explored over the extended area $54 \leqslant Z \leqslant 70$, $63 \leqslant N \leqslant 106$. Figure 41 shows $\delta\langle r^2\rangle$ for Xe, Cs, and Ba (data and references are given in the Appendix) at the low-$Z$ borderline of this region. The curves are similar to those for the Eu case except for the missing shape transition at $N = 88$. The curves for neutron-rich Cs and Ba isotopes run absolutely steadily and parallel to each other until deformation. $\langle\beta_2^2\rangle^{1/2} = 0.29$ is reached at $N = 92$ (Mu 83, We 87). For all three elements the reference point at $N = 82$ is chosen at $\langle\beta_2^2\rangle^{1/2} = 0.09$ which is derived from the $\mathrm{BE2}(0^+ \to 2_1^+)$ value of $^{138}$Ba. The kink at $N = 82$ appears to become stronger with increasing distance from the closed proton shell $Z = 50$. (Owing to the recent measurements on neutron-deficient Nd and Sm isotopes (Al 87, Le 87) these systematics can now be followed up to Eu.) The phenomenon reflects the well-known fact that deformation of the core increases with the total number of valence nucleons. This has been formulated in a very simple way by Casten's empirical rule which states that the value of a collective quantity is a function of the product $N_\pi \cdot N_\nu$—the number of proton and neutron pairs (or holes) in a shell (Ca 85c).

The staggering of the Cs radii for very neutron-deficient isotopes and isomers is very interesting (Th 81b, Co 87). The spins of $^{119,121,122}$Cs are $\frac{9}{2}$, $\frac{3}{2}$, and 1, while those of their isomers are $\frac{3}{2}$, $\frac{9}{2}$, and 8. The three high spin states come from the $\pi[404]\frac{9}{2}$ Nilsson level out of the $1g_{9/2}$ proton shell which is already closed at $Z = 50$. But this level rises very steeply with deformation, passes the Fermi surface around $\beta_2 = 0.3$, and becomes, beyond the $\pi[422]\frac{3}{2}$ level, the ground state (Ek 81). This interchange happens between

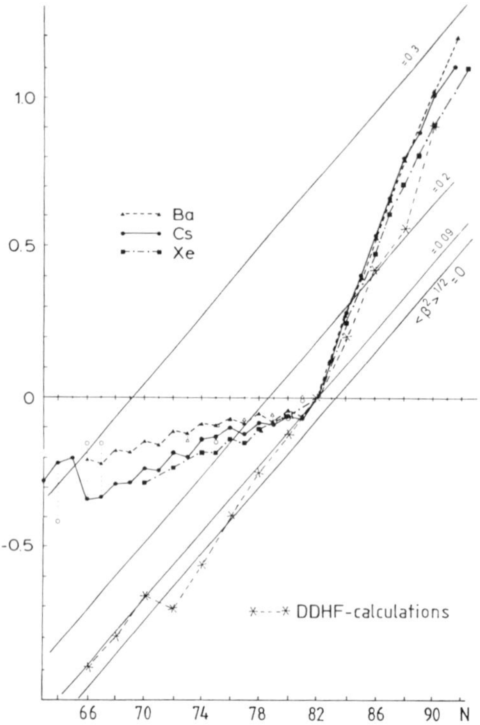


**Figure 41.** Plots of $\delta\langle r^2\rangle^{N,82}$ for Xe, Cs, and Ba (data and references are given in the Appendix). The open circles refer to isomers of Cs. Also shown are equideformation lines given by the droplet model and the results of a density-dependent Hartree–Fock (DDHF) calculation (Ca 75, Ep 81).

mass numbers 121 and 119. It is accompanied by a distinct increase of deformation as shown by the strong isomer shift.

Figure 42, taken from Thibault *et al.* (Th 81b), compares deformation parameters for Cs isotopes derived from $\delta\langle r^2\rangle$ and $Q_s$ values with those of Xe and Ba. The latter have been calculated from energies and transition strengths of the first $2^+$ states. Similar systematic studies for off-line results in this region had previously been conducted by Ullrich *et al.* (Ul 75). The more-recent on-line optical experiments have expanded the area of investigation by a factor of two on the neutron-deficient side, and have also added the neutron-rich branch $N > 82$. With these results we have now a much more complete view of the development of nuclear structure as a function of $N$. The consistency of the data appears convincing on the neutron-deficient side. In particular, the $\langle\beta_2\rangle_{Q_s}$ values of the lightest isotopes constitute the first and only proof of a static and prolate deformation in this region: $\beta$-values for Cs are framed between those of Xe and Ba, as was observed in the case of the isotope shift (Figure 41). On the neutron-rich side, $\langle\beta_2\rangle_{Q_s}$ and $\langle\beta_2\rangle_{BE2}$ are much smaller than is expected from IS. The reasons for this are difficult to trace: certainly a static deformation is not established there. $Q_s$ values of Ba isotopes do not provide a clear explanation either (Ek 81, Mu 83). (Compare also sections 4.6.2 and 4.8.)

### 4.5.3. *Heavier Rare Earths*

The even-heavier rare earths, Gd, Dy, Er, and Yb, have been investigated systematically by Neugart and collaborators (Bu 82, Ne 82, 83, 85) and quite recently by the

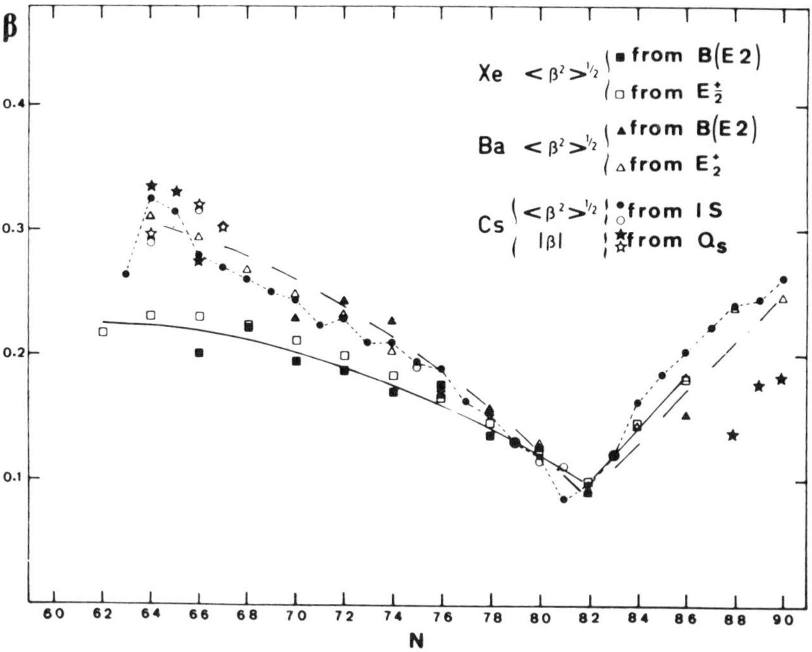


**Figure 42.** Values of the deformation parameter $\langle\beta^2\rangle^{1/2}$ for Xe and Ba, deduced from BE2 and $E_2^+$ and for Cs isotopes and isomers, deduced from IS. Values of $\langle\beta\rangle$ deduced from $Q_s$ are also given for some Cs isotopes (solid symbols) and isomers (open symbols) (Th 81b).

Leningrad–Troitz collaboration (Al 87, Le 87). Although not all of the vast amount of data has yet been published in its final form, the essential features are very clear and have been discussed at several conferences (Ne 82, Ne 85, Ot 85c). Figure 43 shows the ms deformation of some even elements in the range $78 \leqslant N \leqslant 106$, derived from $\delta\langle r^2\rangle$ as well as from $BE2(0^+ \rightarrow 2_1^+)$ values. The agreement is generally good. We again note the sharp minimum at the shell closure, the symmetric rise on both sides of it,

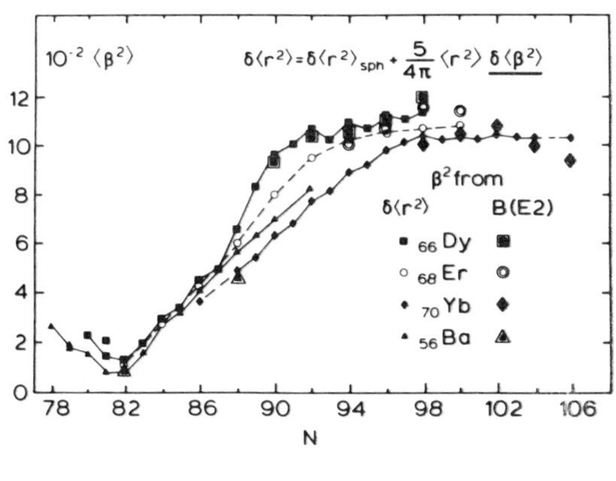


**Figure 43.** Loci of $\langle\beta_2^2\rangle$ in the isotopic sequences of Ba, Dy, Er, and Yb, as determined from isotope-shift data. Values from BE2 measurements are given for comparison (double symbols) (Ot 85c).

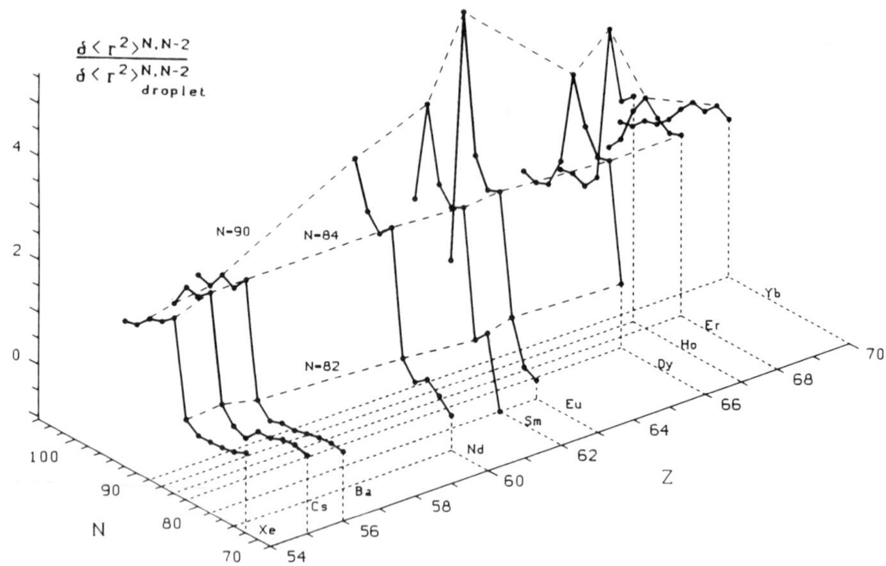


**Figure 44.** Brix–Kopfermann plot of differential IS for even-$N$ isotopes in the region of the rare earths, normalized to the droplet value (My 83). (Data taken from the Appendix.)

and the flat maximum in the deformation around $N = 102$. We further observe that the drastic jump into deformed shapes beyond $N = 88$ (which is strongest for Eu) becomes less marked for heavier elements. It vanishes completely for Yb at $Z = 70$ as well as for Ba at $Z = 56$ (which has been added for comparison). Nevertheless, the heavy Yb isotopes are highly deformed nuclei, although the transition from spherical to deformed shape is smooth. Thus the sharp shape transition is confined to a few elements around $Z = 64$. [The recently measured odd element Tm ($Z = 69$) (Al 87) fits into the picture well.]

Local changes in the course of the IS can be accentuated in a differential plot $\delta\langle r^2\rangle^{N,N-2}$, the so-called Brix–Kopfermann plot (Ko 58). In such a plot the kink at the magic number is transformed into a step, and the step at the shape transition into a cusp, as shown in Figure 44 which is for the region $54 \leqslant Z \leqslant 70$. The cusp is highest for Eu and decays gradually for the neighboring elements. The step at the magic number $N = 82$ is a very clear and regular phenomenon. It increases from Xe to Eu according to the number of valence protons. It is tempting to compare these systematics with corresponding plots of two-neutron separation energies which show similar ridges at shell closures. Indeed, Gerstenkorn found a correlation between nuclear radii and binding energies near shell closures many years ago, when experimental results were still quite limited (Ge 69). Angeli and Lombard recently critically reexamined the theoretical grounds of this correlation and showed the precise range and limits of its applicability (An 86e).

### 4.5.4. *Shape Transition and Subshell Closure at Z = 64*

The localization of a shape transition around $Z = 64$ has been interpreted by Casten *et al.* in terms of the systematics of energies of the $4^+$ and $2^+$ states in the ground-state vibrational/rotational bands; it has been associated with a subshell closure at $Z = 64$ (Ca 81). Although that subshell closure lies in the middle of the major proton shell $50 \leqslant Z \leqslant 82$, it is still able to keep the rate of increase of deformation above $N = 82$ to a moderate value that compares to the one observed for isotones further away from midshell. This continues until the $f_{7/2}$ neutron orbits are filled at $N = 88$. Beyond that point, $\nu h_{9/2}$ orbits, in cooperation with intruding $\pi h_{11/2}$ orbits, abruptly drive all the nuclei into a highly deformed shape and eradicate the $Z = 64$ subshell gap (Fe 77, Fe 78). As soon as the proton subshell disappears into the major one for $N > 88$, the nuclei around $Z = 64$ experience their midshell situation with a maximum number of active valence protons. Therefore, they attain a more-deformed shape than their isotones further away from midshell which need still more neutrons to reach maximum deformation [see also Casten *et al.* (Ca 85c)]. All of these features can be discerned in Figures 43 and 44.

## 4.6. The Rb Region

Figure 45 shows $\delta\langle r^2\rangle$ over a range of 22 Rb isotopes (Th 81, Kl 79b) together with data on Sr (Bu 85, Ea 87, Si 88). Comparison with, for example, the Eu plot, Figure 40, shows that all the characteristics are repeated one shell lower. The kink at the magic number $N = 50$ is still more accentuated, to the extent that the IS actually reverses sign below it. The magic isotopes $^{87}$Rb and $^{88}$Sr have the very smallest ms charge radius

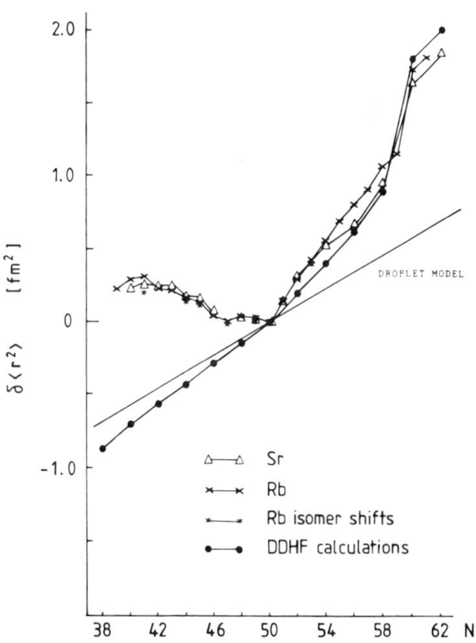


**Figure 45.** $\delta\langle r^2\rangle^{N,50}$ for Rb and Sr isotopes together with equideformation lines given by the droplet model (data and references in the Appendix). Theoretical results from a density-dependent Hartree–Fock (DDHF) calculation are also shown (Ca 80).

in their respective chains, even when compared to nuclei ten mass numbers lower. This strong kink was previously known from classical optical spectroscopy (He 61), but only close to stability.

### 4.6.1. *The Shape Transition at $N = 59$*

At $N = 59$ another shape transition, to strong stable deformation, occurs. Far off stability one has here reached a region of shape coexistence around $Z = 38$, $N = 60$ which has in recent years been extensively studied by nuclear spectroscopy [reviewed, e.g., by Hamilton *et al.* (Ha 85)]. Similarly to the rare-earths region, the shape transition is triggered by the competition of a spherical subshell gap at $Z = 40$ with a deformed shell gap at $Z = 38$. The latter is the preferred one when neutrons start to fill the $\nu g_{7/2}$ orbits around $N = 60$ and interact with intruding $\pi g_{9/2}$ orbits. Nuclear spectroscopy has shown the coexistence of spherical and deformed bands at almost degenerate positions in nuclei such as $^{100}_{38}\mathrm{Sr}^{62}$ and $^{100}_{40}\mathrm{Zr}^{60}$ (Ha 85).

Optical spectroscopy has so far only touched this region. In addition to the Rb chain, the Sr chain has been measured up to $A = 100$ (just at the time of completion of the manuscript for this chapter); its isotope shift runs parallel to that of Rb (Si 88). Beyond Sr, a region of refractory elements starts, which cannot really be coped with in on-line optical experiments for the time being.

Direct measurement of shape coexistence by the isomer-shift would be extremely desirable but requires a sufficiently long isomeric lifetime. In odd nuclei around $A = 100$, one such isomer might possibly exist (see the Hg case below).

Although optical spectroscopy on Rb has so far covered only a few nuclear ground states in the critical region, it has yielded precise information on the shape transition: its size has been fixed from $\delta\langle r^2\rangle^{97,96}$ to be $\delta\langle\beta^2\rangle^{97,96}_{\mathrm{IS}} = 0.08$; and from the spin and moments of $^{97}$Rb ($I = \frac{3}{2}$, $\mu_I = 1.84$, and $Q_s = 0.6$ b) a prolate static deformation $\langle\beta_2\rangle_{Q_s} = +0.33(3)$ has been obtained and the Nilsson state identified to be either $[301]^{3}_{2}$ with $\mu_I = 1.9$ or $[431]^{3}_{2}$ with $\mu_I = 1.99$. These magnetic moments, which were calculated with a particle-plus-rotor model, fit the experiment; but the calculation of energies using a shell-correction method strongly favors the latter one (Ra 80).

### 4.6.2. *The General Course of $\delta\langle r^2\rangle$ and Deformation Parameters*

On the neutron-deficient side, nuclear spectroscopy has not confirmed a clear-cut, stable deformation, although the collectivity of transitions is high. However, $Q_s$ values of $^{77}$Rb and $^{76}$Rb indicate strong prolate static deformation of $\langle\beta_2\rangle_{Q_s} = 0.31(5)$ and 0.52(20), respectively.

The evaluation of the total rms deformation from $\delta\langle r^2\rangle^{N,50}$ under the assumption $\langle\beta_2^2\rangle^{N=50} = 0$ leads to surprisingly high values, namely, 0.38 for $^{76}$Rb and 0.39 for $^{97}$Rb (Th 81). The BE2($0^+ \to 2_1^+$) value for $^{98}$Sr yields $\beta_2 = 0.34$ (Wo 77); this is reproduced by Strutinsky-type calculations (0.35) (Ra 80) and by density-dependent Hartree–Fock calculations (0.36) (Ep 81). Closer to $N = 50$ the Rb isotope shift also indicates a considerably stronger deformation than do BE2 values from neighboring elements. Gerhardt *et al.* observed and discussed this discrepancy already in the case of Kr isotopes and introduced a variable skin thickness in order to remove it (Ge 79, Ge 81). Buchinger *et al.* discussed the same phenomenon in Sr nuclei close to $N = 50$ (Bu 85).

The problem is again met in the Cd–Sn region and we will discuss it there in greater detail (see Section 4.8).

### 4.6.3. Comparison to Microscopic Calculations of $\delta\langle r^2\rangle$

The Hartree–Fock calculation by Campi and Epherre (Ca 80) reproduce the Rb shape transition at the right place reasonably successfully but appreciably overestimates its size. The kink at the magic number is virtually absent in the calculated line. The calculation for Ba also fails to show the kink (Ca 75, Ep 81) (see Figure 41). Campi and Epherre explain this deficiency by the fact that their calculation includes only static deformations, and not dynamic ones, from the zero-point motion of the collective modes. Therefore, the calculated $\delta\langle r^2\rangle$ line lies below the experimental one except for the highly deformed $N = 60$ case where the agreement is fairly good. Thus we can again conclude that the deformation is static beyond the shape transition and that dynamic effects are small both there and at the magic number.

## 4.7. The Mercury Region and the Problem of Shape Coexistence in Near-Closed Shells

### 4.7.1. Observation of Shape Staggering and Shape Isomerism

Historically, the Hg chain was the first one investigated by optical means far off stability at an on-line mass separator. Unexpectedly, a large, sudden increase in $\langle r^2\rangle$ was discovered between $^{187}$Hg and $^{185}$Hg (see Figure 46); it was tentatively interpreted as a transition from a weakly deformed ($\beta_2 \approx 0.15$) to a strongly deformed ($\beta \approx 0.27$) shape (Bo 72). In-beam $\gamma$-spectroscopy experiments on neighboring even Hg isotopes, which followed soon after, did not confirm this finding with respect to the ground states; they continue to be the heads of quasi-spherical bands which are observed with remarkably constant parameters, in all the heavier isotopes. Instead, they found a coexisting, excited band whose energies and transition strengths fitted well with the large deformation claimed from the isotope-shift data (Pr 73, Ru 73, Co 76). The experimental and theoretical work on the problem published up to 1976 is summarized elsewhere (Bo 76).

Later, laser spectroscopic experiments corroborated the quasi-spherical shape of the even isotopes down to $^{182}$Hg by the observation of an extremely large odd-even staggering (Kü 77, Ul 86). In addition, a shape isomerism was discovered in $^{185}$Hg in terms of a huge isomer shift between the deformed ground state and a quasi-spherical $i_{13/2}$ isomer (Da 79). Figure 46 shows the course of $\delta\langle r^2\rangle$ for Hg and the neighboring elements Au and Pb.

The deformed states in $^{181,183,185}$Hg are clearly identified by their spin $I = \frac{1}{2}$ and moment $\mu_I \approx 0.5$ n.m. as members of the $[521]\frac{1}{2}$ Nilsson orbit. Unfortunately a $Q_s$ value has thus far not been measured in the deformed region, neither for Hg because $I = \frac{1}{2}$, nor for $^{185,186}$Au because $J = \frac{1}{2}$, for the atomic states involved (Wa 87). In a recent review of the Hg data (Ul 86), reanalyzed rms deformation parameters for the isotopes $^{181,183,185}$Hg are given as $\langle\beta_2^2\rangle_{\text{IS}} = 0.28$, 0.27, and 0.25, respectively, in accordance with the original numbers quoted above.

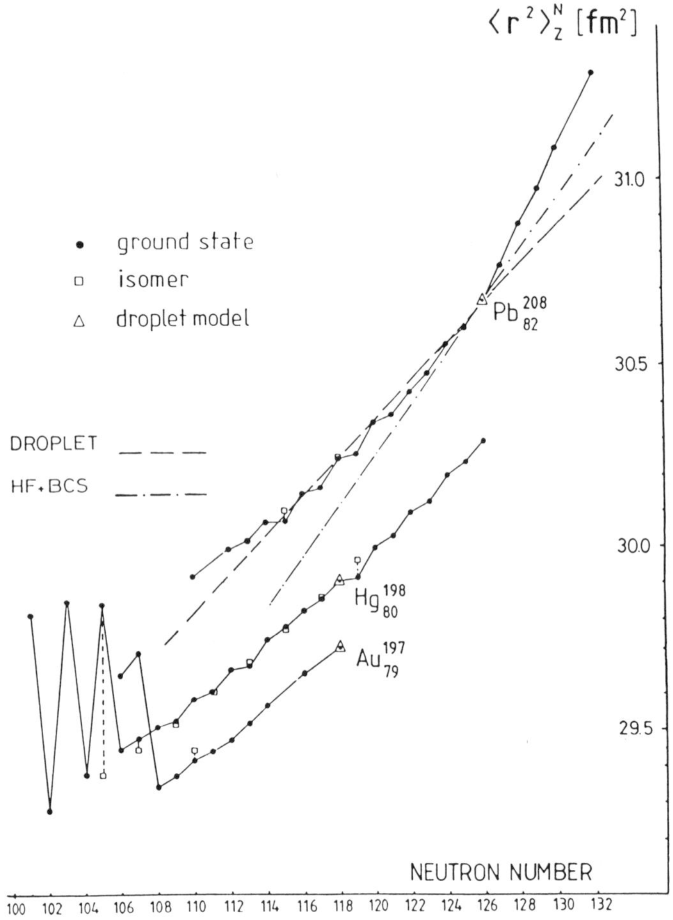


**Figure 46.** Change of $\langle r^2 \rangle$ in the isotopic sequences of Au, Hg, and Pb. The scales have been fixed at $^{208}$Pb, $^{198}$Hg, and $^{197}$Au by the droplet model (for simplicity). Isomers are indicated by open squares (data and references are given in the Appendix).

### *4.7.2. Interpretation of Shape Coexistence*

In the first few years after the discovery of a shape transition so close to the magic proton number $Z = 82$, various theoretical explanations were offered. Most calculations were performed with a Strutinsky or a density-dependent Hartree–Fock (DDHF) method. Based on a paper by Wong (Wo 72) and the observation of missing rotational fine structure in $\alpha$ decay, Hornshøj *et al.* (Ho 73) speculated on the existence of a central depression in the charge density in light Hg isotopes, so-called bubble nuclei. The bubble hypothesis was disproved by Beiner and Lombard (Be 73) and by Nilsson *et al.*; the latter authors concluded that the increase in $\langle r^2 \rangle$ is only partially owing to deformation and predominantly results from an increase in the proton surface density

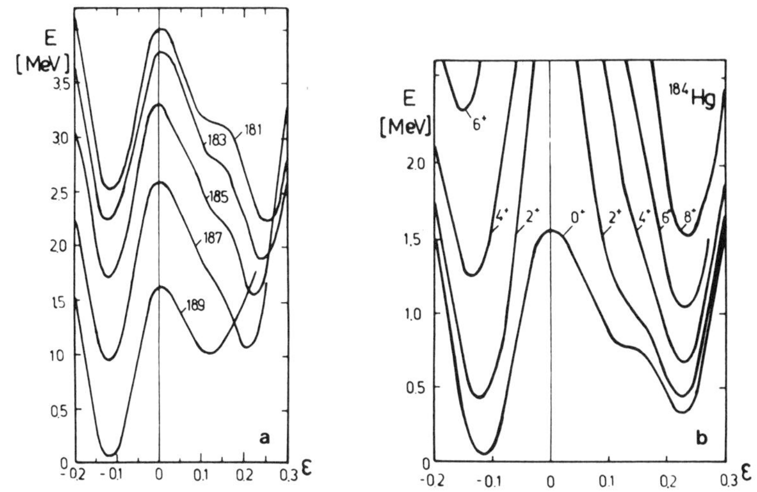


**Figure 47.** Deformation energy of (a) the odd-mass Hg isotopes and (b) the rotational states in $^{184}$Hg. Hexadecapole deformation $\alpha_4$ is equal to $-0.02$ (Fr 75).

as a consequence of the occupation of large (n, l) orbitals (Ni 74). A similar interpretation was recently proposed by Krygin and Mitroshin (Kr 85). On the other hand, Faessler *et al.* (Fa 72), Dickmann and Dietrich (Di 73, 74), Cailliau *et al.* (Ca 73b), Kolb and Wong (Ko 75), and Frauendorf and Pashkevich (Fr 75) have all explained the effect as a transition from small oblate to strong prolate deformation.

Frauendorf and Pashkevich were also the first to give a satisfactory explanation of the odd–even shape staggering. Figure 47 shows their potential-energy surfaces for Hg isotopes. All cases shown exhibit a nearly-spherical oblate minimum and a more-deformed prolate one which compete with each other. The former minimum is caused by the almost-closed proton shell and the latter one by the open neutron shell. $^{184}$Hg and odd masses with $A \geqslant 187$ favor the oblate minimum, while odd masses with $A \leqslant 185$ favor the prolate one, in agreement with experiment. The physical explanation of the odd–even staggering is found in the difference in the neutron pairing energy. On the oblate side it is strong, because of the high level density in the open spherical shell; on the prolate side it is lower, since the level density at the Fermi surface has relaxed owing to the presence of the gap of the deformed neutron shell. Thus, in cooperation with the preferred proton configuration, pairing is able to stabilize the almost-spherical shape of the even nuclei, even up to the middle of the neutron shell at $N \approx 104$. For odd neutron numbers, however, pairing is weakened by the blocking effect; from a certain point onward, these nuclei therefore move over to the prolate side. The balance between the two potential minima is obviously very delicate. This is also seen from the low-lying $0_2^+$-deformed band head in $^{184}$Hg at 375 keV (Co 76). Already in the neighboring $^{185}$Au, neutron pairing is no longer sufficient to stabilize a spherical ground state which again underlines the critical energy balance. But also in Au the deformed region starts with an odd neutron number (107 instead of 105 in Hg) (Wa 87).

Ragnarsson treated the problem of the $^{185}$Hg isomer in a triaxial calculation of the potential-energy surface in the $\beta,\gamma$-plane (see Figure 48). He was able to locate a

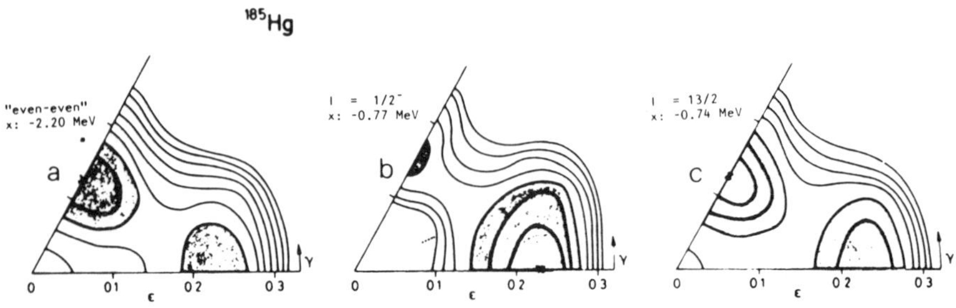


**Figure 48.** Potential surfaces in the $\varepsilon,\gamma$-plane for $^{185}$Hg: (a) the $^{184}$Hg core, calculated with a Strutinsky-like method; (b) the odd particle added in a $1/2^-$ state; and (c) in a $\frac{13}{2}^+$ state, calculated by the particle-plus-rotor model. Equipotential lines are plotted with a spacing of 0.2 MeV (Ra 80).

minimum for the $i_{13/2}$ state on the oblate side, and for the $[521]\frac{1}{2}$ state on the prolate side, at almost degenerate positions, in agreement with experiment (Ra 80). Moreover, Figure 48 shows that the two minima are separated by only a rather shallow saddle, while the path along the $\gamma = 0°$ and $60°$ axes through the point of sphericity at the origin has to surmount a substantially higher barrier. This is a characteristic feature of all calculations in this region that include the $\gamma$-degree of freedom to allow for triaxiality. Also, it is well known that the spectra of heavier Hg isotopes correspond well to $\gamma$-deformed or $\gamma$-unstable nuclei (Ha 85). However, in the context of shape coexistence it is difficult to understand why the $\gamma$ instability—sometimes called the $\gamma$ river—does not apparently cause any substantial shape mixing. This is seen, e.g., from the isotope shift, which distinguishes two absolutely separate classes of nuclei. Also, the monopole strength between the two $0^+$ band heads in $^{184}$Hg is weak (Co 76) and implies a mixing amplitude of only 8% (Kü 77). If the potential-energy surfaces have been calculated correctly, then the solution to the problem could lie with the effective masses of the $\gamma$ vibration which, for some reason, might become very high. Girod and Reinhard have investigated the problem more recently using an elaborate triaxial Hartree–Bogolyubov calculation using the Gogny force. Collective motion was treated using a cranking model. They found the potential-energy surfaces of even isotopes as well as the effective masses, as functions of $\gamma$, to be smooth. They predicted that the ground state of $^{184}$Hg and $^{182}$Hg should lie on the prolate rather than on the oblate side; but the energy differences are marginal. Furthermore, they found that an odd neutron substantially reinforces the minima on either side—depending on its particular state—thus they were able to explain shape isomerism in $^{185}$Hg (Gi 82).

Nuclear spectroscopy has traced the proton configurations that trigger the shape transition in cooperation with the large number of valence neutrons [reviewed, e.g., by Heyde *et al.* (He 83) and Hamilton *et al.* (Ha 85)]. In Tl and Au spectra it has been observed that the spherical $\pi h_{9/2}$ orbit, from across the $Z = 82$ shell gap, moves systematically downward in energy with decreasing $N$ and finally forms a $\frac{5}{2}^-$ ground state in $^{185}$Au in a complicated coupling to the deformed core (Po 83, Bo 82). The downward-sloping intruder state reveals that it takes only a little energy to promote a proton pair from the full $h_{11/2}$ to the empty $h_{9/2}$ shell. In the even Hg core, for example, this mechanism would produce six active valence protons in total (two particles and four holes) which, in cooperation with the valence neutrons, drives the core to deformation. Isotope and isomer shift measurements on light Tl isotopes have in fact revealed that the $\pi h_{9/2}$ intruder isomer is distinctly more deformed than the corresponding $\pi s_{1/2}$ ground state (Bo 85b). In $^{193}$Tl the isomer shift corresponds to $\delta\langle\beta^2\rangle^{9/2,1/2} = 0.015$.

### 4.7.3. *The Isotope Shift around the Double Shell Closure at* $^{208}$*Pb*

The course of $\delta\langle r^2\rangle$ for the heavier Au, Hg, and Pb isotopes below the shell closure runs remarkably straight and, moreover, fairly parallel to the droplet line, especially for Pb and for Hg in the range $120 \leqslant N \leqslant 126$. Again, Pb data show a kink at the magic number but a considerably smaller one that that observed for lighter elements (see the figures above). Clearly, the presence (or neighborhood) of the proton shell closure plays a role in all this. But still in Ra, which already lies six protons above shell closure, the kink is fairly flat (see Figure 60 below). In general it seems that the magnitude of the kink decreases from one major shell to the next. That is to say, a small core polarizes more easily than does a bigger one.

Another interesting difference relative to the situation at lighter $N$-shell closures is the asymmetry of the two $\delta\langle r^2\rangle$ branches for Pb with respect to the droplet line. An interpretation in terms of no core polarization below $N = 126$ and a rather strong one above it seems strange. One could argue that, in addition to the quadrupole mode, higher-multipole deformations contribute and that these contributions steadily increase across the shell closure. This would help to restore the symmetry around $N = 126$ with respect to the quadrupole polarization but shifts the problem of asymmetry to the higher-order polarizations. Andl *et al.* (An 82) and Barranco and Broglia (Ba 85) have discussed ground-state correlations due to octupole vibrations and their influence on the variation of ms charge radii of Ca isotopes. They found them to be very important in $^{40}$Ca and to decrease for the heavier isotopes. The question of higher-multipole contributions in the $N \leqslant 126$ region was also discussed by Ulm *et al.* (Ul 86) in the analysis of heavy Hg isotopes; they explained the impossibility of giving a conclusive answer to that question at the present level of experimental information. For $N > 126$, on the other hand, it is known that the octupole strength increases strongly because of the $g_{9/2}$ valence neutrons which have large octupole matrix elements connecting to $\nu j_{15/2}$ states (Bo 75, Kl 82).

Thompson *et al.* (Th 83) have performed HF and BCS calculations of ms charge radii of Pb isotopes, using Skyrme III forces. They calculated a slope of $\delta\langle r^2\rangle$ that was about 15% steeper than that obtained from the droplet model (My 83). The distance of the calculated line from the experimental data is, in this new calculation, more symmetric for the even isotopes but cannot be bridged by the $\langle\beta_2^2\rangle$ contribution as calculated from the (few) measured BE2 values; these are too small, as already observed in Rb. Also, the microscopic calculation gives practically no evidence for a kink at the magic number, again as already observed for Rb and Ba.

An increase in the standard droplet slope of $\delta\langle r^2\rangle_{sD}$ by about 10–15% in the lead region was also suggested by Berdichevsky and Tondeur on the basis of another microscopic calculation (Be 85) and by Möller *et al.* who modified the droplet model to include finite-range forces (Mö 84).

Ulm *et al.* (Ul 86) applied the standard droplet model to evaluate $\langle\beta_2^2\rangle_{IS}$ for Hg isotopes and found good agreement with $\langle\beta_2^2\rangle_{BE2}$ below $^{198}$Hg. If, instead, the higher slope was adapted then the former would be raised over the latter. In view of the same discrepancy having been observed elsewhere this would not, in fact, come unexpectedly. In the next section this problem will be treated in greater detail.

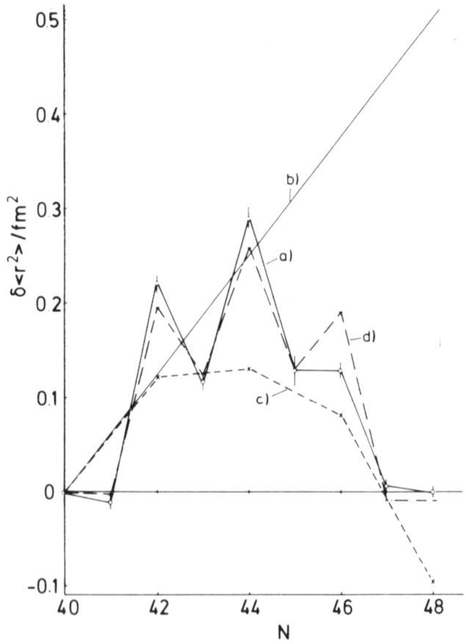


**Figure 49.** Plots of $\delta\langle r^2\rangle^{A,40}$ in the Ca series (data and references are given in the Appendix): (a) experiment; (b) droplet model; (c) droplet model plus contributions from deformation (Ba 85); and (d) fit of the Zamick–Talmi model to the data (Ta 84).

## 4.8. Parabolic Isotope Shifts and the Question of Large Core Polarization in Regions of Closed Proton Shells

### *4.8.1. Isotope Shift and Core Polarization in Ca Nuclei*

Ca is the only element (besides its neighbor $K$) for which IS measurements have been performed over a full neutron shell, namely, the $\nu$ $1f_{7/2}$ shell between $^{40}$Ca and $^{48}$Ca—a small one, admittedly. The even isotopes in the series, which are all stable, have also been investigated by many other methods which have yielded information on the charge and mass distributions: Coulomb excitation, muon spectroscopy, electron scattering, hadron scattering, and so on. This treasure of data therefore comprises not only precise ms radii but also form factors and their isotopic change [for a synopsis see elsewhere (Re 79)]. The isotope shift (An 82), shown in Figure 49, is characterized first of all by the astonishing fact that the two double-magic nuclei $^{40}$Ca and $^{48}$Ca have the same ms charge radius. Electron-scattering experiments have, however, revealed that the charge distributions $\rho(r)$ are not identical. Their difference was shown in Figure 5 in the form $r^2\Delta\rho(r)^{48,40}$. Charge must flow from the center and the outer skin into a region around the half-density radius $c$; but since the total area under this curve is zero, this flow leaves $\langle r^2\rangle$ unchanged (Em 83).

One must bear in mind that (elastic) electron scattering measures the monopole charge distribution averaged over all possible multipole deformations. The latter manifest themselves by simulating an increased skin thickness of $\rho(r)$ which cannot be distinguished from a change in the intrinsic charge distribution. The same is true of the isotope shift. In this picture the two-parameter model [equation (54)] would therefore be changed to

$$\delta\langle r^2\rangle = \delta\langle r^2\rangle_s + 3\delta\sigma^2 \tag{67}$$

The change in skin thickness may be calculated from equation (67) by inserting experimental values on the left-hand side and the droplet expectation $\delta\langle r^2\rangle_{sD}$ on the right-hand side. The result is a shrinking of $\sigma^2$ by 0.166 fm$^2$ between $^{40}$Ca and $^{48}$Ca; or, with the assumption $\sigma^{40} = 0.99$ fm, $\sigma^{48}$ drops to 0.90 fm.

Andl *et al.* (An 82) have pointed out that at least the major part of this effect can be attributed to collective $\langle\beta_2^2\rangle$ and $\langle\beta_3^2\rangle$ contributions [see also the general theoretical treatment of the problem for nuclei in this region by Reinhard and Drechsel (Re 79b)]. The octupole zero-point motion, as calculated from the transition strength to the first $3^-$-state, is very large in $^{40}$Ca and decreases to a small value in $^{48}$Ca. This decrease just compensates the increase in $\langle r^2\rangle_{sD}$. The quadrupole strength is small at both shell closures and maximizes itself in between; this accounts for part of the bump in $\delta\langle r^2\rangle$ in the middle of the shell. Barranco and Broglia (Ba 85) tackled the problem anew and performed a full theoretical calculation of ms deformations up to $\langle\beta_5^2\rangle$ for the even isotopes. For $^{40}$Ca they calculated a total ms deformation $\sum_{i=1}^{5}\langle\beta_i^2\rangle$ of as high as 0.30 and for $^{48}$Ca a value of 0.16. Adding these deformation corrections to the droplet value $\delta\langle r^2\rangle_{sD}$ according to equation (54) yields curve (c) in Figure 49. It lies about 0.1 fm$^2$ below the experimental line but reproduces its structure fairly well, at least in comparison to the pure droplet model. The remaining difference may or may not be interpreted as the residual "noncollective" change of skin thickness. It could also be attributed to deficiencies in the collective model or in the droplet model. After all, one cannot expect gross models to reproduce the exact nuclear wave function in any detail. It should be noted here that the isotope shift of K (To 82b) shows a similar shape to that of Ca (see the Appendix); *grosso modo* it may therefore be interpreted on the same bases.

### *4.8.2. The Occurrence of Parabolic Shapes of $\delta\langle r^2\rangle$ and the Zamick–Talmi Model*

Apart from its odd–even staggering, the Ca isotope shift has a parabolic shape with a maximum in the middle of the shell. Parabolic shapes for $\delta\langle r^2\rangle$ are quite common in regions where abrupt shape transitions do not occur. Isotonic shifts can also exhibit this behavior, as shown in Figure 50 where differential isotone shifts $\delta R^{Z,Z-2} \approx \frac{1}{2}\langle r^2\rangle^{-1/2}\delta\langle r^2\rangle^{Z,Z-2}$ are plotted for stable nuclei in the $\pi 1f_{7/2}$ shell and the next one, up to $Z = 40$; they were obtained from electron scattering data and from muonic spectra (Fr 81, Fr 86). The data are consistent with a downward-sloping straight line in each shell. This corresponds to an integral shift of parabolic shape with negative curvature, as observed in Ca. The step at $Z = 28$ corresponds to the well-known kink of the integral shift, discussed above, with which the parabola of the next shell starts.

The ms deformation $\langle\beta_2^2\rangle^{N,82}$ of Yb, which lies just outside the shape transition region of the rare earths, also has a parabolic shape (see Figure 43 above). The best examples of this kind are for the isotope shifts of Cd (Bu 87), In (Ul 85, Eb 87b), and Sn (An 86b, Lo 85, Eb 87) which are shown in Figure 51. They can be perfectly fitted

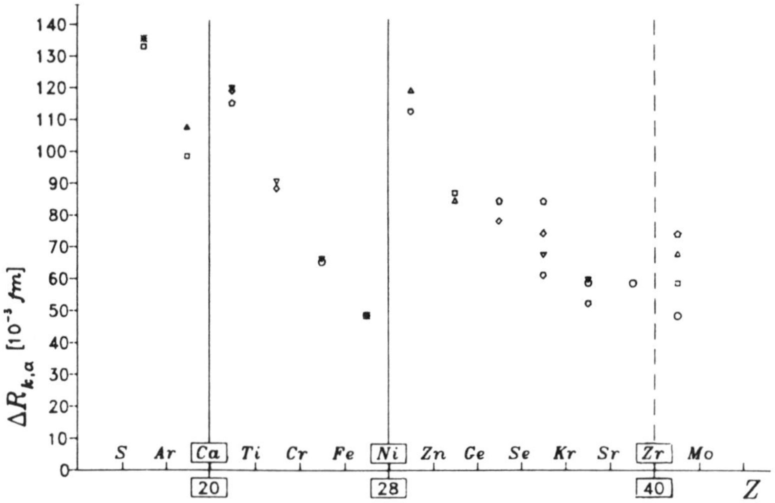


**Figure 50.** Differential isotone shifts $\delta R^{Z,Z-2}$ for even elements within the shells $20 \leqslant Z \leqslant 28$ and $28 \leqslant Z \leqslant 40$ (Fr 81, Fr 86).

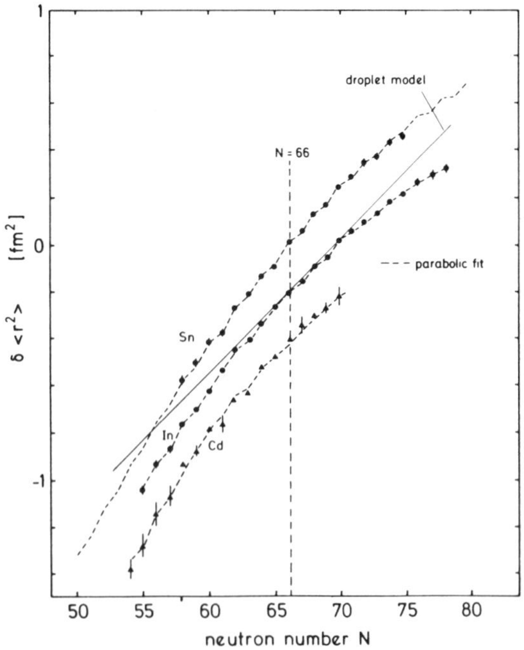


**Figure 51.** Plots of $\delta\langle r^2 \rangle$ for Cd, In, and Sn (data and references are given in the Appendix). The Cd and In curves have been lowered at the reference point $N = 66$ by $0.4\,\text{fm}^2$ and $0.2\,\text{fm}^2$, respectively, for convenience. The broken lines are fitted parabolas with odd-even staggering as given by equation (68). Also shown is the droplet line.

by a polynomial of second order. It may be written in the form

$$\delta\langle r^2\rangle^{N,50} = a(N-50) + b(N-50)(82-N) + \tfrac{1}{2}c[1-(-1)^N] + d \qquad (68)$$

such that the quadratic term maximizes at midshell. The parameter $c$ describes the odd–even staggering—assumed constant over the shell. (The offset $d$ is introduced so as to optimize the fit and has no further physical relevance.)

A parabolic model which includes odd–even staggering was introduced by Zamick to fit the Ca charge radii (Za 71), in formal analogy to an identical equation derived by Talmi and Thieberger (Ta 56) for the isotopic dependence of the nuclear binding energy. When, at a later date, information on the isotope shifts of odd Ca isotopes became available, Träger showed that they also fitted nicely into the model (except for $^{46}$Ca) (Tr 81).

Talmi has recently taken up the problem again and has given an explicit derivation of equation (68) in terms of core polarization of nuclei with closed proton shells by valence neutrons in a pure $\nu j^n$ shell-model configuration and in states of lowest seniority (Ta 84). For the contribution of an even-multipole polarization of order $2l > 0$, the dependence of $\langle r^2\rangle$ on the occupation number $n$ is given by

$$\delta\langle r^2\rangle^{(n)} = \frac{2|\langle j\| T_\nu^{(2l)}\|\rangle|^2}{(2j+1)(2j-1)}\{n(2j+1-n) - \tfrac{1}{2}[1-(-1)^n](j+\tfrac{1}{2})\} \qquad (69)$$

The factor in front of the right-hand side of equation (69) is the reduced matrix element of the multipole operator acting on a single neutron in a state $\nu_j$. The first, quadratic term in the bracket is proportional to the number of particles times the number of holes and is symmetric in both, as in the ansatz [equation (68)]. The second term describes a constant odd–even staggering with a negative sign.

Monopole polarization gives rise to a $n^2$-dependence of $\langle r^2\rangle$ with positive curvature. The data shows no evidence for the presence of such a term. Odd multipoles contribute a positive, constant term for all odd $n$ but do not contribute at even $n$. Therefore, they also give rise to an odd–even staggering but with opposite sign to that of the even multipole case. An expansion of the ms core radius proportional to $A^{2/3}$ is not considered in the model.

Andl *et al.* (An 82) have fitted equation (69) to $\delta\langle r^2\rangle$ values of Ca and obtained a qualitatively correct shape and an indication of odd–even staggering. Talmi (Ta 84) performed a fit with the more general Zamick formula which has three free parameters $\varepsilon$, $\alpha$, and $\beta$ (Za 71):

$$\delta\langle r^2\rangle^{40+n,40} = \varepsilon n + \frac{\alpha}{2}n(n-1) + \left\{\frac{n}{2} - [1-(-1)^n]/4\right\}\beta \qquad (70)$$

Naturally, he got a better fit but it was at the cost of a stringent interpretation; his fit is included in Figure 49.

In a major neutron shell the $n$-dependence of core polarization of semimagic nuclei is still given by equation (70), at least for even $n$, and under certain conditions also for odd ones (Ta 84, Ta 71). The odd–even staggering would again be constant over the shell. Equation (70) preserves the quadratic character of equations (68) and (69).

### 4.8.3. *Interpretation of Parabolic Shapes in Terms of the Two-Parameter Model in the Cd, In, and Sn Region*

Since the parameters of equation (70) are not related to one another, it gives no information about the form of the parabola. Physically, it is clear that it should still have its vertex in the middle of the shell. If the polarization amplitude is made up by some coherent, residual interaction of the pairs in the valence shell, then it should be proportional to their number. This explains why $\delta\langle r^2\rangle$ is a quadratic function of that number. Moreover, since this interaction is limited to the valence space of the major shell in question, and since it obeys the Pauli principle, it should be symmetric in particles and holes; hence it should maximize in the middle of the shell in the same way as equation (69) did. With these assumptions, the quadratic term of equation (68) is most naturally identified with the ms deformation $\langle\beta^2\rangle$. Under the additional assumption $\langle\beta^2\rangle^{N=50} = \langle\beta^2\rangle^{N=82} = 0$, it follows that

$$\langle\beta^2\rangle^N_{\text{model}} = \frac{4\pi}{5} b(N-50)(82-N) \tag{71}$$

where any residual, shell-independent deformation at the shell closures has been neglected (Bu 87). Application of equation (71) implies that the fit can be extrapolated to the shell closures, which have not yet been investigated experimentally in the Cd–Sn region. But there is no reason to expect any strange bending or discontinuity in the locus of $\delta\langle r^2\rangle$ on its final, short path to the shell closures.

Since the explicit formulas for multipole core polarization by a single $j$ shell, discussed above, did not exhibit any linear $n$-dependence, the $a$-parameter in equation (68) might be entirely attributable to a quantity outside the core-polarization model, namely, to the volume effect $\delta\langle r^2\rangle_s$. Fitting equation (68) to the data yields $a(\text{Cd}) = 0.047\ \text{fm}^2$, $a(\text{In}) = 0.060\ \text{fm}^2$, and $a(\text{Sn}) = 0.065\ \text{fm}^2$; the droplet model predicts $a \approx 0.055\ \text{fm}^2$, in fair agreement with the average of the three values. By fitting the parameter $b$ one obtains from equation (71) the ms deformation at midshell as $\langle\beta^2\rangle^{N=66}_{\text{model}} = 0.078$, 0.048, and 0.032 for Cd (Bu 87), In (Eb 87b), and Sn (Eb 87), respectively. Note that the model yields the center deformation from the curvature of the experimental $\delta\langle r^2\rangle$ curve only, without explicit reference to any other input data.

Figure 52 shows $\langle\beta^2\rangle_{\text{IS}}$ for Cd calculated from $\delta\langle r^2\rangle^{N,66}_{\text{exp}}$ by use of the two-parameter model [equation (54)] with $\delta\langle r^2\rangle_s$ taken from the droplet model and using the above calculated $\langle\beta^2\rangle^{N=66}_{\text{model}} = 0.078$ as a reference point (Bu 87). Alternatively, one could adapt $\langle\beta^2\rangle^N_{\text{model}}$ from equation (71) as the final result of the analysis for all $N$. They differ from one another systematically if the fitted parameter $a$ in front of the linear term of equation (68) does not agree with the droplet slope $\delta\langle r^2\rangle^{N,N-1}_{\text{sD}}$. The question of which one comes closer to truth is difficult to answer since in both approaches there is the risk of having introduced systematic errors.

Surprisingly enough, the rms deformation reaches a maximum of as high as 0.28 for Cd at midshell. In fact, $\langle\beta^2\rangle_{\text{IS}}$ values are about a factor of two larger than those calculated from BE2$(0^+ \to 2^+_1)$ values, as shown in Figure 52. The same proportion is observed in In between $\langle\beta^2\rangle^{1/2}_{\text{IS}}$ and $\langle\beta\rangle_{Q_s}$, where the latter is derived from $Q_s$ values in the strong-coupling limit $\Omega = I$ using equation (26) (see Figure 53) (Eb 87b).

In both cases shown, and also in Sn, the IS reveals a much stronger collective contribution to $\langle r^2\rangle$ than is expected from a simple model of vibrational or static quadrupole deformation. In the latter case one could argue that the contribution of

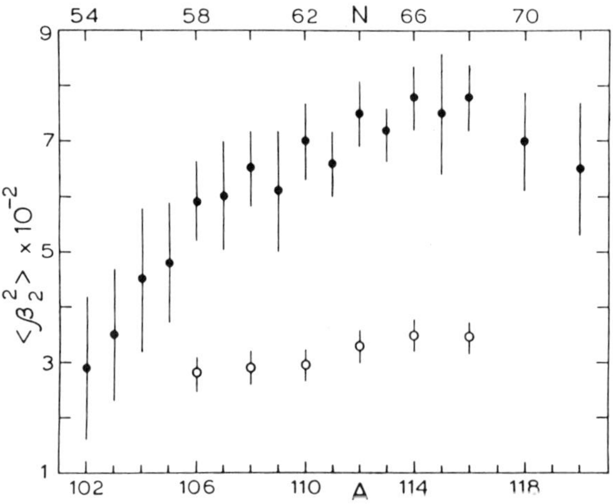


**Figure 52.** $\langle\beta^2\rangle$ of Cd ground states from $\delta\langle r^2\rangle$ (solid circles) compared to those derived from $BE2(0^+ \to 2_1^+)$ values (open circles) (Bu 87).

the zero-point motion to $Q_s$, and hence to $\langle\beta\rangle_{Q_s}$, cancels. Regarding the missing strength in $\langle\beta_2^2\rangle_{BE2}$, admittedly the truncation of equation (63) is certainly unjustified in the case of anharmonic vibration. Strong anharmonicity of a soft mode is postulated, e.g., in the Jahn–Teller model of nuclear deformation (Re 84). Moreover, the question remains open as to what extent multipoles other than the quadrupole contribute to the ground-state collectivity of the Cd–In–Sn region; but they are expected to be much

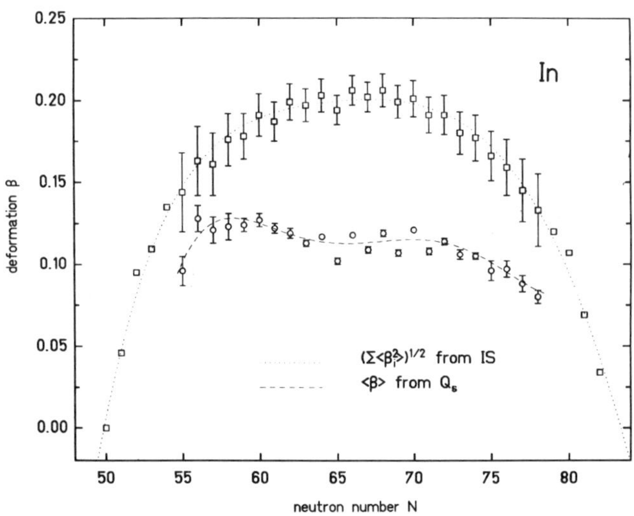


**Figure 53.** $\langle\beta^2\rangle_{IS}^{1/2}$ of In isotopes in comparison with $\langle\beta\rangle_{Q_s}$ values. The error bars on $\langle\beta^2\rangle_{IS}^{1/2}$ increase toward the wings of the curve owing to systematic scaling errors (Eb 87b).

weaker than was observed in the Ca case. An explanation may be sought, at least in part, in a shell dependence of the intrinsic skin thickness, as was advocated by Friedrich and Voegler (Fr 81b, 82) and for Cd, in particular, by Wenz *et al.* (We 81). None of these explanations can at present be excluded.

In summary, the analyses in this section have led us to the following conclusions: Isotope shifts for proton-magic, or nearly proton-magic, elements have a very regular, parabolic shape and include a much larger contribution from (probably) collective core polarization than those offered by $BE2(0^+ \to 2_1^+)$ values alone. The same phenomenon is observed near closed neutron shells, as we discussed earlier. A satisfactory theoretical explanation for this using phenomenological or microscopic model calculations is still awaited. On the other hand, it is to be emphasized that the isotope shift is well described by the droplet model plus the traditional $\beta_2$ contribution whenever it connects rigid spherical with rigid deformed nuclei.

## 4.9. Isotope Shifts in Very Light Nuclei

In light elements, muonic spectra and electron scattering data yield model-independent rms radii; but only very few, and not very precise, measurements of isotope shifts have been reported for elements lighter than Ca (Ja 74, En 74). The situation is no better in the case of optical spectroscopy, except for Na. There the reasons are obvious: (1) the field shift is so small that experimental error matters even in the most precise methods; and (2) the mass shift exceeds the field shift by orders of magnitude (about $10^2$ for Na). The field effect can be traced only from tiny deviations of the total IS from the well-defined $A$-dependence of the mass shift [equations (2) and (3)]. This is possible if a sufficiently large number of isotopes have been measured, as is the case for Na (Hu 78). Still, the quantitative evaluation of the field shift, and especially its scale, depend strongly on the choice of the specific mass shift constant $S$; it has to be chosen such that the $\delta\langle r^2\rangle$ scale is compatible with some model prediction. Figure 54 shows $\delta\langle r^2\rangle$ values for Na isotopes resulting from this type of evaluation (Hu 78) in comparison with deformed Hartree–Fock calculations (Ca 75). In the vicinity of $A = 25$

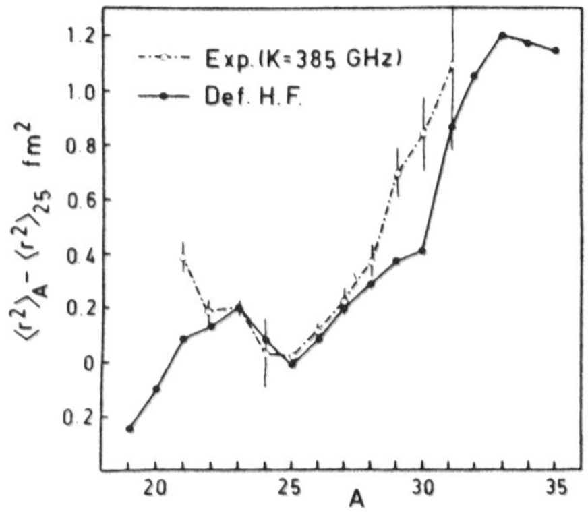


**Figure 54.** Experimental $\delta\langle r^2\rangle^{A,25}$ values for a series of Na isotopes (Hu 78) in comparison with those from a deformed Hartree–Fock calculation (Ca 75). The total mass-shift constant $K = N + S$ was chosen to be 385 GHz.

the curves agree very well; but experiment does not confirm the sharp edges at $A = 23$ and 30.

The example of Na proves that optical spectroscopy of light exotic nuclei can still provide useful data on their charge distribution, although the accuracy is much more limited by experimental and systematic errors than is the case for heavier elements.

## 4.10. Isotone and Isobar Shifts and the Question of a Neutron Skin

The question of whether the distribution of neutrons $\rho_n(r)$ in nuclei differs from that of protons $\rho_p(r)$ was already addressed in Section 4.2. Only hadronic probes have direct access to the former, and electrons only to the latter. Nevertheless, the neutron distribution can be indirectly sensed by electrons through its reaction on the proton distribution. This should be reflected by, for example, differences between isotope and isotone shifts and in isobar shifts of the ms charge radius $\langle r^2\rangle_p$. Isotone shifts between stable nuclei were shown in Figure 50; they are largely limited to $\Delta Z = 2$. Since measurements of isotope shifts off stability have extended our knowledge of ms charge radii over large, compact regions of the nuclear chart, it is now possible to extract long chains of isotone and isobar shifts from the data. So far, this has not been systematically done except for the Cd–Sn region which has been connected toward lower $Z$, up to Rb, and toward higher $Z$, up to Cs and Ba (Bu 87).

For evaluating isotone and isobar shifts one needs, besides the optical isotope shifts, an absolute $\langle r^2\rangle$ value for at least one isotope in each chain, as an additional input parameter that serves as a link between the different elements. It can be taken from muonic spectroscopy or from electron scattering on stable nuclei. As an example, the isotone shift at $N = 66$, i.e., in the middle of the neutron shell, is shown in Figure 55a. As in the case of isotope shift, the shell closure at $Z = 50$ is reflected in a distinct kink in the $\delta\langle r^2\rangle$ curve, which indicates a minimum in the nuclear deformation. After subtracting the deformation contribution in an appropriate way, the $\delta\langle r^2\rangle$ line is straightened and is reproduced by the droplet prediction (Figure 55b).

An example of an isobar shift is given in Figure 56 for the case of $A = 118$; it covers the range from $^{118}_{48}\text{Cd}^{70}$ to $^{118}_{55}\text{Cs}^{63}$ (with seven nucleons exchanged). Corrected for contributions from deformation, it is also reproduced by the prediction obtained

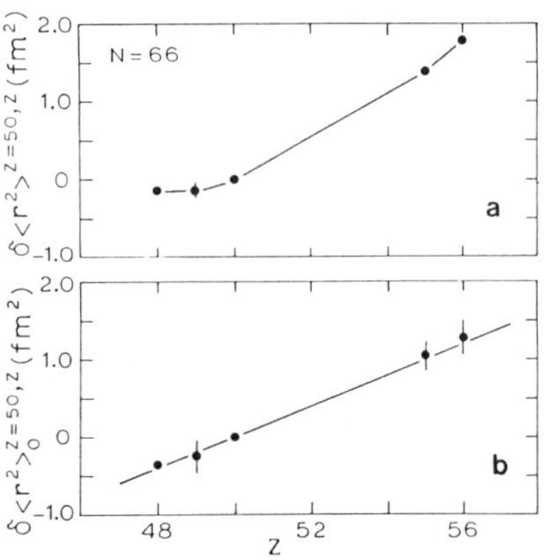


**Figure 55.** (a) $\delta\langle r^2\rangle^{Z,50}$ in the isotonic series $N = 66$ relative to the closed proton shell isotone $^{116}_{50}\text{Sn}$. The line connecting the data points is for guiding the eye. (b) Same as diagram (a), but corrected for the contribution of quadrupole deformation to $\langle r^2\rangle$ (Bu 87). The line indicates the prediction of the spherical-droplet model.

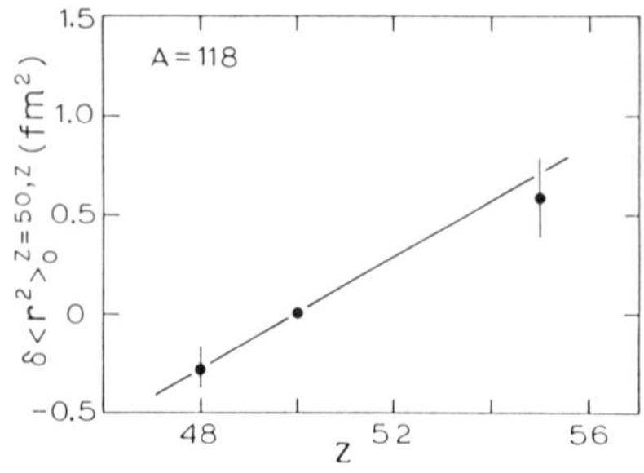


**Figure 56.** $\delta\langle r^2\rangle$ in the isobaric series $A = 118$ corrected for quadrupole deformation (Bu 87). The prediction of the spherical droplet model is shown as a solid line.

from the droplet model. Note that in a naive sense, in a homogeneously charged liquid-drop model of radius

$$R_p = R_n = R_0 = r_0 A^{1/3} \tag{72}$$

the isobar shift would vanish identically. Hence these comparisons demonstrate, in a very simple and clear manner, the capacity of the droplet model to describe neutron and proton radii separately and correctly.

The connection between isobar shift and ms neutron radii may be further illustrated in the derivation of a relation between $\delta\langle r^2\rangle_p$ and $\delta\langle r^2\rangle_n$ under the assumption that the ms nuclear matter radius

$$\langle r^2\rangle_m = \frac{Z}{A}\langle r^2\rangle_p + \frac{N}{A}\langle r^2\rangle_n \tag{73}$$

is constant in a chain of isobars. This assumption is more flexible than equation (72) but still essentially implies the incompressibility of nuclear matter. With this constraint one finds from equation (73) the relation

$$\delta\langle r^2\rangle_n^{N_1,N_0} = -\frac{Z_1}{N_1}\,\delta\langle r^2\rangle_p^{Z_1,Z_0} - \frac{A}{N_0 N_1}(N_1 - N_0)(\langle r^2\rangle_m^A - \langle r^2\rangle_p^{Z_0}) \tag{74}$$

and hence, to first order,

$$\delta\langle r^2\rangle_n = -\frac{Z}{N}\,\delta\langle r^2\rangle_p \tag{75}$$

which is what was sought. For the case considered it follows from the data that

$$\langle r^2\rangle_N({}^{118}_{48}\mathrm{Cd}^{70}) - \langle r^2\rangle_N({}^{118}_{55}\mathrm{Cs}^{63}) = 0.59\ \mathrm{fm}^2 \tag{76}$$

In terms of rms radii, equation (76) implies a swelling of the neutron radius by $\delta\langle r^2\rangle_N^{1/2} = 0.063$ fm at the cost of a shrinking of the charge radius by $\delta\langle r^2\rangle_Z^{1/2} = -0.092$ fm (Bu 87). Angeli *et al.* have performed systematic calculations of proton and neutron radii using a BCS Hartree–Fock code (An 80). Figure 57, taken from their paper, shows differences between neutron and proton rms radii in isotopic sequences with, e.g., $\delta\langle r^2\rangle_n^{1/2} - \delta\langle r^2\rangle_p^{1/2} = 0.12$ fm for the pair ${}^{118}_{48}\mathrm{Cd}^{70}$, ${}^{118}_{52}\mathrm{Te}^{66}$ (with four nucleons exchanged). Extrapolation of this number to exchange of seven nucleons raises it to

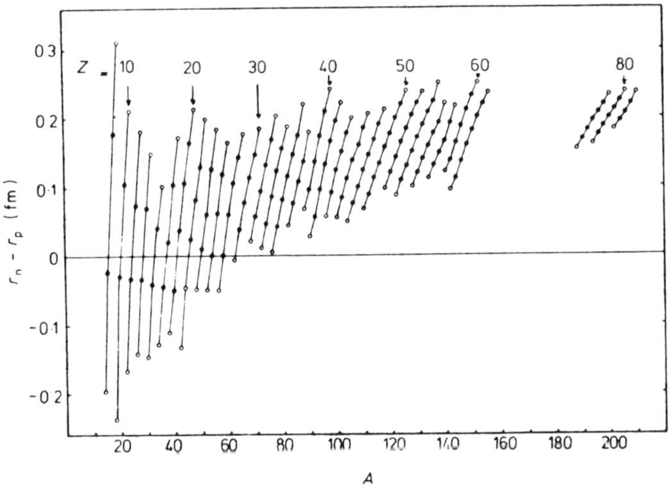


**Figure 57.** Calculated differences between neutron and proton rms radii. Full circles refer to stable nuclei, open circles to unstable nuclei (An 80).

0.196 fm, which is in fair agreement with the value of 0.155 fm estimated from the crude model established above.

## 4.11. Differential Isotope Shifts, Subshell Effects, and Odd–Even Staggering

The experimental accuracy of measurements on integral optical isotope shifts far exceeds the present limits of their interpretation, which are set by systematic errors in the evaluation of mass and field shift and by deficiencies of models of gross nuclear structure. It therefore makes sense to also examine differential shifts between neighboring isotopes; by that means systematic errors are scaled down and even small local changes in the locus of $\delta\langle r^2\rangle$ are sufficiently accentuated to be clearly seen.

### *4.11.1. Indications of Subshell Closures*

Such an accentuation is illustrated by the sharply structured $\delta\langle r^2\rangle^{A,A-2}$ plot for Sn isotops in Figure 58 (An 86b), as compared to the smooth integral isotope shift plotted in Figure 51. The interpretation of such differential structures is not obvious. The authors have attributed the drop at $N = 64$, for example, to a subshell closure, dividing the $d_{5/2}$ and $g_{7/2}$ neutron orbits from the $s_{1/2}$, $d_{3/2}$, and $h_{11/2}$ ones. The same correlation was previously observed in the isotones of Cd (We 81). In other regions of the nuclear chart as well, the appearance of discontinuities in the differential isotope shift could be correlated with subshell closures; but the signature of these effects does not resemble the clear and unique one at major shell closures. Thus far, a clear explanation of the meaning of these structures eludes us. Be that as it may, the Sn structure is reproduced qualitatively by a microscopic Hartree-Fock-Bogolyubov (HFB) calculation, although the dip is shifted relative to the experiment by one neutron number (Do 84).

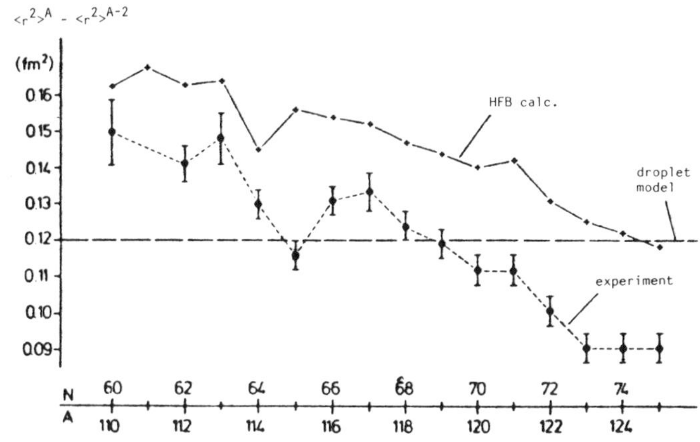


**Figure 58.** Differential isotope shifts $\delta\langle r^2\rangle^{A,A-2}$ of Sn (An 86b), in comparison with theoretical results (Do 84).

### *4.11.2. Odd-Even Staggering and its Interpretation as a Collective Effect*

One famous differential effect is the odd-even staggering of nuclear radii. It has been known since the early days of isotope-shift measurements and has been observed in all regions of nuclei. With a few exceptions (noted in the next section) the effect exhibits a very regular signature, namely, $\langle r^2\rangle$ of an odd-$N$ nucleus is always smaller that the average of its even-$N$ neighbors, which is expressed by the staggering parameter

$$\gamma^{(N)} = 2\delta\langle r^2\rangle^{N,N-1}/\delta\langle r^2\rangle^{N+1,N-1} \qquad N \text{ odd} \tag{77}$$

being smaller than one. The neutron-deficient isotopes of Xe, Cs, and Ba exhibit an especially clear example of the regularity and persistence of this effect, as can be seen in Figure 41.

It is usual in physics to find a clear-cut explanation for a regular phenomenon; but this has not been the case for odd-even staggering. At present, no quantitatively successful theory of this odd-even staggering is available. But qualitatively all explanations that have been offered agree in that the effect is a reflection of nuclear deformation and that $\langle\beta_2^2\rangle$ should be smaller for an odd-$N$ nucleus than for its even neighbor. One explanation along these lines has already been presented in Section 4.9 within the framework of the Zamick-Talmi model. The algebra of core-polarization matrix elements led to a dependence on particle number in a $(j^n)$ shell which included odd-even staggering. Inspection of equation (69), which holds for even multipoles, shows that the sign of the effect is given correctly.

In the framework of a boson picture one could argue, naively, that to a first approximation only the number of pairs of particles or holes counts with respect to the deformation. This mechanism would automatically produce odd-even staggering in the right sense.

The standard explanation of odd-even staggering is, however, given in terms of a pairing-plus-quadrupole model. Reehal and Sørensen formulated the theory and performed extensive calculations of staggering parameters, obtaining results in qualitative agreement with experiment (Re 71). According to their theory the BCS mechanism

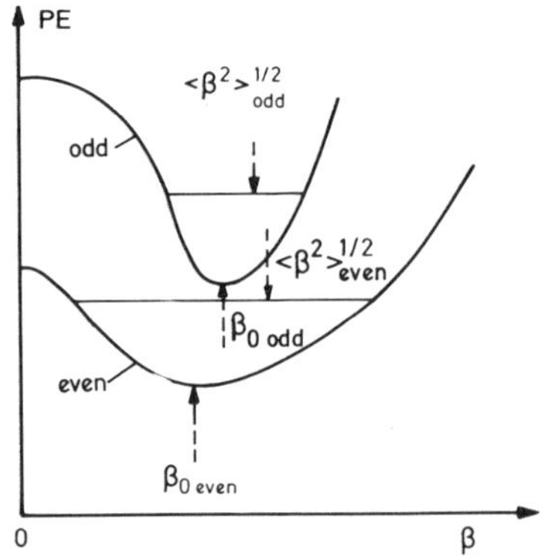


**Figure 59.** Schematic drawing of the potential energy (PE) of an odd and an even nucleus as a function of quadrupole deformation. The PE minima and the rms deformations of the ground level are indicated.

scatters pairs in levels above the Fermi surface giving rise to an enlarged collective vibrational $\beta_2$-amplitude of the zero-point motion which contributes to $\langle r^2 \rangle$ via equation (53). In odd nuclei these vibrations are reduced by blocking. The effect may be visualized graphically by Figure 59, which shows schematically the potential-energy surface of an odd and an even nucleus. Both nuclei may fill the ground level in a deformed minimum created by shell corrections to the liquid-drop energy. The pairing interaction counteracts the shell corrections; it thus reduces the depth of the minimum, which causes the collective wave function to spread more widely. Consequently, the rms deformation shifts to a larger value because of its quadratic weight. In odd nuclei, however, this effect is blocked [see also Ahmad *et al.* (Ah 88)]. It is interesting to note that the same picture is used to explain why spontaneous fission rates are slower in odd than in even nuclei (Bj 80).

The systematics of odd–even staggering that are currently available show variation with occupation number of a particular subshell. In the case of the $\nu i_{13/2}$ isomers in Hg, for example, the effect is very strong for the light isotopes ($\gamma \approx 0$ in $^{185}$Hg) and vanishes for the heavy ones ($\gamma \approx 1$ in $^{197}$Hg) in which the shell is full except for a single hole (St 79, Ul 86). Similar behavior is apparently observed in the $\nu f_{7/2}$ shell of the rare earths. The theoretical literature relevant to the extensive new results of odd–even staggering obtained off stability is sparse. Zawischa *et al.* have again recently taken up the problem; they have included a cluster interaction between neutron and proton pairs in order to transfer the pairing effect of the neutrons, which is the primary cause of the odd–even staggering, to the charge distribution. They were able to reproduce experimental data for stable Mo, Cd, and Ba isotopes with reasonable accuracy by their model, but were unable to produce any satisfactory result by the ansatz of Reehal and Sorensen (Re 71, Za 85, 87).

## 4.12. Evidence for Octupole Shapes in the Ra Region

An interesting feature of nuclear structure discovered in recent years is that intrinsic reflection symmetry is spontaneously broken in the ground states of a group of heavy nuclei around $^{225}$Ra. Nuclear spectroscopy has traced several characteristic signatures of this asymmetry, for example: (1) almost degenerate doublets of the same spin but opposite parity, in odd nuclei; (2) $I^p$ sequences of $0^+$, $1^-$, $2^+$, $3^-$, $4^+$, ... in bands of even nuclei; (3) unusually large E1 matrix elements in $\alpha$ decay and $\gamma$ decay. Most theories ascribe the broken symmetry to an intrinsic octupole deformation of the matter distribution, but others invoke clustering $\alpha$-particle in the nuclear surface. Both explanations would also imply the existence of an intrinsic electric dipole moment, *a priori* in the latter case and *a posteriori* in the former case. The field has been summarized by, e.g., Hamilton *et al.* and Meyer *et al.* (Ha 85, Me 86).

### 4.12.1. Octupole Effects in the Isotope Shift

We may ask what optical spectroscopy can contribute to the resolution of this problem. Systematic measurements of spins, moments, and isotope shifts have recently been performed along extended isotope chains of Rn, Fr, and Ra (data and references are given in the Appendix) that cross the critical region of neutron numbers $130 \leqslant N \leqslant 140$. Figure 60 shows the locus of $\delta\langle r^2\rangle$ for Ra, spanning the range $206 \leqslant A \leqslant 232$. There is a gap between $^{214}$Ra and $^{220}$Ra, since $\alpha$-particle decay lifetimes drop to the microsecond range in this region—too short to be separable with present on-line mass-separation techniques. Nevertheless, by extrapolation from heavier masses, one can discern the characteristic kink at the magic neutron number, $N = 126$.

One may also ask whether it is possible to identify an octupole term $\langle\beta_3^2\rangle$ from an analysis of the integral isotope shift $\delta\langle r^2\rangle^{N,126}$ using equation (54) under the assumption

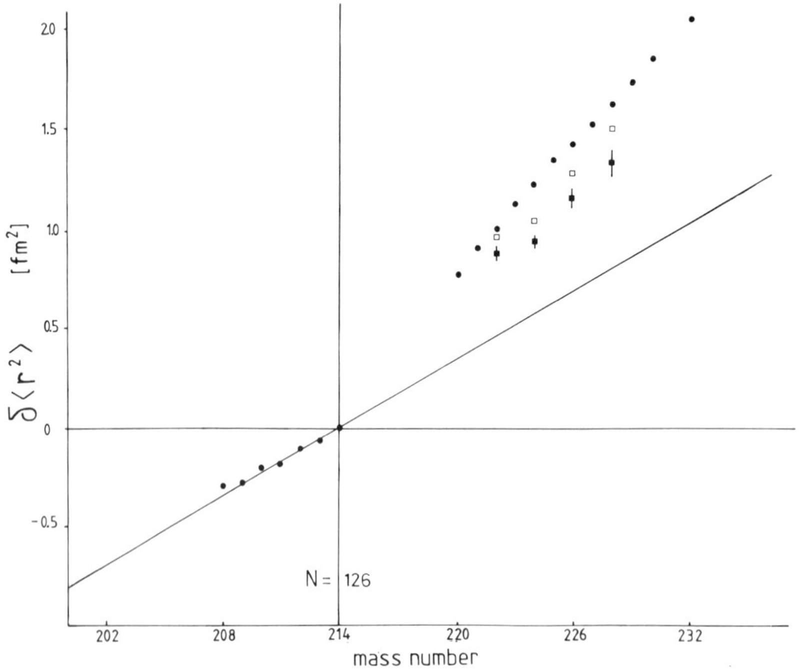


**Figure 60.** Plot of $\delta\langle r^2\rangle^{A,214}$ (full dots) for Ra (Ah 88), in comparison with the revised, steeper droplet line (Mö 84). The shape effect, as calculated from experimental BE2 values, is added to the latter for some of the heavier isotopes (full squares). Also shown is further correction with theoretical $\beta_4$-values (Na 84) (open squares).

that $\langle\beta_i^2\rangle^{N=126} \approx 0$. A $\beta_3$ value of about 0.1—as calculated by the majority of current theories [see, e.g., Leander *et al.* (Le 84b)]—would have an effect of the order of 0.13 fm$^2$ on the ms radius. Since it is competing with a total shift of about 1.5 fm$^2$ over that range, the effect is actually too small to be determined unambiguously in view of the systematic errors in the evaluation of the field shift factor $F$ [equation (13)] and of the volume effect $\delta\langle r^2\rangle_s$ (Ah 88). Alternatively, one could start from the heavy-mass side of the isotope chain, for example from $^{229}$Ra, where octupole deformation should have already vanished. $Q_s$ and BE2 values are known in this region and can be used in order to correct $\delta\langle r^2\rangle$ for quadrupole deformation. This procedure seems to be reliable at the heavy end of an isotopic chain since deformation is static there, as has been proven by comparing $Q_s$ to $Q_0$ by the strong-coupling rule [equation (26)]. But for smaller $N$ the static character of the deformation disappears, thus inhibiting a safe determination of $\langle\beta_2^2\rangle$ from BE2 or $Q_s$ values, as has been pointed out in preceding sections. Moreover, a $\beta_4$-deformation must also be taken into account. In summary, isotope shift analysis is not presently up to the task of determining the size of the octupole deformation reliably.

Nevertheless, the isotope shift reveals a very valuable qualitative indication of the presence of octupole deformation. It does so in connection with the irregular, inverted odd-even staggering observed for $^{221,223,225}$Ra. The ms radius of these isotopes is bigger than the average of their even neighbors, indicating a larger ms deformation. The inversion of the staggering is more clearly seen in the derivative of the integral isotope shift, which is plotted in Figure 61 as $\delta\langle r^2\rangle^{N,N-1}$. The slope always staggers from one isotope to the next but its sense is inverted in the above-mentioned region. There, the odd ones are up and the even ones down, in contrast to the regular staggering below $N = 126$. Moreover, it is very conspicuous that the inverted staggering is not confined to Ra but is repeated in the isotones of Rn (Bo 87) and Fr (Co 87), as shown in this figure.

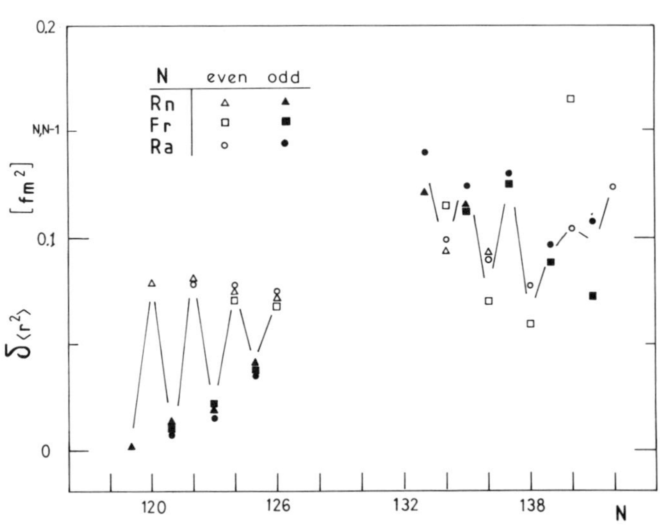


**Figure 61.** Differential isotope shift $\delta\langle r^2\rangle^{N,N-1}$ for Rn, Fr, and Ra isotopes showing regular odd-even staggering below $N = 126$, and with inverted sense for neutron numbers $N = 133$, 135, and 137 (Ah 88).

The observation of inverted staggering came unexpectedly, but it corroborates calculations by Chasman (Ch 80) and by Leander and Sheline (Le 84b), which suggest that octupole deformation should be more pronounced in odd than in even nuclei. In this context it is worth remembering that the core polarization model of Zamick and Talmi (see Section 4.9) also predicted regular odd–even staggering for even multipoles, and inverted ones in the case of odd multipoles. Although the two theories do not have very much in common, their agreement on that point is not necessarily accidental.

It has already been mentioned that octupole deformation can also bring about an intrinsic electric dipole moment. The order of magnitude observed experimentally is $p \approx 1$ $e$fm. Its effect on the ms radius may be estimated as follows: The dipole may be created by moving an amount of charge $q = p/R_0$ from its quadratically weighted average position at $\langle r^2\rangle^{1/2} = \sqrt{\frac{3}{5}R_0^2}$ to the surface at $R_0$. Then $\langle r^2\rangle$ increases by

$$\delta\langle r^2\rangle_{\text{e.D.}} = \frac{q}{Ze}\left(R_0^2 - \frac{3}{5}R_0^2\right) = \frac{2}{5}\frac{p}{Ze}R_0 \tag{78}$$

The effect amounts to 0.033 fm$^2$ in Ra for $p$ as given above; it is small but not completely negligible.

### 4.12.2. *Single-Particle States of Octupole-Deformed Nuclei*

Crucial information on octupole shapes is also provided by single-particle properties, such as spins and magnetic moments. Before explaining these details it may be advisable to review briefly a few basic relationships between geometry and dynamics in an asymmetric potential. By definition, the nuclear potential-energy surface must be symmetric in terms of the collective octupole coordinate $\beta_3$, since change of sign is equivalent to a rotation of the nucleus by 180° (see Figure 62). If the central barrier at $\beta_3 = 0$ is high enough, the collective wave function localizes around the two potential minima at $\pm\beta_{3\text{min}}$ and forms a symmetric (positive-parity) and an antisymmetric (negative-parity) state with respect to reflection. These two states form an (almost) degenerate doublet of opposite parity, which is the characteristic spectroscopic feature of asymmetric shapes—as is well known from molecular physics. The famous maser transition in $NH_3$, for instance, is just this doublet transition. If the doublet splitting is very small compared to other excitations, then one may say that the octupole

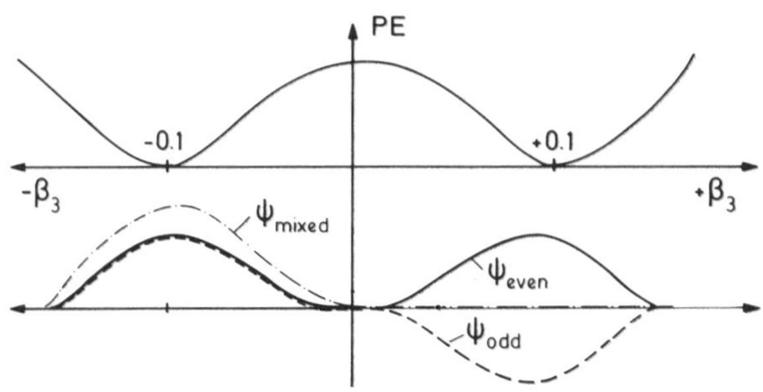


**Figure 62.** Schematic diagram of an octupole potential-energy surface (upper curve). The symmetric and antisymmetric collective wave functions and their superposition are also shown (lower curves).

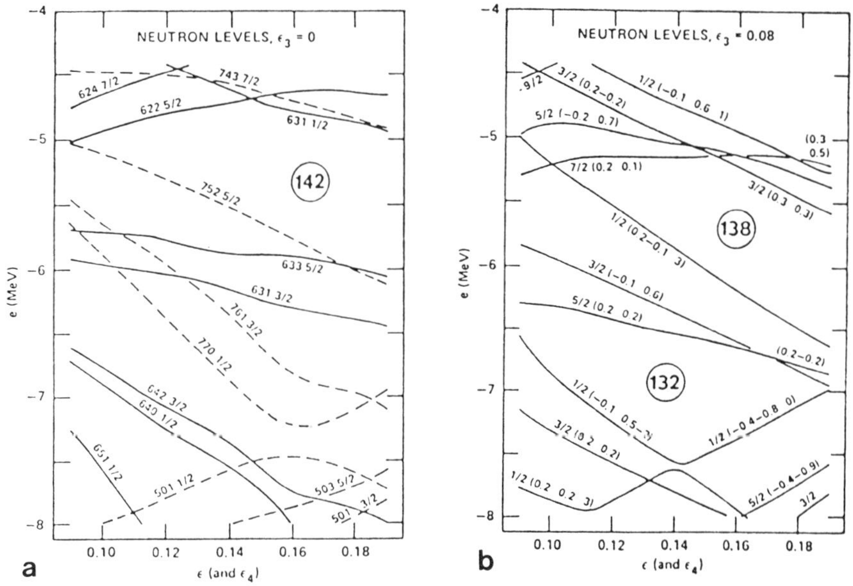


**Figure 63.** Nilsson diagram of neutron levels in Ra as a function of $\beta_2$: (a) without an octupole degree of freedom ($\beta_3 = 0$) and (b) with $\beta_3$ fixed at 0.08 (Le 84b).

deformation is "stable" because the length of time for which the collective wave function can be localized at one side of the barrier by superposition of the doublet states is given by the reciprocal of the transition energy. This doublet structure is quite well established in a number of cases in the Ra region.

In an octupole potential, single-particle levels of opposite intrinsic parity will become mixed—especially degenerate ones. Figure 63a shows Nilsson levels for Ra as a function of the quadrupole deformation $\beta_2$, where the octupole deformation has been held at zero (Le 84b). It can be seen that a number of levels of states with the same spin but opposite parity cross each other. After the turning on of an octupole deformation, $\beta_3 = 0.08$ (Figure 63b), these crossings are removed by the formation of new single-particle states, which have mixed intrinsic parity. In fact, these crossings of levels are the ones which trigger the breakdown from quadrupole to octupole symmetry by residual octupole interactions with the core [reviewed, e.g., by Ragnarsson *et al.* (Ra 78b)]. It has been shown (Re 84) that this mechanism is the same as the one which breaks the symmetry of molecular configurations, namely, the Jahn–Teller effect (Ja 37).

For comparison with experiment it is important that the reshuffled level scheme shows a different spin sequence as a function of neutron number than that characteristic of pure quadrupole deformation. For neutron numbers $133 \leq N \leq 141$ spins $\frac{1}{2}, \frac{3}{2}, \frac{3}{2}, \frac{5}{2}$, and $\frac{5}{2}$ appear consecutively at the Fermi surface in the quadrupole case, while calculations that include octupole deformation reproduce the correct experimental sequence $\frac{5}{2}, \frac{3}{2}, \frac{1}{2}, \frac{3}{2}, \frac{5}{2}$. ($\beta_2$ increases over this range from 0.15 to 0.2.)

Moreover, the magnetic moment is very sensitive to the degree of intrinsic parity mixing in the hybridized single-particle wave functions. This is particularly evident

**Table 1. Spins and Magnetic Moments of Some Ra Isotopes in Comparison with Theoretical Calculations With and Without the Octupole Degree of Freedom (Le 84b)**

| Nucleus | $I^{p}_{exp}$ | $\mu_{exp}$ (n.m.) | $I^{p}_{th}$ | $\mu_{th}$ (n.m.) | $\beta_3$ | Dominant configuration |
|---|---|---|---|---|---|---|
| $^{221}$Ra | $\frac{5}{2}$ | −0.18 | $\frac{5}{2}^{+}$ | −0.13 | 0.1 | $[642]\times[752]\frac{5}{2}$ |
| | | | $\frac{5}{2}^{+}$ | 0.39 | 0 | $[633]\frac{5}{2}$ |
| $^{223}$Ra | $\frac{3}{2}$ | 0.27 | $\frac{3}{2}^{+}$ | 0.50 | 0.1 | $[642]\frac{3}{2}$ |
| | | | $\frac{3}{2}^{+}$ | 0.03 | 0 | $[631]\frac{3}{2}$ |
| | | | $\frac{3}{2}^{-}$ | −0.06 | 0 | $[761]\frac{3}{2}$ |
| $^{225}$Ra | $\frac{1}{2}^{+}$ | −0.73 | $\frac{1}{2}^{+}$ | −0.75 | 0.1 | $[640]\frac{1}{2}$ |
| | | | — | — | 0 | — |
| $^{227}$Ra | $\frac{3}{2}^{+}$ | −0.40 | $\frac{3}{2}^{+}$ | −0.41 | 0.08 | $[631]\frac{3}{2}$ |
| | | | $\frac{3}{2}^{+}$ | −0.06 | 0 | $[631]\frac{3}{2}$ |
| $^{229}$Ra | $\frac{5}{2}$ | 0.50 | $\frac{5}{2}^{+}$ | 0.38 | 0.04 | $[633]\frac{5}{2}$ |
| | | | $\frac{5}{2}^{+}$ | 0.61 | 0 | $[633]\frac{5}{2}$ |
| | | | $\frac{5}{2}^{-}$ | −0.29 | 0 | $[752]\frac{5}{2}$ |

when states with opposite nucleon spin projection, such as $|n, n_z, \Lambda\downarrow\rangle$ and $|n+1, n_z + 1, (\Lambda - 1)\uparrow\rangle$ mix, because the nucleon $g$-factor is canceled in taking the sum. Table 1, taken from Leander *et al.* (Le 84b), is a comparison of experimental magnetic moments with those calculated with and without octupole deformation. It is clear from the table that only in the octupole case is satisfactory agreement between theory and experiment achieved [see also Ragnarsson (Ra 83)].

## 5. Concluding Remarks

What then has been achieved by optical spectroscopy far off stability and what is still to be done? Charge radii, spins, and moments have been measured along extended isotopic chains of about 30 elements, covering more than 500 nuclear ground states and isomers. From a purely numerical point of view we are not yet midway in our task; but experiments are currently under design and construction so as to complete it.

Will greater numbers give us important new insights into nuclear structure and stimulate further theory or will they simply fill tables, like a collection of precious stones on an amateur's shelves? The truth of the matter is that the complexity of nuclear structure cannot be adequately understood and analyzed without detailed and complete experimental information. Critical phenomena, such as shape transitions, provide only one example, proving this. The charge radii of transitional nuclei show surprisingly large contributions from core polarization; but this can only be seen by analyzing the isotope shift along extended chains that cover major parts of a shell. The systematic IS and $Q_s$ measurements over compact regions have also corroborated the traditional gross nuclear models for nuclei with stabilized spherical or deformed shells.

Generally speaking, optical spectroscopy and nuclear spectroscopy are playing complementary roles in the study of the fundamental questions of nuclear size, shape, and symmetry: the former on the side of the geometry, the latter on the side of the dynamics.

On the other hand, there remain a number of open questions in the field of atomic physics: how to evaluate precisely the electric field gradient, the specific mass, and the field shift in electronic spectra. In many-body theory of electronic shells, these problems have an importance and weight in their own right. The present uncertainties—of about 10%, or even more in some cases—are currently impeding, but by no means preventing, the interpretation of the results in nuclear physics. Progress toward answering these atomic-physics questions can soon be expected in theory and also in experiment, from the results of new calibration measurements on muonic spectra.

On-line laser spectroscopy, in combination with sophisticated isotope production and separation, and signal detection techniques will push the frontiers of this entire field to higher sensitivity and enable measurements to be made on refractory elements. The new techniques will enrich the scope of experiments at mass separators on-line with both proton and heavy-ion accelerators. But on-line facilities at high-current accelerators could also lead to important new progress in this field.

## Appendix

Tables 2–31 summarize the present (06/88) status of our knowledge of spins $I$, magnetic moments $\mu_I$, spectroscopic quadrupole moments $Q_s$, and changes of ms charge radii $\delta\langle r^2\rangle$, as measured in extended chains of isotopes by off-stability techniques of optical spectroscopy. In most cases explicit reference is made to publications in this field only. We use these publications also as references to data on stable isotopes, measured by different techniques, if they were remeasured or if such data was compiled. Other data are taken from published material (Le 78). Comments are made with regard to the experimental method used, the optical transitions investigated, and the evaluation of data. In most cases $\mu_I$ is calculated from the ratio of $A$-factors through a known and diamagnetically corrected $\mu_I$ of a reference isotope, and without correction for hfs anomaly, using the equation

$$\mu_I^{\mathrm{X}} = \mu_I^{\mathrm{ref}}(A^{\mathrm{X}} I^{\mathrm{X}} / A^{\mathrm{ref}} I^{\mathrm{ref}})$$

With regard to $Q_s$, it is stated whether or not a Sternheimer correction is applied.

With regard to $\delta\langle r^2\rangle$ values: the method used for the evaluation of the field-shift factor $F$, and its value, are given in each case, along with a reference. Also given is the specific mass shift constant $S$ used in the calculation.

Where more than one data set is available, generally the most precise value that has been published or the most complete data set that was compiled is presented without averaging or further reanalysis. In these cases the corresponding reference is given without brackets, while bracketed references refer to the other measurements in order to point out priorities. In view of the different evaluation methods used, this was felt to be the only viable procedure within the framework of this review. Errors are given as published; mostly they do not include systematic uncertainties. Unless explicitly mentioned, competing measurements agree within their experimental errors. An update of complete, critical tables of nuclear moments (Ra 88) and isotope shifts (Au 87) is published appear in *Atomic Data and Nuclear Data Tables.*

**Table 2. Nuclear Moments and Changes of Charge Radii of Long Isotopic Chains**

| Elements included: Abbreviation | Element | Status: 01/88 Proton number $Z$ |
|---|---|---|
| Li | Lithium | 3 |
| Na | Sodium | 11 |
| K | Potassium | 19 |
| Ca | Calcium | 20 |
| Rb | Rubidium | 37 |
| Sr | Strontium | 38 |
| Ag | Silver | 47 |
| Cd | Cadmium | 48 |
| In | Indium | 49 |
| Sn | Tin | 50 |
| Xe | Xenon | 54 |
| Cs | Caesium | 55 |
| Ba | Barium | 56 |
| Nd | Neodymium | 60 |
| Sm | Samarium | 62 |
| Eu | Europium | 63 |
| Dy | Dysprosium | 66 |
| Ho | Holmium | 67 |
| Er | Erbium | 68 |
| Tm | Thulium | 69 |
| Yb | Ytterbium | 70 |
| Pt | Platinum | 78 |
| Au | Gold | 79 |
| Hg | Mercury | 80 |
| Tl | Thallium | 81 |
| Pb | Lead | 82 |
| Rn | Radon | 86 |
| Fr | Francium | 87 |
| Ra | Radium | 88 |

**Table 3. Element: Lithium, $Z = 3$**

| $A$ | $\tau_{1/2}$ | $I$ | $\mu_I$ (n.m.) | $Q_s$ (b) | $\delta\langle r^2\rangle^{A,6}$ (fm$^2$) | Ref. No. $\mu_I$ | $Q_s$ | $\delta\langle r^2\rangle$ |
|---|---|---|---|---|---|---|---|---|
| 6 | stable | 1 | 0.8220467 (6) | −0.000644 (7) | | 1 | 2 | |
| 7 | stable | $\frac{3}{2}$ | 3.256424 (2) | −0.0370 (8) | | 1 | 3 | |
| 8 | 0.84 s | 2 | 1.653347 (9) | 0.0283 (6) | | 4 | 5 | |
| 9 | 178.3 ms | $\frac{3}{2}$ | 3.4391 (6) | −0.0253 (9) | | 6 | 8 | |
| 11 | 8.5 ms | $\frac{3}{2}$ | 3.76673 (25) | | | 7 | | |

*References*: 1. (Be 74), 2. (Wh 64), 3. (We 85b), 4. (Wi 78), 5. (Du 77), 6. (Co 83), 7. (Ar 87b), 8. (Ar 88).
*Exp. method*: 1–2: ABMR and NMR.
3: Coulomb excitation of aligned nuclei.
4–8: $\beta$ NMR.
*Transitions*: 7–8: LiI D1 and D2 line (670 nm).
$\mu_I$*-eval.*: reference isotope $^6$Li from Ref. 1.
$Q_s$-eval.: direct measurement in Ref. 3.
$\delta\langle r^2\rangle$*-eval.*: not given, only possible under extremely restrictive assumptions about the dominating mass shift.
*Remarks*: Ref. 6 gives a different value for $Q_s(^9\text{Li}) = 0.032\ (6)$.

**Table 4. Element Sodium, $Z = 11$**

| $A$ | $\tau_{1/2}$ | $I$ | $\mu_I$ n.m. | $Q_s$ (b) | $\delta\langle r^2\rangle^{A,23}$ fm$^2$ | Ref. No. $\mu_I$ | $Q_s$ | $\delta\langle r^2\rangle$ |
|---|---|---|---|---|---|---|---|---|
| 20 | 446 ms | 2 | 0.3694 (2) | | −0.13 (7) | 5 | | 2 (1) |
| 21 | 22.5 s | $\frac{3}{2}$ | 2.38612 (10) | 0.050 (37) | 0.12 (5) | 1 | 2 | 2 (1) |
| 22 | 2.60 y | 3 | 1.746 (3) | | −0.05 (4) | 1 | | 2.(1) |
| 23 | stable | $\frac{3}{2}$ | 2.217520 (2) | 0.101 (8) | 0 | 3 | 4 | |
| 24 | 15.02 h | 4 | 1.6902 (5) | | −0.12 (4) | 1 | | 2 (1) |
| 25 | 60.0 s | $\frac{5}{2}$ | 3.683 (4) | −0.095 (50) | −0.10 (3) | 1 | 2 | 2 (1) |
| 26 | 1.07 s | 3 | 2.851 (2) | −0.079 (54) | −0.005 (18) | 1 | 2 | 2 (1) |
| 27 | 290 ms | $\frac{5}{2}$ | 3.895 (5) | −0.058 (54) | 0.12 (4) | 1 | 2 | 2 (1) |
| 28 | 30.5 ms | 1 | 2.426 (3) | −0.017 (37) | 0.28 (5) | 1 | 2 | 2 (1) |
| 29 | 43 ms | $\frac{3}{2}$ | 2.449 (8) | 0.025 (54) | 0.60 (8) | 1 | 2 | 2 (1) |
| 30 | 53 ms | 2 | 2.083 (10) | | 0.76 (12) | 1 | | 2 (1) |
| 31 | 17 ms | $\frac{3}{2}$ | 2.283 (38) | | 1.09 (7) | 1 | | 2 (1) |

*References*: 1. (Hu 78) and references cited therein, 2. (To 82), 3. (Be 74), 4. (Le 78), 5. (Sc 75).
*Exp. method*: 1–2: LIOP and laser rf double resonance.
4: AMBR–NMR.
5: RADOP.
*Transitions*: 1–2: NaI D1 and D2 line (589 nm, 590 nm).
$\mu_I$*-eval.*: reference isotope $^{23}$Na.
$Q_s$*-eval.*: HS calculation from (Ro 72).
$\delta\langle r^2\rangle$*-eval.*: semiempirical calculation, $F = -47$ MHz/fm$^2$, $N + S = 385.5$ GHz amu as discussed in Ref. 2, $\delta\langle r^2\rangle$ values extremely sensitive to $S$!

### Table 5. Element: Potassium, $Z=19$

| $A$ | $\tau_{1/2}$ | $I$ | $\mu_I$ (n.m.) | $Q_s$ (b) | $\delta\langle r^2\rangle^{A,39}$ ($fm^2$) | Ref. No. $\mu_I$ | Ref. No. $Q_s$ | Ref. No. $\delta\langle r^2\rangle$ |
|---|---|---|---|---|---|---|---|---|
| 36 | 342 ms | 2 | ±0.548 (1) | | | 1 | | |
| 37 | 1.226 s | $\frac{3}{2}$ | 0.20321 (6) | | | 4 | | |
| 38 | 7.63 m | 3 | 1.371 (6) | | −0.058 (41) | 3 | | 2 |
| 39 | stable | $\frac{3}{2}$ | 0.3914658 (4) | 0.049 (4) | 0 | 3 | 3 | |
| 40 | stable | 4 | −1.2982 (4) | | 0.0222 (20) | 3 | | 2 |
| 41 | stable | $\frac{3}{2}$ | 0.2148699 (2) | −0.067 (8) | 0.117 (6) | 3 | 3 | 2 |
| 42 | 12.36 h | 2 | −1.1425 (6) | 0.060 (5) | 0.116 (15) | 3 | 3 | 2 |
| 43 | 22.3 h | $\frac{3}{2}$ | 0.1633 (8) | | 0.143 (9) | 2 | | 2 |
| 44 | 22.1 m | 2 | −0.856 (4) | | 0.148 (11) | 2 | | 2 |
| 45 | 17.3 m | $\frac{3}{2}$ | 0.1734 (8) | | 0.176 (13) | 2 | | 2 |
| 46 | 107 s | 2 | −1.051 (6) | | 0.143 (12) | 2 | | 2 |
| 47 | 17.5 s | $\frac{1}{2}$ | 1.933 (9) | | 0.126 (13) | 2 | | 2 |

*References*: 1. (Sc 75), 2. (To 82b), 3. (Le 78), 4. (Kö 69).
*Exp. method*: 2: LIOP.
1, 4: RADOP.
*Transitions*: 1, 2: KI D1 and D2 line (764 nm, 770 nm).
$\mu_I$-*eval.*: reference isotope $^{39}$K.
$Q_s$-*eval.*: see Ref. 3 and references cited therein.
$\delta\langle r^2\rangle$-*eval.*: semiempirical calculation, $F=-128$ MHz/fm$^2$, $S=200(4)$ GHz amu from muonic X-rays according to Ref. 2.

### Table 6. Element: Calcium, $Z=20$

| $A$ | $\tau_{1/2}$ | $I$ | $\mu_I$ (n.m.) | $Q_s$ (b) | $\delta\langle r^2\rangle^{A,40}$ ($fm^2$) | Ref. No. $\mu_I$ | Ref. No. $Q_s$ | Ref. No. $\delta\langle r^2\rangle$ |
|---|---|---|---|---|---|---|---|---|
| 39 | 0.86 s | $\frac{3}{2}$ | 1.02168 (12) | | | 4 | | |
| 40 | stable | 0 | | | 0 | | | |
| 41 | $1\times10^5$ y | $\frac{7}{2}$ | −1.59478 (1) | −0.062 (12) | 0.0012 (25) | 4 | 2 | 3 (2) |
| 42 | stable | 0 | | | 0.2209 (60) | | | 3 (1, 2) |
| 43 | stable | $\frac{7}{2}$ | −1.31726 (6) | −0.040 (8) | 0.1221 (45) | 4 | 2 | 3 (1, 2) |
| 44 | stable | 0 | | | 0.2908 (75) | | | 3 (1, 2) |
| 45 | 165 d | $\frac{7}{2}$ | −1.316 (16) | 0.043 (9) | 0.1272 (89) | 3 | 2 | 3 (2) |
| 46 | stable | 0 | | | 0.1290 (63) | | | 3 (1, 2) |
| 47 | 4.536 d | $\frac{7}{2}$ | −1.380 (24) | | 0.007 (10) | 3 | | 3 (2) |
| 48 | stable | 0 | | | 0.0032 (67) | | | 3 (1, 2) |

*References*: 1. (Br 78), 2. (Be 80), 3. (An 82), 4. (Le 78).
*Exp. method*: 1: Classical atomic-beam spectroscopy, 2–3: Atomic-beam laser fluorescence spectroscopy.
*Transitions*: 1–3: CaI 4s $^1S_0 \rightarrow 4s4p\,^{1,3}P_1$ (422.7 nm and 657.3 nm).
$\mu_I$-*eval.*: reference isotope $^{43}$Ca.
$Q_s$-*eval.*: semiempirical calculation from fs in Ref. 2, Sternheimer corrected.
$\delta\langle r^2\rangle$-*eval.*: weighted-mean values from Refs. 1–3 calibrated by muonic X-ray isotope shift according to (He 85b).

**Table 7. Element: Rubidium, $Z = 37$**

| $A$ | $\tau_{1/2}$ | $I$ | $\mu_I$ (n.m.) | $Q_s$ (b) | $\delta\langle r^2\rangle^{A,87}$ (fm$^2$) | Ref. No. $\mu_I$ | Ref. No. $Q_s$ | Ref. No. $\delta\langle r^2\rangle$ |
|---|---|---|---|---|---|---|---|---|
| 76 | 39.1 s | 1 | −0.376 (9) | 0.38 (15) | 0.220 (27) | 2 | 2 | 2 |
| 77 | 3.70 m | $\frac{3}{2}$ | 0.6568 (41) | 0.695 (32) | 0.2831 (69) | 2 | 2 | 2 |
| 78 | 17.66 m | 0 | | | 0.3060 (23) | | | 2 |
| 78$^m$ | 5.74 m | 4 | 2.5485 (21) | 0.814 (39) | 0.1912 (26) | 2 | 2 | 2 |
| 79 | 22.9 m | $\frac{5}{2}$ | 3.3579 (12) | −0.098 (22) | 0.2249 (23) | 2 | 2 | 2 |
| 80 | 30.34 s | 1 | −0.0834 (3) | 0.348 (20) | 0.2166 (68) | 3 | 2 | 2 |
| 81 | 4.58 h | $\frac{3}{2}$ | 2.0595 (14) | 0.398 (23) | 0.1698 (22) | 2 | 2 | 2 |
| 81$^m$ | 32 m | $\frac{9}{2}$ | 5.5980 (17) | −0.743 (57) | 0.1398 (22) | 2 | 2 | 2 |
| 82 | 1.25 m | 1 | 0.5536 (54) | 0.190 (72) | 0.1322 (65) | 2 | 2 | 2 |
| 82$^m$ | 6.2 h | 5 | 1.6434 (12) | 1.01 (12) | 0.1319 (58) | 3 | 2 | 2 |
| 83 | 86.2 d | $\frac{5}{2}$ | 1.4249 (8) | 0.196 (22) | 0.0512 (15) | 2 | 2 | 2 |
| 84 | 32.87 d | 2 | −1.3246 (16) | −0.015 (35) | 0.0078 (32) | 2 | 2 | 2 |
| 84$^m$ | 20.49 m | 6 | 0.220 (15) | 0.57 (27) | 0.004 (16) | 2 | 2 | 2 |
| 85 | stable | $\frac{5}{2}$ | 1.3533505 (9) | 0.274 (2) | 0.0355 (22) | 3 | 3 | 2 (1) |
| 86 | 18.65 d | 2 | −1.6977 (16) | 0.193 (32) | 0.0271 (31) | 2 | 2 | 2 |
| 86$^m$ | 1.018 m | 6 | 1.8150 (10) | 0.369 (95) | 0.0060 (35) | 2 | 2 | 2 |
| 87 | stable | $\frac{3}{2}$ | 2.751816 (2) | 0.127 (1) | 0 | 3 | 3 | |
| 88 | 17.8 m | 2 | 0.5117 (26) | −0.012 (96) | 0.1364 (78) | 2 | 2 | 2 |
| 89 | 15.2 m | $\frac{3}{2}$ | 2.3836 (7) | 0.144 (26) | 0.3051 (31) | 2 (1) | 2 (1) | 2 (1) |
| 90 | 153 s | 0 | | | 0.4289 (80) | | | 2 (1) |
| 90$^m$ | 4.26 m | 3 | 1.61598 (64) | 0.204 (45) | 0.4149 (31) | 2 | 2 | 2 (1) |
| 91 | 58.4 s | $\frac{3}{2}$ | 2.1815 (15) | 0.154 (26) | 0.5580 (35) | 2 (1) | 2 (1) | 2 (1) |
| 92 | 4.50 s | 0 | | | 0.6963 (80) | | | 2 (1) |
| 93 | 5.85 s | $\frac{5}{2}$ | 1.4095 (16) | 0.177 (40) | 0.8130 (32) | 2 (1) | 2 (1) | 2 |
| 94 | 2.70 s | 3 | 1.4984 (18) | 0.163 (50) | 0.9167 (35) | 2 | 2 | 2 |
| 95 | 0.38 s | $\frac{5}{2}$ | 1.3336 (34) | 0.211 (65) | 1.0781 (56) | 2 | 2 | 2 |
| 96 | 0.20 s | 2 | 1.4658 (17) | 0.246 (56) | 1.1637 (74) | 2 | 2 | 2 |
| 97 | 0.17 s | $\frac{3}{2}$ | 1.8410 (21) | 0.581 (44) | 1.7391 (62) | 2 | 2 | 2 |
| 98 | 0.11 s | (0) | | | 1.821 (15) | 2 | 2 | 2 |

*References*: 1. (Kl 79b), 2. (Th 81), 3. (Le 78).
*Exp. method*: 1: Standard collinear fast-beam laser spectroscopy.
2: LIOP.
*Transitions*: 1: RbI D2′ line (420.3 nm).
2: RbI D2 line (780.0 nm).
$\mu_I$*-eval.*: reference isotope $^{87}$Rb.
$Q_s$*-eval.*: HS calculation from (Ro 72), Sternheimer corrected.
$\delta\langle r^2\rangle$*-eval.*: semiempirical approach, $F_{780} = -650$ MHz/fm$^2$, $S_{780} = 0$ Ghz amu.

**Table 8. Element: Strontium, $Z=38$**

| $A$ | $\tau_{1/2}$ | $I$ | $\mu_I$ (n.m.) | $Q_s$ (b) | $\delta\langle r^2\rangle^{A,88}$ (fm$^2$) | Ref. No. $\mu_I$ | $Q_s$ | $\delta\langle r^2\rangle$ |
|---|---|---|---|---|---|---|---|---|
| 78 | 30.6 m | 1 | | | 0.237 (8) | | | 2 (6) |
| 79 | 2.25 m | $\frac{3}{2}$ | −0.474 (2) | 0.744 (54) | 0.260 (6) | 8 | 8 | 2 |
| 80 | 106.3 m | 0 | | | 0.249 (4) | | | 2 (3, 5, 6) |
| 81 | 22.2 m | $\frac{1}{2}$ | 0.5440 (4) | | 0.238 (4) | 8 (5) | | 2 (3, 5) |
| 82 | 25.6 d | 0 | | | 0.182 (4) | | | 2 (3, 5, 6) |
| 83 | 32.4 h | $\frac{7}{2}$ | −0.8289 (7) | 0.823 (50) | 0.163 (4) | 5 | 5 | 2 (3, 5) |
| 83$^m$ | 4.95 s | $\frac{1}{2}$ | 0.582 (1) | | | 8 | | |
| 84 | stable | 0 | | | 0.110 (2) | | | 1 (2, 3, 6) |
| 85 | 64.84 d | $\frac{9}{2}$ | −0.9994 (7) | 0.323 (20) | 0.047 (14) | 5 | 5 | 5 |
| 85$^m$ | 67.66 m | $\frac{1}{2}$ | 0.599 (2) | | 0.070 (9) | 5 | | 5 |
| 86 | stable | 0 | | | 0.047 (2) | | | 1 (2, 3, 6) |
| 87 | stable | $\frac{9}{2}$ | −1.093602 (1) | 0.335 (20) | 0.006 (2) | 4 | 7 | 1 (2, 3) |
| 87$^m$ | 2.81 h | $\frac{1}{2}$ | 0.6282 (6)* | | −0.001 (10) | 8 (5) | | 5 |
| 88 | stable | 0 | | | 0 | | | |
| 89 | 50.55 d | $\frac{5}{2}$ | −1.147 (2) | −0.274 (19) | 0.147 (8) | 5 | 8 (5) | 1 (2, 3) |
| 90 | 28.6 y | 0 | | | 0.277 (4) | | | 6 (1, 2, 3) |
| 91 | 9.52 h | $\frac{5}{2}$ | −0.8868 (6) | 0.044 (5) | | 8 | 8 | |
| 92 | 2.71 h | 0 | | | 0.512 (5) | | | 6 |
| 93 | 7.6 m | $\frac{5}{2}$ | −0.7942 (5) | 0.265 (19) | | 8 | 8 | |
| 94 | 75.1 s | 0 | | | 0.716 (6) | | | 6 |
| 95 | 25.1 s | $\frac{1}{2}$ | −0.5379 (4) | | | 8 | | |
| 96 | 1.06 s | 0 | | | 0.968 (6) | | | 6 |
| 97 | 0.42 s | $\frac{1}{2}$ | −0.500 (1) | | | 8 | | |
| 98 | 0.65 s | 0 | | | 1.624 (6) | | | 6 |
| 100 | 0.2 s | 0 | | | 1.830 (15) | | | 6 |

*References*: 1. (Bu 85), 2. (Ea 87), 3. (An 86), 4. (Le 78), 5. (An 87), 6. (Si 88), 7. (He 77), 8. (Bu 87b, Bu 88).
*Exp. method*: 1, 3, 5: atomic beam laser fluorescence spectroscopy.
2: collinear fast-beam laser coincidence spectroscopy (FACS).
6, 8: collinear fast-beam laser spectr. with state selective neutralization.
*Transitions*: 1, 2: $5s^2\ ^1S_0 \rightarrow 5s5p\ ^1P_1$ (460.7 nm).
3: $5s^2\ ^1S_0 \rightarrow 5s6p\ ^1P_1$ (293.2 nm).
6, 8: $5s\ ^2S_{1/2} \rightarrow 5p\ ^2P_{1/2,3/2}$ (421.7 nm, 407.9 nm).
*$\mu_I$-eval.*: reference isotope $^{87}$Sr.
*$Q_s$-eval.*: reference isotope $^{87}$Sr, Sternheimer corrected.
*$\delta\langle r^2\rangle$-eval.*: semiempirical approach from (Be 84d). $F_{460} = -896\ (130)$ MHz/fm$^2$, $S_{460} = -0.12\ (8)$ N. $F_{407} = -1582\ (49)$ MHz/fm$^2$, $S_{407} = -0.12\ (13)$ N.
*Remark*: evaluation of IS of neutron-rich odd-*A* isotopes under progress.
* Ref. 5 quotes $\mu_I$ (Sr$^{87m}$) = 0.787 (9) n.m. instead.

## Table 9. Element: Silver, $Z=47$

| $A$ | $\tau_{1/2}$ | $I$ | $\mu_I$ (n.m.) | $Q_s$ (b) | $\delta\langle r^2\rangle^{A,109}$ (fm$^2$) | Ref. No. $\mu_I$ | Ref. No. $Q_s$ | Ref. No. $\delta\langle r^2\rangle$ |
|---|---|---|---|---|---|---|---|---|
| 101 | 10.8 m | $\frac{9}{2}$ | 5.627 (11) | 0.33 (5) | −0.676 (3) | 1 (2) | 1 | 1 |
| 103 | 1.1 h | $\frac{7}{2}$ | 4.432 (2) | 0.80 (9) | −0.486 (2) | 1 (7) | 1 | 1 |
| 104 | 69 m | 5 | 3.914 (3) | 1.02 (11) | −0.419 (2) | 3 (1) | 1 | 1 |
| 105 | 41.3 d | $\frac{1}{2}$ | −0.1014 (10) | | −0.298 (2) | 7 | | 1 |
| 105$^m$ | 7.2 m | $\frac{7}{2}$ | 4.414 (13) | 0.82 (11) | −0.324 (2) | 1 | 1 | 1 |
| 106$^m$ | 8.5 d | 6 | 3.709 (4) | 1.11 (11) | −0.274 (4) | 4 | 5 (1) | 1 |
| 107 | stable | $\frac{1}{2}$ | −0.1136796 (2) | | −0.149 (1) | 7 | | 1 (6) |
| 108$^m$ | 127 y | 6 | 3.580 (20) | 1.32 (7) | −0.122 (12) | 7 | 5 (6) | 6 |
| 109 | stable | $\frac{1}{2}$ | −0.1306905 (2) | | 0 | 7 | | |
| 110$^m$ | 252 d | 6 | 3.607 (4) | 1.44 (10) | 0.029 (2) | 7 | 5 (6) | 6 |

*References*: 1. (Di 88), 2. (Va 83), 3. (Va 86), 4. (Ed 84), 5. (Be 84e), 6. (Fi 75), 7. (Le 78).
*Exp. method*: 1: standard collinear fast-beam laser spectroscopy.
6: classical optical spectroscopy.
*Transitions*: 1: AgI $4d^9 5s^2\ ^2D_{5/2} \rightarrow 4d^{10}\ 6p\ ^2P_{3/2}$ (547.7 nm).
6: AgI D1, D2 line (328 nm, 338 nm).
$\mu_I$-*eval*: reference isotope $^{109}$Ag.
$Q_s$-*eval.*: semiempirical and *ab-initio* calculations according to Ref. 1., Sternheimer corrected.
$\delta\langle r^2\rangle$-*eval.*: see the extensive discussion in Ref. 1., $F_{547} = -12.070\ (966)$ GHz/fm$^2$, $S_{547} = 14\ (9)$ N.

## Table 10. Element: Cadmium, $Z=48$

| $A$ | $\tau_{1/2}$ | $I$ | $\mu_I$ (n.m.) | $Q_s$ (b) | $\delta\langle r^2\rangle^{A,114}$ (fm$^2$) | Ref. No $\mu_I$ | Ref. No $Q_s$ | Ref. No $\delta\langle r^2\rangle$ |
|---|---|---|---|---|---|---|---|---|
| 102 | 5.5 m | 0 | | | −0.982 (43) | | | 1 |
| 103 | 7.3 m | $\frac{5}{2}$ | −0.81 (3) | −0.79 (66) | −0.876 (41) | 1 | 1 | 1 |
| 104 | 58 m | 0 | | | −0.745 (47) | | | 1 |
| 105 | 56.0 m | $\frac{5}{2}$ | −0.7393 (2) | 0.43 (4) | −0.673 (34) | 1 | 1 | 1 |
| 106 | stable | 0 | | | −0.5345 (2) | | | 1 |
| 107 | 6.50 h | $\frac{5}{2}$ | −0.615055 (1) | 0.68 (7) | −0.481 (29) | 2 | 2 | 1 |
| 108 | stable | 0 | | | −0.3920 (2) | | | 1 |
| 109 | 453 d | $\frac{5}{2}$ | −0.827846 (2) | 0.69 (7) | −0.377 (40) | 2 | 2 | 1 |
| 110 | stable | 0 | | | −0.2555 (2) | | | 1 |
| 111 | stable | $\frac{1}{2}$ | −0.5948856 (9) | | −0.2348 (2) | 2 | | 1 |
| 111$^m$ | 48.6 m | $\frac{11}{2}$ | −1.1051 (4) | −0.85 (9) | −0.246 (100) | 2 | 2 | 1 |
| 112 | stable | 0 | | | −0.1206 (220) | | | 1 |
| 113 | stable | $\frac{1}{2}$ | 0.6223005 (9) | | −0.0989 (2) | 2 | | 1 |
| 113$^m$ | 14 y | $\frac{11}{2}$ | 1.087783 (2) | −0.71 (7) | −0.100 (68) | 2 | 2 | 1 |
| 114 | stable | 0 | | | 0 | | | |
| 115 | 53.4 h | $\frac{1}{2}$ | −0.648425 (1) | | 0.023 (45) | 2 | | 1 |
| 115$^m$ | 44.8 d | $\frac{11}{2}$ | −1.041034 (2) | −0.54 (5) | 0.030 (12) | 2 | 2 | 1 |
| 116 | stable | 0 | | | 0.0941 (2) | | | 1 |
| 118 | 50.3 m | 0 | | | 0.125 (22) | | | 1 |
| 120 | 50.8 s | 0 | | | 0.169 (42) | | | 1 |

*References*: 1. (Bu 87) and references cited therein, 2. (Le 78).
*Exp. method*: Laser-excited fluorescence in resonance cells.
*Transitions*: 1: CdI $5s^2\ ^1S_0 \rightarrow 5s5p\ ^3P_1$ (326.1 nm).
$\mu_I$-*eval.*: reference isotope $^{109}$Cd.
$Q_s$-*eval.*: see Ref. 2.
$\delta\langle r^2\rangle$-*eval.*: see extensive discussion in Ref. 1. $F_{326} = 3.91\ (46)$ GHz/fm$^2$, $S = 376\ (230)$ GHz amu.

## Table 11. Element: Indium, $Z=49$

| $A$ | $\tau_{1/2}$ | $I$ | $\mu_I$ (n.m.) | $Q_s$ (b) | $\delta\langle r^2\rangle^{A,115}$ ($fm^2$) | Ref. No. $\mu_I$ | Ref. No. $Q_s$ | Ref. No. $\delta\langle r^2\rangle$ |
|---|---|---|---|---|---|---|---|---|
| 104 | 1.7 m | (5) | 4.435 (22) | 0.659 (106) | −0.842 (15) | 4 | 4 | 4 |
| 105 | 4.9 m | $\frac{9}{2}$ | 5.675 (5) | 0.829 (50) | −0.733 (14) | 2 | 2 | 2 |
| 106 | 6.2 m | 7 | 4.916 (7) | 0.972 (61) | −0.678 (12) | 2 | 2 | 2 |
| 107 | 32.4 m | $\frac{9}{2}$ | 5.585 (8) | 0.807 (52) | −0.575 (10) | 1 | 1 | 1 |
| 108 | 58 m | 7 | 4.561 (3) | 1.005 (7) | −0.509 (5) | 3 | 3 | 3 |
| 108 | 40 m | 2 | 4.935 (5) | 0.467 (14) | −0.501 (5) | 3 | 3 | 3 |
| 109 | 4.2 h | $\frac{9}{2}$ | 5.538 (4) | 0.841 (27) | −0.410 (7) | 1 | 1 | 1 |
| 110 | 4.9 h | 7 | 4.713 (8) | 1.000 (22) | −0.361 (8) | 1 | 1 | 1 |
| 111 | 2.83 d | $\frac{9}{2}$ | 5.503 (7) | 0.804 (22) | −0.262 (5) | 1 | 1 | 1 |
| 112 | 20.9 m | 4 | 5.277 (5) | 0.714 (10) | −0.218 (6) | 4 | 4 | 4 |
| 112 | 14.4 m | 8 | 3.080 (32) | −0.093 (6) | | 4 | 4 | |
| 113 | stable | $\frac{9}{2}$ | 5.5289 (2) | 0.799 | −0.1280 (3) | 5 | 4 | 4 |
| 113 | 99.5 m | $\frac{1}{2}$ | −0.21074 (2) | | | 5 | | |
| 114 | 49.51 d | 5 | 4.653 (5) | 0.739 (12) | −0.087 (2) | 5 | 4 | 4 |
| 115 | stable | $\frac{9}{2}$ | 5.5408 (2) | 0.810 | 0 | 5 | | |
| $115^m$ | 4.49 h | $\frac{1}{2}$ | −0.24398 (5) | | 0.014 (3) | 5 | | 4 |
| 116 | 54.1 m | 5 | 4.235 (15) | 0.802 (12) | 0.048 (1) | 4 | 4 | 4 |
| $116^m$ | 2.18 s | 8 | 3.215 (11) | 0.310 (9) | 0.043 (3) | 4 | 4 | 4 |
| 117 | 43.8 m | $\frac{9}{2}$ | 5.519 (4) | 0.829 (10) | 0.119 (4) | 4 | 4 | 4 |
| 117 | 1.93 h | $\frac{1}{2}$ | −0.25174 (3) | | 0.125 (6) | 4 | | 4 |
| 118 | 4.45 m | 5 | 4.231 (9) | 0.796 (8) | 0.157 (2) | 4 | 4 | 4 |
| 118 | 8.5 s | 8 | 3.321 (11) | 0.441 (7) | 0.153 (2) | 4 | 4 | 4 |
| 119 | 2.4 m | $\frac{9}{2}$ | 5.515 (10) | 0.854 (7) | 0.220 (4) | 4 | 4 | 4 |
| 119 | 18.0 m | $\frac{1}{2}$ | −0.319 (5) | | 0.222 (3) | 4 | | 4 |
| 120 | 44.4 s | 5 | 4.295 (5) | 0.810 (17) | 0.252 (2) | 4 | 4 | 4 |
| 120 | 47.3 s | 8 | 3.692 (4) | 0.530 (10) | 0.243 (6) | 4 | 4 | 4 |
| 121 | 23.1 s | $\frac{9}{2}$ | 5.502 (5) | 0.814 (11) | 0.306 (3) | 4 | 4 | 4 |
| 121 | 3.8 m | $\frac{1}{2}$ | −0.355 (4) | | 0.305 (3) | 4 | | 4 |
| 122 | 9.2 s | 5 | 4.318 (5) | 0.806 (20) | 0.332 (4) | 4 | 4 | 4 |
| 122 | 10.5 s | 8 | 3.781 (6) | 0.592 (21) | 0.330 (5) | 4 | 4 | 4 |
| 123 | 6.0 s | $\frac{9}{2}$ | 5.491 (7) | 0.757 (9) | 0.385 (3) | 4 | 4 | 4 |
| 123 | 48 s | $\frac{1}{2}$ | −0.400 (4) | | 0.379 (4) | 4 | | 4 |
| 124 | 3.2 s | 3 | 4.043 (11) | 0.614 (68) | 0.412 (6) | 4 | 4 | 4 |
| 124 | 2.4 s | 8 | 3.888 (9) | 0.664 (9) | 0.405 (8) | 4 | 4 | 4 |
| 125 | 2.32 s | $\frac{9}{2}$ | 5.502 (9) | 0.710 (36) | 0.452 (5) | 4 | 4 | 4 |
| 125 | 12.2 s | $\frac{1}{2}$ | −0.433 (4) | | 0.446 (5) | 4 | | 4 |
| 126 | 1.53 s | 3 | 4.034 (11) | 0.494 (49) | 0.480 (7) | 4 | 4 | 4 |
| 126 | 1.45 s | 8 | 4.061 (4) | 0.683 (12) | 0.480 (5) | 4 | 4 | 4 |
| 127 | 1.15 s | $\frac{9}{2}$ | 5.522 (8) | 0.589 (29) | 0.508 (7) | 4 | 4 | 4 |

*References*: 1. (Ul 85), 2. (Lo 85), 3. (Eb 86), 4. (Eb 87b) and references cited therein for some $\mu_I$ and $Q_s$ values, 5. (Le 78).
*Exp. method*: 1–4: Standard collinear fast-beam laser spectroscopy.
*Transitions*: 1–4: InI $5p\,^2P_{1/2} \rightarrow 6s\,^2S_{1/2}$ line (410 nm)
and $5p\,^2P_{3/2} \rightarrow 6s\,^2S_{1/2}$ line (451 nm).
*$\mu_I$-eval.*: reference isotope $^{115}$In.
*$Q_s$-eval.*: reference isotope $^{115}$In as discussed in Ref. 1, Sternheimer corrected.
*$\delta\langle r^2\rangle$-eval.*: RCDF calculation from (Fr 85), $F=2.070\,(10)\ GHz/fm^2$.
$S$ from $K_\alpha$ X-ray and King plot as discussed in Refs. 1–4.

**Table 12. Element: Tin, $Z = 50$**

| $A$ | $\tau_{1/2}$ | $I$ | $\mu_I$ (n.m.) | $Q_s$ (b) | $\delta\langle r^2\rangle^{A,116}$ (fm$^2$) | Ref. No. $\mu_I$ | Ref. No. $Q_s$ | Ref. No. $\delta\langle r^2\rangle$ |
|---|---|---|---|---|---|---|---|---|
| 108 | 10.3 m | 0 | | | −0.578 (4) | | | 2 |
| 109 | 18.0 m | $\frac{5}{2}$ | −1.079 (6) | 0.31 (10) | −0.506 (4) | 2 | 2 | 2 |
| 110 | 4.1 h | 0 | | | −0.417 (3) | | | 2 (1) |
| 111 | 35 m | $\frac{7}{2}$ | 0.608 (4) | 0.18 (9) | −0.378 (3) | 2 | 2 | 2 (1) |
| 112 | stable | 0 | | | −0.269 (2) | | | 2 (1) |
| 113 | 115.1 d | $\frac{1}{2}$ | −0.8791 (6) | | −0.212 (1) | 1 | | 2 (1) |
| 114 | stable | 0 | | | −0.136 (3) | | | 2 (1) |
| 115 | stable | $\frac{1}{2}$ | −0.91883 (7) | | −0.096 (1) | 3 | | 2 (1) |
| 116 | stable | 0 | | | 0 | | | |
| 117 | stable | $\frac{1}{2}$ | −1.00104 (7) | | 0.044 (1) | 3 | | 2 (1) |
| 117$^m$ | 14.0 d | $\frac{11}{2}$ | −1.3955 (10) | −0.42 (5) | 0.045 (7) | 1 | 1 | 2 (1) |
| 118 | stable | 0 | | | 0.128 (1) | | | 2 (1) |
| 119 | stable | $\frac{1}{2}$ | −1.04728 (7) | | 0.165 (1) | 3 | | 2 (1) |
| 120 | stable | 0 | | | 0.241 (2) | | | 2 (1) |
| 121 | 27.1 h | $\frac{3}{2}$ | 0.6978 (10) | −0.02 (2) | 0.281 (1) | 1 | 1 | 2 (1) |
| 121$^m$ | 55 y | $\frac{11}{2}$ | −1.3877 (9) | −0.14 (3) | 0.279 (3) | 1 | 1 | 1 (1) |
| 122 | stable | 0 | | | 0.342 (3) | | | 2 (1) |
| 123 | 129 d | $\frac{11}{2}$ | −1.3700 (9) | 0.03 (4) | 0.367 (1) | 1 | 1 | 2 (1) |
| 124 | stable | 0 | | | 0.428 (4) | | | 2 (1) |
| 125 | 9.6 d | $\frac{11}{2}$ | −1.348 (2) | 0.09 (17) | 0.455 (3) | 1 | 1 | 2 (1) |

*References*: 1. (An 86b), 2. (Eb 87), 3. (Le 78).
*Exp. method*: 1: Atomic-beam laser fluorescence spectroscopy.
2: Standard collinear fast-beam laser spectroscopy.
*Transitions*: 1: SnI $5p^2\,{}^3P_0 \rightarrow 5p6s\,{}^3P_1$ (286.3 nm).
2: SnI $5p^2\,{}^1S_0 \rightarrow 5p6s\,{}^{1,3}P_1$ (452.5 nm and 563.2 nm).
$\mu_I$-*eval.*: reference isotopes ${}^{115,117,119}$Sn.
$Q_s$-*eval.*: semiempirical calculation from (De 84) and semiempirical calculation in Ref. 2, not Sternheimer corrected.
$\delta\langle r^2\rangle$-*eval.*: 1: semiempirical approach from (Ba 83):
$F_{286} = 3.3\,(5)$ GHz/fm$^2$, $S_{286} = -1.2$ N.
2: RDF calculation and $K_\alpha$ X-rays according to Ref. 2:
$F_{452} = 2.24\,(27)$ GHz/fm$^2$, $S_{452} = -1.02$ N.

**Table 13. Element: Xenon, $Z=54$**

| $A$ | $\tau_{1/2}$ | $I$ | $\mu_I$ (n.m.) | $Q_s$ (b) | $\delta\langle r^2\rangle^{A,136}$ (fm$^2$) | Ref. No. $\mu_I$ | Ref. No. $Q_s$ | Ref. No. $\delta\langle r^2\rangle$ |
|---|---|---|---|---|---|---|---|---|
| 116 | 56 s | 0 | | | −0.599 (9) | | | 1 |
| 118 | 4 m | 0 | | | −0.460 (7) | | | 1 |
| 120 | 40 m | 0 | | | −0.363 (7) | | | 1 |
| 122 | 20.1 h | 0 | | | −0.299 (6) | | | 1 |
| 124 | stable | 0 | | | −0.242 (5) | | | 1 |
| 126 | stable | 0 | | | −0.193 (7) | | | 1 |
| 128 | stable | 0 | | | −0.152 (4) | | | 1 |
| 129 | stable | $\frac{1}{2}$ | −0.77976 (9) | | −0.151 (1) | 2 | | 1 |
| 130 | stable | 0 | | | −0.117 (3) | | | 1 |
| 131 | stable | $\frac{3}{2}$ | 0.691861 (4) | −0.120 (12) | −0.125 (1) | 2 | 2 | 1 |
| 132 | stable | 0 | | | −0.0844 (17) | | | 1 |
| 134 | stable | 0 | | | −0.0518 (13) | | | 1 |
| 136 | stable | 0 | | | 0 | | | |
| 138 | 14.08 m | 0 | | | 0.254 (3) | | | 1 |
| 140 | 13.60 s | 0 | | | 0.486 (2) | | | 1 |
| 141 | 1.73 s | $\frac{5}{2}$ | 0.010 (3) | −0.60 (6) | 0.590 (2) | 1 | 1 | 1 |
| 142 | 1.22 s | 0 | | | 0.710 (9) | | | 1 |
| 143 | 0.30 s | $\frac{5}{2}$ | −0.460 (2) | 0.96 (10) | 0.792 (3) | 1 | 1 | 1 |
| 144 | 1.15 s | 0 | | | 0.908 (5) | | | 1 |
| 146 | ? | 0 | | | 1.100 (5) | | | 1 |

*References*: 1. (Bo 88), 2. (Le 78).
*Exp. method*: 1. Collinear fast-beam laser spectroscopy with collisional ionization.
*Transitions*: 1: XeI $5p^5 6s\,[\frac{3}{2}]_2 \rightarrow 5p^5 6p\,[\frac{3}{2}]_2$ (823.2 nm).
$\mu_I$-*eval.*: reference isotope $^{131}$Xe.
$Q_s$-*eval.*: reference isotope $^{131}$Xe, not Sternheimer corrected.
$\delta\langle r^2\rangle$-*eval.*: semiempirical approach, $F_{823} = -2.32$ GHz/fm$^2$, $S = 0\,(200)$ GHz amu.
*Comment*: Ref. 1 will also include missing data on odd-$A$ isotopes, currently under evaluation.

## Table 14. Element: Cesium, $Z=55$

| $A$ | $\tau_{1/2}$ | $I$ | $\mu_I$ (n.m.) | $Q_s$ (b) | $\delta\langle r^2\rangle^{A,133}$ (fm$^2$) | Ref. No. $\mu_I$ | $Q_s$ | $\delta\langle r^2\rangle$ |
|---|---|---|---|---|---|---|---|---|
| 118 | 16.4 s | 2 | 3.876 (5) | 1.4 (2) | −0.197 (2) | 5 | 5 | 5 |
| 119 | 37.7 s | $\frac{9}{2}$ | 5.46 (3) | 2.8 (1) | −0.136 (6) | 5 | 5 | 5 |
| $119^m$ | 28 s | $\frac{3}{2}$ | 0.838 (5) | 0.9 (1) | −0.334 (4) | 5 | 5 | 5 |
| 120 | 64 s | 2 | 3.87 (2) | 1.45 (2) | −0.1185 (14) | 5 | 5 | 5 |
| 121 | 2.09 m | $\frac{3}{2}$ | 0.770 (4) | 0.838 (9) | −0.2555 (9) | 5 | 5 | 5 |
| $121^m$ | 2.9 m | $\frac{9}{2}$ | 5.41 (3) | 2.69 (5) | −0.0702 (8) | 5 | 5 | 5 |
| 122 | 21 s | 1 | −0.1333 (9) | −0.19 (1) | −0.2524 (15) | 4 | 4 | 4 |
| $122^m$ | 4.5 m | 8 | 4.77 (2) | 3.29 (8) | −0.0681 (18) | 4 | 4 | 4 |
| 123 | 5.8 m | $\frac{1}{2}$ | 1.377 (7) | | −0.2078 (6) | 4 | | 4 |
| 124 | 31 s | 1 | 0.673 (3) | −0.74 (3) | −0.2008 (12) | 4 | 4 | 4 |
| 125 | 45 m | $\frac{1}{2}$ | 1.409 (7) | | −0.1517 (6) | 4 | | 4 |
| 126 | 1.64 m | 1 | 0.777 (4) | −0.68 (2) | −0.1586 (9) | 4 | 4 | 4 |
| 127 | 6.25 h | $\frac{1}{2}$ | 1.459 (7) | | −0.0985 (7) | 4 | | 4 (7) |
| 128 | 3.62 m | 1 | 0.974 (5) | −0.570 (8) | −0.1131 (4) | 4 | 4 | 4 |
| 129 | 32.06 h | $\frac{1}{2}$ | 1.491 (8) | | −0.0561 (11) | 4 | | 4 (7) |
| 130 | 29.2 m | 1 | 1.460 (7) | −0.059 (6) | −0.0465 (10) | 4 | 4 | 4 |
| $130^m$ | 3.7 m | 5 | 0.629 (4) | 1.45 (5) | −0.0628 (14) | 4 | 4 | 4 |
| 131 | 9.688 d | $\frac{5}{2}$ | 3.543 (2) | −0.620 (6) | −0.0141 (7) | 6 (8) | 8 | 4 (7) |
| 132 | 6.47 d | 2 | 2.222 (7) | 0.508 (7) | −0.0369 (6) | 6 (8) | 8 | 4 (7) |
| 133 | stable | $\frac{7}{2}$ | 2.582023 (9) | −0.003 (1) | 0 | 6 (8) | 6 (8) | |
| 134 | 2.062 y | 4 | 2.9937 (9) | 0.389 (3) | −0.0096 (11) | 6 (8) | 8 | 4 (7) |
| $134^m$ | 2.90 h | 8 | 1.0978 (2) | 0.98 (8) | 0.0037 (16) | 4 | 4 | 4 |
| 135 | $3\times10^6$ y | $\frac{7}{2}$ | 2.7324 (2) | 0.050 (2) | 0.0250 (9) | 4 | 6 | 4 (7) |
| $135^m$ | 53 m | $\frac{19}{2}$ | 2.18 (1) | 0.89 (7) | 0.0169 (10) | 4 | 4 | 4 |
| 136 | 13.10 d | 5 | 3.711 (15) | 0.225 (10) | 0.0168 (14) | 4 (8) | 8 | 4 (7) |
| $136^m$ | 19 s | 8 | 1.319 (7) | 0.74 (10) | 0.0743 (15) | 4 | 4 | 4 |
| 137 | 30.17 y | $\frac{7}{2}$ | 2.8413 (4) | 0.051 (1) | 0.0821 (11) | 3 | 6 | 4 (1) |
| 138 | 32.2 m | 3 | 0.700 (4) | 0.125 (18) | 0.2023 (8) | 4 | 3 | 4 (1, 3) |
| $138^m$ | 2.9 m | 6 | 1.713 (9) | −0.40 (3) | 0.2022 (7) | 4 | 4 | 4 |
| 139 | 9.27 m | $\frac{7}{2}$ | 2.696 (4) | −0.075 (11) | 0.3604 (12) | 3 | 3 | 4 (1, 3) |
| 140 | 63.7 s | 1 | 0.134 (1) | −0.112 (7) | 0.4869 (13) | 4 | 3 | 4 (3) |
| 141 | 24.94 s | $\frac{7}{2}$ | 2.438 (10) | −0.36 (4) | 0.6159 (14) | 3 | 3 | 4 (3) |
| 142 | 1.8 s | 0 | | | 0.7454 (7) | | | 4 (3) |
| 143 | 1.78 s | $\frac{3}{2}$ | 0.870 (4) | 0.47 (3) | 0.8798 (5) | 4 | 4 | 4 |
| 144 | 1.00 s | 1 | −0.546 (3) | 0.30 (1) | 0.9669 (7) | 5 | 5 | 5 |
| 145 | 0.59 s | $\frac{3}{2}$ | 0.784 (4) | 0.62 (6) | 1.0953 (10) | 5 | 5 | 5 |
| 146 | 0.34 s | 1 | −0.515 (2) | 0.22 (3) | 1.185 (2) | 5 | 5 | 5 |

*References*: 1. (Sc 78b), 2. (Hu 78b), 3. (Bo 79), 4. (Th 81b), 5. (Co 87), 6. (Le 78), 7. (Ul 75), 8. (Ac 75).
*Exp. method*: 1, 3: Standard collinear fast-beam laser spectroscopy.
2, 4, 5: LIOP.
7, 8: Optical level crossing technique.
*Transitions*: 1, 3: CsI D2′ line (455.5 nm).
2, 4, 5, 7, 8: CsI D2 line (852.1 nm).
$\mu_I$-*eval.*: reference isotope $^{133}$Cs.
$Q_s$-*eval.*: 1+3, 8: $\langle r^{-3}\rangle_{6p}$ from semiempirical calculation in Ref. 8.
2, 4, 5: HS calculation from (Ro 72).
Sternheimer corrected in both cases.
$\delta\langle r^2\rangle$-*eval.*: semiempirical calculation in Ref. 2.
$F_{852}=-2313$ MHz/fm$^2$, $S=0$ GHz amu.

**Table 15. Element, Barium, $Z = 56$**

| $A$ | $\tau_{1/2}$ | $I$ | $\mu_I$ (n.m.) | $Q_s$ (b) | $\delta\langle r^2\rangle^{A,138}$ (fm$^2$) | Ref. No $\mu_I$ | Ref. No $Q_s$ | Ref. No $\delta\langle r^2\rangle$ |
|---|---|---|---|---|---|---|---|---|
| 120 | 32 s | 0 | | | −0.262 (2) | | | 6 |
| 121 | 29.7 s | $\frac{5}{2}$ | 0.659 (2) | 1.78 (21) | −0.184 (2) | 6 | 6 | 6 |
| 122 | 2.0 m | 0 | | | −0.204 (3) | | | 2 |
| 123 | 2.7 m | $\frac{5}{2}$ | −0.687 (18) | 1.52 (13) | −0.220 (3) | 2 | 2 | 2 |
| 124 | 11.9 m | 0 | | | −0.175 (2) | | | 3 (2) |
| 125 | 3.5 m | $\frac{1}{2}$ | 0.177 (12) | | −0.182 (2) | 2 | | 2 |
| 126 | 100 m | 0 | | | −0.142 (2) | | | 1 (2) |
| 127 | 13 m | $\frac{1}{2}$ | 0.089 (12) | | −0.158 (2) | 2 | | 2 |
| 128 | 2.43 d | 0 | | | −0.111 (1) | | | 1 (2) |
| 129 | 2.23 h | $\frac{1}{2}$ | −0.298 (16) | | −0.118 (1) | 1 | | 1 (2) |
| 129$^m$ | 2.17 h | $\frac{7}{2}$ | 0.930 (17) | 1.60 (13) | −0.130 (1) | 1 | 1 | 1 (2) |
| 130 | stable | 0 | | | −0.086 (1) | | | 1 (2) |
| 131 | 11.8 d | $\frac{1}{2}$ | −0.709 (16) | | −0.093 (1) | 1 | | 1 (2) |
| 131$^m$ | 14.6 m | $\frac{9}{2}$ | −0.870 (18) | 1.46 (13) | −0.096 (1) | 2 | 2 | 2 |
| 132 | stable | 0 | | | −0.068 (1) | | | 1 (2) |
| 133 | 10.74 y | $\frac{1}{2}$ | −0.769 (3) | | −0.084 (1) | 5 | | 1 (2) |
| 133$^m$ | 38.9 m | $\frac{11}{2}$ | −0.910 (51) | 0.89 (7) | −0.072 (1) | 1 | 1 | 1 (2) |
| 134 | stable | 0 | | | −0.049 (1) | | | 4 (1, 2) |
| 135 | stable | $\frac{3}{2}$ | 0.837943 (17) | 0.160 (3) | −0.074 (1) | 5 | 4 (2) | 4 (2) |
| 135$^m$ | 28.7 h | $\frac{11}{2}$ | −1.001 (15) | 0.96 (8) | −0.053 (1) | 1 | 1 | 4 (1, 2) |
| 136 | stable | 0 | | | −0.039 (1) | | | 4 (1, 2) |
| 137 | stable | $\frac{3}{2}$ | 0.937365 (20) | 0.245 (4) | −0.057 (1) | 5 | 4 (2) | 2 |
| 137$^m$ | 2.551 m | $\frac{11}{2}$ | −0.992 (26) | 0.78 (9) | 0.004 (1) | 2 | 2 | 2 |
| 138 | stable | 0 | | | 0 | | | |
| 139 | 84.63 m | $\frac{7}{2}$ | −0.973 (5) | −0.573 (13) | 0.123 (1) | 4 (2) | 4 (2) | 4 (2) |
| 140 | 12.75 d | 0 | | | 0.278 (1) | | | 4 (1, 2) |
| 141 | 18.27 m | $\frac{3}{2}$ | −0.337 (5) | 0.454 (10) | 0.391 (1) | 4 (2) | 4 (2) | 4 (2) |
| 142 | 10.6 m | 0 | | | 0.525 (2) | | | 4 (2) |
| 143 | 14.5 s | $\frac{5}{2}$ | 0.443 (11) | −0.879 (22) | 0.648 (2) | 4 (2) | 4 (2) | 4 (2) |
| 144 | 11.4 s | 0 | | | 0.786 (2) | | | 4 (2) |
| 145 | 4 s | $\frac{5}{2}$ | −0.285 (7) | 1.225 (21) | 0.886 (2) | 4 (2) | 4 (2) | 4 (2) |
| 146 | 2.2 s | 0 | | | 1.011 (2) | | | 4 (2) |
| 148 | 0.64 s | 0 | | | 1.245 (3) | | | 4 |

*References*: 1. (Be 79), 2. (Mu 83), 3. (Re 80), 4. (We 88), 5. (Le 78), 6. (Ea 88).
*Exp. method*: 1, 3: Atomic-beam laser fluorescence spectroscopy.
2, 4: Standard collinear fast-beam laser spectroscopy.
*Transitions*: 1, 3: $6s^2\ {}^1S_0 \to 6s6p\ {}^1P_1$ (553.6 nm).
4: BaII D2 line (455.4 nm).
*$\mu_I$-eval.*: reference isotopes $^{135,137}$Ba from (Ol 72).
*$Q_s$-eval.*: *ab initio* calculation from (He 85).
*$\delta\langle r^2\rangle$-eval.*: 1–3: semiempirical approach, $F_{554} = -3929$ MHz/fm$^2$,
$S = 0$ GHz amu, from King plot to BaII D1 [$F = -4980$ Mhz/fm$^2$ (Fi 74b)]
4: semiempirical approach, $F_{455} = -5120$ MHz/fm$^2$, $S = 254$ (160) GHz amu.

**Table 16. Element: Neodymium, $Z = 60$**

| $A$ | $\tau_{1/2}$ | $I$ | $\mu_I$ (n.m.) | $Q_s$ (b) | $\delta\langle r^2\rangle^{A,142}$ ($fm^2$) | Ref. No. $\mu_I$ | $Q_s$ | $\delta\langle r^2\rangle$ |
|---|---|---|---|---|---|---|---|---|
| 132 | 1.8 m | 0 | | | 0.047 (28) | | | 1 |
| 134 | 8.5 m | 0 | | | 0.005 (20) | | | 1 |
| 135 | 12 m | $\frac{9}{2}$ | −0.782 (28) | 2.05 (41) | −0.035 (31) | 1 | 1 | 1 |
| 136 | 50.65 m | 0 | | | −0.011 (25) | | | 1 |
| 137 | 38.5 m | $\frac{1}{2}$ | −0.632 (5) | | −0.040 (15) | 1 | | 1 |
| 138 | 5.04 h | 0 | | | −0.006 (20) | | | 1 |
| 139 | 29.7 m | $\frac{3}{2}$ | 0.905 (7) | 0.30 (9) | −0.044 (13) | 1 | 1 | 1 |
| 140 | 3.37 d | 0 | | | −0.021 (25) | | | 1 |
| 141 | 2.49 h | $\frac{3}{2}$ | 1.011 (9) | 0.34 (13) | −0.062 (13) | 1 | 1 | 1 |
| 142 | stable | 0 | | | 0 | | | |
| 143 | stable | $\frac{7}{2}$ | −1.065 (5) | −0.56 (6) | 0.122 (5) | 2 | 3 | 4 |
| 144 | stable | 0 | | | 0.278 (3) | | | 4 |
| 145 | stable | $\frac{7}{2}$ | −0.661 (3) | −0.33 (3) | 0.385 (5) | 2 | 3 | 4 |
| 146 | stable | 0 | | | 0.537 (4) | | | 4 |
| 148 | stable | 0 | | | 0.823 (4) | | | 4 |
| 150 | stable | 0 | | | 1.205 (5) | | | 4 |

*References*: 1. (Al 87), 2. (Le 78), 3. (Sm 65), 4. (Ki 73).
*Exp. method*: 1: Resonance ionization spectroscopy.
*Transitions*: 1: NdI 588.8 nm line.
$\mu_I$-*eval.*: reference isotope $^{143}$Nd.
$Q_s$-*eval.*: semiempirical calculation in Ref. 1, Sternheimer corrected.
$\delta\langle r^2\rangle$-*eval.*: see Ref. 4 and semiempirical calculation in Ref. 1, $F_{588} = -5.50$ GHz/fm$^2$, $(S+N)_{588} = -4550$ GHz amu.

**Table 17. Element: Samarium, $Z = 62$**

| $A$ | $\tau_{1/2}$ | $I$ | $\mu_I$ (n.m.) | $Q_s$ (b) | $\delta\langle r^2\rangle^{A,144}$ (fm$^2$) | Ref. No. $\mu_I$ | $Q_s$ | $\delta\langle r^2\rangle$ |
|---|---|---|---|---|---|---|---|---|
| 138 | 3.0 m | 0 | | | 0.073 (11) | | | 1 |
| 139 | 2.57 m | $\frac{1}{2}$ | −0.533 (17) | | 0.031 (11) | 1 | | 1 |
| 140 | 14.82 m | 0 | | | −0.040 (11) | | | 1 |
| 141 | 10.2 m | $\frac{1}{2}$ | −0.735 (23) | | −0.043 (18) | 1 | | 1 |
| $141^m$ | 22.6 m | $\frac{11}{2}$ | −0.83 (2) | 1.63 (44) | 0.007 (11) | 1 | 1 | 1 |
| 142 | 72.49 m | 0 | | | −0.006 (9) | | | 1 |
| 143 | 8.83 m | $\frac{3}{2}$ | 1.01 (2) | 0.41 (21) | −0.043 (11) | 1 | 1 | 1 |
| 144 | stable | 0 | | | 0 | | | |
| 145 | 340 d | $\frac{7}{2}$ | −1.11 (6) | −0.59 (17) | 0.123 (7) | 1 | 1 | 1 |
| 146 | $10^8$ y | 0 | | | 0.266 (10) | | | 2 |
| 147 | stable | $\frac{7}{2}$ | −0.8148 (7) | −0.259 (26) | 0.365 (28) | 3 | 4 | 5 |
| 148 | stable | 0 | | | 0.517 (27) | | | 5 |
| 149 | stable | $\frac{7}{2}$ | −0.6717 (7) | 0.075 (8) | 0.609 (29) | 3 | 6 | 5 |
| 150 | stable | 0 | | | 0.820 (31) | | | 5 |
| 151 | 90 y | $\frac{5}{2}$ | −0.362 (2) | 0.65 (14) | 0.981 (40) | 1 | 1 | 2 |
| 152 | stable | 0 | | | 1.243 (38) | | | 5 |
| 153 | 46.7 h | $\frac{3}{2}$ | −0.026 (14) | 1.27 (13) | 1.339 (2) | 2 | 2 | 2 |
| 154 | stable | 0 | | | 1.473 (40) | | | 5 |

*References*: 1. (Al 87), 2. (Ea 84), 3. (Le 78), 4. (Ba 81), 5. (Br 80), 6. (Ch 72).
*Exp. method*: 1: Resonance ionization spectroscopy.
2: Atomic-beam laser fluorescence spectroscopy.
*Transitions*: 1: SmI 600.4 nm.
2: SmI 570.7 nm.
$\mu_I$-*eval.*: reference isotope $^{147}$Sm.
$Q_s$-*eval.*: reference isotope $^{147}$Er from Ref. 4.
$\delta\langle r^2\rangle$-*eval.*: 1: according to Ref. 5 via King plot, $F_{600} = -5.5$ GHz/fm$^2$, $(S+N)_{600} = 100$ GHz amu.

**Table 18. Element: Europium, $Z = 63$**

| $A$ | $\tau_{1/2}$ | $I$ | $\mu_I$ (n.m.) | $Q_s$ (b) | $\delta\langle r^2\rangle^{A,145}$ (fm$^2$) | Ref. No. $\mu_I$ | $Q_s$ | $\delta\langle r^2\rangle$ |
|---|---|---|---|---|---|---|---|---|
| 138 | 35 s | | | | 0.166 (57) | | | 8 |
| 139 | 22 s | | | | 0.109 (33) | | | 8 |
| 140 | 1.3 s | 1 | 1.365 (13) | 0.31 (4) | 0.033 (12) | 5 | 5 | 5 |
| 141 | 40 s | $\frac{5}{2}$ | 3.494 (8) | 0.85 (4) | 0.035 (9) | 5 | 5 | 5 (2) |
| 142 | 2.4 s | 1 | 1.536 (19) | 0.12 (5) | −0.054 (10) | 5 | 5 | 5 (2) |
| $142^m$ | 73 s | 8 | 2.978 (11) | 1.41 (6) | 0.003 (6) | 5 | 5 | 5 |
| 143 | 2.6 m | $\frac{5}{2}$ | 3.673 (8) | 0.51 (3) | −0.026 (5) | 5 | 5 | 5 (2) |
| 144 | 10 s | 1 | 1.893 (13) | 0.10 (3) | −0.050 (6) | 5 | 5 | 5 (2) |
| 145 | 5.9 d | $\frac{5}{2}$ | 3.993 (7) | 0.29 (2) | 0 | 5 | 5 | 5 (1, 2) |
| 146 | 4.6 d | 4 | 1.425 (11) | −0.18 (6) | 0.124 (9) | 5 | 5 | 5 (1, 2) |
| 147 | 24.6 d | $\frac{5}{2}$ | 3.724 (8) | 0.55 (3) | 0.271 (17) | 5 | 5 | 5 (1, 2, 4) |
| 148 | 54.5 d | 5 | 2.340 (10) | 0.35 (6) | 0.376 (24) | 5 | 5 | 5 (1, 2) |
| 149 | 93 d | $\frac{5}{2}$ | 3.565 (6) | 0.75 (2) | 0.532 (33) | 5 | 5 | 5 (1, 2, 4) |
| 150 | 35.8 y | 5 | 2.708 (11) | 1.13 (5) | 0.625 (39) | 5 | 5 | 5 (2) |
| $150^m$ | 12.6 h | 0 | | | 0.634 (40) | | | 5 |
| 151 | stable | $\frac{5}{2}$ | 3.4717 (6) | 0.903 (10) | 0.851 (51) | 6 | 7 | 5 (2, 3, 4) |
| 152 | 13.33 y | 3 | −1.9414 (13) | 2.54 (22) | 1.396 (81) | 6 | 5 | 5 (2, 3, 4) |
| $152^m$ | 9.30 h | 0 | | | 1.112 (53) | | | 4 |
| 153 | stable | $\frac{5}{2}$ | 1.5330 (8) | 2.412 (21) | 1.453 (83) | 6 | 7 | 4 (3) |
| 154 | 8.8 y | 3 | −2.00 | 3.04 | 1.540 (64) | 4 | 4 | 4 (3) |
| 155 | 4.96 y | $\frac{5}{2}$ | 1.53 | 2.51 | 1.528 (61) | 4 | 4 | 4 |
| 156 | 15.19 d | 0 | | | 1.565 (64) | | | 4 |

*References*: 1. (Al 83), 2. (Fe 84), 3. (Dö 83), 4. (Dö 84), 5. (Ah 85), 6. (Le 78), 7. (Ta 83), 8. (Al 87).
*Exp. method*: 1–2, 8: Resonance ionization spectroscopy.
3–5: Standard collinear fast-beam laser spectroscopy.
*Transitions*: 1–2, 8: EuI 576.5 nm line.
3–4: EuII 604.9 nm line.
5: EuI 459.4 nm and 462.7 nm line.
$\mu_I$-*eval.*: reference isotope $^{151}$Eu.
$Q_s$-*eval.*: reference isotope $^{153}$Eu from Ref. 7.
$\delta\langle r^2\rangle$-*eval.*: 4, 8: semiempirical approach according to (He 74).
5: semiempirical approach, $F_{ss-sp} = -6.55$ GHz/fm$^2$, $S = 0$ GHz amu.

**Table 19. Element: Dysprosium, $Z = 66$**

| $A$ | $\tau_{1/2}$ | $I$ | $\mu_I$ (n.m.) | $Q_s$ (b) | $\delta\langle r^2\rangle^{A,165}$ (fm$^2$) | Ref. No. $\mu_I$ | Ref. No. $Q_s$ | Ref. No. $\delta\langle r^2\rangle$ |
|---|---|---|---|---|---|---|---|---|
| 146 | 29 s | 0 | | | −0.017 (2) | | | 5 |
| 148 | 3.1 m | 0 | | | 0 | | | |
| 149 | 4.23 m | $\frac{7}{2}$ | −1.119 (9) | −0.62 (5) | 0.113 (12) | 1 | 1 | 5 |
| 150 | 7.17 m | 0 | | | 0.254 (25) | | | 5 |
| 151 | 16.9 m | $\frac{7}{2}$ | −0.945 (7) | −0.30 (5) | 0.350 (35) | 1 | 1 | 5 |
| 152 | 2.38 h | 0 | | | 0.502 (50) | | | 5 |
| 153 | 6.4 h | $\frac{7}{2}$ | −0.782 (6) | −0.02 (5) | 0.588 (59) | 1 | 1 | 5 |
| 154 | $10^7$ y | 0 | | | 0.799 (80) | | | 5 |
| 155 | 10.0 h | $\frac{3}{2}$ | −0.385 (4) | 1.04 (3) | 1.02 (10) | 1 | 1 | 5 |
| 156 | stable | 0 | | | 1.19 (12) | | | 5 |
| 157 | 8.1 h | $\frac{3}{2}$ | −0.301 (2) | 1.30 (2) | 1.28 (13) | 1 | 1 | 5 |
| 158 | stable | 0 | | | 1.39 (14) | | | 5 |
| 159 | 144.4 d | $\frac{3}{2}$ | −0.354 (3) | 1.37 (2) | 1.40 (14) | 1 | 1 | 5 |
| 160 | stable | 0 | | | 1.53 (15) | | | 5 |
| 161 | stable | $\frac{5}{2}$ | −0.4805 (51) | 2.51 (2) | 1.56 (16) | 2 | 3 | 5 |
| 162 | stable | 0 | | | 1.66 (17) | | | 5 |
| 163 | stable | $\frac{5}{2}$ | −0.6726 (35) | 2.648 (21) | 1.71 (17) | 2 | 4 | 5 |
| 164 | stable | 0 | | | 1.80 (18) | | | 5 |
| 165 | 2.334 h | $\frac{7}{2}$ | ±0.51 | 3.49 (10) | | 2 | 3 | |

*References*: 1. (Ah 88b), 2. (Le 78), 3. recalibration of (Le 78) with Ref. 4, 4. (Ta 83), 5. (Ne 88b).
*Exp. method*: 1, 5: Collinear fast-beam laser spectroscopy.
*Transitions*: 1: DyI 421.2 nm line.
$\mu_I$-*eval.*: reference isotope $^{163}$Dy.
$Q_s$-*eval.*: reference isotope $^{165}$Dy from muonic X-rays.
$\delta\langle r^2\rangle$-*eval.*: preliminary semiempirical evaluation, $F_{421} = -7.26$ GHz/fm$^2$, $S_{421} = 0$ GHz amu.

### Table 20. Element: Holmium, $Z = 67$

| $A$ | $\tau_{1/2}$ | $I$ | $\mu_I$ (n.m.) | $Q_s$ (b) | $\delta\langle r^2\rangle^{A,165}$ (fm$^2$) | Ref. No. $\mu_I$ | Ref. No. $Q_s$ | Ref. No. $\delta\langle r^2\rangle$ |
|---|---|---|---|---|---|---|---|---|
| 151$^m$ | 42 s | $\frac{1}{2}$ | 6.93 (2) | | −1.643 (9) | 3 | | 3 |
| 151 | 35.6 s | $\frac{11}{2}$ | 1.49 (1) | −0.73 (8) | −1.617 (8) | 3 | 3 | 3 |
| 152$^m$ | 52.3 s | 9 | 5.92 (5) | −1.3 (8) | −1.402 (6) | 1 (3) | 1 (3) | 1 (3) |
| 152 | 2.35 m | 2 | −1.02 (2) | 0.08 (24) | −1.406 (5) | 1 (3) | 1 (3) | 1 (3) |
| 153$^m$ | 2.0 m | $\frac{11}{2}$ | 6.79 (5) | −1.1 (5) | −1.252 (5) | 1 (3) | 1 (3) | 1 (3) |
| 153 | 9.3 m | $\frac{1}{2}$ | 1.19 (1) | | −1.271 (7) | 1 (3) | | 1 (3) |
| 154$^m$ | 3.2 m | 8 | 5.63 (6) | −1.0 (5) | −1.182 (5) | 1 (3) | 1 (3) | 1 (3) |
| 154 | 11.8 m | 2 | −0.641 (6) | 0.19 (10) | −1.167 (3) | 1 (3) | 1 (3) | 1 (3) |
| 155 | 48 m | $\frac{5}{2}$ | 3.50 (3) | 1.52 (12) | −0.949 (2) | 1 (3) | 1 (3) | 1 (3) |
| 156 | 2 m | 4 | 2.99 (2) | 2.43 (30) | −0.907 (5) | 3 | 3 | 3 |
| 156$^m$ | 55.6 m | 1 | | | −0.894 (5) | | | 3 |
| 157 | 12.6 m | $\frac{7}{2}$ | 4.34 (3) | 2.97 (13) | −0.490 (2) | 1 (3) | 1 (3) | 1 (3) |
| 158$^m$ | 27 m | 2 | 2.43 (3) | 1.62 (17) | −0.612 (3) | 1 (3) | 1 (3) | 1 (3) |
| 158 | 11.3 m | 5 | 3.76 (3) | 4.10 (40) | −0.454 (4) | 1 (3) | 1 (3) | 1 (3) |
| 159 | 33 m | $\frac{7}{2}$ | 4.27 (3) | 3.19 (13) | −0.350 (2) | 1 (3) | 1 (3) | 1 (3) |
| 160$^m$ | 5.02 h | 2 | 2.51 (3) | 1.78 (17) | −0.446 (5) | 1 (3) | 1 (3) | 1 (3) |
| 160 | 25.6 m | 5 | 3.70 (3) | 3.95 (23) | −0.363 (2) | 1 (3) | 1 (3) | 1 (3) |
| 161 | 2.48 h | $\frac{7}{2}$ | 4.24 (3) | 3.22 (11) | −0.240 (2) | 1 (3) | 1 (3) | 1 (3) |
| 162$^m$ | 15 m | 1 | 2.37 (5) | 0.54 (12) | −0.231 (5) | 3 | 3 | 3 |
| 162 | 67 m | 6 | 3.59 (4) | 3.9 (7) | −0.207 (7) | 1 (3) | 1 (3) | 1 (3) |
| 163 | 33 y | $\frac{7}{2}$ | 4.22 (4) | 3.6 (6) | −0.117 (6) | 1 (3) | 1 (3) | 1 (3) |
| 165 | stable | $\frac{7}{2}$ | 4.173 (27) | 3.49 (3) | 0 | 2 | 2 | |

*References*: 1. (Al 87), 2. (Le 78), 3. (Ne 88c).
*Exp. method*: 1: Resonance ionization spectroscopy.
3: Collinear fast-beam laser spectroscopy.
*Transitions*: 1: HoI 592.2 nm line.
3: HoI 410.1 nm line.
$\mu_I$-*eval.*: reference isotope $^{165}$Ho.
$Q_s$-*eval.*: reference isotope $^{165}$Ho, Sternheimer-corrected.
$\delta\langle r^2\rangle$-*eval.*: semiempirical calculation, $F_{592} = -8.41$ GHz/fm$^2$, $(S+N)_{592} = 282$ GHz amu.

## Table 21. Element: Erbium, $Z = 68$

| $A$ | $\tau_{1/2}$ | $I$ | $\mu_I$ (n.m.) | $Q_s$ (b) | $\delta\langle r^2\rangle^{A,166}$ (fm$^2$) | Ref. No. $\mu_I$ | Ref. No. $Q_s$ | Ref. No. $\delta\langle r^2\rangle$ |
|---|---|---|---|---|---|---|---|---|
| 150 | 18.5 s | 0 | | | 0 | | | |
| 152 | 10.3 s | 0 | | | 0.253 (25) | | | 4 |
| 153 | 37.1 s | $\frac{7}{2}$ | −0.939 (7) | −0.42 (2) | | 1 | 1 | |
| 154 | 3.75 m | 0 | | | 0.499 (50) | | | 4 |
| 155 | 5.3 m | $\frac{7}{2}$ | −0.671 (5) | −0.27 (2) | | 1 | 1 | |
| 156 | 20 m | 0 | | | 0.760 (76) | | | 4 |
| 157 | 25 m | $\frac{3}{2}$ | −0.414 (3) | 0.92 (1) | | 1 | 1 | |
| 158 | 2.25 h | 0 | | | 1.05 (11) | | | 4 |
| 159 | 36 m | $\frac{3}{2}$ | −0.305 (2) | 1.17 (1) | | 1 | 1 | |
| 160 | 28.59 h | 0 | | | 1.30 (13) | | | 4 |
| 161 | 3.21 h | $\frac{3}{2}$ | −0.367 (3) | 1.35 (2) | | 1 | 1 | |
| 162 | stable | 0 | | | 1.48 (15) | | | 4 |
| 163 | 75.0 m | $\frac{5}{2}$ | 0.560 (4) | 2.56 (2) | | 1 | 1 | |
| 164 | stable | 0 | | | 1.63 (16) | | | 4 |
| 165 | 10.36 h | $\frac{5}{2}$ | 0.646 (4) | 2.71 (3) | | 1 | 1 | |
| 167 | stable | $\frac{7}{2}$ | −0.5665 (24) | 3.565 (29) | | 2 | 3 | |

*References*: 1. (Ah 85b) and (Ah 88b), 2. (Le 78), 3. (Ta 83), 4. (Ne 88b).
*Exp. method*: 1, 4: Collinear fast-beam laser spectroscopy.
*Transitions*: 1: ErI 415.1 nm, 582.6 nm, and 440.9 nm line.
$\mu_I$-*eval.*: reference isotope $^{167}$Er.
$Q_s$-*eval.*: reference isotope $^{167}$Er from Ref. 3.
$\delta\langle r^2\rangle$-*eval.*: preliminary semiempirical evaluation, $F_{583} = -8.08$ GHz/fm$^2$, $S = 0$ GHz amu.

## Table 22. Element: Thulium, $Z = 69$

| $A$ | $\tau_{1/2}$ | $I$ | $\mu_I$ (n.m.) | $Q_s$ (b) | $\delta\langle r^2\rangle^{A,169}$ (fm$^2$) | Ref. No. $\mu_I$ | Ref. No. $Q_s$ | Ref. No. $\delta\langle r^2\rangle$ |
|---|---|---|---|---|---|---|---|---|
| 156 | 80 s | 2 | 0.40 (3) | −0.48 (11) | −1.284 (6) | 1 | 1 | 1 |
| 157 | 3.5 m | $\frac{1}{2}$ | 0.457 (15) | | −1.093 (8) | 1 | | 1 |
| 158 | 4.02 m | 2 | 0.042 (20) | 0.74 (15) | −1.002 (8) | 1 | | 1 |
| 159 | 9 m | $\frac{5}{2}$ | 3.41 (3) | 1.93 (7) | −0.850 (4) | 1 | 1 | 1 |
| 160 | 9.2 m | 1 | 0.156 (18) | 0.58 (4) | −0.741 (4) | 1 | 1 | 1 |
| 161 | 38 m | $\frac{7}{2}$ | 2.39 (2) | 2.90 (7) | −0.632 (3) | 1 | 1 | 1 |
| 162 | 22.0 m | 1 | 0.068 (8) | 0.69 (3) | −0.537 (5) | 1 | 1 | 1 |
| 163 | 1.81 h | $\frac{1}{2}$ | −0.082 (1) | | −0.404 (2) | 1 | | 1 |
| 164 | 2.0 m | 1 | 2.37 (2) | 0.71 (5) | −0.347 (6) | 1 | 1 | 1 |
| 165 | 30.06 h | $\frac{1}{2}$ | −0.139 (2) | | −0.250 (2) | 1 | | 1 |
| 166 | 7.70 h | 2 | 0.092 (1) | 2.14 (3) | −0.209 (3) | 1 | 1 | 1 |
| 167 | 9.24 d | $\frac{1}{2}$ | −0.197 (2) | | −0.126 (4) | 1 | | 1 |
| 168 | 93.1 d | 3 | 0.226 (11) | 3.23 (7) | −0.084 (5) | 1 | 1 | 1 |
| 169 | stable | $\frac{1}{2}$ | −0.231 (2) | | 0 | 2 | | |
| 170 | 28.6 d | 1 | 0.247 (4) | 0.74 (2) | 0.048 (5) | 1 | 1 | 1 |
| 171 | 1.92 y | $\frac{1}{2}$ | −0.230 (4) | | 0.131 (6) | 1 | | 1 |
| 172 | 2.65 d | 2 | 0.8 | | 0.154 (30) | 1 | | 1 |

*References*: 1. (Al 87), 2. (Le 78).
*Exp. method*: 1: Resonance ionization spectroscopy.
*Transitions*: 1: TmI 589.6 nm and TmI 597.1 nm.
$\mu_I$-*eval.*: reference isotope $^{169}$Tm.
$Q_s$-*eval.*: semiempirical evaluation in Ref. 1, Sternheimer corrected.
$\delta\langle r^2\rangle$-*eval.*: semiempirical evaluation from (He 74), $F_{597} = -10.3$ GHz/fm$^2$, $(S + N)_{597} = 268$ GHz amu.

**Table 23. Element: Ytterbium, $Z = 70$**

| $A$ | $\tau_{1/2}$ | $I$ | $\mu_I$ (n.m.) | $Q_s$ (b) | $\delta\langle r^2\rangle^{A,168}$ ($fm^2$) | Ref. No. $\mu_I$ | $Q_s$ | $\delta\langle r^2\rangle$ |
|---|---|---|---|---|---|---|---|---|
| 156 | 24 s | 0 | | | −1.160 (58) | | | 2 |
| 158 | 1.1 m | 0 | | | −0.9554 (58) | | | 2 |
| 159 | 4.6 m | $\frac{5}{2}$ | −0.366 (8) | −0.22 (2) | | 3 | 3 | |
| 160 | 4.8 m | 0 | | | −0.7319 (38) | | | 2 |
| 161 | 4.2 m | $\frac{3}{2}$ | −0.327 (8) | 1.03 (2) | −0.6472 (32) | 3 | 3 | 2 |
| 162 | 18.9 m | 0 | | | −0.5157 (28) | | | 2 |
| 163 | 11.0 m | $\frac{3}{2}$ | −0.374 (8) | 1.24 (2) | −0.4340 (23) | 3 | 3 | 2 |
| 164 | 76 m | 0 | | | −0.3135 (27) | | | 2 |
| 165 | 10 m | $\frac{5}{2}$ | 0.478 (8) | 2.48 (4) | −0.2409 (16) | 3 | 3 | 2 |
| 166 | 56.7 h | 0 | | | −0.1397 (10) | | | 2 |
| 167 | 17.5 m | $\frac{5}{2}$ | 0.623 (8) | 2.70 (4) | −0.0640 (5) | 3 | 3 | 2 |
| 168 | stable | 0 | | | 0 | | | |
| 169 | 32.0 d | $\frac{7}{2}$ | −0.635 (8) | 3.54 (6) | 0.0517 (14) | 3 | 3 | 2 |
| 170 | stable | 0 | | | 0.1173 (12) | | | 2 (1) |
| 171 | stable | $\frac{1}{2}$ | 0.49367 (1) | | 0.1573 (14) | 4 | | 2 (1) |
| 172 | stable | 0 | | | 0.2279 (19) | | | 2 (1) |
| 173 | stable | $\frac{5}{2}$ | −0.67989 (3) | 2.80 (4) | 0.2664 (23) | 4 | 4 | 2 (1) |
| 174 | stable | 0 | | | 0.3144 (27) | | | 2 (1) |
| 176 | stable | 0 | | | 0.3968 (34) | | | 2 (1) |

*References*: 1. (Cl 79), 2. (Bu 82) and (Ah 88b), 3. (Ne 83), 4. (Le 78).
*Exp. method*: 1: Atomic-beam laser fluorescence spectroscopy.
2, 3: Collinear fast-beam laser spectroscopy.
*Transitions*: 1: YbI $6s^2\ {}^1S_0 \rightarrow 6s6p\ {}^3P_1$ (555.6 nm).
2, 3: YbI $6s^2\ {}^1S_0 \rightarrow 6s6p\ {}^3P_1$ (555.6 nm) and YbI $6s6p\ {}^3P_2 \rightarrow 6s7{,}8s\ {}^3S_1$ (769.9 nm, 456.4 nm).
*$\mu_I$-eval.*: reference isotope ${}^{173}Yb$.
*$Q_s$-eval.*: reference isotope ${}^{173}Yb$.
*$\delta\langle r^2\rangle$-eval.*: preliminary semiempirical evaluation (He 85b) of data from (Bu 82) in agreement with recent evaluation from (Au 87), $F_{555} = -11.8\ GHz/fm^2$, $(N+S)_{555} = 296\ (148)$ GHz amu.

## Table 24. Element: Platinum, $Z=78$

| $A$ | $\tau_{1/2}$ | $I$ | $\mu_I$ (n.m.) | $Q_s$ (b) | $\delta\langle r^2\rangle^{A,190}$ ($fm^2$) | Ref. No. $\mu_I$ | $Q_s$ | $\delta\langle r^2\rangle$ |
|---|---|---|---|---|---|---|---|---|
| 186 | 2.0 h | 0 | | | −0.0134 (11) | | | 1 |
| 188 | 10.2 d | 0 | | | −0.0055 (7) | | | 1 |
| 190 | $10^{12}$ y | 0 | | | 0 | | | |
| 192 | stable | 0 | | | 0.0052 (7) | | | 1 |
| 194 | stable | 0 | | | 0.0098 (9) | | | 1 |
| 196 | stable | 0 | | | 0.0141 (9) | | | 1 |
| 198 | stable | 0 | | | 0.0181 (10) | | | 1 |

*References*: 1. (Sa 88).
*Exp. method*: 1: Resonance ionization spectroscopy.
*Transitions*: 1: PtI line (266 nm).
$\delta\langle r^2\rangle$*-eval.*: according to (Au 87): $F_{266}=-28$ GHz/fm$^2$, $S_{266}=0$ GHz amu.

## Table 25. Element: Gold, $Z=79$

| $A$ | $\tau_{1/2}$ | $I$ | $\mu_I$ (n.m.) | $Q_s$ (b) | $\delta\langle r^2\rangle^{A,197}$ ($fm^2$) | Ref. No. $\mu_I$ | $Q_s$ | $\delta\langle r^2\rangle$ |
|---|---|---|---|---|---|---|---|---|
| 185 | 4.3 m | $\frac{5}{2}$ | 2.170 (17) | | −0.075 (9) | 3 | | 4 |
| 186 | 10.7 m | 3 | −1.263 (29) | | −0.014 (8) | 3 | | 4 |
| 187 | 8.0 m | $\frac{1}{2}$ | 0.535 (15) | | −0.385 (9) | 3 | | 4 |
| 188 | 8.8 m | 1 | −0.067 (28) | | −0.347 (9) | 3 | | 4 |
| 189 | 28.7 m | $\frac{1}{2}$ | 0.494 (14) | | −0.309 (7) | 3 | | 4 |
| $189^m$ | 4.6 m | $\frac{11}{2}$ | 6.186 (20) | | −0.283 (9) | 3 | | 4 |
| 190 | 42.8 m | 1 | −0.068 (26) | | −0.285 (6) | 3 (2) | | 4 (2) |
| 191 | 3.18 h | $\frac{3}{2}$ | 0.140 (7) | | −0.242 (5) | 2 | | 4 (2) |
| 192 | 5.0 h | 1 | ±0.0081 (11) | −0.0076 (21) | −0.208 (5) | 2 | 6 | 4 (2) |
| 193 | 17.65 h | $\frac{3}{2}$ | 0.142 (7) | | −0.157 (4) | 2 | | 4 (2) |
| 194 | 39.5 h | 1 | 0.076 (4) | | −0.130 (4) | 4 | | 6 |
| 195 | 186 d | $\frac{3}{2}$ | 0.151 (7) | | −0.076 (6) | 2 | | 2 (1) |
| 196 | 6.18 d | 2 | 0.5914 (14) | | 0.043 (4) | 5 | | 6 |
| 197 | stable | $\frac{3}{2}$ | 0.148158 (8) | 0.594 (10) | 0 | 5 | 5 | |

*References*: 1. (Kl 83), 2. (St 85c), 3. (Wa 87), 4. (Wa 87b), 5. (Le 78), 6. (Le 87b).
*Exp. method*: 1, 2: Fluorescence laser spectroscopy in cells.
3, 4, 6: Resonance ionization spectroscopy.
*Transitions*: 1, 4: AuI D1 line (267.6 nm).
$\mu_I$*-eval.*: reference isotope $^{197}$Au.
$Q_s$*-eval.*: reference isotope $^{197}$Au.
$\delta\langle r^2\rangle$*-eval.*: MCDF calculation from (Ro 84), $F_{268}=-43.07$ GHz/fm$^2$, $S_{268}=0$ GHz amu.

## Table 26. Element: Mercury, $Z = 80$

| $A$ | $\tau_{1/2}$ | $I$ | $\mu_I$ (n.m.) | $Q_s$ (b) | $\delta\langle r^2\rangle^{A,198}$ (fm$^2$) | Ref. No. $\mu_I$ | Ref. No. $Q_s$ | Ref. No. $\delta\langle r^2\rangle$ |
|---|---|---|---|---|---|---|---|---|
| 181 | 3.6 s | $\frac{1}{2}$ | 0.5071 (7) | | −0.0953 (33) | 1 | | 3 (1) |
| 182 | 11 s | 0 | | | −0.6384 (20) | | | 3 |
| 183 | 8.8 s | $\frac{1}{2}$ | 0.524 (5) | | −0.0527 (16) | 1 | | 3 (1) |
| 184 | 31 s | 0 | | | −0.5337 (13) | | | 3 (6) |
| 185 | 50 s | $\frac{1}{2}$ | 0.509 (4) | | −0.0622 (6) | 3 (1) | | 3 (1) |
| 185$^m$ | 27 m | $\frac{13}{2}$ | −1.017 (9) | 0.20 (33) | −0.5357 (22) | 3 (2) | 3 (2) | 3 (2) |
| 186 | 1.4 m | 0 | | | −0.4643 (10) | | | 3 (6) |
| 187 | 2.4 m | $\frac{3}{2}$ | −0.594 (4) | −0.75 (25) | −0.4324 (40) | 3 (1) | 3 (1) | 3 (1) |
| 187$^m$ | 1.6 m | $\frac{13}{2}$ | −1.044 (11) | 0.45 (33) | −0.4630 (24) | 3 (2) | 3 (2) | 3 (2) |
| 188 | 3.3 m | 0 | | | −0.3944 (10) | | | 3 (6) |
| 189 | 7.7 m | $\frac{3}{2}$ | −0.6086 (8) | −0.76 (35) | −0.3793 (24) | 3 (1) | 3 (1) | 3 (1) |
| 189$^m$ | 8.7 m | $\frac{13}{2}$ | −1.058 (6) | 0.66 (26) | −0.3877 (13) | 2 | 3 (2) | 3 (2) |
| 190 | 20 m | 0 | | | −0.3188 (8) | | | 3 (1) |
| 191 | 50 m | $\frac{3}{2}$ | −0.618 (11) | −0.80 (25) | −0.3041 (15) | 3 (1) | 3 (1) | 3 (1) |
| 191$^m$ | 50.8 m | $\frac{13}{2}$ | −1.068 (5) | 0.64 (25) | −0.3037 (13) | 2 | 3 (2) | 3 (2) |
| 192 | 4.9 h | 0 | | | −0.2405 (14) | | | 3 (1, 5) |
| 193 | 6 h | $\frac{3}{2}$ | −0.62757 (2) | −0.72 (38) | −0.2340 (80) | 3 (5) | 3 (5) | 3 (5) |
| 193$^m$ | 11.1 h | $\frac{13}{2}$ | −1.058429 (3) | 0.916 (97) | −0.2160 (24) | 1 | 3 (2) | 3 (2, 5) |
| 194 | 260 y | 0 | | | −0.1607 (7) | | | 3 (5) |
| 195 | 9.5 h | $\frac{1}{2}$ | 0.541475 (1) | | −0.1234 (22) | 4 | | 3 (5) |
| 195$^m$ | 40 h | $\frac{13}{2}$ | −1.044647 (3) | 1.08 (11) | −0.1315 (22) | 1 | 3 (2) | 3 (2, 5) |
| 196 | stable | 0 | | | −0.0809 (3) | | | 3 |
| 197 | 64.1 h | $\frac{1}{2}$ | 0.5273741 (9) | | −0.0532 (30) | 4 | | 3 |
| 197$^m$ | 23.8 h | $\frac{13}{2}$ | −1.027684 (3) | 1.24 (14) | −0.0427 (16) | 1 | 3 (2) | 3 (2) |
| 198 | stable | 0 | | | 0 | | | |
| 199 | stable | $\frac{1}{2}$ | 0.5058851 (9) | | 0.0119 (2) | 4 | | 3 |
| 199$^m$ | 42.6 m | $\frac{13}{2}$ | −1.014703 (3) | 1.18 (45) | 0.0623 (30) | 1 | 3 (1) | 3 |
| 200 | stable | 0 | | | 0.0935 (2) | | | 3 |
| 201 | stable | $\frac{3}{2}$ | −0.560225 (1) | 0.385 (40) | 0.1245 (2) | 4 | 3 (1) | 3 |
| 202 | stable | 0 | | | 0.1970 (2) | | | 3 |
| 203 | 46.6 d | $\frac{5}{2}$ | 0.84895 (13) | 0.343 (36) | 0.2288 (36) | 4 | 3 (1) | 3 |
| 204 | stable | 0 | | | 0.2988 (3) | | | 3 |
| 205 | 5.2 m | $\frac{1}{2}$ | 0.6009 (1) | | 0.3333 (20) | 1 | | 3 (1) |
| 206 | 8.2 m | 0 | | | 0.3987 (17) | | | 3 (2) |

*References*: 1. (Bo 76) and references cited therein, 2. (Da 79), 3. (Ul 86) and references cited therein (updates also earlier measurements), 4. (Le 78), 5. (To 64), 6. (Kü 77).

*Exp. method*: 1: RADOP and nuclear magnetic resonance.
2, 6: Fluorescence laser spectroscopy in cells.
3: Standard collinear fast-beam laser spectroscopy (applied only to even isotopes).
5: Classical high-resolution spectroscopy.

*Transitions*: 1, 2, 5, 6: HgI $6s^2\,{}^1S_0 \rightarrow 6s6p\,{}^3P_1$ (253.7 nm).
3: $6s6p\,{}^3P_2 \rightarrow 6s7s\,{}^3S_1$ (546.1 nm).

*$\mu_I$-eval.*: reference isotope ${}^{199}$Hg.

*$Q_s$-eval.*: semiempirical calculation of $\langle r^{-3}\rangle_{6p}$, see Ref. 3, Sternheimer corrected.

*$\delta\langle r^2\rangle$-eval.*: MCDF calculation from (To 85), $F_{254} = -55.36$ GHz/fm$^2$, $S_{254} = 0$ GHz amu.

**Table 27. Element: Thallium, $Z = 81$**

| $A$ | $\tau_{1/2}$ | $I$ | $\mu_I$ (n.m.) | $Q_s$ (b) | $\delta\langle r^2\rangle$ (fm$^2$) | Ref. No. $\mu_I$ | Ref. No. $Q_s$ | Ref. No. $\delta\langle r^2\rangle$ |
|---|---|---|---|---|---|---|---|---|
| 189$^m$ | 1.4 m | $\frac{9}{2}$ | 3.8776 (63) | −2.29 (4) | | 4 | 2 | |
| 191$^m$ | 5.22 m | $\frac{9}{2}$ | 3.9034 (48) | −2.27 (3) | | 4 | 2 | |
| 193 | 21.6 m | $\frac{1}{2}$ | 1.5912 (22) | | | 4 | | |
| 193$^m$ | 2.11 m | $\frac{9}{2}$ | 3.9482 (39) | −2.20 (2) | | 4 | 2 | |
| 194 | 33.0 m | 2 | ±0.14 (1) | | | 3 | | |
| 195 | 1.16 h | $\frac{1}{2}$ | 1.58 (4) | | | 3 | | |
| 196 | 1.84 h | 2 | ±0.07 (12) | | | 3 | | |
| 197 | 2.84 h | $\frac{1}{2}$ | 1.58 (2) | | | 3 | | |
| 198 | 5.3 h | 2 | ±0.00 (1) | | | 3 | | |
| 198$^m$ | 1.87 h | 7 | ±0.640 (74) | | | 3 | | |
| 199 | 7.42 h | $\frac{1}{2}$ | 1.60 (2) | | | 3 | | |
| 200 | 26.1 h | 2 | ±0.04 (1) | | | 3 | | |
| 201 | 73.1 h | $\frac{1}{2}$ | 1.61 (2) | | | 3 | | |
| 202 | 12.23 h | 2 | ±0.06 (1) | | | 3 | | |
| 203 | stable | $\frac{1}{2}$ | 1.622257 (1) | | | 3 | | |
| 204 | 3.78 y | 2 | ±0.0908 | | | 3 | | |
| 205 | stable | $\frac{1}{2}$ | 1.6382134 (7) | | | 3 | | |
| 207 | 4.77 m | $\frac{1}{2}$ | 1.876 (5) | | | 1 | | |

*References*: 1. (Ne 85b), 2. (Bo 85b), 3. (Le 87), 4. (Bo 87b).
*Exp. method*: 1, 2, 4: Standard collinear fast-beam laser spectroscopy.
*Transitions*: 1, 2: TlI 6p $^2P_{3/2} \rightarrow$ 7s $^2S_{1/2}$ line (535.0 nm).
$\mu_I$-*eval.*: reference isotope $^{205}$Tl.
$Q_s$-*eval.*: *ab initio* calc. from (Li 74), not Sternheimer corrected applied.
$\delta\langle r^2\rangle$-*eval.*: not given, although IS has been measured in Ref. 2.
*Comment*: $\mu_I$ of the neutron-deficient isomers $^{189m,191m,193m}$Tl has been measured but no clear spin assignment is given.

## Table 28. Element: Lead, $Z = 82$

| $A$ | $\tau_{1/2}$ | $I$ | $\mu_I$ (n.m.) | $Q_s$ (b) | $\delta\langle r^2\rangle^{A,208}$ (fm$^2$) | Ref. No. $\mu_I$ | Ref. No. $Q_s$ | Ref. No. $\delta\langle r^2\rangle$ |
|---|---|---|---|---|---|---|---|---|
| 190 | 1.2 m | 0 | | | −0.838 (12) | | | 5 |
| 191 | 1.3 m | $\frac{13}{2}$ | 1.176 (2) | 0.130 (77) | −0.831 (12) | 5 | 5 | 5 |
| 192 | 3.5 m | 0 | | | −0.758 (12) | | | 5 (4) |
| 193$^m$ | 4.0 m | $\frac{13}{2}$ | 1.156 (2) | 0.165 (71) | 0.747 (12) | 5 | 5 | 5 |
| 194 | 10 m | 0 | | | −0.683 (10) | | | 5 |
| 195$^m$ | 15.8 m | $\frac{13}{2}$ | −1.1318 (13) | 0.286 (95) | −0.664 (13) | 4 | 4 | 4 |
| 196 | 37 m | 0 | | | −0.6057 (52) | | | 2 |
| 197 | 8 m | $\frac{3}{2}$ | −1.0753 (22) | −0.078 (174) | −0.6038 (50) | 2 | 2 | 2 |
| 197$^m$ | 42 m | $\frac{13}{2}$ | −1.1045 (27) | 0.512 (135) | −0.5739 (47) | 2 | 5 | 2 |
| 198 | 2.4 h | 0 | | | −0.5214 (49) | | | 2 (1) |
| 199 | 1.5 h | $\frac{5}{2}$ | −1.0742 (12) | 0.078 (86) | −0.5163 (42) | 2 (1) | 2 (1) | 2 (1) |
| 200 | 21.5 h | 0 | | | −0.4286 (35) | | | 2 (1) |
| 201 | 9.4 h | $\frac{5}{2}$ | 0.6753 (5) | −0.009 (43) | −0.4093 (34) | 2 (1) | 2 (1) | 2 (1) |
| 202 | $5.3 \times 10^4$ y | 0 | | | −0.3280 (27) | | | 2 (1) |
| 202$^m$ | 3.62 h | 9 | −0.2276 (7) | 0.581 (86) | −0.3299 (27) | 2 (1) | 2 (1) | 2 (1) |
| 203 | 52.0 h | $\frac{5}{2}$ | 0.6864 (5) | 0.095 (52) | −0.3045 (25) | 2 (1) | 2 (1) | 2 (1) |
| 204 | stable | 0 | | | −0.2231 (18) | | | 2 (1) |
| 205 | $1.5 \times 10^7$ y | $\frac{5}{2}$ | 0.7117 (4) | 0.226 (37) | −0.1967 (16) | 2 (1) | 2 (1) | 2 (1) |
| 206 | stable | 0 | | | −0.1179 (10) | | | 2 (1) |
| 207 | stable | $\frac{1}{2}$ | 0.592582 (9) | | −0.0737 (6) | 3 | | 2 (1) |
| 208 | stable | 0 | | | 0 | | | |
| 209 | 3.25 h | $\frac{9}{2}$ | −1.435 (16) | −0.269 (165) | 0.0937 (9) | 2 (1) | 2 (1) | 2 (1) |
| 210 | 22.3 y | 0 | | | 0.2107 (18) | | | 2 (1) |
| 211 | 36.1 m | $\frac{9}{2}$ | −1.4037 (8) | 0.087 (62) | 0.2995 (25) | 2 | 2 | 2 |
| 212 | 10.64 h | 0 | | | 0.4144 (38) | | | 2 (1) |
| 214 | 26.8 m | 0 | | | 0.6099 (52) | | | 2 |

*References*: 1. (Th 83), 2. (An 86c), 3. (Le 78), 4. (Di 87), 5. (Me 88).
*Exp. method*: 1, 2: Atomic-beam laser spectroscopy.
3, 4: Standard collinear fast-beam laser spectroscopy.
*Transitions*: 1, 2: PbI $6p^2\,{}^3P_0 \rightarrow 6p7s\,{}^3P_1$ (283.3 nm).
3: PbI $6p^2\,{}^1D_2 \rightarrow 6p7s\,{}^3P_1$ (723 nm).
$\mu_I$-*eval.*: reference isotope $^{207}$Pb.
$Q_s$-*eval.*: semiempirical calculation from (De 84), Sternheimer-corrected.
$\delta\langle r^2\rangle$-*eval.*: combined analysis of muonic X-ray and electron scattering (Fr 81c), $F_{283} = 20.26\,(18)$ GHz/fm$^2$, $S = -0.81\,(25)$ N from (Ki 85).

**Table 29. Element: Radon, $Z = 86$**

| $A$ | $\tau_{1/2}$ | $I$ | $\mu_I$ (n.m.) | $Q_s$ (b) | $\delta\langle r^2\rangle^{A,212}$ ($fm^2$) | Ref. No. $\mu_I$ | $Q_s$ | $\delta\langle r^2\rangle$ |
|---|---|---|---|---|---|---|---|---|
| 202 | 9.9 s | 0 | | | −0.4382 (4) | | | 1 |
| $203^m$ | 28 s | $\frac{13}{2}$ | −0.9601 (60) | 1.28 (13) | −0.4099 (3) | 1 | 1 | 1 |
| 204 | 1.24 m | 0 | | | −0.3860 (3) | | | 1 |
| 205 | 2.83 m | $\frac{5}{2}$ | 0.8018 (36) | 0.062 (6) | −0.3849 (3) | 1 | 1 | 1 |
| 206 | 5.7 m | 0 | | | −0.3058 (3) | | | 1 |
| 207 | 9.3 m | $\frac{5}{2}$ | 0.8157 (30) | 0.220 (22) | −0.2926 (2) | 1 | 1 | 1 |
| 208 | 24 m | 0 | | | −0.2125 (2) | | | 1 |
| 209 | 29 m | $\frac{5}{2}$ | 0.83881 (39) | 0.311 (31) | −0.1917 (1) | 2 | 1 | 1 |
| 210 | 2.4 h | 0 | | | −0.1143 (1) | | | 1 |
| 211 | 14.6 h | $\frac{1}{2}$ | 0.6013 (41) | | −0.0735 (1) | 1 | | 1 |
| 212 | 24 m | 0 | | | 0 | | | |
| 218 | 35 ms | 0 | | | 0.7000 (3) | | | 1 |
| 219 | 3.96 s | $\frac{5}{2}$ | −0.4421 (45) | 1.15 (12) | 0.8212 (3) | 1 | 1 | 1 |
| 220 | 55.6 s | 0 | | | 0.9151 (3) | | | 1 |
| 221 | 25 m | $\frac{7}{2}$ | −0.0202 (30) | −0.474 (48) | 1.0320 (4) | 1 | 1 | 1 |
| 222 | 3.8235 d | 0 | | | 1.1236 (4) | | | 1 |
| 223 | 43 m | $\frac{7}{2}$ | −0.776 (8) | 0.80 | | 3 | 3 | |
| 225 | 4.5 m | $\frac{7}{2}$ | −0.696 (8) | 0.84 | | 3 | 3 | |

*References*: 1. (Bo 87), 2. (Ki 88), 3. (Ne 88d).
*Exp. method*: 1: Standard collinear fast-beam laser spectroscopy.
2: RADOP.
*Transitions*: 1: RnI 7s—7p (745 nm).
$\mu_I$-*eval.*: reference isotope $^{209}$Rn from Ref. 2.
$Q_s$-*eval.*: semiempirical calculation in Ref. 1, not Sternheimer corrected.
$\delta\langle r^2\rangle$-*eval.*: semiempirical approach in Ref. 1, $F_{745} = -22.1$ GHz/fm$^2$, $S_{745} = 0$ GHz amu.

## Table 30. Element: Francium, $Z = 87$

| $A$ | $\tau_{1/2}$ | $I$ | $\mu_I$ (n.m.) | $Q_s$ (b) | $\delta\langle r^2\rangle^{A,213}$ (fm$^2$) | Ref. No. $\mu_I$ | Ref. No. $Q_s$ | Ref. No. $\delta\langle r^2\rangle$ |
|---|---|---|---|---|---|---|---|---|
| 207 | 14.8 s | $\frac{9}{2}$ | 3.89 (8) | −0.16 (5) | −0.21794 (16) | 2 | 2 | 3 (2) |
| 208 | 58.6 s | 7 | 4.75 (10) | 0.004 (38) | −0.20804 (12) | 1 | 2 | 3 (1, 2) |
| 209 | 50 s | $\frac{9}{2}$ | 3.95 (8) | −0.24 (2) | −0.13043 (8) | 1 | 2 | 3 (1, 2) |
| 210 | 3.2 m | 6 | 4.40 (9) | 0.19 (2) | −0.10831 (4) | 1 | 2 | 3 (1, 2) |
| 211 | 3.1 m | $\frac{9}{2}$ | 4.00 (8) | −0.19 (3) | −0.03757 (4) | 5 | 2 | 3 (1, 2) |
| 212 | 19.3 m | 5 | 4.62 (9) | −0.10 (1) | 0 | 1 | 2 | |
| 213 | 34.7 s | $\frac{9}{2}$ | 4.02 (1) | −0.14 (2) | 0.06829 (8) | 1 (4) | 2 (4) | 3 (1, 2, 4) |
| 220 | 27.4 s | 1 | −0.67 (1) | 0.482 (2) | 0.86725 (45) | 2 (4) | 2 (4) | 3 (1, 2, 4) |
| 221 | 4.8 m | $\frac{5}{2}$ | 1.58 (3) | −1.00 (1) | 0.98269 (33) | 2 (4) | 2 (4) | 3 (1, 2, 4) |
| 222 | 14.4 m | 2 | 0.63 (1) | 0.51 (4) | 1.09543 (12) | 2 | 2 | 3 (2) |
| 223 | 21.8 m | $\frac{3}{2}$ | 1.17 (2) | 1.17 (1) | 1.16507 (8) | 2 | 2 | 3 (2) |
| 224 | 2.7 m | 1 | 0.40 (1) | 0.517 (4) | 1.28937 (4) | 2 | 2 | 3 (2) |
| 225 | 3.9 m | $\frac{3}{2}$ | 1.07 (2) | 1.32 (1) | 1.34862 (22) | 2 | 2 | 3 (2) |
| 226 | 48 s | 1 | 0.071 (2) | −1.35 (2) | 1.43700 (4) | 2 | 2 | 3 (2) |
| 227 | 2.4 m | $\frac{1}{2}$ | 1.50 (3) | | 1.60249 (8) | 2 | | 3 (2) |
| 228 | 39 s | 2 | −0.76 (2) | 2.38 (5) | 1.67522 (20) | 2 | 2 | 3 (2) |

*References*: 1. (Li 80), 2. (Co 85), 3. (Co 87), 4. (Du 87), 5. (Ek 86).
*Exp. method*: 1, 3: LIOP.
4: Standard collinear fast-beam laser spectroscopy.
*Transitions*: 1, 3: FrI D1 and D2 line.
4: FrI D1′ and D2′ line.
$\mu_I$-*eval.*: reference isotope $^{211}$Fr.
$Q_s$-*eval.*: RMBPT calculation of $\langle r^{-3}\rangle_{7p}$ from (He 83b), Sternheimer correction $R_{6p} = -0.23$ (2).
$\delta\langle r^2\rangle$-*eval.*: semiempirical approach in Ref. 3.

**Table 31. Element: Radium, $Z=88$**

| $A$ | $\tau_{1/2}$ | $I$ | $\mu_I$ (n.m.) | $Q_s$ (b) | $\delta\langle r^2\rangle^{A,214}$ (fm$^2$) | Ref. No. $\mu_I$ | Ref. No. $Q_s$ | Ref. No. $\delta\langle r^2\rangle$ |
|---|---|---|---|---|---|---|---|---|
| 208 | 1.3 s | 0 | | | −0.256 (27) | | | 3 |
| 209 | 4.7 s | $\frac{5}{2}$ | 0.865 (13) | 0.38 (4) | −0.253 (25) | 3 (1) | 2 (1) | 3 |
| 210 | 3.7 s | 0 | | | −0.182 (19) | | | 3 |
| 211 | 13 s | $\frac{5}{2}$ | 0.8780 (38) | 0.46 (5 | −0.168 (17) | 3 (1) | 2 (1) | 3 |
| 212 | 13.0 s | 0 | | | −0.099 (10) | | | 3 |
| 213 | 2.7 m | $\frac{1}{2}$ | 0.6133 (18) | | −0.066 (7) | 4 (1) | | 3 |
| 214 | 2.46 s | 0 | | | 0 | | | |
| 220 | 23 ms | 0 | | | 0.679 (69) | | | 3 |
| 221 | 30 s | $\frac{5}{2}$ | −0.1799 (17) | 1.90 (20) | 0.805 (82) | 3 (1) | 2 (1) | 3 |
| 222 | 38 s | 0 | | | 0.894 (91) | | | 3 |
| 223 | 11.435 d | $\frac{3}{2}$ | 0.2750 (19) | 1.19 (12) | 1.007 (102) | 3 (1) | 2 (1) | 3 |
| 224 | 3.66 d | 0 | | | 1.090 (110) | | | 3 |
| 225 | 14.8 d | $\frac{1}{2}$ | −0.7338 (15) | | 1.208 (122) | 4 (1) | | 3 |
| 226 | $1.6\times10^3$ y | 0 | | | 1.277 (129) | | | 3 |
| 227 | 42.2 m | $\frac{3}{2}$ | −0.4038 (24) | 1.50 (15) | 1.356 (138) | 3 (1) | 2 (1) | 3 |
| 228 | 5.76 y | 0 | | | 1.459 (148) | | | 3 |
| 229 | 4.0 m | $\frac{5}{2}$ | 0.5025 (27) | 2.96 (30) | 1.556 (158) | 3 (1) | 2 (1) | 3 |
| 230 | 93 m | 0 | | | 1.667 (169) | | | 3 |
| 232 | ? | 0 | | | 1.854 (188) | | | 3 |

*References*: 1. (Ah 83), 2. (We 87) and (Ne 88), 3. (Ah 88), 4. (Ar 87).
*Exp. method*: 1, 3: Standard collinear fast-beam laser spectroscopy.
4: Collinear laser spectroscopy with spin rotation.
*Transitions*: RaII D1 line (468.3 nm).
RaI $7s^2\ ^1S_0 \rightarrow 7s7p\ ^{1,3}P_1$ (482.7 nm, 714.1 nm).
RaI $7s7p\ ^3P_2 \rightarrow 7s7d\ ^3D_3$ (468.3 nm).
$\mu_I$-*eval.*: reference isotopes $^{213,225}$Ra from Ref. 4.
$Q_s$-*eval.*: semiempirical calculation in Ref. 2, Sternheimer correction $R_{7p} = -0.24\ (5)$.
$\delta\langle r^2\rangle$-*eval.*: semiempirical calculation $F_{468} = -49.6$ GHz/fm$^2$, $S = 0$ GHz amu.

## Acknowledgments

Having finished this treatise at a meeting point between quantum optics and atomic and nuclear physics, the author feels very grateful to his collaborators and colleagues, J.Bonn, G. Huber, H.-J. Kluge and R. Neugart with whom he has had the pleasure of working in this field over the years and whose ideas and results have been merged into this article. He acknowledges the most valuable help of Mrs. A. Hermanni and Mr. K. Wendt who arranged the figures, references, and the tables of results. Sincere thanks are also due to Mrs. E. Stuck-Kerth for her diligent typing of the manuscript.

## References

(Ab 61) A. Abragam, *The Principle of Nuclear Magnetism*, Oxford University Press, London (1961).

(Ac 75) F. Ackermann, E. W. Otten, G. zu Putlitz, A. Schenck, and S. Ullrich, *Nucl. Phys. A* **248**:157 (1975).

(Ah 83) S. A. Ahmad, W. Klempt, R. Neugart, E. W. Otten, K. Wendt, and C. Ekström, *Phys. Lett.* **133B**:47 (1983).

(Ah 85) S. A. Ahmad, W. Klempt, C. Ekström, R. Neugart, and K. Wendt, *Z. Phys. A* **321**:35 (1985).

(Ah 85b) S. A. Ahmad, C. Ekström, W. Klempt, R. Neugart, and K. Wendt, Proceedings of the "Symposium on Quantum Electronics," Bombay (1985).

(Ah 88) S. A. Ahmad, W. Klempt, R. Neugart, E. W. Otten, K. Wendt, and C. Ekström, *Nucl. Phys. A* **483**:244 (1988).

(Ah 88b) S. A. Ahmad, W. Klempt, R. Neugart, K. Wendt, and C. Ekström, to appear.

(Al 83) G. D. Alkhazov, A. E. Barzakh, E. I. Berlovich, V. P. Denisov, A. J. Dernyatin, V. S. Ivanov, A. N. Zherikhin, O. N. Kompanets, V. S. Letokhov, V. I. Mishin, and V. N. Fedoseyev, *JETP Lett.* **37**:274 (1983).

(Al 85) C. F. Alonso, J. M. Arias, and F. Iachello, *Phys. Lett.* **164B**:241 (1985).

(Al 87) G. D. Alkhazov, A. E. Barzakh, V. N. Buyanov, V. P. Denisov, V. S. Ivanov, V. S. Letokhov, V. I. Mishin, S. K. Sekatsky, V. N. Fedoseyev, and I. Y. Chubukov, Preprint 1309, Institute of Nuclear Physics of the Academy of Sciences USSR, Leningrad (1987).

(An 78) K. R. Anton, S. L. Kaufman, W. Klempt, G. Moruzzi, R. Neugart, E. W. Otten, and B. Schinzler, *Phys. Rev. Lett.* **40**:642 (1978).

(An 80) I. Angeli, M. Beiner, R. J. Lombard, and D. Mas, *J. Phys. G* **6**:303 (1980).

(An 82) A. Andl, K. Bekk, S. Göring, A, Hanser, G. Nowicki, H. Rebel, G. Schatz, and R. C. Thompson, *Phys. Rev. C* **26**:2194 (1982).

(An 86) M. Anselment, S. Chongkum, K. Bekk, S. Göring, A. Hanser, G. Meisel, and H. Rebel, *Z. Phys. D* **3**:421 (1986).

(An 86b) M. Anselment, K. Bekk, A. Hanser, H. Hoeffgen, G. Meisel, S. Göring, H. Rebel, and G. Schatz, *Phys. Rev. C* **34**:1052 (1986).

(An 86c) M. Anselment, W. Faubel, S. Göring, A. Hanser, G. Meisel, H. Rebel, and G. Schatz, *Nucl. Phys. A* **451**:471 (1986).

(An 86d) S. V. Andreev, V. I. Mishin, and V. S. Letokhov, *Opt. Commun.* **57**:317 (1986).

(An 86e) I. Angeli and R. J. Lombard, *Z. Phys. A* **324**:299 (1986).

(An 87) M. Anselment, K. Bekk, S. Chongkum, S. Göring, A. Hanser, H. Hoeffgen, W. Kälber, G. Meisel, and H. Rebel, *Z. Phys. A* **326**:493 (1987).

(Ar 71) L. Armstrong, *Theory of the Hyperfine Structure of Free Atoms*, Wiley, New York (1971).

(Ar 87) E. Arnold, W. Borchers, M. Carre, H. T. Duong, P. Juncar, J. Lerme, S. Liberman, W. Neu, R. Neugart, E. W. Otten, M. Pellarin, J. Pinard, G. Ulm, J. L. Vialle, and K. Wendt, *Phys. Rev. Lett.* **59**:771 (1987).

(Ar 87b) E. Arnold, J. Bonn, R. Gegenwart, W. Neu, R. Neugart, E. W. Otten, G. Ulm, and K. Wendt, *Phys. Lett. B* **197**:311 (1987).

(Ar 88) E. Arnold, J. Bonn, W. Neu, R. Neugart, and E. W. Otten, to appear.

(Au 85) P. Aufmuth and E. Wöbker, *Z. Phys. A* **321**:65 (1985).

(Au 86) P. Aufmuth and E. Wöbker, *Z. Phys. D* **2**:131 (1986).

(Au 87) P. Aufmuth, K. Heilig, and A. Steudel, *At. Data Nucl. Data Tables* **37**:455 (1987).

(Ba 73) R. Bauer, J. Speth, V. Klemt, P. Ring, W. Werner, and T. Yamazaki, *Nucl. Phys. A* **209**:535 (1973).

(Ba 74) R. C. Barret, *Rep. Prog. Phys.* **37**:1 (1974).

(Ba 81) P. Barreau, L. Roussel, and R. J. Powers, *Nucl. Phys. A* **364**:446 (1981).

(Ba 83) P. E. G. Baird, S. A. Blundell, G. Burrows, C. J. Foot, G. Meisel, D. N. Stacey, and G. K. Woodgate, *J. Phys. B* **16**:2485 (1983).

(Ba 85) F. Barranco and R. A. Broglia, *Phys. Lett.* **151B**:90 (1985).

(Ba 86) H. Backe, private communication, Mainz.

(Be 73) M. Beiner and R. J. Lombard, *Phys. Lett* **47B**:399 (1973).

(Be 74) A. Beckmann, K. D. Böklen, and D. Elke, *Z. Phys.* **270**:173 (1974).

(Be 79) K. Bekk, A. Andl, S. Göring, A. Hanser, G. Nowicki, H. Rebel, and G. Schatz, *Z. Phys. A* **291**:219 (1979).

(Be 79b) C. E. Bemis, Jr., J. R. Beene, J. P. Young, and S. D. Kramer, *Phys. Rev. Lett.* **43**:1854 (1979).

(Be 80) E. Bergmann, P. Bopp, C. Dorsch, J. Kowalski, F. Träger, and G. zu Putlitz, *Z. Phys. A* **294**:319 (1980).

(Be 82) C. E. Bemis, Jr. and H. K. Carter (eds.), *Lasers in Nuclear Physics*, Harwood Academic Publishers, New York, (1982).

(Be 84) C. Bengtsson, C. Ekström, L. Robertsson, and J. Heinemeier, *Phys. Scr.* **30**:164 (1984).

(Be 84b) N. Bendali, H. T. Duong, P. Juncar, S. Liberman, J. Pinard, J. M. Saint-Jalm, J. L. Vialle, S. Büttgenbach, C. Thibault, F. Touchard, A. Pesnelle, and A. C. Mueller, *C. R. Acad. Sci. Paris* **299**:1157 (1984).

(Be 84c) N. Bendali, H. T. Duong, J. M. Saint-Jalm, and J. L. Vialle, *J. Phys.* **45**:421 (1984).

(Be 84d) D. Bender, H. Brand, and V. Pfeufer, *Z. Phys. A* **318**:291 (1984).

(Be 84e) I. Berkes, B. Hlimi, G. Marest, E. H. Sayouty, R. Coussement, F. Hardeman, P. Put, and G. Scheveneels, *Phys. Rev. C* **30**:2026 (1984).

(Be 85) D. Berdichevsky and F. Tondeur, *Z. Phys. A* **322**:141 (1985).

(Bj 80) S. Bjørnholm and J. E. Lynn, *Rev. Mod. Phys.* **52**:725 (1980).

(Bl 85) S. A. Blundell, P. E. G. Baird, C. W. P. Palmer, D. N. Stacey, G. K. Woodgate, and D. Zimmermann, *Z. Phys. A* **321**:31 (1985).

(Bo 59) R. A. Bodmer, *Nucl. Phys.* **9**:371 (1959).

(Bo 69) A. Bohr and B. R. Mottelson, *Nuclear Structure*, Vol. 1, Benjamin, New York (1969).

(Bo 71) J. Bonn, G. Huber, H. J. Kluge, U. Köpf, L. Kugler, and E. W. Otten, *Phys. Lett.* **36B**:41 (1971).

(Bo 72) J. Bonn, G. Huber, H. J. Kluge, L. Kugler, and E. W. Otten, *Phys. Lett.* **38B**:308 (1972).

(Bo 74) F. Boehm and P. L. Lee, *At. Data Nucl. Data Tables* **14**:605 (1974).

(Bo 75) A. Bohr and B. R. Mottelson, *Nuclear Structure*, Vol. 2, Benjamin, New York (1975), pp. 556 ff.

(Bo 76) J. Bonn, G. Huber, H. J. Kluge, and E. W. Otten, *Z. Phys. A* **276**:203 (1976).

(Bo 79) J. Bonn, W. Klempt, R. Neugart, E. W. Otten, and B. Schinzler, *Z. Phys. A* **289**:227 (1979).

(Bo 82) C. Bourgeois, P. Kilcher, B. Roussiere, J. Sauvage-Letessier, and M. G. Porquet, *Nucl. Phys. A* **386**:308 (1982).

(Bo 83) G. L. Borchert, O. W. B. Schult, J. Speth, P. G. Hansen, B. Jonson, H. L. Ravn, and J. B. McGrory, *Nuovo Cimento* **73A**:273 (1983).

(Bo 85) J. Bonn, *Hyperfine Interact.* **22**:57 (1985).

(Bo 85b) J. A. Bounds, C. R. Bingham, P. Juncar, H. K. Carter, G. A. Leander, R. L. Mlekodaj, E. H. Spejewski, and W. M. Fairbank, Jr., *Phys. Rev. Lett.* **55**:2269 (1985).

(Bo 87) W. Borchers, R. Neugart, E. W. Otten, H. T. Duong, G. Ulm, and K. Wendt, *Hyperfine Interact.* **34**:25 (1987) and to appear.

(Bo 87b) J. A. Bounds, C. R. Bingham, H. K. Carter, G. A. Leander, R. L. Mlekodaj, E. H. Spejewski, and W. M. Fairbank, Jr., *Phys. Rev. C* **36**:2560 (1987).

(Bo 88) W. Borchers, E. Arnold, W. Neu, R. Neugart, K. Wendt, and G. Ulm, to appear.

(Br 52) P. Brix, *Z. Phys.* **132**:579 (1952).

(Br 77) P. Brix, *Naturwissenschaften* **64**:293 (1977).

(Br 78) H. W. Brandt, K. Heilig, H. Knöckel, and A. Steudel, *Z. Phys. A* **288**:241 (1978).

(Br 80) H. Brand, B. Seibert, and A. Steudel, *Z. Phys. A* **296**:281 (1980).

(Br 85) P. Brix, *Z. Naturforsch.* **41a**:3 (1986).

(Bu 81) F. Buchinger, P. Dabkiewicz, H. J. Kluge, A. C. Müller, and E. A. Otten, *Hyperfine Interact.* **9**:165 (1981).

(Bu 82) F. Buchinger, A. C. Mueller, B. Schinzler, K. Wendt, C. Ekström, W. Klempt, and R. Neugart, *Nucl. Instrum. Methods* **202**:159 (1982).

(Bu 85) F. Buchinger, R. Corriveau, E. B. Ramsay, D. Berdichevsky, and D. W. L. Sprung, *Phys. Rev. C* **32**:2058 (1985).

(Bu 87) F. Buchinger, P. Dabkiewicz, H. J. Kluge, A. C. Mueller, and E. W. Otten, *Nucl. Phys. A* **462**:305 (1987).

(Bu 87b) F. Buchinger, E. B. Ramsay, R. E. Silverans, P. Lievens, E. Arnold, W. Neu, R. Neugart, K. Wendt, and G. Ulm, *Z. Phys. A* **327**:361 (1987).

(Bu 88) F. Buchinger, E. B. Ramsay, R. E. Silverans, P. Lievens, E. Arnold, W. Neu, R. Neugart, K. Wendt, D. Berdichevsky, R. Fleming, D. W. L. Sprung, and G. Ulm, *The Fifth International Conference on Nuclei Far From Stability* (I. S. Towner, ed.), American Institute of Physics, New York (1988), Conf. Proc. 164, p. 197.

(Bü 82) S. Büttgenbach, *Hyperfine Structure in 4d and 5d Shell Atoms*, Springer-Verlag, Berlin (1982).

(Ca 35) H. Casimir, *Physics* **2**:719 (1935).

(Ca 73) U. Cappeler and W. Mazurkewitz, *J. Magn. Reson.* **10**:15 (1973).

(Ca 73b) M. Cailliau, J. Letessier, H. Flocard, and P. Quentin, *Phys. Lett.* **46B**:11 (1973).

(Ca 75) X. Campi, H. Flocard, A. Kerman, and S. Koonin, *Nucl. Phys. A* **251**:193 (1975).

(Ca 80) X. Campi and M. Epherre, *Phys. Rev. C* **22**:2605 (1980).

(Ca 81) R. F. Casten, D. D. Warner, D. A. Brenner, and R. L. Gill, *Phys. Rev. Lett.* **47**:1433 (1981).

(Ca 85) F. Calaprice, *Hyperfine Interact.* **22**:83 (1985).

(Ca 85b) F. Calaprice, W. Happer, D. F. Schreiber, M. M. Lowry, E. Miron, and X. Zeng, *Phys. Rev. Lett.* **54**:174 (1985).

(Ca 85c) R. F. Casten, *Phys. Rev. Lett.* **54**:1991 (1985).

(Ca 86) F. Calaprice, private communication, Princeton and to appear.

(Ca 86b) M. Carre, J. Lerme, and J. L. Vialle, *J. Phys. B* **19**:2853 (1986).

(Ch 72) W. J. Childs and L. S. Goodman, *Phys. Rev. A* **6**:2011 (1972).

(Ch 80) R. R. Chasman, *Phys. Lett.* **96B**:7 (1980).

(Cl 79) D. L. Clark, M. E. Cage, D. A. Lewis, and G. W. Greenlees, *Phys. Rev. A* **20**:239 (1979).

(Co 76) J. D. Cole, J. H. Hamilton, A. V. Ramayya, W. G. Nettles, H. Kawakami, E. H. Spejewski, M. A. Ijaz, K. S. Toth, E. L. Robinson, K. S. R. Sastry, J. Lin, F. T. Avignone, W. H. Brantley, and P. V. G. Rao, *Phys. Rev. Lett.* **37**:1185 (1976).

(Co 83) F. D. Correll, L. Mandansky, R. A. Hardekopf, and J. W. Sunier, *Phys. Rev. C* **28**:862 (1983).

(Co 85) A. Coc, C. Thibault, F. Touchard, H. T. Duong, P. Juncar, S. Liberman, J. Pinard, J. Lerme, J. L. Vialle, S. Büttgenbach, A. C. Mueller, and A. Pesnelle, *Phys. Lett.* **163B**:66 (1985).

(Co 87) A. Coc, C. Thibault, F. Touchard, H. T. Duong, P. Juncar, S. Liberman, J. Pincard, M. Carre, J. Lerme, J. L. Vialle, S. Büttgenbach, A. C. Mueller, and A. Pesnelle, *Nucl. Phys. A* **468**:1 (1987).

(Da 78) P. Dabkiewicz, C. Duke, H. Fischer, T. Kühl, H. J. Kluge, H. Kremmling, E. W. Otten, and H. Schüssler, *J. Phys. Soc. Jpn.* **44**, Suppl.:503 (1978).

(Da 79) P. Dabkiewicz, F. Buchinger, H. Fischer, H. J. Kluge, H. Kremmling, T. Kühl, A. C. Müller, and H. A. Schuessler, *Phys. Lett.* **82B**:199 (1979).

(Da 86) T. P. Das, private communication, Albany.

(De 84) J. Dembczynski and H. Rebel, KFK-Rep. 3815:64 (1984).

(De 84b) J. Dembczynski and H. Rebel, *Z. Phys. A* **315**:137 (1984).

(Di 73) F. Dickmann and K. Dietrich, *Z. Phys.* **263**:211 (1973).

(Di 74) F. Dickmann and K. Dietrich, *Z. Phys.* **271**:417 (1974).

(Di 87) U. Dinger, J. Eberz, G. Huber, R. Menges, S. Schröder, R. Kirchner, O. Klepper, T. Kühl, D. Marx, and G. D. Sprouse, *Z. Phys. A* **328**:253 (1987).

(Di 88) U. Dinger, doctoral thesis, Mainz (1988) and to appear.

(Do 84) J. Dobaczewski, M. Flocard, and J. Treiner, *Nucl. Phys. A* **422**:103 (1984).

(Dö 83) K. Dörschel, W. Hedderich, H. Hühnermann, E. W. Peau, H.Wagner, G. D. Alkhazov, E Y. Berlovich, V. P. Denisov, V. N. Panteleev, and A. G. Polyakov, *Z. Phys. A* **312**:269 (1983).

(Dö 84). K. Dörschel, W. Hedderich, H. Hühnermann, E. W. Peau, H. Wagner, G. D. Alkhazov, E. Y. Berlovich, V. P. Denisov, V. N. Panteleev, and A. G. Polyakov, *Z. Phys. A* **317**:233 (1984).

(Du 77) D. Dubbers, K. Dörr, H. Ackermann, F. Fujura, H. Grupp, M. Grupp, P. Heitjans, A. Körblein, and. H. J. Stöckmann, *Z. Phys. A* **282**:243 (1977).

Du 82) H. T. Duong, P. Juncar, S. Liberman, J. Pinard, J. L. Vialle, F. Touchard, P. Gimbal, S. Büttenbach, R. Klapisch, M. de Saint-Simon, J. M. Serre, and C. Thibault, *Phys. Rev. C* **25**:2725 (1982).

(Du 86) H. T. Duong, S. Liberman, J. Pinard, C. Thibault, F. Touchard, M. Carre, J. Lerme, J. L. Vialle, P. Juncar, S. Büttgenbach, and A. Pesnelle, *J. Phys. (Paris)* **47**:1903 (1986).

(Du 87) H. T. Duong, P. Juncar, S. Liberman, A. C. Mueller, R. Neugart, E. W. Otten, B. Peuse, J. Pinard, H. H. Stroke, C. Thibault, F. Touchard, J. L. Vialle, and K. Wendt, *Europhys. Lett.* **3**:175 (1987).

(Dz 83) V. A. Dzuba, V. V. Flambaum, and O. P. Sushkov, *Phys. Lett.* **95A**:230 (1983).

(Dz 85) V. A. Dzuba, V. V. Flambaum, and O. P. Sushkov, *Phys. Scr.* **32**:507 (1985).

(Ea 84) D. A. Eastham, P. M. Walker, J. A. R. Griffith, D. E. Evans, I. S. Grant, J. G. England, and M. J. Fawcett, Proc. 7th Int. Conf. on Atomic Masses and Fundamental Constants AMCO-7 (O. Klepper, ed.), p. 322, Darmstadt-Seeheim (1984).

(Ea 86) D. A. Eastham, P. M. Walker, J. R. H. Smith, J. A. R. Griffith, D. E. Evans, S. A. Wells, M. J. Fawcett, and I. S. Grant, *J. Phys. G* **12**:205 (1986).

(Ea 87) D. A. Eastham, P. M. Walker, J. R. H. Smith, D. D. Warner, J. A. R. Griffith, D. E. Evans, S. A. Wells, M. J. Fawcett, and I. S. Grant, *Phys. Rev. C* **36**:1583 (1987).

(Ea 88) D. A. Eastham, J. R. H. Smith, J. Groves, D. D. Warner, D. W. L. Tolfree, S. A. Wells, J. A. R. Griffith, D. E. Evans, M. J. Fawcett, I. S. Grant, J. Billowes, and P. M. Walker, *The Fifth International Conference on Nuclei far from Stability* (I. S. Towner, ed.), American Institute of Physics, New York (1988), Conf. Proc. 164, p. 136.

(Eb 86) J. Eberz, U. Dinger, T. Horiguchi, G. Huber, H. Lochmann, R. Menges, R. Kirchner, O. Klepper, T. Kühl, D. Marx, E. Roeckl, D. Schardt, and G. Ulm, *Z. Phys. A* **323**:119 (1986).

(Eb 87) J. Eberz, U. Dinger, G. Huber, H. Lochmann, R. Menges, G. Ulm, R. Kirchner, O. Klepper, T. Kühl, and D. Marx, *Z. Phys. A* **326**:121 (1987).

(Eb 87b) J. Eberz, U. Dinger, G. Huber, H. Lochmann, R. Menges, R. Neugart, R. Kirchner, O. Klepper, T. Kühl, D. Marx, G. Ulm, and K. Wendt, *Nucl. Phys. A* **464**:9 (1987).

(Ed 84) R. Eder, E. Hagn, and E. Zech, *Phys. Rev. C* **30**:676 (1984).

(Eg 78) T. v. Egidy, W. Kaiser, W. Mampe, C. Hillebrand, W. Stöffl, R. G. Lanier, K. Mühlbauer, O. W. B. Schult, H. R. Koch, H. A. Baader, R. L. Mlekodaj, R. K. Sheline, E. B. Shera, J. Ungrin, P. T. Prokofjev, L. I. Simonova, M. K. Balodis, H. Seyfarth, B. Kardon, W. Delang, P. Göttel, D. Breitig, W. R. Kane, R. F. Casten, H. J. Scheerer, P. Glässl, E. Huenges, M. Löffler, H. Rösler, and H. K. Vonach, *Z. Phys. A* **286**:341 (1978).

(Ei 70) J. M. Eisenberg and W. Greiner, *Nuclear Models*, Vol. 1, North-Holland, Amsterdam (1970).

(Ek 77) C. Ekström, S. Ingelman, G. Wannberg, and M. Skarestad, *Nucl. Phys. A* **292**:144 (1977).

(Ek 78) C. Ekström, S. Ingelman, G. Wannberg, and M. Skarestad, *Nucl. Phys. A* **311**:269 (1978).

(Ek 79) C. Ekström, L. Robertsson, G. Wannberg, and J. Heinemeier, *Phys. Scr.* **19**:516 (1979).

(Ek 80) C. Ekström, L. Robertsson, S. Ingelman, G. Wannberg, and I. Ragnarsson, *Nucl. Phys. A* **348**:25 (1980).

(Ek 81) C. Ekström, Proc. 4th Int. Conf. on Nuclei Far From Stability, Helsingör 1981, CERN Rep. 81-09, Geneva (1981), p. 12.

(Ek 84) C. Ekström, L. Robertsson, and J. Heinemeier, *Phys. Scr.* **30**:164 (1984).

(Ek 86) C. Ekström, L. Robertsson, and A. Rosen, *Phys. Scr.* **34**:624 (1986).

(Em 83) H. J. Emrich, G. Fricke, G. Mallot, H. Miska, H. G. Sieberling, J. M. Cavedon, B. Frois, and D. Goutte, *Nucl. Phys. A* **396**:401c (1983).

(En 74) H. Engfer, H. Schneuwly, J. L. Vuilleumier, H. K. Walter, and A. Zehnder, *At. Data Nucl. Data Tables* **14**:509 (1974).

(Ep 81) M. Epherre, G. Audi, and X. Campi, Proc. 4th Int. Conf. on Nuclei Far From Stability, Helsingör 1981, CERN Rep. 81-09, Geneva (1981), p. 62.

(Fa 72) A. Faessler, U. Götz, B. Slavov, and T. Ledergerber, *Phys. Lett.* **39B**:579 (1972).

(Fa 81) A. Faessler, *Hyperfine Interact.* **9**:3 (1981).

(Fe 77) P. Federmann and S. Pittel, *Phys. Lett.* **69B**:385 (1977).

(Fe 78) P. Federmann and S. Pittel, *Phys. Lett.* **77B**:29 (1978).

(Fe 84) V. N. Fedoseyev, V. S. Lethokov, V. I. Mishin, G. D. Alkhazov, A. E. Barzakh, V. P. Denisov, A. G. Dernyatin, and V. S. Ivanov, *Opt. Commun.* **52**:24 (1984).

(Fi 74) W. Fischer, H. Hühnermann, G. Krömer, and H. J. Schäfer, *Z. Phys.* **270**:113 (1974).

(Fi 74b) W. Fischer, M. Hartmann, H. Hühnermann, and H. Vogg, *Z. Phys.* **267**:209 (1974).

(Fi 75) W. Fischer, H. Hühnermann, and T. Meyer, *Z. Phys.* **274**:79 (1975).

(Fr 75) S. Frauendorf and V. V. Pashkevich, *Phys. Lett.* **55B**:365 (1975).

(Fr 81) G. Fricke, Proc. 4th Int. Conf. on Nuclei Far From Stability, Helsingör 1981, CERN Rep. 81-09, Geneva (1981).

(Fr 81b) J. Freidrich and N. Voegler, *Phys. Rev. Lett.* **47**:1385 (1981).

(Fr 81c) G. Fricke, H. Miska, and D. Rychel, private communication, Mainz, cited in (An 86c).

(Fr 82) J. Freidrich and N. Voegler, *Nucl. Phys. A* **373**:192 (1982).

(Fr 83) G. Fricke, private communication, Mainz, cited in (Bo 83).

(Fr 85) B. Fricke, private communication, Kassel.

(Fr 86) G. Fricke, private communication, Mainz.

(Ge 69) S. Gerstenkorn, *C. R. Acad. Sci. Paris, Ser. B* **268**:1636 (1969).

(Ge 79) H. Gerhardt, E. Matthias, H. Rinneberg, F. Schneider, A. Timmermann, R. Wenz, and P. J. West, *Z. Phys. A* **292**:7 (1979).

(Ge 81) H. Gerhardt, F. Jeschonnek, W. Makat, E. Matthias, H. Rinneberg, F. Schneider, A. Timmermann, R. Wenz, and P. J. West, *Hyperfine Interact.* **9**:175 (1981).

(Ge 86) R. Gegenwart, doctoral thesis, Mainz (1986) and R. Gegenwart, E. Arnold, D. Bauer, J. Bonn, E. W. Otten, and K. P. C. Spath, to appear.

(Gi 82) M. Girod and P. G. Reinhard, *Phys. Lett.* **117B**:1 (1982).

(Gr 60) D. P. Grechukhin, *Sov. Phys. JETP* **11**:1359 (1960).

(Gr 71) S. Green, *Phys. Rev. A* **4**:251 (1971).

(Ha 73) R. C. Haskell and L. Mandansky, *Phys. Rev. C* **7**:1277 (1973).

(Ha 85) J. H. Hamilton, P. G. Hansen, and E. F. Zganjar, *Rep. Prog. Phys.* **48**:631 (1985).

(Ha 86) F. Hannachi, G. Bastin, M. G. Porquet, C. Schück, J. P. Thibaud, C. Bourgeois, L. Hildingsson, D. Jerrestam, N. Perrin, H. Sergolle, F. A. Beck, T. Byrski, and J. C. Merdinger, *Z. Phys. A* **325**:371 (1986).

(He 56) R. H. Helm, *Phys. Rev.* **104**:1466 (1956).

(He 61) K. Heilig, *Z. Phys.* **161**:252 (1961).

(He 71) P. Heinecke, *Z. Phys.* **245**:411 (1971).

(He 74) K. Heilig and A. Steudel, *At. Data Nucl. Data Tables* **14**:613 (1974).

(He 77) S. M. Heider and G. O. Brink, *Phys. Rev. A* **16**:1371 (1977).

(He 83) K. Heyde, P. van Isacker, M. Waroquier, J. L. Wood, and R. A. Meyer, *Phys. Rep.* **102**:291 (1983).

(He 83b) J. L. Heully and A. M. Martensson-Pendrill, *Phys. Rev. A* **27**:3332 (1983).

(He 85) J. L. Heully and A. M. Martensson-Pendrill, *Phys. Scr.* **31**:169 (1985).

(He 85b) K. Hellig, *Hyperfine Interact.* **24**–26:349 (1985).

(Ho 73) P. Hornshoj, P. G. Hansen, B. Jonson, A. Lindahl, and O. B. Nielsen, *Phys. Lett.* **43B**:377 (1973).

(Ho 84) B. Hoffmann, G. Baur, and J. Speth, *Z. Phys.* **315**:57 (1984).

(Hu 75) G. Huber, C. Thibault, R. Klapisch, H. T. Duong, J. L. Vialle, J. Pinard, P. Juncar, and P. Jaquinot, *Phys. Rev. Lett.* **34**:1209 (1975).

(Hu 76) G. Huber, J. Bonn, H. J. Kluge, and E. W. Otten, *Z. Phys. A* **276**:187 (1976).

(Hu 78) G. Huber, F. Touchard, S. Büttenbach, C. Thibault, R. Klapisch, H. T. Duong, S. Liberman, J. Pinard, J. L. Vialle, P. Juncar, and P. Jaquinot, *Phys. Rev. C* **18**:2342 (1978).

(Hu 78b) G. Huber, F. Touchard, S. Büttgenbach, C. Thibault, R. Klapisch, S. Liberman, J. Pinard, H. T. Duong, P. Juncar, J. L. Vialle, P. Jacquinot, and A. Pesnelle, *Phys. Rev. Lett.* **41**:459 (1978).

(Hu 79) G. S. Hurst, M. G. Payne, S. D. Kramer, and J. P. Young, *Rev. Mod. Phys.* **51**:767 (1979).

(Hy 80) H. Hyuga, A. Arima, and K. Shimizu, *Nucl. Phys. A* **336**:363 (1980).

(Ia 81) F. Iachello, *Nucl. Phys. A* **358**:89c (1981).

(Ia 83) F. Iachello, *Hyperfine Interact.* **15/16**:11 (1983) and references cited therein.

(Ja 37) H. A. Jahn and E. Teller, *Proc. R. Soc. London, Ser. A* **161**:220 (1937).

(Ja 74) C. W. de Jager, H. de Vries, and C. de Vries, *At. Data Nucl. Data Tables* **14**:479 (1974).

(Ja 76) P. Jaquinot, in *High Resolution Laser Spectroscopy* (K. Shimoda, ed.), *Top. Appl. Phys.* **13**:51, Springer-Verlag, Heidelberg (1976).

(Ja 79) P. Jaquinot and R. Klapisch, *Rep. Progr. Phys.* **42**:773 (1979).

(Ka 76) S. L. Kaufman, *Opt. Commun.* **17**:309 (1976).

(Ka 81) G. Kaindl and H. Haas (eds.), Proc. 5th Int. Conf. on Hyperfine Interactions, Berlin, 1980, *Hyperfine Interact.* **9/10** (1981).

(Ki 73) W. H. King, A. Steudel, and M. Wilson, *Z. Phys.* **265**:207 (1973).

(Ki 84) W. H. King, *Isotope Shifts in Atomic Spectra*, Plenum Press, New York (1984).

(Ki 85) W. H. King and H. Wilson, *J. Phys. G* **11**:L43 (1985).

(Ki 86) W. H. King and W. W. Palmer, *Commun. At. Mol. Phys.* **18**:11 (1986).

(Ki 88) M. Kitano, F. P. Calaprice, M. L. Pitt, J. Clayhold, W. Happer, M. Kadar-Kallen, M. Musolf, G. Ulm, K. Wendt, T. Chupp, J. Bonn, R. Neugart, E. W. Otten, and H. T. Duong, *Phys. Rev. Lett.* **60**:2133 (1988).

(Kl 79) H.-J. Kluge, in *Progress in Atomic Spectroscopy*, Part B (W. Hanle and H. Kleinpoppen, eds.), Plenum Press, New York (1979), p. 727.

(Kl 79b) W. Klempt, J. Bonn, and R. Neugart, *Phys. Lett.* **82B**:47 (1979).

(Kl 82) P. Kleinheinz, J. Styczen, M. Piiparinen, J. Blomquist, and M. Kortelahti, *Phys. Rev. Lett.* **48**:1457 (1982).

(Kl 83) H. J. Kluge, H. Kremmling, H. A. Schuessler, J. Streib, and K. Wallmeroth, *Z. Phys. A* **309**:187 (1983).

(Kl 85) H.-J. Kluge, *Hyperfine Interact.* **24–26**:69 (1985).

(Ko 58) H. Kopfermann, *Nuclear Moments*, Academic Press, New York (1958).

(Ko 75) D. Kolb and C. Y. Wong, *Nucl. Phys. A* **245**:205 (1975).

(Kö 69) U. Köpf, H. J. Besch, E. W. Otten, and C. von Platen, *Z. Phys.* **226**:297 (1969).

(Kr 79) H. Kremmling, P. Dabkiewicz, H. Fischer, H. J. Kluge, T. Kühl, and H. A. Schüssler, *Phys. Rev. Lett.* **43**:1376 (1979).

(Kr 85) G. B. Krygin and V. E. Mitroshin, *Sov. J. Nucl. Phys.* **41**:193 (1985).

(Ku 69) H. G. Kuhn, *Atomic Spectra*, Longmans, London (1969).

(Ku 72) K. Kumar, *Phys. Rev. Lett.* **28**:249 (1972).

(Ku 83) W. Kunold, M. Schneider, L. M. Simons, J. Wüst, and R. Abela, *Z. Phys. A* **313**:11 (1983).

(Kü 77) T. Kühl, P. Dabkiewicz, C. Duke, H. Fischer, H. J. Kluge, H. Kremmling, and E. W. Otten, *Phys. Rev. Lett.* **39**:180 (1977).

(La 78) S. E. Larsson, G. A. Leander, and I. Ragnarsson, *Nucl. Phys. A* **307**:189 (1978).

(Le 78) C. M. Lederer and V. S. Shirley, *Table of Isotopes*, 7th edn., Wiley, New York (1978).

(Le 84) V. S. Letokhov, *Nonlinear Laser Chemistry*, Springer Series Chem. Phys. 22, Springer-Verlag, Heidelberg, Berlin (1984).

(Le 84b) G. A. Leander and R. K. Sheline, *Nucl. Phys. A* **413**:375 (1984).

(Le 86) V. S. Letokhov, private communication, Moscow.

(Le 87) V. S. Letokhov and V. I. Mishin, in *Laser Spectroscopy VIII* (W. Persson and S. Svanberg eds.), Springer Series in Optical Series, Springer-Verlag, Berlin (1987), Vol. 55, p. 167.

(Le 87b) J. K. P. Lee, private communication, Montreal.

(Li 74) I. Lindgren and A. Rosen, *Case Stud. At. Phys.* **4**:93 (1974).

(Li 78) S. Liberman, J. Pinard, H. T. Duong, P. Juncar, J. L. Vialle, P. Jaquinot, G. Huber, F. Touchard, S. Büttgenbach, A. Pesnelle, C. Thibault, and R. Klapisch, *C. R. Acad. Sci. Paris, Ser. B* **286**:253 (1978).

(Li 80) S. Liberman, J. Pinard, H. T. Duong, P. Juncar, P. Pillet, J. L. Vialle, P. Jaquinot, F. Touchard, S. Büttgenbach, C. Thibault, M. de Saint Simon, R. Klapisch, A. Pesnelle, and G. Huber, *Phys. Rev. A* **22**:2732 (1980).

(Lo 85) H. Lochmann, U. Dinger, J. Eberz, G. Huber, R. Menges, G. Ulm, R. Kirchner, O. Klepper, T. Kühl, D. Marx, and D. Schardt, *Z. Phys. A* **322**:703 (1985).

(Lö 70) K. E. G. Löbner, M. Vetter, and W. Hönig, *Nucl. Data Tables A* **7**:495 (1970).

(Ma 71) E. Matthias, B. Olsen, D. A. Shirley, J. E. Templeton, and R. M. Steffen, *Phys. Rev. A* **4**:1626 (1971).

(Ma 76) I. J. Ma and G. zu Putlitz, *Z. Phys. A* **277**:107 (1976).

(Me 77) T. Meier, H. Hühnermann, and H. Wagner, *Opt. Commun.* **20**:397 (1977).

(Me 86) R. A. Meyer and V. Paar (eds.), *Proc. Int. Conf. on Nuclear Structure, Reactions and Symmetries, Dubrovnik* (1986), World Scientific Publishers, Singapore, Vols. 1 and 2.

(Me 88) R. Menges, doctoral thesis, Mainz (1988) and to appear.

(Mi 75) T. Minamisono, J. W. Hugg, D. G. Mavis, T. K. Saylor, S. M. Lazarus, H. F. Glavish, and S. S. Hanna, *Phys. Rev. Lett.* **34**:1465 (1975).

(Mö 84) P. Möller, W. D. Myers, W. J. Swiatecki, and J. Treiner, Proc. 7th Int. Conf. on Atomic Masses and Fundamental Constants, AMCO-7 (O. Klepper, ed.), Darmstadt-Seeheim (1984), p. 457.

(Mu 83) A. C. Mueller, F. Buchinger, W. Klempt, E. W. Otten, R. Neugart, C. Ekström, and J. Heinemeier, *Nucl. Phys. A* **403**:234 (1983).

(My 69) W. D. Myers and W. J. Swiatecki, *Ann. Phys.* (*N.Y.*) **55**:395 (1969).

(My 77) W. D. Myers, *Droplet Model of Atomic Nuclei*, IFI/Plenum Data Co., New York (1977).

(My 83) W. D. Myers and K. H. Schmidt, *Nucl. Phys. A* **410**:61 (1983).

(Na 84) W. Nazarewicz, P. Olanders, I. Ragnarsson, J. Dudek, G. A. Leander, P. Möller, and E. Ruchowska, *Nucl. Phys. A* **429**:269 (1984).

(Ne 78) W. Neuhauser, M. Hohenstatt, P. E. Toschek, and H. G. Dehmelt, *Phys. Rev. Lett.* **41**:233 (1978).

(Ne 82) R. Neugart, in *Lasers in Nuclear Physics* (C. E. Bemis, Jr. and H. K. Carter, eds.), Harwood Academic Publishers, New York (1982).

(Ne 83) R. Neugart, K. Wendt, S. A. Ahmad, W. Klempt, and C. Ekström, *Hyperfine Interact.* **15/16**:181 (1983).

(Ne 85) R. Neugart, *Hyperfine Interact.* **24–26**:159 (1985).

(Ne 85b) R. Neugart, H. H. Stroke, S. A. Ahmad, H. T. Duong, H. L. Ravn, and K. Wendt, *Phys. Rev. Lett.* **55**:1559 (1985).

(Ne 86) R. Neugart, W. Klempt, and K. Wendt, *Nucl. Instrum. Methods B* **17**:354 (1986).

(Ne 87) R. Neugart, in *Progress in Atomic Spectroscopy* (H. J. Beyer and H. Kleinpoppen, eds.), Plenum Press, New York (1987), Part D, Ch. 2, p. 75.

(Ne 88) W. Neu, R. Neugart, B. Fricke, E. Arnold, H. J. Kluge, E. W. Otten, G. Passler, G. Ulm, and K. Wendt, to appear.

(Ne 88b) R. Neugart, F. Buchinger, C. Ekström, W. Klempt, A. C. Mueller, and K. Wendt, to appear.

(Ne 88c) W. Neu, E. Arnold, C. Ekström, R. Neugart, K. Wendt, and G. Ulm, to appear.

(Ne 88d) R. Neugart, E. Arnold, W. Borchers, W. Neu, G. Ulm, and K. Wendt, in *The Fifth International Conference on Nuclei far from Stability* (I. S. Towner, ed.), American Institute of Physics, New York (1988), Conf. Proc. 164, p. 126.

(Ni 57) W. A. Nierenberg, *Annu. Rev. Nucl. Sci.* **7**:349 (1957).

(Ni 74) S. G. Nilsson, J. R. Nix, P. Möller, and I. Ragnarsson, *Nucl. Phys. A* **222**:221 (1974).

(Ni 83) L. Niesen, F. PLeiter, and H. de Waard (eds.), Proc. 6th Int. Conf. on Hyperfine Interactions, Groningen, 1983, *Hyperfine Interact.* **15/16** (1983).

(No 78) G. Nowicki, K. Bekk, S. Göring, A. Hanser, H. Rebel, and G. Schatz, *Phys. Rev. C* **18**:2369 (1978).

(Ol 72) L. Olschewski, *Z. Phys.* **249**:205 (1972).

(Ot 81) E. W. Otten, *Nucl. Phys. A* **354**:471c (1981).

(Ot 85) E. W. Otten, Proc. of Symp. Alfred Kastler, Paris 1985, *Ann. Phys. Fr.* **10**:659 (1985).

(Ot 85b) E. W. Otten, *Hyperfine Interact.* **21**:43 (1985).

(Ot 85c) E. W. Otten, *J. Phys. Soc. Jpn.* **54**, suppl. I:59 (1985).

(Po 83) M. G. Porquet, C. Bourgeois, P. Kilcher, and J. Sauvage-Letessier, *Nucl. Phys. A* **411**:65 (1983).

(Pr 73) D. Proetel, R. M. Diamond, P. Kienle, J. R. Leigh, K. H. Maier, and F. S. Stephens, *Phys. Rev. Lett.* **31**:896 (1973).

(Ra 78) R. S. Raghavan and D. E. Murnick (eds.), Proc. 4th Int. Conf. on Hyperfine Interactions, Madison, New Jersey, 1977, *Hyperfine Interact.* **4** (1978).

(Ra 78b) I. Ragnarsson, S. G. Nilsson, and R. K. Sheline, *Phys. Rep.* **45**:1 (1978).

(Ra 79) H. L. Ravn, *Phys. Rep.* **54**:201 (1979).

(Ra 80) I. Ragnarsson, in *Future Directions in Studies of Nuclei Far From Stability* (J. H. Hamilton, E. H. Spejewski, C. R. Bingham, and E. F. Zganjar, eds.), North-Holland, Amsterdam (1980), p. 367.

(Ra 83) I. Ragnarsson, *Phys. Lett.* **130B**, 353 (1983).

(Ra 85) G. N. Rao (ed.), Proc. Int. Workshop on Hyperfine Interactions, Kanpur, India, 1984, *Hyperfine Interact.* **24–26** (1985).

(Ra 88) P. Raghavan, *At. Data Nucl. Data Tables*, to appear.

(Re 62) A. S. Reiner and L. Wilets, *Nucl. Phys.* **36**:457 (1962).

(Re 71) B. S. Reehal and R. A. Sorensen, *Nucl. Phys. A* **161**:385 (1971).

(Re 79) H. Rebel, H. J. Gils, and G. Schatz (eds.), Proc. Int. Discussion-Meeting on "What do we know about the radial shape of nuclei in the Ca region?" KFK-Rep. 2830, Karlsruhe (1979).

(Re 79b) P. G. Reinhard and D. Drechsel, *Z. Phys. A* **290**:85 (1979).

(Re 80) H. Rebel, K. Bekk, G. Nowicki, and G. Schatz, *Nukleonika* **25**:145 (1980).

(Re 84) P. G. Reinhard and E. W. Otten, *Nucl. Phys. A* **420**:173 (1984).

(Ro 72) A. Rosen and I. Lindgren, *Phys. Scr.* **6**:109 (1972).

(Ro 84) A. Rosen, B. Fricke, and G. Torbohm, *Z. Phys. A* **316**:157 (1984).

(Ru 73) N. Rud, D. Ward, H. R. Andrews, R. L. Graham, and J. S. Geiger, *Phys. Rev. Lett.* **31**:1421 (1973).

(Sa 65) P. G. H. Sandars and J. Beck, *Proc. R. Soc. London, Ser. A* **289**:97 (1965).

(Sa 88) G. Savard, J. K. P. Lee, J. E. Crawford, G. Thekkadath, H. T. Duong, J. Pinard, S. Liberman, F. Le Blanc, P. Kilcher, J. Obert, J. Oms, J. C. Putaux, B. Roussiere, and J. Sauvage, Preprint Institut de Physique Nucleaire, Orsay, France IPNO-DRE-88-10, to appear.

(Sc 35) H. Schüler and T. Schmidt, *Z. Phys.* **94**:457 (1935).

(Sc 55) C. Schwartz, *Phys. Rev.* **97**:380 (1955).

(Sc 57) C. Schwartz, *Phys. Rev.* **105**:173 (1957).

(Sc 75) H. Schweickert, J. Dietrich, R. Neugart, and E. W. Otten, *Nucl. Phys. A* **246**:187 (1975).

(Sc 78) F. Scheck, *Phys. Rep.* **44**:187 (1978).

(Sc 78b) B. Schinzler, W. Klempt, S. L. Kaufman, H. Lochmann, G. Moruzzi, R. Neugart, E. W. Otten, J. Bonn, L. von Reisky, K. P. C. Spath, J. Steinacher, and D. Weskott, *Phys. Lett.* **79B**:209 (1978).

(Se 69) E. C. Seltzer, *Phys. Rev.* **188**:1916 (1969).

(Sh 76) K. Shimoda (ed.), *High Resolution Laser Spectroscopy*, *Top. Appl. Phys.* **13**:51, Springer-Verlag, Heidelberg (1976).

(Sh 82) E. B. Shera, H. D. Wohlfahrt, M. V. Hoehn, and Y. Tanaka, *Phys. Lett.* **112B**:124 (1982).

(Si 80) R. F. Silverans, G. Borghs, G. Dumont, and J. M. van den Cruyce, *Z. Phys. A* **295**:311 (1980).

(Si 85) R. F. Silverans, G. Borghs, P. de Bishop, and M. van Hove, *Hyperfine Interact.* **24**:181 (1985).

(Si 86) R. F. Silverans, private communication.

(Si 88) R. F. Silverans, P. Lievens, V. L. Vermeeren, F. Buchinger, E. B. Ramsay, E. Arnold, W. Neu, R. Neugart, G. Ulm, and K. Wendt, *Phys. Rev. Lett.* **60**:2607 (1988).

(Sm 65) K. F. Smith and P. J. Unsworth, *Proc. R. Soc. London* **86**:1249 (1965).

(Sp 85) K. P. C. Spath, Doctoral thesis, Mainz (1985).

(St 50) R. M. Sternheimer, *Phys. Rev.* **80**:102 (1950).

(St 79) H. H. Stroke, D. Proetel, and H. J. Kluge, *Phys. Lett.* **82B**:204 (1979).

(St 85) R. M. Steffen, *Hyperfine Interact.* **24**:223 (1985).

(St 85b) N. Stone, *Hyperfine Interact.* **22**:3 (1985).

(St 85c) J. Streib, H. J. Kluge, H. Kremmling, R. B. Moore, H. W. Schaaf, and K. Wallmeroth, *Z. Phys. A* **321**:537 (1985).

(Ta 56) I. Talmi and R. Thieberger, *Phys. Rev.* **103**:718 (1956).

(Ta 71) I. Talmi, *Nucl. Phys. A* **172**:1 (1971).

(Ta 83) Y. Tanaka, R. M. Steffen, E. B. Shera, W. Reuter, M. V. Hoehn, and J. D. Zumbro, *Phys. Rev. Lett.* **51**:1633 (1983).

(Ta 84) I. Talmi, *Nucl. Phys. A* **423**:189 (1984).

(Th 81) C. Thibault, F. Touchard, S. Büttgenbach, R. Klapisch, M. de Saint Simon, H. T. Duong, P. Jaquinot, P. Juncar, S. Liberman, P. Pillet, J. Pinard, J. L. Vialle, A. Pesnelle, and G. Huber, *Phys. Rev. C* **23**:2720 (1981).

(Th 81b) C. Thibault, F. Touchard, S. Büttgenbach, R. Klapisch, M. de Saint Simon, H. T. Duong, P. Jaquinot, P. Juncar, S. Liberman, P. Pillet, J. Pinard, J. L. Vialle, A. Pesnelle, and G. Huber, *Nucl. Phys. A* **367**:1 (1981).

(Th 83) R. C. Thompson, M. Anselment, K. Bekk, S. Göring, A. Hanser, G. Meisel, H. Rebel, G. Schatz, and B. A. Brown, *J. Phys. G* **9**:443 (1983).

(To 64) W. J. Tomlinson and H. H. Stroke, *Nucl Phys.* **60**:614 (1964).

(To 82) F. Touchard, J. M. Seere, S. Büttgenbach, P. Guimbal, R. Klapisch, M. de Saint Simon, C. Thibault, H. T. Duong, P. Juncar, S. Liberman, J. Pinard, and J. L. Vialle, *Phys. Rev. C* **25**:2756 (1982).

(To 82b) F. Touchard, P. Guimbal, S. Büttgenbach, T. Klapisch, M. de Saint Simon, J. M. Serre, C. Thibault, H. T. Duong, P. Juncar, S. Liberman, J. Pinard, and J. L. Vialle, *Phys. Lett.* **108B**:169 (1982).

(To 85) G. Torbohm, B. Fricke, and A. Rosen, *Phys. Rev. A* **31**:2038 (1985).

(Tr 81) F. Träger, *Z. Phys. A* **299**:33 (1981).

(Ul 75) S. Ullrich and E. W. Otten, *Nucl. Phys. A* **248**:173 (1975).

(Ul 85) G. Ulm, J. Eberz, G. Huber, H. Lochmann, R. Menges, R. Kirchner, O. Klepper, T. Kühl, P. O. Larsson, D. Marx, D. Murnick, and D. Schardt, *Z. Phys. A* **321**:395 (1985).

(Ul 86) G. Ulm, S. K. Bhattacherjee, P. Dabkiewicz, G. Huber, H. J. Kluge, T. Kühl, H. Lochmann, E. W. Otten, K. Wendt, S. A. Ahmad, W. Klempt, and R. Neugart, *Z. Phys. A* **325**:247 (1986).

(Va 83) D. Vandeplassche, E. van Walle, C. Nuytten, and L. Vanneste, *Nucl. Phys. A* **396**:115c (1983).

(Va 85) D. Vandeplassche, E. van Walle, J. Wouters, N. Severijns, and L. Vanneste, *Hyperfine Interact.* **22**:483 (1985).

(Va 86) D. Vandeplassche, E. van Walle, J. Wouters, N. Severijns, and L. Vanneste, *Phys. Rev. Lett.* **57**:2641 (1986).

(Vo 77) E. I. Volmyanskii and V. G. Dubro, *Bull. Acad. Sci. USSR, Phys. Ser.* **41**, No. 6:124 (1977); Engl. transl. *Izv. Akad. Nauk SSSR, Ser. Fiz.* **41**, No. 6:1252 (1977).

(Wa 85) E. van Walle, J. Wouters, D. Vandeplassche, N. Severijns, and L. Vanneste, *Hyperfine Interact.* **22**:83 (1985).

(Wa 87) K. Wallmeroth, G. Bollen, M. J. G. Borge, J. Campos, A. Dohn, P. Egelhof, J. Grüner, H. J. Kluge, U. Krönert, F. Lindenlauf, R. B. Moore, A. Rodriguez, A. Venugopalan, and J. Wood, *Hyperfine Interact.* **34**:21 (1987).

(Wa 87b) K. Wallmeroth, G. Bollen, A. Dohn, P. Egelhof, J. Grüner, J. Campos, A. Rodriguez, M. J. G. Borge, A. Venugopalan, J. L. Wood, R. B. Moore, and H. J. Kluge, *Phys. Rev. Lett.* **58**:1516 (1987).

(We 81) R. Wenz, A. Timmermann, and E. Matthias, *Z. Phys. A* **303**:87 (1981).

(We 84) K. Wendt, S. A. Ahmad, F. Buchinger, A. C. Mueller, R. Neugart, and E. W. Otten, *Z. Phys. A* **318**:125 (1984).

(We 85) K. Wendt, Doctoral thesis, Mainz (1985).

(We 85b) A. Weller, P. Egelhof, R. Caplar, O. Karban, D. Krämer, K. H. Möbius, Z. Moroz, K. Rusek, E. Steffens, G. Tungate, K. Blatt, I. König, and D. Fick, *Phys. Rev. Lett.* **55**:480 (1985).

(We 87) K. Wendt, S. A. Ahmad, W. Klempt, R. Neugart, E. W. Otten, and H. H. Stroke, *Z. Phys. D* **4**:227 (1987).

(We 88) K. Wendt, S. A. Ahmad, C. Ekström, W. Klempt, R. Neugart, and E. W. Otten, *Z. Phys. A* **329**:407 (1988).

(Wh 64) L. Wharton, L. P. Gold, and W. Klemperer, *Phys. Rev.* **133B**:270 (1964).

(Wi 76) W. H. Wing, G. A. Ruff, W. E. Lamb, Jr., and J. J. Spezewski, *Phys. Rev. Lett.* **36**:1488 (1976).

(Wi 78) A. Winnacker, D. Dubbers, F. Fujara, K. Dörr, H. Ackermann, H. Grupp, P. Heitjans, A. Körblein, and H. J. Stöckmann, *Phys. Lett.* **67A**:423 (1978).

(Wo 72) C. Y. Wong, *Phys. Lett.* **41B**:446 (1972).

(Wo 77) H. Wollnik, F. K. Wohn, K. D. Wünsch, and G. Jung, *Nucl. Phys. A* **291**:355 (1977).

(Za 71) L. Zamick, *Ann. Phys. (N.Y.)* **66**:784 (1971).

(Za 85) D. Zawischa, *Phys. Lett.* **155B**:309 (1985).

(Za 87) D. Zawischa, U. Regge, and R. Stapel, *Phys. Lett.* **185B**:299 (1987).

# 8

WALTER GREINER (far left) is Professor and Director of the Institute for Theoretical Physics at the University of Frankfurt. He received his undergraduate education at that university and his graduate training at the University of Freiburg, receiving the Ph.D. degree in 1961. During the 1962–64 period, he held Assistant Professorships at the University of Maryland and at Florida State University of Freiburg before moving to the Theoretical Physics Chair at Frankfurt in 1964. In 1974 he was awarded the Max Born Prize and Medal jointly by the Institute of Physics in London and the German Physical Society. In 1982 he received the Otto-Hahn-Prize of Germany and an honorary D.Sc. degree from the University of the Witwatersrand in Johannesburg, South Africa.

MARIN IVASCU (middle left) is currently General Director of the Central Institute of Physics and Vice President of the State Committee for Nuclear Energy, Bucharest, Romania. He has been a member of the staff of the Institute for Atomic Physics since 1956. He received BS and Ph.D. degrees from Bucharest University in 1955 and 1968 respectively, and before his appointment as Director in 1977, he was for many years head of its Nuclear Structure Department.

DORIN N. POENARU (middle right) is a Senior Researcher at the Institute of Physics and Nuclear Engineering, Bucharest, Romania. He has been a member of the staff of this Institute, formerly the Institute for Atomic Physics, since 1958. After receiving BE and BS degrees at the Polytechnique Institute and at Bucharest University, respectively, he received the Ph.D. degree in Electronics in 1968 and again in Theoretical Physics in 1979, both in Bucharest. In 1969–70 he was a member of the research staff at the University of Strasbourg and during recent years he is also, temporarily, Professor at the Bucharest University and a frequent visiting professor at JINR in Dubna and at the Institute for Theoretical Physics of the Johann-Wolfgang-Goethe University at Frankfurt a.M.

AUREL SANDULESCU (far right) is Vice Director of the Joint Institute for Nuclear Research, Dubna, USSR, where he was previously Vice Director of its Laboratory for Nuclear Reactions. He received the BE and Ph.D. degrees in Physics from Bucharest University in 1955 and 1962, respectively. Sandulescu has been a member of the staff of the Institute for Atomic Physics in Bucharest since 1956. He is a frequent visiting professor at the Institute for Theoretical Physics in Frankfurt a.M., at GSI in Darmstadt, and at JINR at Dubna, USSR. He has held visiting professorships as well at the University of Helsinki (1965–66), the University of Mainz (1967–68), and at the Technische Hochschule, Darmstadt (1970) and was a visiting scientist at the Niels Bohr Institute in Copenhagen (1968–69).

8

# Cluster Radioactivities

Walter Greiner, Marin Ivascu, Dorin N. Poenaru, and Aurel Sandulescu

## 1. Introduction

Induced fission (Ha 39, Me 39) and spontaneous fission (Pe 40) were discovered long after $\alpha$-particle decay (Be 96, Ru 08) of the nucleus. One of the first successes of quantum mechanics was the explanation of $\alpha$-particle tunneling through the Coulomb barrier (Ga 28, Co 28, Co 29). Details about the historical development of other kinds of radioactivities are given elsewhere (Po 84b). Advances in the study of many nuclear-decay modes, especially in the region far from $\beta$ stability, were reviewed recently by Hamilton *et al.* (Ha 85).

The similarity between fission and $\alpha$-particle decay was recognized in the early stages of fission theory (Bo 39, Fr 39, Fr 46). Nevertheless, the theories of these two phenomena have been developed on essentially different grounds. Nuclear-reaction microscopic methods have been used in the theory of $\alpha$ decay, but for many years a phenomenological liquid-drop model (LDM) dominated nuclear fission theory.

The asymmetric distribution of the fragment masses (Va 73) from spontaneous or low excitation energy-induced fission was a longstanding puzzle in the theory of fission. The first attempt to consider both the collective nature of the nucleonic motion as well as the single-particle effects by adding the shell corrections to the LDM energy (My 66) led to a good estimate of nuclear ground-state (gs) deformations. The elementary calculation of potential-energy curves simply by summing single-particle Nilsson levels

---

Walter Greiner • Institute for Theoretical Physics, University of Frankfurt, Frankfurt am Main, Federal Republic of Germany Marin Ivascu and Dorin N. Poenaru • Central Institute of Physics and Nuclear Engineering, Bucharest, Romania. Aurel Sandulescu • Joint Institute for Nuclear Research, Dubna, USSR. (*Present affiliation*: Central Institute of Physics, Bucharest, Romania.)

as a function of deformation, of course, preceded this method. The next important step, producing a renewed interest in the development of fission theory, was the idea of deformed nuclear shells and the microscopic shell correction method (St 66, St 67, St 68) strongly stimulated by the experimental discovery (Po 62) of fission isomers (Me 80, Bj 80). Secondary minima in the collective potential (shape isomers) and their effect on low-energetic collective nuclear structure were already being considered by Greiner (Gr 65) in 1965.

By considering shell effects it was shown that in the potential-energy surfaces for fission there appears a valley corresponding to mass $A \sim 140$ and to the emission of strongly elongated fission fragments. In this way it was possible to explain, qualitatively, the fission asymmetry (Mö 70, Pa 71, Ad 70, Br 72, Mö 72, Mu 72, Mu 73) as being essentially dominated by shell structure. Also, it was noted (De 86, De 86b) that for very heavy elements around $^{264}$Fm there appears in the potential-energy surface for fission, in addition to the above valley, another one corresponding to symmetric fragmentation wherein both fragments have nearly spherical shapes—the bimodal symmetric fission modes (the usual spontaneous fission and the Sn-decay modes). Such breakups were already suggested in early calculation of potential-energy surfaces based on the two-center shell model (see, e.g., Mo 71). Significant progress was achieved with the development of the two-center shell model (TCSM) by the Frankfurt school (Mo 69, Ma 72) and its extension to asymmetric breakups via an asymmetric two-center shell model (ATCSM). This ATCSM made it possible to follow the shell structure all the way from the original nucleus, over the fission barrier, and into the individual fission fragments. Up to this point all theoretical arguments were based on the Nilsson model, which possessed the wrong asymptotic behavior for fission processes and for heavy-ion collisions. The ATCSM can now be considered as the basic shell model for all fission and heavy-ion (quasi-molecular) phenomena. Fragmentation theory (Fi 74, Ma 74, Ma 80), based on the two-center shell model, was successful in describing both regions of low and high mass asymmetry (Ma 76, Sa 78, Sa 78b, Sa 80). It was shown, for example, that in the potential-energy surfaces for fission, in addition to the above two valleys there exists yet another valley at much higher mass asymmetry corresponding to the double magic nucleus $^{208}$Pb with fragments having nearly spherical shapes. This new valley leads to additional peaks in the fission mass distribution—the superasymmetric fission modes—and constitutes an important step forward in the quantitative understanding and theoretical foundations of mass and charge fragmentation in the fission process.

Mass and charge asymmetry have been considered as dynamical degrees of freedom $[\eta_A = (A_1 - A_2)/A,\ \eta_Z = (Z_1 - Z_2)/Z]$. A Schrödinger equation was formulated for the system wave function $\Psi(\mathbf{R}, \eta_A, \eta_Z)$. Later, also extensions of the theory to describe more than two-center (i.e., three- and more center) breakup followed. Phenomenological shell corrections were extended (Sc 70) to high deformations encountered in the fission process. New variants of the LDM [such as the folded Yukawa model (Sc 68, Sc 69), the finite range of nuclear forces model (FRNFM) (Kr 74), and the Yukawa-plus-exponential folding model (Y + EM) (Kr 79)] were especially developed to describe quasi-molecular fission and fusion processes appearing in heavy-ion reactions. The extension of these models to systems with different charge densities (Po 79e, Po 80c, Po 80d) offered a good basis for the calculation of potentials needed in the fission theory of $\alpha$-particle decay (Po 79b, Po 79c, Po 79d, Po 80, Po 82)—an early attempt toward a unified treatment of these only apparently distinct processes.

Another model describing, in a continuous way, the transition from light-particle emission to fission in the decay of the compound nucleus at higher excitation energies well above the barrier was developed by Moretto (Mo 75) and by Swiatecki (Sw 83). It was recently tested (So 83) by studying the production of $^{4}$He, Li, $^{7,9}$Be, B, C, N, O, and F products in the reaction of 90 MeV $^{3}$He with $^{nat}$Ag. This model is based on a statistical transition-state formalism used in fission and allows computation of product angular and energy distributions. Unlike this model—dealing with excited states above the barrier—we are concerned in this chapter with ground or lower excited states, below the barrier where quantum-mechanical tunneling and shell effects are extremely important.

The purpose of this article is then to present both theory and experiments concerning the new decay modes involving spontaneous emission of clusters heavier than the $\alpha$ particle and lighter than the usual fission fragments. The classical theoretical paper of Sandulescu, Poenaru and Greiner entitled "New type of decay of heavy nuclei intermediate between fission and $\alpha$-decay" (Sa 80) initiated this field of cluster radioactivity in nuclei. Its predictions were based on fragmentation theory and it discussed all relevant aspects of the new phenomenon. Four theoretical models have been developed (Sa 80, Po 80) since 1980—assuming for the mechanism of these new cluster radioactivities either emission of a preformed cluster or a very asymmetric fission process. A new peak reflecting $^{48}$Ca emission appeared in the fission fragment mass distributions calculated using fragmentation theory (FT). Eight even–even (e–e) clusters ($^{14}$C, $^{24}$Ne, $^{28}$Mg, $^{32,34}$Si, $^{46}$Ar, and $^{48,50}$Ca) were suggested as potential candidates for radioactive emission from 16 e–e nuclides (various isotopes of the elements Ra, Th, Pu, Cm, Cf, Fm, and No). A numerical superasymmetric fission model (NSAFM), derived for binary systems with charge asymmetry different from the mass asymmetry, was developed to permit calculation of the half-life, an important measurable quantity, and it was used to describe both the $\alpha$-particle decay process and the new cluster decay modes.

In a systematic search for cluster radioactivity it is necessary to take into consideration a very large number of combinations of parents and emitted clusters. For example, if we were to attempt to study the stability of the 2200 nuclides, listed by Wapstra and Audi (Wa 85) in their new mass tables, relative to the emission of up to 200 different nuclides (various isotopes of the elements with atomic numbers $Z_2$ = 2–26), this number of combinations is about $10^5$. Analytical relationships are therefore needed to limit the computations to reasonable time. For this purpose an analytical superasymmetric fission model (ASAFM) was derived (Po 80) and extended (Po 83c, Po 84) to account for angular momentum and small excitation energy effects. It was shown that all nuclides of known masses are stable (negative $Q$-values) with respect to the emission of $^{2,3}$H, $^{3,6-9}$He, $^{4}$Li, $^{7}$B, and $^{9}$C, but that $^{9}$Be, $^{12}$C, $^{16}$O, etc. are good candidates for emission as clusters in the new radioactive-decay modes. When $^{5}$He radioactivity was predicted, a simple rule governing these processes was found: for a given emitted cluster, the maximum value of the emission rate (minimum half-life) occurs for the daughter nucleus having magic neutron and proton numbers ($N_1$ = 126, $Z_1$ = 82) or for nuclei in the immediate neighborhood of these magic numbers. This rule was confirmed when other emitted clusters with increasing $Z_2$ were considered.

A comprehensive listing (Po 86), containing more than 140 predicted new decay modes, was reported (Po 84b, Po 85c, Po 85e) and multiple heavy-ion radioactivity predicted (Po 85b). The branching ratios relative to $\alpha$-particle decay, or the absolute values of the partial half-lives, determined experimentally in 1984, are reproduced (within 1.5 orders of magnitude) by estimates made using the ASAFM. Since the calculated half-life is extremely sensitive to the various parameters of the potential and of the nuclear shapes, this result seems very promising. Recently Martin Greiner and Werner Scheid (Gr 86b) have calculated new radioactive decays to excited states; here both the daughter nucleus as well as the emitted cluster may be excited.

This model (ASAFM) was also applied to cold fission (Po 84b, Po 86b) in the uranium region and in the region of the very heavy elements around $^{264}$Fm (Sa 85c), and the general trends of the half-lives of the predicted new symmetric fission mode—with nearly spherical fragments—were obtained allowing a prediction of the probability of this process—the so-called Sn decay—relative to the known spontaneous fission with strongly elongated fragment shapes.

Shi and Swiatecki (Sh 85, Sh 85b) also developed a superasymmetric fission model based on a proximity plus Coulomb potential, allowing computation of the fission branching ratio relative to $\alpha$-particle decay, with results that reproduce recent experimental results.

There now exist experimental data concerning two of the predicted new decay modes: $^{14}$C spontaneous emission from $^{223}$Ra (Ro 84, Al 84, Ga 84, Ga 84b, Pr 85, Ku 85) and from $^{222,224}$Ra (Pr 85, Ho 85) and $^{226}$Ra (Ho 85), and $^{24}$Ne radioactivity of $^{232}$U (Ba 85), $^{231}$Pa (Sa 84), $^{233}$U (Tr 85), and $^{230}$Th (Tr 85b). Preliminary data also show that $^{28}$Mg is emitted from $^{234}$U and $^{236}$Pu and this is also predicted by our model (Po 86d). We have also interpreted the experimentally observed (Hu 86) total kinetic energy (TKE) distributions of the fission fragments for $^{258}$Fm, $^{259}$Md, $^{260}$Md, and $^{258}$No peaked either at low TKE (about 195 MeV) or high TKE (about 235 MeV) together with the very narrow symmetric mass distribution for high TKE and broad symmetric mass distribution for low TKE as evidence of the new decay mode—Sn decay. The prediction of cluster radioactivity has received worldwide attention and has led to extensive research activity [see also (Br 85, Ha 85, Gr 85, Sa 85, Sh 85, Iv 88)]. Its principal importance lies in the unification of nuclear-decay modes that it provides and in the possibility of studying nuclear structure effects on radioactive-decay modes over a very broad scale.

## 2. Nuclear Stability

### 2.1. Metastability

In order to see whether a nucleus $^{A}Z$ is stable or not with respect to its fission into two nuclei $^{A_1}Z_1$ and $^{A_2}Z_2$, one can use (Bl 56) the deformation energy of the system as a function of the separation distance between fragments (Figure 1). If the energy of the two nuclei at infinite separation is taken as the origin of the potential, the initial energy

$$E(R)|_{R=0} = Q(A_2, Z_2) = M(A, Z) - [M(A_1, Z_1) + M(A_2, Z_2)] \tag{1}$$

is the $Q$-value (the energy release), which can be computed easily from the well-known experimental masses (Wa 77, Wa 85).

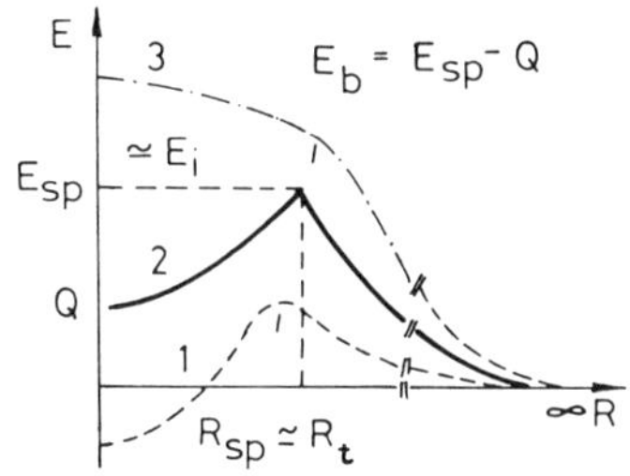


**Figure 1.** Stability (1), metastability (2), and instability (3) of the nuclear systems with respect to the fission process.

The fission barrier height is the difference

$$E_b = E_{sp} - Q \tag{2}$$

where $E_{sp}$ denotes the saddle-point energy.

Three distinct cases are shown in Figure 1:

1. *Stability.* $Q < 0$, $E_b > 0$. The nucleus $^AZ$ is in the position of minimum potential energy; its decay is prevented by energy conservation (an infinitely thick barrier).
2. *Metastability.* $Q > 0$, $E_b > 0$. The fragments are held together temporarily by a potential barrier. Owing to the quantum-mechanical tunneling effect, there is a finite probability $P$ per unit time for penetration through this barrier which decreases with an increase in $E_b$ and in the reduced mass $\mu = (A_1A_2/A)m$, where $m$ is the nucleon mass. The radioactive nuclei are metastable.
3. *Instability.* $Q > 0$, $E_b < 0$. The compound nucleus is unstable mainly due to the spontaneous fission.

One has to consider also $\beta$ instability as a nuclear-decay mode, but this is beyond the scope of this review.

If a given nucleus is metastable with respect to a particular fragmentation (binary or multicenter breakup) and if this decay mode has a larger probability than its neighbors, one expects to find a corresponding peak in the fission fragment mass distribution or a deep minimum in the lifetime spectrum. This can be connected with a local maximum in the barrier penetrability spectrum of the parent nucleus or with a maximum in the preformation probability for a particular fragmentation (decay mode).

## 2.2. Shell Model Valleys in the Potential-Energy Surfaces. Mass Distributions. Fragmentation Theory

The theory of fragmentation (Fi 74, Ma 80) is a consistent method permitting treatment of the two-body and many-body breakup channels in fission, fusion, and heavy-ion scattering. The unifying aspect of these phenomena is that the nuclear system transforms and disintegrates into other nuclear systems via collective mass transfer.

The collective description of a nuclear system has proven to be extremely successful (Ei 74). Consequently, the basic idea of the theory of fragmentation involves the introduction of collective coordinates such as the mass asymmetry

$$\eta = (A_1 - A_2)/A \tag{3}$$

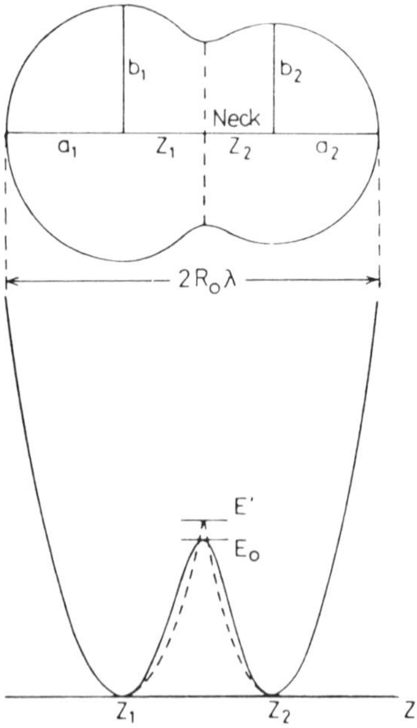


**Figure 2.** Asymmetric two-center potential along the $z$-axis and the associated nuclear shape. Barrier heights for the definition of the neck parameter $\varepsilon = E_0/E'$ are indicated.

and the distance between centers, $R$, or the elongation (see Figure 2)

$$\lambda = \frac{a_1 + a_2 + z_2 - z_1}{2R_0} \tag{4}$$

expressing the length of the nucleus along the symmetry axis in units of the radius $R_0$ of a spherical nucleus with mass number $A$. For a heated nucleus, one takes into account also the excitation energy $E^*$ or the nuclear temperature $\Theta$:

$$E^* = (A/10)\Theta^2 \qquad \text{(in MeV)} \tag{5}$$

If it is assumed that the shell corrections $\delta U$ to the liquid drop model (LDM) deformation energy $E_{\text{LDM}}$ are smoothed by means of a Gaussian factor

$$V(\lambda, \eta, \Theta) = E_{\text{LDM}}(\lambda, \eta) + \delta U(\lambda, \eta)\exp\left(-\frac{\Theta^2}{\Theta_0^2}\right) \tag{6}$$

one obtains for each elongation $\lambda$ and nuclear temperature $\Theta$, taken as parameters, a wave function $\Phi_\lambda^{(\nu)}$ as a solution of the Schrödinger equation

$$\left\{\left[-\frac{\hbar^2}{2(B_{\eta\eta})^{1/2}}\cdot\frac{\partial}{\partial\eta}\frac{1}{(B_{\eta\eta})^{1/2}}\cdot\frac{\partial}{\partial\eta} + V(\lambda, \eta, \Theta)\right] - E_\lambda^{(\nu)}(\lambda, \eta)\right\}\Phi_\lambda^{(\nu)}(\lambda, \eta, \Theta) = 0 \tag{7}$$

The fission-fragment mass yield normalized to 200% is given by

$$Y(A_2) = |\psi(\lambda, \eta, \Theta)|^2(B_{\eta\eta})^{1/2}(400/A) \tag{8}$$

where

$$|\psi(\lambda, \eta, \Theta)|^2 = \sum_{\nu=0}^{\infty} |\Phi_\lambda^{(\nu)}(\lambda, \eta, \Theta)|^2 \exp(-E_\lambda^{(\nu)}/\Theta) \tag{9}$$

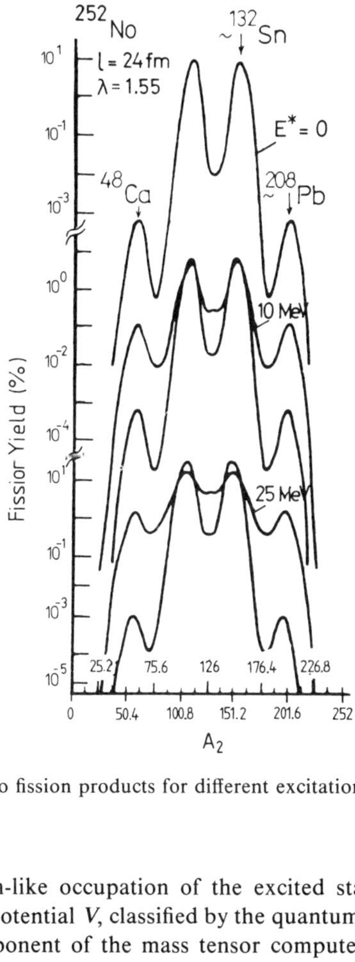


**Figure 3.** Yields of $^{252}$No fission products for different excitation energies and $l = 24$ fm.

expresses the Boltzmann-like occupation of the excited states. The states $\Phi_\lambda^{(\nu)}$ are vibrational states in the potential $V$, classified by the quantum numbers $\nu = 0, 1, 2, \ldots$, and $B_{\eta\eta}$ is the $\eta\eta$-component of the mass tensor computed in the framework of a cranking model within a two-center shell model.

Figure 3 represents the fission yield of $^{252}$No calculated (Sa 78, Sa 78b, Sa 80) for different excitation energies and a nuclear elongation parameter equal to 24 fm. Besides the well-known mass asymmetry due to "daughter" fragments in the neighborhood of the double magic $^{132}$Sn, there appears a new peak corresponding to fragments in the vicinity of the next double magic daughter, $^{208}$Pb, hence $^{48}$Ca for the emitted cluster. With increasing excitation energy, the mass distribution becomes more symmetric. A second peak in the fission-fragment mass distributions is predicted for Fm and Cf isotopes (Lu 80). The mass distributions resulting from the fission of excited medium heavy nuclei have been discussed by Depta *et al.* (De 85).

In fact, in the above equations the potential $V(\lambda, \eta, \Theta) = V(R, \eta, \Theta)$ is obtained numerically by a many-dimensional minimization procedure from a general shape potential $V(R, \eta, \eta_z, \varepsilon, \beta_1, \beta_2, \Theta)$ depending also on the deformation parameters of the two fragments $\beta_1$ and $\beta_2$, the neck parameter $\varepsilon$ defined in Figure 2, and the charge asymmetry $\eta_z = (Z_1 - Z_2)/Z$. The main result of these time-consuming calculations is that for each compound nucleus deep valleys could exist in each coordinate. For example, in the case of the mass-asymmetry degree of freedom, deep valleys in the fragmentation potential occur at only a few $\eta$ values and these valleys are not only stable in $\eta$ for large ranges of $R$ but also no new valleys appear after the two nuclei overlap to form a compound system. For example, the potential-energy surface for $^{252}$No, in the length ($L = 2R$) and mass asymmetry ($\eta$) coordinates minimized relative to the deformations of each fragment ($\beta_1, \beta_2$) and the neck parameter $\varepsilon$, shows two shell-model valleys primarily reflecting the double magic structure of $^{132}$Sn and of $^{208}$Pb, first with strongly elongated fragments and second with nearly spherical fragments (Sa 78, Sa 78b, Sa 80).

We should like to mention that, in the fission processes starting from inside the potential, both of these mass-asymmetry valleys have been observed experimentally—the most symmetric one through the mass distribution of fission fragments and the most asymmetric one through the new $^{14}$C and $^{24}$Ne radioactivities. As an example, Figure 4 presents a sample calculated PES (De 86b) for $^{224}$Ra. This shows only the results of a semirealistic calculation because the two-center shell model has not yet been developed to the point where it can describe very asymmetric shapes reasonably. However, it should suffice for emphasizing some features of this collective approach. The major structure in this PES is the broad minimum near the ground state determining the observable lowest odd multipole moments. A nonvanishing octupole moment does not require an octupole-deformed equilibrium deformation: softness in the octupole direction $\eta$ is completely sufficient. Another interesting feature of the ground-state wave function follows from the quantity $P(\eta) = \int dR|\psi_0(R, \eta)|^2$. Obviously one might interpret $P(\eta)$ as the preformation probability $P_{\text{cluster}}$ for the cluster configuration characterized by $\eta$.

Now the asymmetric valley also has an influence on the fission probability. Softness along a particular valley in the surface will lead to an enhanced probability density at that point and yield a large preformation probability for that specific breakup. In this sense, we see a direct linkage between static multipole moments and highly asymmetric fission: both may be related to valleys in the $(R, \eta)$-plane.

A similar situation exists for the shape degree of freedom including the neck. For very heavy elements around $^{264}$Fm, we expect to find in the potential-energy surface—minimized with respect to the shape coordinates at symmetric fragmentation ($\eta = 0$)

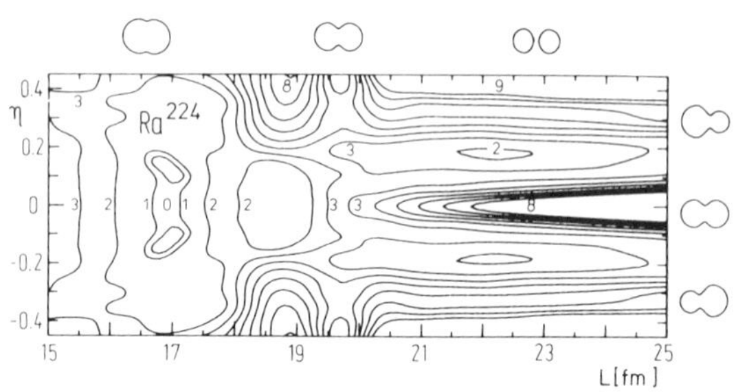


**Figure 4.** Minimized PES of a typical heavy nucleus as a function of $L = 2R$ and $\eta$. Embedded are asymmetric local minima which indicate nonvanishing static octupole deformation, while for larger values of $L$ the valleys crossing the fission barrier appear.

close to the scission configuration—two valleys in the shape degree of freedom. One of these corresponds to the usual spontaneous fission with strongly elongated fragments, and the other to a new type of fission with nearly spherical fragments. There are two minima, the first quite deep and the second very shallow. The deep minimum corresponds to nearly spherical fragments and the shallow minimum to strongly elongated ones. A relatively high barrier between them allows us to define two quite distinct types of fission.

Of course, we also expect other shape valleys at larger mass-asymmetry values when one of the fragments is close to a double magic nucleus.

A complete description of these processes implies the dynamical treatment of few collective coordinates. The other degrees of freedom which are not included explicitly, such as the intrinsic degrees of freedom, can be treated by opening the system. Finally, for the description of the above decay modes we need quantum mechanics for open systems (Sa 85b). We should like to mention that other processes, such as charge equilibration, fast fission, and deep inelastic reactions, can be described in the above framework as quantum trajectories on many-dimensional potential-energy surfaces.

## 2.3. Barrier Penetrability and Half-Life

From the energetic point of view all nuclides with $Z > 40$, as tabulated by Wapstra and Audi (Wa 85), are metastable with respect to several decay modes involving spontaneous emission of heavy clusters (Po 85). If the partial half-life, $T$, for a particular breakup is very long in comparison with the same quantity related to the others, this emission cannot be detected experimentally; only the faster processes are observed. The half-life of a metastable system is given by

$$T = \hbar \ln 2/\Gamma \tag{10}$$

where $\Gamma$ is the partial disintegration width. Two different mechanisms have been considered in the paper of Sandulescu *et al.* (Sa 80) for the new radioactivity phenomena: either emission of a preformed cluster or a superasymmetric fission process. In the first case, $\Gamma$ is calculated from the equation

$$\Gamma = 2\gamma^2 \cdot P' \tag{11}$$

where the reduced width $\gamma^2$ is proportional to the cluster preformation probability. In the framework of $R$-matrix theory of $\alpha$-particle decay one can compute $\gamma^2$ by using microscopically determined wave functions, but for heavier clusters, only recently have corresponding methods been developed (Iv 86). For this reason, only penetrabilities $P'$—or Gamow factors $G' = 1/P'$—have been computed by Sandulescu *et al.* (Sa 80), Rose and Jones (Ro 84), Alexandrov *et al.* (Al 84), and Price *et al.* (Pr 85). One assumes a cluster reduced to point dimensions, moving in a potential $V(r)$, which is primarily of electrostatic origin. The semiclassical Wentzel–Kramers–Brillouin (WKB) method leads to

$$P' = \exp(-K') \qquad \text{where} \quad K' = \frac{2}{\hbar}\int_{R_t}^{R_{out}} \{2\mu[V(r) - Q]\}^{1/2}\, dr \tag{12}$$

$V(R_{out}) = Q$, $\mu = (A_1A_2/A)m$ is the reduced mass, and $m$ is the nucleon mass. The "channel radius" $R_t$ is of the order of the touching-point distance $R_t \simeq R_1 + R_2$ where

$R_j = r_0 A_j^{1/3}$ $(j = 1, 2)$. It depends on the radius constant $r_0$. For this reason Rose and Jones (Ro 84) have used three different values: $r_0 = 1.15$, 1.20, and 1.25 fm. This means that both $P'$ and $\gamma^2$ are model-dependent quantities and only their product, $\Gamma$, can be experimentally determined. From the measured branching ratio relative to $\alpha$-particle decay

$$b = \frac{\Gamma}{\Gamma_\alpha} = \frac{T_\alpha}{T} = \left(\frac{\gamma^2}{\gamma_\alpha^2}\right) \cdot \frac{P'}{P'_\alpha}$$

which was of the order of $10^{-9}$ for $^{14}C$ radioactivity of $^{223}Ra$, while the calculated ratio of penetrabilities $P'/P'_\alpha \simeq 10^{-3}$ (Ro 84). The preformation probability ratio is $\gamma^2/\gamma_\alpha^2 \simeq 10^{-6}$ and it is expected that this ratio will be much smaller for emitted clusters heavier than $^{14}C$. But if we use a more realistic interaction barrier, including the overlapping region of the two fragments, we obtain larger values for $\gamma^2/\gamma_\alpha^2$.

Penetrability spectra for various even–even heavy clusters emitted from 16 even–even parent nuclides were presented by Sandulescu *et al.* (Sa 80). In this way eight new decay modes have been predicted: $^{14}C$, $^{24}Ne$, $^{28}Mg$, $^{32,34}Si$, $^{46}Ar$, and $^{48,50}Ca$ were the predicted clusters expected to be emitted in new radioactive processes. Some of these calculated spectra are shown in Figure 5, which shows that frequently the ratio $P'/P'_\alpha$ is larger than unity. Of course, this does not mean that $b > 1$. There is also a trend of increasing $P'$ when $A_2$ increases, reflecting the fact that the finite size of the fragments—including their deformations—were not taken into account; the neglected contribution for $R < R_t$ increases with $A_2$.*

In fission theory (Pa 75) the disintegration width

$$\Gamma = h\nu \cdot P = 2E_v \cdot P$$

where $\nu = \omega/2\pi = 2E_v/h$ represents the number of assaults on the barrier per second (the characteristic frequency of the collective mode) and $E_v \doteq \hbar\omega/2$ is the zero-point vibration energy (Mo 69, Jo 70, Fr 71, Ma 73); $E_v$ is obtained by fitting the experimental half-life data. The WKB penetrability

$$P = [1 + \exp(K)]^{-1} \simeq \exp(-K) \tag{13}$$

can be expressed as a product of two terms, because in the action integral

$$K = K_{ov} + K_s = \frac{2}{\hbar}\left(\int_{R_a}^{R_t} dr + \int_{R_t}^{R_b} dr\right)\{2\mu[E(r) - Q']\}^{1/2} \tag{14}$$

one has not only the term $K_s$ corresponding to the Coulomb interaction of the separated fragments but also $K_{ov}$, the contribution of the strong interaction in the overlapping region, which was not taken into account in $P'$. In equation (14) one has

$$Q' = Q + E_v \qquad \text{and} \qquad E(R_a) = E(R_b) = Q' \tag{15}$$

The energy barrier $E(r)$ can be computed using the Strutinsky (St 66, St 67, St 68) macroscopic–microscopic method adapted for superasymmetric breakup (Po 79b,

* Unfortunately, in the English-language version of the paper by Sandulescu *et al.* (Sa 80) the quantity plotted in Figure 7 was $\frac{1}{2}\log P'$ instead of $\frac{1}{2}\ln P'$. Due to this fact and to the comparison of the measured $b$ with $P'/P'_\alpha$, a disagreement of 13 orders of magnitude (which in fact does not exist) was mentioned by Bromley (Br 85).

Po 79c, Po 79d) with liquid-drop-model parameters obtained from a fit to experimental data on masses, fusion and fission barriers, electron-scattering experiments, etc. The emitted cluster is no longer treated as a point in a potential; it is taken as having a spatial extension with radius $R_2$ and $E(r)$ is the deformation energy of the two bodies. For $\alpha$-particle decay, $K_{ov}$ is only a few percent of $K_s$, but it increases with increasing mass number $A_2$. Consequently, $E_v$ does not play the role of a preformation probability. Shi and Swiatecki (Sh 85, Sh 85b) use the equation $T = \tau_0 G$ to compute half-lives, where the Gamow factor $G = 1/P$, hence the frequency factor $\tau_0 = h \ln 2/(2E_v)$. The fact that in equation (14) we have $Q'$ instead of $Q$ helps to compensate for a too high barrier, reflecting the cusp of saddle-point shapes approximated by a touching-point configuration. In fact the energy released in the process is $Q$, and not $Q'$, because the zero-point vibration energy is already included in the experimental mass values used to calculate $Q$. By using equations (14) and (15) one gets, from a fit to the experimental half-lives of 379 $\alpha$-emitters (Po 83b, Po 84d), the variation of $E_v$ with neutron number $N_1$ of the daughter nucleus, which is much smaller than that of $\gamma^2$. In a similar way, there is stability with respect to different emitted clusters. For example, $E_v(^{14}C)/E_v(\alpha)$ is as low as 3.5 but $\gamma^2(^{14}C)$ differs by six orders of magnitude from $\gamma_\alpha^2$ (Ro 84). This property of the parameter $E_v$ (Po 84h) is very useful when the method is employed to predict the lifetime for a new decay mode, because one can use an already known $E_v$ value from a neighboring decay mode.

A comparison between $P'$ and $T$ spectra for the spontaneous emission of various even–even clusters from some even–even parent nuclei [calculated by using equation (12) for $P'$] is illustrated in Figure 5. Here the $P'$ spectra are those presented by Sandulescu *et al.* (Sa 80) while the half-life spectra are calculated using the ASAFM. The vertical scales of log $P'$ and log $T$ are chosen in such a way that the experimental $\alpha$-particle decay lifetime, $T_\alpha^{\text{exp}}$, is at the same level as the corresponding penetrability $P'_\alpha$ corrected for $2\gamma_\alpha^2 \neq 1$ MeV:

$$\log T_\alpha^{\text{exp}} = -21.34 - \log P'_\alpha - \log(2\gamma_\alpha^2)$$

The argument used by some experimentalists, that they found a maximum in the ratio $P'/P'_\alpha$, does not imply necessarily that the branching ratio

$$b = \frac{\Gamma}{\Gamma_\alpha} = \frac{T_\alpha}{T} = \left(\frac{E_v}{E_{v\alpha}}\right)\frac{P}{P_\alpha} = \left(\frac{\tau_{0\alpha}}{\tau_0}\right)\frac{G_\alpha}{G} \tag{16}$$

will also have a maximum value. From the cases shown in Figure 5 it is shown that sometimes the penetrabilities, as functions of the emitted cluster, are maximal for the same cluster as in the reciprocal of the lifetime spectra (for example, $^{14}C$ from $^{222}Ra$, $^{34}Si$ from $^{238}U$, and $^{48}Ca$ from $^{254}Fm$). But there are many cases in which one obtains different results from the two spectra, as, for example, in the case of $^{244}Pu$: instead of $^{46}Ar$ and $^{50}Ca$ being maximal in the penetrability spectra, $^{40}S$ has a minimum in the lifetime spectra.

At the same time, the trend of increasing values of $P'$ when $A_2$ increases is not present in the lifetime spectra. In fact, frequently one finds a situation in which the penetrability $P'$—for a given heavy-ion emission—is even larger than that for $\alpha$-particle decay, from which (of course) one cannot deduce that the branching ratio $b = \Gamma/\Gamma_\alpha$ should be larger than unity.

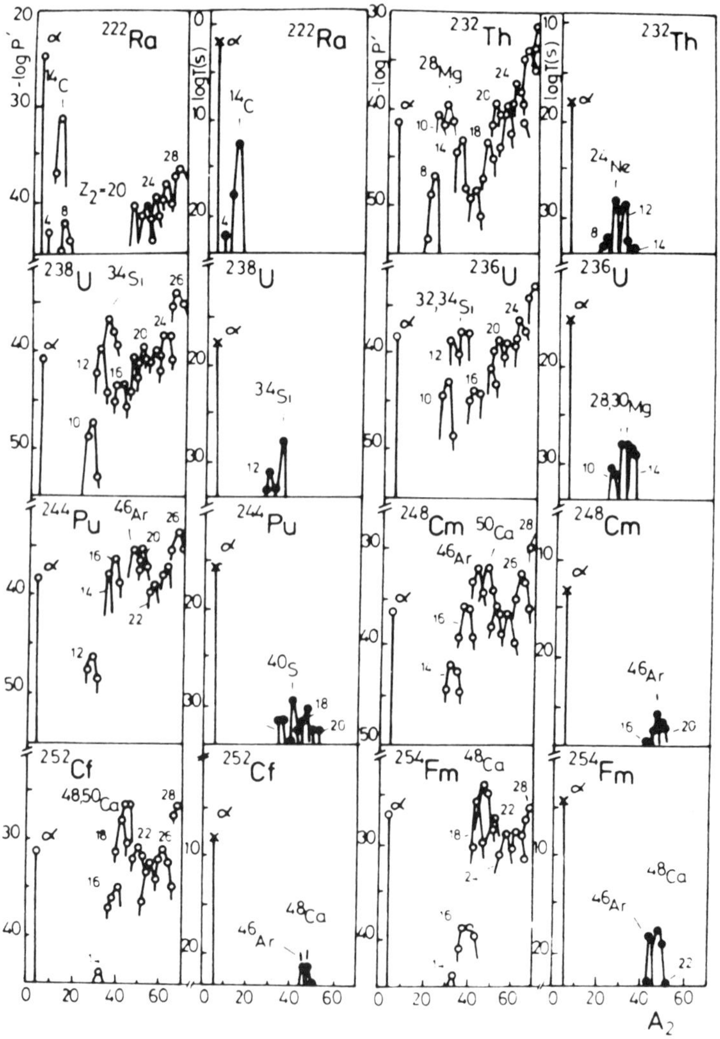

**Figure 5.** Penetrability $P'$ and half-life $T$ spectra for the emission of even–even heavy ions from some even–even nuclei.

In conclusion, the penetrabilities do not hinder the spontaneous emission of heavy clusters and, over a small region of emitted nuclei, i.e., in the vicinity of a certain reference decay, they could be used to obtain rough information about the new decay modes. In order to estimate the branching ratio directly measured in the experiment, one has to use the lifetime spectra. The good selectivity of these spectra is well illustrated for all parent nuclei.

## 3. Deformation Energies for Binary Systems with Charge Asymmetry Different from the Mass Asymmetry

The LDM of Myers and Swiatecki (My 66, My 67), and its generalization replacing the surface energy by a double-folded Y + EM over a sharp surface density distribution (Kr 79), are used in fission theory and in heavy-ion physics. The Y + EM parameters have been determined by fitting nuclear ground-state masses, fission barriers, heavy-ion elastic scattering, and elastic electron-scattering data. New physical effects have been introduced (Mö 81) in the mass formulas, among them a proton form factor, an exact diffuseness correction, a charge asymmetry term, and microscopic zero-point energies.

In order to reduce the number of independent parameters, some authors (Kr 76, Mö 77) have assumed the same charge-to-mass ratio ($Z/A$) by choosing one of the reaction partners to be off the line of $\beta$ stability. On the other hand, Zohni and Blann (Zo 78) considered realistic combinations but ignored the different charge-to-mass ratios. The difference between the mass- ($\eta_A$) and charge- ($\eta_Z$) asymmetry parameters has been considered only by some authors (Gu 75, Ad 76, Gu 77) when potential-energy surfaces (PES) have been calculated, though asymmetric mass and charge fragmentation (Fi 74) is a case frequently met in heavy-ion collisions, because for light beta-stable nuclei $Z_2 = N_2$, but for the heavy ones $N_1 > Z_1$.

It was shown (Iv 78, Po 79) both for the LDM and the FRNFM that the deformation energy can be underestimated if this difference is ignored. An extension to systems with charge asymmetry different from the mass asymmetry of these two models (Po 79e) and of Y + EM (Po 80d, Po 80c) has been developed.

It is obvious that the condition

$$\eta_A \equiv (A_1 - A_2)/A \neq \eta_z \equiv (Z_1 - Z_2)/Z \tag{17}$$

for mass- and charge-asymmetry coordinates, where $A = A_1 + A_2$ and $Z = Z_1 + Z_2$, is equivalent to

$$I \equiv (N - Z)/A \neq I_i \equiv (N_i - Z_i)/A_i \qquad (i = 1, 2) \tag{18}$$

for the nuclear composition, where $N = A - Z$ and

$$\rho_{1e} \neq \rho_{2e} \neq \rho_{0e} \equiv eZ/(\tfrac{4}{3}\pi r_0^3 A) \tag{19}$$

for the charge densities; $e$ is the electron charge and $r_0$ the radius constant.

If one ignores the difference $\eta_Z - \eta_A$, one has actually the pair of nuclei ${}^{A_1}Z_a$, ${}^{A_2}Z_b$, instead of ${}^{A_1}Z_1$, ${}^{A_2}Z_2$, where

$$Z_b = Z/(1 + q), \qquad Z_a = Z - Z_b, \qquad q = A_1/A_2 \tag{20}$$

We shall take into consideration the difference in charge densities for fusion or binary fragmentation and assume uniformity in each of the two fragments. In an intermediate stage of a fusion (or fission) reaction, when the two fragments are close together, as in Figure 6, we assume that the nuclear volume $V$ is divided in two parts, $V_1$ and $V_2$, and that each fragment is homogeneously charged with density

$$\rho_e(\mathbf{r}) = \begin{cases} \rho_{1e}, & \mathbf{r} \in V_1 \\ \rho_{2e}, & \mathbf{r} \in V_2 \end{cases} \tag{21}$$

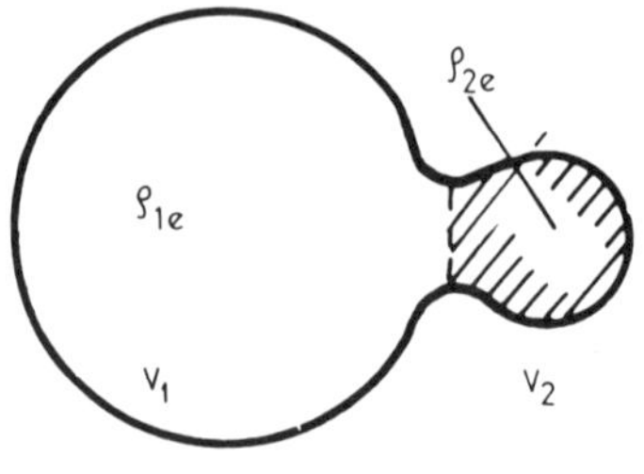


**Figure 6.** Binary system with different charge densities.

The deformation-dependent terms of the potential energy $E_d$, where

$$E_d = E_s + E_c + E_v \tag{22}$$

are the surface energy ($E_s$), Coulomb energy ($E_c$), and volume energy ($E_v$).

### 3.1. Liquid Drop Model (LDM) Surface Energy

The proportionality between surface energy $E_s$ and surface area is expressed by

$$E_s = c_{s1}A^{2/3}B_{s1} + c_{s2}A^{2/3}B_{s2}$$
$$B_s \equiv E_s/E_s^0 = (c_{s1}/c_s)B_{s1} + (c_{s2}/c_s)B_{s2} \tag{23}$$

where

$$B_{si} = S_i/S^0, \qquad c_{si} = a_s(1 - \kappa I_i^2) \qquad (i = 1, 2)$$
$$a_s = 17.9439 \text{ MeV}, \qquad \kappa = 1.7826 \tag{24}$$

$S_i$ is the surface area of fragment $i$ (without the separation area dividing $V_1$ from $V_2$) and $S$ is the surface area of the spherical fused nucleus ${}^AZ$, for which

$$E_s^0 = c_s A^{2/3} \tag{25}$$

with $c_s$ calculated as in equation (24) but with $I_i$ replaced by $I$.

### 3.2. Coulomb Energy for Various Macroscopic Models

The general expression for the Coulomb energy can be divided into three parts, by taking into account the two integration domains. Hence

$$E_c = \tfrac{1}{2}\rho_{1e}^2 \int_{V_1} d^3r_1 \int_{V_1} r_{12}^{-1}\, d^3r_2 + \tfrac{1}{2}\rho_{2e}^2 \int_{V_2} d^3r_1 \int_{V_2} r_{12}^{-1}\, d^3r_2$$
$$+ \rho_{1e}\rho_{2e} \int_{V_1} d^3r_1 \int_{V_2} r_{12}^{-1}\, d^3r_2, \qquad r_{12} = |\mathbf{r}_1 - \mathbf{r}_2| \tag{26}$$
$$B_c = \frac{E_c}{E_c^0} = (\rho_{1e}/\rho_{0e})^2 B_{c1} + (\rho_{2e}/\rho_{0e})^2 B_{c2} + (\rho_{1e}\rho_{2e}/\rho_{0e}^2) B_{c12}$$

where the first two terms represent the self-energies of the two fragments and the last one, their interaction energy. For a spherical homogeneously charged nucleus

$$E_c^0 = 3e^2Z^2/5r_0A^{1/3} \tag{27}$$

with $r_0 = 1.2249$ fm for LDM, $r_0 = 1.16$ fm for FRNFM, and $r_0 = 1.18$ fm for Y + EM. Möller and Nix (Mö 81) also used $r_0 = 1.16$ fm for Y + EM.

### 3.3. Surface Energy within a Finite Range of Nuclear Forces Model (FRNFM)

In the FRNFM (Sc 69, Kr 74) the surface energy is replaced by a folded Yukawa potential energy $E_n$, from which a spurious contribution to the volume energy, $E_{nV}$, has to be subtracted. This type of potential was used previously by Holm and Greiner (Ho 70). Recently Seiwert *et al.* (Se 85) developed a model for heavy-ion potentials, equivalent to the folding model for large separation and to the LDM in the limit of spherical composite systems. If the nuclear matter is homogeneously distributed in the two fragments ($\rho_1 = \rho_2 = \rho_0$) one has, similar to the case of Coulomb energy,

$$E_n = -\frac{1}{4\pi a^3}\left( V_{01}\int_{V_1} d^3r_1 \int_{V_1} d^3r_2\, g(x) + V_{02}\int_{V_2} d^3r_1 \int_{V_2} d^3r_2\, g(x) + 2(V_{01}V_{02})^{1/2}\int_{V_1} d^3r_1 \int_{V_2} d^3r_2\, g(x)\right)$$

$$g(x) = \exp(-x)/x, \qquad x = r_{12}/a \tag{28}$$

$$B_n \equiv (E_n - E_{nV})/(E_n^0 - E_{nV}^0) = \frac{c_{s1}}{c_s}B_{n1} + \frac{c_{s2}}{c_s}B_{n2} + \frac{\sqrt{c_{s1}c_{s2}}}{c_s}B_{n12}$$

where

$$V_0 = [V_0(r_1)V_0(r_2)]^{1/2}, \qquad V_0(r) = V_{0i} = c_{si}/2\pi a r_0^2 \qquad r \in V_i \quad (i = 1, 2) \tag{29}$$

$a = 1.4$ fm, $a_s = 24.7$ MeV, and $\kappa = 4$ as usual (Kr 74). For separated nuclei

$$E_{n\tilde{V}} = E_{nV_1} + E_{nV_2} = (2r_0/3a)(c_{s1}A_1 + c_{s2}A_2) \tag{30}$$

and for a spherical (fused nucleus

$$E_n^0 - E_{nV}^0 = c_s[A^{2/3} - (a/r_0)^2 + (A^{1/3} + a/r_0)^2 \exp(-2R_0/a)], \qquad R_0 = r_0A^{1/3} \tag{31}$$

### 3.4. Yukawa + Exponential Model (Y + EM) Surface Energy

The double-folded Yukawa-plus-exponential potential energy is also divided into three parts:

$$E_Y = -\frac{1}{8\pi^2 r_0^2 a^4}\left[ c_{s1}\int_{V_1} d^3r_1 \int_{V_1} f(x)\, d^3r_2 + c_{s2}\int_{V_2} d^3r_1 \int_{V_2} f(x)\, d^3r_2 + 2(c_{s1}c_{s2})^{1/2}\int_{V_1} d^3r_1 \int_{V_2} f(x)\, d^3r_2\right]$$

$$f(x) = (x-2)g(x), \qquad x = r_{12}/a \tag{32}$$

$$B_Y \equiv E_Y/E_Y^0 = (c_{s1}/c_s)B_{Y1} + (c_{s2}/c_s)B_{Y2} + [(c_{s1}c_{s2})^{1/2}/c_s]B_{Y12}$$

$$E_Y^0 = c_sA^{2/3}\{1 - 3(a/R_0)^2 + (1 + R_0/a)[2 + 3a/R_0 + 3(a/R_0)^2]\, e^{-2R_0/a}\}$$

In spite of using the same notations, for Y + EM one finds $a_s = 21.7$ MeV, $\kappa = 3$, and $a = 0.65$ fm. Möller and Nix (Mö 81) obtained—in a new fit to nuclear masses, fission barriers, and electron-scattering experiments—smaller RMS errors for the ground-state masses, without microscopic zero-point corrections, with the following values: $a_s = 21.14$ MeV, $\kappa = 2.4$, and $a = 0.68$ fm.

### 3.5. Volume Energy

In the LDM the symmetry energy is included in the volume energy $E_V$, so for $\eta_Z \neq \eta_A$ one also has to add the volume contribution to the deformation energy (although volume conservation is assumed):

$$E_{V_1} + E_{V_2} - E_V^0 \neq 0, \qquad E_{V_i} = -c_{V_i}A_i, \qquad c_{V_i} = a_V(1 - \kappa_V I_i^2) \tag{33}$$

where $a_V = 15.4941$ MeV, $\kappa_V = \kappa$ for LDM, and $a_V = 16.4696$, $\kappa_V = 2.31015$ for FRNFM (Mö 76). These values are only slightly different from those adopted by Möller *et al.* (Mö 74). For Y + EM, the corresponding quantities are $a_V = 16.012$ MeV, $\kappa_V = 2.64$, and the new ones $a_V = 16.0053$ MeV, $\kappa_V = 1.959$.

Equations (23), (26), (28), and (32) are generally valid for binary fragmentation even in the absence of axial symmetry. They can be easily generalized to ternary fragmentation (Di 73, Wu 85, Ha 77b, De 79).

## 4. Computation of the Deformation-Dependent Terms of the Potential Energy

### 4.1. Reducing the Order of Integration for General Shapes

Only for very simple nuclear surface shapes (see Section 4.5 below) can equations (24), (26), (28), and (32) be integrated in order to obtain analytical results. Usually, for a given parametrization $\rho = \rho(z, \phi)$ of the nuclear shape, this is not possible; one has to apply appropriate methods of numerical quadrature. Unfortunately, achievement of the required accuracy (more than five significant digits) in the case of sixfold integrals like those in equations (26), (28), and (32) is expensive with respect to computer time. Consequently, it is very important to try to reduce the complexity of the problem analytically.

If $(\rho, \phi, z)$ are the cylindrical coordinates of a point on the nuclear surface, the surface area is given by (Ju 74)

$$S = \int_{z_1}^{z_2} dz \int_0^{2\pi} d\phi \left\{ \rho^2(z, \phi) \left[ 1 + \left( \frac{\partial \rho}{\partial z} \right)^2 \right] + \left( \frac{\partial \rho}{\partial \phi} \right)^2 \right\}^{1/2} \tag{34}$$

where $z_1$ and $z_2$ are the intersection points of the $z$-axis with the nuclear surface. This double integral requires negligible computer time for its evaluation in comparison to all other operations. Of course, the calculation of the volume energy by equation (33) is even faster, once the quantities $A_i$ and $I_i$ have been determined for a given deformation.

The double-volume integral expressing the Coulomb energy of a uniform charge distribution, or the gravitational potential energy of a uniform mass distribution, can be converted (Da 75) in various ways into a double-surface integral by two applications of the Gauss divergence theorem. In this way one obtains

$$E_c = -\frac{1}{12}\oint\oint \rho_e^2 \frac{(d\mathbf{S}_1 \cdot \mathbf{r}_{12})(d\mathbf{S}_2 \cdot \mathbf{r}_{12})}{r_{12}} \tag{35}$$

or

$$E_c = \frac{1}{10}\oint\oint \rho_e^2 \frac{(d\mathbf{S}_1 \cdot \mathbf{r}_1)(d\mathbf{S}_2 \cdot \mathbf{r}_{12})}{r_{12}} \tag{36}$$

or

$$E_c = -\frac{1}{2}\oint\oint \rho_e^2 \frac{(d\mathbf{S}_1 \cdot \mathbf{r}_2)(d\mathbf{S}_2 \cdot \mathbf{r}_{12})}{r_{12}} \tag{37}$$

The computer time can be further reduced by exploiting the symmetry of equation (35) subject to the interchange of $\mathbf{r}_1$ and $\mathbf{r}_2$. In cylindrical coordinates, equation (36) assumes the form

$$E_c = \frac{1}{20}\int_{z_1}^{z_2} dz \int_{z_1}^{z_2} dz' \int_0^{2\pi} d\phi \int_0^{2\pi} d\phi' \, \rho_e^2 \times \frac{\left\{\left(\rho'^2 - \frac{z'}{2}\frac{\partial \rho'^2}{\partial z'}\right)\left[2\rho^2 - 2\rho'\rho\cos(\phi' - \phi) - (z - z')\frac{\partial \rho^2}{\partial z} - 2\rho'\frac{\partial \rho}{\partial \phi}\sin(\phi - \phi')\right]\right\}}{[(z - z')^2 + \rho^2 + \rho'^2 - 2\rho\rho'\cos(\phi - \phi')]^{1/2}} \tag{38}$$

where $\rho' = \rho'(z', \phi')$ and $\rho = \rho(z, \phi)$ denote the equations for the nuclear surface. Similarly, quantities $E_n$ and $E_Y$ are given, respectively, by the double-surface integrals

$$E_n = \frac{1}{8\pi^2 r_0^2 a^4}\oint\oint c_s (d\mathbf{S}_1 \cdot \mathbf{r}_{12})(d\mathbf{S}_2 \cdot \mathbf{r}_{12})\left(\frac{a}{r_{12}}\right)^4 \times \left[\frac{r_{12}}{a} - 2 + \left(\frac{r_{12}}{a} + 2\right) e^{-r_{12}/a}\right] \tag{39}$$

and

$$E_Y = \frac{1}{8\pi^2 r_0^2 a^4}\oint\oint c_s (d\mathbf{S}_1 \cdot \mathbf{r}_{12})(d\mathbf{S}_2 \cdot \mathbf{r}_{12})\left(\frac{a}{r_{12}}\right) \times \left\{\left[\frac{r_{12}}{a}\left(\frac{r_{12}}{a} + 2\right) + 2\right] e^{-r_{12}/a} - 2\right\} \tag{40}$$

To pass from Cartesian $(x, y, z)$ to cylindrical $(\rho, \phi, z)$ coordinates [as from equation (36) to equation (38)] one uses the following expressions:

$$\mathbf{r} = x\mathbf{i} + y\mathbf{j} + z\mathbf{k}, \qquad x = \rho\cos\phi, \qquad y = \rho\sin\phi$$
$$d\mathbf{S} = \left(\frac{\partial \mathbf{r}}{\partial \phi}\right) \times \left(\frac{\partial \mathbf{r}}{\partial z}\right) dz\, d\phi \tag{41}$$

where **i**, **j**, **k** are the unit vectors of the Cartesian system of coordinates.

The general relationships given in this section are not particularized for binary systems with different charge densities. Such expressions can be obtained by applying the method used in the previous section. If the nuclear shape has symmetry with respect to the $z$-axis, the order of integration in equations (34)–(40) can be reduced further by one.

### 4.2. Davies–Sierk Method for the Coulomb Energy of Axially Symmetric Nuclei

Three methods were frequently used in the past to compute the electrostatic energy $E_c$ of an axially symmetric nucleus: Lawrence's (La 65) method in which $E_c$ is expressed as a threefold integral; Hill and Wheeler's (Hi 53) method derived from equation (38), in which $E_c$ is a twofold integral; and Beringer's (Be 63) method containing only a sum of terms. These methods were compared by Poenaru and Galeriu (Po 75), who also extended Beringer's formula for reflection symmetric bodies (Ga 76).

The integrals in the above formula are computed by using Gauss–Legendre numerical quadrature. It is convenient (Po 79) to express the equation of the nuclear surface in cylindrical coordinates $y = y(x)$ with a properly selected scale factor [$y = \rho/(z_2 - z_1)$; $x = z/(z_2 - z_1)$] leading to $(-1, +1)$ instead of $(z_1, z_2)$ intercepts of the surface with the symmetry axis. For axially symmetric shapes one can reduce further the order of integration in equation (38). Integration over one of the angles $\phi$ gives a factor of $2\pi$, and from the remaining angular integration one obtains an integrand containing complete elliptic integrals (Hi 53). With a reliable method for evaluating these elliptic integrals (Co 65), the Hill–Wheeler integral should be faster to compute than equation (47) involving an extra numerical integration (Po 78). All three methods based on equations (35)–(37) are much more accurate in a given time than the methods of Beringer or of Lawrence. The symmetry of equation (35) for interchange of $\mathbf{r}_1$ to $\mathbf{r}_2$ makes the Davies–Sierk formula faster than computation using equation (36) or (37).

This formula adapted to the case of a binary system with charge asymmetry different from the mass asymmetry (Po 80d) leads to the following relationships for the quantities in equation (26):

$$\begin{aligned}
B_{c1} &= b_c \int_{-1}^{x_3} dx \int_{-1}^{x_3} dx' F(x, x'), \\
B_{c2} &= b_c \int_{x_3}^{1} \int_{x_3}^{1} dx' F(x, x'), \\
B_{c12} &= b_c \int_{-1}^{x_3} dx \int_{x_3}^{1} dx' F(x, x')
\end{aligned} \tag{42}$$

where $b_c = 5d^5/(8\pi)$, $x_3$ is the position of the separation planes between fragments (see, for example, $x_c$ in Figure 7 below), and

$$\begin{aligned}
F(x, x') = \Bigg\{ & yy_1[(K - 2D)/3]\left[2(y^2 + y_1^2) - (x - x')^2 + 1.5(x - x')\left(\frac{dy_1^2}{dx'} - \frac{dy^2}{dx'}\right)\right] \\
& + K\left\{y^2 y_1^2/3 + \left[y^2 - 0.5(x - x')\frac{dy^2}{dx}\right]\left[y_1^2 + 0.5(x - x')\frac{dy_1^2}{dx'}\right]\right\}\Bigg\} \Big/ a_\rho
\end{aligned} \tag{43}$$

in which $y_1 = y(x')$ is the equation of the nuclear surface with $(-1, +1)$ intercepts on the symmetry axis and

$$a_\rho^2 = (y + y_1)^2 + (x - x')^2, \qquad D = (K - K')/k^2, \qquad k^2 = 4yy_1/a_\rho^2 \tag{44}$$

Here $K$ and $K'$ are complete elliptic integrals of the first and second kind, respectively, with argument $k$:

$$K(k) = \int_0^{\pi/2} (1 - k^2 \sin^2 t)^{-1/2}\, dt, \qquad K'(k) = \int_0^{\pi/2} (1 - k^2 \sin^2 t)^{1/2}\, dt \tag{45}$$

These elliptic integrals are computed using a Chebyshev polynomial approximation. By removing the indeterminacy arising for $x = x'$, the function to be integrated takes the value $F(x, x) = 4y^3/3$.

### 4.3. Surface Energies

The expressions for the $B$-coefficients in equations (23), (28), and (32) will be developed explicitly in this section.

#### 4.3.1. Liquid Drop Model

Quantities $B_{s1}$ and $B_{s2}$ in equation (23) are given by

$$B_{s1} = 0.5d^2 \int_{-1}^{x_3} dx \left[ y^2 + 0.25 \left( \frac{dy^2}{dx} \right)^2 \right]^{1/2},$$
$$B_{s2} = 0.5d^2 \int_{x_3}^{1} dx \left[ y^2 + 0.25 \left( \frac{dy^2}{dx} \right)^2 \right]^{1/2} \tag{46}$$

for axially symmetric nuclei. The same notation is used as in the preceding section.

#### 4.3.2. Finite Range of Nuclear Force Model

For quantities $B_{n1}$, $B_{n2}$, and $B_{n12}$ in equation (28) one obtains similarly

$$B_{n1} = b_n \left( -0.5d^3 \int_{-1}^{x_3} dx \int_{-1}^{x_3} dx' \int_0^1 dw\, F_1 F_2 Q + 2A_1/3A \right)$$
$$B_{n2} = b_n \left( -0.5d^3 \int_{x_3}^{1} dx \int_{x_3}^{1} dx' \int_0^1 dw\, F_1 F_2 Q + 2A_2/3A \right) \tag{47}$$
$$B_{n12} = -0.5 b_n d^3 \int_{-1}^{x_3} dx \int_{x_3}^{1} dx' \int_0^1 dw\, F_1 F_2 Q$$

where

$$F_1 = y^2 - yy_1 \cos\phi - 0.5(x - x') \frac{dy^2}{dx}, \qquad F_2 = y_1^2 - y_1 y \cos\phi + 0.5(x - x') \frac{dy_1^2}{dx'} \tag{48}$$

$$b_n = c_s A r_0 / [a(E_n^0 - E_{nV}^0)] \tag{49}$$

$$Q = [P^{1/2} - 2a/R_0 d + (P^{1/2} + 2a/R_0 d) \exp(-R_0 d P^{1/2}/a)]/P^2 \tag{50}$$

$$P = y^2 + y_1^2 - 2yy_1 \cos\phi + (x - x')^2, \qquad \phi = 2\pi w \tag{51}$$

When $x = x'$, one has $y = y_1$ and the integrand becomes

$$(F_1F_2Q)_{ii} = 0.5\{y\sin(\phi/2) - a/R_0d + [y\sin(\phi/2) + a/R_0d]\exp[-(2R_0d/a)y\sin(\phi/2)]\} \tag{52}$$

Unlike $B_{ci}$, which depends only on the nuclear shapes and allows rapid computation for many nuclei simultaneously, $B_{ni}$ depends also on the nuclear mass. Hence the computation for this model has to be carried out for each nucleus separately. The same is true for the $B_Y$ quantities in the Y + EM.

### 4.3.3. Yukawa + Exponential Model

Quantities $B_{Y1}$, $B_{Y2}$, and $B_{Y12}$ in equation (32) are expressed as follows:

$$\begin{aligned} B_{Y1} &= b_Y \int_{-1}^{x_3} dx \int_{-1}^{x_3} dx' \int_0^1 dw\, F_1F_2Q_Y \\ B_{Y2} &= b_Y \int_{x_3}^{1} dx \int_{x_3}^{1} dx' \int_0^1 dw\, F_1F_2Q_Y \\ B_{Y12} &= b_Y \int_{-1}^{x_3} dx \int_{x_3}^{1} dx' \int_0^1 dw\, F_1F_2Q_Y \end{aligned} \tag{53}$$

where

$$\begin{aligned} b_Y &= -d^4(r_0/2a^2)c_sR_0A/E_Y^0 \\ Q_Y &= \{(P^{1/2}(P^{1/2} + 2a/R_0d) + 2a^2/(R_0d)^2]\exp(-R_0dP^{1/2}/a) - 2a^2/(R_0d)^2\}/P^2 \end{aligned} \tag{54}$$

while $F_1$, $F_2$, and $P$ remain unchanged.

## 5. α-Particle Decay as a Fissionlike Process

We shall primarily present in this section the numerical superasymmetric fission model (NSAFM) applied to $\alpha$-particle decay and a semiempirical formula for the $\alpha$-particle decay half-life.

### 5.1. Nuclear Shape Parametrization

If a simple parametrization of the nuclear shape is used, one has to use numerical, time-consuming procedures only in the overlapping region of the two fragments; for the touching point configuration and the separated objects one can obtain simple analytical relationships.

The ground-state shapes of the parent, the daughter nuclei, and the emitted cluster are approximated by spheres of radii $R_0$, $R_1$, and $R_2$, respectively. In an intermediate stage during the deformation process (see Figure 7), the distance between centers is $R$. It increases from $R_i = R_0 - R_2$ to $R_t = R_1 + R_2$ at the touching point, in such a way that $R_2$ is kept constant and the volume $V_1 + V_2$ is conserved. The surface equation

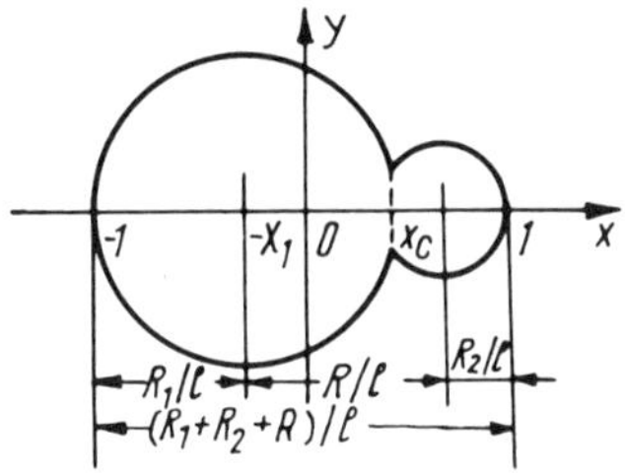


**Figure 7.** Spherical two-center parametrization of the nuclear surface.

$y(x)$, in cylindrical coordinates, with $(-1, +1)$ intercepts on the symmetry axis is

$$y^2 = \begin{cases} R_1^2/l^2 - (x + x_1)^2, & x \leqslant x_c \\ R_2^2/l^2 - (x + x_1 - R/l)^2, & x > x_c \end{cases} \tag{55}$$

where $2l = R + R_1 + R_2$, $x_1 = (l - R_1)/l$, and $x_c = z_c/l - x_1$. For a given deformation parameter $R$, the position of the separation plane between fragments relative to the heavy-fragment center, $z_c$, and the radius of the large fragment, $R_1$, are determined by solving iteratively the nonlinear system of equations expressing the matching and volume conservation conditions:

$$\begin{aligned} &R_1^2 - 2Rz_c + R^2 - R_2^2 = 0 \\ &2R_1^3 + 3R_1 z_c - 3Rz_c^2 + 3(R^2 - R_2^2)z_c + R_2^2(2R_2 + 3R) - R^3 - 4 = 0 \end{aligned} \tag{56}$$

with a first guess $R_{10} = 1$, $z_{c0} = 0.5[(1 - R_2^2)/R + R]$. In this section all the lengths $R$, $R_1$, $R_2$, and $z_c$ are expressed in units of $R_0$. We presume a proportionality between the fragment mass numbers $A_1$ and $A_2$ and their volumes $V_1$ and $V_2$. At the touching point $A_{1f} = A_1(R_t) = A - A_2 = A - 4$ and $A_{2f} = A_2(R_2) = 4$ for $\alpha$ decay.

For deformed nuclei the ground-state shapes of the parent and daughter are approximated by two spheroids of semiaxes $a_0$, $c_0$ and $a_1$, $c_1$, respectively (see Figure 8), the ratio of which can be determined from tables of nuclear deformations (Se 75, Be 84, Mö 74, Mö 76, Mö 81b):

$$\beta_0 \equiv c_0/a_0 = (1 + \alpha_{20} + \alpha_{40})/(1 - 0.5\alpha_{20} + 0.375\alpha_{40}) \tag{57}$$

One assumes a spherical shape with constant radius $R_2 = r_0 A_{2f}^{1/3}$ for the emitted particle. During the deformation process, the distance between centers, $R$, is increased from $R_i = c_0 - R_2$ to $R_t = c_1 + R_2$ at the touching point, then to $\infty$ in such a way that the volume is conserved and the ratio of semiaxes of the large fragment, $\beta \equiv c/a$, is a linear extrapolation of the ratio between that of the parent $\beta_0$ and that of the daughter $\beta_1$:

$$\beta = \beta_0 - (\beta_0 - \beta_1)(R - R_i)/(R_t - R_i), \qquad R_i \leqslant R \leqslant R_t \tag{58}$$

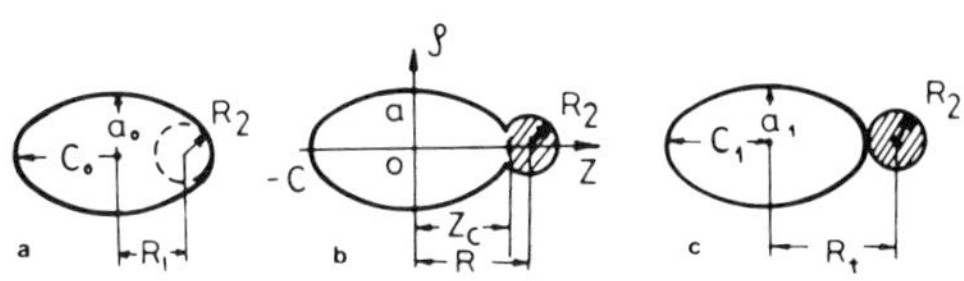


**Figure 8.** Spheroid intersected by a sphere: (a) initial configuration; (b) intermediate position; (c) touching point.

In cylindrical coordinates $(\rho, z)$ the surface equation is given by

$$\rho^2 = \begin{cases} a^2(1 - z^2/c^2), & z \leqslant z_c \\ R_2^2 - (z - R)^2, & z > z_c \end{cases} \tag{59}$$

The volume conservation condition can be written as

$$4\pi R_0^3/3 = 4\pi a_0^2 c_0/3 = 4\pi(a_1^2 c_1 + R_2^3)/3 = V_1 + V_2 \tag{60}$$

where $R_0 = r_0 A^{1/3}$,

$$\begin{aligned} 3V_1/\pi &= c^2[2c + z_c(3 - z_c^2/c^2)]/\beta^2 \\ 3V_2/\pi &= z_c^3 - 3Rz_c^2 + 3(R^2 - R_2^2)z_c + R_2^2(2R_2 + 3R) - R^3 \end{aligned} \tag{61}$$

We assume proportionality between the fragment mass numbers $A_1$ and $A_2$ (at the touching point $A_{1f}$ and $A_{2f}$) and their volumes $V_1$ and $V_2$. From the matching condition at the separation plane $z = z_c$, we obtain

$$c^2 - z_c^2 = \beta^2[R_2^2 - (z_c - R)^2] \tag{62}$$

In this way, for a given value of one independent deformation parameter $R$, the geometry of the system is perfectly determined. The two unknown quantities $x = c$ and $y = z_c$ are the simultaneous solutions of the following nonlinear system of equations:

$$x^2 + (\beta^2 - 1)y^2 - 2R\beta^2 y + \beta^2(R^2 - R_2^2) = 0 \tag{63}$$

$$\begin{aligned} 2x^3 + 3x^2 y + (\beta^2 - 1)y^3 - 3R\beta^2 y^2 + 3(R^2 - R_2^2)\beta^2 y \\ + \beta^2[R_2^2(2R_2 + 3R) - R^3 - 4R] = 0 \end{aligned}$$

This can be solved by refining an initial guess

$$y_0 = c_1, \qquad x_0 = \{y_0^2 + \beta^2[R_2^2 - (y_0 - R)^2]\}^{1/2} \tag{64}$$

with Newton's numerical iteration method.

## 5.2. Variation of the Charge Density

It is known from heavy-ion collision studies that the charge density equilibration process is very fast. This, in fact, has been anticipated in fragmentation theory because, basically, the large symmetry energy is responsible for the driving force of this process (Gu 75). The charge densities of the two fragments are considered to be the same until $R = R_7 = R_i + 0.7(R_t - R_i)$; then the charge number $Z_2$ of the small fragment increases linearly with $R$ up to $Z_{2f} = 2$:

$$Z_2 = Z_{27} + (Z_{2f} - Z_{27})(R - R_7)7/[3(R_7 - R_i)], \qquad R \geqslant R_7 \tag{65}$$

As mentioned above, only deformation-dependent terms are taken into account when the macroscopic potential energy is calculated. In the LDM, FRNFM and Y + EM, one has

$$\begin{gathered} E_d(R) = E_{LD}(R) - E_{LD}(\infty), \qquad E_d(R) = E_{FR}(R) - E_{FR}(\infty), \\ E_d(R) = E_{Y+E}(R) - E_{Y+E}(\infty) \end{gathered} \tag{66}$$

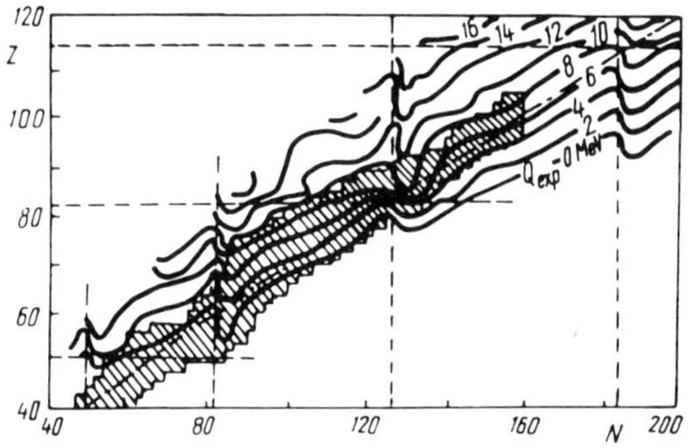


**Figure 9.** Contour maps for the $\alpha$-particle decay $Q_{exp}$-values for even–even nuclei with $Z$ protons and $N$ neutrons. The dashed area indicates measured masses as tabulated by Wapstra and Bos (Wa 77).

where the potential $E(\infty) = E_1 + E_2$ of the two fragments at infinite separation distance $(R \to \infty)$ is taken as the origin of the energy.

## 5.3. Fission $Q$ Values

For $R = R_i$ one obtains the macroscopic model, deformation-dependent part of the $Q$-value of the fission reaction,

$$E_d(R)|_{R=R_i} = Q_d \tag{67}$$

This quantity is different from $Q_{exp}$, the experimental value, because the ground-state deformations of the parent and daughter nuclei cannot be reproduced by the chosen parametrization, other terms of the macroscopic-model mass formula were ignored, and the shell correction energy $\delta E$ has not yet been introduced.

For each of the three models mentioned above, one has

$$Q_{exp} = Q = M - (M_1 + M_2) = Q_M + \delta Q, \qquad \delta Q = \delta E - (\delta E_1 + \delta E_2) \tag{68}$$

where $Q_M$ is the contribution of the macroscopic-model terms of the mass formula and $\delta Q$ is the shell correction part.

If $Q_{exp} < 0$ for a particular nucleus, it is stable with respect to the spontaneous emission of an $\alpha$ particle. This is the case for light nuclei, and for most of the neutron excess nuclei far off the line of $\beta$ stability. The $Q$-value contour lines for even–even nuclei with $Z$ protons and $N$ neutrons are given in Figure 9. The dashed area corresponds to nuclei with experimentally measured masses tabulated by Wapstra and Bos (Wa 77). Calculated values are obtained with a code kindly supplied by Dr. Jänecke (Jä 80); the results of various authors (Ma 76b, My 77) and the Myers–Swiatecki (My 67) mass formula for spherical shapes were used beyond this area.

The point-dotted curve is the Green approximation of the line of $\beta$ stability. From this figure one can see the general trend of increased $\alpha$-particle instability toward neutron-deficient and heavy nuclei—well explained by the macroscopic part of the mass formula—and the important variation from magic parent to magic daughter nuclei, reflecting strong increase in the shell effects.

The pronounced instability corresponding to daughter nuclei with magic neutron number $(N - 2 = 126)$ is clearer in Figure 10.

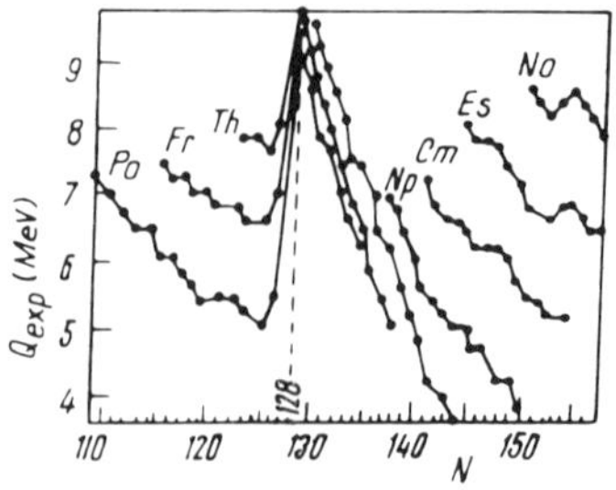


**Figure 10.** $Q_{\text{exp}}$-value systematics for selected $\alpha$-particle emitters.

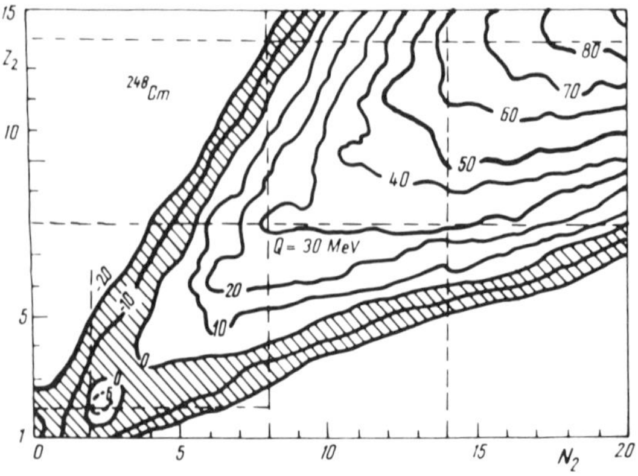


**Figure 11.** Contour maps for the $Q$-values for various asymmetric splittings of the $^{248}$Cm nucleus. The dashed area characterizes stability against cluster emission.

For a given nucleus (such as $^{248}$Cm) it is interesting to consider other possible, very asymmetric spontaneous fission activities that may be in competition with $\alpha$-particle decay. As is shown in Figure 11 the neighboring nuclei are not emitted spontaneously because the corresponding processes have negative $Q$-values, but for emission of many other nuclei with $N_2$, $Z_2 \geqslant 4$, $^{248}$Cm has $Q > 0$.

The shell effects, $\delta Q$ favoring $\alpha$-particle decay of this nucleus with respect to that of its neighbors, are presented in Figure 12. These were computed with a Myers–Swiatecki (My 66) formula, equation (69), for spherical nuclei. At each of the double shell closures ($N_2 = Z_2 = 2, 8$, and 14 in Figure 12), $\delta Q$ has a maximum value (3, 7, and 12 MeV, respectively). In spite of the increased value of $Q$ for $^{16}$O and $^{28}$Si in comparison with $^4$He, the emission of these clusters (Sa 80) is strongly hindered with respect to $\alpha$-particle decay. One must also consider the macroscopic-model contribution to the fission barrier, the vibration frequency, and the inertial mass parameters, all playing an important part in this competition. The $\delta$-function-like mass and charge distributions in this region of asymmetry reflect the fact that asymmetric fission leading to a light fragment different from $^4$He is not possible ($Q < 0$) or has a very low probability (lower than $10^{-9}$ relative to $\alpha$-particle decay).

The barrier shapes for this very asymmetric fission are plotted in Figure 17 below. The barrier height $E_b$ is given by $E_b = E_d(R_m) - Q = E_i - Q$. For the same nuclei as in Figure 10, we have plotted, in Figure 13a, the LDM touching-point interaction

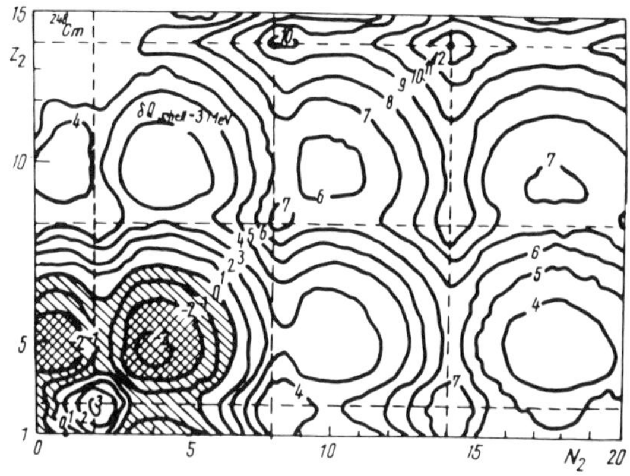


**Figure 12.** Shell-effect contribution $\delta Q$ to the $Q$-values for the same cluster decays as in Figure 11.

energy $E_i = E_d(R_t)$. In Figure 13b the shell effects $\delta Q$, and in Figure 13c the barrier heights $E_b = E_i - Q_{exp}$, are also shown. Figure 13c is almost the same as Figure 10, but instead it is upside down, because $E_i$ is a smoothly varying quantity. The pronounced reduction of the barrier height for magic neutron number, reflecting the steep variation in $\delta Q$ (Figure 13b), is very clear in Figure 13c.

## 5.4. Shell Corrections and Barrier Shape

Usually the macroscopic–microscopic method (Br 72) is applied to heavy nuclei. For example, the mass table of Seeger and Howard (Se 75) refers to $Z$, $N \geq 22$ and that of Möller and Nix (Mö 81) to $Z$, $N \geq 8$. The Strutinsky prescription for light nuclei or for very large mass-asymmetry has not yet been reliably developed.

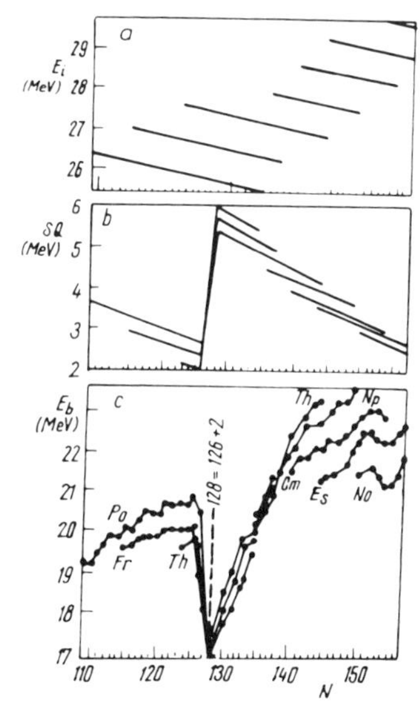


**Figure 13.** Interaction barriers (a), shell effects (b), and fission barriers (c) for $\alpha$-particle decay of the same nuclei as in Figure 10.

Nevertheless, one can use a phenomenological shell-correction method—sometimes called a cluster prescription—extending for large deformations the Myers–Swiatecki (My 66) formula for the ground state of spherical nuclei:

$$\begin{aligned} S(N,Z) &= C\{[F(N)+F(Z)]/(0.5A)^{2/3} - cA^{1/3}\} \\ F(m) &= 0.6[(N_i^{5/3} - N_{i-1}^{5/3})(m - N_{i-1})/(N_i - N_{i-1}) - m^{5/3} + N_{i-1}^{5/3}] \end{aligned} \tag{69}$$

where $m \in (N_{i-1}, N_i)$, $N_i$ being the spherical magic numbers $N, Z = 2, 8, 14, 28, 50, 82$; $N = 126, 184$; $Z = 114$, while parameters $C = 5.8$ MeV and $c = 0.325$. This equation was used to calculate $\delta E$, $\delta E_1$, $\delta E_2$, and $\delta Q$ plotted in Figures 12 and 13b. According to Schultheis and Schultheis (Sc 73) the damping terms of the Myers–Swiatecki shell-correction formula for deformed nuclei were replaced by a curvature-dependent integration usually performed numerically:

$$\delta S = \frac{C}{4\pi r_0^2 A^{4/3}} \int k\{[F(N/k^3) + F(Z/k^3)]2^{2/3}k^3 - cA\}\, d\sigma \tag{70}$$

where

$$k = 2R_0\left(\frac{1-\alpha}{\mathcal{R}_1 + \mathcal{R}_2} + \frac{\alpha}{|S|}\int \frac{d\sigma}{\mathcal{R}_1 + \mathcal{R}_2}\right), \qquad \alpha = 0.7 \tag{71}$$

$\mathcal{R}_1$ and $\mathcal{R}_2$ are the curvature radii and $|S|$ is the surface area. For two intersected spheres (Sc 70) one has a simple formula:

$$\delta S(R) = S(N_1, Z_1)\frac{2\pi R_1(R_1 + z_c)}{4\pi R_1^2} + S(N_2, Z_2)\frac{2\pi R_2(R - z_c + R_2)}{4\pi R_2^2} \tag{72}$$

During the deformation, the variation in $R$ induces the variation in $z_c$, $R_1$, $V_1$, $V_2$, $A_1$, $A_2$ and consequently in $Z_1$, $N_1$, $Z_2$, and $N_2$. Each time one of the nucleon numbers reaches a magic value, the correction energy reaches a minimum (negative value) and has a maximum at a mid-shell number.

For $\alpha$-particle decay, the variation in the fission fragment nucleon numbers during the deformation from $R = R_i$ to $R = R_t$ is as low as two units. Hence the shell-correction energy is a smooth function of $R$ (see Figure 14).

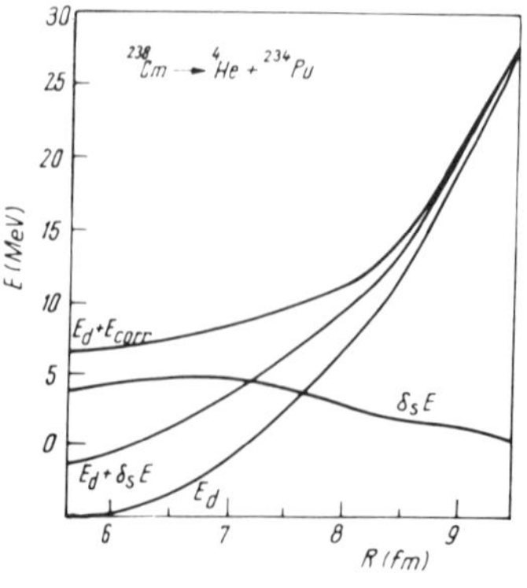


**Figure 14.** The rising part of the LDM, Schultheis's shell corrections, and the total fission barrier for the $\alpha$-particle decay of $^{238}$Cm.

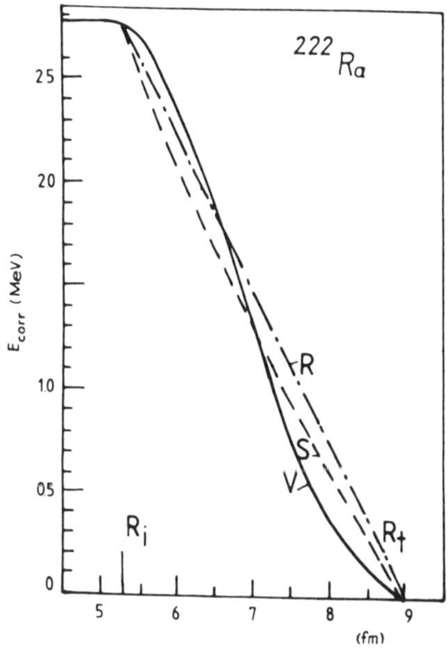


**Figure 15.** Variation in the $^{222}$Ra correction energy for volume ($V$), surface ($S$), and separation distance ($R$) scaling parameters in case of Y + EM.

By assuming that the top of the barrier is unaffected by the shell correction—which agrees with experimental data on fusion interaction barriers (Po 79e)—the shell correction

$$\delta_s E(R) = \delta S(R) - \delta S(R_t) \tag{73}$$

when added to the deformation energy $E_d$ of the LDM, raises the theoretical $Q_d$-values for heavy nuclei by 3–6 MeV. In spite of this improvement, there still remains a smoothly varying shift mentioned above. In order to reproduce the experimental $Q$-value exactly, we have introduced a phenomenological correction energy $E_{corr}(R)$, containing both the shell correction and the smooth shift:

$$E(R) = E_d(R) + E_{corr}(R), \qquad E_{corr}(R) = Q_{corr}[1 - V_2(R)/V_{2f}], \\ Q_{corr} = Q_{exp} - Q_d \tag{74}$$

where $V_{2f} = V_2(R_t)$ is the $\alpha$-particle volume. It seems reasonable to scale the shell correction with $V_2(R)/V_{2f}$, by relating its variation to the bulk properties of the nascent $\alpha$-particle fragment. Another choice could be the ratio $S_2(R)/S_{2f}$ of the surface areas of the small fragment to $S_{2f} = S_2(R_t)$, or the ratio $(R - R_i)/(R_t - R_i)$ as shown in Figure 15 for the example of $^{222}$Ra. A best choice for the scaling parameters $S_2$ and $R$ (Po 79c) has yielded a negligible difference of log $T(S_2)$ and log $T(R)$ with respect to log $T(V_2)$; hence we decided to use equation (74).

The experimental corrections $Q_{corr}$ defined by equation (74) are plotted in Figure 16a for the LDM, in Figure 16b for the FRNFM, and in Figure 16c for the Y + EM. As expected, the very strong shell effects at the magic neutron number in the daughter nucleus are the same, but the deformation-dependent part $Q_d$ of the Y + EM and FRNFM give a good approximation of the $\alpha$-particle $Q$-values. On the other hand the

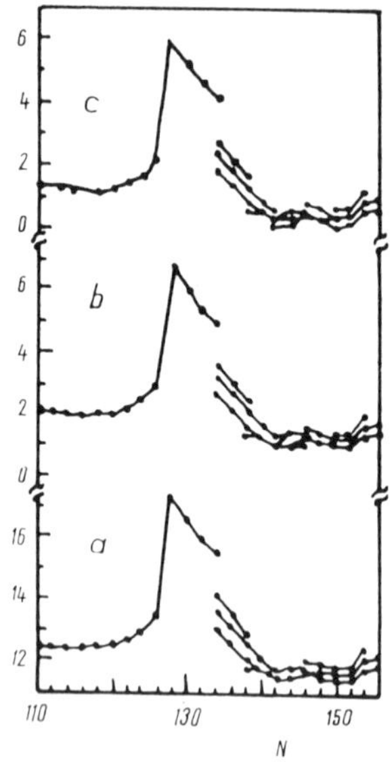


**Figure 16.** Correction energies vs. neutron number $N$ of parent nucleus for various $\alpha$-particle emitters (Po, Ra, Th, U, Pu, Cm, Cf, Fm, and No): (a) LDM, (b) FRNFM, (c) Y + EM.

LDM, giving a large error for the $\alpha$-particle mass, leads to large discrepancies (even negative $Q_d$-values).

The corrections of the barrier for the $\alpha$-particle decay of nonmagic nuclei are very small in the framework of FRNFM and Y + EM. As an example, in Figure 17 the barrier shapes for the $\alpha$ decay of $^{238}$Cm are shown. The LDM (a), FRNFM (b), and Y + EM (c) potential energies without (dashed lines) and with (full line) corrections are plotted. As mentioned in the preceding section, one can see that for the last two models $R_m > R_t$. In the LDM, $R_m = R_t$; the top of the barrier is very sharp and higher

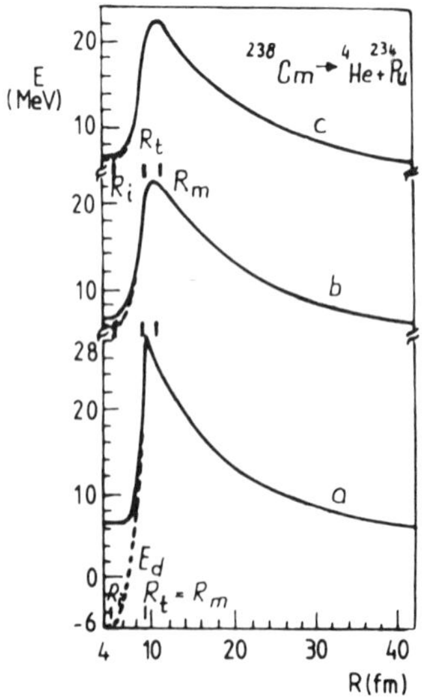


**Figure 17.** Barrier shapes for the $\alpha$ decay of $^{238}$Cm, without (dashed line) and with (full line) shell correction: (a) LDM, (b) FRNFM, (c) Y + EM.

in energy. Because of the different values of the radius constant $r_0$, the initial separation distance $R_i$ (in fm) is also different from model to model.

### 5.5. $\alpha$-Decay Lifetime of Heavy and Superheavy Nuclei

After replacing the numerical values of the parameters, equations (10), (13), and (14) yield

$$T = \frac{1.4333 \times 10^{-21}}{E_{vib}} \exp\left(0.4239 \int_{R_a}^{R_b} \{\mu_A(R)[E(R) - Q']\}^{1/2} \, dR\right) \tag{75}$$

where $T$ is expressed in s, while $R$, $R_a$, and $R_b$ are in fm, and $E_{vib}$, $E$, and $Q'$ in MeV. The penetrability integral is computed by a Gauss-Legendre numerical quadrature, dividing the whole range $(R_a, R_b)$ into two subintervals $(R_a, R_m)$ and $(R_m, R_b)$. For $R \in (R_i, R_t)$ the potential energy $E(R)$ is calculated numerically. The lower limit of the integral, $R_a$, the solution of the equation $E(R_a) - Q'$ in this interval, is found by a searching computer code. The upper limit, $R_b$, is given by the Coulomb interaction energy: $R_b = Z_1 Z_2 e^2/Q$, because the nuclear interaction term is vanishingly small at large distances.

For the mass parameter $\mu_A(R) \equiv \mu(R)/m$, where $m$ is the nucleon mass, we have tried three kinds of variations:

1. Instantaneous reduced mass

$$\mu_i(R) = A_1(R)A_2(R)/A \tag{76}$$

which is an increasing function of $R$.

2. Final reduced mass

$$\mu_f(R) = A_{1f}A_{2f}/A = 4(A-4)/A = \text{const} \tag{77}$$

3. The semiempirical relationship suggested by Randrup *et al.* for almost symmetrical fission:

$$\mu_s(R) = \mu_f\left[1 + 7.37 \exp\left(-2.452 \frac{R - R_i}{R_t - R_i}\right)\right] \tag{78}$$

The zero-point vibration energy $E_{vib}$ was adjusted in each case, in order to obtain the best fit to experimental half-lives: 0.37 MeV for $\mu_i$, 0.4 MeV for $\mu_f$, and 0.63 MeV for $\mu_s$ in the LDM. The difference between the maximum and minimum value of $\log(T/T_{exp})$ for the seven isotopes of U is a good measure of the deviation of our theoretical values from the experimental ones. This is practically the same (1.4 and 1.5) for $\mu_i$ and $\mu_f$ but much higher (2.2) for $\mu_s$. Hence we decided to use for the moment the simple law of $\mu_f$. It can be seen from Figure 18 with the examples of $^{230}$Th (1), $^{248}$Cf (2), and $^{252}$No (3) that the half-life is very sensitive to the variation of $E_{vib}$. In order to obtain a good overall fit to the experimental data, the values $E_{vib} = 0.37$, 0.37, and 0.30 MeV were chosen for LDM, FRNFM, and Y + EM, respectively, with $\mu_i$.

Experimental half-lives (solid points) after Rytz (Ry 79) and Nuclear Data Sheets (Vol. 5, Nos. 3 and 6; Vol. 6, No. 4; Vol. 7, No. 2, and Vol. 8, No. 2) are used in Figure 19. By comparing this with Figure 13c one can see an almost identical trend, revealing the importance of the barrier height in this process. The good agreement with

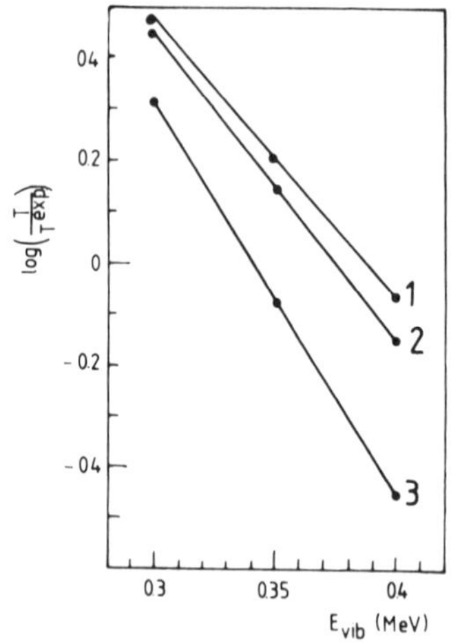


**Figure 18.** Half-life sensitivity to the variation of zero-point vibration energy: (1) $^{230}$Th, (2) $^{248}$Cf, (3) $^{252}$No.

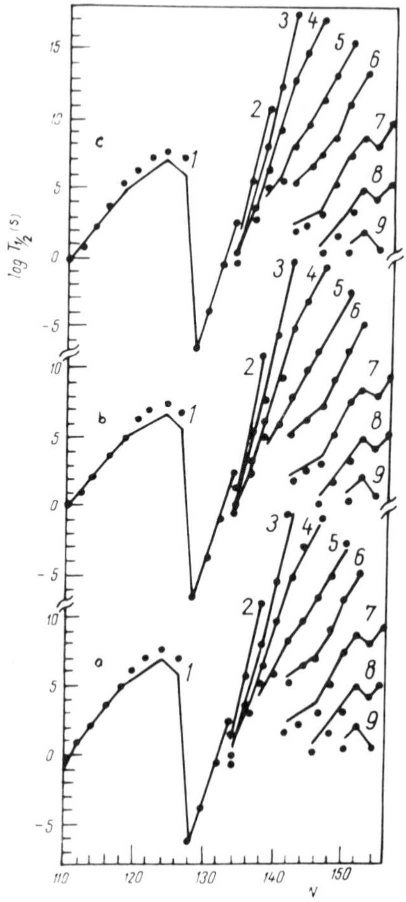

**Figure 19.** Experimental (points) and theoretical (lines) $\alpha$-decay half-lives for the same even-even nuclei as in Figure 17: (a) LDM, (b) FRNFM, (c) Y + EM, (1) Po, (2) Ra, (3) Th, (4) U, (5) Pu, (6) Cm, (7) Cf, (8) Fm, (9) No.

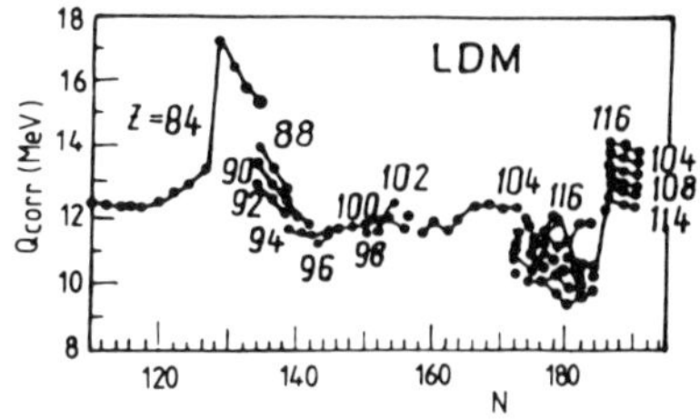


**Figure 20.** Correction energies vs. neutron number of parent nucleus for $\alpha$ emitters including superheavies.

experimental data, irrespective of the model used for the macroscopic energy $E_d$, in the range of $T$ varying over 24 orders of magnitude, suggests that the $\alpha$-decay process can indeed be interpreted as a fission phenomenon. This conclusion is also supported by the fact that a new semiempirical relationship (Po 80b, Po 82b, Po 83) for $T_{1/2}(Q)$, derived on the basis of fission theory of $\alpha$ decay, gives the best agreement with experimental data. In the following we will use the LDM (plus shell corrections) in order to exploit its simplicity.

When compared with other theories of $\alpha$-particle decay, these computations are faster and more accurate. To calculate the lifetime for one nucleus, the IBM 370/135 computer running time is of the order of 45 s for LDM and 2 min 20 s for FRNFM or Y + EM.

On the basis of our method one can predict reliable values for the $\alpha$-particle decay lifetimes in the new region of nuclei including the superheavies. $\alpha$-Particle disintegration of these nuclei is particularly important, because in many cases it puts a limit on their survival.

As an example we have studied (Po 81c) the even–even isotopes of the elements with $Z = 106$–$120$ and $N = 172$–$190$, where there are, as yet, no experimental data. In order to examine the complete systematic trend with $N$ between 110 and 190 (see Figure 20) the isotopes of element 104 with $N = 154$–$190$ have also been considered. The $Q$-values for energy correction were taken from Fiset and Nix (Fi 72). As is shown in Figure 20, the shell effects of the daughter nuclei with magic numbers of neutrons, for which $N = 128$ and 186, induce a steep increase in $Q_{corr}$. The same happens at the magic proton number 144 ($Z = 116$).

A corresponding increase in the decay probability is shown in Figure 21. One can see that nuclei with $Z > 114$ decay considerably faster. For some isotopes of elements

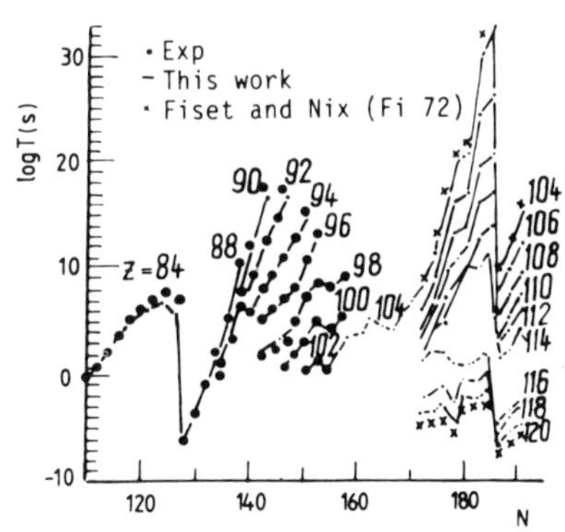


**Figure 21.** $\alpha$-Decay half-lives of some heavy and superheavy even–even nuclei.

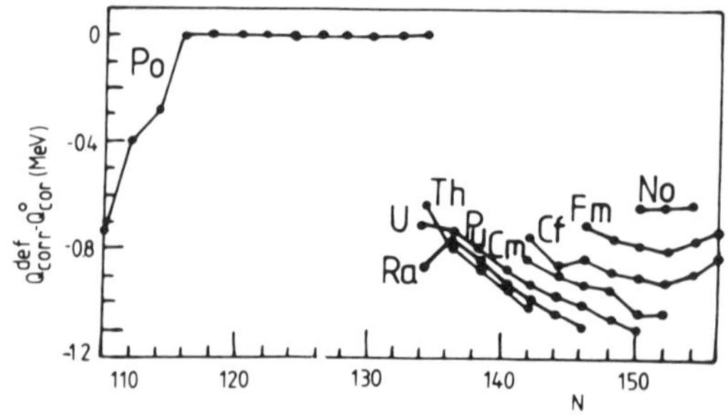


**Figure 22.** Nuclear deformation contributions to the $\alpha$-decay $Q$-values.

104, 106, and 108 we predict lower half-lives than Fiset and Nix (Fi 72), but in general the present results are more optimistic: about an order of magnitude larger. When other competing modes of decay (fission and $\beta$ decay) are also taken into account, the nucleus $^{294}110$, for example, has the longest calculated total half-life. Of course, the centrifugal barrier (Mö 80) can change the situation drastically. $\alpha$-Particle decay is used to identify new elements (such as 107 and 108) both in Dubna (Flerov and co-workers) and in Darmstadt (Armbruster and co-workers)).

## 5.6. Transitions between Deformed Ground States and Fission Isomers in $\alpha$-Particle Decay

In the last section the ground-state nuclear deformations of the parent and daughter nuclei (Se 75) were not taken into consideration explicitly; they were introduced only through the experimental $Q$-value in the correction energy. The parametrization of a spheroid intersected with a sphere, presented in Section 5.1, makes it possible to consider the deformations (Po 81b) and at the same time preserve the advantage of the analytical relationship for $E_d(R)$, in the LDM, when the fragments are separated ($R \geqslant R_t$).

Because of the deformation, $Q_d$ increases thus improving the agreement with experimental data. The reduction of $Q_{corr}$ by 0.6–1.2 MeV is evident in Figure 22.

Owing to the increase in $R_t$, the LDM barrier height (Figure 23) and consequently the zero-point vibration energy are smaller ($E_{vib} = 0.2$ MeV for the case of constant reduced mass $\mu_f$).

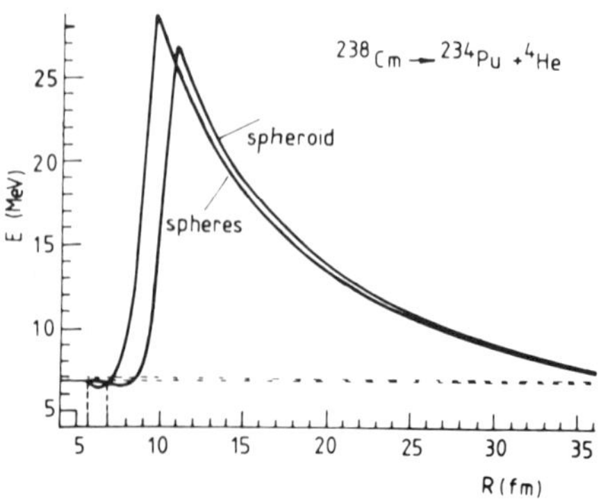


**Figure 23.** Influence of the deformation on LDM barrier shape for the $\alpha$ decay of $^{238}$Cm.

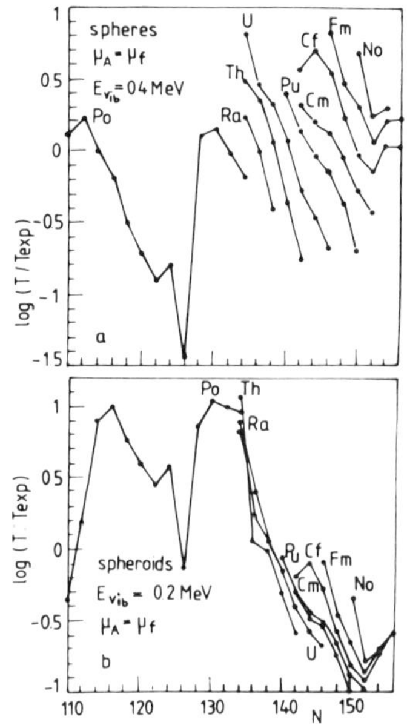


**Figure 24.** Residual discrepancies between the theoretical and experimental lifetimes without deformation (a) and with deformation (b) taken into account.

Another interesting result is illustrated in Figure 24, where an enlarged view of the residual discrepancies of the calculated lifetimes as compared to experimental ones are displayed, both without and with deformations taken into account. It is obvious that the dependence of the discrepancies on $Z$ is reduced in the presence of deformation. This is an argument that the neutron-shell effects shadow the proton ones in this region. The same conclusion can be drawn from Figure 25, where the individual zero-point vibration energies of each nucleus, determined from the condition $T = T_{\text{exp}}$, were plotted, assuming a constant inertial mass parameter $\mu_A = \mu_{\text{f}}$.

The parametrization of a spheroid for each of the parent and daughter nuclei and of a sphere for $\alpha$-particle fragments allowed us to study the $\alpha$ transition $Q_{\text{is,g}}$ from a shape isomeric state (Va 77, Po 77, Me 80, Bj 80) of a parent nucleus to the ground state of the daughter. In this case

$$Q_{\text{is,g}} = Q_{\text{gg}} + E_{\text{IIp}} \tag{79}$$

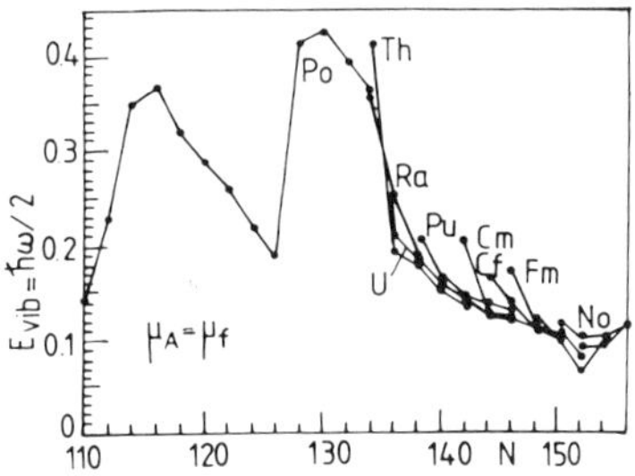


**Figure 25.** Zero-point vibration energy of some spherical and deformed nuclei.

where $E_{IIp}$ is the fission isomer excitation energy (Br 73, Iv 81) of the parent nucleus and $Q_{gg}$ is the $Q$-value for the $\alpha$ decay between ground states (called $Q_{exp}$ above). The shape isomeric state has a large deformation $\beta_p \simeq 2$ (Bj 80), but the ground-state deformation of the daughter nucleus is of the order of $\beta_d = 1.24$.

The results obtained for the known even-$N$ isotopes of U, Pu, and Cm are plotted in Figure 26. The same figure shows the measured lifetime for the decay—by spontaneous fission—from the isomeric state $f_{is}$. This process is much faster. The branching ratio for the $\alpha$-particle decay is lower than $10^{-8}$—a figure which explains why the $\alpha$ particles from the fission isomers could not be found experimentally (Le 66, Be 73).

Another type of $\alpha$-particle transition $\alpha_{is,is}$ between the shape isomeric states of the parent and daughter nuclei has a lower $Q$-value,

$$Q_{is,is} = Q_{is,g} - E_{IId} = Q_{gg} + (E_{IIp} - E_{IId}) \tag{80}$$

and the shape of the daughter nucleus is also well deformed, $\beta_d \simeq 2$. It is shown in Figure 26 that the probability of this process is at least five orders of magnitude lower than that of $\alpha_{is,g}$.

The measured lifetime of other processes occurring in these nuclei are shown in Figure 26: the spontaneous fission of the ground state $f_{gg}$ and the ground state–ground

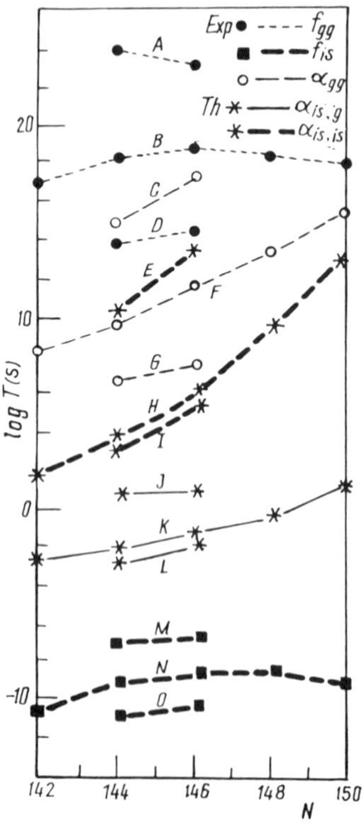


**Figure 26.** Half-lives for some fission (f) and $\alpha$-decay ($\alpha$) processes for the ground state and shape isomeric states of some U (curves A, C, E, J, and M), Pu (curves B, F, H, K, and N), and Cm (curves D, G, I, L, and O) even-$N$ isotopes.

state $\alpha$ decay $\alpha_{gg}$. For low neutron number $\alpha$-particle decay dominates, but for large $N$ values the fission competition becomes stronger and stronger (Po 81b).

## 5.7. Time-Dependent Hartree–Fock Study of $\alpha$ Decay

It was shown (Sa 83) that in the TDHF approximation, the collision of an $\alpha$ particle with a lead nucleus leads to nuclear dynamical effects during the capture process and to periodic oscillations of the compound system. The oscillations were interpreted as zero-point motions associated with the $\alpha$-particle decay collective mode. The computed frequency corresponds exactly with the experimentally deduced value based on the $\alpha$-particle decay description as a fission process.

Weiss (We 81) has suggested that the TDHF approximation may also include some high-energy collective excitations of the residual nuclei after a grazing collision. It was shown that the high-frequency structures of the Fourier transform of the Cartesian moments of the density of the residual nuclei for several moments are correlated. This may explain the observed maxima in the cross-section plotted versus the final kinetic energy, for different species in the system $^{40}Ca + {}^{40}Ca$ at $E_{lab} = 100$ MeV (Ro 81). From these calculations it is not clear whether the frequency components represent non-linearity effects, multiple phonon or primary collective excitations.

Below we present the results of a study, in the TDHF approximation, of another collective mode, which cannot be simply described by multipole expansion of the surface, i.e., the collective mode associated with $\alpha$-particle decay. In order to excite such a collective mode, the central collision of an $\alpha$ particle with a lead nucleus, at relative kinetic energy just above the Coulomb barrier, was considered.

A simplified Skyrme plus Yukawa interaction was used, the small Coulomb exchange correction energy was neglected, and no spin–orbit interaction was included. The effective charge quartet model with a filling approximation for the outermost shells (Cu 80, St 80) has been used.

The calculations were conducted over a long period of time (3.315 fm/c). From the total number of pictures taken in steps of 19.5 fm/c only a few, which were considered to be characteristic of the $\alpha$- particle decay process, were selected.

In Figure 27 the first 10 shapes, which illustrate capture of the $\alpha$ particle by the lead nucleus, have been plotted. It is evident that if we assume that the emission process is just the capture process reflected in time, the $\alpha$-decay process implies many dynamical effects, such as polarization of the heavy nucleus in the vicinity of the $\alpha$ particle and formation of a long neck before the $\alpha$ particle is emitted.

From the TDHF calculations it was concluded that the whole process looks like a fission process. This gives full support to the theory of $\alpha$ decay as a fission process based on the liquid drop model with phenomenological shell corrections.

After the capture process, the compound system undergoes complicated shape oscillations. Looking at the time evolution of these shapes one can clearly see some shapes which appear periodically. First, the compound system reaches a more or less symmetric shape, similar to a sphere with two bumps on opposite sides and, second, an asymmetric shape which can be approximately described by a sphere with only one bump on the side on which the two partners first come into contact. These shapes, which appear periodically with half-period $T/2 = 780$ fm/c $= 2.6 \times 10^{-21}$ s, are shown in Figure 28.

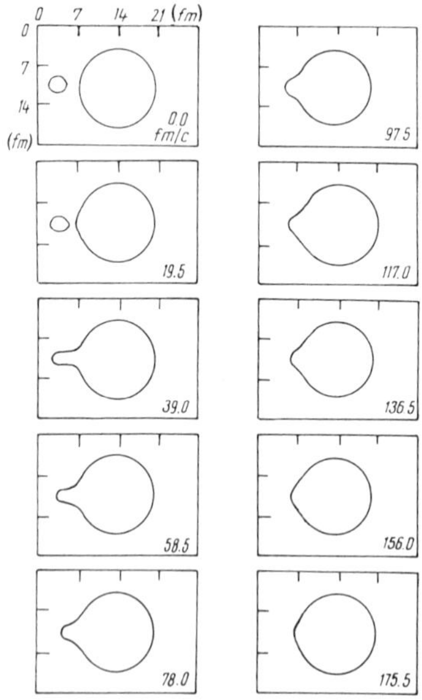


**Figure 27.** Capture of an $\alpha$ particle by the $^{208}$Pb nucleus illustrated in time-steps of 19.5 fm/c.

These oscillations have been interpreted as zero-point oscillations of the collective mode associated with $\alpha$-particle decay $E_{\text{vib}} = \hbar\omega/2 = 0.4$ MeV. We stress that this value corresponds quite well with the experimentally deduced value based on the description of $\alpha$-particle decay as a fission process (see Section 5.5).

## 5.8. Semiempirical Formulas for $\alpha$-Particle Decay Half-Lives

During the last few years, the number of known $\alpha$-particle emitters has increased mainly through measurements on new neutron-deficient nuclei (Ha 85) produced in

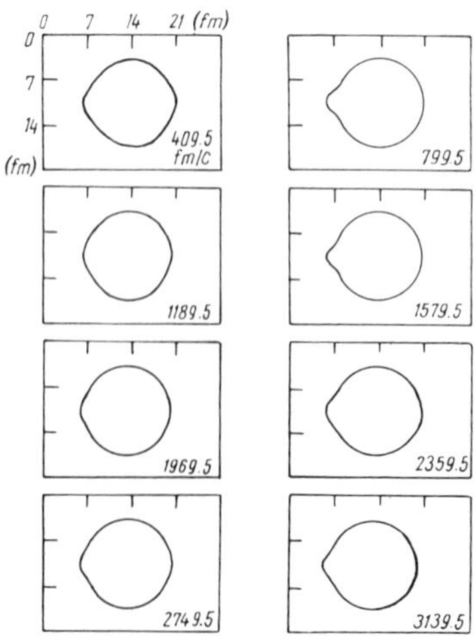


**Figure 28.** Shapes which appear periodically in the time evolution of the compound system $^{212}$Po.

heavy-ion reactions (Ga 75, Ca 78, Ha 79, Ho 79, Ri 81, Sc 79), and in reactions induced by high-energy protons (Ha 77, Al 83). A new island of $\alpha$-particle activity in the neighborhood of the double magic nucleus $^{100}Sn$ was also studied (Sc 79, Sc 81).

$\alpha$-Particle decay usually competes with fission and $\beta$ decay in the disintegration of the heaviest nuclei synthesized thus far (Be 81), and it is expected to be met frequently in the superheavy region (Ni 72).

As long ago as 1911, Geiger and Nuttall found a simple dependence of the $\alpha$-particle decay partial half-life $T$ on the $\alpha$-particle range in air. The disintegration period can be estimated if the kinetic energy of the emitted particle, $E_\alpha$, is known, by using any one of a number of semiempirical relationships (Fr 57, Wa 59, Ta 61, Vi 66, Ke 72, Ho 74). Some of these formulas were derived only for a limited region of parent proton and neutron numbers, $Z$ and $N$. Their parameters have been determined by fitting a given set of experimental data selected by the various authors from the available data on strong collectivity in even–even, even–odd, and odd–odd nuclides.

More recently, the precision of some measurements has been increased and new $\alpha$-particle emitters were discovered. This process of improving both the quality and quantity of the data is expected to continue in the future.

In an attempt to improve the description of data even in the neighborhood of the magic neutron and proton numbers, where the errors of the other semiempirical relationships are large, a new formula with six parameters $B_k$, based on the fission theory of $\alpha$-particle decay, has been derived (Po 80b, Po 80, Po 83). This formula takes into explicit consideration not only the dependence on the proton number, but also that on the neutron number and on the difference of both from magicity. A corresponding computer program (Po 82b) allows us to automatically improve the parameters of the formulas whenever a better set of experimental data is available (Po 84c).

A set of 376 ($Q$, $T$) data on the most probable (ground state to ground state or favored) $\alpha$-particle transitions in 123 even–even, 111 even–odd, 83 odd–even, and 59 odd–odd nuclei can be found in the literature (Po 83b, Po 84c). They were selected to meet, in each of four groups, the criteria of best fitting the systematics of the $Q$-value, log $T$, and the quantity $\chi_{\text{exp}}$ defined below (see Figure 29), versus neutron number and the Geiger–Nuttall plot (Figure 30).

The parameters of our formula were obtained from these data. Many of the $\alpha$ active nuclei also display other competing decay modes ($\beta$ decay, spontaneous fission, etc.), hence the partial $\alpha$-particle lifetimes represent only a fraction (given by the branching ratio $b_\alpha$) of the total disintegration periods $T_t$. From all the $\alpha$-particle transitions we have selected the strongest ones: ground-state to ground-state transitions in even–even nuclei and the so-called favored transitions in odd–odd and odd-mass nuclei. The ratio of the intensity of this selected strong $\alpha$ line to the total $\alpha$ strength will be called $i_p$ in the following. Consequently the partial decay lifetime $T_\alpha$ of the most probable $\alpha$ transition is given by

$$T_\alpha = \frac{100}{b_\alpha} \cdot \frac{100}{i_p} T_t \tag{81}$$

where $b_\alpha$ and $i_p$ are expressed in percent.

The released energy $Q$ is related to the $\alpha$-particle kinetic energy $E_\alpha$ by the relationship

$$Q = E_\alpha A / A_1 \tag{82}$$

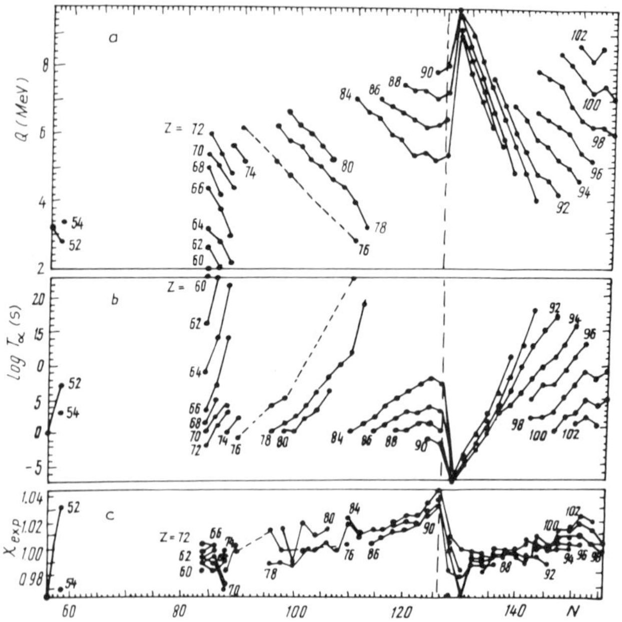


**Figure 29.** Systematics of $Q$ (a), log $T_\alpha$ (b), and $\chi_{\text{exp}}$ (c) values of the even–even basic set of nculei.

where $A$ and $A_1 = A - 4$ are the mass numbers of the parent and daughter nucleus, respectively. For $\alpha$ transitions from the ground state of the parent nucleus to the ground state of the daughter, $Q$ is given by the mass difference $Q = M(A, Z) - M(4, 2) - M(A_1, Z_1)$; otherwise (for favored transitions) one must also add the difference of the excitation energies.

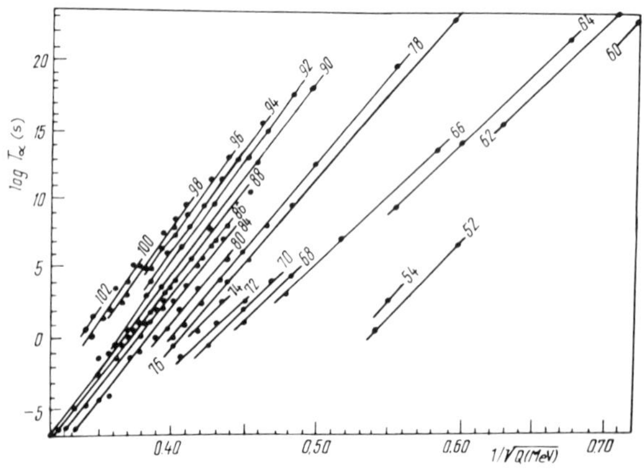


**Figure 30.** Geiger–Nuttall plot for even–even nuclei.

Information concerning the quantities $T_t$, $b_\alpha$, and $i_p$ of equation (81) were compiled by Rytz (Ry 79). Our basic set of experimental data was selected from these tables and from the publications by Gauvin *et al.* (Ga 75), Gorbachev *et al.* (Go 75), Cabot *et al.* (Ca 78), Lederer and Shirley (Le 78), Hagberg *et al.* (Ha 79), Hofmann *et al.* (Ho 79), Schardt *et al.* (Sc 79), and Schmidt *et al.* (Sc 79b).

For $\alpha$ decay, the value of $K_{ov}$ from equation (14) is always only a small percent of $K_s$ and one can write

$$\log T_\alpha = 0.43429K - 20.446 \tag{83}$$

$$K = \chi K_s \tag{84}$$

$$\begin{aligned} K_s &= 2.52956 Z_1 (A_1/AQ)^{1/2}[\arccos\sqrt{x} - \sqrt{x(1-x)}], \\ x &= 0.423Q(1.5847 + A_1^{1/3})/Z_1 \end{aligned} \tag{85}$$

The quantity $\chi$ is close to unity and shows a "sawtooth" variation versus the neutron number $N = A - Z$ of the parent nucleus, maximum values being attained at the magic numbers for even-$N$ and at the magic-plus-one numbers for odd-$N$ nuclides. This quantity is expressed as a second-order polynomial

$$\chi = B_1 + B_2 y + B_3 z + B_4 y^2 + B_5 zy + B_6 z^2 \tag{86}$$

in the reduced variables $y$ and $z$ expressing the distance from the closest magic-plus-one numbers $N_i$ and $Z_i$ in the neighborhood of $N$ and $Z$:

$$y \equiv (N - N_i)/(N_{i+1} - N_i), \qquad N_i < N \leqslant N_{i+1}, \qquad N_i = \ldots, 29, 51, 83, 127, 185, \ldots \tag{87}$$

$$z \equiv (Z - Z_i)/(Z_{i+1} - Z_i), \qquad Z_i < Z \leqslant Z_{i+1}, \qquad Z_i = \ldots, 29, 51, 83, 115, \ldots \tag{88}$$

with parameters $\{B_k\}$, given in Table 1, obtained from a fit to the experimental data (Po 83b, Po 84c) on 125 even-even, 111 even-odd, 84 odd-even, and 60 odd-odd nuclides. A computer program allows us to automatically improve the parameters $B_k$ whenever a new set of experimental data (completed with new $\alpha$ emitters or more accurate measurements) is available. The quantity $\chi_{exp}$ is given by

$$\chi_{exp} = (\log T_{exp} + 20.446)\ln 10/K_s \tag{89}$$

Details concerning different semiempirical formulas in various groups of nuclei are given elsewhere (Po 83b, Po 83e). It was shown that the present relationship, in which shell effects are taken into account, is able to reproduce experimental data even in the

**Table 1. Values of Parameter $B_k$ for the Semiempirical Formula of $\alpha$-Decay Half-Life**

| Group of nuclei | $B_1$ | $B_2$ | $B_3$ | $B_4$ | $B_5$ | $B_6$ |
|---|---|---|---|---|---|---|
| even–even | 0.985 911 | 0.022 841 | 0.024 584 | 0.023 279 | −0.000 716 | −0.022 562 |
| odd–even | 1.000 560 | 0.010 783 | 0.050 671 | 0.013 919 | 0.043 657 | −0.079 999 |
| even–odd | 1.017 560 | −0.113 054 | 0.019 057 | 0.147 320 | 0.230 300 | −0.101 523 |
| odd–odd | 1.007 740 | −0.184 136 | 0.260 268 | 0.231 900 | 0.326 025 | −0.407 280 |

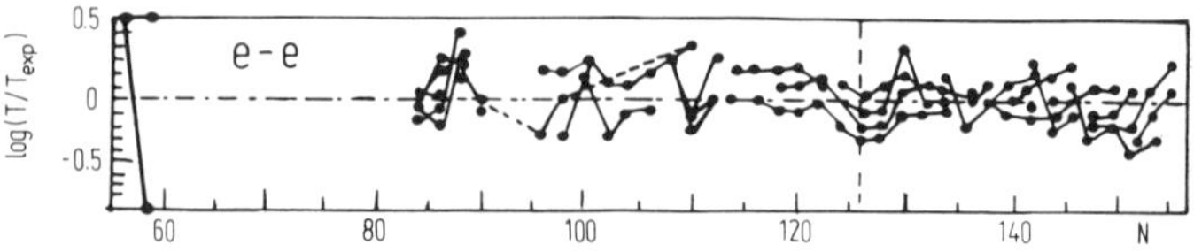


**Figure 31.** Errors of $\alpha$-decay lifetime predictions with our semiempirical formula.

regions of the magic proton and neutron numbers (see Figure 31) where the use of other relationships has always resulted in large errors. Many predictions for $\alpha$-decay half-lives have been made (Po 81, Po 84c, Ru 82, Po 84i, Po 86), including the island of $\alpha$ activity close to the double magic $^{100}Sn$, studied intensively by Schardt *et al.* (Sc 79, Sc 81). [See also (Ti 81, Ma 65, Ka 67).] The need for improved mass formulas has been stressed by Schardt *et al.* (Sc 81), Plochocki *et al.* (Pl 82), Epherre (Ep 83), and Poenaru and Ivascu (Po 84c).

## 6. Analytical Superasymmetric Fission Models

When performing a systematic investigation of the stability of about 2200 nuclides, $^{A}Z$, with known masses (Wa 85) subject to the possible emission of 379 nuclides with $Z_2 \leqslant 28$ and $A_2 < 70$, the number of parent-emitted cluster combinations is larger than $10^5$. One could consider this large number of cases by using ASAFM, a model leading to analytical results both for the static (potential-energy barrier) and dynamic (half-life) properties of the process. This model was developed by Poenaru and Ivascu (Po 80) and improved (Po 83c, Po 84) to account for angular momentum and small nuclear excitation effects. It was used to predict new decay modes $^{A_2}Z_2$ in several steps. At first (Po 80) it was shown that all nuclides tabulated by Wapstra and Bos (Wa 77) are stable against spontaneous emission of $^{2,3}H$, $^{3}He$, $^{4}Li$, $^{7}B$, and $^{9}C$ and it was mentioned that $^{8}Be$, $^{12}C$, $^{16}O$, $^{22}Ne$, $^{26}Mg$, $^{30}Si$, $^{36}S$, $^{40}Ar$, and $^{48}Ca$ have a relatively high probability of being observed as emitted clusters. As an example of our use of ASAFM, spontaneous and $\beta$-delayed $^{5}He$ radioactivities have been predicted (Po 83c, Po 84) by performing a systematic investigation of nuclear stability against He-isotope emission ($Z_2 = 2$, $A_2 = 3$–$10$). It was shown that, for a given cluster, the maximum emission rates are obtained when the daughter has a magic neutron and proton number or is not too far removed from $N_1 = 126$, $Z_1 = 82$. After that, by increasing $Z_2$, in the region 2–10 (Po 84d, Po 84f, Po 84g, Po 84i, Po 85d) the list of possible new decay modes was enlarged, including those with odd $Z_2$ and $N_2$ which had not been considered previously. A very comprehensive list of more than 140 possible decay modes was presented by Poenaru *et al.* (Po 85c). $\beta$-Enhanced emission of heavy clusters was predicted and the region of potential parent nuclides was extended (Po 84b) for $Z < 82$ showing (Po 85) that even the "stable" nuclides with $Z \geqslant 40$ are metastable with respect to heavy-cluster emission. The partial half-lives, and branching ratios relative to $\alpha$-particle decay, kinetic energies, and $Q$-values for the eight most probable clusters emitted from 766 nuclides with $Z = 47$–$106$ were tabulated by Poenaru *et al.* (Po 86). Double and multiple heavy-cluster radioactivities have been predicted (Po 84e, Po 85b). Recently, ASAFM was extended by Martin Greiner and Werner Scheid (Gr 86b, Gr 86c) to describe transitions leading to excited final states. Nuclear deformation was taken into account

(Po 86c) and the odd–even staggering of heavy-cluster emission rates explained (Po 86d).

A superasymmetric fission model using a proximity-plus-Coulomb potential was developed by Shi and Swiatecki (Sh 85, Sh 85b). This model was used to predict branching ratios with respect to $\alpha$ decay for several parent-emitted ion combinations expected to have a reasonable probability of being observed experimentally.

## 6.1. Parabolic Approximation of the Interaction Potential Containing a Centrifugal Term

Within the framework of an LDM two-center spherical parametrization (see Sections 5.1, 5.3, and 5.4) for separated fragments, $R \geqslant R_t$, only the Coulomb interaction energy $Z_1Z_2e^2/R$ has been considered for $\alpha$-particle decay of even–even nuclei. The maximum of the potential energy at $R = R_t$ was

$$E_c = Z_1Z_2e^2/R_t \tag{90}$$

where $e$ is the electron charge. In this way, a barrier shape similar to that shown in Figure 32 was obtained.

For a heavy cluster which possesses spin $\mathbf{I}_2$, spin ($\mathbf{I}$) and parity ($\pi$) conservation must be fulfilled:

$$\mathbf{I} = \mathbf{I}_1 + \mathbf{I}_2 + \mathbf{l}, \qquad \pi = \pi_1\pi_2(-1)^l \tag{91}$$

To the Coulomb interaction $E_c$ one must also add the contribution of angular momentum $l\hbar$—the centrifugal term

$$E_l = \hbar^2 l(l+1)/(2\mu R_t^2) \tag{92}$$

Substitution of numerical values yields, for $R = R_t$ in fm, the total interaction energy at $R = R_t$, in MeV:

$$E_i = E_c + E_l = 1.43998 Z_1Z_2/R_t + 20.735 l(l+1)A/(A_1A_2R_t^2) \tag{93}$$

In the overlapping region, a convenient analytical approximation for the potential energy $E(R)$, leading from $E(R_i) = Q$ to $E(R_t) = E_i$, suggested by the potential barrier shape (Figure 32) and allowing us to obtain a better expression for the lifetime $T$ (see the next section), is a second-order polynomial in $R$. Finally, therefore, one has

$$E(R) = \begin{cases} Q + (E_i - Q)[(R - R_i)/(R_t - R_i)]^2, & R \leqslant R_t \\ Z_1Z_2e^2/R + \hbar^2 l(l+1)/(2\mu R^2), & R \geqslant R_t \end{cases} \tag{94}$$

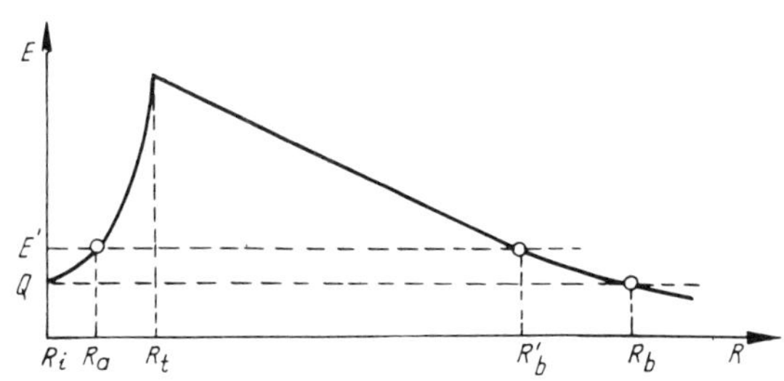


**Figure 32.** Barrier shape for heavy cluster emission.

For a nonzero excitation energy (a nuclear temperature $\tau$), the Coulomb energy is only slightly reduced (Sa 76), being multiplied by the factor $(1\text{–}10^{-3} \times \tau^2)$, where $\tau$ is expressed in MeV. Consequently, at relatively small excitation energies, this Coulomb energy remains practically unchanged.

## 6.2. Closed Formula for the Lifetime

As in fission, the half-life of a metastable system is given by equations (10)–(15), in which $Q' = Q + E_{\text{vib}} + E^*$ and $E^* < E_{\text{i}} - (Q + E_{\text{vib}})$ is the fraction of the excitation energy concentrated in the separation degree of freedom. The barrier transmission model, used in the present work, describes only low excitations (Bj 80). Statistical equilibrium among all the degrees of freedom of the nucleus is only reached for high excitation energies $U$. In this case $E^* = U - (E^{\text{i}} + E_{\text{rot}})$, where $U = a\tau^2$ ($a \simeq A/10\,\text{MeV}^{-1}$ is the level density parameter), $E^{\text{i}}$ is the internal energy available for other degrees of freedom, and $E_{\text{rot}}$ is the rotational energy. In a hot nucleus, the energy concentrated on motion in the deformation mode is of the order of the nuclear temperature. Hence the residual nucleus is usually also left in a highly excited state. According to a statistical model (Mi 81) the probability of exciting a collective state of $E^*$ energy is roughly proportional to $\exp(-E^*/\tau)$. Since $E(R_{\text{a}}) = E(R_{\text{b}'}) = Q' = E'$ it follows that

$$R_{\text{a}} = R_{\text{i}} + (R_{\text{t}} - R_{\text{i}})[(E_{\text{v}} + E^*)/E_{\text{b}}^0]^{1/2} \tag{95}$$

$$E_{\text{b}}^0 = E_{\text{i}} - Q \tag{96}$$

$$R'_{\text{b}} = (R_{\text{t}}E_{\text{c}}/Q')[0.5 + (0.25 + Q'E_l/E_{\text{c}}^2)^{1/2}] \tag{97}$$

According to equation (94), one can divide the action integral into two terms, $K = K_{\text{ov}} + K_{\text{s}}$, by integrating from $R_{\text{a}}$ to $R_{\text{t}}$ in the overlapping region and from $R_{\text{t}}$ to $R'_{\text{b}}$ for separated fragments. By expressing the time in seconds, the energies in MeV, and the lengths in fm, one obtains (after introducing the numerical constants) the following relationships:

$$T = \frac{1.4333 \times 10^{-21}}{E_{\text{v}}}[1 + \exp(K)], \qquad K = K_{\text{ov}} + K_{\text{s}} \tag{98}$$

$$K_{\text{ov}} = 0.1296(E_{\text{b}}^0 A_1 A_2/A)^{1/2}\left(\sqrt{b^2 - a^2} - \frac{a^2}{b}\ln\frac{b + \sqrt{b^2 - a^2}}{a}\right] \tag{99}$$

$$K_{\text{s}} = 0.4392(Q'A_1A_2/A)^{1/2}R_{\text{b}}\mathcal{J}_{\text{mc}} \tag{100}$$

$$\begin{aligned}\mathcal{J}_{\text{mc}} = {} & (c + m - 1)^{1/2} - [r(c - r) + m]^{1/2} \\ & + \frac{c}{2}\left(\arcsin\frac{c - 2r}{(c^2 + 4m)^{1/2}} - \arcsin\frac{c - 2}{(c^2 + 4m)^{1/2}}\right) \\ & + \sqrt{m}\ln\left\{\frac{2m^{1/2}[r(c - r) + m]^{1/2} + cr + 2m}{r[2m^{1/2}(c + m - 1)^{1/2} + c + 2m]}\right\}\end{aligned} \tag{101}$$

where

$$a = R_{\text{a}} - R_{\text{i}} = b[(Q' - Q)/E_{\text{b}}^{\text{c}}]^{1/2}, \qquad b = R_{\text{t}} - R_{\text{i}} \tag{102}$$

$$c = rE_{\text{c}}/Q', \qquad m = r^2E_l/Q', \qquad r = R_{\text{t}}/R_{\text{b}'} \tag{103}$$

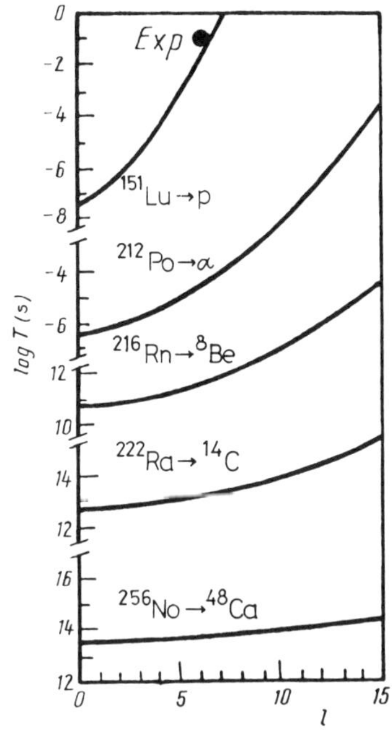


**Figure 33.** The centrifugal barrier effect on the lifetime for emission of various charged particles: protons from $^{151}$Lu ($E_v = 0.51$ MeV), $\alpha$ from $^{212}$Po, $^{8}$Be from $^{216}$Rn, $^{14}$C from $^{222}$Ra, and $^{48}$Ca from $^{256}$No.

If $l = 0$, one has $c = 1$, $m = 0$, and the well-known formula

$$\mathscr{J}_{01} = \arc\cos r^{1/2} - [r(1-r)]^{1/2} \tag{104}$$

is obtained. From Figure 33, we can see that a small angular momentum—up to five units of $\hbar$—makes an important contribution to the lifetime only for light emitted particles (especially for protons). For heavier ions, one can neglect small angular-momentum effects; hence, in the following, we shall neglect spin and parity conservation.

## 6.3. Even–Odd Effect in $^{14}$C Radioactivity

Earlier in this chapter the zero-point vibration energy was obtained by fitting to a selected set of experimental data (Po 82b, Po 83b) $T_i^{\exp}$ on 376 $\alpha$ emitters. Now there are 380 $\alpha$-decay nuclides (Po 84c) and, as is shown by curve a in Figure 34 for $E_v = 0.51$ MeV, the standard deviation of log $T$

$$\sigma = \left[\frac{1}{379}\sum_{i=1}^{380} [\log/(T_i - T_i^{\exp})]^2\right]^{1/2} \tag{105}$$

takes the minimum value $\sigma = 0.96$.

When experimental data for $^{14}$C radioactivity became available (Ro 84), we saw that our predictions made with $E_v = 0.51$ MeV had been too pessimistic by three orders of magnitude. Agreement has been obtained subsequently (Po 84f, Po 84g) by using the law $E_v = 0.1275A_2$.

Using a semiempirical formula (Po 80, Po 83, Po 80b, Po 82b) for $\alpha$ decay, in which shell effects have been taken into account, one gets $\sigma = 0.41$. The variation in

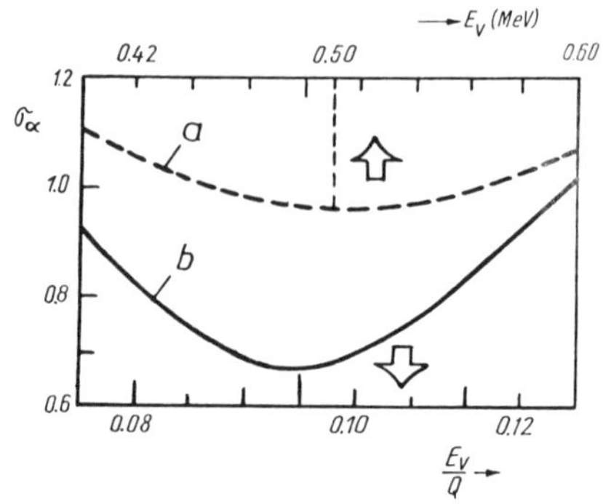


**Figure 34.** Choosing the zero-point vibration energy $E_v$ by minimization of the standard deviation of log $T$ values for 380 $\alpha$ emitters: $E_v$ = constant (a) and $E_v$ proportional to $Q_\alpha$ when $Q_\alpha > 0$ (b).

$E_v$, calculated to reproduce exactly the lifetime for each $\alpha$ emitter, suggests a simple way of introducing shell effects into $E_v$ by making it proportional to $Q$. Curve b of Figure 34 shows the optimum value $0.095Q$ for which $\sigma = 0.73$. A convenient law was found (Po 85c) to be

$$E_v = Q\left[0.056 + 0.039 \exp\left(\frac{4 - A_2}{2.5}\right)\right], \qquad Q > 0 \quad \text{and} \quad A_2 \geqslant 4 \qquad (106)$$

One has $E_v = 0.095Q$ for $\alpha$ decay and $E_v = 31.8 \times 0.056 = 1.77$ MeV for $^{14}C$ radioactivity of $^{223}Ra$.

This equation will be used below. But we require further experimental data in order to check and improve the accuracy of our predictions. Owing to the exponential dependence of $T$ on $K$, the estimated lifetime is very sensitive to the potential barrier shape and parameters. An error of one or even two orders of magnitude is not unacceptable at the present stage of this theory. We can obtain only rough estimates based on this simple model, but the most interesting results obtained merit further study experimentally or with more refined theoretical models.

An odd–even effect with respect to the parent nuclei has been observed for $\alpha$ decay (Po 83). When the 380 $\alpha$ emitters were grouped according to even and odd parent proton and neutron numbers, it was shown (Po 84b) that agreement with experimental results is improved if $E_v/Q$ takes different values in different regions of nuclides, instead of only one (0.095) for all $\alpha$ parents regardless of their even-odd character. The following $E_v/Q$ optimum values (giving minimum $\sigma$) have been obtained: 0.105 for 125 even–even emitters ($\sigma = 0.39$), 0.090 for 111 even–odd ($\sigma = 0.73$), 0.095 for odd–even ($\sigma = 0.48$), and 0.075 for odd–odd ($\sigma = 0.87$). One can see, for example, that $\alpha$-particle emission rates from even–odd parents are somewhat hindered (smaller $E_v/Q$ leading to larger $T$) with respect to that from even–even nuclei or, equivalently, emission rates from even–even parents are enhanced.

Half-life $T$ and $T'$ estimates were made by Poenaru *et al.* (Po 85, Po 86) using both equation (106) for $E_v$ (without the even-odd effect taken into account) and

$$E'_v = E_v \times \begin{cases} 1.105 & \text{for even–even parent} \\ 0.947 & \text{for even–odd parent} \\ 1.000 & \text{for odd–even parent} \\ 0.789 & \text{for odd–odd parent} \end{cases} \qquad (107)$$

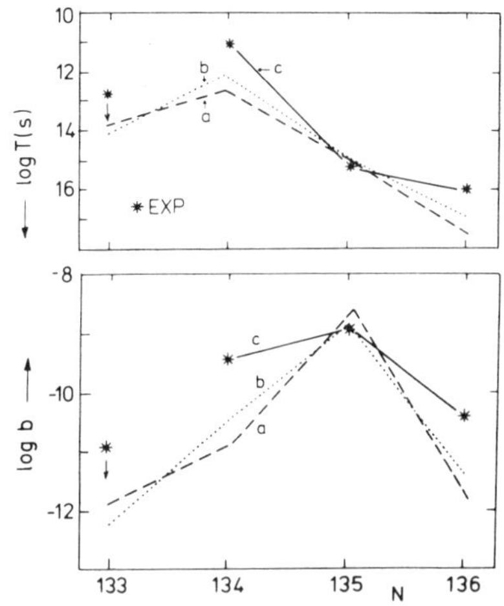


**Figure 35.** Even-odd effect in $^{14}C$ radioactivity of Ra isotopes. * denotes experimental value; - - - - - - $E_v/Q = 0.056$; $\cdots\cdots$ $E_v/Q = 0.062$ and 0.053 for even-even and even-odd parents, respectively; ——— $E_v/Q = 0.075$ and 0.048 for even even and even-odd, respectively.

A comparison of calculated $T$ (dashed line) and $T'$ (dotted line) values with experimental results obtained by Price *et al.* (Pr 85) for $^{14}C$ radioactivity of Ra isotopes is presented in Figure 35a; the corresponding branching ratios are given in Figure 35b. The agreement improves when equation (107) is used, but the observed even-odd effect (Ba 85) is more pronounced than that expected in analogy with $\alpha$ decay: one needs $E_v/Q = 0.075$ for even-even Ra isotopes and 0.048 for even-odd ones to reproduce (solid line) the experimental data. One has to compare these numbers with the values of 0.062 and 0.053, respectively, used to find $T'$ values and with 0.056 for $T$. Recently, the even-odd effect was studied (Po 86d) in detail.

It would be interesting to see whether a similar even-odd effect in the zero-point vibration energy is also present for proton ($Z_2$) and neutron ($N_2$) numbers for the emitted heavy clusters of different parities (for example, to compare $^{24}Ne$ emission with that of $^{25}Ne$).

Martin Greiner and Werner Scheid (Gr 86b) extended ASAFM in order to calculate partial decay rates for transitions into excited states of the daughter nuclei. The corresponding model was called the excitation model (EXM). The result obtained with it for $^{14}C$ and $^{24}Ne$ radioactivities of some nuclei is shown in Figure 36, where the logarithmic difference between the total half-life and the partial half-life of the ground-state transition is shown. One can see that in some cases the total decay rate for cluster emission might be greater by half an order of magnitude as a result of transitions into excited states. When the daughter nucleus is $^{208}Pb$, which has a high-lying first excited state, this effect is, of course, very small.

## 6.4. Branching Ratios Relative to $\alpha$-Particle Decay Estimated Using a Proximity plus Coulomb Potential

Shi and Swiatecki (Sh 85, Sh 85b) used a proximity potential and improved values of nuclear radii to calculate ratios of penetrabilities on the assumption of an extremely asymmetric spontaneous fission mechanism. Equation (16) in Section 2.3 shows that the half-life is expressed as a product of a frequency factor $\tau_0$ and a Gamow penetrability

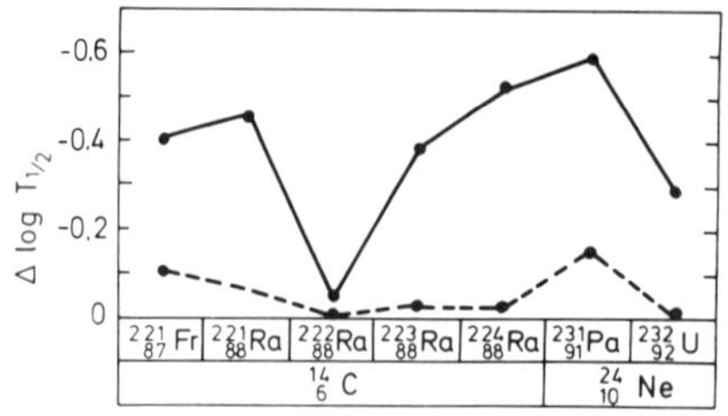


**Figure 36.** Logarithmic difference between the total half-life and that of the ground-state transition. Solid line: excitation radius $R_{ex}$ = exit point; dashed line: $R_{ex}$ = touching point.

factor $G$, hence the branching ratio relative to $\alpha$ decay is given by

$$b = \frac{\tau_0^\alpha}{\tau_0} \cdot \frac{G_\alpha}{G} \tag{108}$$

A comparison of the estimated $G_\alpha/G$ ratio with the experimentally determined branching ratios, $b$, assumes implicitly that $\tau_0 = \tau_0^\alpha$.

The fission barrier is constructed by adding to the Coulomb repulsion between the fragments, the nuclear proximity attraction (Bl 77, Bl 81):

$$E(L) = \begin{cases} -Q + Z_1 Z_2 e^2/R + V_p(z), & L > L_t \\ a(L - L_0)^\nu, & L_0 < L < L_t \end{cases} \tag{109}$$

where $z = b\xi$ is the distance between the near surfaces, $b$ is the diffuseness ($b \simeq 1$ fm), $L$ is the overall length of the configuration ($L_0$ is the diameter of the parent nucleus and $L_t$ is the sum of the fragment diameters at the touching-point configuration), while $R = L - (C_1 + C_2)$ is the separation between fragment centers. The central radii $C_i$ ($i = 0, 1, 2$) are related to the effective sharp radii $R_i$ by

$$C_i = R_i - b^2/R_i \tag{110}$$

where the semiempirical formula (Bl 77)

$$R_i = 1.28A_i^{1/3} - 0.76 + 0.8A^{-1/3} \tag{111}$$

is used.

The proximity interaction potential

$$V_p(z) = K\Phi(\xi), \qquad K = 4\pi \frac{C_1 C_2}{C_1 + C_2} \gamma b$$

$$\gamma = 0.9517[1 - 1.7826((N - Z)/A)^2] \text{ MeV/fm}^2 \tag{112}$$

$$\Phi(\xi) \simeq \begin{cases} -4.41 \exp(-\xi/0.7176) & \text{for } \xi \geqslant 1.9475 \\ -1.7817 + 0.9270\xi + 0.01696\xi^2 - 0.05148\xi^3 & \text{for } 0 \leqslant \xi \leqslant 1.9475 \end{cases}$$

is used only for separated fragments ($L > L_t$). In the overlapping region, the approximation of a $\nu$-power law is adopted (for example, $\nu = 4.678$ for $\alpha$ decay and 3.366 for $^{14}$C decay of $^{222}$Ra).

The coefficients $a$ and $\nu$ are determined from a smooth continuity condition on $V$ at $z = 0$:

$$\nu = -\xi_0 V_t'/V_t, \qquad a = V_t/(-\xi_0)^{\nu}$$
$$V_0' = \left.\frac{dV}{d\xi}\right|_{\xi=0} = -Z_1 Z_2 e^2 b/R_t^2 + K\phi'(0) \tag{113}$$

The Gamow penetrability factor is expressed as

$$G = \exp[(0.43749/(\mathrm{MeV})^{1/2})(b/\mathrm{fm})(A_1 A_2/A)^{1/2} S$$
$$S = \int_{\xi_0}^{\xi_{\mathrm{exit}}} (V(\xi))^{1/2}\, d\xi, \qquad \xi_0 = 2(C - C_1 - C_2)/b, \qquad V(\xi_{\mathrm{exit}}) = 0 \tag{114}$$

The integral $S$ may be evaluated analytically in the interval $\xi_0 < \xi < 0$ and for $\xi > 6$. In the range $0 < \xi < 6$, one can use an approximation (Sh 85b) based on Simpson's rule with seven ordinates.

Tables containing penetrability ratios $G_\alpha/G$ for the spontaneous emission of $^{14}C$, $^{22}O$, $^{24,25,26}Ne$, and $^{28}Mg$ from various parent isotopes of Fr, Ra, Ac, Th, Pa, U, and Np, leading to daughters Hg, Tl, Pb, and Bi, are given by Shi and Swiatecki (Sh 85b).

## 7. New Decay Modes Involving Spontaneous Emission of Heavy Nuclei

In the early stages of the present work we have investigated the region of parent nuclides heavier than lead, because of the enhanced emission rates for cluster decays leading to a daughter nucleus with a magic number of neutrons ($N_1 = 126$) and protons ($Z_1 = 82$). Poenaru *et al.* (Po 84b, Po 85) have shown that, from an energy point of view, heavy-cluster emission is allowed in a region of the nuclide chart larger than that for $\alpha$ decay (see Figure 37).

The importance of the two magic numbers, $Z = 50$ and $N = 82$, for the explanation of this new mode of fission with compact shapes has been discussed recently (De 86b, De 86, Po 84b, Po 86b, Sa 85c). In the following we shall discuss these questions in some detail.

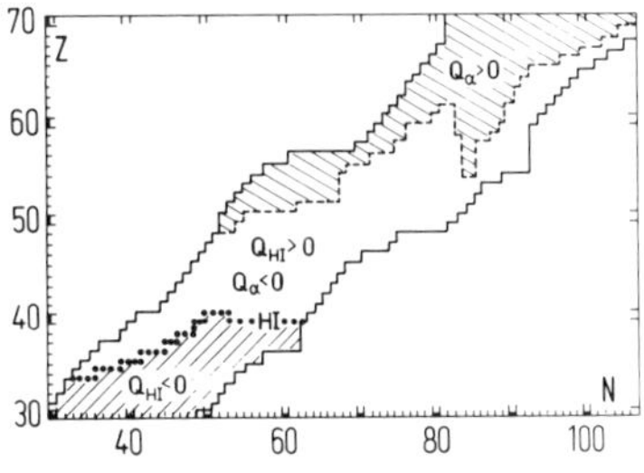


**Figure 37.** Lower limits for the regions where $\alpha$ decay (dashed line) and various cluster radioactivities (dotted line) are allowed from an energy point of view.

### 7.1. Parent Nuclides with Atomic Number Smaller than 84. Decay of "Stable" *trans*-Zirconium Nuclides

Figure 37 shows a part of the map of the nuclides tabulated by Wapstra and Audi (Wa 85) indicating the lower limits of the regions where $\alpha$ decay and various cluster radioactivities are allowed. For example, the neutron-deficient nucleus $^{67}_{34}$Se—which is stable relative to $\alpha$ decay—can be split into $^{27}$Si + $^{40}$Ca ($Q = 0.37$ MeV), $^{28}$Si + $^{39}$Ca ($Q = 1.91$ MeV), $^{28}$Si + $^{39}$Ca ($Q = 1.91$ MeV), $^{31}$S + $^{36}$Ar ($Q = 2.42$ MeV), and $^{32}$S + $^{35}$Ar ($Q = 2.20$ MeV). For $Z > 40$, *all* nuclides, including the so-called "stable" ones, are metastable with respect to the new decay modes. However, if the lifetime is long enough, $T > T_{max}$, one can—from a practical point of view—consider the nuclides as being stable. The question then is: What sensible limit $T_{max}$ should be chosen? Which decay channel determines it? And can it be measured? Indeed, measurements of lifetimes have reached new sensitivity. For example, half-lives of the order of $10^{30}$ s have been measured for the spontaneous fission of some actinides and are expected to be found for double beta decay (Ch 84, Kl 84).

Attempts are made to measure the proton half-life, estimated to be about $10^{38}$–$10^{40}$ s. Hence we shall consider $T_{max} \leq 10^{50}$ s.

There are 1741 nuclides with $Z \geq 34$ and total lifetime longer than 1 $\mu$s (Se 81), of which almost 950 have lifetimes shorter than $10^{50}$ s for several decay modes. For example, $^{151}_{63}\text{Eu} \to {}^{12}_{6}\text{C}_6 + {}^{139}_{57}\text{La}_{82}$ with $Q = 12.6$ MeV and $T = 10^{42.7}$ s and $^{202}_{82}\text{Pb} \to {}^{66}_{28}\text{Ni}_{38} + {}^{136}_{54}\text{Xe}_{82}$ with $Q = 126.5$ MeV and $T = 10^{44.4}$ s. Even $^{208}$Pb has a chance to fission either into $^{74}_{30}\text{Zn}_{44} + {}^{134}_{52}\text{Te}_{82}$ ($Q = 133.4$ MeV, $T = 10^{52.6}$ s) or into $^{80}_{32}\text{Ge}_{48} + {}^{128}_{50}\text{Sm}_{78}$ ($Q = 138.2$ MeV, $T = 10^{52.7}$ s).

Only very few of the above-mentioned nuclides have partial lifetimes short enough (see Table 2) to permit possible observation. These are obtained usually for daughters with magic number of neutrons $N_1 = 50$ or $N_1 = 82$. The double magic nucleus $^{132}_{50}\text{Sn}_{82}$ is too neutron-rich (far off the $\beta$-stability line) to play the role of a very stable daughter (like $^{208}$Pb for *trans*-lead nuclei) for parent-emitted heavy-ion combinations lying not very far off the $\beta$-stability line. A more important role could be played by the $^{100}$Sn double magic nucleus because, as in the case of $\alpha$ decay, neutron-deficient parents are preferred. One cannot investigate this region in which one also has to consider the possible competition of proton radioactivity, owing to the present lack of relevant mass measurements.

The branching ratio with respect to $\alpha$ decay here is typically much smaller than $10^{-10}$, but there are some cases (such as emission of $^{32}$Si from $^{169}$Tm, of $^{48}$Ca from $^{176}$Yb, $^{180}$Hf, $^{181}$Ta, $^{183,184}$W, or of $^{49}$Ca from $^{185}$W) for which one might expect that nuclear-cluster emission would be more probable than $\alpha$ decay!

In Table 3 only some of the "stable" parent nuclei with $T < 10^{50}$ s are listed for cluster emission with $Z_2 \leq 28$. A more complete table containing also $^{162}$Er, $^{171,172,174,176}$Yb, $^{175}$Lu, $^{176-179}$Hf, $^{180}$Ta, $^{190}$Os, $^{193}$Ir, $^{194-196}$Pt, $^{198-201}$Hg, and $^{203}$Tl in addition to many other radioactive nuclei is published elsewhere (Po 86). Data relevant to the decay of some neutron-deficient nuclides in this region have also been published (Po 84b). $\alpha$-Decay half-lives $T_\alpha$ have been estimated with our semiempirical formula. One can see that $T < 10^{30}$ s is expected for $^{151}$Eu, $^{176}$Hf, $^{180}$W, and $^{184,187}$Os. One also has $T < 10^{42}$ s for $^{16}$O emission from $^{156}$Dy, for $^{48}$Ca emission from $^{184}$W, $^{185}$Re, and $^{184}$Os, and for $^{49}$Ca emission from $^{187}$Os.

In conclusion, according to the mentioned estimates obtained within the framework of ASAFM, even the so-called "stable" nuclei with $Z > 60$ are expected to decay

**Table 2. Some Emitters ($Z < 84$) of Heavy Ions with Short Lifetimes $T$ or Large Branching Ratio Relative to $\alpha$ Decay**

| Emitted heavy ion | Emitter Nuclide | Emitter log $T_t$(s) | log $T_\alpha$(s) | Daughter $Z_1$ | Daughter $N_1$ | Q (MeV) | log $T$(s) |
|---|---|---|---|---|---|---|---|
| ${}^{8}_{4}Be_4$ | ${}^{110}Xe$ | −0.40 | −0.4 | 50 | 52 | 8.11$^c$ | 17.2 |
| | ${}^{155}Tm$ | 1.59 | 3.6 | 65 | 82 | 9.2 | 26.6 |
| | ${}^{158}Hf$ | 0.51 | 0.8 | 68 | 82 | 10.8$^c$ | 22.1 |
| | ${}^{160}W$ | −1.09 | −1.1$^b$ | 70 | 82 | 12.0$^c$ | 18.9 |
| | ${}^{161}Re$ | −2.00 | −2.0$^b$ | 71 | 82 | 12.7$^c$ | 17.2 |
| | ${}^{186}Bi$ | 0.30 | 0.3$^b$ | 79 | 99 | 19.6$^c$ | 4.1 |
| ${}^{12}_{6}C_6$ | ${}^{113}Cs$ | 0.40 | 1.4$^b$ | 49 | 52 | 16.80 | 16.3 |
| | ${}^{114}Ba$ | −0.4 | −0.4$^b$ | 50 | 52 | 20.07$^d$ | 9.3 |
| | ${}^{154}Dy$ | 13.5 | 13.5 | 60 | 82 | 15.6 | 33.7 |
| | ${}^{160}Hf$ | 1.08 | 2.7 | 66 | 82 | 21.9 | 20.9 |
| | ${}^{161}Ta$ | 0.70 | 1.9$^b$ | 67 | 82 | 22.9$^c$ | 19.4 |
| | ${}^{162}W$ | 0.14 | 0.3 | 68 | 82 | 23.8$^c$ | 18.3 |
| ${}^{16}_{8}O_8$ | ${}^{110}Te$ | 1.27 | 6.4 | 44 | 50 | 15.16$^c$ | 43.4 |
| | ${}^{156}Dy$ | *a* | 32.3 | 58 | 82 | 22.3 | 41.0 |
| | ${}^{164}W$ | 0.81 | 2.4 | 66 | 82 | 34.4 | 21.9 |
| ${}^{28}_{14}Si_{14}$ | ${}^{126}Nd$ | 0.30 | 23.8$^b$ | 46 | 52 | 48.47$^d$ | 28.6 |
| ${}^{32}_{14}Si_{18}$ | ${}^{172}Hf$ | 7.77 | 20.59$^b$ | 58 | 82 | 55.8 | 42.0 |
| ${}^{48}_{20}Ca_{28}$ | ${}^{184}W$ | *a* | 43.9$^b$ | 54 | 82 | 84.9 | 40.9 |
| | ${}^{186}Os$ | 22.80 | 22.8 | 56 | 82 | 89.5 | 39.5 |
| | ${}^{188}Pt$ | 5.95 | 12.5 | 58 | 82 | 94.5 | 37.5 |
| ${}^{50}_{22}Ti_{28}$ | ${}^{188}Pt$ | 5,95 | 12.5 | 56 | 82 | 101.9 | 37.7 |
| ${}^{56}_{24}Cr_{32}$ | ${}^{194}Hg$ | 10.06 | 28.5$^b$ | 56 | 82 | 111.3 | 41.1 |

[a] "Stable."
[b] Estimated with semiempirical relationship. If $T_\alpha < T$ one takes $T_\alpha = T$.
[c] Mass of the daughter or of the parent from systematics (Wa 85).
[d] Estimated mass of the daughter or of the parent.

spontaneously, by emission of clusters such as ${}^{12}C$, ${}^{16}O$, ${}^{30,32}Si$, ${}^{48,50}Ca$, and ${}^{68}Ni$ and with half-lives $T > 10^{40}$ s, leading to daughters with $Z_1 = 50$–58 and $N_1 \simeq 78$–82. This means that in nature there is a very slow trend toward continuously increasing the abundance of lighter nuclei because of the spontaneous decay of heavier ones by cluster emission.

## 7.2. Lifetimes of *trans*-Lead Nuclei

A part of the nuclear chart, with $Z \geqslant 60$ and $N \geqslant 80$, is shown in Figure 38, where contour plots of $Q$-values for ${}^{14}C$ radioactivity are drawn. For a given mass number, $A = Z + N$, there is a smooth (liquid-drop model) trend of increase in $Q$ from the neutron-rich toward the neutron-deficient side, similar to that for $\alpha$-decay $Q$-values.

**Table 3. Some "Stable" Nuclides Possessing Half-Life $T$ with Respect to Heavy Cluster Emission Shorter than $10^{50}$ s**

| Nuclide | Emitted heavy ion | Daughter $Z_1$ | Daughter $N_1$ | $Q$ (MeV) | $Q_\alpha$ (MeV) | $\log T_\alpha$ (s) | $\log T$ (s) | $\log T'$ (s) | $\log\left(\frac{T}{T_\alpha}\right)$ | $\log\left(\frac{T'}{T'_\alpha}\right)$ |
|---|---|---|---|---|---|---|---|---|---|---|
| $^{150}$Sm | $^{12}$C | 56 | 82 | 11.21 | 1.45 | 35.8 | 48.8 | 48.3 | 13.0 | 12.5 |
| $^{151}$Eu | | 57 | 82 | 12.57 | 1.96 | 25.7 | 42.7 | 42.7 | 17.0 | 17.0 |
| $^{154}$Gd | $^{16}$O | 56 | 82 | 19.29 | 0.92 | 60.4 | 48.5 | 48.0 | −11.9 | −12.4 |
| $^{156}$Dy | | 58 | 82 | 22.29 | 1.76 | 32.2 | 41.1 | 40.5 | 8.8 | 8.3 |
| $^{169}$Tm | $^{32}$Si | 55 | 82 | 49.36 | 1.20 | 54.7 | 48.3 | 48.3 | −6.4 | −6.4 |
| $^{168}$Yb | $^{30}$Si | 56 | 82 | 51.13 | 1.95 | 32.1 | 45.5 | 44.6 | 13.3 | 12.5 |
| $^{170}$Yb | $^{32}$Si | 56 | 82 | 51.58 | 1.74 | 37.1 | 45.9 | 45.0 | 8.8 | 7.9 |
| $^{180}$Hf | $^{48}$Ca | 52 | 80 | 79.64 | 1.28 | 54.3 | 44.0 | 42.8 | −10.3 | −11.5 |
| $^{181}$Ta | | 53 | 80 | 81.68 | 1.52 | 47.6 | 43.6 | 43.6 | −3.9 | −3.9 |
| $^{180}$W | | 54 | 78 | 83.86 | 2.51 | 25.9 | 43.3 | 42.1 | 17.4 | 16.2 |
| $^{182}$W | | 54 | 80 | 84.09 | 1.77 | 40.7 | 42.6 | 41.4 | 1.9 | 0.7 |
| $^{183}$W | | 54 | 81 | 84.35 | 1.68 | 46.2 | 42.0 | 42.7 | −4.2 | −3.6 |
| $^{184}$W | | 54 | 82 | 84.94 | 1.66 | 43.9 | 40.9 | 39.7 | −3.0 | −4.2 |
| $^{186}$W | $^{50}$Ca | 54 | 82 | 83.48 | 1.12 | 64.9 | 43.9 | 42.7 | −21.0 | −22.2 |
| $^{185}$Re | $^{48}$Ca | 55 | 82 | 86.95 | 2.19 | 32.8 | 40.7 | 40.7 | 7.8 | 7.8 |
| $^{184}$Os | | 56 | 80 | 88.87 | 2.97 | 21.2 | 40.8 | 39.6 | 19.5 | 18.3 |
| $^{187}$Os | $^{49}$Ca | 56 | 82 | 88.34 | 2.72 | 27.2 | 41.6 | 42.3 | 14.5 | 15.1 |
| $^{188}$Os | $^{52}$Ti | 54 | 82 | 94.75 | 2.14 | 34.4 | 42.8 | 41.5 | 8.4 | 7.1 |
| $^{189}$Os | $^{53}$Ti | 54 | 82 | 94.27 | 1.97 | 41.9 | 43.8 | 44.5 | 1.9 | 2.5 |
| $^{192}$Os | $^{58}$Cr | 52 | 82 | 98.57 | 0.36 | 161.6 | 47.4 | 46.0 | −114.2 | −115.6 |
| $^{191}$Ir | $^{56}$Cr | 53 | 82 | 102.39 | 2.08 | 37.1 | 44.4 | 44.4 | 7.4 | 7.4 |
| $^{192}$Pt | $^{56}$Cr | 54 | 82 | 105.41 | 2.41 | 31.2 | 43.2 | 41.8 | 12.0 | 10.6 |
| $^{198}$Pt | $^{68}$Ni | 50 | 80 | 113.74 | 0.09 | 399.3 | 48.3 | 46.7 | −351.0 | −352.6 |
| $^{197}$Au | $^{62}$Fe | 53 | 82 | 111.54 | 0.95 | 83.6 | 47.1 | 47.1 | −36.5 | −36.5 |
| $^{196}$Hg | $^{60}$Fe | 54 | 82 | 115.99 | 2.04 | 40.6 | 43.7 | 42.2 | 3.1 | 1.6 |
| $^{202}$Hg | $^{68}$Ni | 52 | 82 | 118.52 | 0.13 | 317.5 | 48.7 | 47.1 | −268.7 | −270.4 |

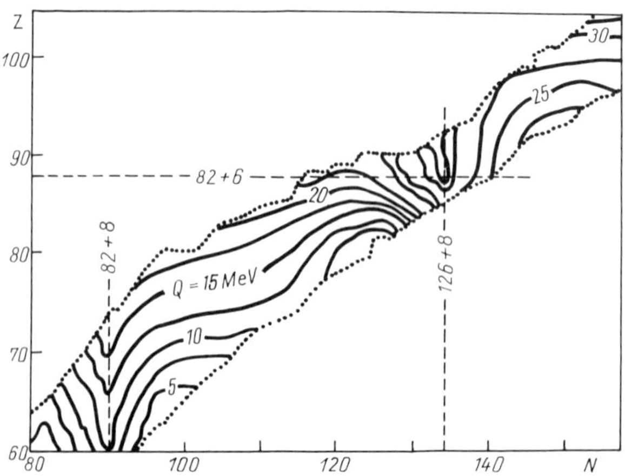


**Figure 38.** Contour plots of $Q$-values for $^{14}$C emission from various nuclides with $Z \geq 60$ and $N \geq 80$, having masses tabulated by Wapstra and Bos (Wa 77).

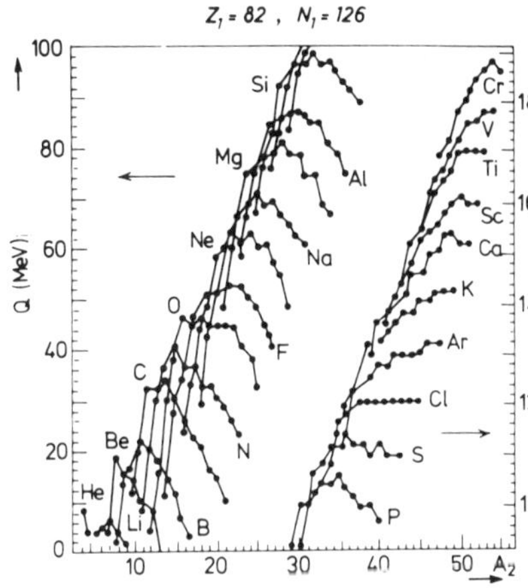


**Figure 39.** Released energy for the emission of various isotopes of the specified elements from the parents leading to the double magic $^{208}$Pb daughter nucleus.

The reverse is true for very neutron-rich, emitted heavy ions such as $^{8,9}$He. The shell effects for the daughter magic numbers $Z_1 = 82$, $N_1 = 82$ and 126 are also very clearly evident, pushing down the maximum $Q$-value for $^{14}$C emission to $^{222}_{88}$Ra (33.05 MeV) and to $^{222}_{89}$Ac (33.08 MeV).

The very strong shell effect at the double magic daughter $^{208}$Pb in Figure 38 is also observed for all other emitted heavy ions. Hence we can focus, in Figure 39, on those $Q$-values for heavy-ion emission in which the daughter is $^{208}$Pb. The released energy increases with $Z_2$, and for $Z_2 > 16$ one has $Q > 100$ MeV. This energy is shared between the emitted particle and the daughter. As for $\alpha$ decay, one can assume that the kinetic energy of the emitted heavy ion is given by $E_k = QA_1/A$ and that the recoil energy of the daughter is $Q - E$. But it can happen that for heavier emitted particles as in fission, a part of this energy is lost in exciting or deforming one or both of the fragments.

The corresponding predicted lifetimes are plotted in Figure 40, displaying the best values expected for almost all new decay modes. To guide the eye, various isotopes of a given emitted element are joined by a solid line if $Z_2$ is even, or by a dashed line if $Z_2$ is odd. The specified mass numbers of some emitted heavy ions are used as reference

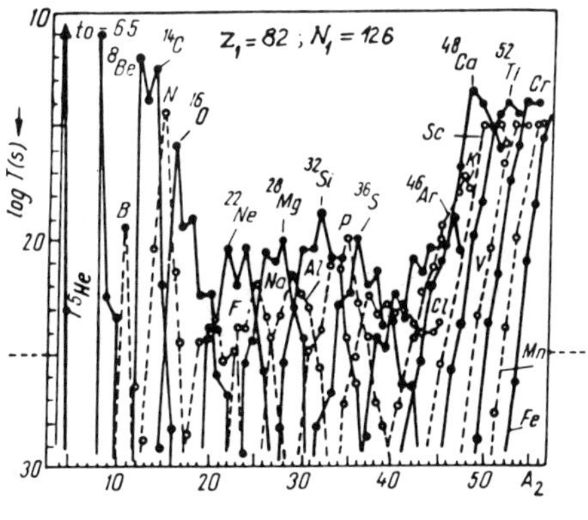


**Figure 40.** Decimal logarithm of the lifetimes for the same processes as in Figure 39.

points. For a practical reason, $Z_2$ is restricted to values less than 26 (parents with $Z < 108$), but of course we could continue even to the superheavy nuclei.

Almost all nuclei are metastable with respect to several decay modes, but if the lifetimes are very long, they are effectively stable. One must consider all the competing decay modes, from which the most important one is that having the minimum partial half-life. Available experimental techniques now permit measurement of very slow processes ($T \sim 10^{22}$ yr), as for example the spontaneous fission of some actinides. Consequently, we cut Figure 40 at $T = 10^{30}$ s. The time scale on many figures of this chapter is reversed, because we wished to create an impression of the relative intensity which is proportional to $T^{-1}$.

The most important shell effect exists in the region of the double magic ${}^{208}_{82}$Pb daughter. Similar predictions for lighter parent nuclei show longer lifetimes. In Figure 40 we also show a pairing effect: for odd $N_2$ or $Z_2$ emitted heavy ions, the lifetime is longer than in the cases of their even neighbors. There are a few exceptions to this rule concerning $Z_2$, such as ${}^{15}_{7}N_8$ and ${}^{35}_{15}P_{20}$, but in these cases the neutron numbers $N_2$ are magic, and the proton numbers are almost magic.

For each $A_2$ in Figure 40 one has a $Z_2$ giving a minimum lifetime. When $A_2$ is increased beyond $A_2 = 16$, the most probable emitted cluster for a given $A_2$ becomes more and more neutron-rich and the corresponding parent becomes the more neutron-deficient nucleus. For $Z_2 \geqslant 17$ there is a cut in the curves at higher values of $A_2$, because the corresponding masses (on the neutron-rich side of the emitted particles or the neutron-deficient side of the parents) are not yet available.

The list of all new decay modes given in this figure is very large: ${}^{5}$He, ${}^{8-10}$Be, ${}^{11,12}$B, ${}^{12-16}$C, ${}^{13-17}$N, ${}^{15-22}$O, ${}^{18-23}$F, ${}^{20-26}$Ne, ${}^{23-28}$Na, ${}^{23-30}$Mg, ${}^{27-32}$Al, ${}^{28-36}$Si, ${}^{31-39}$P, ${}^{32-42}$S, ${}^{35-45,\ldots}$Cl, ${}^{37-47,\ldots}$Ar, ${}^{40-49,\ldots}$K, ${}^{42-51,\ldots}$Ca, ${}^{44-53,\ldots}$Sc, ${}^{46-53,\ldots}$Ti, ${}^{48-54,\ldots}$V, ${}^{49-55,\ldots}$Cr, etc. There are more than 140 kinds of possible exotic radioactivities with $Z_2 = 2$–24 and they continue for $Z_2 > 24$. From these one can select a list of the most interesting processes for a given $A_2$: ${}^{5}$He, ${}^{8}$Be, ${}^{11}$B, ${}^{12-14}$C, ${}^{15}$N, ${}^{16-20}$O, ${}^{21}$F, ${}^{22-24}$Ne, ${}^{25}$Na, ${}^{26-28}$Mg, ${}^{29}$Al, ${}^{30-34}$Si, ${}^{35}$P, ${}^{36-38}$S, ${}^{39,41}$Cl, ${}^{40,42-44}$Ar, ${}^{45}$K, ${}^{46-50}$Ca, ${}^{51-53}$Ti, ${}^{54,55}$Cr, etc. For each element one or two isotopes are most probably emitted: ${}^{8}$Be, ${}^{11}$B, ${}^{12,14}$C, ${}^{15}$N, ${}^{16}$O, ${}^{21}$F, ${}^{22,24}$Ne, ${}^{25}$Na, ${}^{26,28}$Mg, ${}^{29}$Al, ${}^{32}$Si, ${}^{35}$P, ${}^{36}$S, ${}^{37}$Cl, ${}^{46}$Ar, ${}^{47}$K, ${}^{48}$Ca, ${}^{49}$Sc, ${}^{52}$Ti, ${}^{53,54}$V, ${}^{54,55}$Cr.

A detailed illustration of the $N_1 = 126$ neutron and $Z_1 = 82$ proton shell effects in the daughter nucleus is given in Figure 41, where the lifetime for the 12 decay modes (${}^{8}$Be, ${}^{12-14}$C, ${}^{15}$N, ${}^{23}$F, ${}^{24,25}$Ne, ${}^{28}$Mg, ${}^{32}$Si, ${}^{46}$Ar, and ${}^{48}$Ca) are plotted vs. the daughter neutron number $N_1$ for various daughter proton numbers $Z_1 = 80$–87. The minimum value of the lifetime for a given $Z_1$ is almost always obtained when $N_1 = 126$. For some light emitted heavy ions (Like ${}^{8}$Be and ${}^{12}$C), however, $Z_1$ for the minimum lifetime is not 82 as usual. Nevertheless, the improvement with respect to the values given in Figure 40 is not larger than one order of magnitude. The even–odd effect can also be seen better in Figures 41(c), (e), (f), and (h). For even–odd emitted heavy ions (${}^{13}_{6}C_7$ and ${}^{25}_{10}Ne_{15}$), even $N_1$ (odd $N$) are preferred. For odd–even (${}^{15}_{7}N_8$ and ${}^{23}_{9}F_{14}$), even $Z_1$ (odd $Z$) are the most probable.

One can display, simultaneously, the information about $Q$-values and half-lives in a diagram (Figure 42) similar to the Geiger–Nuttall (GN) plot for $\alpha$ decay. The difference is that $1/\sqrt{Q}$ is usually used in a GN plot instead of $Q$ as here.

In Figure 43, we compare the results obtained using numerical methods (the points) with those obtained from our analytical formula (the curves). One can see that, not far from the magic neutron number of the daughter, the agreement is good, but in

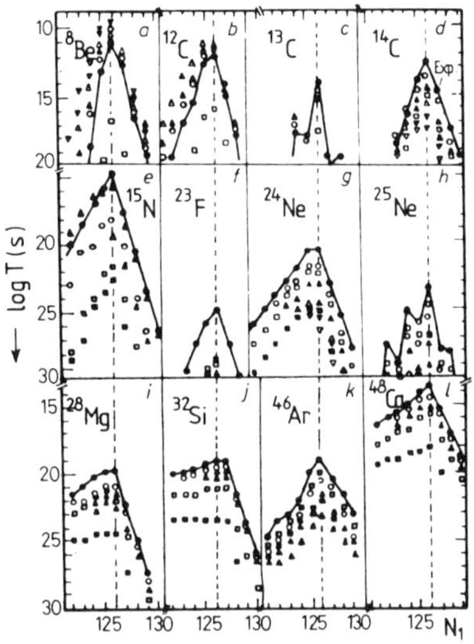


**Figure 41.** Lifetimes vs. daughter neutron number $N_1$ for various decay modes by spontaneous emission of the following clusters: $^{8}$Be (a), $^{12}$C (b), $^{13}$C (c), $^{14}$C (d), $^{15}$N (e), $^{23}$F (f), $^{24}$Ne (g), $^{25}$Ne (h), $^{28}$Mg (i), $^{32}$Si (j), $^{46}$Ar (k), and $^{48}$Ca (l). The daughter atomic numbers are the following: ■—80; □—81; ●—82; ○—83; ▲—84; △—85; ▼—86, and ▽—87.

general the results of the analytic formula are more optimistic than those of the numerical method.

## 7.3. Branching Ratios Relative to $\alpha$ Decay

The most important competitor of the new decay modes is usually $\alpha$-particle decay. Consequently, for some experiments, the absolute value of the lifetime $T$ is not as important as the branching ratio relative to $\alpha$ decay, $T/T_{\alpha}$. We have shown that from

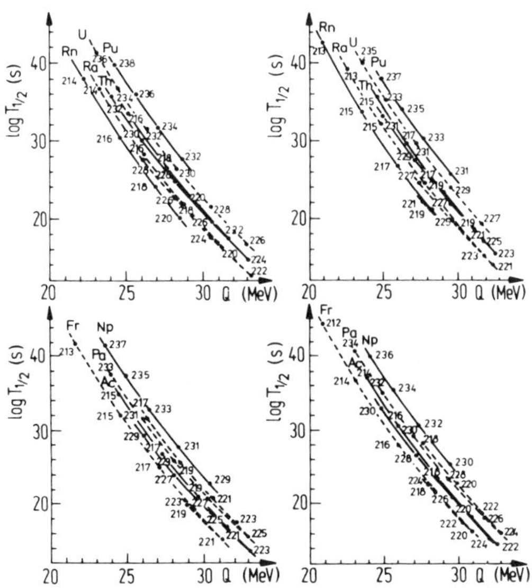


**Figure 42.** Half-lives and $Q$-values for $^{14}$C radioactivity of even–even, even–odd, odd–even, and odd–odd nuclides with $Z = 87$–$94$.

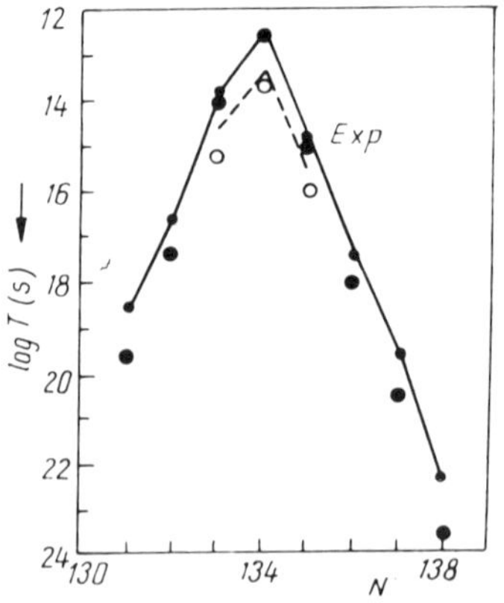


**Figure 43.** Comparison of lifetimes for $^{14}C$ emission from various isotopes of Ra and Ac, calculated by a numerical method (●—Ra, ○—Ac) and the analytical formula (——, Ra, - - -, Ac).

this standpoint, Rose and Jones (Ro 84) discovered the best emitter of $^{14}C$, namely $^{223}Ra$, with a branching ratio $b = \Gamma/\Gamma_\alpha = T_\alpha/T = (8.5 \pm 2.5) \times 10^{-10}$. Even if the $\alpha$ particles are deflected in a magnetic field, the lifetime is determined from the measured branching ratio, $b$, because it is not possible to measure directly an exponential decay curve for these very slow processes.

Many of the $\alpha$-radioactive nuclides also have other competing decay modes ($\beta$ decay, spontaneous fission, etc.). Hence the partial $\alpha$ lifetime $T_\alpha$ is longer than or equal to the total disintegration period $T_t$. Usually there are several $\alpha$ transitions belonging to a parent nucleus, because the process does not take place only between ground states and different excited states of the daughter are usually populated.

In the table given by Poenaru and Ivascu (Po 83b, Po 84c) form all $\alpha$ transitions, only the strongest have been selected: ground-state to ground-state transitions in even–even nuclides or so-called "favored" transitions in odd–odd or odd-mass nuclides, because the partial half-life $T_\alpha$ calculated with our semiempirical formula refers also to these single transitions having an intensity $i_s$ expressed in percent, sometimes different from 100%. Consequently, the total $\alpha$-decay half-life $T_\alpha^t$ can be found by using the relationship

$$T_\alpha^t = \frac{i_s}{100} T_\alpha \tag{115}$$

In experiments and in work by Greiner and Scheid (Gr 86b), the branching ratio relative to the total number of $\alpha$-decay lines is determined as

$$b_t = T_\alpha^t / T \tag{116}$$

but in our table (and in similar tables computed with other models) only one of the $\alpha$ transitions is taken into account:

$$b_s = T_\alpha / T \tag{117}$$

Hence in order to compare the measured $b_t$ with calculated $b_s$ values, one must apply a correction $b_s = (100/i_s)b_t$, where $i_s$ is given in tables for $\alpha$-decay (Le 78, Ry 79, Po 83b) half-lives.

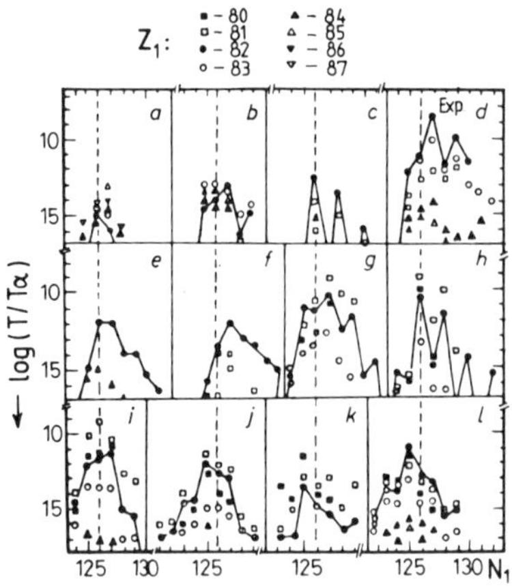


**Figure 44.** Branching ratios relative to $\alpha$ decay for the same processes as in Figure 41.

For $^{24}Ne$ radioactivity our estimates appear to be more optimistic than is shown by experiment, a trend also observed for heavier emitted clusters. If more experimental data are available in future, it will be possible to modify equations (106) and (107) [see (Po 86d)] in order to obtain a better fit as, e.g., carried out for the 379 $\alpha$ emitters. Comparing the results of Shi and Swiatecki (1985b) with ours, it is seen that generally estimates for C and Ne radioactivities are in agreement, but for $^{22}O$ radioactivity the former are more optimistic and for $^{28}Mg$ more pessimistic.

We have demonstrated, in the case of $^{14}C$ radioactivity, the simple rule that the largest decay probability is obtained for a doubly magic daughter or neighboring final product. In Figure 44 many more decay modes are presented.

Only for a small number of cases ($^{13}C$, $^{15}N$, $^{25}Ne$, and $^{28}Mg$) are the most probable branching ratios obtained at $N_1 = 126$. Usually $N_1 = 127$ is preferred for clusters lighter than $^{25}Ne$ and $N_1 = 125$ for emitted ions heavier than $^{28}Mg$. The value $Z_1 = 82$ is selected for $^{13,14}C$, $^{15}N$, $^{23}F$, and $^{48}Ca$, but one can have $Z_1 = 85$ for $^{8}Be$, $Z_1 = 80$ for $^{46}Ar$, and $Z_1 = 81$ for $^{24,25}Ne$, $^{28}Mg$, and $^{32}Si$. These calculated numbers suggest that it will be difficult to detect these modes experimentally because of their weak predicted intensity, but it is known that small branching ratios relative to $\alpha$ decay—in the range of $10^{-13}$–$10^{-10}$—have been determined in the spontaneous fission of some Th, U, Np, and Am isotopes. In the experimental search for new decay modes one should use the extended tables (Po 86).

Some examples are given in Table 4. For each parent nucleus one has a line with a code number $C$ for the mass, the decimal logarithm of its total half-life (Se 81) $T_t$ (in s), the decimal logarithm of its partial half-life $T_\alpha$ for its strongest $\alpha$ transition, and the corresponding code $C_\alpha$. For $C_\alpha = 2$ we have used an experimental value for $T_\alpha$ selected by us; for $C_\alpha = 1$, reflecting the paucity of data in our table we have used our semiempirical relationship to estimate $T_\alpha$. The following lines contain information on the emitted cluster (its mass $A_2$ and atomic number $Z_2$ and the code number of its mass $C_2$), the daughter nucleus (proton number $Z_1$, neutron number $N_1$, and code $C_1$), the released energy $Q$, the cluster kinetic energy $E_k = Q \cdot A_1/A$, the partial lifetime for the cluster emission calculated by using the present relationship for $E_v$, $T$, and by using the old one ($E_v = 0.1275\ A_2$), $T'$, and the branching ratio $T/T_\alpha$. The mass code numbers $C$, $C_1$, and $C_2$ show the source for the masses used to find the $Q$-value: 0

**Table 4. Some of the Cluster Emissions from Different Parent Nuclei**

| Parent | | $C$ | log $T_t$ (s) | | | | | log $T_\alpha$ (s) | | $C_\alpha$ |
|---|---|---|---|---|---|---|---|---|---|---|
| $A_2$ | $Z_2$ | $C_2$ | $Z_1$ | $N_1$ | $C_1$ | $Q$ (MeV) | $E_k$ (MeV) | log $T$ (s) | log $T'$ (s) | $\log\left(\frac{T}{T_\alpha}\right)$ |
| $^{223}$Ra | | 0 | | | 5.99 | | | 6.27 | | 2 |
| 14 | 6 | 0 | 82 | 127 | 0 | 31.85 | 29.85 | 14.8 | 14.8 | 8.5 |
| 13 | 6 | 0 | 82 | 128 | 0 | 28.86 | 27.18 | 19.7 | 19.7 | 13.4 |
| $^{226}$Ra | | 0 | | | 10.73 | | | 10.73 | | 2 |
| 14 | 6 | 0 | 82 | 130 | 0 | 28.21 | 26.47 | 22.4 | 21.9 | 11.7 |
| 20 | 8 | 0 | 80 | 126 | 0 | 40.83 | 37.22 | 27.6 | 26.8 | 16.8 |
| $^{224}$Ac | | 0 | | | 4.02 | | | 5.61 | | 2 |
| 14 | 6 | 0 | 83 | 127 | 0 | 32.0 | 30.00 | 15.4 | 15.5 | 9.8 |
| 15 | 7 | 0 | 82 | 127 | 0 | 37.74 | 35.21 | 17.7 | 18.2 | 12.1 |
| $^{229}$Th | | 0 | | | 11.36 | | | 11.61 | | 2 |
| 20 | 8 | 0 | 82 | 127 | 0 | 43.42 | 39.63 | 24.8 | 24.5 | 13.2 |
| 24 | 10 | 0 | 80 | 125 | 0 | 57.84 | 51.78 | 24.8 | 23.5 | 13.2 |
| 21 | 8 | 0 | 82 | 126 | 0 | 43.22 | 39.26 | 26.2 | 25.5 | 14.6 |
| $^{230}$Th | | 0 | | | 12.39 | | | 12.51 | | 2 |
| 24 | 10 | 0 | 80 | 126 | 0 | 57.78 | 51.75 | 24.9 | 25.3 | 12.4 |
| 22 | 8 | 0 | 82 | 126 | 0 | 43.19 | 39.06 | 27.1 | 26.0 | 14.6 |
| $^{232}$Th | | 0 | | | 17.64 | | | 17.76 | | 2 |
| 24 | 10 | 0 | 80 | 128 | 10 | 55.62 | 49.87 | 28.4 | 28.5 | 10.6 |
| 28 | 12 | 0 | 78 | 126 | 10 | 69.43 | 61.05 | 28.5 | 29.3 | 10.8 |
| 26 | 10 | 0 | 80 | 126 | 0 | 55.97 | 49.70 | 29.4 | 28.9 | 11.6 |
| $^{231}$Pa | | 0 | | | 12.01 | | | 12.01 | | 1 |
| 24 | 10 | 0 | 81 | 126 | 0 | 60.42 | 54.14 | 22.0 | 22.7 | 10.0 |
| 23 | 9 | 0 | 82 | 126 | 0 | 51.84 | 46.68 | 24.8 | 24.7 | 12.8 |
| $^{232}$U | | 0 | | | 9.56 | | | 9.52 | | 2 |
| 24 | 10 | 0 | 82 | 126 | 0 | 62.31 | 55.86 | 20.4 | 21.3 | 10.9 |
| 28 | 12 | 0 | 80 | 124 | 0 | 74.32 | 65.35 | 24.5 | 25.7 | 15.0 |
| $^{233}$U | | 0 | | | 12.70 | | | 12.78 | | 2 |
| 24 | 10 | 0 | 82 | 127 | 0 | 60.50 | 54.27 | 23.1 | 23.9 | 10.3 |
| 25 | 10 | 0 | 82 | 126 | 0 | 60.85 | 54.32 | 23.3 | 23.8 | 10.5 |
| 28 | 12 | 0 | 80 | 125 | 0 | 74.24 | 65.32 | 24.5 | 25.7 | 11.7 |
| $^{234}$U | | 0 | | | 12.88 | | | 13.02 | | 2 |
| 28 | 12 | 0 | 80 | 126 | 0 | 74.13 | 65.26 | 24.6 | 25.8 | 11.5 |
| 24 | 10 | 0 | 82 | 128 | 0 | 58.84 | 52.81 | 25.7 | 26.3 | 12.7 |
| 26 | 10 | 0 | 82 | 126 | 0 | 59.47 | 52.87 | 26.2 | 26.2 | 13.1 |
| $^{235}$U | | 0 | | | 16.33 | | | 16.57 | | 2 |
| 20 | 12 | 0 | 80 | 127 | 0 | 72.20 | 63.60 | 27.3 | 28.4 | 10.8 |
| 29 | 12 | 0 | 80 | 126 | 0 | 72.61 | 63.65 | 27.4 | 28.2 | 10.8 |
| 32 | 14 | 0 | 78 | 125 | 10 | 85.74 | 74.06 | 27.9 | 29.4 | 11.3 |
| 25 | 10 | 0 | 82 | 128 | 0 | 57.83 | 51.68 | 28.1 | 28.3 | 11.6 |
| 24 | 10 | 0 | 82 | 129 | 0 | 57.36 | 51.50 | 28.1 | 28.5 | 11.6 |
| 26 | 10 | 0 | 82 | 127 | 0 | 58.11 | 51.68 | 28.4 | 28.3 | 11.8 |
| 30 | 12 | 0 | 80 | 125 | 0 | 72.33 | 63.09 | 28.4 | 28.9 | 11.9 |
| $^{237}$Np | | 0 | | | 13.82 | | | 14.16 | | 2 |
| 32 | 14 | 0 | 79 | 126 | 10 | 88.41 | 76.48 | 25.8 | 27.6 | 11.6 |
| 30 | 12 | 0 | 81 | 126 | 0 | 75.02 | 65.52 | 25.8 | 26.6 | 11.6 |

**Table 4.** *(Continued)*

| Parent | | C | log $T_t$ (s) | | | | log $T_\alpha$ (s) | | | $C_\alpha$ |
|---|---|---|---|---|---|---|---|---|---|---|
| $A_2$ | $Z_2$ | $C_2$ | $Z_1$ | $N_1$ | $C_1$ | $Q$ (MeV) | $E_k$ (MeV) | log $T$ (s) | log $T'$ (s) | $\log\left(\frac{T}{T_\alpha}\right)$ |
| 28 | 12 | 0 | 81 | 128 | 0 | 73.55 | 64.86 | 26.8 | 28.0 | 12.6 |
| 31 | 13 | 0 | 80 | 126 | 0 | 80.93 | 70.34 | 27.0 | 28.2 | 12.9 |
| $^{236}$Pu | | 0 | | | 7.96 | | | 8.10 | | 2 |
| 28 | 12 | 0 | 82 | 126 | 0 | 79.67 | 70.22 | 19.8 | 21.6 | 11.7 |
| 32 | 14 | 0 | 80 | 124 | 0 | 91.67 | 79.24 | 23.3 | 25.4 | 15.2 |
| $^{239}$Pu | | 0 | | | 11.88 | | | 12.01 | | 2 |
| 34 | 14 | 0 | 80 | 125 | 0 | 90.76 | 77.85 | 25.2 | 26.7 | 13.1 |
| 33 | 14 | 0 | 80 | 126 | 0 | 90.12 | 77.68 | 25.5 | 27.2 | 13.5 |
| $^{240}$Pu | | 0 | | | 11.31 | | | 11.45 | | 2 |
| 34 | 14 | 0 | 80 | 126 | 0 | 90.95 | 78.07 | 24.8 | 26.3 | 13.3 |
| 32 | 14 | 0 | 80 | 126 | 10 | 88.43 | 76.64 | 27.3 | 29.1 | 15.8 |
| $^{242}$Pu | | 0 | | | 13.08 | | | 13.19 | | 2 |
| 34 | 14 | 0 | 82 | 128 | 10 | 88.80 | 76.33 | 27.6 | 28.9 | 14.4 |
| 36 | 14 | 13 | 80 | 126 | 0 | 88.44 | 75.29 | 29.1 | 29.9 | 15.9 |
| $^{239}$Cm | | 13 | | | 4.02 | | | 7.03 | | 2 |
| 32 | 14 | 0 | 82 | 125 | 0 | 97.65 | 84.57 | 18.9 | 21.5 | 11.8 |
| 31 | 14 | 0 | 82 | 126 | 0 | 95.81 | 83.88 | 20.7 | 23.4 | 13.7 |
| $^{251}$Cf | | 0 | | | 10.45 | | | 10.91 | | 2 |
| 48 | 20 | 0 | 79 | 125 | 10 | 139.09 | 112.49 | 21.9 | 25.1 | 11.0 |
| 46 | 18 | 0 | 80 | 125 | 0 | 126.16 | 103.4 | 22.6 | 24.9 | 11.6 |
| 49 | 20 | 0 | 78 | 124 | 10 | 138.66 | 111.59 | 22.7 | 25.6 | 11.8 |
| $^{252}$Cf | | 0 | | | 7.91 | | | 8.00 | | 2 |
| 48 | 20 | 0 | 78 | 126 | 10 | 139.21 | 112.70 | 21.6 | 24.8 | 13.6 |
| 46 | 18 | 0 | 80 | 126 | 0 | 126.72 | 103.59 | 21.8 | 24.1 | 13.8 |
| 50 | 20 | 0 | 78 | 124 | 10 | 138.84 | 111.29 | 22.5 | 25.2 | 14.5 |

and 13 means Wapstra and Audi (Wa 85) from experiment (0) and from the systematics (13), while 10 refers to extrapolations made by us using a code supplied by J. Jänecke. It is very important to have these masses as accurately as possible, because a small difference in $Q$-value produces a large variation in $T$. By giving $Z_1$ and $N_1$ for the daughter one determines the distance from the magic numbers $Z_1 = 82$, $N_1 = 126$.

In Table 4 one finds several modes which are missing in the tables of our older publications, such as $^{24}$Ne from $^{231}$Pa, $^{24}$Ne and $^{28}$Mg from $^{232}$Th, $^{28,29}$Mg from $^{235}$U, $^{34}$Si from $^{238}$U, and $^{48}$Cs from $^{252}$Cf. For $A_2 > 24$ calculations are overly optimistic.

Another important problem in planning an experiment is to know the potential spectrum of the emitted particles from a given source $A$ and $Z$. The $Q$-value variation with the neutron and proton numbers $N_2$ and $Z_2$ of the emitted cluster, as is shown in Figure 45 for the $^{233}$U parent nucleus, does not enable one to answer the question. There are many emitted particles with positive $Q$-values. But the ridge shape appearing in this figure indicates that, for each $A_2$, there is usually one combination of $Z_2$ and $N_2$ giving the maximum $Q$, which increases for increasing $A_2$.

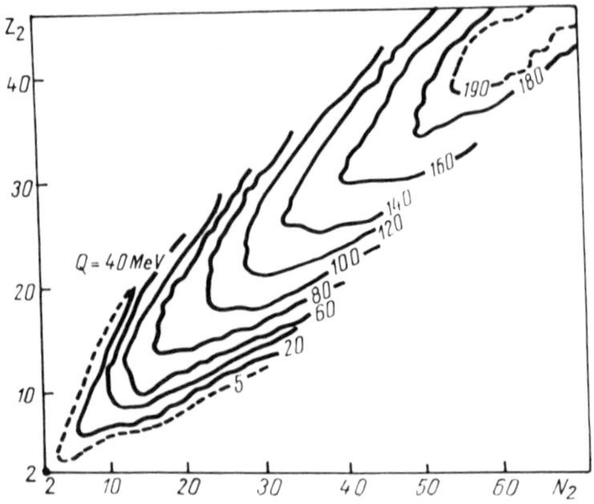


**Figure 45.** $Q$-values for the emission of heavy ions with $Z_2$ protons and $N_2$ neutrons from a source of $^{233}$U.

Figure 46 presents the predicted spectrum of the emitted particles from various sources: $^{223}$Ra, $^{233}$U, $^{238}$U, and $^{254}$Fm. In the first case one can see that the other clusters ($^{12,13,15}$C) emitted from $^{223}$Ra are more than five orders of magnitude less probable than is $^{14}$C. The latter has already been measured. We expect for $^{233}$U that the most probable emission will be that of $^{24,25}$Ne ($T \simeq 10^{23}$ s and $T/T_\alpha \simeq 10^{10.5}$). Then follows $^{28}$Mg with $T = 10^{24.5}$ and $T/T_\alpha \simeq 10^{12}$.

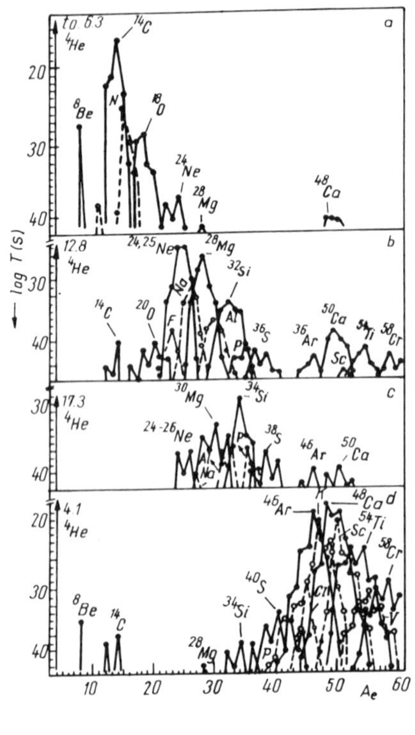


**Figure 46.** Lifetime spectra for various modes of decay by heavy-ion emission of $^{223}$Ra (a), $^{233}$U (b), $^{238}$U (c), and $^{254}$Fm (d).

$^{238}$U is predicted to have $T \simeq 10^{28}$ s for the emission of $^{34}$Si and $T/T_\alpha \simeq 10^{11}$; $^{30}$Mg is predicted to be four orders of magnitude less probable. But the case of $^{34}_{14}\mathrm{Si}_{20}$ deserves further examination in order to check the validity of the approximation for the mass value of the daughter ($^{204}_{78}\mathrm{Pt}_{126}$). For this reason, $^{34}$Si is not included in our tables and there are numerous other similar cases. Better accuracy in mass predictions and more experimental data are needed to increase confidence in our estimates. From $^{254}$Fm, the most probable emitted cluster is $^{48}$Ca with $T \simeq 10^{18}$ s and $T/T_\alpha \simeq 10^{14}$.

By comparing the spectra of all sources in Figure 46, the trend toward heavier emitted clusters with increasing mass number of the parent can be recognized. This trend is similar to that observed in the light-fragment part of the asymmetric fission-fragment mass distribution. In our case it is explained by the shell effect at $N_1 = 126$:

$$^{223}\mathrm{Ra} \to {}^{14}\mathrm{C} + {}^{209}_{82}\mathrm{Pb}_{127}, \qquad {}^{233}\mathrm{U} \to {}^{24,25}\mathrm{Ne} + {}^{209,208}_{82}\mathrm{Pb}_{127,126}$$

$$^{238}\mathrm{U} \to {}^{34}\mathrm{Si} + {}^{204}_{78}\mathrm{Pt}_{126}, \qquad {}^{254}\mathrm{Fm} \to {}^{48}\mathrm{Ca} + {}^{206}_{80}\mathrm{Hg}_{126}$$

Again we find that the neutron shell effects at $N_1 = 126$ are usually much stronger than the proton ones at $Z_1 = 82$.

Fission of some heavy Fm isotopes has a symmetric mass distribution, because the light fragment is approximately a magic one (neutron number 82 or not too far from it). It can be also interpreted as heavy cluster emission.

## 7.4. $^5$He and $^8$Be Radioactivities

Like the diproton (Go 60, Go 80, Ca 83, Ca 84, Ja 85), dineutron (Be 70, Az 79), and trineutron (Az 80), $^5$He and $^8$Be are unstable (Aj 74), but one can assume a mechanism in which the components leave the parent nucleus together and, after tunneling through the potential barrier, are moving apart. Hence $^5$He radioactivity could be experimentally determined by observation of the $\alpha$ particles or the neutrons produced in its own disintegration, as well as by the presence of the requisite daughter nucleus. For $^8$Be the situation seems to be more complicated, because one has to distinguish among the $\alpha$ particles coming from $\alpha$ decay, from $^8$Be fission, and from double $\alpha$ radioactivity (Po 84c, Po 85b). Nevertheless, in some plastic detectors $^8$Be is very easily identified by its specific T-shaped track.

Figure 47 illustrates the fact that for some Po isotopes, besides the well-known $\alpha$ decay, exotic $^5$He spontaneous emission from the ground state is also energetically possible. All other He isotopes ($A_2 = 3, 6, 7, 8, 9, 10$) can be emitted only from excited Po states, because $Q < 0$ (Po 83d).

The variation in $Q$-values for the emission of $^5$He from the ground states of nuclei with masses tabulated by Wapstra and Bos (Wa 77) is plotted in Figure 48. One can see the pronounced odd–even effect, which is present for all odd-mass He isotopes (see also Figure 47).

There are two islands (in fact, two archipelagos due to odd–even effects) of $^5$He radioactivity. Figure 48 shows the detailed position of $^5$He emitters, relative to the Green approximation for the line of $\beta$ stability. The main archipelago formed from two islands involves the medium-mass nuclei with $Z = 83$–$92$ and $N = 127$–$137$. The enhanced $Q$ of the $N = 129$, $Z = 84$ nucleus, leading to the double magic daughter

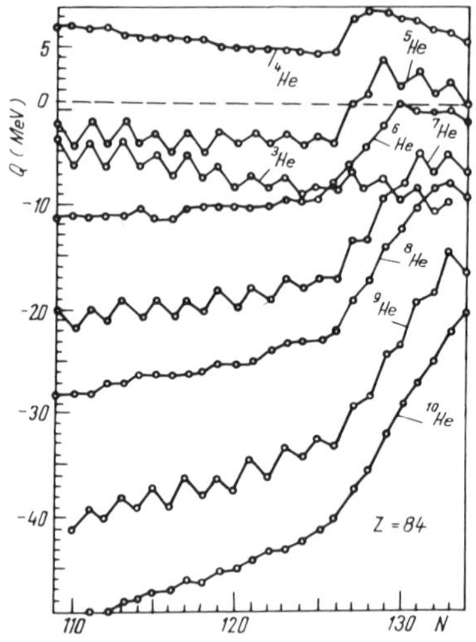


**Figure 47.** $Q$-values for the emission of various He isotopes from $Z = 84$ nuclides with different neutron numbers.

$N_1 = 126$, $Z_1 = 82$, stems from a strong shell effect disturbing the smooth LDM-like trend toward larger $Q$-values of the neutron-deficient nuclei. This trend manifests itself in the second archipelago of four islands of heavy *trans*-curium nuclei with $Z = 97$–$105$, $N = 145$–$157$. Similarly, the neutron subshell $N_1 = 152$ explains the larger $Q$-values for $N = 155$.

Unfortunately, the lifetimes for $^5$He spontaneous emission from the ground states of the above-mentioned nuclei are very long. Only 15 of all 110 emitters have a disintegration period $T$ smaller than $10^{45}$ s (see Figure 49). Such a small emission probability, as in the case of spontaneous fission of some actinides, makes it difficult to observe this phenomenon experimentally, particularly in the presence of other competing decay modes. For example, the partial lifetime for $^5$He radioactivity of $^{213}$Po is $10^{20.9}$ s, but its total half-life due to $\alpha$ emission is only 4.2 $\mu$s.

One also has to consider the contribution of angular momentum $l\hbar$ raising the potential barrier. This is determined from the spin and parity conservation condition. For example, if $^5$He is emitted from $^{213}$Po, one obtains $l = 3$ or 5, because $I^\pi = 9/2^+$,

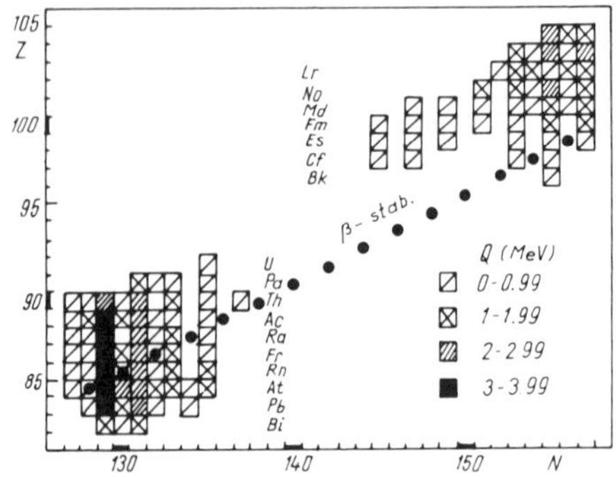


**Figure 48.** $^5$He emitters. Solid dots correspond to the Green approximation for the line of $\beta$ stability.

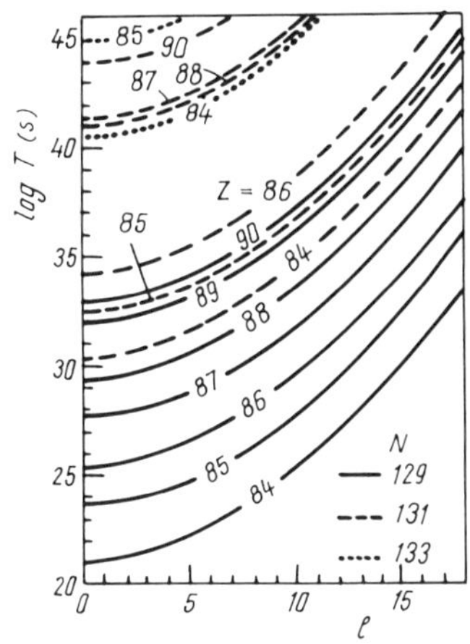


**Figure 49.** Variation in the lifetimes of some $^5$He emitters with angular momentum.

$I_1^{\pi_1} = 0^+$, and $I_2^{\pi_2} = 3/2^-$. Five units of angular momentum produce an increase of about an order of magnitude of the lifetime, as is shown in Figure 49.

A convenient method of obtaining shorter lifetimes is to excite the cluster emitters (Po 83c, Po 84, Po 84e). A $\beta$-decaying precursor far off the $\beta$-stability line could populate the excited levels of the emitter. $^5$He decay of $^9$Be excited states fed by the $\beta$ decay of a $^9$Li precursor has already been experimentally observed by Nyman *et al.* (Ny 81). Alternatively, the feeding of excited levels could be supplied by thermal neutron capture as in $(n, \alpha)$ reactions or, in analogy to Coulomb fission (Ob 85), by Coulomb cluster emission. Such studies are being carried out by Martin Greiner and Werner Scheid (Gr 86b).

Figures 40 and 41a indicate that, after $\alpha$ decay, the shortest half-lives are obtained for $^8$Be spontaneous emission. Nevertheless, the branching ratio (see Figure 44a) relative to the $\alpha$ decay is smaller than for other kinds of radioactivity. The largest values of the branching ratio (less than $10^{-13.6}$) for $^8$Be radioactivity are obtained for some short-lived nuclei such as $^{216}$Rn, $^{217,218}$Fr, $^{218}$Ra, and $^{220}$Ac.

## 7.5. Sn Radioactivity

We have shown that a major role in cluster radioactivities is played by the shell effects in the final fragments. Consequently, heavy nuclei have high emission rates when one of the fragments is close to the magic numbers $N = 126$ and $Z = 82$ and, for the medium nuclei, when one of the fragments is close to the magic numbers $N = 82$, 50, 28 and $Z = 50$, 28, 20.

It is quite natural to ask if similar decay modes do not exist for double magic numbers $Z = 50$, $N = 82$, i.e., for very heavy Fm isotopes or, close to this extreme case, when both partners are strongly bound. As we have already shown in our introductory section, microscopic calculations indicate the possibility of such a decay. If we look, as before, from the standpoint of mass distributions, such a process would be very similar to spontaneous fission, but there are other characteristics of this decay mode—let us call it Sn radioactivity for reasons that will become clear and from the fact that at least one of the fragments must be an isotope of Sn or not far from it ($Z \simeq 50$, $N \simeq 82$), that make this process different from spontaneous fission.

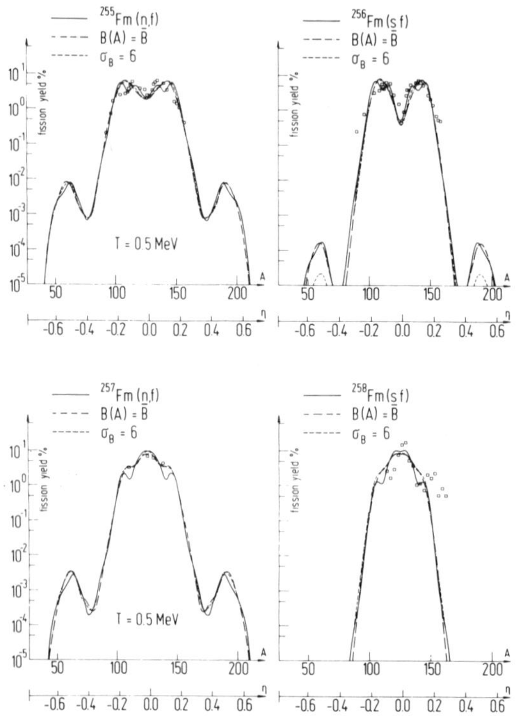


**Figure 50.** Mass yields for various Fm isotopes using $B_{\eta\eta}(\eta)$ (full curves), the average mass ($\bar{B}$, chain curves), and the smoothed masses (broken curves). Experimental values are denoted by points (Lu 80).

The example of Fm fission (see Figure 50) shows the dominant influence of shell effects (Lu 80). A transition from asymmetric fission of the lighter isotopes, such as $^{255}$Fm, to pure symmetric fission for the heavier ones, such as $^{260}$Fm, was confirmed in the experimental data of Hoffman *et al.* (Ho 80). It is important to stress that the dominant symmetric decay of $^{258}$Fm, measured recently (Hu 86), is caused not only by liquid-drop effects but also by the double magic shell closure of the $^{132}_{50}Sn_{82}$ fragment.

Spontaneous fission is different from Sn radioactivity. In both processes the total available energy is given by the difference between masses in the initial channel (fissioning nucleus) and in the final channel (fission fragments).

In fission, the energy balance includes the TKE of the fragments and their initial excitation energies. These initial excitation energies are obtained as a result of the deformation of the fragments at scission and the dissipative forces that acted up to this point. This energy is eventually released by particle evaporation and $\gamma$-ray emission. The deformation of the fissioning nucleus and the corresponding appearance of the neck ensures, in the framework of the liquid drop model, lowering of the barrier height and hence penetration of this barrier with relatively low total kinetic energy (TKE).

In the other process, Sn radioactivity, the scission point is at a more compact configuration since the fragments are considered to be spherical. Inclusion of the finite range forces (Yukawa + exponential or proximity forces) ensures a drastic decrease in the Coulomb forces at short distances, so that the barrier is considerably lowered.

At this point we should mention that Royer and Remaud (Ro 84b), trying to show that spontaneous fission could also proceed through compact shapes as well as elongated

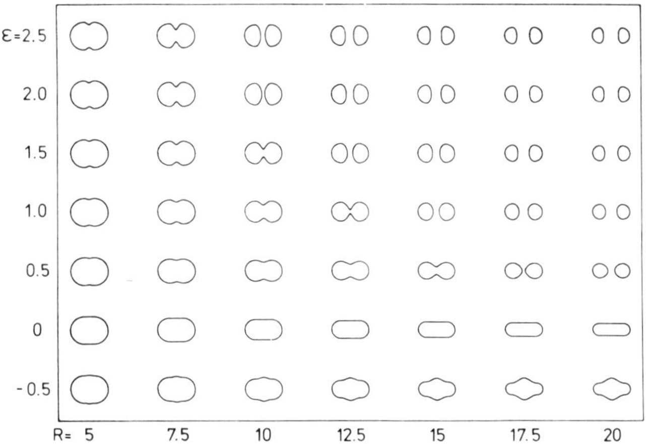


**Figure 51.** Family of shapes in the $R$–$\varepsilon$ plane for symmetric $^{258}$Fm fission ($\eta = 0$, $\beta_1 = \beta_2 = 1$).

ones, have already succeeded in showing that the fission barrier obtained by including proximity forces between the two compact shapes in addition to Coulomb and surface forces results in a barrier that is not significantly higher than that given by the liquid drop model with elongated shapes. Of course, they started from a hypothesis of degeneracy between the two mechanisms and not from the idea that Sn emission is a new process which, for some nuclei, could become competitive with spontaneous fission. Another characteristic of the new process is that the fragments are expected to be almost completely unexcited, since dissipative forces could act only along the new path with compact shapes and thus is much shorter than the spontaneous fission path with elongated shapes. Consequently, the compactness of the scission configurations, together with the absence of fragment excitation, will lead to values of the kinetic energy of the fragments much higher than in a fission process, and should almost approach the reaction $Q$-value. Microscopically, the introduction of the potential-energy surfaces in the neck and the relative distance between the fragments will allow us to describe the TKE distributions and the two main paths corresponding to compact and enlongated shapes. The dynamical treatment must give the distribution of the scission point shapes, i.e., the TKE distributions in a way similar to the treatment of the mass distributions (Sa 80).

Figure 51 shows (Gr 65), in the $R$–$\varepsilon$ plane, some of the shapes which can be used to study the various processes discussed above. The neck parameter $\varepsilon$ describes not only the necking-in of the nucleus, but also its higher moments of deformation.

The liquid-drop part of the potential-energy surfaces, in the case of $^{258}$Fm, is unable to reproduce the experimentally observed double-humped TKE distribution for the fragments. On adding to the liquid-drop part the corresponding shell corrections (De 86b), the resulting total potential offers the system two possible decay channels, separated by a ridge. Following these two paths (sketched in Figure 52) along the two valleys of the potential-energy surface, the nucleus fissions at two different points on the scission line. This corresponds to two different energies in the exit channel, as is measured in the experiments. The different paths reflect, of course, different shape configurations: The path belonging to higher energies corresponds to two touching

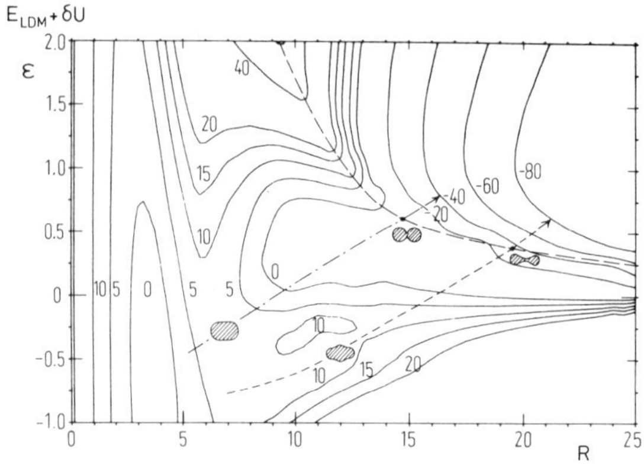


**Figure 52.** Total potential energy as a sum of the liquid-drop $E_{\mathrm{LDM}}$, and the shell corrections $\delta U$ depending on $R$ and $\varepsilon$ for the symmetric division of $^{258}$Fm. The dashed curve indicates the separation line, the two paths correspond to the possible decay channels. The nuclear shapes at saddle and scission points of the two paths are indicated.

spheres, and the path at lower energies to a deformed configuration, as indicated in the figure. For simplicity we neglected, in our considerations, the pairing correction energy, which gives a minor contribution of around 1–2 MeV to these surfaces.

Figure 53 shows the half-life spectrum for emission of various clusters from $^{233}$U and $^{234}$U (Po 86b), in agreement with experimental results [see (Cl 86) and the references cited therein]. This simple static picture gives us a natural explanation of the experimental data. The shell effects lead to a kind of "fission-path isomer" causing "bimodal fission." Dynamical calculations describing the precise details of the distribution will

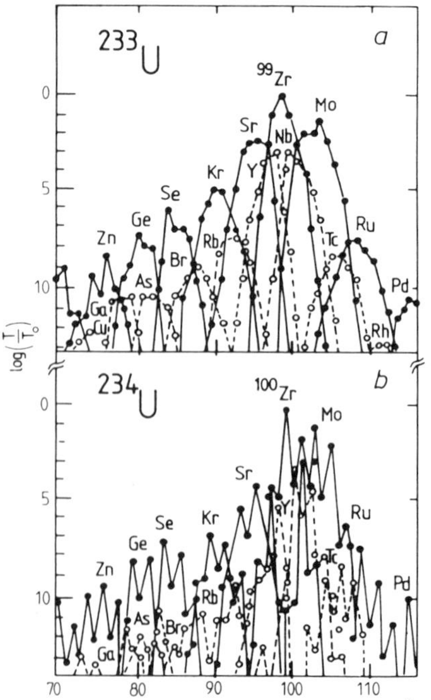


**Figure 53.** Half-life spectrum for the emission of various clusters from $^{233}$U (a) and $^{234}$U (b) with $A_e > 70$.

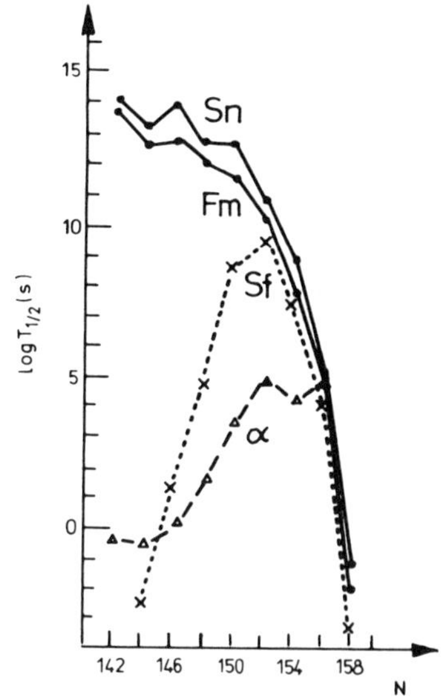


**Figure 54.** Logarithm of half-lives for spontaneous fission ( × ), $\alpha$ decay (△), and Sn emission ( · ) of Fm isotopes. For Sn emission only the first two combinations of clusters with shortest half-lives are given. It is seen that half-lives for all three processes become comparable for the heavier isotopes:
$^{242}$Fm($^{120}$Sn + $^{122}$Sn, $^{104}$Ru + $^{138}$Ba),
$^{244}$Fm($^{122}$Sn + $^{122}$Sn, $^{120}$Sn + $^{124}$Sn),
$^{246}$Fm($^{122}$Sn + $^{124}$Sn, $^{120}$Sn + $^{126}$Sn),
$^{248}$Fm($^{124}$Sn + $^{124}$Sn, $^{122}$Sn + $^{126}$Sn),
$^{250}$Fm($^{124}$Sn + $^{126}$Sn, $^{122}$Sn + $^{128}$Sn),
$^{252}$Fm($^{126}$Sn + $^{126}$Sn, $^{124}$Sn + $^{128}$Sn),
$^{254}$Fm($^{126}$Sn + $^{128}$Sn, $^{124}$Sn + $^{130}$Sn),
$^{256}$Fm($^{128}$Sn + $^{128}$SN, $^{126}$Sn + $^{130}$Sn),
$^{258}$Fm($^{128}$Sn + $^{130}$Sn, $^{126}$Sn + $^{132}$Sn).

follow this schematic explanation, but should not change the principal underlying mechanism.

In Figure 54, the systematics of the spontaneous half-lives of even Fm isotopes are presented together with calculated half-lives for Sn emission from the same isotopes. The $\alpha$-emission half-lives are also presented. For light Fm isotopes the probability of Sn emission is many orders of magnitude below that for spontaneous fission. For the heavier isotopes the two curves come closer. We expect the shortest half-lives for $^{264}$Fm—which could decay into two double magic nuclei $^{132}_{50}$Sn$_{82}$—after which the half-lives must increase, again reaching the ms region at $N \geqslant 168$–$170$.

In Figure 55, we show that the same behavior remains valid for the other very heavy elements. Because we expect that the half-lives will decrease up to $N = 164$ (double $N = 82$ closed shells) we conclude that the half-lives of the very heavy elements with $158 \leqslant N \leqslant 170$ are shorter than the ms range. That can explain the lack of success thus far in producing superheavy elements with longer half-lives.

Up to now the two fission components, with elongated and with nearly spherical fragments, have been observed only for four nuclei: $^{258}$Fm, $^{259}$Md, $^{260}$Md, and $^{258}$No (Hu 86). New data, especially for $^{256}$Fm and $^{256}$No, are essential to test the predicted systematic behavior.

## 7.6. Experimental Evidence

Four years after our predictions, experimental evidence for one of the new decay modes, $^{14}$C radioactivity of $^{223}$Ra, was published by Rose and Jones (Ro 84) of Oxford University at the beginning of 1984. This experiment was confirmed in Moscow (Al 84), Orsay (Ga 84, Ga 84b), Berkeley-Geneva (Pr 85), and Argonne (Ku 85). Other experimental discoveries refer to the emission of $^{14}$C from $^{222,224}$Ra (Pr 85) and to $^{24}$Ne radioactivity of $^{232}$U (Ba 85) and of $^{231}$Pa (Sa 84). Either semiconductor detector

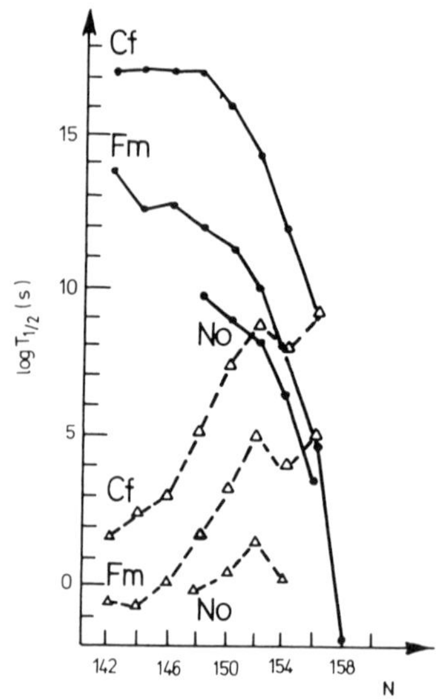


**Figure 55.** Logarithm of half-lives for the new process—cold fission with compact shapes or Sn emission—for even-even elements with $98 \leq Z \leq 102$. Only the first combination of clusters with the shortest half-life is given. We conclude that the half-lives of the very heavy elements with $158 \leq N \leq 170$ are shorter than ms: $^{242}Cf(^{106}Ru + ^{136}Xe)$, $^{244}Cf(^{108}Ru + ^{136}Xe)$, $^{246}Cf(^{118}Cd + ^{128}Sn)$, $^{248}Cf(^{120}Cd + ^{128}Sn)$, $^{250}Cf(^{120}Cd + ^{130}Sn)$, $^{252}Cf(^{122}Cd + ^{130}Sn)$, $^{254}Cf(^{122}Cd + ^{132}Sn)$, $^{248}No(^{122}Sn + ^{126}Te)$, $^{250}No(^{122}Sn + ^{128}Te)$, $^{252}No(^{122}Sn + ^{130}Te)$, $^{254}No(^{124}Sn + ^{130}Te)$, $^{256}No(^{126}Sn + ^{130}Te)$. (Symbols as in Figure 54.)

telescopes (Po 69) or solid-state track-recording films (Fl 75) are used in these experiments. A summary of experimental results is given in Table 5.

### *7.6.1. $^{14}C$ Radioactivity*

An $^{227}Ac$ source ($T_t = 22$ y) with which $^{223}Ra$ ($T_t = 11.4$ days) is in secular equilibrium was employed by Rose and Jones (Ro 84). $^{223}Ra$ is a member of the "actinium" ($A = 4n + 3$) natural radioactive family (the decay chain of $^{235}U$) with $^{227}Ac$ being a convenient precursor: $^{227}Ac \xrightarrow{\beta^-} {^{227}Th} \xrightarrow{\alpha} {^{223}Ra}$.

A protective layer of gold (20 $\mu$g cm$^{-2}$) evaporated on the surface and a 15 $\mu$g cm$^{-2}$ carbon foil inserted between the source and the semiconductor telescope prevented source material sputtering onto the $\Delta E$ counter as a consequence of the daughter recoil after $\alpha$ decay.

A counting rate of about 4000 s$^{-1}$ was obtained in an 0.3 sr effective solid angle of detection from the 3.3 $\mu$Ci source. The active area of the $\Delta E$ and $E$ detectors were 200 mm$^2$ and 300 mm$^2$, respectively. The sensitive thickness of the $\Delta E$ detector was 8.2 $\mu$m.

In spite of the pile-up rejection systems used, quadruple pile-up events were observed and a large number of double and triple $\alpha$-pulse pile-ups were recorded. A two parameter $\Delta E$, $E$ spectrum was used for $^{14}C$ identification. The runs were conducted automatically. Every 12 hours the results were read out and updated; the 12 h run was rejected if any detectable error occurred during that time. In a total run of 189 days 11 events were recorded, and in another 194-day run, 8 events. The $^{14}C$ kinetic energy was $QA_1/A = 29.8$ MeV and the branching ratio relative to $\alpha$ decay, $b = \Gamma/\Gamma_\alpha = (8.5 \pm 2.5)10^{-10}$. A $^{241}Am$ $\alpha$ source (5.48 MeV) was used for energy calibration.

Using a similar technique, but a stronger source, Alexandrov *et al.* (Al 84) have obtained 7 events in only 30 days. A heavy-ion cyclotron beam (70 MeV $^{14}N$) has been used to obtain the position of the $^{14}C$ hyperbola in the $\Delta E$-$E$ spectrum. An elastically scattered 29.2 MeV $^{12}C$ ion beam on $^{197}Au$ and $^{12}C$ targets was used for energy calibration in the range 27–30 MeV. The branching ratio $b = (7.6 \pm 3.0)10^{-10}$ was in good agreement with the Oxford measurement.

A second experimental confirmation was obtained by Gales *et al.* (Ga 84). They used a very large solid angle (0.1 sr) superconducting solenoidal magnetic spectrometer (SOLENO) at the Orsay MP tandem accelerator, with a $\Delta E$-$E$ semiconductor detector telescope placed at its focal plane, 1.53 m from the source. The $^4He^{++}$ and $^4He^+$ particles were deflected by the magnetic field. The magnetizing current in SOLENO ($I = 285$ A, $B = 2.43$ T) was preset to focus selectively only the expected (most probable charge state $6^+$) $^{14}C^{6+}$ at 29.7 MeV on the $\Delta E$-$E$ telescope. In this way the very strong source (247 $\mu$Ci) made it possible to obtain 11 events in a run of only 5 days.

By using energy-analyzed $^{12}C$, $^{14}C$, and $^{16}O$ particle beams elastically scattered at a laboratory angle of $8.3 \pm 0.2°$, an unambiguous calibration was obtained. The measured branching ratio $b = (5.5 \pm 2.0)10^{-10}$ was also in very good agreement with the result obtained by Rose and Jones (Ro 84).

Kutschera *et al.* (Ku 85) at Argonne used an Enge split-pole magnetic spectrograph to study the energy and mass of carbon nuclei emitted in the decay of $^{223}Ra$. A $^{227}Th$ ($T_t = 18.7$ day) source was used, to prevent long-term contamination of the spectrograph. The $\alpha$ activity (about 9 m Ci obtained from a thickness of 100 $\mu g/cm^2$) led to the detection of 24 $^{14}C$ events in only 6 days, despite the small solid angle of the spectrograph ($4.4 \times 10^{-4}$ of $4\pi$). A mixed beam of $^{12,13,14}C$ ions provided by the Argonne tandem-superconducting linac operated in a mode used for very low concentrations of long-lived radioisotopes permitted an accurate calibration of the magnetic spectrograph.

A gas ionization detector in the focal plane determined the magnetic rigidity $B\rho$, and the specific energy loss $\Delta E$ and total energy $E$ of the particles simultaneously. It was established beyond doubt that the detected events are indeed $^{14}C$. The charge state fractions of the $^{14}C$ ions were 58.1% ($6^+$), 36.7% ($5^+$), and 5.2% ($4^+$). Agreement with other measurements of the branching ratio was obtained and a weighted mean of the five measurements (see Table 5) was determined as $(6.0 \pm 1.0)10^{-10}$.

By using polycarbonate track-recording detectors—which are not sensitive to $\alpha$ particles—Price *et al.* (Pr 85) not only confirmed the branching ratio for $^{223}Ra$ with the best statistics (56 events) to date, but also discovered $^{14}C$ radioactivity of the neighboring Ra isotopes, namely $^{222,224}Ra$. This technique is the most efficient way to detect the rare decay modes and was used also to discover $^{24}Ne$ radioactivity of $^{232}U$ and of $^{231}Pa$. Fr and Ra isotopes with mass numbers 221, 222, 223, and 224 were produced in spallation reactions by bombarding a 55g/cm$^2$ Th $C_2$ powder target, maintained at approximately 2100 °C, with 2.6 $\mu$A of 600 MeV protons from the CERN synchrocyclotron. The ISOLDE electromagnetic isotope separator selected various masses from the beam extracted from the ion source. Each of these four mass-separated beams was implanted in the base of a cylindrical aluminum cup having a 125 $\mu$m Tuffak sheet mounted on its interior wall. For $A = 221$ both Fr and Ra are $\alpha$ emitters, but for the other three masses, Fr isotopes $\beta$-decayed to Ra. A detector area corresponding to a solid angle of about 2 sr was scanned after etching 8 h at 70 °C in a NaOH solution.

From the etched-track diameters and lengths, the charge and range of the emitted clusters have been determined. The charge calibration of the films was made with beams of B, C, and Ne at the Lawrence Berkeley Laboratory Superhilac.

**Table 5. Summary of Experiments on $^{14}C$ and $^{24}Ne$ Decay Modes**

| Emitted heavy ion | Parent and its $T_t$ | Source | Kinetic energy (MeV) | Branching ratio $T_\alpha/T$ | Number of events | Run (days) | Detector | Reference |
|---|---|---|---|---|---|---|---|---|
| $^{14}C$ | $^{221}Ra$ 28 s | mass sep. ISOLDE (p + Th) | 30.3 | $<4.4 \times 10^{-12}$ | 0 | | track-recording not sensitive to $\alpha$ | (Pr 85) (Berkeley and CERN) |
| | $^{222}Ra$ 38 s | mass sep. ISOLDE (p + Th) | 31.0 | $(3.7 \pm 0.5) \times 10^{-10}$ | 52 | | track-recording not sensitive to $\alpha$ | (Pr 85) (Berkeley and CERN) |
| | $^{223}Ra$ | $^{227}Ac$ | 29.8 | $(8.5 \pm 2.5) \times 10^{-10}$ | 11 + $\frac{8}{19}$ | 189 $\frac{194}{383}$ | semiconductor telescope $\Delta E$–$E$ ident. | (Ro 84) (Oxford) |
| | 11.4 d | 21.7 y | | $(7.6 \pm 3.0) \times 10^{-10}$ | 7 | 30 | semiconductor telescope $\Delta E$–$E$ ident. | (Al 84) (Moscow) |
| | | | | $(5.5 \pm 2.0) \times 10^{-10}$ | 11 | 5 | superconducting magnetic spectrometer and $\Delta E$–$E$ ident. | (Ga 84) (Orsay) |
| | | mass sep. ISOLDE | | $(6.1 \pm 0.8) \times 10^{-10}$ | 56 | | track-recording not sensitive to $\alpha$ | (Pr 85) (Berkeley and CERN) |

| | | | | | | | |
|---|---|---|---|---|---|---|---|
| | | $^{227}$Th 18.7 d | | $(4.7 \pm 1.3) \times 10^{-10}$ | 24 | 6.09 | Enge split-pole magnetic spectrograph $\Delta E-E$ gas ionizer | (Ku 85) (Argonne) |
| | $^{224}$Ra 3.7 d | mass sep. ISOLDE | 28.6 | $(4.3 \pm 1.1) \times 10^{-11}$ | 22 | | track-recording not sensitive to $\alpha$ | (Pr 85) (Berkeley and CERN) |
| | $^{226}$Ra 1600 y | $^{226}$Ra 1600 y | 26.5 | $(3.2 \pm 1.6) \times 10^{-11}$ | 4 | 9.3 | superconducting magnetic spectrometer and $\Delta E-E$ ident. | (Ho 85) (Orsay) |
| | | | | $(2.9 \pm 1.0) \times 10^{-11}$ | 23 | 1 | track-recording not sensitive to $\alpha$ | (Ba 86) (Berkeley and Orsay) |
| $^{24}$Ne | $^{231}$Pa $3.3 \times 10^4$ y | $^{231}$Pa | 54.1 | $(4.3 \pm 0.9) \times 10^{-12}$ | 25 | | track-recording not sensitive to $\alpha$ | (Sa 84) (Dubna and Bucharest) |
| | $^{232}$U 72 y | $^{232}$U | 55.9 | $(1.0 \pm 0.25) \times 10^{-12}$ | 31 | | track-recording not sensitive to $\alpha$ | (Ba 85) (Berkeley) |
| | $^{233}$U $1.59 \times 10^5$ y | $^{233}$U | 54.3 | $(7.5 \pm 2.5) \times 10^{-13}$ | 16 | 28 | track-recording not sensitive to $\alpha$ | (Tr 85) (Dubna and Bucharest) |
| | $^{230}$Th $7.54 \times 10^4$ y | $^{230}$Th | 51.75 | $(5.6 \pm 1.0) \times 10^{-13}$ | 165 | 61 | track-recording not sensitive to $\alpha$ | (Tr 85b) (Dubna and Bucharest) |

Only a limit could be determined for $^{221}$Fr and $^{221}$Ra, because no $^{14}$C events were observed during the measurement. A comparison between estimates and the experimental results is given in Figure 35.

### *7.6.2. $^{24}$Ne Radioactivity*

Toward the end of 1984 a second decay mode, $^{24}$Ne, was discovered by Barwick *et al.* (Ba 85) for $^{232}$U. They determined a maximum tolerable $\alpha$-particle dose of $6 \times 10^{10}\ \text{cm}^{-2}$ incident on Tuffak polycarbonate detectors used for $^{14}$C radioactivity and sensitive to particles with $Z > 2$. For Cronar film (polyethylene terephthalate) sensitive only to $Z > 6$ ions, the limiting $\alpha$-particle dose was $2 \times 10^{11}\ \text{cm}^{-2}$ permitting detection of a branching ratio $\gtrsim 10^{-14}$.

A hemispherical array of Cronar film was exposed for one month to the 0.5 mCi $^{232}$U source in a vacuum chamber at a pressure of 0.01 torr.

From the two detectors scanned in transmitted light (200 cm$^2$ surface area), $24 \pm 7$ tracks of $^{24}$Ne were found using a silicon replica (Fl 75) technique. The Cronar film was calibrated with $^{20}$Ne and $^{18}$O ion beams. Simultaneously (Ba 85) a polyethylene terephthalate track detector (170 $\mu$m thick) was used by Sandulescu *et al.* (Sa 84) to discover $^{24}$Ne radioactivity of $^{231}$Pa.

In Dubna, plane sources were prepared on a 0.1 mm nickel backing in the form of the oxides of the isotopes being investigated. The special-purpose technology used for the preparation of the sources eliminated contamination of the detectors by the very active specimens under study during long exposures.

For studying the emitted clusters, 175 $\mu$m Melinex detectors were irradiated in air using Pa, U, Th, and Np sources at a distance of 1 mm and an Am source in vacuum at a distance of 1 cm. The detectors allowed one to detect clusters with atomic numbers $6 \leqslant Z \leqslant 20$ and separate them from fission fragments.

The exposure time was determined taking into account the $\alpha$ activity of each source. For accurate spectrometric measurements in polyethylene terephthalate detectors, the integral $\alpha$-particle flux has to be lower than $5 \times 10^{11}\ \alpha/\text{cm}^2$.

Inside the polymer material irradiated in air, some processes occur along the particle trajectory leading to a change in the material structure, and this affects the measured geometrical parameters of the track. It was shown experimentally that this process, i.e., the change in the etched track length, in the case of 46 MeV $^{20}$Ne ions, is finished at most one month after the irradiation. Moreover, the high $\alpha$-particle flux makes this process faster. Nevertheless, in order to ensure that all the recorded cluster tracks, in the case of a very long exposure time, have the same etching characteristics, the detectors were stored 30 days in air prior to the chemical processing, which was carried out in a 20% NaOH solution at a temperature of 60 °C.

Together with the detector exposed to the radioactive sources, there were etched calibration samples irradiated with $^{18}$O, $^{20}$Ne, and $^{26}$Mg ions with energies of 1.6–3.0 MeV/nucleon and with dip angles of 30° and 45°; these calibration samples were exposed to sources together with the detectors. The geometric efficiency of cluster detection was equal to $0.66(2\pi)$.

Since the identification method chosen involves variation of $V_t/V_b$ with residual range, the detectors were etched two or three times for different time intervals. After a total time of four hours, the etching reached the cluster stopping point. The track

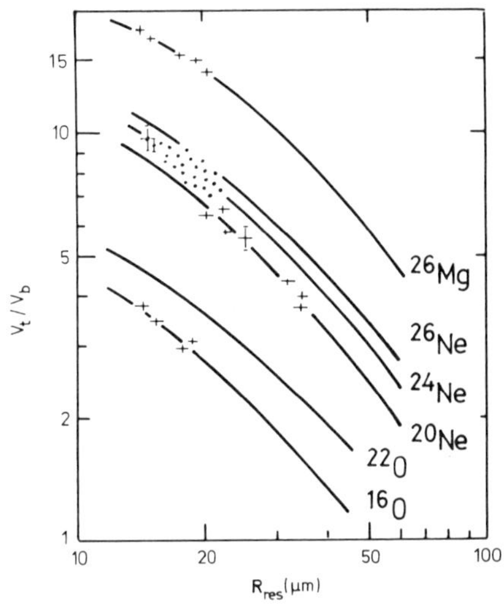


**Figure 56.** Calculated etching selectivity $V_t/V_b$ (where $V_t$ and $V_b$ are the rates of polymer etching along the track and the bulk etch rate for the detector material, respectively) vs. residual range $R_{res}$. Dots indicate experimental data for clusters; crosses denote results of measurements with control detectors exposed to $^{16}$O, $^{20}$Ne, and $^{26}$Mg ions with energies of 1.6-3.0 MeV/nucleon.

length and $V_t/V_b$ were calculated from the cone geometrical parameters, measured under optical microscopes with a magnification of $16 \times 25 \times 40$.

Calibrations with $^{16}$O, $^{20}$Ne, and $^{26}$Mg ions enabled the $V_t/V_b = f(dE/dx)$ dependence to be obtained for the $\alpha$-particle fluxes of $2 \times 10^{11}$ $\alpha/\text{cm}^2$: $V_t/V_b = 6.3 \times 10^{-3}$ $(dE/dx)^{2.1}$ where $dE/dx$ is expressed in MeV cm$^2$/mg.

By taking into account the $dE/dx$-range relationship, the $V_t/V_b$ variation with the residual range was obtained for different ions (Figure 56). As shown, the experimental points measured with the Th source cluster around the curve for $^{24}$Ne ions. The data treatment using a minimum $\chi^2$ method leads to the conclusion that $^{24}$Ne is the most probable decay cluster. The $V_t/V_b$ values have been obtained only for a fraction of the cluster tracks obtained, as the corresponding measurements were too time-consuming.

The track-length distribution for the decay products of Th, given in Figure 57, has a maximum at about 29 $\mu$m, which is the range of the $^{24}$Ne ions corresponding to the calculated value of the kinetic energy of the emitted heavy cluster.

The distribution of the full ranges, $R$, is rather broad. This is due to experimental shortcomings, namely, a large source thickness and a rather high recoil nuclei background, which have intorudced ambiguities in the estimate of the cluster mass number because of difficulties either in the estimation of the self-absorption process in the source or in the measurements of the geometrical parameters of the track.

By taking into account the theoretical predictions as well as the experimental results, one can conclude that the new decay mode of $^{230}$Th takes place most probably by emission of a $^{24}$Ne ion.

No clusters have been detected from the Np and Am sources, this providing additional evidence for the absence of a background in our experiment.

It is noteworthy that from the experimental data (Ho 85) as well as from the $^{226}$Ra quantity contained in the $^{230}$Th + $^{232}$Th source, we would have expected about 40 tracks from $^{226}$Ra decay through emission of $^{14}$C. Due to the sensitivity of the detector the maximum etched track length for $^{14}$C ions is equal to about 10 $\mu$m. Since the C and

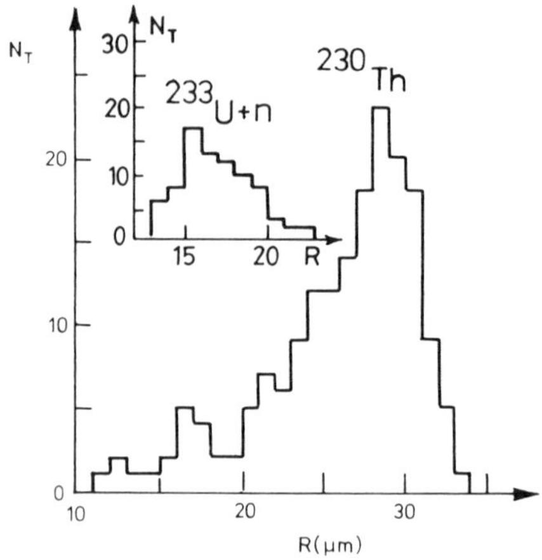


**Figure 57.** Distribution of the full ranges of the $^{230}$Th decay products and those of fragments from the neutron-induced fission of $^{233}$U in polyethylene terephthalate. The thickness of the $^{230}$Th and $^{233}$U sources was equal to 0.3 mg/cm$^2$.

O recoil nuclei have lower energies, they are etched faster than the $^{14}$C cluster tracks. After a longer etching time the background becomes so high, that it is impossible to discriminate the $^{14}$C cluster tracks against the recoil nuclei tracks.

These experiments were not aimed at studying spontaneous fission. Nevertheless, in experiments with thorium we have detected 25 spontaneous fission fragments which differ from clusters in both their range (see Figure 57) and shape. Bearing in mind the possible sources of background, such as thorium fission induced by neutrons from the $(\alpha, n)$ reaction on light nuclei in the detector material, the effect obtained can be used only to provide a lower limit on the spontaneous-fission half-life of $^{230}$Th. According to our data, this limit is set at $\geqslant 2 \times 10^{18}$ years, that is, one order of magnitude higher than the previously known value of $1.5 \times 10^{17}$ years. The partial half-life for $^{24}$Ne emission is $(1.3 \pm 0.3)10^{17}$ years. The results for $^{231}$Pa and $^{233}$U are presented in Table 5.

The estimated half-life spectra from $^{231}$Pa and $^{232}$U and the measurements are compared in Figure 58. One can see that our estimates are about one order of magnitude too optimistic.

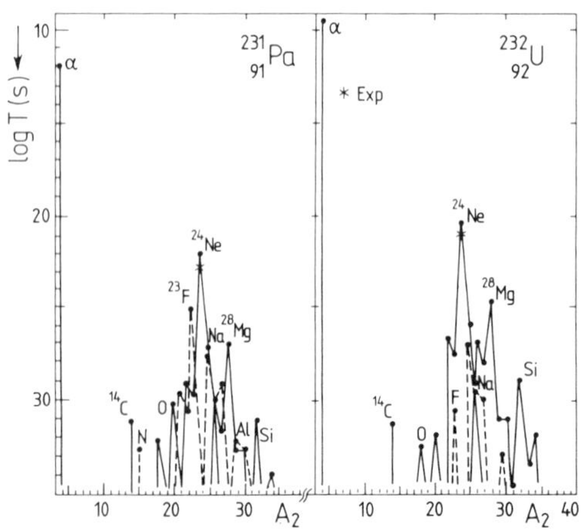


**Figure 58.** Lifetime spectra for the emission of various heavy clusters from $^{231}$Pa and $^{232}$U compared with experimental points.

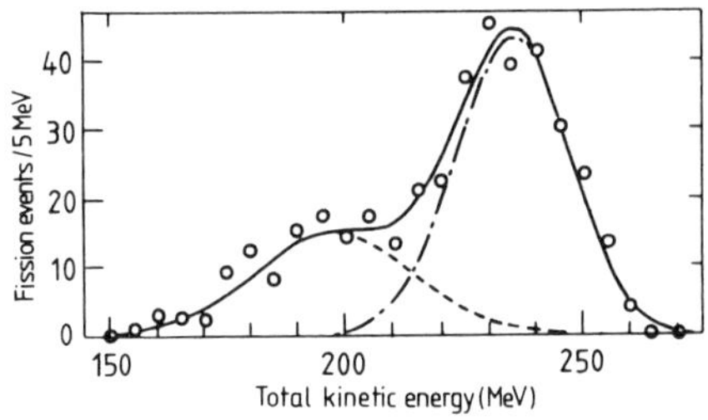


**Figure 59.** Fission events (○) for $^{260}$Md as a function of total kinetic energy (TKE). Clearly two components are seen. Tentatively two Gaussians, one for the high TKE component (-··-) and another for the low TKE component (- - -) are fitted with the experimental data (○).

### *7.6.3. Sn Radioactivity*

The recently measured TKE distribution for four very heavy nuclei—$^{258}$Fm, $^{259}$Md, $^{260}$Md, and $^{258}$No—indicate a composite of two energy distributions, one peaked at 200 MeV and another at 235 MeV (Hu 86). This was interpreted as reflecting a mixture of liquid-drop-like and fragment-shell-directed symmetric fission and not as evidence for Sn emission. We have concluded above that for this region the two processes, Sn emission and spontaneous fission, are of comparable probability. It was also shown experimentally that events with very high TKE could always be associated with the symmetric mode and that the asymmetric fission mass fragments were always associated with the low TKE distributions, facts which support our interpretation (see Figures 59 and 60).

The other characteristics of the new process, the absence of neutrons and $\gamma$ rays or low-energy and low-multiplicity $\gamma$ rays, make the new mechanism experimentally distinguishable from spontaneous fission. Indeed, coincident measurements of mass and kinetic energy of the fragments associated with no neutrons, and especially with low-energy and low-multiplicity $\gamma$ rays, will allow the determination of a small decay component in a large background of spontaneous fission fragments, because in spontaneous fission we see high-multiplicity $\gamma$ rays.

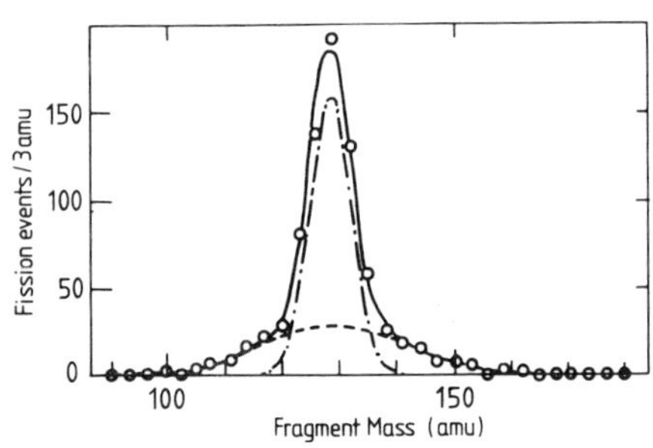


**Figure 60.** Fission events (○) for $^{260}$Md as a function of the fragment mass. We can see that the high TKE component corresponds to a very narrow mass distribution and the low TKE component to a very large mass distribution. (Symbols as in Figure 59.)

The observed TKE distributions for $^{298}$Fm, $^{299}$Md, $^{260}$Md, and $^{258}$No skewed upward or downward from the peak in each case could be easily explained if we assume the existence of the new decay mode—Sn emission. New data close to the present ones on $^{256,254}$Fm or $^{256,254}$No could clarify this situation. Also, the large change observed in the TKE distributions from $^{258}$Fm, which is mainly peaked at high TKE, to $^{259}$Md, which is mainly peaked at low TKE (nuclei which differ only by one proton), could be explained simply by larger effects of the odd proton on the fission barrier for compact shapes with few levels at the Fermi surface than for elongated shape with a large density of levels at this surface.

## 8. Conclusions

Based on the existence of shell-model valleys in the potential-energy surfaces along mass asymmetry and shape (including the neck) degrees of freedom, a unified description of $\alpha$ decay, heavy-cluster radioactivities, and a new type of fission with compact shapes is given, as are spontaneous fission modes with compact shapes in which one or both fragments have magic or almost magic numbers.

Promising results have been obtained by applying fission theory to the calculation of $\alpha$-disintegration half-lives between ground states or low excited states below the barrier. Consequently, the traditional picture of $\alpha$-particle preformation, followed by tunneling through the potential barrier, can be replaced by the larger amplitude oscillations (asymmetric fission model) of the nuclear surface favored energetically by shell effects and leading to a very asymmetric fission as a consequence of a quantum-mechanical tunnel effect.

For a very large mass asymmetric fission, best described in terms of fragmentation theory based on an asymmetric two-center shell model, the charge-density asymmetry plays an important part. Three macroscopic models have been developed for nuclear systems with different charge densities. A phenomenological correction was introduced, accounting for shell effects, nuclear deformation, and neglected terms in the mass formulas. The shell-correction part could be also described by Schultheis' cluster prescription.

The fact that the macroscopic models, with parameters obtained from a fit to experimental data for masses and almost symmetrical fission and fusion barriers, with a suitable phenomenological correction, can reproduce the potential-barrier shape is attested by the good agreement of the calculated lifetimes for $\alpha$-particle decay with the experimental values, over a range of some 24 orders of magnitude. The time-dependent Hartree–Fock method gives the same zero-point vibration frequency as that obtained from a fit to experimental data.

By taking into account the deformation of the parent and daughter nuclei, the agreement of $Q$-values with experimental results was improved and it was possible to first predict the lifetime of $\alpha$ decay from a shape isomeric state. This is many orders of magnitude larger than that of isomeric spontaneous fission, which explains the unsuccessful experimental research for the latter.

In principle, this theory can reproduce both the $Q$-values and the half-lives. It has to be improved by extending the microscopic shell correction method for light nuclei or very asymmetric fission. Also, the mass parameters and the zero-point vibration energy of this mode need further investigation (Sc 85).

The new semiempirical relationship for the $\alpha$-decay lifetime takes into consideration explicitly the dependence not only on the proton but also on the neutron number, and their departure from magicity. In comparison with other formulas for the known regions of $\alpha$ emitters, it gives the best estimates.

The estimate of the general trend of nuclear stability with respect to the emission of charged particles revealed many interesting facts. For the spontaneous emission of $^{2,3}$H, $^{3,6-10}$He, $^{4}$Li, $^{7}$N, $^{9}$C, and some other clusters, all nuclei with masses tabulated by Wapstra and Bos are stable ($Q < 0$) in their ground states. $^{5}$He radioactivity is energetically possible for some 110 nuclides grouped in two archipelagos with $Z = 83$–92, $N = 127$–137 and $Z = 97$–105, $N = 145$–157. The last region extends also to superheavy nuclei.

In the framework of ASAFM one can obtain reliable values for the nuclear lifetimes relative to the new decay modes, if the masses of the three partners (parent and two fragments) are well known and if a suitable value of the zero-point vibration energy is used. By introducing shell effects in $E_v$ one obtains better agreement with experimental data on 380 $\alpha$ emitters and the $^{14}$C radioactivity of $^{223}$Ra is reproduced. We need, of course, more experimental data (new decay modes with emitted ions heavier than $^{14}$C) to further improve our model, because the lifetime is very sensitive to the details of the barrier shape and parameters. Hence, the present calculations should be taken only as an orientation toward what can be accomplished with such an approach.

From a systematic search across the nuclear table, we have shown the great complexity and diversity of the new phenomena: there are more than 140 possible emitted clusters, various isotopes of the elements with $Z_2 = 2$–24, for which the lifetimes are in the range $10^{10}$–$10^{30}$ s. For a given decay mode $(A_2, Z_2)$, the minimum lifetime is obtained when the daughter has a magic number of neutrons $N_1 = 126$ and a magic number of protons $Z_1 = 82$, or not too far from it.

In the *trans*-tin region, where $N_1 = 82$ plays the major role, the lifetimes are much longer. In fact not only the shell effects for $N = 126$, $Z = 82$ are stronger than for $N = 50, 82$, $Z = 50$, but also $^{280}$Pb is lying on the $\beta$-stability line, while $^{100}$Sn or $^{132}$Sn are both far off stability. A pairing effect is also present: usually for odd $N_2$ or $Z_2$ emitted heavy ions, the lifetime is longer than for their even neighbors. Also for even–even, even–odd, odd–even, and odd–odd parent nuclides, as in the case of $\alpha$ decay, $E_v/Q$ should have different values in various groups of nuclei. Almost all nuclides including the "stable" ones with $Z > 40$ are metastable with respect to several decay modes, but if the lifetime is very long one can say, from any practical point of view, that the corresponding nucleus is stable. When the mass $A_2$ increases, for $Z_2 > 16$ the position of the optimum-optimorum emitted heavy ion moves in the neutron-rich side of the nuclear chart and the parent becomes more and more neutron-deficient.

According to our calculations, the largest value of the branching ratio for $^{14}$C emission relative to $\alpha$ decay ($\sim 10^{-9}$) is obtained for the parent $^{223}$Ra and has already been measured. In this case the absolute value of the lifetime ($\sim 10^{15}$ s) is by two orders of magnitude longer than that of $^{222}$Ra for $^{14}$C emission and by five orders of magnitude longer than the minimum value for $^{8}$Be radioactive decay.

A systematic investigation of a new type of fission with nearly spherical fragments, for the very heavy elements around $^{264}$Fm, and showing the influence of two magic numbers with $Z = 50$, will complete our picture. New branching ratios, not only the two fission modes, but also relative to the cluster decays, are necessary to clarify the simplifying assumptions in the present phenomenological theories.

The principal importance of magic radioactivity rests on the following facts:

1. It is a new phenomenon unifying fission and $\alpha$ radioactivity.
2. It is a molecular phenomenon, very well described within fragmentation theory (and its derivative models) and based on a two-center shell model.
3. Through it, nuclear structure can be tested in a comprehensive and sensitive manner. The various emission rates are strongly dependent on the structure of parent and daughter fragments and their barrier.
4. Coulomb-induced cluster emission, as proposed by Martin Greiner and Werner Scheid, will furnish a promising tool for further and more detailed investigations of the new radioactivities. The same is true for cluster radioactivity in excited states and has been studied by the same authors.
5. There is an inherent connection between cluster radioactivities and the observed dipole, quadrupole, octupole, etc., moments of heavy nuclei. Fragmentation theory yields the most natural way to join these apparently unconnected phenomena together (De 86, He 86).

## Acknowledgments

The first, third, and fourth authors of this chapter were supported in part by Bundesministerium für Forschung und Technologie and by Internationales Bureau KFK Karlsruhe, and the second, third, and fourth authors in part by the Romanian State Committee for Nuclear Energy.

## References

(Ab 64) M. Abramowitz and I. A. Stegun (eds.), *Handbook of Mathematical Functions*, National Bureau of Standards, Washington (1964).

(Ad 70) G. D. Adeev, I. A. Gamalya, and P. A. Cherdantsev, *Yad. Fiz.* **12**:272 (1970).

(Ad 76) G. D. Adeev, E. A. Filipenko, and P. A. Cherdantsev, *Yad. Fiz.* **23**:30 (1976).

(Aj 74) P. Ajzenberg-Selove and T. Lauritzsen, *Nucl. Phys. A* **227**:1 (1974).

(Al 83) G. D. Alkhazov, K. A. Mezilev, Yu. N. Novikov, A. M. Murmukhamedov, N. Ganbaatar, K. Ya. Gromov, V. G. Kalinnikov, A. Potempa, and F. Tarkanyi, *Z. Phys. A* **311**:245 (1983).

(Al 84) D. V. Alexandrov, A. F. Belyatsky, Yu. A. Gluhov, E. Yu. Nikolsky, B. V. Novatsky, A. A. Ogloblin, and D. N. Stepanov, *Pis'ma JETP* **40**:152 (1984).

(Az 79) R. E. Azuma, L. C. Carraz, P. G. Hansen, B. Jonson, K. L. Kratz, S. Mattsson, G. Nyman, H. Ohm, H. L. Ravn, A. Schroder, and W. Ziegert, *Phys. Rev. Lett.* **43**:1652 (1979).

(Az 80) R. E. Azuma, T. Bjornstadt, H. A. Gustafsson, P. G. Hansen, B. Jonson, S. Mattsson, G. Nyman, A. M. Poskanzer, and H. L. Ravn, *Phys. Lett.* **96B**:31 (1980).

(Ba 85) S. W. Barwick, P. B. Price, and J. D. Stevenson, *Phys. Rev. C* **31**:1984 (1985).

(Ba 86) S. W. Barwick, P. B. Price, H. L. Ravn, E. Hourani, and M. Hussonnois, private communication (1986).

(Be 61) R. Beringer and W. J. Knox, *Phys. Rev.* **121**:1195 (1961).

(Be 63) R. Beringer, *Phys. Rev.* **131**:1402 (1963).

(Be 70) E. E. Berlovich, O. M. Golubev, and Ju. N. Novikov, *Pis'ma JETP* **12**:289 (1970).

(Be 73) A. G. Below, Yu. P. Grangrsky, B. Dalkhsuren, A. M. Kucher, and Nguen Kong Khan, *Yad. Fiz.* **17**:942 (1973).

(Be 81) C. R. Bemis (Jr), P. F. Dittner, R. L. Ferguson, D. C. Hensley, F. Plasil, and F. Pleasonton, *Phys. Rev. C* **23**:555 (1981).

(Be 84) R. Bengtsson, P. Möller, J. R. Nix, and Jing-ye Zhang, *Phys. Scr.* **29**:402 (1984).

(Be 96) H. Becquerel, *Compt. Rend.* 122 (1896).

(Bj 80) S. Bjornholm and J. E. Lynn, *Rev. Mod. Phys.* **52**:725 (1980).

(Bl 56) J. M. Blatt and V. F. Weisskopf, *Theoretical Nuclear Physics*, Wiley, New York (1956).

(Bl 77) J. Blocki, J. Randrup, W. J. Swiatecki, and C. F. Tsang, *Ann. Phys.* **105**:427 (1977).

(Bl 81) J. Blocki and W. J. Swiatecki, *Ann. Phys.* **132**:53 (1981).

(Bo 39) N. Bohr and J. A. Wheeler, *Phys. Rev.* **56**:426 (1939).

(Br 72) M. Brack, J. Damgaard, A. S. Jensen, H. C. Pauli, V. M. Strutinsky, and C. Y. Wong, *Rev. Mod. Phys.* **44**:320 (1972).

(Br 73) H. C. Britt, *At. Data Nucl. Data Tables* **12**:408 (1973).

(Br 85) D. A. Bromley, in *Clustering Aspects of Nuclear Structure*, invited papers at the 4th Int. Conf. Chester 1984 (J. S. Lilley and M. A. Nagarajan, eds.), Reidel, Dordrecht (1985), p. 1.

(Ca 61) B. C. Carlson, *J. Math. Phys.* **2**:441 (1961).

(Ca 78) C. Cabot, S. Della Negra, C. Deprun, H. Gauvin, and Y. Le Beyec, *Z. Phys. A* **287**:71 (1978).

(Ca 83) M. D. Cable, J. Honkanen, R. F. Parry, S. H. Zhou, Z. Y. Zhou, and J. Cerny, *Phys. Rev. Lett.* **50**:404 (1983).

(Ca 84) M. D. Cable, J. Honkanen, E. C. Schloemer, M. Ahmed, J. E. Reiff, Z. Y. Zhou, and J. Cerny, *Phys. Rev. C* **30**:1276 (1984).

(Ch 84) Ching Chen-rui and Ho Tso-hsiu, *Phys. Rep.* **112**:1 (1984).

(Cl 86) H. G. Clerc, W. Lang, M. Mutterer, C. Schmitt, J. P. Theobald, V. Quade, K. Rudolph, P. Armbruster, F. Gönnenwein, H. Schrader, and D. Engelhardt, *Nucl. Phys. A* **452**:227 (1986).

(Co 28) E. U. Condon and R. W. Gurney, *Nature* **122**:439 (1928).

(Co 29) E. U. Condon and R. W. Gurney, *Phys. Rev.* **33**:127 (1929).

(Co 62) S. Cohen and W. J. Swiatecki, *Ann. Phys.* (*N.Y.*) **19**:67 (1962).

(Co 65) W. J. Cody, *J. Math. Phys.* **2**:441 (1965).

(Cu 80) R. Y. Cusson, J. A. Maruhn, and H. Stöcker, *Z. Phys. A* **294**:245 (1980).

(Da 75) K. T. R. Davies and A. J. Sierk, *J. Comput. Phys.* **18**:311 (1975).

(Da 75b) P. J. Davis and P. Rabinowitz, *Methods of Numerical Integration*, Academic Press, New York (1975).

(De 79) A. R. Degheidy and J. A. Maruhn, *Z. Phys. A* **290**:205 (1979).

(De 85) K. Depta, R. Herrmann, W. Greiner, W. Scheid, and A. Sandulescu, *J. Phys. G.* **11**:1087 (1985).

(De 85b) K. Depta, R. Herrmann, V. Schneider, J. A. Maruhn, W. Greiner, D. N. Poenaru, M. Ivascu, and A. Sandulescu, Proc. Am. Chem. Soc. Symp. on Recent Advances in the Study of Nuclei Off the Line of Stability, Chicago (1985).

(De 86) K. Depta, J. A. Maruhn, W. Greiner, W. Scheid, and A. Sandulescu, *Mod. Phys. Lett.* **A1**:377 (1986).

(De 86b) K. Depta, R. Herrmann, J. A. Maruhn, W. Greiner, M. Greiner, and W. Scheid, Proc. Symp. on Nuclear Fission, Rochester (1986).

(Di 73) H. Diehl and W. Greiner, *Phys. Lett.* **45B**:35 (1973).

(Ei 74) J. M. Eisenberg and W. Greiner, *Nuclear Theory*, Vol. 1, North-Holland, Amsterdam (1974).

(Ep 83) M. Epherre, in Proc. 7e session d'études biennale de Physique Nucléaire Aussois IPN Lyon (1983), c. 4.1.

(Fi 72) E. O. Fiset and J. R. Nix, *Nucl. Phys. A* **193**:647 (1972).

(Fi 74) H. J. Fink, J. Maruhn, W. Scheid, and W. Greiner, *Z. Phys.* **268**:321 (1974).

(Fl 75) R. L. Fleischer, P. B. Price, and R. M. Walker, *Nuclear Tracks in Solids, Principles and Applications*, University of California Press, Berkeley (1975).

(Fr 39) J. Frenkel, *Phys. Rev.* **53**:987 (1939).

(Fr 46) J. Frenkel, *J. Phys (Paris)* **10**:533 (1946).

(Fr 57) P. O. Fröman, *Mat. Fys. Sk. Dan. Vidensk. Selsk.* **1**:3 (1957).

(Fr 71) R. Fraser and W. Greiner, *Nucl. Phys. A* **177**:174 (1971).

(Ga 28) G. Gamow, *Z. Phys.* **52**:510 (1928).

(Ga 75) H. Gauvin, Y. Beyec, J. Livet, and J. L. Reyss, *Ann. Phys. (Paris)* **9**:241 (1975).

(Ga 76) D. Galeriu and D. N. Poenaru, *Rev. Roum. Phys.* **21**:109 (1976).

(Ga 84) S. Gales, E. Hourani, M. Hussonnois, H. P. Shapira, L. Stab, and M. Vergnes, *Phys. Rev. Lett.* **53**:759 (1984).

(Ga 84b) S. Gales, E. Hourani, M. Hussonnois, H. P. Shapira, L. Stab, and M. Vergnes, in *Atomic Masses and Fundamental Constants*, Proc. 7th Int. Conf. Darmstadt (O. Klepper, ed.), Techn. Hochschule, Darmstadt (1984), p. 279.

(Go 49) M. Goepert-Mayer, *Phys. Rev.* **75**:1969 (1949).

(Go 60) V. I. Goldansky, *Nucl. Phys.* **19**:482 (1960).

(Go 75) V. M. Gorbachev, Yu. S. Zamiatnin, and A. A. Lbov, *Fundamental Characteristics of Elementary Isotopes* (in Russian), Atomizdat, Moscow (1975).

(Go 80) V. I. Goldansky, *Pis'ma JETP* **32**:554 (1980).

(Gr 65) W. Greiner, in Proceedings of the Fourth International Conference on Neutron Capture, Jülich (1965).

(Gr 85) W. Greiner, M. Ivascu, D. N. Poenaru, and A. Sandulescu, *Z. Phys. A* **320**:347 (1985).

(Gr 86) W. Greiner, *Past and Future of Heavy Ion Research*, GSI, Darmstadt (1986).

(Gr 86b) M. Greiner and W. Scheid, *J. Phys. G* **12**:L229 (1986).

(Gr 86c) M. Greiner, Diploma Thesis, University of Giessen (1986).

(Gu 75) R. K. Gupta, W. Scheid, and W. Greiner, *Phys. Rev. Lett.* **35**:353 (1975).

(Gu 77) R. K. Gupta, *Z. Phys. A* **281**:159 (1977).

(Ha 39) O. Hahn and F. Strassmann, *Naturwissenschaften* **27**:11, 89 (1939).

(Ha 77) E. Hagberg, P. G. Hansen, J. C. Hardy, P. Hornshoj, B. Jonson, S. Mattsson, and P. Tiedemand-Petersson, *Nucl. Phys. A* **293**:1 (1977).

(Ha 77b) J. Hahn, H. S. Lustig, and W. Greiner, *Z. Naturforsch.* **32a**:215 (1977).

(Ha 79) E. Hagberg, P. G. Hansen, P. Hornshoj, B. Jonson, S. Mattsson, and P. Tiedemand-Petersson, *Nucl. Phys. A* **318**:29 (1979).

(Ha 85) J. H. Hamilton, P. G. Hansen, and E. F. Zganjar, *Rep. Prog. Phys.* **48**:631 (1985).

(He 86) R. Herrmann, J. A. Maruhu, and W. Greiner, *J. Phys. G* **12**:L285 (1986).

(Hi 53) D. L. Hill and J. A. Wheeler, *Phys. Rev.* **89**:1102 (1953).

(Ho 70) G. Holm and W. Greiner, *Phys. Rev. Lett.* **24**:404 (1970).

(Ho 74) P. Hornshoj, P. G. Hansen, B. Jonson, H. L. Ravn, L. Westgaard, and O. N. Nielsen, *Nucl. Phys. A* **230**:365 (1974).

(Ho 79) S. Hofmann, W. Faust, G. Münzenberg, W. Reisdorf, P. Armbruster, K. Güttner, and H. Ewald, *Z. Phys. A* **291**:53 (1979).

(Ho 80) D. C. Hoffman, J. B. Wilhelmy, J. Weber, W. R. Daniels, E. K. Hulet, R. W. Lougheed. J. H. Landrum, J. F. Wild, and R. J. Dupzyk, *Phys. Rev. C* **21**:972 (1980).

(Ho 85) E. Hourani, M. Hussonnois, L. Stab, L. Brillard, S. Gales, and J. P. Schapira, *Phys. Lett.* **160B**:375 (1985).

(Hu 86) E. K. Hulet *et al.*, *Phys. Rev. Lett.* **56**:313 (1986).

(Iv 78) M. Ivascu and D. N. Poenaru, in Proceedings of the Predeal International School on Heavy Ion Physics (A. Berinde, V. Ceausescu, and I. A. Dorobantu, eds.), Central Institute of Physics, Bucharest (1978), p. 1119.

(Iv 81) M. Ivascu and D. N. Poenaru, *Nuclear Deformation Energy and Shape Isomers*, Romanian Academy Publ. House, Bucharest (1981).

(Iv 86) M. Ivascu and I. Silisteanu, *Nucl. Phys.* **A485**:93 (1988).

(Jä 80) J. Jänecke, private communication (1980).

(Ja 85) R. Jahn, R. L. McGrath, D. M. Moltz, J. E. Reiff, X. J. Xu, J. Äystö, and J. Cerny, *Phys. Rev. C* **31**:1576 (1985).

(Jo 70) T. Johansson, S. G. Nilsson, and Z. Szymansky, *Ann. Phys. (Paris)* **5**:377 (1970).

(Ju 74) K. Junker, *Acta Phys. Austriaca* **40**:335 (1974).

(Ka 67) V. A. Karnaukhov and G. M. Ter-Akopyan, *Ark. Fiz.* **36**:419 (1967).

(Ke 72) K. A. Keller and H. Munzel, *Z. Phys.* **255**:419 (1972).

(Kl 84) H. V. Klapdor and K. Grotz, *Phys. Lett.* **142B**:323 (1984).

(Kr 74) H. J. Krappe and J. R. Nix, in *Proceedings of the Symposium on Physics and Chemistry of Fission*, Vol. 1, IAEA, Vienna (1974), p. 159.

(Kr 76) H. J. Krappe, *Nucl. Phys. A* **269**:493 (1976).

(Kr 79) H. J. Krappe, J. R. Nix, and A. J. Siek, *Phys. Rev. C* **20**:992 (1979).

(Ku 85) W. Kutschera, J. Ahmad, S. G. Armato III, A. M. Friedman, J. E. Gindler, W. Henning, T. Ishii, M. Paul, and K. E. Rehm, *Phys. Rev. C* **32**:2036 (1985).

(La 65) J. N. P. Lawrence, *Phys. Rev.* **139B**:1227 (1965).

(Le 66) R. M. Leachman and B. H. Erkkila, *Bull. Am. Phys. Soc.* **10**:1204 (1966).

(Le 78) C. M. Lederer and V. S. Shirley, *Table of Isotopes*, Wiley, New York (1978).

(Lu 80) H. J. Lustig, J. A. Maruhn, and W. Greiner, *J. Phys. G* **6**:L-25 (1980).

(Ma 65) R. D. Macfarlane and A. Siivola, *Phys. Rev. Lett.* **14**:114 (1965).

(Ma 72) J. A. Maruhn and W. Greiner, *Z. Phys.* **251**:431 (1972).

(Ma 73) J. A. Maruhn and W. Greiner, *Phys. Lett.* **44B**:9 (1973).

(Ma 74) J. A. Maruhn and W. Greiner, *Phys. Rev. Lett.* **32**:548 (1974).

(Ma 76) J. A. Maruhn and W. Greiner, *Phys. Rev. C* **13**:2404 (1976).

(Ma 76b) S. Maripuu (ed.), *At. Nucl. Data Tables*, **17**:411 (1976).

(Ma 80) J. A. Maruhn, W. Greiner, and W. Scheid, in *Heavy Ion Collisions* (R. Bock, ed.), North-Holland, Amsterdam (1980), p. 397.

(Me 39) L. Meitner and O. R. Frisch, *Nature* **143**:239 (1939).

(Me 76) W. D. Meyers, *At. Data Nucl. Data Tables* **17**:411 (1976).

(Me 80) V. Metag, D. Habs, and H. F. Specht, *Phys. Rep.* **65**:1 (1980).

(Mi 81) A. Michaudon, *Nuclear Fission and Neutron-Induced Fission Cross-Sections*, Pergamon Press, Oxford (1981).

(Mö 70) P. Möller and S. G. Nilsson, *Phys. Lett.* **31B**:283 (1970).

(Mö 72) P. Möller, *Nucl. Phys. A* **192**:529 (1972).

(Mö 74) P. Möller, S. G. Nilsson, and J. R. Nix, *Nucl. Phys. A* **229**:292 (1974).

(Mö 76) P. Möller and J. R. Nix, *Nucl. Phys. A* **269**:493 (1976).

(Mö 77) P. Möller and J. R. Nix, *Nucl. Phys. A* **281**:354 (1977).

(Mö 80) P. Möller and S. Aberg, *Phys. Lett.* **92B**:223 (1980).

(Mö 81) P. Möller and J. R. Nix, *Nucl. Phys. A* **361**:117 (1981).

(Mö 81b) P. Möller and J. R. Nix, *At. Data Nucl. Data Tables* **26**:165 (1981).

(Mo 69) U. Mosel and W. Greiner, *Z. Phys.* **222**:261 (1969).

(Mo 71) U. Mosel and H. W. Schmitt, *Phys. Rev. C* **4**:2185 (1971).

(Mo 75) L. G. Moretto, *Nucl. Phys. A* **247**:211 (1975).

(Mu 72) M. G. Mustafa, U. Mosel, and H. W. Schmitt, *Phys. Rev. Lett.* **28**:1536 (1972).

(Mu 73) M. G. Mustafa, U. Mosel, and H. W. Schmitt, *Phys. Rev. C* **7**:1519 (1973).

(My 66) W. D. Myers and W. J. Swiatecki, *Nucl. Phys.* **81**:1 (1966).

(My 67) W. D. Myers and W. J. Swiatecki, *Ark. Fys.* **36**:343 (1967).

(My 77) W. D. Myers, *Droplet Model of Atomic Nuclei*, Plenum Press, New York (1977).

(Ni 72) J. R. Nix, *Ann. Rev. Nucl. Sci.* **22**:65 (1972).

(Ny 81) G. Nyman, R. E. Azuma, B. Jonson, K. L. Kratz, P. O. Larsson, S. Mattsson, and W. Ziegert, CERN 81-09 (1981), p. 312.

(Ob 85) V. Oberacker, T. Pinkston, and H. Kruse, *Rep. Prog. Phys.* **48**:31 (1985).

(Pa 71) V. V. Pashkevich, *Nucl. Phys. A* **169**:275 (1971).

(Pa 75) H. C. Pauli and T. Ledergerber, *Proceedings of the Symposium on Physics and Chemistry of Fission, Vienna 1974,* Vol. 1, IAEA, Vienna (1975), p. 46.

(Pe 40) K. A. Petrjak and G. N. Flerov, *JETP* **10**:1013 (1940).

(Pe 57) J. Perlman and J. O. Rasmussen, in *Handbuch der Physik,* Band XLII, Springer-Verlag, Berlin (1957), p. 109.

(Pe 71) L. K. Peker, E. I. Volmyansky, V. E. Bumakov and S. G. Ogloblin, *Phys. Lett.* **36B**:547 (1971).

(Pl 82) A. Plochocki, J. Zylicz, R. Kirchner, O. Klepper, E. Roeckl, P. Tidemand-Pettersson, I. S. Grant, and P. Misaelides, *Nucl. Phys. A* **388**:91 (1982).

(Po 62) S. M. Polikanov, V. A. Druin, V. A. Karnaukhov, V. M. Mikheev, A. Pleve, N. K. Skobelev, V. G. Subotin, G. M. Ter-Akopyan, and V. A. Fomichev, *JETP* **42**:1464 (1962).

(Po 69) D. N. Poenaru and N. Vîlcov, *Measurement of Nuclear Radiations with Semiconductor Detectors,* Chemical, New York (1969).

(Po 75) D. N. Poenaru and D. Galeriu, Report CRD-60, Institute for Atomic Physics, Bucharest (1975).

(Po 77) D. N. Poenaru, *Ann. Phys. (Paris)* **2**:133 (1977).

(Po 78) D. N. Poenaru and M. Ivascu, *Comput. Phys. Commun.* **16**:85 (1978).

(Po 79) D. N. Poenaru and M. Ivascu, *Rev. Roum. Phys.* **24**:119 (1979).

(Po 79b) D. N. Poenaru, M. Ivascu, and A. Sandulescu, *J. Phys. (Paris) Lett.* **40**:L-465 (1979).

(Po 79c) D. N. Poenaru, M. Ivascu, and A. Sanduleseu, *J. Phys. G* **5**:L-169 (1979).

(Po 79d) D. N. Poenaru, M. Ivascu, and A. Sandulescu, *Rev. Roum. Phys.* **24**:917 (1979).

(Po 79e) D. N. Poenaru, D. Mazilu, and M. Ivascu, *J. Phys. G* **5**:1093 (1979).

(Po 80) D. N. Poenaru and M. Ivascu, *Proceedings of the International Summer School Critical Phenomena in Heavy Ion Physics,* Central Institute of Physics, Bucharest (1980), p. 743.

(Po 80b) D. N. Poenaru, M. Ivascu, and D. Mazilu *J. Phys. (Paris) Lett.* **41**:L-589 (1980).

(Po 80c) D. N. Poenaru, M. Ivascu, D. Mazilu, and A. Sandulescu, *Rev. Roum. Phys.* **25**:55 (1980).

(Po 80d) D. N. Poenaru, M. Ivascu, and D. Mazilu, *Comput. Phys. Commun.* **19**:205 (1980).

(Po 81) D. N. Poenaru and M. Ivascu, *Proceedings of the EPS Division Conference "Nuclear and Atomic Physics with Heavy Ions"* Central Institute of Physics, Bucharest (1981), p. 681.

(Po 81b) D. N. Poenaru and M. Ivascu, *J. Phys. G* **7**:965 (1981b).

(Po 81c) D. N. Poenaru, M. Ivascu, and A. Sandulescu, *Rev. Roum. Phys.* **26**:253 (1981).

(Po 82) D. N. Poenaru and M. Ivascu, *Rev. Roum. Phys.* **27**:129 (1982).

(Po 82b) D. N. Poenaru, M. Ivascu, and D. Mazilu, *Comput. Phys. Commun.* **25**:297 (1982).

(Po 83) D. N. Poenaru and M. Ivascu, *J. Phys.* **44**:791 (1983).

(Po 83b) D. N. Poenaru and M. Ivascu, *Rev. Roum. Phys.* **28**:309 (1983).

(Po 83c) D. N. Poenaru and M. Ivascu, Report NP-27, Central Institute of Physics, Bucharest (1983).

(Po 83d) D. N. Poenaru, M. Ivascu, and A. Sandulescu, *Proceedings of the International Conference on Nuclear Physics, Florence,* Vol. 1 (R. A. Ricci and R. Blasi, eds.), Tipografia Compositori, Bologna (1983), p. 662.

(Po 83e) D. N. Poenaru, M. Ivascu, and A. Sandulescu, JINR Dubna Report E4-83-858 (1983).

(Po 84) D. N. Poenaru and M. Ivascu, *J. Phys. (Paris)* **45**:1099 (1984).

(Po 84b) D. N. Poenaru and M. Ivascu, *Proceedings of the International Summer School on Atomic and Heavy Ion Physics, Brasov-1984,* Central Institute of Physics, Bucharest (1984).

(Po 84c) D. N. Poenaru and M. Ivascu, *Rev. Roum. Phys.* **29**:587 (1984).

(Po 84d) D. N. Poenaru and M. Ivascu, *Rev. Roum. Phys.* **29**:623 (1984).

(Po 84e) D. N. Poenaru and M. Ivascu, Report NP-40, Central Institute of Physics, Bucharest (1984).

(Po 84f) D. N. Poenaru, M. Ivascu, A. Sandulescu, and W. Greiner, *J. Phys. G* **10**:L-183 (1984).

(Po 84g) D. N. Poenaru, M. Ivascu, A. Sandulescu, and W. Greiner, *Izv. Akad. Nauk SSSR, Ser. Fiz.* **48**:1897 (1984).

(Po 84h) D. N. Poenaru, M. Ivascu, A. Sandulescu, and W. Greiner, *JINR Dubna Rapid Communications* **5**:29 (1984).

(Po 84i) D. N. Poenaru, M. Ivascu, A. Sandulescu, and W. Greiner, JINR Dubna Report E4-84-446 (1984).

(Po 85) D. N. Poenaru, W. Greiner, M. Ivascu, and A. Sandulescu, *Phys. Rev. C* **32**:2198 (1985).

(Po 85b) D. N. Poenaru and M. Ivascu, *J. Phys. (Paris) Lett.* **46**:L-591 (1985).

(Po 85c) D. N. Poenaru, M. Ivascu, A. Sandulescu, and W. Greiner, *Phys. Rev. C* **32**:572 (1985).

(Po 85d) D. N. Poenaru, M. Ivascu, A. Sandulescu, and W. Greiner, *Rev. Roum. Phys.* **30**:173 (1985).

(Po 85e) D. N. Poenaru, M. Ivascu, A. Sandulescu, and W. Greiner, Report Np-46, Central Institute of Physics, Bucharest (1985).

(Po 86) D. N. Poenaru, W. Greiner, K. Depta, M. Ivascu, D. Mazilu, and A. Sandulescu, *At. Data Nucl. Data Tables* **34**:423 (1986).

(Po 86b) D. N. Poenaru, M. Ivascu, and W. Greiner, *Proc. Int. Conf. on SSNTD*—Rome 1985, in *Nucl. Tracks and Radiation Measurements* **12**:313 (1986).

(Po 86c) D. N. Poenaru, M. Ivascu, D. Mazilu, and I. H. Plonski, *Rev. Roum. Phys.* **31**:125 (1986).

(Po 86d) D. N. Poenaru, M. Ivascu, W. Greiner, D. Mazilu, and I. H. Plonski, *Z. Phys. A* **325**:435 (1986).

(Pr 85) P. B. Price, J. D. Stevenson, S. W. Barwick, and H. L. Ravn, *Phys. Rev. Lett.* **54**:297 (1985).

(Pr 85b) P. B. Price, *Phys. Bull.* **36**:489 (1985).

(Ri 81) B. G. Ritchie, K. S. Toth, H. K. Carter, R. J. Mlekodaj, and E. H. Spejevski, *Phys. Rev. C* **23**:2342 (1981).

(Ro 39) R. B. Roberts, R. C. Meyer, and P. Wang, *Phys. Rev.* **55**:510 (1939).

(Ro 81) J. C. Roynette, N. Frascaria, Y. Blumenfeld, J. C. Jacmart, E. Plagnol, J. P. Garron, A. Camp, and H. Fuchs, *Z. Phys. A* **299**:73 (1981).

(Ro 84) H. J. Rose and G. A. Jones, *Nature* **307**:245 (1984).

(Ro 84b) G. Royer and B. Remaud, *J. Phys. G* **10**:1057 (1984).

(Ru 08) E. Rutherford and H. Geiger, *Proc. R. Soc. London Ser. A* **81**:141 (1908).

(Ru 16) E. Rutherford and A. B. Wood, *Phil. Mag.* **31**:379 (1916).

(Ru 82) E. Rurarz, Report INR 1950/IA/PL/A, Warszawa Institute of Nuclear Research (1982).

(Ry 79) A. Rytz, *At. Data Nucl. Data Tables* **23**:507 (1979).

(Sa 76) G. Sauer, H. Chandra, and U. Mosel, *Nucl. Phys. A* **264**:221 (1976).

(Sa 78) A. Sandulescu, H. J. Lustig, J. Hahn, and W. Greiner, in *Proceedings of the Predeal International School on Heavy Ion Physics* (A. Berinde, V. Ceausescu, and I. A. Dorobantu, eds.), Central Institute of Physics, Bucharest (1984), p. 691.

(Sa 78b) A. Sandulescu, H. J. Lustig, J. Hahn, and W. Greiner, *J. Phys. G* **4**:L-279 (1978).

(Sa 80) A. Sandulescu, D. N. Poenaru, and W. Greiner, Sov. J. Part. Nucl. **11**(6):528 (1980).

(Sa 83) A. Sandulescu, R. Y. Cusson, and W. Greiner, *Lett. Nuovo Cimento* **36**:321 (1983).

(Sa 84) A. Sandulescu, Yu. S. Zamiatnin, J. A. Lebedev, B. F. Myasoedov, S. P. Tretyakova, and D. Hasegan, *JINR Dubna Rapid Communications* **5**:5 (1984); *Izv. Akad. Nauk SSSR, Ser. Fiz.* **41**:2104 (1985).

(Sa 85) A. Sandulescu, D. N. Poenaru, W. Greiner, and J. H. Hamilton, *Phys. Rev. Lett.* **54**:490 (1985).

(Sa 85b) A. Sandulescu and H. Scutaru, *JINR Dubna Rapid Communications* **10**:5 (1985).

(Sa 85c) A. Sandulescu, *Proceedings of the International School of Nuclear Structure, Alushta* (V. G. Soloviev and Yu. P. Popov, eds.), D4-85-851, Dubna (1985).

(Sc 68) W. Scheid, R. Ligensa, and W. Greiner, *Phys. Rev. Lett.* **21**:1479 (1968).
(Sc 69) W. Scheid and W. Greiner, *Z. Phys.* **226**:364 (1969).
(Sc 70) H. Schultheis, R. Schultheis, and G. Süssmann, *Nucl. Phys. A* **144**:545 (1970).
(Sc 73) H. Schultheis and R. Schultheis, *Nucl. Phys. A* **215**:329 (1973).
(Sc 79) D. Schardt, R. Kirchner, O. Klepper, W. Reisdorf, E. Roeckl, P. Tidemand-Petersson, G. T. Ewan, E. Hagberg, B. Jonson, S. Mattsson, and G. Nyman, *Nucl. Phys. A* **326**:65 (1979).
(Sc 79b) K. H. Schmidt, W. Faust, G. Münzenberg, H. G. Clerc, W. Lang, K. Pielenz, D. Vermeulen, H. Wohlfarth, H. Ewald, and K. Güttner, *Nucl. Phys. A* **218**:353 (1979).
(Sc 81) D. Schardt, T. Batsch, R. Kirchner, O. Klepper, W. Kurcewicz, E. Roeckl, and P. Tidemand-Pettersson, *Nucl. Phys. A* **368**:153 (1981).
(Sc 85) V. Schneider, J. A. Maruhn, and W. Greiner, Frankfurt University Report UFTP 155/198 (1985).
(Se 75) P. A. Seeger and W. M. Howard, *Nucl. Phys. A* **238**:491 (1975).
(Se 81) W. Seelmann-Eggebert, G. Pfennig, H. Müntzel, and H. Klewe-Nebenius, *Nuklidkarte*, Kernforschungszentrum, Karlsruhe (1981).
(Se 85) M. Seiwert, W. Greiner, and W. T. Pinkston, *J. Phys. G* **11**:L-21 (1985).
(Sh 85) Shi Yi-Jin and W. J. Swiatecki, *Phys. Rev. Lett.* **54**:300 (1985).
(Sh 85b) Shi Yi-Jin and W. J. Swiatecki, *Nucl. Phys. A* **438**:450 (1985).
(So 83) L. G. Sobotka, M. L. Padgett, G. J. Wozniak, G. Guarino, A. J. Pacheco, L. G. Morett, Y. Chan, R. G. Stokstad, J. Tserruya, and S. Wald, *Phys. Rev. Lett.* **51**:2187 (1983).
(St 66) V. M. Strutinsky, *Yad. Fiz.* **3**:614 (1966).
(St 67) V. M. Strutinsky, *Nucl. Phys. A* **95**:420 (1967).
(St 68) V. M. Strutinsky, *Nucl. Phys. A* **122**:1 (1968).
(St 80) H. Stöcker, R. Y. Cusson, J. A. Maruhn, and W. Greiner, *Z. Phys. A* **294**:125 (1980).
(Sw 83) W. J. Swiatecki, *Aust. J. Phys.* **36**:641 (1983).
(Ta 61) R. Taagepera and M. Nurmia, *Ann. Acad. Sci. Fenn., Ser. A* 78 (1961).
(Ti 81) P. Tidemand-Petersson, R. Kirchner, O. Klepper, E. Roeckl, A. Ptochocki, J. Zylicz, and D. Schardt, CERN 81-09, Vol. 1 (1981), p. 205.
(Tr 85) S. P. Tretyakova, A. Sandulescu, Yu. S. Zamiatnin, Yu. S. Korotkin, and V. L. Micheev, *JINR Dubna Rapid Communications* **7**:23 (1985).
(Tr 85b) S. P. Tretyakova, A. Sandulescu, V. L. Micheev, D. Hasegan, I. A. Lebedev, Yu. S. Zamyatnin, Yu. S. Kortkin, and B. F. Myassoedov, *JINR Dubna Rapid Communications* **13**:34 (1985).
(Va 73) R. Vandenbosch and J. R. Huizenga, *Nuclear Fission*, Academic Press, New York (1973).
(Va 77) R. Vandenbosch, *Annu. Rev. Nucl. Sci.* **27**:1 (1977).
(Vi 66) V. E. Viola (Jr.) and G. T. Seaborg, *J. Inorg. Nucl. Chem.* **28**:741 (1966).
(Wa 59) A. H. Wapstra, G. J. Nijgh, and R. Van Lieshout, in *Nuclear Spectroscopy Tables*, North-Holland, Amsterdam (1959).
(Wa 77) A. H. Wapstra and K. Boss, *At. Data Nucl. Data Tables* **19**:215 (1977).
(Wa 85) A. H. Wapstra and G. Audi, *Nucl. Phys. A* **432**:1 (1985).
(We 81) M. S. Weiss, in *Dynamics of Heavy Ion Collisions* (N. Cindro, R. A. Ricci, and W. Greiner, eds.), North-Holland, Amsterdam (1981), p. 115.
(Wu 85) X. Z. Wu, K. Depta, R. Herrmann, J. A. Maruhn, and W. Greiner, *Nuovo Cimento* **87A**:309 (1985).
(Zo 78) O. Zohni and M. Blann, *Nucl. Phys. A* **297**:170 (1978).

# Index

LM269033
Coutts
9-5-91